Introduction to Multidisciplinary Science with Artificial Intelligence

The book is about multidisciplinary science education. The challenges of our time, such as improving the length and quality of lives on Earth and short- and long-distance communication and transportation. In this book, we provide readers with the multidisciplinary education necessary to meet the scientific and technological challenges of our time while optimizing the college experience for students.

The fundamental notions addressed in this book include gravitational forces and energy; dark matter and dark energy; heat transfer in solid Earth, stars' interiors, and human bodies; electromagnetic radiation and spectroscopy; quantum entanglement and computing; accretion disks; matter in plasma state; and exoplanets. We illustrate the importance of these notions with applications across disciplines, including monitoring the deformation of the solid Earth's surface using satellite measurements, unusual gravity anomalies in Antarctica, a view and characterization of the far side of our Moon, Earth's climate, Titan's anti-greenhouse effect, long-distance communication between Earth and the planets and exoplanets, etc. Finally, the book contains analytical and computational problems, including MATLAB® software developed especially for the classes associated with this book.

Key Features:

- Contains multiple analytic and computational (MATLAB®) exercises
- Explores applications related to space programs' discoveries
- Provides an accessible introduction and response to growing Multidisciplinary Science programs

Introduction to Multidisciplinary Science with Artificial Intelligence

Geodesy, Geotherms, Quantum Entanglement, and Spectroscopy

Luc Thomas Ikelle

CRC Press
Taylor & Francis Group
Boca Raton London New York

CRC Press is an imprint of the
Taylor & Francis Group, an **informa** business

Designed cover image: https://www.eso.org/public/news/eso1029/ (Creative Commons Attribution 4.0 International License); Mauk, B., Haggerty, D., Paranicas, C. et al. Discrete and broadband electron acceleration in Jupiter's powerful aurora. Nature 549, 66–69 (2017). https://doi.org/10.1038/nature23648 (With Permission from Springer Nature); and Duncan, F., Que, E., Zhang, N. et al. The zinc spark is an inorganic signature of human egg activation. Sci Rep 6, 24737 (2016). https://doi.org/10.1038/srep24737 (Open Access).

First edition published 2024
by CRC Press
2385 NW Executive Center Drive, Suite 320, Boca Raton FL 33431

and by CRC Press
4 Park Square, Milton Park, Abingdon, Oxon, OX14 4RN

CRC Press is an imprint of Taylor & Francis Group, LLC

© 2024 Luc Thomas Ikelle

Library of Congress Cataloging-in-Publication Data

Names: Ikelle, L. (Luc), author.
Title: Introduction to multidisciplinary science with artificial intelligence : geodesy, geotherms, quantum entanglement, and spectroscopy / Luc Thomas Ikelle.
Description: First edition. | Boca Raton, FL : CRC Press, 2024. | Includes bibliographical references and index. | Summary: "Technological developments and advances in scientific research have enabled the cross-fertilization of scientific disciplines. This book aims to provide readers with the tools necessary to address scientific and technological challenges in astronomy and space exploration"--
Provided by publisher.
Identifiers: LCCN 2023048003 | ISBN 9781032617794 (hardback) | ISBN 9781032620596 (paperback) | ISBN 9781032620619 (ebook)
Subjects: LCSH: Geophysics--Data processing. | Artificial intelligence--Geophysical applications. | Astrophysics--Data processing. | MATLAB.
Classification: LCC QC808.6 .I34 2024 | DDC 502.85/63--dc23/eng/20240129
LC record available at https://lccn.loc.gov/2023048003

ISBN: 978-1-032-61779-4 (hbk)
ISBN: 978-1-032-62059-6 (pbk)
ISBN: 978-1-032-62061-9 (ebk)

Typeset in CMR10
by KnowledgeWorks Global Ltd.

DOI: 10.1201/9781032620619

Publisher's note: This book has been prepared from camera-ready copy provided by the authors.

This book is dedicated to Larissa, Kevin-Luc Matiba, and Darrell-Thomas Udoh, my gifts of 4D life.

Luc Thomas Ikelle

Contents

Preface   xiii

Author Biography   xvii

1   **Gravitational Forces and Geodesy**   1
   1.1. Newton's Universal Law of Gravitation 1
      1.1.1. The Gravitational Field 1
      1.1.2. Kepler's Three Laws of Planetary Motion 14
      1.1.3. Gravity of Dark Matter 25
      1.1.4. The Gravitational Potential and Geoid and Gravitational Anomalies 33
      1.1.5. Satellite Gravity Missions 54
      1.1.6. Lagrange Points: Asteroids, Satellites, and Telescopes in Space   59
      1.1.7. Application: Gravity Anomalies Versus Topography in Antarctica 64
   1.2. Application: Tidal Forces 69
      1.2.1. The Barycenters of Earth-Moon and Earth-Sun Systems . . . 70
      1.2.2. Tide-Generating Forces 72
      1.2.3. Tide Occurrence 76
      1.2.4. Tide Range 77
      1.2.5. Far Side of the Moon 83
      1.2.6. The Titan Tides 86
      1.2.7. High Tides May Be the Origin of Life on Earth 92
   1.3. Application: Isostasy of Crustal Thickness 94
      1.3.1. Archimedes' Principle 94
      1.3.2. Isostasy 96
      1.3.3. Crustal Thickness: an Explanation of Thick-Continental and Thin-Oceanic Crusts 97
   1.4. Application: Tectonic Geodesy 100
      1.4.1. The Global Positioning System: Relative Motion 101
      1.4.2. The Absolute Motion: the Hotspot Model 109
      1.4.3. The Global Positioning System: Absolute Motion Using the NNR Frame 114
      1.4.4. The Global Positioning System: Absolute Motion Using the ITR Frame 117
      1.4.5. The Interferometric Synthetic Aperture Radar 130

1.5. Application: The Human Trip to Mars and Beyond 130
        1.5.1. The Bio-Physics of Space: NAD^+ Reduction and Bone and
                 Muscles Loss 134
        1.5.2. The Moon Colony and the Upcoming Mars Trip 138
1.6. Essay: Estimation of Tectonic Plate Velocity 142
1.7. Essay: Earthquakes in California and Volcanic Eruption of Yellow-
        stone National Park 145
1.8. Conclusions 148

2 Heat Radiation, Conduction, and Convection   149
2.1. Heat Transfer and Entropy 149
        2.1.1. Conservation of Energy and the First Law of Thermodynamics 149
        2.1.2. Entropy, Equilibrium, and the Second Law of Thermodynamics 159
        2.1.3. Entropy: Public Policies 165
        2.1.4. Heat-Transfer Mechanisms 170
        2.1.5. The Heat-Conductive Equation 173
        2.1.6. The Convective Heat-Diffusion Equation 177
2.2. Application: Heat Radiation and Greenshouse Effects 181
        2.2.1. Blackbodies, the Four Laws of Radiation, and Planck's Law . 181
        2.2.2. The Climate Model Without the Atmosphere: Mercury . . . 190
        2.2.3. The Climate Model with the Atmosphere: Venus, Earth,
                 Mars, and Jovian Planets 192
        2.2.4. The Climate Model with the Greenhouse Effect: Earth, Mars,
                 and Venus 194
        2.2.5. Absorption of Electromagnetic Radiation by the Atmosphere 199
        2.2.6. Temperature of TRAPPIST-1 Exoplanets 201
        2.2.7. Magnetic Fields of TRAPPIST-1 Exoplanets 203
2.3. Application: Geothermics and Energy Transport in the Stars 204
        2.3.1. The Conductive Heat Transport in the Lithosphere 205
        2.3.2. Mantle Heat Convection 216
        2.3.3. Conduction and Convection in the Core 226
        2.3.4. Energy Transport in Stars 229
2.4. Bioheat Transfer 237
        2.4.1. Fundamental Aspects of Bioheat 239
        2.4.2. Bioheat Transfer Equations: Thermal Diagnostics and Ther-
                 apies 243
2.5. Heat Engines and Pumps 256
        2.5.1. Heat Engines and the Second Law of Thermodynamics . . . 257
        2.5.2. Heat Pumps 265
        2.5.3. Is the Heart a Pump Engine? 267
2.6. Essay: Entropy and the Arrow of Time: The Irreversible Law 269
2.7. Essay: Cloud Formations on Earth 271
        2.7.1. Water Vapor and Aerosols in the Earth's Atmosphere 272
        2.7.2. Cloud Formations 274
        2.7.3. Some Weird Cloud Formations 276
2.8. Essay: Cloud Formations on Titan 279
        2.8.1. Titan's Landscape 279

2.8.2. Titan's Atmosphere Composition and Haze-Layer Formation 281
2.8.3. Anti-Greenhouse Effect 288
2.8.4. Cloud Formation 290
2.9. Essay: From Antarctica's Lakes to Europa's Subsurface Ocean . . . 291
2.9.1. Heat Sources of Antarctica's Sub-glacial Lakes 292
2.9.2. Oxygenic Photosynthesis in Antarctica's Sub-glacial Lakes . . 293
2.9.3. Europa's Subsurface Ocean 294
2.10. Essay: Magnetic Reversal May Not Be Catastrophic for Living Things 295
2.10.1. Earth's Magnetosphere 295
2.10.2. Geomagnetic Major Coordinates 296
2.10.3. Magnetic Polar Reversals 298
2.10.4. How Dangerous is the Coming Magnetic Reversal to Our Civilization? 300

3 Electromagnetic Radiation and Spectroscopy **305**
3.1. Microscopic and Macroscopic Maxwell's Equations 306
3.1.1. Microscopic and Macroscopic Scales 306
3.1.2. Lorentz Force Law at the Macroscopic Scale 307
3.1.3. Microscopic and Macroscopic Maxwell's Equations 312
3.1.4. Linear Constitutive Equations 321
3.1.5. Conductors, Semiconductors, Insulators, and Dielectrics . . . 329
3.1.6. Nonlinear Constitutive Equations 334
3.1.7. Electric Dipole Moments 342
3.1.8. Magnetic Dipole Moments 355
3.1.9. Ground Penetration Radar 366
3.2. Application: Infrared Spectroscopy Technique 379
3.2.1. Molecular vibrations 380
3.2.2. Formulation of the Infrared Spectroscopy Technique 381
3.2.3. Infrared spectroscopy of organic molecules 384
3.2.4. Infrared Spectroscopy: Inorganic Molecules 388
3.3. Application: Raman Spectroscopy Technique 392
3.3.1. Formulation of Raman Spectroscopy 392
3.3.2. Raman versus IR 394
3.4. pplication: Terahertz Spectroscopy and Unconscious Awareness . . . 396
3.4.1. Anesthetic Drugs and Unconsciousness 397
3.4.2. A Quantification of Consciousness; Conscious Awareness . . . 399
3.4.3. Terahertz Frequencies: the Transition from Quantum to Classical Mechanisms 401
3.4.4. Loss Memory, Enhanced Memory, and Missing Time 401
3.4.5. Can A Machine Be Conscious? 402
3.5. Application: NMR Spectroscopy 402
3.5.1. Quantum Spin 402
3.5.2. Nuclear Magnetic Resonance (NMR) 403
3.5.3. Chemical Shift 404
3.6. Application: Spectroscopy of Human Eyes 407
3.6.1. Cornea, Pupil, and Lens 408
3.6.2. Retina 409

   3.6.3. Sunlight and Earth's Atmosphere Filtering 410
  3.7. Application: Spectroscopy of Stars 412
   3.7.1. Optical Telescopes 412
   3.7.2. The Birth of the Famous Hubble Telescope 414
   3.7.3. James Webb Space Telescope and Search of Exoplanets . . . 416
   3.7.4. Seeing into the Past and at What Scale? 418
   3.7.5. Radio Telescopes 420
   3.7.6. Composition of Stars 423
  3.8. Essay: Spectroscopy of Cancer-causing Chemicals: Polycyclic Aromatic Hydrocarbons 429
  3.9. Essay: Some Biosignature Gases and Light Signatures 433
   3.9.1. Some Biosignature Gases 433
   3.9.2. Light Signatures 435
  3.10. Essay: Decoding of Molecule, Mineral, and Star Spectroscopy 437

4 A View of our Universe from Quantum Computing   439
  4.1. Schrödinger Equation 440
   4.1.1. Wave Functions 440
   4.1.2. Quantum Tunneling 453
   4.1.3. Heisenberg's Uncertainty Principle 464
  4.2. Quantum Entanglement: Nature Inseparability 468
   4.2.1. Dirac Notations 473
   4.2.2. A Formulation of Quantum Entanglement 480
   4.2.3. Density Matrix and Mixed States 485
   4.2.4. Source of Quantum Entanglements 488
  4.3. Application: Quantum Computing 493
   4.3.1. Classical Mechanics-Based Computers 493
   4.3.2. Quantum Mechanics-Based Computers 494
   4.3.3. Shor's Quantum Factorization Algorithm 496
   4.3.4. Will Quantum Computing Work? 503
   4.3.5. The Future Is Not Fixed 504

5 Spontaneous Technology   507
  5.1. Electricity in the Human Body 507
   5.1.1. Brain and Mind Interaction 508
   5.1.2. The Brain: An Electrochemical Network 511
   5.1.3. Limits of Humanoid's Brains 522
   5.1.4. Membrane Potential: Signaling within the Body and with the Environment 526
   5.1.5. Electrophysiological Activity 535
   5.1.6. Analogy between Cell Membranes and Electric Capacitors . . 544
  5.2. Electricity from the Ground 552
   5.2.1. Electromagnetic Induction 552
   5.2.2. Resistivity Well-Logging 560
   5.2.3. Spontaneous Electricity from the Well-logging 565
   5.2.4. Application: Artificial Geothermal Energy from Dry Oil Reservoirs 572

5.2.5. Application: Spontaneous Nuclear Fission Reactors,
the Gabon Case. 573
5.3. Essay: Casimir Effect and Ultracapacitors 579
5.3.1. Ultracapacitors: Energy Storage 579
5.3.2. Casimir Effect: Generating Energy from Vacuum 580

6 Plasma, Magnetohydrodynamics, Accretion, and Exoplanets 583
6.1. Solar Structure and Wind 585
6.1.1. Sun Structure 585
6.1.2. From Solar Plasma to Earth Atmosphere 590
6.1.3. The Sun's Magnetic Field: A Dynamo 594
6.2. Motion of Charged Particles in Collisionless Plasmas 595
6.2.1. Gyromotion (Larmor Motion) 598
6.2.2. Drift Motion 602
6.2.3. Magnetic Mirrors 610
6.2.4. Charged Particles in Earth-like Magnetic Fields and Van
Allen Radiation Belts 619
6.2.5. Auroras on Jupiter 624
6.2.6. Fusion Reactors by Magnetic Confinement 626
6.3. Magnetized Fluids: Magnetohydrodynamics 631
6.3.1. MHD Equations 631
6.3.2. MHD Equations Are Scale-Independent 641
6.3.3. Application: Parker's Solar Wind Model 642
6.3.4. Application: Magnetohydrodynamic Waves 650
6.4. MHD Instabilities and Accretion 657
6.4.1. Eulerian versus Lagrangian Formalisms 658
6.4.2. Gravitational Collapse 668
6.4.3. Kelvin-Helmholtz and Rayleigh-Taylor Instabilities 677
6.4.4. Convective Instability and Energy Transport in the Stars . . 684
6.4.5. Accretion Disks and Magnetorotational Instability (MRI) . . 687
6.4.6. Star Formation: Gravitational Collapse and Accretion Disk . 692
6.4.7. Accretion of Planets and Oumuamua 697
6.5. Generalized Ohm's Law and Magnetic Reconnection 700
6.5.1. Microscopic Interpretation of Electrical Conductivity 701
6.5.2. Three-fluid Ohm's Law 703
6.5.3. Magnetic Reconnection: A Departure from the Ideal MHD . 704
6.6. Application: Exoplanets 707
6.6.1. Exploration of Exoplanets 710
6.6.2. Why is the Exoplanet Radius Distribution Bimodal? 712
6.6.3. Exoplanets with Their Own Stars: Spontaneous Fusion . . . 715
6.6.4. Earth-like Exoplanets with Potential Life 717
6.6.5. Drake's Equation 721

A Advanced Numerical Modeling 723
A.1. Difference Operators 723
A.2. Finite-Difference Modeling of the Heat-Diffusion Equation 724
A.2.1. The Heat (or Diffusion) Equation 725

A.2.2. The Explicit Finite-Difference Method 725
A.2.3. The Implicit Finite-Difference Method with Backward Differ-
       ences in Time 727
A.2.4. The Implicit Finite-Difference Method with Forward Finite
       Differences in Time 728
A.2.5. The Case of Variable Conductivity 729
A.2.6. Multidimensional Thermal Calculation 729
A.3. Nonlinear Heat-Transfer Modeling 730
A.4. Madariaga-Virieux Staggered-grid Scheme 732
A.4.1. Stability and Dispersion Conditions 741
A.4.2. Free-surface Boundary Conditions 746
A.4.3. Absorbing Boundary Conditions 750
A.5. Modeling of the Schrödinger Equation 756
A.5.1. Fimite-Difference Scheme 756
A.5.2. Time-Space Solution in 1D Case 758
A.5.3. Time-Independent Solution in the 1D Case 759

B Answers to Some of the Quizzes and Exercises   761

References   781

Index   797

Preface

Higher science education is about preparing people to address the scientific and technological challenges of our times. The list of these challenges includes, but is not limited to, feeding the world's growing population, improving the length and quality of lives on Earth, improving short- and long- distance communication and transportation, predicting of, adapting to, and mitigating of natural hazards, and exploring space to better our lives on Earth and for the survival of our civilization. This book series prepares students to understand and contribute, to address these challenges. Because these challenges, even in most of their narrow focuses, transcend current college curriculums or programs. Earth's climate change is an example of the narrow focus of the large topic of natural hazards. It spans almost all aspects of the modern sciences, including biology, chemistry, and physics. Moreover, the span of these challenges may also be telling us that we have to rethink higher sciences education because the current splintering of science education into endless disciplines may not be the best or a unique way to prepare minds to address these challenges. We may be locking talented young minds to a certain viewpoint forever, an unintended indoctrination. At least some universities and colleges must start moving away from the monolithic way of delivering higher science education or to create a multidisciplinary science as a separate program and hopefully unleash a new generation of super engineers and scientists who are speaking scientific language. Moreover, this approach may optimize higher education time for some people.

Nature does not differentiate between physics, chemistry, and biology. Perhaps modern humans separate them for convenience's sake. The separation of sciences in biology, chemistry, and physics began in the 1800's primarily as recognition of expertise. Gradually, the split moved into basic science education in the 1900's and the boundaries between them artificially hardened in the late 1900's as college politics took center stage. People started to specialize more, and there was not any formal agreement on science's divisions. It just happened and spread from school to school, as adequate educators became available. Moreover, in the 1900's, terminology, techniques, and tools used in different subjects studied were sometimes wildly different. Even though specialization is still appropriate, there is significant overlap and cross-fertilization today. The cross-fertilization of three of these scientific areas of knowledge will continue to increase as we ascend to higher levels of knowledge.

The splintering of science education may not even be the most effective way of preparing young minds for the jobs of tomorrow. With the huge progress being made in the field of artificial intelligence (AI), the new jobs are about innovation. Broadly speaking, AI is a set of techniques that are aimed at approximating some aspects of human or animal cognition by inserting computer software into robots and other related systems. AI systems can nowadays perform many traditionally human tasks with high efficiency and without errors. Yet they are very far from being able to create. Very soon AI systems will become central to all science disciplines and the central platform of multidisciplinary studies. They are modern versions of computers, which extend beyond programming a set of scientific rules for a discipline-specific dataset. The AI systems can automatically learn these rules, improve them, and even create new ones. We can easily input different kinds of datasets to them and even introduce the rules for outcomes. So, while specialization and expertise are especially critical in graduate schools and adult life, a broad horizon is critical for expertise and innovation and to command and create AI systems. Moreover, how can one teach AI without having a background in the biology of life, chemistry, and physics? Likewise, how can one teach the biology of life, chemistry, or physics without incorporating AI?

Therefore, the objective of this book series is to provide an example of a text for undergraduate science education that prepares students to contribute to the challenges of improving the length and quality of lives on Earth, improving short- and long-distance communication and transportation, predicting of, adapting to, and mitigating natural hazards, and exploring space, to improve our lives on Earth and for the survival of our civilization. The main topics covered here are vacuum and matter, heat, chemical and nuclear reactions, porosity, permeability, viscosity, immersibility, wettability, gravity, electricity, and magnetism, acoustic and elastic waves, the biology of lives, and tomography. A basic background in matter, heat, chemical and nuclear reactions, gravity, electricity, and magnetism, and the biology of lives is central for present and future jobs in healthcare. Along with electromagnetic wave propagation in a vacuum and matter and elastic wave propagation in matter, one can develop a solid background for understanding energy flow, which is an essential tool in healthcare, communication, and transportation, natural hazard prediction and mitigation, and space exploration. Tomography and spectroscopy experiments are the basis of most technologies in medicine and mineral exploration. We demonstrate in the last part of this book series that the basic equations of physics and chemistry are the same everywhere in our solar system, and probably in our entire universe. Biological equations, including equations of lives, are probably also the same everywhere in the universe, although some of them are still unknown. Because of the multidisciplinary approach adopted in this text, the description of the atmosphere, hydrosphere, biosphere, and the lithosphere, along with solar energy, which is essential to the function of each of these spheres and their interactions, readily follows from these equations. The interconnectivity and interactions of these spheres are difficult to be taught in current programs, yet they are central to our livelihoods and come naturally in this text.

This text covers the materials of multiple current programs in a short space because these programs include huge duplications that we have avoided here. They also unnecessarily over-elaborate on some materials, which are actually just examples, exercises, or applications of a given topic, another issue avoided in this text. No chapter is divided into biology, chemistry, and physics. We define and motivate the importance of a topic or problem. Then we address it, both qualitatively and quantitatively, starting from the basic principles followed by a series of applications. I also mentioned various terminologies of the same quantity to accommodate readers with a fixed background from many current sciences and engineering disciplines. Finally, the book series contains analytical problems, as well as computational problems, that include MATLAB software and Mathematica software developed especially for the classes associated with this book series. The computational exercises focus on numerical modeling, imaging, and statistics. In some chapters, essays on topics of considerable broad interest and debate are discussed at the end of these chapters based on the background developed in this book series. These essays may help students see how the background that they have acquired plays a central role in understanding and addressing the major scientific and technological questions of our times.

The book series is intended for multi-semester course work. The first two chapters of the first book are broad. They intend to provide a broad perspective on some basic notions that science students are expected to be familiar with. They also motivate the rest of the book series and provide a clear indication of cross-fertilization of disciplines in the 21st century. In the other chapters, we motivate the topic, followed by a focus presentation of the topic, and then a broad set of examples of applications. For example, in Chapter 2 of the fourth book (i.e., this book), we discuss heat radiation, conduction, and convection. We then describe their applications in climate change, in determining the solid Earth temperature gradient. In addition, we describe the Earth's mantle convection, radiation of energy from the core of the Sun to its surfaces, and bioheat transfer. We also introduce the second law of thermodynamics and describe its application to efficiently develop heat and pump engines, including wind turbines.

The chapters are also closed related with a clear progression. For example, in Chapter 1 of the first book, we introduced oxygen in the atmosphere as essential for some lives on Earth. In Chapter 2, we described oxygen isotopes. In Chapter 3, we introduce subsurface oxides and minerals. In Chapter 1 of the second book, we distinguish living things that require atmospheric oxygen for their survival. We also present photosynthesis, rock burial in the ground, and other examples of chemical reactions involving oxygen and oxygen compounds, and so on. It does not make sense for people to call themselves engineers or scientists if they are not familiar with the basic features of oxygen gas and electromagnetic waves around us.

The book series is divided into five volumes. This fourth book has six chapters. We continue to introduce, in significant detail, some of the core fundamental notions that underpin modern science and engineering education and scientific training. These notions help address some major challenges of our time. These notions are (1) gravitational forces, (2) Kleeper's laws, (3) Lagrangian points, (4) the Geoid, (5) dark matter and dark energy, (5) Archimedes' principle, (6) entropy, (7) the two laws of thermodynamics, (8) diffusion equations, (9) electromagnetic radiation, (10) quantum entanglement, (11) spectroscopy, (12) terahertz manipulation, (13) radio and optical telescopes, (14) chemical composition of stars, and (15) Bell's inequality. We then illustrate, with applications across disciplines, the importance of these notions in our lives and in understanding the world around us. These include monitoring the deformation of the solid Earth's surface using satellite measurements, ocean tides, monitoring earthquakes and volcanic eruptions, unusual gravity anomalies in Antarctica, a view of the far side of our Moon, geotherms, energy transfer in the stars, Earth's climate, Titan's anti-greenhouse effect, spontaneous electricity in human body, spontaneous electricity in solid Earth, spontaneous electricity in atmosphere and vacuum, long-distance communication between Earth and the planets and exoplanets, space travel, quantification of the notion of conscious awareness, characterizing the far side of the moon, finding life on the TRAPPIST-1e, TRAPPIST-1f, and TRAPPIST-1g planets, thermal medical diagnostics and treatment, terahertz spectroscopy, finding the ideal location on Earth for a human body with slow muscle loss with age, listening for extraterrestrial signals, quantum computing, spontaneous production electricity by nuclear fission in Gabon, bio- and light-signatures of extraterrestrial civilizations, and ways of attracting extraterrestrial civilizations to Earth.

Author Biography

Dr. Luc Thomas Ikelle is a scientist working for Imode Energy. He is also an adjunct professor at the Department of Geology and Geophysics at Texas A&M University. Previously, he worked at Cray Research Inc. in Minneapolis, developing 3D seismic inversion algorithms for CRAY Y-PM. From 1988 to 1997, he was employed as a scientist by Schlumberger Geco-Prakla, Schlumberger Doll Research, and Schlumberger Cambridge Research. From 1997 to 2014, he was Robert R. Berg Professor at the Department of Geology and Geophysics at Texas A&M University. He earned a PhD in geophysics and geochemistry from Paris 7 University in France. He was awarded Le Prix de Thesis du CNRS in 1986 for the PhD thesis. He received an SEG award in 2010 for his contribution to the creation of Geoscientists Without Borders. He was also awarded a Texas A&M University award as an outstanding scientist in 2012.

He received the 2023 Reginald Fessenden Award from the Society for Exploration Geophysics. The Reginald Fessenden Award is awarded to a person who has made a specific contribution to exploration geophysics, such as invention or theoretical or conceptual advancement, which merits special recognition. He is the co-author of *Introduction to Petroleum Seismology [2005 edition (an SEG high-selling seller) and 2018 edition]*, author of *Coding and Decoding: Seismic Data, The Concept of Multishooting* (2010 and 2017 editions), and *Introduction to Earth Sciences: A Physics Approach* (2017 and 2020 editions). He is also a co-founder of Geoscientists Without Borders and Imode Energy Research. He is a member of SEG, AGU, and APS. He worked as a DOE (US Department of Energy) special employer from 2005 to 2012 and was a member of the Ultra-Deepwater Advisory Committee (an advisory committee to the Secretary of Energy) from 2005 to 2012. In most seismology studies, the cost of data collection far outweighs the cost of data analysis. Therefore, it is important to use the most efficient and accurate techniques to collect seismic and EM data. In 1998, when he set up the CASP consortium at Texas A&M University, he started working on the topic of near simultaneous multiple shooting (or multishooting). This is based on the idea of generating waves from several source locations that are nearly simultaneous instead of from one single-source location at a time. He filed the first US patent on multiple shot acquisition and processing, granted in 2001, and the first book on the topic in 2010.

Chapter 1

Gravitational Forces and Geodesy

There are four fundamental forces—that is, gravity, electromagnetism, and the weak and strong nuclear forces—responsible for every event, action, and reaction that occurs in our ordinary matter universe. Gravity is the dominant force in our universe because it shapes the large-scale structure of galaxies, stars, planets, exoplanets, and black holes.

In addition to being an extremely powerful force, gravity is an invisible force. It keeps us at a comfortable distance to enjoy the Sun's light, keeps the planets in orbit around the Sun, and holds our world together (see Figure 1.1). Earth's gravity keeps us on the ground and makes objects fall. Its spatial variations have played a central role in the study of dynamic processes in the Earth's interior and in the exploration and production of the Earth's interior. Our objective here is to introduce gravitational forces and laws. These forces and laws are covered in the first section of this chapter. The later sections discuss applications of these laws, including satellite-based gravity measurements.

Key sentence: *It looks like there is no fixed reference on Earth to which all points on Earth can be defined: It looks more and more that the increasing amount of geodesy data will simply help to refine the motion of the center of the Earth, but it will not help to define a time-independent fixed reference on Earth. We are stuck with relative motions.*

1.1. Newton's Universal Law of Gravitation

1.1.1. The Gravitational Field

In 1687, Isaac Newton discovered that masses attract one another. He formulated his discovery into an equation called Newton's law of gravitation. This equation explains that the Earth's pull or attractive force that we are all familiar with is the force that causes objects to fall. For any two objects with masses m_1 at position x_0 (i.e., the center of mass m_1) and m_2 at position x (i.e., the center of mass m_2)

DOI: 10.1201/9781032620619-1

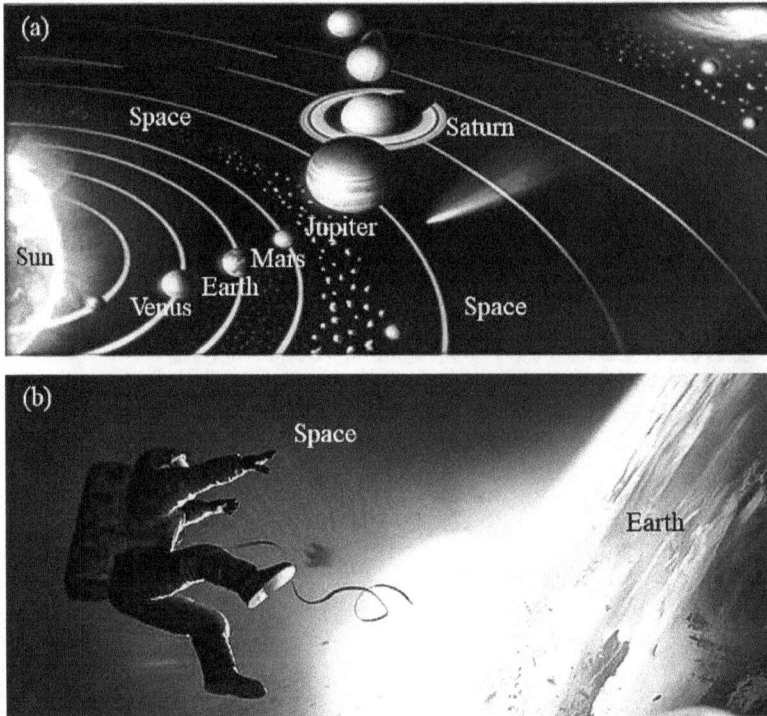

Figure 1.1. (a) Here is a partial illustration of the solar system. It is the Sun's gravity that keeps the Earth in orbit around the Sun. So gravity helps the Earth to stay just the right distance from the Sun and therefore it is not too cold or hot. Venus is too hot and Mars is too cold. Note that there is no gravity in space. (b) Here is an illustration of an astronaut in space. The gravity of planets is so small that they appears weightless and float in space instead of falling down. The weightless effect in space is such that astronauts can move equipment weighing hundreds of kilograms with their fingertips.

the gravitational force, $\boldsymbol{F}$, between them is

$$\boldsymbol{F} = -G\frac{m_1 m_2}{|\boldsymbol{x} - \boldsymbol{x}_0|^3}(\boldsymbol{x} - \boldsymbol{x}_0) \; , \tag{1.1}$$

with a magnitude

$$F = |\boldsymbol{F}| = G\frac{m_1 m_2}{|\boldsymbol{x} - \boldsymbol{x}_0|^2} \; , \tag{1.2}$$

where $G = 6.673 \times 10^{-11}$ m^3 kg^{-1} s^{-2} (or N m^2 kg^{-2}) is the universal gravitational constant; G has the same value anywhere in our universe for all pairs of objects and it does not vary with time. The minus sign accounts for the fact that the force vector $\boldsymbol{F}$ points inward (i.e., toward m_1) whereas the vector $\boldsymbol{x} - \boldsymbol{x}_0$ points outward (away from m_1) as illustrated in Figure 1.2a. So Newton's law of universal gravitation in (1.1) tells us that every object in our universe attracts every other object with a force directed along the line of centers for the two objects. This force is proportional to the product of their masses and inversely proportional to the square of the separation between the two objects because of the $1/|\boldsymbol{x} - \boldsymbol{x}_0|^2$ term.

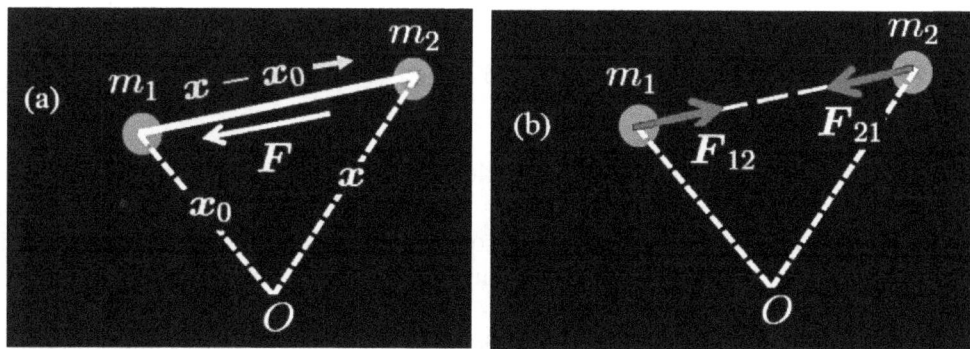

Figure 1.2. (a) Newton's law of gravitation in vector form. Mass m_2 is attracted to mass m_1. (b) The gravitational forces on m_1 and m_2. $\boldsymbol{F}_{12}$ is the force that m_1 exerts on m_2, F_{21} is the force that m_2 exerts on m_1, and $\boldsymbol{F}_{12} = -\boldsymbol{F}_{21}$ according to the third Newton's law of motion. Basically, the two forces are an action/reaction pair. r is the distance between the centers of masses.

If we define the distance between the centers of the two masses as $r = |\boldsymbol{x} - \boldsymbol{x}_0|$, then

$$F = G\frac{m_1 m_2}{r^2} \, . \tag{1.3}$$

Before we continue our discussion on gravitational forces, let us recall that there are three additional well-known Newton's laws of motion that are different from Newton's universal law of gravitation in equation (1.1). The first law of motion, also known as inertia, says that a body will continue in its state of rest or uniform motion unless acted upon by a net force. The second law states that the net force acting on a body equals mass times acceleration (i.e., $\boldsymbol{F} = m \times \boldsymbol{\gamma}$, where m is the mass of the object and $\boldsymbol{\gamma}$ is its acceleration). The third law states that for every force there is an equal and opposite force; in other words, for every action, there is an equal and opposite reaction, as illustrated in Figure 1.2b.

Example #1: Consider two 1-kg objects that are 1 m apart in space. The gravitational force between the two objects is

$$F \approx \frac{\left(6.67 \times 10^{-11}\,\mathrm{N\,m^2\,kg^{-2}}\right)(1\,\mathrm{kg})\,(1\,\mathrm{kg})}{(1\,\mathrm{m})^2} = 6.67 \times 10^{-11}\,\mathrm{N} \, . \tag{1.4}$$

This force is very, very small indeed. It will take a long time for these two masses to come together, but they will eventually come together. You may be wondering if this example can be carried out with the two masses on the Earth's surface. The answer is no because the static friction forces are several billions of times greater than this gravitational force.

The reason why the gravitational force between the 1-kg masses is so small is that G is very small compared to these masses. Instead of considering two 1-kg masses, consider, for example, one object with 1-kg mass at 20 m above the Earth's surface and the second object to be the Earth ($m_1 = 6 \times 10^{24}$ kg). The gravitational force between this object and the Earth is now

$$F \approx \frac{\left(6.67 \times 10^{-11}\,\mathrm{N\,m^2\,kg^{-2}}\right)\left(6 \times 10^{24}\,\mathrm{kg}\right)(1\,\mathrm{kg})}{\left(6.371 \times 10^6\,\mathrm{m}\right)^2} = 9.82\,\mathrm{N}. \qquad (1.5)$$

If you let go of the 1-kg mass, the gravitational force acting on it immediately pulls it toward the surface because this force is large enough. Note that the separation between the two masses is almost equal to the mean radius of the Earth which is $r = |\boldsymbol{x} - \boldsymbol{x}_0| = 6.371 \times 10^6$ m and the 20-m distance above the surface of the Earth is negligible compared to the radius of the Earth.

Let us make this example more practical by using a 300-kg satellite above the surface of the Earth at a height equal to the Earth's radius (i.e., $r = |\boldsymbol{x} - \boldsymbol{x}_0| = 2 \times 6.371 \times 10^6$ m) for mass m_2. The magnitude of the gravitational force exerted on the satellite by the Earth is

$$F \approx \frac{\left(6.67 \times 10^{-11}\,\mathrm{N\,m^2\,kg^{-2}}\right)(300\,\mathrm{kg})\left(6 \times 10^{24}\,\mathrm{kg}\right)}{\left(2 \times 6.371 \times 10^6\,\mathrm{m}\right)^2} = 7.42 \times 10^2\,\mathrm{N} \ .$$

Remember that when an object is moving in a circular path, even at a constant speed like the satellite one in orbit, the object is accelerating because its velocity is changing direction. The acceleration of the object is $\gamma = v^2/r$, where r is the radius of the circular path and $v = |\boldsymbol{v}|$ is the (constant) speed of the object. Since gravitational force is always directed toward the center of the Earth (see Figure 1.3) and since this is the only force that acts on the satellite, we can deduce the satellite's orbital speed:

$$\frac{m_2 v^2}{r} = F = G\frac{m_1 m_2}{r^2} = 7.42 \times 10^2\,\mathrm{N}$$

or

$$v = \sqrt{\frac{Gm_1}{r}} = 5.6\,\mathrm{km/s} \ . \qquad (1.6)$$

And we can deduce the period of the satellite's orbit as

$$T = \frac{2\pi r}{v} = 1.42 \times 10^4\,\mathrm{s} = 3.96\,\mathrm{hours}. \qquad (1.7)$$

Remember that $2\pi r$ is the circumference of a circle. We have assumed here that the orbital motion of a satellite is circular. Actually, it is elliptic. But in most cases, the elliptic orbit differs only slightly from being circular.

Example #2: Let us look at the attractive force between the Earth and the Sun. The magnitude of the gravitational force exerted on the Earth by the Sun given the mass of the Earth ($m_2 = 6 \times 10^{24}$ kg), the mass of the Sun ($m_1 = 2 \times 10^{30}$ kg), and the distance between them ($r = |\boldsymbol{x} - \boldsymbol{x}_0| = 1.5 \times 10^{11}$ m) is

$$F \approx \frac{\left(6.67 \times 10^{-11}\,\mathrm{N\,m^2\,kg^{-2}}\right)\left(6 \times 10^{24}\,\mathrm{kg}\right)\left(2 \times 10^{30}\,\mathrm{kg}\right)}{\left(1.5 \times 10^{11}\,\mathrm{m}\right)^2}$$

$$= 3.5 \times 10^{22}\,\mathrm{N} \ . \qquad (1.8)$$

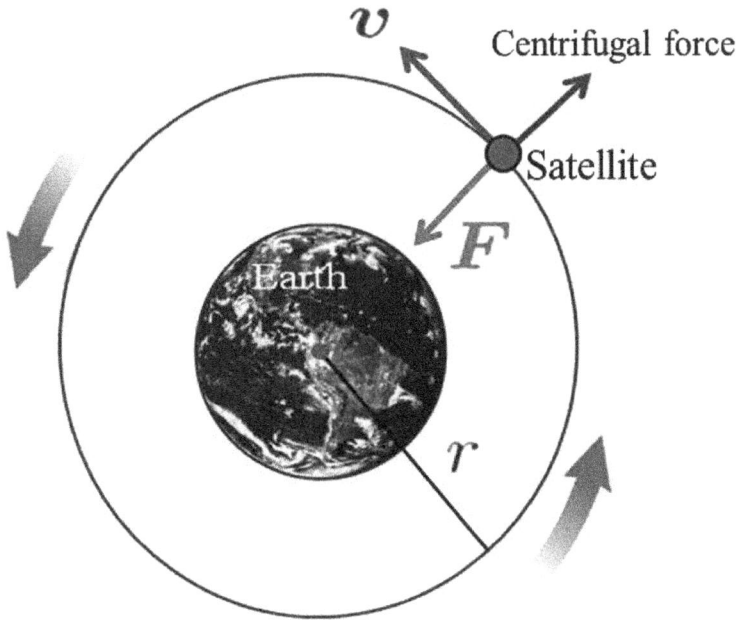

Figure 1.3. We consider the orbital motion of a satellite due to the force of gravity. The satellite must have speed $\sqrt{Gm_1/r}$.

Therefore, the magnitude of the gravitational force exerted on the Earth by the Sun is 3.5×10^{22}; a huge force. Note that formulas for the satellite's orbital speed in (1.6) and for the period of the satellite's orbit in (1.7) are also valid for orbiting planets (i.e., mass m_2) around the Sun. We just replace the orbiting satellite with an orbiting planet, and the planet (like the Earth) with the sun.

Let us now return to the discussion of Newton's law universal gravitation in (1.1). This law has been shown to be experimentally valid at a wide range of distances about 100 years since its formulation. This validation was carried out by Henry Cavendish. The key stumbling block that caused this delay was the measurement of the universal gravitational constant, G. In 1779, Cavendish published a paper titled "Weighing the Earth" in which he reports the first measurements of G obtained by measuring the tiny attractive forces between lead spheres of known masses and known distances between masses. Decades later, many measurements were carried out of the universal gravitational constant at different locations on the Earth and the invariance of the universal gravitational constant was established.

> **Quiz:** Suppose it takes a planet 3.09 days to make one orbit (assumed to be circular) around the Sun. The distance of this planet from our Sun is 6.433×10^6 km. (1) How fast is this planet moving? (2) What is the mass of the planet?

Cavendish realized that by knowing the value for G, the mass of the Earth can be determined. That is why he titled his 1779 paper "Weighing the Earth." He found that the Earth's mean density, $\rho \approx 5,500 \, \text{kg/m}^3$, is much larger than the density of rocks at the Earth's surface, which have densities between 1,500 and 3,500 kg/m^3 with extreme values up to 4,000 kg/m^3 in massive ore deposits.

This observation was one of the first strong indications that density must increase substantially toward the center of the Earth.

Constant G must not be confused with the acceleration vector $\boldsymbol{g}$, also known as the gravity field or simply as gravity, which is gravitational acceleration, or force on a unit of mass due to gravity. The mathematical expression for $\boldsymbol{g}$ can be obtained by using the Newton's second law of motion. If m_1 is the mass of the Earth, the gravitational acceleration, or force of a unit of mass due to gravity, can be obtained by equating the gravitational force exerted on mass m_2 by the Earth with mass m_2 times gravitational acceleration; i.e.,

$$\boldsymbol{F} = m_2\boldsymbol{g} = -G\frac{m_1 m_2}{|\boldsymbol{x} - \boldsymbol{x}_0|^3}(\boldsymbol{x} - \boldsymbol{x}_0) \implies \boldsymbol{g} = -G\frac{m_1}{|\boldsymbol{x} - \boldsymbol{x}_0|^3}(\boldsymbol{x} - \boldsymbol{x}_0),$$

with

$$g = |\boldsymbol{g}| = G\frac{m_1}{|\boldsymbol{x} - \boldsymbol{x}_0|^2} = G\frac{m_1}{r^2}. \tag{1.9}$$

So $\boldsymbol{g}$ points toward the center of the Earth, i.e., in the $-(\boldsymbol{x} - \boldsymbol{x}_0)$ direction. The gravitational acceleration was first determined by Galileo (1589). He was the first to realize that a falling body picked up speed at a constant rate (i.e., it had constant acceleration). The unit of gravity, *Gal*, was adopted to honor him. The magnitude of $\boldsymbol{g}$ varies over the surface of the Earth, but on average it is $g = 9.8\,\text{m/s}^2 = 9.8 \times 10^2\,\text{Gal}$. We obtained this value in the first example of this section as a gravitational force between one object with 1-kg mass (i.e., a unit mass) and the Earth.

Let us make some observations about gravitational acceleration. We can compute the mass of the Earth, Venus, Mars, or asteroids if their radius and gravitational acceleration are known. For example, we can calculate the Earth's mass, assuming that the gravitational field strength on the surface of the Earth is known (e.g., $g = 9.8\,\text{m/s}^2$) and the distance from the center of the Earth to the Earth's surface is approximately 6371 kilometers therefore

$$m_1 = \frac{gR^2}{G} \approx 5.964 \times 10^{24}\,\text{kg}, \tag{1.10}$$

and we can deduce the Earth's density as

$$\rho = \frac{m_1}{\frac{4}{3}\pi R^3} = 5493\,\text{kg/m}^3. \tag{1.11}$$

Remember that the volume of the sphere is $V = (4/3)\pi R^3$.

Another significant observation is that mass does not depend on gravity. It is a fundamental quantity throughout our universe, and it is the same everywhere at a fixed time. However, the weight of an object depends on the force of gravity at that point and the force of gravity varies with elevation, rock densities, latitude, and topography. Remember that weight is the force of gravity exerted on a body (i.e., $w = mg$, where w is the weight and g is the gravitational acceleration). If you weigh 600 Newtons on Earth, you will weigh 150 Newtons on planet B which has one-tenth of the mass of Earth and half the radius of Earth.

$$\text{On Earth}: g_e = G\frac{m_e}{\|\boldsymbol{x}_e\|^2}$$

On planet B : $g_b = G \dfrac{m_b}{\|\boldsymbol{x}_b\|^2} = G \dfrac{0.1\,m_e}{(0.5\|\boldsymbol{x}_e\|)^2} = G \dfrac{m_e}{4\|\boldsymbol{x}_e\|^2} = \dfrac{1}{4} g_e$.

The gravitational acceleration of planet B is one-fourth of the gravitational acceleration of the Earth. As the mass is unchanged, your weight on planet B (i.e., $w_b = m_e g_b = 0.25\,m_e g_e$) is one-fourth of your weight on Earth (i.e., $w_e = m_e g_e$).

Again the force of gravity varies with elevation, rock densities, latitude, and topography. In life, we often express weight in kilograms and pounds because we assume that g is constant and we implicitly use w/g instead of w. An 80-kg man weighs less than 80 kg Cameroon (19 m above sea level) and more than 80 kg in Mexico (10 m below sea level). Therefore, the farther from the center of the Earth, the less one weighs, because w/g is changing but the actual mass is unchanged.

Our Moon has about 1/81 of the mass of the Earth. Therefore, you might think its gravity would be 81 times less. But its radius is only 1/3.7 times that of the Earth. Remember, gravity, as defined in (1.9) is an inverse square law because of the $1/r^2$ relationship, so from the small radius, you would expect the Moon's gravity to be $(3.7)^2 = 13.7$ times larger. By combining these two effects, we can verify that the Moon's surface gravity is about 1/6 times that of the Earth ($g_m = g_e/6$). When astronauts landed on the Moon for the first time, they realized how weak the gravity was and they could bounce around in slow motion. When they jumped, they went high, and when they came back down, they came down slowly. It is this gravity that keeps the Moon in orbit around the Earth, and the Earth in orbit around the Sun (see Figure 1.4). If gravity was turned off, satellites would move in straight lines, rather than in circles. Even at long distances from the Earth or on other planets, the force of gravity never drops down to zero. The sensation of weightlessness felt by astronauts in very low-gravity areas of outer space (see Figure 1.1) is really the sensation of continuous falling.

Let us close this subsection with examples of calculations of the gravitational field $\boldsymbol{g}$ near and inside spherical bodies as such bodies are useful when dealing with the Earth, other planets, as well as the Sun and the stars, which can all be approximated as being spherically symmetric.

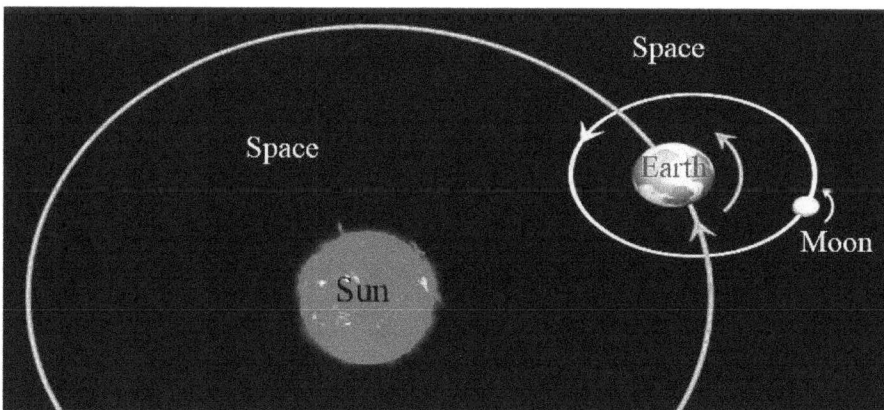

Figure 1.4. We consider Mars and Venus, Earth orbits around the Sun. The Moon orbits around the Earth and not the Sun.

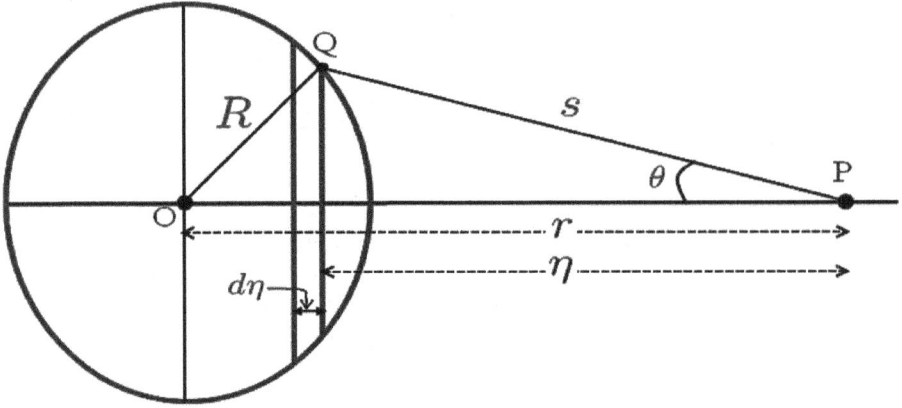

Figure 1.5. We consider the gravitational field outside a hollow spherical shell of radius R, with a surface density sigma, and a point P at a distance r from the center of the sphere. We can compute the gravitational force using the elemental zone of thickness $d\eta$.

Example #3: Consider a hollow spherical shell of radius R with surface density σ and a point P at a distance r from the center of the sphere, as illustrated in Figure 1.5. The mass of the spherical shell is $m_1 = \sigma 4\pi R^2$; remember that the area of a sphere is $A = 4\pi R^2$. We divided the spherical shell into narrow zones of thickness $d\eta$. The mass of this element is $\delta m = \sigma 2\pi R d\eta$, where $2\pi R d\eta$ is the area of the rectangle describing the elementary zone. The field due to this zone, in the direction PO is

$$dg = G\frac{\delta m}{s^2}\cos\theta \ . \tag{1.12}$$

Let us express dg in terms of a single variable, s. By using the law of cosines of a triangle, which describes OQP in Figure 1.5, we have

$$R^2 = r^2 + s^2 - 2rs\cos\theta = r^2 + s^2 - 2r\eta \tag{1.13}$$

and

$$\cos\theta = \frac{r^2 - R^2 + s^2}{2rs} \ . \tag{1.14}$$

We can differentiate (1.13) with respect to η to obtain

$$\frac{ds}{d\eta} = \frac{r}{s} \ . \tag{1.15}$$

By substituting (1.13), (1.14), and (1.15) into (1.12), the field at P due to the elementary zone, in Figure 1.5 is

$$dg = G\frac{\delta m}{s^2}\cos\theta = G\frac{\sigma 2\pi R d\eta}{s^2}\cos\theta = \frac{\pi R G\sigma}{r^2}\left(1 + \frac{r^2 - R^2}{s^2}\right)ds \ .$$

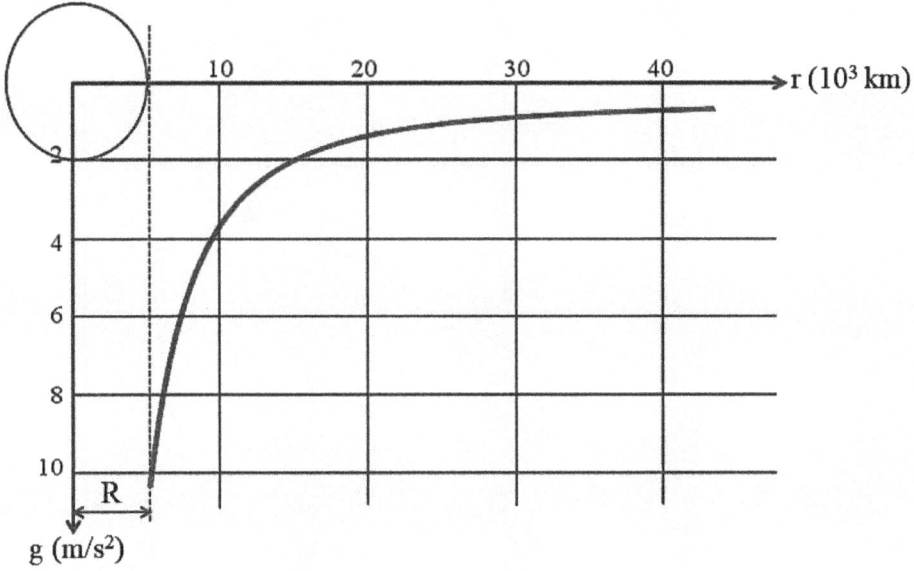

Figure 1.6. We consider the gravitational field of a hollow shell. Inside the hollow sphere, the gravitational field is zero. The gravitational field outside a hollow spherical shell is $-Gm_1/r^2 \mathbf{i}_r$ toward the center. It can be treated as if the entire mass of the sphere is concentrated at its center.

If P is an external point, to find the field due to the entire spherical shell, we integrate dg from $r - R$ to $r + R$; i.e.,

$$g(r) = \frac{Gm_1}{4Rr^2} \int_{r-R}^{r+R} \left(1 + \frac{r^2 - R^2}{s^2}\right) ds \quad \text{by using} \ \ m_1 = \sigma 4\pi R^2$$

$$= \frac{Gm_1}{4Rr^2}\left[s - \frac{r^2 - R^2}{s}\right]_{r-R}^{r+R} = \frac{Gm_1}{r^2} . \tag{1.16}$$

The gravitational attraction is as if the whole sphere were one particle at the center of the sphere and of the same mass as the sphere. Therefore, outside a hollow sphere, we can treat the sphere as if its entire mass was concentrated at the center (see Figure 1.6). But if P is an internal point, to find the field, due to the entire spherical shell, we integrate from $R - r$ to $R + r$; i.e.,

$$g(r) = \frac{Gm_1}{4Rr^2} \int_{R-r}^{R+r} \left(1 + \frac{r^2 - R^2}{s^2}\right) ds$$

$$= \frac{Gm_1}{4Rr^2}\left[s - \frac{r^2 - R^2}{s}\right]_{R-r}^{R+r} = 0 . \tag{1.17}$$

Inside a hollow sphere, the gravitational field is zero (see Figure 1.6). Figure 1.7 illustrates this result. For a given point P inside the spherical shell, there are two small areas, A_1 and A_2, on opposite sides, which exert gravitational attraction on

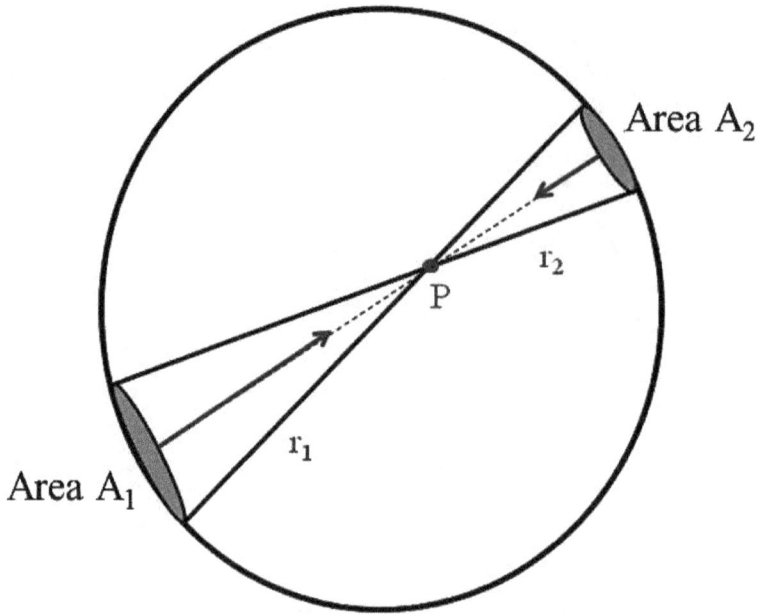

Figure 1.7. We consider the gravitational force inside a hollow spherical shell of radius R, with a surface density sigma, and a point P at a distance r from the center of the sphere. We can compute the gravitational force using the elemental zone of thickness. Two small areas of the shell on opposite sides exert gravitational force attraction on a mass at P in opposite directions which cancel each other out so that the force of gravity inside a spherical shell is zero.

a mass at P in opposing directions. It turns out that they cancel because the ratio of the areas A_1 and A_2 at distances r_1 and r_2 are proportional to r_1^2 and r_2^2, respectively, therefore the two opposite areas have equal magnitude gravitational force,

$$\frac{G\sigma A_1}{r_1^2} = \frac{G\sigma A_2}{r_2^2} \ , \tag{1.18}$$

but opposite gravitational fields at P. In fact, the gravitational pull from every small part of the shell is balanced by a part on the opposite side.

Let us now consider a solid sphere. A solid sphere is just lots of hollow spheres nested together like layers of an onion. Therefore, the field at an external point is just the same as if all the mass was concentrated at the center (see Figure 1.8). This is true not only for a sphere of uniform density, but for any sphere in which the density depends only on the distance from the center, i.e., any spherically symmetric distribution of matter. This justifies Newton's treatment of the Sun and planets as point masses when analyzing planetary motion.

Inside the solid sphere (i.e., $r \leq R$), the mass m_{1r} contained in the volume V_r with radius r is (see Figure 1.9)

$$m_{1r} = \rho V_r = \rho \frac{4}{3}\pi r^3 = \rho \left(\frac{4}{3}\pi R^3\right) \left(\frac{r}{R}\right)^3 = m_1 \left(\frac{r}{R}\right)^3 \ , \tag{1.19}$$

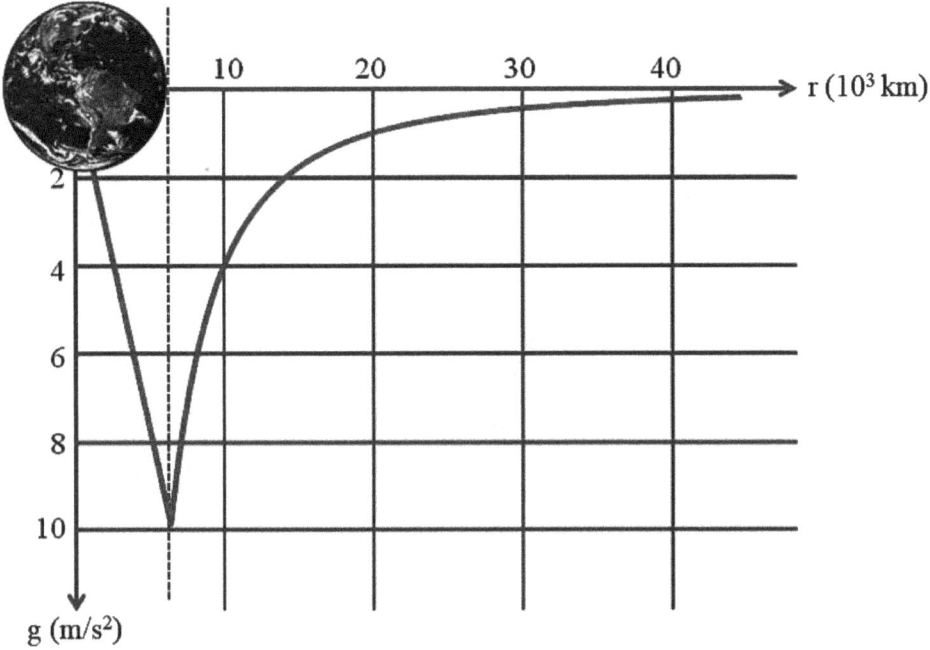

Figure 1.8. We consider the gravitational field of a dense sphere. The gravitational field outside a uniform solid sphere is $-Gm_1/r^2 \boldsymbol{i}_r$ toward the center and linearly proportional to the distance from the center of the sphere. Outside a solid sphere, we can also treat the sphere as if all the mass is at the center of the sphere. Inside the sphere, only the mass inside the radius of interest counts.

where m_1 is the mass of the entire solid sphere. The magnitude of the gravitational force inside the sphere is then

$$g = \frac{Gm_{1r}}{r^2} = \frac{Gm_1 r}{R^3} \; ; \tag{1.20}$$

that is, the gravitational field inside the Earth, for example, is linearly proportional to the distance from the center (see Figure 1.8). Therefore, there is no gravitational field or force at all at the center of the Earth as $r = 0$. In all directions, the masses are attracted equally.

The acceleration due to gravity at radius $r < R$ inside the Earth (assuming that it is spherically symmetric) can alternatively be written as

$$g(r) = \frac{G}{r^2} \int_0^r 4\pi\eta^2 \rho(\eta) d\eta = \frac{G}{r^2} M(r) \; , \tag{1.21}$$

where

$$M(r) = \int_0^r 4\pi\eta^2 \rho(\eta) d\eta \; . \tag{1.22}$$

By assuming that the density inside the Earth is uniform (i.e., $\rho = $ constant), we arrive at (1.20).

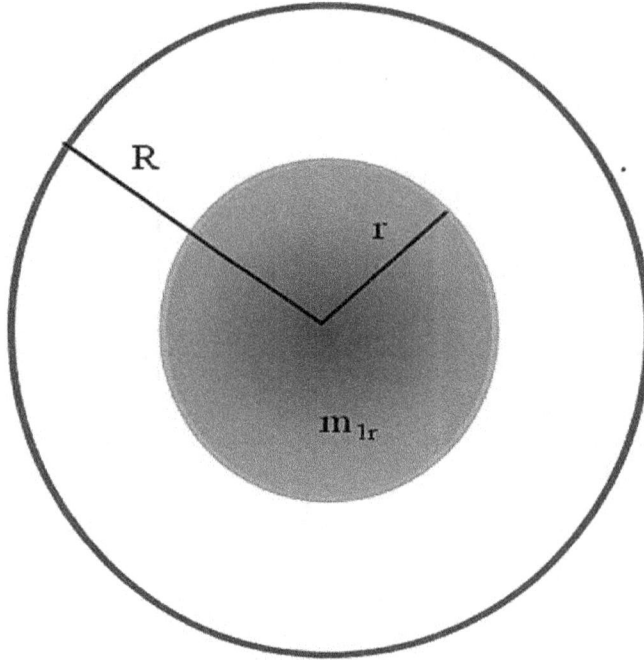

Figure 1.9. An illustration of the fraction of the solid-sphere mass used in the calculations of the gravitational field inside a solid sphere. Inside the sphere, only the mass inside the radius of interest m_{1r}, counts.

Quiz: What is the force of Earth's gravity on a 100-kg man at the to the top of Mount Everest ($h = 8850$ m)?

Example #4: Free-fall time and cloud collapse. Consider a sphere made of a cloud of hydrogen gases. We assume that the cloud is incompressible. As described in Ikelle (2023c), the motion of the gas can be expressed by the Navier Stoke equation as follows:

$$\rho \frac{\partial \boldsymbol{v}}{\partial t} = \underbrace{-\rho(\boldsymbol{v} \cdot \boldsymbol{\nabla})\boldsymbol{v}}_{\text{Convection}} + \underbrace{-\boldsymbol{\nabla} p}_{\text{Gas pressure}} + \underbrace{\eta \nabla^2 \boldsymbol{v}}_{\text{Viscosity}} + \underbrace{\rho \boldsymbol{g}}_{\text{Gravity}} + \underbrace{\frac{1}{c}(\boldsymbol{j} \times \boldsymbol{B})}_{\text{Lorentz}} + \underbrace{\rho \boldsymbol{F}}_{\text{External force}} \quad (1.23)$$

The main force here is the attractive gravitational forces and the opposite forces to it are gas pressure, radiation pressure, magnetic force (molecular clouds are magnetized), and fluid-flow forces due to turbulence. Suppose that the attractive gravitational force is greater than the sum of opposite forces (gas cloud internal pressure, magnetic force, external pressure, and other external forces). At some point, the gas cloud will collapse at the center of the cloud. How long does it take for most particles to collapse at the center? We can get a rough idea of this time, which is generally called free-fall time, by ignoring all the opposite forces; i.e., (1.23) reduces to

$$\rho \frac{\partial \boldsymbol{v}}{\partial t} = \rho \boldsymbol{g} \; . \quad (1.24)$$

Our objective in this example is to estimate the free-fall time in this case. The gravity in this case varies only radially, as defined in (1.21). Equation (1.24) can be rewritten, as follows:

$$\frac{d^2r}{dt^2} = -\frac{GM(r)}{r^2} \iff \frac{1}{2}\frac{d(v^2)}{dr} = -\frac{GM(r)}{r^2} \quad \text{or} \quad \frac{1}{2}d(v^2) = -\left(\frac{GM(r)}{r^2}\right)dr \;, \quad (1.25)$$

using the fact that

$$\frac{d^2r}{dt^2} = \frac{d}{dt}\left(\frac{dr}{dt}\right) = \frac{dv(r)}{dt} = \frac{dr}{dt}\frac{dv(r)}{dr} = \frac{1}{2}\frac{d(v^2)}{dr} \;. \quad (1.26)$$

By integration, using the last equation, in (1.25), we obtain

$$v^2 = 2GM(R)\left(\frac{1}{r} - \frac{1}{R}\right) \iff v = \frac{dr}{dt} = \pm\left[2GM(R)\left(\frac{1}{r} - \frac{1}{R}\right)\right]^{1/2} \;.$$

The constant of integration was chosen based on the initial conditions $v(t = 0) = 0$, $r(t = 0) = R$. By choosing the solution with $v < 0$ with the substitutions $\zeta = r/R$ and $d\zeta = dr/R$, we arrive at

$$dt = -\left[\frac{2GM(R)}{R^3}\right]^{-1/2}\left(\frac{\zeta}{1-\zeta}\right)^{1/2}d\zeta \;. \quad (1.27)$$

Integrating the right-hand side from $\zeta = 1$ to $\zeta = 0$ (i.e. $r = R$ to $r = 0$) and using the fact that $\rho_0 = 3M(R)/(4\pi R^3)$, we obtain the so-called free-fall time t_{ff} of the cloud,

$$\begin{aligned} t_{ff} &= -\left[\frac{3}{8\pi G\rho_0}\right]^{1/2}\int_1^0\left(\frac{\zeta}{1-\zeta}\right)^{1/2}d\zeta \\ &= -2\left[\frac{3}{8\pi G\rho_0}\right]^{1/2}\int_{\pi/2}^0 \sin^2\theta d\theta \\ &= \left(\frac{3\pi}{32G\rho_0}\right)^{1/2} \;. \end{aligned} \quad (1.28)$$

That is, the time it takes until the (pressureless) cloud shrinks to a point at $r = 0$. The integral can be calculated analytically by using the substitution $\zeta = \sin^2\theta$. Again, note that in the actual interstellar space, the gas pressure is not zero. However, t_{ff} gives a rough time-scale for a gas cloud to collapse and to form stars in it. For $\rho_0 = 4.7 \times 10^{-19}$ g/cm^{-3}, the free-fall time is

$$\begin{aligned} t_{ff;sun} &= \left(\frac{3\pi}{32 \times 6.7 \times 10^{-8} \times 4.7 \times 10^{-19}}\right)^{1/2} \\ &= 3.058 \times 10^{12} \text{ seconds} \approx 98,315 \text{ years} \;. \end{aligned} \quad (1.29)$$

This time is very fast by astronomical standards. We should therefore have to be very lucky to catch a star in the act of formation. In other words, without the opposite forces, especially gas pressure, this cloud would collapse very quickly! Note also that the free-fall time depends only on the average density of the cloud, ρ_0, and not on the actual distribution of particles in the cloud.

Box 1.1: Fusion by Gravitational Confinement

As discussed in Ikelle (2023b), nuclear fusion powers the stars, including our Sun. It collides two lightweight nuclei to create two new particles with the release of energy. Here is one example discussed in Ikelle (2023b):

$$^2_1\text{D} + ^3_1\text{T} \rightarrow ^4_2\text{He} + ^1_0\text{n} + \gamma\,(17.6\,\text{Mev})\,, \tag{1.30}$$

where D is deuterium and T is tritium, and $^2_1\text{D} = ^2_1\text{H}$ and $^3_1\text{T} = ^3_1\text{H}$. Tritium decays by emitting an electron (beta radiation) and has a half-life of 12.3 years. Tritium occurs both naturally and as a by-product in nuclear fission reactors. Deuterium and tritium are also known as the heavy isotopes of hydrogen. The 17.6 MeV of energy is split between the kinetic energy of the neutron (14.1 MeV) and the helium nucleus (3.5 MeV). In the stars, the fusion is accomplished by the increasing gravitational pressure ($p = F/A$, where F is the magnitude of the gravitational force and A is the area on which the gravitational force is applied; and pressure, p, is proportional to the mass of the star just like F), as they burn their hydrogen, to 2 or more billion atmospheres. Remember that this process takes place in the vacuum and at very high pressures, which imply at very high temperatures (i.e., more than 10 million degrees Kelvin). The challenges of accomplishing fusion on Earth include the construction of a very-high-temperature environment and the confinement of the very hot plasma in vacuum—that is, the plasma cannot be in direct contact with any solid material. Note that, if the object were completely composed of deuterium, the minimum mass needed to generate gravitational pressures sufficient to initiate fusion would be equivalent to the mass of the planet Jupiter. This type of mass is only found in the stars. Realistic fusions are discussed in Ikelle (2024).

1.1.2. Kepler's Three Laws of Planetary Motion

In the early 1600's, Johannes Kepler (1571–1630) formulated three laws of celestial mechanics that revolutionized the scientific world and that led Newton to the law of gravity, which we described in the previous subsection. Pondering over data collected by Tycho Brahe (1546–1601), the first experimental physicist Kepler identified three laws that govern the dynamics of planetary orbits.

The first law, also known as the law of ellipses, states that the planets all move in elliptical orbits with the Sun as one of the focuses. *The second law*, which is known as the law of equal areas, says that, as a planet moves in its orbit, the line from the center of the Sun to the center of the planet sweeps out equal areas in equal time. *The third law*, also referred to as the harmonic law, states that the time, T, it takes a planet to make one complete orbit around the Sun (one planet year) is related to its average distance from the Sun, r—that is, the square of the orbital period is proportional to the cube of the average distance between the Sun and the orbiting planet (i.e., $T^2 \propto r^3$).

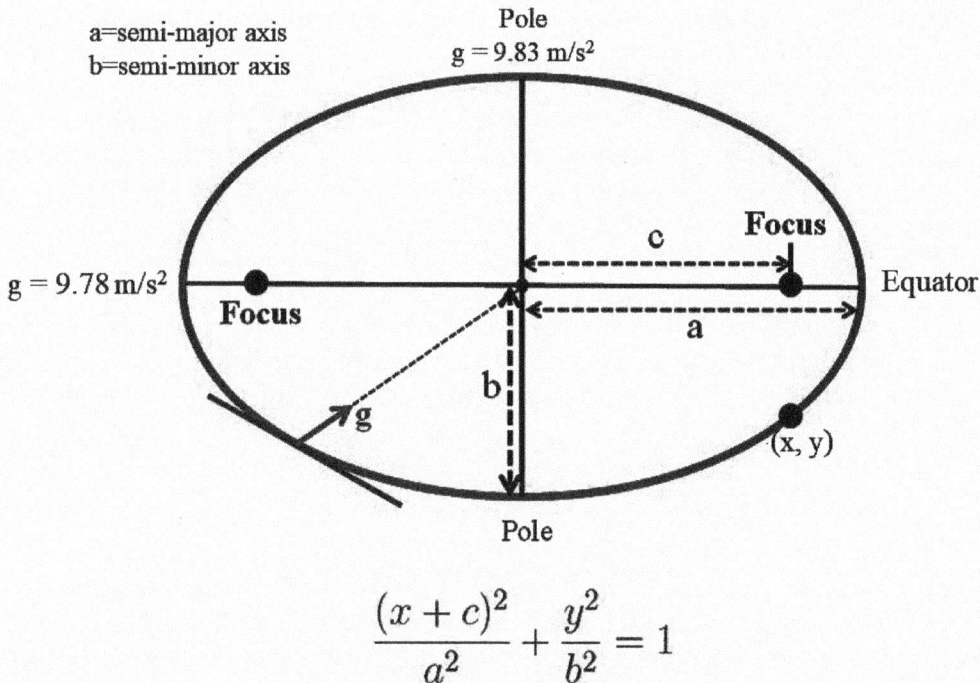

Figure 1.10. Here are some basic facts about ellipses and ellipsoids. An ellipsoid is defined by its semi-major axis (equatorial radius), a, its semi-minor axis (polar radius), b, its flattening (the relationship between equatorial and polar radius), $f = (a - b)/a$, and its eccentricity, $e^2 = 2f - f^2$. For the Earth, $a = 6378.136$ km and $b = 6356.751$ km. The Earth is fatter at the equator and thinner at the pole. The further you are from the rotation axis the greater is the gravitational acceleration, g.

As we progress through this subsection, you will find that your old basic materials on ellipses and ellipsoids will come in handy. Figure 1.10 recalls some of these facts. An ellipsoid is defined by its semi-major axis (equatorial radius), a, and its semi-minor axis (polar radius), b. We will also use its flattening (the relationship between equatorial and polar radius), $f = (a - b)/a$, and its eccentricity, $e^2 = 2f - f^2 = (a^2 - b^2)/a^2$. Solid Earth is not a sphere. It has the shape of an oblate ellipsoid. The amount of flattening is $f = 1/298$. This shape was one of the first clues that the Earth's mantle behaves like a fluid. Notice also that the acceleration of the Earth due to gravity is less at the equator ($g = 9.78\,\text{m/s}^2$) than at the Poles ($g = 9.83\,\text{m/s}^2$). This happens because the Earth's rotation causes a centrifugal force ($F_c = mv^2/r$) that is in proportion to the distance away from the axis of rotation, as the equator is further from the Earth's center ($r = 6,378$ km) than the Poles ($r = 6,357$ km).

Proof of the third law using circular orbits. Kepler's laws, as we call them today, are precise, but they are only descriptive, as Kepler did not understand why planets behave in this way. Newton's most significant achievement was to prove that all planetary behaviors were the consequence of the simple law of attraction, which we described in the previous subsection.

Surprisingly, the first of Kepler's laws, that the planetary paths are elliptical, is the toughest to prove beginning with Newton's assumption of inverse-square gravitation. Newton himself did it with an ingenious geometrical argument, famously difficult to follow. It can be demonstrated easily using calculus, but even this is non-trivial, and we shall work through it later. The most effective strategy is to attack the laws in reverse order. Thus, we start with the third law under the assumption of circular orbits.

Again, formulas of the satellite's orbital speed, in (1.6), and of the period of the satellite's orbit in (1.7) are also valid for the planets (i.e., mass m_2) around the Sun. We simply replace the orbiting satellite with the orbiting planet (i.e., mass m_2) and the Earth with the Sun (i.e., mass m_1). Therefore, by substituting (1.6), in (1.7), we obtain the ratio of the square of the period over the cube of the radius r, thus, Kepler's third law; i.e.,

$$\frac{T^2}{r^3} = \frac{4\pi^2}{Gm_1} = k = \text{constant} . \tag{1.31}$$

The coefficient of proportionality $k = T^2/r^3$ is known as Kepler's constant. It has the same numerical value for all of the Sun's planets (the Earth, Mars, Venus, Mercury, Jupiter, etc.), as illustrated in Table 1.1. The Moon has a different Kepler's constant because it orbits a different object, the Earth and not the Sun. In more general terms, k has the same numerical value if the object being orbited stays the same. Note that we will generalize the proof of this third law to elliptic orbits later by replacing the orbital radius, r, with the semi-major axis of the ellipse (see Figure 1.10) with a.

We can write out Kepler's third law a couple of different ways by rearranging parameters and variables so that we can cancel out some terms and facilitate some calculations. For example, we know that Kepler's constant is the same for the planets orbiting the same object, so we can equate the formulas of the Earth and Mars together, as follows:

$$\frac{T^2_{\text{earth}}}{r^3_{\text{earth}}} = \frac{T^2_{\text{mars}}}{r^3_{\text{mars}}} , \tag{1.32}$$

Table 1.1. Different Kepler's constants, k. Other parameters are the length of year T for each planet in the solar system, and its average distance from the Sun r. The Earth, Mars, and Venus give approximately the same value for Kepler's constant, confirming that Kepler's third law is correct in predicting that there is a constant ratio between the squared period and cubed radius of objects orbiting the same object. The Moon has a different Kepler's constant because it orbits a distinct object, the Earth and not the Sun.

Planet & satellite	Orbit	T (days)	r (km)	k (days^2/km^3)
Earth	Sun	365	149×10^6	4.03e-20
Mars	Sun	684	228×10^6	3.95e-20
Venus	Sun	225	108×10^6	4.01e-20
Moon	Earth	27.32	384400	1.31e-14

and eliminate Kepler's constant.

> **Quiz:** The distance between the Sun and the planets and their orbital periods with respect to the Sun are the most significant parameters for analyzing our solar system. These parameters increase with the planet distance to the Sun (e.g., orbital periods of Mercury, Venus, Mars, Jupiter, Saturn, Uranus, and Neptune compared with that of the Earth are 0.241, 0.615, 1.88, 11.9, 29.4, 83.80, 164, 248, respectively). However, the gravities, masses, and radii of the planets do not increase or decrease with the planet distance to the Sun (e.g., gravities at the surface of Mercury, Venus, Mars, Jupiter, Saturn, Uranus, and Neptune compared with that of the Earth are 0.378, 0.894, 0.379, 2.54, 1.07, 0.8, 1.2, and 0.059, respectively), why?

Proof of the second law. Let us recall some basic equations for angular motion before we continue with this proof. Recall that anything that is in motion has momentum and momentum is the quantity of motion. If something is moving in a straight line it has linear momentum. Anything that is moving around a curve or in a circle has an angular momentum. Ikelle (2023b) shows the analogy between linear motion (translational motion) and angular motion (e.g., rotational motion). The analogy needs to be treated with caution. For example, the moment of inertia, I is not a constant property of the body but its mass is a constant property. Newton's law of linear motion,

$$\boldsymbol{F} = \frac{d\boldsymbol{p}}{dt} = m\frac{d\boldsymbol{v}}{dt} = m\boldsymbol{\gamma} \, , \tag{1.33}$$

becomes

$$\boldsymbol{\tau} = \frac{d\boldsymbol{L}}{dt} = I\frac{d\boldsymbol{\omega}}{dt} = I\boldsymbol{\alpha} \, , \tag{1.34}$$

for angular motion. Inertia is the resistance to change and rotational inertia (resistance to change in rotation) is called *moment of inertia*. The object located by its position vector $\boldsymbol{x}$ rotates about an axis given by the rotation vector $\boldsymbol{\omega}$ (also termed an angular velocity vector). The magnitude of this vector, $|\boldsymbol{\omega}|$, is the angular velocity giving the speed of the rotation in units of angle/time. The rotating object moves at a linear velocity $\boldsymbol{v} = \boldsymbol{x} \times \boldsymbol{\omega}$. In other words, the linear velocity vector, $\boldsymbol{v}$, is perpendicular to both the angular velocity vector, $\boldsymbol{\omega}$, and the position vector, $\boldsymbol{x}$. As the object moves in the direction of the linear velocity, its position vector moves with it, so the object moves in a circle about the rotation axis if the angular velocity does not change with time.

Suppose that the area swept out by a planet orbiting the Sun between times t_1 and t_2 is A_{12} and the area swept out by the same planet between times t_3 and t_4 is A_{34}, as illustrated in Figure 1.11. According to Kepler's second law, if $t_2 - t_1 = t_4 - t_3$ then $A_{12} = A_{34}$. In other words, the ratio of the area swept out by this planet over time is constant. For the elementary angle $d\theta$, the infinitely thin triangular area swept out by the orbiting planet is (see Figure 1.11)

$$dA = \frac{1}{2}(\text{height} \times \text{base}) = \frac{1}{2}r \times (rd\theta) \implies \frac{dA}{dt} = \frac{1}{2}r^2\frac{d\theta}{dt} \, . \tag{1.35}$$

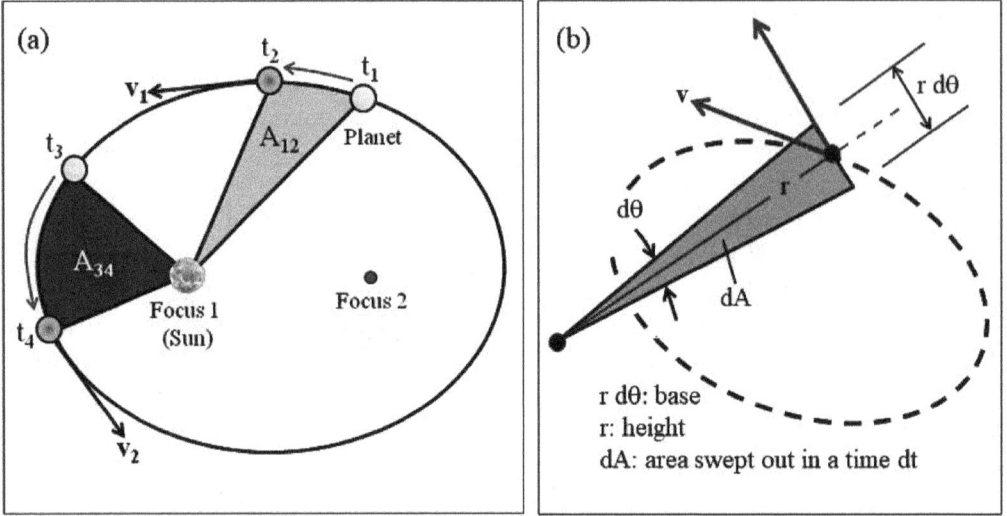

Figure 1.11. We consider the area swept by a planet orbiting the Sun in a given time interval. (a) In its orbit around the Sun, the planet sweeps out equal areas at equal time. They travel faster and faster as they get closer to the Sun. (b) The triangular area dA is the area swept as the planet travels an infinitesimal arc $rd\theta$. This area is one-half the product of the radius r and the arc $rd\theta$.

Therefore, the proof of the second law is equivalent to showing that dA/dt is constant; which is our task here.

Let us start by introducing a physical quantity known as angular momentum. It measures the quantity of rotation of a rotating body just like linear momentum (simply known as momentum) measures the quantity of motion of a moving body. The angular momentum of a mass m_2 with position vector $\boldsymbol{x}$ and velocity $\boldsymbol{v}$ is $\boldsymbol{L} = \boldsymbol{x} \times (m_2\boldsymbol{v})$, where $\times$ is the cross product defined in Ikelle (2023b). The derivative of $\boldsymbol{L}$ with respect to time, is

$$\frac{d\boldsymbol{L}}{dt} = \underbrace{\frac{d\boldsymbol{x}}{dt} \times (m_2\boldsymbol{v})}_{=0} + \boldsymbol{x} \times \left(m_2 \frac{d\boldsymbol{v}}{dt} \right)$$

$$= \boldsymbol{x} \times (m_2\boldsymbol{\gamma}) = \boldsymbol{x} \times \boldsymbol{F} = 0 . \tag{1.36}$$

In these derivations we used the fact that the cross product of two parallel vectors is zero. Because $\boldsymbol{F}$ is collinear with $\boldsymbol{x}$ (i.e., the force is along the direction of the line joining the centers of the Sun and the orbiting planet), $d\boldsymbol{L}/dt$ is zero and therefore $L = |\boldsymbol{L}|$ is a constant with respect to time, which means in physics that angular momentum is conserved. The other implication of this result is that the orbiting planet always moves in the plane (two-dimensional space) spanned by its velocity, $\boldsymbol{v}$, and the position vector, $\boldsymbol{x}$, as described in Figure 1.12.

In this plane, the position vector of the orbiting planet is $\boldsymbol{x} = r\,\boldsymbol{i}_r$ with $\boldsymbol{i}_r = [\cos\theta, \sin\theta]^T$ being the unit vector in the direction of the vector position of the orbiting planet (see Figure 1.12). The vector $\boldsymbol{i}_\theta = [-\sin\theta, \cos\theta]$ is the unit vector orthogonal to the direction of the vector position of the orbiting planet. The velocity

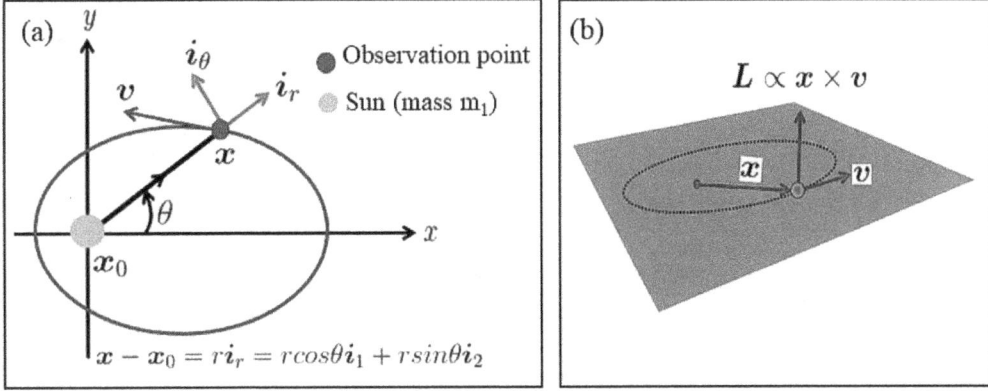

Figure 1.12. (a) We consider the position vector and velocity vector, v, in a plane containing an orbiting planet and the Sun. We also introduce polar coordinates: $i_r = [\cos\theta, \sin\theta]^T$ is the unit vector in the direction of the vector position of the orbiting planet. The vector $i_\theta = [-\sin\theta, \cos\theta]^T$ is the unit vector orthogonal to the direction of the vector position of the orbiting planet. (b) Another significant vector associated with this plane is the angular momentum vector, L, which is the scaled cross-product of the position vector and the velocity of the orbiting planet. The position vector is orthogonal to the angular momentum vector.

vector of the orbiting planet, v, is the time derivative of the vector position and can be written as

$$v = \frac{dx}{dt} = \frac{dr}{dt}i_r + r\frac{di_r}{d\theta}\frac{d\theta}{dt} = \frac{dr}{dt}i_r + r\frac{d\theta}{dt}i_\theta \tag{1.37}$$

and the magnitude of the angular momentum vector of the orbiting planet is

$$\frac{L}{2m_2} = \frac{1}{2}|x \times v| = \frac{1}{2}r^2\frac{d\theta}{dt} = \frac{dA}{dt} \tag{1.38}$$

and therefore dA/dt is constant because L is constant. We can now deduce that the area swept out by the planet's orbit between time t_1 and t_2 is

$$A_{12} = \int_{t_1}^{t_2} \frac{dA}{dt} = \frac{1}{2}\int_{t_1}^{t_2} r^2\frac{d\theta}{dt}dt = \frac{1}{2m_2}L(t_2 - t_1) \tag{1.39}$$

and the area swept out by the same planet between time t_3 and t_4 is

$$A_{34} = \int_{t_3}^{t_4} \frac{dA}{dt} = \frac{1}{2}\int_{t_3}^{t_4} r^2\frac{d\theta}{dt}dt = \frac{1}{2m_2}L(t_4 - t_3) \ . \tag{1.40}$$

If $t_2 - t_1 = t_4 - t_3$, then $A_{12} = A_{34}$ because the magnitude of the angular momentum of the orbiting planet, L, is constant. This conclusion proves Kepler's second law. Also we can clearly see that Kepler's second law is really a statement of the conservation of angular momentum. Note that the key assumption made in these derivations is that the Sun is not moving and lies at the origin of our coordinate system.

Quiz: (a) What is the eccentricity of the Earth's orbit? (b) Show that the equation of an ellipse can also be written as

$$r = \frac{a(1 - e^2)}{1 + e\cos\theta} \ , \tag{1.41}$$

where r and θ are the distance and angle, respectively, as defined in Figure 1.12.

Proof of the first law. Our next task is to prove the first law, which says that the planets move in elliptical orbits with the Sun being at one of the focuses. By using the heliocentric diagram with the polar coordinates, in Figure 1.12a, and the definition of the velocity vector of the orbiting planet, in (1.37), we can define the acceleration vector, $\gamma = d\boldsymbol{v}/dt$, which is the second time derivative of the vector position, as

$$\gamma = \left[\frac{d^2r}{dt^2} - r\left(\frac{d\theta}{dt}\right)^2\right]\boldsymbol{i}_r + \left[2\frac{dr}{dt}\frac{d\theta}{dt} + r\frac{d^2\theta}{dt^2}\right]\boldsymbol{i}_\theta$$

$$= -\frac{Gm_1}{r^2}\boldsymbol{i}_r \qquad\qquad + \qquad 0\ \boldsymbol{i}_\theta \ . \tag{1.42}$$

Because the gravity force $\boldsymbol{F} = m_2\gamma = -Gm_2m_1/r^2\boldsymbol{i}_r$ is collinear with the position vector of the orbiting planet, $\boldsymbol{x} = r\,\boldsymbol{i}_r$, the θ-component of the acceleration vector of the orbiting planet is zero and the radial component of the acceleration vector of the orbiting planet is nonzero; i.e.,

$$\frac{d^2r}{dt^2} = r\left(\frac{d\theta}{dt}\right)^2 - \frac{Gm_1}{r^2} = \frac{L^2}{m_2r^3} - \frac{Gm_1}{r^2} \tag{1.43}$$

$$2\frac{dr}{dt}\frac{d\theta}{dt} + r\frac{d^2\theta}{dt^2} = \frac{d}{dt}\left(r^2\frac{d\theta}{dt}\right) = 0 \implies \frac{L}{m_2} = r^2\frac{d\theta}{dt} \ . \tag{1.44}$$

We can see from equation (1.43) that θ-component of the acceleration vector provides an alternative way of showing that the magnitude of the angular momentum vector of the orbiting planet is constant. Our focus here is on equation (1.44). As we will see later the solution to this equation is an ellipse and therefore useful in the proof of the first Kepler's law.

Quiz: The apparent weight w of an object depends on the latitude θ at which it is measured. In contrast, the true weight is entirely due to the Earth's gravity and its magnitude equals $w = mg_e$ and is the same everywhere (m is the mass of the object). (a) If the tangential speed of a point on the surface of the Earth at the equator is v_θ, then find the apparent weight of the object at the equator. (b) What is the apparent gravitational acceleration at the equator?

By using the following identity

$$\frac{d}{dt} = \frac{d}{d\theta}\frac{d\theta}{dt} = \frac{LB^2}{m_2}\frac{d}{d\theta} \ , \tag{1.45}$$

with $B = 1/r$, we can rewrite equation (1.43)

$$\frac{d^2r}{dt^2} = \frac{L^2}{m_2 r^3} - \frac{Gm_1}{r^2} \iff \frac{d^2B}{d\theta^2} + B = \frac{Gm_1 m_2^2}{L^2} . \tag{1.46}$$

We can verify that the solution of this equation is

$$r = \frac{1}{B} = \frac{k}{(1 - \varepsilon \cos\theta)} \quad \text{or} \quad r = k - \varepsilon x , \tag{1.47}$$

where

$$k = \frac{L^2}{Gm_1 m_2^2} \quad \text{and} \quad \varepsilon = \frac{L^2}{Gm_1 m_2^2} C , \tag{1.48}$$

and where $x = r\cos\theta$ and C is a constant of integration. We can also verify that the solution in (1.47) is an ellipse; i.e.,

$$\frac{(x+c)^2}{a^2} + \frac{y^2}{b^2} = 1 . \tag{1.49}$$

by setting

$$c = \frac{k\varepsilon}{1 - \varepsilon^2} , \quad a = \frac{k}{1 - \varepsilon^2} , \quad \text{and } b = \frac{k}{\sqrt{1 - \varepsilon^2}} \quad . \tag{1.50}$$

This equation and associated parameters are illustrated in Figure 1.10. Semi-major and semi-minor axes are a and b, respectively.

Proof of the third law with elliptic orbits. We have already shown that the time it takes a planet to make one complete orbit around the Sun (one planet year) is related to the radius of circular orbits. We now extend this proof to elliptical orbits. The area traced out in one period of the orbit, which is the time to travel around the entire ellipse is,

$$A = \int_0^T \frac{dA}{dt} = \frac{1}{2m_2} LT \implies T = \left(\frac{2m_2}{L}\right) A = \left(\frac{2m_2}{L}\right) \pi ab , \tag{1.51}$$

where $A = \pi ab$ is the area of the ellipse bounded by the planetary orbit. We now know that $b^2 = a^2(1 - \varepsilon^2)$ and $a = k/(1 - \varepsilon^2)$. We can deduce the third law;

$$\frac{T^2}{a^3} = \frac{4\pi^2}{Gm_1} = \text{constant} . \tag{1.52}$$

Example #5: We can use Kepler's third law to find how long it will take for the spacecraft to arrive on Mars. The major axis of a spacecraft's orbit is $2a_S = r_E + r_M$, assuming the Earth, designed by r_E and T_E, and Mars, designed by r_M and T_M, move in nearly circular orbits (see Figure 1.13). Using Table 1.2, the semi-major axis is therefore

$$a_S = \frac{r_E + r_M}{2} = \frac{149 + 228}{2} \times 10^6 = 189 \times 10^6 \, \text{km} . \tag{1.53}$$

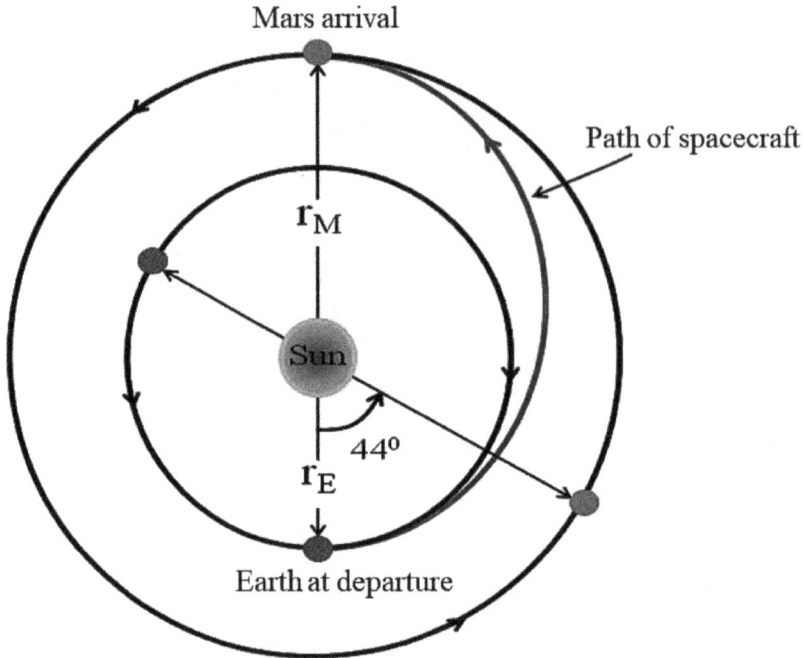

Figure 1.13. Journey to Mars. We blast off from Earth heading in the same direction as Earth and Mars are both traveling around the Sun, and by adding a little speed using the spacecraft's engines, we eventually match up with Mars's orbit and catch up to Mars itself. Remember that Mars is the fourth planet from the Sun, and Earth is the third planet from the Sun (see Figure 1.16).

The spacecraft travels through half of this complete elliptical orbit on its way out to Mars, so the travel time is

$$T = \frac{1}{2} \left[\frac{a_S}{r_E} \right]^{\frac{3}{2}} T_E = \frac{1}{2} \left(\frac{1.89}{1.49} \right)^{\frac{3}{2}} (1 \text{ year}) = 258 \text{ days} . \qquad (1.54)$$

NASA launched the InSight (Interior Exploration using Seismic Investigations, Geodesy and Heat Transport) mission from the central coast of California on May 5, 2018. InSight landed on Mars on November 26, 2018. The trip took almost 205 days. The difference between the actual traveltime of the InSight mission and the predicted result, in (1.54), are due to the fact that we did not take into account the speed of the launch and the alignment of Earth and Mars because both Mars and Earth's orbits are not perfectly circular.

Notice also the large gravity of Jupiter, especially when compared to that of its neighbor, Mars. This large gravity of Jupiter stirred up most asteroids in the asteroid belt between Mars and Jupiter (see Figure 1.14). This region is known as the *main asteroid belt*, with more than 90 percent of asteroids in our solar system. About 100,000 of them are more than one kilometer across. Two groups of asteroids, known as the Trojan asteroids, precede and follow Jupiter on its orbit. The combined gravitational pull of the Sun and Jupiter prevents Trojan asteroids from straying far from Jupiter's orbit. There is also the Kuiper belt, from about

Table 1.2. A gravity fact sheet. T is the orbital period, a is the orbital semi-major axis, m_2 is the mass of the orbiting planet, e is the orbital eccentricity, and g/g_e is the comparison of g at the surface of other planets to the Earth's gravity ($g_e = 9.81$ m/s^2). Note that this table includes the Moon of Earth which is a natural satellite and not a planet and the Galilean satellites of Jupiter. It also includes the Sun, which is a star and not a planet. Because there are many perturbations in the motion of the Sun around the center of the Milky Way, the notions of orbital semi-major axis and eccentricity do not apply to the orbit of the Sun around the center of our galaxy.

Planet	Orbit	T	a	m_2	e	g/g_e
		(days)	(10^6 km)	(kg)	$\times 0.01$	
Earth	Sun	365	149	5.97×10^{24}	1.67	1
Mars	Sun	684	228	6.42×10^{23}	9.34	0.38
Venus	Sun	225	108	4.87×10^{24}	0.677	0.89
Mercury	Sun	88	58	3.30×10^{23}	20.6	0.38
Jupiter	Sun	11.9	778	1.90×10^{27}	4.84	2.54
Saturn	Sun	29.4	1430	5.69×10^{26}	5.42	1.07
Uranus	Sun	83.8	2870	8.69×10^{25}	4.72	0.8
Neptune	Sun	164	4500	1.02×10^{26}	0.86	1.2
Moon	Earth	27.32	0.3484	7.35×10^{22}	5.49	0.167
Ganymede	Jupiter	7.2	0.5262	1.48×10^{23}	0.15	0.146
Callisto	Jupiter	16.7	0.4821	1.07×10^{23}	0.7	0.126
Io	Jupiter	1.8	0.3643	8.93×10^{22}	0.4	0.184
Europa	Jupiter	3.6	0.3122	48.0×10^{22}	1.01	0.134
Sun	Milky Way	83×10^9	N/A	1.99×10^{30}	N/A	27.96

30 to 50 AU from the Sun (see Figure 1.14). It stretches past Neptune. Actually, Neptune's gravity stirred up the Kuiper belt. Following up from our discussion in Ikelle (2023b), it is actually the asteroids of the Kuiper belt that may contain the ingredients for life on Earth and even the sperms and eggs of animal life. The asteroids of the main-asteroid belt generally contain inorganic matter and thus are unlikely to contain the ingredient for life on Earth.

Quiz: (1) As we can see in Table 1.2, Mars and Mercury have nearly identical gravitational pull at the surface, despite the fact that Mars is larger and more massive than Mercury. Explain why?
(2) Why is it helpful to have the mass and radius of a planet?
(3) If you weigh 100 kg on Earth, how much would you weigh on the other planets of our solar system?

As described in Ikelle (2023a and 2023b), TRAPPIST-1 is another star system with a star of mass of about 9 percent the mass of our Sun (i.e., 0.089 m_{sun}; only slightly larger than Jupiter) and an effective temperature of 2511 degrees Kelvin, which is much cooler than the Sun. Seven of the star's explonates have the potential to support carbon-based life due to their cool temperatures and gravity that approaches that of Earth. However, humans and/or humanoids on these planets are

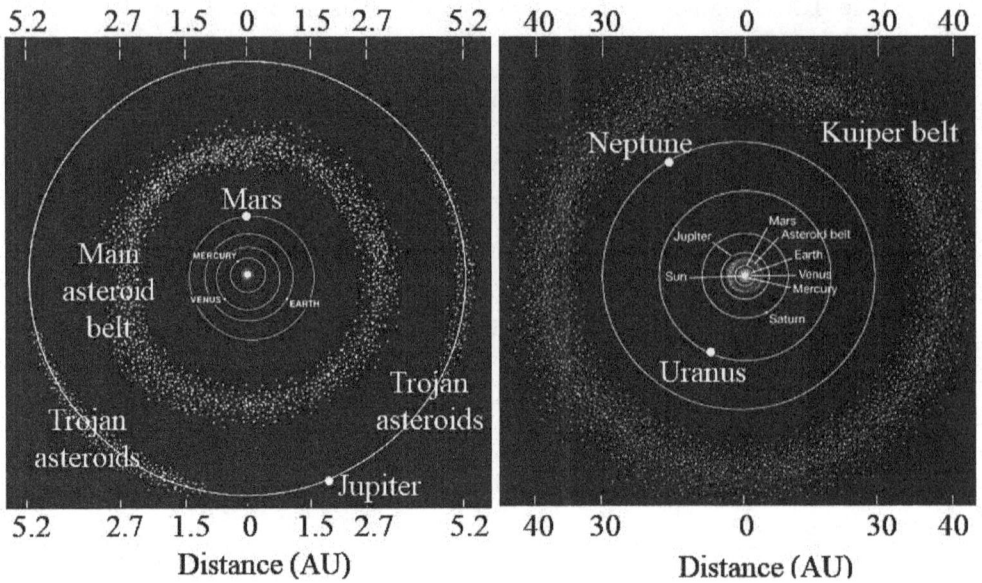

Figure 1.14. Measuring the Sun's orbital motion (radius and velocity) gives us the mass inside the Sun's orbit. Note that we cannot measure the mass outside of the Sun's orbit in this fashion.

likely to be shorter, on average, than us because their gravity is smaller than that of Earth. The seven exoplanets are labeled b through h. Table 1.3 captures the properties of these seven planets that have been observed by using NASAs Spitzer, Kepler, and TRAPPIST space telescopes or inferred from these observations. All these planets are smaller or equal to Earth in terms of mass and radius, but larger than Mars in the same terms. In other words, they vary in a narrower band than even the rocky planets of our solar system. One of the key missing properties in Table 1.3 is the exoplanet temperatures. They are difficult to infer from the other parameters in Table 1.3 or from telescope observations. They require flybys of probes near the exoplanets, as we will discuss in Ikelle (2024). But because we know the distance of the TRAPPIST-1 exoplanets to TRAPPIST-1, and the temperature of TRAPPIST-1, TRAPPIST-1 exoplanets receive comparable levels of light and heat to Earth and its neighboring planets.

By comparing the solar and TRAPPIST-1 systems, and using the fact that luminosity, mass, temperature, and radius of stars are all related, we see a trend which suggests that smaller stars in mass produce less Jovian exoplanets because of their limited temperature and hence luminosity. TRAPPIST-1 system has no Jovian exoplanets while our hotter sun has five Jovian planets. Despite this, TRAPPIST-1 and the solar system are quite similar in terms of the size and mass distribution of planets compared to the mass of the star. The massive exoplanet g is far away from TRAPPIST-1 just like Jupiter is far away from the Sun. One can expect to discover a Pluto version of the TRAPPIST-1 system if this trend is correct. Additionally, TRAPPIST-1 does not seem to be connected to any icy asteroid belt, which is in line with the rocky planets of our solar system. However, the bombardment of planets by the icy asteroid was central to the evolution of our tectonics on Earth and even

Table 1.3. A gravity fact sheet of the seven planets of TRAPPIST-1 with their orbital periods, distances from TRAPPIST-1, their radii, masses, densities and surface gravity as compared to those of Earth. T is the orbital period, a is the orbit's semi-major axis, m_2 is the mass of the orbiting planet, ρ is the planet density, and g is the gravity at the surface of the planet. (Source of data: Delrez et al., 2018.)

Exoplanet	Orbit	T (days)	Distance to star	a/r_e	m_2/m_e	ρ/ρ_e	g/g_e
T-1b	T-1	1.51	0.0115 AU	1.12	1.37	0.99	1.10
T-1c	T-1	2.42	0.0158 AU	1.10	1.31	0.99	1.09
T-1d	T-1	4.05	0.0223 AU	0.79	0.39	0.79	0.62
T-1e	T-1	6.10	0.0293 AU	0.92	0.69	0.89	0.82
T-1f	T-1	9.21	0.0385 AU	1.05	1.04	0.91	0.95
T-1g	T-1	12.36	0.0469 AU	1.13	1.32	0.92	1.04
T-1h	T-1	18.76	0.0619 AU	0.76	0.33	0.75	0.57

Star	Orbit	T	Distance to Earth	a/r_s	m_2/m_s	ρ/ρ_s	g/g_e
T-1	Milky Way	N/A	41 light-years	0.121	0.089	51.1	172.15

life on Earth. So the idea that a star system with limited or no bombardments by icy asteroids can produce a planet with plate tectonic—that is, planets with large accumulations of water on their surfaces—is yet to be formulated. Trappist-1 may have an asteroid belt similar to the Kuiper belt in our solar system that is still to be discovered. There is a possibility that a belt such as this could also contain life and Pluto-like planets that are still to be discovered.

Because the average densities of TRAPPIST-1's exoplanets are quite close to those of Earth and Venus, we can expect typical layered structures of these exoplanets—this is, crust, mantle, and core. However, because the thermonuclear reactions of the Sun are different to those of red stars like TRAPPIST-1, we can expect the composition of rocky planets and TRAPPIST-1's exoplanets to be different from those of our solar system.

Note that TRAPPIST-1's exoplanets were discovered in 2016. From 2016 to 2022, more than 5,000 new exoplanets have been discovered, but almost none with Earth-like properties. In other words, exoplanets with Earth-like properties are so far not common in our Universe, as pointed out in Ikelle (2023b).

1.1.3. Gravity of Dark Matter

How do we measure the mass of the galaxy? As shown in (1.6) and (1.7), we can measure an object's mass from the orbital period and the distance of the centers of bodies in orbit around it. For example, we can determine the mass of the Earth based on its moon's orbits, as follows:

$$v^2 = \frac{Gm_1}{r} \iff m_1 = \frac{rv^2}{G} \, , \tag{1.55}$$

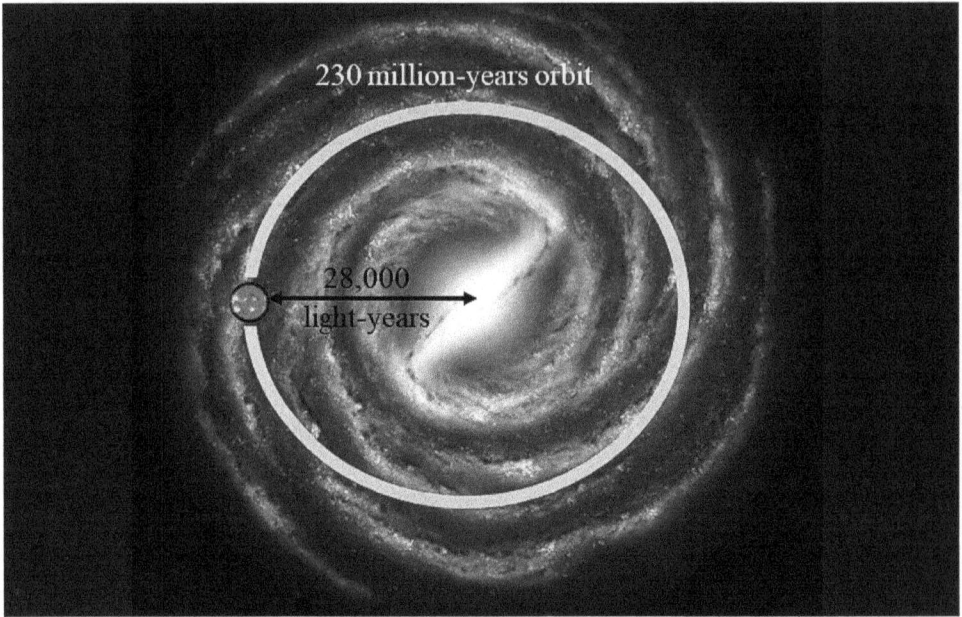

Figure 1.15. Galaxy rotation Curves: observed rotation speeds and expected rotation speeds. The visible mass is insufficient to account for the observed rotation speeds. Dark Matter dominates at large radii.

v is the orbital velocity, r is the orbital separation, m_1 is the mass enclosed by the orbit, and G is the universal gravitational constant, as defined earlier. The mass enclosed by the orbit here is the mass of the Earth because we assume the only mass enclosed inside the orbit is the Earth and the remaining space is vacuum (see Figure 1.4). As illustrated in Figure 1.15, we can also use the Sun's orbital motion (radius and velocity) in the Milky Way to estimate the mass of matter enclosed in this orbit, as first proposed by Rubin et al. (1980). For $v = 280$ km/s and $r = 280,000$ light-years, this mass is $1 \times 10^{11} \, m_{\mathrm{sun}}$. Actually, our Milky Way's total mass is at least $5 \times 10^{11} \, m_{\mathrm{sun}}$. The fact that we cannot measure the mass outside of the Sun's orbit in this fashion cannot compensate for this deficit. The unexpected mass is dark matter. More than 85 percent of this mass is opaque. In other words, most of our galaxy's mass is invisible to us and even to the entire electromagnetic spectrum, as we know it today—that is, *neutrons and protons of which living things, Earth, and the stars are made of, do not account for most of the mass in our Milky Way, and in our entire Universe*. It is our objective here to prove the existence of this mass.

Evidence of dark matter. In a spherical concentration of mass $m(r)$, the circular gravitational orbit from balancing the centripetal force with gravity is again

$$v(r)^2 = \frac{Gm(r)}{r} \implies v(r) \propto \frac{1}{\sqrt{r}} \, , \tag{1.56}$$

where $v(r)$ is the orbit velocity and $m(r)$ is the total mass interior to r. Figure 1.16 shows deviations from predictions of orbital speed based on Newtonian gravity

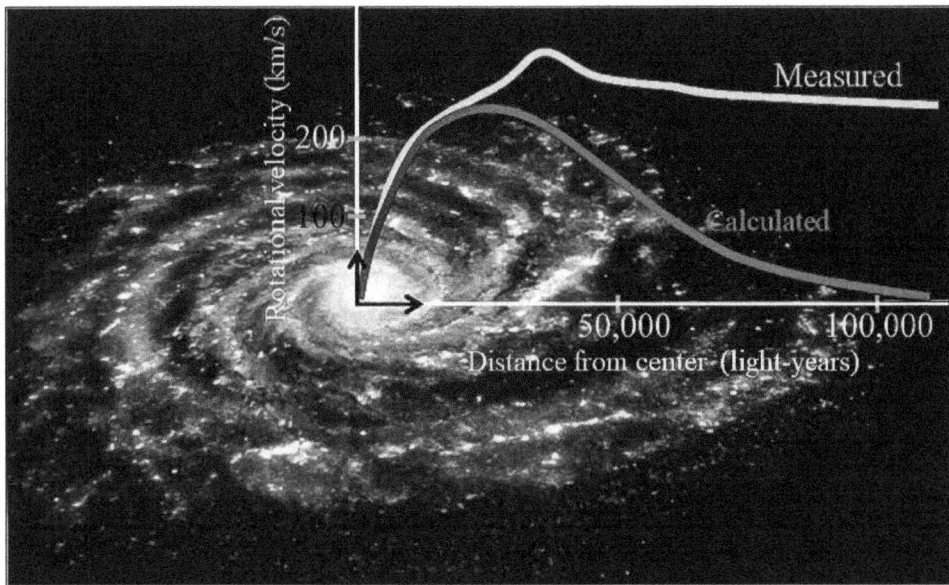

Figure 1.16. An illustration of evidence for dark matter based on orbiting objects' velocity. Stars in the outer areas orbit much faster than expected. So far, our galaxies models do not predict these deviations. These deviations are due to dark matter. Moreover, we need dark matter to be about five times as massive as ordinary matter to compensate for these deviations.

in (1.56) and the observed orbital speed with the radius describing r. The two orbital speed curves align until 4,000 light years. Beyond this limit, the observed orbital speed remains relatively constant suggesting that objects orbiting the galaxy at larger distances from the galactic center move around at more or less the same velocity as objects much closer to the center, contrary to what is expected (Rubin et al., 1980). The rotational velocity is expected to drop as $v(r) \propto 1/\sqrt{r}$, like in our solar system [see (1.55) and (1.56)]. Instead, at large distances, the rotational curve becomes flat. The deviations between two sets of velocities in Figure 1.16, are then used to determine dark matter amounts.

Let us add some mathematical description of the curves in Figure 1.16. By plugging a mass distribution with constant density $m(r) \approx 4\pi\rho r^3/3$ in (1.56) we obtain $v(r) \propto r$. Assuming galaxies are axisymmetric objects of constant density, we can see that a linear growth in $v(r)$, followed by $v(r) \propto 1/\sqrt{r}$ decay after leaving the region where most ordinary matter lives. However, the actual measurements in galaxies suggest that $v(r)$ is nearly flat after leaving the region where most ordinary matter lives. In other words, the matter model $m(r) \approx 4\pi\rho r^3$ works very well at small distances. One can reproduce a nearly-flat velocity rotation curve at larger r by assuming $\rho(r) \propto r^{-2}$ which corresponds to $m(r) \propto r$. In other words, a non-visible mass component, which increases linearly with radius, must exist; that is dark matter (a mass in a non-luminous form). It is distributed in a spherically symmetric halo about the center of the Galaxy, which means that galaxies are thus surrounded by a halo of dark matter. Also, additional invisible mass is needed to prevent peripheral stars from flying away and the galaxies from breaking apart. Note that in the constant velocity portion of the galactic curve, we can obtain from (1.56) the following:

$$m(r) = \frac{v_c^2}{G}r \implies \frac{dm(r)}{dr} = \frac{v_c^2}{G} . \tag{1.57}$$

where v_c is constant circular speed. By substituting the equation for the conservation of mass in a spherically symmetric system (see Chapter 2)

$$\frac{dm(r)}{dr} = 4\pi r^2 \rho(r) \tag{1.58}$$

in (1.57), we obtain for the density in the outer parts of the galaxy as

$$\rho(r) = \frac{v_c^2}{4\pi r^2 G} \propto r^{-2} , \tag{1.59}$$

which is a reasonable approximation of dark matter density. A better form of the density of dark matter covering the entire galactic curve is

$$\rho(r) = \frac{\rho_0}{1 + (r/r_0)^2} \implies \rho(r) \approx \begin{cases} \rho_0 & r \ll r_0 \\ \rho_0(r_0/r)^2 & r \gg r_0 \end{cases} . \tag{1.60}$$

The constants ρ_0 (the density at $r = 0$) and r_0 (the core radius) can be obtained by fitting the observable galactic rotation curves.

The first historical evidence for non-luminous matter in the universe is made by Fritz Zwicky (1933) from observations of the dynamics of clusters of galaxies known as Coma cluster.[1] This cluster of galaxies contains 1 percent luminous matter, 9 percent intergalactic gas (i.e., intracluster medium), and 90 percent dark matter. Zwicky (1937) deduced that galaxies had masses much larger than he could observe from their movements. His calculations suggest that 100 times more luminous mass is needed to account for galaxies' speed (Sanders, 2010); a tad less than expected. Zwicky (1937) concluded that most matter does not radiate. Later observations and studies confirm that even individual galaxies like our Milky Way are dominated by dark matter (Rubin and Ford, 1970; and Rubin et al., 1980).

One word that we have used multiple times through this book series that we have not properly defined so far is *galaxy*. This incompleteness is due to not discussing dark matter so far. Actually, a galaxy is a gravitationally bound system composed of dark matter, stars, gas, and dust. Stars include stellar remnants (black holes to white dwarfs), gas includes ions, neutrals, and molecules, and supermassive black holes are located at the centers of most, if not all, galaxies. Dust is listed here as a separate component because of its key role in redistributing electromagnetic radiation emitted by galaxies, radiation which constitutes most of the information on galaxies available to us. Roughly half of all stars are in spiral galaxies, and half are in elliptical galaxies. In spiral galaxies such as the Milky Way, most stars are in a thin

---

[1]The Coma cluster consists of approximately 1,000 galaxies spread over two degrees in the sky. Gravity binds these galaxies together into a cluster. However, galaxies do not orbit a central heavy object like the Sun.

disc. The Milky Way disc has a radius of about 15 kpc, and contains about 6×10^6 m_{sun} of stars. The disc stars rotate around the center of the galaxy in nearly circular orbits, with a rotation speed of approximately 220 km/s almost independent of the radius r. Such rotation curves require a very extended mass distribution, much more extended than the observed light distribution. This indicates the presence of large amounts of invisible dark matter. Curiously, the spiral arms do not rotate at the same speed as the stars, hence stars may move in and out of spiral arms. Elliptical galaxies are roundish objects that appear yellow. They have lower star formation rates than spirals of the same mass. The low star formation rate is because the central supermassive black hole prevents star formation.

Quiz: Black hole. Given a massive star of mass m_1, calculate the radius (called the Schwarzschild Radius) at which not even light emanating from the star's surface can escape. Determine the Schwarzschild radius of the Sun and the mean mass density it would have if its mass were contained in the corresponding spherical volume (the spherical surface bound to this volume is called the event horizon). Note that the current understanding of our universe is that new black holes will continue to form in the next 1,000 years or so.

Some have considered black holes as a possible explanation for dark matter. As described in Ikelle (2023b), a black hole is an astronomical object with a gravitational pull so strong that nothing, not even light, can escape it. Matter and radiation fall in, but cannot leave (Wald, 1997). They form as massive stars collapse at the end of their life cycle. The presence of a black hole can be inferred through its interaction with electromagnetic radiation such as visible light. So far, black holes have been detected by gravitational lensing[2] (Carr et al., 2016). Unfortunately, there may not be enough black holes in our Universe to account for dark matter in our universe. Also, we still do not really understand how galaxies form and evolve, an imperative step to explaining dark matter.

One can describe dark matter as made of slow moving, uncharged, and collisionless heavy particles based on the difference between ordinary matter and dark matter in terms of mass and their invisibility with respect to the electromagnetic spectrum. It may also be possible that our descriptions of nucleosynthesis, electromagnetism, gravity, and/or particle physics are incomplete. We will discuss possible forms of dark matter and energy in Ikelle (2024).

The virial theorem applies to dark matter. Let us start by recalling that the equation of hydrostatic equilibrium of an object in space with a spherical mass distribution:

$$\frac{dP(r)}{dr} = -\frac{Gm(r)\rho(r)}{r^2} ,$$

---

[2]Cluster dark matter creates a cosmic mirage known as gravitational lensing. Albert Einstein theorized this phenomenon. The gravitational pull of dark matter in a cluster bends light emitted by distant galaxies passing by. Near the center of the cluster, this effect can be strong enough to produce multiple images of distant galaxies. Farther out, it distorts the shape of each and every distant galaxy. Gravitational lensing is routinely used to map dark matter in clusters and individual galaxies. This method shows that clusters of galaxies have 100 times more mass than can be accounted for in stars.

where $P(r)$ is the local pressure at a radius r, $m(r)$ is the mass enclosed in sphere with a radius r, and $\rho(r)$ is the density enclosed in sphere with a radius r. It expresses the balance between internal pressure and gravity. The virial theorem is another hydrostatic equilibrium equation in energy terms rather than forces. Claussius obtained it in 1851. It is the relation between the total kinetic energy, K, and the total potential energy, W, of a system in equilibrium. It allows us to analyze systems of millions or billions of stars and clusters of galaxies that may have thousands of galaxies. This theorem avoids solving the million-body problem. Moreover, sometimes we do not know much about these large systems. In this theorem, we also assume that there are no other sources of internal support against gravity in the system apart from its internal kinetic energy (i.e., there is no magnetic field at the source or rotation). Note that, although we focus on dark matter in our discussion here, the virial theorem has a wide range of applications. These applications range from star formation to the formation of the largest structures in the universe. These structures include galaxies and clusters of galaxies. Galaxy clusters are the most massive gravitationally bound systems in the universe.

Let us consider a system of N bodies with mass m_i, position vector $\boldsymbol{x}_i$, velocity vector $\boldsymbol{v}_i$ and momentum $\boldsymbol{p}_i = m_i \boldsymbol{v}_i$ for body i. The time derivative of the moment of inertia, called the *virial*, of this system is

$$Q = \frac{1}{2}\frac{dI}{dt} = \sum_{i=1}^{N} \boldsymbol{p}_i \cdot \boldsymbol{x}_i \, ,$$

where

$$I = \sum_{i=1}^{N} m_i |\boldsymbol{x}_i|^2 \, , \tag{1.61}$$

is the moment of inertia. To deduce the virial theorem we need to take the time derivative of the virial

$$
\begin{aligned}
\frac{dQ}{dt} &= \sum_{i=1}^{N} \frac{d\boldsymbol{p}_i}{dt} \cdot \boldsymbol{x}_i + \sum_{i=1}^{N} \boldsymbol{p}_i \cdot \boldsymbol{v}_i \\
&= \sum_{i=1}^{N} \boldsymbol{F}_i \cdot \boldsymbol{x}_i + \sum_{i=1}^{N} m_i v_i^2 \qquad \text{with } \boldsymbol{F}_i = \frac{d\boldsymbol{p}_i}{dt} \\
&= \underbrace{-\frac{1}{2}\sum_{i=1}^{N}\sum_{j\neq i} \frac{Gm_im_j}{x_{ij}}}_{\text{Potential energy }(W)} + \underbrace{\sum_{i=1}^{N} m_i v_i^2}_{\text{Kinetic energy }(2K)} = 0 \, , \tag{1.62}
\end{aligned}
$$

where $x_{ij} = |\boldsymbol{x}_i - \boldsymbol{x}_j|$ is the distance between body i and body j and the factor $1/2$ corrects for double counting the number of pairs in the total potential energy. For a system in equilibrium, $dQ/dt = d^2I/dt^2 = 0$ is zero. Thus, the *virial theorem* can be expressed as

$$2K + W = 0 \, . \tag{1.63}$$

Consider a gravitational system consisting of N galaxies. For simplicity, let us consider the case of a perfectly spherically symmetric elliptical galaxy composed of stars which all have the same mass $m_i = m$. In virial equilibrium, we then can rewrite the mass of our cluster of galaxies as

$$m_{\text{eff}} = \frac{2}{G} \frac{\langle v^2 \rangle}{\langle 1/r \rangle} = \frac{1}{G} r_h \langle v^2 \rangle \, , \tag{1.64}$$

where

$$m_{\text{eff}} = \sum_{i=1}^{N} m_i = Nm \, , \quad \langle 1/r \rangle = \frac{1}{N(N-1)} \sum_{i=1}^{N} \sum_{j \neq i} \frac{1}{x_{ij}} \, , \quad \langle v^2 \rangle = \frac{1}{N} \sum_{i=1}^{N} v_i^2 \, ,$$

$r_h = 2/\langle 1/r \rangle$ is the radius of a sphere centered on the center of mass of the cluster and within which half of the cluster mass is contained. Unfortunately, we cannot infer m_{eff} from $\langle v^2 \rangle$ and $\langle 1/r \rangle$ because they are not observable. Zwicky (1937), who inferred that the total mass in the Coma cluster is much larger than the sum of the masses of its galaxies, used instate

$$m_{\text{eff}} = \frac{3}{\alpha G} r_h \langle v_{\text{los}}^2 \rangle \, , \tag{1.65}$$

v_{los} is the line-of-sight velocity and α is a parameter of order unity that describes our ignorance of the radial distribution of galaxies. Note that, under the assumption of isotropy, $\langle v_{\text{los}}^2 \rangle = \langle v^2 \rangle /3$ (i.e., the relationship between radial velocity relative to the galactic center and orbital velocity) and one can also determine the mean reciprocal pair separation from the projected pair separations. Therefore, determining the mass of a galaxy cluster via virial arguments has reduced to determining $\langle v^2 \rangle$ and the half mass radius r_h.

Consider the Coma cluster of galaxies studied by Fritz Zwicky (1937). The measured line-of-sight velocity by telescopes is $v_{\text{los}} = 880$ km/s. Assuming the galaxy orbits are isotropic then $\langle v^2 \rangle = 2.323 \times 10^{12}$ m^2/s^2. By also assuming that the unknown dark matter component displays the same distribution as the order matter (i.e., visible matter of the galaxies), we can take

$$r_h \approx 1.5 \text{ Mpc} \approx 4.6 \times 10^{22} \text{ m} \, . \tag{1.66}$$

We can then deduce the mass of the Coma cluster as

$$m_{\text{coma;virial}} = \frac{\langle v^2 \rangle \, r_h}{\alpha G} \approx 1 \times 10^{45} \text{ kg} = 2 \times 10^{15} \, m_{\text{sun}} \, , \tag{1.67}$$

when $\alpha = 1.2$. For comparison, the total stellar mass of Coma is $m_{\text{coma},*} \approx 3 \times 10^{13} \, m_{\text{sun}}$; i.e., stars only make 1.5 percent of the cluster's mass. The intracluster gas mass observed at X-ray wavelengths is $m_{\text{coma,icm}} \approx 2 \times 10^{14} \, m_{\text{sun}}$; i.e., intracluster gas makes 10 percent of the cluster mass. Dark matter represents 88.5 percent of the cluster's mass. In other words, the Coma cluster is dominated by dark matter.

Since we know the existence of galaxies mostly through the light they emit, it makes sense to use the mass-to-light ratio (i.e., the ratio of the total mass of an object to its total luminosity) to characterize dark matter present in a galaxy or in a cluster of galaxies. It is a convenient measure for comparing gravitational mass with luminous mass. Let $m = \alpha' m_{\text{sun}}$ be the mass of a given galaxy and $L = \beta' L_{\text{sun}}$ be its luminosity. Its mass-to-light ratio is measured in solar system units:

$$\Upsilon = \frac{\alpha' \, m_{\text{sun}}}{\beta' \, L_{\text{sun}}} \implies \Upsilon_{\text{coma}} = \frac{2 \times 10^{15} \, m_{\text{sun}}}{8 \times 10^{12} \, L_{\text{sun}}} \approx 250 \frac{m_{\text{sun}}}{L_{\text{sun}}} \; . \tag{1.68}$$

So the mass-to-light ratio of Coma is 250, that of the Sun is 1, and that of the Milky Way is 27.1 assuming that $m_{mw} = 3.8 \times 10^{11} m_{\text{sun}}$ and $L_{mw} = 1.4 \times 10^{10} m_{\text{sun}}$ for the Milky Way. Most extragalactic objects have mass-to-light ratios greater than 1, indicating that not all their mass is in visible stars. There is a high value inferred for galaxies up to 30 and for clusters of galaxies up to 300. These numbers indicate the existence of considerable amounts of dark matter. In more general terms, the mass-to-light ratio depends on the number and types of stars in the galaxy, and the amount of dark matter.

Other examples of the virial theorem applications: Estimation of the Sun's mass-averaged temperature. Let us determine the Sun's mass-averaged temperature. Assuming that the sun is filled with a monatomic ideal gas, the internal energy per particle is $(3/2)k_B T$. So the Sun's internal energy is

$$K = \frac{3}{2} m_{\text{sun}} \frac{\mathcal{R} \langle T \rangle}{\mu} \; , \tag{1.69}$$

where μ is the mean mass per particle in the gas and $\mathcal{R}$ is the ideal gas constant. The Sun's gravitational energy is

$$W = -\alpha \frac{G m_{\text{sun}}^2}{r_{\text{sun}}} \tag{1.70}$$

where α is a constant of order unity that describes our ignorance of the internal density structure. By applying the virial theorem, we obtain

$$2K + W = 0 \implies \langle T \rangle = \frac{\alpha}{3} \frac{\mu}{\mathcal{R}} \frac{G m_{\text{sun}}}{r_{\text{sun}}} \tag{1.71}$$

If $\mu = 1/2$ (for a gas of pure, ionized hydrogen), and $\alpha = 2/5$ (for a uniform sphere), we need $\langle T \rangle = 1.53 \times 10^6$ degrees Kelvin to keep the Sun in hydrostatic equilibrium. The Sun's surface temperature is about 6000 degrees Kelvin and its core is 1 million degrees Kelvin. It is remarkable that we can obtain the average internal temperature of the Sun to within a factor from nothing more than its bulk characteristics, and without any knowledge of the Sun's internal fusion. That is the power of the virial theorem.

Other examples of the virial theorem applications: the Sun has a gaseous center and Earth has a liquid or solid center. We can also use the virial theorem to prove that the Sun has a gaseous center and Earth has a liquid or solid center. Consider a celestial object of density ρ and temperature T, consisting of atoms with atomic mass A and atomic number Z. The number density of the particles is $n = \rho/(Am_H)$, so the typical distance between them must be

$$d = n^{-1/3} = \left(\frac{Am_H}{\rho}\right)^{-1/3} . \tag{1.72}$$

The electromagnetic potential energy is

$$K \approx \frac{Z^2 e^2}{d} = Z^2 e^2 \left(\frac{\rho}{Am_H}\right)^{1/3} , \tag{1.73}$$

where e is the electron charge. We neglected any cancellation from electrons of opposite charges screening the nuclear charges. For a spherical object of mass m and radius r, $\rho = 3m/(4\pi r^3)$, the virial theorem tells us that the mean temperature will be

$$\langle T \rangle = \frac{\alpha}{3} \frac{\mu}{\mathcal{R}} \frac{Gm}{r} = \frac{1}{3} \frac{A}{\mathcal{R}} \frac{Gm}{r} \tag{1.74}$$

or

$$\frac{K}{k_B \langle T \rangle} \approx \frac{Z^2 e^2}{G A^{4/3} m_H^{4/3} m^{2/3}} = 0.011 \frac{Z^2}{A^{4/3}} \left(\frac{m}{m_{sun}}\right)^{-2/3} , \tag{1.75}$$

where $\mu \approx A$ is the mean atomic mass per particle. Even for a pure iron composition, $Z = 26$ and $A = 56$, we have

$$\frac{K}{k_B \langle T \rangle} = 0.035 \left(\frac{m_{earth}}{m_{sun}}\right)^{-2/3} \approx 168.15 \implies \frac{K}{k_B \langle T \rangle} \gg 1 . \tag{1.76}$$

For the hydrogen composition, $Z = A = 1$, we have

$$\frac{K}{k_B \langle T \rangle} = 0.035 \left(\frac{m_{jupiter}}{m_{sun}}\right)^{-2/3} \approx 3.61 . \tag{1.77}$$

and $K/(k_B \langle T \rangle) - 0.035$ for the Sun. Thus solid or liquid phases should be absent in bodies substantially larger than Jupiter, which includes the Sun. They begin to appear once the mass drops to that of Jupiter or less, which includes Earth. It is remarkable that we can deduce the phase of celestial bodies based on their masses even when mass estimates are not significantly accurate. Again, that is the power of the virial theorem.

1.1.4. The Gravitational Potential and Geoid and Gravitational Anomalies

Spatial variations of gravity have played a central role in the study of dynamic processes in the Earth's interior and in the exploration and production from the Earth's interior. Our objective here is to introduce gravitational potential, to describe how gravity measurements are used in the study of dynamic processes in the Earth's interior and in exploration and production of minerals.

The calculations of gravitational potentials allow us to introduce equipotential surfaces and later on two important equipotential surfaces in geophysics studies and applications, which are the geoid and reference ellipsoid. Again, the force exerted by mass m_1 (Earth) on mass m_2 is

$$\boldsymbol{F}(r) = -\frac{Gm_1m_2}{r^2}\boldsymbol{i}_r = -m_2\frac{dU(r)}{dr}\boldsymbol{i}_r \ , \qquad (1.78)$$

where

$$dU(r) = \frac{Gm_1}{r^2}dr$$

or

$$U(r) = Gm_1 \int_\infty^r \frac{1}{\eta^2}d\eta = Gm_1 \left[\frac{1}{\infty} - \frac{1}{r}\right] = -\frac{Gm_1}{r} \ . \qquad (1.79)$$

Centers of masses m_1 and m_2 are separated by a distance r. So we have introduced a new scalar field $U(r)$ (has no directional properties) that is known as the gravitational potential. It is actually the potential difference between two points because when going from an elementary potential $dU(r)$ to the total potential $U(r)$, we need two points (or two limits of integration). We define the potential at a particular point $r = \infty$ to be zero with respect to a reference point so that the gravitational potential generated by m_1 depends on the distance r only.

Using the definition of gravitational potential, in (1.79), the gravitational acceleration of mass m_2 toward m_1 can be written as

$$\boldsymbol{g}(r) = -\frac{Gm_1}{r^2}\boldsymbol{i}_r = -\frac{dU(r)}{dr}\boldsymbol{i}_r \ . \qquad (1.80)$$

In general terms,

$$\boldsymbol{g}(\boldsymbol{x}) = -\mathrm{grad}[U(\boldsymbol{x})] = -\boldsymbol{\nabla}U(\boldsymbol{x}) = -\left[\frac{\partial U}{\partial x}\boldsymbol{i}_1 + \frac{\partial U}{\partial y}\boldsymbol{i}_2 + \frac{\partial U}{\partial z}\boldsymbol{i}_3\right] \ . \qquad (1.81)$$

Because potential is a scalar rather than a vector, it is sometimes easier first to calculate the potential and then to calculate the gradient of the potential. A surface, S, on which the gravitational potential is constant (i.e. $U = $ constant hence $dU = 0$) is known as an equipotential surface. Thus $\boldsymbol{g} = -\boldsymbol{\nabla}U$ has no components along S, being perpendicular to the equipotential surface. However $\boldsymbol{g}$ is not necessarily constant on equipotential surfaces. For $U = -Gm_1/r$, $U = $ constant implies $r = $ constant (i.e., equipotential surfaces are spheres centered on m_1) for a non-rotating homogeneous Earth (see Figure 1.17a).

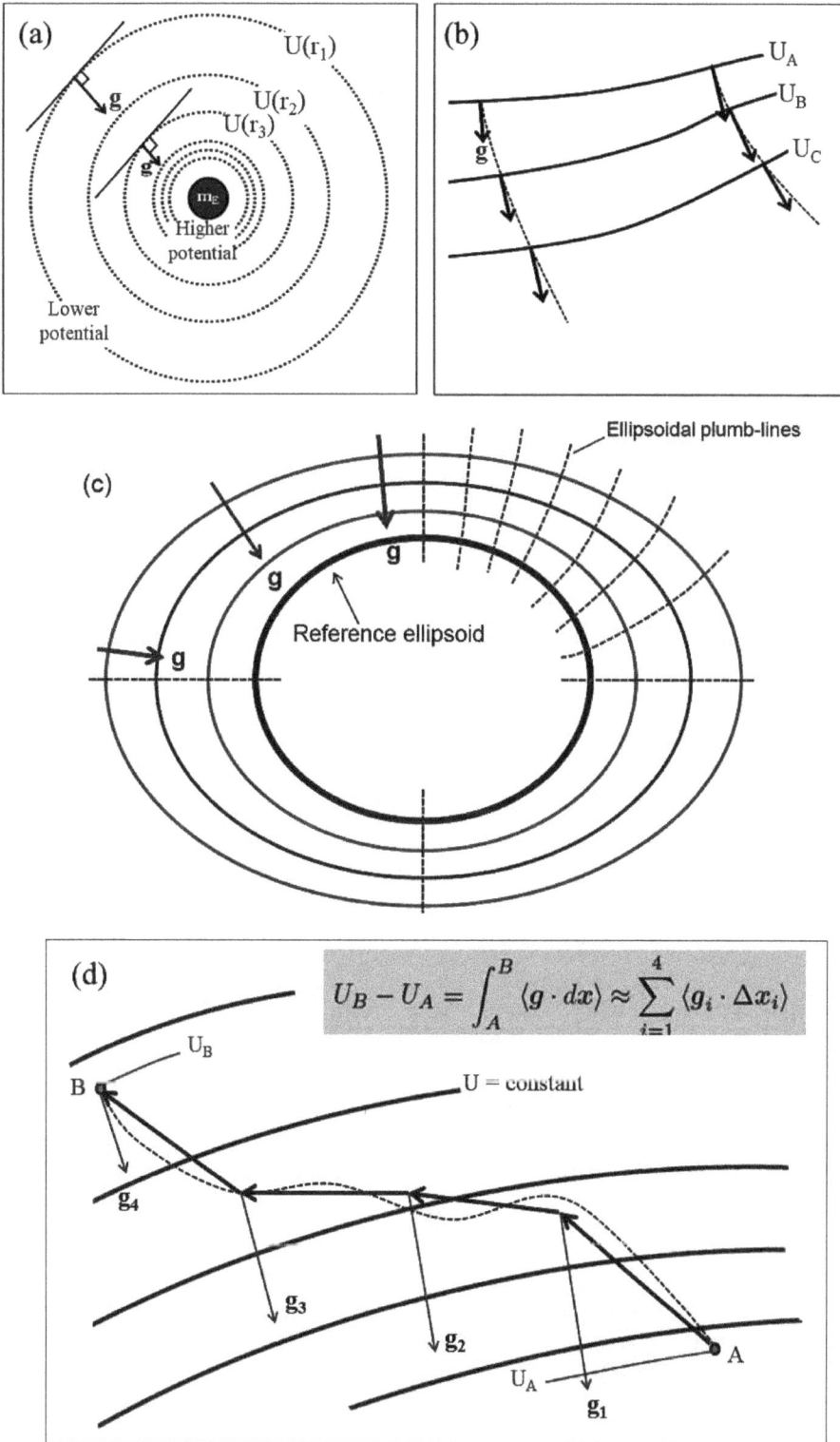

Figure 1.17. (a) Each of the circles represents an equipotential surface. (b) Equipotential surfaces with local undulations due to density variations and/or topographic variations. (c) Ideal reference ellipsoid (thick black line), level surfaces of the gravity potential (thin lines), and plumb lines (dashed lines). (d) The integral path of work.

Let us connect the gravitational potential with the notion of work done by the gravitational force on the object of mass m_2. That is, when the particle moves from a point $\boldsymbol{x}_A = r_A \boldsymbol{i}_r$ to another point at $\boldsymbol{x}_B = r_B \boldsymbol{i}_r$ in the Earth's gravitational field we have

$$
\begin{aligned}
W &= \int_{\boldsymbol{x}_A}^{\boldsymbol{x}_B} -m_2 \left[\boldsymbol{g}(\boldsymbol{x}) \cdot d\boldsymbol{x} \right] \\
&= m_2 \int_{r_A}^{r_B} G \frac{m_1}{r^3} \left(r \boldsymbol{i}_r \right) \cdot \left(dr \boldsymbol{i}_r \right) = m_2 \int_{r_A}^{r_B} \frac{Gm_1}{r^2} dr \ .
\end{aligned}
\tag{1.82}
$$

Using the definition gravitational potential in equation (1.79) we then write

$$
W = -Gm_2 m_1 \left[\frac{1}{r_B} - \frac{1}{r_A} \right] = m_2 \left[U(\boldsymbol{x}_B) - U(\boldsymbol{x}_A) \right] \ .
\tag{1.83}
$$

This result shows that the work required to move a mass, m_2, from point $\boldsymbol{x}_A$ to point $\boldsymbol{x}_B$ is the gravitational potential difference between A and B, with B being at the higher potential. It also shows that in moving a point mass from A to B, it does not matter what route is taken. All that matters is the potential difference between $\boldsymbol{x}_A$ and $\boldsymbol{x}_B$. So the gravitational force is indeed conservative, since the change in potential energy is not a function of the path taken between the initial and final points.

Although the results in this subsection were derived using the gravitational interaction with the Earth as an example, they apply to any body.

Example #6: Let us now consider the work involved in lifting a mass m_2 from the Earth's surface, r_{earth}, to r under the assumption that $r = r_{\text{earth}} + h$ with $h \ll r_{\text{earth}}$ (e.g., $r_{\text{earth}} = 6400$ km and $h = 100$ km). The total work needed to lift a mass m_2 from the Earth's surface to a point at r is

$$
W = -Gm_1 m_2 \left[\frac{1}{r} - \frac{1}{r_{\text{earth}}} \right] = -Gm_1 m_2 \left[\frac{h}{r_{\text{earth}}^2} \right] = -m_2 gh \ ,
$$

where

$$
g = |\boldsymbol{g}| = \frac{Gm_1}{r_{\text{earth}}^2} \ .
\tag{1.84}
$$

Example #7: Using the concept of gravitational potential, we can determine the minimum initial velocity an object (e.g., a 1000-kg rocket) must have to escape the gravitational field of the Earth under the assumption of a non-rotating Earth. *Escape* means it is able to move infinitely far from the Earth's gravitational field. For example, the Moon is *infinitely* far away from the Earth, so the Apollo spacecraft needed to escape from the Earth's gravitational field to get there. Because the gravitational force is conservative, the mechanical energy of the system must be conserved. The total mechanical energy for an object in orbit is the sum of its kinematic and potential energy. Remember that kinetic energy is the energy due to the movement of an object with mass. If we denote by "1" and "2," the conditions at the moment the mass is ejected and when it is away from the gravitational attraction of the Earth, respectively, we then write

$$E_{K1} + m_2 U_1 = E_{K2} + m_2 U_2$$

or

$$\frac{1}{2} m_2 v_1^2 - \frac{G m_1 m_2}{r_1} = \frac{1}{2} m_2 v_2^2 - \frac{G m_1 m_2}{r_2} \,, \qquad (1.85)$$

where E_{K1} and E_{K1} represent kinematic energies and U_1 and U_2 are potential energy. The minimum launch speed for escape, which is called the escape speed, will cause the rocket to stop ($v_2 = 0$, hence, kinematic energy becomes zero) only as it reaches $r_2 = \infty$ (i.e., gravitational energy becomes zero). Then we find the escape speed is

$$E_{K1} + m_2 U_1 = 0 \iff \frac{1}{2} m_2 v_1^2 = \frac{G m_1 m_2}{r_1}$$

and therefore

$$v_1 = \sqrt{\frac{2 G m_1}{r_1}} = 11.12 \, \text{km/s} \,.$$

Remember that m_1 and r_1 are the mass and radius of solid Earth, respectively. For Mars, Venus, Mercury, Moon, Jupiter, Saturn, Uranus, Neptune, and Sun, the escape velocities are 5.0 km/s, 10.3 km/s, 4.2 km/s, 2.3 km/s, 61 km/s, 37 km/s, 22 km/s, 25 km/s, and 618 km/s, respectively. We simply replace m_1 and r_1 with the mass and radius of the source under consideration, accordingly, to obtain these speeds.

Let us recall that kinetic energy is associated with the speed of an object. If a body has mass m and speed v, the kinetic energy is $E_K = \frac{1}{2} m v^2$. Note that while the velocity of an object may be positive or negative (depending on direction), the kinetic energy is always positive. When the object motion is rotational, the kinetic energy is $E_R = I \omega^2$, where I is the moment of inertia of the object and ω is its angular velocity (see also Ikelle, 2023b). For a solid cylinder of radius R, mass m, spinning around its symmetry axis, $I = m R^2$. Again the gravitational potential energy is $E_g = mgh$ for a mass m that is at a height h above the ground.

The Poisson and Laplace equations: The gravitational field of a planet is caused by its density. The mass distribution of a planet is inherently three-dimensional, but the most we can do is measure the gravitational acceleration at the Earth's surface. Can we reconstruct the mass distribution of the planet from the surface measurements? The answer is yes, thanks to a fundamental relationship known as Gauss's theorem (i.e., the surface integral of a vector field, g, over a closed surface is equal to the volume integral of its divergence, $\text{div}(g) = \nabla \cdot g$), described in Ikelle (2020), which links surface measurements and the properties of the whole planet, if we can solve the resulting equations.

Suppose we measure g everywhere at the surface ∂D bounding the domain D with an outwardly pointing unit vector n. In the following we assume that the origin of our coordinate system is at the planet's center, that $r = |\boldsymbol{x} - \boldsymbol{x}_0| = |\boldsymbol{x}|$. Gauss's theorem tells us that:

$$\int_{\partial D} [\boldsymbol{g}(\boldsymbol{x}) \cdot \boldsymbol{n}(\boldsymbol{x})] \, dS(\boldsymbol{x}) = \int_D [\boldsymbol{\nabla} \cdot \boldsymbol{g}(\boldsymbol{x})] \, dV(\boldsymbol{x})$$

$$= -\int_D [\boldsymbol{\nabla}^2 U(\boldsymbol{x})] \, dV(\boldsymbol{x}) \ . \tag{1.86}$$

We also used (1.81) in the derivation of (1.86). By assuming that ∂D is a spherical surface, we can alternatively link surface measurements and the density of the whole planet as follows:

$$\int_{\partial D} [\boldsymbol{g}(\boldsymbol{x}) \cdot \boldsymbol{n}(\boldsymbol{x})] \, dS(\boldsymbol{x}) = -\int_{\partial D} g_n(\boldsymbol{x}) dS(\boldsymbol{x}) = -4\pi R^2 \left(\frac{G m_1}{R^2} \right)$$

$$= -4\pi G \int_D \rho(\boldsymbol{x}) dV(\boldsymbol{x}) \ , \tag{1.87}$$

where $g_n(\boldsymbol{x})$ being the component of $\boldsymbol{g}(\boldsymbol{x})$ normal to $dS(\boldsymbol{x})$. We use the fact that, within a spherically symmetric body, the gravitational acceleration $\boldsymbol{g}$ is determined only by the mass between the observation point at $r = |\boldsymbol{x}|$ and the center of mass. Equating the two ways of going from surface measurements to potential and density inside the planet, we obtain the Poisson's equation:

$$\boldsymbol{\nabla}^2 U(\boldsymbol{x}) = 4\pi G \rho(\boldsymbol{x}) \ . \tag{1.88}$$

We use the fact that the equality of (1.87) and (1.88) hold for their integrands everywhere. Poisson's equation is fundamental. It requires no information about exactly how the density is distributed within D. The homogeneous case (i.e., no material in a region of space; density is zero) of the Poisson's equation gives the Laplace's equation:

$$\boldsymbol{\nabla}^2 U(\boldsymbol{x}) = 0 \ . \tag{1.89}$$

This result is just a statement that field lines cannot start or end in a region where there is no mass. Since all of the planet's mass lies inside the surface, the gravitational potential exterior to the planet satisfies Laplace's equation.

Spherical harmonics: We are now left with the challenge of solving Poisson's equation. We present here the solution to this equation without proof because of the lengthy derivations associated with this solution. Furthermore, these derivations require a mathematics background beyond that of the intended readers of this book.

For the Earth, it is advantageous to use spherical coordinates because of its shape. They are related to Cartesian, as follows:

$$\begin{cases} x = r \sin \theta \cos \phi \\ y = r \sin \theta \sin \phi \\ z = r \cos \theta \end{cases} \ , \tag{1.90}$$

θ is the co-latitude (θ varies between 0 and π) and ϕ is the longitude (ϕ varies between 0 and 2π). See Figure 1.18 for more details. A very important thing to

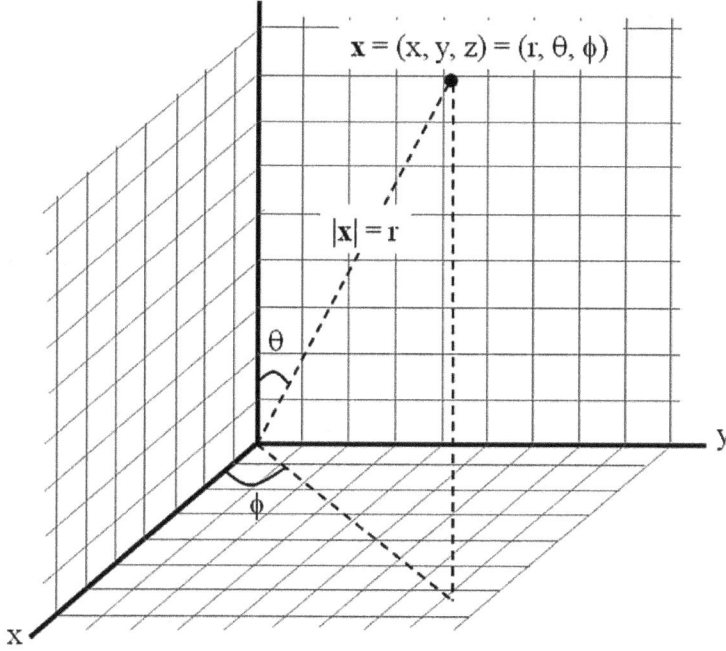

Figure 1.18. Spherical and Cartesian coordinates with $r = |\boldsymbol{x}| \in [0, \infty]$, $\theta \in [0, \pi]$, and $\phi \in [0, 2\pi]$.

realize is that, while the Cartesian frame is described by immobile unit vectors, the spherical frame is described by unit vectors that depend on where the point is located. After some algebra, we can write the spherical Laplacian as

$$\boldsymbol{\nabla}^2 U(\boldsymbol{x}) = \frac{1}{r^2} \frac{\partial}{\partial r} \left(r^2 \frac{\partial U}{\partial r} \right) + \frac{1}{r^2 \sin \theta} \frac{\partial}{\partial \theta} \left(\sin \theta \frac{\partial U}{\partial \theta} \right) + \frac{1}{r^2 \sin^2 \theta} \left(\frac{\partial^2 U}{\partial \phi^2} \right) = 0. \quad (1.91)$$

The generic solution for U is inside the mass distribution (e.g., Hubbard, 1984)

$$U^{(in)}(r, \theta, \phi) = -\frac{Gm_1}{R} \sum_{n=0}^{\infty} \sum_{m=0}^{n} \left(\frac{r}{R} \right)^n \left[A_n^{m,i} \cos m\phi + B_n^{m,i} \sin m\phi \right] P_n^m(\cos \theta) \quad (1.92)$$

and outside we have

$$U^{(out)}(r, \theta, \phi) = -\frac{Gm_1}{R} \sum_{n=0}^{\infty} \sum_{m=0}^{n} \left(\frac{R}{r} \right)^{n+1} \left[A_n^{m,o} \cos m\phi + B_n^{m,o} \sin m\phi \right] P_n^m(\cos \theta). \quad (1.93)$$

A pair (n, m) describes a solid spherical harmonic of degree n and order m. $P_n^m(x)$ are associated Legendre functions (see Ikelle, 2020). The process of determining $A_n^{m,i}$, $B_n^{m,i}$, $A_n^{m,o}$, and $B_n^{m,o}$ is just like determining the components of a vector in a given orthogonal basis. Actually, the notion of orthonormal basis representation is not limited to vector spaces. It can also be used for spaces made of functions as long as an appropriate scalar product can be defined (see Ikelle and Amundsen, 2018, and Ikelle, 2023b). By multiplying, say, $U^{(out)}(r, \theta, \phi)$, by $\cos m'\phi P_1^m(\cos \theta)$, integrating over θ, and using the orthogonality of associated

Legendre functions (see Ikelle, 2020), we obtain $A_l^{m,o}$. In theory an infinite series is necessary to fully characterize the gravitational field, but practically the double sum is truncated to a maximum of about n between 100 and 200.

Gravitational gradiometry principle. In addition (or as an alternative) to measuring the gravity vector from one accelerometer, we can measure the gradiometry tensor (or gradiometry matrix), which is defined as follows (e.g., Baur and Grafarend, 2006):

$$\mathbf{\Gamma} = \begin{pmatrix} \frac{\partial g_x}{\partial x} & \frac{\partial g_x}{\partial y} & \frac{\partial g_x}{\partial z} \\ \frac{\partial g_y}{\partial x} & \frac{\partial g_y}{\partial y} & \frac{\partial g_y}{\partial z} \\ \frac{\partial g_z}{\partial x} & \frac{\partial g_z}{\partial y} & \frac{\partial g_z}{\partial z} \end{pmatrix} = - \begin{pmatrix} \frac{\partial^2 U}{\partial x^2} & \frac{\partial^2 U}{\partial y \partial x} & \frac{\partial^2 U}{\partial z \partial x} \\ \frac{\partial^2 U}{\partial x \partial y} & \frac{\partial^2 U}{\partial y^2} & \frac{\partial^2 U}{\partial z \partial y} \\ \frac{\partial^2 U}{\partial x \partial z} & \frac{\partial^2 U}{\partial y \partial z} & \frac{\partial^2 U}{\partial z^2} \end{pmatrix} , \qquad (1.94)$$

from an array of accelerometers, especially when taking measurements using satellites or airplanes. This tensor is a symmetric tensor because the order of differentiation of a scalar quantity is irrelevant and it has only five independent components because the sum of the diagonal components is zero (Laplace's equation in free space); i.e.,

$$\mathbf{\nabla}^2 U(\boldsymbol{x}) = \frac{\partial^2 U}{\partial x^2} + \frac{\partial^2 U}{\partial y^2} + \frac{\partial^2 U}{\partial z^2} = - \left(\frac{\partial g_x}{\partial x} + \frac{\partial g_y}{\partial y} + \frac{\partial g_z}{\partial z} \right) = 0 .$$

These five components can, for example, be $(\partial g_x/\partial x,\ \partial g_x/\partial y,\ \partial g_x/\partial z,\ \partial g_z/\partial y,\ \partial g_z/\partial z)$. The practical implementation of the gradiometry tensor consists of taking differences of the accelerometers as follows:

$$\Gamma_{jk} = \frac{g_j(x_k + \Delta x_k) - g_j(x_k)}{\Delta x_k} , \qquad (1.95)$$

where $j, k \in \{x_1, x_2, x_3\} = \{x, y, z\}$ and Δx_k is the short distance between accelerometers in the x_k-direction. By measuring higher derivatives of gravity, small scale features can be accurately determined.

Geoid.[3] There is an infinite number of equipotential surfaces and a particular equipotential surface of the Earth known as geoid. It coincides with the mean sea level (ignoring tides and other dynamical effects in oceans); that is, the geoid is the equipotential surface that best fits the mean sea level. Geoid would be the Earth's surface if the oceans were to flow freely and cover the entire solid Earth's surface (in the absence of winds, tides and currents). The surface of the oceans would align with the surface of the geoid, so the geoid is a theoretical shape of the Earth. The choice of mean sea level is totally arbitrary but it makes sense because the surface of a fluid in equilibrium must follow an equipotential surface. So over the oceans, the geoid is the ocean surface. Over the continents, the geoid is not the topographic surface (its location can be calculated from gravity measurements). The geoid is

---

[3]Representing the shape of the Earth with all the variation of its complex surface topography (e.g. mountains and valleys) would be an impossible task. An entire branch of science, called geodesy, is dedicated to the study of the shape and size of the Earth. Geodesists have defined two main models or reference surfaces, namely geoid and ellipsoid, to approximate the Earth's shape.

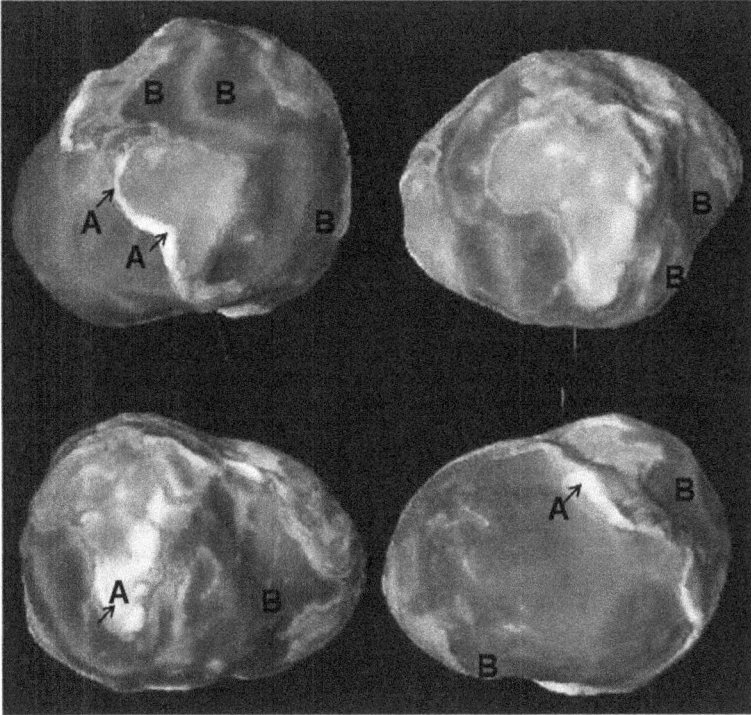

Figure 1.19. Four photos of the geoid (also known as *The figure of the Earth*) recorded by satellites. Geoid is fatter at the equator and thinner at the poles. The "B" areas represent areas where the geoid is below the reference ellipsoid and the "A" areas are areas above the reference ellipsoid. The Indian Ocean has the largest negative geoid anomalies [Credit: GOCE (ESA)].

recorded by satellites that use radar technology to measure the elevation of the sea surface. As illustrated in Figure 1.19, the geoid is fatter at the equator and thinner at the Poles. Remember that the geoid is an equipotential surface therefore it must remain perpendicular to the direction of the gravity field. The bumps and hollows of geoid inform us of the submarine topography and the composition of the subsurface.

The reference ellipsoid is another (theoretical) shape of the Earth. It is the ellipsoid that most closely fits the geoid. It can be shown (Clairaut and Colson, 1739; and Blakely, 1995) that the (theoretical) value of gravity on this reference ellipsoid is:

$$
\begin{aligned}
g(\theta) &= g_{\text{eq}} \left(\frac{1 + k \sin^2 \theta}{\sqrt{1 - e^2 \sin^2 \theta}} \right) \\
&\approx g_{\text{eq}} \left(1 + k \sin^2 \theta \right) \left(1 + \frac{1}{2} e^2 \sin^2 + \frac{3}{8} e^4 \sin^4 \theta \right) \\
&= g_{\text{eq}} \left(1 + k_1 \sin^2 \theta + k_2 \sin^4 \theta \right) \\
&= g_{\text{eq}} \left[1 + (k_1 + k_2) \sin^2 \theta - \frac{k_2}{4} \sin^2(2\theta) \right],
\end{aligned}
\tag{1.96}
$$

where θ is the latitude, g_{eq} is the gravity at the equator, $e^2 = (a^2 - b^2)/a^2$ (a is the semi-major axis and b is the semi-minor axis) is the eccentricity of the ellipsoid, $k = (b\, g_{\text{pole}} - a\, g_{\text{eq}})/(a\, g_{\text{eq}})$ is a constant that depends on the shape and the rotation of the Earth, and g_{pole} is the gravity at the poles. The other two parameters in the last equality are related to k and e as follows: $k_1 = k + \frac{1}{2}e^2$ and $k_2 = \frac{1}{2}ke^2 + \frac{3}{8}e^4$. The typical values of g_{eq}, k, and e are $g_{\text{eq}} = 978031.846$ m/s^2, $k = 0.001931851$, and $e^2 = 0.006694379$ and we can deduce $k_1 = 0.005278895$ and $k_2 = 0.000023462$. These values are more accurately determined by the increasing amounts of satellite data. We used Taylor's expansion in (1.96) and limited the calculations to the third term of the Taylor's series to obtain the final equality. So if the Earth were an homogeneous ellipsoid, the value of gravity at the Earth's surface would be given by (1.96).

Quiz: What is the farthest location on the Earth's surface from the center of the Earth?

Geoids are used in oceanography to determine (mean) sea surface topography. In polar regions where oceans are partly covered by sea ice, icebergs, or ice shelves, the geoid also provides a link between the surface ellipsoidal height and the freeboard height, which in turn, can be used to infer the thickness of the floating ice, including Antarctic subglacial lakes (Ewert et al., 2012). In geophysics, analysis of gravitational anomalies yields insight into the thicknesses of the lithosphere and into the tectonic and geodynamic processes that shape the continents and surrounding oceans.

Again, note that we cannot directly see the geoid surface and reference ellipsoid; therefore, we cannot actually measure the heights above or below the geoid surface. We must infer where the geoid surface is by making gravity measurements and where the reference ellipsoid is by modeling it mathematically or numerically.

Geoid height anomalies: Geoid undulations are differences, in meters, between the geoid and the reference ellipsoid. Geoid undulations are also known as geoid height anomalies. They are caused by the distribution of mass in the solid Earth, the topography of the seafloor and the Earth's surface, and mantle processes. They are generally denoted by N which is the ellipsoidal height with respect to the geoid (see Figure 1.20). In absolute values, N, is less than 100 m always everywhere—that is, about $10^{-5} \times r_e$, where r_e is the Earth radius. A map of global geoid height anomalies is shown in Figure 1.21. The width of the geoid height anomalies indicates the depth of the density anomalies that cause them. When underground density anomalies are near the surface, the geoid height anomalies are narrow. When density anomalies are very deep, the geoid height anomalies are very broad, sometimes thousands of kilometers across. These large-scale geoid height anomalies are a result of the large upwellings (light color) and downwellings (dark color) of mantle convection, which we will discuss in later chapters.

As illustrated in Figure 1.21, there are significant geoid height anomalies over the ocean due to bathymetry (i.e., the topography of the ocean floor). For example, over the oceans, there are seamounts, which lead to water being replaced by rock. The rock increases the gravitational pull in that area and consequently changes the

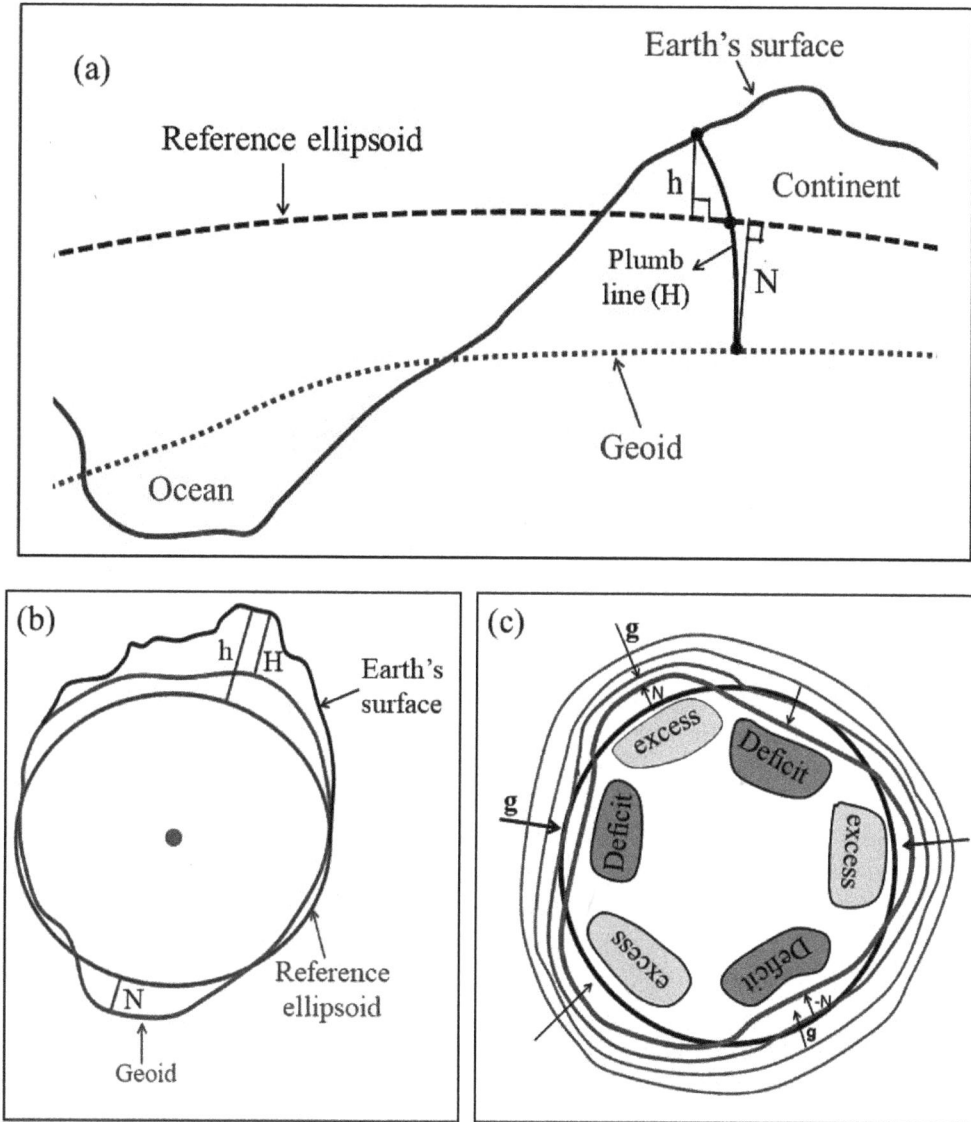

Figure 1.20. (a) Geoid and its relationship with the reference ellipsoid and Earth's surface topography. Differences between geoid and reference ellipsoid along the normal of the ellipsoid are known as geoid height (or undulations) anomalies, N. The geoid height is negative everywhere in the continental US except in Alaska. h is the difference between the Earth's surface and the reference ellipsoid along the normal of the ellipsoid. It is known as ellipsoid height. When the geoid does not coincide with the reference ellipsoid, $h \approx H + N$ where H is the elevation above sea level along the plumb line. H is also known as the orthometric height. In Houston (Texas), we have $N = -27.5$ meters, $h = 13$ meters, and $H = 40.5$ meters. (b) An illustration of a horizontal slice of geoid, reference ellipsoid, and Earth's surface topography. (c) The variations of gravity versus the uneven distribution of masses inside the Earth. The thick red curve is the geoid and the thick black curve is the reference ellipsoid.

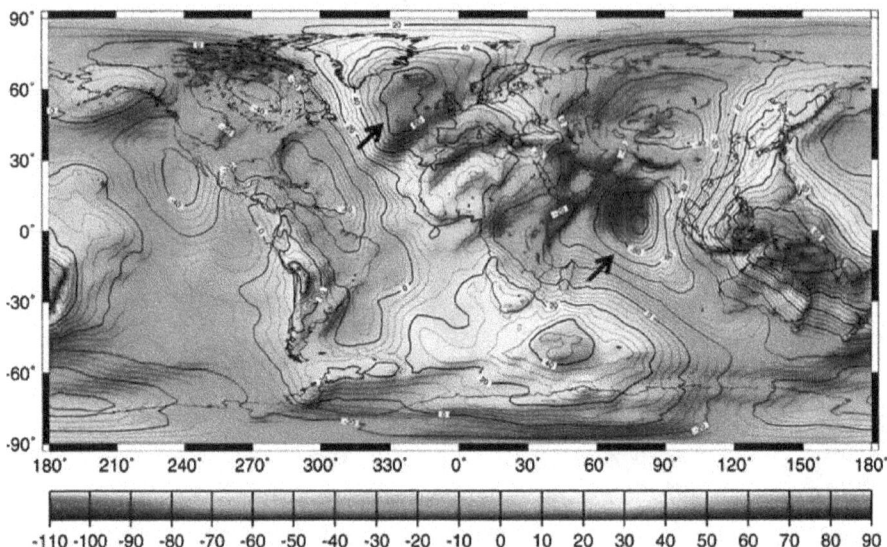

Figure 1.21. Here is a world map of the differences between the geoid and the reference ellipsoid. The heights above this reference are positive (toward +90) and below (toward −110) are negative. In absolute values, these geoid undulations are less than 100 meters everywhere. These variations are very broad, sometimes thousands of kilometers across as indicated here with two dark arrows. They are the result of broad differences in density within the Earth's deep interior. Largest anomalies are associated with oceanic trenches.

level of the sea surface. Note that the sea surface is elevated above seamounts, ridges, and underwater plateaus. Largest negative anomalies are associated with the deepest trenches (10 km or deeper and filled with water rather than rock, as illustrated in Ikelle, 2023a), large negative amplitudes, and large positive values over the landward island arcs.

Gravitational anomalies and corrections. If the Earth were a perfectly homogeneous sphere with no lateral inhomogeneities and did not rotate, g would be the same everywhere. Unfortunately this is not the case. The Earth is inhomogeneous and it rotates. Rotation causes the Earth to be an oblate spheroid with an eccentricity of 1/298. The polar radius of the Earth is 20 km less than the equatorial radius, which means that g is 0.4 percent less at the equator than at the Poles. At the equator, $|g|$ is about 5300 mGal (milliGals). Gravity measurements at the Earth's surface show that there are more gravitational variations than the slight changes in gravity, as the elliptical nature may suggest. These variations of $|g|$ depend on two factors: topography of the seafloor and the Earth's surface (e.g., mountains, valleys, plains, ocean ridges, and deep ocean trenches) and density variations in the subsurface (e.g., density variations in crust and mantle, salt domes, sediment basins, and ores). These variations are known as gravity anomalies. They are local variations and have amplitudes of several mGal.

There are also variations of $|g|$ with time due to tides, atmospheric pressure, ocean topography, water masses, and polar motion. These variations are important for studies of weather, climate, and water storage over continents but not for studies of gravity anomalies related to density changes. For the most part, the variations of the subsurface density are constant over thousands and even millions of years.

Terrestrial gravity measurements are taken at almost constant periods of time with respect to these temporal gravity variations and therefore their effects are negligible. However, these temporal variations to $|g|$ are important in satellite gravity measurements because the satellite recordings are almost continuous with time, as we will describe later. In this case, averages of the gravity field over a period of weeks or months are made to cancel out temporal gravity variations before using them for estimating variations in subsurface density.

Let us reiterate that density (mass per unit volume) is one of the most important parameters of the solid Earth. All processes inside solid Earth are in one way or another related to the non-equilibrium distribution of masses. Masses of the major features of solid Earth cannot be determined directly. However, density can be measured in the upper part of the Earth's crust (less than 8 km) with the use of core-rock samples from drilled boreholes. In deeper parts of the Earth, density can be determined from velocities of seismic waves (see Ikelle, 2023b) and gravity anomalies.

Quiz. (1) What is the variation in range of g on the Earth's surface? (2) What are the differences between topographic surface, ellipsoidal surface, and the geoid?

Free-air correction. As we have seen from the Newton's law of gravity in Ikelle (2023b), the gravity decreases with the square of distance. Hence, if we lift the gravimeter from the point R0 on the surface (or any other datum) to M0, as illustrated in Figure 1.22, gravity will change. To be able to compare data measured at different elevations, we have to reduce them to a common datum (i.e., we have to regularize the data). This correction for the topographic effects has several steps, free-air correction being the first step.

As we discussed in this chapter so far, subsurface mass distribution is directly related to the density of rocks which is different at different locations (see Ikelle,

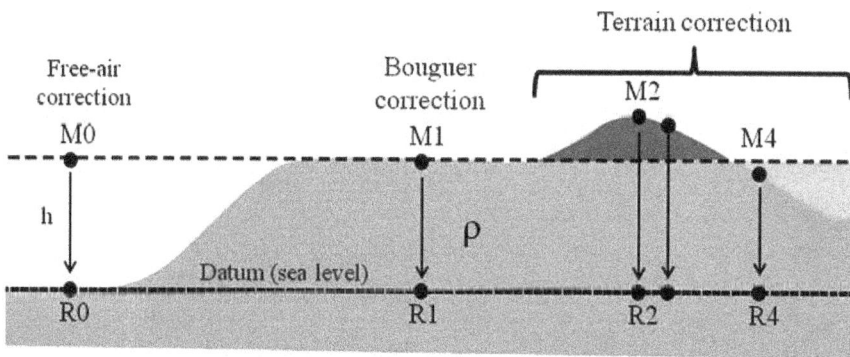

Figure 1.22. Topographic and bathymetric corrections. On the leftmost side, we have the free-air correction. It is mostly used for marine surveys and for investigating deep mass distribution. In the center, we have the Bouguer correction. This correction strips away everything above sea level. It is the correction most commonly made in geophysical prospecting. It requires the knowledge of the density, ρ of the slab. On the rightmost side, we have the terrain correction for mountain belts and valleys on the Earth's surface. M0, M1, M2, and M4 are the actual measurement points (i.e., locations of gravity stations) and R0, R1, R2, and R4 are the extrapolated measurement points after topographic and bathymetric corrections.

2023a). If there are denser rocks underground, the resulting increased mass will cause a greater gravitational force. In other words, the density of rocks, which is the only physical property that causes variations in the gravitational field, accounts for the majority of gravity anomalies (differences from the expected). Another significant cause of gravity variations at the surface is due to topography on land and bathymetry (the topography of the ocean floor). These variations are not directly related to any physical property in the subsurface. Consequently, in geophysics studies, especially in the prospect for mineral deposits, we are only interested in the density variations in the subsurface. Hence, the removal of unwanted components of $|\boldsymbol{g}|$ related to topography of land and bathymetry are needed. These removals are often referred to as reduction or correction. We here describe some of the key corrections (see also Figure 1.22) assuming that the geoid surface and reference ellipsoid are known.

Suppose that a given gravity measurement was made at an elevation h (point M0) above sea level. Free-air correction corrects for the change in the elevation of the gravimeter, considering only air (leftmost part of Figure 1.22) being between the gravimeter and selected datum point R0. The free air anomaly is then determined as follows:

$$\Delta g_{\text{fa}}(\theta) = g_{\text{obs}} - g_0(\theta) + \delta g_{\text{fa}} = g_{\text{obs}}(\theta) - g_0(\theta)\left(1 - \frac{2h}{r_e}\right), \qquad (1.97)$$

with

$$\delta g_{\text{fa}} = \frac{2h}{r_e}g_0(\theta) \approx 0.308769 \times h , \text{ (in milliGals if } h \text{ is in m)} \qquad (1.98)$$

where g_{obs} is the measured gravity at a particular location, g_0 is the theoretical gravity given by a reference ellipsoid Earth model (i.e., the value of gravity at the Earth's surface if the Earth were an homogeneous ellipsoid), r_e is the Earth radius, and δg_f is the free-air correction. When the geoid does not coincide with the reference ellipsoid, one can simply replace h by $h' = h + N$, where N is the geoid anomaly (i.e., the geoid height or undulations). Note that if the gravimeter is at a higher elevation than the reference ellipsoid, gravity at the recording station will increase after the free-air correction is applied.

Box 1.2: Gravity and Inertia

With a few exceptions, celestial bodies moving in space do not stop if there is no atmosphere or ground resistance to slow them down. Space itself does not have gravity–that is, its gravity is zero. However, celestial bodies generally have gravity since density directly affects gravity. This is why larger planets have stronger gravity and our Sun has the greatest mass, the strongest gravitational pull. Planetary bodies are made in a variety of ways, but various forces get them in motion. No matter how slowly they move, they will not stop. When the Sun and a smaller celestial body, such as a comet, come within range of each other, the greater gravitational pull wins. Either the celestial body will hit the Sun straightaway or it will eventually hit the Sun after moving in an ellipse. Alternatively, it will take many circles around the sun before it is pulled in.

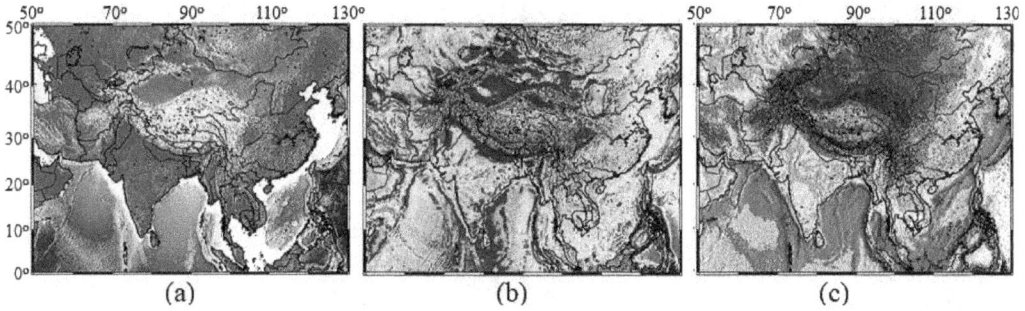

Figure 1.23.   Section of the gravity anomaly map of the Tibet region. (a) Topography and bathymetry of the Tibet region. (b) The free-air-gravity anomaly in Tibet. (c) Bouguer gravity anomaly map of the Tibet region. (Adapted from Fowler, 1990 and https://wiki.tfes.org/Gravimetry).

In general, the free-air-gravity anomaly map of a given region correlates well with its topography in most regions and continents of the Earth, including the Tibet region (see Figure 1.23). Even on extraterrestrial planets like Mars, we have a strong correlation between the topography and the free-air gravity profile, as illustrated in Figure 1.24.

The Bouguer and terrain corrections. The Bouguer correction removes from the data an effect of rocks lying between the measurement point M1 and reference datum point R1, as illustrated in Figure 1.22 (the center). Ignoring the terrain correction, the Bouguer gravity anomaly is then given by

$$\Delta g_B = g_{\text{obs}} - g_0(\theta) + \delta g_{\text{fa}} + \delta g_B$$

$$= g_{\text{obs}} - g_0(\theta) \left[1 - \frac{2h}{r_e}\right] \underbrace{-2\pi G \rho h}_{\text{Bouguer}}, \tag{1.99}$$

with

$$\delta g_B = 2\pi G \rho h \approx -0.04193 \times 2.67 \times h, \ (\text{in mG if } h \text{ is in m}) \tag{1.100}$$

where G is the universal gravitational constant, ρ is the assumed mean density of crustal rock and h is again the height above sea level. The density of the topography, ρ, is generally taken to be 2.67 g/cm^3.

The Bouguer correction assumes the slab to be infinite in the horizontal direction. This assumption is not always true due to the Earth's surface topography (terrain). It is straightforward to apply the terrain correction if one has access to digital topography/bathymetry data. We simply divide the surrounding of the gravity stations into zones, estimate average altitude in every zone and compute the gravity effect of the zones based on the Bouguer correction. Knowledge of the density of subsurface rocks is essential for the Bouguer and terrain corrections.

Gravity prospecting. One use of gravity is that gravity measurements help map the buried craters. One of the most fascinating discoveries is the buried crater on the Yucatan peninsula in Mexico (e.g., Alvarez et al., 1980). The crater was filled

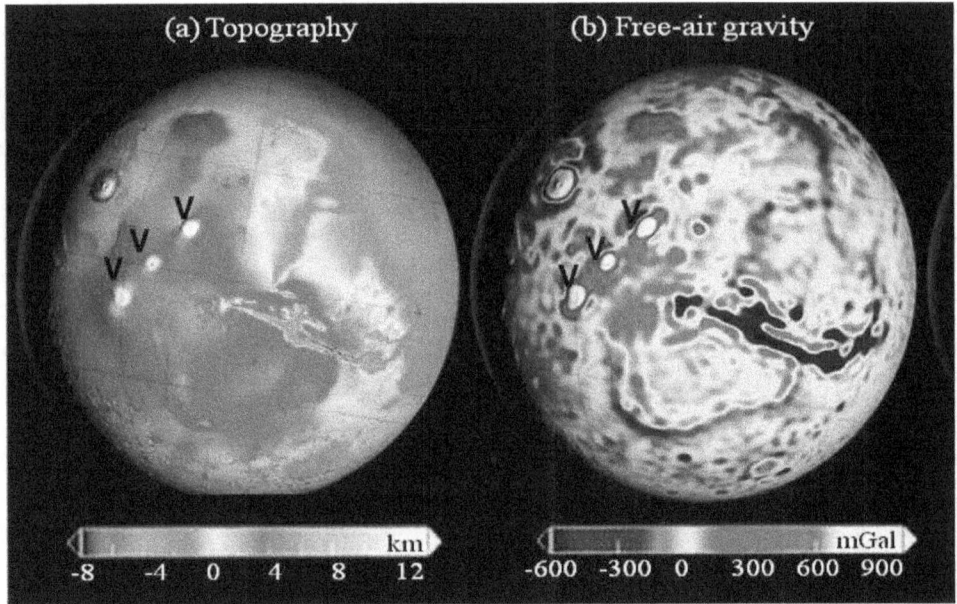

Figure 1.24. (a) MOLA (Mars Orbiter Laser Altimeter)-based topography of Mars showing highlands (2 kilometers) and lowlands (< -4 km). Tall volcanoes are indicated by "V." (b) Free-air gravity. Notice the correlation between the topography and the free-air-gravity anomaly map. Because Mars has no oceans and hence no "sea level," zero altitude is defined as the height at which there is 610.5 Pa (6.105mbar) of atmospheric pressure. (Credit: NASA's Scientific Visualization Studio.)

in by sedimentary rock that was lighter than the original rock, so even though it is filled, it shows a gravity anomaly. An airplane flying back and forth over this region made sensitive measurements of the strength of gravity, and they produced the map shown in Figure 1.25. In this map, the crater is estimated to be 150 kilometers in diameter and 20 km in depth, well into the continental crust of the region of about 10-30 km depth. It is believed the crater was left behind when an asteroid or comet killed the dinosaurs and many animal species about 66 million years ago. The impact is also believed to have delivered an estimated energy of about 21 billion Hiroshima A-bombs.

Let us briefly describe the key features of gravimeters. Modern gravimeters are designed to measure the free fall of a test mass (see Figure 1.26a). The free-fall motion is carried out in a vacuum to avoid air resistance (see Figure 1.26b). The parameters of this motion measured by components of the gravimeters are the pair of heights and times (z_i, t_i') along the vertical axis, as illustrated in Figure 1.26a, with i taking values from 0 to, say, 1,000 at a distance of about 20 cm and a time of about 0.2 seconds. Laser interferometers are used to determine (z_i, t_i'). The gravity is deduced from these pairs.

The motion equation of a free-falling mass in the gravity field in the rotating coordinate system on the Earth is (e.g., Nagornyi, 1995)

$$\frac{\partial^2 z}{\partial t^2} = g_0 + \gamma_z z \,, \tag{1.101}$$

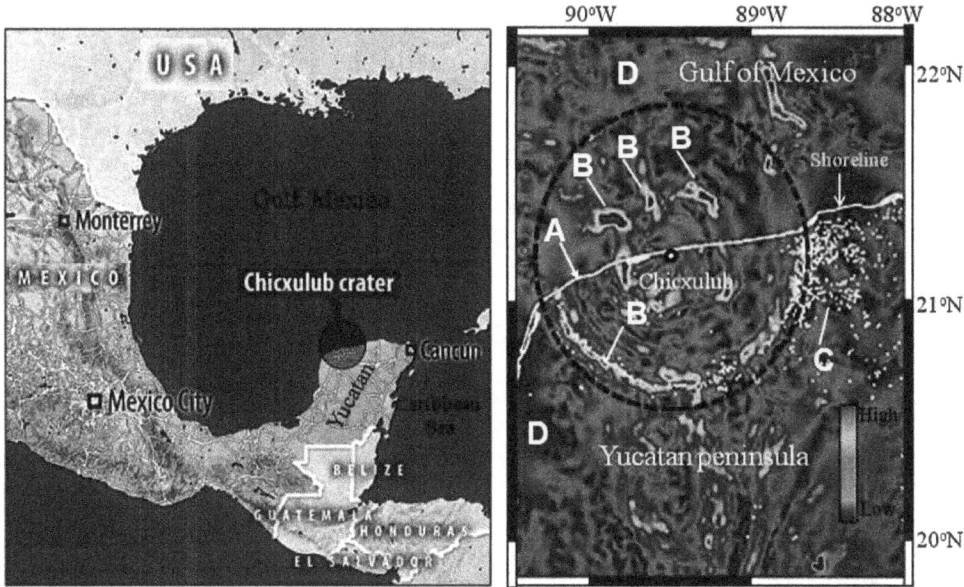

Figure 1.25. (a) Location of Mexico's Yucatán Peninsula, including the Chicxulub impact crater. This crater was discovered by Antonio Camargo and Glen Penfield, geophysicists who were looking for hydrocarbons in the Yucatán in the late 1879s. (b) Gravity anomaly map of the Chicxulub impact area. The coastline is shown as a "A" line. "C" dots represent cenotes (water-filled sinkholes); "B" structures are gravity highs; and "D" structures are gravity lows.

with

$$\gamma_z = \frac{\partial g_z}{\partial z} \, , \tag{1.102}$$

where g_0 is the gravity value at $z = 0$ and γ_z is the vertical gradient of the gravity field. An approximate solution of equation (1.101) for $\gamma_z \ll 1$ is (e.g., Timmen, 2003)

$$z_i = z_0 + v_o t_i + \frac{1}{2} \left(g_0 + z_0 \gamma_z \right) t_i^2 + \frac{1}{6} v_0 \gamma_z t_i^3 + \frac{1}{24} g_0 \gamma_z t_i^4 \, , \tag{1.103}$$

with

$$t_i = t_i' + \frac{z_i}{c_0} \, , \tag{1.104}$$

where the initial coordinate and velocity of the test mass are $z_0 = z(t = 0)$ and $v_0 = v(t = 0)$, respectively, and c_0 is the speed of light. Equation (1.101) is related to the instrumentation of the gravimeter. Often there is a time delay between the time of the actual interaction of the light with the falling test mass and the time registered by the photodetector of the laser interferometer. Equation (1.101) corrects for this delay. For a significant number of pairs (z_i, t_i'), we can deduce g_0, v_0, z_0, and γ_z by least-squares optimization or statistical regression.

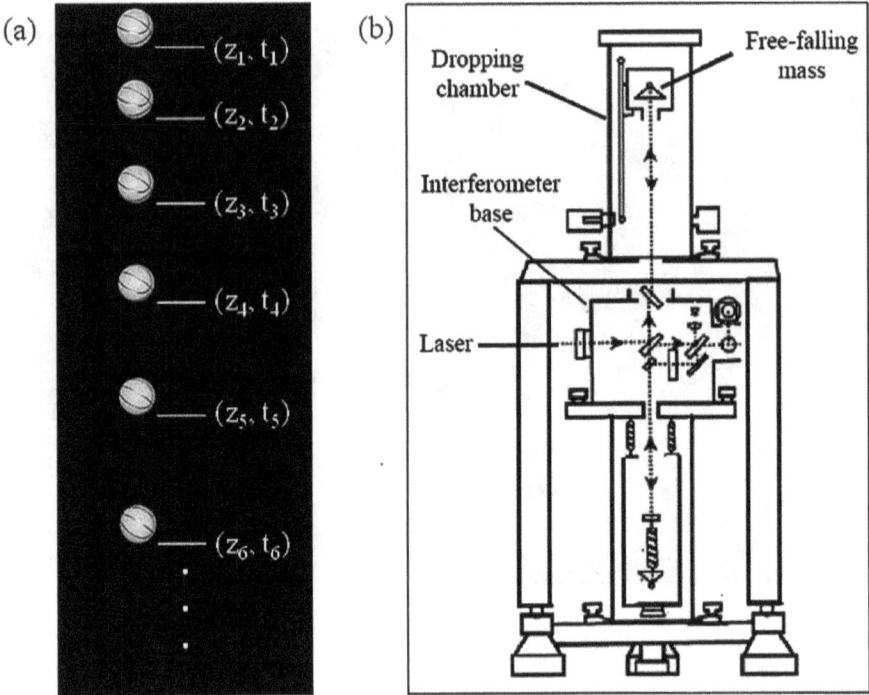

Figure 1.26. (a) An illustration of test masses in free fall. The concept of a gravimeter is based on the story of the falling test masses. Gravity can be measured between the various position-time pairs; (z_i, t_i) with $i = 1, 2, 3, \ldots$. To measure one $/mu$gal variation, time must be kept to within a nanosecond. (b) A schematic diagram of a modern gravimeter. This apparatus consists of (i) a vacuum dropping chamber with mechanics to let the test mass fall freely, and (ii) beam laser displacement interferometers to measure position-time pairs. (Source: Micro-g Solutions, Inc.)

Example #8: A buried uniform sphere. Simple geometrical bodies are often used for the gravity modeling before the survey is carried out. The aim is to get a rough estimate of the anomaly effect of the target structure. If we, for example, find that the amplitude of a modeled anomaly is lower than the sensitivity of our instrument. Then there is no need to do any measurements at all. There is a small number of simple basic bodies; however, combining them together can build up even a complex model.

The gravity measurements can be written as

$$g_{\text{pred}} = g_0 - \delta g_{\text{fa}} + \delta g_B + \delta g_z , \tag{1.105}$$

where g_0 is the gravity response of the homogeneous reference ellipsoid, δg_{fa} is the free-air contribution above the reference ellipsoid, δg_B is the contribution of excess mass above the reference ellipsoid, and δg_z is the contribution of the excess mass in the subsurface. We focus on the calculation of δg_z.

A sphere is the most basic body and can be used as a part of other models or could approximate symmetrical bodies. Let us describe how geological structures can cause gravity anomalies. We consider surface gravity measurements across a buried sphere in a homogeneous medium. The sphere has an excess mass m_s and

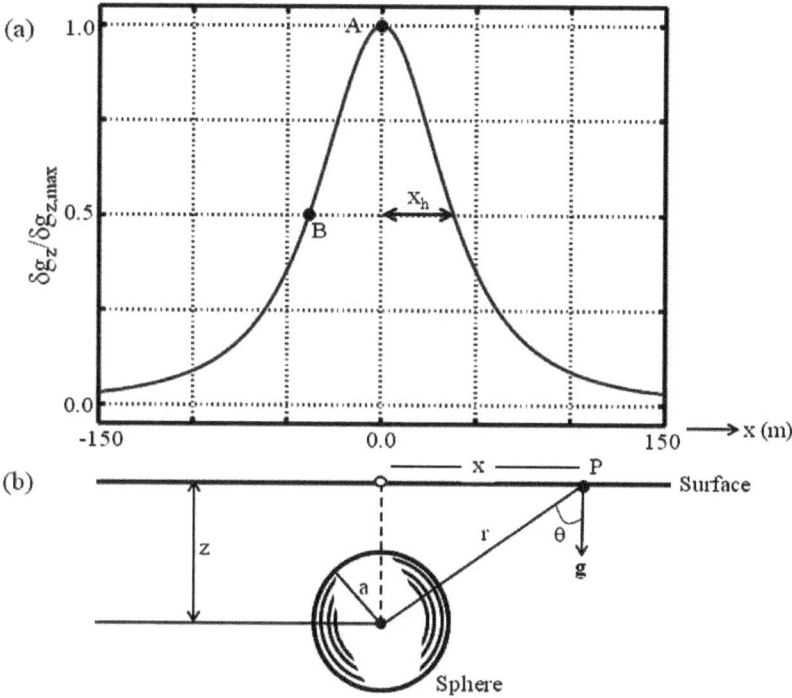

Figure 1.27. (a) The variation in δg_z plotted on a profile (anomaly of a sphere) with $z = 50$ m. (b) Gravity survey parameters.

the center is at a depth z. We want to determine the excess mass by using surface gravity measurements.

As described earlier, a sphere has the same gravitational pull as a point mass located at its center. Simple mathematics can be used to show that at point P, the vertical component of Δg is given by the equation below. The gravity effect of a sphere at point P (Figure 1.27b) is:

$$\delta g_z = |g| \cos \theta = \frac{Gm_s}{r^2} \cos \theta = \frac{Gm_s}{r^2} \frac{z}{r} = \frac{Gm_s z}{(x^2 + z^2)^{3/2}} \, , \tag{1.106}$$

with

$$m_s = \frac{4}{3} \pi a^3 \Delta \rho \tag{1.107}$$

where δg_z is the vertical gravitational effect of the buried sphere, x is the horizontal distance from the center, z is the depth of the center of the sphere, a is the radius of the cylindrical cross section, $\Delta \rho = \rho_1 - \rho_1$ is the density contrast between the homogeneous medium and the buried sphere; ρ_0 is the density of the homogeneous medium and ρ_1 is the density of the buried sphere. We can see that the gravity anomaly due to this buried sphere is symmetrical about the center of the sphere and essentially is confined to a width of about two to three times the depth of the sphere (see Figure 1.27a). Therefore, different depths of the mass m_s causes variations in the gravitational field. Mass near the ground produces high gravity. Likewise, a large mass produces high gravity. Notice that Δg_z has its maximum value, denoted here as $\Delta g_{z,\max}$, at $x = 0$ with

$$\Delta g_{z,\max} = \frac{Gm_s}{z^2} . \tag{1.108}$$

At point B, in Figure 1.27a, g_z has fallen to half the peak value. The distance x_h is called the half-width of the curve; i.e.,

$$\delta g_z = \frac{1}{2}\delta g_{z,\max} \iff x_h = \sqrt{(2^{2/3} - 1)}\, z = 0.7664\, z , \tag{1.109}$$

or $z = 1.3048 \times x_h$. In other words, we can determine z from the value of x_h and deduce the excess mass as

$$m_s = \frac{g_{z,\max} z^2}{G} . \tag{1.110}$$

The depth z of the sphere can be estimated from the half-width of the measured anomaly at half of its value.

Example #9: A buried infinite cylinder. When gravity measurements are made across a buried cylinder, we have

$$\delta g_z = \frac{2G\pi a^2 z \Delta \rho}{(x^2 + z^2)} . \tag{1.111}$$

where a is the radius of the cylindrical cross-section, $\Delta \rho = \rho_0 - \rho_1$ is the density contrast between the homogeneous medium and the buried cylinder; ρ_0 is the density of the homogeneous medium, and ρ_1 is the density of the buried cylinder. Note that the cylinder anomaly falls off more slowly than the sphere, $1/z$ rather than $1/z^2$, and that the anomaly is broader than that of the sphere. While it might be difficult to distinguish between these two bodies in a survey profile (see Figure 1.29), a map would show the parallel linear contours of the cylinder versus the circular contours of the sphere.

Quiz: Despite technological advances in the last 100 years or so, gravity measurements reflect a perceptible displacement of a mass relative to some reference (e.g., test mass). The first gravity data collected in the United States was obtained by G. Putman around 1890. Putnam used a pendulum gravity meter based on the relation between gravity and the period of a pendulum:

$$g = \frac{4\pi^2 l}{T^2} \tag{1.112}$$

where l is the length of the pendulum and T is the pendulum period. Assuming the period of a pendulum is known to be 1 s, find the pendulum displacement for a gravity anomaly of 10 mGal.

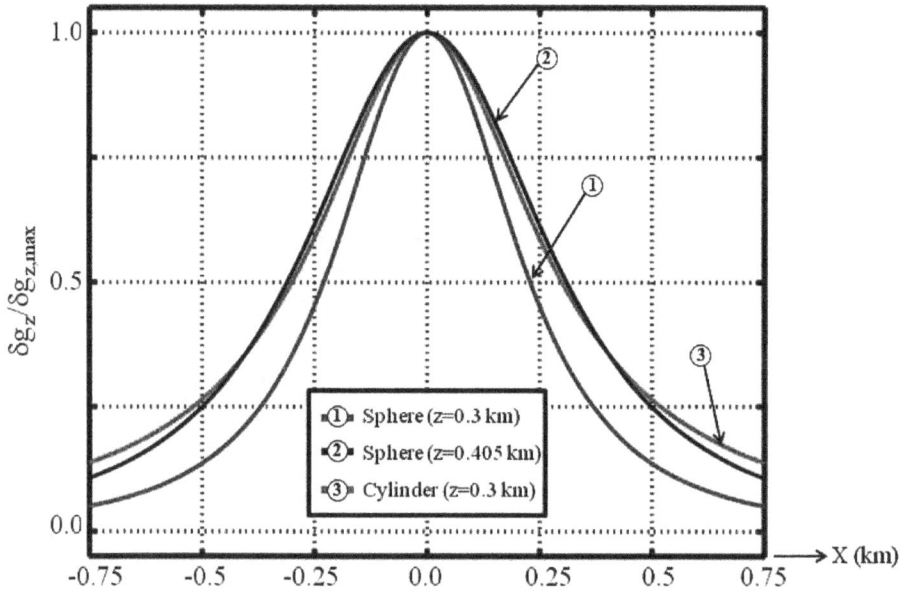

Figure 1.28. An illustration of the fact that the interpretation (inversion) of gravity data is non-unique. Spherical and cylindrical bodies can produce similar gravity profiles.

Again, notice that g_z has its maximum value, denoted here as $g_{z,\max}$, at $x = 0$ with

$$g_{z,\max} = \frac{2G\pi a^2 \Delta\rho}{z} \ . \tag{1.113}$$

The half-width distance, x_h, is now

$$g_z = \frac{1}{2}g_{z,\max} \Longleftrightarrow x_h = z \tag{1.114}$$

As illustrated in Figure 1.28, the cylinder anomaly falls off more slowly than the sphere, $1/z$ rather than $1/z^2$, and that the anomaly is broader than that of the sphere. We can see that it is difficult to distinguish a buried sphere from a cylinder with just a single profile. We need to collect gravity on a grid to detect gravitational changes in directions.

In summary, the two fundamental characteristics of the Earth's gravity field are geoid and gravity anomalies. Gravity anomalies can be used to locate regions with large mass concentrations and regions containing depressions of the landmass. The geoid is the equipotential surface of the Earth's gravity field that coincides with the mean sea level from which the anomalies are derived.

Table 1.4. The Matlab code used to produce Figure 1.28

```
clear all; G=6.67e-3; rho=1; a=1; x = -800:0.2:800; z=300;
g=((4*pi*G*rho*a∧3)/3).*z./(x.∧2+z.∧2).∧1.5; % sphere
gm = ((4*pi*G*rho*a∧3)/3)/z∧2; plot(x',g/gm); hold % blue line
g1=((2*pi*G*rho*a∧2)/1).*z./(x.∧2+z.∧2).∧1; % cylinder
g1m = ((2*pi*G*rho*a∧2)/1)/z; plot(x',g1/g1m,'r'); % blue line
z=405; g=((4*pi*G*rho*a∧3)/3).*z./(x.∧2+z.∧2).∧1.5; % another sphere
gm = ((4*pi*G*rho*a∧3)/3)/z∧2; plot(x',g/gm,'k'); % black line
```

Computational exercise: (1) Use the Matlab code in Table 1.4 to reproduce the results in Figure 1.28.

(2) Show that the gravity response of excess spherical mass over a $(x\text{-}y)$ surface can be written as

$$\delta g_z = \frac{G m_s z}{(x^2 + y^2 + z^2)^{3/2}} \,, \tag{1.115}$$

(3) Use the Matlab code in Table 1.5 to reproduce the results in Figure 1.29.
(4) Modify the code in Table 1.5 to predict the gravity response for the case in which the $(x\text{-}y)$ surface coincides with the surface of the reference ellipsoid.

1.1.5.   Satellite Gravity Missions

Producing an accurate and detailed model of the geoid has proven very challenging by using only surface gravity measurements over land, gravity measurements at sea obtained by using ships, and gravity measurements obtained by using airplanes. It is difficult to combine these gravity measurements accurately over long and global distances for the construction of the geoid, as we cannot make measurements everywhere. The resulting data were far from being continuous or globally distributed,

Table 1.5. The Matlab code used to produce Figure 1.29.

```
clear all; close all; G=6.67e-3; rho=1; a=40; z=200;
[x,y]=meshgrid(-400:0.5:400,-400:0.5:400); r=(x.∧2+y.∧2).∧0.5;
gm=((4*pi*G*rho*a∧3)/3).*z./(r.∧2+z.∧2).∧1.5;
[n,m] = size(gm); delgz1 = zeros(n,m); gfa = zeros(n,m);
r=((x-300).∧2+y.∧2).∧0.5;
gn=((4*pi*G*rho*a∧3)/3).*z./(r.∧2+z.∧2).∧1.5;
r=((x-300).∧2+(y+200).∧2).∧0.5;
gp=((4*pi*G*rho*a∧3)/3).*z./(r.∧2+z.∧2).∧1.5;
r=((x+300).∧2+(y+200).∧2).∧0.5;
gq=((4*pi*G*rho*a∧3)/3).*z./(r.∧2+z.∧2).∧1.5;
g0= 9.78032677; z=100; gfa(100:1100,600:1100) = g0*((2*z)/(1000*6370.));
z1=2.5; delrho=2.67; h=5;
delgz1=G*0.001*delrho*h*(pi/2+ -atan(x./z1)); delgz1(1200:n,:) = 0;
g = g0 -gfa - delgz1 + 0.6*(gm+(0.8*gn)+gp+gq);
contourf(x,y,g,10);
```

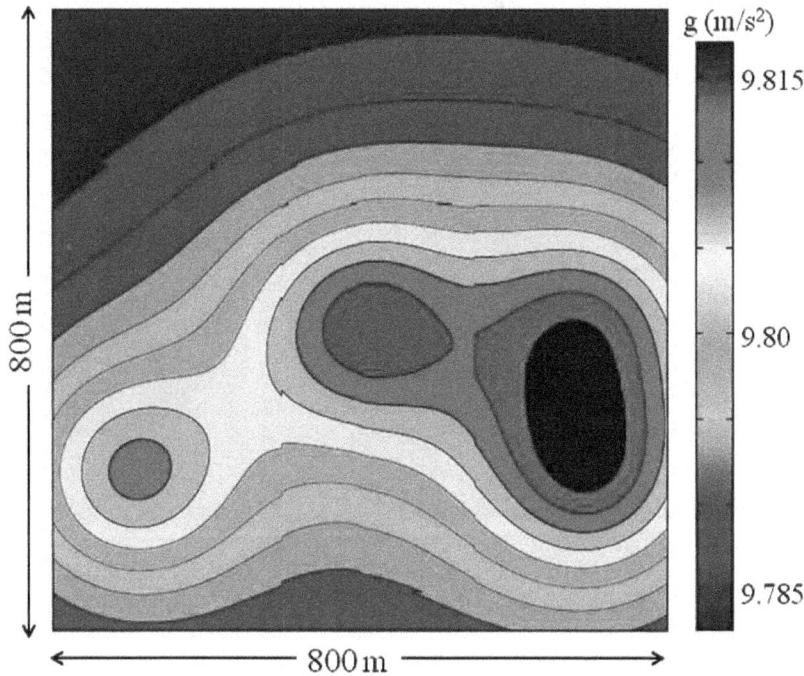

Figure 1.29. A gravity map obtained by using the code in Table 1.5. Average gravity on the Earth's surface is about $g = 9.8$ m/s^2, and varies by about 0.2% of g.

and it was only possible to achieve a rather unsatisfying geoid. A remedy was found in 1957 with the advent of artificial (man-made) satellites in low-Earth orbit (LEO[4]). Nowadays, there are satellites in orbit dedicated to scientific measurements, including gravity measurements. For example, GRACE (Gravity Recovery and Climate Experiment) satellites, which were launched by NASA in 2002 (2002-2017), have delivered us global and continuous high quality gravity data (e.g., Tapley et al., 2004). These data allow us to construct detailed and accurate geoids such as the ones shown in Figures 1.18 and 1.20, among other features of the Earth's surface and atmosphere. The second generation of GRACE satellites, known as GRACE-FO (Gravity Recovery and Climate Experiment Follow-On) satellites, was launched by NASA in 2018.

GRACE satellites measure the long wavelength (i.e., 40000 km down to 400 km) structure of gravity and geoid with very high precision. They even detect temporal changes in the Earth system due to mass redistribution (ice, sea level, continental hydrology) with details of how water is moving around the planet. Temporal change in gravity was observed by GRACE after the 2004 Sumatra earthquake and tsunami (Han et. al., 2006). These changes were due to mass redistribution of the plates and were around 15 μgal. GRACE data are also used to create revised average gravity fields, which contain less of the effects of short-time variations of gravity fields associated with atmosphere and ocean surfaces and glaciers. These data are relevant for the studies of the structure of the solid Earth.

---

[4]Low-Earth orbit is the altitude between 160 km and 2,000 km above the Earth's surface. It is here that the Earth's communications, navigation, and satellites reside. It is also here that the International Space Station (ISS) conducts its operations.

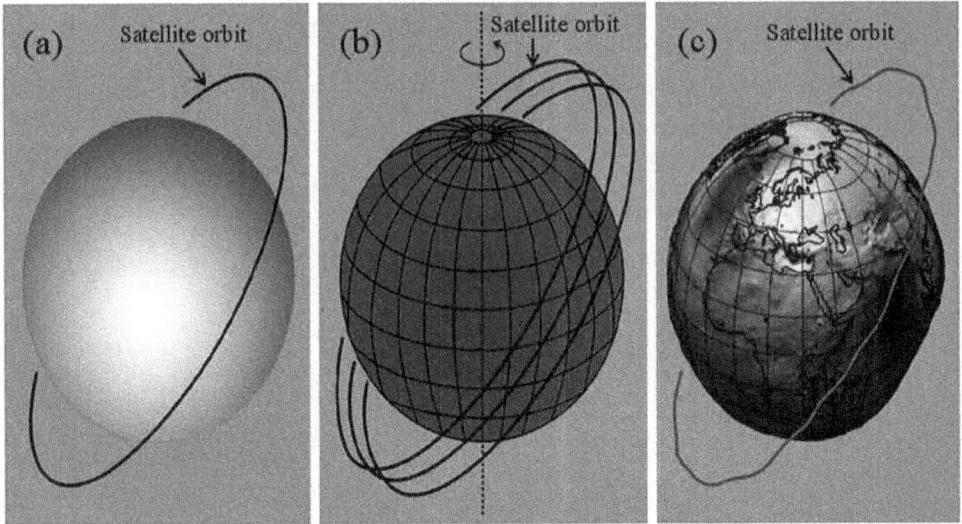

Figure 1.30. Satellite orbital paths around two idealized Earth models and the actual Earth. (a) If the Earth was a homogeneous sphere, the paths of satellites would be elliptical as predicted by the first Kepler's law. (b) If the Earth were a homogeneous ellipsoid, the paths of satellites would also be elliptical as predicted by the first Kepler's law. (c) The paths of satellites orbiting the actual Earth are distorted ellipses (orbit perturbations from a reference ellipse) due to gravitation. As shown in Figure 1.25, the orbit perturbation varies with position-time pairs. (Adapted Rummel et al., 2002)

The orbit motion of planets obeys the same law as a free-falling test mass. Measuring positions and times of the falling test mass are replaced by measuring satellite-orbit perturbations (see Figure 1.30). With GRACE, we have two identical satellite orbits (Wolf, 1969): one behind the other in the same orbital plane at an approximate distance of 220 km from the Earth's surface (see Figure 1.31a) at an altitude of about 500 km above the Earth's surface. Both GRACE satellites are tracked by GPS satellites. GPS receivers are used to determine the exact positions of the GRACE satellites over the Earth to within a centimeter or less and their relative motion is measured precisely by a highly sensitive microwave link to within a few micrometers. As the satellite pair circles the Earth, areas of slightly stronger gravity (excess mass concentration, fixed or or moving) affect them as follows: consider the case in which GRACE satellites are approaching a gravity anomaly. As the lead satellite approaches the anomaly, the separation between the two satellites increases—when the trailing satellite approaches the anomaly, it accelerates and catches up with the lead satellite and their separation decreases.

Let us now consider the case in which the GRACE satellites are moving far away from this particular anomaly. As the lead satellite moves far from the anomaly and of its influence, the trailing satellite may still be tugged backwards by the anomaly, and we have an increasing separation. The trailing satellite will also move under the influence of this anomaly, but the variations in the separation of GRACE satellites have captured the signature of this anomaly. Therefore, huge progress in mapping the global Earth gravity field has been made in recent years aided in particular by the satellite gravity missions GRACE and GOCE (Gravity field and steady state Ocean Circulation Explorer). GOCE was launched in March 2009 and fell from orbit in November 2013 (Floberghagen et al., 2011). GOCE was also a low-orbit satellite, streamlined to cut through the uppermost atmosphere. It measures two independent gravity gradients (gradiometers), as described in (1.94), in addition to the actual gravity. The two gradients are $\partial g_x/\partial y$ and $\partial g_z/\partial y$. However, GOCE

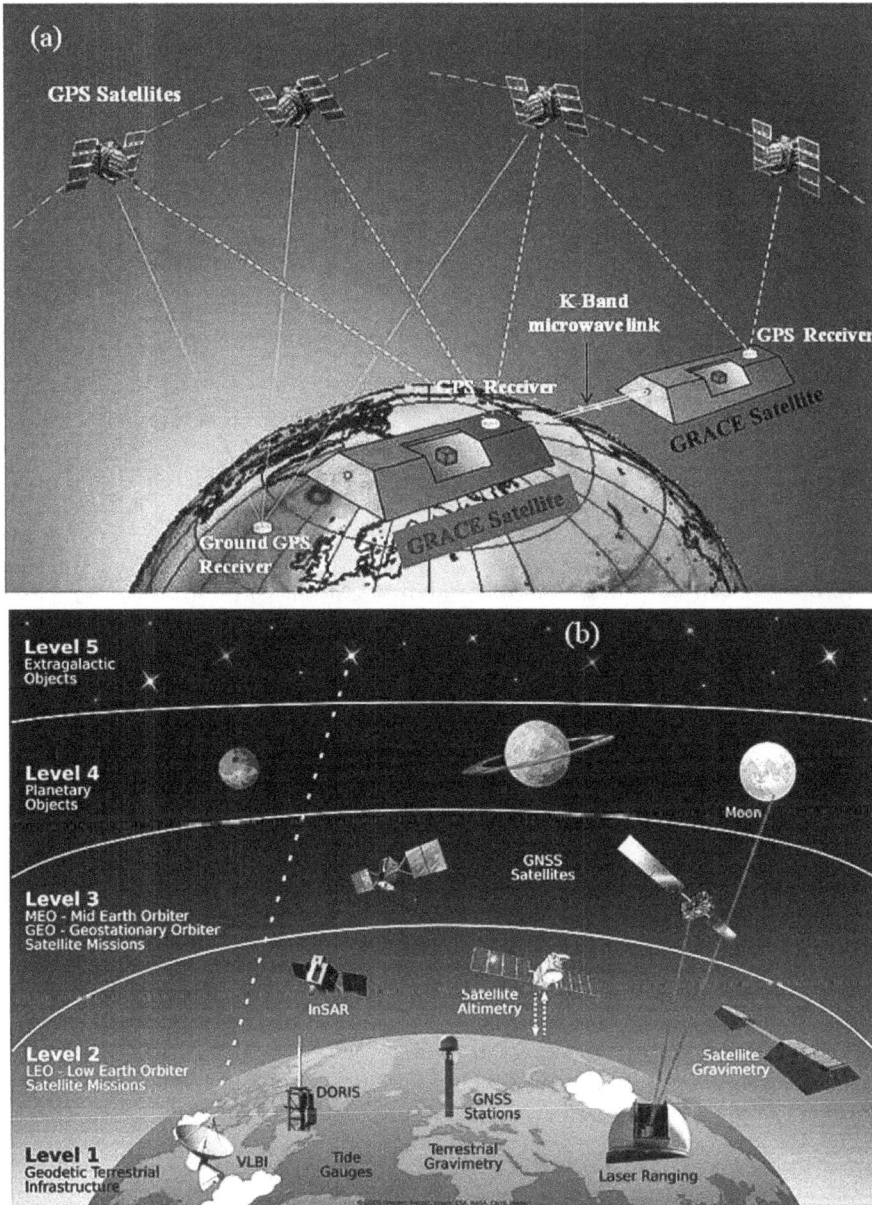

Figure 1.31. (Continued.)

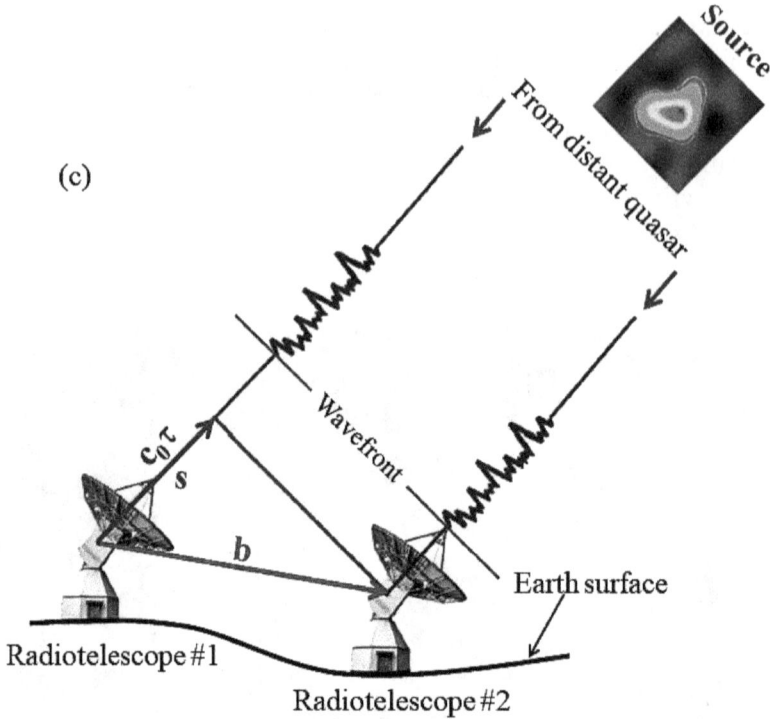

Figure 1.31. (a) An illustration of the GRACE mission with its two satellites. The two satellites at an altitude of about 500 km are flying one after another at a distance of about 220 km. The inter-satellite measurement device, called K-Band microwave link, determines relative distances between the two satellites (i.e., differential orbit perturbations between the satellites). The on-board GPS receivers of the twin GRACE satellites measure their absolute positions. GRACE is sensitive to large scale temporal variations in the gravity field and enables monitoring of mass redistributions on the Earth surface including glacier melting and ground water storage variations. (b) Space geodesy techniques include Very Long Baseline Interferometry (VLBI), Lunar Laser Ranging (LLR), Satellite Laser Ranging (SLR), DORIS, GNSS (Global Navigation Satellite Systems), InSAR, satellite altimetry, and satellite gravimetry. These satellites are used in determining positions on Earth. This geodesic infrastructure depends on the distance to the Earth's surface. At the Earth'surface (level 1), we have a ground networks of in situ instruments and space-geodetic tracking stations as well as the data and analysis centers. Level 2 includes the Low Earth Orbit (LEO) satellites. InSAR, satellite altimetry, and satellite gravimetry are located at this level. Most GNSS are located at Level 3. Level 4 comprises the Moon and planets. The quasars are found at Level 5, where they provide the inertial reference frame fixed in space.

data have a polar data gap larger than that of GRACE with a diameter of approximately 1,400 km. Therefore, in order to use the GRACE and GOCE data type, one has to apply a certain type of regularization (Pail et al., 2011) or include terrestrial gravity data.

GRACE and GOCE enable a coherent coverage and consistent accuracy of gravity measurements up to an unprecedented resolution of 130 km and 90 km, respectively. The gravity fields resulting from these experiments allow us to study deeper lithospheric features and large-scale regional-to-continental-scale geoid patterns. Basically, the higher you are from the Earth, the more the gravity field resembles an ellipsoid, because all these humps and bumps attenuate. Hence, GRACE and GOCE's estimates for the geoid and gravity field changes are only sufficiently accurate for large regions, such as the entire Colorado Plateau. For higher-resolution (up to 10 km or below) data, we also need to include terrestrial data over most continents and oceans to satellite-based data.

Quiz: You are in a city that is 9184 km from New York City, 6,904 km from Paris, and 11800 km from Shanghai. What city are you in?

Over the last decade, in particular, the international scientific community has deployed aircraft and ships equipped with gravimeters to collect a huge amount of new terrestrial gravity data. Yet, gravity surveying is still challenging in Tibet and Antarctica because of their hostile environments. For example, the available gravity anomaly data about the Antarctic continent in 2018 cover 73 percent of the Antarctic continent (equal to the entire area of Europe). In other words, Antarctica remains the most difficult-to-access region on Earth and, therefore, still suffers from considerable gravity data coverage gaps (see Ikelle, 2020).

Quiz: Satellite systems for gravity measurements can be divided into four categories: (1) displacements of a satellite in orbit relative to tracking stations on the ground; (2) displacements of a satellite and the sea surface with respect to each other; (3) displacements of two satellites with respect to each other; and (4) relative displacements of two masses within a single satellite. Use this grouping to categorize GRACE and GOCE.

1.1.6. Lagrange Points: Asteroids, Satellites, and Telescopes in Space

So far, we have limited ourselves to the analysis of two-body gravity. This analysis can be extended to a three- or more-body gravity system. The typical three-body example is Earth, Sun, and a human satellite. In this case we ideally want the satellite to be stationary with respect to Earth and the Sun so that we can monitor dynamic changes on Earth and Sun, including tectonic changes on Earth, and atmospheric, gravity, and magnetic changes both on Earth and Sun. We are basically dealing with two massive objects that we denote m_1 (e.g., Sun) and m_2 (e.g., Earth) in circular orbits around their common center of mass (CM) and the motion of a third body whose mass m (e.g., satellite or an asteroid) is essentially negligible compared to m_1 and m_2—that is, $m \ll m_1$ and $m \ll m_2$. The *Lagrange points*, also known as *L*-points, are places in space where the gravity of m_1 and m_2 cancel each other out, so that m, such as a satellite, stays exactly where you put it (i.e., stays stationary). In other words, the gravitational pull of m_1, m_2, and m zeros itself out at five Lagrange points. These points are named in honor of the French-Italian mathematician Joseph Lagrange, who discovered them in 1772 while studying the gravity of restricted three bodies. The term *restricted* refers to the condition that two of the masses are very much heavier than the third. Today we know that the full three-body problem is chaotic, and so cannot be solved in a closed form. Therefore, Lagrange had good reason to make this assumption. Note that very little energy is needed to maintain telescopes and satellites at the Lagrange points.

Coordinates of Lagrange points. Let us now derive the coordinates of Lagrange points. Our derivations significantly borrowed from that of Cornish (1988). Let x_1 and x_2 be the respective positions of m_1 and m_2, the total force exerted on a third mass m, at a position x is

$$F(x) = -\frac{Gm_1 m}{\| x - x_1 \|^3} (x - x_1) - \frac{Gm_2 m}{\| x - x_2 \|^3} (x - x_2) \ . \tag{1.116}$$

The easiest approach to finding the Lagrange points of the equation of motion of the mass m is to transform the three-body problem in a coordinate system whose origin is the center of mass of m_1 and m_2 and that rotates with an angular frequency ω equals the orbital angular frequency of m_1 and m_2 (i.e., a co-rotating frame in which m_1 and m_2 hold fixed positions). This angular frequency is given by the third Kepler's law: $\omega^2 r^3 = G(m_1 + m_2)$, $R = |x_1 - x_2|$ is the separation of m_1 and m_2 and $\omega = |\omega|$. In the rotating coordinate system the positions of m_1 and m_2 are fixed and the equation of motion for m now includes the fictitious centrifugal and Coriolis forces (see Ikelle, 2023c, for the description of Coriolis forces). The force F_m acting on m in the rotating reference frame is then

$$F_m(x) = \underbrace{F(x)}_{\text{gravity}} - \underbrace{m\Omega \times (\omega \times x)}_{\text{centrifugal force}} - \underbrace{2m\left(\omega \times \frac{dx}{dt}\right)}_{\text{Coriolis force}}$$

$$= -\nabla U(x) = 0 \ , \tag{1.117}$$

where

$$U(x) = -G\left(\frac{m_1 m}{|x - x_1|} + \frac{m_2 m}{|x - x_2|}\right) - \frac{1}{2} m\omega^2 |x|^2 \tag{1.118}$$

is the generalized potential and F is the gravitational force given in (1.116). The second term on the right side of equation (1.117) is the centrifugal force and the third term is the Coriolis force. The Lagrange points (i.e., stationary points) correspond to x for which $F_m = -\nabla U = 0$. Consider Cartesian coordinates originating from the center of mass with the z-axis aligned with the angular velocity as follows:

$$\omega = \omega i_z \, , \quad x = x(t)i_x + y(t)i_y \, , \quad x_1 = -\alpha R i_x \, , \quad x_2 = -\beta R i_x \, , \tag{1.119}$$

where

$$\alpha = \frac{m_2}{m_1 + m_2} \quad \text{and} \quad \beta = \frac{m_1}{m_1 + m_2} \ . \tag{1.120}$$

Equation (1.117) reduces to

$$\frac{F_m}{\omega^2} = \left(x - \frac{\beta(x + \alpha R)R^3}{[(x + \alpha R)^2 + y^2]^{3/2}} - \frac{\alpha(x - \beta R)R^3}{[(x - \beta R)^2 + y^2]^{3/2}}\right) i_x$$

$$+ \left(y - \frac{\beta y R^3}{[(x + \alpha R)^2 + y^2]^{3/2}} - \frac{\alpha y R^3}{[(x - \beta R)^2 + y^2]^{3/2}}\right) i_y = 0 \, . \tag{1.121}$$

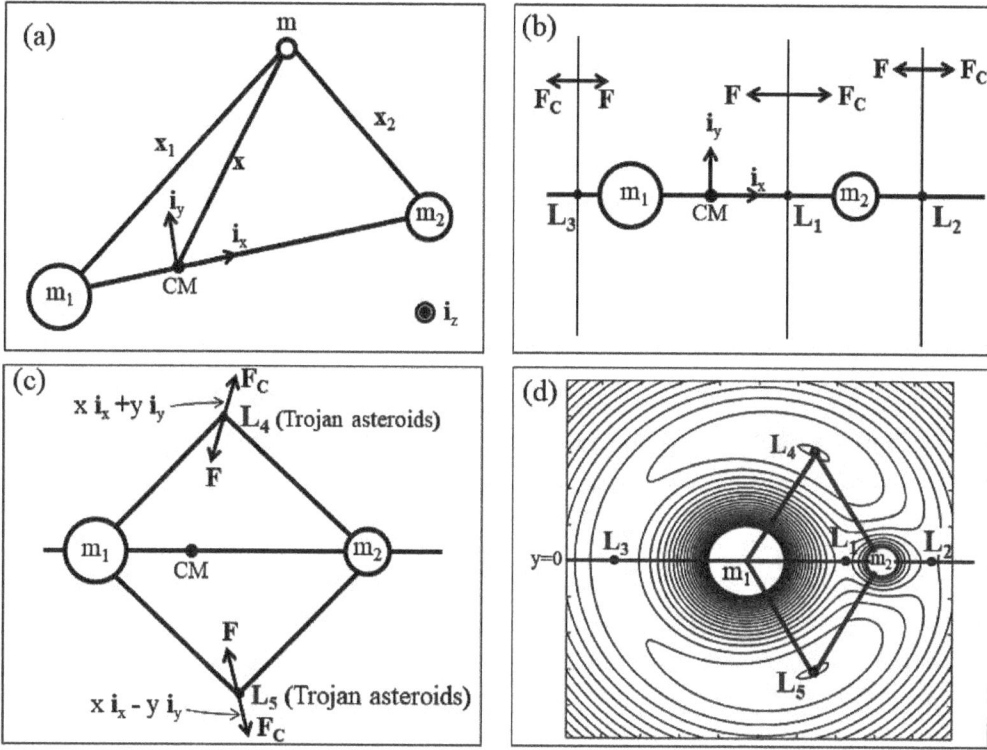

Figure 1.32. (a) A reference frame with its origin at the common center of mass and the x-axis along the reference line of the two heavy bodies, m_1 and m_2, which rotates uniformly with the angular velocity ω (the mean motion of the heavy bodies) around the z-axis. In this reference frame the positions of the two heavy masses remain fixed. (b) Positions of the Lagrangian points L_1, L_2, and L_3. $\boldsymbol{F}_C$ denotes centrifugal force and $\boldsymbol{F}$ denotes gravitational forces. (c) Positions of the Lagrangian points L_4, and L_5. $\boldsymbol{F}_C$ denotes centrifugal force and $\boldsymbol{F}$ denotes gravitational forces. (d) A contour plot of the gravitational potential on the orbital plane of the Sun-Earth system showing the locations of the five Lagrangian points.

Here the mass m has been set equal to unity without loss of generality. We can solve numerically for these conditions to obtain the five Lagrange points, as shown in Figure 1.32d. We can alternatively seek a closed-form solution to (1.121)—that is our next task.

Let us first focus on determining L_1, L_2, and L_3. As illustrated in Figures 1.31b and 1.31d, these points are along the x-axis. By setting $y = 0$ and $x = R(u + \beta)$ (so that u measures the distance from m_2 in units of R), the condition in (1.121) reduces to

$$u^2 \left[(1 - s_1) + 3u + 3u^2 + u^3 \right] = \alpha \left[s_0 + 2s_0 u + (1 + s_0 - s_1) u^2 \right.$$

$$\left. + \ 2u^3 + u^4 \right] , \tag{1.122}$$

where $s_0 = \text{sign}(u)$ and $s_1 = \text{sign}(u + 1)$. There are only three possible sets of (s_0, s_1), which are $(-1, 1)$, $(1, 1)$ and $(-1, -1)$. The case $(1, -1)$ cannot occur because gravitational forces are only attractive forces. In each case there is one real root to the quintic equation, giving us the positions of the first three Lagrange points. We are unable to find closed-form solutions to equation (1.122) for general values of α, so instead we seek approximate solutions valid in the limit $\alpha \ll 1$. To the lowest order in α, we find the first three Lagrange points to be positioned at

$$L_1 : \left[R \left(1 - \left(\frac{\alpha}{3} \right)^{1/3} \right), 0 \right]^T , \quad L_2 : \left[R \left(1 + \left(\frac{\alpha}{3} \right)^{1/3} \right), 0 \right]^T ,$$

$$\text{and} , \quad L_3 : \left[-R \left(1 + \left(\frac{5\alpha}{12} \right) \right), 0 \right]^T .$$

For the earth-sun system, $\alpha \approx 3 \times 10^{-6}$, $R = 1$ AU $\approx 1.5 \times 10^8$ km, and the first and second Lagrange points, L_1 and L_2, are located approximately 1.5 million kilometers from the Earth. The third Lagrange point, L_3, is located about 1 AU from the Sun.

For determining the other two Lagrange points, we need to balance the centrifugal force, which acts in a direction radially outward from the center of mass, with the gravitational force exerted by the two masses, as illustrated in Figures 1.32c and 1.32d. Clearly, force balance in the direction perpendicular to centrifugal force will only involve gravitational forces. This suggests that we can resolve the force into directions parallel and perpendicular to $\boldsymbol{x}$. The appropriate projection vectors are $x\boldsymbol{i}_x + y\boldsymbol{i}_y$ and $y\boldsymbol{i}_x - x\boldsymbol{i}_y$. The perpendicular projection of $\boldsymbol{F}_m$ is

$$F_m^\perp = \alpha\beta y \omega^2 R^3 \left(\frac{1}{[(x + \beta R)^2 + y^2]^{3/2}} - \frac{1}{[(x + \alpha R)^2 + y^2]^{3/2}} \right) .$$

Setting $F_m^\perp = 0$ and $y \neq 0$ tells us that the Lagrange points must be equidistant from two masses. The parallel projection of $\boldsymbol{F}_m$ is

$$F_m^\| = \omega^2 \frac{x^2 + y^2}{R} \left(\frac{1}{R^3} - \frac{1}{[(x - \beta R)^2 + y^2]^{3/2}} \right) . \tag{1.123}$$

Requiring that the parallel component of the force vanish leads to the condition that the Lagrange points are at a distance R from each mass. In other words, L_4 is situated at the vertex of an equilateral triangle, with the two masses forming the other vertices (see Figure 1.32c). L_5 is obtained by a mirror reflection of L_4 with respect to the x-axis. Explicitly, the fourth and fifth Lagrange points have the following coordinates

$$L_4 : \left[\frac{R}{2} \left(\frac{m_1 - m_2}{m_1 + m_2} \right), \frac{\sqrt{3}}{2} R \right]^T , \quad L_5 : \left[\frac{R}{2} \left(\frac{m_1 - m_2}{m_1 + m_2} \right), -\frac{\sqrt{3}}{2} R \right]^T .$$

Stability analysis. L_1, L_2, and L_3 are simply the spots in space and L_4 and L_5 are little oval-shaped hollows that form independent, as illustrated in Figure 1.32d— that is, L_1, L_2, and L_3 are unstable points and L_4 and L_5 are stable points. Small departures from L_1, L_2, or L_3 will grow exponentially with a time of $\tau \approx 2/5\omega$. For the Earth-Sun system $\omega = 2\pi$ year^{-1} and $\tau \approx 23$ days. In other words, a satellite parked at L_1 or L_2 will wander off after a few months unless course corrections are made.

Consider, for example, a Sun-Earth system and a satellite in space. Just about anywhere you choose to put a satellite, the gravity pull means that it will end too close to either the Sun or the Earth. If a satellite is too close to the Sun, it will end up either crashing into the Sun, or going into an orbit around it. Alternatively if your satellite is too close to the Earth it will crash back down into our atmosphere, or go into orbit around the planet. A satellite can stay stable only at Lagrange points. A satellite put at L_1, L_2, or L_3 will eventually move away because these three Lagrange points are unstable. At L_4 and L_5, even if a satellite is not perfectly placed, the gravity will push it into the ideal position and it will stay there forever because L_4 and L_5 are stable. Note that L_4 and L_5, in the Sun-Earth system, follow the planet in its orbit by about 60 degrees. They are being considered as potential locations for extraterrestrial space colonies.

The stability of the first and second Lagrange points of the Earth-Sun system is an important consideration in space missions. Currently the Solar and Heliospheric Observatory (SOHO) satellite is parked at Earth-Sun L_1 and the James Webb Space Telescope is at Sun-Earth L_2, located about 1.5 million km from Earth. As all objects located at L_1 orbit the Sun at Earth's angular frequency and SOHO has an uninterrupted view of the Sun. Since L_1 is an unstable point, occasional course adjustments are necessary in order to keep the satellite stationary. From L_2, the Sun, Earth, and Moon are huddled up in a tiny location in the sky, leaving the rest of the universe free for observation by the James Webb Space Telescope. This telescope has 10 years of fuel.

Note also that, if m_1 is our Sun, five Lagrange points are associated with every planet and with every moon of our solar system. The stronger the planet's gravity (i.e., the more massive the planet), the more stable the Lagrange points become. In the Sun-Jupiter system, L_4 and L_5 are locations of about 1691 asteroids called *Trojans*—asteroids are mostly in the asteroid belt between Mars and Jupiter; they are made of rock, metal, or a mixture of both. This name was taken from mythological characters appearing in Homer's Iliad, an epic poem set during the Trojan War. Asteroids at the L_4 point, ahead of Jupiter, are named after Greek characters in the Iliad and referred to as the *Greek camp*. Those at the L_5 point are named after Trojan characters and referred to as the *Trojan camp*. Both camps are considered to be types of Trojan bodies. The Sun-Earth system has no Trojan satellites.

Quiz: (1) Determine the Lagrange points, L_1, L_2, and L_3, for the Earth–Moon-satellite system. (2) Determine the Lagrange points L_1, L_2, and L_3, for the Sun–Mars-satellite system. L_1, L_2, and L_3 of the Earth–Moon-satellite and Earth–Moon-satellite systems.

Halo orbit or orbital control. As we discussed earlier, L_1, L_2, and L_3 Lagrange are unstable points. So satellites are in practice equipped with systems for keeping them around these points. In other words, a satellite is actually placed in a known orbit around, say, L_2. When the satellite begins to slowly drift back toward Earth, in the case of the Sun-Earth system, an electronic system within the satellite pushes the satellite back toward L_2. If the satellite was actually parked at L_2, there would be no way to predict which direction it would eventually drift off to. If it drifted further from the sun the whole mission would be over. Thus satellites are kept in orbits around L_1, L_2, or L_3, where their motions are easily monitored and corrected if necessary. These orbits are characterized as halo orbits. Robert Willard Farquhar (1932-2015) came out with the idea of halo orbit (Farquhar, 1966) and was involved in a number of spaceflight missions with NASA.

The first mission to use a halo orbit was a joint ESA and NASA spacecraft launched in 1978. It traveled to the Sun–Earth L_1 point and remained there for several years. The joint ESA/NASA SOHO mission also uses a halo orbit around the Sun–Earth L_1 point (Howell, 1984). In May 2018, the Chinese space program placed the first communications relay satellite, Queqiao, into a halo orbit around the Earth-Moon L_2 point (Li et al, 2021). On January 3, 2019, the Chang'e 4 spacecraft landed in the Von Kármán crater on the far side of the Moon, using the Queqiao relay satellite to communicate with the Earth, as we will discuss later.

1.1.7. Application: Gravity Anomalies Versus Topography in Antarctica

Modern Antarctic-wide free-air-gravity-anomaly map is shown in Figure 1.33b. This map includes both strong negative, ranging from -30 mGal to -60 mGal, and strong positive free-air gravity anomalies of about $+50$ mGal anomaly. These strong free-air gravity anomalies are found to occur mainly on the Antarctic continent, in particular, in the Wilkes Land, Princess Elizabeth Land, Ross Sea, central continental, and Weddell Sea sectors. Some of these anomalies are very large and comparable to gravity anomalies in other regions on Earth, but some are surprisingly different. Moreover, these anomalies are distributed almost randomly. Clearly, the present

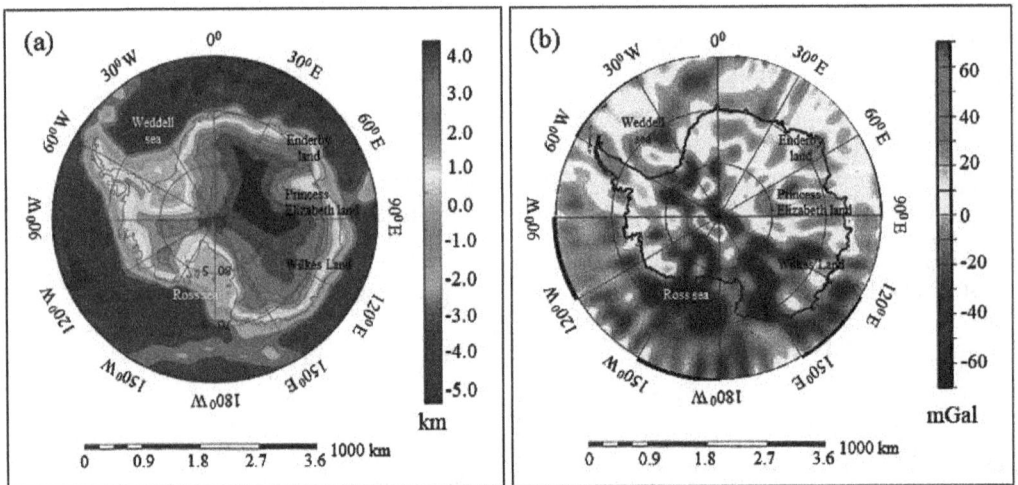

Figure 1.33. (a) Topography and bathymetry of Antarctica. (b) The free-air gravity anomaly of Antarctica.

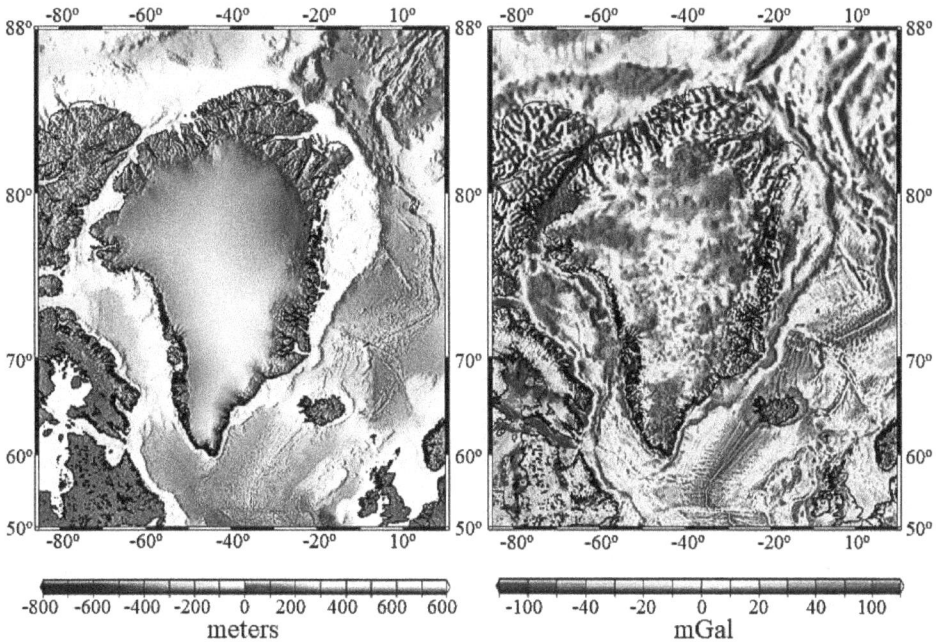

Figure 1.34. (a) Topography and bathymetry of Arctic, including Greenland. (b) The free-air gravity anomaly of Arctic. (adapted from https://bgi.obs-mip.fr/grids-and-models-2/).

Antarctica free-air-gravity anomaly map in Figure 1.33b does not correlate with its topography (see Figure 1.33a) in contrast with most regions and continents of the Earth, including Greenland and Australia (see Figures 1.34 and 1.35).

In these two examples, we have a high correlation between free-air anomalies with the topography of area—that is, elevation above sea level create positive free-air gravity anomalies and valleys below sea create negative gravity anomalies, as described in (1.97). Even on extraterrestrial planets like Mars, we have a strong correlation between the topography and the free-air gravity profile, as illustrated in Figure 1.24.

We can see that the central continental of Antarctica has some negative free-air anomalies despite the fact that the whole region is well above sea level. How can one interpret these large and almost randomly distributed free-air gravity anomalies in Antarctica (see Figure 1.33). It turns out the ice covering the continent of Antarctica has its own topography. In other words, the ice covering the Antarctica continent has a non-uniform distribution. As we can see in Figure 1.36, when the overlaying ice is removed, the topography and free-air anomaly correlate more effectively.

Note how the Bouguer anomaly profile of Antarctica is quite similar to those of Africa, Australia, and Greenland in Figures 1.37 and 1.38. Remember that the Bouguer anomaly reflects density and topography. If we ignore the topography or assume the topographic changes are small compared to the density changes, the excess lighter rocks create negative gravity anomalies and denser rocks will tend to form positive gravity anomalies, as described in Figure 1.20c and in equation (1.99). Notice also that the ice sheet has no noticeable effect on the free-air anomalies of

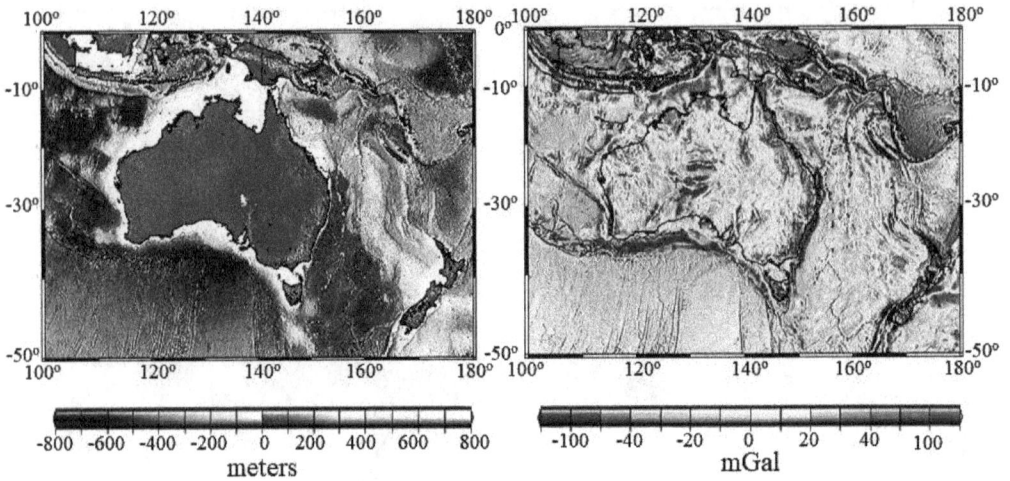

Figure 1.35. Section of the gravity anomaly map of Australia. (a) Topography and bathymetry of Australia. (b) The free-air gravity anomaly of Australia. (adapted from https://bgi.obs-mip.fr/grids-and-models-2/).

the Arctic in Figure 1.34—that is, there is an unconformity between the Antarctica ice-sheet and the strata below it and not in the Arctic.

So what causes different ice-sheet behavior in the Arctic and Antarctica? There is a lake on the East Antarctic Ice Sheet known as Lake Vostok. It is the largest lake in Antarctica. Located approximately 4 km beneath Russia's Vostok Station on the East Antarctic Ice Sheet, the water body is also the largest subglacial lake known. It is approximately 240 km long with a maximum width of about 50 km. It holds nearly 5,400 cubic km of water (see BBC news, Lake Vostok drilling in Antarctic running out of time By Katia Moskvitch, Science reporter, 27 January

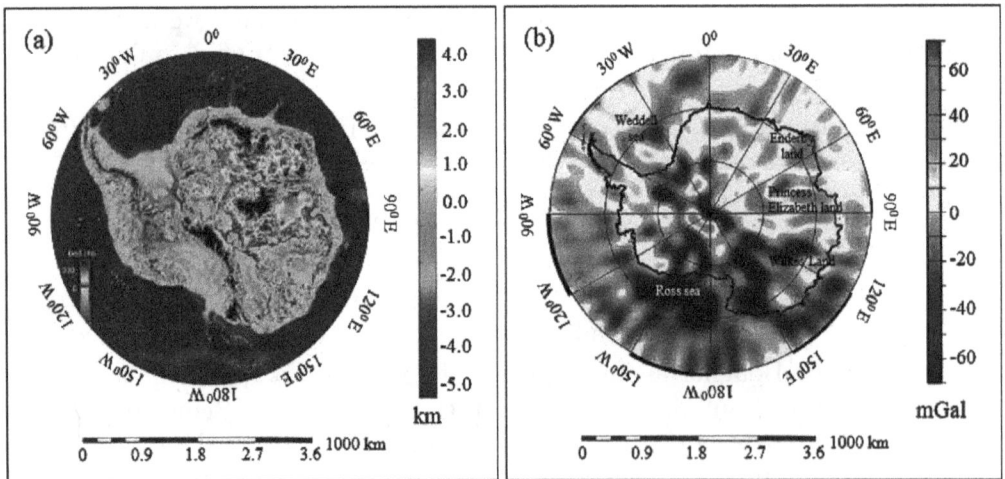

Figure 1.36. (a) Topography and bathymetry of Antarctica under the ice sheet. (Credit: Mathieu Morlighem/UCI.) (b) The free-air gravity anomaly of Antarctica.

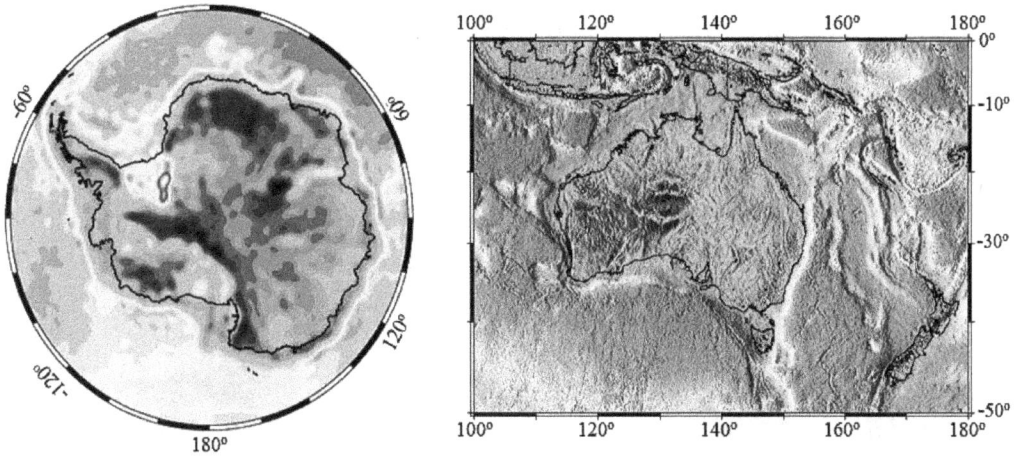

Figure 1.37. The Bougeur gravity anomaly profiles of (a) Antarctica (adapted from Tenzer et al., 2008) and (b) Australia (adapted from https://bgi.obs-mip.fr/grids-and-models-2/) with the effects of topography removed to reveal density variations underneath the surface and variations in crustal thickness. Low (negative) values indicate lower density beneath the measurement point. High (positive) values indicate higher density beneath the measurement point.

2011). In the Antarctic Ice Sheet there are more than 70 subglacial lakes. These lakes suggest that gravity anomalies of the Antarctic Ice Sheet can be calculated in two steps: a free-air correction for the ice followed by a Bouguer correction for the lakes. It is the result of this two-step process that must be compared to the topography of Antarctica rather than the free-air correction alone.

The discovery and/or confirmation of the 70 or so lakes beneath the Antarctic ice sheet lakes have been made possible by GPR data-acquisition surveys and GPR data imaging (see Chapter 4), especially that water from none of the lakes has not been sampled directly yet because the lakes are subject to high pressure (about 355 bars), low temperatures (about $-3°$ degrees Celsius), and permanent darkness. GPR is very useful for distinguishing between warm ice and cold ice based on their temperatures. Temperature ice is ice which is at or close to its pressure melting point and cold ice is below this temperature. Water is formed between grain boundaries, cavities and channels within temperature ice. This type of ice is found in temperature glaciers and polythermal glaciers. Polythermal glaciers also contain cold ice which usually contains no water. Cold ice is transparent to EM signals whereas warm ice shows multiple reflections due to the water and ice interface. Thus they are easily distinguishable by GPR (Moran et al., 2003). Polythermal and temperature glaciers have veins, moulins, channels and conduits through which meltwater flows. In a cold glacier, water produced at the surface will drain through the surface channels only and not penetrate the ice.

Note that the mass density of the Antarctic ice sheet increases with depth from approximately 300 kg/m^3 to 900 kg/m^3 while the mass density of lake water is about 1000 kg/m^3. These differences in density are significant enough that if they are ignored, the Bouguer anomalies of Antarctica are likely to be incorrect. Furthermore, if the lakes beneath Antarctica's ice sheet are not taken into account, the

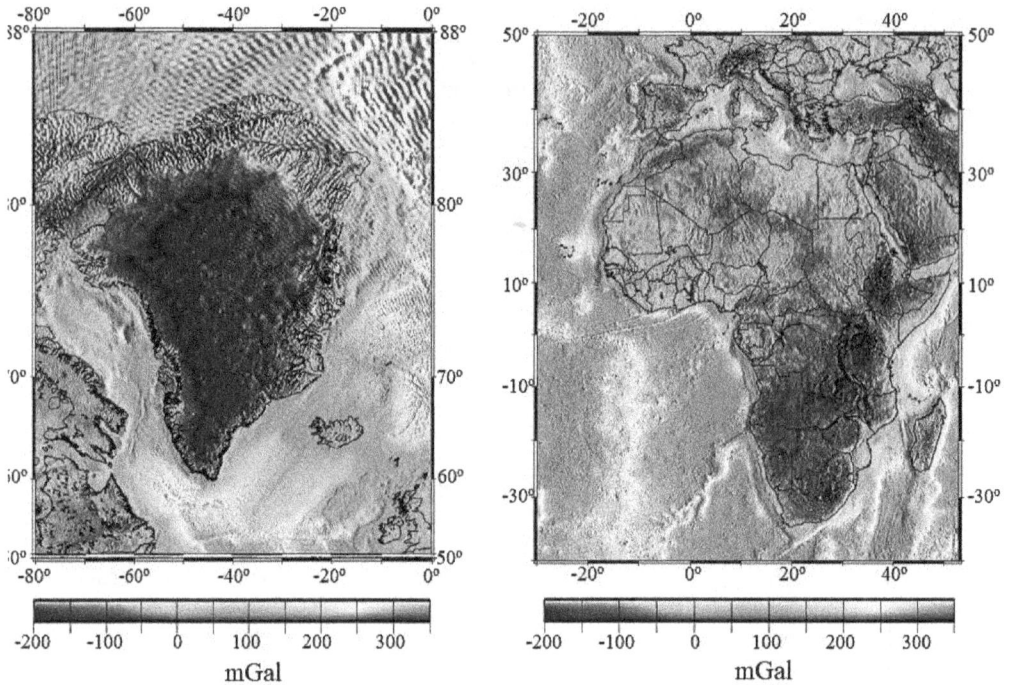

Figure 1.38. The Bougeur gravity anomaly profiles of (a) the Arctic and (b) Africa with the effects of topography removed to reveal density variations underneath the surface and variations in crustal thickness. Low (negative) values indicate lower density beneath the measurement point. High (positive) values indicate higher density beneath the measurement point. (adapted from https://bgi.obs-mip.fr/grids-and-models-2/).

topography of the bedrock will be erroneous, and the free-air anomaly of Antarctica will be erroneous. Remember that mass density of underground matter and the topography on land and bathymetry (the topography of the ocean floor) are the only causes of variations in the gravitational fields. The lakes beneath the Antarctic ice sheet also increase the possibility of biological life under the ice sheet.

There are unconfirmed suggestions that subglacial lakes in Antarctica date back over 35 million years old. There is a possibility that they contain unique life forms, including ancient microbes. The dark, nutrient-deprived environment of the lakes could resemble conditions on Jupiter's moon Europa, which is assumed to hold a large ocean beneath its frozen surface. The Russian Antarctic programme drilled into the Antarctic ice sheet in 2008-2009. Their drill bit got stuck 80 meters above the estimated surface of Lake Vostok, which is about 4 km. There are suggestions that the lake surfaces are constantly falling and rising, and the lakes occasionally drain and then refill. Technical modifications have been made. Currently, plans to drill into subglacial lakes are under consideration. These include Lake Vostok, Lake Ellsworth, a relatively small lake in western Antarctica, and Lake Whillans near Antarctica's Ross Ice Shelf. Anyway we still have a lot to learn and educate about

Figure 1.39. These two photos show the same area on the beach of the Bay of Fundy roughly twelve hours apart. We can see that the water level has risen and covered the beach entirely. Assuming that the people in these photos are between 1.5 m and 1.8 m tall, the water has therefore risen at least 5 to 7 m in 12 hours [courtesy Ocean Drilling Program (ODP)].

Antarctica, including how Antarctica moved from being a free-ice area 100 million years ago to present-day Antarctica covered by a thick ice sheet.

1.2. Application: Tidal Forces

The rise and fall in sea levels with respect to land on a daily basis is known as tides. Tides result from a combination of two basic forces: (1) the force of gravitation exerted by the Moon upon the Earth and (2) the centrifugal forces produced by the revolutions of the Earth and the Moon around their common center of gravity (mass). The most familiar evidence of tides along coastlines is high and low water—usually twice daily. Tides typically have ranges (vertical, high-to-low) of two meters, but there are regions of the world where various conditions conspire to produce virtually no tides at all (e.g., in the Mediterranean Sea, the range is 2 to 3 cm) and others where the tides are greatly amplified (e.g., on the northwest coast of Australia, the range can be up to 12 m, and in the Bay of Fundy, Nova Scotia, it can rise up 15 m). Figure 1.39 illustrates an example of high and low tides in the Bay of Fundy. This figure shows the same location about 12 hours apart, with a range of about 6 m.

Understanding tides is not just an illustration of the gravitational forces that keep the Moon in its orbit, they have a profound effect on coastal marine life. Actually, coastal life is sorted into zones and subzones, depending on the amount of emergence and submergence the organisms can tolerate. Besides solving the Kepler problems, Newton also was able to explain the ocean tides as a gravitational effect caused by the Moon and Sun.

The tides of the ocean are affected by celestial bodies other than the Earth, including the Moon and Sun. However, the tidal phenomenon is dominated by the Moon and Sun. The tide-producing forces due to the Sun are about 0.45 times that of the Moon and that of Venus is 0.0054 times that of the Moon—for most places, tides related to Venus are less than 0.1 mm. So we first limit our discussion to the Earth-Moon system.

1.2.1. The Barycenters of Earth-Moon and Earth-Sun Systems

The Earth-Moon system is a twin-planet system moving around the common center of gravity, known as the barycenter. In other words, the Earth and Moon are locked in motion, rotating for a period of one month. Let r_c, r_1, and r_2 be the distances between a reference point and the barycenter, the center of the Earth, and the center of the Moon, respectively. We can define the barycenter as

$$r_c(m_e + m_m) = r_1 m_e + r_2 m_m \ , \tag{1.124}$$

where m_e is the mass of the Earth and m_m is the mass of the Moon. Setting the reference point at the center of the Earth (see Figure 1.40a), we have $r_1 = 0$ and

$$r_c = \frac{m_m}{m_e + m_m} r \ , \tag{1.125}$$

where $r_2 = r$ is the distance between the center of the Earth and the center of the Moon. The ratio of the distance between the center of the Earth and the center of the Moon and the Earth's radius is about 60, $r/R \approx 60$, and $m_e/m_m \approx 81.53$ yield $r_c \approx 60/82R \approx 4500$ km (R is the Earth's radius). So the barycenter, due to the difference in the mass of the Earth and Moon, is located approximately $3/4$ times the Earth's radius from the center of the Earth (see Figure 1.40a).

We can alternatively introduce the barycenter by examining the balance of forces at the center of rotation of the Earth-Moon system. Again, we assume that the Earth and Moon rotate about one another and that under the action of gravity, they are in a stable binary orbit. The center of rotation is located on the line connecting the center of the Earth and the center of the Moon, as illustrated in Figure 1.40a. Two types of forces are in play here: centrifugal forces and forces of gravitational attraction. Centrifugal forces are derived from the inertial tendency of the Earth (or the Moon) to go flying off into space tangentially to its orbit. Since the Earth does not fly off into space, the gravitational force must be balanced by the centrifugal force (equal but oppositely directed force vectors) at the center of mass of the Earth. This balance of forces can be expressed in magnitude as follows:

$$\underbrace{\frac{G m_e m_m}{r^2}}_{\text{gravity}} = \underbrace{\frac{m_e v_e^2}{r_c}}_{\text{centrifuge}} = \underbrace{\frac{m_m v_m^2}{r - r_c}}_{\text{centrifuge}} \ , \tag{1.126}$$

where v_e is the velocity of the Earth, and v_m is the velocity of the Moon. The first term in (1.126) is the gravitational force of attraction; that is, $\boldsymbol{F}_g = G m_e m_m / r^2 \, \boldsymbol{i}_1$ is the force that m_e exerts on m_m, $-\boldsymbol{F}_g$ is the force that m_m exerts on m_e. These two gravitational forces are directed toward the centers of the Moon and the Earth, as illustrated in Figure 1.40a. The second and third terms in (1.126) are the centrifugal forces of the Earth and Moon. Centrifugal forces are everywhere equal (but of opposite sign) to the gravitational force at the center of mass of the Earth. They are directed away from the centers of the Moon and the Earth, as illustrated in Figure 1.40a. Again, equation (1.126) reflects the fact that the Earth-Moon system is kept in dynamic equilibrium by the balance between centrifugal forces and gravitational

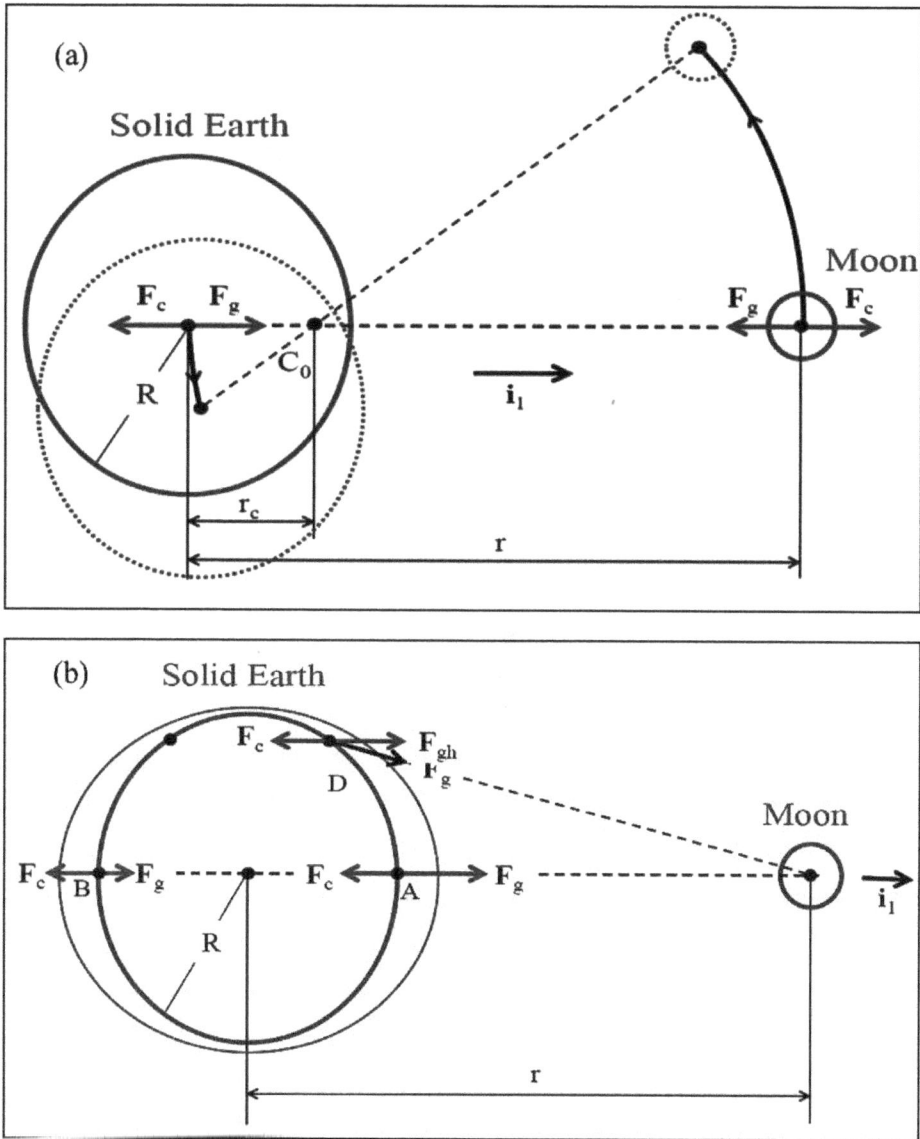

Figure 1.40. (a) The Earth and Moon are locked in the motion, rotating within the period of one month. C_0 is the location of their common center of mass which is known as the Earth-Moon barycenter. The Earth-Moon system is kept in a dynamical equilibrium by their centrifugal forces, $\boldsymbol{F}_c$, and their forces of gravitational attraction, $\boldsymbol{F}_g$. (b) We here consider gravitational forces due to the Moon acting on the particles A, B, and D with unit mass each. These particles are located on the surface of the Earth. The tide-generating forces are the sums of $\boldsymbol{F}_g$ and $\boldsymbol{F}_c$. Force $\boldsymbol{F}_g$ at D can be split into horizontal ($\boldsymbol{F}_{gh}$) and vertical components. The centrifugal force is the same for all particles (points) in and on Earth.

forces. The rotation rate of the body's system can be defined as $\omega = v_e/r_c = v_m/(r - r_c)$. Equation (1.126) can then be rewritten as

$$\frac{Gm_e m_m}{r^2} = m_e \omega^2 r_c = m_m \omega^2 (r - r_c) \implies m_e r_c = m_m (r - r_c)$$

or

$$r_c = \frac{m_m}{m_e + m_m} r , \tag{1.127}$$

and thus we can deduce the barycenter location. Note that the Earth is a rigid body, so every medium point in it executes an identical orbit, and therefore all its points are subject to the same centrifugal force. Gravitational forces, however, will vary because they vary with the distance between points on Earth and the center of the Moon, as described in equation (1.3) and illustrated in Figure 1.40b.

Let us now consider the Sun-Earth system. The mean distance between the center of the Sun and the center of the Earth is about 1.5×10^8 km. Their center of mass is located about 450 km (mean) from the center of the Sun. The Sun and the Earth revolve around this common center of mass in one year.

1.2.2. Tide-Generating Forces

Figure 1.40b shows two particles with unit mass located at points A and B on opposite sides of the Earth. Point A is the point of the Earth's surface closest to the Moon and point B is the point of the Earth's surface furthest from the Moon. We will analyze how the Moon alters the forces on these two particles. At the center of the Earth, the gravitational force of the Moon is

$$\boldsymbol{F}_{g,O}^{(m)} = \frac{Gm_m}{r^2} \boldsymbol{i}_1 . \tag{1.128}$$

Because of the balance of forces, the centrifugal force at the Earth's center is directed away from the Moon; i.e.,

$$\boldsymbol{F}_c^{(m)} = -\frac{Gm_m}{r^2} \boldsymbol{i}_1 , \tag{1.129}$$

At point A on the side of the Earth facing the Moon, the gravitational force of the Moon is

$$\boldsymbol{F}_{g,A}^{(m)} = \frac{Gm_m}{(r - R)^2} \boldsymbol{i}_1 . \tag{1.130}$$

This force is directed toward the Moon, of course and its magnitude is greater than at the center of the Earth because it is closer to the Moon. At point A (see Figure 1.40b), the net Moon's gravitational force is

$$\boldsymbol{F}_A^{(m)} = \boldsymbol{F}_{g,A}^{(m)} + \boldsymbol{F}_c^{(m)} = \boldsymbol{F}_{g,A}^{(m)} - \boldsymbol{F}_{g,O}^{(m)} \tag{1.131}$$

or

$$\begin{aligned}
\boldsymbol{F}_A^{(m)} &= \left[\frac{Gm_m}{(r - R)^2} - \frac{Gm_m}{r^2} \right] \boldsymbol{i}_1 \approx \left[\frac{Gm_m}{r^2} \left(1 + \frac{2R}{r} \right) - \frac{Gm_m}{r^2} \right] \boldsymbol{i}_1 \\
&= \frac{2Gm_m R}{r^3} \boldsymbol{i}_1 . \tag{1.132}
\end{aligned}$$

This force is the sum of the attractive force by the Moon and the centrifugal force (or the differences between the gravitational attraction at the center of the Earth and the gravitational attractions at point A), which corresponds to a force directed toward the Moon, and the "tide-generating" force is $F_A^{(m)}$. We used the fact that $(R/r) \ll 1$ as $R/r \approx 1/60$ to carry out the expansion $1/(1-\alpha)^2 \approx 1 + 2\alpha$ for small $\alpha = R/r$ to obtain the net force toward the Moon in (1.132). The gravitational force of the Earth is not included in (1.132).

At point B, the gravitational field of the moon is

$$F_{g,B}^{(m)} = \frac{Gm_m}{(r+R)^2} i_1 \qquad (1.133)$$

while the centrifugal force given in (1.129) is the same at B. At this point, the magnitude of the gravitational force is less than at the center of the Earth because it is further away from the Moon. The net Moon's gravitational force at point B is

$$
\begin{aligned}
F_B^{(m)} &= F_{g,B}^{(m)} - F_{g,O}^{(m)} = \left[\frac{Gm_m}{(r+R)^2} - \frac{Gm_m}{r^2} \right] i_1 \\
&\approx \left[\frac{Gm_m}{r^2} \left(1 - \frac{2R}{r} \right) - \frac{Gm_m}{r^2} \right] i_1 \\
&= -\frac{2Gm_m R}{r^3} i_1.
\end{aligned}
\qquad (1.134)
$$

So the gravitational force prevails over the centrifugal force on point A closest to the Moon and the centrifugal force prevails on point B further from it. In other words, the opposite (points) hemispheres have net forces in opposite directions, causing the ocean to bulge on both sides, as illustrated in Figure 1.40b. The net attraction at point A causes the water on this "near side" of the Earth to be pulled toward the Moon. At point B, the centrifugal force exceeds the gravitational force, and the water tries to keep going in a straight line, moving away from the Earth, also forming a bulge. Thus, the combination of gravity and inertia creates two bulges of water. The bulges on opposite sides of the Earth are proportional to the ratio m_m/r^3.

The Sun also affects the size and position of the two tidal bulges. These effects on opposite sides of the Earth, which are approximately proportional to the ratio m_s/r^3, also give rise to solar tidal forces; i.e.,

$$F_A^{(s)} \approx \frac{2Gm_s R}{r_s^3} i_1 , \qquad (1.135)$$

$$F_B^{(s)} \approx -\frac{2Gm_s R}{r_s^3} i_1 , \qquad (1.136)$$

where m_s is the mass of the Sun and r_s is the distance between the center of the Earth and the center of the Sun. But because of the much greater distance of the Sun to the Earth, the ratio m_s/r_s^3 for the Sun is only about half of that for the Moon [$m_s = 27.07 \times 10^6 \, m_m$ and $r_s^3 = (389)^3 r^3 = 58.8 \times 10^6 \, r^3$]. Basically, the

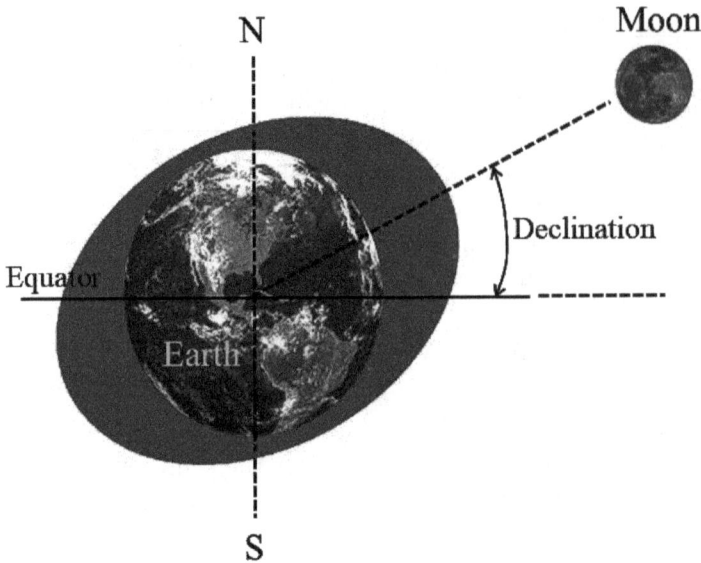

Figure 1.41. An illustration of the declination which is the angular distance of the Moon above or below Earth's equator. The two tidal bulges track the changes in lunar declination.

gravitational force varies inversely as the square of the distance between the two bodies and the tide-generating force varies inversely as the cube of the distance. This latter observation accounts for the Moon being 2.18 times more influential in causing tides than the Sun.

For simplicity, we have presented the tide-generating forces for selected points A and B (see Figure 1.40b), which are positioned on a line connecting the centers of the Earth and the Moon (or Sun). This representation is incomplete. Each point of the Earth, including the one labeled D, experiences the radial gravity of the Moon and Sun, and therefore the tidal forces, differently. However, only the tidal forces' horizontal components actually tidally accelerate the water particles since there is almost no resistance. In other words, horizontal components are more effective at moving water than the vertical components that try to lift water. Thus, on the surface of the Earth, the lateral tide-generating forces are more significant than the vertical forces in generating the tide bulges.

Over time, the positions of the Moon and Sun change relative to the Earth's equator. These changes have a direct effect on daily tidal heights and intensity. As the moon revolves around the Earth, its angle increases and decreases in relation to the Earth's equator. This angle is known as its declination (see Figure 1.41). Similarly, the Sun's relative position to the Earth's equator changes over the course of a year as the Earth rotates around it. The Sun's declination also affects the tides. Let us look at specific examples of the declination effects of the Moon and Sun on tides. When the Sun, Moon, and Earth are in alignment (as at the times of the new Moon or full Moon), the two tidal effects reinforce one another, and the solar tide has an additional effect on the lunar tide, creating extra-high tides and very low tides (shown in Figure 1.42a). These tides are called spring tides.

One week later, when the Sun and Moon are at right angles to each other (i.e., as we go from the new Moon to the first quarter Moon or the full Moon to the third quarter Moon), the low solar tides occur at the times of the high lunar tides and vice versa (shown in Figure 1.42b). Therefore, smaller-than-average tidal ranges occur. These tides are called neap tides. As the moon goes from the new Moon to the first quarter Moon or from the full Moon to the third quarter Moon, the solar tides are behind the lunar tides such that the composite is retarded. Thus, high and low tides of the composite occur earlier than with just the lunar tides (shown in Figure 1.42c). This scenario is called priming. Similarly, as the moon goes from the first quarter Moon to a full Moon or the third quarter Moon to a new Moon, the solar tides are ahead of the lunar tides such that the high and low tides of the composite occur later than with just the lunar tides (shown in Figure 1.42d). This scenario is called lag.

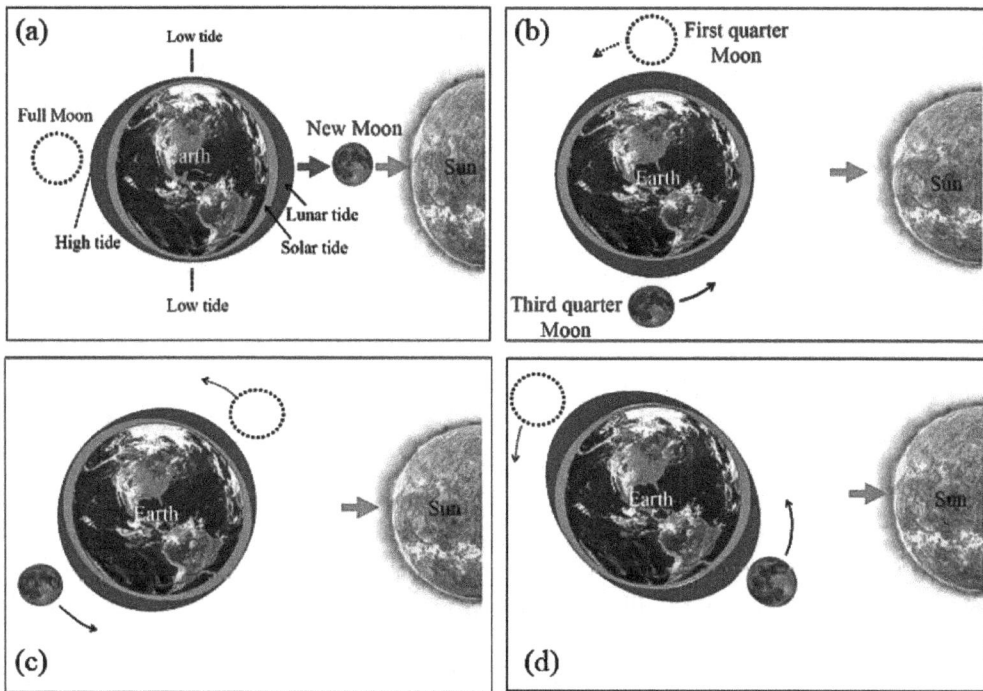

Figure 1.42. (a) Spring tides. At times, the Sun and the Moon pull together on Earth's waters in the same direction [i.e., the Moon is new (on the side of the earth toward the Sun) or full (on the side of the earth away from the Sun). The high solar tides occur at the same time as the high lunar tides, likewise, the low solar tides occur at the same time as the low lunar tides. (b) Neap tides. When the Moon is at right angle to the Sun (i.e., the moon is in its first or third quarters), the low solar tides occur at the times of the lunar high tides and the high solar tides occur at the low lunar tides of the lunar. (c) At the first quarter Moon coming from new Moon and third quarter Moon coming from full Moon, the high and low tides of the composite occur earlier than with just the lunar tides. (d) At the first quarter Moon going to full Moon and third quarter Moon going to full Moon, the high and low tides of the composite occur later than with just the lunar tides.

1.2.3.   Tide Occurrence

Tide-producing forces set the waters of the ocean basins into periodic motion, which can be written as a series of harmonic components or partial tides (Doodson, 1921); i.e.,

$$h(t) = \sum_i H_i \cos\left(\frac{v_i}{L_i}t + \xi_i + \kappa_i\right) , \tag{1.137}$$

where $h(t)$ is the height at time t, H_i is the height of mean water level above a selected datum, v_i is the wave-tide speed, L_i is the wave-tide wavelength, ξ_i is the phase lag relative to equilibrium tide, and κ_i is the local phase of the i-th harmonic. Parameters H_i, v_i, L_i, ξ_i, v_i, and κ_i are associated with the i-th harmonic component. Equation (1.137) can be used to model almost all of the perturbations in the relative motions of the Sun, Moon, and Earth, including their relative distances, their declinations, and cases where lunar and solar tides get into phase (maximal lined up) and out of phase (maximum of one lined up with the minimum of the other), to produce spring and neap tides, respectively (see Figure 1.42). Note that the study of tide height by harmonic analysis was begun by Laplace, William Thomson (Lord Kelvin), and George Darwin. Doodson (1921) extended their work. The underlying assumption in (1.137) is that, no matter how complex they appear, the tidal variations at any location can be represented by the sum of a finite number of simple harmonics.

Figure 1.43a illustrates a tidal wave for a single harmonic. The crest is a high tide and the trough is a low tide. The height of a tidal wave is the range (or tidal range) and the period of a tide is the time interval between the occurrences of two successive crests at a fixed point. The tide wave speed is approximately given by

$$v = \sqrt{gH} , \tag{1.138}$$

where v is the wave speed of the tide, H is the water depth, and g is the acceleration of gravity. Note that the wave speed depends only on water depth as g is almost constant. As a tide wave enters coastal areas (shallower waters with decreasing H), it slows down. The wave then gets taller [i.e., $h(t)$ increases] and the wavelength gets significantly shorter, as illustrated in Figure 1.43b, in order to maintain the principle of conservation of energy.

The ocean responses to tide-producing forces are generally grouped in three ways: semidiurnal, diurnal, and mixed tides. When the two high tides and two low tides of each tidal day are approximately equal in height, the tide is semidiurnal (as shown in Figure 1.44a). The time period between successive high waters of semidiurnal tides is about 12.42 hours. When there is a relatively large difference in the two high and/or low tides of each tidal day (i.e., large diurnal inequality), the tide is mixed (shown in Figures 1.44c and 1.44d). Typical examples of mixed tides are found on the Pacific coast of the United States. When there is only one high tide and one low tide each tidal day, the tide is diurnal (shown in Figure 1.44b). In this case, the time period between successive high waters is about 24.84 hours. Diurnal tides are the least common. Typical examples are the Java Sea, the Gulf of Tonkin, and Pei-Hai (China).

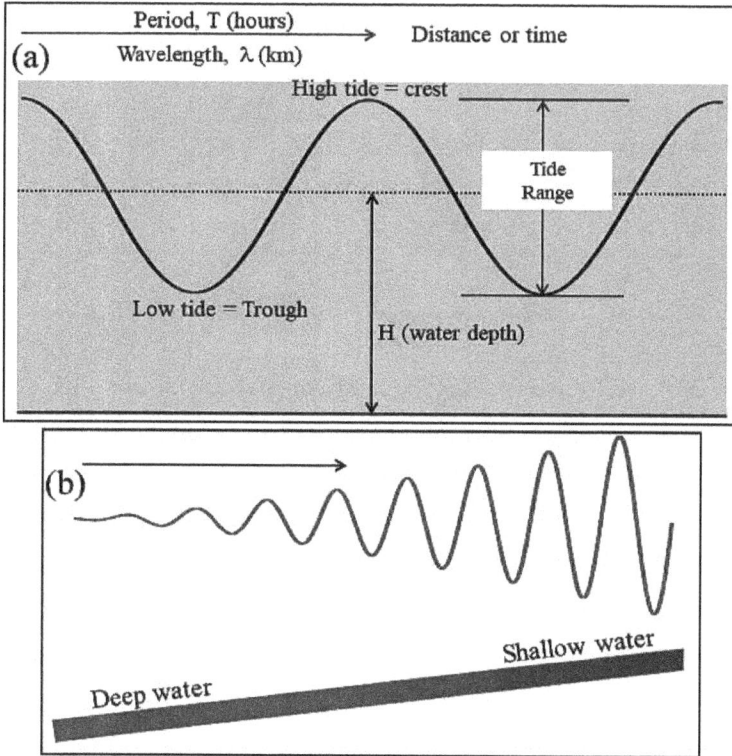

Figure 1.43. (a) Characteristics of a tidal wave. (b) Propagation of a tidal wave toward shallow waters.

At most places (e.g., the Africa coast, except the Somalia coast, which is mixed semidiurnal, the coast of West Europe, and many areas on the East Coast of North America), the tidal changes are semidiurnal. The genesis of semidiurnal tides is the fact that, unlike a 24-hour solar day, which we are well familiar with, a lunar day lasts 24 hours and 50 minutes. This difference occurs because the Moon revolves around the Earth in the same direction that the Earth is rotating on its axis. Therefore, it takes the Earth an extra 50 minutes to "catch up" to the Moon. Since the Earth rotates through two tidal bulges every lunar day, we experience two high and two low tides every 24 hours and 50 minutes. High tides occur 12 hours and 25 minutes apart, taking six hours and 12.5 minutes for the water at the shore to go from high to low, and then from low to high.

Note that the mean sea level that we will use in the next chapter is the average height of the sea surface over a longer period of time (usually a month or year) in which the short-term variations associated with tides and storm surges average out.

1.2.4. Tide Range

If the Earth were a perfect sphere without large continents, all areas would experience two equally proportioned high and low tides every lunar day. In fact, in the open ocean, the actual height of the tidal wave crest is relatively uniform but quite small (usually 1 m or less). It is only when the tide waves move into shallow waters

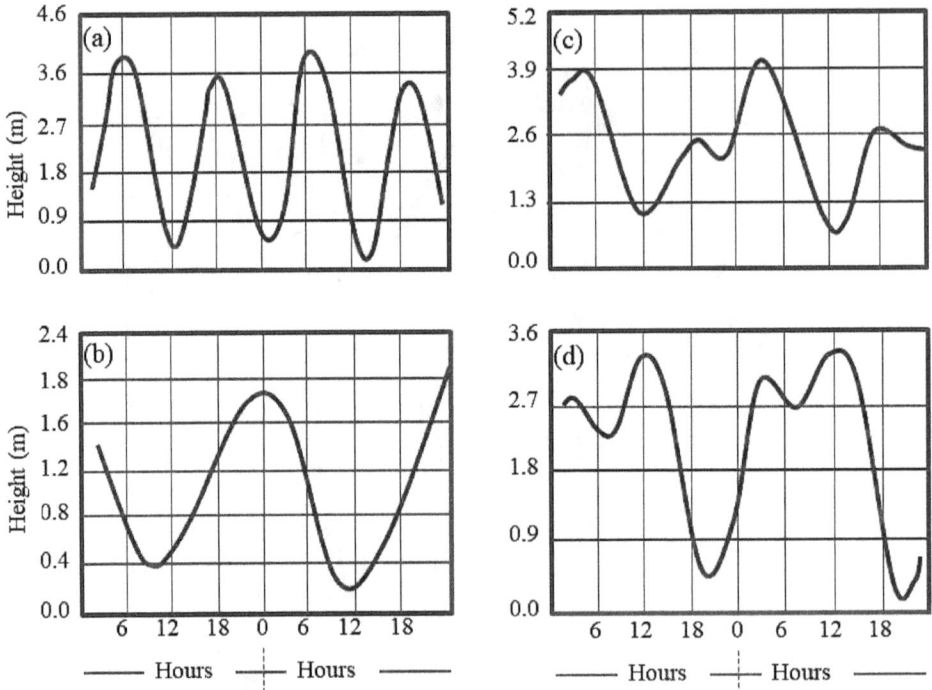

Figure 1.44. Types of tides: semidiurnal, diurnal, and mixed. (a) Tides are most commonly semidiurnal (two high waters and two low waters each day) in cycles of 12 hours 25 minutes, or (b) diurnal (one tidal cycle per day) in cycles of 24 hours 50 minutes. (c) and (d) are mixed tides. There is a large inequality in the high water heights, low water heights, or in both.

with reflections on land masses and into confining channels that large tidal ranges occur. Because land masses, channels, and water depths vary considerably from one coastline to another, the tide range has a complex pattern around the world, as we can see in Figure 1.45.

Every location has a unique coastline and a unique bathymetry (i.e., submarine topography) that gives each location its unique tide ranges. For example, when oceanic tidal bulges hit wide continental margins, the height of the tides can be magnified. Conversely, on mid-oceanic islands not near continental margins, the height of the tides are typically very small tides, less than 1 meter (Thurman, 1994). The shape of bays and estuaries can also magnify the intensity of tides. Funnel-shaped bays, in particular, can dramatically alter tidal magnitude. The Bay of Fundy in Nova Scotia is one of the most notable examples of this effect. It has the highest tides in the world, 15 meters (Thurman, 1994). The amplification of the tide occurs from the mouth to the head of the bay (see Figure 1.46). The narrowing of the bay causes tide wave reflections on the coast and consequently the amplification of the tidal range. This phenomenon is known as tide resonance. Further amplification is added, in this case, because the natural period of oscillations of the bay is close to the tidal period. Constructive interference increases tide ranges.

Regions where a continental shelf has the right combination of depth and width for tidal resonance to occur, such as on the northwest Australian shelf (shown in

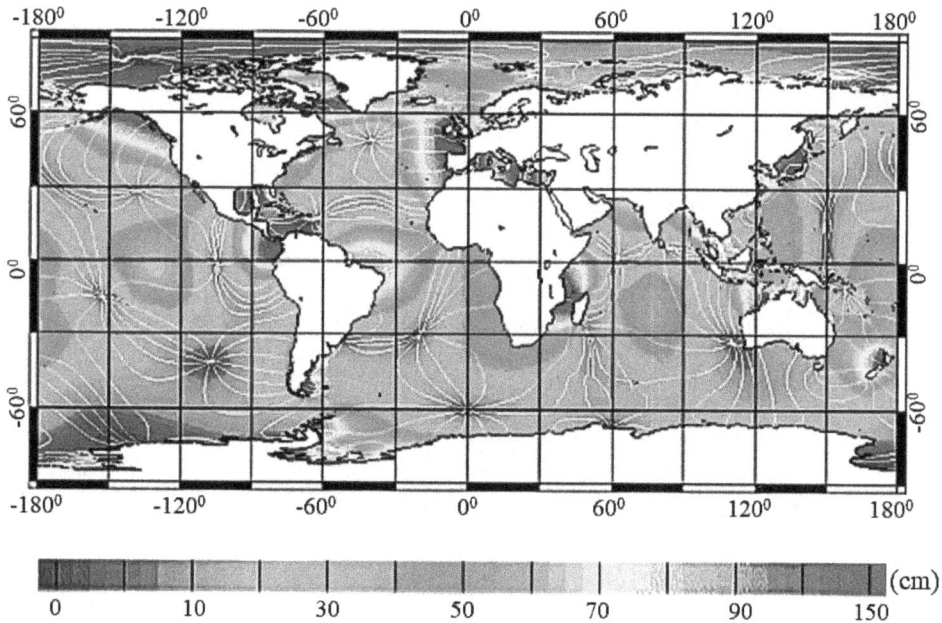

Figure 1.45. Global tide ranges. (Adapted from Egbert and Erofeeva, 2002).

Figure 1.47), can also exhibit large tide ranges. In general, tides are usually largest in semi-enclosed seas and funnel-shaped entrances to bays and estuaries, such as Penzhinskaya Guba bay in Russia (12 m) and the Bristol Channel in the United Kingdom (12 m). Other areas with high tide ranges include Granville in France (11 m), Rio Gallegos in Argentina (10 m), St. Helier in the Channel Islands (9.5 m), and Cook Inlet in Alaska (9 m).

Quiz: Are there always two high tides a day?

Local wind and weather patterns can also affect coastal tides. Strong offshore winds and other patterns can move water away from coastlines, exaggerating low-tide exposures, as in the Mediterranean Sea and the Baltic Sea. Basically, the oscillation modes of the Mediterranean Sea (2- to 3-cm tides) and the Baltic Sea (less than 10-cm tides) associated with local wind and weather patterns cancel out tide-producing forces in these seas. In general, tidal ranges are typically smallest in the open ocean, on mid-oceanic islands not near continental margins, along open ocean coastlines, and in almost fully enclosed seas. Note that the Mediterranean and Baltic Seas are almost fully enclosed seas (see Figure 1.48).

To conclude the above discussion on ocean tides, let us make very brief remarks on frictions, climate change, Roche limit, and cohesiveness of celestial associated with tides. The Earth's daily tides cause loss of energy from its rotation due to friction which slows down the Earth's rotation rate. Currently this rotation rate is about 0.0016 seconds per 100 years. At the same time the Moon moves away from us at about 3.8 meters per 100 years, so its orbital period increases. This process will carry on until a stable situation is reached when the Earth's rotation period equals the Moon's revolving period. In such a scenario, the tidal forces

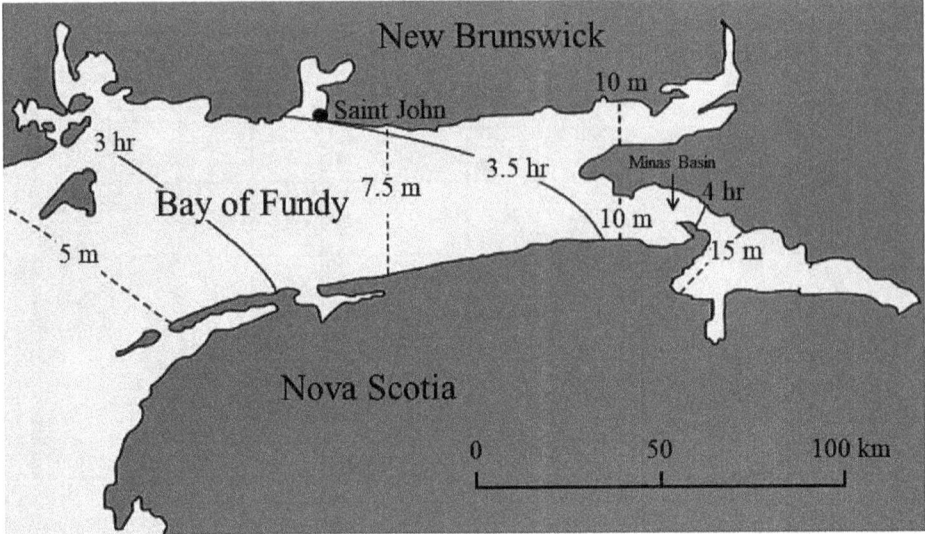

Figure 1.46. The Bay of Fundy has a tidal range of 15 m. The bay is located on the Atlantic coast of Canada. Tides in the bay are dominated by the semi-diurnal oscillations with high waters and low waters of approximately of the same range. Tide range changes from approximately 2 m at the Bay entrance to more than 15 m in Minas Basin.

by the Moon on the Earth will remain steady because the Earth will be tidally locked to the Moon, always keeping the same face toward it, only solar tides will

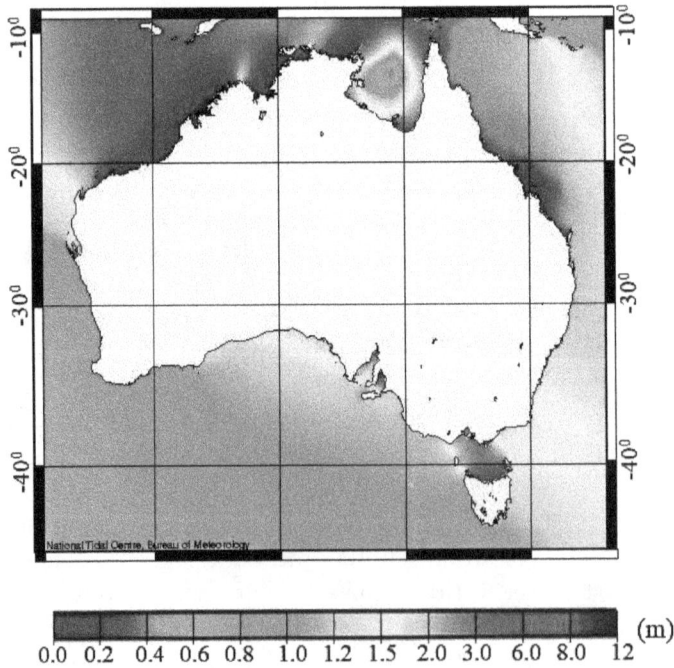

Figure 1.47. Tide ranges of Australian coastlines. [Adapted from the 2018-tidal map of NTF (National Tidal Facility) of Australia].

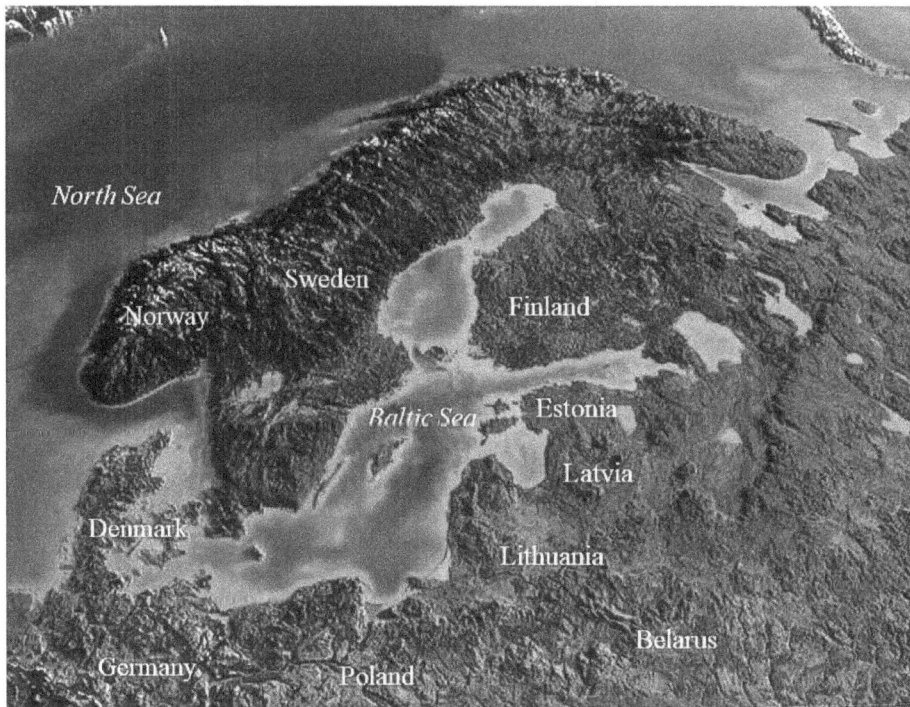

Figure 1.48. (Top) A map of the Mediterranean Sea, showing how it is nearly completely enclosed by land masses.(Bottom) A Baltic-Sea map showing that it is also an almost fully enclosed sea by land masses.

slow the Earth's rotation. Note also that cyclical changes in Earth's obliquity and eccentricity, ultimately due to Jupiter's tidal forces, account for some climate changes that have happened. These changes include continental glaciers' advances and retreats.

Consider a planet and a moon orbiting it. The Roche limit, also known as the Roche radius, is essentially the shortest distance these two celestial bodies can come together without the moon getting pulled apart by the planet's tidal and gravitational forces. Basically, closer to the Roche limit the moon is deformed by tidal forces. Within the Roche limit the moon's own gravity can no longer withstand tidal forces, and the moon disintegrates. For large moons, particles closer to the planet move faster than particles farther away. The material's orbital speed eventually forms a ring. In other words, inside the Roche limit, the orbiting body disperses and forms rings, whereas outside the limit, the material tends to coalesce. This limit is named after Édouard Roche who first calculated this theoretical limit in 1849 (Roche, 1849a, 1849b, and 1850). Between a rigid spherical moon and a planet, the Roche limit is

$$
d_{\text{roche}} \approx 2.423 \; r_{\text{planet}} \left(\frac{\rho_{\text{planet}}}{\rho_{\text{moon}}} \right)^{1/3} = 2.423 \; r_{\text{moon}} \left(\frac{m_{\text{planet}}}{m_{\text{moon}}} \right)^{1/3} , \qquad (1.139)
$$

where where r_{planet} is the radius of the planet, ρ_{planet} is the density of the planet, ρ_{moon} is the density of the moon, m_{planet} is the mass of the planet, and m_{moon} is the mass of the moon. For Earth and Moon, the Roche limit is about 18,237 kilometers, assuming that the mean Earth radius is 6371 kilometers, the mean density of the Earth is 5.51 g/cm^3, and the mean density of the Moon is 3.34 g/cm^3. So the Roche limit distance is approximately equal to 2.863 r_{earth}. The Moon has to orbit at least 2.863 r_{earth} (from Earth's center to Moon's center) away from Earth's center to avoid being disintegrated or getting torn apart by Earth's tidal forces.

Currently the Moon is about 60.336 r_{earth} away from Earth. The Roche limit is not an issue. Actually, as pointed out earlier, the Moon will never reach Earth's Roche limit because the Moon is slowly moving away from Earth. Also, the Roche limit suggests that an object the size of our Moon does not form within the Roche limit. Instead, it would have been captured and our Earth would have a ring system. So our Earth has captured our Moon from somewhere or it has been placed at its currents by forces that have yet to be determined.

Humans are within the Roche limit of the Earth. But the limit does not affect us because we are gravitationally tied to Earth. Also human satellites in orbit are affected by the Roche limit because they are held together by mechanical strength, not their own gravity. As opposed to this, an asteroid will break apart if its orbit reaches a planet's Roche limit. Comet Shoemaker–Levy 9 broke apart on its first pass by Jupiter because it passed too close or entered inside Jupiter's Roche limit.

The Sun also has a Roche limit. We simply have to replace planet by the Sun and moon by planet in (1.139) to obtain the Roche limit for the planets in our solar system. There is a Roche limit for rocky planets and moons within the sun itself, as it turns out. For a fluid body, say Jupiter, it is about 1.17 million kilometers outside the Sun. However, it is insignificant since the distance between the Sun and Jupiter is 633 times Roche limit, about 741.08 million km.

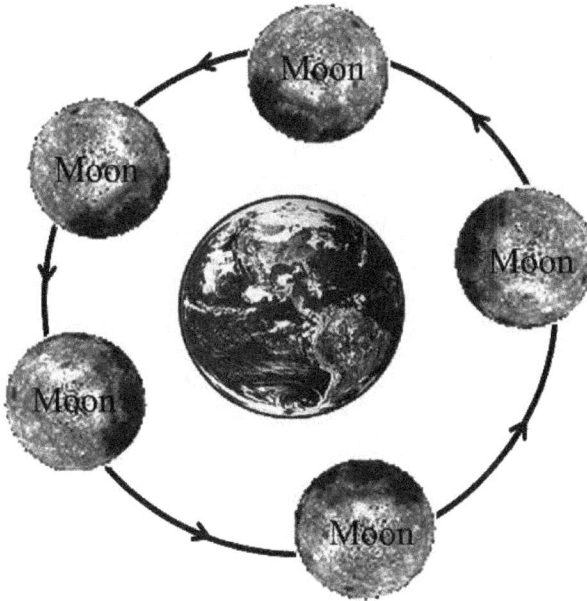

Figure 1.49. An illustration of how Moon always has one face facing us. The Moon orbits Earth once every 27.3 days and spins on its axis once every 27.3 days (i.e., synchronous rotation). So, it always faces us.

A human falling into a black hole will also experience tidal forces, which are lethal! The difference in acceleration between the head and feet could be many thousands of times Earth's gravity. People would literally be pulled apart! Similarly, if the tide force is stronger than a celestial body's cohesiveness, the body will be disrupted. As we will discuss later, the minimum distance a moon comes to an exoplanet before it is shattered this way is called the Roche distance from the exoplanet. As space exploration continues, we will come across a moon tidally pulled apart by an exoplanet's strong gravity one day.

1.2.5. Far Side of the Moon

Yes, there is a lunar hemisphere that we have never seen through naked eyes and even from a telescope until recently in 2019. This hemisphere of the Moon always faces away from the Earth (see Figure 1.49). Every time we see the Moon, we are only seeing one side of the Moon. The visible side of the Moon to us is known as the *near side of the Moon*. The other side of the Moon, known as the *far side of the Moon*, is hidden to us. This phenomenon is due to the fact that Earth and Moon are tidally locked, as described above—that is, the Moon rotates around its axis and at the same speed that it rotates around the Earth. The Moon completes one full rotation on its axis in the time it takes to orbit the Earth, approximately 27.3 days. That means the same side is always turned toward us. However, it is significant to note that the far side of the Moon receives equal amounts of sunlight as the near side of the Moon (i.e., the Earth-facing side of the Moon).

Note that the rotation of the Moon around Earth in Figure 1.49 is sometimes called as synchronous rotation. Many moons in the Solar System (including Mars's

Phobos and Deimos, Jupiter's Io, Europa, Ganymede and Callisto, Saturn's Enceladus, Tethys, Dione, and Titan, Uranus' Miranda, Ariel, Umbriel, Titania and Oberon, and Neptune's Triton) are also in synchronous rotation; that is, these moons always show the same face to their planet while revolving in a nearly circular orbit. In other words, their rotational periods have an exact relation to their period of revolution. These relations are a direct consequence of the tidal forces generated by the planets. Note also that the Mercury, instead of always keeping the same face toward the Sun, makes three complete rotations for every two complete orbits as a direct consequence of its rather eccentric orbit ($e = 0.206$). TRAPPIST-1 b, the closest planet to TRAPPIST-1 star, is tidally locked by the TRAPPIST-1 star, according to recent observations from NASA's James Webb Space Telescope. Again, that means that one side of TRAPPIST-1 b faces the star at all times and the other is in permanent darkness.

Landing process. On January 3, 2019, the China National Space Administration's Chang'e 4 probe made the first ever landing on the lunar far side and provided us with high-resolution images of both sides of the Moon. One may wonder why there have not been probes and rovers around the far side of the Moon earlier. This landing is a long way from being a mundane event. US and former Soviet Union missions to the moon, including the Apollo and Luna (Lunik) programs, never landed on the far side of the Moon. The challenges of landing on the far side of the Moon include:

- ensuring communication between the far side and the Earth for the safety of the operations. Communication with Earth is impossible from the far side of the Moon. Spacecrafts can neither send to nor receive a radio signal from its control center on Earth; a dangerous proposition for a manned, even unmanned, landing, and/or malfunctioning probes and scientific equipments.

- ensuring spacecrafts/rovers can have revolution trips around the Moon, including travel over the far side during their revolution period. It is the difficulty of performing these revolution trips that has made taking photographs and/or seeing the far side of the Moon impossible until recently.

- and making sure of soft landing of spacecraft on the far side of the Moon to ensure that the required data can be recorded and returned to Earth. Essentially, the recorded data, including photographs, are the proofs of landing. We need to avoid crash landing to ensure cameras, radio astronomy, geophysical, and geochemical instruments can function after landing. In other words, these cameras and instruments must be deployable on the Moon's surface, and the continuous recording and/or transmitting of data to Earth must occur.

The China's Chang'e-4 (CE-4) mission, composed of a lander, rover, and relay satellite overcame these challenges. It places a relay satellite, named Queqiao, in a halo orbit of the Earth–Moon L_2 Lagrange point. Queqiao was successfully launched on May 21, 2018, and entered the halo orbit of the L_2 point on June 14, becoming the first satellite connecting the Earth and the Moon's far side, circumventing one of the major challenges of visiting the Moon's far side. From the Earth–Moon L_2 Lagrange point, Queqiao can communicate with both the far side of the Moon and the Earth. Queqiao became the hub of three-way communication: from the Moon's

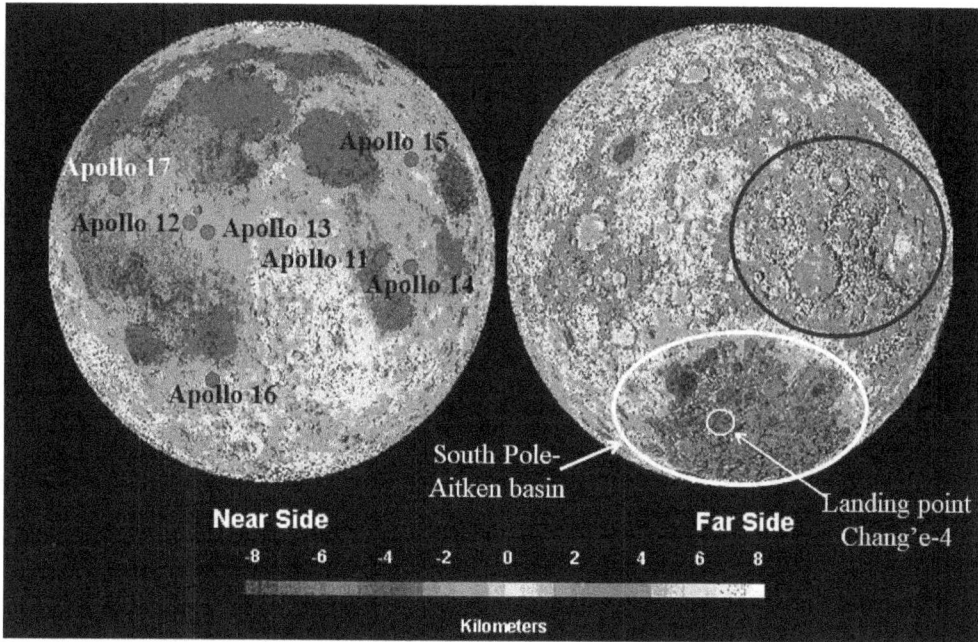

Figure 1.50. The topography of the near side and far side of the Moon. The South Pole-Aitken basin is about 25,000 km across. (Adapted from the Clementine-Mission Images).

far side to Earth–Moon L_2 Lagrange point and from Earth–Moon L_2 to Earth and vice-versa.

The next step was the launch of the CE-4 probe composed of a lander, a rover, and relay satellite on December 8, 2018. It entered lunar orbit on December 30, 2018 after several path adjustments and corrections. The last step was the soft landing; the make-or-break step. Communication between the probe and relay satellite is essential for monitoring the descent, as before. There are techniques for performing such descent remotely. The oil and gas industry, for example, performs such descent routinely today to place sensors more than 1.5 kilometers below the sea surface, at the sea floor, through radio communication between remotely operated vehicles, which actually place sensors on the sea floor instead of humans, and the control center on the surface.

The rover landed on one of the flattest spots in the South pole Aitken basin known as von Karman crater since that is a simpler place to land a rover. This crater is about 2500 km wide and 12 km deep. Due to the thin crust of the Moon, the impact that hit this area may have brooked through the crust into the Moon's mantle.

Quiz: The distance between the lunar far side and the Earth is 500,000 km. Define the Earth-Moon L_2 point.

Comparison of near and far sides of the Moon. Figure 1.50 shows the topography of the near and far sides of the Moon. It shows that the far side is significantly different from the near side despite the fact that both receive almost equal amounts of sunlight. The circled black line in Figure 1.50 captures some of these differences. The near side has an area covered by dark lava flows (the lunar maria or *seas*). The

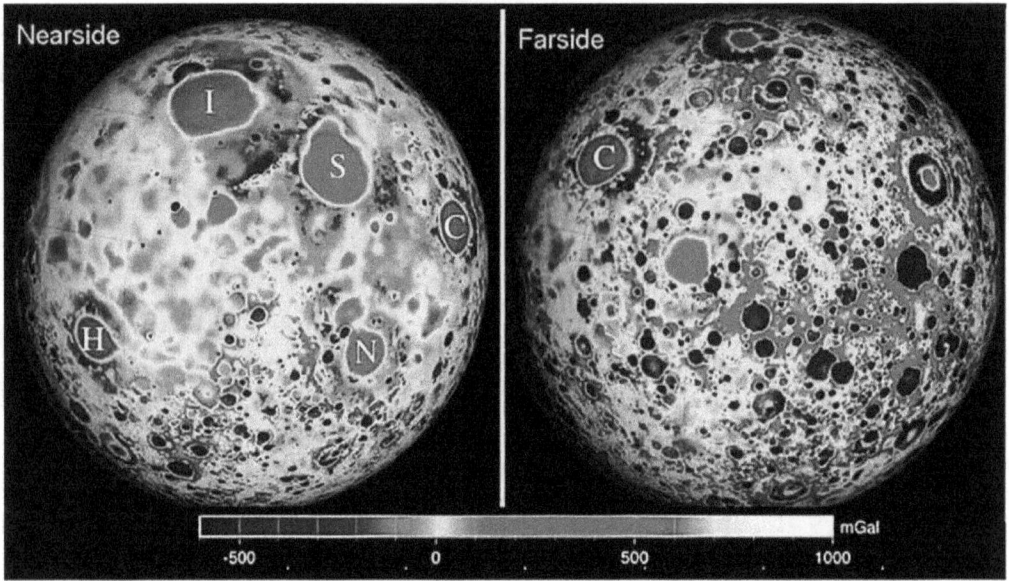

Figure 1.51. The Free-air anomaly of the near side and far side of the Moon. (Image credit: NASA/JPL-Caltech/CSM).

far side's terrain is rugged, with a multitude of impact craters and relatively few flat. These impacts give the far side, in some parts, a heavily-cratered appearance closer to that of Mercury (see Ikelle, 2023a). It has one of the largest craters in the Solar System, the South Pole–Aitken basin, which is a 2,500 kilometers wide depression and occupies much of the southern part of the far side. No seas are apparent on the far side.

The differences between the near and far sides of the Moon suggest that the Moon was formed from mixtures of matter of two totally different compositions. Perhaps, Earth may have shielded the near side from a significant amount of bombardments. As a result, their differences may be caused by materials deposited on the far side by colliding objects. The free-air anomalies of both sides in Figure 1.51 enhance and further clarify their topographic differences. The Bouguer anomalies in Figure 1.52 show typical blue, yellow, and red colors corresponding to negative, near-zero, and positive values of anomalies, respectively. By comparing these results with Figures 1.37 and 1.38, one can surmise that the area in the circled line has light rocks just like many continents on Earth, and therefore, a likely suitable area for a future Moon colony. Note, however, the strange texture of this area (inside the black circle) in Figure 1.50. We have not found this texture on Earth's maps so far.

Quiz: Are there tides on Jupiter?

1.2.6.   The Titan Tides

Our discussion so far has focused on water tides, thus on a planet with surface water. Actually, celestial bodies' solid parts are also subject to tides. In more general

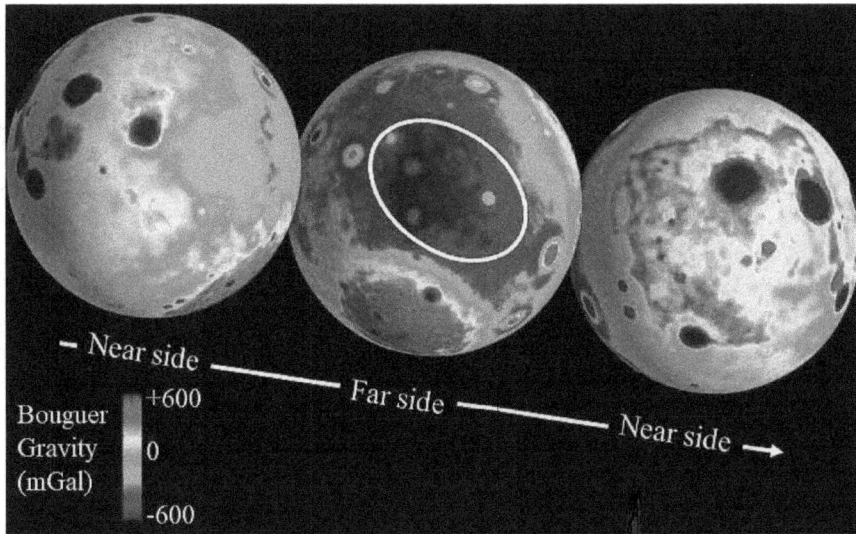

Figure 1.52. The Bouguer anomaly of the near side and far side of the Moon with the effects of topography removed to reveal density variations underneath the surface and variations in crustal thickness. (Image credit: NASA/JPL-Caltech/CSM).

terms, any celestial body placed in another's gravitational field will experience tidal forces that distort it. This is because attraction is stronger on one side than the other. For example, solid body tides on the Moon caused by Earth vary by about ±0.1 m each month. In addition to changes in shape, the Moon's gravity field and orientation in space are affected by tides. In other words, tidal forces have reshaped star system orbits, determining the location and orbital parameters of the moon system around planets and exoplanets. Due to the work done per period in a steady state, tides also produce heat in celestial bodies. Moreover, information about the internal structure, composition, and evolution of exoplanets may well come first from tide measurements because current exoplanetary mass and radius cannot be used directly to identify the interior structure of exoplanets, as different composition and density profiles can lead to the same solution. Direct subsurface exploration of solid exoplanets and even Jovian moons are long off.

Titan, the largest of Saturn's 82 moons, is another celestial object with significant tides. Titan has a diameter of 5,150 kilometers, an eccentricity of 0.0288, and a distance of 1.4 billion kilometers from the Sun. Titan completes its elliptical orbit around Saturn every 16 days. Saturn's gravity causes tidal forces during that time (see Figure 1.53). At its farthest point from Saturn, Titan is almost perfectly spherical. However, at its closest point where Saturn's gravitational pull is the strongest Titan grows flatter, its diameter alters by 10 meters (Castillo-Rogez and Lunine, 2010).

A huge leap forward in our current understanding of Titan took place in 2005, when NASA–ESA–ASI's Cassini–Huygens unmanned spacecraft entered orbit around Saturn on July 1, 2004 and landed on Titan on January 14, 2005 after traveling for seven years (from October 15, 1997 to January 14, 2005) in space with flybys of Earth, Venus, and Jupiter. This mission is named after the astronomer Giovanni

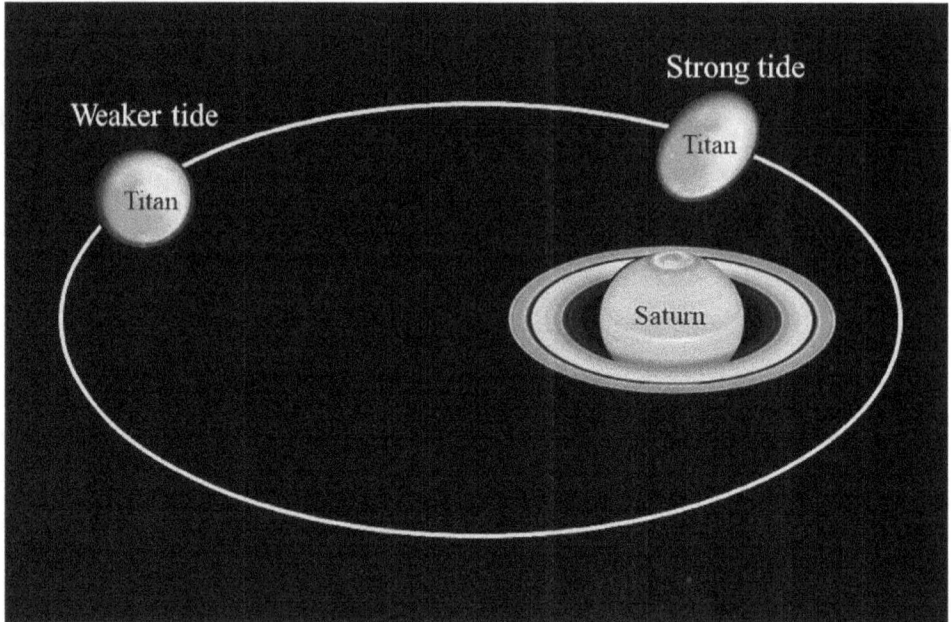

Figure 1.53. The distance between Saturn and Titan varies with time, the tidal force on Titan varies with time, and the amplitude variation of body response has a lag along with what amounts to energy dissipation.

Domenico Cassini, who discovered Saturn's ring divisions and four moons of Saturn. It is also named after the astronomer, mathematician and physicist Christiaan Huygens, who discovered Titan in 1655. The Cassini-Huygens mission ended in 2017 due to dwindling rocket fuel resources. As of April 2023, Titan is the only known celestial object with liquid on its surface other than Earth. Moreover this liquid is methane (CH_4), the simplest hydrocarbon, and not water (Stofan et al., 2007). Most people know methane as the gas we use when cooking on a gas stove. Strangely enough there are lakes, and even oceans, of liquid methane on Titan. Tidal forces in these lakes and oceans are almost zero. However, precise measurements of the Cassini spacecraft's acceleration during six close flybys between 2006 and 2011 have revealed that Titan responds to Saturn's variable tidal field with periodic shape changes. The tidal field requires that Titan's interior is deformable over orbital period time scales, in a way that is consistent with a global ocean at depth. To establish this link, we need measurements of a parameter directly related to tides, known as the Love number. It describes the amount of distortion. Our next task is to introduce this parameter and define its values for Titan's tides obtained from Cassini–Huygens data.

Tidal potential and Love number. A planet and its moons, the Sun and its planets, a star and its exoplanets, binary star systems, triple star systems, and quadruple star systems, for example, are subject to mutual gravitational interactions. When one goes beyond the point-mass approximation for a given body (called the primary), the gravitational force exerted by another body (called the secondary or the companion) is different at its surface and at its center of mass. A differential gravitational force is applied, which is by definition the tidal force

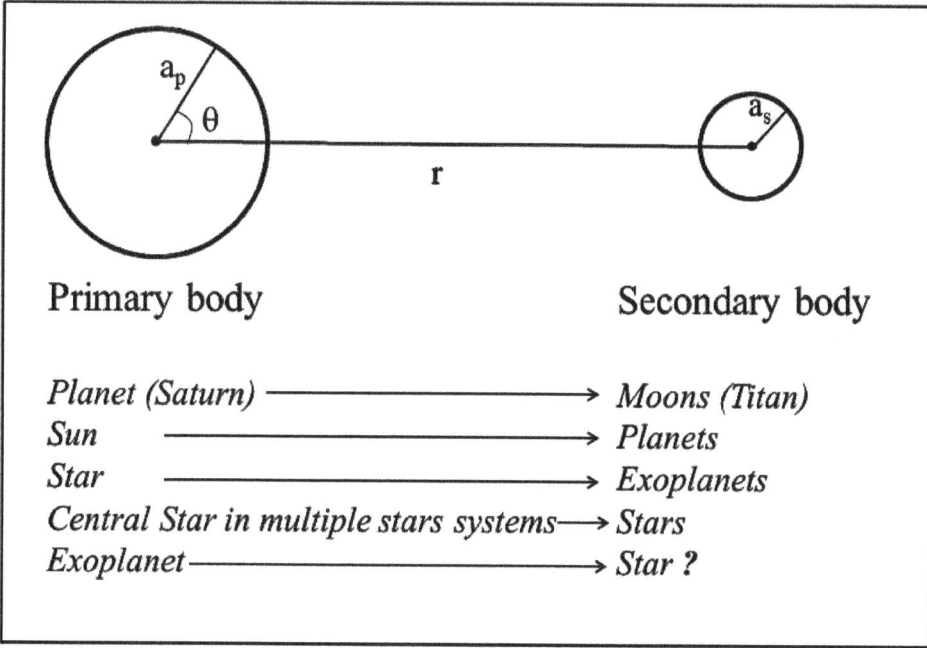

Figure 1.54. The geometry of the problem for computing tidal potential and surface deformations.

$(\boldsymbol{F}_{\text{tides}} \approx 4GMma/r^3$ for $r \gg a)$, also known as the force producing diurnal ocean tides. We can derive the tidal potential (W) from it as follows:

$$\boldsymbol{F}_{\text{tides}} = \boldsymbol{\nabla}W \implies W = \frac{Gm_p}{r} \sum_{n=0}^{\infty} \left(\frac{a_s}{r}\right)^n P_n(\lambda) , \qquad (1.140)$$

or

$$W = \frac{Gm_p}{r} \left[1 + \left(\frac{a_s}{r}\right) P_1(\lambda) + \left(\frac{a_s}{r}\right)^2 P_2(\lambda) + ...\right] , \qquad (1.141)$$

where G is the universal gravity constant, P_n is the Legendre polynomial of degree n, m_p is the mass of the primary body treated here as a point-like body with a negligible angular momentum, $\lambda = \sin\theta\cos\phi$ as defined in (1.90) and Figure 1.18, r is the distance from the center of the secondary body to the primary's center, a_p is the primary body radius with $a_p < r$. Distances a_p, a_s, and r are also defined in Figure 1.54. Assuming that $a_s/r \ll 1$, we alternatively wrote tidal potential as a sum of Legendre polynomials in (1.141). The $n = 0$ term is constant in space, so its gradient (the force) is zero, and it can be discarded. The gradient (the force) of the $n = 1$ term is a constant force along the direction of the tidal companion; but we can subtract it to get the tidal force. The higher terms are the ones that are tide-raising of the potential at point r. By neglecting fourth- and higher-degree tides (i.e., terms corresponding to $n > 2$), the tide-raising potential reduces to the largest of the tide-raising potential term; i.e.,

$$W_{\text{tides}}^{(2)}(r, \lambda, t) \approx \frac{Gm_p}{r(t)} \left(\frac{a_s}{r(t)}\right)^2 P_2(\lambda(t)) , \qquad (1.142)$$

where we have made r and θ, as they actually are functions of time t. The primary body (e.g., Saturn)gravity causes tides both in the secondary (e.g., Titan) interior as well as its surface. In the interior of a secondary object (e.g., Titan), matter is redistributed. This redistribution of matter further modifies the potential, exaggerating the effect. For $r = a_s$ (i.e., the surface of the secondary body) we have

$$V_{\text{tides}}^{(2)}(a_s, \lambda, t) = (1 + k_2) W_2(a_s, \lambda, t) = h_2 W_2(a_s, \lambda, t) , \qquad (1.143)$$

where $V_{\text{tides}}^{(2)}(a_s, \theta, t)$ is the new potential that takes into account the redistribution of matter inside secondary object (e.g., Titan) through k_2, which is known as the Love number of degree 2, and $h_2 = 1 + k_2$. The Love number is named after Love (1911) who introduced it. When higher-order terms of Legendre polynomials are used, we have multiple Love numbers with k_j as Love numbers of degree j. From equations (1.141) and (1.143) we obtain the tidal surface deformations

$$\Delta r_{\text{tides}} = \frac{V_{\text{tides}}^{(2)}}{g} = a_s \frac{m_p}{m_s} h_2 \left(\frac{a_s}{r}\right)^3 P_2(\lambda) . \qquad (1.144)$$

Rotational potential and Love number. We consider the case where the secondary object's (e.g., Titan) spin axis is not necessarily perpendicular to the orbital plane. We define the obliquity, Θ, as the angle between the arbitrary radius vector $\boldsymbol{x} = [r \sin\theta \cos\phi, r \sin\theta \sin\phi, r \cos\theta]^T$ and the spin axis. Kopal (1959) and Hellar et al. (2019) showed that the rotational potential in this case can be written as follows:

$$W_{\text{rotation}}^{(2)} \approx -\frac{1}{3} \frac{Gm_s}{r} \left(1 + \frac{m_p}{m_s}\right) F_s^2 \left(\frac{a_s}{r}\right)^2 P_2(\cos\Theta) , \qquad (1.145)$$

where $F_s = \tau_{\text{orbit}}/\tau_{\text{rotation}}$ is the ratio between the orbital and rotational periods. Rotational surface deformations are

$$\Delta r_{\text{rotation}} = \frac{h_2 W_{\text{rotation}}^{(2)}}{g} = -\frac{a_s}{3} \left(1 + \frac{m_p}{m_s}\right) F_s^2 h_2 P_2(\cos\Theta) . \qquad (1.146)$$

Surface deformations. Assuming that the surface deformations are simply additive, the total surface shape is defined by

$$\begin{aligned}
\Delta a_s(\theta, \phi) &= a_s + \Delta r_{\text{tides}} + \Delta r_{\text{rotation}} \\[2mm]
&= a_s \left[1 + \frac{m_p}{m_s} h_2 \left(\frac{a_s}{r}\right)^3 P_2(\lambda)\right. \\[2mm]
&\quad \left. -\frac{1}{3} h_2 \left(1 + \frac{m_p}{m_s}\right) \left(\frac{a_s}{r}\right)^3 F_s^2 P_2(\cos\Theta)\right] . \qquad (1.147)
\end{aligned}$$

We can determine the value of k_2 (or h_2) by fitting $\Delta a_s(\theta, \phi)$ to a three-degree polynomial for different flybys of the secondary planet using for example, least squares fitting.

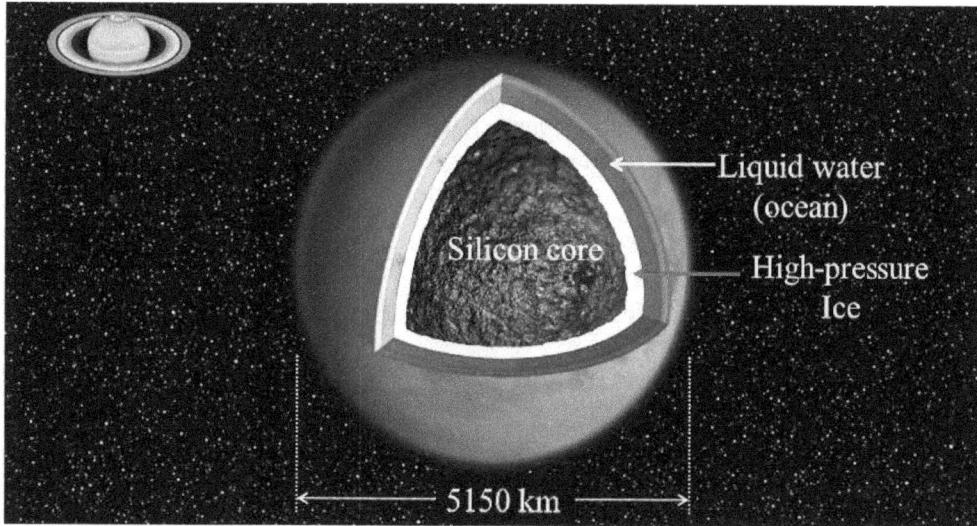

Figure 1.55. A Titan's interior mostly based on the value of k_2.

Table 1.6 shows some values for various secondary bodies. It is clear that k_2 is influenced by celestial body deformability (i.e., elastic shear modulus which is also known as rigidity) and uniformity (i.e., variation of mass density) which are two significant indicators of the interior of a body's interior. Actually, we can see from (1.147), the tide effect of the interior body is entirely enclosed in the Love number k_2, or equivalently h_2. It measures, for example, the level of central condensation of an object. The more homogeneous the celestial body is in its mass distribution, the bigger the Love number k_2. Maximum homogeneity is represented by a body of constant density and zero rigidity, yielding the maximum value of $k_2 = 1.5$. If the planet is more centrally condensed, the Love number decreases. Planets with a core generally have a smaller Love number due to stronger central condensation. As an example, Saturn with a core model of 10 m_{earth} has $k_2 = 0.39$. In other words, k_2 infers the presence of a massive core. Note that a planet with a dense core has a smaller Love number than a uniform body with equal rigidity.

The Love number $k_2 = 0$ corresponds to rigid bodies because matter is not redistributed. We can see in Table 1.6 that most planets and moons have low Love numbers because their interiors are nonuniform and significantly rigid as one may expect. Love number $k_2 = 3/2$ corresponds to zero rigidity (i.e., shear modulus is zero) like a perfect liquid.

For $k_2 > 0.4$, it is generally extrapolated that an ocean may be present in the interior of the celestial body. The idea that Titan hosts a global subsurface ocean (see Figure 1.55) is primary based on the fact that its Love number is significantly greater than 0.4 as we can see in Table 1.6. Such a large response to the tidal field requires that Titan's interior is deformable over time scales of the orbital period, in a way that is consistent with a global ocean at depth.

For the most part, the Moon's interior density is constant, as we discussed in Ikelle (2023c). Its low Love number in Table 1.6 indicates that the Moon's core is also very rigid and *its ratio of core mass over its total mass is greater than 0.85*.

Note that k_2 may not be sufficient to determine the interior of the celestial body. However it provides an excellent indication of its matter distribution and

Table 1.6. Some Love numbers, k_2, and their interpretation and examples.

k_2	Interpretation and examples
0	Rigid body (no redistribution of matter) and constant density.
0.0198 ± 0.0025	Moon (Williams, 2007) from lunar laser ranging (LLR) data.
0.27 ± 0.02	Neptune (extrapolated based on core mass/planet mass=0.45).
0.307 ± 0.02	Solid earth without ocean masses.
0.390 ± 0.024	Saturn (Lainey et al., 2017) from Cassini observations.
0.464 ± 0.023	Mercury (Verma and Margot, 2016).
0.411	Venus.
0.425 ± 0.005	Neptune (extrapolated based on core mass/planet mass=0.27).
0.45	Mars.
0.464 ± 0.023	Mercury (Verma and Margot, 2016).
0.535 ± 0.006	Jupiter (Ni, 2018) from Juno observations.
0.616 ± 0.067	Titan (Durante et al., 2019) from Cassini observations.
0.94	A fluid body with the earth's density profile.
1.20	Tethys (Tethy's density is 0.973g/cc).
1.5	Uniform density fluid body.

overall rigidity. Thus it is possible to have several acceptable micro-models of a celestial for a given k_2 value.

> **Quiz:** There is no physics that suggests an exoplanet cannot have its own star (see Figure 1.54). A Jupiter-like exoplanet can have its own small star. After all, a fusion reactor is designed to build our own Sun on Earth. It will turn out that nature already does so spontaneously. This is what we need to locate and duplicate. Discuss.

1.2.7.   High Tides May Be the Origin of Life on Earth

Earth's daily rotation around its axis has an angle to its orbital plane. Earth's orbital plane is the path it takes around the sun as it orbits the sun. This plane is called the ecliptic. The equator is defined as a plane perpendicular to the earth's rotation axis. Therefore, the two planes are independent of each other. The angle between these two planes represents Earth's tilt. This angle is called axial tilt or obliquity. It is defined as

$$\theta \approx 23.45396° - 0.01301346° \, t - 0.000002° \, t^2 + 0.000001° \, t^3 \,, \qquad (1.148)$$

where θ is the obliquity and t is the time the sun takes to return to the same position in the cycle of seasons, as seen from Earth. The date of start is January 1, 1900. Because of t variations, the Earth's axial tilt varies between 22.1 and 24.5 degrees every 41,000 years. This cycle time of 41,000 years of oscillations between 22.1 degrees and 24.5 degree is called Milankovitch Cycle.[5] The Earth's tilt relative to its orbital plane is not an assumption. Instead, it is perfectly measurable if you

---

[5]Earth's orbital variations are characterized by eccentricity, obliquity, and precession. The elliptical shape of Earth's orbit around the sun is characterized by its eccentricity—that is, a measure of the deviation of an orbit from a perfect circle; i.e., $e^2 = 1 - (P^2/a^2)$, where e is the eccentricity, P is the closest distance to the sun known as the perihelion, and a is the furthest

chart the angle of sunlight incidence at different points on the planet's surface. The most direct way of observing this is to look at the angles made by the ecliptic and the equator. Once a year, on the summer solstice (on or about June 21), the North Pole points directly toward the sun, and the South Pole is entirely hidden from incoming radiation. Half a year later, on the winter solstice (on or about December 21), the North Pole points away from the sun and does not receive sunlight.

Because there are no two planets in our solar system with the same tilt angle (Mercury, 0.1 degrees; Venus, 177 degrees; Earth, 23 degrees; Mars, 25 degrees, Jupiter, 3 degrees, Saturn 27 degrees, Uranus, 97 degrees; Neptune, 30 degrees; and Pluto, 120 degrees), and no obvious correlations between the tilt angles of the planets, the tilt angles are generally viewed as the results of random processes. No planet had a rotational axis perpendicular to its orbit. The early (less than a billion years old) inner solar system was home to a myriad of young planets with overlapping and unstable orbits. One possible random reason for Earth tilting is a giant whack the planet got billions of years ago, when another smaller planet collided with it. Over four or so billion years ago, a celestial object called Theia, about the size of Mars, collided with Earth to create a tilted axis. Prior to this event, the Earth's rotation axis was straight at 180 degrees.

The most obvious consequence of the current Earth's tilt is that it has made Earth suitable for biological life and produces seasons; it is tilt that gives us spring, summer, fall, and winter. Seasonal changes are vital to agriculture, plant growth, and many species. The length of the day and night changes with latitude and time of the year. Without the axial tilt the relationship of the Earth to the Sun would not exist, seasonality would not exist and the Earth's biosphere would be significantly different. The tropics could be covered in ice and Antarctica transformed into a vast desert. There would be 12 hours of light and 12 hours of darkness everywhere. The poles would be entombed in eternal freezing twilight while the equator was baked in endless heat.

It is the Moon that maintains Earth's tilt between 22.1 and 24.5 degrees. The Moon has a tidal effect on Earth, as discussed in the previous subsections. It is also this tidal effect that prevents the Earth's tilt from changing too much—that is, the moon is tidally locked to the Earth, so their rotation and revolution become synchronized. [6] Without the Moon, the Earth's rotation axis would wobble violently between 0 and 90 degrees and the days would be five to six hours long. These shifts in the Earth's tilt would lead to rapid climate change. It is the Moon that saves us from such disasters and allows life to exist. In fact, when the Moon was formed, the tides were 1,000 times higher than today. They would have moved inland like a water wall, 4 kilometers high like a mountain range. They covered

---

distance to the sun. Earth behaves like a wobbling top, and its precession—the alignment of its diurnal axis of diurnal its distance from the sun—oscillates with an average period of about 21,000 years.

[6]The Moon shows us the same face all the time today because it is tidally locked to Earth. The moon is spinning, but it has slowed down significantly today compared to when it started spinning. Eventually, we would see the two Moon sizes if the Moon did not rotate. Because the Moon rotates at the same rate that it orbits the Earth—synchronuous rotation—we always see the same Moon's side on Earth. The far side of the Moon is not always dark. If you were out in space you could observe the Sun hitting all sides of the Moon as the moon makes its month-long orbit around the Earth

hundreds of kilometers, taking minerals and nutrients from the land into the oceans. They created different combinations of minerals bound together and torn apart. There is a possibility that the right combination of minerals can be forged into life in this violent environment. The changes in chemical concentrations as high moon-induced tides flow in and out may have helped DNA evolve. This may also have caused DNA to split and replicate.

Also billions of years ago our planet rotated four times faster than today. Huge hurricane-force winds and four-kilometer-high tides were almost constant in the Earth's atmosphere. The environment was too hostile for life. The tides slowly slow the speed of rotation of our planet eventually lengthening a day from 6 to 24 hours. Even two billion years ago, days were 20 hours long; a year lasted 438 days. The moon was 1.6 kilometers closer to Earth than today. It had a greater effect on ocean tides than today. As the Earth's rotation slows the atmosphere ceases to whip around the globe. Hurricane strength winds are no longer the norm and more complex life forms evolve in our planet's relative peace and calm. Today, we still witness hurricanes, monsoons, and cyclones, 150 kilometers an hour winds, houses losing their roofs, and tremendous flooding. However, after a few days, they are history and things settle back to normal. People get on with their lives. The moon's influence has waned but not disappeared. The gravitational relationship between the Earth and the Moon has never ceased. Our existence on this planet is the result of many things coming together at the right time. The moon moving away from Earth and the tides' strength and height coming down are two examples. Along with planet tectonics and the liquid core, they are responsible for our existence today.

1.3.   Application: Isostasy of Crustal Thickness

Again, solid Earth is composed principally of the solid central core, the fluid peripheral core, and viscous mantle (see Ikelle, 2023a). Crust slides over the asthenosphere (upper mantle). There is a system of forces that allows these loads, including the low standing oceans and the high standing continents, to be in equilibrium. Isostasy is a concept of physical equilibrium of vertical motions that allows us to describe the vertical dynamics of the lithosphere, including the creation of the low-standing oceans and high-standing continents, the forces that hold up the mountains or conversely hold down basins, the causes of stranded shorelines, and the fact that oceanic crust thicknesses are less than 10 km and continental crust thicknesses are between 35 and 70 km.

We will start this section by recalling Archimedes' principle because the equilibrium of low-standing oceans and high-standing continents is based on this principle, along with Newton's gravity. In other words, the equilibrium of ocean and mountains is conceptually similar to that of large, heavy ships.

1.3.1.   Archimedes' Principle

Archimedes' principle is central to the concept of isostasy. This physics' principle was discovered by Archimedes millennia ago. Consider pushed up by a buoyant force (uplift), F_{up}; that is the weight of the fluid displaced by the body, as illustrated in

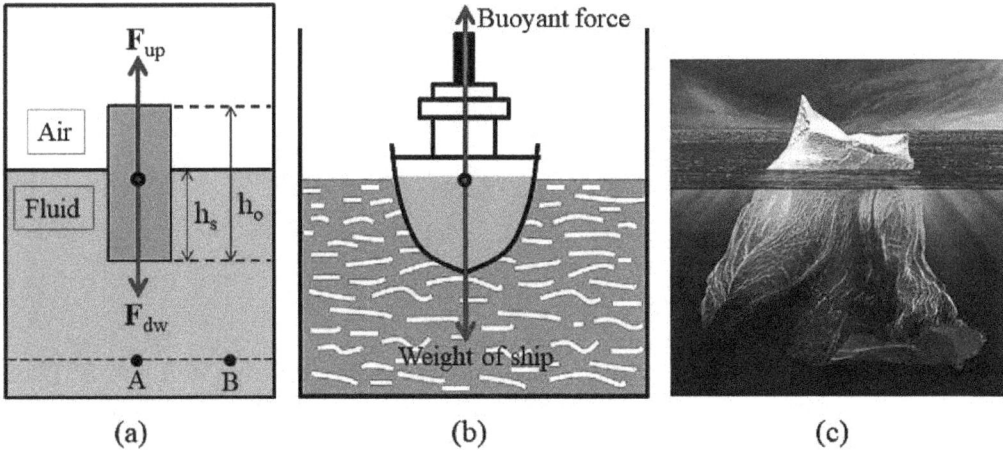

Figure 1.56. Examples of buoyancy. (a) An object can change its density to change its buoyancy force and move up or down in a gravity field. Archimedes' principle states that the weight of a block that floats in a liquid is equal to the volume of liquid that is displaced. (c) An example of icebergs. Icebergs are also called *ice mountains*. They are huge pieces of freshwater ice that are floating in open water. They are formed after breaking off of continental ice shelves or glaciers. (adapted from https://www.theactivetimes.com/users/hbyrnes).

Figure 1.56a, and the downward gravitational force on the object, $\boldsymbol{F}_{\mathrm{dw}}$, that acts in the upward direction at the center of mass of the displaced fluid. Archimedes' law of buoyancy, or simply Archimedes' principle, states that a body, whether fully or partially submerged, in a fluid (i.e., liquids and gasses including air, water, and oil) will experience an upward force equal in magnitude to the weight of the fluid displaced by the body (Heath, 2010; Halliday et al., 2011). In other words, the weight of a floating solid is supported by the weight of the fluid that it displaces. Thus, the magnitude of the net vertical force on the object, $\Delta F = |\boldsymbol{F}_{\mathrm{up}}| - |\boldsymbol{F}_{\mathrm{dw}}|$, is the difference between the magnitudes of the buoyant force and its weight. If this magnitude of the net force is positive, the object rises; if negative, the object sinks; and if zero, the object is neutrally buoyant—that is, it remains in place without either rising or sinking. In other words, Archimedes' principle helps to determine whether an object placed in a fluid will float or it will sink.

Equating downward gravitational force acting on the object and upward buoyant forces we will get

$$|\boldsymbol{F}_{\mathrm{dw}}| = |\boldsymbol{F}_{\mathrm{up}}| \iff m_o g = m_d g$$

or

$$\rho_o V_o g = \rho_f V_d g \iff \frac{\rho_o}{\rho_f} = \frac{V_d}{V_o}, \tag{1.149}$$

where $m_o = \rho_o V_o$ is the mass of the object, ρ_o is the density of the object, V_o is the volume of the submerged portion, $m_d = \rho_f V_d$ is the mass of fluid that is displaced, ρ_f is the density of the fluid, and V_d is the displaced fluid volume. We can see that $\rho_o < \rho_f$, then $V_d < V_o$ and the object floats. If $\rho_o > \rho_f$, then $V_d > V_o$ and the object sinks. The sign of the density contrast between the object and the liquid determines whether the object floats (block density less than liquid) or sinks (block density greater than liquid). By using the fact that

$$\text{volume} = \text{length} \times \text{width} \times \text{height} \tag{1.150}$$

and assuming that for each column, the width and length are both equal to one unit, we can deduce from (1.149) that

$$h_s \rho_f = h_o \rho_o \iff \frac{h_s}{h_o} = \frac{\rho_o}{\rho_f} \iff h_s = h_o \frac{\rho_o}{\rho_f} \tag{1.151}$$

where h_s is the height of the part of the object submerged in the fluid and h_o is the height of the entire object. Remember that the density of water is 1 g/cm^3 and air at normal atmospheric pressure is about 0.0012 g/cm^3. Objects that are less dense than water will float, whereas objects that are denser than water will sink. The density of ice is about 0.9 g/cm^3; therefore, the submerged portion of the iceberg is about 90 percent of its volume. Only 10 percent is above water and visible; for example, if h_o is 100 m then h_s is 90 m or 100,000 kg of water is displaced by 90,000 kg of ice, and the remaining 10,000 kg of the iceberg stands above the surface. In other words, the iceberg's density is lower, but close to the density of water, it only floats with a significant part submerged, as illustrated in Figure 1.56c.

Notice that ships (see Figure 1.56b) and submarines are designed on the basis of the Archimedes' principle. A ship is hollow at the bottom to allow its average density to be less than that of water, so the ship is able to displace water equal to its weight and hence it floats. A submarine is a vessel that can change its buoyancy that enables it to sail on the surface (i.e., when part of the submarine is above the water surface) or run submerged (i.e., when the submarine is completely submerged, and no part or appendages are above the water surface). It includes ballast tanks, which can be filled with water to increase its weight, so that it sinks into the sea and sinks lower. At a desired depth, it uses trim tanks to maintain neutral buoyancy so that they don't sink or rise (i.e., $\Delta F = |\boldsymbol{F}_{\text{up}}| - |\boldsymbol{F}_{\text{dw}}| = 0$). When the submarine is ready to return to the surface of the water, compressed air is forced into the ballast tanks to drive the water out so that the submarine rises. The operating depth of most modern submarines is between 300 and 500 meters deep.

1.3.2. Isostasy

An expansion of Archimedes' principle is the concept of isostasy that tells us where the displaced water will flow. If you measured the hydrostatic pressure [i.e., force per square meter in every direction; pressure = force/unit area = (mass $\times g$)/unit area] at directly beneath the floating object at point A, P_A, and at the adjacent point B to the floating object, P_B, as illustrated in Figure 1.57a, the two pressures would be equal—that is, these points are in isostatic equilibrium. Any change in P_A is instantly compensated by some redistribution of the fluid so that P_A becomes again equals to P_B. For example, by increasing the bulk density of the floating object by adding to its weight, the pressure under the object, P_A, becomes greater than the pressure on the adjacent point, P_B. However, this potential difference in pressure does not occur because the fluid spontaneously flows from higher to lower

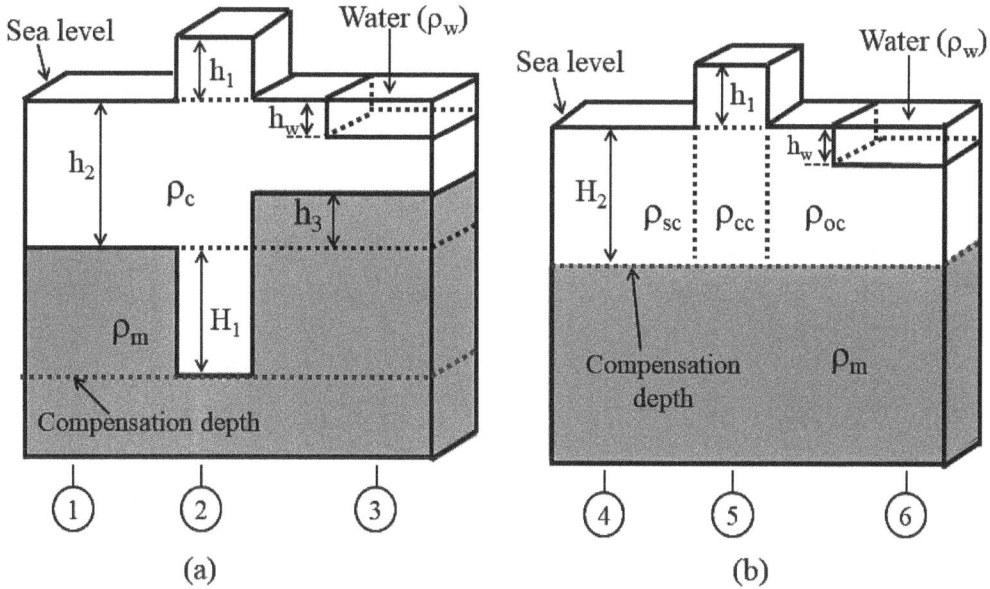

Figure 1.57. (a) The Airy model of isostatic compensation. The crust as a whole is floating in a semi-layer according to Archimedes' principle. The compensating masses lie directly beneath mountains and oceans. The crust has a constant density and the depth of the interface between the crust and the mantle varies. (b) The Pratt model of isostatic compensation. The density of the crust below sea level varies laterally, being less below mountains than below oceans. The depth of compensation is the same everywhere.

pressure until the pressure across the surface containing the points A and B is again equal—that is, why we have isostatic equilibrium. In short, depths larger than a certain *compensation depth* no longer contribute to lateral variations in pressure. This concept of fluid flow in response to uneven pressure applies to all sorts of phenomena, including the circulation of the atmosphere and ocean on larger spatial scales, and even to the state of gravitational equilibrium between the Earth's crust and mantle, as if the (lighter) crust were floating on a (heavier) mantle.

1.3.3. Crustal Thickness: an Explanation of Thick-Continental and Thin-Oceanic Crusts

In analogy to an iceberg floating on water, Fisher (1881) described the Earth's crust as an iceberg and the underlying asthenosphere as water. In other words, he considers that the crust can be approximated as being in isostasic equilibrium—the mantle is not a liquid but behaves like a viscous liquid (a very critical condition for isostasy to work); the analogy is about the fact that the crust slides over the mantle and the pressure in the asthenosphere is invariant at a certain depth. Fisher (1881) actually used the term *hydrostatical equilibrium* rather than *isostasic equilibrium*. The word *isostasy* was introduced by Dutton in 1882 and his complimentary review of Fisher's book and in one of his subsequent papers (Dutton, 1889).

Crustal rocks with densities between 2.6 and 3.0 g/cc are less dense than the upper mantle, which has a density of 3.3 g/cc. Therefore, crustal rocks can float in the mantle in a similar way as icebergs float in the ocean. In other words, Archimedes' principle can be used to explain why continental and oceanic crusts sit on top of the mantle at a height determined by densities and thicknesses.

As in the example, in Figure 1.57a, there is a common depth (*compensation depth*) within the mantle where pressures exerted by the overlying materials are equal (i.e., where there is isostatic equilibrium), the mass of each of the vertical columns of materials over this depth is the same. Any deviation from this equilibrium will provoke an isostatic adjustment in the form of a vertical motion whose speed is determined by the viscosity of the mantle and the magnitude of the pressure difference within the mantle. In other words, if one column is deficient in mass, then the mantle in that region is at lower than normal pressure, and this deficiency initiates magma flow in the mantle to fill in this low pressure area. If a column has excess mass, the mantle below is higher than normal pressure, and magma flows in the mantle away from that region to reinstate the isostatic equilibrium.

Again, the concept of isostatic equilibrium can be used to answer a number of questions about the vertical dynamics of the lithosphere. For instance, we can figure out the thickness of the oceanic crust, as shown in the example below, by equating two lithospheric columns in isostatic equilibrium. By equating the masses in vertical columns 1 and 2 above the compensation depth, as illustrated in Figure 1.57a, if these two columns of rock are in isostatic equilibrium, they are neither rising nor subsiding, so the masses of these two columns must be equal. We can figure out the thickness of the root, H, underlining the mountain with height h_1:

$$h_2\rho_c + H_1\rho_m = (h_1 + h_2 + H)\rho_c \Longleftrightarrow H_1 = \left(\frac{\rho_c}{\rho_m - \rho_c}\right)h_1 . \qquad (1.152)$$

Similarly, by equating the masses in vertical columns 1 and 3, we can figure out the thickness of anti-root, h_3, underlining the ocean basin with depth, h_w:

$$h_2\rho_c + H\rho_m = h_w\rho_w + (h_2 - h_w - h_3)\rho_c + (H + h_3)$$

or

$$h_3 = \left(\frac{\rho_c - \rho_w}{\rho_m - \rho_c}\right)h_w . \qquad (1.153)$$

The model in Figure 1.57a is known as Airy's model. It accounts for the mass deficiency with two densities, that of the rigid upper layer, ρ_c, and that of the mantle, ρ_m. Mountains therefore have deep roots. Basically, lateral variations in crustal thickness allow surface topography to be compensated by a deep crustal root. Plugging in realistic values for the different parameters including $h_2 = 30$ km (sea-coast crust thickness), $h_1 = 6$ km (mountain elevation), $h_w = 5$ km (ocean depth), $\rho_c = 2.8$ g/cc (crust density), $\rho_m = 3.3$ g/cc (mantle density), and $\rho_w = 1.0$ g/cc (water density), we find that $H_1 = 33.6$ km, $h_3 = 18$ km and therefore the continental crust beneath the mountain has $h_{cc} = h_1 + h_2 + H_1 = 69.6$ km and the ocean crust has $h_{oc} = h_2 - h_w - h_3 = 7$ km. So Airy's model explains why the thicker, less dense, and more buoyant continental crust is topographically higher than the ocean crust, which is thinner and denser. It also explains why mountains at high elevations extend deep into the upper mantle.

Figure 1.57b shows an alternative model to account for the mass deficiency beneath mountains, known as Pratt's hypothesis. The depth of the base of the upper layer is constant in this model and constant pressure is achieved by lateral variations in density. By equating the masses in vertical columns 4 and 5 above the compensation depth, as illustrated in Figure 1.57a, we can figure out the low density material, ρ_1, underlining the mountain with height h_1:

$$H_2\rho_{sc} = (h_1 + H_2)\rho_{cc} \iff \rho_{cc} = \left(\frac{H_2}{h_1 + H_2}\right)\rho_{sc} , \tag{1.154}$$

where ρ_{cc} is the density of the crust above sea level. Similarly, by equating the masses in vertical columns 4 and 6, we can figure out the high density material, ρ_{oc}, underlining the ocean basin with depth, h_w:

$$H_2\rho_{sc} = h_w\rho_w + (H_2 - h_w)\rho_{oc} \iff \rho_{oc} = \left(\frac{H_2\rho_{sc} - h_w\rho_w}{H_2 - h_w}\right) . \tag{1.155}$$

If we assume $\rho_w = 1.0$ g/cc, $\rho_{sc} = 2.67$ g/cc, $H_2 = 113.7$ km, $h_1 = 6$ km, and $h_w = 5$ km, we find $\rho_{cc} = 2.56$ and $\rho_{oc} = 2.78$ g/cc. Due to this, the density under the mountains is lower than that beneath the sea coast and the ocean.

Quiz: (1) What crustal density (ρ_{cc}) is needed to explain the 5 km high Tibetan Plateau? Assume $\rho_{sc} = 2.8$ g/cc, $\rho_m = 3.1$ g/cc, and $H_2 = 30$ km.
(2) How deep are the roots needed to support the Tibetan plateau?

The Airy and Pratt models are not the only possible models. A combination of density changes and crust thickness may occur. In general, most data supports the Airy model for continental mountain ranges because continental mountain ranges have thick crustal roots and data supports the Pratt model for mid-ocean ridges because most mid-ocean ridges have topography that is supported by density changes.

Note that all these examples confirm that Earth's surface is in isostatic equilibrium. Continents rise high relative to ocean basins because the underlying rock differs in density and thickness. Earth has an equatorial "bulge."

So the Earth's lithosphere is composed of roughly 12 large distinct fragments that, owing to their stiffness, that behave as rigid plates. The description of the motions and interactions of these lithospheric fragments is called plate tectonics, as described in Ikelle (2023a). The interaction between plates at their shared boundaries determines the overall pattern of mantle convection, and accounts for most of the earthquakes and volcanic activity on Earth. Continental crust is thicker (25-70 km) and less dense, on average, than the oceanic crust (8 km). This difference leads to greater buoyancy for the continental lithosphere. The principle of isostasy predicts that the continental lithosphere (root) will float higher on the viscous asthenosphere, and the oceanic lithosphere lower. Note that only vertical movements of solid blocks have been used here to determine the crust's thicknesses. These results are consistent with the idea that the crust is formed from solidification of lava coming from magma chambers in ascendant motions, as described in Ikelle (2023a). Moreover, our formulation of vertical movements of solid blocks is based on two well verified laws: Newton's law of gravity and Archimedes's principle of buoyancy. In contrast, we have seen, in Ikelle (2023a), that the tectonic plate theory includes both vertical and lateral movements of the crust. The lateral movements are associated with continental drifts. The forces behind the horizontal motions are not settled. Let us emphasize that isostasy is fundamentally different from other tectonic processes described in Ikelle (2023a) such as continental drift and seafloor spreading. It is essentially about vertical motions and not horizontal motions like continental drift. The process of crust subduction in the mantle is not invoked here. Yet, isostasy allows us to predict quite well the crustal thickness; therefore isostasy is one of the key constraints that any description/model of the horizontal plate motion must include (see Ikelle, 2024).

1.4.   Application: Tectonic Geodesy

The science of determining the size, shape, orientation, and gravitational field of the Earth and the variations of these quantities over time is sometimes called geodesy (e.g., Lambeck, 1988). The gravity field and gravity potential of the Earth, which we discussed in the first section of this chapter, are considered physical geodesy. The other branch of geodesy, known as geometric geodesy, is primarily concerned with measurements of dynamical (4D = space and time) phenomena, such as ground subsidence, sea level, and climate change. As described in the previous section, in the last 60 years or so, geodetic measurements have become so precise (to the millimeter level), thanks to measurements from space, that measurements of plate motions on Earth and deformations over inter-continental distances and over very short and very long periods of time are now possible. Today's geodetic technologies are so advanced that the motion of every point on the Earth's surface is measurable. We here focus on one application of geometric geodesy known as *tectonic geodesy*. Again, Earth is the only planet of our solar system on which plate tectonics has influenced the presence of our form of life here.

Tectonic geodesy uses 3D positions and displacements in space and time on the Earth's surface (i.e., surface deformation) to quantify changes due to active tectonic processes such as plate motion, plate boundary deformation, intraplate deformation (i.e., deformation inside the plates), volcanic processes, and land subsidence. These surface and subsurface deformations help us better understand the growth of mountains, rifting of continents, evolution of the geomorphic landscape, and processes that lead to earthquakes (e.g., Segall and Davis, 1998; and Bock and Melgar, 2016). For example, the *elastic rebound theory* (Reid, 1910), which is one of the most profound geophysical discoveries, was established by analyzing surface-deformation measurements before and after the 1906 magnitude-7.8 San Francisco earthquake. Reid (1910) concluded that the 1906 earthquake resulted from the *elastic rebound* of strain energy stored in the crust over a wide region surrounding the San Andreas Fault in the 100 to 1000 years prior to the earthquake. Eventually, the accumulated strain energy becomes larger than the rocks on the fault can withstand, and the fault slips to cause the earthquake. Nowadays, quantifying the crustal deformation cycle and its transients in both space and time, including the time between earthquakes (known as interseismic), the time following an earthquake (known as postseismic), and the earthquake phase itself (known as coseismic)—is the purview of tectonic geodesy. In more general terms, modern tectonic geodesy, with its spatial measurements, has confirmed that plate tectonics, including continent drifts,

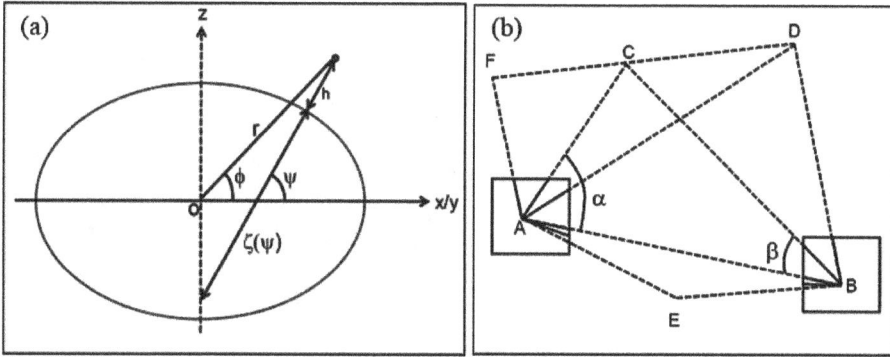

Figure 1.58. (a) An alternate representation of Cartesian (x, y, z) is spherical (r, ϕ, θ), and geodetic coordinates. (b) An illustration of triangulation. We assume that the horizontal coordinates of A and B are known. That is, at point A, all directions to points B, C, D, and E are measured using a theodolite. Then we move to point B, and directions to targets A, C, D, and E are measured. We can determine the horizontal coordinates of C, D, and E by intersection using these horizontal angles and distances.

across the face of our planet, are observations.

Our objective in this section is to describe the basic principles behind space tectonic geodetic methods. We then illustrate how these methods are applied to identify stable blocks of the lithosphere and measure their motions. In addition, we map surface deformation over large intra-continental regions.

1.4.1. The Global Positioning System: Relative Motion

Terrestrial geodesy. Geodesy addresses the basic problem of determining $\boldsymbol{x}(t) = [x(t), y(t), z(t)]^T$. Until the 1980s, tectonic geodesy mainly relied on ground-based measurements from a network of terrestrial stations and calculation techniques such as triangulation and leveling. Figure 1.58b shows a typical illustration of a network of terrestrial stations, in which the horizontal coordinates of points $\boldsymbol{x}_A$ and $\boldsymbol{x}_B$ are known and those of $\boldsymbol{x}_C$, $\boldsymbol{x}_D$, and $\boldsymbol{x}_E$ are unknown. By using simple trigonometric formulae, we can obtain the horizontal coordinates, say $\boldsymbol{x}_C$, as follows:

$$x_C = \frac{x_A \cot \beta + x_B \cot \alpha}{\cot \alpha + \cot \beta} + \frac{y_B - y_A}{\cot \alpha + \cot \beta} , \tag{1.156}$$

$$y_C = \frac{y_A \cot \beta + y_B \cot \alpha}{\cot \alpha + \cot \beta} - \frac{x_B - x_A}{\cot \alpha + \cot \beta} . \tag{1.157}$$

Similarly, one can deduce the horizontal coordinates of $\boldsymbol{x}_D$ and $\boldsymbol{x}_E$. Therefore, we can determine the horizontal coordinates of a point, say $\boldsymbol{x}_C$, by measuring angles α and β to it from two other points, say $\boldsymbol{x}_A$ and $\boldsymbol{x}_B$, and the distance between $\boldsymbol{x}_A$ and $\boldsymbol{x}_B$. These angular measurements are carried out with rotable telescopes known as theodolites. Moving the theodolite through a widespread network of stations provides the angles of many such inter-station triangles. The key assumption here is that there are lines of sight (LOS) between measurement points. This assumption

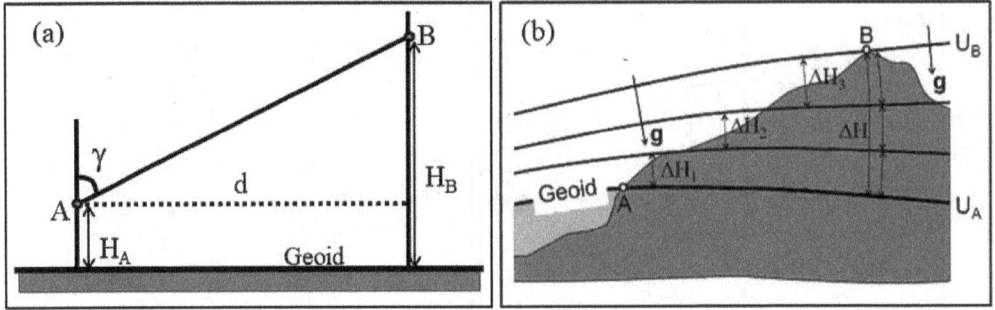

Figure 1.59. (a) An illustration of how to determine the height difference between two points A and B using zenith-angle measurement. (b) An illustration of height differences estimation using leveling methods. Note: Height differences obtained from leveling may not just be added together to get a point's height.

limits theodolites to distances less than 200 km at most. We can reach 200 km or so because the points in Figure 1.58b are actually observation towers typically 20-30 meters in height. The heights improve the line-of-sight of neighboring towers.

The technique used to determine vertical coordinates (i.e., height deformation) is known as leveling. Again the vertical coordinate here is the height of a given point above the geoid. As illustrated in Figure 1.59a, we can determine the height between x_A and x_B as follows:

$$\Delta H = d \sin \gamma \quad \text{or} \quad H_B = H_A + d \sin \gamma , \tag{1.158}$$

where d is the horizontal distance between the two points, γ is the vertical angle, H_A is the orthometric height of x_A, H_B is the orthometric height of x_B, and $\Delta H = H_B - H_A$. Therefore, measuring the vertical angle, γ, using for example a theodolite, and the distance between with x_A and x_B allows us to determine the elevation difference between the two points, and even the elevation of x_B if that of x_A is known.

In areas with complex surface topography, equation (1.158) is not valid, especially when the distance between the two points is large, as the orthometric heights of points on a given equipotential surface generally vary with distance. One solution, in this case, is to make a series of small vertical height measurements. We assume that vertical height differences are so small that we consider gravity constant for each height difference. The technique can be described using equation (1.80). In discrete form, equation (1.80) can be written as follows:

$$U(x_B) - U(x_A) = \sum_{i=1}^{N} g_i \Delta H_i = \bar{g} H , \tag{1.159}$$

with

$$\bar{g} = \frac{\sum_{i=1}^{N} g_i \Delta H_i}{\sum_{i=1}^{N} \Delta H_i} , \tag{1.160}$$

and

$$H \neq \sum_{i=1}^{N} \Delta H_i \ , \tag{1.161}$$

where H is the orthometric height at B and ΔH_i represents the separation between the equipotential surfaces on the plumb-line of the point, and $\bar{g}$ is the mean gravity along the plumb-line. As illustrated in Figure 1.59b, height differences obtained from leveling may not just be added together to get a point's height. Because $\boldsymbol{x}_A$ is located on the geoid, (1.159) reduces to

$$U(\boldsymbol{x}_B) = \sum_{i=1}^{N} g_i \Delta H_i = \bar{g} H \ . \tag{1.162}$$

The objective is to reconstruct H. We first compute $U(\boldsymbol{x}_B)$ using ΔH_i and g_i and then deduce $H = U(\boldsymbol{x}_B)/\bar{g}$.

In the 1900s, terrestrial geodesy was already used. The geodesic measurements of crust deformation before and after the 1906 San Francisco earthquake were probably one of the first major contributions of tectonic geodesy to earthquake studies. The triangulation measurement of deformation showed that the land to the west of the San Andreas Fault moved northwestward by 2-6 m relative to stations well to the east of the fault, based on the difference in horizontal coordinates derived (Hayford and Baldwin, 1908). Again, these results were of fundamental importance in the recognition of elastic rebound (Reid, 1910) which is still useful for crustal deformation cycle studies and earthquake forecasting through estimation of fault slip. More recent studies of the same geodesic measurements largely confirm these results. In Hayford and Baldwin's (1908) measurements of elevation change during the earthquake, deformation was almost completely horizontal, confirming the strike-slip (transform fault) nature of this event—that is, this event glided horizontally along the other (respecting the wrench and transform boundaries).

The global positioning system. Measuring and/or monitoring plate tectonic far-field motions or two points (stations) over long distances, say 200 km, is beyond the reach of ground-based geodetic techniques briefly described above, because they require line of sight between stations. The deployment of dedicated Earth-orbiting satellite constellations, in particular the Global Positioning System (GPS), in the 1980s, and Interferometric Synthetic Aperture Radar (InSAR) methods, in the 1990s have made such measurements possible and hence, revolutionized tectonic geodesy. Space geodetic methods are cheaper, faster, and easier to use than ground-based geodetic methods. Once the relevant satellites are launched, the measurements are accurate, relatively easy to make, and are sometimes publicly available. Precise relative positioning can be obtained over very large distances between the two measurement points without line of sight between stations. It yields 3D positions with respect to an Earth-fixed and Earth-centered reference frame. However, we are still at the early stage of space tectonic geodesy. Only four giant earthquakes have occurred in the era of satellite geodesy: (1) the 2004 M_W-9.1 and 2005 M_W-8.6 both in Sumatra, Indonesia, (2) the 2010 M_W-8.8 Maule, Chile, and (3) the 2011 M_W-9.1 Tohoku-Oki, Japan. Analyses of these events, especially the postseismic (the time following an earthquake) deformation associated with these events, are still inconclusive in most cases due to the limited geodetic network near seas and oceans. Here are a few review articles, books, and reports on space geodesy and tectonic geodesy: Hofmann-Wellenhof, 1992; Leick, 2003; and Segall and Davis, 1998.

Let us now focus on the GPS method. At present there is a constellation of more than 30 GPS satellites orbiting around Earth in 12 hours, on six different orbital planes at an altitude of about 20,000 km. On the ground, a world-wide network of GPS receivers solidly coupled to the ground is used to control and monitor the constellation. GPS satellites transmit two signals at 19 cm (1.575 GHz) and 24 cm (1.227 GHz) wavelengths. GPS satellites launched since 2010 transmit a third signal at 25.5 cm wavelengths. We need two signals because their weighted sum allows us to remove the effects of the passage of GPS radio signals through the ionosphere. Most GPS receivers contain the electronics, software, and computation power to determine the travel time between satellites and ground stations and deduce the distance between the satellites and the ground stations by using the speed of light, c_0 (i.e., $z = c_0 \Delta t$, where c_0 is the light speed). The addition of a third signal emitted by GPS satellites increases these calculations' accuracy (Hatch et al., 2000).

Quiz and Computational exercise:: Three points fixed relative to each other are sufficient to define a reference frame in three-dimensional space.

(1) Consider $x_1 = (0,0,0)$, $x_2 = (x_{21},0,0)$, and $x_3 = (x_{31},x_{32},0)$ are the centers of three spheres with radii r_1, r_2, and r_3, respectively. Determine the point of intersection, (x,y,z), of these three spheres.

(2) Use this case to explain why we used four spheres in Figure 1.60 instead of three.

(3) Use the following Matlab script for the case in which $x_1 = (x_{11},x_{12},x_{13})$, $x_2 = (x_{21},x_{22},x_{23})$, and $x_3 = (x_{31},x_{32},x_{33})$ are the centers of three spheres with radii r_1, r_2, and r_3, respectively, to determine the condition under which the point of intersection of these three spheres has components with imaginary numbers. Here is the Matlab script:

```
A = [1 1 1; 1 0 1; 1 1 0]'; %A = [x11 x12 x13; x21 x22 x23; x31 x32 x33]'
r = [1.5; 2; 1]; %r = [r1; r2; r3]
ii1 = (A(:,2) − A(:,1))/norm((A(:,2) − A(:,1)),'fro');
iia = ii1' * (A(:,3) − A(:,1));
ii2 = (A(:,3) − A(:,1) − iia * ii1)/norm((A(:,3) − ...
    A(:,1) − iia * ii1),'fro');
B = [ii1, ii2, cross(ii1, ii2)]; cc = B' * (A − repmat(A(:,1), 1, 3));
x = (r(1)^2 − r(2)^2 + d^2)/(2 * cc(1, 2));
y = (r(1)^2 − r(3)^2 + iia^2 + cc(2, 3)^2)/(2 * cc(2, 3)) − iia/cc(2, 3) * x;
z = sqrt(r(1)^2 − x^2 − y^2); xx = B * [x; y; z] + A(:,1);
```

From the signal of four or more satellites measured at the same time, and the position of each satellite, we can compute the three-dimensional station position to sub-centimeter precision just like locating earthquake epicenters from arrivals at multiple seisometers (e.g., Ikelle, 2017b and 2020). If we know the distance from GPS satellite A to the ground receiver, d_A, we are somewhere on a sphere whose

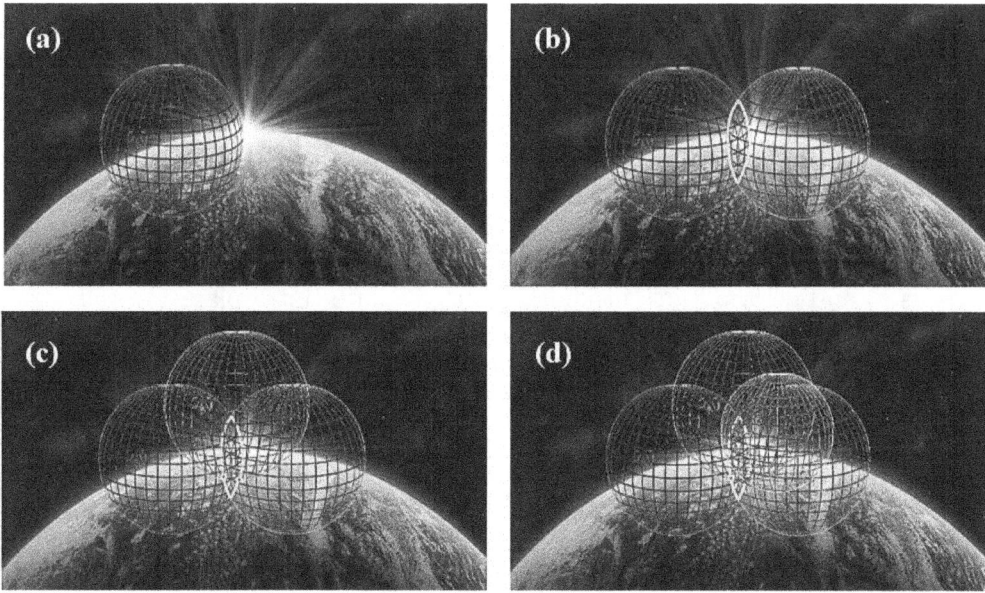

Figure 1.60. We need four GPS satellites to calculate a precise location on Earth using the Global Positioning System. (a) From only one satellite, we can be anywhere on an imaginary sphere with a radius equal to the distance between the GPS receiver and the satellite. (b) If we add a second satellite, we can only be where these two imaginary spheres intersect. (c) If we add a third satellite, there are only two points where we can possibly be. (d) From four satellites, the exact location of the Earth can be determined just like locating earthquakes from arrivals at multiple seismometers.

middle is satellite A and the diameter is d_A (see Figure 1.60). If we also know the distance from GPS satellite B to the same ground receiver, d_B, we are now on the circle where two spheres intersect. If we also know the distance from GPS satellite C to the same ground receiver, d_C, we are now on either of two points where three spheres intersect (see Figure 1.60). A fourth satellite is needed to determine which one of the two positions is not correct. That is the basic principle of determining the three-dimensional station position by using satellite triangulation. Note again that, unlike terrestrial methods, GPS methods do not require line-of-sight between stations in a network. Note also that GPS measures vertical positions with respect to an idealized ellipsoidal reference (i.e., ellipsoidal heights) described in the section of this chapter, as opposed to the Earth's geoid. Thus, knowledge of the local gravity field is needed when converting GPS vertical positions to orthometric heights, like the one obtained from leveling measurements. Remember that surfaces of constant ellipsoidal height are not necessarily equipotential surfaces. Hence, in an ellipsoidal height system, it is possible that water flows from a point with low "height" to a point with a higher "height."

Until the early 1990s only episodic GPS field surveys (sGPS)—that is, GPS field campaigns in which GPS receivers were planted in the ground for a short period of time and then removed just like classical seismic surveys for oil and gas exploration—were used to monitor the Earth's surface deformation. These surveys were generally repeated on average every two years in seismically active regions like the Chile-Peru region. Nowadays the sGPS have been replaced and augmented by the permanent network GPS (cGPS) approach, thanks to a drop in equipment costs. In the permanent network GPS approach, continuous recording is made by GPS receivers permanently coupled to the ground. They give us daily positions and observe transient effects. There are now thousands of permanently installed GPS receivers around the world. These permanent GPS receivers are having a profound impact on the study of Earth dynamics. Networks of these permanent GPS networks (also known as continuous GPS) detect deformations between plates (inter-plate) and deformations inside the tectonic plate (intra-plate) as well. They have become an essential component of geophysical research for studying fundamental solid Earth processes that drive natural earthquakes and volcanic eruptions. Continuously recording also allows us to measure the relative velocity of plates under this approximation:

$$\boldsymbol{x}(t) \approx \boldsymbol{x}_0 + \boldsymbol{v}_R(t - t_0) + \sum_i \Delta \boldsymbol{x}_i(t) \ , \tag{1.163}$$

where $\boldsymbol{x}(t)$ is the instantaneous position of the GPS receiver at time t, $\boldsymbol{x}_0$ is the point position at a reference epoch t_0, $\boldsymbol{v}_R$ is the relative velocity of plates, and $\Delta \boldsymbol{x}_i$ are corrections due to various time changing effects. Basically $\boldsymbol{v}_R$ is estimated by a weighted least squares line fit to positions, $\boldsymbol{x}(t)$, versus the time plot. The accuracy estimation of $\boldsymbol{v}_R$ increases with time. The corrections, $\Delta \boldsymbol{x}_i$, include earthquakes, solid Earth tide displacement, ocean loading, pole tide, atmospheric loading, and geocenter motion (geocenter is the center of mass (CM) of the whole Earth, including oceans and the atmosphere).

Quiz: The point vector can be defined at time t as follows:

$$\boldsymbol{x}(t) = r(t) \begin{bmatrix} \sin\theta(t)\cos\phi(t) \\ \sin\theta(t)\sin\phi(t) \\ \cos\theta(t) \end{bmatrix} = \begin{bmatrix} [\zeta(\psi) + h(t)]\sin\theta(t)\cos\psi(t) \\ [\zeta(\psi) + h(t)]\sin\theta(t)\sin\psi(t) \\ [(b^2/a^2)\zeta(\psi) + h(t)]\cos\theta(t) \end{bmatrix}$$

with $\zeta(\psi) = a^2 \left(a^2\cos^2\psi + b^2\sin^2\psi\right)^{-1/2}$. Draw a diagram of this configuration including a, b, h, ψ, θ, $\zeta(\psi)$.

Relative motion. Let us emphasize that $\boldsymbol{x}(t)$ and $\boldsymbol{v}_R$ are fundamentally relative quantities. As the plates move, the positions of GPS receivers change relative to one another. In fact, Figure 3.23a of Ikelle (2023a) and Table 1.7 of this book show relative motions between plates. Actually, most evidence for plate motions is actually measures of relative plate motion—motion of plate B relative to plate A; one fault side is "fixed" so that we can see what the other side is doing relative to it. Nowadays, most of these relative plate motions are obtained by GPS measurements and we can use the fact that plate motions are generally very steady to extrapolate the plate motions of the last millions or so years. The 10-year time history of these measurements shows, for example, that the separation of North America and Europe is proceeding at a steady rate of about 1.7 centimeters per year (see Figure 3.23a of Ikelle (2023a)) and the Pacific Plate slips past the North American Plate at the rate of about 5 centimeters per year. However, motions due to ruptures, like those associated with earthquakes, are non-linear motions.

Table 1.7. Plate motions are relative to North American (first five rows) and Pacific Plates (last eight rows). Notice that we also included the relative motion of the NNR frame to North American. These motions are averaged over the past three million years. (adapted from Demets et al., 1994).

Plate	Latitude	Longitude	$|\omega|(°/\text{Ma})$
Antarctica (AN)	60.511	119.619	0.2540
Australian (AU)	29.112	49.006	0.7579
African (AF)	78.807	38.279	0.2380
Pacific (PA)	−48.709	101.833	0.7486
Eurasian (EU)	62.408	135.831	0.2137
North American (NA)			
Nazca (NZ)	61.544	−109.781	0.6362
Cocos (CO)	27.883	153.892	1.3572
Caribbean (CA)	74.346	−0.11384	0.1031
Arabian (AR)	44.132	25.586	0.5688
Indian (IN)	43.281	29.570	0.5803
South American (SA)	−16.290	121.876	0.1465
Juan de Fuca (JF)	−22.417	67.203	0.8297
Philippine (PH)	−43.986	−19.814	0.8389
Rivera (RI)	22.821	−109.407	1.8032
Scotia (SC)	−43.459	123.120	0.0925
NNR Frame	2.429	93.965	0.2064

Because the Earth is approximately a sphere made of moving fragments (plates), let us also recall some basic quantities associated with equations for angular motion (e.g., rotational motion). Recall that anything that moves has momentum and momentum is the quantity of motion. If something moves in a straight line, it has linear momentum. Anything that moves around a curve or in a circle, like solid Earth, has angular momentum. Table 2.2 of Ikelle (2023b) shows the analogy between linear motion translational motion) and angular motion. The analogy needs to be treated with caution. For example, the moment of inertia, I is not a constant property of the body, but its mass is a constant property. The physical quantity known as angular momentum measures the quantity of rotation of a rotating body just like linear momentum (simply known as momentum) measures the quantity of motion of a moving body. The angular momentum of a mass m with position vector $\boldsymbol{x}$ and linear velocity $\boldsymbol{v}$ is $\boldsymbol{L} = \boldsymbol{x} \times (m\boldsymbol{v})$, where $\times$ is the cross product. The linear velocity is related to the angular velocity vector (also known as the Euler vector), $\boldsymbol{\omega}$, as follows:

$$\boldsymbol{v} = \boldsymbol{\omega} \times \boldsymbol{x} \iff \begin{bmatrix} v_1 \\ v_2 \\ v_3 \end{bmatrix} = \begin{bmatrix} z\omega_2 - y\omega_3 \\ x\omega_3 - z\omega_1 \\ y\omega_1 - x\omega_2 \end{bmatrix} \tag{1.164}$$

or

$$\boldsymbol{v} = \begin{bmatrix} 0 & z & -y \\ -z & 0 & x \\ y & -x & 0 \end{bmatrix} \begin{bmatrix} \omega_1 \\ \omega_2 \\ \omega_3 \end{bmatrix} = \begin{bmatrix} 0 & -\omega_3 & \omega_2 \\ \omega_3 & 0 & -\omega_3 \\ -\omega_2 & -\omega_1 & 0 \end{bmatrix} \begin{bmatrix} x \\ y \\ z \end{bmatrix}.$$

Computational exercise: In this exercise, we compute the velocity at a given point (λ, ϕ) on the Earth's surface on Plate B, with respect to plate A (i.e., plate A is the reference frame). Use the script below to associate the arrows in Figure 3.23a of the first volume of this book with specific values of velocity and to verify the accuracy of relative motion direction for Antarctica, Eurasia, North America, and South America Plates with respect to the Africa Plate, which is here the reference frame, using the following values Africa-Antarctica ($\lambda = 5.6°$ N, $\phi = 39.2°$ W, $\omega = 0.13°$/Ma), Africa-Eurasia ($\lambda = 21°$ N, $\phi = 20.6°$ W, $\omega = 0.13°$/Ma), Africa-North America ($\lambda = 78.8°$ N, $\phi = 38.3°$ E, $\omega = 0.24°$/Ma), and Africa-South America ($\lambda = 62.5°$ N, $\phi = 39.4°$ W, $\omega = 0.32°$/Ma).
Here is the Matlab script:

```
lam = 34.75; phi = −116.5; lamr = 48.7; phir = −78.2;
om = 0.78; R = 6371; rad = pi/180.; lam = lam * rad;
phi = phi * rad; lamr = lamr * rad; phir = phir * rad;
xx = [cos(lam) * cos(phi), cos(lam) * sin(phi), sin(lam)]';
omv = [cos(lamr) * cos(phir), cos(lamr) * sin(phir), sin(lamr)]' * om;
B = [−sin(lam) * cos(phi), −sin(lam) * sin(phi), cos(lam); ...
−sin(phi), cos(phi), 0; ...
−cos(lam) * cos(phi), −cos(lam) * sin(phi), −sin(lam)];
v = R * rad * B * cross(omv, xx); vv = norm(v);
psi = atan(v(2)/v(1)) * 180/pi;
```

For latitude, north is positive and south is negative. For longitude, east is positive and west is negative.

There are three parameters in the plate angular velocity vector and often, two data from each site velocity corresponding to the horizontal velocity components are available. So we need at least two sites to determine the plate angular velocity vector. In general, the more sites, and the farther apart, the more accurate the angular velocity can be determined.

Using the latitude, λ, and the longitude, ϕ, we can alternatively write (1.164) for point on the Earth's surface as

$$
\begin{bmatrix} v_1 \\ v_2 \\ v_3 \end{bmatrix} = R \begin{bmatrix} \omega_3 \sin \lambda - \omega_2 \cos \lambda \sin \phi \\ \omega_3 \cos \lambda \cos \phi - \omega_1 \sin \lambda \\ \omega_1 \cos \lambda \sin \phi - \omega_2 \cos \lambda \cos \phi \end{bmatrix}
$$

$$
= R\omega \begin{bmatrix} \sin \lambda \cos \theta \sin \varphi - \sin \theta \cos \lambda \sin \phi \\ \sin \theta \cos \lambda \cos \phi - \cos \theta \cos \varphi \sin \lambda \\ \cos \lambda \cos \theta \sin(\phi - \varphi) \end{bmatrix} , \qquad (1.165)
$$

where

$$
x = R[\sin(\pi/2 - \lambda) \cos \phi, \sin(pi/2 - \lambda) \sin \phi, \cos(pi/2 - \lambda)]^T
$$

$$
= R[\cos \lambda \cos \phi, \cos \lambda \sin \phi, \sin \lambda]^T ,
$$

R is the radius of the Earth, $\boldsymbol{\omega} = \omega[\cos \theta \cos \varphi, \cos \theta \sin \varphi, \sin \theta]^T$ is the angular velocity (the Euler vector), θ and ψ describe the Euler Pole at the Earth's surface, and $\omega = |\boldsymbol{\omega}|$ is the scalar angular velocity or rotation rate. The conversion of $\boldsymbol{v}$ to north, east, and down coordinates is

$$
\begin{bmatrix} v_{\text{north}} \\ v_{\text{east}} \\ v_{\text{down}} \end{bmatrix} = \begin{bmatrix} -\sin\lambda\cos\phi & -\sin\lambda\sin\phi & \cos\lambda \\ -\sin\phi & \cos\phi & 0 \\ -\cos\lambda\cos\phi & -\cos\lambda\sin\phi & -\sin\phi \end{bmatrix} \begin{bmatrix} v_1 \\ v_2 \\ v_3 \end{bmatrix} . \qquad (1.166)
$$

Note that we can also define the angular velocity (Euler vector) $\boldsymbol{\omega} = \omega[\cos\theta\cos\psi, \cos\theta\sin\psi, \sin\theta]^T$, where θ (latitude) and ψ (longitude) describe the Euler Pole at the Earth's surface and $\omega = |\boldsymbol{\omega}|$ is the scalar angular velocity or rotation rate. In plate tectonics, it describes the rotation of a plate about a fixed axis originating at the center of the sphere. The Euler Pole is the intersection of the Euler vector $\boldsymbol{\omega}^{(n)}$ and the sphere surface. The Earth's center of mass is the origin of the coordinate system. Let us now define the relative velocity of tectonic plates. We denote linear relative velocity with two superscripts; i.e.,

$$
\boldsymbol{v}^{(rs)} = \boldsymbol{\omega}^{(rs)} \times \boldsymbol{x} , \qquad (1.167)
$$

where $\boldsymbol{v}^{(rs)}$ is the relative linear velocity of plate r with respect to plate s. In other words, plate s here is the reference frame; $\boldsymbol{v}^{(rs)}$ is defined relative to this frame.

1.4.2. The Absolute Motion: the Hotspot Model

Plate motions can be relative or absolute. Again, relative plate motion describes the motion of one plate relative to another; that corresponds to selecting a reference frame whose position can vary with time independently of the other plates. Relative motion occurs at plate boundaries (i.e., ridges, trenches, and transform faults; breaks in the Earth's lithosphere), as illustrated in Figure 3.23a of Ikelle (2023a). Absolute plate motions describe the motion of plates relative to a fixed reference system such as the Earth's mesosphere (the relatively strong mantle beneath the weak asthenosphere) and the center of the Earth. Our main focus in the remainder of this section is on calculations of absolute plate motions. These calculations start by determining or choosing a fixed reference system with respect to all plate motions.

It is critical to determine the (absolute) coordinates of plates with respect to a global unique (fixed) reference frame because such coordinates allow us to know what is actually moving and what is fixed (zero motion). Global unique reference frames are very convenient for visualizing and modeling tectonic deformation, including small deformations inside the large plates, as depicted in Figure 1.61. Such visualizations improve our understanding of the evolution of the systems and processes of the Earth's surface and near subsurface. Absolute coordinates also ensure interoperability and, as we will see in Ikelle (2024), they are critical when looking at Earth's tectonics in the context of our solar system. Note that plate tectonic theory was developed from relative measurements between plates. These measurements are actually local and in general easier to perform than absolute measurements, which are global. While the relative motion between two plates on the spherical earth can be described as $\boldsymbol{v}^{(rs)} = \boldsymbol{\omega}^{(rs)} \times \boldsymbol{x}$, where $\boldsymbol{v}^{(rs)}$ is the relative linear velocity of plate r with respect to plate s, the absolute linear velocity can be described with one

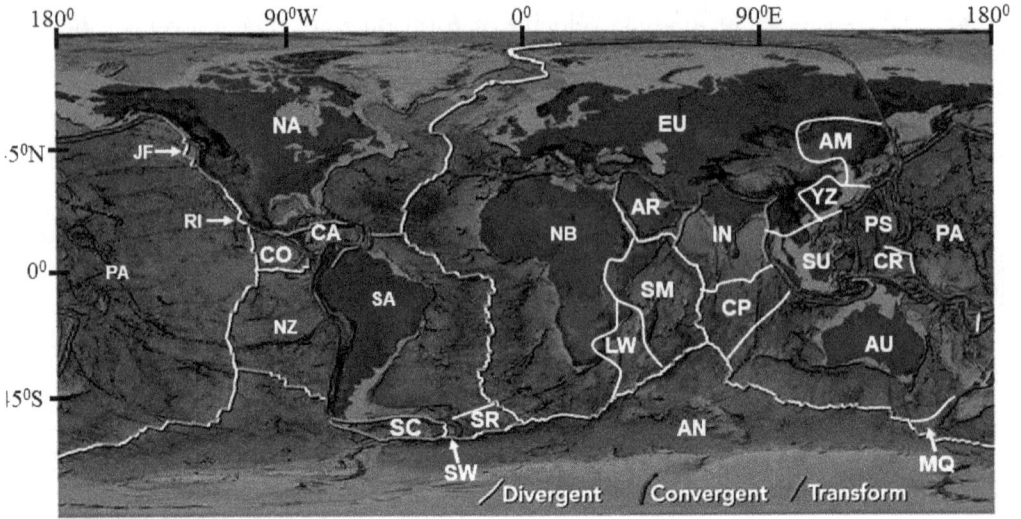

Figure 1.61. Another global plate tectonics map shows that many major plates, as defined in Figure 1.1a, are not so "rigid." GPS measurements have revealed some active deformations. It is now considered that the Indo-Australian plates are made up of Indian, Australian, and Capricorn plates. The eastern part of Africa is made up of the Lwandle and Somalia plates. (adapted from DeMets et al., 1994).

superscript as $\boldsymbol{v}^{(r)} = \boldsymbol{\omega}^{(r)} \times \boldsymbol{x}$ because s is no longer a variable, it is fixed, thus $\boldsymbol{v}^{(r)}$ is the linear velocity of plate r with respect to a fixed global frame.

If the Earth was a non-deforming radially-symmetric sphere (like present-day Mercury, Mars, and Venus), the definition of a global unique reference frame would be straightforward—since the plates move in relative motion over the entire Earth's surface, the Earth rotates, and we have ocean tides, there is no obvious or definitive Earth-fixed reference frame that co-rotates with the Earth in its diurnal motion in space. One approach is to identify rigid blocks or points (i.e., stationary blocks or points), as such blocks or points co-rotate with the Earth about a geocentric axis. The continental plates are not the solution because they are not necessarily completely rigid or in one piece; rather, there are all kinds of faults inside them along which tectonic movements have taken place—and surely still take place. The active deformations within plate interiors observed predominantly from GPS measurements have produced the global plate motion model with more than 14 plates (e.g., DeMets et al., 1990). In other words, some of the large "rigid" (major) plates, in Figure 3.23a of Ikelle (2023a), actually include blocks that move sufficiently independently from the major plates that they were formally considered part of. These blocks are called microplates or simply plates. The seismically active localities defining microplates within the "rigid" plates are also termed diffuse plate boundaries or diffuse plate boundary zones (e.g., Gordon et al., 1999 and Argus et al., 2010). For example, the model, in Figure 3.23a of Ikelle (2023a), combines India and Australia into a single plate, the Australian-Indian Plate, while new plate models distinguish Indian and Australian Plates through the deformations observed in the Central Indian Ocean related to Himalayan uplift (see Figure 1.61).

The model, in Figure 1.61, also slanders the North American Plate at its western boundary with Eurasia to include the Okhotsk and Bering microplates. Overall,

Figure 1.61 contains 38 small plates (Okhotsk, Amur, Yangtze, Okinawa, Sunda, Burma, Molucca Sea, Banda Sea, Timor, Birds Head, Maoke, Caroline, Mariana, North Bismarck, Manus, South Bismarck, Solomon Sea, Woodlark, New Hebrides, Conway Reef, Balmoral Reef, Futuna, Niuafoou, Tonga, Kermadec, Rivera, Galapagos, Easter, Juan Fernandez, Panama, North Andes, Altiplano, Shetland, Scotia, Sandwich, Aegean Sea, Anatolia, Somalia). Again, all these regions are considered microplates (or simply plates) because GPS measurements show that their movements are sufficiently independent of the major plates they were formally considered part of. Note, however, that the idea that plates are rigid and do not deform internally, but they can move relative to one another along plate boundaries, is a reasonable approximation for explaining most global geological phenomena and processes. However, the plates are not rigid enough to be used as Earth-fixed reference frames.

Early absolute plate-motion calculation attempts have used some hotspots (i.e., locations of intraplate and along plate boundaries with especially vigorous volcanism) that originate in the lower mantle and that are fixed with respect to each other as Earth-fixed reference frames (Wilson, 1963; Minster and Jordan 1978; Gripp and Gordon, 1990; and Morgan and Morgan, 2007)—the basic idea is that some hotspots are the results of plumes of hot rock rising up from deep in the mantle and that the lower mantle has very slow relative motion to the lithosphere. Furthermore, absolute plate motions can be specify the lithosphere's motion relative to the lower mantle. This is because the lower mantle accounts for 70 percent of the solid Earth's mass and deforms more slowly than the asthenosphere above and the outer core below. Therefore, hotspots from the deep mantle can be used as an Earth reference frame over geologic times. Remember that three points fixed relative to each other are sufficient to define a reference frame in three-dimensional space. Thus, two hotspots fixed relative to each other at a constant angular distance over a significant amount of time, say several million years, along with the center of mass of the Earth (the Earth's geocenter) as the third point, provide us with the desired reference frame. The next step is to select the two hotspots. To start, one has to exclude hotspots on plate boundaries and other deforming zones (e.g., solid Earth tides, tidal ocean loading deformations, and post-glacial rebound) such as Iceland, which is located at the Mid-Atlantic Ridge, Réunion Island, which is located at a transform fault, and Afar, which is located at the East African continental rift. Pacific hotspots (Hawaii, Louisville, Samoa, Tahiti, and Marquesas) located within the Pacific plate are virtually the only ones that can be considered reliable for the construct of a hotspot-reference frame. Two suitable examples of hotspots are Hawaii (latitude: 19.4° N, longitude. 155.3° W, and age: about 120 million years) and Louisville (latitude: 50.9° S, longitude: 138.1° W, and age: about 120 million years), since they are both on the Pacific plate, and reliable age data are available from both chains. Hawaiian hotspot is found in the middle of the Pacific Plate more than 3,200 km from the nearest plate boundary. In the South Pacific Ocean, the Louisville hotspot is located north of Tonga. The analysis of Norton (2000) suggests that Pacific hotspots have been reasonably well fixed relative to one another during the last 80 million years and independent of the moving Pacific plates. These two hotspots along with the geocenter suffice to define a reference frame in three-dimensional space for the last 80 million years or so. Let us denote $\omega^{(\text{plate,hotspot})}$ as the angular velocity of a given plate with respect to this frame. We can then obtain the absolute linear velocity of this plate as

$$v^{(\text{plate})} = \omega^{(\text{plate,hotspot})} \times x , \qquad (1.168)$$

where x is the vector from the center of the Earth to the point of interest on Earth's surface of this plate. The Euler vector of the hotspot frame relative to a plate is $\omega^{(\text{hotspot,plate})}$ and $\omega^{(\text{plate,hotspot})} = -\omega^{(\text{hotspot,plate})}$—that is, the absolute Euler vector for the plate in the hotspot reference frame. Suppose we have the absolute Euler vector of plate r. Assuming that the plates are rigid, we can obtain the absolute Euler vector for s plate by using readily available relative Euler vectors as follows:

$$\omega^{(r,\text{hotspot})} = \omega^{(r,s)} + \omega^{(s,\text{hotspot})}$$

or

$$\omega^{(s,\text{hotspot})} = \omega^{(r,\text{hotspot})} - \omega^{(r,s)} . \qquad (1.169)$$

Once the absolute angular velocity vector for each plate is known, we can deduce their absolute linear velocities using (1.168). Suppose that (r) is the Pacific Plate, (s) is the African Plate. We can determine the absolute motion of the African plate relative to a supposed fixed Hawaiian-Emperor hotspot by using the motion of the Pacific plate relative to the African plate and the absolute motion of the Pacific plate relative to the Hawaiian-Emperor hotspot. So we simply need one measurement of the angular velocity of a fixed hotspot relative to any plate and relative measurements of all planets relative to one plate. For example, the relative angular velocity between the "fixed" Hawaiian hotspot and the rigid Pacific Plate is 0.7895 degrees per million years and unchanged for the last 25 million years. We assume that the motion of the Pacific Plate relative to the Hawaiian hotspot is limited to Earth's surface and the angular-velocity vector is perpendicular to this surface.

Figure 1.62 shows absolute plate motion obtained assuming the Hawaiian hotspot is a fixed global reference. We can see that (i) plates move faster near the equator than near the poles, (ii) the Pacific and Australian plates are the fastest moving plates on Earth, (iii) North America seems to rotate around a point in the Pacific off South America, (iv) Antarctica is moving very little, (v) a general westward motion of most plates, although with different velocities, with Nazca plate, which moves eastward, being one of the few exceptions, (vi) at a divergent plate boundary, a plate moves faster toward west than its adjacent plate behind, to the east, and (vii) conversely, at a convergent plate boundary, a plate moves faster to the west than its preceding plate, to the west. Westward drift of most plates implies that plates have a general sense of motion, and are not moving randomly.

Note that we have excluded some hotspots from consideration as points in the construction of a fixed reference frame based on their proximity to plate boundaries. A hotspot located along or near a plate boundary cannot be used for a fixed reference frame. It is also important to exclude hotspots whose sources are not close to the mantle because the assumption in the computations of absolute plate motions is that these motions are relative to the mantle. For this purpose, it is also fundamental to

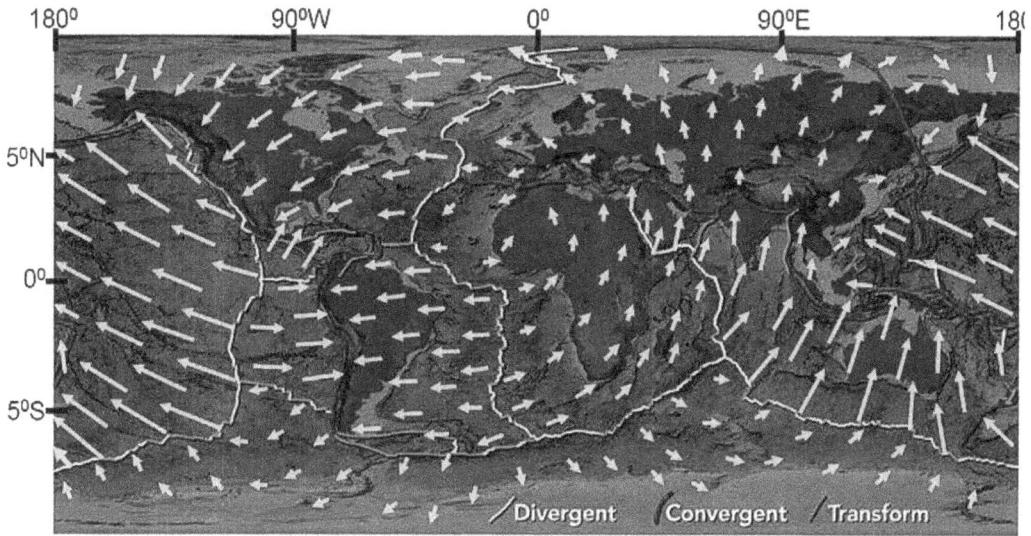

Figure 1.62. Absolute plate motion obtained assuming the deep Hawaiian hotspot is a fixed global reference. The longer the arrows (vectors) the higher the plate velocity at that geographical point. (Adapted from Cox and Hart, 1986).

know whether hotspots are fixed relative to the mantle. As we can see in Table 1.8, shallow and deep intraplate hotspots of the Pacific Plate give significantly different absolute plate angular velocities, as pointed out by Cuffaro and Doglioni (2007). The shallow hotspot here is the Society hotspot located in the south Pacific Ocean and the deep hotspot is the Hawaii hotspot located in the northern Pacific Ocean. Plates move faster relative to the mantle if the source of hotspots is taken to be the middle-upper asthenosphere and therefore are significantly erroneous compared to plate motions measured relative to hotspots with sources in the lower asthenosphere. Pacific hotspots are located within the plate and are probably the only ones to be considered for a hotspot reference frame.

In summary, most hotspots are fixed, thus they do not represent an absolute reference frame. This is because they are located on plate margins such as moving ridges (e.g., Galapagos and Iceland), transform faults (e.g., Reunion), or continental rifts (e.g., Afar). These are locations of features that move relative to one another and the mantle. The only hotspots relevant to a fixed-hotspot reference frame are those located within plates. Norton (2000) analyzes that the Pacific hotspots (the fastest plate on Earth) have been reasonably well fixed relative to one another during the last 80 My. Deep and shallow hotspot interpretations generate two hotspot reference frames. In the case of a deep mantle source for hotspots, a few plates still move eastward relative to the mantle (Figure 1.62), whereas in the In the case of shallow sources, all plates have a westward component, although with different velocities. Note that this westward drift persists even when plate motions are computed relative to Antarctica (Le Pichon, 1968; Knopoff and Leeds, 1972). Antarctica lies on a plate with almost no subducting slab. It is thus often considered stationary, or moving only slowly, with respect to the underlying mantle.

Table 1.8. Global plate motions with respect to the deep (Hawaii) and shallow (Society) hotspot reference frames. (data: Gripp and Gordon, 2002). The second, third, and fourth columns are related to the deep hotspot and the last three columns are related to the shallow hotspot.

Plate	Lat	Long	$\|\omega\|$ °/Ma	Lat	Long	$\|\omega\|$ °/Ma
Antarctic	−47.34	74.51	0.202	−59.38	86.98	1.256
Australian	−0.091	44.48	0.747	−38.86	62.78	1.488
African	−43.39	21.14	0.199	−61.75	76.73	1.213
Pacific	−61.47	90.33	1.061	−61.47	90.33	2.123
Eurasian	−61.90	73.47	0.205	−62.35	87.51	1.265
North American	−74.70	13.40	0.383	−67.52	79.79	1.409
Nazca	35.88	−90.91	0.323	−71.73	91.65	0.782
Cocos	13.17	−116.0	1.162	−42.84	−135.9	0.982
Caribbean	−73.21	25.92	0.283	−65.54	82.59	1.322
Arabian	2.95	23.17	0.508	−46.99	56.73	1.239
Indian	3.069	26.47	0.521	−46.05	57.93	1.256
South American	−70.58	80.40	0.436	−64.18	88.12	1.492
Juan de Fuca	−39.21	61.63	1.012	−51.45	72.84	2.010
Philippine	−53.88	−16.67	1.154	−68.89	25.66	1.999
Scotia	−76.91	52.23	0.445	−66.65	84.27	1.488
Lithosphere	−55.90	69.93	0.436	−60.24	83.66	1.490

1.4.3. The Global Positioning System: Absolute Motion Using the NNR Frame

No-net-rotation (NNR) frame (also known as the mean-lithosphere reference frame) is another example of an approximate Earth-fixed frame. It is based on the assumption that the sum of lithosphere movements is zero; that is, there is no net rotation of the lithosphere as a whole, and therefore there is no net rotation of its N plates (e.g., Solomon and Sleep, 1974; Minster and Jordan, 1978; and Argus et al., 2010). In this frame, plates are imagined to move relative to a fixed Earth center, and the lithosphere does not move relative to the underlying mantle. So, the sum of the absolute motion of all plates weighted by their area is zero; i.e.,

$$\boldsymbol{L} = \sum_{n=1}^{N} \int_{D_n} \left[\boldsymbol{x} \times \boldsymbol{v}^{(n)}(\boldsymbol{x}) \right] dS(\boldsymbol{x}) = \sum_{n=1}^{N} \int_{D_n} \left[\boldsymbol{x} \times \left(\boldsymbol{\omega}^{(n)} \times \boldsymbol{x} \right) \right] dS(\boldsymbol{x})$$

$$= \sum_{n=1}^{N} \int_{D_n} \left[\boldsymbol{\omega}^{(n)} (\boldsymbol{x} \cdot \boldsymbol{x}) - \boldsymbol{x} \left(\boldsymbol{\omega}^{(n)} \cdot \boldsymbol{x} \right) \right] dS(\boldsymbol{x}) = 0, \qquad (1.170)$$

where $\boldsymbol{L}$ is the total angular momentum summation of the Earth's lithosphere, $\boldsymbol{v}^{(n)} = \boldsymbol{\omega}^{(n)} \times \boldsymbol{x}$ is the linear horizontal velocity vector of the corresponding surface position vector $\boldsymbol{x}$, $dS(\boldsymbol{x})$ is an infinitesimal element of surface area at $\boldsymbol{x}$, $\boldsymbol{\omega}^{(n)} = [\omega_1^{(n)}, \omega_2^{(n)}, \omega_3^{(n)}]^T$ is the angular velocity of the n-plate, which is also known as Euler vector, and D_n is the surface of the n-plate. We used the identity $\boldsymbol{a} \times (\boldsymbol{b} \times \boldsymbol{a}) = (\boldsymbol{a} \cdot \boldsymbol{a})\boldsymbol{b} - (\boldsymbol{a} \cdot \boldsymbol{b})\boldsymbol{a}$ in the derivation of (1.170) and define the radius of the Earth to equal one so that $\boldsymbol{x} \cdot \boldsymbol{x} = 1$. Note that the condition in (1.170) is nowadays used to test the accuracy of global reference frames. The closer to zero $|\boldsymbol{L}|$ is, the more accurate the global frame is.

In a compact notion, (1.170) can be written as

$$L_k = \sum_{n=1}^{N} \sum_{l=1}^{3} I_{kl}^{(n)} \omega_l^{(n)} = 0$$

or

$$\left(\sum_{n=1}^{N} I_{kl}^{(n)} \right) \omega_l^{(r)} = -\sum_{n=1}^{N} \sum_{p=1}^{3} I_{kp}^{(n)} \omega_p^{(nr)} \ , \tag{1.171}$$

the quantities $I_{kl}^{(n)}$, which is known as inertial tensor, is

$$\begin{aligned} I_{kl}^{(n)} &= \int_{D_n} (\delta_{kl} - x_k x_l) \, dS(\boldsymbol{x}) \\ &= D^{(n)} \delta_{kl} - \int_{D^{(n)}} x_k x_l \, dS(\boldsymbol{x}) = D^{(n)} \delta_{kl} - \int_{D^{(n)}} x_k x_l \sin\theta \, d\theta \, d\phi \end{aligned} \tag{1.172}$$

where $\boldsymbol{\omega}^{(n)} = \boldsymbol{\omega}^{(nr)} + \boldsymbol{\omega}^{(r)}$ is the absolute Euler vector of the n-th plate, $\boldsymbol{\omega}^{(r)}$ is the Euler vector of the reference plate, δ_{kl} is the Kronecker delta ($\delta_{kl} = 0$ if $k = l$ and zero otherwise), and $D^{(n)}$ is the area of the n-th plate. In (1.171), we assume that the Earth's radius is normalized to unity (i.e., $|\boldsymbol{x}| = 1$ and the sphere area is $4\pi|\boldsymbol{x}|^2 = 4\pi$). The first equation in (1.171) can then be written as a linear system of three equations with respect to the three unknown components of $\boldsymbol{\omega}^{(r)}$, as described in (1.171). The tensors $\boldsymbol{I}^{(n)}$ are generally evaluated over small areas such as (0.6° by 0.5°) (Schettino, 1999). Equation (1.172) can be reduced to the following expression

$$I_{kl}^{(n)} = \sum_{p=1}^{K^{(n)}} \left[D_p^{(n)} \delta_{kl} - x_{k,p}^{(n)} x_{l,p}^{(n)} \right] \tag{1.173}$$

where $K^{(n)}$ is the number of small areas on the n-plate. Table 1.9 shows the components of $\boldsymbol{I}^{(n)}$ of major plates. The total $\boldsymbol{I}$ tensor for the whole lithosphere can be obtained simply by summating the tensors $\boldsymbol{I}^{(n)}$ associated with each plate. It results in:

$$I_{kl} = \sum_{n=1}^{N} I_{kl}^{(n)} = \frac{8\pi}{3} \delta_{kl} \ . \tag{1.174}$$

Using (1.171) and (1.174), we obtain $\boldsymbol{\omega}^{(r)}$ in terms of the relative Euler vectors $\boldsymbol{\omega}^{(nr)}$ as follows:

$$\omega_k^{(r)} = -\frac{3}{8\pi} \sum_{n=1}^{N} \sum_{p=1}^{3} I_{kp}^{(n)} \omega_p^{(nr)} \ . \tag{1.175}$$

Table 1.9. The components of the inertial I-tensor of some major modern plates and the entire Earth's lithosphere. Schettino (1999) obtained similar results.

Plate (n)	$I_{11}^{(n)}$	$I_{22}^{(n)}$	$I_{33}^{(n)}$	$I_{12}^{(n)}$	$I_{13}^{(n)}$	$I_{23}^{(n)}$
Africa	0.6606	1.5415	1.6385	-0.2514	0.0543	0.1094
Antarctica	1.3303	1.1769	0.3653	-0.0509	0.0522	0.0788
Arabia	0.0790	0.0692	0.1039	-0.0503	-0.0311	-0.0334
Australia	0.8365	0.6313	0.9430	0.2103	-0.2251	0.2902
Caribbean	0.0969	0.0139	0.0995	0.0235	-0.0058	0.0211
Cocos	0.0726	0.0035	0.0718	-0.0062	0.0013	0.0108
Eurasia	1.4305	1.0959	0.9048	0.1081	-0.1547	-0.4144
India	0.2545	0.0400	0.2523	-0.0546	-0.0136	-0.0580
Nazca	0.3902	0.0699	0.3442	-0.0139	-0.0036	-0.1152
N. America	1.2695	0.9950	0.5798	0.0752	0.0189	0.3848
S. America	0.6779	0.6458	0.8554	0.3512	0.2054	-0.1911
Juan de Fuca	0.0061	0.0051	0.0036	-0.0017	0.0022	0.0029
Philippine	0.0776	0.0771	0.1239	0.0606	0.0283	-0.0279
Pacific	1.1948	2.0117	2.0909	-0.3997	0.0711	-0.0581
Earth	8.3775	8.3774	8.3775	-0.0000	-0.0001	0.0001

The angular velocities of major plates as solutions of (1.175) are shown in Table 1.10. Therefore, the absolute velocity of one plate can be determined solely from the geometry (i.e., the tensors $I^{(n)}$) and relative motions of the plates. The tensors $I^{(n)}$ are generally evaluated over finite areas such as (0.6° by 0.5°) (Schettino, 1999). There are three main problems with this method: (1) it depends on choosing the number of plates (N), (2) the location of plate boundaries in determining the area of plates, and (3) the accuracy of the computation inertial I-tensors which depends on the GPS-stations coverage of the Earth's surface.

The absolute velocities in Figure 1.63, relative to the NNR frame, confirm that the Pacific Plate moves faster than every other plate, with the Australia-Indian Plate close behind. Absolute velocities in equatorial regions tend to be significantly higher than absolute velocities in Polar regions—in other words, the features of absolute plate motions in Figure 1.62, obtained using the hotspot reference frame, are broadly similar to those obtained using the NNR reference frame. These two sets of absolute plate motions also have a good correlation between their vector directions with the exception that Eurasia and Africa appear more clearly to have very similar motions in Figure 1.63.

East-moving plates (e.g., Pacific Plate), and subducting plates move three to five times faster than their overriding plates. However, the magnitudes of plate motions associated with the NNR

reference frame differ significantly from those of the hotspot reference frame. A comparison of the relative velocities derived from absolute plate motions in Figures 1.62 and 1.63 (i.e., $\omega^{(nm)} = \omega^{(n)} - \omega^{(m)}$) with the actual relative velocities directly derived from GPS data shows that the absolute plate motions related to the NNR reference frame are significantly more accurate than those related to the Hawaii hotspot.

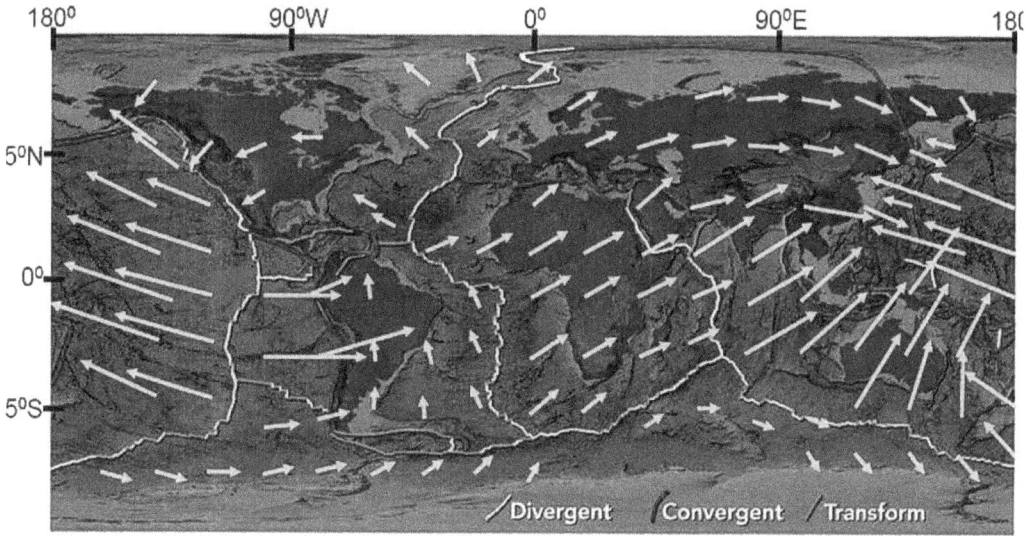

Figure 1.63. Present-day absolute plate velocities relative to NNR reference frame; that is, we assume no-net rotation of the lithosphere as a whole.

1.4.4. The Global Positioning System: Absolute Motion Using the ITR Frame

ITRF: mathematical model. As discussed above, motions and positions can be described with respect to a reference $(O, \mathbf{i}_x, \mathbf{i}_y, \mathbf{i}_z)$, where the origin point O is the center of mass (CM) of the whole Earth, including oceans and atmosphere, and is close to the Earth's center of mass. The vector set, $(\mathbf{i}_x, \mathbf{i}_y, \mathbf{i}_z)$, is a

Table 1.10. NNR Euler vectors of some major modern plates. Similar results were obtained by Schettino (1999).

Plate	Latitude	Longitude	$|\omega|(°/\text{Ma})$
Africa	50.640	-73.591	0.291
Antarctica	63.222	-115.261	0.238
Arabia	44.935	-4.437	0.544
Australia	33.866	33.154	0.647
Caribbean	25.135	92.627	0.214
Cocos	24.526	-115.758	1.509
Eurasia	50.806	-111.889	0.234
India	45.432	0.348	0.546
Nazca	47.850	-99.962	0.743
North America	-2.299	-85.540	0.207
South America	-25.341	-123.920	0.115
Juan de Fuca	-27.4	58.1	0.64
Philippine	-37.948	-35.319	0.900
Pacific	-62.990	107.041	0.641

right-handed, orthogonal basis. We assume that the basis vectors have the same length (i.e., $|e_x| = |e_y| = |e_z| = \lambda$). This length is known as scale. The orientation of the basis is also assumed to be equatorial—that is, e_z is the direction of the pole, e_y is aligned Greenwich meridian, and e_x is aligned with the equator. So a terrestrial fixed reference frame is entirely characterized by its origin, O, its scale, λ, and the orientation of its basis vectors. By definition, the time evolution of the reference fixed frame orientation obeys the no-net-rotation (NNR) condition over the whole Earth's crust.

The satellite system in the sky has allowed, and continues to allow us, to record a large amount of 3D vector positions and velocities on the surface of the Earth from several hundreds of geodetic stations on the surface of the Earth. The velocities are needed to account for tectonic plate motions. These positions and velocities change with time. If all these positions are standardized to a fixed frame, the problem of determining that frame from the large amount of 3D vector positions and velocities is well-posed and even overdetermined, so long as the distribution of geodetic stations on Earth is dense and not over-sparse (see Figure 1.64). The fact that the Northern Hemisphere has about four times more geodetic stations than that in the Southern Hemisphere and the distribution of geodetic stations is still very sparse, especially in the Southern Hemisphere, limits our capabilities to accurately represent physical Earth and its rotation, and therefore, to determine the global fixed global frame. We can only hope that this other indicator of wealth distribution will be overcome in coming years because understanding Earth tectonics is probably one of the most important factors in our lives, and even in the survival of living things on Earth. Not all stations are GNSS (GPS, GLONASS, Galileo, ...) systems. Other stations are based on very-long-baseline-interferometry (VLBI), satellite-laser-ranging (SLR), and Doppler-orbitography-and-radiopositioning-integrated-by-satellite (DORIS) systems, and these varieties of recording further constraints the frame estimation (see Figure 1.31). We will describe these systems later on. Anyway, the fixed unique reference frame can be related to 3D vector positions and velocities as follows (Altamimi et al., 2016 and 2017):

$$x_s^{(i)}(t) - \delta x_{PSD}^{(i)}(t) = x_c^{(i)}(t_0) + (t - t_0)\frac{dx_c^{(i)}}{dt} + \Delta x_f^{(i)}(t)$$

$$+ \quad T_k + D_k x_c^{(i)} + R_k x_c^{(i)} + [t - t_0]$$

$$\left[\frac{dT_k}{dt} + \frac{dD_k}{dt}x_c^{(i)} + \frac{dR_k}{dt}x_c^{(i)}\right] , \qquad (1.176)$$

$$\frac{dx_s^{(i)}}{dt} = \frac{dx_c^{(i)}}{dt} + \frac{dT_k}{dt} + \frac{dD_k}{dt}x_c^{(i)} + \frac{dR_k}{dt}x_c^{(i)}, \qquad (1.177)$$

with

$$R_k = \begin{bmatrix} 0 & -\Omega_z & \Omega_y \\ \Omega_z & 0 & -\Omega_x \\ -\Omega_y & \Omega_x & 0 \end{bmatrix} , \qquad (1.178)$$

where $\boldsymbol{x}_s^{(i)}(t)$ and $(d\boldsymbol{x}_s^{(i)}/dt)(t)$ are the various 3D vector-positions and velocity at the various stations denoted here by upperscipt i and $\boldsymbol{x}_c^{(i)}$ are the desired coordinates of the reference frame. The coordinates of the reference system vary with time because it can be improved as more data are collected and new installations of geodetic

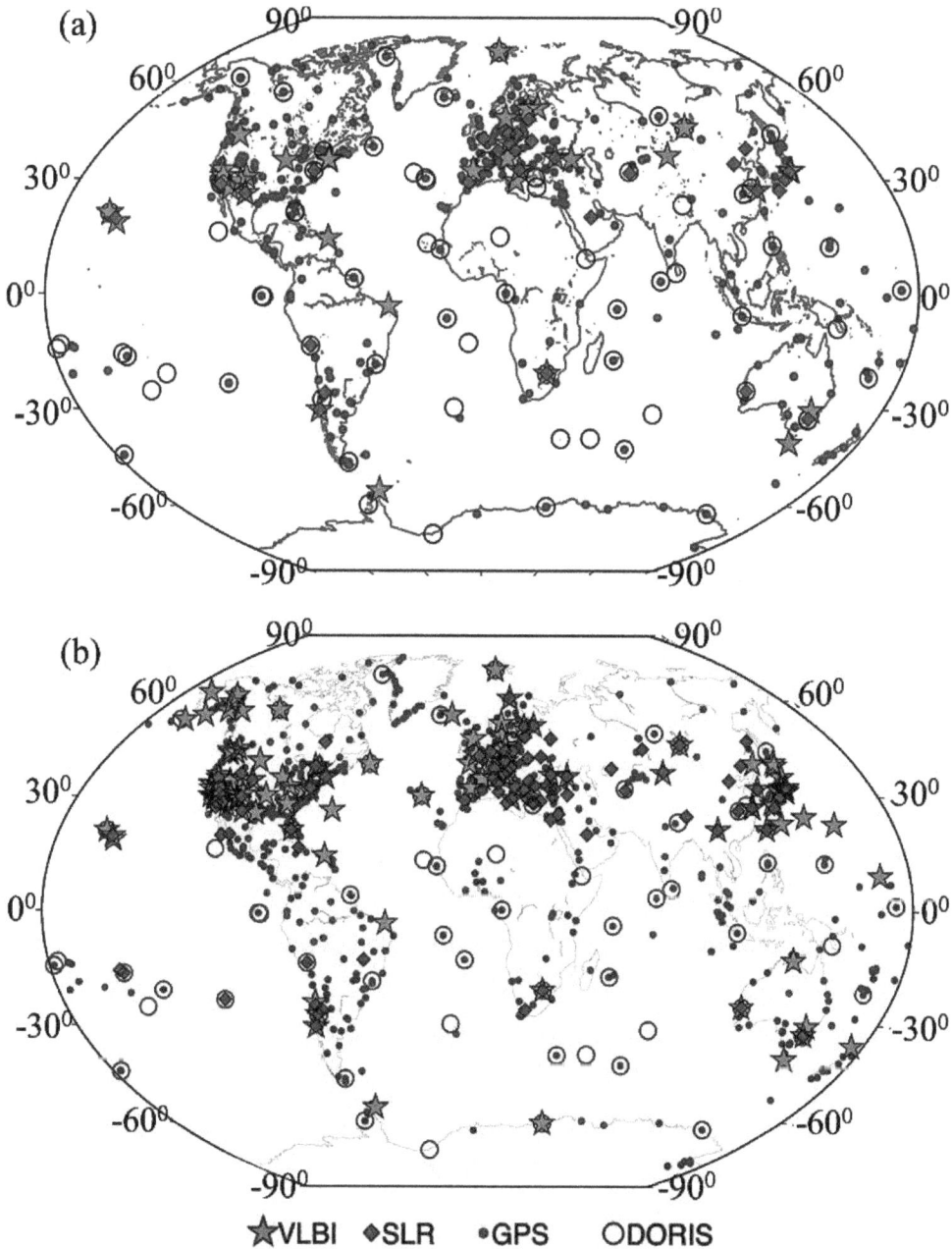

Figure 1.64.   Global distribution of stations used for computation of ITRFs: (a) ITRF2008 and (b) ITRF2014. Notice how bad the global distribution of SLRs is: no site at high latitudes and very few Southern sites. (Adapted from Altamimi et al., 2016).

stations become available. Vector $\boldsymbol{T}_k$ describes the potential translation between the fixed frame and the stations; $\boldsymbol{R}_k$ describes the rotation matrix (Ω_x, Ω_y, and Ω_z are the rotation angles about x-, y-, and z-axes, respectively) between the fixed frame and the stations; D_k expresses the small scale difference between geodetic station systems, $\Delta\boldsymbol{x}_f^{(i)}$ is the correction due to various time changing effects such as solid Earth tide displacement, ocean loading, post-glacial rebound, and atmospheric loading; and $\delta\boldsymbol{x}_{PSD}^{(i)}$ is the total sum of post-seismic deformation corrections which compensate for the fact that some recording sites are subject to major earthquakes and volcanic eruptions, in particular the devastating recent ones in Chile (2010) and Japan (2011). About 117 sites (including 10 GPS-SLR, 13 GPS-VLBI, and 7 GPS-DORIS co-location sites) were significantly impacted by 59 or so large earthquakes in the last decade.

Quiz: What is the difference between linear momentum and angular momentum? Provide mathematical definitions of these two quantities.

In theory $\boldsymbol{x}_s^{(i)}(t)$ and $\boldsymbol{x}_c^{(i)}$ are nonlinearily related. Several terrestrial reference frames have been constructed in the past. Because we are not starting from scratch and experience suggests that previous frames are close enough to each other so we here adopted the linearized version of the relationship between $\boldsymbol{x}_s^{(i)}(t)$ and $\boldsymbol{x}_c^{(i)}$—that is, equation (1.176) is a linearized version of this relationship under the assumption that the changes from $\boldsymbol{x}_c^{(i)}$ to $\boldsymbol{x}_s^{(i)}(t)$, as illustrated in Table 1.11. Equation (1.177) is obtained by differentiating equation (1.176) with respect to time. The scale factors, D_k, and the rotation matrices $\boldsymbol{R}_k$ are in the order of 10^{-9} and $d\boldsymbol{x}_c^{(i)}/dt$ is about 10 cm per year, so the rates $Dd\boldsymbol{x}_c^{(i)}/dt$ and $\boldsymbol{R}_k d\boldsymbol{x}_c^{(i)}/dt$ are of the order of 10^{-4} mm over 100 year, therefore the are negligible and that is why these rates are not included in equation (1.177). The transformation between two ITRFs is generally a function of time to reflect temporal variations between their origins, scales, and orientations.

ITRF is improved regularly as more data becomes available and more geodetic stations become online. Thus, there are various ITRFs. A new ITRF version is

Table 1.11. Transformation parameters from ITRF2020 to ITRF2014 and ITRF2020 to ITRF2008 at epoch 2015, and from ITRF2014 to ITRF2008 at epoch 2000. These parameters are taken from the IERS Annual Reports, and from the transformation parameters between ITRF2020, ITRF2014 and ITRF2008. The matrices $\boldsymbol{R}_k$ are zero and $d\boldsymbol{T}_1/dt$ is also zero.

ITRFs	T_1 (mm)	T_2 (mm)	T_3 (mm)	D (ppb)	$d\boldsymbol{T}_2/dt$ (mm/y)	$d\boldsymbol{T}_3/dt$ (mm/y)	dD/dt (ppb/y)
20 to 14	-1.4	-0.9	1.4	-0.42	-0.1	0.2	0.00
20 to 08	0.2	1.0	3.3	-0.29	-0.1	0.1	0.03
14 to 08	1.6	1.9	2.4	-0.02	0.0	-0.1	0.03

released every few years (e.g., ITRF2000, ITRF2005, ITRF2008, ITRF2014, and ITRF2020). ITRF2014 realizations consist of coordinates and velocities of 1,499 globally distributed geodetic stations. The 1,499 stations of ITRF2014 are located at 975 sites, with 91 sites (co-location sites) in which two or more space geodesy close instruments (less than hundred meters) are operating. In addition to GPS, there are 33 SLR, 40 VLBI, and 46 DORIS sites. GPS data dominate the ITRF computations, however. Notice that, in this context, a site can contain several stations. For example, ITRF in 2014 was based on 975 sites with a total of 1,499 stations.

Earth orientation parameters (EOPs). An ITRF consists of positions and velocities for a set of geodetic markers and reference points of geodetic instruments, as indicated above. Nowadays, it also includes daily series of Earth orientation parameters. The parameters are obtained by the combined use of the Geocentric Celestial Reference System (GCRF) and ITRF.

GCRF can be viewed as the counterpart of the ITRF, which has the same origin as the ITRF—that is, close to the Earth's center of mass—whose axes have a fixed orientation with respect to distant celestial objects. The relative orientation between the axes of the ITRF and those of the GCRS provides Earth's rotation parameters, including motion of the Earth's rotation axis in the GCRS (in space), the difference between the Universal Time and the Coordinated Universal Time, and motion of the Earth's rotation axis in the ITRS (with respect to the Earth's crust). More specifically, this information is captured in the following additional equations:

$$
\begin{cases}
x_s^{(p)} = x_c^{(p)} + R_{yk} \\
y_s^{(p)} = y_c^{(p)} + R_{xk} \\
UT_s = UT_c - \frac{1}{f}R_{zk} \\
\frac{dx_s^{(p)}}{dt} = \frac{dx_c^{(p)}}{dt} \\
\frac{dy_s^{(p)}}{dt} = \frac{dy_c^{(p)}}{dt} \\
LOD_s = LOD_c
\end{cases}
\tag{1.179}
$$

where $x_s^{(p)}$, $y_s^{(p)}$, $x_c^{(p)}$, and $y_c^{(p)}$ are pole coordinates, UT_s and UT_c are universal times, LOD_s and LOD_c are length of days rates of UT_s and UT_c, respectively, and $f = 1.002737909350795$ is the conversion factor from UT into sidereal time. LOD_s is uniquely determined by VLBI. The link between the combined frame and the EOPs is ensured via the three rotation parameters appearing in the first three lines of (1.179). So the term EOPs thus refers to daily the parameters on the left-side of The term EOPs will thus refer to daily pole coordinates.

The advantage of including EOPs in the ITRF computation is that they improve the consistency of the data of the four techniques (GPS, SLR, VLBI, and DORIS) from which ITRF is made of because EOPs are common to all terrestrial data-acquisition techniques.

Computational exercise: Here is a Matlab version of equation (1.177) transforming velocity of plates between ITRF2008 and ITRF2014. Use this code to verify the velocity of plate in 2008.

Here is the Matlab script:

vel2014 = [−8.3985; 19.1556; 13.5900]; % mm/year
T = [0.0; 0.0; −0.1]; % mm/year
D = 0.03 * (10⁻9); % ppb/year
XYZ = [5022375141.73; −289813708.24; 3908939819.91]; % mm
RR = [0 − 0.000.00; 0.000 − 0.00; −0.000.000]; % rad/year
vel2008 = vel2014 + T + D * XYZ + RR * XYZ;

SLR, VLBI, and DORIS. As illustrated in Figure 1.31, SLRs simultaneously determine the distance from the ground to a satellite and satellite orbits by measuring the round trip time of a light pulse that is sent to the satellite. The laser instrument is pointed at a satellite in space. The laser beam is reflected at the satellite by special reflectors and the returning light is detected at the ground. SLRs are placed into a nearly circular orbit at high altitude (about 6,000 km) to minimize orbit estimation errors, while ensuring strong signal returns to the laser systems. There were about 60 SLR stations operational around the world in 2020. This limited number is due to the fact that SLRs are costly to operate. Moreover, they require clear skies for precise measurements. This requirement also limits the utility of the method. The main contribution of SLRs to ITRF is that it provides a precise location of the Earth's geocenter based on the fact that satellite orbits are naturally related to the geocenter.

The basic geometry of VLBIs is illustrated in Figure 1.31c. VLBIs measure the difference in arrival times of microwave signals from radio sources by cross-correlating the two received signals. The sources, which are extragalactic objects called quasars or radio galaxies, are observed by globally distributed antenna networks. Due to the large distance of billions of light years between the Earth and these extragalactic sources, the arriving wavefronts can be considered as plane waves. Hence, the directions to the sources s_0 can be considered identical for all antennas on Earth. If the reception times at the stations 1 and 2 are denoted as t_1 and t_2, the observed delay time (also known as geometric delay), $\tau = t_2 - t_1$, is

$$\tau(t) = \frac{\boldsymbol{b} \cdot \boldsymbol{s}_0}{c_0} , \tag{1.180}$$

where $\boldsymbol{b}$ is the baseline vector, as defined in Figure 1.31c, and c_0 is the speed of light. Here s_0 is the unit vector pointing toward the source. Delays estimated for many radio sources and over a period of time produce a dataset large enough to over-determine the baseline vector $\boldsymbol{b}$, the coordinates of the observed radio sources, and

Earth orientation parameters (precession[7], nutation[8], polar motion[9], and universal time). In fact VLBI is the only geodetic technique which is able to measure nutation parameters and the Earth rotation angle directly. Hence, it is the only technique which is able to provide a full set of EOPs to ITRF.

Remember that the Doppler effect (or Doppler shift) is the change in frequency of a wave in relation to an observer who is moving relative to the wave source; i.e.,

$$f' = f_0 \left(\frac{1}{1 + \frac{v_0}{c_0}} \right), \tag{1.181}$$

where f' is the frequency of the moving object with speed v_0 with respect to the stationary observer and f_0 is the source frequency. So, the light from any part of the electromagnetic spectrum can shift up and down in frequency depending upon your relative motion to the emitting source. The DORIS system is based on accurate measurements of the Doppler shifts on a radio-frequency signal emitted by ground stations and received on-board different satellites, carrying DORIS receivers. Measurements of Doppler shifts are made as the satellite's orbit moves over the ground-based bacons. The emitted radio-frequency signal is dual-frequency signals, at 400 MHz and 2 GHz. DORIS receivers on board the satellites compute the position and the velocity of the satellite at a real time on-board. This velocity is used for orbit determination. So the main contributions of DORIS to ITRFs are its very precise and real-time orbit determination, precise ground beacon position determination, Earth-rotation parameters, and measurement of Earth-center position.

Table 1.12 indicates how each of these geodetic techniques contributes to the realization of ITRF2014. So far none of the four space geodetic techniques (GPS, SLR, VLBI, and DORIS) has been able to provide a frame as accurate as the ITRF—for example, it is still difficult to recover the Earth center of mass position from GPS and DORIS data. ITRF gathers its strengths from the four space geodesy techniques and compensates for their weaknesses, systematic and random errors, including effects of sparse and non uniform distribution of stations, and noise. So the origin of ITRF and its rate are essentially based on SLR data, the scale and the rate of scale of ITRF are essentially based on VLBI, SLR, and DORIS data, and the orientation of ITRF is based on all the four geodetic data types.

---

[7]These orbital variations are characterized by their eccentricity, obliquity, and precession. Eccentricity is a measure of the deviation of an orbit from a perfect circle. Earth's daily rotation around its axis has an angle with respect to its orbital plane. This angle is called the axial tilt or obliquity. Earth behaves like a wobbling top, and its precession is the measure of the alignment of its diurnal axis of diurnal its distance from the Sun.

[8]Nutation is a phenomenon that causes the orientation of the axis of rotation of a spinning object to vary over time. Nutation is caused by the gravitational forces of other nearby bodies acting upon the spinning object.

[9]While precession and nutation are motions of the Earth's spin axis viewed by an observer in space, outside of the Earth, the Pole of the Earth's spin rotation also changes with time with respect to an observer on Earth.

Table 1.12. Contributions of various geodetic techniques to ITRF.

	VLBI	SLR	GPS	DORIS
Origin and rate	No	Yes	No	Yes
Scale	Yes	Yes	No	Yes
Orientation and rate	Yes	Yes	Yes	Yes
Satellite orbits	No	Yes	Yes	No
Quasar coordinates	Yes	No	No	No
Polar motion and rates; EOP	Yes	Yes	Yes	No
Universal Time (dUT); EOP	Yes	No	No	No
Length of Day (LOD); EOP	Yes	Yes	Yes	No
Nutation (and nutation rates); EOP	Yes	Yes	Yes	No
Earth's gravity field	No	No	Yes	No
Troposphere	Yes	No	Yes	No
Ionosphere	Yes	No	Yes	No

In practice, the ITRF (say, ITRF2014) origin is designed in such a way that there are zero translation parameters at a given epoch, t_0 (say, 2010), and zero translation rates between the ITRF and the ILRS and SLR time series. In the ITRF scale, there is zero scale factor at a given epoch, t_0, and null scale rate between the ITRF and VLBI time series. The ITRF orientation is defined in such a way that there are zero rotation parameters at a given epoch, t_0, and zero rotation rates between the ITRF (say, ITRF2014) and the previous ITRF (say, ITRF2008) version.

Data processing. Numerical codes for solving the system of equations in (1.176), (1.177), (1.178), and (1.179) are available on the International Terrestrial Reference Frame (ITRF) website. The input data are $x_s^{(i)}$, their time derivatives, and EOPs (i.e., $x_s^{(p)}$, $y_s^{(p)}$, rates of $x_s^{(p)}$ and $y_s^{(p)}$, UT_s, and LOD_s) and the outputs are $x_c^{(i)}$, D_s, $\boldsymbol{T}_s$, $\boldsymbol{R}_s$ and their time derivatives, and EOPs ($x_c^{(p)}$, $y_c^{(p)}$, rates of $x_c^{(p)}$ and $y_c^{(p)}$, R_{xk}, R_{zk}, UT_c, and LOD_c). In other words, the ITRFs are constructed from the time series of station positions and velocities and EOPs. As we can see in (1.176), (1.177), (1.178), and (1.179), the resulting frame is not a tectonic-centered frame and does not require the definitions of plate tectonics or the number of plate tectonics in its designs. It is used in multiple applications, including aviation, terrestrial and maritime navigation, cartography, geographic boundary disputes, and the quantification of changes that are affecting the Earth system. The ITRF requires very accurate and very long-term stability in order to determine, for example, global sea-level rise which can be as small as 1 mm/year. So, the ITRF is fundamental to monitoring Earth dynamics, including volcanic eruptions and earthquakes (see Ikelle, 2023c), and to determining the orbits of artificial satellites. Figure 1.65 shows examples of input times series of the different techniques used in ITRF2014 and ITRF2020. The data time span of ITRF2014 is 1980.0-2015.0 for VLBIs, 1993.0-2015.0 for SLRs, 1993.0–2015.0 for DORISs, and 1994.0–2015.1 for GPSs. The time series represent the variations of the positions of the GPS sites with time. These time series are actually weekly solutions in case of GPS, SLR and DORIS and session-wise VLBI normal equation systems (one session comprises about 24 hours). The positions can be represented in three dimensions as motion in a north-south direction (i.e., changes in latitude and longitude), motion in an east-west direction (i.e., changes in longitude), and motion in an up-down (vertical height) direction. The trend of the dots shows the direction of the GPS site's movement. If a line has a slope of zero, the site is not moving in that dimension.

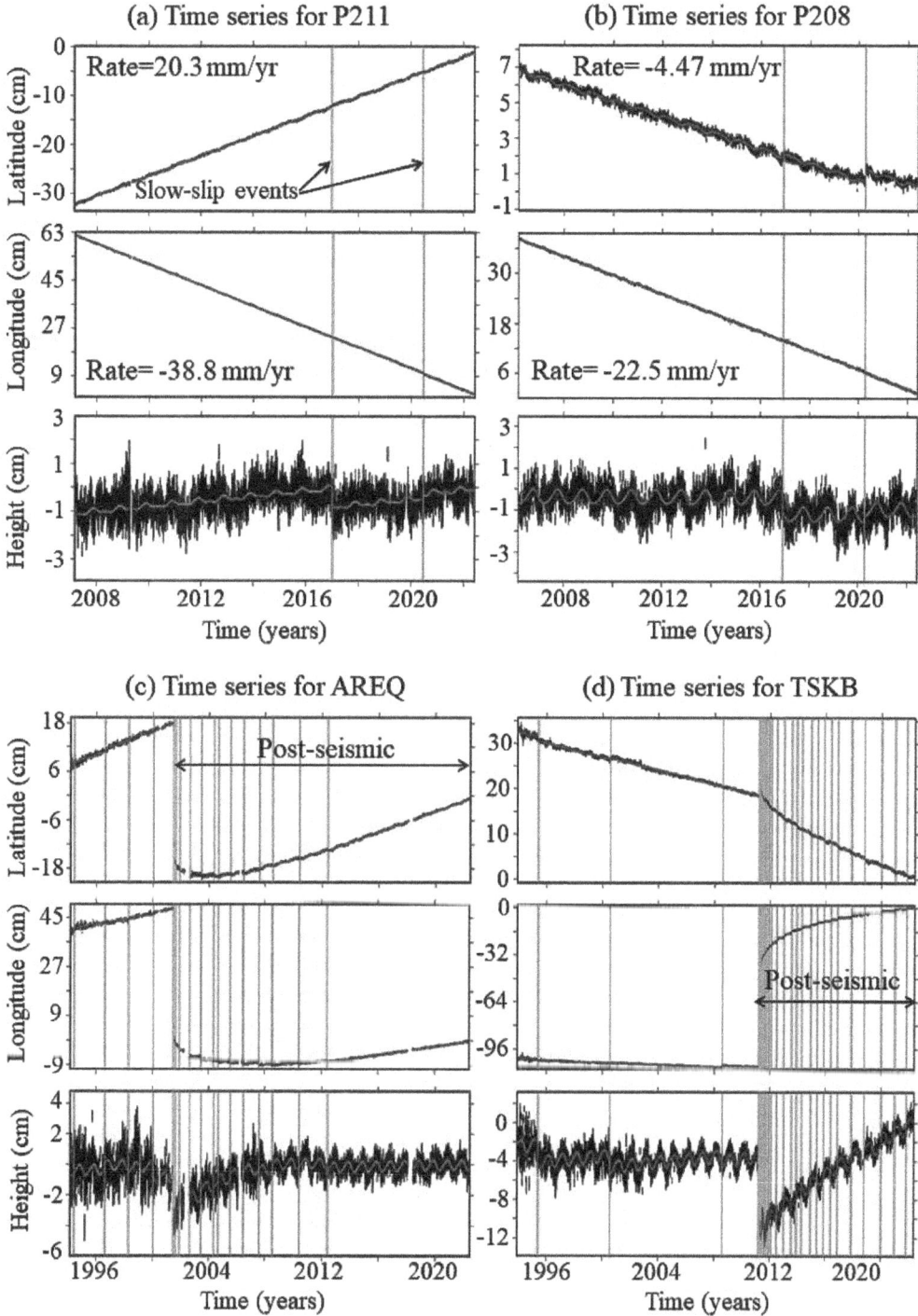

Figure 1.65. *(Continued.)*

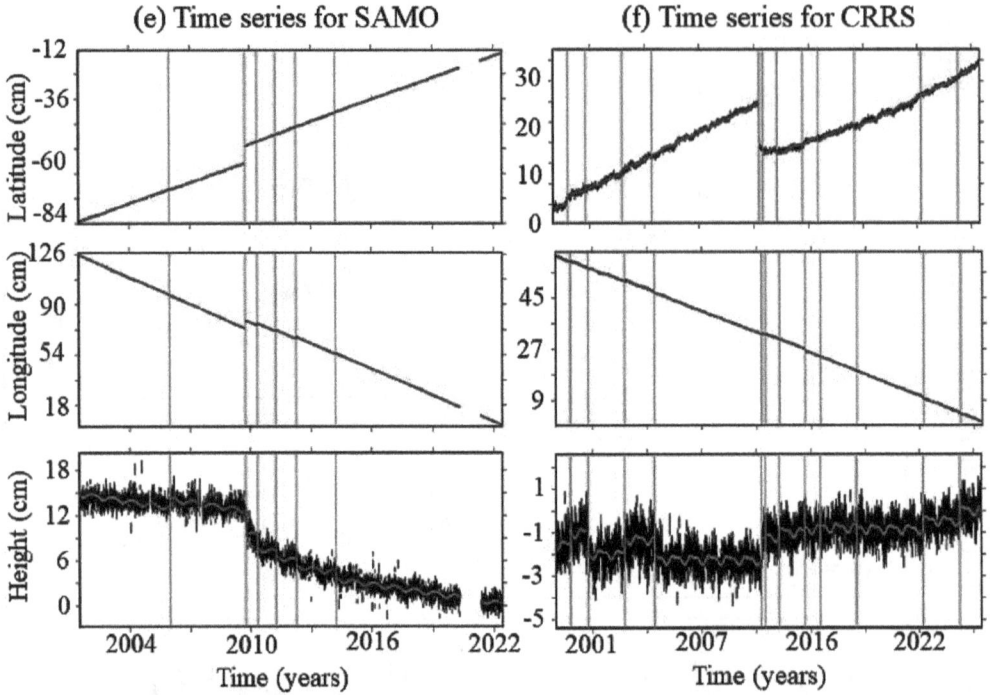

Figure 1.65. Examples of time series from various stations used in the computation of ITRF2014. We are showing the motions of latitude, longitude, vertical height as functions of time. (a) Station P211 in Aburatsu, Nichinan, Japan; about 31.5 degrees latitude and 131.4 degrees longitude. (b) Station P208 in Kushimoto-Wakayama, Japan, about 33.45 degrees latitude and 135.7 degrees longitude. (c) Station AREQ in Arequipa, Peru; about -16.5 degrees latitude and -71.5 degrees longitude. (d) Station TSKB in Tsukuba, Japan; about 36.1 degrees latitude and 140.1 degrees longitude. (e) Station SAMO in Salmon Island; about -13.8 degrees latitude and 188.1 degrees longitude. (f) Station CRRS in United States of America; about -13.8 degrees latitude and 188.1 degrees longitude. (Adapted from time series of the Jet Propulsion Laboratory, California Institute of Technology; https://sideshow.jpl.nasa.gov/post/series.html).

Notice that some time series include discontinuities and nonlinear movements. Discontinuities are mainly caused by equipment changes and geophysical effects such as earthquakes. VLBI and SLR time series have few discontinuities compared to GPS and DORIS because VLBI and SLR equipment changes are rare due to their complexity and cost. Figure 1.65c shows a time series affected by the land movement due to the 2001 Arequipa-Camana-Tacna area ($M_w = 8.4$) earthquake. Arequipa (AREQ) GPS station is located in southwestern Peru, in the subduction zone where the oceanic Nazca plate is subducted under the South American plate along the Peru-Chile trench (see Ikelle, 2023a). It is interesting to note that the GPS time recorded the earthquake on June 23, 2001. Before the event, the trend was -2.6 ± 0.5 mm/year, changing direction to a positive 4.2 ± 0.1 mm/year uplift trend afterward. From 2005 on, a slight trend reduction can be observed. This is consistent with exponential behavior that can be expected for a post-seismic event, as we will describe later. Figure 1.65d illustrates the land displacement of Tsukuba VLBI station caused by the magnitude-9.1 Tohoku (Japan) earthquake on March

Table 1.13. The motions of GPS stations as functions of R_N, the rate of the GPS latitude-time-series plot, and R_E, the rate of the GPS-longitude-time-series plot.

$R_N = R_E = 0$	GPS station is not moving
$R_N > 0 \; R_E > 0$	GPS station is moving to the Northeast
$R_N > 0 \; R_E < 0$	GPS station is moving to the Northwest (Fig. 1.65a)
$R_N > 0 \; R_E = 0$	GPS station is moving to the North
$R_N < 0 \; R_E = 0$	GPS station is moving to the South
$R_N = 0 \; R_E > 0$	GPS station is moving to the East
$R_N = 0 \; R_E < 0$	GPS station is moving to the West
$R_N < 0 \; R_E > 0$	GPS station is moving to the Southeast
$R_N < 0 \; R_E < 0$	GPS station is moving to the Southwest (Fig. 1.65b)

11, 2011 (see Ikelle, 2023a). Unfortunately, the seismic-related time series do not contain any obvious pre-seismic behavior that can be used to predict or monitor earthquakes. More than 100 sites used in the computations of ITRF2014 are subject to post-seismic deformation due to major earthquakes. Additional examples are shown in Figures 1.65e and 1.65f, including the magnitude-7.8 Samoa earthquake and tsunami that took place on September 29, 2009 in the southern Pacific Ocean adjacent to the Kermadec-Tonga subduction zone and recorded by SAMO station, and the magnitude-7.2 El Mayor-Cucapah earthquake on April 20, 2010 recorded by CRRS station. The El Mayor-Cucapah earthquake ruptured 120 km of upper crust in Baja California to the US-Mexico border. Table 1.13 shows the motions of GPS stations as functions of R_N, the rate of the GPS latitude-time-series plot, and R_E, the rate of the GPS longitude-time-series plot.

In data analysis, all time series exhibit periodic signals with some of them having a very low amplitude that is not discernible to naked eyes. So the correction term Δx_f in (1.176) can be modeled as a Fourier series; i.e.,

$$\Delta x_f = \sum_{q=1}^{Q} a^{(q)} \cos(\omega_q t) + b^{(q)} \sin(\omega_q t) \; , \tag{1.182}$$

for $a^{(q)}$ and $b^{(q)}$ are Fourier coefficients of the vector Δx. The post-seismic trajectory, δx_{PSD} in (1.176), at an epoch t can be written for each component as follows (Altamimi et al., 2016):

$$\delta L(t) = \sum_{i=1}^{N_l} A_i^{(l)} \log\left(1 + \frac{t - t_i^{(l)}}{\tau_i^{(l)}}\right) + \sum_{i=1}^{N_e} A_i^{(e)} \left[1 - \exp\left(-\frac{t - t_i^{(e)}}{\tau_i^{(e)}}\right)\right] \; ,$$

where N_l is number of logarithmic terms of the parametric model; N_e is number of exponential terms of the parametric model; $A_i^{(l)}$ is amplitude of the ith logarithmic term; $A_i^{(e)}$ is amplitude of the ith exponential term; $\tau_i^{(l)}$ is relaxation time of the ith logarithmic term; $\tau_i^{(e)}$ is relaxation time of the ith exponential term; $t_i^{(l)}$ is earthquake time (date) corresponding to ith logarithmic term; and $t_i^{(e)}$ is earthquake time (date) corresponding to ith logarithmic term. With these two models, we can fit time series for velocities, offsets, and amplitudes of annual and semi-annual periodic functions. So the computations of ITRF2014 and the upcoming ITRF2020 use linear motions along with nonlinear motions like effects like discontinuities, periodic signals, post-seismic deformation, velocity changes due to current ice melting, and some other unidentified behaviors of station trajectories.

As illustrated in Figure 1.66, the ITRF construction is a two-step procedure. (1) The first step is to estimate local (station) transformation parameters, which include scale variations, translation variations, and rotation variations with time and their time derivatives (rates). More precisely, we estimate $\boldsymbol{T}_k$, D_k, and $\boldsymbol{R}_k$ and their time derivatives. It also takes into account nonlinear motions (e.g., discontinuities, periodic signals, post-seismic deformation, velocity changes, and some other unidentified behaviors of station trajectories) and identifies and removes outliers. This analysis is based on multivariable regression with least-square optimization (see Ikelle, 2024). The estimations of $x_{m,s}^{(i)}(t)$ are based on (1.176) under the assumption that the rates are zeros and their times derivatives. (2) The second step is a stacking (accumulating) of the time series of the four-technique solutions to generate long-term cumulative solutions of station positions and velocities, and EOPs; and combining the resulting stacked solutions together with local ties at co-location sites. The inputs of this step are new $x_{m,s}^{(i)}(t)$, also known as

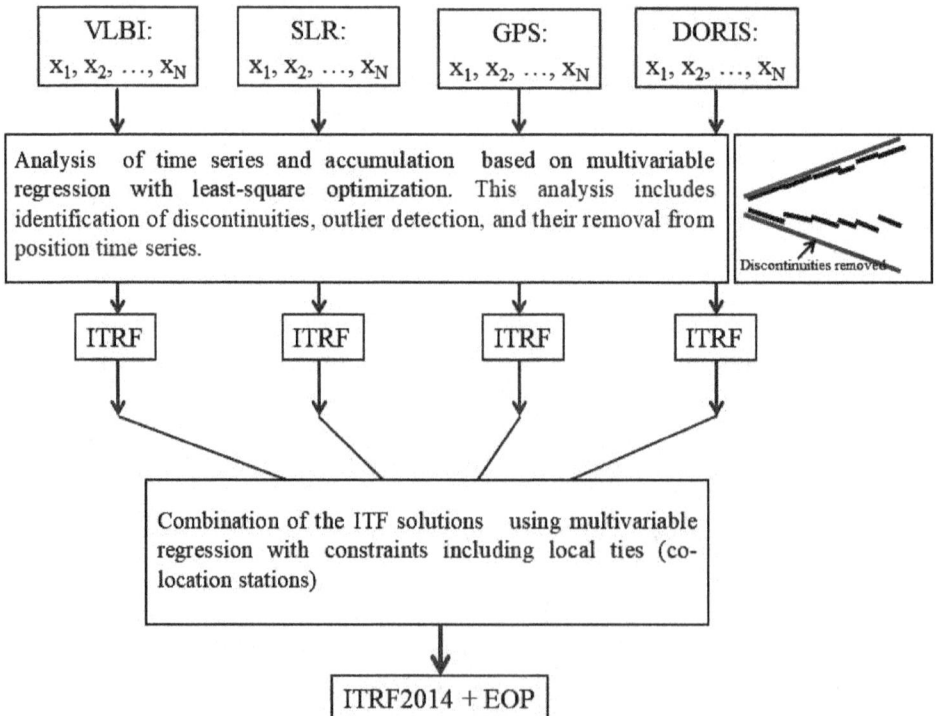

Figure 1.66. Key steps of the computation of the ITRF2014 solution and other similar frame such as DTRF2014.

Table 1.14. Comparison of estimated absolute plate rotation poles defined with respect to the Hawaii hotspot, NNR frame, ITRF2008, and ITRF2014. The unit is degree per million years or (°/Ma).

	Hotspot	NNR	ITRF2008	ITRF2014
Antarctic (AN)	0.2024	0.238	0.214	0.216
Australian (AU)	0.7467	0.647	0.633	0.630
African (AF)	0.1987	0.291	0.268	0.267
Pacific (PA)	1.0613	0.641	0.634	0.680
Eurasian (EU)	0.2047	0.234	0.256	0.255
North American (NA)	0.3835	0.207	0.189	0.193
Nazca (NZ)	0.3231	0.743	0.682	0.632
Caribbean (CA)	0.2827	0.214	0.208	0.337
Arabian (AR)	0.5083	0.544	0.570	0.582
Indian (IN)	0.5211	0.546	0.552	0.523
South American (SA)	0.4358	0.115	0.127	0.122

a long-term stacked TRF, for each measurement technique under the assumption of linear motions. The associated EOPs are readjusted at the same time to make them consistent with their stacked TRFs.

Some numerical results. One of the key features of ITRFs is that they readily provide not only horizontal movements but also vertical movements. Figures 1.67 show the ITRF2014 and ITRF2020 absolute velocity fields, respectively—these figures show that horizontal movements can reach up to 20 cm/year and vertical components of up to 40 mm/year. Moreover, the vertical velocity can be negative. In the NNR frame, vertical motion is constrained to zero.

Also these ITRF solutions allow us to quantify plate movements at submillimetre per year. ITRF2014 vertical velocity data confirms the uplift in North America, Greenland, and Fennoscandia. Overall, differences between ITRF2020 and ITRF2014 are small for geo-referencing applications. By examining relative plate motions in Ikelle (2023a) and absolute plate motions in Figures 1.67, we can see that the velocity of plate motion is independent of the size, circumference, and ridge length of a plate, as first noted by Forsyth and Uyeda (1975). The Caribbean plate is obviously not a rigid plate. Table 1.14 provides some specific angular velocities of some plates for hotspot frame, NRR frame, ITRF2008, and ITRF2014.

Due to continuous deformations in the Earth's crust, the accuracy of any particular ITRF will inevitably degrade over time. Thus ITRF realization must then be periodically updated in order to account for newly acquired observations and for upgrades in data analysis procedures and/or combination techniques. A time-independent fixed reference frame for the Earth's surface may be achievable. Note that next generation of laser-ranged satellites known as LARES were launched in 2022 into a high altitude orbit at about 5900 km altitude by ESA. LARES applications include the tests of Einstein's theory of general relativity.

1.4.5.   The Interferometric Synthetic Aperture Radar

GPS are point measurements of surface deformation and InSAR (i.e., a microwave system with an interferometric capability) is a spatial mapping of surface deformation. InSAR illuminates the Earth's surface with their own radiation energy. It is used today in studies of active tectonics and earthquake deformation. The first use of InSAR was to measure the coseismic deformation of the 1992 Landers, California, earthquake. Nowadays the number of earthquakes studied with InSAR is just too large to be listed here.

The current approach is that two images of the same location in space, but at different times, are obtained from InSAR illuminations (see Figure 1.68). In this way, the phase differences between the two images can be measured and analyzed on a pixel-by-pixel basis. These phase differences are translated to the displacement of the ground surface by using this simple formula

$$\Delta\phi(\boldsymbol{x}, t) = \frac{4\pi}{\lambda} \Delta r(\boldsymbol{x}, t) , \qquad (1.183)$$

where $\Delta\phi$ is the phase difference between InSAR images, λ is the wavelength of the radar wave, and Δr is the ground displacement which is typically between 5 and 100 meters. The ground displacements can be used to identify previously unmapped faults, the distribution of slip (i.e., portions of the fault that have failed and that did not), as illustrated in Figure 1.68a. Phase data from two precisely aligned SAR images can be differentiated to produce images which are known as interferograms. Interferograms contain information on minute surface displacements toward or away from the radar between the times of the two image acquisitions, as illustrated in Figure 1.68b.

1.5.   Application: The Human Trip to Mars and Beyond

Again, extraterrestrial travel beyond low-Earth orbit (LEO) and establishing sustained access to destinations such as the Moon (0.384 million kilometers from the Earth's surface on average), asteroids, and Mars (225 million kilometers from Earth's surface on average), is primarily about moving people from one gravity field to another, from an area with photosynthesis to area without photosynthesis (i.e., almost no atmospheric oxygen, and/or from the surface of a planet protected from solar radiation by the atmosphere and magnetosphere to planets or moons with limited atmospheric and magnetosphere protection. For Mars's mission, human bodies will experience three different gravity fields. On the six-month or so journey between Earth and Mars, human bodies will be weightless ($0g_{earth}$ or zero-gravity). The term *microgravity* is often used instead of zero-gravity because gravitational forces are never zero but just very small. On the surface of Mars, people will live and work in approximately one-third of the Earth's gravity ($\frac{1}{3}g_{earth}$), and when you return home, you will have to re-adapt to our gravity ($1g_{earth}$), as illustrated in Figure 1.69. Remember that weight is $W = mg$. So when g is zero we become weightless and when $g > 9.8$ m/s^2, we start putting on weight.

With a communication delay of up to 20 minutes one-way while on Mars and the possibility of equipment failures, you must be able to complete the mission on

(a) ITRF 2014

2 cm/year   Horizontal velocity

(b) ITRF 2020

2 cm/year   Horizontal velocity

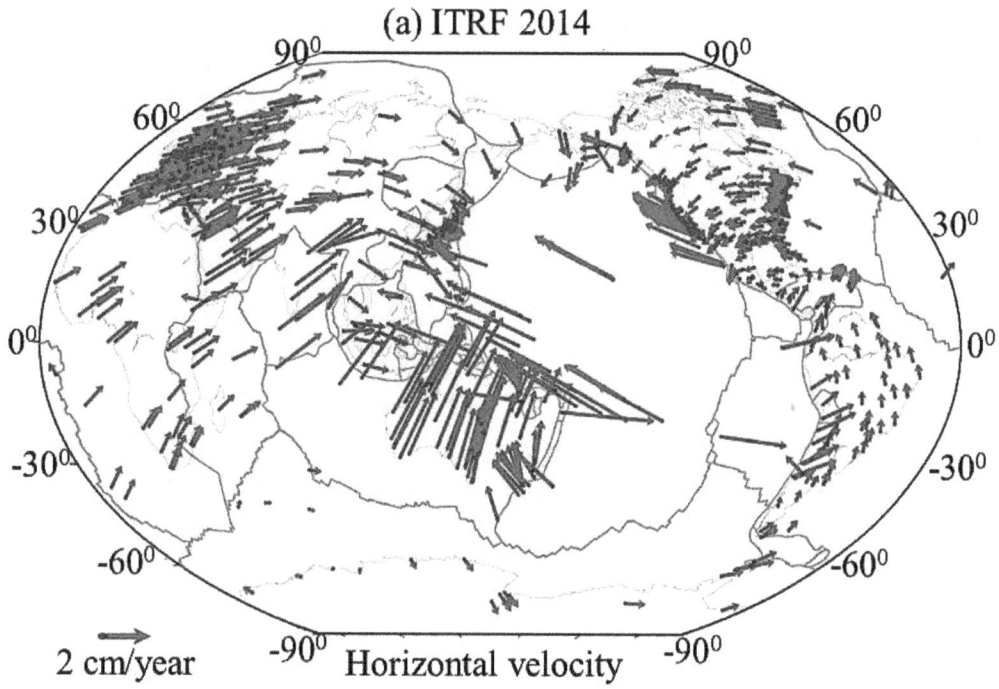

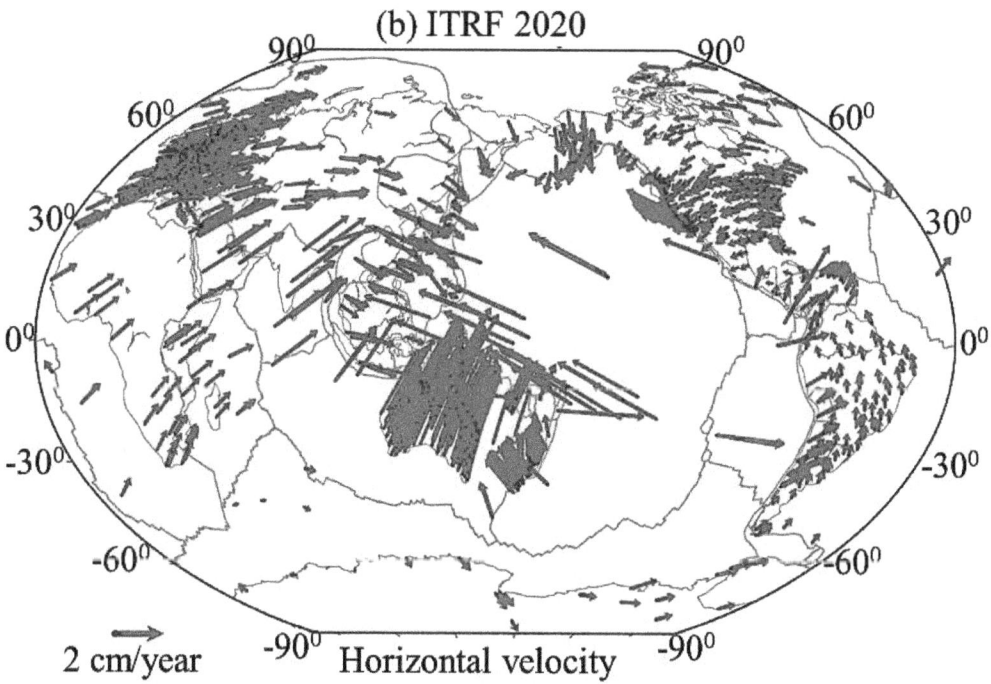

Figure 1.67. *(Continued.)*

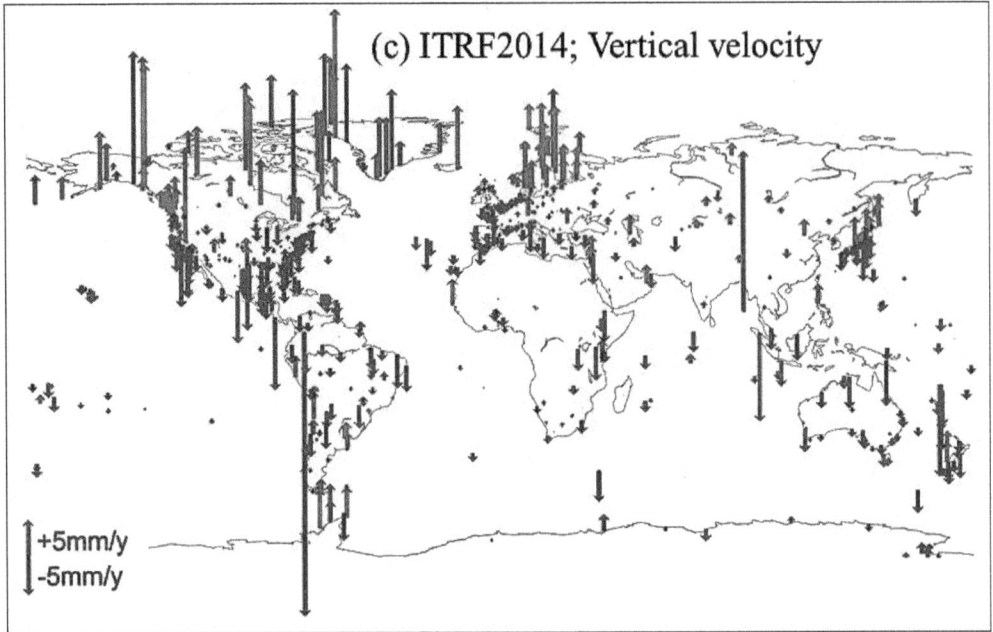

Figure 1.67. (a) ITRF2014 horizontal site velocities with an error less than 0.2 mm/year. The lines indicates plate boundaries. (b) ITRF2020 horizontal site velocities. (c) ITRF2014 vertical of site velocities with an error less than 0.2 mm/year. (Adapted from Altamimi et al., 2016).

your own. Remember that distance communications are carried by electromagnetic radiation—commonly by radio waves. The speed of propagation of these carriers is about 300,000 kilometers per second (the speed of light is the speed of all electromagnetic radiation). For a given point on Earth and another one on Mars, the distance between the two points is between 55 million kilometers and 378 million kilometers. So the delay between Earth and Mars varies between $t_1 = (55 \times 10^6)/(300000 \times 60) = 3$ minutes and 3 seconds and $t_2 = (378 \times 10^6)/(300000 \times 60) = 21$ minutes. For reference, the time delay between Australia and USA is 0.25 second for a distance of about 80,000 kilometers, between Earth and its Moon is 1.3 second for a distance of about 384,000 kilometers, and between Earth and Proxima Centauri (also known as Alpha Centauri C; the nearest star to Earth) is 4 years 3 monthsand 20 days for a distance of about 4.25 light-years (or 4.02×10^{13} kilometers). In addition, the exploration of our universe requires long-distance travel. Ideally, we would like to cover 50 light-years in 12 months or so (i.e., at a velocity of about 60 billion kilometers per hour) in order to conquer our solar system (see Ikelle, 2024).

The pull of gravity governs our lives: *It affects how hard our heart pumps, the density of our bones, cells, and every mineral and every molecule.* Almost all of our entire bodies are uniquely tuned to $1g_{earth}$. Several studies (e.g., Pass, 2008 and Smith and Zwart, 2008) show any gravity field that deviates from $1g_{earth}$, whether the field is greater than $1g_{earth}$, or as more common in human experiences during a space mission, less than $1g_{earth}$ can result in profound changes in the human body. The variable gravity fields and high radiation levels are the primary causes of these

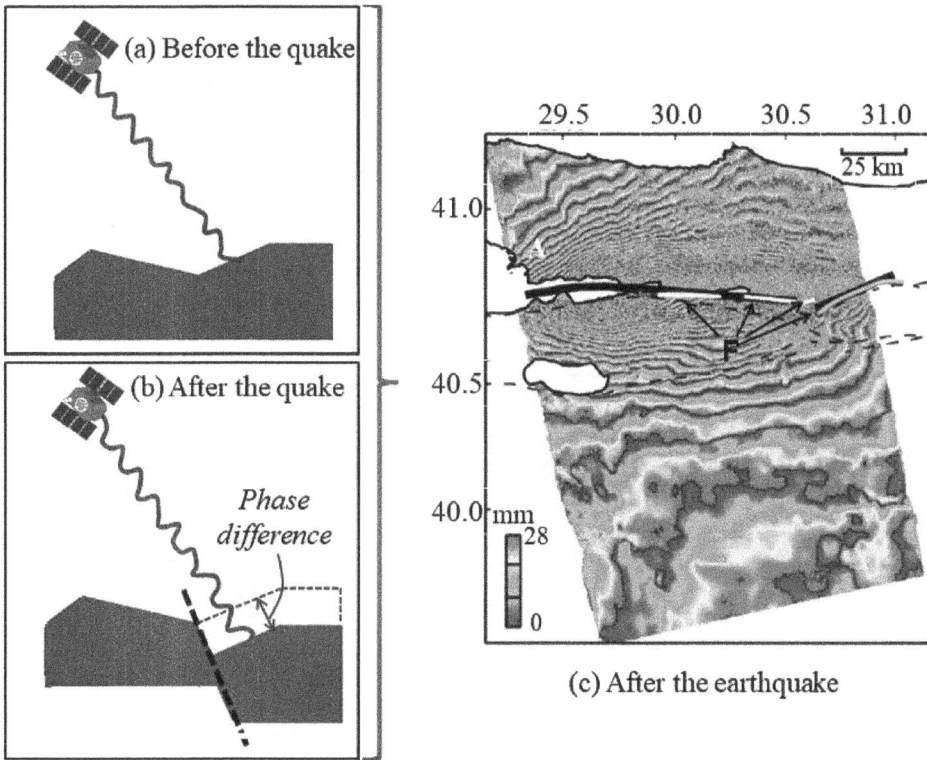

Figure 1.68. An illustration of synthetic aperture radar (InSAR). (a) An image of the Earth's surface before an earthquake. (b) An image of the Earth's surface after the earthquake. (c) The difference in phase between signals before and after earthquakes (or any two successive images) allows us to quantify Earth's surface changes. The "F" lines are faults. (Adapted from NASA/JPL-Caltech).

changes. The list of physiological changes to the functioning of the human body in microgravity environments, as reported by NASA and other researchers, include: (1) dehydration; (2) cardiovascular effects that involve lessened blood flow due to a decreased demand on the heart; (3) bone loss and demineralization (loss of calcium and other minerals) or the deterioration of the skeletal system; (4) alteration of pulmonary function; (5) muscle loss/deconditioned muscles; (6) a decrease in production of red blood cells resulting in mild "space" anemia; (7) loss of blood plasma and other body fluids, (7) balance disorders; (8) a weakened immune system; (9) "space adaptation syndrome" (SAS) characterized by nausea, headaches, vertigo, lethargy, and sweating, affecting half of all astronauts, though only lasting a few days; (10) unknown effects on brain function due to increased blood accumulation; and (11) minor annoyances that include puffiness in the face, increased flatulence, weight and muscle mass during a long space mission, loss, nasal congestion and sleep disturbance. Humans aboard the ISS (International Space Station), who have long-term exposure to microgravity, have harmful physiological effects, even in the LEO environment. There are obviously other changes to human bodies exposure to $0g_{earth}$ over a considerable time period still to be discovered, for example, how infants born in space will differ from infants born on Earth. Yet, the changes that

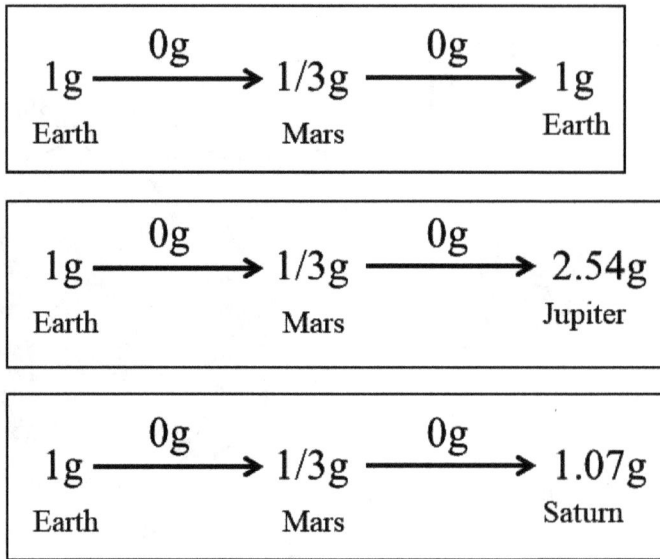

Figure 1.69. Examples of flight paths to space.

we now know off today will have major healthcare hazards if you made it back to Earth and are likely to produce physiological effects, including both transitory and potentially chronic physical body transformations. For those who are not returning and able to survive these new conditions, probably living in caves, the changes can lead to new human species different from what we witness on Earth. Actually, it is possible that we may end with a new branch of human species for each gravity field and even hybrids of Earth humans and computers, because at some gravity maybe the mind will disconnect from the body or some of the body organs can no longer function without computer assistance. Yes, gravity changes can have unimaginable consequences on humanity; as gravity ($1g_{earth}$) is critical to our existence. Note that for trips to places with gravity higher than 1g we may increase muscles and get shorter.

A new bio-physics of space is needed for humans in space. This bio-physics may turn out to be very beneficial to terrestrial human beings as it will potentially improve our understanding of aging, trauma, disease, and how the heart functions. Our objective in this section is to elaborate a little bit more on the reasons for physiological changes during space. We will focus on the production and reduction of NAD^+, blood flow and accumulation, and bone and muscle densities. Note that we discussed the radiation effects for space travelers in Ikelle (2023b) and we expand further in this chapter. Note also that in Chapter 3, we will discuss the challenges in telecommunication between humans in deep space and those on Earth. We will also discuss ways of artificially modifying the gravitational environment.

1.5.1. The Bio-Physics of Space: NAD^+ Reduction and Bone and Muscles Loss

Blood flow is essential for human life. It plays an important role in delivering nutrients and oxygen to tissues. Proper blood flow is required for cardiovascular health. Moreover, our blood flow is designed for $1g_{earth}$ only. Hence, its circulation is greatly affected by significant changes in gravity such as the ones encountered in space travel or space emigration.

As described in the previous chapters, the heart is the main source of blood. It pumps blood in the main arteries (e.g., carotid artery, brachial arteries, thoracic artery, and gastric arteries) that take blood to different organs. The veins then take blood back to the right side of the heart to be deoxygenated, to remove carbon dioxide from it, and to be reoxygenated before the cycle can start again. In this cycle, the blood flow toward the organs is accelerated by gravity for organs located below the heart and is working against gravity for organs located above the heart. Similarly, the flow of blood back to the heart is accelerated by gravity for organs located above the heart and is working against gravity for organs located below the heart. By decreasing or increasing gravity, we can alter this cycle and even shoot it down. For example, contraction of skeletal muscles and one-way valves contained in the veins that assist in the blood flow back to the heart may not be handled in the sudden fast flow in the case of low gravity or in the sudden slow flow in the case of high gravity. The elasticity of the arteries may be reduced and even may stiffen resulting in the reduction of blood flow, to its deregulation, or even to its halt. The blood pressure at the aorta has to be higher than normal to pump the blood to the top of your brain. Backflow of blood can be excluded. The veins in our heads are not designed to pump blood back to the heart (as are those in the lower body). Since almost all the body organs, and even cells, are affected by such changes. One person dies from blood-flow diseases (including an enlarged heart and blocked arteries) every 37 seconds in the United States alone according to the CDC. Because a very large number of deaths are associated with abnormal blood flow, under $1g_{earth}$ it tells us the risks associated with human bodies at other gravities. The specially designed capsules and modules for humans must be handled with special care.

Because of the changes in the functioning of the cardiovascular and pulmonary systems caused by gravity, humans exposed to microgravity exhibit swelling in the face and neck, loss of leg volume, decrease in plasma volume, but an increase in overall cardio-pulmonary blood volume. While standing at $1g_{earth}$ gravity, the blood pressure in our feet is about 200 mmHg and the pressure in our brains is only 60 to 80 mmHg. Take gravity to $0g_{earth}$ and the blood pressure equalizes around 100 mmHg throughout our body; fluids do not move around the body as well as when on Earth. Space travelers urinate more and drink less. They can become dehydrated very easily. Our faces puff up with fluid (really the entire upper body swells) and our legs thin out because the fluid drains out. The shift to higher blood pressure in the head triggers an alarm that the body has too much blood. Blood vessels can bleed when blood pressure is increased. It is possible for the optic nerves to swell and affect vision. In other words, $1g_{earth}$ gravity is an incredibly powerful force that helps to maintain the correct pressure in the right places in our body and even determines our looks. Clearly, advances in bio-physics are needed to ease some of the cardiovascular (pulmonary and systemic circulation) issues of space travelers.

Also indigenous beings of other planets, including standing beings, are likely to have different skin due to thermoregulation (see Chapter 3) and totally different internal organs that allow them to cope with gravity, higher amounts of radiation, and brightness.

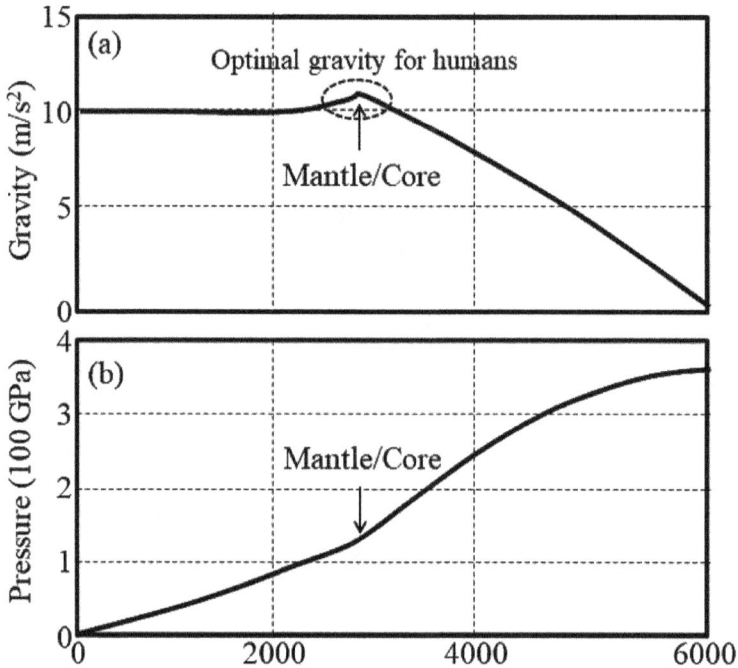

Figure 1.70. (a) Gravity as a function of depth in the solid Earth. (b) Pressure as a function of depth. Pressure increases with depth in the solid Earth due to the increasing mass of the rock overburden.

Bond density and optimal gravity for humans. Living bone is constantly being remodeled. The state of our bones is always close to equilibrium between bone formation and bone resorption (the breakdown and assimilation of old bone in the cycle of bone growth). In childhood and during the teens, bone formation is slightly predominant because we are growing our skeletons. We reach peak bone mass in the twenties, and from then onwards, resorption has the upper hand because the gravity force does not totally balance the stresses that hold our bones. Entering a gravity greater than $1g_{earth}$ after our teens will limit our bone resorption and will therefore limit our bone losses. In contrast, entering a gravity smaller than $1g_{earth}$ after our teens will accelerate our bone resorption and will therefore accelerate our bone losses.

The human body tends to relax in a state of weightlessness because it no longer fights the pull of gravity and body activities are considerably reduced. This lack of the gravitational pull alleviates the mechanical strain otherwise endured by our skeletal system. The reduced stress on bones is responsible for progressive bone loss. Calcium that is normally stored in bones is released into the bloodstream.

Calcium and sodium are naturally extracted from rocks as minerals found in the Earth's surface soil. They are absorbed by crops growing in the soil and find their way to our foods and then to our bones. Also, calcium is added to foods such as milk and orange juice to prevent bone disease. Some medications exist that can reduce the excretion of calcium from our bones. Unfortunately, some reports suggest that they are not effective in space. Nutrition, including eating enough, becomes imperative, otherwise you could compromise your health since nutrients are required for the function of every cell and system in your body.

Mineral loss from a space traveler's bones is estimated at over 1 percent per month. By comparison, the rate of bone loss for elders on Earth is about 1 percent per year (Mazess, 1982; and Nilas and Christiansen, 1987). Assuming the bone losses are linearly related to gravity, we have

$$\text{Bone loses(\% per year)} = -\frac{11}{g_{earth}}g + 12(\%) \tag{1.184}$$

At this rate, a 27-year-old traveler to Mars will return with the bones of a 60-year-old traveler after four years. Note also that there may be an Earth-like planet with a gravity of $g = 10.7\,\text{m/s}^2$ in which humans can exist without bone loss for many years. Alternatively the search for exoplanets with a g of about $10.7\,\text{m/s}^2$ is on (see Ikelle, 2024) or the construction of houses with chambers of $g = 10.7\,\text{m/s}^2$. Perhaps space bio-physics will help people on Earth live longer healthy lives, like Abraham, who apparently lived more than 175 years, or even like Noah, who apparently reached over 950 years, as depicted in the Bible.

The profile of gravity, from the Earth's core to deep in the kilometers deep in the Earth's atmosphere, in Figures 1.71 and 1.72 suggests that the maximum gravity is also about $g = 10.7\,\text{m/s}^2$. In other words, the optimal gravity for homo sapiens sapiens to live longer than their current life expectancy is also the maximum gravity in the Earth's gravity profile. Unfortunately, this optimal gravity is located in the lower mantle, a place we cannot physically access. In some mountains, such as Mount Erebus, there are big holes going into the lower mantle. Mount Erebus is the second-highest volcano in Antarctica, the highest active volcano in Antarctica, and the southernmost active volcano on Earth.

Nicotinamide adenine dinucleotide (NAD^+). Cells are the basic building blocks of tissues. All cells experience changes with aging and with gravity different from $1g_{earth}$. They become larger and are less able to divide and multiply. Among other changes, many cells lose their ability to function along with some of their molecules, and/or they begin to function abnormally.

As described by Fischetti and Christiansen (2021), the human body regularly replaces its own 30 trillion or so cells. About 72 percent of those, by mass, are fat and muscle, which last an average of 12 to 50 years, respectively. The tiny cells in our blood live only three to 120 days, and those lining our gut, typically live less than a week. Those two groups therefore make up the giant majority of the turnover. About 330 billion cells are replaced daily, equivalent to about 1 percent of all our cells. In 80 to 100 days, 30 trillion will have replenished. As we get older, the rate of cell replenishment reduces along with some of the molecules which make up these cells due to the reduced activities in the latter days of our lives. For example, the reproduction of molecules NAD^+ reduces as we get older, along with some blood cells, despite the fact that our immune system consumes more NAD^+ as we age, leading to the depletion of the NAD^+ levels in the body. Similar effects apply to humans in deep-space microgravity.

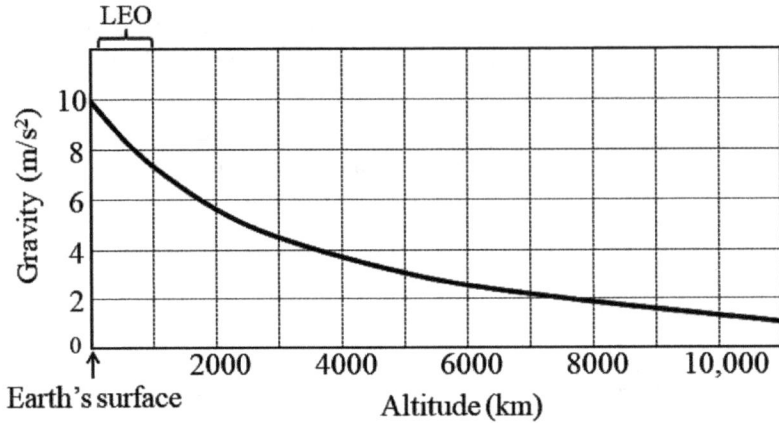

Figure 1.71. Gravity as a function of altitude.

Muscle loss: Like bone density, muscle density adapts to levels of physical activity. Age-related muscle loss is a natural part of aging because of reduced levels of physical activity. After age 30, we begin to lose about 0.4 percent of our muscle per year. Most humans will lose about 30 percent of their muscle mass during their lifetimes. Walking is a great way to improve your overall muscle power, as it is your legs that are responsible for our mobility.

Reduced muscle activities during space microgravity travels leads to a loss of muscle strength and power, and therefore to muscle losses. According to Trappe et al. (2009), space travelers experienced about 2.2 percent loss in calf muscle volume per month and about 5.3 percent decrease in peak muscle power month aboard the ISS despite the use of exercise programs that include treadmill running, cycling exercises, and resistance exercises. Muscle size and strength are generally functions of fibers that support and maintain our posture, endurance activities, shorter-duration bursts of speed, and muscle power and strength. In microgravity, the limited activities are, especially those related to shorter-duration bursts of speed and muscle strength, are responsible for muscle losses. For a Mars trip across several hundred millions of kilometers at near 0_{gearth} and for several months, travelers will experience large losses in muscle mass and strength before reaching Mars. These losses may impede the success of their mission.

1.5.2. The Moon Colony and the Upcoming Mars Trip

The other major problem that makes space unfriendly to human bodies is the high radiation levels. On Earth, we are protected by the atmosphere and its structure from most of the harmful Sun's radiation. In space, we are not. In other words, without a strong magnetic field and atmosphere, such as those provided by the Earth, we cannot escape high levels of radiation in other space environments in our solar system. The radiation is so penetrating that it can reach internal organs, the human's entire body, in a matter of minutes. Consequences include the development of cancer and other illnesses with deaths occurring rather quickly. On the Moon, for example, the lack of an atmosphere exposes human bodies to radiation without mitigation. On Mars, with a thin atmosphere, only about 1 percent of the Earth (see Ikelle, 2023a), the radiation level is so intense that underground habitats (e.g., caves) are being considered, along with strongly shielded vehicles for excursions on Mars's surface. In fact, harmful radiation is more of a deterrent to human space travel than variations in gravity fields.

As discussed in Ikelle (2023b), high-density metals, such as lead, iron, and uranium can be used to shield space travelers from solar radiation. Progress in light-weighted lead and iron fabrics is needed to address this problem.

Moon colony: Probably, the most complex aspect of settlement on the Moon is finding living quarters not affected by earthquakes. As described in Ikelle, 2023a, more than 12,000 seismic events were recorded on the Earth's Moon between 1969 and 1977. There is no indication of settlement in the caves underground because there are earthquakes that originate up to 200 kilometers below the surface, with magnitudes of up to 5. They are regular occurrences, about one a month. There may be a need for settlers to develop multiple living quarters so that they can move around to avoid earthquake effects, especially since moonquakes occur periodically. There are also shallow meteoroid-caused moonquakes that one has to contend with.

The other major issue is the availability of oxygen gas in the underground of the Moon. For the other necessities of life, just imagine that you are going to live on the seafloor, the Arctic, Antarctica, or in the middle of the Sahara desert. The Moon, as well as Mars, contains almost all the materials needed to build solar panels—a limitless and sustainable source of energy.

Note that, based on remote observations by NASA radar instruments aboard Indian Chandrayaan-1 in 2008, there are more than 600 billion kilograms of water ice water ice in the darkest and coldest regions of our Moon's south and north poles where temperatures do not get warmer than about -160 degrees Celsius. Before then, the only evidence of water on the Moon was deep underground. Using ice waters as a source of air, drinking water, propellant, and even entire lunar industries will make it easier to build a permanent base on the Moon. Also note that water in any form has not yet been found on Mars. The reports of potential water are just a misinterpretation of clay.

Mars trip. No earthquake to be worried about but the severe blowing Martian winds (more than 100 kilometers per hour; 62 mph) probably eliminate any visit on the surface of Mars lasting minutes. We have to turn again to the cave's options as resting places.

There are large lava tubes (i.e., conduits formed by flowing lava from a volcanic vent that moves beneath the hardened surface of a lava flow) on Mars. They were first pointed out by Viking orbiter images, and later identified using orbiter imagery from Mars Odyssey, Mars Global Surveyor, Mars Express, and Mars Reconnaissance Orbiter. These tubes are probably the safest places for humans to live on Mars. Some of these lava tubes are located at Hellas Planitia and experience less cosmic and solar radiation than much of the rest of the planet's surface and are not affected by severe wind storms on Mars's surface. They are more than 100 times wider than those on Earth and thus, they can be inhabited by a significant population.

In the minds of many people, permanent bases on the Moon and Mars are often unclear. We propose rotating small teams of scientists and engineers on the Moon and Mars, similar to what we do currently in Antarctica. On Earth, Antarctica cannot yet be permanently inhabited. The fact that we never had indigenous lives in Antartica tells us that our physiology are unlikely to adapt or adjust anywhere else. However, penguins have lived and are living in Antarctica. We first have to determine or find a living thing on Earth that can reside in the Moon and/or Mars. In January 2019, Chinese scientists experimented with growing plants on the Moon through their Chang'e-4 lunar rover. It brought to the lunar surface a mini-biosphere called the Lunar Micro Ecosystem (LME). The LME contained: potato seeds, cotton seeds, rapeseeds, yeast, fruit fly eggs, and *Arabidopsis thaliana*, a common weed. All of these died in few hours, except the cotton plant which grew two leaves before dying due to the cold temperatures after about two weeks. In any case, earthquakes are rare in Antarctica just as they are in Antarctica, and gravity is not a factor in Antarctica. Imagine the changes of physiology needed to permanently live in a hypothetical Antarctica with a gravity different from 1g, compared to us in warmer lands, at 1g.

Radiation-proof fabrics. As discussed in Ikelle (2023b), high-density metals, such as lead, iron, and uranium can be used to shield humans from gamma rays and therefore from solar radiation. Progress in light-weight lead and iron fabrics is needed to address this problem. Grains of basalt can also be included in such fabrics.

Computational exercise: Use the data in Table 1.15 and the following Matlab command to reproduce Figure 1.72:
`quiver(long, lat, veast, vnord, 1.5,' r',' lineWidth', 1.5)`

Table 1.15. Some GPS horizontal velocities; lat stands for latitude, long stands for longitude, vnord stands for GPS site velocities along the nord axis, and veast stands for GPS site velocities along the east axis. The velocities are measured in meters per year.

lat	long	vnord	veast	lat	long	vnord	veast
30.40	268.81	-0.0009	-0.0134	40.23	277.01	0.0018	-0.0153
34.11	242.91	0.0066	-0.0286	34.59	243.57	-0.0035	-0.0186
63.65	189.43	-0.0235	-0.0007	16.74	297.78	0.0140	0.0108
60.38	193.79	-0.0241	-0.0035	32.96	273.99	0.0007	-0.0134
58.95	198.25	-0.0225	-0.0063	31.78	274.03	0.0010	-0.0130
61.03	200.12	-0.0240	-0.0034	45.95	281.92	0.0023	-0.0165
51.86	183.33	-0.0116	-0.0116	36.88	239.33	-0.0116	-0.0339
67.05	203.09	-0.0236	-0.0055	34.01	240.63	0.0184	-0.0415
60.07	217.61	0.0144	-0.0301	39.01	283.39	0.0040	-0.0148
66.55	214.78	-0.0211	-0.0092	33.34	242.67	0.0170	-0.0380
58.19	223.35	-0.0047	-0.0183	36.11	265.82	-0.0021	-0.0142
56.24	225.35	-0.0083	-0.0137	36.18	266.96	-0.0006	-0.0138
58.41	225.45	-0.0132	-0.0127	35.20	241.08	-0.0003	-0.0281
16.26	298.47	0.0147	0.0101	43.66	247.37	-0.0100	-0.0161
59.77	208.13	-0.0280	-0.0061	43.82	238.63	-0.0089	-0.0229

58.92	206.35	-0.0221	-0.0073	39.39	242.69	-0.0090	-0.0168
54.83	200.41	-0.0120	-0.0130	63.48	197.99	-0.0227	-0.0114
60.48	210.27	-0.0046	-0.0185	59.37	206.64	-0.0319	-0.0035
58.92	207.75	-0.0251	-0.0071	54.14	194.23	-0.0222	-0.0066
55.92	200.87	-0.0139	-0.0144	53.87	193.45	-0.0226	-0.0076
58.21	205.84	-0.0193	-0.0072	54.21	194.10	-0.0193	-0.0054
59.42	213.66	0.0156	-0.0256	59.38	206.46	-0.0296	-0.0050
61.47	209.26	-0.0232	-0.0092	54.84	195.61	-0.0211	-0.0062
59.37	209.20	-0.0109	-0.0140	54.81	196.00	-0.0198	-0.0099
57.75	206.65	-0.0141	-0.0141	43.06	277.31	0.0014	-0.0162
55.90	199.59	-0.0177	-0.0124	32.98	248.48	-0.0071	-0.0140
61.24	210.43	-0.0153	-0.0175	33.05	249.08	-0.0075	-0.0134
60.08	207.37	-0.0300	-0.0045	34.60	247.53	-0.0018	-0.0203
61.49	208.16	-0.0272	-0.0066	34.26	248.75	-0.0192	-0.0070
61.76	209.93	-0.0215	-0.0117	31.95	249.22	-0.0038	-0.0188
61.76	209.93	-0.0215	-0.0117	49.18	291.73	0.0058	-0.0161
57.15	189.78	-0.0236	-0.0039	48.83	234.86	-0.0060	-0.0042
64.02	217.92	-0.0188	-0.0081	33.48	240.97	0.0227	-0.0411
62.83	216.29	-0.0091	-0.0172	18.20	288.90	0.0093	0.0084
63.11	217.97	-0.0179	-0.0036	42.60	245.12	-0.0083	-0.0195
63.69	214.11	-0.0205	-0.0057	34.58	240.01	0.0206	-0.0413
63.04	216.74	-0.0153	-0.0047	50.54	233.15	-0.0094	-0.0116
59.99	212.59	0.0273	-0.0273	13.08	300.39	0.0148	0.0131
9.371	280.05	0.0118	0.0162	48.75	237.52	-0.0009	-0.0019
44.05	238.68	-0.0110	-0.0149	47.61	237.87	0.0120	-0.0145
33.96	241.84	0.0158	-0.0385	45.97	252.00	-0.0081	-0.0152
47.53	235.74	0.0134	0.0045	35.28	243.91	-0.0078	-0.0157
42.76	250.44	-0.0075	-0.0150	33.61	245.28	-0.0076	-0.0143
33.96	243.01	0.0087	-0.0270	9.76	274.79	0 .0146	0.0083
37.91	237.84	-0.0060	-0.0384	27.94	278.21	0 .0014	-0.0119
44.73	247.46	-0.0094	-0.0152	34.61	239.80	0 .0303	-0.0458
35.15	240.65	0 .0042	-0.0289	42.83	235.43	0 .0014	-0.0066
41.72	272.46	0.0003	-0.0156	35.88	239.56	0.0142	-0.0330
42.81	253.61	-0.0059	-0.0154	46.19	236.63	-0.0041	-0.0154
19.73	280.24	0.0038	-0.0067	34.80	240.98	0.0088	-0.0334
34.25	253.03	-0.0062	-0.0136	12.97	272.44	0.0051	0.0061
45.65	275.53	0.0015	-0.0167	43.22	238.21	-0.0100	-0.0202
38.77	284.91	0.0040	-0.0149	37.08	239.78	-0.0098	-0.0360
45.48	236.02	-0.0032	-0.0056	34.27	242.57	0.0070	-0.0268
34.53	241.85	0.0104	-0.0304	42.97	237.91	-0.0026	-0.0155
40.55	245.12	-0.0095	-0.0166	40.44	235.60	0.0234	-0.0332
17.04	298.23	0.0157	0.0101	18.56	291.64	0.0104	0.0058
17.90	288.32	0.0097	0.0088	17.02	282.21	0.0075	0.0090
20.97	286.32	0.0058	-0.0080	17.40	276.05	0.0032	-0.0026
13.40	272.57	0.0057	0.0087	19.57	271.94	0.0006	-0.0081
19.66	290.06	0.0067	-0.0058	11.99	276.22	0.0073	0.0125
13.37	278.63	0.0062	0.0105	12.22	288.01	0.0128	0.0108
8.94	291.95	0.0112	-0.0050	12.48	298.57	0.0133	0.0116
18.58	296.57	0.0161	0.0103	8.54	281.98	0.0182	0.0097
48.53	238.25	-0.0052	-0.0184	32.86	244.87	-0.0078	-0.0134
34.08	278.87	0.0042	-0.0137	35.70	278.76	0.0029	-0.0138
38.20	282.62	0.0036	-0.0148	33.56	248.11	-0.0065	-0.0132
33.73	243.61	0.0075	-0.0270	36.90	284.28	0.0042	-0.0145

46.84	237.74	-0.0077	-0.0098	35.79	239.24	0.0221	-0.0374
35.24	240.27	0.0079	-0.0313	17.75	295.41	0.0128	0.0101
18.76	290.95	0.0083	0.0035	33.86	241.74	0.0172	-0.0391
34.40	240.62	0.0160	-0.0400	41.28	287.33	0.0051	-0.0153
34.12	243.62	0.0004	-0.0187	7.93	287.48	0.0144	0.0014
47.42	236.78	-0.0053	-0.0005	45.61	237.50	-0.0105	-0.0258
37.74	241.01	-0.0018	-0.0184	39.67	284.25	0.0041	-0.0148
37.87	238.08	-0.0036	-0.0350	33.43	242.69	0.0173	-0.0384
36.42	279.28	0.0029	-0.0141	49.32	240.37	-0.0104	-0.0130
70.33	211.52	-0.0200	-0.0088	33.73	243.28	0.0092	-0.0287
36.18	284.24	0.0047	-0.0130	47.65	241.85	-0.0129	-0.0155
39.65	253.15	-0.0086	-0.0161	37.91	245.73	-0.0090	-0.0159
34.94	241.16	0.0055	-0.0279	16.91	270.13	0.0005	-0.0073
33.58	241.87	0.0192	-0.0398	33.64	242.57	0.0163	-0.0371
34.59	241.54	0.0117	-0.0295	45.43	242.71	-0.0086	-0.0148
41.25	248.07	-0.0087	-0.0149	10.14	274.43	0.0118	0.0100
34.10	242.47	0.0122	-0.0341	64.80	212.15	-0.0213	-0.0057
34.06	241.55	0.0027	-0.0216	34.09	243.06	0.0053	-0.0255
34.35	241.11	0.0184	-0.0389	34.73	241.75	0.0068	-0.0257
35.19	276.60	0.0021	-0.0141	46.20	236.04	-0.0087	-0.0176
33.95	276.67	0.0022	-0.0125	32.38	276.65	0.0018	-0.0129
35.31	278.81	0.0027	-0.0138	47.74	269.65	-0.0023	-0.0171
49.68	233.87	-0.0079	-0.0072	34.29	243.61	-0.0009	-0.0174
34.30	245.81	-0.0071	-0.0142	45.83	239.18	-0.0089	-0.0122
35.42	243.11	-0.0030	-0.0169	39.91	248.10	-0.0079	-0.0162
12.22	298.35	0.0104	0.0127	63.83	211.02	-0.0192	-0.0103
45.36	239.21	-0.0080	-0.0194	39.54	252.67	-0.0172	-0.0167
43.24	246.75	-0.0093	-0.0162	58.41	224.30	-0.0140	-0.0116
34.28	272.14	0.0007	-0.0146	30.51	269.53	-0.0005	-0.0125
34.46	239.31	0.0256	-0.0412	33.78	241.71	0.0190	-0.0375
34.75	243.56	-0.0054	-0.0184	66.40	321.78	0.0185	-0.0233
34.75	242.20	0.0048	-0.0230	44.68	296.38	0.0084	-0.0153
73.67	331.87	-0.0113	-0.0221	38.58	283.86	0.0040	-0.0148
34.45	242.15	0.0076	-0.0290	70.73	242.23	-0.0123	-0.0170
37.65	241.17	-0.0044	-0.0132	49.66	276.48	0.0003	-0.0182
44.11	238.15	-0.0165	0.0153				

1.6. Essay: Estimation of Tectonic Plate Velocity

Anecdotally, some ideas are only theories because some of their answers or observations are of limited use or of limited interest. That is not the case in the sciences, in which a theory is a well-tested and widely accepted view that, scientists agree, most accurately describes their observations. A theory becomes a paradigm because it describes many interrelated observations. The tectonic plate theory is one such paradigm. It explains the correlations between earthquakes and volcanoes, and sometimes their noncorrelations, continental drift, large mid-ocean ridge systems, deep oceanic trenches, seamount chains, and rift valleys within ridges. The tectonic plate theory explains all these observations. Tectonic plate theory relies heavily on the computation of plate velocity vectors. Our objective is to revisit

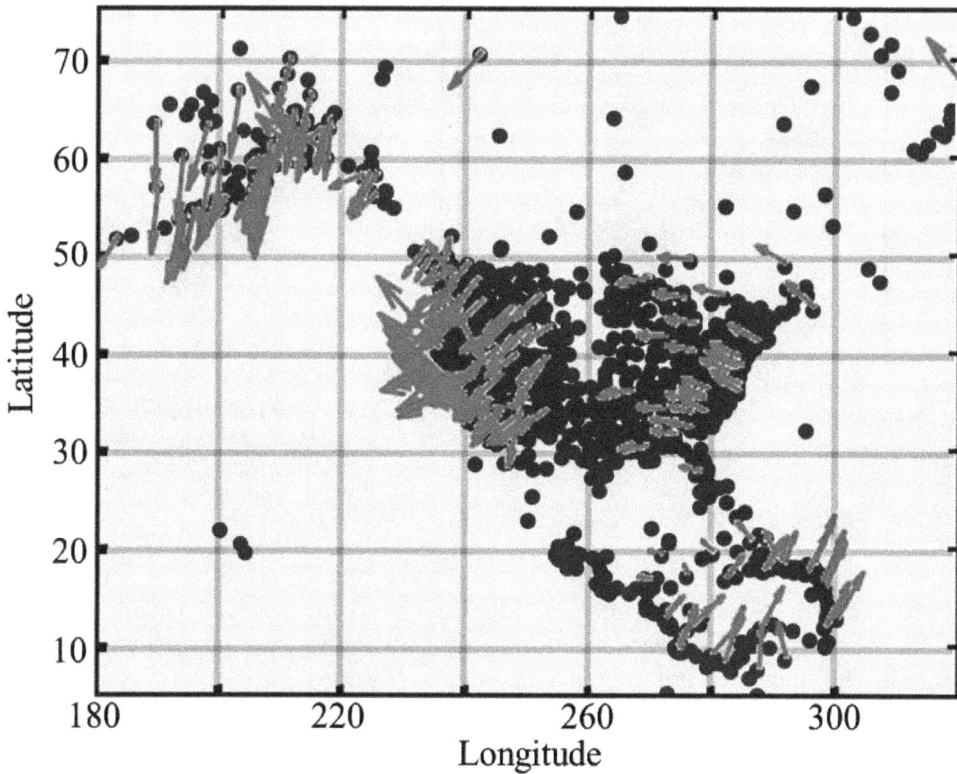

Figure 1.72. A graphic representation of the velocity vectors is in Table 1.15. The dots represent the GPS stations in the region, which includes North America, Mexico, the Caribbean, and a small portion of South America.

these computations and propose an alternative based on satellite measurements of ground deformations.

Consider the Pacific plate. The plate motion is calculated by using the Hawaiian island chain of seamounts under the assumption that seamounts result from the Pacific plate motion over a fixed Hawaiian hotspot. The basic idea is that the distance between two seamounts, say δd, over their age difference, δt, provides us with the velocity of the Pacific, including its direction because δd is a vector. Defining points that describe seamounts to determine the distance between mounts, determining the age of lavas and thus the age of seamounts, and the assumption that the hotspot is fixed are potential sources of significant errors in such calculations. For a plate without mountain chains and a hotspot, the classical method is seafloor spreading (e.g., Ikelle, 2020; Chapter 12). Our position is at the divergent boundary, where a new floor is created. Basalt with iron has different magnetic polarities when new floors are created. This iron is polarized by the Earth's magnetic field. The magnetic field of the Earth will switch when the northern and southern poles switch. By looking at the polarity of the iron that comes out of the seafloor spreading we can determine how old the material is at the bottom of the ocean is along with its distance to the divergent boundary to determine the plate velocity.

With multiple stations, we nowadays have access to several vectors of ground deformations and therefore several particle velocities. We can see in Figures 1.67

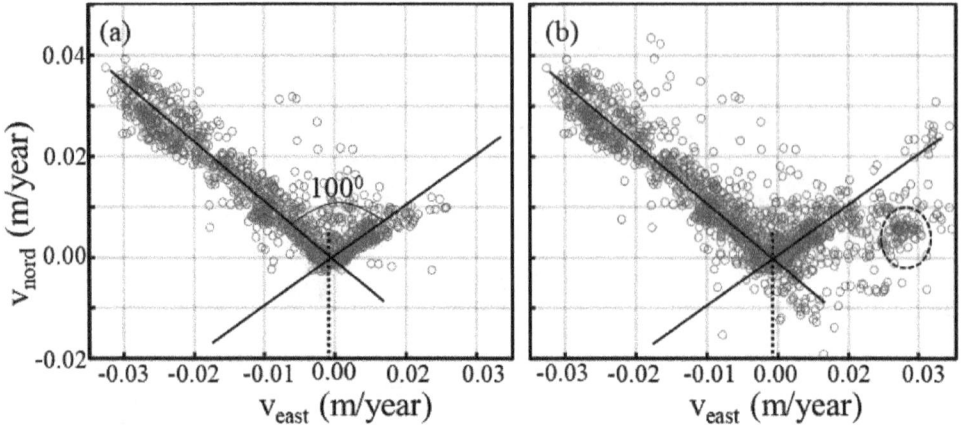

Figure 1.73. (a) A scatterplot of v_{nord} versus v_{east} for all the data points in Figure 1.72 in the interval $[30 - 50^o$ N, $180 - 260^o$ E]. (b) A scatterplot of v_{nord} versus v_{east} for all the data points in Figure 1.72. We circle some totally outlier datapoints.

and 1.72 that these velocities are not identical even for stations within the same plate. In other words the plates are not as rigid as one way have expected, thus the stations in the plate are not moving in unison. We may be tempted to define the place velocity vector the average of vectors associated with each station as

$$v = \frac{1}{N} \sum v_i \, , \tag{1.185}$$

where v_i is the particle velocity at the ith station and N is the number of stations. By using the data points in Figure 1.72, we have obtain $v = [0.0097, -0.0076]^T$. The angle between vv and $[1,0]^T$ is about 38.078^o. Because we dealing with multiple plates in Figure 1.72, we decided to restrict our computations of the resulting to interval in which latitudes are between 30° N and 50° N and longitudes are between 180° E to 360° E. In this interval we are limited to North America without Canada and Alaska where data coverage is anyway negligible. The resulting vector is $v = [0.0108, -0.0085]^T$ is different from the one corresponding to whole region as one may expect and the angle is 38.20^o. However, we further restrict ourselves to the east coast, the resulting vector changes again to $v = [0.0126 - 0.0100]^T$ and the angle is 38.43^o. The variation in magnitude of these vectors are just too large in the same plate. We again used the data recorded by GSSN (https://www.unavco.org/data/gps-gnss/gps-gnss.html).

The scattering plot of v_{east} and v_{nord} in Figure 1.73 for the data points in Figure 1.72 shows significant scattering of data points. This spread indicates that significant data corrections must be made. The variations in the resulting vectors are also due to data errors. The datapoints in the circle are totally erroneous. First of all, if our basis vectors are orthogonal, we should expect perpendicular narrow lines in Figures 1.73a and 1.73b. Moreover by using the medians over two scattering lines we must end up with 90° angle instead of 100°. Also the reference origin must coincide with $[0,0]^T$. In Figure 1.73, we used medians (ℓ_1-norm) instead of averages (ℓ_2-norm). As described in Ikelle (2024), all these errors can be fixed and must be corrected for effective data interpretation. Advanced statistical tools for

Table 1.16. Some recent eartquakes in California; lat stands for latitude, long stands for longitude.

Date	Name	County	Lat (^{o} N)	Long (^{o} E)	Mag
2022-12-20	Humboldt	North Coast	40.745	236.14	6.4 Mw
2019-07-05	Ridgecrest	Eastern	35.766	242.395	7.1 Mw
2019-07-04	Ridgecrest	Eastern	35.766	242.395	6.4 Mw
2014-08-24	South Napa	North Bay	38.22	237.69	6.0 Mw
2014-03-28	La Habra	LA Area	34.037	241.73	5.1 Mw
2010-04-04	Baja California	Baja California	32.13	244.7	7.2 Mw
2010-01-09	Eureka	North Coast	40.65	235.24	6.5 Mw
2008-07-29	Chino Hills	LA Area	33.947	242.284	5.4 Mw
2007-10-30	Alum Rock	Bay Area	37.26	238.074	5.6 Mw

making these corrections, including nonlinear independent component analysis are described and applied in Ikelle (2024).

1.7. Essay: Earthquakes in California and Volcanic Eruption of Yellowstone National Park

In this chapter, we focus on tectonic geodesy to obtain reference frames. Along with classical earthquake data described in Ikelle (2023a), we will also use tectonic geodesy to monitor Earth's surface deformations associated with natural hazards, such as earthquakes, volcanic eruptions, and landslides as described in Ikelle (2024). As we can see in Figure 1.74, tectonic geodesy contains a wealth of information that can be employed to monitor natural hazards through their variations in directions and time at a given GPS station and through the use of multiple stations available nowadays. We used the GSSN data (https://www.unavco.org/data/gps-gnss/gps-gnss.html). In Figure 1.74a, we show data from station P580 (35.62^o N, 242.80^o E) near the epicenter of the July 4 and 5, 2019 earthquakes in Ridgecrest, California. Ridgecrest is located in Kern County and west of Searles Valley (approximately 200 km north-northeast of Los Angeles). This seismic event included a magnitude 6.4 foreshock in Searles Valley on July 4, 2019. It also included a magnitude 7.1 mainshock event in Ridgecrest on July 5, about 34 hours later. There were also many aftershocks, mainly near Naval Air Weapons Station China Lake. Eleven months later, a Mw 5.5 aftershock occurred (the largest aftershock of the sequence) east of Ridgecrest. The other recent significant earthquakes in California are listed in Table 1.16 and indicated in Figures 1.74a and 1.74b

A station OFW2 (latitude 44.451^o N, longitude 249.168^o) near Yellowstone National Park shows ground deformations before and after the 2016 volcanic eruption. On February 16, 2016, a magnitude 6.3 earthquake ruptured in Yellowstone National Park, Wyoming. Its epicenter was only 2 kilometers NNW of Old Faithful. It was felt 140 kilometers from the epicenter. Being one of the strongest quakes in almost 10 years in the continental US, it caught most people off guard. On February 16, 2016 an eruption occurred. The signatures of these events are obvious in Figure 1.74. Notice the height difference between the scale of ground deformation due to transform-fault earthquake in 1.74a and that of Yellowstone in Figure 1.74b. Also notice how data at station P580 vary differently from station OFW2. Table 1.16 contains most significant events that have occurred in California since 2007. However, there are no obvious signatures that will allow us to predict these earthquakes and eruptions months in advance of these events. In Ikelle (2024), we used advanced statistics to construct machine learning schemes and AI systems that will help us make these predictions.

Basically, there is no system of equations that allows us to make these predictions months in advance from ground motions obtained from GNSS, SLR, DORIS, and VLBI. However, we have enough data to learn how such equation systems work and/or what additional information we may need to obtain such systems. We can do so with statistical tools such as data decomposition, clustering, and neuronetworks, which have now been integrated with machine learning and artificial intelligence. The key component of these tools is determining the features that distinguish between deformations leading to earthquakes and not earthquakes and between volcanic eruptions and non-volcanic eruptions, as described in Ikelle (2023a). In Ikelle (2024) we describe some of these features.

After all, earthquakes are about fracturing, just like a bridge collapsing in your area. Rocks contain cracks/fractures. Ultimately, rock failure occurs through the growth of fractures, the creation of new fractures, the reactivation of existing cracks (fractures), and their propagation through rock formation. All the processes affect

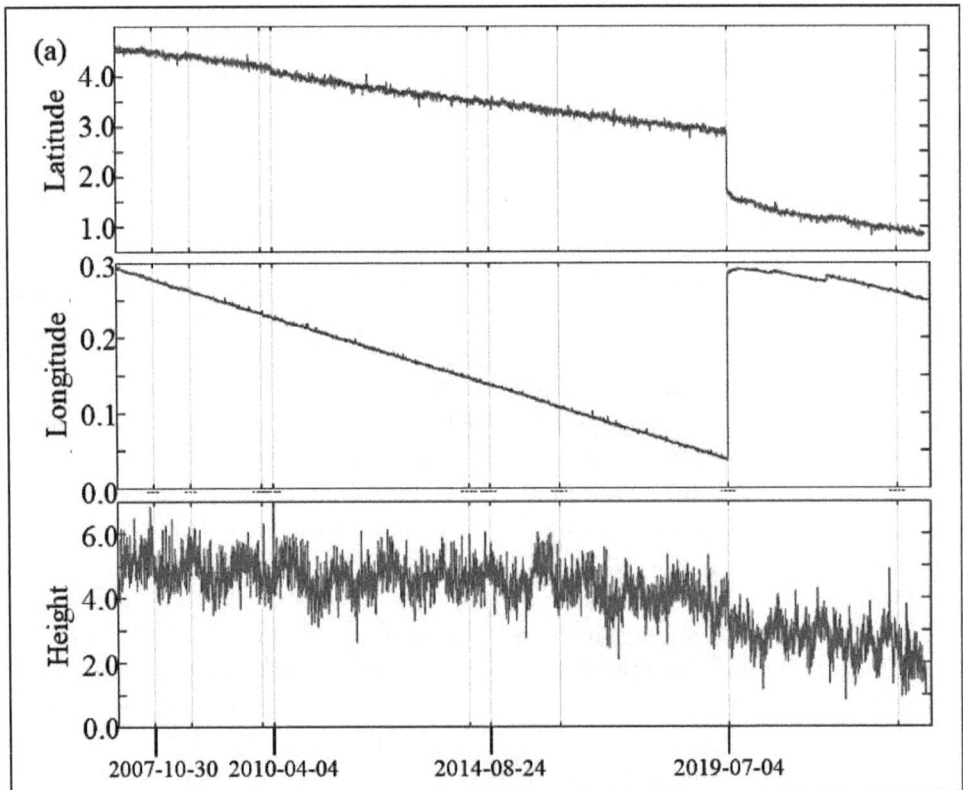

Figure 1.74. *(Continued.)*

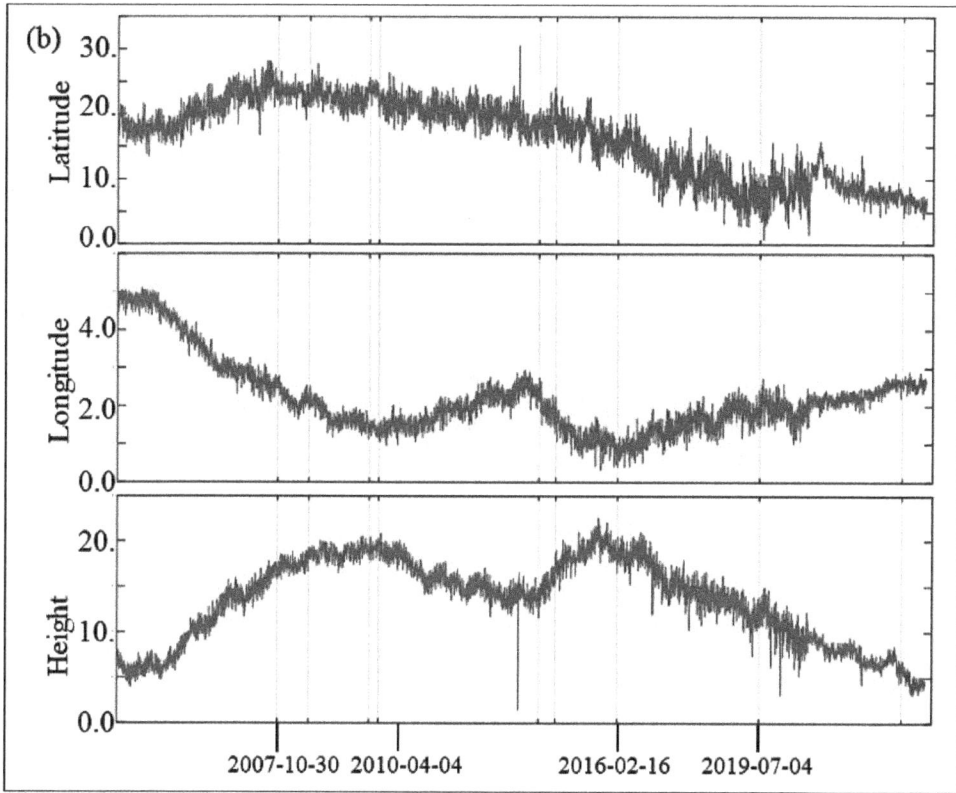

Figure 1.74. (a) Latitude, longitude, and height variations with time at station P580 ($35.62°$ N, $242.80°$ E) near the epicenter of the July 4 and 5, 2019 earthquakes in Ridgecrest, California. The vertical lines indicate the time the earthquakes in Table 1.16 occurred. Latitude here is essentially the actual latitude minus a reference latitude which is 356209466×10^{-7} decimal degrees. Similary the long is the actual longitude minus the reference longitude, which is 24280777×10^{-5} decimal degrees and the height is the actual height minus the reference height which is 166959 cm. (b) Latitude, longitude, and height variations with time at the station OFW2 (latitude $44.451°$ N, longitude $249.168°$) near Yellowstone National Park station, USA. Latitude here is actually the actual latitude minus a reference latitude which is 356209466×10^{-7} decimal degrees. Similarly the long is the actual longitude minus the reference longitude, which is 24280777×10^{-5} decimal degrees and the height is the actual height minus the reference height which is 166959 cm.

the Earth's surface deformations at a scale, most of the time, not visible with the naked eyes, until a significant earthquake, volcanic eruption, or landslide takes place. Finding the precursory processes behind major changes in rock formations is challenging. However, we know that these precursory mechanisms exist because foreshocks explicitly occur for a number of earthquakes. Similar mechanisms exist implicitly for earthquakes and landscapes without foreshocks. Also the fact that small earthquakes increase dramatically before volcanic eruptions confirms the existence of precursory mechanisms.

1.8.   Conclusions

It is easy to see the wide range of potential applications and narrow expertise that can be developed once one has mastered gravity basics. By studying space programs, we can realize that China, the USA, Russia, and India are the world's most advanced economies nowadays, since space science and technology determine their development. Modern medical treatments, extending human lifespans, new materials, transportation, etc., are all related to space program advances. Despite their low budgets, China and Indian programs are making remarkable progress due to their lower footprints of social scientists who lack understanding and rigor for sciences and engineering. As an example, the Indian's Chandrayaan-3 mission to the Moon's pole is a strategic mission. Just like Earth's poles, it is becoming clear that the poles of many planets, exoplanets, and moons have large potential water reserves. This serves as the central point of human exploration and rocket fuel. We can see this phenomenon on Mercury and Mars, as well as on our Moon and Enceladus.

Chapter 2

Heat Radiation, Conduction, and Convection

Heat is the measure of the motions of atoms, which compose the matter and whose temperature we measure. Heat moves. In this chapter, we describe the heat movements, also known as heat-transfer mechanisms. They are radiation, conduction, convection, and latent heat. Understanding heat-transfer mechanisms is crucial to understanding Earth's climate, solid Earth's heating and cooling, mass extensions on Earth, star evolutions in the sky, human thermotherapy, heat engines, and heat pumps. Some of these applications of heat transfer mechanisms are central to modern human lives. Heat transfer mechanisms also explain why a cold glass of water in a room warms up and a warm glass of water in a refrigerator cools down. To cook delicious food, it is necessary to have a good understanding of heat transfer. We start this chapter by describing the laws of thermodynamics of conservation of energy and non-conservation of entropy.

Key sentences: *Increasing entropy in our lives describes the winding down of our energy as we get older. For molecular living things, the ultimate end is death. For machines, they will one-day fatigue, stop working, and be decommissioned. One of the challenges of our times is to reduce the rate of entropy increase. Such a reduction leads to healthier and longer lives, and better mankind.*

2.1. Heat Transfer and Entropy

2.1.1. Conservation of Energy and the First Law of Thermodynamics

Terminology. Let us start by defining the terminology of heat transfer adopted in this discussion. Heat, like work, is energy in transit. The transfer of energy as heat, however, occurs at the molecular level as a result of a temperature difference. The symbol Q is used to denote heat. As with work, the amount of heat transferred depends upon the path and not simply on the initial and final conditions of the system. The unit of heat is Joule (J) since heat is a form of energy. The heat in many systems, including the solid Earth, varies over time and space, thus, $Q = Q(\boldsymbol{x}, t)$.

DOI: 10.1201/9781032620619-2

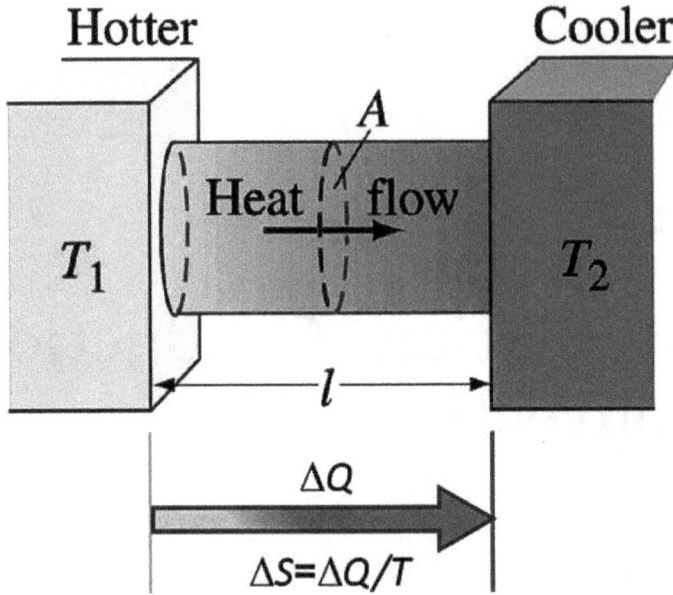

Figure 2.1. An illustration of heat conduction—that is, how heat transfers through direct contact with objects (solids and non-moving fluids) that are touching. Any time that two objects or substances touch, the hotter object passes heat to the cooler object ($T_1 > T_2$). A represents the area normal to the direction of heat transfer. Entropy (S) is another thermodynamic variable. As a process of a system goes from one equilibrium state to another, the entropy of the system plus the environment increases if the process is irreversible and remains constant for a reversible process.

The symbol $Q' = \frac{dQ}{dt}$ is the rate of heat flow—that is, the amount of heat transferred per unit time. The unit of heat flow is Joules per second (J/s) or Watts (W). This quantity reiterates that the heat here varies with time, t.

The symbol $\boldsymbol{q} = \frac{dQ'}{dA}\,\boldsymbol{n}$ (W/m^2) is heat flux (also known as heat flow-rate density). It describes the rate of heat transfer per unit area along the normal to the direction of heat transfer. Vector $\boldsymbol{n} = \boldsymbol{n}(\boldsymbol{x})$ is this normal to the direction of heat transfer. The unit of heat flux is W/m^2). Figure 2.1 depicts heat transfer through direct contact with objects (solids and non-moving fluid) that are touching along with area A. Any time that two objects or substances touch, the hotter object passes heat to the cooler object. Figure 2.2 shows the planetary vertical component of heat flux (also known as heat-flow-rate density, the rate of heat transfer per unit area normal to the direction of heat transfer). In this map, small heat-flow areas are hotter, midle heat-flow areas cooler. There is high heat flow at mid-ocean ridges (large heat flow) and low heat flow on the continents (small heat flow). Most heat losses occur in the oceans.

The symbol q' (W/m^3) is heat production (volumetric heat release)—that is, the source of heat from, say, radioactive decay of long-lived radioactive elements, such as uranium-238 (see Ikelle, 2023a and 2023b).

The theory of classical thermodynamics and its laws was predominantly developed by Carnot (1824), Clausius (1850 and 1852), Kelvin (1851), Planck (1897),

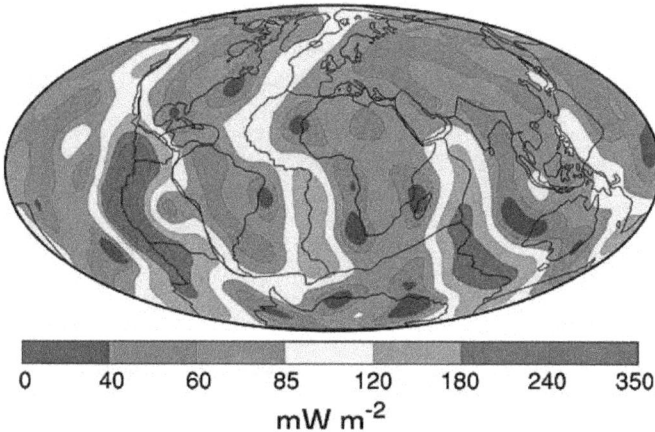

Figure 2.2. The planetary vertical component of heat flux (also known as heat flow-rate density, the rate of heat transfer per unit area normal to the direction of heat transfer). We have high heat flow at mid-ocean ridges (large values) and low on the continents (small values). Most heat loss in the oceans takes place during the creation and cooling of oceanic plates. Basically the Earth's heat budget drives plate tectonics. (Source: Pollack, Hurter, and Johnson (1993) Rev. Geophys. 31, 267-280.)

Gibbs (1957), and Carathéodory (1925), We now recall these laws, starting with the first law of thermodynamics. Suppose that a closed system, like the solid Earth, receives a certain amount of thermal energy (heat), Q, say, from radiation. Heat is a spontaneous flow of energy due to temperature differences. As a result, the system may do a certain amount of external or internal work, W—that is, the flow of energy in or out of the system not caused by temperature differences. Heat is negative if the system is hotter than the surroundings (i.e., heat leaves the system) and positive if the surroundings are hotter than the system (i.e., heat enters the system). Work is positive when work is done on the system and negative when work is done by the system. A thermal gradient, a gradient in the electrical potential, a gravitational potential that causes a mass to fall are examples of conditions under which work may be produced by a system such as Earth. The principle of conservation of energy states that the excess of the energy supplied to the system over and above the external work done by the system is $Q - W$; i.e.,

$$\Delta U = Q - W \Longrightarrow dU = dQ - dW \iff \frac{dQ}{dt} = P\frac{dV}{dt} + \frac{dU}{dt} \,, \qquad (2.1)$$

or

$$\left(\begin{array}{c} \text{Change in the} \\ \text{total energy of} \\ \text{the system} \end{array} \right) = \left(\begin{array}{c} \text{Total energy} \\ \text{entering the} \\ \text{system} \end{array} \right) - \left(\begin{array}{c} \text{Total energy} \\ \text{leaving the} \\ \text{system} \end{array} \right) \,,$$

where ΔU is the change in internal energy (related to temperature) of the system, Q is heat added to the system, and W is work done by the system (see Figure 2.3). So the first law of thermodynamics, also known as the conservation of energy principle, states that the net change (increase or decrease) in the total energy of the

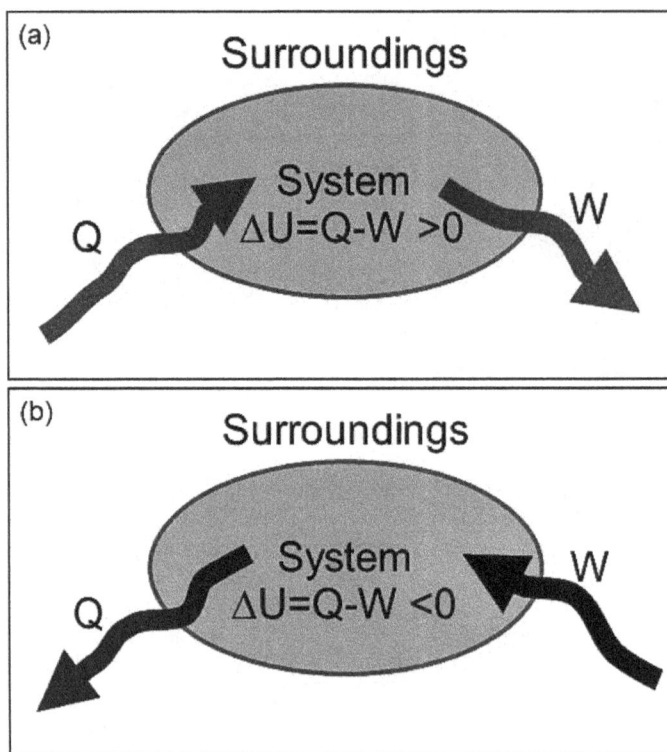

Figure 2.3. An illustration of the first law of thermodynamics. ΔU is the internal energy (i.e., the total energy both kinetic and potential energy) in a system; Q is the heat and W is the work. A system only contains internal energy. A system does not contain energy in the form of heat or work. Heat and work only exist during a change in the system. Any energy entering the system is positive; that is, if the heat *is absorbed* by the system (heat transfer into the system), then $Q > 0$ and work is done *on* the system, $W > 0$. Any energy leaving the system is negative; that is, if heat *is given off* by the system, then $Q < 0$, and work is done *by* the system, $W < 0$. (a) A total of Q (e.g., $Q = 16$ Joules) of heat transfer occurs in the system, while work takes out W (e.g., $W = 6$ Joules). The change in internal energy is $\Delta U = Q - W = 16 - 6 = 10$ Joules. (b) Heat Q removes from the system (e.g., $Q = -100$ Joules) while work W puts into it (e.g., $W = -110$ Joules), producing an increase of $\Delta U = Q - W = 10 = -100 - (-110) = 10$ Joules in internal energy.

system during a process is equal to the difference between the total energy entering and the total energy leaving the system during that process. In other words, energy is not created or destroyed; it can be transformed from one form to another. Note again that internal energy is a form of energy; it is a property of a system. Work is also a form of energy, but it is energy in transit (e.g., compressing a gas). Work is not a property of a system. Work is a process done by or on a system, but a system contains no work. This distinction between the forms of energy that are properties of a system and the forms of energy that are transferred to and from a system is important in the applications of thermodynamics, which we will discuss in the next sections. Heat, like work, is also energy in transit. However, this transfer of energy as heat occurs at the molecular level as a result of a temperature difference (e.g., heating a gas).

The second equation in (2.1) is the differential form of the principle of conservation of energy, where dQ is the differential increment of heat added to the system, dW the differential element of work done by the system, and dU the differential increase in internal energy of the system. In the third equation of (2.1), it follows that the heat-transfer rate is equal to the work transfer rate plus the rate of change in internal thermal energy. We used the fact that the differential work is a force of moving material, F, times the distance (or depth), dz, moved in the forced direction; i.e.,

$$dW = F dz = P A dz = P dV . \tag{2.2}$$

Most of the time, we are concerned with pressure on materials. Since $F = PA$, where P is the pressure and A is the cross-sectional area, and since $dV = A dz$ is the differential volume, we get the last equality in (2.2)—that is, $dW = P dV$ (work equals applied pressure $\times$ the displaced volume).

Two well-known special cases of (2.1) are

$$\text{Constant volume} : Q' = \frac{dQ}{dt} = \frac{dU}{dt} = \frac{dU}{dT}\frac{dT}{dt} = m c_V \frac{dT}{dt} , \tag{2.3}$$

$$\text{Constant pressure} : Q' = \frac{dQ}{dt} = \frac{dH}{dt} = \frac{dH}{dT}\frac{dT}{dt} = m c_P \frac{dT}{dt} , \tag{2.4}$$

with

$$c_V = \frac{1}{m}\left(\frac{dU}{dT}\right)_V \text{ and } c_P = \frac{1}{m}\left(\frac{dH}{dT}\right)_P , \tag{2.5}$$

where $H = U + PV$ is known as the enthalpy, m is the mass of the body, and c_V and c_P are the specific heat capacities at constant volume and constant pressure, respectively. At constant volume dW is zero (i.e., $dQ = dU$) and the specific heat c_V is the amount of heat needed to raise the temperature of a unit mass of material by 1 degree Kelvin. By using the fact that

$$dQ = d(U + PV) - V dP = dH - V dP , \tag{2.6}$$

at constant pressure $dQ = d(U + PV) = dH$ and the specific heat c_P is the amount of heat needed to raise the temperature of a unit mass of material by 1 degree Kelvin; specific heat c_P is a measure of a material's ability to store thermal energy. For example, $c_P = 4.18\,\text{kJ/kg}^\circ\text{C}$ for water and $c_P = 0.45\,\text{kJ/kg}^\circ\text{C}$ for iron at room temperature, which indicates that water can store almost 10 times the energy that iron can per unit mass. Note that enthalpy of a substance cannot be measured directly, and it is given with respect to some reference value. For example, the specific enthalpy of water or steam is given using the reference that the specific enthalpy of water is zero at 0.01 degree Celsius and normal atmospheric pressure.

Most liquids and solids in the lithosphere under moderate-to-low pressure conditions ($P < 2$ GPa) can be modeled as incompressible—that is, volume V is constant for any pressure or temperature variation—materials, and the two specific heats are equal, $c_V = c_P = c$; i.e.,

$$Q' = \frac{dQ}{dt} = m c \frac{dT}{dt} . \tag{2.7}$$

When a body cools, energy is released according to this equation. In fact, this equation is fundamental to the derivation of geothermics of the lithosphere in the third section of this chapter. Note that when air parcels move upward in the atmosphere, their sizes (volumes) change. The air pressure inside parcels always equals the air pressure immediately surrounding the parcel.

Example: what causes climate change? Let solid Earth be the system and the energy radiated by the sun that reached the Earth be Q (see Figure 2.4). Some of the energy reaching the Earth is reflected, W, while some is absorbed and emitted back by the Earth, ΔU; i.e.,

$$\left[\text{ Emitted radiation by Earth } (\Delta U) \right] = \left[\begin{array}{c} \text{Incoming radiation} \\ \text{from the sun } (Q) \end{array} \right] \qquad (2.8)$$
$$- \left[\begin{array}{c} \text{Reflected radiation} \\ \text{by Earth } (W) \end{array} \right].$$

There is a balance between (1) the incoming radiation from the sun, (2) the radiation reflected by the Earth, and (3) the radiation emitted by the Earth. We have an example of the first law of thermodynamics in action. Note that changes in internal energy, ΔU, involve changes in the temperature of the system. Also, these changes in internal energy depend only on the initial and final states of the system, and are therefore independent of the manner by which the system is transferred between these two states.

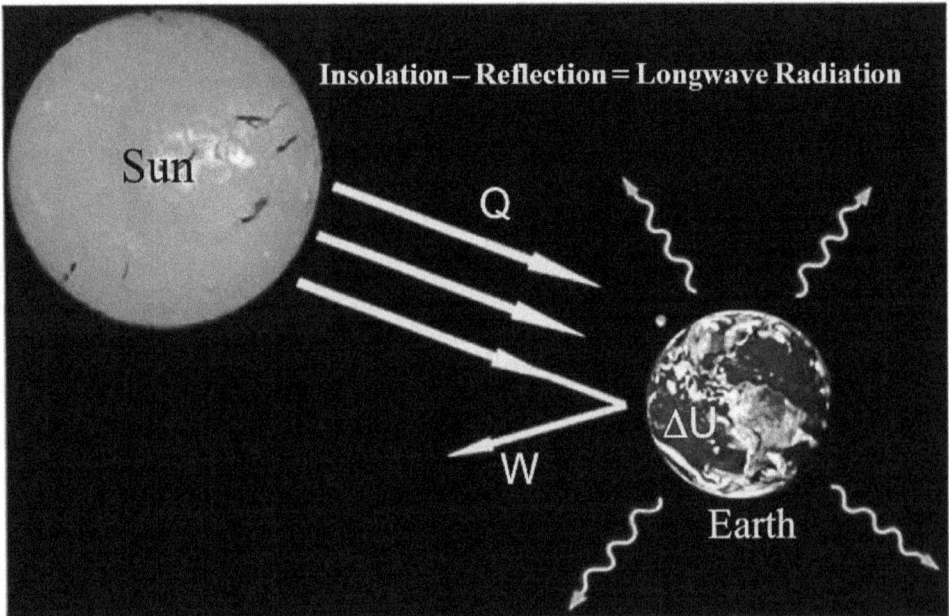

Figure 2.4. Here is a simplified climate system. The sun radiates energy in all directions. Some of the energy reaching the earth is reflected, whereas some is absorbed and emitted back by the earth.

One consequence of radioactive decays of the major long-lived radioactive elements such as ^{238}U, ^{235}U, ^{232}Th, and ^{40}K (Ikelle, 2023a and 2023b), is that the Earth is cooling. When a body cools, energy is released according to the first law of thermodynamics. The system here is the solid Earth and the process here is the cooling. We have an example of the first law of thermodynamics in action.

In terms of climate systems, we can interpret this law as follows: A certain amount of solar energy falls on the Earth's surface at any given place and time, creating heat. Unless this same amount of heat energy is dissipated back into space, the Earth will warm. Conversely, if more heat is reradiated back into space more than it is received, the planet cools. Also any change in any of the components of the climate system, in (2.9), causes climate change. For example, anything that alters the amount or distribution of solar energy arriving on the Earth (solar variations, orbital variations) can cause climate change. As humans cannot change the amount of solar radiation, this change is natural, and therefore the corresponding climate change is natural. A change in the amount of reflected radiation of the Earth can produce climate change. Reflected radiation by the Earth is caused by clouds, ice, deserts, etc. In fact, anything that alters the flow of energy in and/or out of the Earth's surface or changes its distribution (desertification, continental drift) can result in climate change. For example, humans can affect the radiation reflected by the Earth by cutting or planting trees. A change in the amount of radiation emitted by the Earth can also cause climate change. Humans can also affect this radiation by introducing particles and gases into the atmosphere, possibly blocking the longwave radiation from escaping into space. Anything that changes the radiative properties of the atmosphere (e.g., volcanic aerosols) can cause climate change.

Example: The first law of thermodynamics explains human metabolism. Human metabolism (e.g., oxygen consumption, carbon dioxide, waste production, glucose conversion to pyruvate, the breakdown of macromolecules to form high-energy compounds such as ATP and ADP) is the conversion of food (into heat, work, and fat). The human body is here the system. Its temperature is kept constant by its surroundings, thus Q is negative. Moreover, if you exercise, the body is doing work, thus, W is also negative. Internal energy is taken out of the body, thus, ΔU is negative. If you eat just the right amount of food, then your average internal energy remains constant $\Delta U = 0$; food puts back the losses. If you overeat repeatedly, then, ΔU is always positive, and your body stores this extra internal energy as flat. You can also end with ΔU being negative if you eat too little. If ΔU is negative for a few days, then the body metabolizes its own fat to maintain body temperature and do work that takes energy from the body. This process is how dieting produces weight loss. Note that, aside from heat and work, there is excretion and change in body weight.

Example: Enthalpy and energy changes in chemical reactions. As described in Ikelle (2023b), the bonds between atoms may break, combine, reform, or both, to either absorb or release energy during chemical reactions. So, the heat absorbed by or released from a system (i.e., the chemical reaction). Under constant pressure this heat is known as enthalpy, as described in (2.6), and the change in

enthalpy that results from a chemical reaction is the enthalpy of reaction, often denoted as ΔH_{rxn}. This enthalpy of reaction is actually the difference of the enthalpy
of the products and the enthalpy of the reactants; i.e.,

$$\Delta H_{\text{rxn}} \;=\; \sum_{i} n_i \Delta H^{(i)}_{\text{products}} - \sum_{k} n_k \Delta H^{(k)}_{\text{products}}$$

$$=\; \sum \text{energy required to brake bonds}$$

$$-\; \sum \text{energy released by bond formations},\qquad (2.9)$$

where n_i are the stoichiometric coefficients. It is the same whether the actual
reaction occurs in one step or in multiple intermediate steps—that is, the total
enthalpy change for a reaction is independent of the path by which the reaction
occurs. When $\Delta H_{\text{rxn}} < 0$, we say that the reaction is exothermic, $\Delta H_{\text{rxn}} > 0$ the
reaction is endothermic, and $\Delta H_{\text{rxn}} = 0$ the reaction is thermoneutral.

The quantification of the enthalpy of a reaction is not straightforward because it
cannot be measured directly. However, from equation (2.4), the change in enthalpy
can be expressed as follows:

$$\Delta H_{\text{rxn}} = m \times c_P \times \Delta T\ ,\qquad (2.10)$$

where m is the mass of the reactants, c_P is the specific heat of the product, and ΔT is
the change in temperature from the reaction. Therefore, we can indirectly measure
the enthalpy of reaction trough the measurements of m, c_P, and ΔT. Consider, for
example, the following reaction:

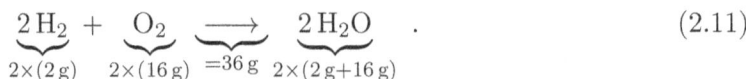

$$\underbrace{2\,H_2}_{2\times(2\,g)} + \underbrace{O_2}_{2\times(16\,g)} \xrightarrow{\;=36\,g\;} \underbrace{2\,H_2O}_{2\times(2\,g+16\,g)}\ .\qquad (2.11)$$

The mass is 36 grams and the product, which is water, has a specific heat of about
4.2 Joule/(gramK). Let us say that our reaction was 185 K at its very start but
had cooled to 95 K. By the time the reaction is finished $\Delta T = 95\,\text{K} - 185\,\text{K} = -90$
K. We can deduce that $\Delta H_{\text{rxn}} = (36\,\text{g}) \times (4.2\,\text{J K}^{-1}\,\text{g}^{-1}) \times (-90\,\text{K}) = -13,608\,\text{J}$.
This reaction is exothermic.

One can alternatively use known bond energies (e.g., the N–N bond has an
energy of about 160 kJ/mol, the N–F bond has an energy of about 272 kJ/mol, and
the N–Br bond has an energy of about 243 kJ/mol; see Table 2.7 of Ikelle (2023a)
for more examples) to estimate the enthalpy of the reaction. Consider for example,
the following reaction:

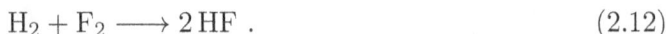

$$H_2 + F_2 \longrightarrow 2\,HF\ .\qquad (2.12)$$

The energy required to break the H atoms in the H_2 molecule apart is 436 kJ/mol,
while the energy required to break F atoms in F_2 is 158 kJ/mol. The energy needed
to form HF from H and F is -568 kJ/mol. We can add all these quantities based on
(2.9) to obtain $\Delta H_{\text{rxn}} = 436 + 158 - 2 \times 568 = -542$ kJ/mol.

Quiz: Calculate the enthalpy of reaction for the following reactions:

(a) $3\,Fe_2O_3(s) + 2\,NH_3(aq) \longrightarrow 6\,FeO(s) + 3\,H_2O(l) + N_2(aq)$

(b) $2\,N_2O_5 \longrightarrow 4\,NO_2 + O_2$.

Example: Ideal gas. Gases are full of billions of energetic gas molecules that can collide and possibly interact with each other. The concept of an *ideal gas* allows us to approximate the behavior of real gases. Ideal gas molecules do not attract or repel each other; they can only elastically collide with the walls of their container. Also, they are modeled as particles with no volume in a gas. There are no real gases that are exactly ideal, but there are plenty of gases that are close enough to an ideal gas, especially at temperatures near room temperature and pressures near atmospheric pressure.

We consider a particle with mass m and velocity $[v_x, v_y, v_z]^T$. It strikes once a unit area of the yz-wall of a cube volume L^3 per time interval $\Delta t = 2L/v_x$. After the reflection at the wall, the velocity of the particles is changed into $[-v_x, v_y, v_z]^T$. The change of the linear momentum gives a force on the plane normal to the x axis. We then have $\Delta p_x = p_{x,\text{final}} - p_{x,\text{initial}} = -2mv_x = F'_x \Delta t$. As a result of Newton's third law (the action-reaction; in Ikelle (2023b)), the particle exerts a force $F_x = -F'_x = p_x v_x/L$ on the wall, with $p_x = mv_x$. The pressure produced by N particles is then $P = F/A = N p_x v_x/(LA) = n p_x v_x$ with $n = N/(LA)$ or for an isotropic distribution (and isotropic motion of particles) with

$$\langle v_x^2 \rangle + \langle v_y^2 \rangle + \langle v_z^2 \rangle = v^2 \quad \text{or} \quad \langle p_x^2 \rangle + \langle p_y^2 \rangle + \langle p_z^2 \rangle = p^2 \ , \tag{2.13}$$

(i.e., $v_x = v/\sqrt{3}$, $p_x = p/\sqrt{3}$), $p = \frac{1}{3}nvp = \frac{1}{3}nmv^2$. We assume that the number of particles per unit volume that have velocity components of $v_x - (v_x + dv_x)$, $v_y - (v_y + dv_y)$, and $v_z - (v_z + dv_z)$ is $n_v(v)dv$ and the corresponding pressure is $(\frac{1}{3}mv^2 n(v)dv)$ so that the number of particles and the total pressure can be written as

$$N = V \int_0^\infty n_v(v)dv$$

or

$$P = \frac{1}{3}\int_0^\infty mv^2 n_v(v)dv = \frac{1}{3}\int_0^\infty n_p(p)vpdp \ , \tag{2.14}$$

where $n_v(v)$ and $n_p(p)$ describe the particle's velocity and momentum distributions, respectively. The last equation here is known as the pressure integral. Maxwellian velocity distribution of ideal gases can be written as follows (Dunbar, 1982):

$$n_v(v)dv = n\left(\frac{m}{2\pi k_B T}\right)^{3/2} \exp\left(\frac{-mv^2}{2k_B T}\right) 4\pi v^2 dv \ , \tag{2.15}$$

where $k_B = 1.381 \times 10^{-23}$ J/K is the Boltzmann constant. We can insert now $n_v(v)dv$ into the pressure integral (2.14) to arrive at the gas equation of state; i.e.,

$$P = \frac{1}{3}\int_0^\infty mv^2 n_v(v)dv = n\left(\frac{m}{2\pi k_B T}\right)^{3/2}$$

$$\int_0^\infty dx\; x^4 \exp\left(-\frac{m}{2k_B T}x^2\right) = nk_B T \;. \tag{2.16}$$

Here are a variety of equivalent forms of this ideal gas equation of state:

$$P = nk_B T = \frac{N}{V}k_B T = \frac{N}{N_{Av} V}RT = \rho\frac{k_B T}{\mu}$$

or

$$PV = nRT = Nk_B T = \rho\frac{RT}{M} \;, \tag{2.17}$$

where n is the number density of gas particles, V is the volume, $N = nV$ is the number of particles contained in a volume V, N/N_{Av} is the number of moles in the gas volume V, $R = k_B N_{Av} = 8.31$ Joule/(Kelvin × mole) is known as the universal gas constant, $N_{Av} \approx 6.02 \times 10^{23}$ is Avogadro's number (see Ikelle, 2023a), $\mu = \rho N_{Av}/n$ is the average molecular weight for the gas particles, ρ is the mass density, and M is the molecular mass. Alternatively, $R = 0.082$ (L × atom)/(Kelvin × mole) if the pressure is in atmospheres and volume is liters. Therefore, for pressure, P (in pascals), volume, V (in m^3), and temperature, T (in Kelvin), the ideal gas law equation is $PV = nRT$. If we want to use N number of molecules instead the number of moles, n, we can write the ideal gas law as,

$$PV = Nk_B T \;, \tag{2.18}$$

$k_B = 1.381 \times 10^{-23}$ J/K is the Boltzmann constant.

For the case of an ideal gas, the thermodynamic equation in (2.6) can be written as

$$dQ = mc_V dT + d(PV) - VdP$$

$$= mc_V dT + d(nRT) - VdP$$

$$= m(c_V + R_d)dT - VdP \;, \tag{2.19}$$

by using the fact that the mass of gas is moles × molecular mass ($m = n \times M$) and $R_d = R/M$. At constant pressure, the last term vanishes; therefore $c_P = c_V + R_d$. This relation is sometimes known as Mayer's relation. Then, $R_d = c_P - c_V = 287$ J/(kg K). For an ideal monatomic gas (e.g., Ar, O, and H),

$$c_V = \frac{3}{2}R_d \quad\text{and}\quad c_P = \frac{5}{2}R_d \;, \tag{2.20}$$

and for diatomic gases at moderate temperature, $50 < T < 600$ K,

$$c_V = \frac{5}{2}R_d \text{ and } c_P = \frac{7}{2}R_d . \tag{2.21}$$

For low temperatures, the specific heat is well modeled as a constant; here the internal energy change is strictly proportional to the temperature change. For moderate to high temperatures, a temperature-variation of the specific heat is no longer constant, hence changes in internal energy are no longer strictly proportional to changes in temperature.

The ideal-gas equation can be adjusted to take these deviations from ideal behavior into account. The corrected ideal-gas equation is known as the van der Waals equation;

$$\left(P + \frac{n^2 a}{V^2}\right)(V - nb) = nRT , \tag{2.22}$$

where a and b are empirical constants that differ for each gas; for example, $a = 0.0341$ (L$^2 \times$ atm)/mol^2 and $b = 0.0237$ L/mol for helium, $a = 0.211$ (L$^2 \times$ atm)/mol^2 and $b = 0.0171$ L/mol for neon, $a = 1.34$ (L$^2 \times$ atm)/mol^2 and $b = 0.0322$ L/mol for argon, $a = 0.244$ (L$^2 \times$ atm)/mol^2 and $b = 0.0266$ L/mol for hydrogen gas, $a = 1.39$ (L$^2 \times$ atm/mol^2) and $b = 0.0391$ L/mol for hydrogen gas, and $a = 1.36$ (L$^2 \times$ atm)/mol^2 and $b = 0.0381$ L/mol for oxygen gas.

2.1.2. Entropy, Equilibrium, and the Second Law of Thermodynamics

Suppose that we take a hot material from a high pressure environment (e.g., deep in the Earth's mantle) and bring it to a low pressure environment (e.g., the Earth's crust) without letting it exchange any heat with the surrounding medium. Because the pressure drops, the volume expands and its temperature decreases. Thus, we have changes in thermodynamics variables that allow us to describe this process— that is, changes in temperature, pressure, and volume. However, there is no variable that describes gain or loss of heat. That is a weakness of the first law of thermodynamics that the second law of thermodynamics corrected. The second law tells us that there is another useful thermodynamics variable called entropy, S, just like pressure, temperature, and volume, but entropy cannot be measured directly and the entropy of a substance is given with respect to some reference value. Also, the second law states that the change in entropy, dS, is equal to the heat transfer, dQ, divided by the temperature T (Clausius, 1865)

$$dS = \frac{dQ}{T} \implies \Delta S = S_B - S_A = \int_A^B \frac{dQ}{T} , \tag{2.23}$$

and

$$\frac{dS}{dt} > 0 \quad \text{(entropy always increases)}. \tag{2.24}$$

More precisely, dS is the change in entropy of a system during some process, dQ is the amount of heat transferred to or from the system during the process, and T is

Table 2.1. Some groups of systems and processes: A thermodynamic system is a collection of matter and space with its boundaries. A thermodynamic process is the succession of states that a system passes through. Surroundings are everything not in the system being studied.

Process/system	Definition
Reversible process	The change of entropy of the system
Irreversible process	If reversed, the entropy of the system and/or surroundings will increase.
Isentropic process	The entropy of the system remains unchanged
Adiabatic process	There is no heat transfer across the system boundaries
Isobaric process	The pressure, P, of the system is constant
Isochoric process	The volume, V, of the system is constant
Isotherm process	The temperature, T, of the system is constant
Isolated system	Neither mass nor energy can cross the boundaries
Closed system	Only energy can cross the boundaries
Open system	Both mass and energy can cross the boundaries

the absolute temperature at which the heat was transferred. The unit of entropy is J/K (Joule per Kelvin). For a given physical process, if the combined entropy of the system and the environment remains a constant, the process can be reversed—that is, the change in entropy of the system is zero. Engineers call such a process an *isentropic* process, as isentropic means constant entropy. If the physical process is irreversible[1] (i.e., occurs in one direction only), the combined entropy of the system and the environment must increase—the final entropy must be greater than the initial entropy for an irreversible process. And an *adiabatic* process is one in which there is no heat transfer into or out of the system. The adiabatic system can be considered to be perfectly insulated (see also Table 2.1).

As we will in the next sections, and actually in the entire book, most processes occur spontaneously in one direction only—that is, they are irreversible. For example, a broken glass does not recover its original state; and heat involves the transfer of energy from higher to lower temperature objects but the reverse is impossible. Therefore, an irreversible process can go in only one direction. The reverse path differs fundamentally when it possible. In other words, the second law of thermodynamics forbids these processes and that is why it is stated in many ways that may seem different, but that in fact are equivalent. Other ways to express the second law include the following:

- The entropy of a physical system left to itself will increase or, if the system is already at its maximum entropy, the entropy will remain the same.

- Any system, when left to itself, tends toward equilibrium with its surroundings. For example, hot tea cools and ice cream melts, at room temperature.

---

[1]A reversible process is a process in which, both the system and the surroundings return to their initial states at the end of the reverse process. A reversible process is possible only if the net heat and net work exchanged between the system and the surroundings are zero for the combined (original and reverse) process.

- The entropy of a system that is in equilibrium with its surroundings remains constant. In other words, an object in equilibrium with its surroundings has the temperature of its surroundings.

An important engineering implication of the second law is that it limits the efficiency of heat engines (e.g., power plants and internal combustion engines of automobiles), as we will discuss later. Thus, the second law leads to the definition of entropy. But what exactly is this new fascinating word *entropy* along with the familiar concepts of heat, work, energy, and temperature? And why does it always increase?

Entropy is basically a macroscopic (an effective) description of the microscopic (molecular or atomic) state of a system. It recognizes that matter is made of a huge number of particles and captures in a macroscopic scale the state of these particles (a kind of average statistic value of loss of heat). On that basis, it is similar to electromagnetic and elastic parameters (Ikelle, 2023a and 2023b), which are also averages that capture microscopic aspects of heterogeneous media. These parameters are allowed to increase and decrease depending of the medium under consideration. In contrast, the entropy cannot decrease in most realistic cases,

---

Box 2.1: Statistical Physics Distribution Functions

In the description of an ideal gas, we have introduced a distribution function n_p (or n_v) that gives us the number density of particles in the gas with momentum p such that $n_p dp$ is the number of particles with momentum between p and $p + dp$. This distribution function is known as the Maxwell-Boltzmann distribution. It is a particular probability distribution named after James Clerk Maxwell and Ludwig Boltzmann. In momentum terms, the Maxwell-Boltzmann distribution function for an ideal gas is

$$n_p = 4\pi n \left(\frac{m}{2\pi k_B T} \right)^{3/2} \frac{p^2}{m^2} \exp \left(\frac{-p^2}{2mk_B T} \right) = nf(p) , \qquad (2.25)$$

with

$$f(p) = 4\pi \left(\frac{m}{2\pi k_B T} \right)^{3/2} \frac{p^2}{m^2} \exp \left(\frac{-p^2}{2mk_B T} \right) , \qquad (2.26)$$

In the description of an ideal gas, we have introduced a distribution function n_p (or n_v) that gives us the number density of particles in the gas with momentum p such that $n_p dp$ is the number of particles with momentum between p and $p + dp$. This distribution function is known as the Maxwell-Boltzmann distribution. It is a particular probability distribution named after James Clerk Maxwell and Ludwig Boltzmann. In momentum terms, the Maxwell-Boltzmann distribution function for an ideal gas is

$$n_p = 4\pi n \left(\frac{m}{2\pi k_B T} \right)^{3/2} \frac{p^2}{m^2} \exp \left(\frac{-p^2}{2mk_B T} \right) = nf(p) , \qquad (2.27)$$

with

$$f(p) = 4\pi \left(\frac{m}{2\pi k_B T} \right)^{3/2} \frac{p^2}{m^2} \exp \left(\frac{-p^2}{2mk_B T} \right) , \qquad (2.28)$$

where n is the total number of particles per volume, n_p is the density of particles with momentum p, m is the mass of a gas molecule, k_B is Boltzmann's constant, T is the absolute temperature, and $f(p)$ is the fundamental expression that describes the distribution of momentum in n gas molecules. This distribution function describes particle momentum in idealized gases, where the particles move freely inside a stationary container without interacting with one another, except for very brief collisions in which they exchange energy and momentum with each other or with their thermal environment. The most probable momentum (i.e., the momentum at which the distribution curve reaches a peak—that is, $df(p)/dp = 0$) is $p_m = \sqrt{2k_B mT}$ or the most probable velocity $v_m = \sqrt{(2k_B T)/m}$. Also the average speed increases with increasing temperature because this distribution has an asymmetric shape. The lowest possible speed is 0 and the highest is infinity. Physically, equation (2.27) describes the fact that the air molecules surrounding us are not all moving at the same speed, even when all molecules are at the same temperature. Some of the air molecules will be traveling extremely fast, some will be traveling at moderate speeds, and some of the air molecules will hardly be moving. So the speed of an air molecule in a gas does not make any sense. There is also a second statistical physics distribution known as *Planck distribution*, which we will introduce later. The Planck distribution is the number density of photons within a given frequency range.

because the averages of ratios dQ/T are always increasing, despite the fact that some of them can be negative, whereas the averages of electromagnetic and elastic parameters of heterogeneous media can be positive as well as negative.

In 1877, Boltzmann (Schrödinger, 1952; Ben-Naim, 2007) stated that every spontaneous change in nature tends to occur in the direction of higher disorder, and entropy is the measure of that disorder or randomness. In other words, ever-increasing entropy can be interpreted as nature's way of proceeding from order to disorder. For example, current rapid glacier melting (we are losing about 30 percent more snow and ice per year than we did 15 years earlier) is accompanied by a large increase in entropy because a block of ice is a highly structured and orderly system of water molecules and melting converts it into a disorderly liquid in which molecules have no fixed positions (see Ikelle 2023a).

The use of the intangible concepts of order and disorder in this definition has meant that some have associated entropy and the second law of thermodynamics with our own lives, far away from in terms of heat and temperature. The fact that things inevitably become less organized in our lives and we gradually age and die is viewed as a statement of the second law of thermodynamics (Schrödinger, 1945; and Lindley, 2001). Obviously, one must consider this statement for large ensembles rather than one person or one family. The second law is also viewed as the fundamental law of our universe. Basically, our universe is at a higher-entropy state today than, say, 4.5 billion years ago, despite the fact that things look well organized and all the planets are almost spherical and ordered with solar systems, galaxies, and intricate cosmic structure. The ever-increasing entropy state of our universe is primarily attributed to the increasing formation of black holes in our universe and the fact that our universe is filled with newly generated photons. Note that the interpretation of the ever-increasing entropy as the measure of disorder or randomness holds quite well experimentally despite some "exotic" counter examples. Bridgman (1941) showed that the crystallization of some super-cooled melt (blinded fluids that remain fluids rather than becoming solid below their freezing points) can increase their entropy. McGlashan (1966) showed that the crystallization of some supersaturated solutions (solutions that have a tendency to crystallize and preciptate) can also increase their entropy. As a result, the change in entropy of our universe must be greater than zero for an irreversible process and equal to zero for a reversible process. Maybe, at some point, the entropy of our universe will reach a maximum value (which is known as the heat death of our universe)—that is, our universe will be in a state of uniform temperature and density. All physical, chemical, and biological processes will cease as the state of complete disorder implies that no energy is available for doing work because, as entropy increases, less and less energy in our universe is available to do work. Also, at the beginning of our universe, entropy must have been very small or near zero if the estimations that entropy can continue for at least 10^{100} years are correct. Near zero entropy implies that matter was almost homogeneous at the beginning of our universe. Notice as well that human brain-imaging studies have found a rapid rise in entropy production from early life, throughout childhood, and into early adulthood is also seen, followed by a slow production in entropy production thereafter (Aoki, 1991; and Shyu et al., 2011).

On Earth, we still have large stores of chemical, gravitational, nuclear, solar, and wind energies. One day, all fuels producing these energies will be exhausted, all temperatures will equalize, and hence, heat engines will no longer function.

Equation (2.23) allows us to write the first law for reversible processes solely in terms of state variables as

$$dU \;=\; TdS - PdV = \left(\frac{\partial U}{\partial S}\right)_V dS - \left(\frac{\partial U}{\partial V}\right)_S dV, \qquad (2.29)$$

from which we can see that

$$\left(\frac{\partial U}{\partial S}\right)_V = T \ \text{ and } \ \left(\frac{\partial U}{\partial V}\right)_S = -P. \qquad (2.30)$$

Equation (2.29) is usually referred to as the combined first- and second-law equation, and serves as the starting point for discussing many heat-related problems. The partial derivatives of the internal energy in (2.30) can be utilized to find one of Maxwell's thermodynamic relations; i.e.,

$$\left(\frac{\partial}{\partial V}\right)_S \left(\frac{\partial U}{\partial S}\right)_V = \left(\frac{\partial}{\partial S}\right)_V \left(\frac{\partial U}{\partial V}\right)_S$$

or

$$\left(\frac{\partial T}{\partial V}\right)_S = -\left(\frac{\partial P}{\partial S}\right)_V. \qquad (2.31)$$

We used the fact that the second partials of exact differentials must be the same, no matter what the order of differentiation. The other three Maxwell's thermodynamic relations are (VanWylen and Sonntag, 1993):

$$\left(\frac{\partial V}{\partial S}\right)_P = \left(\frac{\partial T}{\partial P}\right)_S \;, \; \left(\frac{\partial S}{\partial V}\right)_T = \left(\frac{\partial P}{\partial T}\right)_V \;,$$

$$\text{and} \;\; \left(\frac{\partial S}{\partial P}\right)_T = -\left(\frac{\partial V}{\partial T}\right)_P \;. \tag{2.32}$$

They are obtained by replacing changes in the internal energy, $dU = TdS - PdV$, in (2.30), with changes in the enthalpy, $dH = TdS + VdP$, changes in the Helmholtz free energy, $dF = -SdT - PdV$, and changes in the Gibbs free energy, $dG = -SdT + VdP$, respectively. So we have four thermodynamics variables: T, P, V, and S. They are related to the four Maxwell relations in (2.31) and (2.32). Notice that by using the combined first- and second-law equation in (2.29), we can alternatively write the specific volume as

$$c_V = \frac{1}{m}\left(\frac{dU}{dT}\right)_V = \frac{1}{m}T\left(\frac{dS}{dT}\right)_V \;\; \text{and}$$

$$c_P = \frac{1}{m}\left(\frac{dH}{dT}\right)_P = \frac{1}{m}T\left(\frac{dS}{dT}\right)_P \;. \tag{2.33}$$

Example: Derivation of the speed of sound in an ideal gas. Our goal in this example is to determine the velocity of a sound wave that is excited in an ideal gas, say air, by a large membrane like a drum, a speaker, or an explosion. The generated waves in the air are just like compressional waves in a solid or a fluid (liquid or gas) described in Ikelle (2023a). The air will compress and then expand as the wave passes by. Imagine an elementary tube containing air. The velocity of the sound entering the tube is c_s and the sound leaves with velocity $c_s + dc_s$. The density of the gas on the entry face of the tube is ρ and on the other side it is $\rho + d\rho$. Applying the principle of conservation of mass and ignoring the product of two infinitesimals (i.e., $d\rho dc_s \approx 0$), we have

$$(\rho + d\rho)A(c_s + dc_s) - \rho A c_s = 0 \iff c_s d\rho = -\rho dc_s \;, \tag{2.34}$$

where A is the area of the tube faces. Assuming that the tube can be approximated by a rectangular parallelepiped, the front face experiences a force of $(P + dP)dydz$, where $P + dP$ is the pressure experienced by the front face along the x-axis, and the back face experiences a front pressure of P, hence a force of $Pdydz$. The net force on the tube is $-dPdydz$ and it is subjected to an acceleration of

$$\frac{dc_s}{dt} = \frac{\text{force}}{\text{mass}} = \frac{-dPdydz}{\rho dxdydz} = \frac{-dP}{\rho dx}$$

or

$$dc_s = \frac{-dP}{\rho dx} \times dt = \frac{-dP}{\rho c_s} \;. \tag{2.35}$$

We used the fact that the total time it takes sound to cross the elementary tube is $dt = dx/c_s$. Furthermore, by substituting mass conservation in (2.34) into (2.35), we arrive at

$$\rho c_s dc_s = -dP \Longrightarrow c_s^2 d\rho = dP \Longleftrightarrow c_s = \left(\frac{dP}{d\rho}\right)^{1/2} . \tag{2.36}$$

The sound enters and leaves the tube in the same state, without increasing the entropy; i.e., $PV^\gamma = \text{constant}$ (where $\gamma = c_P/c_V$) which implies that

$$\ln P + \gamma \ln V = \text{constant} \Longrightarrow \ln P - \gamma \ln \rho = \text{constant}$$

or

$$\frac{dP}{d\rho} = \gamma \frac{P}{\rho} = \gamma \frac{RT}{M} , \tag{2.37}$$

where M is the molecular mass of the gas, as defined earlier. Combining (2.36) and (2.37), we obtain the velocity of sound passing through air as

$$c_s = \left(\gamma \frac{RT}{M}\right)^{1/2} . \tag{2.38}$$

Note that the hotter the gas, the faster the sound is. Also, the heavier the gas, the slower the sound is. For air, $\gamma = 1.4$, $M = 0.029$ Kg/mole, $R = 8.314$ Joule/(Kelvin × mole), and $T = 300$ K. We then obtain

$$c_s = \left(\frac{1.4 \times 8.314 \times 300}{29}\right)^{1/2} = 347 \text{ m/s} . \tag{2.39}$$

Quiz: (1) The amount of heat needed to change one gram of ice at at 0 degree Celsius to one gram of liquid water at 0 degree Celsius, 3.35×10^7 Joules. What is the change in entropy when 100 grams of ice at 0 degree Celsius melt into 100 grams of liquid water at 0 degree Celsius?
(2) Suppose that two reservoirs are brought into direct contact. An amount of heat, Q, is transferred from the high temperature reservoir (heat source) to the low temperature reservoir (heat sink). Determine the entropy change and show that heat transfer is irreversible.
(3) No heat is supplied in adiabatic expansion, but work is done. Show that, for an ideal gas, this work is $W = c_V(T_B - T_A)$, where T_A is the temperature at the beginning of the adiabatic process and T_B is the temperature at the end of the adiabatic process.

2.1.3. Entropy: Public Policies

The use of entropy is not limited to thermodynamics. It extends to public policies and the field of economics. However, the key parameter in these cases is no longer the temperature. It is now the statistical distribution, including random variables.

In other words, in public policies and the economy, entropy is essentially viewed as a statistical measurement. Here, we examine entropy from this perspective.

Examples of random variables. The building block of statistics is random variables. If the outcome of an experiment depends upon a chance mechanism, then the proper description of the experiment is stated in terms of the different possible outcomes and the possibility of each outcome. Suppose the experiment consists of tossing a coin that has sides marked -1 and 1. The outcome of the experiment is called a random variable, X, which may take on either of these values with a probability of $1/2$. The probabilities are assigned by the experimenter, based upon the state of her/his knowledge. Here we consider that the coin does not contain any deformation that can affect the outcome of the experiment; in other words, we believe the coin to be fair. The possible values of X form a discrete set of values, $x_1 = -1$ and $x_2 = 1$, and X is therefore called a discrete random variable. A discrete random variable is completely described by a listing of its possible values $\{x_1, x_2, ...\}$ and the associated probabilities $\{p(x_1), p(x_2), ...\}$. Each of the probabilities must lie in a range between $0 \to 1$, and the sum of all probabilities must equal 1. That is,

$$0 \le p(x_i) \le 1 \quad \text{for} \quad i = 1, 2, 3, ... \quad \text{with} \quad \sum_i p(x_i) = 1 . \qquad (2.40)$$

As a second example of a random variable, let us consider a roulette wheel that has a uniform scale, $0 \to a$, marked on the circumference. The experiment consists of spinning the wheel and observing the value under the pointer when it comes to rest. The outcome is a random variable that may have any value between $0 \to a$ with equal probability. It is impossible to list all the possible values of X and their probabilities. In this case X is called a continuous random variable. It is completely described by the probability-density function (PDF) of the random variable, which is written $p(x)$. It is customary to use capital letters X, Y, etc., for random variables and corresponding lowercase letters x, y, etc., for the range of values that the random variables assume. Probability-density functions are non-negative and are normalized so that the area under the curve equals unity:

$$p(x) \ge 0 \quad \text{for all} \quad x \quad \text{with} \quad \int_{-\infty}^{+\infty} p(x)dx = 1 . \qquad (2.41)$$

The probability that X takes on a value in the interval $x \le X \le x + dx$ is defined by the incremental area $p(x)dx$. As we can see in these two examples, random variables are typically characterized by their probability-density functions (PDFs). We gain insight of a PDF of a random variable X through its histogram, which are graphic representations of a set of counts of each possible outcome of the random variable. When there are large number of outcomes, PFD of a random variable can be accurately represented the histogram of the random variable. A random variable with a bell-shaped-curve histogram is known as a random variable with the Gaussian PDF also referred as a Gaussian random variable. A random variable with any other shape of PDF is considered a nonGaussian random variable.

Gaussianity and non-Gaussianity are two of the most important characteristics of random variables. A random variable—say, X—is Gaussian if it has a Gaussian

PDF (also called a normal density function), which is defined as follows:

$$p_X(x) = \frac{1}{\sqrt{2\pi\sigma^2}} \exp\left[-\frac{(x-\eta)^2}{2\sigma^2}\right] , \tag{2.42}$$

where σ is the standard deviation, which is a measure of the variability of x and η is the mean of X. This function has a bell-shaped curve, and it is symmetrical about $x = \eta$. The first term $(1/\sqrt{2\pi\sigma^2})$ is a normalization constant which ensures that the area under the normal PDF sums to unity. A random variable is non-Gaussian (non-normal) if its PDF is non-Gaussian. Laplace and uniform PDFs are defined, respectively, as follows:

$$p_X(x) = \frac{\lambda}{2} \exp\left(-\lambda|x|\right) , \ \lambda > 0 , \text{ and } p_X(x) = \begin{cases} 1/2c & -c \leq x \leq c \\ 0 & \text{elsewhere} \end{cases}$$

are two examples of non-Gaussian PDFs. The Gaussian, Laplace, and uniform PDFs can be described as particular cases of the so-called generalized Gaussian PDF, which is defined as follows:

$$p_X(x) = \frac{\alpha}{2\beta\Gamma(1/\alpha)} \exp\left[-\left(\frac{|x|}{\beta}\right)^\alpha\right] , \tag{2.43}$$

where $\Gamma(.)$ denotes the Gamma function [i.e., $\Gamma(x) = \int_0^\infty t^{x-1} \exp(-t)dt$], $\alpha > 0$ describes the shape of the distribution, and $\beta > 0$ describes its scale.

One can easily verify that (2.43) reduces to the Gaussian shape if $\alpha = 2$ and $\beta = \sqrt{2}$, to the Laplacian shape if $\alpha = 1$ and $\beta = 1/\sqrt{2}$, and to the uniform PDF shape when $\alpha \to \infty$. Figure 2.5 shows the generalized Gaussian distribution for $\alpha = 0.5, 1, 2,$ and 6. Notice that these curves go from a peaky shape for values of α smaller than 2 (super-Gaussian) and progressively move toward a uniform distribution as α increases (sub-Gaussian). In other words, the generalized Gaussian PDF is a useful representation of almost the entire spectrum of symmetric PDFs. In term of public policies, distributions with α greater than 2 trend toward equitable use of public public wealth and those with α smaller than 2 trend toward looting of public wealth (the so-called *biens mal acquis*) like the worldwide looting associated with Covid-19 that took place in 2020, 2021, and even 2022.

Entropy. Let us start by defining entropy H of a random variable X with probability density function $p_X(x)$ (Shannon, 1948); i.e.,

$$H[X] = -\int p_X(x) \log\left[p_X(x)\right] dx = E\left[\log\left(\frac{1}{p_X(x)}\right)\right] , \tag{2.44}$$

where the symbol $E[.]$ stands for expectation. We have introduced square brackets in our notation of entropy to emphasize that entropy is a statistical quantity. Notice that entropy depends on the probability density function (PDF) of X instead of X itself. Depending on what the base of the logarithm is, different units of entropy are obtained. Usually the logarithm with base 2 is used, in which case the unit is called a bit.

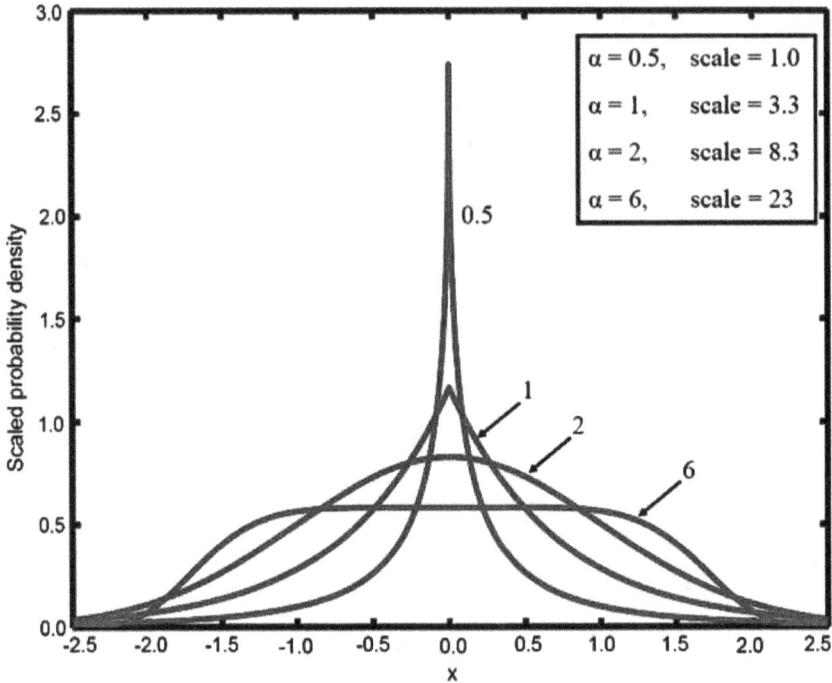

Figure 2.5. Generalized Gaussian distribution for $\alpha = 0.5$, 1 (Laplace), 2 (Gaussian), and 6. For the scale parameter β, we have used the following expression: $\beta^2 = \Gamma\left(\frac{1}{\alpha}\right)/\Gamma\left(\frac{3}{\alpha}\right)$. Notice that we have scaled up the curves corresponding to $\alpha = 1$, 2, and 6 to facilitate their display. The scaling factors are shown in the top right corner of this picture.

The fundamental question at this point is: How to interpret entropy? To answer this question, let us return to the example of coin tossing that we discussed earlier. Our discussion of this example was carried out under the assumption that the equiprobability of the outcomes; i.e., the probability of observing the head is 0.5 as well as that of observing the tail. Suppose that the coin has been modified such that the coin-tossing experiment becomes biased. Such modifications can lead to the production of more heads than tails or to a deterministic experiment in which the outcomes are always heads, for instance. These possible modifications of the coin show the need for an additional statistical quantity which characterizes the degree of randomness of the experiment itself. The entropy provides this characterization. If the experiment is really as unpredictable as coin tossing, with two equiprobable outcomes, the entropy $H[X]$ has its maximal value. This maximal value is 1 in the case of the coin-tossing experiment. If the experiment is totally biased toward either heads or tails, the entropy is zero.

So the entropy is a function of the probability-density functions, p_X, for the coin-tossing experiments:

$$H[X] = -\left[p_X \log p_X + (1 - p_X) \log(1 - p_X)\right] . \tag{2.45}$$

In this case, the PDF can be represented by a single value. Figure 2.6 shows this entropy as a function of p_X. Maximum entropy is achieved for equiprobable outcomes; minimum entropy is achieved for deterministic outcomes $p_X = 0$ and

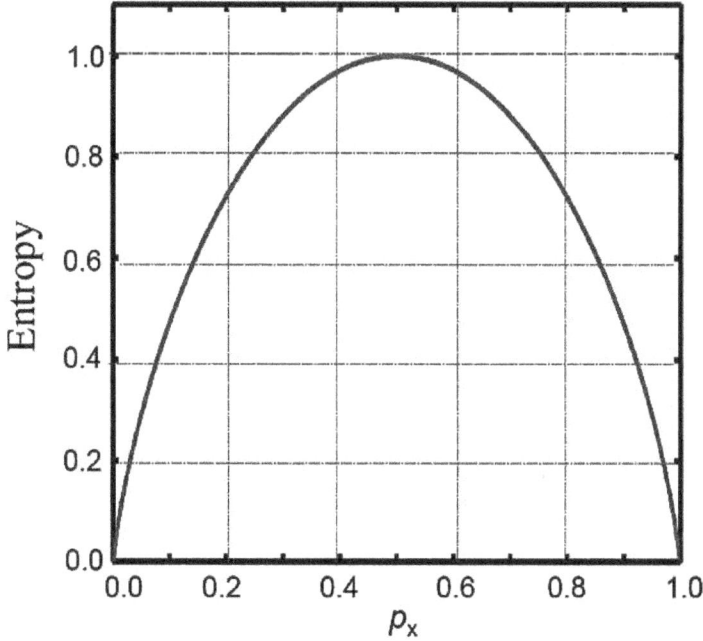

Figure 2.6. Entropy as a function of the probability density function (p_X) in the coin-tossing experiment. Maximum entropy is achieved for equiprobality, and minimum entropy is achieved for $p_X = 0$ and $p_X = 1$. Entropy as a function of the probability density function (p_X) in the coin-tossing experiment. Maximum entropy is achieved for equiprobality, and minimum entropy is achieved for $p_X = 0$ and $p_X = 1$.

$p_X = 1$. Entropy is always positive; i.e., $H[X] \geq 0$.

Let us also look at the entropy of the random variables for the generalized Gaussian PDF in (2.43) under the normalization condition that the random variable has unit variance. A simple calculation shows that the differential entropy

$$
\begin{aligned}
H[X] &= -\int_{-\infty}^{+\infty} p_X(x) \log\left[p_X(x)\right] dx \\
&= -\left\{ \frac{\alpha}{2\beta\Gamma(1/\alpha)} \log\left[\frac{2\beta\Gamma(1/\alpha)}{\alpha}\right] \int_{-\infty}^{+\infty} dx \exp\left[-\left(\frac{|x|}{\beta}\right)^{\alpha}\right] \right\} \\
&\quad + \left\{ \frac{\alpha}{2\beta\Gamma(1/\alpha)} \int_{-\infty}^{+\infty} dx \left(\frac{|x|}{\beta}\right)^{\alpha} \exp\left[-\left(\frac{|x|}{\beta}\right)^{\alpha}\right] \right\} .
\end{aligned}
\tag{2.46}
$$

For the particular case in which

$$
\beta = \left[\frac{\Gamma\left(\frac{1}{\alpha}\right)}{\Gamma\left(\frac{3}{\alpha}\right)}\right]^{1/2} ,
\tag{2.47}
$$

we arrive at

$$
H[X] = \frac{1}{\alpha} - \log\left[\frac{\alpha}{2\beta\Gamma(1/\alpha)}\right] .
\tag{2.48}
$$

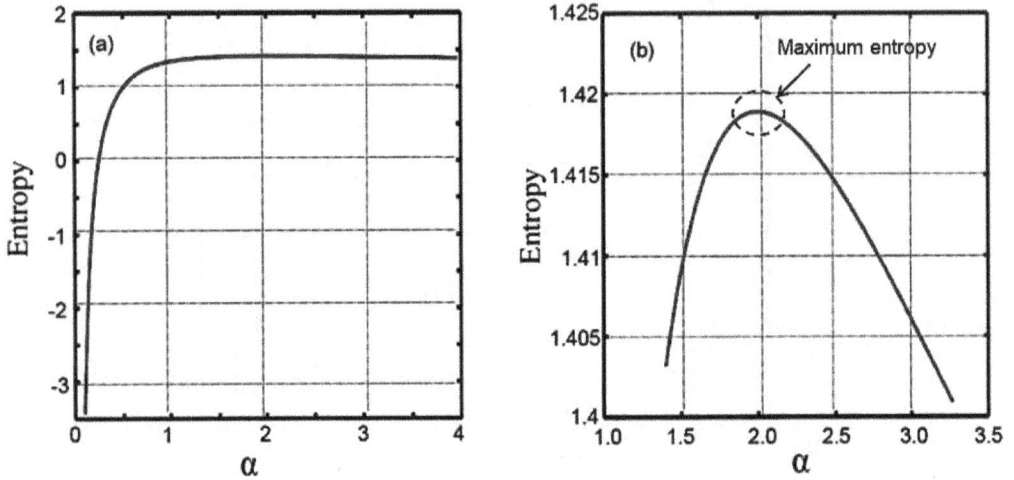

Figure 2.7. (a) Entropy of the generalized Gaussian PDFs as a function of α, which is a positive real number and describes the sharpness of the distribution. (b) We zoomed the plot of entropy versus α around $\alpha = 2$ to point out that the maximum entropy is achieved for $\alpha = 2$.

Figure 2.7 captures this description of entropy as a function of α. Notice that the Gaussian PDF (i.e., $\alpha = 2$) has the largest entropy. This result shows that the Gaussian distribution is the "most random" or the least structured of all symmetric distributions. We can also notice that the entropy is as small for "spiky" PDFs.

2.1.4.   Heat-Transfer Mechanisms

Heat can move in four ways: conduction, convection, latent heat, and radiation (Figure 2.8). Heat energy is most intense in substances whose molecules are moving rapidly in a very disorderly way. Such substances will give up some of their heat to another substance whose molecules are less agitated. When this happens, heat is said to "flow" from one substance to another (or from one body to another). The energy transfer is indicated by a change in temperature. This transfer of heat is known as conduction, which requires the presence of matter and therefore can occur only in solids, liquids, and gases (see Figures 2.1 and 2.8). So conduction is a good way of moving heat around when dealing with objects (solids, liquids, and gases) in contact with each other. But it is not useful in a vacuum (regions where there is no matter), such as space. In other words, conduction is not important for heat exchange between the sun and the Earth (or for climate sciences) because there is nothing to conduct heat between Earth and space.

Heat can be transferred from hot places to cold places by convection. Convection occurs when warmer areas of a liquid or gas rise to cooler areas in a liquid or gaseous form. The cooler liquid or gas then takes the place of the warmer areas, which have risen. This exchange results in a continuous circulation pattern known as convection. Water boiling in a pan is a good example of a convection current (see Figure 3.33 of Ikelle, 2023a and Figure 2.8). Convection is important for describing the motion of the atmosphere and the ocean on both local and global scales.

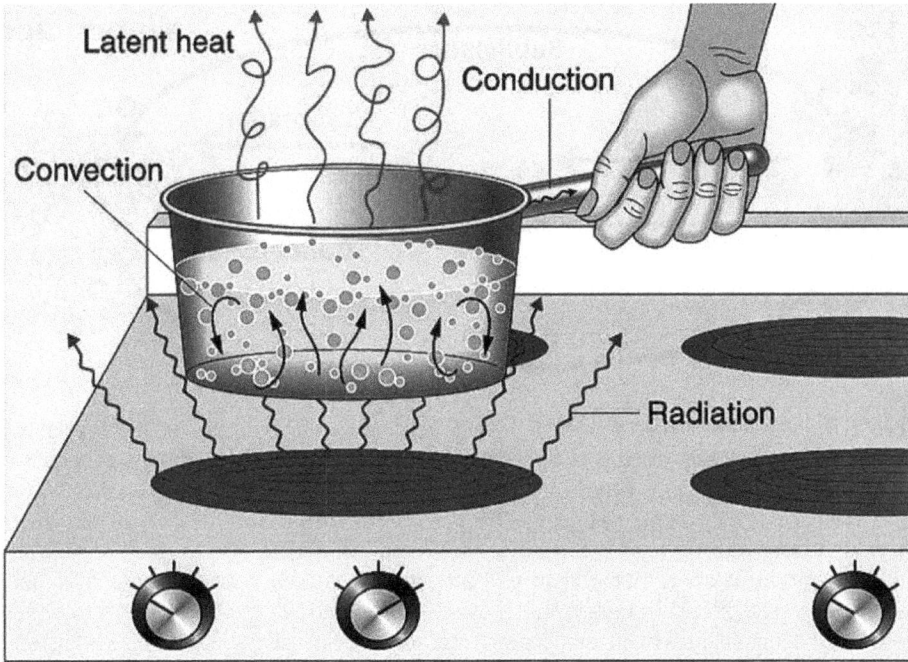

Figure 2.8. Illustrations of the four heat transfer mechanisms. The pan is held above the burner and is not in direct contact with it. Yet energy travels from the burner to the pan via the process of radiation. The pan absorbs much of this radiation and becomes hot. In turn, the water nearest the bottom of the pan takes on more heat energy than the water above. The hotter water at the bottom rises and water not as hot at the top descends, setting up vertical circulation currents known as convection. The convection currents bring new but cooler water to the bottom of the pan to absorb heat energy from the pan's bottom. Once the water reaches the boiling point, some of the liquid water will change to gas bubbles (latent heat). Heat in the lower part of the pan migrates, molecule to molecule, up the side of the pan to its handle, and then out along the handle, all of this by the process of conduction.

As a substance changes phase, such as from a solid to a liquid or a liquid to a gas, it requires latent heat energy. For example, heat is needed to transform water into vapor. This way of moving heat is used in the climate sciences in relationship with the water cycle between the oceans and the atmosphere. In more general terms, the internal energy associated with the phase of a system is latent energy or latent heat (see Figures 2.8 and 2.9). It is associated with the intermolecular forces (forces that bind molecules to each other) between the molecules of a system. As described in Ikelle (2023a), these forces are strongest in solids and weakest in gases.

Radiation, which is important and even central to climate-science studies, is a method of heat transfer that does not rely on any contact between the heat source and the heated object, as is the case with conduction and convection (see Figure 2.4 and 2.8). Heat can be transmitted though empty space by thermal radiation, often called infrared radiation. Radiation-absorption is accomplished directly through electromagnetic interactions. The process involves the motion of atoms and molecules at the beginning (radiation) and the end (absorption) of the

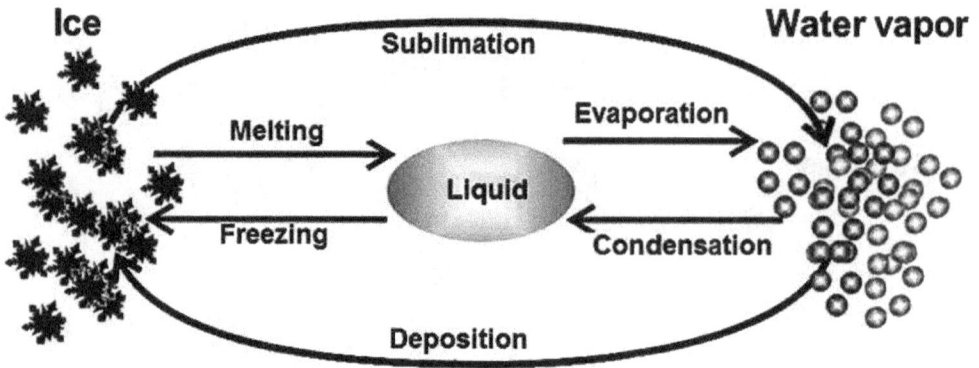

Figure 2.9. Another illustrations of latent heat (i.e., the amount of energy that is either absorbed or released during a phase transition) through components of water cycle, described in Ikelle (2023a). Freezing (from liquid to solid) and condensation (from gas to liquid) release latent heat. Melting (from solid to liquid) and evaporation (from liquid to gas) absorb latent heat. Sublimation (from solid directly to gas) absorbs latent heat as The potential energy stored in the interatomic forces between molecules needs to be overcome by the kinetic energy the motion of the particles before the substance can change phase. Deposition (from gas directly to solid) released latent heat. (adapted from https://arenahanna.wordpress.com/).

process, but not during the actual transfer through space. Basically, the thermal energy present in the vibrations of the molecules in the energy source is transferred to the thermal energy in the vibrations of molecules in the energy receiver. Electrical forces are responsible for the entire process. The energy travels through space in the form of electromagnetic waves. Examples of radiation include the heat from the sun. Energy travels through empty space from the sun to the Earth and the other planets in the solar system. Radiation also occurs between the Earth's surface and the atmosphere, and within the atmosphere and the ocean.

Radiation is the mechanism that allows us to probe a significant portion of our universe. It allows us to describe the light emitting astronomical objects in our universe. The most distant astronomical objects humans have traveled to and brought back samples, so far, are our Moon and a single asteroid. As described in Ikelle (2023a), meteorites on earth and unmanned missions provide insight into our solar system planets asteroids, and comets. Otherwise, most knowledge of our universe that we have gained in the past centuries is primarily based on studies of radiation. As a result, understanding the properties of radiation is a fundamental part of modern science and technology.

In summary, all four forms (or mechanisms) of heat transfer move heat from one body to another or from one part of a system (such as the climate system) to another—heat always moves from a place where the temperature is relatively high to a place where it is relatively low. Latent describes heat transfer between phases (solid, liquid, and vapor); radiation describes heat in empty space (e.g., sunlight that heats up the Earth and infra-red radiation allowing us to see in the dark); conduction describes heat transfer within materials; and convection describes heat transfer within fluids. Conduction, convection, latent heat, and radiation can be written as follows:

$$\begin{cases} q = -k\nabla T & \text{conduction} \\ q = h\left(T - T_\infty\right) & \text{convection} \\ q = \epsilon\sigma\left(T^4 - T_0^4\right) & \text{radiation} \\ q = L_f\, m & \text{latent} \end{cases} , \qquad (2.49)$$

where k is the thermal conductivity, T is the temperature of the object, T_∞ is the temperature of the environment, h is the convection-heat-transfer coefficient, T_0 is the temperature of surroundings, ϵ is emissivity (it varies between 0 and 1), L_f is the latent heat of fusion, and σ is Stefan-Boltzmann constant. For heat conduction, the heat-flux equation has a vector form, stating that heat flux is a vector aligned with the temperature gradient field. For heat convection, the heat-flux equation has a scalar form.

2.1.5. The Heat-Conductive Equation

Convection is a very efficient means of transporting heat but can only operate if material can flow. The strong (rigid) solid materials cannot convect and instead they only conduct heat. In other words, only conduction is possible in the strong materials through the combination of vibrations of the molecules and the energy transport by free electrons. Conduction can take place in liquids and gases. In gases and liquids, conduction is due to the collisions and diffusion of molecules during their random motion.

The conduction is described by Fourier's law; i.e.,

$$\begin{aligned} q &= -k\nabla T = -k\left[\frac{\partial T}{\partial x}i_1 + \frac{\partial T}{\partial y}i_2 + \frac{\partial T}{\partial z}i_3\right] \\ &= -k\left[\frac{\partial T}{\partial x}, \frac{\partial T}{\partial y}, \frac{\partial T}{\partial z}\right]^T . \end{aligned} \qquad (2.50)$$

This law is named after Jean Fourier, who expressed it first in his heat transfer text in 1822. It is a relationship between the heat flux and the temperature gradient. If the temperature is constant in a given region of interest, there is no heat flow—of course! The negative sign ensures that heat transfer in the positive direction is a positive quantity.

The rate of heat conduction through a medium depends on the geometry of the medium, its thickness, and the material of the medium, as well as the temperature difference across the medium. We know that wrapping a hot water tank with glass wool (an insulating material) reduces the rate of heat loss from the tank, and the thicker the insulation, the smaller the heat loss. We also know that a hot water tank will lose heat at a higher rate when the temperature of the room housing the tank is lowered, further, the larger the tank, the larger the surface area and thus, the rate of heat loss.

Table 2.2. The thermal conductivity of some common materials: Conduction in solids and liquids works by transferring energy through bonds between atoms or molecules. Metals have generally high thermal conductivity (conductors). Styrofoam, wool fabric, and fiberglass have very low thermal conductivity (insulators).

Matter	k [W/(m K)]	Matter	k [W/(m K)]
Basalt	2.09	Diamond	1000
Average Crust	2.51	Silver	406
Ice	1.6	Copper	385
Heart	0.59	Gold	314
Liver	0.57	Aluminum	205
Kidney	0.54	Iron	73.2
water at 273 K	0.57	Lead	35
water at 647 K	0.24	Mercury	8.3
Skin	0.5	Salt	5.4 - 7.2
Muscle	0.38 - 0.54	Sandstone	1.5 - 4.2
Brain	0.16 - 0.57	Granite	2.4 - 3.8
Fat	0.19 - 0.20	Periodite	3.35
fiberglass	0.038	Limestone	2.15
Styrofoam	0.025	Coal	0.21
Air	0.024	Shale gas	0.049

For many problems of heat conduction, like in solid Earth, horizontal heat flow is not important and we can reduce the problem to a one-dimensional one along the depth axis, which is the first implication here:

$$\boldsymbol{q} = -k\boldsymbol{\nabla}T \Longrightarrow q \approx -k\frac{dT}{dz} \Longrightarrow q \approx -k\frac{\Delta T}{h} \ . \tag{2.51}$$

We can only access a thin region near the surface of the Earth (e.g., the ocean crust, which is only 0 to 10 kilometers thick), so the temperature gradient can be assumed to be roughly linear; which is the second implication in (2.51). To measure heat flow, all we have to do is measure a temperature difference, ΔT, over some depth range, h, in the Earth, measure the thermal conductivity of the material, and use the third formula in (2.51).

The quantity k is referred to as the thermal conductivity of the medium and is measured in Watts per meter times Kelvin degrees (W/(m K)). It is defined as the heat flow per unit area per unit time when the temperature decreases by 1 degree in unit distance. In other words, it measures the ability of a material to conduct heat, reflecting the relative ease or difficulty of the transfer of energy through the material. And it depends on the bonding and structure of the material. Table 2.2 indicates that iron conducts heat more than a hundred times faster than water. Thus, water is a poor heat conductor relative to iron, although water is an excellent medium to store thermal energy. In general, solids are better conductors than liquids or gases because their molecules are very tightly packed together, so it is much easier for the molecules to pass the heat along. The molecules in liquids and gases are spread further apart, so they are not touching as much.

The heat-diffusion equation. The objective of this subsection is to predict the distribution of temperature and heat transfer within the strong materials. The fundamental physical principle we will employ to meet this objective is conservation of energy, also referred to as the first law of thermodynamics, in (2.7).

We begin with a restatement of the first law of thermodynamics in terms of the local value of density:

$$Q' = mc_P \frac{\partial T}{dt} = \int_D \left(\rho c_P \frac{\partial T}{dt} \right) dV , \tag{2.52}$$

where D is a fixed region in space (see Figure 2.10a), representing a portion of a given medium, with a surface ∂D describing the boundaries of this domain. An element of the surface ∂D, dA, is shown in Figure 2.10a, with two vectors: a normal vector $\boldsymbol{n}$ and the heat-flux vector $\boldsymbol{q} = -k\boldsymbol{\nabla} T$.

We can alternatively predict the rate of heat flow (second) by using the heat conducted (out) of ∂D plus the heat generated (or consumed) within the region D, by a nuclear reaction, for example. In accordance with Figure 2.10a, we can write the heat flux, dQ', out of dA, as

$$dQ' = \boldsymbol{q} \cdot (\boldsymbol{n} dA) = [(-k\boldsymbol{\nabla} T) \cdot \boldsymbol{n}] \, dA . \tag{2.53}$$

We are now going to derive the heat rate over D. Note that the mathematics of this domain is slightly more complicated than those described in the previous sections. However, a quick review of Appendix A of Ikelle (2020) will allow you to go to these three or so equations. Anyway, using Gauss's theorem (see Appendix A of Ikelle, 2020), the heat conducted (out) of ∂D plus the heat generated (or consumed) within

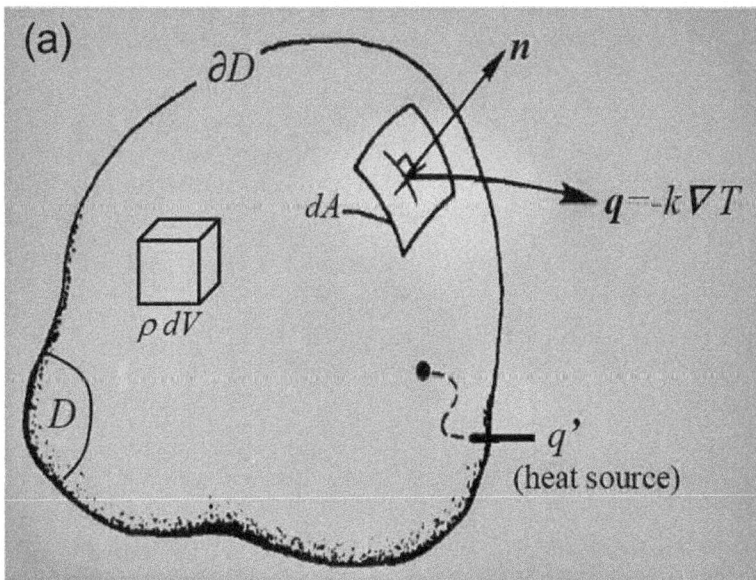

Figure 2.10. *(Continued.)*

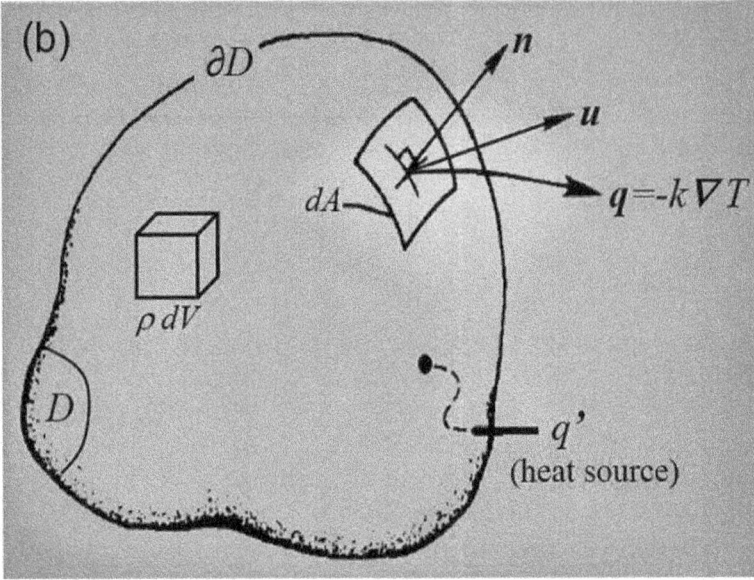

Figure 2.10. (a) Solid domain D in a heat-flow field with a surface ∂D. Infinitesimal mass element is ρdV and infinitesimal surface element is dS with unit normal $\boldsymbol{n}$ ($|\boldsymbol{n}| = 1$) and heat-flux vector is $\boldsymbol{q} = -k\boldsymbol{\nabla}T$. (b) Now this volume contains fluid. So D includes heat-flow and fluid-flow fields with at the surface ∂D. Heat flux vector is $\boldsymbol{q} = -k\boldsymbol{\nabla}T$ and fluid particle velocity is $\boldsymbol{u}$.

the region D can be written as

$$
\begin{aligned}
Q' &= -\int_{\partial D} [(-k\boldsymbol{\nabla}T) \cdot \boldsymbol{n}]\, dA + \int_D q'dV \\
&= \int_D (\boldsymbol{\nabla} \cdot k\boldsymbol{\nabla}T + q')\, dV \ .
\end{aligned}
\tag{2.54}
$$

Notice that we have included the possibility that a volumetric heat release can be distributed through the region, q' (W/m^3). By equating equations (2.52) and (2.54), we obtain

$$
\int_D \left(\boldsymbol{\nabla} \cdot k\boldsymbol{\nabla}T - \rho c_P \frac{\partial T}{dt} + q' \right) dV = 0
$$

or

$$
\boldsymbol{\nabla} \cdot k\boldsymbol{\nabla}T + q' = \rho c_P \frac{\partial T}{\partial t} \ .
\tag{2.55}
$$

The implication here is based on the fact that, since the domain D is arbitrary the integrand of this integral equation must vanish identically. This implication is the desired heat-diffusion equation of the strong materials or block of materials, such as lithosphere, under the assumptions that the material or the block of materials is incompressible. At the end, the diffusion equation relates the rate of temperature change and the temperature gradient. It is obtained as a combination of the first law of thermodynamics and Fourier's law. The rate of heat production per unit volume of a rock, q', can be the sum of the products of the decay energies of radioactive isotopes in D.

Thermal diffusivity. If the variations of thermal conductivity (k) are small, thermal conductivity can be factored out from temperature in the diffusion equation, we obtain

$$\frac{1}{\alpha}\frac{\partial T}{\partial t} = \nabla^2 T + \frac{q'}{k} \tag{2.56}$$

with the thermal diffusivity defined as

$$\alpha = \frac{\text{Heat conducted}}{\text{Heat stored}} = \frac{k}{\rho c_P} \ (\text{m}^2/s) , \tag{2.57}$$

where

$$\nabla^2 T = \nabla \cdot (\nabla T) = \frac{\partial^2 T}{\partial x^2} + \frac{\partial^2 T}{\partial y^2} + \frac{\partial^2 T}{\partial z^2} . \tag{2.58}$$

As the name implies, thermal diffusivity can be viewed as a measure of the rate at which heat "diffuses" through the material. Its dimension is length^2/time. Thermal diffusivity ranges, α, ranges from 0.14×10^{-6}m^2/s for water to 174×10^{-6}m^2/s for silver, which is a difference of more than a thousand times. The larger the thermal diffusivity, the faster the propagation of heat into the medium—a small value of thermal diffusivity means that heat is mostly absorbed by the material and a small amount of heat will be conducted further. The possible boundary condition of the thermal diffusivity are captures in Figure 2.11. Applications of the heat diffusion equation to many types of systems and processes are discussed in the next sections.

2.1.6. The Convective Heat-Diffusion Equation

Convection is how heat passes through fluid [see Figure 3.33 of Ikelle (2023a) and Figure 2.8]. One important property of fluid is that it rises when heated. The hot fluid becomes less dense and rises up. Cooler fluid is less dense and so it sinks down. In other words, the fluid-density differences are a direct result of temperature differences between the fluid and the solid wall surface. This up-and-down motion creates what are called convection currents. Convection currents are circular movements of heated fluid that help spread the heat. So, as illustrated in Figure 3.33 of Ikelle (2023a), we need a fluid layer bounded by two surfaces and a heater to produce convection. The Rayleigh number is a dimensionless number that describes the vigor of convection within a convecting layer. It is defined as follows

$$R_a = \frac{\rho g \beta \Delta T d^3}{\eta \alpha} , \tag{2.59}$$

with

$$\beta = \frac{1}{V}\left(\frac{\partial V}{\partial T}\right)_P , \tag{2.60}$$

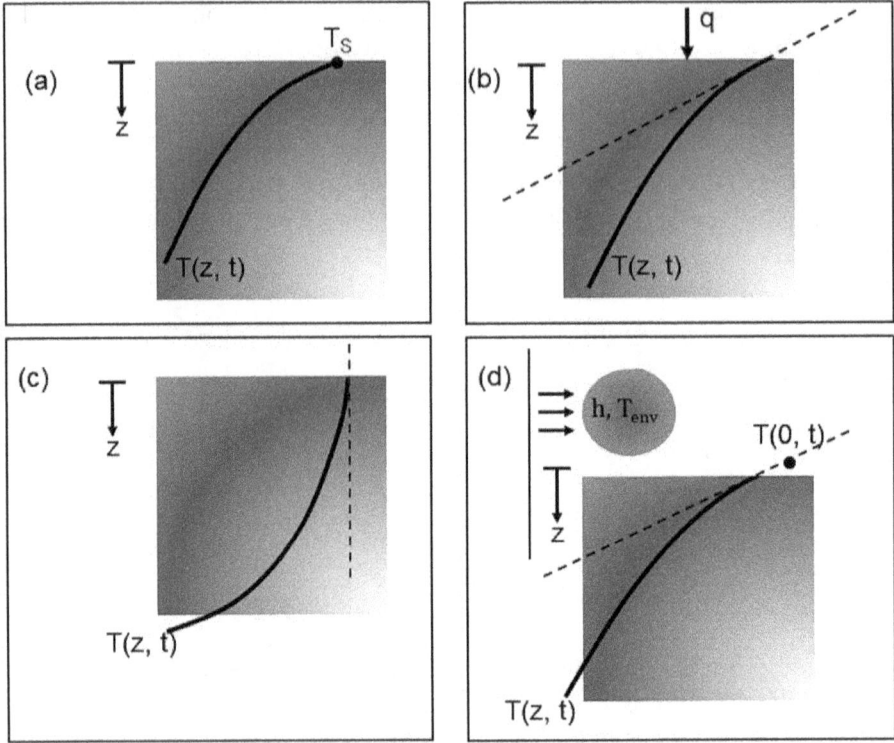

Figure 2.11. Boundary conditions of the heat diffusion at the surface, $z = 0$. (a) Constant surface temperature, $T(0,t) = T_S$; (b) Constant surface heat flux with finite heat flux, $-k[\partial T/pz](z = 0) = q$; (c) Adiabatic or insulated surface, $[\partial T/pz](z = 0) = 0$; and (d) convection surface condition, $-k[\partial T/pz](z = 0) = h[T_{env} - T(0,t)]$.

where g is gravity, ρ is fluid density, β is thermal expansivity, d is the thickness of the fluid layer, α is the fluid thermal diffusivity, η is fluid viscosity, and ΔT is the difference in temperature between the bottom and top surfaces (see Figure 3.33 of Ikelle (2023a)). The Rayleigh number is essentially the ratio of the parameters that increase the vigor of convection to those that inhibit convection. It tells us what makes convection likely. An increasing fluid viscosity retards convection. Large values of thermal diffusivity, α, also hinder convection. In contrast, a large temperature difference, ΔT, fluid layer thickness, d, and fluid density, ρ, aid convection. In other words, the Rayleigh number represents the competition between forcing by thermal buoyancy, damping by viscosity, and thermal diffusion. Note that, on occasion, you will see the Rayleigh number defined with no density term; this is because the kinematic viscosity is used, η/ρ.

Experiments (laboratory and computer) show that the Rayleigh number, R_a, must exceed the critical Rayleigh number R_{ac}, which is between 200 and 1000 (dependent on geometry), for convection to occur. For $R_a < R_{ac}$ the material is stable and transports heat by conduction. For $R_a > R_{ac}$ the material will be convectively unstable and transport heat more rapidly via conduction. Because the equations that govern fluid flow are very complicated and required numerical solutions as we will see later, the Rayleigh number gives us useful insight into whether convection can occur and how vigorous it might be in a given material.

Let us turn to the derivation of the convection-diffusion equation. In convection heat transfer, the transfer of heat is between a surface and a moving fluid (liquid or gas), when they are at different temperatures. The rate of heat transfer is given by Newton's law of cooling; i.e.,

$$q = h(T_s - T_\infty) = -k \left. \frac{dT}{dz} \right|_{z=z_0} , \tag{2.61}$$

T_∞ is fluid temperature and T_s is surface temperature, with $T_s > T_\infty$. The temperature gradient here is normal to the fluid-solid interface and is evaluated at the fluid-solid interface, and k is the conductivity of the fluid. Typical values of the convection heat transfer coefficient are between 50 and 1,000 W/(m^{2}K) for free convection with liquid, 35 and 250 W/(m^{2}K) for forced convection with gas, and 50 and 20,000 W/(m^{2}K) for forced convection with liquid. In forced convection, the fluid motion is produced by mechanical means, such as fans blowing air in domestic fan-coolers and heaters. Natural convection occurs when heat is added to fluid and fluid density varies with temperature. Flow is induced by force of gravity acting on density variations.

To derive the convective-heat-transfer equation, we will also employ the first law of thermodynamics in (2.52), as we did for the derivation of the conductive-heat-transfer equation, in the previous section, to meet this objective. The key difference between the two derivations is that we now allow the domain D to contain fluids. So the element of surface ∂D, dA, shown in Figure 2.10b, is now represented by three vectors, instead of two, as shown in Figure 2.10a. We have a normal vector, $\boldsymbol{n}$, the heat-flux vector, $\boldsymbol{q} = -k\boldsymbol{\nabla}T$, and the fluid particle velocity, $\boldsymbol{u}$. We can predict the rate of heat flow by using the heat conducted (out) of ∂D, the fluid flow (out) of ∂D, and the heat generated (or consumed) within the region D by a nuclear reaction, for example. In accordance with Figure 2.10b, we can write the heat flux, dQ', out of dA as

$$dQ' = [(-k\boldsymbol{\nabla}T) \cdot \boldsymbol{n}]\,dA + \rho c_P T (\boldsymbol{u} \cdot \boldsymbol{n}) dA . \tag{2.62}$$

This equation is actually a version of conduction in (2.53), albeit enhanced by fluid flow, which is the second term on the right-hand side of (2.62). This term describes the energy of the mass crossing surface dA in an interval of time dt. This mass is

$$dm = \rho dV = \rho dt\,(\boldsymbol{u} \cdot \boldsymbol{n})\,dA \quad \text{or} \quad \frac{dm}{dt} = \rho\,(\boldsymbol{u} \cdot \boldsymbol{n})\,dA . \tag{2.63}$$

Energy crossing the boundary associated with dm is

$$dQ'_{\text{adv}} - \rho c_P T \frac{dm}{dt} - \rho c_P T (\boldsymbol{u} \cdot \boldsymbol{n}) dA . \tag{2.64}$$

Let us return to equation (2.62). By using Gauss's theorem (see Appendix A of Ikelle (2020)) the heat conducted (out) and fluid out of ∂D plus the heat generated (or consumed) within the region D can be written as

$$\begin{aligned}
Q' &= -\int_{\partial D} [(-k\boldsymbol{\nabla}T) \cdot \boldsymbol{n}]\,dA - \int_{\partial D} [\rho c_P\,(\boldsymbol{u} \cdot \boldsymbol{n})]\,dA + \int_D q'dV \\
&= \int_D \left\{ (\boldsymbol{\nabla} \cdot k\boldsymbol{\nabla}T) - \rho c_P\,[\boldsymbol{\nabla} \cdot (\boldsymbol{u}T)] + q' \right\} dV .
\end{aligned} \tag{2.65}$$

By equating equations (2.52) and (2.65), we obtain

$$\int_D \left(\nabla \cdot k\nabla T - \rho c_P \left(\frac{\partial T}{dt} + [\nabla \cdot (\boldsymbol{u}T)] \right) + q' \right) dV = 0 \ . \tag{2.66}$$

Since the region D is arbitrary, the integrand of this equation must be zero. Thus, we obtain the convective-heat-diffusion equation; i.e.,

$$\nabla \cdot k\nabla T + q' = \rho c_P \left(\frac{\partial T}{dt} + [\nabla \cdot (\boldsymbol{u}T)] \right) \ . \tag{2.67}$$

Let us break down $\nabla \cdot (\boldsymbol{u}T)$ as follows:

$$\nabla \cdot (\boldsymbol{u}T) = (\nabla \cdot \boldsymbol{u})T + \boldsymbol{u} \cdot \nabla T \ . \tag{2.68}$$

If we assume that the flow is incompressible then $\nabla \cdot \boldsymbol{u} = 0$. We can rearrange the convection diffusion equation, in (2.67), and use the thermal diffusivity term $[\alpha = k/(\rho c_P)]$ to arrive at

$$\underbrace{\frac{\partial T}{\partial t}}_{\text{storage}} + \underbrace{(\boldsymbol{u} \cdot \nabla T)}_{\text{advection}} = \underbrace{\alpha \nabla^2 T}_{\text{conduction}} + \underbrace{\frac{q'}{\rho c_P}}_{\text{generation}} \ . \tag{2.69}$$

The first term of the new equation describes energy storage, the second term describes the heat transport due to bulk fluid motion, the third term describes the heat conduction, and the last term describes the heat production. It is the same equation as the corresponding conductive diffusion equation for a solid body obtained in (2.56), except for the second term, which is known as advection. In other words, convection is a conduction, plus advection. Again, the heat produced by the body under consideration is included in the convection-diffusion equations through q'. Keep in mind the assumptions used to arrive at this equation are that properties (thermal conductivity, k, the density, ρ, the specific heat, c_P) are constant and flow is incompressible ($\nabla \cdot \boldsymbol{u} = 0$). For steady state (no variation of temperature with time), the storage term in (2.69) can be ignored and the advection term can be ignored in a solid with no bulk flow through it (see also Table 2.3). As we will discuss later, boundary conditions (BCs) are the conditions at the surfaces of a body and initial conditions are the conditions at time $t = 0$. Boundary and initial conditions are needed to solve convective-diffusion h and heat-diffusion equations for any specific physical situation. The three typical of heat-transfer-boundary conditions are (1) temperature at a specific surface, (2) heat flux at a specific surface, and (3) convective- and/or radiative-heat-transfer conditions at a specific surface.

To summarize, convection heat transfer differs from conduction heat transfer in that bulk fluid motion augments the overall heat transfer by an advection mechanism; i.e.,

$$\text{Convection} = \text{Conduction} + \text{Advection}$$

where advection is a heat transport due to bulk fluid motion. Basically, we have simultaneously fluid-flow fields, which is the transfer of heat by transfer of the material that the system contains (here the flowing fluid), and heat-flow by conduction

Table 2.3. Meaning of terms of the convective heat-diffusion equation.

Term	Meaning	When can we gnore it?
• Storage	Rate of change of stored energy	No variation of temperature with time (steady state)
• Advection	Energy transport due to bulk flow	In a solid with no bulk flow
• Conduction	Energy transport due to conduction	Slow thermal conduction
• Generation	Generation of energy	No internal heat generation

(i.e., the diffusion of heat in atom-to-atom collisions). Applications of the convective heat-diffusion equation to many types of systems and some of the processes are discussed in the next sections.

Note that, in Chapter 1, we introduced wave equations, and in this chapter, we just introduced diffusion equations. These two groups of equations are important and classical in the sciences. Both are second partial differential equations with one key difference: a wave equation is always a second order in time, while the diffusion equation or heat equation is always a first order in time. For the particular case of homogeneous isotropic media and propagation along one direction, these equations are

$$\frac{\partial^2 u_z(z,t)}{\partial t^2} - c_0^2 \frac{\partial^2 u_z(z,t)}{\partial z^2} = f(z,t) \quad \text{(wave equation; hyperbolic)}$$

and

$$\frac{\partial c_z(z,t)}{\partial t} - k \frac{\partial^2 c_z(z,t)}{\partial z^2} = g(z,t) \quad \text{(diffusion equation; parabolic)} ,$$

where u_z can be a displacement of the surface in the vertical direction c_0 is the wave speed, $f(z,t)$ is the source of wave propagation, c_z can be the concentration of a chemical at position $x = [0,0,z]^T$, and time t, k the diffusion, and $g(z,t)$ is the source of chemical production.

2.2. Application: Heat Radiation and Greenshouse Effects

The exchange of heat between the sun and the Earth is the main driver of the climate model. As we described in the previous section, this heat-exchange mechanism is the radiation. It is carried out through an electromagnetic radiation exchange, so we will start by describing the fundamental laws of radiation exchange before using them to predict the ground temperature and to describe the greenhouse effect.

2.2.1. Blackbodies, the Four Laws of Radiation, and Planck's Law

The first law of radiation. Let us start with the phenomenon of light traveling in time through space. At some point, light hits matter. Let us idealize the problem by assuming that this matter can vibrate. If the wavelength of the light is equal to the wavelength of the matter, i.e.,

$$\lambda_{\text{matter}} = \lambda_{\text{light}} , \tag{2.70}$$

the matter can absorb the light and thus warm. But this matter can also emit light. Basically, any form of matter that has a temperature greater than absolute zero can emit light. So we have our first law of radiation, which states that all bodies (or all matter) with temperatures above 0 K (absolute zero) emit electromagnetic radiation.

Quiz: (1) Is a human body a blackbody? (2) Are stars blacbodies?

Blackbody. If an object has oscillations at all frequencies (i.e., all wavelengths), we call it a blackbody.[2] A blackbody is an ideal body that absorbs all radiation incident to it, and it reflects none. It is also the most efficient emitter of radiation. The light that comes from a blackbody is called the blackbody spectrum, and is represented by the intensity of radiation emitted by the blackbody versus the wavelength (or frequency). The solid earth and the sun are good examples of blackbodies, as they can oscillate at any frequency. Because of the importance of these two blackbodies in the climate system, we have found it useful to familiarize ourselves with the physical properties of blackbodies. Actually, the blackbody spectrum is a cornerstone in the study of quantum mechanics. The analysis of blackbody spectra led to the discovery of the field of quantum mechanics, which provides a complete explanation of the mechanism of radiation from very long wavelengths to very short wavelengths, whereas continuum mechanics (classical mechanics) is correct only for long wavelengths, such as infrared and visible radiation.

Planck's law. Consider a source of radiation in vacuum (e.g., the sun) to be a blackbody at a given temperature, T (in Kelvin), with surface brightness $I(\lambda, T)$ over a solid angle $d\Omega$ and a wavelength range $[\lambda, \lambda + d\lambda]$, as depicted in Figure 2.12. The energy at a point in space from the source of radiation received from the source is

$$dE = I(\lambda, T) \cos\theta dA d\Omega d\lambda dt \quad \text{or} \quad I(\lambda, T) = \frac{dE}{\cos\theta dA d\Omega d\lambda dt} , \tag{2.71}$$

where dA is area in space that the radiation passes through, θ is the angle away from the normal to that surface, and dt is the window of time of detecting of the radiation. To obtain the corresponding formula as a function of the frequency, f, we use the fact that $\lambda = c/f$ and $d\lambda = -c/f^2 df$, where c is the speed of light.

As we can see, in Figure 2.12, the variation of dE with distance from a source is included in solid angle $d\Omega$, which decreases proportionally to the inverse of the source-detector distance. As a result, the surface brightness, which is generally known as specific intensity, is independent of this distance. It is a superposition of plane waves of different wavelengths. In other words, $I(\lambda, T)$ is conserved through the vacuum and hence, it is conserved along the flow of photons. By using the

---

[2]The blackbody concept was introduced by Kirchhoff in 1860.

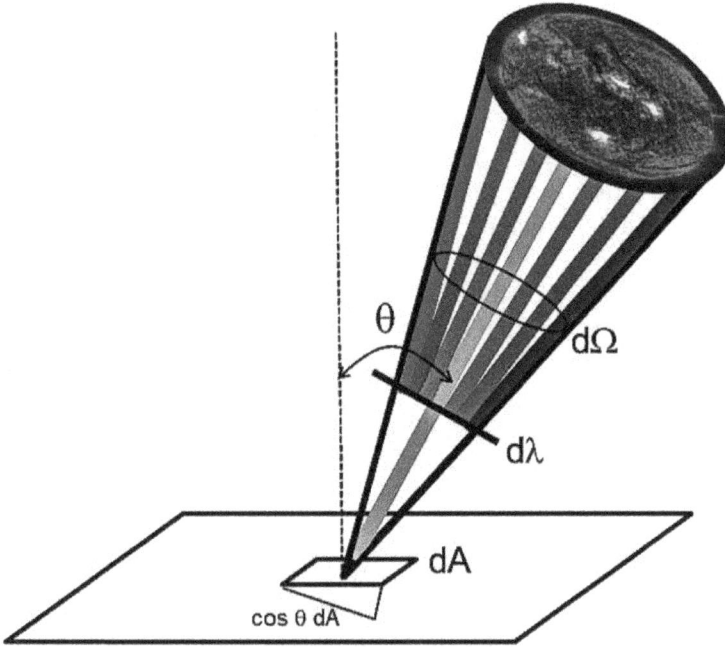

Figure 2.12. Description of the energy detected at a location in space for a period of time dt over a solid angle $d\Omega$ and an area dA arriving at an angle θ from an object with intensity I_0, a solid angle $d\Omega$, through a wavelength range $d\lambda$.

Planck's distribution function of photons, as predicted in quantum mechanics, the specific intensity can be written as (Planck, 1897) :

$$I(\lambda, T) = \frac{2hc^2}{\lambda^5} n_{\text{photon}}$$

$$= \frac{2hc^2}{\lambda^5 \left[\exp\left(\frac{hc}{\lambda k_B T}\right) - 1\right]} = \frac{a_1}{\pi \lambda^5 \left[\exp\left(\frac{a_2}{\lambda T}\right) - 1\right]}, \qquad (2.72)$$

with

$$n_{\text{photon}} = \frac{1}{\left[\exp\left(\frac{hc}{\lambda k_B T}\right) - 1\right]}, \qquad (2.73)$$

where n_{photons}, which is Planck's distribution, provides the average number of photons in a single wavelength (or frequency), c is the speed of light, $h = 6.626 \times 10^{-34}$ J·s $= 4.136 \times 10^{-15}$ eV s is the Planck constant, k_B is the Boltzmann constant, $a_1 = 2\pi hc^2 = 3.74 \times 10^{-16}$ W/m^{-2}, and $a_2 = hc/k = 1.44 \times 10^{-2}$ m K. This formula is known as Planck's law. By using the Taylor series of $\exp(x)$; i.e.,

$$\exp(x) = 1 + x + \frac{x^2}{2!} + \frac{x^3}{3!} + ... = 1 + x + \frac{x^2}{2} + \frac{x^3}{6} + ... , \qquad (2.74)$$

and limiting ourselves to the first two terms of this series, we can approximate equation (2.72) as follows:

$$I(\lambda, T) \approx \frac{2ck_B T}{\lambda^4} , \qquad (2.75)$$

and Planck's constant h disappears from the expression for $I(\lambda, T)$. Equation (2.75) is known as the Rayleigh-Jeans' law. It agrees with experimental results at large wavelengths (low frequencies) but strongly disagrees with experimental results at short wavelengths (high frequencies). Figure 2.13 illustrates the spectrum-intensity distribution of blackbody radiation as a function of wavelength at different temperatures. The 3000-K curve is close to what the sun emits. Increasing the temperature also increases the radiation intensity at all wavelengths. The total area under the curve corresponds to the energy emitted by the blackbody. Therefore, the energy emitted (the area below the curve) by a blackbody under the Rayleigh-Jeans law goes to infinity as the wavelengths decrease, whereas the total area under Planck's curve increases as the temperature increases, corresponding to increased emission at each wavelength and in total as the object becomes hotter. This is the prediction of classical mechanics (continuum mechanics). The actual measurements show that this prediction is very wrong, and this wrong prediction is known as the ultraviolet catastrophe.

Quantum mechanics corrects this problem by saying that energy is not continuous. It comes in chunks. At long wavelengths, these chunks are very small and close together, so we can describe the energy of radiation at these wavelengths as continuous. This is the case for both infrared and visible radiation. In these cases, continuum mechanics offers a good approximation. When we get to X-rays and gamma rays, the energy chunks are wide apart, and continuum mechanics is no longer valid. Notice also that the maximum of the intensity shifts to shorter wavelengths as blackbody temperatures increase. In other words, the wavelength at which this spectrum peaks is inversely proportional to temperature. These two observations yield the so-called Stefan-Boltzmann law and Wien's law, which will be mathematically described next. Before we do so, let us draw our attention to Figure 2.14. This figure also shows the blackbody spectra in Figure 2.13, but with wavelengths in log scale. Notice the change in the shape of the Planck's and Rayleigh-Jeans' curves.

The second law of radiation. The Stefan-Boltzmann law tells us that the total emission over all wavelengths is the area under Planck's curve. Therefore, we can obtain the Stefan-Boltzmann law by integrating the Planck law; i.e.,

$$E(T) = \int_{\lambda=0}^{\lambda=\infty} I(\lambda, T)d\lambda = \frac{2k^4\pi^2}{15c^2h^3}T^4 = \sigma T^4 , \qquad (2.76)$$

with

$$\sigma = \frac{2k^4\pi^2}{15c^2h^3} , \qquad (2.77)$$

where $\sigma = 5.67 \times 10^{-8}$ Wm^{-2}K^{-4} is the Stefan-Boltzmann constant and $E(T)$ is the energy radiated by the blackbody in W/m^2 (also known as the bolometric flux or

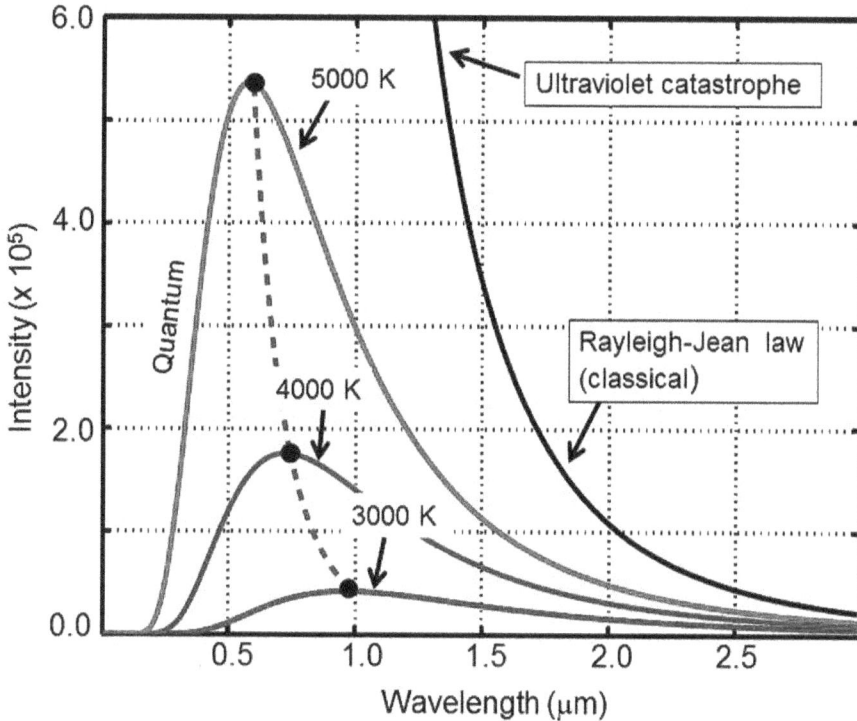

Figure 2.13. The blackbody emission (the Planck function) for absolute temperatures, as indicated, plotted as a function of wavelength on a linear scale. Making the object hotter causes it to emit more radiation across the entire spectrum. Notice also that increasing the temperature causes the peak intensity to shift to a shorter wavelength. The higher the temperature, the shorter the wavelength of the peak of the spectrum.

the total flux at all wavelengths). This formula is known as the Stefan-Boltzmann law. It tells us that as temperature increases, emitted energy increases. It is a very important formula for our discussion here because we will use it in the Earth energy-balance equation. Notice that we have used the Kelvin temperature scale instead of the Celsius or Fahrenheit scales in these derivations and throughout this subsection so that the temperature at power four, for example, can preserve negative temperatures on the Celsius and Fahrenheit scales in our formulation. This is the second law of radiation, which states that a blackbody at temperature T (K) emits radiation from its surface at the rate of $E = \varepsilon \sigma T^4$, where ε is the emissivity of the surface of the object under consideration and $\varepsilon = 1$ for blackbodies.

Note that the luminosity of a blackbody is the total energy emitted per unit time. The SI unit of luminosity is just Watts. Luminosity can be determined from the energy radiated by the blackbody by integrating over its entire surface:

$$L = \int_S E \, dA \, . \tag{2.78}$$

For a spherical blackbody with radius R and temperature T, the luminosity density is

$$L = 4\pi R^2 E = 4\pi \sigma R^2 T^4 \, . \tag{2.79}$$

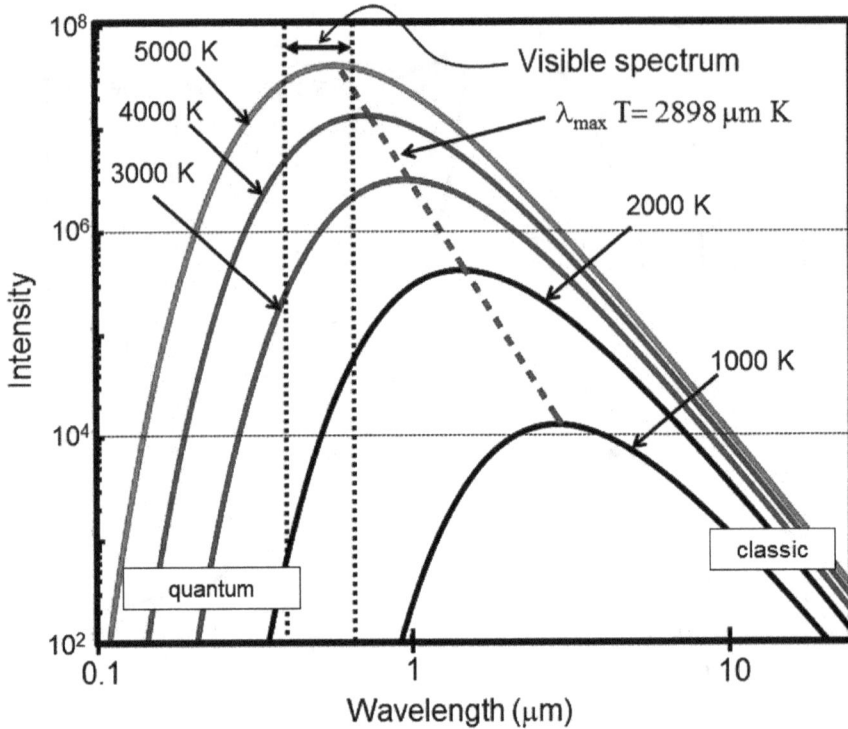

Figure 2.14. The blackbody emission (the Planck function) at absolute temperatures, as indicated, is plotted as a function of wavelength on a log scale.

This result leads to the common definition of an effective temperature of stars:

$$T_{\text{eff}} = \left(\frac{L}{4\pi\sigma R^2} \right)^{1/4} . \tag{2.80}$$

Other important related quantities to the Planck blackbody function are energy intensity (i.e., the energy density of photons),

$$U(T) = \frac{4\pi}{c} E(T) , \tag{2.81}$$

and the radiation pressure,

$$P(T) = \frac{4\pi}{3c} E(T) = aT^4 , \tag{2.82}$$

where $a = 4/3c$. Note also that $P(T) = U(T)/3$ is valid whenever $I(\lambda, T)$ is isotropic, regardless of whether we have blackbody radiation.

The third law of radiation. Differentiating Planck's intensity function in (2.72) and setting the derivative equal to zero, i.e.,

$$\left. \frac{\partial I(\lambda, T)}{\partial \lambda} \right|_{\lambda_{\max}} = 0 , \tag{2.83}$$

yields the wavelength of the peak emission for a blackbody at temperature T, which is

$$\lambda_{max} \approx \frac{2978}{T} , \tag{2.84}$$

where λ_{max} is expressed in microns and T is expressed in degrees Kelvin. This equation represents the third law of radiation, which states that the wavelength at which this spectrum peaks is inversely proportional to the temperature—that is, the peak intensities of thermal radiation coming from blackbodies occur at shorter wavelengths because as temperature increases, wavelength of maximum emission decreases. The third law is also known as the Wien's displacement law. On the basis of this equation, it is possible to estimate the temperature of a radiation source from knowledge of its emission spectrum, as illustrated in Figures 2.13 and 2.14.

Exercise: (1) A proof of Wien's displacement law. For values of interest to climate science, the exponential term is much larger than unity. By using this assumption, verify that (2.72) can alternatively be written as follows:

$$\log I(\lambda, T) = \log a_1 - 5 \log \lambda - \frac{a_2}{\lambda T} . \tag{2.85}$$

Verify also that

$$\frac{\partial \left[\log I(\lambda, T)\right]}{\partial \lambda} = -\frac{5}{\lambda} + \frac{a_2}{\lambda^2 T} . \tag{2.86}$$

Deduce (2.84). (2) Estimate the surface temperature of from the sensitivity of the human eye to light at 550 nm. We assume that the atmosphere exhibits all frequencies in the visible range similarly transparent.

Sun and Earth radiation. Figure 2.15 illustrates the blackbody spectra of the sun and the Earth. Clearly the sun emits much more radiation than the Earth, with very little overlap between the Sun's shortwave radiation and the Earth's longwave radiation. Note that 99 percent of sunlight has wavelengths of less than 4.0 μm, and sunlight essentially known as shortwave radiation, yet we consider the sun to be a blackbody because the other 1 percent covers the remainder of the electromagnetic spectrum. Note also that 99 percent of earthlight has wavelengths of more than 5.0 μm, which essentially represent longwave radiation, yet we consider the Earth to be a blackbody because the other 1 percent covers the remaining electromagnetic spectrum.

The fourth law of radiation. The atmosphere is a blackbody, but individual gases in the atmosphere are not blackbodies because they absorb and emit only at a limited wavelength range, which is the infrared wavelength range. In other words, greenhouse gases, which we will describe later, are not blackbodies, but their behavior can nonetheless be understood by applying the radiation laws derived for blackbodies. For this purpose it is useful to define the emissivity and the absorptivity of a body. At any given wavelength, λ, emissivity is defined as the ratio of

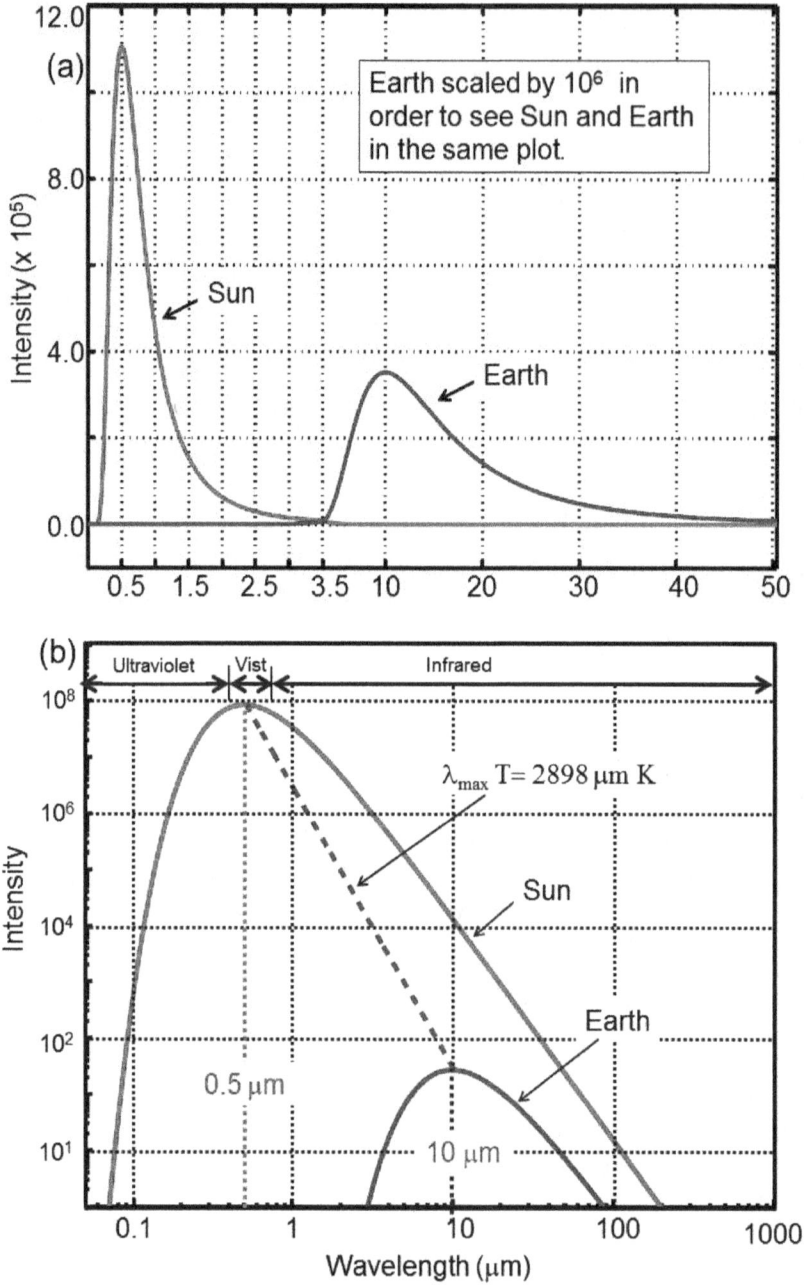

Figure 2.15. (a) A blackbody spectrum of the sun and the Earth in a linear scale. The earth spectrum is scaled by a factor of 10^6 because the sun emits much more radiation than the earth does. Also there is very little overlap between the two spectra: solar (shortwave) and Earth (longwave). The temperature of the sun is 5778 degrees Kelvin, and the temperature of the earth is 290 degrees Kelvin. (b) A blackbody spectrum of the sun and the Earth in log scale.

the intensity of the radiation emitted by the body under consideration, E_λ, to that of an ideal blackbody, $I(\lambda, T)$:

$$\varepsilon_\lambda = \frac{E_\lambda(\text{emitted})}{I(\lambda, T)} \; . \tag{2.87}$$

So emissivity is a measure of how strongly a body radiates at a given wavelength; it ranges between 0 and 1 for all real substances. A gray body is defined as a substance whose emissivity is independent of wavelength. The absorptivity, a_λ, is the fraction of the incident monochromatic intensity that is absorbed by a body under consideration; i.e.,

$$a_\lambda = \frac{E_\lambda(\text{absorbed})}{E_\lambda(\text{emitted})} \; . \tag{2.88}$$

A body that is not a blackbody (any gas) will absorb a fraction, a_λ, of the radiation incident upon it; this usually varies (strongly) with the wavelength, λ. It will also emit a fraction, ε_λ, of the radiation that a blackbody would emit at this wavelength. The fourth law of radiation states that the fractional absorptivity equals the fractional emissivity at all wavelengths; i.e.,

$$\varepsilon_\lambda = a_\lambda \; . \tag{2.89}$$

In other words, materials that are strong absorbers, at a given wavelength, are also strong emitters at that wavelength; similarly, weak absorbers are also weak emitters. This law is known as Kirchhoff's law.

In summary, Planck's law describes the intensity as a function of wavelength for each temperature; Stefan-Boltzmann's law describes the total emission at all wavelengths. Wien's displacement law estimates the peak emission wavelength, and Kirchhoff's law describes the absorption and emission of non-blackbodies. Also it is important to remember that thermal radiation is one of the fundamental mechanisms of heat transfer. The characteristics of thermal radiation depend on the absorptivity and emissivity of the surfaces. Moreover, the radiation does not consist of just a single frequency, but comprises a continuous dispersion of photon energies.

Some blackbody examples. (1) A spherical object with no internal heat source but shining is a blackbody because of stored up heat; i.e.,

$$\frac{d}{dt}(\text{heat capacity} \times \text{mass} \times T_{\text{eff}}) = 4\pi r^2 \sigma T^4 \; . \tag{2.90}$$

White dwarf stars (see Ikelle, 2023b) are typical examples of such backbody. Despite their lack of nuclear fusion, white dwarfs are very hot. Neutron stars cool are also blackbodies. (2) A main sequence star of Herzsprung-Russell diagram in thermal equilibrium is a blackbody; i.e.,

$$\int_0^M \varepsilon_{\text{nuc}} dm = 4\pi r^2 \sigma T^4 \; . \tag{2.91}$$

(3) Planets are blackbodies. A planet absorbs some small fraction of the power emitted by a star, reprocesses it into heat and radiates as a black body with a temperature quite different from that of the star. The interior of the planet does not participate on relevant time scales but the surface and atmosphere quickly acquire a temperature as needed to satisfy energy balance. (4) Stars are good blackbodies. Their luminosities in (2.79) provides both a way of measuring stellar radii (if we can measure the temperature and luminosity) and another way of understanding the Herzsprung-Russell diagram (i.e., diagram of temperature versus luminosity of stars). For example, TRAPPIST-1 has a mass of only 0.09 m_{sun}, a radius of 0.11 r_{sun}, and an effective temperature of 2560 degrees Kelvin, and therefore a luminosity of only 5.25×10^{-4} L_{sun}.

2.2.2.   The Climate Model Without the Atmosphere: Mercury

Our objective in this subsection is to explain how the climate system works by considering the simplified climate model shown in Figures 2.4 and 2.16. Here is the question we are trying to answer in regard to climate modeling: How does the flow of energy in and out of the Earth affect the temperature of the Earth?

Imagine that planets take in light only at short wavelengths of the electromagnetic spectrum from the sun (mostly in the visible region) and that planets emit light into space only at longer wavelengths (i.e., in the thermal infrared part of the electromagnetic spectrum). In other words, sunlight is the source of the energy coming into the planet. When the sun's energy reaches the Earth, it is partially absorbed by different parts of the climate system. The absorbed energy is converted back to heat, which causes the Earth to warm and become habitable. The heat is infrared light shining in all directions, which is the way energy leaves the Earth. That is, the Earth loses energy only through longwave radiation. Note that solar radiation absorption is uneven in both space and time, and this fact gives rise to the intricate patterns and seasonal variations of our climate. However, we will ignore this fact here and focus on global average temperatures.

Most solar radiation comes directly to the Earth's surface: (1) 70 percent is absorbed at the Earth's surface, and then emitted as infrared radiation, and (2) 30 percent is reflected directly back to space (albedo). To establish the Earth's heat budget, we assume that the energy is conserved through the climate system—that is, energy can be moved around but cannot be destroyed. By using the energy-conservation principle, we can write the balancing equation of the climate model shown in Figure 2.4 as follows:

$$S(1 - \xi)\pi R^2 = \sigma T^4 4\pi R^2 \tag{2.92}$$

or

$$T = T_{\text{skin}} = \left[\frac{S(1 - \xi)}{4\sigma} \right]^{\frac{1}{4}}, \tag{2.93}$$

where S is the solar radiative flux[3] (also known as the sun constant), ξ is the albedo (the albedo of the Earth is presently about 0.3), R is the solid-Earth radius, and

---

[3]The sun emits energy at a rate of $Q = 3.87 \times 10^{26}$ W. The solar constant is $S = Q/(4\pi R^2)$, where R is the distance from a given planet to the sun.

T is the ground temperature. In this model, we assume that the energy from the sun is balanced by the outgoing infrared radiation into space. In other words, the energy balance occurs when heating [on the left-hand side of (2.92)] equals cooling [shown on the right-hand side of (2.92)] or equivalently when the shortwave radiation coming in [shown on the right-hand side of (2.92)] equals the longwave radiation going out [shown on the right-hand side of (2.92)]. The term (πR^2) is introduced in this equation to compensate for the fact that planets, including the Earth, are spherical, and their surfaces are tilted with respect to the incoming radiation. In other words, at any given time the sun is shining only on one side of the Earth. From the sun's perspective, the Earth is just a circle, not a sphere, so solar radiation is received by a circle with area πR^2. The term πR^2 is the area of this circle. The term $4\pi R^2$ is the total surface area of the planet. Notice that the temperature resulting from the climate model without the atmosphere in (2.93) is often referred to as the skin temperature, hence the notation T_{skin}.

We now have the answer to the question: How does the flow of energy into and from the Earth affect the temperature of the Earth? The answer is shown in (2.93). By plugging all the relevant parameters into the temperature equation for the terrestrial planets (Venus, Mars, Earth), the average (effective) temperatures of these planets are given in Table 2.4.

The Earth's average temperature of -18.15 degrees Celsius (i.e., 255 degrees Kelvin) is quite low. The actual average temperature of the Earth is 15.85 degrees Celsius (i.e., 289 degrees Kelvin). Therefore, is our climate model totally wrong? No, it is just incomplete because we need to include the atmosphere in this model. Note that the variability of solar output will not help. The solar constant varies over the 11-year sunspot cycle. This variation is about 0.07 percent over an 11-year period. The effect on global temperatures is about 0.2 K, a very negligible effect.

As described in Ikelle (2023a), Mercury has no atmosphere. It is therefore expected that its temperature in a model without an atmosphere coincides with its true effective temperature.

Quiz: (1) Suppose the sun emits energy at a rate of $Q = 3.8736 \times 10^{26}$ W. The solar constant is

$$S - \frac{Q}{4\pi r^2} , \qquad (2.94)$$

where r is the distance of a given planet from the sun. Determine the solar constant of Earth, Venus, Mars, Jupiter, and Mercury. (2) Albedos for oceans are between 0.02 and 0.1; for forests, they are between 0.06 and 0.18; for cities, they are between 0.14 and 0.18; for grasslands, they are between 0.16 and 0.20; for deserts (sand), they are between 0.35 and 0.45; for clouds (thick stratus), they are between 0.60 and 0.70; and for fresh snows, they are between 0.75 and 0.95. Provide explanations for these variations of albedo of clouds with the type of surfaces.

Table 2.4. The average temperature of planets our solar system, as derived from the energy-balancing equations; S is the sun constant, ξ is the albedo, T is the temperature of the planet without the atmosphere, and T_{true} is the actual temperature. Remember that a planet's albedo is the percent of incoming solar radiation that is immediately reflected into space.

Planet	S (W/m^2)	ξ	$T = T_{\text{skin}}$ (K)	T_{true} (K)
Mercury	9121	0.12	435	440
Venus	2600	0.71	240	733
Earth	1370	0.3	255	289
Mars	600	0.16	216	220
Jupiter	51	0.34	110	165
Saturn	15	0.34	81	133
Uranus	4	0.30	59	78
Neptune	1.5	0.29	47	72
Pluto	0.9	0.4	39	47
Titan (moon)	15	0.3	82	94

2.2.3. The Climate Model with the Atmosphere: Venus, Earth, Mars, and Jovian Planets

Let us now modify our climate models by asking what effect the atmosphere has on the Earth's temperature. To answer this question, we again consider the one-layer model depicted in Figure 2.16. In this model, the incoming shortwave radiation (after removing the reflected component) is transmitted by the atmosphere and is completely absorbed by the ground. The ground emits as a blackbody. The atmosphere absorbs all of this energy and re-emits energy as a blackbody from both surfaces; i.e., into space and back to the surface. In this one-dimensional model, the emission is shown in Figure 2.16 as only up and down relative to the Earth's surface. Here we have to estimate the average temperature of the ground and that of the atmosphere. We can do so by using the following energy-budget equations in the atmosphere and in the ground:

$$\frac{S(1-\xi)}{4} = \sigma T_{\text{atmos}}^4 , \tag{2.95}$$

$$\frac{S(1-\xi)}{4} + \sigma T_{\text{atmos}}^4 = \sigma T_{\text{ground}}^4 , \tag{2.96}$$

where T_{atmos} is the average temperature of the atmosphere and T_{ground} is the average temperature of the ground. Equation (2.95) shows the energy balance for the atmosphere (infrared light), and (2.96) is the energy balance for the ground (the visible light from the sun and the infrared light from the ground and the atmosphere). When these equations are solved for the two temperatures, we obtain:

$$T_{\text{atmos}} = T_{\text{skin}} = \left[\frac{S(1-\xi)}{4\sigma} \right]^{\frac{1}{4}} \tag{2.97}$$

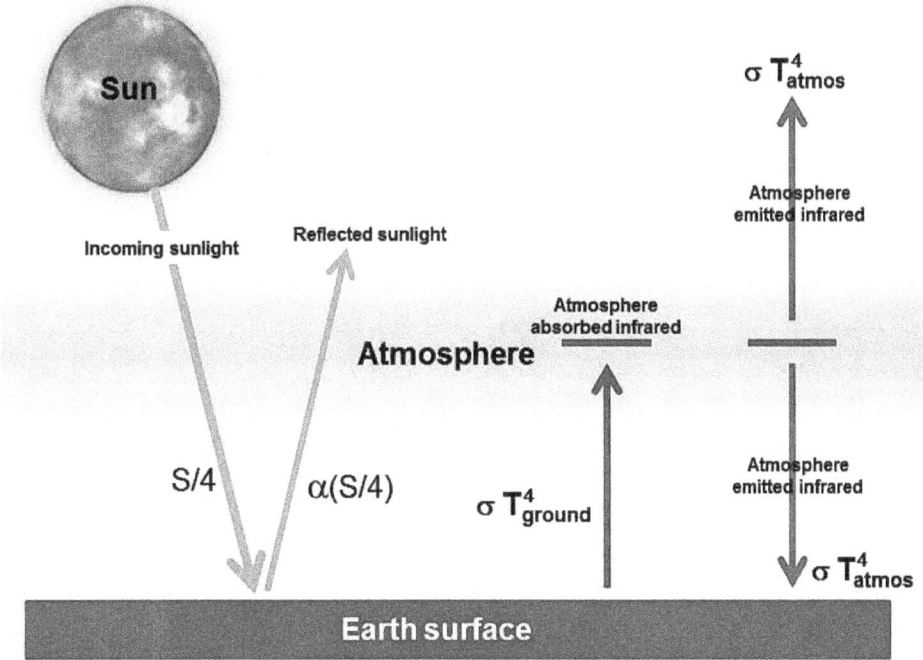

Figure 2.16. Diagram of the energy flows for the sun-ground-atmosphere system in which the atmosphere is modeled as one layer with $\alpha = \xi$.

$$T_{\text{ground}} = 2^{\frac{1}{4}} T_{\text{skin}} \approx 1.189 \, T_{\text{skin}} \; . \tag{2.98}$$

As we described in Ikelle (2023a), the atmosphere has several layers. So for an atmospheric model of n layers, (2.98) becomes

$$T_{\text{ground}} = \left[\frac{S(n+1)(1-\xi)}{4\sigma} \right]^{\frac{1}{4}} . \tag{2.99}$$

Notice that for $n = 1$, (2.99) reduces to (2.98). By plugging all the relevant parameters into the temperature equations [(2.97) and (2.98)] for the terrestrial planets (Venus, Mars, Earth), the average (effective) temperatures of these planets are given in Table 2.5.

Remember that a planet's albedo is the percent of incoming solar radiation that is immediately reflected into space. So this time the Earth's surface is a little too warm. The new ground temperature is about 26.85 degrees Celsius (the actual average temperature on the Earth is about 15.85 degrees Celsius), which is 45 degrees Celsius more than for a planet without an atmosphere. The Earth's surface now receives not only the net solar radiation, but the infrared light from the atmosphere, as well. Because the surface feels more incoming radiation than it would if the atmosphere were not present (or were completely transparent to infrared radiation) the Earth becomes warmer than in the climate model without an atmosphere. This process is valid for the Earth and for other terrestrial planets, Venus and Mars. The result, in Table 2.5, is better than that of a planet without an

Table 2.5. The average temperature of planets of our solar system, as derived from the energy-balancing equations, which take into account their atmospheres. S is the sun constant, ξ is the albedo, n is the numbers of layers in their atmosphere, $T_{\text{ground}}^{(1)}$ is the average temperature of the planet without the atmosphere, $T_{\text{ground}}^{(2)}$ is the average temperature of the planet with the atmosphere, and T_{obs} is the average temperature of the planet.

Planet	S (W/m^2)	ξ	n	T_{obs}(K)	$T_{\text{ground}}^{(1)}$(K)	$T_{\text{ground}}^{(2)}$(K)
Venus	2600	0.71	1	737	240	285
Earth	1370	0.3	1	289	255	303
Mars	600	0.16	1	220	216	256
Jupiter	51	0.34	4	165	110	164.5
Saturn	15	0.34	6	133	81	131.8
Uranus	4	0.30	2	78	59	77.6
Neptune	1.5	0.29	4	72	47	70.3
Pluto	0.9	0.4	1	47	39	46.4
Titan (moon)	15	0.3	1	94	82	97

atmosphere, in Table 2.4, but not yet good. The greenhouse effect[4] is needed. The greenhouse effect is actually less strong than in the model in Figure 2.16, because it permits partial transmission of infrared light through the atmosphere. Notice that Jovian planets have almost no greenhouse effects. It is the numbers of layers of their atmospheres that lead us to their effective temperatures.

> **Quiz:** Consider an atmosphere that is completely transparent to shortwave (solar) radiation but very opaque to infrared terrestrial radiation. Specifically, assume that it can be represented by n layers of atmosphere, each of which completely absorbs infrared radiation. How many layers would be required to obtain the actual average temperatures of Earth, Venus, and Mars? Explain why N is zero for some planets.

2.2.4. The Climate Model with the Greenhouse Effect: Earth, Mars, and Venus

The Earth emits about 390 W/m^2 of longwave radiation, but only 40 W/m^2 make it directly into space. The rest is trapped by clouds and greenhouse gases (gases that absorb some of the infrared radiation, see Ikelle, 2023b) because of the greenhouse effect (see Figure 2.17). A more intuitive way to understand the greenhouse effect is to pay attention to desert weather. In the desert there are no clouds and little water vapor into the atmosphere. During the day, the climate is similar to that of the one-layer model described in the previous subsection; thus it is quite hot. At night, temperatures plunge because there is little greenhouse effect in contrast with non-desert places with a greenhouse effect. These are clear illustrations of the

---

[4]Literally, the greenhouse effect refers to the farming practice of warming garden plots by placing them in a glass (or plastic) enclosure.

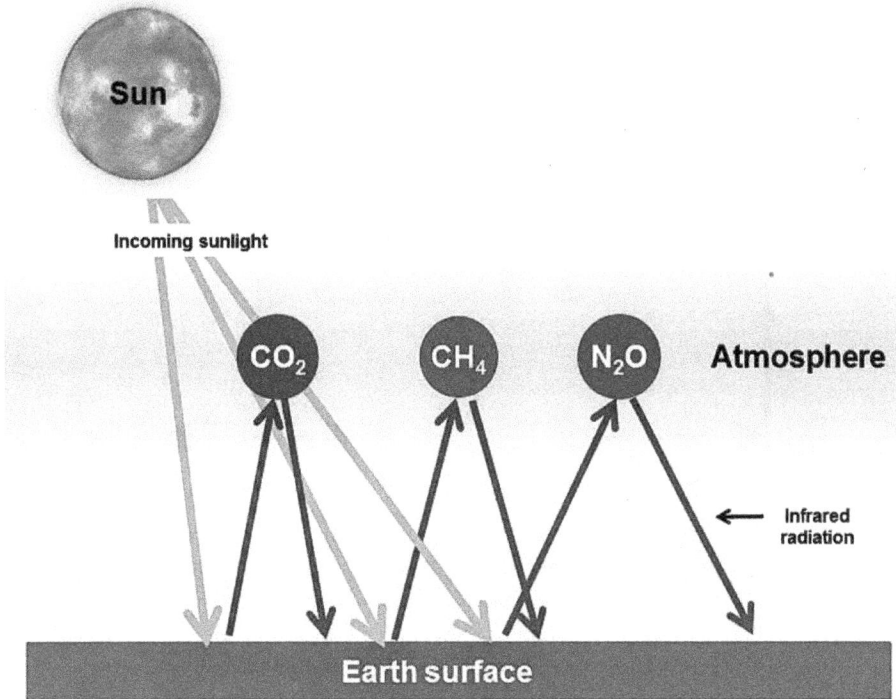

Figure 2.17. An illustration of an idealized model of the greenhouse effect. Sunlight passes through the atmosphere to warm the planetary surface. The planet's surface emits infrared radiation. Infrared radiation is absorbed by greenhouse gases in the atmosphere. These gases also radiate energy, some of which is directed to the surface and the lower atmosphere. This process is called the greenhouse effect; the earth's surface receives not only net solar radiation but also infrared from the atmosphere. So the greenhouse effect is a warming effect.

greenhouse effect. Our objective in this subsection is to describe how the greenhouse effect affects the Earth's temperature.

Understanding why greenhouse gases absorb infrared electromagnetic radiation requires the introduction of bond dipoles, molecular vibrations, and molecular dipole moments. A bond dipole is a vector that represents the degree of charge separation in a bond. For example, in Figure 2.18 we can see the bond dipoles of each of the H-O bonds in water. We can substitute the bond dipoles with vectors along the x- and z-axes. The vector sum of all of the bond vectors in a molecule is the molecular dipole moment. Water has a dipole moment, as shown in Figure 2.18. However, the molecule CO_2 has no dipole moment because the two-bond dipoles cancel each other (see Ikelle, 2023b for more details).

Quiz: As described, in Ikelle (2023a), the surface of Mercury changes in temperature are the most drastic in our entire solar system. Why?

One question that follows from our discussion in the previous subsections concerns the effect of CO_2 and other greenhouse gases on ground temperature. To a

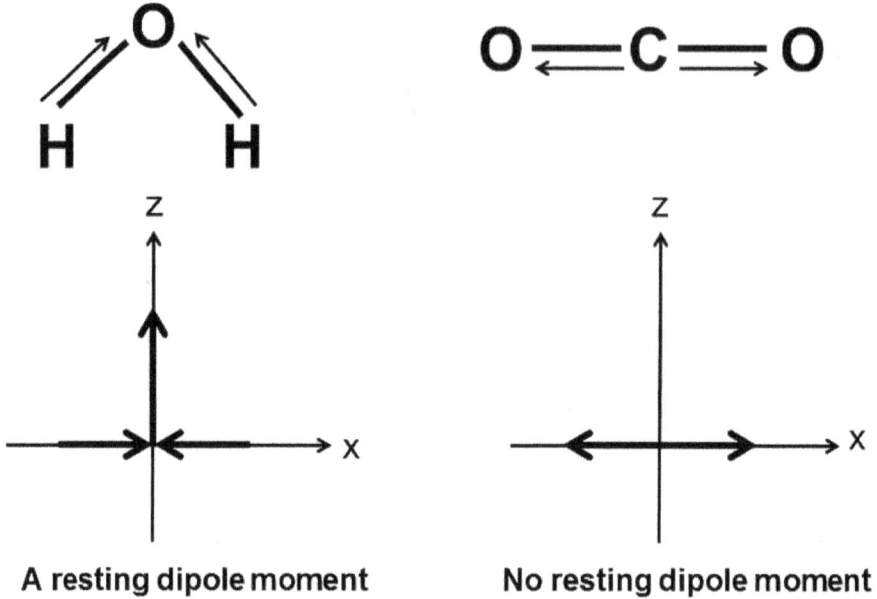

Figure 2.18. The molecules H_2O and CO_2 at their resting states, along with their molecular dipole moments. H_2O has a dipole moment in its resting state, whereas CO_2 does not have a dipole moment in its resting state. CO_2 has no dipole moment because the two bond vectors cancel each other.

certain extent, we can use the one-layer model described in the previous subsection to answer this question. The parameter of the climate model most affected by greenhouse gases effect is the absorptivity/emissivity of the atmosphere. The higher the concentration of CO_2 in the atmosphere, the bigger the emissivity will be, because the atmosphere will be more efficient at absorbing and emitting longwave radiation. That is why the greenhouse gases have a general warming effect. If we denote by a and ϵ absorptivity and emissivity of the atmosphere, respectively, the energy-budget equations in the atmosphere and at the ground can be described as follows:

$$\frac{S(1-\xi)}{4} = \epsilon\sigma T_{\text{atmos}}^4 + (1-\epsilon)\sigma T_{\text{ground}}^4 , \qquad (2.100)$$

$$\frac{S(1-\xi)}{4} + \epsilon\sigma T_{\text{atmos}}^4 = \sigma T_{\text{ground}}^4 . \qquad (2.101)$$

Equation (2.100) shows the energy balance for the atmosphere, and (2.101) shows the energy balance for the ground (see also Figure 2.19). The first term on the left-hand side of (2.100) represents the shortwave radiation absorbed by the atmosphere, the second term represents the longwave radiation power emitted by the ground, and the third term represents the fraction, ϵ, of the longwave radiation from the surface, which is absorbed within the atmosphere (so the model in Figure 2.17 corresponds to $\epsilon = 1$). The right-hand side represents the longwave radiation power emitted by the atmosphere. The first term on the left-hand side of (2.101) represents the shortwave

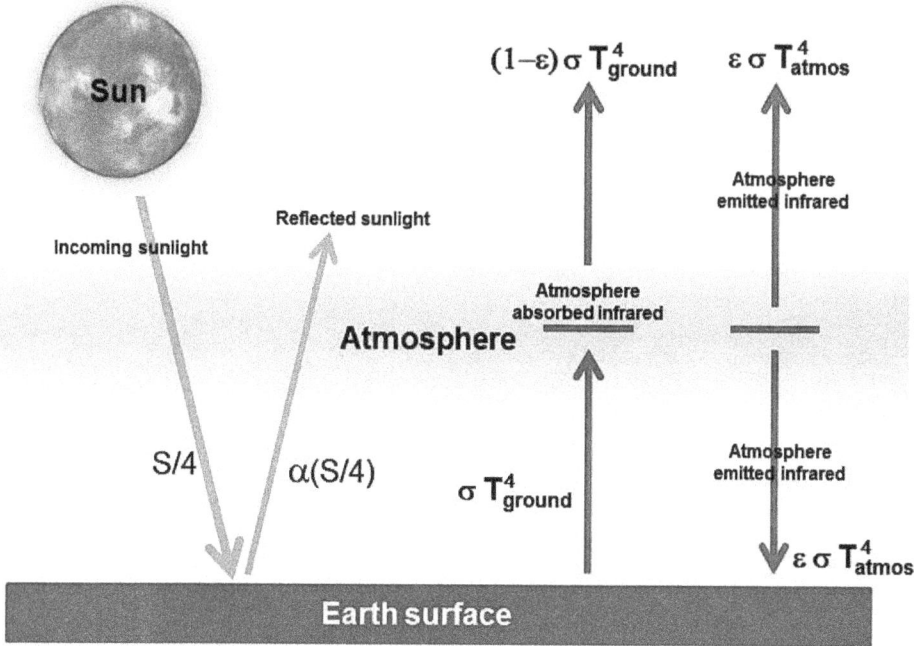

Figure 2.19. Diagram of energy flows for the sun-ground-atmosphere system in which the atmosphere is modeled as an absorbing one-layer model with $\alpha = \xi$.

radiation absorbed by the ground, the second term represents the longwave radiation power emitted by the atmosphere, and the right-hand side represents longwave radiation power emitted by the ground. Notice that we have used the Kirchhoff's law in these equations (i.e., the emissivity of the atmosphere is equal to its absorptivity). When these equations are solved for the two temperatures, we obtain

$$T_{\text{ground}} = \left(\frac{2}{2 - \epsilon}\right)^{\frac{1}{4}} T_{\text{skin}} , \tag{2.102}$$

where

$$T_{\text{skin}} = \left[\frac{S(1 - \xi)}{4\sigma}\right]^{\frac{1}{4}} . \tag{2.103}$$

Thus, when $\epsilon = 0$ (the climate model without the atmosphere, as shown in Figure 2.4), we have $T_{\text{ground}} = T_{\text{skin}}$, and when $\epsilon = 1$ (the climate model with an opaque atmosphere, as shown in Figure 2.16), we have $T_{\text{ground}} = 2^{\frac{1}{4}} T_{\text{skin}}$, as described in the previous section. Because of the greenhouse effect, we have $0 < \epsilon < 1$, which means that the actual effective temperature of a given ϵ is greater than the temperature of the Earth without the atmosphere (i.e., warmer than 255 degrees Kelvin) and smaller than the temperature of the Earth with an opaque atmosphere (i.e., smaller than 300 degrees Kelvin). There is an effective value of ϵ that is greater than 1 but less than 2 (i.e., $1 \leq \epsilon < 2$) for multilayer atmospheres and

Table 2.6. The average temperature of Venus, Earth, and Mars, as derived from the energy-balancing equations which take into account their atmospheres; S is the sun constant, α is the albedo, $T^{(1)}_{\text{ground}}$ is the average temperature of the planet without the atmosphere, $T^{(2)}_{\text{ground}}$ is the average temperature of the planet with the opaque atmosphere, $T^{(3)}_{\text{ground}}$ is the average temperature of the planet with the greenhouse effect, and T_{obs} is the actual average temperature of the planet. Titan's temperature captures greenhouse effects as observed but it is still correct because we must include the anti-greenhouse effects that we will discuss later.

Planet	S (W/m^2)	ξ	ϵ	T_{obs} (K)	$T^{(1)}_{\text{ground}}$ (K)	$T^{(2)}_{\text{ground}}$ (K)	$T^{(3)}_{\text{ground}}$ (K)
Venus	2600	0.71	1.978	737	240	285	732.9
Earth	1370	0.3	0.788	289	255	300	289
Mars	600	0.16	0.01	220	216	259	220.3
Titan (moon)	15	0.3	1.203	94	82	97	103

atmospheres that include the greenhouse effect (warming of the atmosphere) and the anti-greenhouse effect (cooling of the atmosphere). Basically, with appropriate values of ϵ, we can reconstruct average temperatures that are quite close to the actual average temperatures, which is 289 degrees Kelvin for the Earth. Note that the greenhouse model in Figure 2.19 is suitable for Venus.

By plugging all the relevant parameters into the temperature equations [(2.103) and (2.102)] for all the terrestrial planets (Venus, Mars, Earth), the average (effective) temperatures of these planets are given in Table 2.6.

We can use (2.100) and (2.101) to obtain the following atmospheric temperature for the model in Figure 2.16:

$$T_{\text{atmos}} = \left(\frac{1}{2-\epsilon}\right)^{\frac{1}{4}} T_{\text{skin}} = \left(\frac{1}{2}\right)^{\frac{1}{4}} T_{\text{ground}} . \qquad (2.104)$$

Note that $T_{\text{atmos}} < T_{\text{ground}}$: the atmosphere is always cooler than the ground on average.

Two parallel planes emitting black body radiation at each other forms a blackbody. Suppose that both planes emit and absorb radiation perfectly. T_1 and T_2 evolve depending on the situation (sizes, masses, etc). If the surfaces of these planes extend to an infinite extent and the space between the surfaces either is a vacuum, then the net energy flux (W/m^2) transferred by radiation from plane 1 to plane 2 is (Siegel and Howell, 1990)

$$q = \left[\frac{1}{e(T_1)} + \frac{1}{e(T_2)} - 1\right]^{-1} \sigma \left(T_1^4 - T_2^4\right) , \qquad (2.105)$$

where $e(T_1)$ and $e(T_2)$ are the emissivities of planes 1 and 2, respectively. Notice that the flux is independent of the spacing between the surfaces.

2.2.5. Absorption of Electromagnetic Radiation by the Atmosphere

Inspection of high-resolution spectroscopic[5] data reveals that the absorption bands for the 320-380 cm^{-1} region is caused by water vapor, and for the 680-740 cm^{-1} region is caused carbon dioxide. Figure 2.20 shows the absorption spectra for H_2O, CO_2, and CH_4. Notice that even if the atmosphere is saturated with water vapor, a great deal infrared still gets through. CO_2 and CH_4 absorb the infrared wavelengths that H_2O does not. Also, water absorbs the incoming solar infrared radiation because the frequency of the internal vibration of the water molecules is the same frequency of the waves of the solar infrared radiation.

Figure 2.20 also shows an example of an infrared spectrum; clearly, in certain portions of the infrared spectrum, radiation is trapped by various gases in the atmosphere. Among these gases, carbon dioxide, water vapor, and ozone are the most important absorbers. Carbon dioxide absorbs infrared radiation significantly from about 600 to 800 cm^{-1}, which also corresponds to the maximum intensity of the Planck function. Water vapor absorbs thermal infrared light from about 1200 to 2,000 cm^{-1} and in the rotational band below 500 cm^{-1}. Except for ozone, which has an absorption band of about 1,040 cm^{-1}, the atmosphere is relatively transparent from 800 to 1,200 cm^{-1}. This region is referred to as the atmospheric window. Carbon dioxide also has an absorption band in the shorter wavelength of the 4.3 μm region (2,325 cm^{-1}). The global distribution of carbon dioxide is fairly uniform, although there has been observational evidence indicating a continuous global increase over the past century because of the increased combustion of fossil fuels. Unlike carbon dioxide, however, water vapor and ozone are highly variable with respect to both time and the geographical location. These variations are vital to the radiation budget of the Earth-atmosphere system and to long-term climatic changes.

Exercise: White et al. (1997) show that the expected temperature change, ΔT, caused by the 11-year periodic change in solar activity can be described by the following equation:

$$\Delta T(t) = \frac{\Delta S}{\rho h c_P \sqrt{(K^2 + \omega_S^2)}} \cos\left(\omega_S t + \theta_S\right) , \qquad (2.106)$$

where $\Delta S \approx 0.2\,\mathrm{W/m^2}$ describes the change in the sun's constant (nearly all the energy that the Earth receives), $\omega_S = 2\pi/T_S = 2\pi/(11\,\text{years})$ is the angular version of the 11-year period, $\rho = 1.0\,\mathrm{g/cm^3}$ is the average density of ocean water, $K = 1/\text{year}$ is a dissipative timescale for energy loss to the deep ocean and to the atmosphere, $h = 50$ m is the average depth of the mixed layer of the oceans, $c_P = 4.2 \times 10^3$ Ws/(K kg) is the average heat capacity of ocean water at constant pressure, and $\theta_S = \tan^{-1}(\omega_S/K)$.
(1) Determine the range of temperature change during a solar cycle.
(2) Are the changes in solar activity have significant impact on the Earth's surface climate?.

---

[5]Vibrations of molecules can be studied via infrared spectroscopes. By comparing some well-known reference beams with the one passing through the sample, we can deduce the frequencies absorbed by the excitation of molecules at their vibrational energy levels.

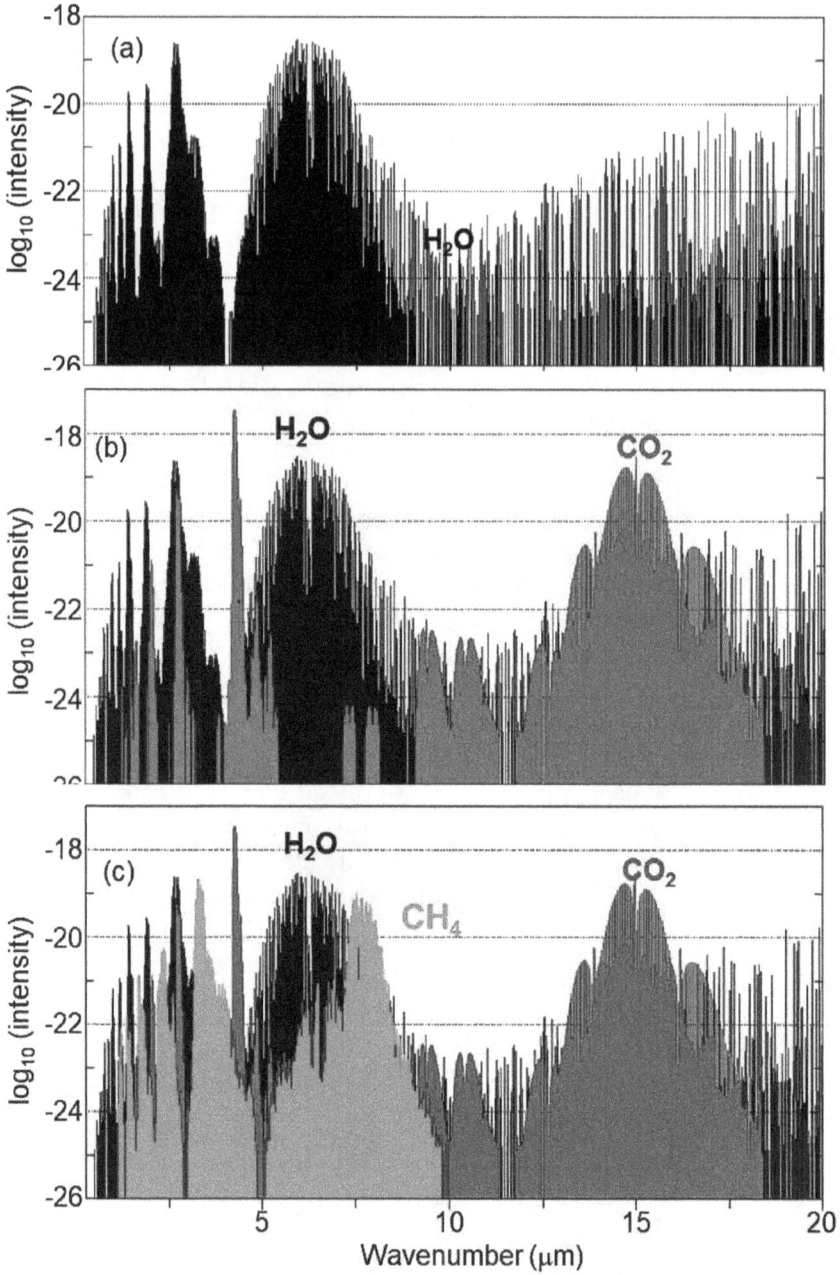

Figure 2.20. (a) Absorption spectrum of water vapor, (b) A superposition of the absorption spectra of water vapor and carbon dioxide, (c) A superposition of the absorption spectra of water vapor, carbon dioxide, and methane. Water vapor does most of the absorption of shortwave radiation, which means that even if the atmosphere is saturated with water vapor, a lot of infrared radiation still gets through; generally, CO_2 and CH_4 absorb infrared wavelengths that H_2O does not. (Adapted from spectralcal.com).

Table 2.7. (top) Matlab commands for computing blackbody spectra. (bottom) Matlab commands for computing the Planck Radiation law and the Rayleigh-Jeans law.

```
>> h = 6.6261*10^-34; c = 2.9979*10^8; k = 1.3807*10^-23;
>> lambda = 1e-9:10e-9:3000e-9; T1 = 4500; T2 = 6000; T3 = 7500;
>> A1=(h.*c)./(k.*T1.*lambda); A2=(h.*c)./(k.*T2.*lambda);
>> A3=(h.*c)./(k.*T3.*lambda); B=(8.*pi.*h.*c)./lambda.^5;
>> BB1=B.*(1./(exp(A1)-1)); BB2=B.*(1./(exp(A2)-1));
>> BB3=B.*(1./(exp(A3)-1));
>> figure; clf; plot(lambda,BB1,lambda,BB2,lambda,BB3); grid on
```

Computational exercise: (1) Use Matlab commands in Table 2.7 to reproduce the spectra in this section. (2) The Planck's law can be derived from classical physics by assuming that radiation is emitted and absorbed by oscillators in the wall, each with a characteristic frequency, f. Use Matlab commands for these derivations in Table 2.8 to reproduce Figure 2.21.

```
>> h = 6.626e-34; % Planck's constant in J-sec
>> c = 3.00e8; % Speed of light in a vacuum (m/sec)
>> k = 1.381e-23; % Boltzmann constant in J-K-1
>> freqrj = .015e14:.015e14:1.5e14; % R-J over limited frequencies.
>> prj = (8*pi*k.*6000/c^3).*freqrj.^2; % R-J law for T = 6000 K.
>> tvec = 3000:1000:6000; %the temperature vector
>> freq = .015e14:.015e14:15e14; % All frequencies
>> [F,T] = meshgrid(freq,tvec);
>> pnu = (8*pi*h*F.^3/c^3).*(1./(exp(h*F./(k*T))-1));
>> figure(1); plot(freqrj,prj,freq,pnu)
>> title('Comparison of Planck and Rayleigh-Jeans Laws')
>> ylabel('Energy Density, J/m^3')
>> xlabel('Frequency, Hz')
>> legend('R-J Law (6000 K)', 'T = 3000 K', 'T = 4000 K', ...
         'T = 5000 K','T = 6000 K')
```

2.2.6. Temperature of TRAPPIST-1 Exoplanets

As discussed in Chapter 2 of Ikelle (2023b), the recent discovery of seven Earth-sized, terrestrial exoplanets around TRAPPIST 1, with masses ranging from 0.33 to 1.37 m_{earth} and radii ranging from 0.755 to 1.05 r_{earth}, may be suitable for Earth-like lives because their gravities are similar to Earth gravity and their distances to TRAPPIST-1 suggest that their surface temperatures are below 400 degrees Kelvin, well within the temperature window of liquid water, which is between 273 and 373 degrees Kelvin. This estimate of temperatures of TRAPPIST-1's exoplanets is based on how much light the exoplanet receives. To get a better estimate of the real temperature of these planet—and therefore an improved indication of their suitability to Earth-like lives—you need to know the greenhouse effect and planetary albedo, as the climate models above suggest. For a definitive answer, we need observations

Table 2.8. Matlab commands for computing and plotting Stefan-Boltzmann law. The Stefan-Boltzmann law is written in terms of the fundamental physical constants h, c, and k.

```
>> h = 6.626e-34; % Planck's constant in J-sec
>> k = 1.381e-23; % Boltzmann constant in J-K-1
>> c0=2.997*10.^8; % m/s speed of light in vacuum
>> n=1; %refravtive index of the medium
>> step= 2.8429e-007; lambda=[10.^-7:step:2* ...
          10.^-5]; T=1;
>> figure(1);
for i=300:50:800
   VD = 2*pi.*h.*(c0.^2);
   Mi=VD./((n.^2).*(lambda.^5).*(exp((h.*c0)./(n.*k.*T.*lambda.*i))-1));
plot(lambda,Mi); hold on; end; grid; xlabel('lambda'); ylabel('M(lambda)')
```

that can confirm existence of some present or past oxygenic and/or anoxygenic photosynthesis in these exoplanets, through, for example, the amount of N_2, O_2, H_2O, CO_2, and Ar in their atmospheres, the existence of present or past plate tectonics in the interiors of these planets, and the existence of magnetosphere and moons,

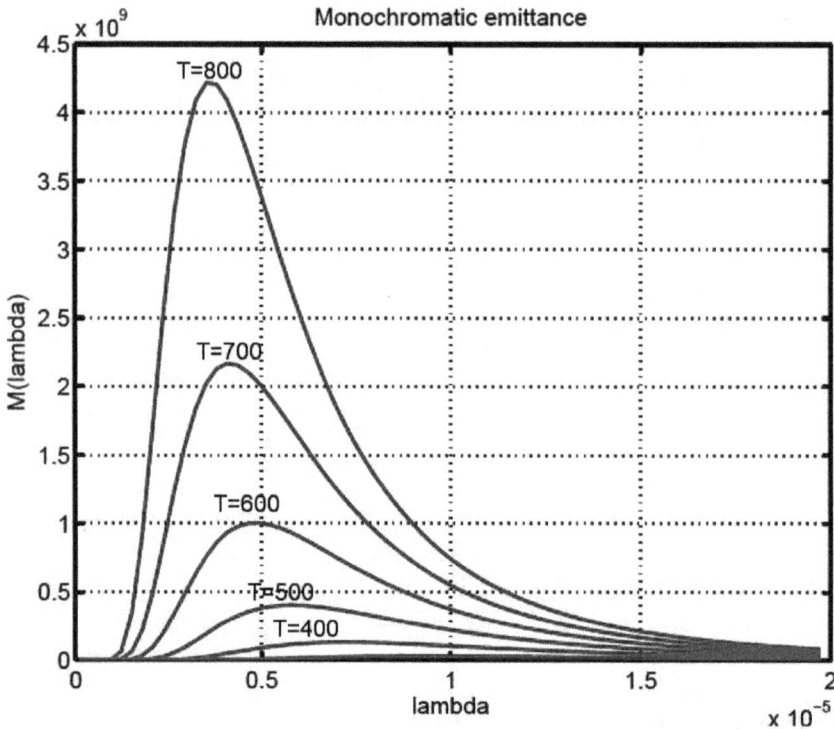

Figure 2.21. An illustration of results of Stefan-Boltzmann law in terms of the fundamental physical constants h, c, and k. The mathematical formula illustrated is derived from classical physics (statistical thermodynamics) instead of quantum mechanics.

including icy moons that can contain/preserve ingredients of biological life. It is also important to keep in mind the cyclical nature of our universe between creation and mass expansion when determining how to find Earth-like planets with organic life. Note also that, as discussed in Ikelle (2023a), all living things on Earth are built from the same six essential elemental ingredients: carbon, hydrogen, nitrogen, oxygen, phosphorus, and sulfur (CHNOPS). Taking into account that phosphorus production is associated with supernovae, which are not a part of the life cycle of small stars like TRAPPIST-1, it is possible that potential life in TRAPPIST-1 may differ significantly from our own due to the absence of phosphorus. On Earth, some microbes do not contain phosphorus.

Because TRAPPIST-1 planets are quite similar to Earth in mass, density, and size, it is very likely their albedo is between 0.7 and 0.2 and their stellar constants between 1000 and 3000 W/m^2 based on the calculations in the previous subsections. These assumptions imply that the ground temperatures of TRAPPIST-1 planets are likely to vary between 215 and 500 degrees Kelvin. In other words, the possibility of liquid water in some of the TRAPPIST-1 planets cannot be discounted based on their planet temperatures. Note that the above temperature window of TRAPPIST-1 planets does not include greenhouse effects.

TRAPPIST-1 b, the closest exoplanet to TRAPPIST-1 star, is tidally locked by the TRAPPIST-1 star, according to recent observations from NASA's James Webb Space Telescope. Again, that means that one side of TRAPPIST-1 b faces the star at all times and the other (the night side) is in permanent darkness and turned forever toward space. Its dayside temperature is about 500 degrees Kelvins (227 degrees Celsius); any water at such a temperature will turn into vapor. According to these temperatures, TRAPPIST-1 b has no significant atmosphere, just like Mercury, our star's closest planet. Note that the temperatures of TRAPPIST-1 b are not too dissimilar to those of Mercury in Table 2.4. The fact that TRAPPIST-1 b is nearer to TRAPPIST-1 (1.72 million kilometers) compared to the distance between Mercury and the Sun (67.5 million kilometers) and the Sun is more luminous than TRAPPIST-1 ($0.055\,L_{sun}$, mostly as infrared radiation) may compensate for the small differences in temperature between TRAPPIST-1 b and Mercury. Although TRAPPIST-1 exoplanets all orbit much closer to their star than any of our planets orbiting the Sun (all TRAPPIST-1 exoplanets could fit comfortably within Mercury's orbit around the Sun) they receive comparable amounts of energy from their tiny star. Exoplanets TRAPPIST-1 d, e, f, and g are likely to have surfaces with liquid, including water, since temperatures decrease with increasing distance from their star in general (see Table 2.5). The relationship between star types and their exoplanets and the modeling of star systems are discussed in significant detail in Chapter 6 and Ikelle (2024).

2.2.7. Magnetic Fields of TRAPPIST-1 Exoplanets

The minerals called oxides (see Chapter 3 of Ikelle, 2023a) describe rocks' chemical compositions. Ikelle (2023a) shows some of these key oxides along with their mass densities. Let ρ_i be the i-th oxide with i varying from 1 to N. The density of a rocky planet can be described as

$$\rho = \sum_{i=1}^{N} \alpha_i \rho_i \qquad \text{with} \qquad \sum_{i=1}^{N} \alpha_i = 1 \, ,$$

where α_i is the ratio of volume occupied by the i-th oxide of a given TRAPPIST-1 planet over the entire volume of the planet. Based on the average values of mass densities of TRAPPIST-1 exoplanets in Table 1.3 and mass densities of oxides in Ikelle (2023a), it is highly likely that TRAPPIST-1b, TRAPPIST-1c, TRAPPIST-1e, TRAPPIST-1f, and even TRAPPIST-1g have large metallic cores just like Mercury and Earth. As a result, these five planets can have global magnetic fields and significant magnetospheres, which are prerequisites for biological life. Because TRAPPIST-1b and TRAPPIST-1c receive four and three times more solar radiation than Earth, life cannot spread widely on these two worlds. In addition, these planets are quite comparable to Mercury. They have a mass density close to Mercury's and are the closest planets to their star. These planets lack atmospheres due to their close proximity to their stars. They must have the highest possible mass density from the available material they can accrete to them to ensure that they are not broken apart by TRAPPIST-1's gravity pull.

In contrast, the large potential metallic cores in TRAPPIST-1e, TRAPPIST-1f, or TRAPPIST-1g are designed to generate magnetic fields that can produce magnetospheres capable of protecting these planets from dangerous radiation. This is just like Earth. So livelihoods in TRAPPIST-1 are likely located in TRAPPIST-1e, TRAPPIST-1f, and/or TRAPPIST-1g although the low radiation reaching them implies low temperatures for surface life in TRAPPIST-1f and TRAPPIST-1g. Beings in TRAPPIST-1e float due to gravity while in TRAPPIST-1f and TRAPPIST-1g they can walk or float just like on Earth. TRAPPIST-1d and TRAPPIST-1h are hot and cold Mars versions, respectively.

2.3. Application: Geothermics and Energy Transport in the Stars

Volcanoes, earthquakes, plate tectonics, geothermal energy, natural hot springs, black smokers from hydrothermal vents, and the mid-Atlantic ridge are phenomena that show that the heat at the Earth's surface comes from inside the solid Earth. Volcanoes and geothermal activity allow us to say that the temperature inside the solid Earth is higher than the temperature observed on its surface. Thus, it is natural that there is heat transfer from the interior to the surface of the Earth. Moreover, as we can only go from hot to cold, heat flows by conduction, convection, or radiation down a gradient of decreasing temperature; thus, changes in temperature with depth in the atmosphere, glaciers, ocean, and solid earth, and in the stars are evidence that heat transfer is occurring. Our first objective here is to develop geothermic models of the lithosphere, mantle, and core by using the idea of heat transfer from the interior to the surface of the Earth. *Geothermics* is the study of

Table 2.9. An illustration of U, Th, and K concentration (ppm). ρ is the density $(\mathrm{g/m^3})$, q' is the rate of radiogenic heat production in $\mu\mathrm{W/m^3}$, and k is the thermal conductivity in $\mathrm{W/(m\,K)}$.

	U	Th	K	ρ	q'	k
Granite	4.7	20.	36,000	2.65	2.95	2.93
Basalt	0.9	2.2	15,000	2.80	0.56	2.09
Average Crust	1.55	5.75	15,000	2.80	1.00	2.51
Periodite	0.019	0.05	59	3.15	0.01	3.35

the thermal regime of the solid Earth, including its geothermal gradient, as well as the use of the Earth's heat for heating/cooling and generation of electrical power.

Our second objective is to discuss energy transport through the interior of stars, including our sun. Nowadays, this energy is an important source of electricity and plays a central role in the creation of a large number of molecules and minerals found on Earth, as discussed in Ikelle (2023b).

2.3.1.   The Conductive Heat Transport in the Lithosphere

Convection is a very efficient means of transporting heat but can only operate if material can flow. The lithosphere is strong and hence translates as a quasi-rigid block over the asthenosphere; therefore, heat cannot be convected across the lithosphere and instead is conducted across the lithosphere. In other words, only conduction is possible in the lithosphere. Hence we can used equation (2.55) to describe the heat transfer through the lithosphere. Alternatively, we can use equation (2.56) when the variations of thermal conductivity, k, are small. The rate of heat production per unit volume of a rock, q', can be the sum of the products of the decay energies of radioactive isotopes in the lithosphere. If the energies of isotopes are e_i, the concentrations of the isotope in the rock are α_i (ppm) and the density of the rock (ρ) then $q' = \rho \sum_i \alpha_i e_i$ in $\mathrm{W/m^3}$. Rocks are radiogenic to varying degrees. We can use q' to incorporate a heat generation to the diffusion equation. Table 2.9 shows some useful examples of heat production sources.

Example: A one-layer thermal steady state. Let us look at how specific examples of the geothermal gradient can be deduced from the diffusion equation of conduction. Remember that the geothermal gradient is the rate at which the temperature of the Earth's interior increases with increasing depth. In this first example, we assume a thermal steady state (i.e., $\partial T/\partial t = 0$) and that horizontal heat flow is not important, so we can reduce the problem to a one-dimensional one (i.e., 1D assumption). Thus the equation (2.56) becomes

$$\frac{d^2 T}{dz^2} = -\frac{q'}{k} \iff T = -\frac{q'}{2k}z^2 + C_1 z + C_2$$

or

$$T = -\frac{q'}{2k}z^2 + \frac{q_S}{k}z + T_S \ . \tag{2.107}$$

where C_1 and C_2 are the integration constants resulting from the integration of the simplified diffusion equation twice with respect to z. Remember that the depth z coordinate is positive downward. We can determine these two unknown integration constants from geological boundary conditions. The first boundary condition is that the temperature is zero at $z = 0$—that is, surface temperature is $T(z = 0) = C_2 = T_S$. The second boundary condition is that the heat flux at the surface is $q_S = k[dT/dz](z = 0) = kC_1$; an insulated boundary condition corresponds to zero surface heat flux. And the geotherm is the last equation of (2.107). We can see the curvature of the geotherm is dictated by the heat-production rate divided by the thermal conductivity, q'/k. In other words, the geoterm of the lithosphere is just a straight line in the absence of a heat-generation source. That is the case of the oceanic lithosphere. This type of lithosphere is not enriched in radioactive elements and its radioactive sources are negligible. Therefore, the geotherm of the oceanic crust is essentially a straight line.

Let us now calculate a geotherm dictated by a surface temperature and a basal (e.g., the boundary between the Earth's crust and the mantle, also known as Moho or Mohorovicic discontinuity) heat flux at depth z_M—that is, the first boundary condition remains surface temperature and the second is about that of a basal (e.g., Moho) heat flux at depth z_M. In other words, the heat is explicitly coming from the interior of the Earth. The geotherm then becomes

$$T = -\frac{q'}{2k}z^2 + C_1 z + C_2$$

or

$$T = -\frac{q'}{2k}z^2 + \left(\frac{q_M + q' z_M}{k}\right)z + T_S \ , \tag{2.108}$$

with $T(z = 0) = T_S = C_2$. The new boundary condition then is

$$q_M = k \left.\frac{dT}{dz}\right|_{z=z_M} = -q' z_M + kC_1 \text{ or } C_1 = \frac{q_M}{k} + \frac{q'}{k}z_M \ . \tag{2.109}$$

Note that the sign of the derivative of temperature with respect to depth in the second boundary is + at the basal (that is, at depth z_M) to reflect the convection of heat flux into the surface as positive. Note also that this geotherm equation confirms that the basal heat flow contributes, through $q' z_M/k$, to the gradient of the temperature at depth z. By using the following standard model: $k = 2.5\text{W/m}^\circ\text{C}$; $q' = 1.25 \times 10^{-6}\text{W/m}^3$; and $q_M = 21 \times 10^{-3}\text{ W/m}^2$, we can verify that the shallow gradient is about 30° C/km and a deep gradient is about 15° C/km. We can increase these gradients by decreasing conductivity, increasing heat generation, or increasing basal heat flow (see Figure 2.22).

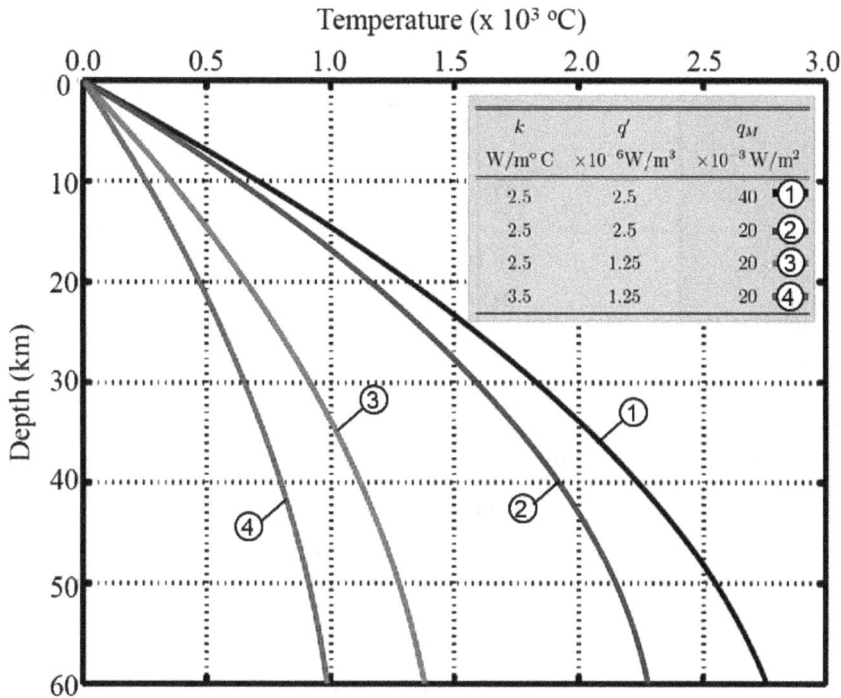

Figure 2.22. Here are lithosphere thermal geotherms under the assumptions of thermal steady state and horizontal heat flow (one-dimensional), as described in (2.108). The curvature in these geotherms is dictated by the heat production rate. Notice how quickly it gets hot with depth in the lithosphere. As we go down through the crust the temperatures at 100 km are somewhere between 1000 and 1800 °C.

Quiz: We should be interested in climate change because climate determines the type and location of human-managed ecosystems, such as agricultural farmlands, and changes in climate will impact crops and livestock. Rising temperatures lead to shorter growing seasons, earlier harvest dates, lower crop quality, and changes in soil temperatures. Climate also determines the quantity and quality of water available for human use. For example, a warming of one degree Celsius is sufficient to move climate zones about 150 km south. A regional temperature change of two degrees Celsius is likely to seriously impact most life forms and most ecosystems and agricultural areas. Climate change is sometime related to extreme weather events. Our objective is to define some of these extreme weather events.
(1) Define monsoons, hurricanes, typhoons, and cyclones.
(2) Define El Niño.
(3) List natural hazards associated with the Earth's atmosphere.

Example: A two-layer thermal steady state. We now consider a more realistic model of the lithosphere by describing it as a two-layer model. Each layer has its own heat production source and thermal conductivity and therefore different thermal geotherms.

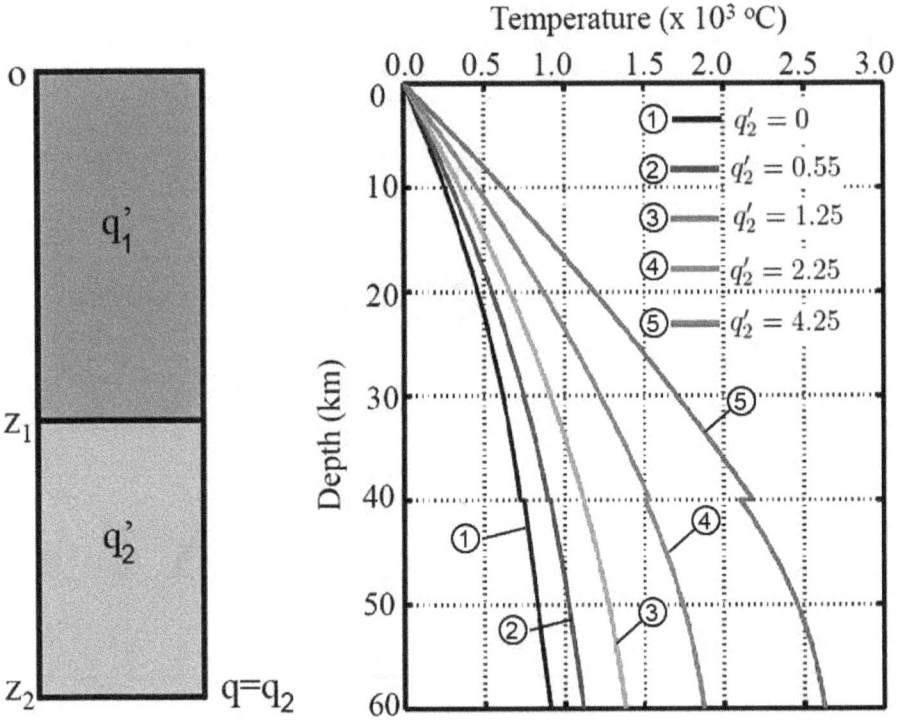

Figure 2.23. (a) A two-layer model with two different heat production rates. (b) Geotherms associated with this model for various heat-production models with $z_1 = 40$ km, $z_2 = 60$ km, $k = 2.5$ W/m$^\circ$ C, $q_2 = 20 \times 10^{-3}$ W/m^2, $q_1' = 1.25 \times 10^{-6}$ W/m^3, and $q_2' = \{0, 0.55, 1.25, 2.55, 4.55\} \times 10^{-6}$ W/m^3.

- Layer #1: $[0, z_1]$

$$\frac{d^2 T}{dz^2} = -\frac{q_1'}{k_1} \Longrightarrow T = -\frac{q_1'}{2k_1} z^2 + \left[\frac{q_2}{k_1} + \frac{q_2'(z_2 - z_1)}{k_1} + \frac{q_1' z_1}{k_1} \right] z .$$

- Layer #2: $[z_1, z_2]$

$$\frac{d^2 T}{dz^2} = -\frac{q_2'}{k_2} \Longrightarrow T = -\frac{q_2'}{2k_2} z^2 + \left(\frac{q_2}{k_2} + \frac{q_2' z_2}{k_2} \right) z + \frac{(q_1' - q_2')}{2k_2} z_1^2 .$$

At the surface, the temperature is zero ($T_S = 0$) and $q = -q_2$ at $z = z_2$. The geotherm is not a straight line as one might expect (see Figure 2.23). Note that this model can be used to describe the "hot dry rock" geothermal energy resource with the second layer describing the hot rock. For an effective geothermal production of energy, the hot rock must be as close to the Earth's surface as possible. Geothermal energy resources are discussed in Ikelle (2017b and 2020).

Example: Cooling of an oceanic lithospheric plate. Let us now consider unsteady conduction-diffusion equations. We begin with the 1-D version of the diffusion equation for temperature in (2.56); i.e.,

$$\frac{1}{\alpha}\frac{\partial T(z,t)}{\partial t} = \frac{\partial^2 T(z,t)}{\partial z^2} + \frac{q'}{\rho c_P} \, , \tag{2.110}$$

where $\alpha = k/(\rho c_P)$ is the thermal diffusivity. We first assume that heat generation is zero. The solution to the heat equation is (Carslaw and Jaeger, 1959)

$$T(z,t) = C_1 + C_2 \, \mathrm{erf}\left(\frac{z}{2\sqrt{\alpha t}}\right) \, , \tag{2.111}$$

where

$$\mathrm{erf}(y) = \frac{2}{\sqrt{\pi}} \int_0^y d\eta \exp\left(-\eta^2\right) d\eta \tag{2.112}$$

is the error function, and C_1 and C_2 are the integration constants. The error function represents the area under the Gaussian function, $\phi(\eta) = (1/\sqrt{2\pi}) \exp\left(\eta^2\right)$, from $\eta = 0$ to $\eta = y$, so that $\mathrm{erf}(0) = 0$, $\mathrm{erf}(\infty) = 1$, and $\mathrm{erf}(-y) = -\mathrm{erf}(y)$. It is a standard function, just like sine, cosine, or log. The Matlab error function is erf. For the initial condition $T(z, t = 0) = T_M$, constant temperature condition, $T(z = 0, t) = T_S$, the solution to the heat equation in (2.110) is then

$$T(z,t) = T_S + (T_M - T_S) \, \mathrm{erf}\left(\frac{z}{2\sqrt{\alpha t}}\right) \, . \tag{2.113}$$

In other words, we have a half-space at temperature T_M that cools at surface T_S. The initial temperature can be due to the initial gravitational collapse (see Ikelle, 2023b), but that the surface is in equilibrium with the solar radiation and stays at a temperature of T_S, so the Earth cools. This solution is a commonly used oceanic lithosphere, as illustrated in Figure 2.24 for $T_S = 0$. At the ridge, the mantle temperature almost reaches the surface ($z = 0$). The materials formed at the ridge move away from the spreading center at a rate equal to the speed of the tectonic plate. The key point in this example, through its solution to the unsteady diffusion-heat equation, is that depth distribution of lithosphere is related to lithosphere age, with greater ages corresponding to greater depths. Also we can see that as the plate cools, thermal boundary layers increase in thickness.

Quiz: Some climate changes are abrupt—that is, they are rapid and unpredictable. In the past, the average global temperature has risen or fallen more than 6° C in less than 10 years more than 10 in a million year span, and at least once in as few as five years. An increase of 6° C in this century would be considered an abrupt climate change. When was the last abrupt climate change?

Quiz: Let us look at how specific examples of the geothermal gradient can be deduced from the diffusion equation of conduction. Remember that the geothermal gradient is the rate at which the temperature of the Earth's interior increases with increasing depth. In this exercise, we assume a thermal steady state and no radioactive heat production rate and that horizontal heat flow is not important. Determine the geotherm of this model.

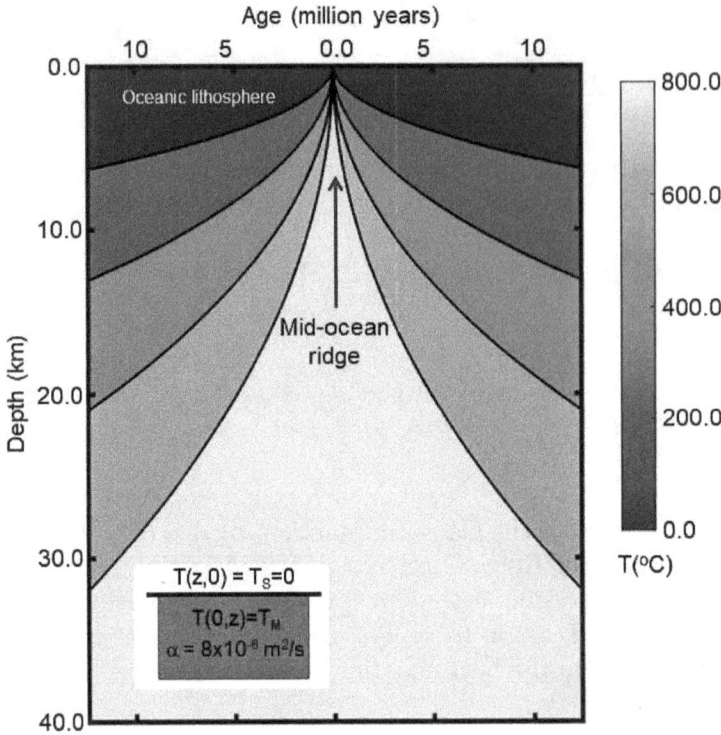

Figure 2.24. Here are lithosphere thermal geotherms under the thermal linear unsteady state and horizontal heat flow (one-dimensional one) with no heat production, as described in (2.113). The thermal diffusivity here is the order of that of granite; that is, $\alpha = 8 \times 10^{-7}$ m^2/s.

By taking the derivative of (2.113) with respect to z, we obtain conductive heat flow versus age; i.e.,

$$q = -k \left. \frac{\partial T(z,t)}{\partial z} \right|_{z=0} = -\frac{k(T_M - T_S)}{\sqrt{\pi \alpha t}} . \tag{2.114}$$

In other words, the heat flow at the surface decreases with increasing age as one might expect.

Let us now add internal heat generation to our unsteady heat-conduction equation. We can use the finite-difference scheme described in Ikelle and Amundsen (2018) to solve this equation. Here is an example of a Matlab script:

```
L = 30 * 10^3; tau = 1 * 10^(15); TS = 0;
TM = 800.00; Ti = 800;
alpha = 8.0645e - 7; q = 15.0e - 6; rhocp = 3.1 * 10^6;
nx = 100; dz = L/(nx - 1); dt = 0.25 * (dz^2/alpha); nt = tau/dt;
r = alpha * dt/dz^2; s = q/(rhocp) * dt; T = zeros(nx + 1, 1) + Ti;
T(1) = TS; T(nx + 1) = TM;
for n = 1 : nt; Told = T; forj = 2 : nx
    T(j) = Told(j) + r * (Told(j + 1) - 2 * Told(j) + Told(j - 1)) + s;
```

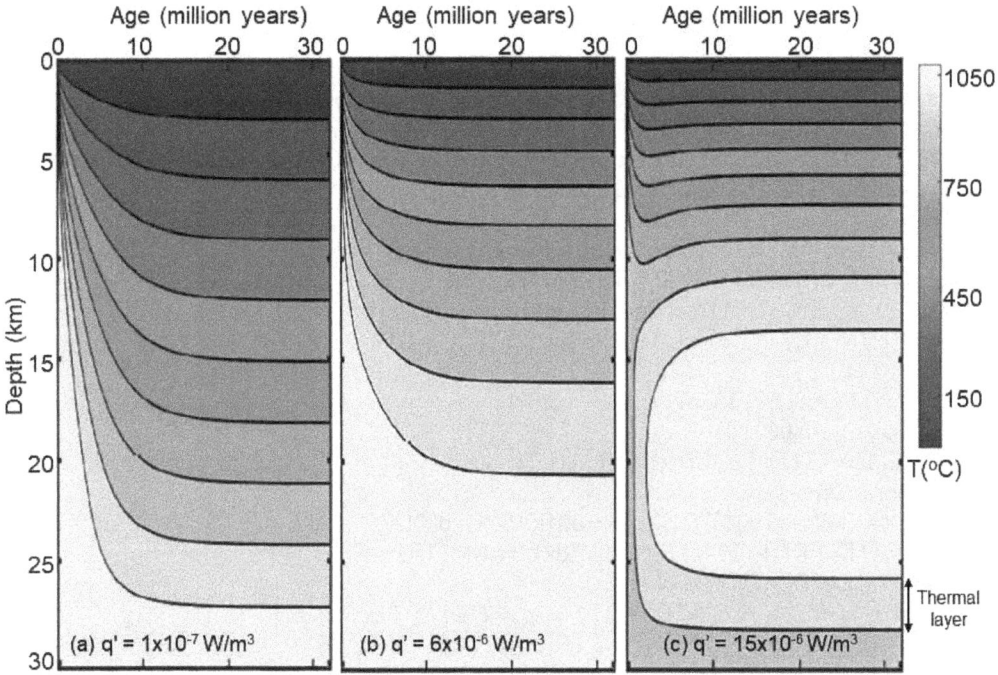

Figure 2.25. Here are lithosphere thermal geotherms under the thermal linear unsteady state and horizontal heat flow (one-dimensional one) with no heat production, as described in (2.113). The thermal diffusivity here is $\alpha = 8 \times 10^{-7}$ m^2/s, the surface heat flux is $q_0 = 8 \times 10^{-2}$ W/m^2, and the thermal conductivity is $k = 2.0$ W/m° C.

```
end; TF(:, n) = T(:); end
[M, c] = contourf(TF, 9); colormap; c.LineWidth = 1.05;
```

The results in Figure 2.25, for $T(z, t = 0) = T_M$ and $T(z = 0, t) = T_S$, show that even uniform heat production has significant effects on the temperature profile. In addition to the creation of the ocean crust, it is possible to model decreasing temperatures with increasing depth when heat production is very high. Because large heat production is rare in the ocean crust, the occurrence of the temperature profile, in Figure 2.25c, is more likely in the continental crust.

Example: Nonlinear heat-conduction equation. Experience shows that the thermal conductivity of the lithosphere is temperature-dependent (e.g., Schatz and Simmons, 1972; and McKenzie et al., 2005). If we assume that horizontal variations of temperature are negligible, we can deduce from (2.55) that the temperature, $T = T(z, t)$, within a cooling plate, satisfies (McKenzie et al., 2005)

$$\rho c_P \frac{\partial T}{dt} = \frac{\partial}{\partial z}\left[k(T) \frac{\partial T}{\partial z} \right] + q' , \qquad (2.115)$$

or

$$\rho c_P \frac{\partial T}{dt} = k(T) \frac{\partial^2 T}{\partial z^2} + \frac{dk(T)}{dT}\left(\frac{\partial T}{\partial z} \right)^2 + q' . \qquad (2.116)$$

Table 2.10. The Matlab commands for predicting the nonlinear Earth's crust geotherm.

```
clear all; close all; rho=3*10^3; cp=2.5*10^3; k0=2.5; b0=0.0022;
nt=91; nx=81; tfinal=15*10^(13); dt=tfinal/(nt-1); A=rho*cp/dt;
x0=0.01*10^3; xL=30*10^3; dx=(xL-x0)/(nx-1); T0=0.75*10^3; TL=1.5*10^3;
x=zeros(nx,1); TT0=zeros(nx,1); t=zeros(nt,1);
TT1=zeros(nx,1); TT=zeros(nx,1);
Tnew=zeros(nx,1); G=zeros(nx,1); TF=zeros(nx,nt);
for i=1:nx; x(i)=x0+(i-1)*dx; TT0(i)=TL; end;
for n=1:nt; t(n)=(n-1)*dt; end;
L=zeros(nx,nx); L(1,1)=1; L(nx,nx)=1; TF(:,1)=TT0;
for n=2:nt
   TT=TF(:,n-1); TT1=TF(:,n-1); eps=1;
   while(eps>0.0001)
     G(1)=TT(1)-T0; G(nx)=TT(nx)-TL;
     for i=2:nx-1
       k=k0*exp(b0*TT(i)); Dk=b0*k; D2k=b0*Dk;
       DT=(TT(i+1)-TT(i-1))/(2*dx); f=(A*(TT(i)-TT1(i))- ...
         Dk*DT*DT)/k;
       q=(-f*Dk+A-D2k*DT*DT)/k; p=-2*Dk*DT/k;
       G(i)=TT(i+1)-2*TT(i)+TT(i-1)-dx*dx*f;
       L(i,i-1)=1+0.5*dx*p; L(i,i)=-2-dx*dx*q; L(i,i+1)=1-0.5*dx*p;
     end
     Tnew=TT-L\G; eps=sqrt(dx*(Tnew-TT)'*(Tnew-TT)); TT=Tnew;
   end
   TF(:,n)=TT;
end; [M,c]= contourf(x,t,TF'); colormap; c.LineWidth = 1.05;
```

Equation (2.115) is generally solved numerically by using the finite-difference technique using a code like the one in Table 2.10 (see also Ikelle and Amundsen, 2018). Figure 2.26 shows the solutions to linear (i.e., $k = k_0 =$ constant, $\rho =$ constant, $c_P =$ constant, and $q' = 0$) and nonlinear conduction-diffusion equations nonlinear conduction-diffusion equations for $k(T) = k_0 \exp(b_0 T)$, $\rho =$ constant, $c_P =$ constant, and $q' = 0$. The temperature increases at a slower rate as one moves deep in the solid Earth for the nonlinear case than in the linear case. The linear solution suggests the lithosphere will not reach a steady state (i.e., a state at which temperatures do not vary time) at an unrealistic age, while in the nonlinear solution steady state, around 2 million years; the thickness of the lithosphere does not grow without limit. The linear and nonlinear solutions in Figure 2.26, along with the linear solutions in Figure 2.25, explain why the thickness of the oceanic crust is smaller than that of the continental crust. Also remember that the most formulas derived in this subsection and the associated figures describe heat flows by conduction down a gradient of decreasing temperature. From the changes in temperatures in Figures 2.24, 2.25, and 2.26, it is clear that thermal conduction in the lithosphere is very slow; about 10 km per a million year or 1 cm/year. The temperature rises more slowly as one moves farther away from the heat source.

Summary. To conclude this subsection, let us make some remarks on lithosphere heat flow on geotherm. The crust of continents and oceans (the continental

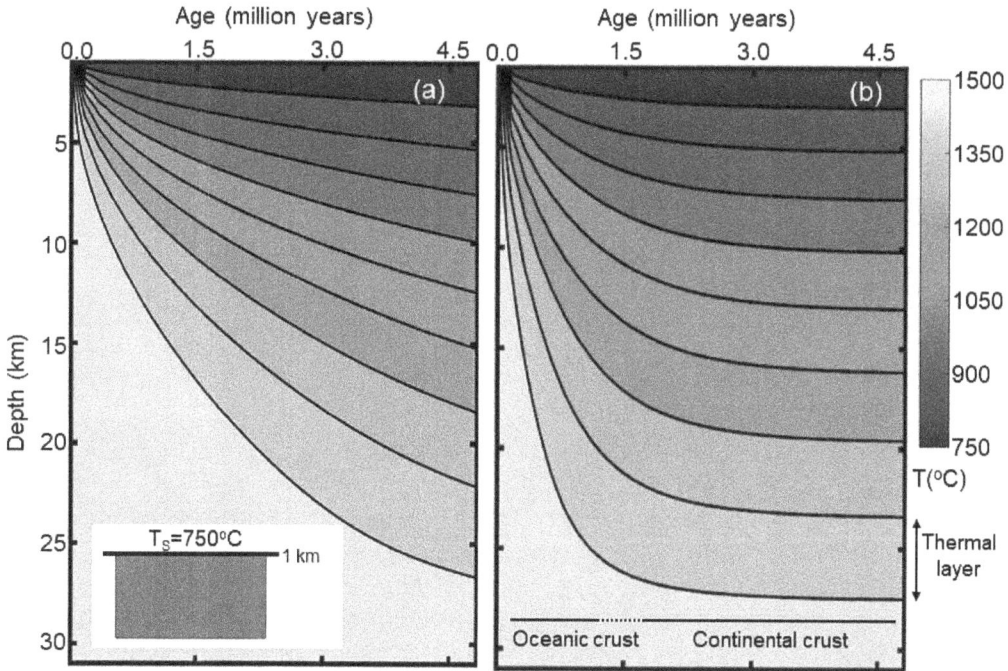

Figure 2.26. Here are lithosphere thermal geotherms under the thermal linear unsteady state and horizontal heat flow (one-dimensional one) with heat production. The thermal diffusivity here is $\alpha = 8 \times 10^{-7}$ m^2/s.

lithosphere and the oceanic lithosphere) are made of different kinds of rocks. They vary in thickness and conduct heat flow differently. The oceanic crust is only 0 to 10 kilometers thick, much thinner than the continental crust. The thinner crust, along with relatively rapid hydrothermal circulation, allows more of the Earth's interior heat to conduct through. The youngest parts of the ocean floor are hot and shallow compared to older parts that are cold and deep. Heat flow is greatest at the very young ocean floor, because it starts out as molten lava and then cools with age. The oceanic lithosphere is not enriched in radioactive elements and its radioactive sources are negligible. Therefore, the geotherm of the ocean crust is essentially a straight line with $dT(z)/dz \approx 30°$ C/km. The key to oceanic heat flow is the process of seafloor spreading whereby the oceanic crust is formed at an ocean ridge and then spreads away until it is eventually subducted back into the mantle at a trench.

The continental crust is 35 to 70 kilometers thick, much thicker than the oceanic crust. The thick crust acts like an insulating blanket, reducing heat flow from the interior to the surface. The heat flow in the continents is controlled partly by the heat supplied to the base of the lithosphere (the reduced heat flow) and partly by radioactive sources in the crust ($q' \neq 0$). Thus, the conductive heat flux in the continental lithosphere increases toward the surface and the geotherm is a curved line, as illustrated in Figure 2.27.

The negative temperature gradient appears because heat flows in the direction of decreasing temperature; that is, heat transfer is in the negative direction. The most important effect causing negative in glaciers is the horizontal motion of the ice.

This horizontal movement of the ice then reduces the vertical conduction. There are some restrictions to the idealized picture glacier model described in this exercise. For example, the free-surface boundary condition is not a perfect sine function and advection is not negligible.

Exercise: As described in Ikelle (2023a), glaciers are very large ice sheets that spread outward in all directions like pancake batter. Arctic and Antarctica are the two most important examples on Earth. The top 15 meters of a glacier are subject to seasonal variations of temperature, which are often described by heat diffusion as follows:

$$\frac{1}{\alpha}\frac{\partial T(z,t)}{\partial t} = \frac{\partial^2 T(z,t)}{\partial z^2} \ , \qquad (2.117)$$

with $T(0,t) = T_0 + \Delta T_0 \sin(wt)$ and $T(\infty,t) = T_0$, where $\alpha = 1/(\rho c_P) = 1.09 \times 10^{-6}$ m^2/s is the thermal diffusivity of ice, T_0 is the mean surface temperature, and ΔT_0 is the amplitude of the periodic changes of the surface temperature, and $2\pi/\omega = 1$ year is duration of a period. Periodically changing boundary conditions at the surface due to winter/summer are here approximated with a sine function. (1) Verify that the solution of equation (2.117) is

$$T(z,t) = T_0 + \Delta T_0 \exp\left(-z\sqrt{\frac{\omega}{2\alpha}}\right) \sin\left(\omega t - z\sqrt{\frac{\omega}{2\alpha}}\right) \ . \qquad (2.118)$$

(2) Here is a Matlab script of this solution:

```
t = 100 * [12096, 38016, 63936, 89856, 115776, 141696, 167616,
193536, 219456, 245376, 271296, 297216]; T = 31104000;
alpha = 1.1e − 6; T0 = −13, DT0 = −6;   yyG = sqrt(pi/(T * alpha)); hold on
for i = 1 : 12;
u(i, :) = T0 + DT0 * exp(−z * yyG). * sin(2 * pi * (t(i)/T)
−z * yyG); plot(z, u(i, :)); end;   grid on; xlabel('z'); ylabel('Temperature')
```

Reproduce the solution in Figure 2.28 for each month of the year using the following parameters: $T_0 = -13°$C, $\Delta T_0 = -6°$C, and $\alpha = 1.1 \times 10^{-6}$ m^2/s. Explain the negative temperature gradient in some cases.

Notice that knowledge of the distribution of temperature in glaciers and ice sheets is of high practical interest because the Antarctic ice sheet, for example, represents a large potential source of sea level rise and glacier lengths are directly related to average air temperature. These glacier lengths also represent an example of proxy measurements before the event of direct temperature measurements. In Europe, for example, we know how they have retreated or advanced since 1850, sometimes even earlier. These measurements are inferred from sketches, paintings, and early maps.

Note also that the geothermal heat flux, which is the key parameter controlling ice sheet dynamics, under Antarctic ice is still significantly unknown. In Ikelle (2024), we discuss how the use of magnetic data can help quantify this parameter.

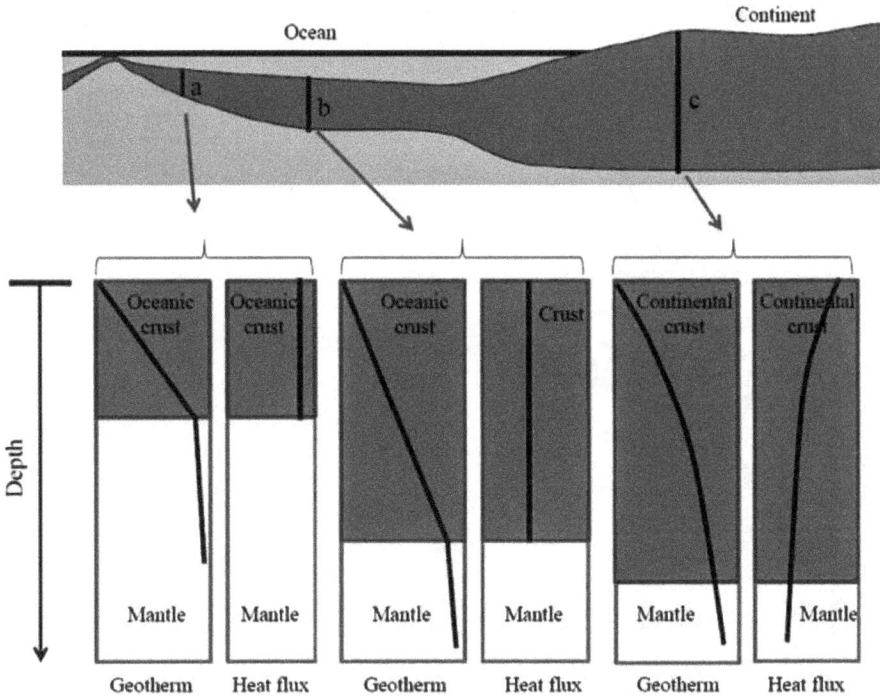

Figure 2.27. Radioactive elements (source of heat) are concentrated in the continental crust and the oceanic crust is depleted of radioactive elements. This is reflected (a) and (b) in a constant heat flux with respect to depth in the oceanic lithosphere and hence the geotherm is a straight line, (c) whereas the continental heat flux increases toward the surface and the geotherm is a curved line. The geotherm in the lithosphere is conductive with a very steep gradient and the geotherm in the convecting mantle is adiabat with a gradient that is much shallower.

Throughout its long geologic history the planet Earth has normally been ice-free. About 50 million years ago, the Earth was free of ice. Giant trees grew on islands near the North Pole, where the annual mean temperature was about 15 degrees Celsius. This period was far warmer than today's mean of 0 degrees Celsius. Remember what the North Pole is today? It is now located in the middle of the Arctic Ocean amid waters that are almost permanently covered with constantly shifting sea ice. So there has been a very large climate change in the last 50 million years. There is also evidence that the earth was almost entirely covered with ice at various times around 500 million years ago; in between, the planet was exceptionally hot. These very large swings in temperature characterize climate change.

We are still in the *Quaternary glaciation Ice Age* which started 2.58 million years ago and continues to the present. This period is characterized by a series of glacial events (the coldest parts of the cycle) separated by interglacial events (the warmest parts of a cycle). At present we are in the warmest part of a cycle. Another cold part of the cycle is coming soon. The change from warm to cool phases in the cycle occurs abruptly with a temperature drop of about 4 to 6 degrees Celsius. Over a period of 85,000 years (on average) the temperature drops to about 9 or 10 degrees Celsius below present. During the last glacial maximum the average global

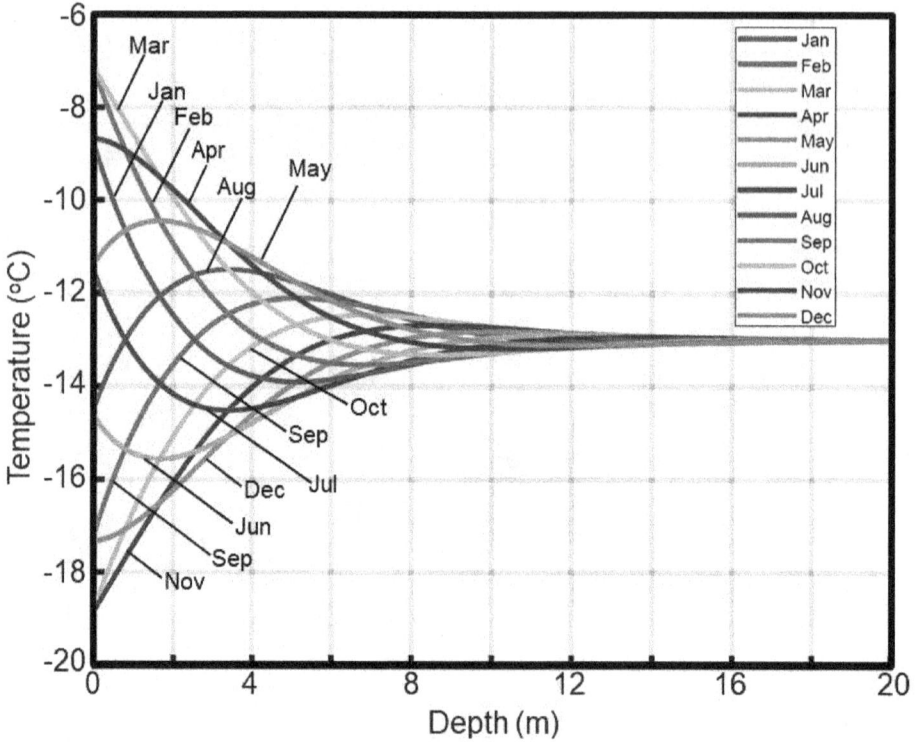

Figure 2.28. Variation of temperature with depth for glaciers. Numbers next to curves indicate months (1 corresponds to January).

temperature was around 6 degrees Celsius cooler than today. For perspective today the average global temperature in the 20th century is 14 degrees Celsius, making the average global temperature back then 8 degrees Celsius.

At best, the CO_2 emission of the industrial era, which began in the late 1700s, have a marginal effect on the Quaternary glaciation Ice Age. So far, it is just like most past ice ages. This is a period when thick ice sheets cover vast areas of land, especially near the poles. An ice age lasts for millions of years. The causes of an ice ages are still unclear. Major causes of ice ages include plate tectonics, ocean currents, and heat distribution. Earth's climate did not stay freezing cold during the entire ice age. Instead, climate changes between glacial and interglacial periods. Glacials are very cold, and last 100,000 years. Interglacials are a much warmer period between two glacials, and they last between 10,000 and 15,000 years. We are currently in the Holocene Interglacial.

2.3.2. Mantle Heat Convection

The mantle starts at a 10 to 70 km depth and goes down to the outer core at a 2,900 km depth. Our objective here is to use the convection-diffusion equation in (2.67) and (2.69), which allows us to determine the geothermic of the mantle. We start by defining mantle convection. As illustrated in Figure 3.33a of Ikelle (2023a), we need a fluid layer bounded by two surfaces and a heater to produce convection. In the interior solid Earth, we have convection between the cooler crust (cooler rocks at shallow depth) and the hotter core (hotter rocks at great depth). In other words, the mantle can be described as a layer of fluid heated from below and cooled from above (see Figure 3.33b of Ikelle (2023a)). The "fluids" (mantle plume) here are the sublithospheric mantle's rocks, which heat up and partially or completely melt in the deeper and hotter mantle. These so-called fluids are viscous (viscosity is the rock's resistance to flow; it decreases with increasing temperature). In any event, we have mantle convection as first pointed out by Bull (1921) and Holmes (1931).

As pointed out in Ikelle (2023a), convection takes place in fluids (liquid or gas) with a constant density before the change in temperature. It is obvious that the mantle is not a fluid and its density is not homogeneous. Yet mantle-wide convection (or partial mantle convection) is a central tenet of plate tectonics. We will revisit this issue in Chapter 6. However, in the context of building an educational background, which is the main goal of this book, the validity of these assumptions has little effect. If these assumptions are far from reality, we still have to demonstrate to our students how to derive and interpret geotherm related to convention. Moreover, this exercise will help them understand and contribute to future developments in this area.

Here are some mantle parameters (Schubert et al., 2001):

$$\left. \begin{array}{l} \rho \approx 4000 \ (\text{kg/m}^3) \\ g = 10 \ (\text{m/s}^2) \\ \beta = 3 \times 10^{-5} \ (\text{K}^{-1}) \\ \nu = 10^{22} \ \text{Pa} \cdot \text{s} \\ \Delta T \approx 3000 \ \text{K}, \\ d = 2900 \ \text{km}, \\ \alpha = 10^{-6} \ \text{m}^2/\text{s} \end{array} \right\} \Longrightarrow R_a \approx 10^7$$

The Rayleigh number is approximately 10^7, which is well beyond critical although the mantle viscosity is extremely high. However, the mantle is also very hot and very large and hence it is convecting vigorously.

Note again that rock expands as its temperature increases, its density thereby decreases slightly. A one percent expansion requires an increase of 300-400 degrees Celsius and leads to a 1 percent decrease in density. Viscosity is the propensity of rock to ductile flow. Rock does not need to melt before it can flow. The presence of H_2O encourages flow in solid rock. Convection currents bring hot rocks upward from Earth's interior. The rock in the lithosphere is too cool for convection to continue. Heat moves through the lithosphere primarily by conduction. The lithosphere-asthenosphere boundary is about 1,300 degrees Celsius.

To summarize, solid Earth moves heat via convection and conduction as follows:

$$\underbrace{\text{Heating} \rightarrow \text{expansion}}_{\text{Mantle}} \rightarrow \text{rise} \rightarrow \underbrace{\text{cooling} \rightarrow \text{contraction}}_{\text{Lithosphere}} \rightarrow \text{sink} \ .$$

Basically, hot material expands and rises, and cold material contracts and sinks through a process of creation and destruction of the oceanic lithosphere. Heat is delivered to the base of the lithosphere by mantle convection that brings hot material to shallower depths, thus releasing pressure, enabling melting of both the rising material and the lithospheric rocks that it contacts.

Thermal geotherm. Deriving the geotherm of the mantle comes down to solving the convection-diffusion equation, in (2.67) or in (2.69). This equation is also known as the advection-diffusion equation. In this first example, we assume the thermal steady state (i.e., $\partial T/\partial t = 0$), and no radioactive heat production rate (i.e., $q' = 0$), and that horizontal heat flow is not important so that we can reduce the problem to a one-dimensional one (i.e., 1D assumption). Equation (2.69) becomes

$$\left(u_z \frac{\partial T}{dz}\right) - \alpha\frac{\partial^2 T}{dz^2} = 0 \iff T(z) = C_1 \left[\exp\left(\text{Pe } z\right) - 1\right] + C_2 , \qquad (2.119)$$

where $\text{Pe} = (u_z \times L)/\alpha$ is the Péclet number (a dimensionless number, named after the French physicist Jean Claude Eugène Péclet), L is the length scale of the convection process (e.g., the width of a fluid duct), u_z is the flow velocity, α is thermal diffusivity, and C_1 and C_2 are the integration constants resulting from the integration of the simplified diffusion equation twice with respect to z. We here assumed that u_z and α are constants. The Péclet number can alternatively be written in term of diffusion and advection times, as follows:

$$\text{Pe} = \frac{t_{\text{diffusion}}}{t_{\text{advection}}} = \frac{L^2 u_z}{L\alpha} = \frac{L u_z}{\alpha} . \qquad (2.120)$$

We can see that if $\text{Pe} \ll 1$ then, advection can be neglected and the processes will be diffusion dominated. Temperatures can be modeled with $u_z = 0$ in (2.119). If $\text{Pe} \approx 1$, then both advection and diffusion terms must be included in the modeling of temperatures. If $\text{Pe} \gg 1$ then advection plays the dominant role in the modeling of temperatures.

To obtain C_1 and C_2 in (2.119), we assume constant temperatures at the top of mantle (i.e., $T = T_S$ at $z = 0$) and the bottom of mantle base mantle ($T = T_M$ at depth $z = z_M$). The analytical solution of the 1D steady advection-diffusion equation in (2.119) is then

$$T(z) = (T_M - T_S) \frac{\exp\left(\text{Pe } z\right) - 1}{\exp\left(\text{Pe } z_{\text{M}}\right) - 1} + T_S . \qquad (2.121)$$

Figure 2.29 shows this geotherm for various values of Pe. When $\text{Pe} = 0$, the advection-diffusion equation in (2.119) reduces to pure diffusion transport; i.e.,

$$\frac{d^2 T(z)}{dz^2} = 0 \iff T(z) = C_1' z + C_2' , \qquad (2.122)$$

where C_1' and C_2' are the integration constants. For $T(z = 0) = 0 = C_2'$ and at $T(z = z_M) = T_M = z_M C_1'$, the geotherm is a straight line between $z = 0$ and z_M like in oceanic crust, as illustrated in Figure 2.29; i.e.,

$$T = T_M \frac{z}{z_M} . \qquad (2.123)$$

When Pe is positive, the flow direction is from the top of the mantle to the base of the mantle. For large Pe values, T_S prevails almost everywhere except the small

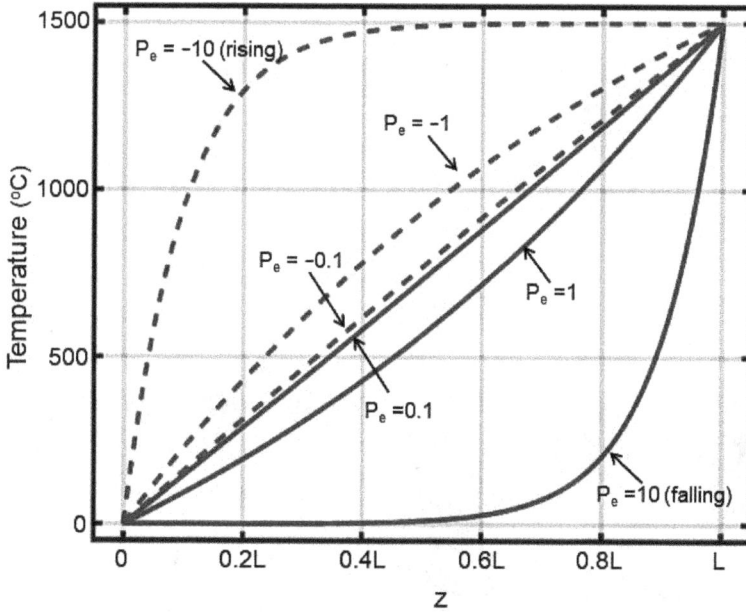

Figure 2.29. Here are lithosphere thermal geotherms under the thermal unsteady state and horizontal heat flow (one-dimensional one) with no heat-production, as described in (2.108). The thermal diffusivity here is the order of that of granite; that is, $\alpha = 8 \times 10^{-7}$ m^2/s.

region near $z = L$ in this case, as illustrated in Figure 2.29. For negative Pe, the flow direction is from the base of the mantle to the Earth's surface. For large Pe values, T_M prevails almost everywhere except the small region near $z = 0$ in this case, as illustrated in Figure 2.29. Basically, the hotter materials are now coming up. Hence, the difference between the top of the mantle and the base of the mantle is not significant. In other words, the temperature does not change in the mantle.

Consider now the one-dimensional time-dependent convection-diffusion equation of the evolution through time of the temperature patterns; i.e.,

$$\rho c_P \left[\frac{\partial T}{\partial t} + \left(u_z \frac{\partial T}{\partial z} \right) \right] - \frac{\partial T}{\partial z} \left[k(T) \frac{\partial T}{\partial z} \right] = q' . \tag{2.124}$$

Again we assume that the thermal conductivity, $k(T)$, is temperature-dependent. We can solve this equation by an implicit finite-difference method as described in Ikelle (2018). Figure 2.30 shows the solution of this equation for $q' = 0$ and $k(T) = k_0 \exp(b_0 T)$, where k_0 and b_0 are constants for a specific material or layer. Generally speaking, b_0 is negative for good conductors and positive for good insulators. For $b_0 = 0$, we have the results, in Figure 2.30b, which are similar to those conduction-heat transfer through lithosphere described in the previous subsection because Pe is close to zero. Also, for $b_0 = 0$, we have the results, in Figure 2.30e, that represent unsteady geotherm in which advection plays the dominant role. We can see that the temperature variations are limited in the base of the mantle. The temperature-dependence of thermal conductivity can cause the steady states of conduction to occur at early age (see Figure 2.30a) compared to the

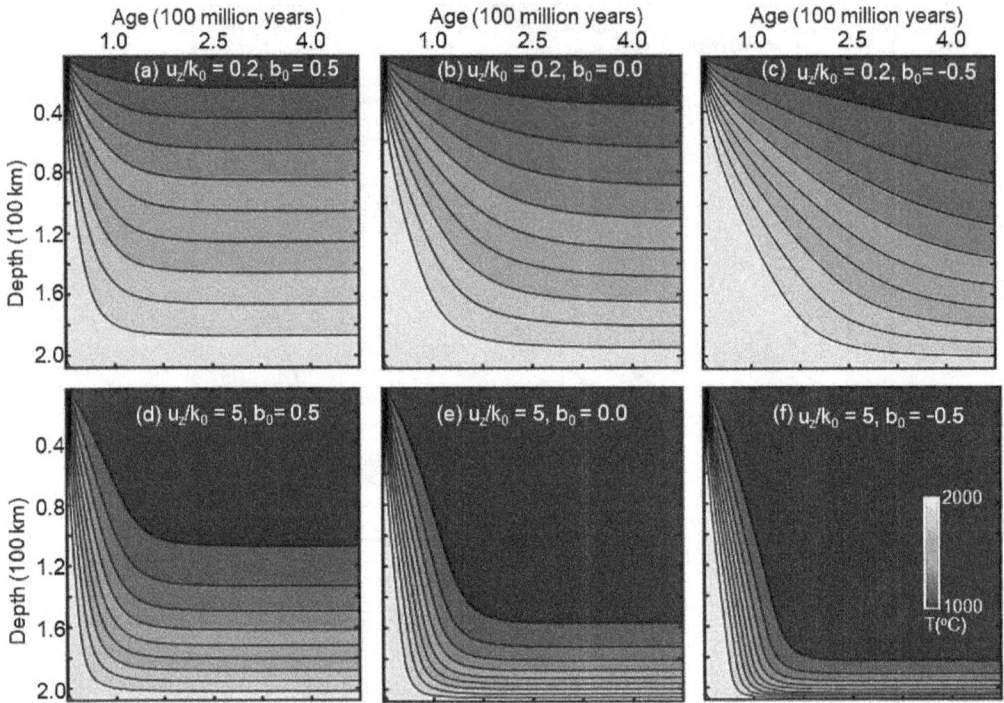

Figure 2.30. The steady advection-convection Peclet number. Pe = 10 indicates fluid advection and Pe = 0.1 indicates thermal conduction.

linear case (see Figure 2.30b), when the exponential factor is positive and quite late age (see Figure 2.30c), compared to the linear case when the exponential factor is positive. Similar, the temperature-dependence of thermal conductivity can cause the steady states of advection to tend toward to those of geothermal conduction, when the exponential factor is positive (see Figure 2.30d) or to further limit the temperature variations to the base of the mantle (see Figure 2.30f). In other words, the temperature-dependence of thermal conductivity is central in fitting observed or inferred temperature distributions to a particular heat model of the subsurface.

Quiz: Is the Péclet number mainly positive or negative for glaciers? Explain your answer.

Example: Adiabatic geotherm. We here derive of the temperature gradient of the mantle under the assumption of no heat exchange occurs during convection (that is constant entropy); i.e.,

$$\left(\frac{dT}{dz} \right)_S .$$

This gradient is known as the adiabatic gradient and the corresponding geotherm as adiabatic geotherm (or adiabat). We refer to the gradient derived in the previous example as thermal geotherms. Remember that constant entropy means that no heat is lost or gained in the process under consideration. So the adiabatic gradient is a limit: It is the temperature gradient that would exist if material rose and fell without losing or gaining heat through conduction. The comparison between the thermal geotherm and adiabatic geotherm provides us with an additional insight into the mantle convection process, as we will see later.

We need these two identities in the derivations of the adiabatic gradient:

$$\left(\frac{\partial z}{\partial T}\right)_S \left(\frac{\partial T}{\partial P}\right)_S \left(\frac{\partial P}{\partial z}\right)_S = 1 \tag{2.125}$$

or

$$\left(\frac{\partial T}{\partial z}\right)_S = \left(\frac{\partial T}{\partial P}\right)_S \left(\frac{\partial P}{\partial z}\right)_S , \tag{2.126}$$

$$\left(\frac{\partial P}{\partial T}\right)_S \left(\frac{\partial T}{\partial S}\right)_P \left(\frac{\partial S}{\partial P}\right)_T = -1 \tag{2.127}$$

or

$$\left(\frac{\partial T}{\partial P}\right)_S = -\left(\frac{\partial T}{\partial S}\right)_P \left(\frac{\partial S}{\partial P}\right)_T . \tag{2.128}$$

Using the coefficient of thermal expansion in (2.60), the specific heat in (2.33), the pressure gradient

$$\frac{dP}{dz} = g\rho , \tag{2.129}$$

and the Maxwell relation in (2.30), we obtain the change of temperature with pressure at constant entropy as

$$\left(\frac{\partial T}{\partial P}\right)_S = -\left(\frac{\partial T}{\partial S}\right)_P \left(\frac{\partial S}{\partial P}\right)_T = \left(\frac{\partial T}{\partial S}\right)_P \left(\frac{\partial V}{\partial T}\right)_P \tag{2.130}$$

or

$$\left(\frac{\partial T}{\partial P}\right)_S = \frac{T\beta}{\rho c_P} . \tag{2.131}$$

We now have the adiabatic gradient as a function of pressure. However, we really want it as a function of depth. By using the hydrostatic equation in (2.129) and the identity in (2.126), we arrive at the adiabatic gradient as a function of depth:

$$\left(\frac{dT}{dz}\right)_S - \left(\frac{dT}{dP}\right)_S \left(\frac{dP}{dz}\right)_S \Longleftrightarrow \left(\frac{dT}{dz}\right)_S = \frac{T\beta g}{c_P} . \tag{2.132}$$

Here is an example of the adiabatic gradient obtained by using some typical values from Turcotte et al. (2001).

$$\left.\begin{array}{l} T = 1600 \text{ K} \\ \beta = 3 \times 10^{-3} \text{ K}^{-1} \\ \rho \approx 4000 \text{ (kg/m}^3) \\ g = 10 \text{ (m/s}^2) \\ c_P = 1 \times 10^3 \text{ J/(kg K)} \end{array}\right\} \Longrightarrow \left(\frac{dT}{dz}\right)_S \approx 0.5°\text{C/km} .$$

The adiabatic gradient in the shallow mantle works out to be 0.5°C/km, in the uppermost mantle it is 0.4°C/km, and at greater depth it is 0.3°C/km. These gradients are very small compared with the lithospheric gradients obtained in the previous section (see also Figures 2.22 and 2.23). They tells us that the temperature does not change much in the mantle. Basically, mantle heat convection is not driven by the temperature itself, but rather the densities of the rocks that are important for convection. Density depends much more strongly on pressure than on temperature. In the Earth's mantle, density increases downward because the increasing pressure has a considerably larger effect on density than does the increasing temperature.

As we mentioned earlier, the adiabatic gradient is a limit. If the thermal gradient is greater than the adiabatic gradient, there is energy available to drive the vertical motion of rocks in the mantle, and we have thermal convection (see Figure 2.31a). In other words, it is the difference in temperature between the thermal geotherm (the actual temperature profile in the Earth as a function of depth) and the adiabatic geotherm (the curve whose slope is the adiabatic gradient at all depths) that drives convection (Figure 2.31b). A mantle with a thermal geotherm at or below the adiabatic geotherm cannot convect and can only lose heat by the slow process of conduction. Another way to grasp these conclusions is to directly relate thermal

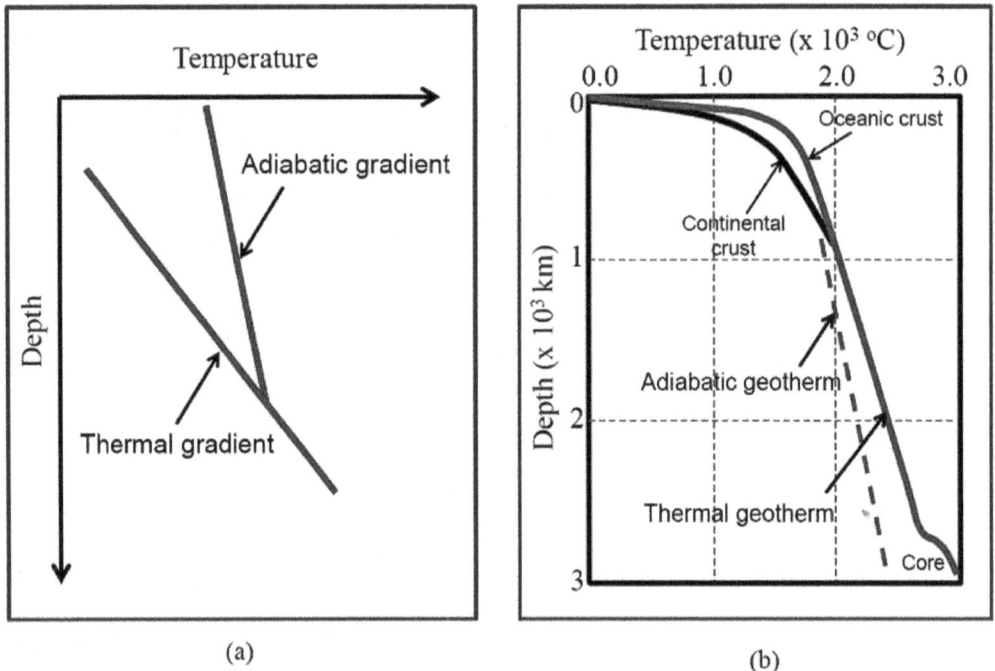

Figure 2.31. (a) An illustration of thermal and adiabatic gradients. If the thermal gradient is greater than the adiabatic gradient, we have thermal convection. (b) Similarly, a mantle convects when the thermal geotherm is above the adiabatic geotherm (adiabat). Modeling results show that thermal and adiabatic geotherms in the upper mantle (below 500 km) are almost identical. However, there are significant differences between the two geotherms in the lower mantle (500 km to 3000 km).

and adiabatic gradients. In general we can write the thermal gradient as follows:

$$\frac{dT}{dz} = \underbrace{\left(\frac{\partial T}{\partial P}\right)_S \frac{\partial P}{\partial z}}_{\text{Adiabat}} + \underbrace{\left(\frac{\partial T}{\partial S}\right)_P \frac{\partial S}{\partial z}}_{\text{Non-adiabat}}. \tag{2.133}$$

Notice that the first term on the right-hand side includes the adiabatic gradient and the second term is obviously zero for adiabatic geotherm (adiabat). The second term characterizes the mantle's convection, including its vigor.

Example: Solidus and liquidus. Along with adiabatic geotherms, solidus and liquidus are other limits of thermal geotherms of solid Earth. Solidus (also known as the melting curve) is the curve above which rocks start to melt. Liquidus is the curve above which rocks become entirely liquid. Both the geotherm and melting-point curves generally increase gradually with depth (see Figure 2.32). Despite the large increase in temperature, materials are not necessarily melting because of the simultaneous increases in temperature and pressure. Materials under increasing temperature tend toward being liquid and under increasing pressure tend toward being solid. This dynamic allows the melting-point curve to also increase with depth. Generally, the melting point curve increases more rapidly with depth than the geotherm, as illustrated in Figure 2.32.

Our knowledge of solidus and liquidus is derived from laboratory experiments of melting key subsurface materials, namely, olivine, periodite, basalt, granite, and iron. Because of the very high in-situ temperatures and very high in-situ pressures of

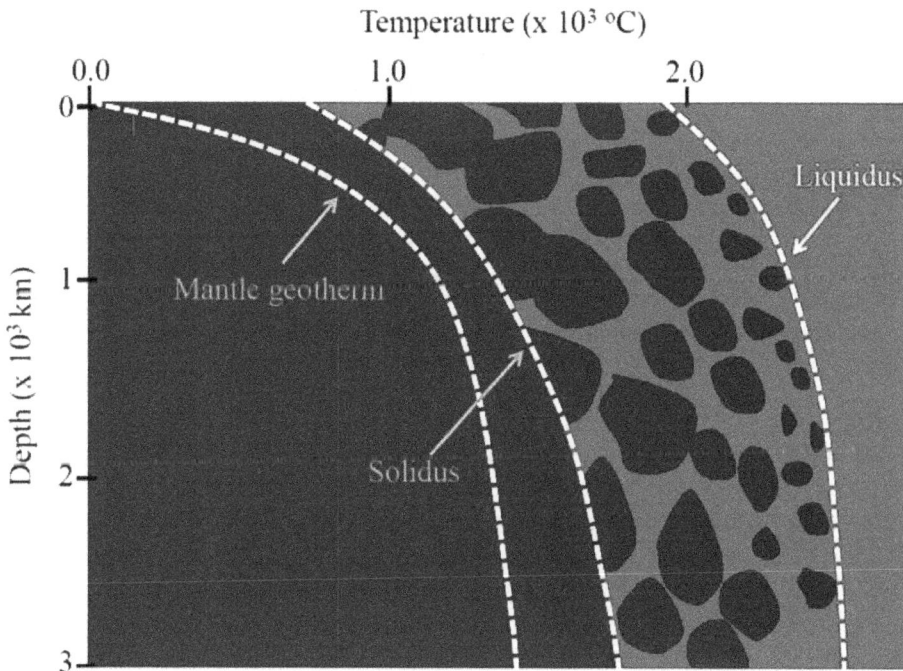

Figure 2.32. A schematic diagram of solidus and liquidus of mantle rocks. The dark color indicates solid rock, and the light color indicates molten rock. At a certain depth, with increasing temperature, rock starts to melt. The first melt appears at solidus temperature. Then the amount of the solid portion reduces gradually with more temperature. The rock completely melts beyond liquidus.

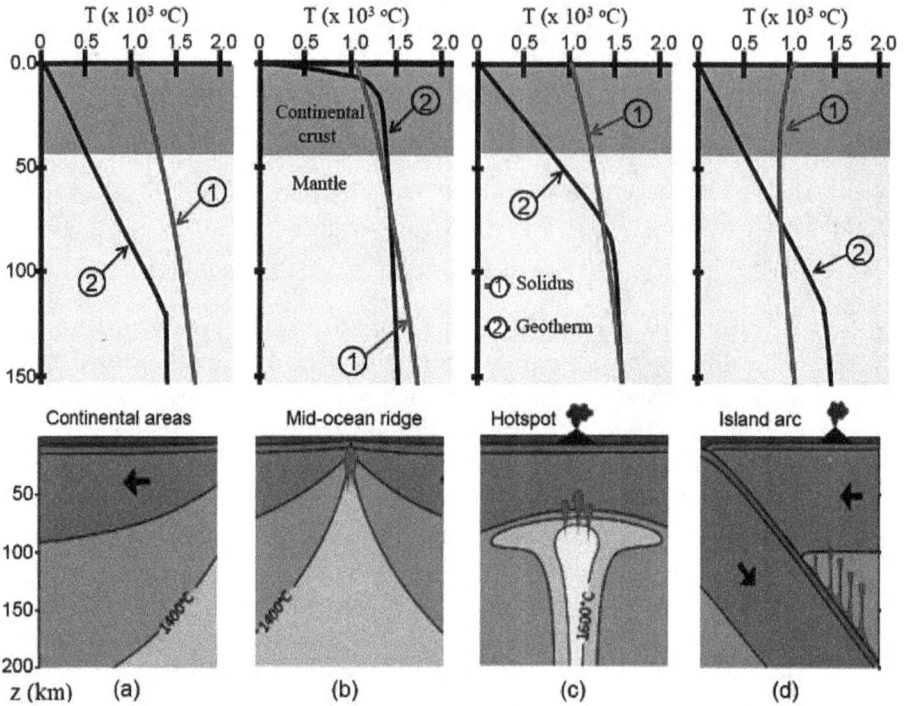

Figure 2.33. (a) In continental areas away from tectonically active zones, geotherm never reaches the solidus (melting points of dry basalt at in-situ pressure; dry granite is also used) instead of the rock; geotherm is always well below mantle solidus temperature, so the mantle does not melt. The melting curves are characterized as dry because we are in a dry environment. (b) At mid-ocean ridges, the temperature increases quickly with depth due to rising hot magma beneath a thin lithosphere. The geotherm crosses the solidus and the mantle starts melting. (c) At hotspots, there is also a zone in which the geotherm exceeds the solidus, leading to melting and magma formation. Mantle plumes also undergo melting as they rise toward the surface. This is known as decompression melting. (d) Temperature increases slowly with depth at subduction zones due to the relatively cool sediments and fluids (i.e., seawater) being subducted along with the old, cold ocean lithosphere. Even though the rock is relatively cool, the solidus (melting point of wet granite at insitu pressure; wet basalt is also used) is depressed by water-bearing rock and sediment, leading to melting and convergent plate boundary volcanism. The melting curves are characterized as wet because we are in a wet environment.

the subsurface, it cannot be replicated in laboratories (e.g., in the core, pressure and temperature conditions can exceed 364 GPa and 5,000 degrees Kelvin, respectively), the solidus and liquidus are obtained by extrapolating the results of experimental data obtained at moderate pressures.

How close the geotherm is to the melting point of a material not only determines whether a material is molten or not, an essential feature for the occurrence of mantle convection, but how stiff it is. Melting begins when the geotherm, including the adiabat, crosses the solidus. In continental areas away from tectonically active zones, the geotherm never reaches the solidus (melting point) of the rock (see Figures 2.33 and 2.34). In mid-ocean ridges, temperature increases quickly due to rising hot

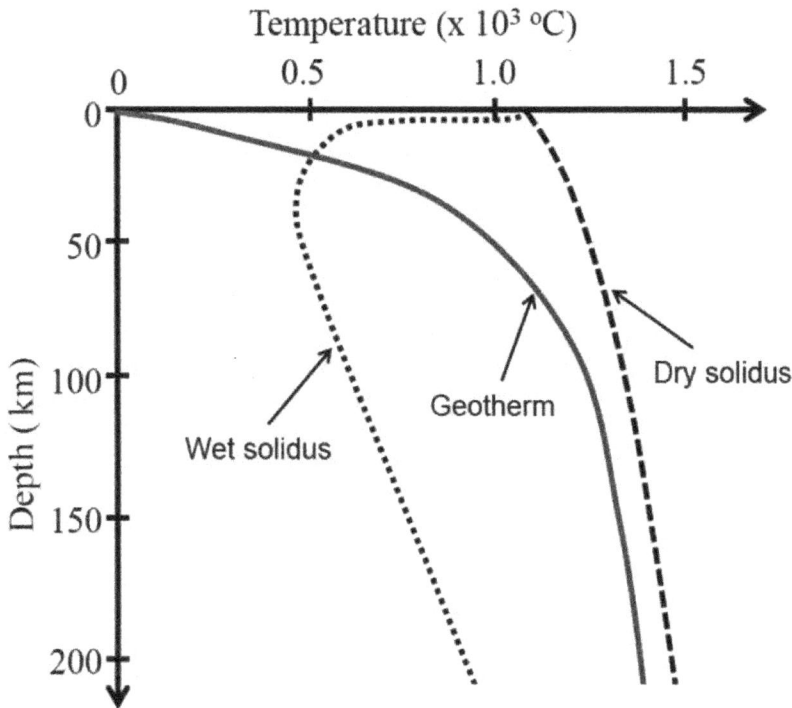

Figure 2.34. A comparison of solidus and geotherm. Melting begins when the geotherm crosses the solidus. In continental areas away from tectonically active zones, the geotherm never reaches the solidus (dry solidus) of the rock. There is a zone in which the geotherm exceeds the wet solidus.

magma beneath a thinner lithosphere (see Figures 2.33 and 2.34). Note that there is a zone in which the geotherm exceeds the solidus, leading to melting and magma formation. In the subduction zones, temperature increases slowly with depth, due to the relatively cool sediments and fluid (i.e., sea water) being subducted along with the old, cold ocean lithosphere that penetrates to great depths. Even though the rock is relatively cool, the solidus is depressed by water-bearing rock and sediment, leading to melting and convergent plate boundary volcanism.

Melting beneath the crust occurs for two basic reasons. The activity at sub-duction is one reason. Subduction zones drive the ocean crust and its blanket of marine sediments down into the mantle. The subducting slabs are relatively cold. They have been at the cold ocean bottom and have cooled since formation at a mid-ocean ridge. Further, the sediments they carry contain minerals that have low melting points, having been formed at the Earth's surface at cool temperatures. Subducting slabs are also accompanied by water, which further depresses melting points for the minerals—water is for rocks what salt is for snow ice; throw some salt on ice and it will melt at a lower temperature than 0 degrees Celsius. Consequently, as the slabs are carried into the deeper, hotter mantle, they heat up and partially or completely melt. This molten rock, or magma, is less dense than the surrounding rock, and rises buoyantly through the lithosphere to the surface, where it makes volcanoes.

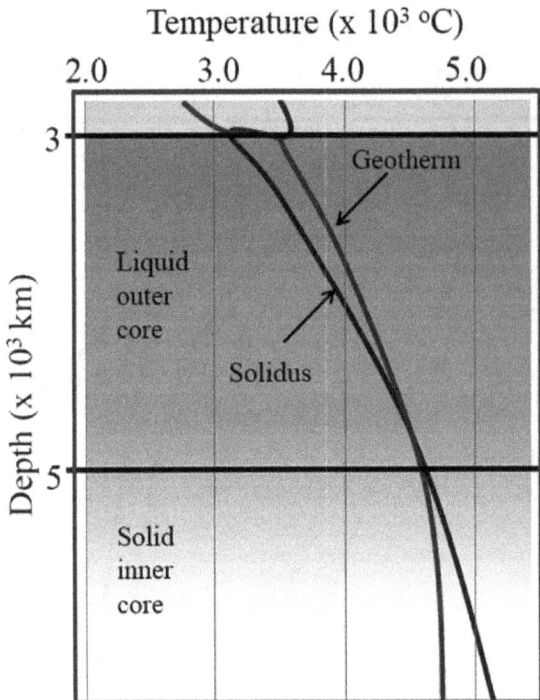

Figure 2.35. A comparison of the geotherm and solidus of the Earth's core. The outer core is liquid because the melting curve lies to the left of the geotherm, so it melts. The inner core is solid because the geotherm is not great enough to melt it at such high pressure.

The other reason melting occurs beneath the crust is mantle convection. Mantle convection, which brings heat to the surface, causes hotspots to occur. Heat is delivered to the base of the lithosphere by mantle convection, which brings hot material to shallower depths, thus, releasing pressure, enabling the melting of both the rising material, and the lithospheric rocks that it contacts. The sites where volcanoes form from this sort of deep-seated heat delivery are called hot-spots.

Quiz: (1) Define a process where a parcels temperature changes due to expansion or compression, but no heat is added or taken away from the parcel. (2) Explain why Africa and Antarctica have numerous hotspots (see Ikelle, 2023a).

2.3.3.   Conduction and Convection in the Core

As illustrated in Figure 2.35, the outer core is liquid because the melting curve lies to the left of its geotherm, so it melts. The inner core is solid because its geotherm is not large enough to melt it due to high pressure.

The Earth's liquid outer core is also a major contributor to heat production on solid Earth and consequently at the surface. Its exact temperature profile is still not certain. It is generally inferred from melting experiments on materials (e.g., iron) that we believe are present in the deep Earth. Some calculations suggested that its temperature is above 6000 degrees Kelvin and its pressure is about

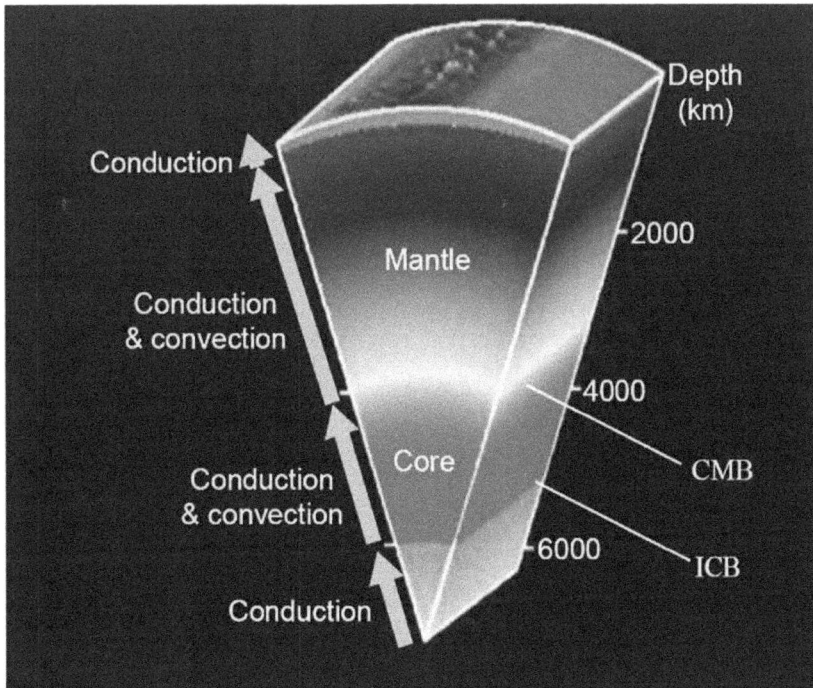

Figure 2.36. A diagram showing the dominant mode of heat transfer in the main layers of the solid Earth. Basically, heat travels from the Earth's interior to the surface, through the processes of convection and conduction. ICB stands for inner core boundary and CMB stands for core-mantle boundary.

330 GPa. This large amount of heat in the inner core is leftover from the accretion of the planet more than 4.5 billion years ago, and the inner core still contains large amounts of this heat. Some of it is conducted from the inner core into the outer core, as illustrated in Figure 2.36. Convection plays a significant role in carrying this heat to the top of the core. Once the heat reaches the lower mantle, it is carried toward the surface through mantle convection, as described in the previous sections. Actually, the thermal evolution of the core is controlled by mantle convection that imposes the total heat loss of the core.

Convection is an important mechanism within the outer core. As iron crystallizes and sinks to form the solid inner core, it leaves behind a melt that contains a higher percentage of lighter elements, so basically we have solidification on one hemisphere of the inner core boundary (ICB) and melting on the other. Because this liquid is more buoyant than the surrounding materials, it rises upward, creating convection. However, heat is only passed from the core to the mantle through conduction, not convection, because iron is much too dense to intrude into the lighter mantle, floating on top.

By assuming that the temperature profile of the Earth's core follows the adiabatic geotherm, we can obtain the geotherms of the Earth's core by solving (2.128) and (2.131), with the boundary condition that the temperature at the ICB is equal to the temperature of crystallization at the ICB. By using these equations and boundary conditions, Labrosse et al. (2003) shows that the temperature profile of

Table 2.11.   The Matlab commands for predicting the Earth's geotherm.

```
>> clear all; close all
>> zl = 190000; k = 3.8; qp = 0.25e-6; qm = 10e-3; A = qp/(2*k);
>> B = (qm+(qp*zl))/k; zz=0:1000:zl; T=-A*(zz.*zz)+B*zz;% Litho
>>
>> ra=2900; rm=191:1:ra; T0 = T(191);
>> T(192:ra+1)=T0+(0.55*rm); % Mantle
>>
>> r0 =3480; r=r0:-1:1; Lp = 7680;
>> r1= 1220; rnm = 1.5*((r1*r1)-(r.*r))/(Lp*Lp); % Core
>> T(ra+2:ra+r0+1) = 5000*exp(rnm); zzkm =0:ra+r0; plot(zzkm,T)
>> grid on
>> axis([0 6500 0 6000])
>> print figure37.pdf -dpdf
>>
```

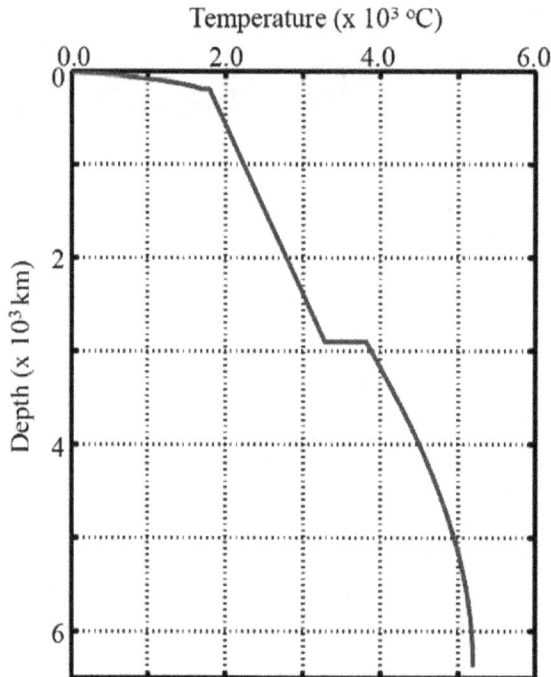

Figure 2.37. Geotherm of the Earth obtained by using the Matlab commands in Table 2.11.

the core can be approximated as follows:

$$T(r) = T_{LI} \exp\left[\frac{3}{2}\left(\frac{r_0^2 - r^2}{L_\rho^2}\right)\right], \tag{2.134}$$

where r is the radial position in the core, $r_0 = 1220$ km is the radius of the inner core, T_{LI} is the liquidus of the outer core (generally the lowest temperature at which iron becomes completely liquid for the pressure at ICB). The quantity L_ρ is known as the density length scale and is defined as $L_\rho = 3K_c/(2\pi G\rho_c^2) \approx 7680$ km, where G is the gravitational constant and K_c and ρ_c are the bulk modulus and the density at the center of solid Earth, respectively.

In summary, there are four main radioactive isotopes that are sources of heat on solid Earth: uranium-235, uranium-238, thorium-232 and potassium-40. The half-lives of these four isotopes are on the order of billions of years. As a result, large quantities of these isotopes remain in solid Earth and have kept mantle convection and plate tectonics active for billions of years. The Earth's liquid outer core is also a major contributor to heat production in solid Earth left over from the accretion of the planet more than 4.5 billion years ago.

> **Computational exercise:** Use the Matlab commands, in Table 2.11 to reproduce the geotherms of the Earth in Figure 2.37.

Heat travels in solid Earth by conduction and convection (see Figure 2.36). Heat conducts from the core into the mantle. Once the heat reaches the lower mantle, it is carried toward the surface through mantle convection. Heat from the mantle moves to the earth's surface through the rocky lithosphere—in hotspots, convection carries heat directly to the surface, in the form of volcanic lava. Everywhere else the heat makes it to the surface by conducting slowly across the strong, rigid lithosphere. The resulting solid Earth's temperature increases from about 0 degree Celsius at its surface to about 6,500 degrees Celsius at its center. Within the earth's crust, the temperature increases rapidly, about 30 degrees Celsius per kilometer of depth. At the base of the lithosphere, about 100 kilometers down, the temperature is roughly 1,400 degrees Celsius. However, you would need to descend to almost the bottom of the mantle before the temperature doubled to 2,800 degrees Celsius. For most of the mantle, the temperature increases very slowly—about 0.5 degree Celsius per kilometer of depth.

> **Computational exercise:** The Matlab code in Table 2.12 contains a numerical solution of the conductive heat diffusion equation with boundary conditions at $z = 0$ and $z = L$ and initial conditions at $t = 0$. Use this code to reproduce the geotherm solution of the lithosphere in Figure 2.38.

Geoscientists determine these temperatures by solving equations (2.55) and (2.67). The boundary and initial conditions needed to solve these equations are provided by the location of boundaries of layers of the solid Earth determined from seismic data and the laboratory measurements of melting curves of subsurface minerals, such as olivine, granite, basalt, periodite, and iron.

2.3.4. Energy Transport in Stars

As described in the previous subsections of this section, solid Earth loses its internal heat through heat conduction and convection. The upper 100-200 km layer of Earth cools and stiffens to form the lithosphere. Internal heat conducts through the lithosphere to the Earth's surface. Beneath the lithosphere, internal heat is transported by moving rock in a convective circulation. What happens in the stars?

Table 2.12. The Matlab code for solving the conductive heat diffusion.

```
L = 100000; % Lithosphere depth in meters (100 km)
dz = 1000; % depth step is meter (1 km)
tmax = 300*5.154*10^13; % duration of the propagation in s (300 MA)
dt = 10*5.154*10^10; % time step in seconds (10 thousand years)
alpha = 8.5*10^-7; % thermal diffusivity in m^2/s
r = alpha*dt/dz^2; r2 = 1 - 2*r; % r must be smaller than 0.5
nz = 1 + floor(L/dz); nt = 1 + floor(tmax/dt);
z = linspace(0,L,nz)'; t = linspace(0,tmax,nt); T = zeros(nz,nt);
T(:,1) = 273*(z/L); T(1,:) = 0; T(nz,:) = 1200; % ic and bc
for m=2:nt
    for i=2:nz-1
        T(i,m) = r*T(i-1,m-1) + r2*T(i,m-1) + r*T(i+1,m-1);
    end
end
nsnap=floor(nt/5); nsnap2=floor(nt/100); plot(z,T(:,1),'k-'); hold;
plot(z,T(:,2*nsnap2),'k'); plot(z,T(:,5*nsnap2),'b');
plot(z,T(:,10*nsnap2),'g'); plot(z,T(:,15*nsnap2),'m'); plot(z,T(:,5*nsnap),'r');
```

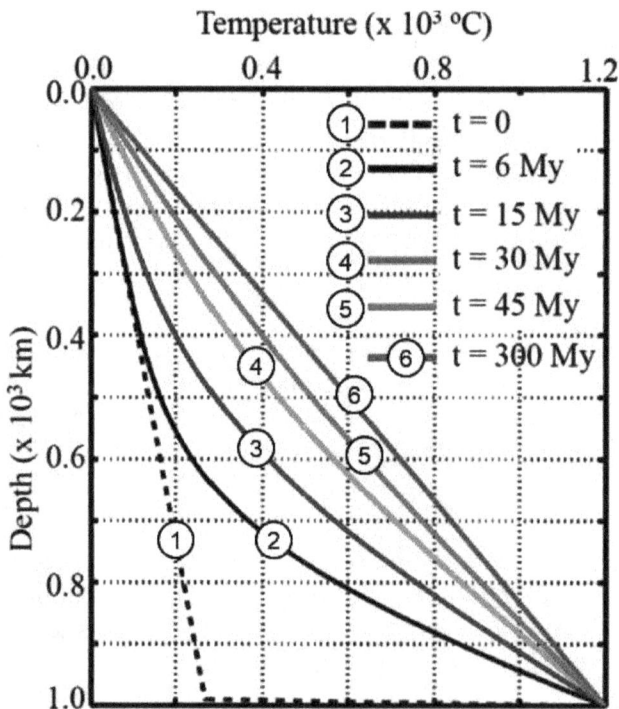

Figure 2.38. Evolution of the lithosphere geotherm is predicted by using equation (2.117). We first constructed a Matlab code of this equation. The code is shown in Table 2.12.

As described in Ikelle (2023b), energy in the sun is generated by nuclear fusion of four hydrogen nuclei at very high temperatures and pressure to one helium

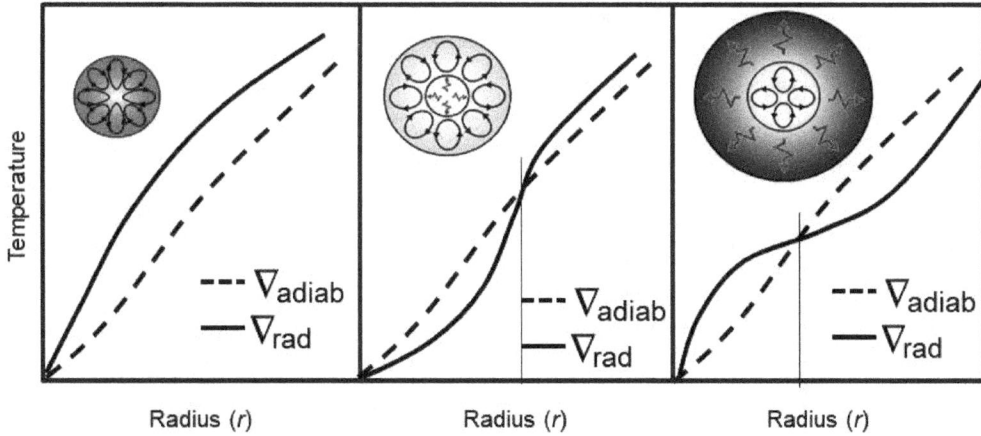

Figure 2.39. A schematic diagram of the actual (radiative) temperature gradient (solid line) and the adiabatic temperature gradient (dashed line) for radiation and convection energy transport in stars. When the radiative gradient is steeper than the adiabatic gradient, convection is the most efficient means of energy transport. The region is then said to be convectively unstable.

nucleus. This energy is transported by radiation through the radiation zone, and then by convection to the sun's surface. Notice that, in contrast to solid Earth, energy transport through the sun does not include conduction because no bulk solid materials are present in the stars, including the sun—interior of a star is a mixture of ions, electrons, and photons. It is, for the most part, transported by radiation and in small part, near its surface, by convection. Also, while energy generated in any other star's hot core is transported outward to its cooler surface, just like in the sun, some stars have different energy transport paths from their core to their surfaces than that of the sun. In some stars the energy is only transported by convection and others include large convection at their cores (see Figures 2.39 and 2.40). Our objective is to determine the temperature gradient for the stars, just like we did in the previous section for the solid Earth and deduce the energy-transport mechanism in the stars.

Energy transport by radiation. Because we are mostly dealing with gases inside the stars with no solid materials like in the Earth's lithosphere, conduction is not an efficient heat-transfer mechanism in stars, and so we can neglect it. We will consider radiation and convection with both energy-transport mechanisms occurring in different regions of a given star depending on its temperature gradient. To obtain the temperature-gradient equation, we consider two blackbody slabs at T and $T+dT$ separated by a photon mean free path $\ell = 1/(\kappa\rho)$—the average distance between collisions is called the mean free path, where κ is the star's opacity (we will further elaborate on the star's opacity later). At every radius, r, in the star, the opacity depends on the density, the temperature, and the chemical composition at that

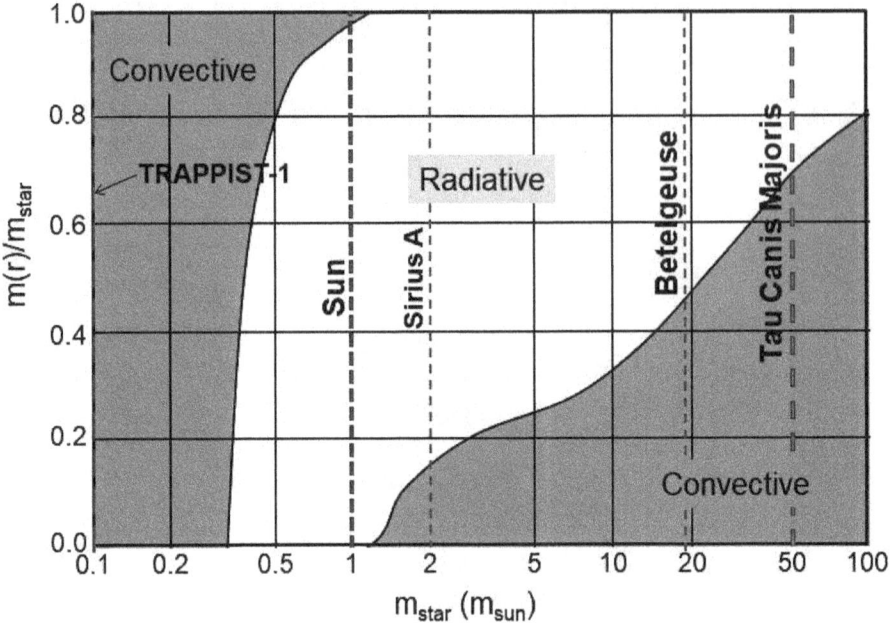

Figure 2.40. Occurrence of convective regions (gray shading) on the stars in terms of fractional mass, $m(r)/m_{\text{star}}$ as a function of stellar mass, m_{star}—the thick dotted line indicates the Sun. (adapted from Kippenhahn and Weigert, 1990).

radius. The net heat flux is

$$F = -[F(T + dT) - F(T)]$$

$$= -\sigma_{SB}T^4\left[\left(1 + \frac{dT}{T}\right)^4 - 1\right]$$

$$\approx -\sigma_{SB}T^4\left[\left(1 + 4\frac{dT}{T}\right) - 1\right]$$

$$\approx -4\sigma_{SB}T^3 dT\,, \tag{2.135}$$

F is the heat flux as defined earlier. The minus sign is introduced here to indicate the fact that temperature in stars increases with decreasing radius (temperatures of stars increase downward). The temperature difference between two slabs is given by the temperature gradient times the mean free path of a photon—that is,

$$dT = \frac{dT}{dr}\ell\,. \tag{2.136}$$

Box 2.2: Radiation Pressure

EM waves carry momentum despite being massless. If an incident EM wave is absorbed by a body, or it reflects from a body, the EM wave will transfer momentum to the body. The longer the EM wave is incident on the body, the more momentum is transferred. This time dependence can be ignored by working with the pressure, known as *radiation pressure*, that is exerted on the surface of the body by the incident EM wave. Radiation pressure is central to the study of stars' interiors in which EM radiation is present along with a mixture of ions, electrons, and photons. The ions and electrons can be treated as an ideal gas and quantum effects can be neglected. Total pressure is $P = P_{gas} + P_{rad}$, where $P_{gas} = P_{ion} + P_{elect}$, P_{ion} is the pressure of the ions, P_{elect} is the electron pressure, and P_{rad} is the radiation pressure. Gas pressure is most important in low mass stars and radiation pressure is most important in high mass stars. In the first section of this chapter, we determined P_{gas}. We used the Maxwell-Boltzmann distribution function of ideal gases to obtain the relationship between P_{gas}, temperature, and volume. Our objective here is to determine P_{rad}.

As we are dealing with radiation, we can similarly use the Planck frequency distribution (number density of photons within a given frequency range)

$$n(f)df = \frac{4\pi}{c} \frac{2f^2 df}{c^2 \left[\exp(hf/k_B T) - 1\right]} , \qquad (2.137)$$

to determine P_{rad} from the integral equation in (2.14). With $p = hf/c$ and $n_f(f)df = n_p(p)dp$, it follows that

$$
\begin{aligned}
P_{rad} &= \frac{1}{3} \int_0^\infty n_f(f) hf\, df \\
&= \frac{4\pi}{3c} \int_0^\infty \frac{2hf^3}{c^2} \frac{1}{\exp(hf/kT) - 1} df \\
&= \frac{4\pi}{3c} \frac{2k_B^4 T^4}{c^2 h^3} \int_0^\infty \frac{x^3}{\exp(x) - 1} dx \\
&= \frac{1}{3} aT^4 , \qquad (2.138)
\end{aligned}
$$

where $a = 4\sigma_{SB}/c = 8\pi^5 k_B^4/(15 c^3 h^3)$ is the radiation density constant, $x = hf/(k_B T)$ and $df = (k_B T/h)dx$. Note that gas pressure is equal radiation to pressure when $T^3 = (3R\rho)/(a\mu)$: i.e.,

$$P_{gas} = P_{rad} \iff \frac{R}{\mu} \rho T = \frac{1}{3} aT^4 \implies T^3 = \frac{3R}{a\mu} \rho . \qquad (2.139)$$

Note also that, at the center of the sun, $P_{gas}/P_{rad} \approx 7 \times 10^{-4}$—that is, radiation pressure is not important in the center of the sun. We used $\rho = 150$ g/cm^3, $T = 1.6 \times 10^7$ K, and $R = 8.31$ J/K/mol. In order to assess the importance of radiation pressure. Hence radiation pressure appears to be negligible at an average point in the sun. Without recourse to our knowledge of how energy is generated in the sun, a value for its internal temperature has been derived and it has been deduced that the material is essentially an ideal gas with negligible radiation pressure. Clearly radiation pressure becomes increasingly significant as stellar mass increases.

Table 2.13. Equations of stellar structure of symmetric stars. ρ is density, m is enclosed mass, L is luminosity, T is temperature, κ is opacity (how transparent the star is to radiation), G is universal constant, q' is rate of energy generation per unit mass, and r is the distance from the star center.

$dm/dr = 4\pi r^2 \rho(r)$	Mass conservation
$dP/dr = -\left(Gm/r^2\right)\rho(r) = g(r)\rho(r)$	Hydrostatic equilibrium
$dL/dr = 4\pi r^2 \rho\, q'$	Energy generation
$dT/dr = -\left[(\kappa\rho)/(16\pi r^2 \sigma_{SB}T^3)\right] L(r)$	Energy transport by radiation

So the net heat flux becomes

$$F = -4\sigma_{SB}T^3 \frac{dT}{dr}\ell = -\frac{4\sigma_{SB}T^3}{\kappa\rho}\frac{dT}{dr}\,. \tag{2.140}$$

We then deduce that temperature gradient equation of radiation

$$L(r) = 4\pi r^2 F = -\frac{16\pi r^2 \sigma_{SB}T^3}{\kappa\rho}\frac{dT}{dr}$$

or

$$\left(\frac{dT}{dr}\right)_{\text{rad}} = -\frac{\kappa\rho}{16\pi r^2 \sigma_{SB}T^3}L(r)\,, \tag{2.141}$$

or when combined with the mass-conservation equation for dm/dr (see Table 2.13) we arrive at

$$\left(\frac{dT}{dm}\right)_{\text{rad}} = -\frac{\kappa}{64\pi^2 \sigma_{SB}r^4 T^3}L(r)\,, \tag{2.142}$$

where $L(r)$ is the luminosity (rate of energy flow) through a shell of thickness dr at a radius r and $\sigma_{SB} = (ac_0)/3$ is the Stefan-Boltzmann constant, c_0 is the speed of light in vacuum and a is the radiation density constant. Equation (2.141) shows that we need to know the luminosity profile to determine the thermal profile. Large luminosity and/or a large opacity κ imply a large (negative) value of dT/dr. In other words, the opacity coefficient κ appearing in (2.141) determines the flux that can be transported by radiation for a certain temperature gradient. Therefore κ is an important quantity in the energy transportation through a star.

In hydrostatic equilibrium, we can combine equation (2.142) and the hydrostatic-equilibrium equation in Table 2.13 as follows:

$$\frac{dT}{dm} = \frac{dT}{dP}\frac{dP}{dm} = -\frac{Gm}{4\pi r^4}\frac{T}{P}\frac{d\log T}{d\log P} \tag{2.143}$$

and we define the dimensionless radiative temperature gradient as follows:

$$\nabla_{\text{rad}} = \left(\frac{d\log T}{d\log P}\right)_{\text{rad}} = \frac{\kappa P}{16\pi\sigma_{SB}GmT^4}L(r)\,,$$

G is universal constant. Remember that $\partial\log A = \partial A/A$.

The opacity has unit of area divided by mass. It indicates how the interior of a star is opaque to electromagnetic radiation, which means that it determines how the flux is transported by radiation, and so it is an important quantity for evaluating the radiative temperature gradient. Solar opacity varies between 1 cm^2/g at the center of the sun, assuming that $T = 1.6 \times 10^7$ K and $\rho = 157$ g/cm^3, to 100 cm^2/g near the sun's surface, assuming that $T = 6 \times 10^5$ K and $\rho = 0.026$ g/cm^3. These opacities indicate that the interior of the sun is extremely opaque to electromagnetic radiation. In general, star's opacities decrease as temperature increases in power laws (Kippenhahn and Weigert, 1990; and Prialnik, 2009), as follows:

$$\kappa = \kappa_0 \rho^r T^{-s}$$

where r and s are positive constants that vary with stars. For $T < 10^4$ K, we have $r = s = 0$ and for the core of the sun, we have $n = 1$ and $s = 3.5$. For reference, the radiation leaving the sun's surface is at about $T_{sun} = 5760$ K and reach the Earth's surface at about $T_{earth} = 255$ K.

Let us use the preceding results to estimate the temperature gradients in the sun assuming that at $r = 0.3\, r_{sun}$ (the interior of the sun), the luminosity is about $L(r) \approx 4 \times 10^{33}$ erg/s, temperature is $T(r) \approx 6.8 \times 10^6$ K, the density is $\rho(r) \approx 12$ cm^3/g, and the opacity is $\kappa(r) \approx 2$ cm^2/g. Then the average solar temperature gradient is

$$\frac{dT}{dr} \approx -1 \times 10^{-4} \text{ K/cm} . \tag{2.144}$$

We can see that the temperature changes very slowly. This result is consistent with the fact that the sun is radiative at this radius.

In Ikelle (2023b), we derived three equations of stellar structure that describe how parameters like the density, pressure, temperature, and luminosity change within the star. The equation for energy transport in (2.141) represents the fourth equation of stellar structure. These four equations are captured in Table 2.13. The boundary conditions associated with these are that the pressure is zero at $r = R$ [i.e., $\rho(R_{star}) = 0$ and $P(R_{star}) = 0$, where R_{star} is the star's radius and m_{star} is the star's mass], the temperature at the star's surface is $T(R_{star}) = T_{eff}$, and the luminosity at the surface is L_{star}. We also assume that these parameters within the star are only functions of the radius r from the center of the star with density. The solutions to these equations allow us to make predictions about the luminosity and evolution of stars. Different classes and ages of stars have different internal structures, reflecting their elemental makeup and energy transport mechanisms.

Adiabatic temperature gradient. As discussed in the previous section, knowledge of the adiabatic (no heat exchange with the environment) temperature gradient is important in the analysis of convection. For this reason, we here derive the equation of the adiabatic temperature gradient specifically for ideal gases although it is not an equation of stellar structure. Ideal gases obey $P = (R/\mu)\rho T$. If we move the gas parcel upward a small distance dr, then the change in pressure is

$$dP = \frac{R}{\mu}\left(\rho\frac{dT}{dr} + T\frac{d\rho}{dr}\right)dr = \left(\frac{P}{T}\frac{dT}{dr} + \frac{P}{\rho}\frac{d\rho}{dr}\right)dr . \tag{2.145}$$

We have assumed that μ is constant. If the gas is adiabatic, we also know that $P = K\rho^\gamma$, where $\gamma = c_P/c_V$ is the ratio of specific heats. Thus, we also have

$$dP = K\gamma\rho^{\gamma-1}\frac{d\rho}{dr}dr = \gamma\frac{P}{\rho}\frac{d\rho}{dr}dr \ . \tag{2.146}$$

By equating (2.145) and (2.146), we arrive at

$$\left(\frac{dT}{dr}\right)_{\text{adiab}} = \left(1 - \frac{1}{\gamma}\right)\frac{T}{P}\frac{dP}{dr} \ . \tag{2.147}$$

This value of dT/dr here is the adiabatic temperature gradient and the value of dT/dr in (2.141) is the star temperature gradient. An alternative form is

$$\left(\frac{dT}{dr}\right)_{\text{adiab}} = \left(1 - \frac{1}{\gamma}\right)\frac{T}{P}\frac{dP}{dT}\frac{dT}{dr}$$

or

$$\nabla_{\text{adiab}} = \left(\frac{d\log T}{d\log P}\right)_{\text{adiab}} = \frac{\gamma}{\gamma - 1} = \frac{T}{P}\frac{dP}{dT} \ . \tag{2.148}$$

For an ideal gas, we have $\gamma = 5/3$, so $\nabla_{\text{adiab}} = [(5/3) - 1]/(5/3) = 0.4$. This value is important in the analysis of energy transport through stars.

Energy transport by radiation and convection. The energy that a star radiates from its surface is generally replenished from the star's core which is the heat source. The outward energy flux at every dr requires an effective means of transporting energy through the star. Radiation heat transports via photon motions is often the most important means of energy transport in the stars. However, convection transports heat via bulk motions of large parcels of gas also occur in the stars. As described earlier for the sun, the radiation mechanism takes 50 million years to transport energy through the radiation zone. In contrast, the convection mechanism takes just hours through the convective regions. So by predicting the time heat takes to pass through various regions of the star's interiors, we can determine region of stars over which the energy transport is carried by radiation or convection. Unfortunately, such prediction is not yet possible because the four equations of stellar structure, in Table 2.13, are time-independent. However, in these equations, temperatures vary with mass $m(r)$ (mass at radius, r) and luminosity, $L(r)$. An alternative way to compare the (radiative) thermal gradient and the adiabatic gradient; i.e.,

$$\left(\frac{dT}{dr}\right)_{\text{star}} > \left(\frac{dT}{dr}\right)_{\text{adiab}} \ . \tag{2.149}$$

The left-hand side is given in (2.141) and the right-hand side is given in (2.147); the star gradient is the radiation gradient. If this inequality holds, there is no convection in the region of stars under consideration. In other words, convection begins if this inequality is violated. Because dT/dr is negative, we can alternatively express this condition with the absolute values as follows:

$$\left| \left(\frac{dT}{dr} \right)_{\text{star}} \right| < \left| \left(\frac{dT}{dr} \right)_{\text{adiab}} \right| \iff \nabla_{\text{rad}} < \nabla_{\text{adiab}} . \qquad (2.150)$$

This inequality is known as the Schwarzschild stability criterion against convection. A schematic illustration of the inequality is displayed in Figure 2.39. Similar to what we described in the previous subsection for Earth's mantle and core, the adiabatic gradient represents an upper limit to the temperature gradient inside a star—if this limit is exceeded, heat is transported by convection.

We need to solve the set of equations, in Table 2.13, to find the gradients of stars. Because this set of equations is highly non-linear and time-dependent, their full solution requires a numerical procedure (e.g., Maeder, 2009; and Kippenhahn and Weigert, 1990). The current numerical solutions of the equations, in Table 2.13, suggest that the occurrence of convective regions on the stars is a function of stellar mass. Here are some key conclusions:

- For $m_{\text{star}} < 0.35\ m_{\text{sun}}$, stars are completely convective (Figure 2.40). The energy to be transported in these cases is too large to be efficiently carried by radiative transport; it is transported by convection. This behavior implies that $\nabla_{\text{rad}} > \nabla_{\text{adiab}}$. The large radiation gradient is due to large opacity, which can reach 10^5 cm^2/g.

- For $0.35\ m_{\text{sun}} < m_{\text{star}} < 1.2\ m_{\text{sun}}$, stars have radiative cores and a convective envelope, as illustrated in Figure 2.40. Energy production is distributed over a larger area, which keeps ∇_{rad} low in the center of the stars and thus, their cores are radiative. By definition, the sun is in this range of stars.

- For $m_{\text{star}} > 1.2\ m_{\text{sun}}$, stars have convective core and radiative zone (Figure 2.40). Energy produced by these stars is very large for their core masses, which are very small. The results are steep increases of ∇_{rad} toward the center and thus convective cores. The size of these convective cores increases with stellar mass (Figure 2.40), and it can encompass up to 80 percent of the mass of a star when the star's mass approaches $100\ m_{\text{sun}}$.

Stars rotate, can convect, and have magnetic fields. In the above derivations, we neglected magnetic fields. We assume that stars are spherically symmetric. In reality stars rotate, convect, and have magnetic fields; these induce deviations from spherical symmetry. However, these deviations are small enough that, for most stars, we can ignore them to first order. In Chapter 6, we include magnetic fields in the derivations of the Schwarzschild stability criterion.

2.4. Bioheat Transfer

So far, most of our examples of heat transfer have been limited to nonliving things, such as the atmospheres of our solar system planets, solid Earth, and stars. In the next section, we will discuss heat transfer through heat engines. However, heat transfers also occur inside living things. Almost all the chemical processes inside living things are temperature-dependent—that is, there are transports of thermal energy in living systems. Moreover, heat transfer is used in diagnostic and therapeutic applications. We are all familiar with heating-cooling pads and may even have used them as therapeutic. Sport activities, especially professional sports, utilize these pads extensively to alleviate pain. They are vivid illustrations of the use of heat transfer in prevention, treatment, preservation, and protection techniques for biological systems. They are used in heat or cold treatments to destroy tumors, to improve patients' outcomes after brain injury, and to protect humans from extreme environmental conditions.

As skin tissues are effective conductors of heat, in extreme heat the surroundings can cause an excess of internal energy in the body that cannot be compensated for by doing work as prescribed by the first law of thermodynamics. Hence extreme heat has direct effects on health and kills more people than any other extreme weather event. Extreme heat events in Europe (2003) and Russia (2017 and 2010) resulted in over 70,000 and 55,000 deaths, respectively, as described in Ikelle (2020), and contributed to many crop and livestock losses. Extreme heat causes death because the body can no longer regulate its temperature by doing some work. Symptoms of extreme heat include sweating, fatigue, and skin burns. Moreover, extreme heat worsens chronic conditions such as cardiovascular disease, respiratory disease, and diabetes. So, heat transfer is an important component of living things.

Our objective here is to link the radiation, conduction, and convection mechanisms of heat, described in the first section of this chapter, to some current diagnostic and therapeutic mechanisms. Bioheat transfer is the generic name used for the study of the transport of thermal energy in living systems. It is generally divided into three groups, based on the cause of temperature alterations on chemical rate processes of living things (see also Table 2.14): hyperthermia (increased temperature to 40 degrees Celsius or more), hypothermia (decreased temperature below 35 degrees Celsius), and cryobiology (subfreezing temperature).

Table 2.14. Temperature (in degrees Celsius) ranges with bioheat processes and biomedical applications.

T (°)	Bioheat processes
< -1	Cryobiology (cells are freezing or frozen)
< 28	Death expected in less 12 hours due ventricular fibrillation
28-32	Loss of consciousness
32-35	Hypothermia
36.5-37.5	Normal body
40-45	Hyperthermia
46-50	Irreversible cellular damage in less than 1 hour
50-52	Coagulation necrosis in less than 6 minutes
60-100	Nearly instantaneous coagulation necrosis
> 100	Tissues vaporize

2.4.1. Fundamental Aspects of Bioheat

One of the most remarkable features of the human system is that we can maintain a core temperature (within the head, thorax, and abdomen) near 37 degrees Celsius (i.e., between 36.2 and 37.8 degrees Celsius) over a wide range of environmental conditions and during thermal stress. Maintaining this temperature is critical to human survival. Whenever our core temperature deviates from this value, our systems automatically trigger reversal mechanisms to bring our system back to 37 degrees Celsius. These reversal mechanisms include heat production through vasoconstriction and shivering (heat generation) that increase temperature toward the core temperature and vasodilation, panting, and sweating (heat dissipation) that reduce temperature toward the core temperature. The mechanisms by which our body temperatures are tightly bound around 37 degrees Celsius, most of the time, are collectively known as thermoregulation. In other words, our systems produce heat, and the dissipation of this heat leads to entropy production.

Let us expand on the notions of heat gain and loses associated with thermoregulation. The body's surface is the main place where heat exchange between the body and the environment takes place. Controlling the flow of blood to the skin is an important way to control the rate of heat loss to or gain from the surroundings. Warm blood from the body's core typically loses heat to the environment (Earth's surface temperatures are on average less 30 degrees Celsius as it passes near the skin. The self-initiated process by the body, known as *vasoconstriction*, allows us to shrink the diameter of blood vessels that supply the skin, thus reduce blood flow and help us retain heat. On the other hand, when we need to get rid of heat, the self-initiated process by the body, known as *vasodilation*, allows us to widen these blood vessels, thus increases blood flow to the skin and helps us lose some of its extra heat to the environment. Sweat production and the subsequent evaporation are the principal modes of heat loss in humans.

Let us also expand on the notion of core temperature. This notion implies the existence of peripheral shell temperatures. The core temperature reflects the temperature within the deep body tissues and organs that keep the body's vital functions going, such as breathing and keeping warm, while at rest or in sleep. These organs include the brain, heart, and liver. The shell temperatures are measured at the skin of hands and feet, parts of the body that are influenced by blood flow to the skin. In a warm environment, this difference decreases (generally 4 degrees Celsius lower than the core temperature, see Figure 2.41) because skin blood flow increases and skin temperature approaches ambient temperature. During cold stress, this difference increases because skin blood flow reduces leading to a decrease in shell temperature and conservation of heat to the core.

Let us now briefly discuss the functioning of human thermoregulation. When your brain, more specifically its region called the hypothalamus, senses your internal temperature rising too high, it initiates heat dissipation responses and heat conservation or heat generation when the temperature is too cold by sending signals to your muscles, organs, glands, and nervous system. They respond in a variety of ways to help return your temperature to normal. These responses include change in behavior, increase in metabolic heat production, and control of the exchange of heat with the environment. (1) Taking more cold water than usual and sitting in

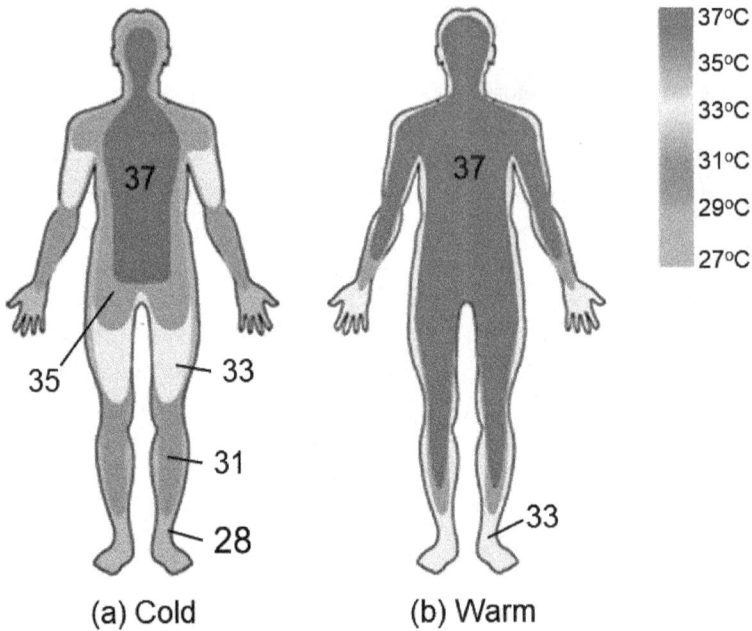

Figure 2.41. Temperature field of the human body. The body core remains largely constant in temperature, which is around 37°C, while the temperature of the body shell is subject to external and internal influences. (a) Human body temperature under cold ambient temperature (around 25°C). (b) Human body temperature under warm ambient temperature (around 30°C). (Adapted from https://www.oilsandplants.com/temperature.htm).

the shade are parts of the change in behavior on a hot day. On a cold day, you might put on a coat, sit in a cozy corner, and eat hot food. (2) The increase in metabolic heat production can be done through muscle contraction—for example, if you shiver uncontrollably when you are very cold. Both deliberate movements, such as rubbing your hands together or jogging for a brisk walk, and shivering increase muscle activity and thus boost heat production; in other words, we produce our own body heat and can regulate our body temperature. (3) Controlling the exchange of heat with the environment include circulatory mechanisms, such as altering blood flow patterns and evaporation mechanisms, panting, and sweating. Therefore, thermoregulation is only possible when the cardiovascular system is working properly. This means that the system delivers oxygen and nutrients to the tissue fluid that surrounds the cells and eliminates metabolic waste, arteries takes blood from the heart, and vessels return blood to the heart. In other words, a systemic transport of blood to all parts of the body, including skin blood flow, is a sinequanon of thermoregulation.

Despite the need for tight regulation of the core temperature, humans can briefly survive in the most inhospitable of places, like the Sahara desert (a place where day-time temperatures can reach 50 degrees Celsius and in freezing Nordic countries—the thermoregulatory capacity can survive these extreme conditions for two or three hours. After that, the thermoregulatory fails. That is why a number of people sleeping on the streets are found dead after very cold winter nights and heat waves kill mostly those who are not capable of escaping affected regions in a matter of hours.

Human thermal balance. Here is the heat-balance equation (Tansey and Johnson, 2015)

$$\text{heat storage} \; = \; \text{metabolism} - \text{work} - \text{evaporation}$$

$$\pm \; \text{radiation} \pm \text{conduction} \pm \text{convection} \, ,$$

where work is the external work done, evaporation is the heat loss to the environment, as water vaporized from the respiratory passages and sweat vaporized from skin surface, radiation is the electromagnetic radiation (heat) transferred to bodies, conduction is the movement of heat to/from the body directly to objects in contact with the body, convection is the transfer of heat to a moving gas or liquid, and metabolism refers to the chemical reactions occurring within the body that produce heat. When heat storage = 0 (i.e., thermoneutral conditions), the body is thermally balanced. The heat gain is balanced by heat loss and the core temperature is maintained at 37 degrees Celsius. However, in more severe exposure to extreme weather, heat storage $\neq$ 0 and heat may be stored or lost. For example, high humidity (which limits evaporation) and/or the wearing of thick clothing (which creates a barrier to effective evaporation of sweat) can yield heat storage > 0 and therefore heat is stored.

Hypothermia, hyperthermia, and cryobiology. Heat-related illness is a spectrum of disease that occurs when the body's thermoregulatory system fails as a result of overheating (hyperthermia) or being too cool (hypothermia). Either state can have deleterious effects on the various body systems, most significantly, reduced blood flow leading to ischemia (inadequate blood flow to body organs, including the heart muscles) and multiple organ failure.

When body temperatures are severely decreased in hypothermia (i.e., below 35 degrees Celsius), the body reduces oxygen demand, increases blood viscosity, diminishes blood flow to the brain, decreases renal plasma flow, and can cause direct tissue injury. Cold-related inflammation can trigger a heart attack by making plaque in our arteries more likely to dislodge and block an artery, cutting off blood flow to the heart.

Again, hyperthermia is defined as a body temperature greater than 40 degrees Celsius and the body becomes unable to properly cool. Skin blood flow can reach 6 to 8 liters per minute during hyperthermia while skin blood flow in thermoneutral environments is approximately 0.250 liters per minute. Several conditions can cause hyperthermia. They include toxic ingestions, certain medication reactions, and exposure to extreme heat. Prompt treatment of heat-related illnesses with aggressive fluid replacement and cooling of the core body temperature is critical to reducing illness and preventing death. Air conditioning is the number-one protective factor against heat-related illness and death, especially during a heat wave.

Note that chemicals called pyrogens released by white blood cells can raise the body's core temperature above 37 degrees Celsius by 2 to 3 degrees Celsius to help kill bacteria and viruses. This elevation of the core temperature can cause the fever and shivering, even though you are hot. Fever is a result of the body releasing pyrogens. Viral illnesses or another infectious disease can cause a person to develop a fever.

Cryobiology, through its application, known as cryosurgery, also referred to as cryoablation, is a technique where undesirable or diseased tissue is frozen down from the exterior, using temperatures between -15 degrees Celsius and -40 degrees Celsius. Cryosurgery is a low-invasive procedure with minimal blood flow, a localized site of surgery, and reduced recovery time and hospitalization time for the patient (Wells et al., 2005).

Mass extension. Thermoregulation is not limited to humans; it is actually valid for most living things, although at different core temperatures. In fact, animals are sometimes divided into endotherms and ectotherms, based on their temperature regulation. Endotherms, such as birds and mammals, use metabolic heat to maintain a stable internal temperature, which is often different from the environment. Ectotherms, like lizards and snakes, do not use metabolic heat to maintain their body temperature, but take on the temperature of the environment. So, it is normal and natural that the thermoregulation of some spices can fail due to significant changes in climate. Polar bears in the Arctic and penguins in Antarctica provide two significant examples of different types of thermoregulation.

Penguins maintain a body temperature around 38 degrees Celsius but live in temperatures that range from 30 degrees Celsius to -60 degrees Celsius on the sea ice of Antarctica. They have thick skin and a lot of fat under their skin to keep warm in cold weather. They also huddle together in groups that may comprise several thousand penguins to keep comfortable. In addition to absorbing heat from the sun, the penguin's dark back surface also helps it to heat up. They lose internal body heat to the surrounding air through thermal radiation, just as our bodies do on a cold day. To maintain their core temperature, while losing heat, penguins, like all warm-blooded animals, rely on the metabolism of food.

Polar bears, with a core temperature around 37 degrees Celsius, have an 11-cm-thick insulating fat layer beneath their skin and 15 cm of fur (thick growth hair) that keeps them warm, even when air temperatures drop to -37 degrees Celsius. They lose so little heat to their environment that they are almost invisible to thermal-imaging cameras.

Given sufficient time through a smooth climate change, some species with dispersal capabilities will migrate to suitable thermal environments. Other organisms may adapt to the new local environment by adapting to plasticity or evolution. Some may just fail to adapt or migrate. In this situation, extinction of local populations or entire species will occur.

As described earlier, some climate changes are abrupt—that is, they are rapid and unpredictable. In the past, the average global temperature has risen or fallen more than 6 degrees Celsius in less than 10 years, and more than 10 degrees Celsius in a million-year span. An increase of 6 degrees Celsius in this century would be considered an abrupt climate change. Adaptation and migration times are too short for many species to adapt to such changes.

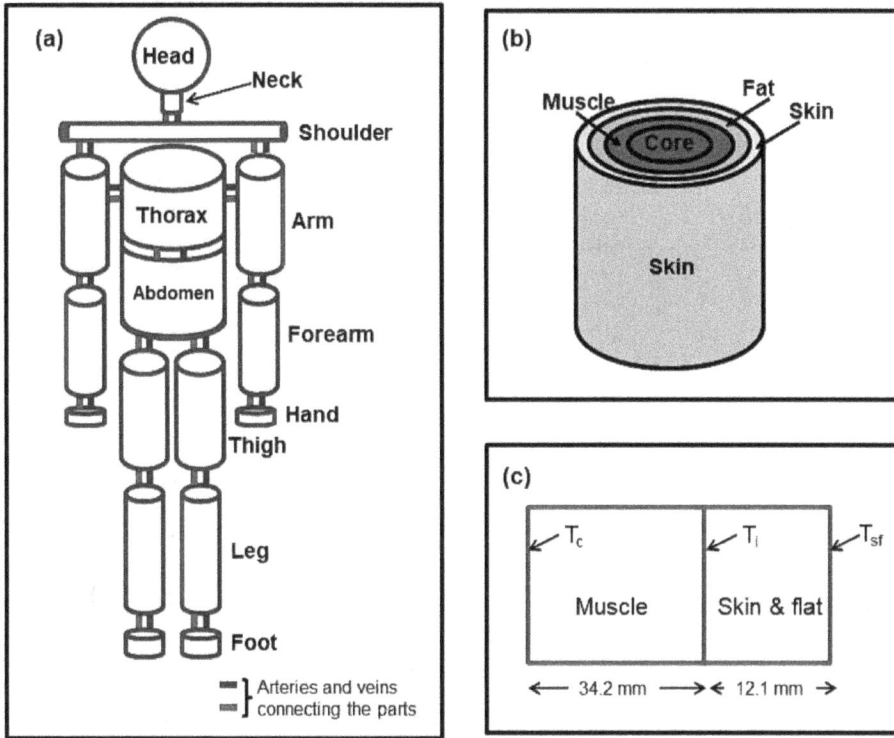

Figure 2.42. (a) Schematic of the multi-segment human anatomy. Segments include the head, neck, upper arms, thighs, forearms, hands, and feet. (b) A detailed representation of a segment as a cylinder. It includes skin, fat, muscle, and the core (bone). The heat transfer model is based on core, skin, artery and vein that exchange heat through convection, conduction and perfusion. (c) Another illustration of a bioheat model of skin tissue. The muscle thickness is about 34.2 mm and the skin fat thickness is about 12.08 mm. The core body temperature, T_c and arterial temperature, T_a are both assumed to be 37°C; T_{sf} is the skin temperature; and T_i the temperature at the interface between muscle and skin.

2.4.2. Bioheat Transfer Equations: Thermal Diagnostics and Therapies

Physical model of the human body. Before we dive into the heat-transfer equations and their solutions, it is useful to recall that a human body can be conceptualized as the multi-segmental model (Fiala, 1999), as illustrated in Figure 2.42. The segments are generally depicted as cylinders, except the head, which is described as a sphere. The segments representing shoulders, legs, thighs, arms, and forearms consist of, from inside to outside, bone, muscle, fat tissue, and skin layers. The head, thorax, and abdomen segments also include the brain, lungs, and viscera, respectively, as illustrated in Table 2.15. Because the head is described as a sphere, the layer radii are re-sized to keep the head volume constant.

Blood circulation through the two principal sets of arteries and veins is the primary mechanism for maintaining body temperature. As discussed in Ikelle (2023a), blood leaves the heart through the aorta (the largest blood vessel with a diameter of about 5000 μm). It is transported to the muscles through the arteries and the

Table 2.15. The key elements of the multi-segmental body model and their sizes. (adapted from Coccarelli et al., 2016). The number in parentheses is the average length in cm.

Cylinder	Tissues	Layer radii (cm)
Head (23.5)	Brain, bone, fat, skin	6.6, 7.6, 7.8, 8.0
Neck (7.9)	Bone, muscle, fat, skin	1.9, 5.4, 5.6, 5.8
Thorax (15.6)	Lung, bone, muscle, fat, skin	7.7, 9, 12.3, 12.6, 13
Abdomen (25)	Viscera, bone, muscle, fat, skin	8, 8.3, 10.9, 12.4, 13
Shoulder (13.4)	Bone, muscle, fat, skin	3.7, 3.9, 4.4, 4.6
Arm (29.6)	Bone, muscle, fat, skin	1.5, 3.4, 4.0, 4.2
Forearm (23.7)	Bone, muscle, fat, skin	1.5, 3.4, 4.0, 4.2
Thigh (58.5)	Bone, muscle, fat, skin	2.2, 4.8, 5.3, 5.5
Leg (34.3)	Bone, muscle, fat, skin	2.2, 4.8, 5.3, 5.5

veins with diameters within the range of 300-1,000 μm. The blood is then delivered to arterioles that are 20-40 μm in diameter and connected to the smallest transport vessels. The returning blood loop to the heart consists of veins.

Skin is the largest organ in the body. It protects us from the surrounding environment and it plays an important role in thermoregulation, sensory, and host defense functions. The thermal energy exchange between the body and the environment occurs at the skin and through the skin. The common thermophysical parameters of human skin are: density ($\rho = 1,000$ kg/m^3), heat conductivity ($k = 0.5 - 0.628$ W/(K $\cdot$ m)), and heat capacity ($c_P = 4.2$ (J $\cdot$ kg)/K). The interior tissue temperature attains a constant value at $L = 2 - 3$ cm from the surface.

The cylindrical body-segment description in Figure 2.42 allows us to model and analyze body segments separately. Furthermore, such modeling and analysis are available in modern medicine and take into account the sensitivity to environmental conditions, the temperature of the media surrounding tissues, the metabolism of muscles, and the flow of blood. Heat exchange between tissues and the arterial wall occurs by convection in large vessels and by perfusion in small arteries. Some heat-transfer analysis of these segments can be conducted under one-dimensional (1-D) Cartesian coordinates, which correspond to surface heating/cooling, 1-D cylindrical coordinate, which is suitable for treatments using heating/cooling probes, and 1-D spherical symmetric coordinate for small heating/cooling sections. The segments are also analyzed in two- and three-dimensional coordinates. As described in Appendix A of Ikelle (2020), mathematical and computational descriptions of one coordinate system can also be applied to another coordinate system.

The bioheat equation. The heat exchanges presented in the previous subsections in this section for humans and their environment are identical to those presented in the previous sections for non-living things. So, the temperature field in biological tissues follows the same heat-diffusion equations, as those of the temperature fields of non-living things with different source terms; i.e., (Pennes, 1948)

$$\underbrace{\rho\, c_P \frac{\partial T(\boldsymbol{x},t)}{\partial t}}_{\text{Storage}} = \underbrace{\boldsymbol{\nabla}\cdot[k\boldsymbol{\nabla}T(\boldsymbol{x},t)]}_{\text{conduction}} + \underbrace{\rho_b c_b w_b(T)\,[T_a - T(\boldsymbol{x},t)]}_{\text{convection}}$$

$$+ \underbrace{q'_m(\boldsymbol{x},t) + q'_e(\boldsymbol{x},t) + q'_r(\boldsymbol{x},t)}_{\text{generation}}\;, \tag{2.151}$$

where $T = T(\boldsymbol{x},t)$, with $t > 0$, stands for the temperature of the tissue in Kelvin that varies with time t [s] and space $\boldsymbol{x}$, ρ is the tissue density [kg/m^3], c_P is the tissue specific heat capacity [J/(kg K)], k is the tissue thermal conductivity [W/(kg K)], $w_b = w_b(T)$ is the blood perfusion rate [kg/(m^3 s)], ρ_b is the blood density [kg/m^3], c_b is the blood specific heat capacity [J/(kg K)], T_a is the arterial temperature K, q'_m is the metabolic heat generation rate [W/m^3], q'_e is the blood perfusion heat generation [W/m^3], and q'_r is the regional heat source [W/m^3]. Equation (2.151) is essentially a conduction heat-diffusion equation process with biological heat-generation terms. In biology, this equation is attributed to Pennes (1948) and is known as the bioheat equation. As long as an appropriate initial condition and boundary conditions are prescribed, this equation allows us to predict the temperature field in the tissue. Many disciplines and industries use this equation to predict human thermal responses in different thermal environments. They include biometeorology, the automobile industry, clothing research, and medical applications. In the following paragraphs, our examples will focus on medical applications.

The non-directional term $q = w_b(T)c_b(T_a - T)$ is known as blood perfusion. It describes the volumetric blood flow at the capillary level in tissue and it is similar to equation (2.50). It carries nutrients through blood exchanges with cells and wastes away. Perfusion is intrinsic to the healthy function of the body and therefore its quantification is important in health care. The parameters in equation (2.151) are usually assumed to be constant except for the blood perfusion, which varies with temperature, T. One of the most widely used applications of blood perfusion effects and measurements is tumor detection. Tumors are known to have a different perfusion rate than normal healthy tissue because blood flows in tumors are quicker than in healthy tissues. Note that in cryobiology, the blood-perfusion term in (2.151) is negligible because rapidly freezing tissues first cause vasoconstriction in the capillaries before freezing all the blood in the capillaries and w_b drops to zero. Also, cells are not able to generate any metabolic heat when frozen, thus q'_m is zero for temperatures below zero Kelvin. Note also that temperature monitoring is difficult during the cryosurgery procedure because the insertion of temperature probes increases the invasiveness of the surgery. Thus, the analytic and/or numerical solutions of (2.151) are generally used to determine the temperature profile of the tissue in this procedure.

Table 2.16. Heat transfer mechanisms in biothermal systems.

	Conduction	Convection	Radiation
Tissues	Significant	Significant	Negligible
Bones	Significant	Negligible	Negligible
Blood vessels	Significant	Significant	Negligible
Skins	Negligible	Significant	Significant

Equation (2.151) is subject to three kinds of BCs (boundary conditions) on the surface of biological tissue or the boundary between the biological tissue and treatment device. They are:

$$(1) \quad T|_{\text{boundary}} = T_S \quad , \quad (2) \quad -k \frac{\partial T}{\partial x}\bigg|_{\text{boundary}} = q_S \ ,$$

$$\text{and} \quad (3) \quad -k \frac{\partial T}{\partial x}\bigg|_{\text{boundary}} = h_f \left(T_f - T|_{\text{boundary}}\right) \ , \qquad (2.152)$$

where boundary is either the whole or a part of the boundary, h_f is the heat transfer coefficient, T_f is the temperature of the surrounding medium (i.e., the exterior ambient temperature). The first kind, BC represents heating/cooling at a constant temperature, T_S, the second kind, BC represents heating/cooling by constant heat flux, q_S, and third kind, BC represents heating/cooling by convective heat transfer, which means heat exchange between the tissue surface and fluid at a constant temperature T_f. Alternatively, heating/cooling can occur by radiation instead of convection; i.e., the third kind of BCs can also be expressed as

$$q = -\sigma_{SB}\varepsilon \left(T_f^4 - T^4|_{\text{boundary}}\right) \ ,$$

where σ_{SB} is the Stefan-Boltzman constant, ε is the radiative interchange factor between the surface and the exterior. Almost all heating/cooling methods include, at least, one of these three kinds of BCs. For example, some heating devices for thermal therapy utilize a heated metal disc and constant temperature BC; other devices utilize a heated metal disc with convective heat transfer BC. See Table 2.16 for the importance of these boundary conditions for biothermal systems.

The complexity of equation (2.151) clearly shows that human body temperature is a complex mechanism. This mechanism is affected by release of hormones and neurotransmitters, redistribution of blood flow to the skin, respiration, evaporation, adjustment of metabolic rate, certain pathologic events (e.g., fever), and external events (e.g., hyperthermia and hypothermia treatments that freeze or superheat tumors in the body, different air temperatures, humidity levels, and wind velocities). Consequently, temperature is a critical parameter in the diagnosis and treatment As an example, elevated local tissue temperatures can indicate high metabolic rates, the body's response to infection, removal of foreign substances, and an increase in temperature of skeletal muscles and joints due to exercise, as well as the presence of malignant tumors. Temperature measurement is also critical in many therapeutic procedures involved in either hyperthermia or hypothermia because heat can be added or removed during these procedures. For example, an artificial elevation of tissue temperature into the 40-42 degrees Celsius range can provide some relief from pain (analgesia) and can enhance wound healing. Note, however, increased temperature does not heal by itself; it simply improves conditions for body self-healing and/or treatments. A typical clinical application of hyperthermia is the treatment of cancer by selectively attacking tumors. This application consists of artificially raising the temperature in the target area to approximately 42 degrees Celsius for up to one hour without excessively heating adjacent tissues. The knowledge of body temperature distribution is critical to these types of applications. This knowledge is readily obtained by solving the bioheat equation in (2.151). In other words, bioheat equation in (2.151) plays a crucial role in many medical treatments. Heat transfer mechanisms in biothermal systems.

Invasive temperature measurements can provide information about the tissue temperature only at discrete points. They have fast response times and periodic temperature monitoring instead of continuous monitoring. The number of probes that can be implanted in the body is restricted by patient tolerance and thus the tissue temperature knowledge is limited—a mathematical model, like the one in equation (2.151), for calculating the tissue temperature during the treatments complements the invasive temperature measurements with space-time distributions of temperature. A treatment that utilizes solutions of (2.151) can be optimized with respect to the heat application geometry by maximizing the therapeutic effect while minimizing unwanted side effects. The outcome of a treatment can be evaluated based on the predictions of equation (2.151). The solutions of this equation can even be used to develop new treatment strategies and to look at more in detail the thermal properties of tissues.

The bioheat equation, in (2.151), distinguishes itself from the heat-diffusion equations, discussed in the previous section, by accounting for the thermal effects due to thermophysical properties of the tissue (heat capacity, thermal conductivity, etc.), geometry of the irradiated organism, heat production, due to absorption of light, heat production, due to metabolic processes, and heat flow due to perfusion of blood. We here discuss analytical solutions to (2.151) for simple geometries and numerical solutions to (2.151) for more realistic, complicated tissue geometries.

Example: 1D steady-state bioheat solution. Heating devices for thermal therapy that utilizes a heated metal disc can be approximated by the 1-D bioheat equation. In addition, body surface cooling with a cooling pad can also be approximated by the 1-D bioheat equation. So we start by solving the 1D bioheat. We assume that the steady-state temperature of biological tissue and the tissue properties are independent of the tissue temperature and constant. Under these assumptions, equation (2.151) can be simplified as follows:

$$k\frac{d^2T(x)}{dx^2} + \rho_b c_b w_b \left[T_B - T(x)\right] = 0 \ , \qquad (2.153)$$

where

$$T_B = T_a + \frac{q'_m}{\rho_b c_b \omega_b} \qquad (2.154)$$

is the equilibrium temperature of each tissue or organ described by the bioheat-transfer model. The local arterial temperature T_a is generally assumed to be constant and equal to the body core temperature $T_c = 37°C$. The solution of equation (2.153) is

$$T(x) = C_1 + C_2 \exp\left(-\frac{x}{\ell_B}\right) , \qquad (2.155)$$

where

$$\ell_B = \left[\frac{k}{\rho_b c_b \omega_b}\right]^{1/2} \qquad (2.156)$$

is the characteristic depth of bioheat transfer, and C_1 and C_2 are the integration constants. The final steady-state solutions are

$$(1): T(x) = T_B + (T_S - T_B)\exp\left(-\frac{x}{\ell_B}\right) , \qquad \text{for}$$

$$T(x = 0) = T_S = C_1 + C_2 \quad \text{and} \quad T(x = \infty) = T_B = C_1 ;$$

$$(2): T(x) = T_B + \frac{q_w \ell_B}{k}\exp\left(-\frac{x}{\ell_B}\right) , \qquad \text{for}$$

$$-k\frac{\partial T}{\partial x}\bigg|_{x=0} = \frac{kC_2}{\ell_B} = q_w \quad \text{and} \quad T(x = \infty) = T_B = C_1 ;$$

$$(3): T(x) = T_B + \frac{T_f - T_B}{1 + k/(h_f \ell_B)}\exp\left(-\frac{x}{\ell_B}\right) , \qquad \text{for}$$

$$-k\frac{\partial T}{\partial x}\bigg|_{x=0} = h_f\left[T_f - T(x = 0)\right] \quad \text{or}$$

$$\frac{kC_2}{\ell_B} = h_f\left[T_f - (C_2 + C_1)\right] \text{ and } \quad T(x = \infty) = T_B = C_1.$$

Notice that the characteristic depth of bioheat transfer, ℓ_B, is a controlling parameter of these solutions. Experiments show that the thermal penetration depth, ℓ_T, which is the deepest depth where the heat effect reaches, is related to the characteristic depth of bioheat transfer, as follows: $\ell_T \approx 2.3\,\ell_B$ (see Table 2.17). This depth is an important parameter in estimating the depth affected by heat transfer during treatment.

Figure 2.43 illustrates these analytical solutions of the bioheat transfer equation for hypothermia (i.e., $T_S < T_a$) and hyperthermia (i.e., $T_S > T_a$) using the typical tissue and blood properties. We can see that the tissue temperature increases with the increase of the thermal conductivity of tissue and it increases with decreasing blood perfusion at the same point x for hyperthermia. The tissue temperature decreases with the increase of the thermal conductivity of tissue and increases

Table 2.17. Some thermophysical properties of organ or tissue for arterial blood temperature, T_a, is constant at around 37 degrees Celsius and the metabolic heat generation rate, q'_m, is constant around 4200 W/m^3. Here we are dealing with adipose tissue and skeletal muscle; k is in [W/(m $\cdot$ K)] and c_b is in [J/(kg $\cdot$ K)].

	k	c_b	ρ_b [kg/(m^3)]	ω_b [1/s]	T_B [°C]	ℓ_B [mm]	ℓ_T [mm]
Kidney	0.54	3700	1050	0.061	37.02	1.5	3.45
Liver	0.52	3600	1060	0.015	37.07	3.0	6.90
Prostate	0.50	3850	1000	0.0061	37.18	4.6	10.58
Brain	0.50	3700	1050	0.0035	37.31	6.1	14.03
Skin	0.45	3300	1200	0.0013	37.82	9.3	21.39
Tissue	0.27	3100	950	0.0005	39.85	13.5	31.05
Muscle	0.50	3465	1050	0.0009	38.28	12.4	28.52

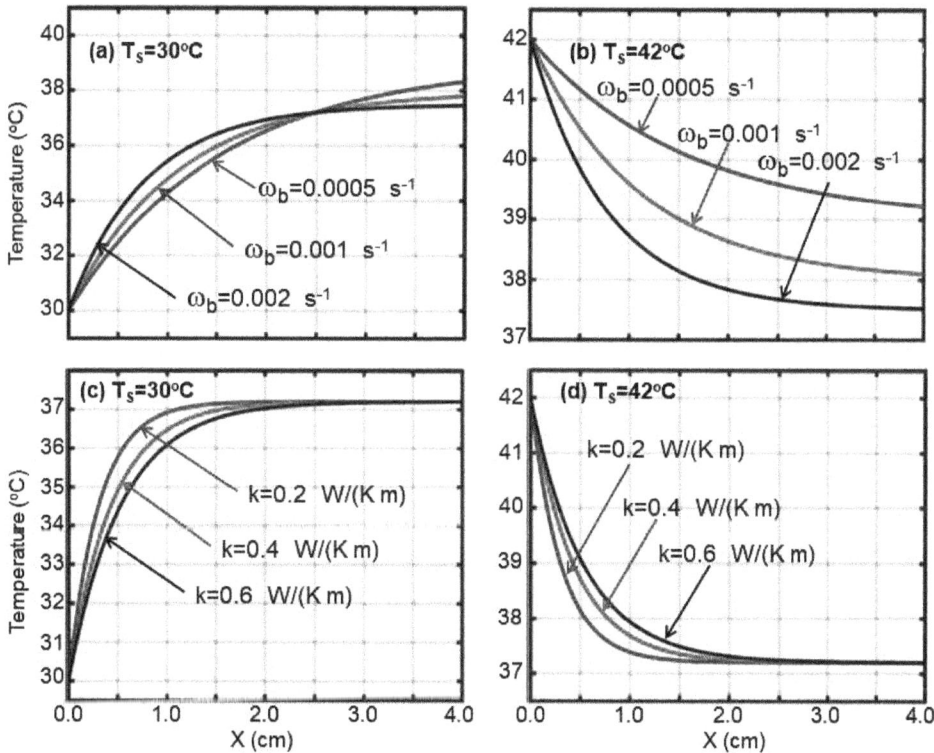

Figure 2.43. Steady-state temperature distributions along the axial direction when temperature T_S is applied to a homogeneous body segment. (a) Temperature distributions for $T_S = 30°C$ (hypothermia) for three blood-perfusion levels. The other parameters are fixed. (b) Temperature distributions for $T_S = 42°C$ (hyperthermia) for three blood-perfusion levels. The other parameters are fixed. (c) Temperature distributions for $T_S = 30°C$ (hypothermia) for three thermal conductivities. The other parameters are fixed. (d) Temperature distributions for $T_S = 42°C$ (hyperthermia) for three thermal conductivities.

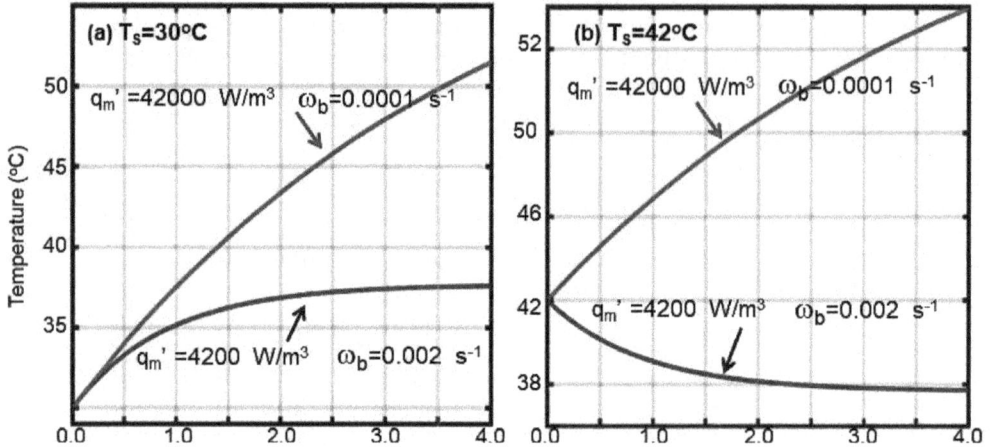

Figure 2.44. The temperature distribution as a function of depth for a steady-state bioheat.

with increasing blood perfusion at the same point x for hypothermia as long as $T(x) < 37.5°C$. Figure 2.44 illustrates a possible way of diagnosing a tumor. According to Deng and Liu (2002), blood perfusion and metabolic rate in the tumors often appear abnormally higher than that in the surrounding healthy tissues. For example, in a tumor we can have $[\omega_b = 0.0001 \, 1/s, q'_m = 42000 \, W/m^3]$ and in a healthy tissues $[\omega_b = 0.002 \, 1/s, q'_m = 4200 \, W/m^3]$. Steady-state temperature distribution at most points clearly show large differences between the healthy skin and the skin with tumor for both, hypothermia and hyperthermia.

Example: 1D unsteady-state linear bioheat solution. For the one-dimensional unsteady-state system, the bioheat equation in (2.151) reduces to

$$\rho c_P \frac{\partial T(x,t)}{\partial t} = \frac{\partial}{\partial x} \cdot (k \nabla T(x,t)) + \rho_b c_b w_b(T) \, (T_a - T(x,t))$$

$$+ \quad q'_m(x,t) + q'_e(x,t) \; . \tag{2.157}$$

We numerically solve this equation by using the implicit finite-difference modeling technique described in Ikelle (2018). Table 2.18 shows a sample code for solving (2.157) for the following boundary and initial conditions:

$$-k \frac{\partial T}{\partial x}\bigg|_{x=0} = q_w \; , \; T(x = L, t) = T_c, \; \text{and} \; T(x, t = 0) = T_c \, .$$

In other words, we assume that the initial temperature at $t = 0$ inside the skin is uniform and equal to the body core temperature $T_c = 37°C$. Figure 2.45a shows the temperature distribution as a function of depth and time. We can see the tissue temperature at the early stage of heating decreases with increasing the tissue depth, like in the steady-state, in Figure 2.45a; the surface temperature of the skin decreases gradually. Note that with the increase of time, the temperature curves do not become steeper, but rather tend to a steady state—that is, there is a time at

Table 2.18. The Matlab commands for predicting the 1D bioheat.

```
nt=100; nx=40; L=0.05; dz=L/nx; time=30; dt=time/nt; wrcp=0.9*3770; cc=4200; kc=0.5;
qo=1400; Tb=37; Tinit=37; Tnew=zeros(nx+2,nt); Tnew(1:nx+2,1)= Tinit;
for n=1:(nt-1)
   atop=zeros(1,nx+1); atop(1)=2*kc/dz; atop(2)=kc/dz; atop(nx+1)=2*kc/dz;
   abot=zeros(1,nx+1); abot(1)=2*kc/dz; abot(nx)=kc/dz;
   a=zeros(1,nx+2); a(1)=atop(1); a(nx+2)=1;
   a(2)=(3*kc/dz)+(cc*dz/dt)+wrcp*dz; a(nx+1)=(3*kc/dz)+(cc*dz/dt)+wrcp*dz;
   b=zeros(nx+2,1); b(1)=qo; b(nx+2)=Tb; b(2)=wrcp*Tb*dz+(cc*dz/dt)*Tnew(2,n);
   b(nx+1)=wrcp*Tb*dz+(cc*dz/dt)*Tnew(nx+1,n); A=zeros(nx+2,nx+2);
   for i=3:nx
      atop(i)=kc/dz; abot(i-1)=kc/dz; a(i)=(2*kc/dz)+(cc*dz/dt)+wrcp*dz;
      b(i)=wrcp*Tb*dz+(cc*dz/dt)*Tnew(i,n);
   end; A=diag(a,0)+diag(-atop,1)+diag(-abot,-1);
   for i=1:nx+2
      D(i,i)=1; U(1,i)=A(1,i);
   end; Y(1,1)=b(1);
   for i=2:nx+2
      D(i,i-1)=A(i,i-1)/U(i-1,i-1); U(i,i)=A(i,i)-D(i,i-1)*A(i-1,i); Y(i,1)=b(i)-D(i,i-1)*Y(i-1);
   end
   for i=1:nx+1
      U(i,i+1)=A(i,i+1);
   end; Tnew(:,n+1)=U\Y;
end; [M, c]= contourf(Tnew);colormap; c.LineWidth = 1.05;
```

which temperature distribution in a biological tissue reaches the steady-state, as we have seen in the previous example for depth. The surface temperature of the skin decreases gradually toward the steady state. Along with the penetration depth, the time to reach steady-state temperature distribution in a biological tissue is also an important parameter in the estimation of treatment time. This time depends on boundary conditions, as we can see in Figure 2.45, through its dependence on the heat flux at surface, q_S. The time to reach steady-state with the first kind boundary condition, in (2.152), is $1.35 \times \ell_b/\alpha$, while that of second kind boundary condition, in (2.152), $2.30 \times \ell_b/\alpha$, where $\alpha = \rho_b c_b/k$. The time of third kind boundary condition, in (2.152), varies from 1.35 to 2.30.

Accurate early assessment of the skin damage in a burn injury can greatly improve post care. It has been found that burns cause changes in blood flow and heat conduction of the skin, depending on the degree of the burn. For example, the normal and burned tissues can be defined with different properties: $k = 0.5$ W/m, $\omega = 0.0005$ ml/s/ml for normal tissue, and $k = 0.2$ W/m/, $\omega = 0.0001$ ml/s/ml for burned tissue. The main reason that the burned tissues have lower values of thermal conductivity and blood perfusion rate is the possible loss of water and the damage of blood vessels during burning. Results in Figure 2.45 demonstrate a very different transient temperature profile at the skin surface for the two types of tissue, so the detection of surface temperature can be used to determine the burned degree.

Exercise: We can alternatively solve (2.153) for these two Bcs:

$$-k\frac{dT}{dx}\bigg|_{x=0} = h_f\,[T_f - T(x=0)] \quad \text{and} \quad T(x=L) = T_c\,,$$

where T_c is the core temperature, h_f is the heat convection coefficient between the skin surface and the surrounding air that describes the overall contribution from natural convection and radiation, and T_f is the surrounding air temperature. Verify that the solution of (2.153) becomes

$$T(x) = T_B \; + \; \frac{(T_c - T_B)\left[\frac{1}{\ell_B}\cosh(x/\ell_B) + \frac{h_f}{k}\sinh(x/\ell_B)\right]}{\frac{1}{\ell_B}\cosh(L/\ell_B) + \frac{h_f}{k}\sinh(L/\ell_B)}$$

$$+ \; \frac{\frac{h_f}{k}\,(T_f - T_B)\sinh\left[(L - x)/\ell_B\right]}{\frac{1}{\ell_B}\cosh(L/\ell_B) + \frac{h_f}{k}\sinh(L/\ell_B)}\,.$$

One of the other major problems that makes space unfriendly to human bodies is the high radiation levels. On Earth, we are protected by the atmosphere and its structure from most of the harmful sun's radiation. In space, we are not. In other words, without a strong magnetic field and atmosphere, such as those provided by the Earth, we cannot escape high levels of radiation in other space environments in our solar system. As illustrated in Figure 2.45c, the high radiation is so penetrating that it can reach internal organs, the human's entire body, in a matter of minutes and even seconds because high radiation is accompanied by significant changes in blood perfusion. As we can see in Figure 2.45d, in a matter of seconds, organs are affected and death may occur nearly instantaneously. On the surface of the Moon, for example, the lack of an atmosphere exposes human bodies to radiation without mitigation. On Mars, with a thin atmosphere, only about 1 percent of the Earth (see Ikelle, 2023a), the radiation level is so intense that underground habitats (e.g., caves/lava tubes) are being considered, along with strongly shielded vehicles for excursions on Mars's surface. In fact, harmful radiation is more of a deterrent to human space travel than variations in gravity fields.

Example: 1D unsteady-state nonlinear bioheat solution. Tumors are known to have a different perfusion rate than normal healthy tissue, as mentioned earlier. Blood flows through them more quickly than healthy tissues. Therefore, the ability to know the effects and measurements of this abnormal perfusion rate helps evaluate the size and severity of a tumor. By varying the blood perfusion, $\omega_b(T)$, it is possible to examine the effect that different volume flow rates of blood have on the heat transfer inside the biological bodies. In a number of hyperthermia treatments, blood perfusion is usually assumed to vary in the following exponential form

$$\omega_b(T) = a_1 \exp(a_2 T)\,, \tag{2.158}$$

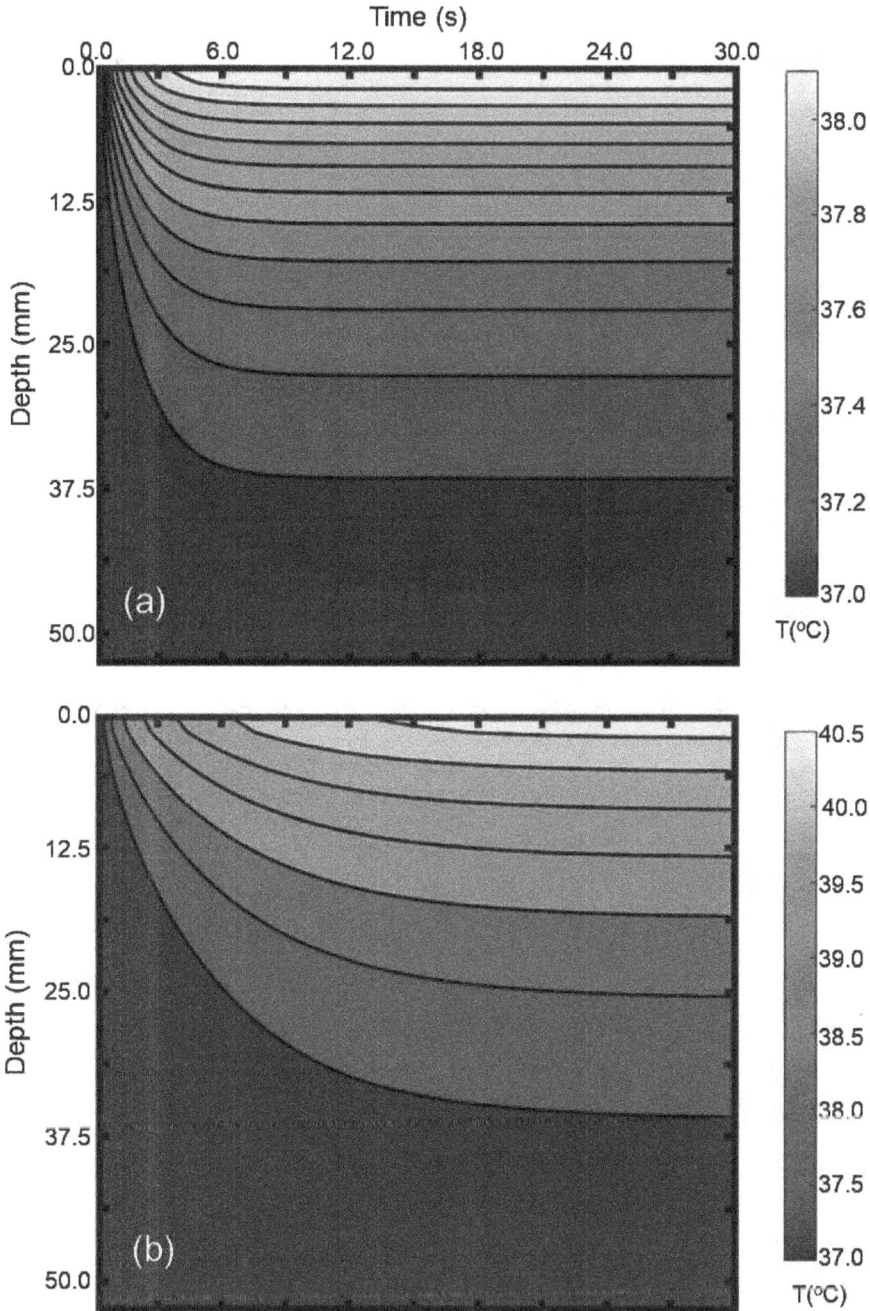

Figure 2.45. *(Continued.)*

where a_1 and a_2 are positive constants with $a_1 = 0.0002$ 1/s and $a_2 = 0.008$ being the typical values for normal healthy tissue and $a_1 = 0.002$ 1/s and $a_2 = 0.008$ for tumors. For small values of a_2 such as $a_2 = 0.001$, the exponential blood-perfusion rate is almost linear because a_2^2 is very small, therefore the quadratic

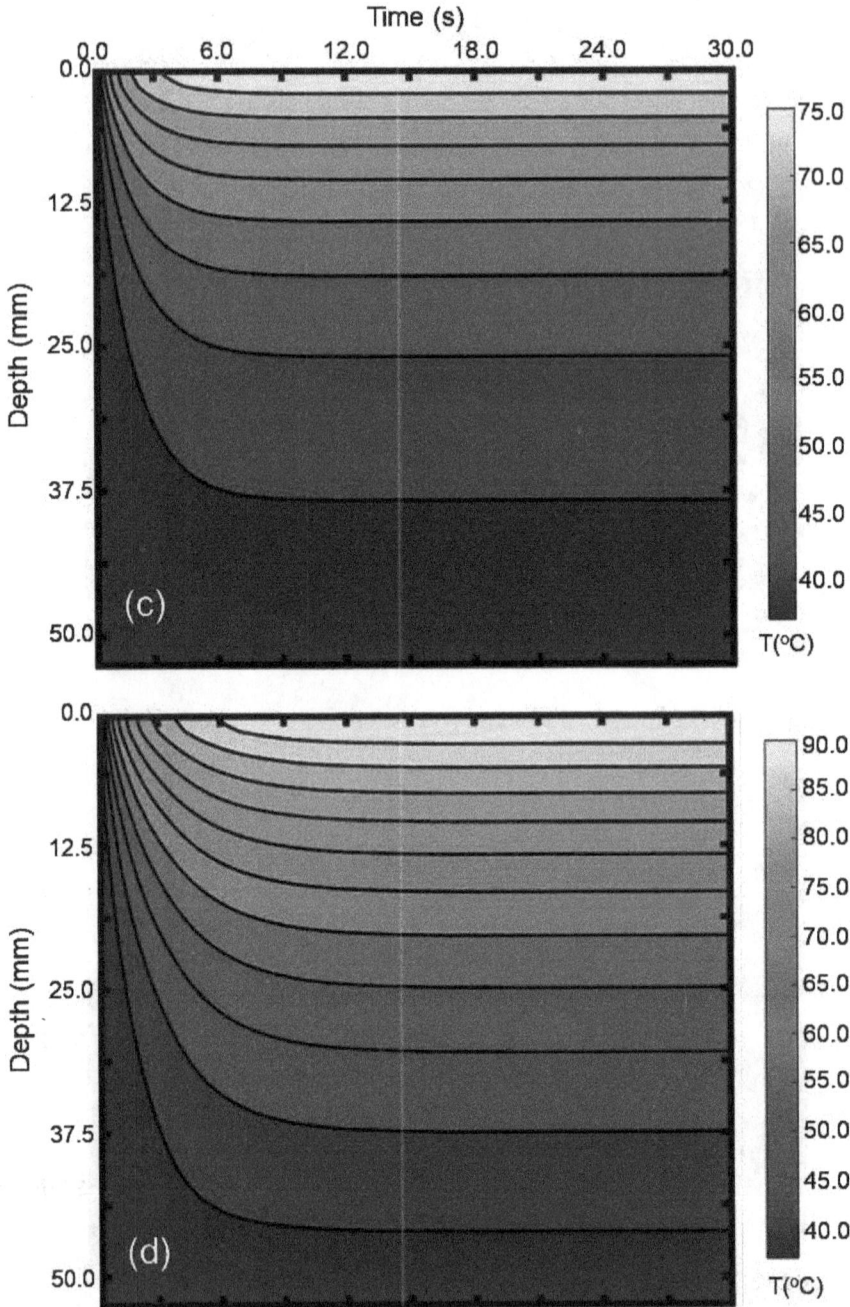

Figure 2.45. The temperature distribution as a function of depth and time for a 1D unsteady-state linear bioheat: (a) $q_w = 40$ W/m^3 and $\omega_b = 0.0007$ 1/s; (b) $q_w = 1400$ W/m^3 and $\omega_b = 0.0007$ 1/s; (c) $q_w = 1000$ W/m^3 and $\omega_b = 0.0003$ 1/s; and (d) $q_w = 1400$ W/m^3 and $\omega_b = 0.0003$ 1/s.

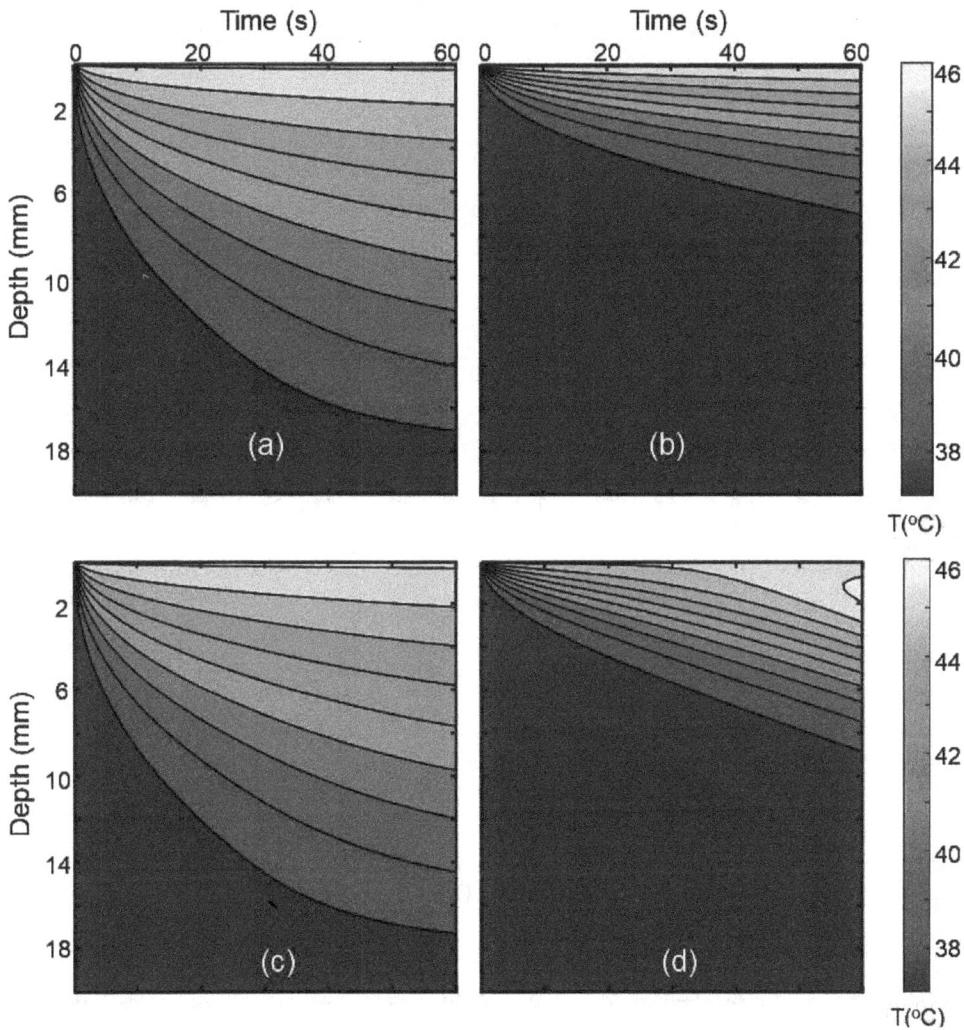

Figure 2.46. The temperature distribution as a function of depth and time for a 1D unsteady-state nonlinear bioheat: (a) $a_1 = 0.0002$ 1/s, $a_2 = 0$, and $c_p = c_b = 4200$ Ws/kg/m^3; (b) $a_1 = 0.002$ 1/s, $a_2 = 0$, and $c_p = c_b = 42000$ Ws/kg/m^3; (c) $a_1 - 0.0002$ 1/s, $a_2 = 0.008$, and $c_p = c_b = 4200$Ws/kg/m^3; and (d) $a_1 = 0.002$ 1/s, $a_2 = 0.008$, and $c_p = c_b = 42000$ kg/s/m^3.

temperature-dependent blood perfusion and the higher terms of the Taylor expansion are negligible. Figure 2.46 shows the temperature distribution as a function of distance and time for exponential and constant blood-perfusion rates. The highest temperature is found to exceed 45 degrees Celsius, due to the nonlinearity of the bioheat-transfer equation.

Computational exercise: Consider the following one-dimensional unsteady bio-heat with radiative heat:

$$\frac{\partial T(x,t)}{\partial t} = \alpha \frac{\partial^2 T(x,t)}{\partial x^2} + \omega_b(T_b - T(x,t)) + h\left[T_\infty^4 - T^4(x,t)\right] \; ,$$

where $T(x,t)$ is the temperature distribution in the body segment, T_∞ is the surrounding temperature. The time step of explicit finite-difference modeling (see Ikelle and Amundsen, 2018) can be written as follows:

$$T_{n+1,j} = \frac{U_n}{D_n} \; ,$$

where

$$U_n = T_{n,j}\left(1 - \frac{\Delta t}{(\Delta x)^2}\right) + \frac{\Delta t}{(\Delta x)^2}\left(T_{n,j+1} + T_{n,j-1}\right) + \frac{h\Delta t}{T_b^4}\left[T_{k,j}^2 + T_\infty^2\right]\left[T_{k,j} + T_\infty\right]T_\infty$$

and

$$D_n = 1 + \frac{h\Delta t}{T_\infty^3}\left[T_{k,j}^2 + T_\infty^2\right]\left[T_{k,j} + T_\infty\right] \; .$$

with the following initial $T_{0,j} = T_0 \sin(\pi \times j \times \Delta x / L)$, and BCs $T_{n,0} = T_1$ and $T_{n,0} = T_2$. Develop a Matlab script for this problem.

2.5. Heat Engines and Pumps

When the two reservoirs are brought into direct contact, an amount of heat is transferred from the high-temperature reservoir to the low-temperature reservoir (Figure 2.1). No work is produced by this process. Also, this process is the reason you cannot cool your kitchen by leaving your refrigerator open! When the two reservoirs are not in direct contact, and an engine is used to transfer heat from the high-temperature reservoir to the low-temperature reservoir by engine doing work; we have the basic principle of design of heat engine. Thermodynamics originally grew from the desire to design and build efficient versions of such heat engines, before it becomes central to all theoretical sciences through the discovery of entropy and the first and second laws of thermodynamics by Carnot, Clausius, Kelvin, Planck, Gibbs, and Carathéodory. So a heat engine is a physical or theoretical device that converts thermal energy (heat) to mechanical output (work). It operates in a cyclic process—that is, it periodically returns to its initial conditions, repeating the same process over and over. Examples of heat engines include power plants that produce electricity, internal combustion engines of automobile, steam engines, and jet engines. We here describe the basic components of heat engines and how they are related to the second law of thermodynamics. We also describe pump engines, which are just reverse-heat engines; they take heat from a cold reservoir and transfer it to a hot reservoir. Refrigerators and air conditioners are typical examples of pump engines.

2.5.1. Heat Engines and the Second Law of Thermodynamics

Let us start by recalling that a cyclical process brings a system, such as heat engines, back to its original state at the end of every cycle. For example, after firing an automobile heat engine, hot gases in the piston cylinder of a car increase their volume, pushing the piston outward. But at the end of the cycle, the piston is restored to its original thermodynamic state so that the work cycle can be repeated. In such a system the internal energy, U, is the same at the beginning and end of every cycle; i.e.,

$$\Delta U = 0 \Longleftrightarrow W = Q , \tag{2.159}$$

in accordance with the first law of thermodynamics. Let us now consider a heat engine operating between two thermal reservoirs. In a full cycle, the engine takes the heat (Q_H from hot reservoir, which is the source), it does work (W) with some of this heat and returns the rest to the cooler reservoir (Q_C). The net heat transfer during the cycle is

$$Q = Q_H - Q_C \Longrightarrow Q_C = Q_H - W . \tag{2.160}$$

In this expression and in many others throughout this section, to be consistent with traditional treatments of heat engines, we take both Q_H and Q_C to be positive quantities, even though Q_C represents the energy leaving the engine. The temperatures of the hot and cold reservoirs are T_{hot} (in Kelvin) and T_{cold} (in Kelvin), respectively.

Because the hot reservoir is heated externally (say, for example, through the use of petroleum), it is important that the work is done as efficiently as possible. In fact, we would like to have $W = Q_H$—that is, no heat is transferred to the environment ($Q_C = 0$). Unfortunately, the second law of thermodynamics tells us it is impossible in any cyclical system to convert heat into work. However, for practical reasons, we would like to have an optimal engine that will do the maximum amount of work with the minimum amount of fuel. We can measure the performance of a heat engine in terms of its thermal efficiency, η, defined as

$$\eta = \frac{W}{Q_H} = 1 - \frac{Q_C}{Q_H} = \frac{\text{work done}}{\text{energy put into the system}} = \frac{\text{what you get out}}{\text{what you put in}} , \tag{2.161}$$

with $\eta < 1$ as 100 percent efficiency is not possible. For example, a good automobile engine has an efficiency of about 20 percent, and diesel engines have efficiencies ranging from 35 percent to 40 percent.

So to summarize, most standard heat engines (steam, gasoline, diesel) work by supplying heat to a gas, the gas then expands in a cylinder and pushes a piston to do its work. The catch is that the heat and/or gas have to somehow be dumped out of the cylinder to get ready for the next cycle.

The Carnot engine: ideal heat engine. How can a heat engine achieve its maximum efficiency? To determine the upper limit of efficiency for heat engines, a hypothetical heat engine, now known as the Carnot engine, was proposed by Carnot (1824). A Carnot engine cannot be realized in reality because it runs reversibly (see Figure 2.47 and Table 2.19). It consists of an ideal gas is confined to a cylinder by a frictionless piston and a 4-step work cycle (two adiabatic processes and two isothermal processes, all reversible):

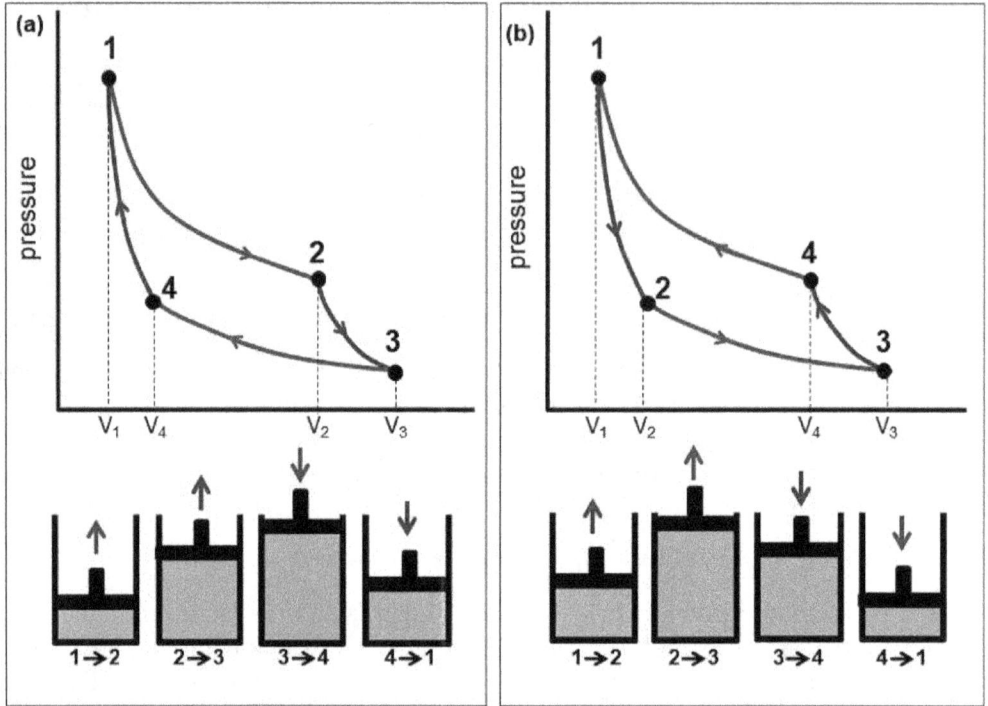

Figure 2.47. (a) A pressure-volume diagram of the Carnot cycle and its corresponding gas-piston model. (b) A pressure-volume diagram of the reversed Carnot cycle and its corresponding gas-piston model.

- Step #1 is an isothermal expansion, $T = T_{\text{hot}}$, from V_1 to V_2. Therefore $\Delta U_1 = 0$ and the work done is

$$W_1 = Q_H = -\int_{V_1}^{V_2} P dV = -\int_{V_1}^{V_2} \frac{RT_{\text{hot}}}{V} dV = -R\,T_{\text{hot}} \ln\left(\frac{V_2}{V_1}\right) \quad (2.162)$$

- Step #2 continues the reversible expansion from V_2 to V_3, but to do so adiabatically (i.e., $Q_2 = 0$). Because the engine is doing work as the gas expands, the energy of the gas must decrease, and because the gas is ideal, this means the temperature must decrease to T_{cold}; i.e.,

$$W_2 = \Delta U_2 = c_V \left(T_{\text{cold}} - T_{\text{hot}}\right) \quad (2.163)$$

- Step #3 reverses the piston movement. The piston compresses the gas isothermally and reversibly from V_3 to V_4, giving off heat Q_C at $T = T_{\text{cold}}$. The work

Table 2.19. The four steps of one Carnot cycle.

$1 \to 2$	isothermal expansion	$T = T_{\text{hot}}$	$V_1 \to V_2$	$Q_1 = Q_H$
$2 \to 3$	adiabatic expansion	$T_{\text{hot}} \to T_{\text{cold}}$	$V_2 \to V_3$	$Q_2 = 0$
$3 \to 4$	isothermal compression	$T = T_{\text{cold}}$	$V_3 \to V_4$	$Q_3 = Q_C$
$4 \to 1$	adiabatic compression	$T_{\text{cold}} \to T_{\text{hot}}$	$V_4 \to V_1$	$Q_4 = 0$

is

$$W_3 = Q_C = -R\,T_{\text{cold}} \ln\left(\frac{V_4}{V_3}\right) \tag{2.164}$$

- Step #4 is diabatic, reversible compression from V_4 back to V_1. Because work is done in compressing the gas with the piston, the temperature of the gas must increase from T_{cold} back to T_{hot}. The system (i.e., the gas in the cylinder and the piston) is restored to its original state. Because this step is adiabatic and reversible, the work is

$$W_4 = \Delta U_4 = -c_V\,(T_{\text{cold}} - T_{\text{hot}}) \ . \tag{2.165}$$

Because steps #1 and #3 are adiabatic and reversible, the volume changes are related by

$$\left(\frac{T_{\text{cold}}}{T_{\text{hot}}}\right)^{c_V} = \left(\frac{V_2}{V_3}\right)^R \quad \text{and} \quad \left(\frac{T_{\text{hot}}}{T_{\text{cold}}}\right)^{c_V} = \left(\frac{V_4}{V_1}\right)^R$$

or

$$\frac{V_1}{V_2} = \frac{V_4}{V_3} \ . \tag{2.166}$$

By using (2.166), the net work is

$$W = W_1 + W_2 + W_3 + W_4 = -R\,(T_{\text{hot}} - T_{\text{cold}}) \ln\left(\frac{V_2}{V_1}\right) \tag{2.167}$$

and the engine efficiency, as defined in (2.161), is

$$\eta_{\text{carnot}} = \frac{W}{Q_H} = \frac{-R\,(T_{\text{hot}} - T_{\text{cold}}) \ln\,(V_2/V_1)}{-R\,T_{\text{hot}} \ln\,(V_2/V_1)} = \frac{T_{\text{hot}} - T_{\text{cold}}}{T_{\text{hot}}} \tag{2.168}$$

or

$$\eta_{\text{carnot}} = 1 - \frac{|Q_C|}{|Q_H|} = 1 - \frac{R\,T_{\text{cold}} \ln\,(V_2/V_1)}{R\,T_{\text{hot}} \ln\,(V_2/V_1)} = 1 - \frac{T_{\text{cold}}}{T_{\text{hot}}} \ , \tag{2.169}$$

where T_{hot} and T_{cold} are in Kelvin, respectively. Efficiency is zero if $TT_{\text{hot}} = T_{\text{cold}}$ and 100 percent only if $T_{\text{cold}} = 0$ Kelvin. Because such reservoirs are not available, $T_{\text{hot}} > T_{\text{cold}}$ and $T_{\text{cold}} \neq 0$ Kelvin, 100 percent efficiency cannot be achieved. In most practical cases, T_{cold} is near room temperature, 300 K. Moreover, no real heat engine operating between two energy reservoirs can be more efficient than a Carnot engine operating between the same two reservoirs.

Quiz: An engine transfers 2×10^3 J of energy from a hot reservoir during a cycle and transfers 1.5×10^3 J to a cold reservoir. (1) Find the efficiency of the engine. (2) How much work does the engine do in one cycle?

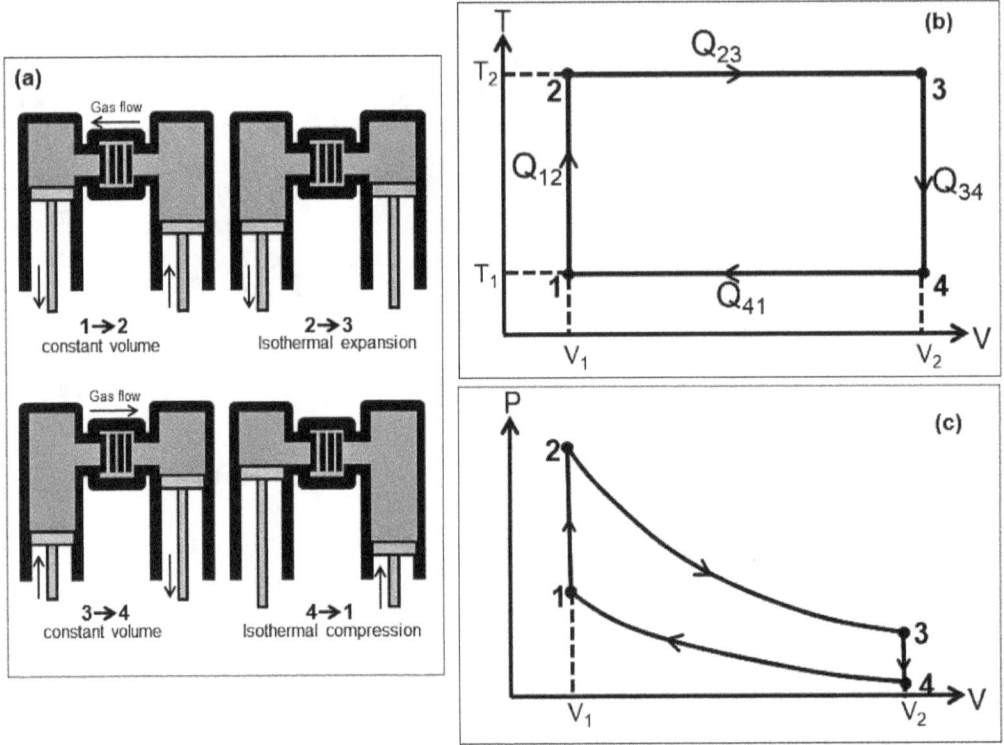

Figure 2.48. (a) The gas-piston model of the Stirling cycle. The model consists of two piston-cylinder systems whose chambers have been filled with gas and are connected by a tube. The process starts at constant volume V_1, the temperature of the gas is increased from $T_1 = T_{cold}$ to $T_2 = T_{hot}$. The gas in the left piston is expanded and the gas in the right piston is compressed so that the total volume remains fixed. The gas in the engine is expanded at a constant temperature T_{hot}. The left piston moves down and the right piston is fixed. At constant volume V_2, the temperature of the gas is reduced from T_{hot} to T_{cold}. The gas in the left piston is compressed and the gas in the right piston is expanded so that the total volume remains fixed. The gas is compressed at constant temperature T_{cold}. The right piston compresses the gas and the left piston is fixed. (b) Temperature-volume and (c) pressure-volume diagrams of the Stirling cycle.

Example: Stirling engines are among the most efficient practical heat-engines ever built. The gasses used inside a Stirling engine never leave the engine. There are no exhaust valves that vent high-pressure gasses, as in a gasoline or diesel engine, and there are no explosions taking place. Because of these characteristics, Stirling engines are very quiet. The cycle of a Stirling engine is shown in Table 2.20 and Figure 2.48. The displacer piston drives gas into the hot zone, where the gas absorbs heat Q_{12} at constant volume (isochoric) and raises the temperature from T_1 to $T_2 = T_{hot}$. Gas absorbs heat Q_{23} isothermally at T_2 and the piston moves to expand the volume from V_1 to V_2. Gas gives off heat Q_{34} and cools to $T_1 = T_{cold}$ at constant volume V_2. Contraction of gas moves the power piston downward decreasing the volume back to T_1 and giving off more heat Q_{41}.

Table 2.20. The four steps of one Stirling-engine cycle.

$1 \to 2$	constant volume (isochoric)	$T_1 \to T_2$	$V_2 = V_1$, Q_{12}, $W_{12} = 0$
$2 \to 3$	isothermal expansion	$T_3 = T_2$	$V_2 \to V_3$, Q_{23}
$3 \to 4$	constant volume (isochoric)	$T_3 \to T_4$	$V_4 = V_3$, Q_{34}, $W_{34} = 0$
$4 \to 1$	isothermal compression	$T_4 = T_1$	$V_4 \to V_1$, Q_{41}

The efficiency of the Stirling engine is

$$\eta = \frac{W}{Q_{12} + Q_{23}} = \frac{Q_{12} + Q_{23} - (-Q_{34} - Q_{41})}{Q_{12} + Q_{23}} = \frac{Q_{23} + Q_{41}}{Q_{12} + Q_{23}}$$

$$= \frac{(T_2 - T_1)) \ln(V_2/V_1)}{\frac{3}{2}(T_2 - T_1) + T_2 \ln(V_2/V_1)} , \tag{2.170}$$

with

$$Q_{12} = -Q_{34} = \Delta U_{12} = n c_V (T_2 - T_1) = \frac{3nR}{2}(T_2 - T_1) , \tag{2.171}$$

$$Q_{23} = -W_{23} = nRT_2 \ln\left(\frac{V_2}{V_1}\right) \tag{2.172}$$

and

$$Q_{41} = -W_{41} = nRT_1 \ln\left(\frac{V_1}{V_2}\right) = -nRT_1 \ln\left(\frac{V_2}{V_1}\right) . \tag{2.173}$$

We can verify that a Stirling Engine is less efficient than a Carnot engine because

$$\frac{(T_2 - T_1)) \ln(V_2/V_1)}{\frac{3}{2}(T_2 - T_1) + T_2 \ln(V_2/V_1)} < \frac{T_2 - T_1}{T_2} \iff \eta_{\text{stirling}} < \eta_{\text{carnot}} .$$

Otto engines: gasoline and diesel engines. The Otto engines are used in automobiles. Gasoline and diesel engines are examples. Table 2.21 and Figure 2.49 illustrate the four-cycle process (intake, compression, power, exhaust) of the Otto-engine cycle, in which the net work of the intake and exhaust cycles can be considered negligible. The input to the cycle consists of increasing volume from V_0 to V_1 by drawing an air-fuel mixture into the cylinder at atmospheric pressure, as the piston moves downward. The intake valve is closed. The piston moves upward, the air-fuel mixture is compressed adiabatically from V_1 to V_2, and the temperature increases from T_1 and T_2. After the piston has reached its highest point, the air-fluid mixture is ignited by a spark plug, and the rapid heating resulting from the combustion process causes the gas to expand and the pressure to increase. This combustion is modeled as quasi-static heat transfer from a series of reservoirs at temperatures ranging from T_2 to T_3. The gas expands adiabatically from V_3 to V_4 and the temperature drops from T_3 to T_4. The exhaust valve opens and the gas is expelled. The volume, however, remains approximately constant because of the

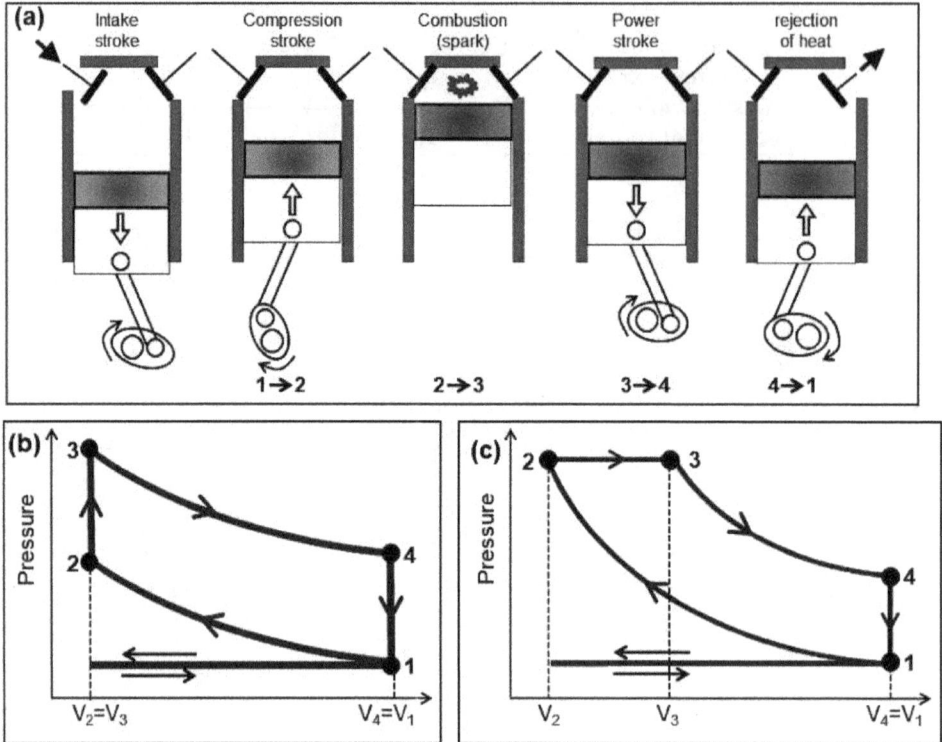

Figure 2.49. (a) Piston and valves in a four-stroke internal combustion engine. (b) Pressure-volume diagrams of the Otto cycle. (c) Pressure-volume diagrams of the diesel cycle. The diesel and Otto (or gasoline) cycles differ only in the $2 \rightarrow 3$ phase; in terms of heat, we have $Q_H = c_V(T_3 - T_2)$ for gasoline engines and $Q_H = c_P(T_3 - T_2)$ for diesel engines. The three other phases are identical in diesel and gasoline cycles.

short time interval and no work is done by the gas. Then the piston moves upward, while the exhaust valve remains open. Residual gases are exhausted at atmospheric pressure and the volume decreases from V_4 to V_1. The cycle then repeats.

Using the expressions, in Table 2.21, we obtain thermal efficiency of Otto engines as

$$\eta_{\text{otto}} = 1 - \frac{Q_C}{Q_H} = 1 - \frac{T_4 - T_1}{T_3 - T_2} = 1 - \frac{1}{(V_1/V_2)^{\gamma-1}} . \tag{2.174}$$

Table 2.21. The four steps of one Otto-engine cycle with $\gamma = c_P/c_V$.

• $1 \rightarrow 2$	isentropic compression	$T_1 \rightarrow T_2$	$V_1 \rightarrow V_2$, $Q_{12} = 0$
			$T_1 = T_2(V_2/V_1)^{\gamma-1}$
• $2 \rightarrow 3$	isochoric heating	$T_2 \rightarrow T_3$	$V_2 = V_3$, $Q_{23} = Q_H$
			$Q_{23} = c_V(T_3 - T_2)$
• $3 \rightarrow 4$	isentropic compression	$T_3 \rightarrow T_4$	$V_3 \rightarrow V_4$, $Q_{34} = 0$
			$T_4 = T_3(V_3/V_4)^{\gamma-1}$
• $4 \rightarrow 1$	isochoric rejection of heat	$T_4 \rightarrow T_1$	$V_4 = V_1$, $Q_{41} = Q_C$
			$Q_{41} = c_V(T_4 - T_1)$

where $\gamma = c_P/c_V$ is the ratio of the molar specific heats and V_1/V_2 is generally known as the compression ratio. We also use the fact that the expressions of adiabatic gas expansions in steps #1 and #3 can be expressed as follows:

$$\frac{T_4 - T_1}{T_3 - T_2} = \left(\frac{V_2}{V_1}\right)^{\gamma - 1} = \frac{T_3}{T_4} = \frac{T_2}{T_1} , \qquad (2.175)$$

since $V_4 = V_1$ and $V_2 = V_3$. The typical compression ratio is 8 and $\gamma = 1.4$, thus, $\eta_{\text{otto}} = 0.56$. Efficiencies of real engines are between 15 percent to 20 percent mainly due to friction, energy transfer by conduction, and incomplete combustion of the air-fuel mixture. By combining (2.174) and (2.175), we arrive at

$$\eta_{\text{otto}} = 1 - \frac{T_2}{T_1} < \eta_{\text{carnot}} , \qquad (2.176)$$

with where the efficiency of a Carnot engine is

$$\eta_{\text{carnot}} = 1 - \frac{T_3}{T_1} \qquad (2.177)$$

because, during the Otto-engine cycle, the lowest temperature is T_3 and the highest temperature is T_1. Therefore, the efficiency of a Carnot engine, which operates between reservoirs at these two temperatures, is more efficient than the Otto engine.

The diesel-engine cycle operates on a cycle similar to the Otto cycle without a spark plug. In this case, higher compression ratios are possible because only air is let into the cylinder during the intake stroke. The fuel is injected into the cylinder at the end of the compression stroke, thus avoiding spontaneous ignition of the fuel-air mixture during the compression stroke. The combustion occurs spontaneously without a spark plug along the constant pressure segment $(2 \rightarrow 3)$ after fuel injection. Because the fuel is injected over a time period in which the piston is moving out of the cylinder, this step can be modeled as a constant-pressure heat intake; i.e.,

$$Q_H = c_P(T_3 - T_2) . \qquad (2.178)$$

In this segment, it is assumed that heat is absorbed from a series of reservoirs at temperatures between T_1 and T_2 in a quasistatic process. In the other segments, the same processes occur, as described for the Otto cycle. Therefore, the efficiency is given by

$$\eta_{\text{diesel}} = 1 - \frac{Q_C}{Q_H} = 1 - \frac{1}{\gamma}\left(\frac{T_4 - T_1}{T_3 - T_2}\right) , \qquad (2.179)$$

where $\gamma = c_P/c_V$. Diesel engines are more efficient than gasoline engines because γ is greater than 1. Real diesel engines used in trucks and passenger cars have efficiencies in the range of 0.30 to 0.35.

Example: Wind-turbine technology. Wind energy is available for electricity generation in locations where the average wind speeds are great enough to drive wind turbines (e.g., Ikelle, 2017 and 2020). As illustrated in Figure 2.50, a wind turbine

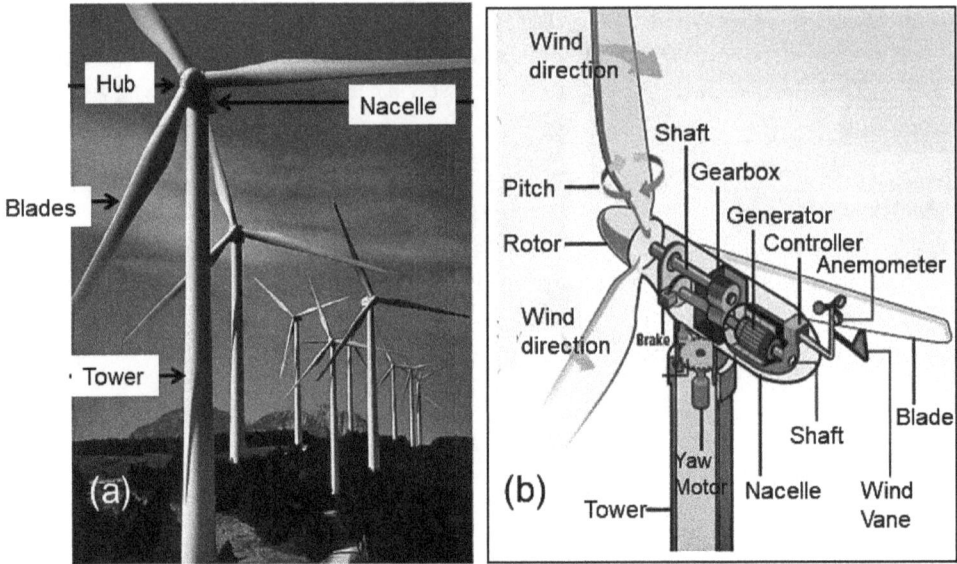

Figure 2.50. Some of the key components of a wind turbine. (a) A modern wind turbine in a wind turbine farm. (b) Basic components of a modern wind turbine with a gearbox.

includes a tower, a nacelle, a rotor (a hub and three blades), and a foundation. The rotating blades are connected to a central hub that rotates with them. Wind passes over the blades, exerting a turning force that depends on wind strength. The rotating blades turn a shaft inside the nacelle, which goes into a gearbox. The gearbox increases the rotation speed of the generator, which uses magnetic fields to convert rotational energy into electrical energy.

Wind or any other moving fluid is practically described by the power rather than energy. Power is kinetic energy per unit time; i.e.,

$$P = \frac{1}{2}\frac{dm}{dt}v_1^2 = \rho A v_1 , \tag{2.180}$$

where m is the mass of the moving object, v_1 is its speed, and dm/dt is the mass flow rate (i.e., density $\times$ volume flux), ρ is the air density and $A = \pi r^2$ is the rotor swept area (turbine size). In this equation, we use the fact that volume moves in 1 second inside a pipe that is the product of the cross-sectional area of the pipe and the speed of the fluid ($dV/dt = Av$). We can estimate wind power as follows:

$$P_{\text{wind}} = \frac{1}{2}\rho A v_1^3 . \tag{2.181}$$

The power available from the wind is a function of the cube of the wind speed. Therefore, if the wind blows at twice the speed of the day before, its energy content will increase eightfold. A turbine cannot necessarily capture all of this power; it can only capture a portion of it. Assume the wind blows axially into the turbine blade (i.e., parallel to the rotor blade shaft) with velocity v_1 and exists in the rotor area with velocity $v_2 < v_1$. Assume no drag exists between the wind and the turbine blades. By the conservation of energy, the power of the turbine is given by

$$P_{\text{turbine}} = \frac{1}{2}\frac{dm}{dt}\left(v_1^2 - v_2^2\right) = \frac{1}{2}Av\left(v_1^2 - v_2^2\right)$$

$$= \frac{1}{2}P_{\text{wind}}\left(1 + \frac{v_2}{v_1}\right)\left(1 - \frac{v_2^2}{v_1^2}\right)\left(1 - \frac{v_2^2}{v_1^2}\right)\,, \qquad (2.182)$$

where $v = (v_1 + v_2)/2$ is the velocity of the wind at the rotor. The efficiency of the wind turbine is the ratio of these two powers:

$$\eta_{\text{turbine}} = \frac{P_{\text{turbine}}}{P_{\text{wind}}} = \frac{1}{2}\left(1 + \frac{v_2}{v_1}\right)\left(1 - \frac{v_2^2}{v_1^2}\right)\,. \qquad (2.183)$$

In analogy to the heat engine, where the efficiency is only a function of T_{hot} and T_{cold}, the efficiency of a wind turbine only depends of the ratio of v_1 and v_2. The maximum turbine efficiency corresponding to the ratio v_2/v_1 that maximizes η_{turbine}; i.e.,

$$\frac{d\eta_{\text{turbine}}}{d(v_2/v_1)} = \frac{1}{2}\left(1 - 2\frac{v_2}{v_1} - 3\frac{v_2^2}{v_1^2}\right) = 0 \implies \frac{v_2}{v_1} = \frac{1}{3}\,. \qquad (2.184)$$

By substituting $v_2/v_1 = 1/3$ in (2.183), we can deduce that the maximum value of η_{turbine} is 0.59. This result is known as Betz' Law. It states that an ideal wind turbine can convert at most 59 percent of the incoming wind power into the rotary motion of the turbine blade. Most current wind turbines operate with efficiencies in the range 10-30 percent. In more general terms, turbine power output can be expressed as follows (Ikelle, 2017b and 2020)

$$P_{\text{turbine}} = C_P P_{\text{wind}}\,, \qquad (2.185)$$

where C_P is the coefficient associated with the turbine. This coefficient depends on the type of turbine, its efficiency, and other operational conditions. Nevertheless, no turbine performing at its best can exceed $C_P = 16/27 = 0.59$.

2.5.2. Heat Pumps

As described above, in a heat engine, energy transfer is from the hot reservoir to the cold reservoir, which is the natural direction. The role of the heat engine is to process the energy from the hot reservoir so as to do useful work. In a refrigerator or a heat pump, we want to transfer energy from the cold reservoir to the hot reservoir. This transfer is not possible between these reservoirs without a device between them. Devices that perform this task are heat pumps and refrigerators. For example, homes in summer are cooled using air conditioners, which transfer energy from the cool room in the home to the warm air outside—that is, in a heat pump, the engine takes in energy Q_C from a cold reservoir and expels energy Q_H to a hot reservoir as follows:

$$Q_H = Q_C + W\,, \qquad (2.186)$$

The work, W, is done on a heat pump. The effectiveness of a heat pump is described by a number called the coefficient of performance (COP), which is

$$\mathrm{COP_R} = \frac{\text{energy transferred at low temperature}}{\text{work done on the pump}} = \frac{Q_C}{W} = \frac{T_{\mathrm{cold}}}{T_{\mathrm{hot}} - T_{\mathrm{cold}}} \; . \; (2.187)$$

As the difference between the temperatures of the two reservoirs approaches zero in this definition, $\mathrm{COP_R}$ tends to ∞. In practice, $\mathrm{COP_R}$ is limited 10. Typical values of $\mathrm{COP_R}$ of a good refrigerator are 5 or 6. We obtained the last equality by using the fact that the hypothetical Carnot heat pump is simply reverses the cycle shown in Figure 2.47.

There are actually two modes of COP. The one in (2.187) is known asthe cooling mode in which you gain energy removed from the cold-temperature reservoir. The other mode, known as heating mode, is defined as follows

$$\mathrm{COP_H} = \frac{\text{energy transferred at high temperature}}{\text{work done on the pump}} = \frac{Q_H}{W} = \frac{T_{\mathrm{hot}}}{T_{\mathrm{hot}} - T_{\mathrm{cold}}} \; . \; (2.188)$$

A comparison of equations (2.187) and (2.188) reveals that

$$\mathrm{COP_H} = \mathrm{COP_R} + 1 \; . \tag{2.189}$$

Example: Refrigerator. The Carnot refrigerator is a reverse of the Carnot cycle, shown in Figure 2.47a. The P-V diagram of the reversed Carnot cycle is shown in Figure 2.47a and Table 2.19. Practical difficulties of the reversed Carnot cycle are a compression of two-phase mixture in step #1 and expansion in step #4. They result in a very wet refrigerant (the working fluid used in the refrigeration cycle is called a refrigerant), causing erosion of turbine blades.

These impracticalities can be overcome by vaporizing the refrigerant before it is compressed. In other words, isothermal processes are replaced by constant pressure processes,as shown in Figure 2.51 and Table 2.22. Therefore, in an actual refrigerator cycle, the fluid passes through a nozzle and expands into a low-pressure area. The cool gas is in thermal contact with the inner compartment of the fridge; it heats up as heat is transferred to it from the fridge. This takes place at constant pressure, so it's an isobaric expansion. The gas is transferred to a compressor, which does most of the work in this process. The gas is compressed adiabatically, heating and turning it back to a liquid. The hot liquid passes through coils on the outside of the fridge, and heat is transferred to the room. This is an isobaric compression process. An air conditioner works the same way; it is a refrigerator whose inside is the room to be cooled (i.e., $T_{\mathrm{room}} = T_{\mathrm{cold}}$) and whose outside is the great outdoors ($T_{\mathrm{outside}} = T_{\mathrm{hot}}$).

Table 2.22. The four steps of one reversed Carnot cycle.

$1 \to 2$	adiabatic compression	$T_{\mathrm{hot}} \to T_{\mathrm{cold}}$	$V_1 \to V_2$, $Q_1 = 0$
$2 \to 3$	isothermal compression	$T = T_{\mathrm{cold}}$	$V_2 \to V_3$, $Q_2 = Q_C$
$3 \to 4$	adiabatic expansion	$T_{\mathrm{cold}} \to T_{\mathrm{hot}}$	$V_3 \to V_4$, $Q_3 = 0$
$4 \to 1$	isothermal expansion	$T = T_{\mathrm{hot}}$	$V_4 \to V_1$, $Q_4 = Q_H$;

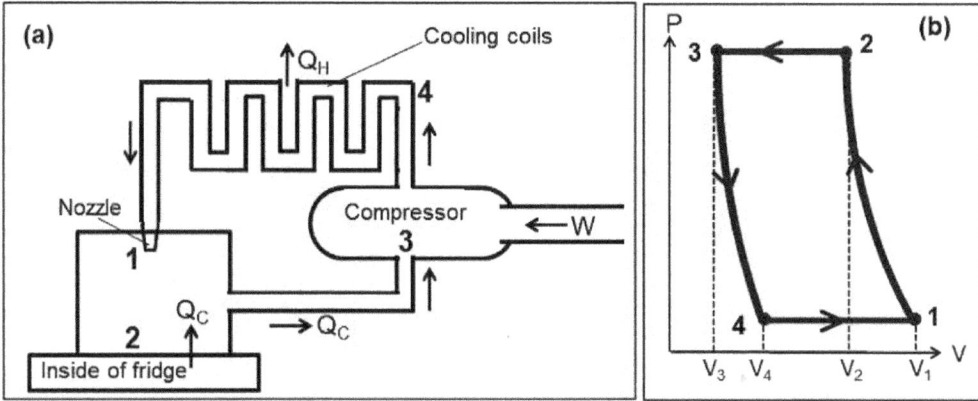

Figure 2.51. (a) A schematic diagram of the four-step process of a refrigerator. Step #1: the fluid passes through a nozzle and expands into a low-pressure area. This step is essentially an adiabatic expansion. Step #2: the cool gas is in thermal contact with the inner compartment of the fridge; it heats up as heat is transferred to it from the fridge. This step is an isobaric expansion. Step #3: The gas is transferred into a compressor, which does most of the work in this process. The gas is compressed adiabatically, heating it and turning it back to a liquid. Step #4: The hot liquid passes through coils on the outside of the fridge, and heat is transferred to the room. This step is an isobaric compression process. An air conditioner works the same way. (adapted from http://physics.bu.edu/ duffy/py105/Heatengines.html) (b) The P-V graph for a refrigerator cycle.

2.5.3.   Is the Heart a Pump Engine?

A heart is an organ that pumps blood through the entire body, including the brain and every cell, nerve, and muscle. It is the source of energy that makes the blood flow. It is composed of two separate pumping systems, the right side and the left side (Figure 2.52). The right-side system receives oxygen-poor blood from the veins and pumps it to the lungs, where it picks up oxygen and gets rid of carbon dioxide. The left-side system receives oxygen-rich blood from the lungs and pumps it; through the arteries, to the rest of your body. Without the heart, the body would not be able to receive the vital oxygen and nutrients that are carried by the blood. The blood also removes waste products. When the heart stops functioning, the body starves of oxygen and quickly dies. Thus, the heart is one of the most important organs in the body. In fact, heart-rate variability, along with respiratory rate, blood pressure, venous oxygen saturation, and oxygen and glucose metabolism, are key indicators of youth, health, aging, illness, and eventually death.

The heart beats unceasingly from birth to death, varying it is rate of activity according to the needs of the body. Its beats are not particularly strong, only

Table 2.23. The four steps of one refrigerator cycle.

$1 \to 2$	isobaric expansion	$T_{\text{hot}} \to T_{\text{cold}}$	$V_1 \to V_2, Q_1 = 0$
$2 \to 3$	adiabatic compression	$T = T_{\text{cold}}$	$V_2 \to V_3, Q_2 = Q_C$
			$Q_2 = c_P(T_2 - T_3)$
$3 \to 4$	isobaric compression	$T_{\text{cold}} \to T_{\text{hot}}$	$V_3 \to V_4, Q_3 = 0$
$4 \to 1$	adiabatic expansion	$T = T_{\text{hot}}$	$V_4 \to V_1, Q_4 = Q_H$
			$Q_4 = c_P(T_2 - T_3)$

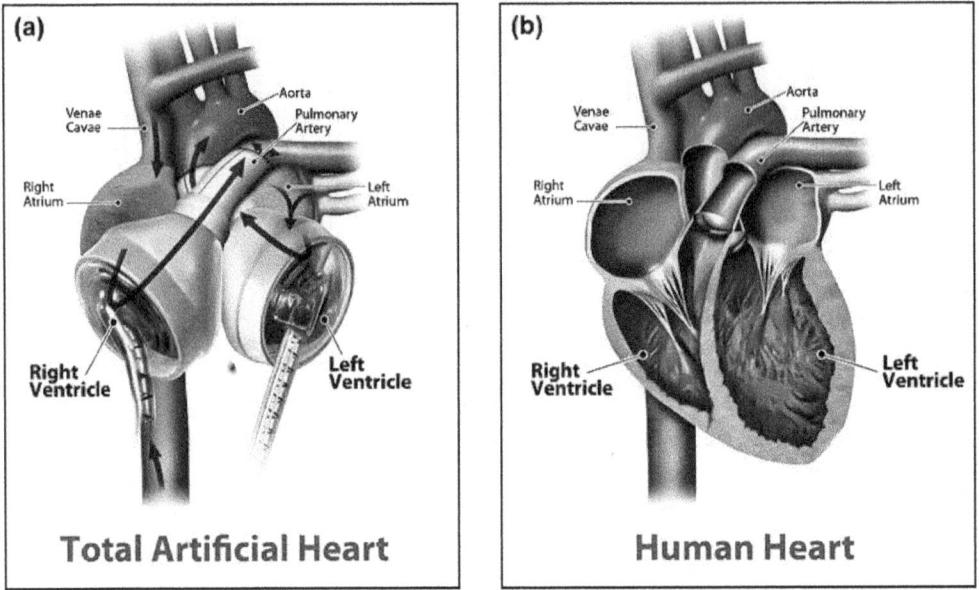

Figure 2.52. (a) Artificial Heart beside (b) a human heart. (Source: SynCardia Systems, LLC.)

strong enough to make the blood flow. Basically, the heart has its own electrical system that coordinates the work of the heart chambers (heart rhythm) and also controls the frequency of beats (heart rate). The arteries help in this regard as well. Heart problems are mostly related to the malfunctioning of this electrical system. These problems include arrhythmia (i.e., the heart is beating in an uncoordinated or random way), tachycardia (i.e., the heart is beating too fast), and bradycardia (i.e., the heart is beating too slowly).

A heart is made of a special type of muscle called myocardium. The myocardium squeezes (contracts) to pump the blood. After pumping, the myocardium relaxes enough so that it can fill with blood properly before it restarts squeezing again to pump the blood. Can this cycle be described in terms of thermodynamically processes such as isentropic, isobaric, and isotherm processes? That is one of the two fundamental questions that we address in this subsection. The other question is: can we mimic the heat and pump engines described in the previous subsections to design artificial mechanical hearts?

The cardiac cycle as a thermodynamic cycle. Figure 2.53 shows a simple diagram of the four phases of the cardiac cycle—that is, the mechanism by which the heart supplies blood to the body. On average, the human heart at rest pumps 5.6 liters of blood per minute—more than 8,000 liters of blood circulate through a healthy human heart per day. The $1 \to 2$ phase, in Figure 2.53, is the relaxation phase in which the heart refills with blood from its contracted state at the end of the previous cycle—this phase is essentially a return stroke volume (the amount of blood pumps out by heart), in which blood slowly fills the heart, increasing its volume without changing the pressure (diastole). In phase $2 \to 3$ the heart begins to contract around the trapped blood (systole). In phase $3 \rightarrow 4$, the heart continues to contract, resulting in a rapid efflux of blood at a constant pressure. In phase $4 \rightarrow 1$, the heart tissue expands, relieving pressure, the blood rushes back in, and the cycle repeats. This figure is based on the measurements of volume and shape of the heart and the volume of blood that passes through the heart during each part of the cardiac cycle. We can see that we are dealing with a pressure-volume diagram of a full cardiac cycle: $1 \rightarrow 2$ filling; $2 \rightarrow 3$: isovolumic contraction; $3 \rightarrow 4$: ejection; $4 \rightarrow 1$: isovolumic relaxation. In other words, the cardiac cycle can be described as a thermodynamic PV cycle that describes pressure generation and blood-volume changes. Moreover, pressure, volume, and flow curves obtained from normal individuals almost perfectly coincide with this cycle.

The artificial mechanical heart. No heat engine or pump engine is comparable to a human heart; the heart is not an engine. Friction or electric resistance are not present in the heart. Hence, it never gets overheated. The muscles, which can be considered the engine of the heart, are automatically repaired or replaced by mitosis (replacing old cells with newly formed cells; see Ikelle, 2023a), as necessary. In contrast, the combustion that takes place inside the engine of a car, for example, produces enormous heat. It is difficult to imagine a physical pump anywhere near as efficient at operating and lasting as a human heart. Nonetheless, there are now multiple artificial mechanical heart designs and even commercial versions being used to help end-stage heart failure patients. These artificial mechanical hearts are not donated hearts or hybrids of mechanics and donated hearts. They are pure mechanical hearts. They can potentially reduce the demand for donated hearts and even eliminate the cumbersome requirements of blood type or antibody level matches between donors and receivers. How these artificial mechanical hearts differ from heat and pump engines was described in the two previous subsections.

Any pump that is expected to move more than 8,000 liters of blood daily requires a lot of power in the form of a big battery, for example. Such a battery is unlikely to fit in a human chest cavity and may require regular replacements. Moreover, several safety issues are associated with the implant of such large batteries in the body. The most common solution today is a pump in the heart, like the one in Figure 2.53b, which includes drivers that can produce pulses of air and vacuum that help pump blood in and out of the ventricles, with a wireless external power system. It is important that this system be wireless instead of a wired system, which can potentially become a conduit for undesired bacteria and other germs.

2.6.   Essay: Entropy and the Arrow of Time: The Irreversible Law

Entropy is always increasing—that is, the second law of thermodynamics is irreversible. Here irreversibility means that we cannot use the second law of thermodynamics to return to a previous state or even to the physical state that led to the present state. The property also describes the fact that the direction of time is unidirectional, always pointing forward. However, all other known laws of classical and quantum physics are reversible. We have yet to determine whether these laws

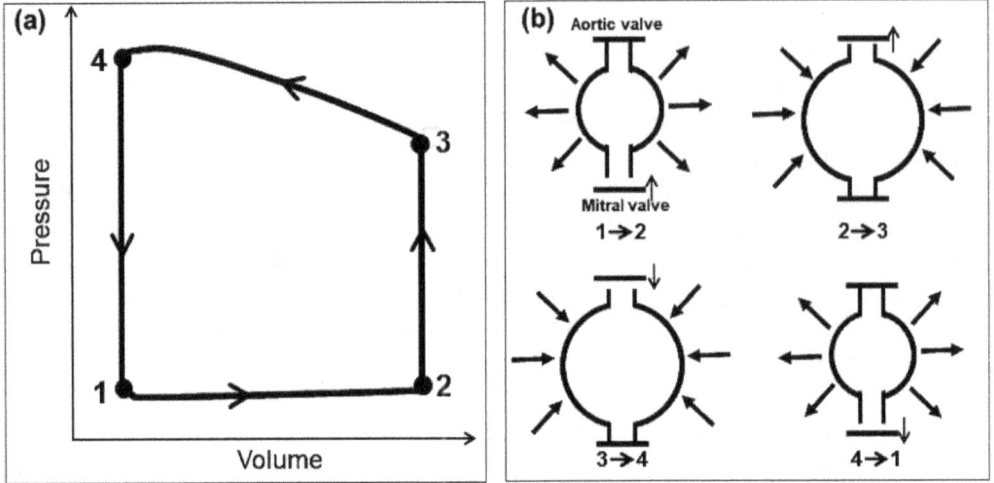

Figure 2.53. (a) Pressure-volume diagram of the cardiac cycle—the cardiac cycle represents all the events that take place during a single heartbeat. The $1 \to 2$ phase is the diastolic filling. Mitral valve closes at the end of this phase. The $2 \to 3$ phase is the isovolumic contraction. It occurs in early systole during which the ventricles contract with no corresponding volume change. This short-lasting portion of the cardiac cycle takes place while all heart valves are closed. Aortic valves, which help keep blood flowing in the correct direction through the heart, open at the end of this phase. The $3 \to 4$ phase is the ejection. Aortic valves closes at the end of this phase.

apply to back holes. In any case, with the current laws of physics you can tell exactly where you are heading and where you came from, when they are properly applied. In other words, these laws tell us that there is no restriction on going into the future and into the past. Let us recall this notion through a simple example of the pressure response of an infinite homogeneous acoustic medium with velocity c and a delta function source. In the frequency domain, the pressure field and its complex conjugate are, respectively,

$$p(\boldsymbol{x}, \omega, \boldsymbol{x}') = \frac{1}{4\pi \, |\boldsymbol{x} - \boldsymbol{x}'|} \exp\left(i\frac{\omega}{c} \, |\boldsymbol{x} - \boldsymbol{x}'|\right) ,$$

$$p(\boldsymbol{x}, -\omega, \boldsymbol{x}') = \frac{1}{4\pi \, |\boldsymbol{x} - \boldsymbol{x}'|} \exp\left(-i\frac{\omega}{c} \, |\boldsymbol{x} - \boldsymbol{x}'|\right) ,$$

where

$$|\boldsymbol{x} - \boldsymbol{x}'| = \sqrt{(x - x')^2 + (y - y')^2 + (z - z')^2}$$

for $|\boldsymbol{x} - \boldsymbol{x}'| \neq 0$. In the time domain, these fields are

$$p(\boldsymbol{x}, t, \boldsymbol{x}') = \frac{1}{4\pi \, |\boldsymbol{x} - \boldsymbol{x}'|} \delta\left(t - \frac{|\boldsymbol{x} - \boldsymbol{x}'|}{c}\right) = \frac{1}{4\pi \, |\boldsymbol{x} - \boldsymbol{x}'|} \delta\left(t - t_o\right) ,$$

$$p'(\boldsymbol{x}, t, \boldsymbol{x}') = \frac{1}{4\pi \, |\boldsymbol{x} - \boldsymbol{x}'|} \delta\left(t + \frac{|\boldsymbol{x} - \boldsymbol{x}'|}{c}\right) = \frac{1}{4\pi \, |\boldsymbol{x} - \boldsymbol{x}'|} \delta\left(t + t_o\right) ,$$

where

$$t_o = \frac{|\boldsymbol{x} - \boldsymbol{x}'|}{c} \ .$$

$p(\boldsymbol{x}, t, \boldsymbol{x}')$ and $p'(\boldsymbol{x}, t, \boldsymbol{x}')$ are the Fourier transforms of $p(\boldsymbol{x}, \omega, \boldsymbol{x}')$ and $p(\boldsymbol{x}, -\omega, \boldsymbol{x}')$, respectively. From (2.190), it follows that $p(\boldsymbol{x}, t, \boldsymbol{x}')$ is the time-retarded field, and its Fourier transform is $p(\boldsymbol{x}, \omega, \boldsymbol{x}')$. Similarly, $p'(\boldsymbol{x}, t, \boldsymbol{x}')$ is the time-advanced field, and its Fourier transform is $p(\boldsymbol{x}, -\omega, \boldsymbol{x}')$. Retarded waves progressively move with increases in time, as visualized in the classical movies of wave propagation (e.g., Ikelle and Amundsen, 2005 and 2018; and Ikelle, 2023c). They are consistent with the way current scattering experiments are carried out; they arrive at receiver locations at some time after they have left the source location. Advanced waves travel backward in time; that is, they arrive at the receivers before they have left the source point. These waves are really an affront to our common sense and our understanding of how the world operates—our ever-aging bodies being an obvious testimony. So despite the fact that advanced waves are valid solutions to the wave equations, they are generally ignored in most scattering studies, at least in part because of their counterintuitive nature. In more general terms, the laws of motions are reversible. They provide the rules for updating the physical state. At a given physical state, they allow us to move to the next state but also where we came from.

Entropy encompasses everything around us, including all the other laws of physics. So as entropy is always increasing, our modern physics must include some irreversible laws of physics that we are still to discover in order for the entropy to still grow. Alternative, at microscopic and macroscopic levels, the world around us sparsely uses our laws of physics backward in time. So The forward time of these laws dominate the entropy and ensure that it is, so far, the only quantity of physics which is always increasing.

From a human perspective, the offspring is an ensemble of molecules stuffed off into a small area and therefore it represents a low entropy entity. It includes too small components that we cannot see (quantum mechanics) and too numerous to take into account, even in modern physics. As we grow we spread in space (molecules diffuse), as no information is lost; our entropy continues to increase until we reach our point of death.

2.7.   Essay: Cloud Formations on Earth

Clouds cover roughly two thirds of the globe. They represent a critical climate variable on Earth. They interact with solar radiation through absorption and scattering and, to a lesser extent with terrestrial radiation through absorption, scattering and emission. Moreover clouds have always fascinated us humans. When we picture heaven, clouds usually make an appearance of some kind. They are always hovering above us, showering us with rainwater, and floating as if by magic. Clouds (including a change of phase from water vapor[6] to liquid droplets or ice crystals) can appear out of nowhere, literally out of the blue, on a perfectly clear day, and cover the whole sky with incredible speed. And, likewise, clouds often dissolve, from a cloudy state, back into perfectly clear air (involving a change of phase from liquid

---

[6]Remember that water vapor is a gas and not a liquid; it is water in gaseous instead of liquid form.

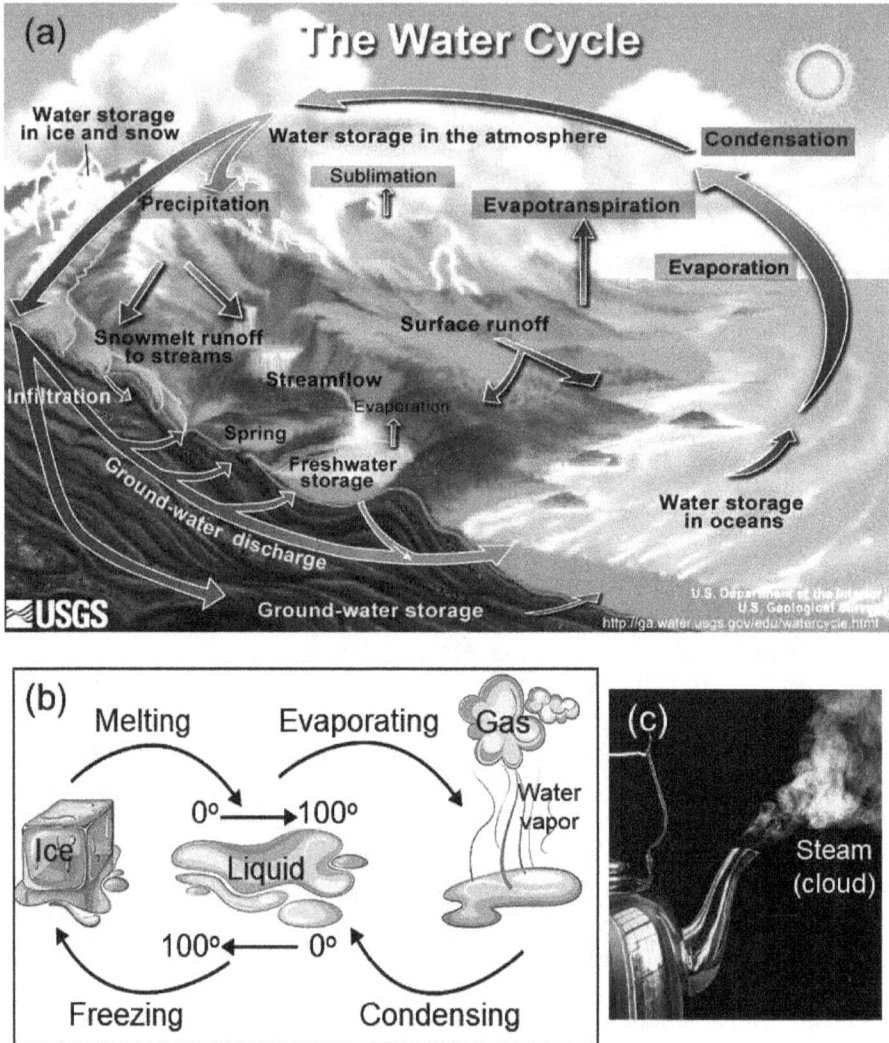

Figure 2.54. (a) The water cycle (also known as the hydrologic cycle). (b) Phase changes in water. (c) An illustration of condensation and the formation of clouds by boiling water in a kettle.

to gas) just as quickly. Also, we would have no rivers, lakes, or fresh water without clouds. Our objective is to examine cloud formations, including their shapes and colors.

2.7.1.   Water Vapor and Aerosols in the Earth's Atmosphere

Water, the only substance that exists in all three phase states (gas, liquid, and solid) at the temperatures found on the Earth, is around us all the time, underground, near the Earth's surface, and even in the atmosphere, as we can learn from the water cycle in Ikelle (2023a) and Figure 2.54. The atmosphere is composed mostly of gases ("air"), but also contains tiny liquids—primarily water—and solid matter—primarily salt and dust—in the form of particles. These particles of water are known as water vapor and the solid particles are known as aerosols. Despite their small mass and small volume in the atmosphere compared to the air, these particles in the atmosphere strongly influence the transfer of radiant energy and the spatial distribution of latent heating through the atmosphere, thereby influencing the weather and climate.

The sources of water in the atmosphere are mostly part of the water cycle (the continuous movement of water on, above, and below the surface of the Earth). Actually, clouds constitute a key step in the water cycle (also known as the hydrological cycle). They include evaporation from the subtropical oceans, evapotranspiration from plants, evaporation from lakes and rivers, and sublimation from snow and ice (see Ikelle, 2023a and Figure 2.54). The sources of aerosols include smoke from natural and man-made fires (biomass burning), mineral soil dust, sea salt, volcanic eruptions, and other pollution processes.

Liquid water droplets generally have the same spherical shape with a radius of about 1 μm and the same composition. The shape of ice particles varies, with an average diameter of 10 μm. The composition and the shape/mixture of aerosols are also variable with characteristic radius between 0.001 μm and 50 μm. Small aerosols with radii < 0.2 μm are referred to as Aitken particles, $0.2 < r < 2$ μm are large aerosols, and $r > 2$ μm are giant aerosols. In other words, taken individually, water vapor drops and aerosols are actually invisible components of our atmosphere.

Besides evaporation and condensation, the physics of these processes in our atmosphere also involves the concept of saturation. Evaporation and condensation are competing processes that occur simultaneously at the liquid water-air interface and saturation, in the context of atmospheric physics, means the capacity or maximum amount of water vapor that can exist in the air at a particular temperature. Warm air can hold more water vapor than cold air.

Let us add to the concepts of evaporation, condensation, and saturation the meteorological notion of relative humidity (RH). The importance of this notion for cloud formation will become clearer later on. Relative humidity refers to the moisture content (i.e., water of the atmosphere, expressed as a percentage of the amount of moisture that can be retained by the atmosphere (moisture-holding capacity) at a given temperature and pressure without condensation.

$$\text{RH} = \frac{\text{water vapor in the air}}{\text{water vapor capacity that the air capable of holding}}$$

$$= \frac{\text{actual mass of water vapor in grams}}{\text{mass of water vapor required for saturation in grams}},$$

at a given temperature. Again, relative humidity is expressed in percentage at that temperature. For example, air with 50 percent relative humidity actually contains one-half the amount of water vapor required for saturation. Air with 100 percent relative humidity is said to be saturated because it is filled to capacity with water vapor. If the relative humidity starts to exceed 100 percent, water vapor in the air immediately turns into liquid. The excess relative humidity over the equilibrium value of 100 percent is generally known as supersaturation. This indicates the amount of excess water vapor available to form rain and snow. Note that this definition makes no reference to any other gases that may be mixed with water vapor.

Table 2.24. An example of the variations of the dew point with altitude of an air parcel.

Altitude (m)	Air parcel (° C)	Dew Point(° C)	Saturation
6000	-15.5	-15.5	Yes
5000	-9.5	-9.5	Yes
4000	-3.5	-3.5	Yes
3000	3.5	3.5	Yes
2000	13.5	3.5	No
1000	23.5	3.5	No
0 (sea level)	33.5	3.5	No

The temperature at which that amount of water vapor makes 100 percent humidity is known as the dew point. It is the temperature at which moisture in the air begins to condense into water. So the dew point is another measure of humidity and it is always at or below the current air temperature. Table 2.24 illustrates the variations of the dew point with altitude of an air parcel. Note that the dew point temperature remains constant until saturation is reached. In this example, the parcel cools to saturation after it has been moved up to 3000 meters above sea level. It is at this altitude that clouds begin to form in the parcel. Once saturation is reached, condensation occurs as the parcel continues to move upward. Due to the release of latent heat by condensation, the rate of temperature changes drops from 10 degrees Celsius per 1000 meters to 6 degrees Celsius per 1000 meters because, once condensation starts, the temperature of the air within a rising parcel is not only affected by cooling but also by the release of latent heat during cloud formation (condensation from water vapor to liquid)—latent heat transfer describes all heat transfer associated with phase changes as discussed earlier. Because the dew point temperature keeps track of the amount of water vapor in the parcel, it must decrease once a cloud begins to form by condensation (water vapor condensing into the tiny liquid droplets that make up a cloud). The decrease in the dew point temperature indicates that there is a decreasing amount of water vapor in the air parcel. The water does not disappear though, it is condensed into the liquid that is the cloud. The dew point temperature in a rising air parcel will remain the same as the air temperature so that the relative humidity in the parcel stays at 100 percent. Note that the rate of cooling of saturated parcels of 6 degrees Celsius per 1000 meters is just a rough approximation for teaching purposes. The actual cooling rate depends on how much water vapor condenses.

2.7.2. Cloud Formations

By evaporating water, a boiling kettle of water produces a lot of water vapor molecules (steam) in the air, as illustrated in Figure 2.54. This steam is basically a cloud. The same process explains why you are able to temporarily "see" your breath when walking outside on cold winter days. In the atmosphere of Earth, most clouds form well above the ground surface. Consider a hypothetical air parcel near the Earth's surface. It contains water vapor, which evaporates from liquid surfaces such as oceans and lakes. As the air parcel moves upward, it takes this water in the form of water vapor with it. As it rises, it expands and cools; that is, the temperature in the air parcel decreases. As the air continues to cool, its relative humidity increases. Once the relative humidity reaches 100 percent, any further cooling results in net condensation and cloud formation; that is, the excess water vapor in the air condenses (or deposits) into liquid water droplets (or ice crystals) and onto microscopic aerosol particles, such as dust, smoke, and salt, which are called cloud condensation nuclei.

Let us expand just a bit more on condensation in the atmosphere. Once the relative humidity reaches 100 percent, any further cooling results in net condensation and cloud formation. In fact, just enough water condenses to keep the relative humidity at 100 percent. The invisible water vapor and the aerosols suspended in the air are constantly bumping into each other. When the air is cooled to its dew point some of the water vapor sticks to the aerosols when they collide; this is condensation. Eventually, bigger visible water droplets form around the aerosol particles or ice crystals, and these water droplets start sticking together with other droplets, forming clouds. Fog is a shallow layer of cloud at or near ground level.

The mechanisms that force air upwards are at the heart of cloud formation. Earth's surface heating, free convection, horizontal air motions that include air inflow and/or outflow of a region, and height structures are some of the major mechanisms that contribute to upward air motion and cloud formation. For example, when air moving along the surface of the Earth encounters a mountain, it is compelled to move up and over the mountain, cooling as it moves upward. If the air cools to its saturation point, water vapor condenses and a cloud forms. Another example is the horizontal net inflow of air into a region from various directions. The converging air is forced to move upward since it cannot go downward. As a third example, heat from the sun causes the air to ascend. Because some surfaces better absorb radiation from the sun and become warmer than surrounding surfaces (e.g., a blacktop surface will typically become warmer than a grass covered surface or a rocky surface will typically be warmer than wet soil). Air forms above these hot surfaces and begins to rise upward. These warmed parcels continue to rise as long as they remain warmer than the air surrounding them.

The mechanisms that force air upwards are at the heart of cloud formation. Earth's surface heating, free convection, horizontal air motions that include air inflow and/or outflow of a region, and height structures are some of the major mechanisms that contribute to upward air motion and cloud formation. For example, when air moving along the surface of the Earth encounters a mountain, it is compelled to move up and over the mountain, cooling as it moves upward. If the air cools to its saturation point, water vapor condenses and a cloud forms. Another example is the horizontal net inflow of air into a region from various directions. The converging air is forced to move upward since it cannot go downward. As a third example, heat from the sun causes the air to ascend. Because some surfaces better absorb radiation from the sun and become warmer than surrounding surfaces (e.g., a blacktop surface will typically become warmer than a grass covered surface or a rocky surface will typically be warmer than wet soil). Air forms above these hot surfaces and begins to rise upward. These warmed parcels continue to rise as long as they remain warmer than the air surrounding them.

One may wonder how massive amounts of water with a density of 1 g/cc, more than 780 times heavier than atmospheric air, stay floating in the sky as clouds. Let us emphasize that we are dealing with water vapor, which is water in gaseous form, and not in liquid form. Actually, water vapor, with a density 4.77×10^{-6} g/cm^3 at zero degree Celsius, is lighter than air, which has a density of 1.275×10^{-3} g/cm^3 at zero degree Celsius. Because liquid water is denser than air, it rains when water vapor condenses into liquid water and/or liquid water breaks up and turns back into water vapor. Also when drops are below a certain size, at which they have no significant weight, they cannot overcome air resistance, and effectively stop dropping. It is a similar reason why aerosols are able to float forever in the atmosphere without sinking.

2.7.3. Some Weird Cloud Formations

Here is one the classifications of cloud formations often adopted by meteorologists. It is illustrated in Figure 2.55.

- *Lenticular clouds* are stationary lens-shaped clouds that form at high altitudes, sometimes aligned perpendicular to the wind direction. In most cases, they are created on or above mountain peaks, downwind of mountains, or even inside a building. When high winds encounter a tall structure, the air is sometimes diverted up and over it. That air cools as it rises, and if it contains enough moisture, it will condense into a flat cloud formation at the crest of the wave. The air will continue to move in waves on the other side of the obstacle. The lenticular clouds do not move because they are created by the combination of winds and a stationary structure.

- *Wave clouds* are created by stable fast air flows over a raised land feature such as a mountain range. Also known as Kelvin-Helmholtz instability. In some ways, they resemble ocean waves. In other words, the atmosphere moves and responds to the environment around it just like. After all, water and air are both fluids and obey the same wave equations.

- *Noctilucent (night shining) clouds* are made of crystals of water ice at high altitudes, around 100 kilometers. They are most commonly observed in the summer months at latitudes between 50 and 70 degrees north and south of the equator. Note that rare clouds can reach an altitude of around 20 km. This is as high as the troposphere, which is the lowest layer of the atmosphere. Noctilucent clouds can sometimes be seen in the mesosphere at an altitude of around 76–85 kilometers.

- *Morning glory clouds* are roll clouds that can be up to 1,000 kilometers long, 1 to 2 kilometers latitude, often only 100 to 200 meters above the ground, and can move at speeds up to 60 kilometers per hour. Sometimes there is only one roll cloud, sometimes there are up to twenty consecutive roll clouds. These clouds are exceedingly rare, and the only place that produces them with any consistency is Australia's Cape York. They do not necessarily coincide

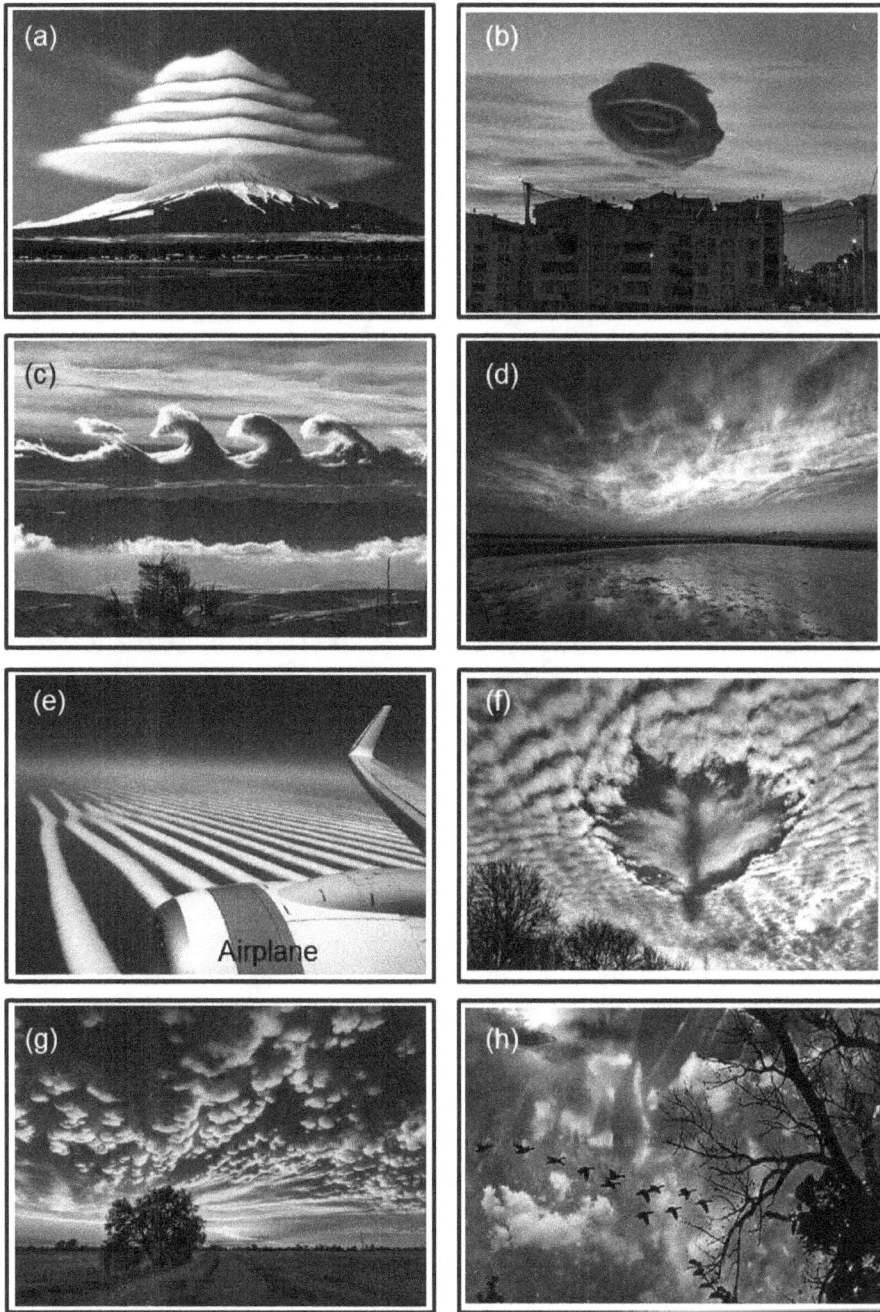

Figure 2.55. (a) A lenticular cloud indicates the height of a mountain. (b) Another lenticular cloud over Bursa, Turkey. (c) A rare wave cloud over Wyoming, USA. (d) A night-shining noctilucent cloud about 80 km over the Northern Hemisphere. (e) A rare morning glory cloud around Burketown, Australia. (f) Fallstreak hole cloud in Midlands, England with a large gap. (g) A mammatus cloud over Illinois, USA. (h) An iridescent cloud caused by diffraction by water droplets and/or ice crystals. Google images.

with storms. Morning glory clouds can often be observed over Burketown in northern Australia, from September to mid-November.

- *Pyrocumuls clouds*, also known as fire clouds, are dense cumuliform clouds associated with fires, volcanic activities, or any other intense heating of the air from the Earth's surface.

- *Fallstreak hole clouds*, also known as hole punch clouds, contain large circular gaps that form when the water temperature in the clouds is below freezing but the water has not frozen yet due to the lack of ice nucleation particles. A fallstreak hole cloud sometimes looks like someone used a massive hole-puncher on a cloud.

- *Mamatus clouds*. Normally, cloud bottoms are flat because moist warm air that rises and cools will condense into water droplets at a specific temperature, which usually corresponds to a very specific height. As water droplets grow, an opaque cloud forms. Under some conditions, however, cloud pockets can develop that contain large droplets of water or ice that fall into clear air as they evaporate. Such pockets may occur in turbulent air near a thunderstorm. The resulting mammatus clouds can appear especially dramatic if lit from the side. Cellular patterns of pouches hanging underneath the base of a cloud, typically a cumulonimbus raincloud, although they may be attached to other classes of parent clouds. These cloud formations are associated with towering cumulonimbus thunderstorms reaching as high as 18 kilometers tall at times. Often, mammatus clouds are thought to be associated with tornadoes, as supercell thunderstorms that produce tornadoes are strong enough to also produce these. As water droplets grow, an opaque cloud forms.

- *Iridescent clouds* are results light scattering by small water droplets or small ice crystals individually scattering light. Larger ice crystals produce halos, which are a refraction phenomenon rather than iridescence. Iridescence should similarly be distinguished from the refraction in larger raindrops that makes a rainbow.

Notice that clouds exhibit various colors, mainly white, blue, and red. These colors are determined by the light illuminating the clouds. Clouds that are directly illuminated by sunlight are seen by us as white because our eyes perceive a mix of all the colors of the rainbow as white. Again, clouds are made up of many small water drops and ice crystals. In a cloud, light can reflect or scatter in many different ways from these particles, and we see the cloud as fairly even and bright white as a result of the mixture of rainbow colors. The non-white clouds are due to shades and/or scattering that take place before the light reaches the cloud such as at sunset or sunrise. The color of sunlight can be yellow to deep red due to the scattering of the blue component of sunlight as the light travels a longer path through the atmosphere. For example, a cloud can look bluish if it is heavily shaded and is illuminated by scattered blue light. Rare green clouds are clouds that contain a large amount of water that absorbs red light. The light falling on the cloud appears red because the blue component has been scattered, and then the red component is absorbed by the water in the cloud, leaving green.

There are many types of weird, fantastic, and stunning cloud formations, as we can see in Figure 2.55. The saucer-like orange-tinted cloud that hovered over Bursa, Turkey on January 26, 2023 was an example of a lenticular cloud and not a flying saucer about to land on Earth. Bursa is nestled in the foothills of 2,500-meter Mount Uludag. One of the key conclusions here, which we will expand on in Chapter 6 and Ikelle (2024), is that cloud formations can be related to surface planetary shape. So we can use clouds to infer the structures on the surface of solid Saturn and Jupiter.

In summary, to form a cloud in the atmosphere of a planet, we need (1) heat to raise the temperature of volatile chemicals to cause them to evaporate into gaseous form. We also need (2) adequate gravity to keep gases close to the planet. We also need (3) a magnetic field to deflect the solar wind and prevent it from stripping the gases off as it passes by. On Earth the volatile chemical molecule is water. Temperatures are high enough to evaporate liquid water into gaseous form (water vapor). The gravity is large enough to keep vapor water close to Earth solid. Solar wind is deflected by the Earth's magnetic field. Jovian planets such as Jupiter and Saturn, have atmospheres of hydrogen and helium, thus evaporation is not necessary. Moreover they have significant gravity and huge magnetic fields.

2.8.   Essay: Cloud Formations on Titan

Cloud formation occurred in at least one other place in our solar system; Titan (with a diameter of 5150 kilometers), the largest moon of Saturn and close second largest in our solar system after Ganymede (with a diameter of 5262 kilometers)of Jupiter. The Cassini-Huygens mission (1997-2017) returned rare images of Titan clouds that we will discuss here. We will start by briefly describing the landscape of Titan's surface, its atmospheric composition, the anti-greenhouse effect on Titan, and the liquid cycle on Titan before describing cloud formation.

2.8.1.   Titan's Landscape

Titan is Saturn's largest moon with a -180 degrees Celsius surface temperature. Figure 2.56 shows its key features. The landscape includes mountains, valleys, dunes, lakes, oceans, and rivers made of water ice frozen harder than granite. However, Titan's topography dynamic range is very narrow (less than 2 kilometers) compared with the 10-30 kilometers range seen on Earth, Mars, the Moon and Venus (e.g., Lorenz et al., 2011). What look like rocks and pebbles whirling across a flat plane are actually chunks of ice. The rocky desert landscape on Titan resembles the Sahara desert on Earth. Sand dunes on Titan, for example, suggest that the winds that move the sand are still active. As stones are worn by rivers on Earth, ice chunks on Titan have sharp edges. Due to the fact that they are rounded like pebbles, Titan's surface is liquid, as described in the previous chapter. In fact, the black spots and blotches visible near Titan's North Pole in Figure 2.56b are

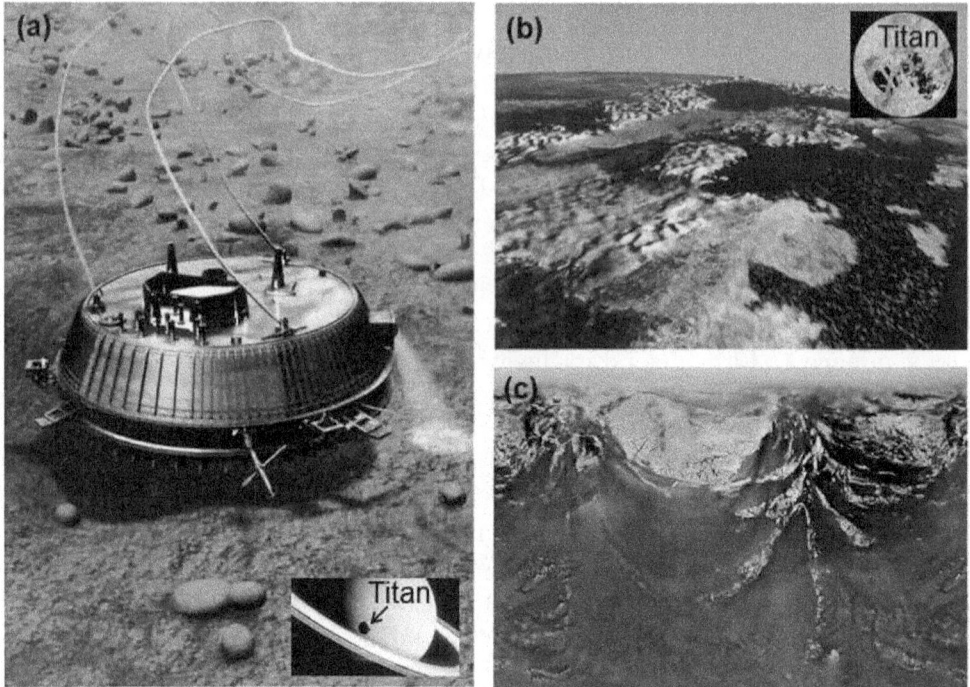

Figure 2.56. (a) The Huygens Probe, a component of the joint ESA-NASA Cassini-Huygens mission to Saturn and its moon, Titan, landed in 2005 on Titan. It pierced the shroud of Titan's thick and massive atmosphere to reveal an earth-like desert world. (b) Methane filled lakes in Titan's north polar region. Dark areas are liquid methane and light color is the solid surface of Titan. This colorized mosaic from NASA's Cassini mission shows the most complete view yet of Titan's northern land of lakes and seas. (c) An aerial view showing some branching channels carved out of Titan's surface by flowing liquid methane (Adapted from images from ESA/NASA/JPL).

lakes filled with liquid methane (CH_4). Methane, which is a vapor on Earth, has a melting point of -182 degrees Celsius and boils at -161 degrees. Between these two extremes, it exists in liquid form. Titan's surface temperatures are just right for liquid methane. Titan has hundreds of lakes and oceans, some huge. It seems they form lakes only in the polar regions. Also, some 20 percent of Titan, all at low latitudes, is covered with giant sand dunes. A striking feature of the equatorial dunes is that they are longitudinal in form, lining up along the mean transport direction. The most spectacular examples of such dunes are found in Namibia (see Chapter 4), where a seasonal change in wind direction forms striking dunes. The reason for these dunes formations on Titan is still unclear.

With a mean density of $\rho_{titan} = 1.88$ g/cm^3 (more dense than water and smaller than dry rocks density), Titan is composed of a mixture of rocks and ice. Like many of the larger moons of the Jovian planets, the rotation of Titan itself is tidally locked to its orbit around Saturn, with a period of 15 days and 23 hours.

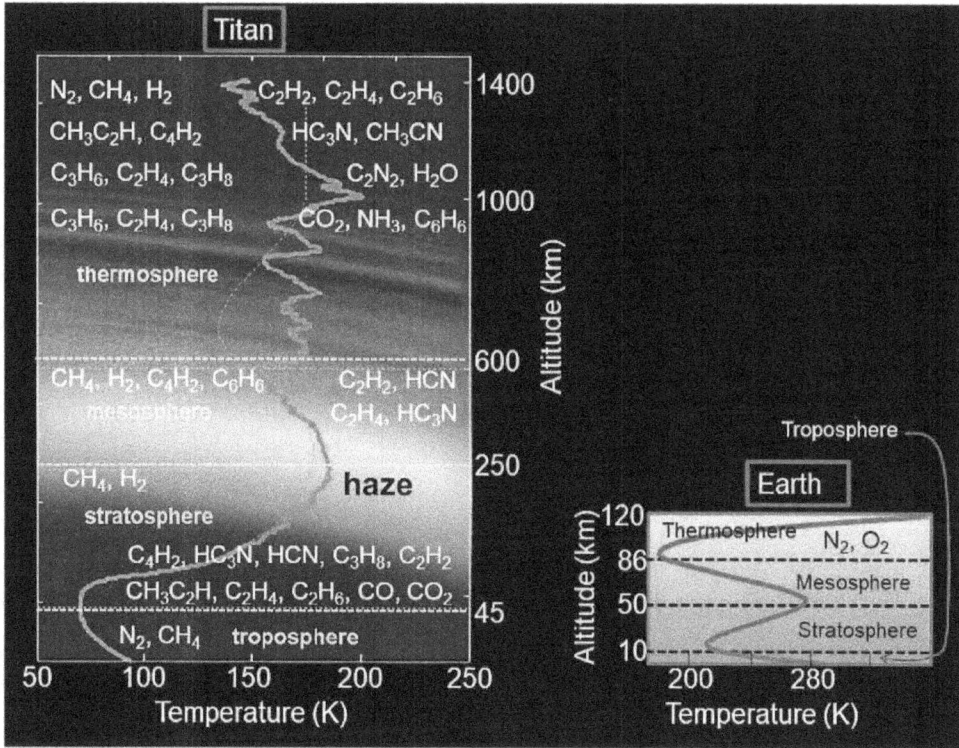

Figure 2.57. Titan's atmosphere and Earth's atmosphere are on the same scale. Earth's atmosphere fits tightly around the planet. Titan has a thick blanket covering the moon. Similar to Earth, Titan's atmosphere is mostly nitrogen. However, it also contains 3.5 percent methane. On Earth and many other planets, chemical processes take place near the surface. On Titan however, its chemical reactor is located above 480 kilometers in the mesosphere and thermosphere. The results are huge amounts of organic compounds that ultimately descend to the surface.

2.8.2. Titan's Atmosphere Composition and Haze-Layer Formation

The discovery of a significant atmosphere sets Titan apart from the other moons in the solar system. As a matter of fact, its atmosphere makes it more planet-like than many other planets (Figure 2.57), the only other thick N_2 atmosphere besides Earth's. Titan's atmosphere is so dense and thick that it is difficult to see its surface from afar, even using ground or space telescopes. Titan's atmosphere appears like a fuzzy orange ball that blocks further views of Titan's surface and troposphere. It is so thick that the Huygens probe from 1,000 kilometers above Titan's surface took more than 2.5 hours to land instead of minutes. Thanks to the Cassini-Huygens mission, carrying a variety of instruments, we now know that Titan's atmosphere extends far further into space than Earth's due to Titan's low gravity ($g_{titan} = 0.143\, g_{earth} = 1.352\,\mathrm{m/s^2}$). For example, Earth's mesosphere/thermosphere boundary is about 86 kilometers in altitude while Titan's is 600 kilometers.

To explain the large Titan atmosphere compared to Earth's atmosphere, let us assume the pressures of Titan and Earth's atmosphere can be roughly expressed as follows:

$$p_{\text{titan}} \approx p_{0,\text{titan}} \left(1 - g_{\text{titan}}\, \rho_{\text{titan}}\, h_{\text{titan}}\right)$$

and

$$p_{\text{earth}} \approx p_{0,\text{earth}} \left(1 - g_{\text{earth}}\, \rho_{\text{earth}}\, h_{\text{earth}}\right),$$

where $p_{0,\text{titan}} = 1.5\, p_{0,\text{earth}} = 1.5\,\text{bar}$ are the surface pressures, h_{titan} and h_{earth} are the thicknesses of Titan and Earth atmospheres, respectively, and ρ_{titan} and ρ_{earth} are the mass densities of Titan and Earth atmospheres, respectively. Because Titan and Earth's atmosphere pressure are the same range as well as their densities, we must have $h_{\text{titan}} \gg h_{\text{earth}}$ to compensate for the fact that $g_{\text{titan}} \ll g_{\text{earth}}$.

Similar to Earth, Titan's atmosphere is mostly nitrogen gas. However, it also contains between 2 and 5 percent methane gas, depending on altitude (5 percent at at surface near equator; 1.4 percent in lower stratosphere; and 2 percent at about 1000 kilometers), and 0.5 percent hydrogen at high altitudes and less than 0.2 percent in lower atmosphere. While the temperature on Titan's surface is -180 degrees Celsius, at 320 kilometers it is -90 degrees Celsius. At the very top the Titan sky turns blue as it fields primarily nitrogen and hydrogen but at the height of about 400 km the aforementioned orange haze begins to form which continues all the way down to 100 km only clearing as you approach the surface. At 30 miles the moon's surface becomes visible. The orange Haze layer is fairly evenly opaque all the way through. Here sunlight never reaches the surface. The top layer is composed of hydrogen and nitrogen.

On Earth and many other planets, chemical processes producing molecules take place near the surface. On Titan, however, its chemical reactor is located above 480 kilometers and produces huge amounts of organic compounds (Figure 2.57). Some of them eventually descend to lower layers to form the haze layer and even down to the surface, rivers, oceans, and dunes. We will next discuss Titan's unexpected atmospheric chemistry.

The most complex atmospheric chemistry in the solar system. Similar to planets in our solar system, the Titan's atmosphere has well-defined troposphere, stratosphere, mesosphere, thermosphere, and exosphere (not shown in Figure 2.57). Titan's thermosphere layer receives about 1 percent of the sun's flux that reaches Earth because it is located farther away from the Sun than Earth. Because Titan does not have a significant magnetic field, some charged particles (e.g., electrons and ions like H^+) from solar radiation also penetrate Titan's atmosphere. In Ikelle (2024), it was explained that the Earth's magnetic field creates an invisible magnetic envelope, known as the magnetosphere, around the planet. This envelope blocks the sun's charged particles from reaching us but allows sunlight to pass through. Saturn has a significant magnetic field and therefore a magnetosphere. However,

Saturn's magnetosphere does not completely cover Titan. About 5 percent of Titan is not protected from the sun's radiation and consequently, some solar-charged particles reach Titan along with sunlight. The interactions of these energetic particles and sunlight on the one hand and the mother molecules of Titan—named nitrogen gas, methane, and hydrogen gas—on the other hand create the most complex atmospheric chemistry in our solar system.

Nitrogen and methane molecules are broken up by sunlight or highly energetic particles entering Titan's atmosphere above 1,000 kilometers. This process results in the formation of massive positive ions and electrons, which trigger a chain of chemical reactions that produce a variety of *hydrocarbons* [acetylene (C_2H_2), ethylene (C_2H_4), ethane (C_2H_6), cyclopropenylidene (c-C_3H_2), propyne and propadiene (C_3H_4), methylacetylene (C_3H_4), propylene (C_3H_6), propane (C_3H_8), propyne (CH_3C_2H), diacetylene (C_4H_2), benzene (C_6H_6), etc.], *nitriles* [hydrogen cyanide (HCN), hydrogen isocyanide (HNC), cyanoacetylene (HC_3N), cyanogen (C_2N_2), acrylonitrile, acetonitrile, and cyanopropyne (C_3H_3N), propionitrile (CH_3CH_2CN), dehydrotetrazine (C_6N_2), dicyanodiacetylene (C_6N_2), etc.], and *oxygen compounds* [water (H_2O), carbon monoxide (CO), and carbon dioxide (CO_2)]. Some of these molecules are isomers (i.e., compounds with the same formula). It is remarkable that even ring-type molecules such as benzene (see Chapter 2 of Ikelle, 2023a) are found in Titan's atmosphere. Benzene is a building block for larger hydrocarbons — two- and three-ring structures known as polycyclic aromatic hydrocarbons (PAHs)—that make up the molecules and particles in Titan's hazy atmosphere. Researchers in this study combined a two-ring PAH with another hydrocarbon and formed a 3-ring PAH. Remember that the carbon chains are the building blocks toward more complex molecules that may be the basis for the earliest forms of life (see Chapter 2 of Ikelle, 2023a). During the Cassini-Huygens mission, all of these molecules were confirmed to be spontaneously produced in Titan's atmosphere (e.g., Waite et al. 2009; Vuitton et al., 2009; Cravens et al., 2009). It detected polycyclic aromatic hydrocarbons (PAHs), a class of chemicals found naturally in coal, crude oil, and gasoline. On Earth, they result from burning coal, oil, gas, wood, garbage, and tobacco. Some PAHs detected in Titan's atmosphere also contain nitrogen atoms. PAHs are the first step in a sequence of increasingly larger compounds. These PAHs can coagulate and form large aggregates, which tend to sink, due to their greater weight, into the lower atmospheric layers.

Chemical reactions occurring in Titan's atmosphere are generally known as photochemistry. Photochemistry is based on light absorption, as illustrated in Figure 2.58a. Photochemical reactions are when sunlight is used to drive chemical reactions and/or generate electricity. These processes begin with the dissociation[7] of N_2 and CH_4 by energetic particles and solar ultraviolet photons. The dissociated and ionized N_2 and CH_4 such N_2^+, N^+, CH_4^+, and CH_5^+ initiate chemical reactions that result in the irreversible destruction of CH_4 and lead to the formation of H_2 and a wealth of other complex organic molecules including the hydrocarbons and

---

[7]In dissociation ions are getting separate. Ionization is a chemical reaction in which some species are transformed in ions.

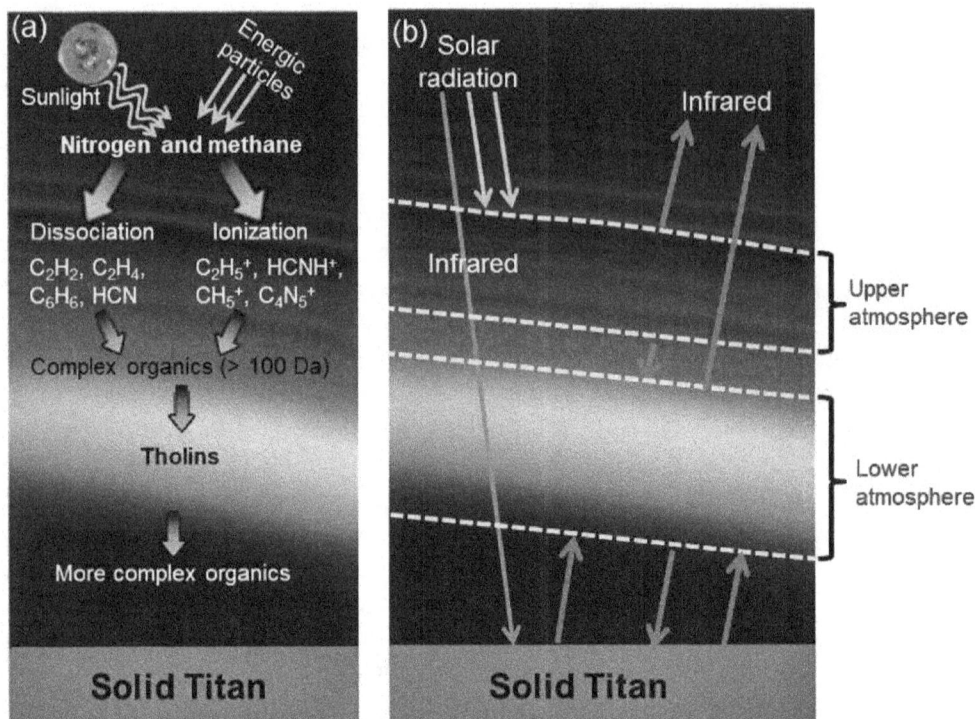

Figure 2.58. (a) Schematic of the organic chemical process leading to tholin formation in Titan's atmosphere. The process begins with free energy from solar UV radiation and energetic particles impinging on Titan's atmosphere. The CH_4, N_2, and H_2 combine through a number of reactions to form larger organic compounds (100 to 350 daltons) that eventually lead to the formation of tholin aerosols (20 to 8,000 daltons) observed at 1000 km. (b) Schematic flow of energy within Titan's climate system. Incoming solar visible and ultraviolet light is absorbed in Titan's atmosphere, mainly due to its haze, is directly scattered back out to space within the stratosphere. The remaining solar energy enters the deep troposphere. Thick arrows denote infrared radiation.

nitriles described above. These gaseous molecules will eventually become heavy enough to become organic aerosols.[8] Thus, molecular growth is driven by gas phase reactions involving radicals and positive and negative ions, all possibly in some excited electronic state, as well as by heterogeneous chemistry on the surface of the aerosols. Some of these reactions are captured in Table 2.25. The efficiency and outcome of these reactions depend strongly on the atmosphere's physical characteristics, namely pressure and temperature. Moreover, species distribution is affected by molecular diffusion and vertical and horizontal winds. Further, species escape from the upper stratosphere and condense in the lower stratosphere.

---

[8]Again, aerosol refers to any particle suspended in a gas and is therefore the most general term used to describe particles in Titan's atmosphere.

Box 2.1: Liquefied Natural Gas (LNG)

On Earth, we also deal with methane in LNG (liquefied natural gas). LNG is mainly made up of methane. As described in this section and illustrated in Figure 2.59b, when natural gas (methane) is cooled at a liquefaction facility to approximately -161.5 degrees Celsius at atmospheric pressure (i.e., at 1.01 bar), it condenses into a liquid. Careful purification and dehydration are necessary to remove impurities that become solids at liquefaction temperatures. When natural gas is turned into LNG, its volume shrinks by approximately 600. Natural gas is transported to the market via pipelines or LNG tankers. In the United States, natural gas is transported by pipelines in the United States and Canada and as LNG outside North America by LNG tankers. As of April 2023, the United States had about 500,000 kilometers of natural gas transmission pipelines. As described in Ikelle (2023a and 2023c), Earth was free of molecular oxygen with some hydrogen and carbon. This was an ideal environment for the formation of methane lakes. If these lakes were formed on Earth billions of years ago, they are now buried underground by years of erosion and diagenesis. Some natural gas fields may have formed through such hydrocarbon lakes when they were buried at pressure and temperatures where they turned from liquid to gas.

As mentioned earlier and demonstrated by Strobel (1974), methane is destroyed by sunlight. Current methane in Titan's atmosphere will last about less than 100 million years, which is still a short time in geologic terms. How is it replenished? That is a basic and fundamental scientific question. The answer to this question will probably require a definite understanding of the composition of and tectonic processes in solid Titan along with the interactions between Titan's interior and atmosphere. It is very likely that the source of methane on Titan is totally different from the ones found in the atmospheres of the Jovian planets and in the atmosphere of Earth. Methane is abundant on the giant planets-Jupiter, Saturn, Uranus and Neptune because these planets are made of the same materials and chemical processes of primordial solar nebula, which include methane, a widely distributed gas in the Solar System. Terrestrial planets are generally methane-poor, with Earth being a exception. Of the 1,750 parts per billion by volume (ppbv) of methane in Earth's atmosphere, 90 to 95 percent is biological in origin, especially the metabolism of many living things, including grass-eating ungulates such as cows, goats, and yaks who belch out one fifth of the annual global methane release (e.g., Keppler and Röckmann, 2007). In other words, life itself replenishes Earth's methane supply. Also, because of the link between methane and living things on Earth, life on Titan is still an open question.

One of the takeaways from Figure 2.58a and Table 2.25 is that aerosols fall down onto Titan's surface for billions of years. If left alone, they could accumulate to hundreds of meters. However, methane rain is expected to wash some of the deposits into lake beds or river basins. Nevertheless, relatively large quantities are expected to survive intact at the surface.

Table 2.25. Some examples of chemical reactions in the Titan atmosphere. Note that many molecules form in a cascade of chemical reactions when solar radiation hits methane, nitrogen and other gases in Titan's atmosphere. The amount of N_2 is 95–98%, CH_4 is 1.5–5%, H_2 is 0.1–0.2%, C_2H_6 is about 20 ppm (parts per million); CO is about 45 ppm; C_2H_2 is 19–3.3 ppm; C_3H_8 is about 700 ppb (parts per billion); HCN is 100–800 ppb; and C_2H_4 is about 160 ppb.

Ions and neutrals	Reaction	Type
N_2^+:	$H^+ + N_2 \rightarrow H + N_2^+$	Capture process
N^+:	$H^+ + N_2 \rightarrow H + N^+ + N$	Dissociative collision
N_2^+:	$H^+ + N_2 \rightarrow H^+ + N_2^+ + e^-$	Ionization
CH_2:	$CH_4 + h\nu \rightarrow CH_2 + H_2$	Dissociation
CH_2:	$\rightarrow CH_2 + 2H$	Dissociation
CH:	$\rightarrow CH + H + H_2$	Dissociation
CH:	$\rightarrow CH_3 + H$	Dissociation
CH_4^+:	$\rightarrow CH_4^+ + e^-$	Ionization
CH_5^+:	$CH_4^+ + CH_4 \rightarrow CH_5^+ + CH_3$	Charge exchange
C_2H_2:	$C_2H_4 + h\nu \rightarrow C_2H_2 + H_2$	Dissociation
$C_2H_2^+$:	$\rightarrow C_2H_2^+ + H_2 + e^-$	Ionization
$C_2H_3^+$:	$\rightarrow C_2H_3^+ + H + e^-$	Ionization
$C_2H_4^+$:	$\rightarrow C_2H_4^+ + e^-$	Ionization
H_3O^+:	$CH_5^+ + H_2O \rightarrow H_3O^+ + CH_4$	Charge exchange
C_2H_2:	$2CH_2 \rightarrow C_2H_2 + H_2$	Neutral-neutral
C_2H_4:	$CH + CH_4 \rightarrow C_2H_4 + H$	Neutral-neutral
CO:	$H_2CO + h\nu \rightarrow CO + H_2$	Dissociation
C_2H_6:	$CH_3 + CH_3 + M \rightarrow C_2H_6 + M$	Condensation
C_4H_2:	$C_2H + C_2H_2 \rightarrow C_4H_2 + H$	Neutral-neutral
CO_2:	$CO + OH \rightarrow CO_2 + H$	Condensation
HCN:	$N + CH_3 \rightarrow HCN + 2H$	Neutral-neutral
HC_3N:	$HCN + C_2H \rightarrow HC_3N + H$	Neutral-neutral
CH_2NH:	$N + CH_4 \rightarrow CH_2NH + H$	Neutral-neutral
CH_3NH:	$CH + CH_2NH \rightarrow CH_3CN + H$	Neutral-neutral
C_6H_6:	$2C_2H_2 + 2CH_4 \rightarrow C_6H_6 + 2H + 2H_2$	Benzene
PAH:	$C_6H_6 + C_2H \rightarrow PAH - polymer$	PAHs

One may wonder about Titan's atmosphere being flammable because of its hydrocarbon content. Carbon and hydrogen atoms are very stable molecules. One part carbon and four parts hydrogen (i.e., methane) is one of the most stable compounds we can make. Oxygen gases are the real cause of methane's flammability on Earth. Oxygen is responsible for the burning of methane and other hydrocarbons as well as the rusting of metals and some rocks. Actually, nothing in our world would work without oxygen driving the reactions including most living things as described in Ikelle (2023a and 2023b). Titan is an entire other world. Titan's atmosphere contains no oxygen gas, only tiny amounts of CO, CO_2, and H_2O. It is therefore impossible to start an inferno on Titan except to send a tank filled with oxygen into the planet. A nitrogen-methane atmosphere is equally stable as Earth's nitrogen-oxygen atmosphere.

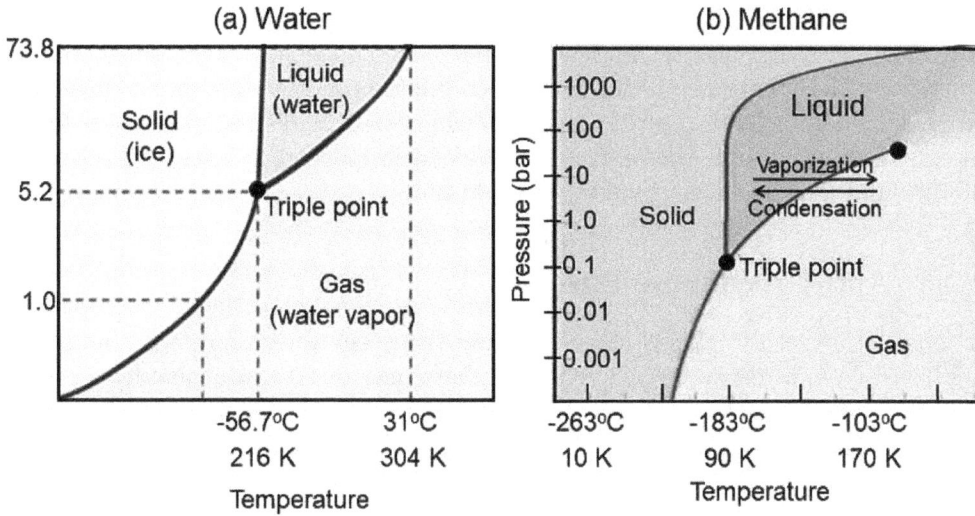

Figure 2.59. (a) Phase diagram of water and (b) phase diagram of methane. Methane is an odorless, colorless, tasteless gas that is lighter than air. When methane burns in the air it has a blue flame. In sufficient amounts of oxygen, methane burns to give off carbon dioxide (CO_2) and water (H_2O). When it undergoes combustion it produces a lot of heat, which makes it very useful as a fuel source. Methane is also a very effective greenhouse gas.

Haze layer: Titan's atmosphere and surface share a unique connection.
The connection between Titan's surface and atmosphere is also unique in our solar system; atmospheric chemistry produces materials that are deposited on the surface and subsequently altered by surface-atmosphere interactions such as aeolian and fluvial processes resulting in the formation of extensive dune fields and expansive lakes and seas. Between 320 and 400 km (see Figures 2.57 and 2.58a), before reaching the surface, Titan's atmosphere is made up of tholin haze. Earth's tholin haze is soot produced by combustion. Tholin is a Greek term coined by Carl Sagan (Sagan and Khare, 1979; Carl Sagan and Dermott, 1984; Dermott and Sagan, 1995) to describe the reddish-brown fog of nitrogen-containing organic substances now known as tholin haze. These suspended solid nanoparticles are one of the most complex organic materials in the Solar System. Pyrolysis coupled to mass spectrometry along with infrared spectrometry measurements (see Chapter 3) by the Cassini-Huygens mission reveal that the tholin haze has a nitrogen-rich core with poly-aromatic hydrocarbon (PAH) signatures (Imanaka et al. 2004; Israël et al. 2005; Szopa et al. 2006; Coates et al. 2007, Vinatier et al. 2012; Waite et al. 2007; Coll et al. 2013; Sciamma-O'Brien et al. 2014; He and Smith 2014; Gautier et al. 2017; He et al. 2017). These molecules form in a cascade of chemical reactions when ultraviolet and cosmic radiation hits the mix of methane, nitrogen, and other gases in atmospheres like Titan's. These observations support a general model of nitrogenated polycyclic aromatic hydrocarbon (N-PAH). Experimental on simulated haze synthesis by Schulz et al. (2021) confirm that tholin haze is very likely composed of N-PAH molecules involving five-membered aromatic rings, including $C_{10}N_3H_9$, $C_9N_2H_7$, C_7NH_9, $C_{23}N_6H_{13}$, $C_8N_2H_{10}$, $C_{21}N_6H_{12}$, $C_8N_3H_7$, $C_{20}N_3H_{13}$, $C_{12}NH_{11}$, C_7H_{15}, and C_7NH_7. These molecules absorb ultraviolet light

as we will discuss in Chapter 3. The haze conceals more complex molecules from view. However, they are an indication that such molecules exist, since tholin is obtained by the reactions of increasingly complex organics. There is a possibility that most macromolecules of life on earth, described in Ikelle (2023a) are formed beneath the haze. Note that infrared wavelengths penetrate Titan's atmosphere, including the haze layer, and visible ones do not.

2.8.3. Anti-Greenhouse Effect

Let us start by recalling that the Sun emits three key radiations, for climate: infrared, visible and ultraviolet. The atmosphere and surface of a planet or a moon transform part of visible and ultraviolet light into infrared. Greenhouse gases reflect infrared light from the planet or moon back to its surface.

In the second section of this chapter, we introduced the greenhouse effect which results in planet and moon surfaces warming. The anti-greenhouse effect, coined by McKay et al. (1991), has the opposite of the greenhouse effect; it results in planet and moon surfaces cooling. The greenhouse gases on Earth—namely carbon dioxide (CO_2), water (H_2O), nitrous oxide (N_2O), and methane (CH_4)—allow solar energy in ordinary visible light to reach the planet's surface, but they block some of the infrared heat energy radiated away by the surface. Titan's greenhouse gases—methane (CH_4), nitrogen (N_2), and hydrogen (H_2)—do the same thing. Atmospheric gases radiate some of this energy back to the ground, warming it. Of the main greenhouse gases on Titan, hydrogen plays a key role. Even though it is present in only small amounts, it absorbs radiation in parts of the spectrum that the more abundant gases methane and nitrogen do not. Thus hydrogen on Titan, like carbon dioxide on Earth, plays a key role in climate evolution. Planetary surface is warmed by Sun energy. This energy shines brightest in light from the electromagnetic spectrum visible to human eyes. The planet cools by emitting infrared radiation, which we sense as warmth. A greenhouse effect arises when certain gases block the escape of the infrared radiation and radiate it back to the surface. This raises the surface temperature even more.

Unlike conventional greenhouse gases, there are not many known anti-greenhouse gases that cool the environment or are still unknown. Sulphuric acid (H_2SO_4) is potential anti-greenhouse gas. When injected into the atmosphere in a sufficient amount, it can reflect the incoming solar radiation and therefore reduce the amount of sunlight reaching the ground. Adding clouds to existing clouds would make them denser, increasing the reflectivity of clouds. Dimethylsulfide (C_2H_6S) can also be used to create clouds that allow heat to escape into space. The current drawback of most known anti-greenhouse gases is that they are short lived and need to be replenished daily or every year, resulting in high maintenance costs. Some of them can cause depletion of the ozone layer or acid rains. Hopefully, we will one day discover an exoplanet with effective anti-greenhouse gases that we can adapt to Earth when necessary. In more general terms, the anti-greenhouse effect occurs when energy from a planet or a moon with an atmosphere absorbs or scatters incoming sunlight in its upper atmosphere, preventing energy from reaching the surface of the celestial object. In Titan's stratosphere, a haze layer absorbs solar radiation and is nearly transparent to infrared energy from Titan's surface as discussed earlier. This absorption reduces solar energy reaching Titan's surface and lets infrared energy escape, cooling Titan's surface. In other words, the haze layer reflects heat from the sun, but allows heat from Titan to radiate into space.

As shown in Table 2.4 and in equation (2.93), the skin temperature T_{skin} of Titan (i.e., temperature in the absence of the atmosphere assuming Titan's albedo is $\xi = 0.3$) is unequal to the observed temperature T_{true}. On all planets in our solar system, T_{true} exceeds T_{skin}. As we discussed in the second section of this chapter, the differences between T_{true} and T_{skin} describe the greenhouse and/or atmospheric effects on all the planets. In addition to the greenhouse and atmospheric effects, these differences also describe the anti-greenhouse effect in Titan's atmosphere (see Figure 2.58b).

Let us now consider the atmosphere, the haze layer, and greenhouse effects in the model for determining Titan's ground temperature. The atmosphere that includes these contributions to climate is best described as a multiple-layer atmosphere. We here used a two-layer model described in Figure 2.58b: the upper atmosphere which behaves like the planet's atmosphere described earlier and includes greenhouse gases and the lower atmosphere which includes anti-greenhouse effects. By using (2.102) and the result in Table 2.6, we can verify that the greenhouse effect increase T_{skin} by $\Delta T_{\text{green}} = 21$ K. We have an increase in temperatures because the greenhouse effect results in warming. The derivations leading to (2.102) can repeated for anti-greenhouse effect to arrive at

$$T_{\text{ground}} = \left(\frac{2}{2-\epsilon}\right)^{-\frac{1}{4}} T_{\text{skin}} , \qquad (2.190)$$

where T_{skin} is given (2.103). For $\epsilon = 0.88$, the average temperature due to anti-greenhouse is $T_{\text{ground}} = 73$ K. So we can see the anti-greenhouse effect decreases the average temperature from T_{skin} by $\Delta T_{\text{anti}} = 9$ K. By assuming the greenhouse and anti-greenhouse effects are additive, we arrive at Titan's temperature:

$$T_{\text{titan}} = T_{\text{skin}} + \Delta T_{\text{green}} - \Delta T_{\text{anti}}$$

$$= (82 + 21 - 9) \text{ K} = 94 \text{ K} . \qquad (2.191)$$

In closing, let's note that the 1991 eruption of Mount Pinatubo in the Philippines, the largest volcanic eruption in over a century, lofted more than twenty tons of sulfur dioxide into the atmosphere. This sulfur dioxide produced a global haze layer that lasted for months. Earth's surface cools in the Northern Hemisphere by 0.5 degrees Celsius and the amount of net radiation reaching Earth's surface decreases. So we witnessed a short-lived haze in 1991 that had an anti-greenhouse effect on the planet. Also two billion year ago, the color of the sky would have appeared more violet and less blue due to absence of oxygen and the presence of more hydrogen and nitrogen, which tend to scatter violet light, as does Titan today. The earth would have no green plants or any other vegetation or flowers to add color to landscapes. Animals, insects, and sea creatures didn't emerge for at least another 1,500 million years. Two billion years ago, Earth appears a lonely place just like Titan today.

2.8.4. Cloud Formation

As discussed in the previous section, clouds hover above us and shower us with rainwater. And we would have no rivers, lakes, or fresh water without clouds. Just like Earth, Titan is the only known place in the solar system that has evidence that rain falls on the surface. Another step in establishing that Titan has oceans, lakes, and rivers of methane on its surface was Cassini-Huygens' cloud detection. Moreover, Titan's surface temperature is close to methane's triple point, just as Earth's is close to water's (see Figures 2.59a and 2.59b). Because of Titan's thick atmosphere and low gravity, rain drops fall at only 1.6 m/s compared with 10 m/s on Earth.

Figure 2.60 shows an example of a stratosphere cloud over Titan's north pole. The Cassini-Huygens mission and intermittently space- and ground-based telescopes have observed clouds in Titan at altitudes ranging from 10 km (mid-altitudes in the troposphere) to 300 km (low mesosphere), well above the haze layer; that is, we have tropospheric and stratospheric clouds, and even mesospheric clouds in Titan's atmosphere. Clouds are primarily found in the polar regions, including the south pole, where no lakes exist. Only very few clouds are present in the equatorial regions.

As we described in the previous section about clouds on Earth, we need some equivalent form of hydrological cycle of hydrocarbons, nitriles, and oxygen compounds to attempt to understand cloud formation. The cascal hydrological cycle involves condensation, evaporation, saturation, and transport of gases from the sur-

Figure 2.60. A stratosphere cloud over Titan's north pole. Image Credit: NASA/JPL-Caltech/Space Science Institute.

Table 2.26.   Some physical characteristics for Ganymede (Jupiter's largest moon), Europa (Jupiter's moon), Enceladus (Saturn's moon), and Titan (Saturn's largest moon). g stands for Earth's gravity.

	Europa	Enceladus	Titan	Ganymede
• Mean surface temperature (K)	130	75	90	110
• Gravity (g)	0.134	0.0116	0.138	0.146
• Radius (km)	1,561	2,521	2,574	2,631
• Density (g/cc)	2.989	1.609	1.879	1.942

face up to the stratosphere. None of these processes can be easily performed on Titan. For example, stratospheric clouds require ten or so organic vapors to condense to form pure and co-condensed ice. Clouds in the stratosphere are highly complex. Also, with Titan's surface getting one-thousandth as much sunlight as Earth, how effective is the evaporation process? More information about cloud composition is needed along with the source of supplying methane to Titan's atmosphere. This will enable us to model Titan's methane cycle, as well as cloud formation. In Chapter 1 of Ikelle (2923d), we describe possible Titan cloud formations based on their temperature, pressure, and gravity profiles.

2.9.   Essay: From Antarctica's Lakes to Europa's Subsurface Ocean

Deep below Vostok station in Antarctica, Russian scientists discovered a pristine body of liquid water sealed from the outside world by four kilometers of ice (see Figure 2.61) and numerous microbes. This has been sealed for some 15 or so million years. It is one of the most significant discoveries of the modern era. The implication of this discovery is that life (not homo sapients as we cannot live in that kind of temperature), especially microbial life, may exist deep in Antarctica. Moreover, this discovery supports the idea of the possible existence of oceans of liquid water and even life under the icy crusts on other planets and moons even without an atmosphere, such as Jupiter's moons Europa, Ganymede and Callisto, and Saturn's moons Enceladus and Titan. It is still unclear why water persists in liquid form beneath ice.

Strangely, the enormous pressure exerted by the four kilometers of ice overhead may prevent the lake from freezing. Ice, which is less dense than water,[9] may act as an insulator for the lake, preventing the ice sheet from sinking. We all know that water freezes on the surface of northern lakes and ponds through winter but not below. This allows plants, microbes, and fish to survive winter and thrive.

---

[9]H_2O has the unusual property that its solid phase is less dense than its liquid phase.

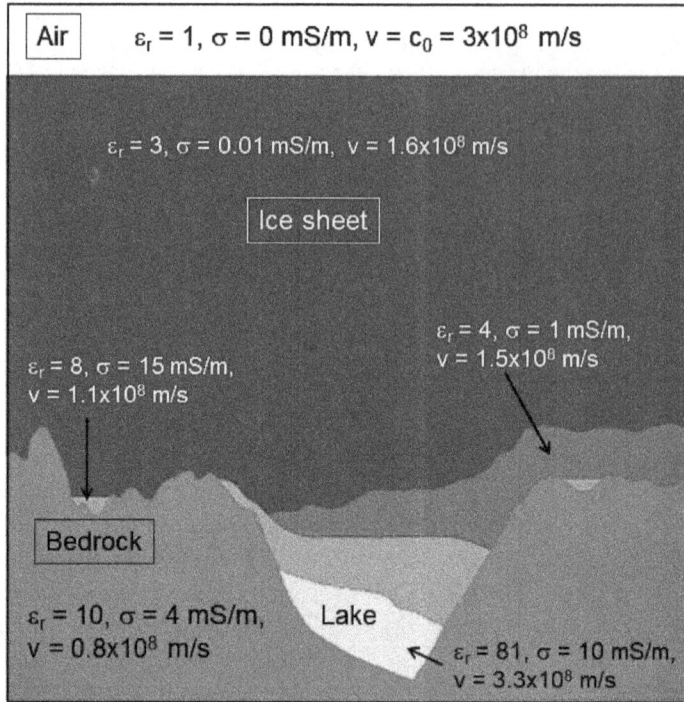

Figure 2.61. A model of a cross-section of Antarctica containing Lake Vostok, one of the 70 or so lakes beneath the Antarctic ice sheet. The relative permeability is $\mu_r = (c_0/v)^2(1/\varepsilon)$. The lakes are subject to high pressure (about 355 bars), low temperatures (about $-3°$ degrees Celsius), and permanent darkness. ε_r is the relative permittivity; σ is the electrical conductivity; and v is the velocity. See also Chapter 3.

2.9.1. Heat Sources of Antarctica's Sub-glacial Lakes

Possible sources of heat that have kept lakes liquid for millions of years deep below the Antarctica ice may include Antarctica volcanoes. Several studies suggest that Antarctica has over 100 volcanoes in the present Holocene period. Some volcanoes are entirely under the ice sheet (Wyk de Vries et al., 2018). There are also ones you can see above ground like Mount Erubus (3,794 m) which throws out a few puffs every now and again just to prove it is active. The majority of volcanoes are around Antarctica's western area with Mt Erebus being situated on Ross Island, not that far from McMurdo Station. Mount Erebus is Antarctica's most active volcano. Moreover Antarctica may well be home to the largest volcano range on Earth, greater than east Africa and the Himalayas, according to Wyk de Vries et al., 2018. Antarctica's largest volcano is Mount Sidley, 4,285 m above sea level. Mount Sidley is located in Marie Byrd Land, a remote and unclaimed region of Antarctica. Mount Sidley's last eruption occurred about 4.7 million years ago.

Also a huge lake (90 and 215 meters in diameter) of sizzling hot lava has been discovered in a volcano on a remote sub-Antarctic island in the South Atlantic Ocean (Gray et al., 2019). While many volcanoes throw out lava when they erupt and form temporary pools and lakes of molten rock, these usually dry into solid rock within a few days or weeks. In contrast, the lava lake remains molten. Lava lakes are rare. It is only the eighth molten rock lake discovered on Earth. The other seven persistent lava lakes are: the volcanoes of Nyiragongo in the Democratic Republic of Congo; Erta Ale in Ethiopia; Mount Erebus beside the Ross Sea in Antarctica; Mount Yasur in Vanuatu; the volcanic island of Ambrym in Vanuatu; Kilauea in Hawaii; and the Masaya caldera in Nicaragua.

Antarctic volcanoes are fueled by a massive upwelling of mantle magma, also known as a hotspot (see Ikelle, 2023a). This is the same mechanism that produces the Hawaiian Islands and fuels volcanic and geothermal activity in Yellowstone (see Ikelle, 2023a). Hotspots coincide with temperature readings above them, as well as surface temperatures. This volcanism is not correlated with subduction zones and is correlated with earthquakes.

2.9.2.   Oxygenic Photosynthesis in Antarctica's Sub-glacial Lakes

In contrast to hydrothermal vents described in Ikelle (2023b), we are not dealing with anoxygenic photosynthesis in Lake Vostok. Because of the high levels of oxygen in the lake water we are rather dealing with an unusual form of oxygenic photo-synthesis. This is because the lake is completely dark and cut off from the outside world making it an unlikely place for photosynthetic organisms to survive. We have a situation akin the Earth when it was 2 billion years old. For 2 billion years of Earth's history, oxygen was scarce in the air. Shang et al. (2019) propose that oxygen accumulated in the atmosphere between 2.1 billion and 2.4 billion years of Earth's history, a period known as the Great Oxidation Event (GOE), as the result of interactions between certain marine microbes (e.g., cyanobacteria) and minerals in ocean sediments. This model can be used to describe oxygenic photosynthesis in sub-glacial lakes of water.

At two billion years in Earth's history, liquid oceans were covered by ice. Ice covered the land too as the last clouds deposited icy loads over high peaks. The Kola Superdeep Borehole (KSDB) scientific drilling project provided us with rock samples containing sulfur isotopes to reconstruct this period. This project drills into the Earth's crust to a depth of 15 km. It was carried out in the former USSR. Solid Earth's high temperature and pressure make drilling difficult. KSDB began digging in 1970. In 1992, it reached a final depth of 12 km (less than 1 percent of the distance to Earth's center) and a temperature of 245 degrees Celsius, a very high temperature compared to the temperature at the surface.

What was next for the planet's surface trapped inside this seemingly eternal icy prison, just like Titan, Europa, and Ganymede's oceans may be trapped? Why and how Earth evolved from the ancient O_2-deficient environment trapped outside the world to the modern O_2-rich world? O_2 is produced by oxygenic photosynthesis and consumed by aerobic respiration or the oxidation of reducing compounds. O_2 accumulates when its production rate exceeds its consumption rate. Through water and the interaction between water and bedrock in the sub-glacial lakes, we have all the ingredients for oxygenic photosynthesis. The real challenge is adding a control mechanism so that oxygen accumulation does not exceed a certain threshold and defining a threshold. GOE modeling faces the challenge of not exceeding 21 percent oxygen gas in the atmosphere. Using partially oxidized organic matter, Shang et al. (2019) propose a way of reaching such a threshold.

Figure 2.62. Montage of Enceladus, Europa, and Ganymede—imaged individually.

2.9.3. Europa's Subsurface Ocean

When NASA's Galileo space probe flew past Jupiter's moon Europa in 1995, it discovered a strange word covered in a vast fractured ice crust. This crust may hide a liquid ocean beneath just like Lake Vostok and other Antarctic sub-glacial lakes. Europa has a global ocean beneath its ice shell that could be up to 150 kilometers deep. It is hidden under a thick layer of ice that could be 15 to 25 kilometers thick. Europa's fractured crust (Figure 2.62) may have been produced by a series of eruptions of warm ice as Europa's crust spread open to expose the warmer layers beneath.

Contrary to Titan, a probe has not yet landed on Europa. Europa is about 630 million kilometers away from Earth, which means it would take at least three years to get there with currently known engine technology. Europa's surface temperature is -143 degrees Celsius, colder than Antarctica. It is exposed to intense radiation from Jupiter's powerful magnetic field. Europa receives 5.4 Sv (540 rem) of radiation per day, which is about 1,800 times the average annual dose for humans on Earth. Europa has very low gravity, about 13.4 percent of Earth's (Europa: 1.315 m/s^2 and Earth: 9.81 m/s^2). A spacecraft, lander, or drone going to Europa would need lots of shielding to survive radiation, less fuel and thrust to land safely. However, it would also have less traction and stability on icy surfaces. To reach the ocean, a spacecraft or lander must drill or melt through the ice. This would require a lot of power and equipment. And once in the ocean, it would face unknown dangers such as pressure, darkness, coldness, and possibly unknown life forms. Europe may have almost 3 times as much water as the Earth. Just like in Antarctica, volcanic activity on Europa's ocean floor which would also provide warmth. The first civilizations on Earth were based around rivers. It is amazing that we continue to observe that everything that we are looking for on other planets and moons of our solar system is here at some scale or was here on Earth at a certain point in time and space at some scale. Everything that happened or is happening to our other planets and moons is happening or happened here on Earth at a certain point in time and space at some scale. Answers are around you and/or in Earth's history.

Note that another remarkable discovery by Cassini occurred in July 2005. Cassini detected water vapor spewing from Enceladus' south polar region. Figure 2.62 shows jets of water vapor spewing from Enceladus' south polar region (see Ikelle, 2024 for more details).

2.10.   Essay: Magnetic Reversal May Not Be Catastrophic for Living Things

More than two thousand years ago, mankind discovered natural ferromagnets that attract iron. They were named lodestones, also known as *the leading stone* in old English because of their importance to early navigation. In other words, the first magnets and compasses were not invented; rather, they were found from a naturally occurring magnetite mineral. By the nineteenth century, and thanks to Gilbert (1600), Bertelli (1868), Peregrinus (1904), and Thompson (1913), it was well-established that all magnets, including lodestones, have a north and south pole, with the poles of two magnets either pulling together (attracting) or pushing apart (repelling). These magnets align consistently north-south; however, compasses did not point exactly to geographical north, as defined by the stars. Magnetic north and geographical north deviations are defined as declination and inclination. These declination and inclination variations are now called secular variations in the field. It was also well established that the solid Earth itself behaves like a large magnet. This is based on the analogy between the attraction properties of the lodestone spheres and the known magnetic properties of the Earth. Based on these series of observations and experiments, a branch of geophysics for the study of the Earth's electromagnetic field, including its spatial and temporal variations, was born. This branch is now known as geomagnetism and is the oldest geophysics branch. However, seismology is nowadays the most widely investigated geophysics branch.

2.10.1.   Earth's Magnetosphere

Geomagnetism is based on electromagnetism physics. It helps us understand the origin, causes, and characteristics of the Earth's magnetic field, the geophysical processes (e.g., polar wandering and global plate tectonics described in Ikelle, 2023a), including the dynamics of the core (e.g., the existence of inner and outer cores), and how the Earth's magnetic field interacts with solar activity to produce the invisible envelop known as *magnetosphere* that protects us on the Earth from Sun radiation.

The Earth's magnetic field source is roughly dipolar and the resulting magnetic field extends into space. This magnetic field is the effect of the liquid outer core rotating around a solid iron-nickel inner core. While the field is strongest at the center of the planet, the magnetic field extends well beyond the Earth's surface. At some point, about 70-500 kilometers away, this magnetic field comes into contact with the solar wind of charged particles traveling outward from the Sun's photosphere. The Earth's magnetic field is generally an impenetrable obstacle to the solar wind (Chapman and Ferraro, 1931; and Dungey, 1961). However, the solar wind pushes and stretches the Earth's magnetic field into a comet-shaped region called the magnetosphere (see Figure 2.63). Despite its name, the magnetosphere is not spherical, but bullet-shaped. In other words, the magnetosphere deflects the solar

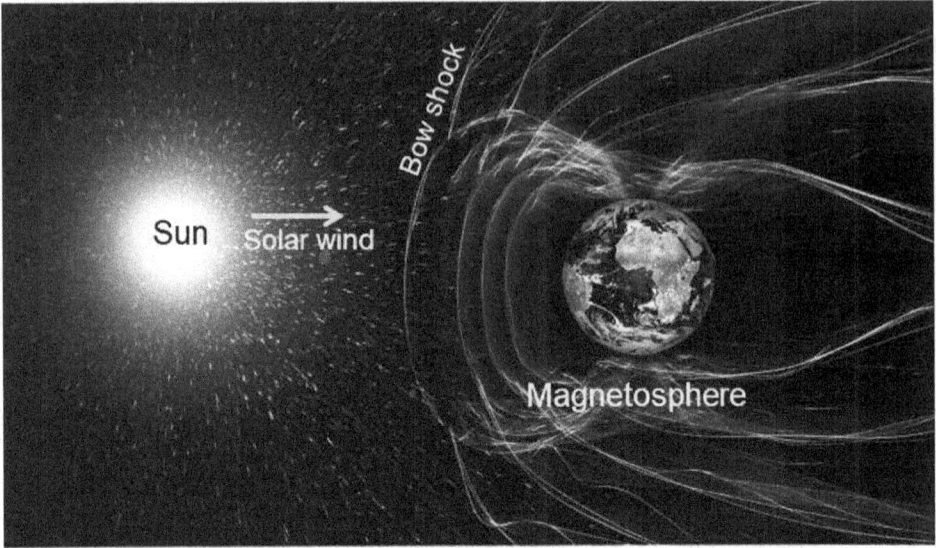

Figure 2.63. Magnetic fields, known as the magnetosphere, surround Earth. Shown here is an artist's conception of the constant stream of particles flowing from the solar wind. (Adapted from NASA Image)

wind and protects us from the solar wind. This includes dangerous charged particle radiation, especially coronal mass ejections from the sun. Dedicated geomagnetic satellites in space can measure the magnetosphere by measuring electromagnetic fields (see, for example, Ikelle, 2020 and 2024).

2.10.2. Geomagnetic Major Coordinates

The key parameters of the geomagnetic field are magnitude (B), inclination (I), and declination (D). As illustrated in Figure 2.64, these parameters are related to the geomagnetic field, $\boldsymbol{B}$, as follows:

$$B_x = B \cos I \cos D \ , \quad B_y = B \cos I \sin D \ , \text{ and } \quad B_z = B \sin D \ , \qquad (2.192)$$

with

$$D = \tan^{-1}\left(\frac{B_y}{B_x}\right) \ , \quad I = \tan^{-1}\left(\frac{B_z}{\sqrt{B_x^2 + B_y^2}}\right) \ , \text{ and } \quad B = \sqrt{B_x^2 + B_y^2 + B_z^2} \ ,$$

where D is the angle between the true north (geographic) and field vector $\boldsymbol{B}_h = B_x \boldsymbol{i}_1 + B_y \boldsymbol{i}_2$, measured positive eastwards (from 180° W to 180° E, from −180° to 180°, or sometimes from 0° to 360°), I is the angle between the horizontal plane and field vector $\boldsymbol{B}$, measured positive downwards (from 90° N to 90° S or from −90° to 90°), B_x is the northerly intensity, B_y is the easterly intensity, and B_z is the vertical intensity with positive downwards, and B is the magnetic intensity. Examples of three independent components are: (B_x, B_y, B_z), (B_h, D, B), and (B_h, D, B_z).

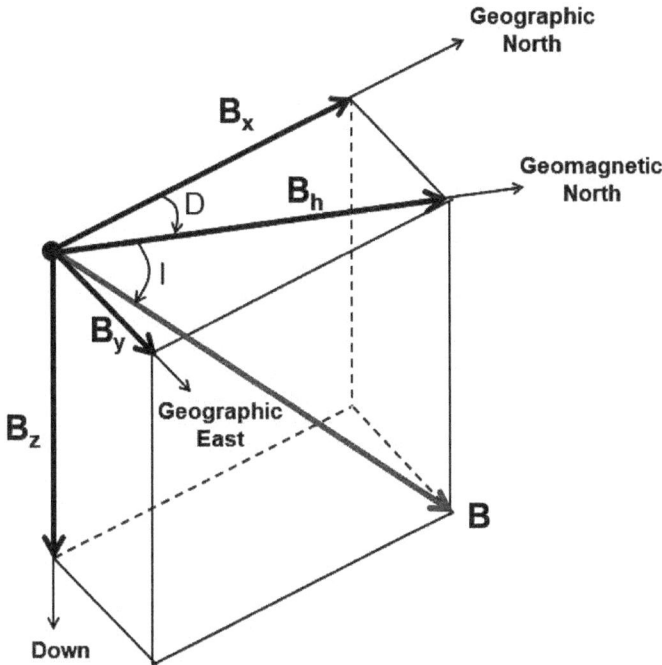

Figure 2.64. B is the total magnetic field, B_h is the horizontal component, B_x is the component in the northern direction, B_y is the component in the east direction, B_z is the vertical component, D is the declination, and I is the inclination.

Table 2.27 shows these parameters at four locations and at various times. The data were obtained using the NOAA (National Oceanic and Atmospheric Administration, United States Department of Commerce) magnetic field calculator. We can see that the Earth's magnetic field changes in intensity and direction with location and time. As one may expect, the intensity near the equator (e.g., Bangui, the capital of the Central African Republic in Africa) is about 30000 nT and smaller than that of locations near the Pole (e.g., Anchorage, Alaska, in the United States). Actually, we can see a steady increase in the intensity field, based on the four locations in Table 2.27, as we move from the equator to the Poles in this order: Bangui, Kursk, Anchorage, and Wilkes Land. These changes in the intensity of the geomagnetic field reflect the fact that the geomagnetic field is not affected by features near the Earth's surface and the ocean floor. These features include mountain ranges, mid-ocean ridges, trenches, etc. They also imply that the main sources of the geomagnetic field at the Earth's surface are located deep within the solid Earth or in deep space. This is another argument for the case that the main source of the geomagnetic field is likely to be located in the Earth's core. Moreover, these changes indicate that Earth's surface temperatures are inversely related to the magnetic field; Earth's surface temperatures decrease from the equator to the magnetic poles, whereas the magnitude of the magnetic field increases from the equator to the magnetic poles.

Table 2.27. Declination (D) [+ east of north, in degrees]; inclination (I) [+ down, in degrees]; north component of horizontal intensity (B_x) [+ north, in nT]; east component of horizontal intensity (B_y) [+ east, in nT]; vertical intensity (B_z) [+ down, in nT]; and total intensity (B) [in nT] of Alaska (61.20 N, 149.96 W), Bangui, the capital of the Central African Republic (4.40 N, 18.56 E), Kursh, in Russia (51.73 N, 36.19 E), Wilkes Land, in Antarctica (68 S, 120 E).

	Anchorage	Bangui	Kursk	Wilkes Land
March 11, 1900				
D (°)	28.041	-11.219	1.376	-75.263
I (°)	74.720	-9.866	65.133	-84.738
B_x (nT)	13334.8	31,723.5	20282.6	1567.5
B_y (nT)	7115.3	-6,292.4	487.3	-5959.5
B_z (nT)	55325.6	-5,625.0	43773.4	-66915.9
B (nT)	57352.9	32,827.0	48246.6	67199.0
March 11, 2000				
D (°)	21.571	-1.286	5.074	-118.060
I (°)	74.152	-16.125	69.396	-83.215
B_x (nT)	14194.3	31922.5	17864.8	-3589.6
B_y (nT)	5611.7	-717.0	1585.9	-6733.9
B_z (nT)	53767.4	-9231.7	47705.2	-64134.6
B (nT)	55891.8	33238.3	50965.2	64586.9
March 11, 2019				
D (°)	15.522	1.200	9.358	-123.722
I (°)	74.185	-16.538	68.686	-83.412
B_x (nT)	14604.3	32193.3	18482.0	-4107.6
B_y (nT)	4056.3	674.7	3045.8	-6154.0
B_z (nT)	53512.7	-9561.2	48008.0	-64066.3
B (nT)	55617.9	33589.9	51532.8	64492.2

2.10.3. Magnetic Polar Reversals

As we can see in Figure 2.65, the intensity and direction of the geomagnetic field, including at the poles, change with time. The changes are such that, from time to time, the geomagnetic north pole moves suddenly to the geomagnetic south, and vice versa (i.e., sometimes a compass pointing north will be aimed at Antarctica rather than the Arctic)—the magnetic poles or dip pole positions are the areas where the field lines penetrate the Earth vertically (see Figure 2.66), that is, the inclination reaches 90. Current estimations are that orientation swaps have occurred more than 183 times over the last 83 million years. The latest reversal occurred 781,000 years ago (Cande and Kent, 1995). There have also been periods as long as 60 million years with no reversals (superchrons are long periods when no reversals took place) and many aborted reversals when the magnetic poles are observed to move equatorward for a while then move back and align closely with the Earth's spin axis. Aborted reversals, which may be considered failed reversals, are called

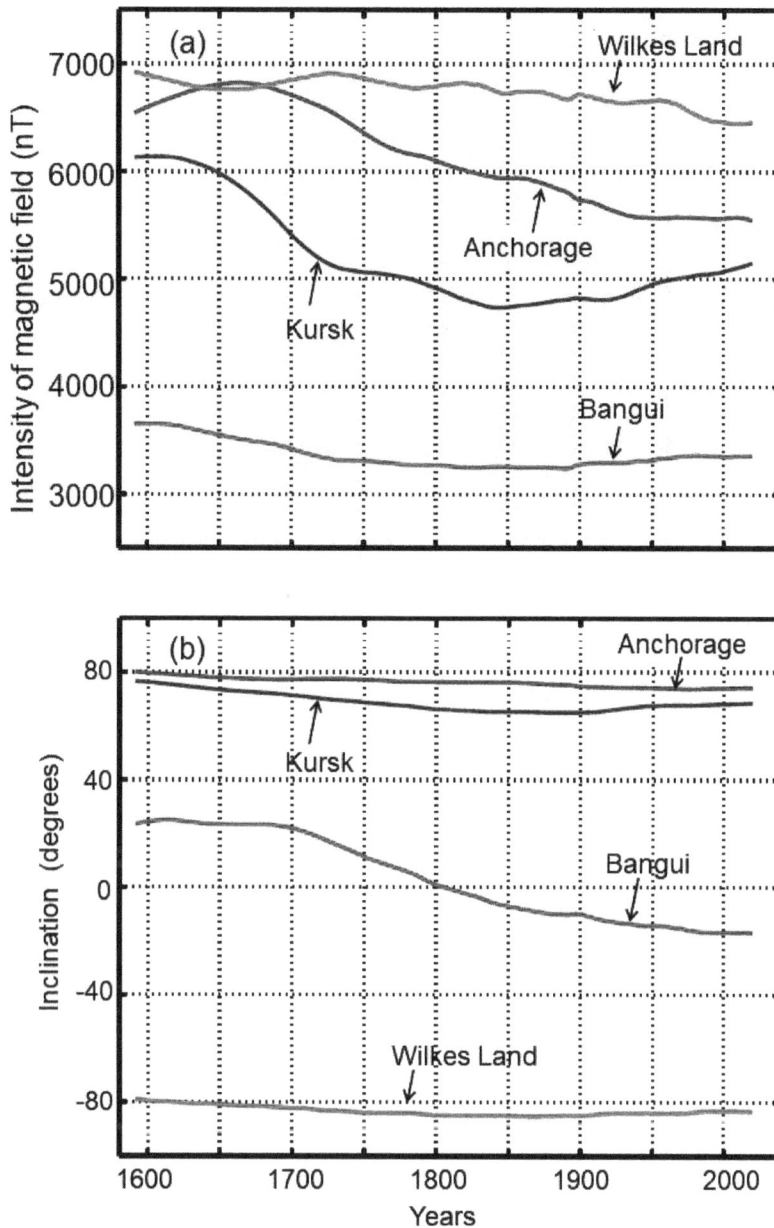

Figure 2.65. *(Continued.)*

excursions. The last excursion took place 40,000 years ago, and evidence suggests we are heading in that direction again. The geomagnetic field has lost 30 percent of its intensity over the last 3,000 years. At this rate, the geomagnetic field will drop to near zero in a few centuries or a millennia for the purpose of calculation only. There is no indication that it will vanish entirely. A 90 percent reduction in the geomagnetic field will generally reverse. The causes of reversals in the Earth's magnetic field are discussed in Ikelle (2024).

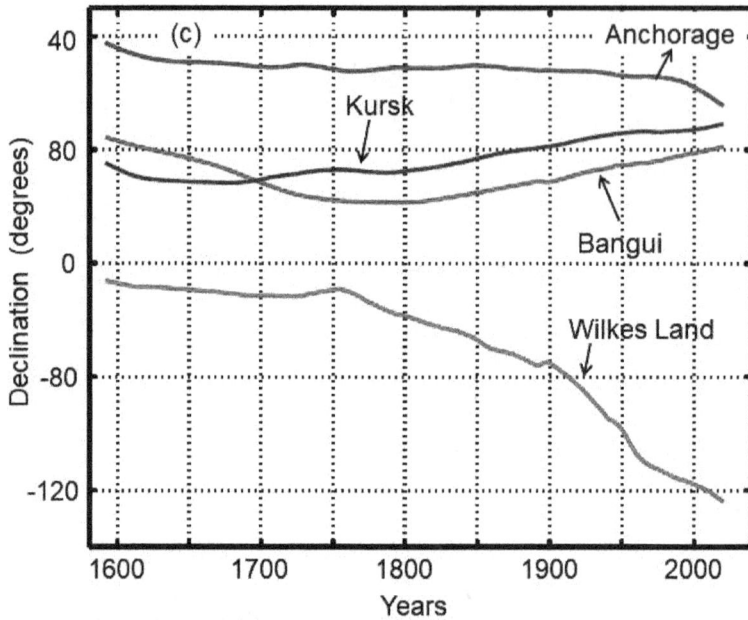

Figure 2.65. Variation with time of the geomagnetic field for the last 400 years at four locations: Anchorage, Alaska, in the United States (61.2° N, 149.96° W), Bangui, the capital of the Central African Republic in Africa (4.4° N, 18.56° E), Kursh in Russia (51.7° N, 36.2° E), and Wilkes land, in Antarctica (69° S, 120° E); (a) is the intensity of the magnetic field, (b) is the declination, and (c) is the inclination.

2.10.4. How Dangerous is the Coming Magnetic Reversal to Our Civilization?

During the magnetic reversal, the magnetosphere is not totally opened because there is no switch-over point between the north and south geomagnetic poles (Ikelle, 2020). The geomagnetic field is not symmetric with respect to the equator and nonsymmetric with respect to the zero Meridian. In other words, it is unlikely that the giant dipole center coincides with solid Earth's center. In fact, during the 100 to 1000 years or so of the magnetic reversal, only a small portion of the atmosphere can be accessed by charged particles from the sun at any given time. In other words, we have a scenario where only about 5 percent or so of the magnetosphere is permeable to charged particles from Sun radiation while the other 95 percent remains impermeable. Just like Titan about 95 percent is protected by Saturn's magnetosphere and about 5 percent is exposed to solar radiation. This window produces all types of reactions that lead to the production of molecules as described in Table 2.28. Most of these reactions are described in Chapter 1 of Ikelle (2023b). Some of these molecules may have contributed to life on Earth and the Great Oxidation Event (GOE). Another complex atmospheric chemistry occurs during magnetic polar reversals just like in Titan's atmosphere.

It is generally assumed that geomagnetic reversal is a very deadly event that can even produce mass extension of some species, including our ancestors who have been around for about six million years—the modern form of humans only evolved about 230,000 years ago and civilization as we know it today is only about 6,000 years old. Because the magnetosphere is only narrowly open at a given time mass extension

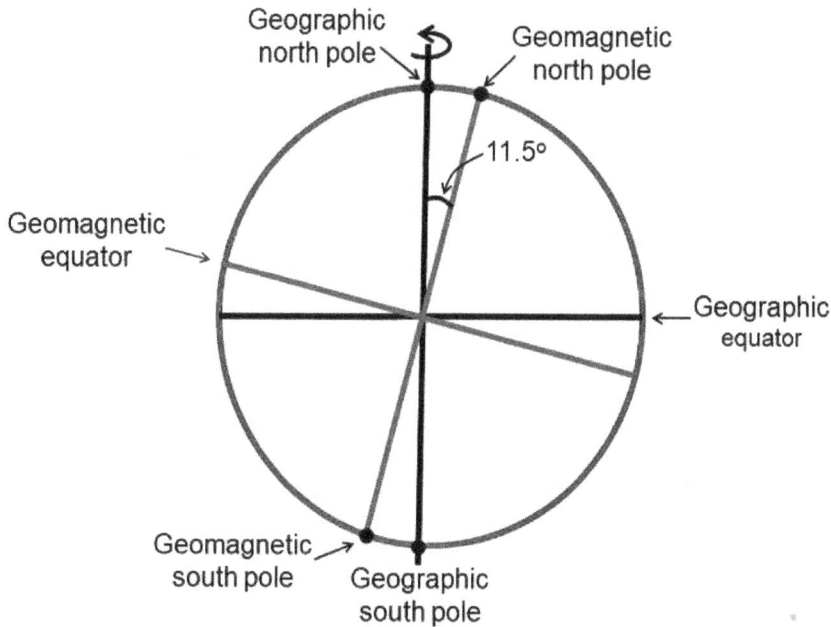

Figure 2.66. An illustration of Earth's dipole model coordinates. The south pole of this dipole is near the geographical South Pole and its north pole is near the geographical North Pole. Best-fit of this dipole to geomagnetic data, in 2018, is 11.5 degrees from the geographic North Pole. The north magnetic pole is currently situated in the icy Arctic Ocean. The south magnetic pole is in the ocean just off the Antarctic coast and south of Australia.

is very unlikely. Also dramatic events like earthquakes are unlikely due to the difference between Earth's liquid core and its mantle. The mantle causes tectonic shifts, which cause earthquakes. Although the mantle and liquid core make contact, it would take millions of years for any changes in the liquid core to affect the mantle and disrupt the tectonic plates. There are no fossils and other records of quakes

Table 2.28. Some examples of chemical reactions in the Earth's atmosphere during geomagnetic polar reversals.

Reaction	Reaction
$O_2 \longrightarrow O + O$	$O + O_2 \longrightarrow O_3$
$O_3 \longrightarrow O_2 + O$	$O + O_3 \longrightarrow 2\,O_2$
$CO_2 + H_2 \longrightarrow H_2O + CO$	$Cl + O_3 \longrightarrow ClO + O_2$
$ClO + O \longrightarrow Cl + O_2$	$Cl + O_3 + ClO + O \longrightarrow ClO + Cl + 2\,O_2$
$CH_4 \longrightarrow C + 2\,H_2$	$ClO + ClO \longrightarrow ClO_2$
$HCl + ClONO_2 \longrightarrow Cl_2 + HNO_3$	$H_2O + ClONO_2 \longrightarrow HOCl + HNO_3$
$2\,O_3 \longrightarrow 3\,O_2$	$O_2 + e^- \longrightarrow O_2^{\bullet -}$
$O_2 + 4\,e^- + 4\,H^+ \longrightarrow 2\,H_2O$	$O_2 + e^- + e^- \longrightarrow O_2^{-2}$
$CO_2 + H_2O \longrightarrow CH_2O + O_2$	$2\,CH_2O \longrightarrow CH_4 + CO_2$
$CO_2 + 2\,H_2O \longrightarrow CH_4 + 2\,O_2$	$CH_4 + OH \longrightarrow CH_3 + H_2O$
$2\,H_2O + 4\,e^- \longrightarrow 2\,H_2 + O_2$	$CO + 3\,H_2 \longrightarrow CH_4 + H_2O$
$2\,CO \longrightarrow C + CO_2$	$CO_2 + H_2 \longrightarrow CO + H_2O$

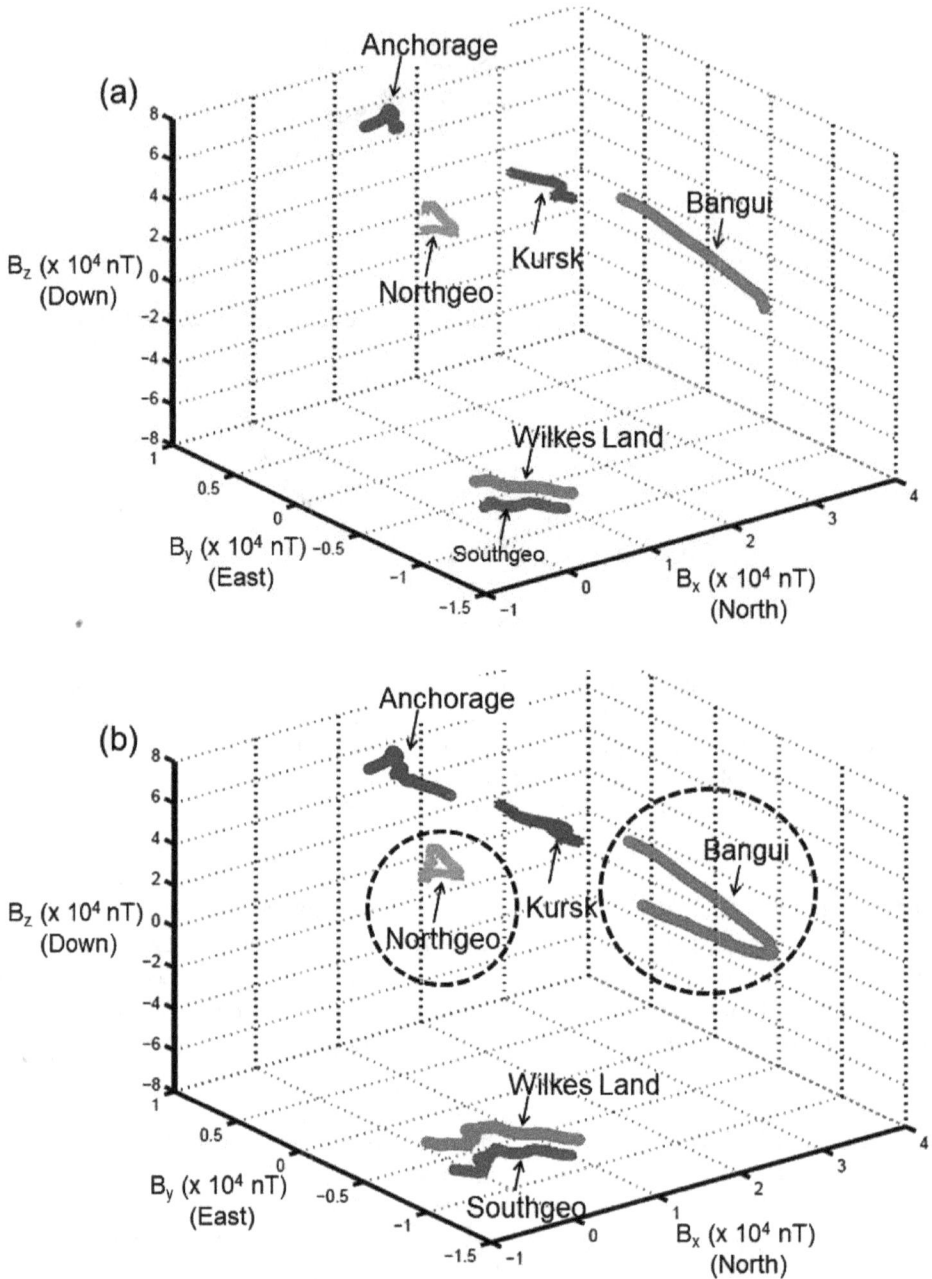

Figure 2.67. (a) 3D scatterplot of B_x, B_y, and B_z based on the values of these quantities from 1590 to 1700 at six locations: Anchorage, Alaska, in the United States (61.2° N, 149.96° W), Northgeo (80.31° N, 72.62° W), Bangui, the capital of the Central African Republic, in Africa (4.4° N, 18.56° E), Kursh in Russia (51.7° N, 36.2° E), Wilkes land in Antarctica (69° S, 120° E), and Southgeo (80.31° S, 107.38° E). The curves here indicate how the poles have changed since 1590. (b) The same scatterplot for a period from 1590 to 2019.

associated with past magnetic reversals. Moreover, there are no geological records or proxy measurements of mass extensions during the past reversals. Also, as we can see in Figure 2.67, there are currently local magnetic reversals that have been completed in the last 150 years (e.g., Northgeo, a location close to the geomagnetic north pole) or about to be completed in the Central African Republic without any known major appeal. Note however that some space human satellites will fall down or malfunction during magnetic reversals due to their exposure to charged particles coming from the Sun. This failure could be one of the few effects on our civilization. This is because of our increasing reliance on space satellites for our communication, electric power grid, electronic devices, and maritime and air navigation. We further discuss the planet's magnetic reversals in Ikelle (2024).

In the Atlantic Ocean, between South America and Africa, there is a vast region known as the South Atlantic Anomaly (SAA). The SAA stretches from Chile to Zimbabwe (a long distance). The Earth's magnetic field strength is three times weaker in SAA than at the poles. Satellites and spacecraft consistently experience electronic failures in this region. This suggests a very narrow part of the atmosphere where the magnetosphere is not protecting us.

SAA was first noticed in the 1950s, and since then, Earth's magnetic field strength in SAA has decreased in strength by 6 percent. In addition, SAA has moved closer to the west. Since 1590, B_z has increased from negative values toward zero. In the last 200 years, B_zs have moved back to negative values. This region illustrates a locally aborted magnetic reversal. In any case, the SAA is likely to contain clues to polar reversal and abortion. This is an excellent place to study hazards, mitigation, and adaptation to magnetic reversals (see Ikelle, 2024).

Chapter 3

Electromagnetic Radiation and Spectroscopy

Electromagnetic (EM) radiation is all around us and in our entire universe. Yet EM radiation is extremely difficult to get a handle on. Although we cannot see with light, which is electromagnetic radiation, we can appreciate the beauty of a Fiji coral reef in blue waters, the warmth of sunshine, the sting of sunburn, the natural electrical phenomenon of lightning, the magnetic forces that can cause lodestones to point north, and the contribution of light to living things' metabolism, as described in Ikelle (2023b). Electromagnetic waves bring us observations. In other words, electromagnetic fields cannot be seen, but they are real and can work (by applying forces) on objects and exchange energy. It took scientists until the 1890s to accept this fact. Today, we harness electromagnetic waves to reveal a broken bone (through X-rays), to cook our food in microwave ovens, to wirelessly communicate around the world instantly, and to construct computers, robotics, and artificial intelligence systems, which are all now central to modern life.

Fundamental questions about electromagnetic waves include how they are created, how they travel, how we can understand and organize their widely varying properties, and what their relationship is to electric and magnetic effects. Our aim in this chapter is to answer one of these questions, light's interaction with matter known as spectroscopy. The main objective is to use light's interaction with matter to determine the molecular and atomic compositions of matter, and sometimes even the electronic and nuclear structures and states of matter. Many phenomena can occur when a beam of light interacts with matter, including the incident radiation being reflected back by matter, the incident radiation being partially or fully absorbed by matter, and matter emitting radiation in the form of heat, light, and even mechanical perturbations. Spectrometers can read and record EM responses. These responses offer insight into the behavior of electrons in atoms and molecules, the composition of living-thing molecules and organs, stars, minerals, and rocks, and help us understand such things as the temperature and pressure of stellar atmospheres, the ages of galaxies, why copper is red, why gold is almost inert, and why certain substances glow in the dark. Spectroscopy and matter energy levels continue to be at the forefront of our technological advances. They will even be central to many future technological advances.

DOI: 10.1201/9781032620619-3

Key sentences: *Electromagnetic forces underlie the mechanisms of DNA, consciousness, chemical composition from molecules to stars and galaxies, the possibility of all photons being linked together in this entire universe, and the possibility that light may be the origin of all quantum entanglements in our universe. These mechanisms suggest that human-like life forms are not possible on exoplanets and exomoons with weak electromagnetic fields. In other words, an electromagnetic field is a prerequisite for life in the matter universe.*

3.1. Microscopic and Macroscopic Maxwell's Equations

In Ikelle (2023b), we describe Maxwell's equations in vacuum. Maxwell's equations govern the electromagnetic field. The solutions to these equations are electromagnetic waves (or photons). These waves cover the entire electromagnetic frequency spectrum, from 0 Hz to infinity Hz, including gamma rays, X-rays, ultraviolet light, visible light, terahertz light, infrared light, and radio waves. We here describe Maxwell's equations for matter (including physical materials) and their solutions, which represent electromagnetic wave propagation in matter.

3.1.1. Microscopic and Macroscopic Scales

When dealing with electromagnetics in matter, special attention must be paid to the microscopic scale (or atomic scale) and the macroscopic scale (scale of human senses) at which experimental measurements are made. This distinction is not necessary in a vacuum because we are dealing with a homogeneous (the same properties at all points) and isotropic medium (the same properties along any direction).

As described in Ikelle (2023a), matter is made of atoms (the size of about 10^{-10} m). We can describe Maxwell's equations, which govern the electromagnetic fields, at the atomic scale by describing the fields entering in these equations at the atomic scale. This choice implies that Maxwell's equations consider all matter to be made of charged and uncharged particles (i.e., point charges). Maxwell's equations that deal with microscopic fields associated with discrete charges are generally known as *microscopic Maxwell's equations*. If the point charges are, say, electrons, then a macroscopic object will consist of an enormous number of electrons, around 10 billion electrons or more for the lower limit of the macroscopic scale cell, of about 10^{-8} m = 100 angstroms. This is why, for many problems, it is not convenient to work with point charges, and therefore, with the microscopic scale. We here primarily describe macroscopic Maxwell's equations that deal with fields that are local spatial averages over microscopic fields associated with discrete charges. Hence, the microscopic nature (the microscopic behavior, at atomic scale) of matter is not explicitly included in the macroscopic fields. In other words, macroscopic Maxwell's equations ignore many details on a fine scale that may not be important to understanding matter on a grosser scale (the scale of the human senses) by calculating fields that are averaged over some suitably sized volume.

The basic assumption in the derivation of macroscopic Maxwell's equations is that the spatial averages of the microscopic quantities lead to auxiliary macroscopic quantities that vary piecewise continuously with position. Matter is considered continuous and completely fills the space it occupies, leaving no holes or empty spaces, and its properties can be described by piecewise continuous functions. It contains no empty spaces—that is, we will disregard the atomic scale (microscopic scale) of matter and envision it without holes or empty spaces. Yet macroscopic Maxwell's equations agree with experimental measurements, and they implicitly contain the fundamental property of conservation of charges—that is, charges are not created or destroyed, just moved, meaning they are preserved by macroscopic Maxwell equations.

So, there are two major variants of Maxwell's equations: microscopic and macroscopic. Microscopic Maxwell's equations include the difficult-to-calculate atomic-level charges. Macroscopic Maxwell's equations include auxiliary fields that allow us to sidestep having to know atomic-sized charges. Our primary focus in this book is on macroscopic Maxwell's equations, although the microscopic scale remains constant to keep in mind for interpretation purposes.

3.1.2. Lorentz Force Law at the Macroscopic Scale

In Ikelle (2023b), we introduced electric and magnetic fields using the Lorentz force for charged particles. Now that we have abandoned the atomic (microscopic) scale for the macroscopic scale, it is necessary to revisit the Lorentz force law and thereby reintroduce the electric and magnetic fields in the context of the macroscopic scale. Let us start again by recalling our description of the Lorentz force at the microscopic scale. A point charge is a charge, q, that occupies a region of space that is negligibly small compared to the distance between the point charge and any other object. Consider a single electric point charge in a vacuum. Experiments show that the force exerted on this point charge is (in vector and subscript notations)

$$\boldsymbol{F} = q\boldsymbol{E} + q\mu_0 \boldsymbol{v} \times \boldsymbol{H} \quad \text{(microscopic)}, \tag{3.1}$$

where $\boldsymbol{F}$ is the Lorentz force, $\boldsymbol{E}$ is the electric field intensity, $\boldsymbol{H}$ is the magnetic field intensity, μ_0 is the magnetic permeability of the vacuum, and $\boldsymbol{v}$ is the velocity of the point charge with respect to the observer in a vacuum domain, as depicted in Figure 1.36 of Ikelle (2023b). Note that electric and magnetic forces are not separate but different manifestations of the same thing: the electromagnetic force. From the Lorentz force law, we can define the electric field as $\boldsymbol{E} = \boldsymbol{F}/q$ for the particular case of a charge q at rest (that is, $\boldsymbol{v} = 0$). Similarly, we can define the magnetic field from the Lorentz force law by assuming that the electric field is zero. Thus, the magnetic field, $\boldsymbol{H}$, is the effect produced by a change in velocity of an electric charge, q. Note that the Lorentz force tells us how the electromagnetic fields affect charged particles and Maxwell's equations tell us how charged particles affect the electromagnetic fields; a typical action-and-reaction phenomenon that ensures conservation of energy. In other words, electromagnetic fields can add energy to charged particles and charged particles can give energy to electromagnetic fields.

For macroscopic-scale problems, we are not dealing with point charges but with macroscopic objects that consist of an enormous number of very tiny charges (e.g., electrons). We thus introduce the concept of volume charge density, $\rho_E(\boldsymbol{x}, t)$, at $\boldsymbol{x}$ and time t, as the ratio of electric charges contained in a domain, $dV(\boldsymbol{x}, t)$, which is a physically infinitesimal volume, bounded by a small closed surface over the volume

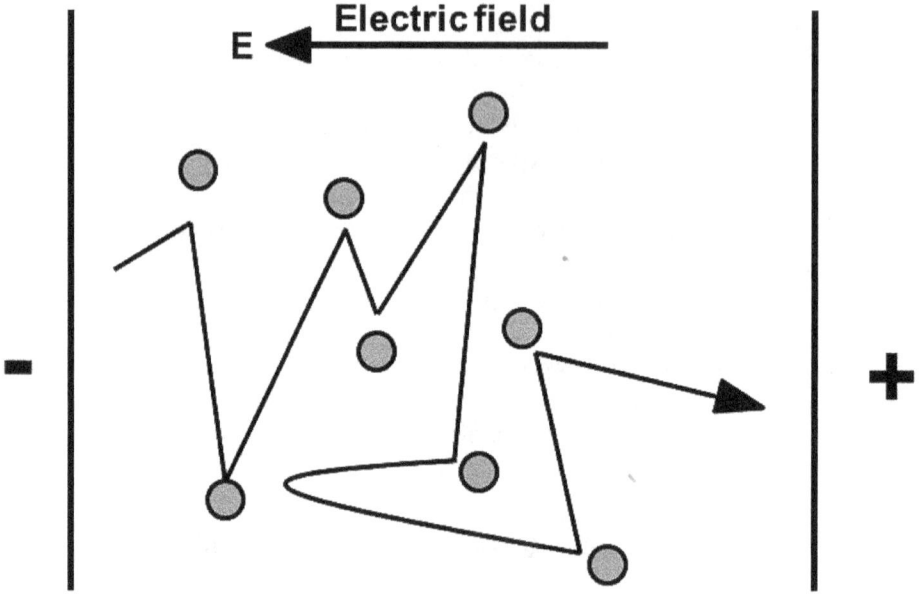

Figure 3.1. An illustration of a charged particle motion in response to an electric field. This motion is known as drift. Particles tend to spread out or redistribute from areas of high concentration to areas of lower concentration. Motion is highly nondirectional on a microscopic scale, but it has a net direction on a macroscopic scale. The average net motion of charged particles gives rise to a current. The drift velocity, $v = -\mu_e E$, is proportional to the applied electric field and μ_e is the electron mobility. The minus sign is related to the fact that the direction of drift velocity is opposite the electric field.

of this same bounded domain, i.e.,

$$\rho_E(\boldsymbol{x}, t) = \frac{dQ(\boldsymbol{x}, t)}{dV(\boldsymbol{x}, t)} , \qquad (3.2)$$

where $dQ(\boldsymbol{x}, t)$ represents electric charges contained in a bounded domain, $dV(\boldsymbol{x}, t)$, at time t. The volume charge density is also subject to the Lorentz force in the presence of fields. Based on the fact that the microscopic scale describes charged particles in motion and the macroscopic scale describes a collection of charged particles in motion, we can define the Lorentz force on the macroscopic scale by analogy to (3.2) as follows:

$$\boldsymbol{F} = \rho_E \boldsymbol{E} + \rho_E \mu_0 \boldsymbol{v} \times \boldsymbol{H} = \rho_E \boldsymbol{E} + \mu_0 \boldsymbol{J} \times \boldsymbol{H} \quad \text{(macroscopic)} , \qquad (3.3)$$

where $\boldsymbol{J} = \rho_E \boldsymbol{v}$ is the flow of the electric charges (i.e., the volume density of electric current) and $\boldsymbol{v}$ is now the average electron velocity, also known as drift velocity (see Figure 3.1). Basically, particle motions in a given volume $dV(\boldsymbol{x}, t)$ are highly nondirectional, as illustrated in Figure 3.1, but have a net direction on a macroscopic scale. That is, we ignore the random motions, and regard the particles as moving through the material with a velocity vector field $\boldsymbol{v}(\boldsymbol{x}, t)$, which is

$$\boldsymbol{v}(\boldsymbol{x}, t) = \frac{1}{N_p(\boldsymbol{x}, t)} \sum_{i=1}^{N_p(\boldsymbol{x}, t)} \boldsymbol{w}^{(i)} , \qquad (3.4)$$

where $N_p(\boldsymbol{x},t)$ is the total number of particles in the bounded domain $dV(\boldsymbol{x},t)$ at time t and $\boldsymbol{w}^{(i)}$ is the velocity of the i-th particle. Note that the chaotic part of the motion of particles, which determines the thermodynamic notion of their temperature, does not contribute to drift velocity. The volume density of electric current is related to individual charges and their velocities, as follows:

$$\boldsymbol{J}(\boldsymbol{x},t) = \frac{1}{dV(\boldsymbol{x},t)} \sum_{i=1}^{N_p(\boldsymbol{x},t)} q_i \boldsymbol{w}^{(i)} . \tag{3.5}$$

Suppose that $q_i = q$ for each i (i.e., a single-species medium), then we arrive at

$$\boldsymbol{J}(\boldsymbol{x},t) = \left[\frac{q\,N_p(\boldsymbol{x},t)}{dV(\boldsymbol{x},t)}\right] \left[\frac{1}{N_p(\boldsymbol{x},t)} \sum_{i=1}^{N_p(\boldsymbol{x},t)} \boldsymbol{w}^{(i)}\right] = \rho_E(\boldsymbol{x},t)\boldsymbol{v}(\boldsymbol{x},t) \tag{3.6}$$

with

$$\rho_E(\boldsymbol{x},t) = \frac{q\,N_p(\boldsymbol{x},t)}{dV(\boldsymbol{x},t)} , \tag{3.7}$$

where $q\,N_p(\boldsymbol{x},t)$ represents the electric charges contained in a bounded domain $dV(\boldsymbol{x},t)$. Note that the electric current density, $\boldsymbol{J}$, is the flow of the amount of charge that flows (perpendicularly) through a unit surface in unit time, as illustrated in Figure 3.2a. Thus, if the amount of charge dQ flows through the surface, dS, in time, dt, we can alternatively define the electric current density, $\boldsymbol{J}$, as follows:

$$J = \rho_E v = \frac{dQ}{dV}\frac{dl}{dt} = \frac{dQ}{dSdl}\frac{dl}{dt} = \frac{dQ}{dS}\frac{dl}{dt} = \frac{dI}{dS} , \tag{3.8}$$

with

$$dI = \frac{dQ}{dt} = \frac{\text{charge}}{\text{time}} , \tag{3.9}$$

where $J = |\boldsymbol{J}|$, $v = |\boldsymbol{v}|$, and dI is the amount of current through the surface, dS. Electric current is the charge flowing through a point per unit of time. The unit of current is Ampères (A). For sphere of radius $r = |\boldsymbol{x}|$, we can verify that

$$I = \int_S J dS = 4\pi r^2 J = 4\pi|\boldsymbol{x}|^2 J . \tag{3.10}$$

Thus, the macroscopic Lorentz force per unit volume, in (3.3), acts on ρ_E and $\boldsymbol{J}$. We can define the electric field as $\boldsymbol{E} = \boldsymbol{F}/\rho_E$ for a static distribution of charges (i.e., $\boldsymbol{v} = \boldsymbol{0}$). Similarly, we can define the magnetic field from the Lorentz force law by assuming that the electric field is zero.

Conservation of the charges on macroscopic scale. Let us make three observations. One is about the total charge, the second is about charged particles whose movement creates electric current, and the third is about the relationship between

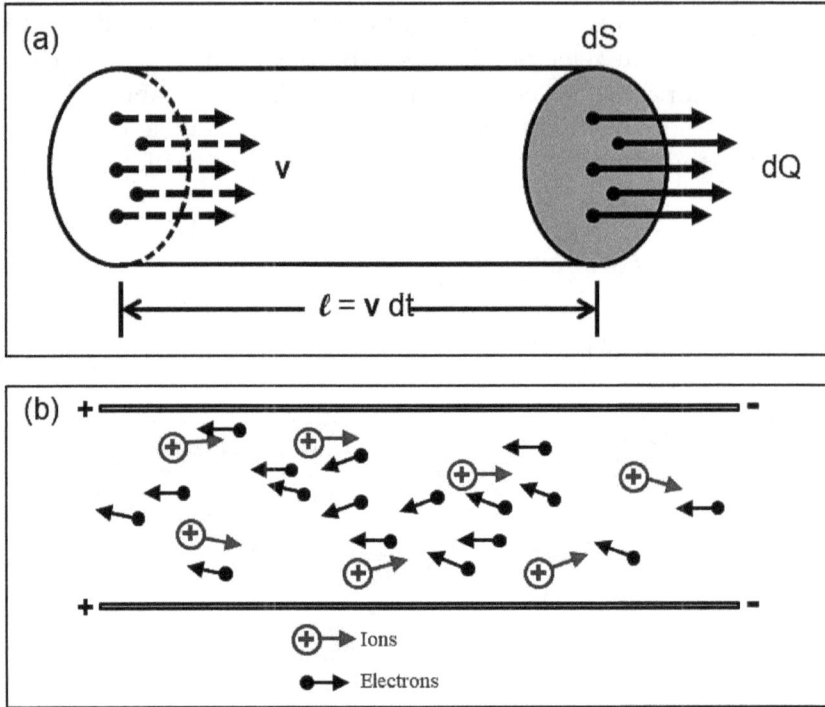

Figure 3.2. (a) Flux of current density. The arrows point in the direction of flow. The flux vector, $\boldsymbol{J}$, is related to the transport velocity $\boldsymbol{v}$ and the volume density of charges, ρ, of the flowing current density. (b) Electrons flow from the negative plate (negative electrode) to the positive plate (positive electrode). Ions move in the opposite direction.

the volume-charge density and the flow of the electric charges. The total charge in the bounded domain, $\mathcal{D}(t)$, is

$$Q(t) = \int_{\mathcal{D}(t)} \rho_E(\boldsymbol{x}, t) dV(\boldsymbol{x}, t) , \tag{3.11}$$

where $Q(t)$ is the total charge at time, t. The net electric charge of any closed system is always conserved; whatever happens, this net charge is not going to change. Therefore, the total charge, Q, describes the conservation of charges on the macroscopic scale and not the volume-charge density, $\rho_E(\boldsymbol{x}, t)$.

Charge particles and movement of ions. Electric charged particles, such as electrons, do not suddenly jump from one point to another; they must flow from one point to another point. In macroscopic scale, the net motion of charged particles gives rise to an electric current. In metals, it is carried by electrons, whereas in liquids, gasses, and plasmas, the current is carried by ions. In other words, ions are also charged particles that give rise to an electric current. Typical examples of ions are sodium (Na^+ has one positive charge), chloride (Cl^- has one negative charge), potassium (K^+ has one positive charge), calcium (Ca^{++} has two positive charge), oxygen (O^+ has one positive charge), helium (He^{++} has two positive charges), and hydrogen (H^+ has one positive charge); H^+ is the proton.

Positively charged ions are called cations and negatively charged ions are called anions. These ions can be part of the flowing of fluid (electrolyte) through a capillary. Note that certain solids such as water-ice are called proton conductors. They contain positive hydrogen ions, which are free to move. In these materials, currents of electricity are composed of moving ions, as opposed to the moving electrons found in metals. If ions cannot move, there will be no current flow. The dominant ions in deep space are H^+, He^{++}, and O^+. The ions that reside in human cells include Na^+, Cl^-, K^+, Ca^{++}, and proteins (A^-).

Consider, for example, a medium with protons and electrons, such as the collisionless media in deep space. We denote by m_e and m_i the masses of electrons and protons, respectively, by n_e and n_i the particle densities (number of particles per unit volume) of electrons and protons, and by q_e and q_i the charges of electrons and protons, and can define the mass density, charge density, and current density of this medium, as follows:

$$\rho = n_e m_e + n_i m_i \ , \ \ \rho_E = q_e n_e + q_i n_i \ , \ \text{ and } \ \boldsymbol{J} = q_e n_e \boldsymbol{v}_e + q n_i \boldsymbol{v}_i \ ,$$

where $\boldsymbol{v}_e$ and $\boldsymbol{v}_i$ are the velocities of electrons and protons. Suppose that $n_i = n_e = n$, and using the fact that $q_i = -q_e = e$, then

$$\boldsymbol{J} = q_e n_e \boldsymbol{v}_e + q n_i \boldsymbol{v}_i = en(\boldsymbol{v}_i - \boldsymbol{v}_e) \tag{3.12}$$

and $\rho_E = 0$. The macroscopic charge density can be zero even when the individual charges of particles are nonzero. Moreover, current can exist in media with zero density of charge as long as $|\boldsymbol{v}_e| \neq |\boldsymbol{v}_i|$ or electrons and ions are moving in opposite directions, as illustrated in Figure 3.2b.

Continuity equation of electrical current. Let us also mention the relationship between the macroscopic quantities $\rho_E(\boldsymbol{x}, t)$ and $\boldsymbol{J}(\boldsymbol{x}, t)$; i.e.,

$$\frac{\partial \rho_E(\boldsymbol{x}, t)}{\partial t} + \boldsymbol{\nabla} \cdot \boldsymbol{J}(\boldsymbol{x}, t) = 0$$

or

$$\rho_E(\boldsymbol{x}, t) = \int_0^t \left[\boldsymbol{\nabla} \cdot \boldsymbol{J}(\boldsymbol{x}, t') \right] \ . \, dt' \ . \tag{3.13}$$

The first equation here is known as the continuity equation for electrical current. This continuity equation is the local form of electric charge conservation, and it always holds true for every physical system of charges and currents, without any exception.

Again, note that an ion can also be defined as an atom that is attracted to another atom because it has an unequal number of electrons and protons. If an atom has more electrons than protons, it is an anion.

Quiz: The microscopic and macroscopic scale distinction is not an issue in vacuum. Why?

3.1.3. Microscopic and Macroscopic Maxwell's Equations

In Ikelle (2023b), we described Maxwell's equations for a vacuum. The simplest way to account for matter in this case is to retain Maxwell's equations for a vacuum and replace the zero terms on the right-hand side with quantities that describe the presence of matter, i.e.,

$$-\nabla \times \boldsymbol{H}(\boldsymbol{x}, t) + \varepsilon_0 \frac{\partial \boldsymbol{E}(\boldsymbol{x}, t)}{\partial t} = -\boldsymbol{J}^{(s)} - \boldsymbol{J}^{(p)} \tag{3.14}$$

$$\nabla \times \boldsymbol{E}(\boldsymbol{x}, t) + \mu_0 \frac{\partial \boldsymbol{H}(\boldsymbol{x}, t)}{\partial t} = -\boldsymbol{K}^{(s)} - \boldsymbol{K}^{(p)} . \tag{3.15}$$

Notice that we described the presence of matter with two parts: $\boldsymbol{J}^{(s)}$ and $\boldsymbol{K}^{(s)}$ describe the external sources of volume density of material electric current and volume density of material magnetic current, respectively, $\boldsymbol{J}^{(p)}$ and $\boldsymbol{K}^{(p)}$ describe the induced part of volume density of material electric current and volume density of material magnetic current, respectively. We here assume that $\boldsymbol{J}^{(s)} = \boldsymbol{K}^{(s)} = \boldsymbol{0}$. Note that the $\boldsymbol{J}$'s and $\boldsymbol{K}$'s are zero in a vacuum because there are no charges of any kind in a vacuum.

Free as well as bound charges contribute to the induced electric current (Figure 3.3). Some materials contain free charges that are able to move around (mostly these

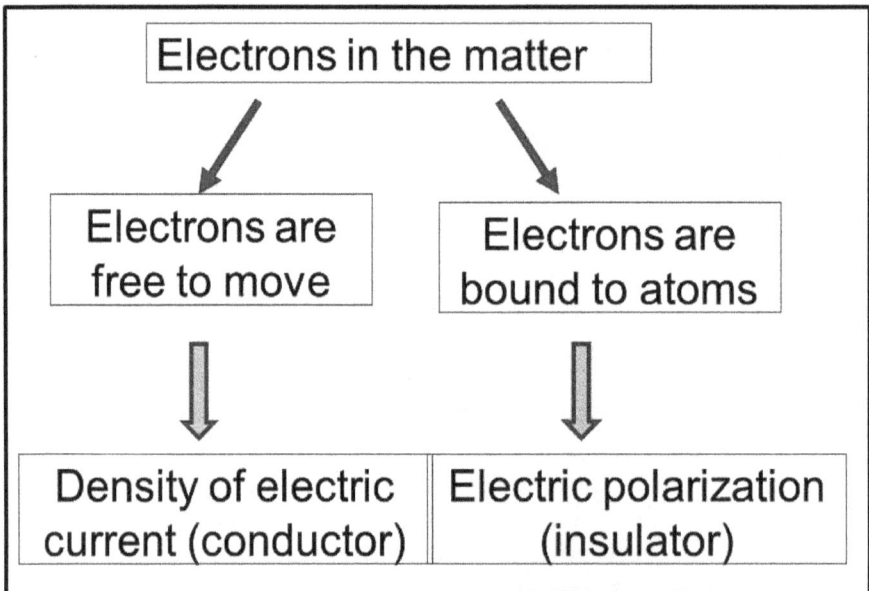

Figure 3.3. The induced volume density of electric current is the sum of the induced volume density of electricity current (free-charge density) and electric polarization (bound current density). The induced volume density of electricity current describes the action of free charges that are able to move around. Usually, these "free charges" are electrons that are not attached to any particular atom. These free charges move under the influence of the external E-field. Electric polarization describes the action of electrons that are bound to atoms and are not free to move about the material.

free charges are electrons that are not attached to any particular atom). These free charges move under the influence of an external E-field. Many materials do not have free electrons that can move around, but have electrons bound to atoms or have bound charges in addition to free charges. Thus the induced volume density of electric current can be written as the sum of the induced volume density of electricity current due to free-charge density and the electric polarization due to bound current density, i.e.,

$$J^{(\mathrm{p})}(\boldsymbol{x},t) = \boldsymbol{J}(\boldsymbol{x},t) + \frac{\partial \boldsymbol{P}(\boldsymbol{x},t)}{\partial t} \ , \tag{3.16}$$

where $\boldsymbol{J}$ is the induced volume density of the electric current due to free-charge density and $\boldsymbol{P}$ is the electric polarization due to bound-current density. A changing state of polarization corresponds to charge movement and so $\partial \boldsymbol{P}/\partial t$ is equivalent to a current. Note that for microscopic Maxwell equations the electric polarization is zero because each particle is treated differently.

In theory, we can also decompose the induced part of volume density of material magnetic current, i.e.,

$$\boldsymbol{K}^{(\mathrm{p})}(\boldsymbol{x},t) = \underbrace{\boldsymbol{K}(\boldsymbol{x},t)}_{=0} + \mu_0 \frac{\partial \boldsymbol{M}(\boldsymbol{x},t)}{\partial t} \ , \tag{3.17}$$

where $\boldsymbol{K}$ is the induced volume density of the magnetic current due to material magnetic monopoles (i.e., isolated magnetic charges), $\boldsymbol{M}$ is the magnetization due to the magnetic dipole, and $\partial \boldsymbol{M}/\partial t$ is the magnetic current density. There are no magnetic monopoles. Unlike electric charges, which can be isolated (i.e., positive and negative electric charges exist separately), the two magnetic poles always come in pairs. When you break the bar magnet, two new bar magnets are obtained, each with a north pole and a south pole. In other words, a magnetic monopole means that there exists an isolated magnet with only one magnetic pole (a north pole without a south pole, or vice versa). No one has ever found an isolated magnetic charge (a magnetic monopole), at least in our universe. So, although in theory, we can also decompose the induced part of the volume density of material magnetic current into two parts, as we did for the induced electric current, only the induced magnetic dipole part matters. Actually, the absence of the magnetic charge is why magnetic-field lines neither begin nor end anywhere (see Ikelle, 2023b). Also, the absence of the magnetic charges is the reason Maxwell's equations in matter will appear non-symmetrical. Note that for microscopic Maxwell equations, magnetism is zero, as each particle is treated differently.

By using the definitions of the induced volume density of material electric and magnetic currents, in (3.16) and (3.17), we obtain this alternative form of Maxwell's equations:

$$-\boldsymbol{\nabla} \times \boldsymbol{H} + \varepsilon_0 \frac{\partial \boldsymbol{E}}{\partial t} + \boldsymbol{J} + \frac{\partial \boldsymbol{P}}{\partial t} = \boldsymbol{0} \tag{3.18}$$

$$\boldsymbol{\nabla} \times \boldsymbol{E} + \mu_0 \frac{\partial \boldsymbol{H}}{\partial t} + \mu_0 \frac{\partial \boldsymbol{M}}{\partial t} = \boldsymbol{0} \ , \tag{3.19}$$

with

$$J^{(\mathrm{p})} = J + \frac{\partial P}{\partial t} \quad \text{and} \quad K^{(\mathrm{p})} = \mu_0 \frac{\partial M}{\partial t} \; . \tag{3.20}$$

Again, the nonsymmetry in these equations is due to the absence of magnetic charges.

Causal microscopic and macroscopic Maxwell's equations. It is customary to introduce electric and magnetic displacement fields in Maxwell's equations, i.e.,

$$-\nabla \times H + J + \frac{\partial D}{\partial t} = 0 \tag{3.21}$$

$$\nabla \times E + \frac{\partial B}{\partial t} = 0 \; , \tag{3.22}$$

where $D = \varepsilon_0 E + P$ and $B = \mu_0 (H + M)$. Note that the definitions of displacement fields do not impose any conditions on the medium and therefore are always valid. Let us now derive the compatibility equations. By differentiating the first and second Maxwell's equations with respect to x and noticing the first terms of these equations are zero because the divergence of the curl of any vector field is always zero (see Ikelle, 2020), we arrive at the following two new relations:

$$\frac{\partial (\nabla \cdot D)}{\partial t} = -\nabla \cdot J \tag{3.23}$$

$$\frac{\partial (\nabla \cdot B)}{\partial t} = 0 \; . \tag{3.24}$$

These relations are known as compatibility equations. We are going to assume that we have a causal action of the source, which has been switched on at $t = 0$ in such a way that the electric and magnetic fields are zero at $t = 0$ and before $t = 0$. We now integrate these compatibility equations with respect to time so that we can eliminate their time derivatives. We thus obtain

$$\nabla \cdot D(x, t) = \int_0^t \left[\frac{\partial [\nabla \cdot D(x, t')]}{\partial t'} \right] dt' = -\int_0^t \left[\nabla \cdot J(x, t') \right] dt' \tag{3.25}$$

$$\nabla \cdot B(x, t) = \int_0^t \left[\frac{\partial [\nabla \cdot B(x, t')]}{\partial t'} \right] dt' = 0 \tag{3.26}$$

because the electric and magnetic displacement fields are zero for $t \leq 0$. By rewriting (3.25) as follows:

$$\nabla \cdot D(x, t) = \rho_E(x, t) \quad \text{Gauss law} \tag{3.27}$$

with

$$\rho_E(x, t) = -\int_0^t \left[\nabla \cdot J(x, t') \right] dt' \tag{3.28}$$

we can now deduce Gauss's law. Quantity $\rho_E(\boldsymbol{x}, t)$ is effectively the volume-charge density, introduced in (3.2) and (3.13). By taking the derivative of (3.28) with respect to time t, we again obtain the continuity equation, i.e.,

$$\frac{\partial \rho_E(\boldsymbol{x}, t)}{\partial t} = -\boldsymbol{\nabla} \cdot \boldsymbol{J}(\boldsymbol{x}, t) \quad \text{Conservation of charge .} \tag{3.29}$$

Again, the continuity equation is the equation of conservation of charges (that is, electrons are not created or destroyed, just moved). It is very interesting to see that the conversation of charges is implicitly contained in Maxwell's equations. Gauss's law for the magnetic displacement field is given in (3.26), i.e.,

$$\boldsymbol{\nabla} \cdot \boldsymbol{B}(\boldsymbol{x}, t) = 0 . \tag{3.30}$$

In summary, whereas the Lorentz force law defines how electric and magnetic fields can be measured and observed, Maxwell's equations, in Table 3.1, explain how these fields can be created directly from charges and currents. Gauss's law for electricity states that the net quantity of electric flux leaving a sample volume is proportional to the charge inside the volume. It shows that the source of static electric fields is electric charges. Gauss's law for magnetism states that it is impossible to produce a magnetic monopole. It shows that the total magnetic flux through a closed surface is zero. Without a magnetic monopole, the only way to generate a magnetic field is by moving an electric field generated by an electrically charged particle. Thus an iron bar magnet, for example, generates its magnetic field by aligning electron motion in the iron atoms. Ampère's law states that currents (moving charges) and electric fields changing over time induce magnetic fields. The law of Faraday states that every time the magnetic field changes, an electric field is created.

In other words, Maxwell's equations also describe how electric and magnetic fields relate to each other. Furthermore, they describe how a time-varying electric field generates a time-varying magnetic field and vice versa. Electric fields arise from volume charge density, according to Gauss's law. Magnetic field lines lack beginnings or ends due to magnetic charge absence. Maxwell's equations describe the electromagnetic field. Maxwell's equations do not directly relate to gravitation, nor describe strong nuclear force or the weak nuclear force. Therefore, we cannot use Maxwell's equations alone to understand planetary motion or atom structure. Note also that Maxwell's equations and the Lorentz force law are independent physical laws. Maxwell's equations tell you what electromagnetic fields are due to a given distribution of charges and currents. The force law tells you the response (motion) of the charges and currents to given electromagnetic fields.

Table 3.1 also shows microscopic Maxwell's equations for matter. Unlike macroscopic equations, microscopic Maxwell's equations do not include the electric polarization, $\boldsymbol{P}$, due to the bound-charge currents and the magnetization, $\boldsymbol{M}$, of the material. These fields are zero because each charge is treated separately on a microscopic scale. Remember that electric polarization and the magnetization of the material are used to capture the very complicated and granular bound charges and granular bound currents by averaging these charges and currents on a sufficiently large scale so as not to see the granularity of individual atoms, but also sufficiently

Table 3.1. Maxwell's equations on the macroscopic and microscopic scales and constitutive equations for the macroscopic scale. The causality of the medium's responses is enforced in these equations: H and E are zeros before $t = 0$. For dielectric media, ε_0 and μ_0 of free space are replaced with ε and μ.

Macroscopic scale	Microscopic scale	Law name
intensity fields		
$\nabla \times H = \varepsilon_0 \frac{\partial E}{\partial t} + J + \frac{\partial P}{\partial t}$	$\nabla \times H = \varepsilon_0 \frac{\partial E}{\partial t} + J$	Ampère
$\nabla \times E = -\mu_0 \frac{\partial H}{\partial t} - \mu_0 \frac{\partial M}{\partial t}$	$\nabla \times E = -\mu_0 \frac{\partial H}{\partial t}$	Faraday
$\nabla \cdot (\varepsilon_0 E + P) = \rho_E$	$\nabla \cdot E = \rho_{E,f}/\varepsilon_0$	Gauss (electric)
$\mu_0 \nabla \cdot (H + M) = 0$	$\nabla \cdot H = 0$	Gauss (magnetic)
and displacement fields		
$\nabla \times H = \frac{\partial D}{\partial t} + J$	$\nabla \times B = \mu_0 \varepsilon_0 \frac{\partial E}{\partial t} + \mu_0 J$	Ampère
$\nabla \times E = -\frac{\partial B}{\partial t}$	$\nabla \times E = -\frac{\partial B}{\partial t}$	Faraday
$\nabla \cdot D = \rho_E$	$\nabla \cdot E = \rho_{E,f}/\varepsilon_0$	Gauss (electric)
$\nabla \cdot B = 0$	$\nabla \cdot B = 0$	Gauss (magnetic)
Constitutive equations		
$D = \varepsilon_0 E + P = \varepsilon E$		
$B = \mu_0 H + \mu_0 M = \mu H$		
$J = \sigma E$		Ohm
$P = \varepsilon_0 \chi^{(e)} E$		
$M = \chi^{(m)} H$		
Maxwell and constitutive		
$\nabla \times H = \varepsilon \frac{\partial E}{\partial t} + \sigma E$		Ampère
$\nabla \times E = -\mu \frac{\partial H}{\partial t}$		Faraday
$\nabla \cdot (\varepsilon E) = \rho_E$		Gauss (electric)
$\nabla \cdot (\mu H) = 0$		Gauss (magnetic)

small that they vary with location in the material. In other words, polarization, P, and magnetization, M, are the distinguishing features of macroscopic Maxwell's equations. Also, one can define electric charge density in macroscopic Gauss's law for electric displacement as the sum of the electric charge density of free electrons, $\rho_{E,f}$, and the electric charge density of bound electrons, $\rho_{E,b}$; i.e.,

$$\nabla \cdot D = \rho_E \iff \varepsilon_0 \nabla \cdot E + \nabla \cdot P = \rho_{E,f} + \rho_{E,b} , \qquad (3.31)$$

with $\varepsilon_0 \nabla \cdot E = \rho_{E,f}$ and $\nabla \cdot P = \rho_{E,b}$. Because each particle is treated separately in microscopic scale, the electric charge density of the bound electrons, $\rho_{E,b}$, associated with the electric polarization is ignored at this scale as P is zero. So, in microscopic scale, Gauss's law for the electric field is

$$\nabla \cdot E = \frac{\rho_{E,f}}{\varepsilon_0} . \qquad (3.32)$$

Table 3.2 lists some of the fundamental EM quantities associated with Maxwell's equations along with their meanings and units. We have also introduced a significant number of new measurements so far. Like all measurements, they are comparisons with standards that are based on a unit. Table 3.3 is a summary of the key units used in this book series.

Table 3.2. A list of some of the fundamental EM quantities associated with Maxwell's equations, along with their meanings and units.

Name	Symbol	Unit (symbol)
Volume density of electric current	$\boldsymbol{J}\,(J_i)$	Ampère/meter^2 (A/m^2)
Electric polarization	$\boldsymbol{P}\,(P_j)$	Coulomb/meter^2 (C/m^2)
Magnetization	$\boldsymbol{M}\,(M_j)$	Ampère/meter (A/m)
Electrical conductivity	σ	siemens/meter (S/m)
(Absolute) permittivity	ε	farad/meter (F/m)
Relative permittivity	ε_r	dimensionless
(Absolute) permeability	μ	henry/meter (H/m)
Relative permeability	μ_r	dimensionless
Electric susceptibility	$\chi^{(e)}$	dimensionless
Magnetic susceptibility	$\chi^{(m)}$	dimensionless

Non-causal microscopic and macroscopic Maxwell's equations. We arrive at the Gauss laws in (3.27) and (3.30) by integrating compatibility equations in (3.23) and (3.24) under the assumption that we have a causal action of the source, which has been switched on at $t = 0$ in such a way that the electric and magnetic fields are zero at $t = 0$ and before $t = 0$. Let us now consider the non-causal action of the source. These integrations can be written as

$$
\boldsymbol{\nabla} \cdot \boldsymbol{D}(\boldsymbol{x},t) = \int_{-\infty}^{t} \left[\frac{\partial\,[\boldsymbol{\nabla} \cdot \boldsymbol{D}(\boldsymbol{x},t')]}{\partial t'} \right] dt'
$$

$$
= -\int_{0}^{t} \left[\boldsymbol{\nabla} \cdot \boldsymbol{J}(\boldsymbol{x},t')\right] dt' - \int_{-\infty}^{0} \left[\boldsymbol{\nabla} \cdot \boldsymbol{J}(\boldsymbol{x},t')\right] dt' , \qquad (3.33)
$$

$$
\boldsymbol{\nabla} \cdot \boldsymbol{B}(\boldsymbol{x},t) = \int_{0}^{t} \left[\frac{\partial\,[\boldsymbol{\nabla} \cdot \boldsymbol{B}(\boldsymbol{x},t')]}{\partial t'} \right] dt' + \int_{-\infty}^{0} \left[\frac{\partial\,[\boldsymbol{\nabla} \cdot \boldsymbol{B}(\boldsymbol{x},t')]}{\partial t'} \right] dt' . \qquad (3.34)
$$

By rewriting (3.33) as follows:

$$
\boldsymbol{\nabla} \cdot \boldsymbol{D}(\boldsymbol{x},t) = \rho_E(\boldsymbol{x},t) + g(\boldsymbol{x}) \qquad \text{Gauss law} \qquad (3.35)
$$

Table 3.3. Some of the key units used in this book.

Name	Abbreviation	Name	Abbreviation
hertz	Hz $= 1/$s	ohm	Ω
newton	N $= (\mathrm{kg\,m})/\mathrm{s}^2$	siemens	S $= 1/\Omega$
pascal	Pa $= \mathrm{N/m}^2 = \mathrm{kg}/(\mathrm{m\,s}^2)$	weber	Wb $= \mathrm{V\,s}$
joule	J $= \mathrm{N\,m} = (\mathrm{kg\,m}^2)/\mathrm{s}^2$	henry	H
tesla	T $= \mathrm{Wb/m}^2 = \mathrm{kg}/(\mathrm{C\,s})$	nanotesla	nT $= 10^{-9}$T
gauss	Gauss $= 10^{-4}$Tesla $= 10^5$T	coulomb	C
watt	W $= (\mathrm{kg\,m}^2)/\mathrm{s}^3$	volt	V
farad	F		

with

$$\rho_E(\boldsymbol{x}, t) = -\int_0^t \left[\boldsymbol{\nabla} \cdot \boldsymbol{J}(\boldsymbol{x}, t')\right] dt' \quad \text{and} \quad g(\boldsymbol{x}) = -\int_{-\infty}^0 \left[\boldsymbol{\nabla} \cdot \boldsymbol{J}(\boldsymbol{x}, t')\right] dt', \quad (3.36)$$

we can now deduce the non-causal version of Gauss's law. By taking the derivative of (3.36) with respect to time t, we can see that the continuity equation in (3.29) is unchanged. In other words, the conservation of charges is unchanged by non-causal source actions.

The non-causal Gauss's law for the magnetic displacement field is given in (3.34), i.e.,

$$\boldsymbol{\nabla} \cdot \boldsymbol{B}(\boldsymbol{x}, t) = f(\boldsymbol{x}) , \qquad (3.37)$$

where

$$f(\boldsymbol{x}) = \int_{-\infty}^0 \left[\frac{\partial \left[\boldsymbol{\nabla} \cdot \boldsymbol{B}(\boldsymbol{x}, t')\right]}{\partial t'}\right] dt' . \qquad (3.38)$$

Basically, in non-causal Maxwell equations, Gauss's law for magnetism can produce a time-independent magnetic monopole because the total magnetic flux through a closed surface can be nonzero. We can generate the magnetic field using a magnetic monopole and by moving an electric field generated by an electricity charged particle.

Non-Causal Maxwell equations vs Maxwell's equations with magnetic monopoles. The idea of magnetic monopoles was first put forward by Paul Dirac in 1931. He came up with a hypothetical elementary particle that is an isolated magnet with only one magnetic pole (a north pole without a south pole or vice versa). Assuming that $\boldsymbol{P} = \boldsymbol{M} = \boldsymbol{0}$, here is a comparative description of non-causal Maxwell's equations and Maxwell's equations with magnetic monopoles.

Causal + magnetic monopoles	Non-Causal only
$\boldsymbol{\nabla} \times \boldsymbol{B}(\boldsymbol{x}, t) = \mu_0 \left(\varepsilon_0 \dfrac{\partial \boldsymbol{E}(\boldsymbol{x}, t)}{\partial t} + \boldsymbol{J}(\boldsymbol{x}, t)\right)$	Unchanged
$\boldsymbol{\nabla} \times \boldsymbol{E}(\boldsymbol{x}, t) = -\dfrac{\partial \boldsymbol{B}(\boldsymbol{x}, t)}{\partial t}$	Unchanged
$\boldsymbol{\nabla} \cdot \boldsymbol{E}(\boldsymbol{x}, t) = \dfrac{\rho_E(\boldsymbol{x}, t)}{\varepsilon_0}$	$\boldsymbol{\nabla} \cdot \boldsymbol{E}(\boldsymbol{x}, t) = \dfrac{\rho_E(\boldsymbol{x}, t)}{\varepsilon_0} + g(\boldsymbol{x})$
$\boldsymbol{\nabla} \cdot \boldsymbol{B}(\boldsymbol{x}, t) = \mu_0 \varepsilon(\boldsymbol{x}, t)$	$\boldsymbol{\nabla} \cdot \boldsymbol{B}(\boldsymbol{x}, t) = f(\boldsymbol{x})$

As we described earlier, magnetic monopoles are still to be discovered or designed. Similarly, white holes have never been observed even though they are Einstein's field equations' solutions. Magnetic monopoles [i.e., $\nabla \cdot B(x, t) \neq 0$] would be a source or sink of the field, where it terminates or emanates from, just as the electric field does for electric charges.

Maxwell's equation system is asymmetric when $\nabla \cdot B(x, t) = 0$ because only electric fields can be created from electrically charged particles. That is not true of magnetic fields. We have to move an electric particle around to produce a magnetic field, as described in (3.1). In other words, Maxwell's equation system is asymmetric because the two Gauss laws do not contain analogous source terms. There is no fundamental law or physical principle that makes these two laws so different. It is simply the fact that we have yet to discover magnetic monopoles. Maybe the magnetic monopole is very large and can only be seen galactically. Perhaps magnetic monopole particles are responsible for cosmic ray events that are still incompletely understood. It is remarkable that non-causal Maxwell's equations have the same structure as Maxwell's equations with magnetic monopoles. It is very difficult to construct a system without causality. The common approach is to include memories in descriptions of permittivity, electromagnetic permeability, and/or conductivity, as we describe in Ikelle (2024).

Exchange of energy in the electromagnetic fields and Poynting vector.
Electromagnetic waves contain energy. The obvious example is summertime. For example, at some locations on the Earth's surface, the Sun's radiation has an intensity of about 1360 Watts/m^2. Energy comes from electromagnetic waves at many frequencies. The total intensity is just the sum of the individual waves' intensities. Using light speed, we can estimate that this energy travels for about eight minutes before reaching Earth. Another less obvious example of energy in electromagnetic waves is the energy of gamma rays. These rays can destroy living cells and/or interact with rocks to reveal rock properties. Electromagnetic waves can bring energy into a system (e.g., rock formation) through their electric and magnetic fields. These fields can exert forces and move charges in the system and, thus, work on them. Our objective here is to answer the following question: How much energy is carried by an electromagnetic wave of a given amplitude? From the answer to this question, the Poynting[1] vector is the quantity that characterizes the area density of power flow in electromagnetic fields. We will address this question by calculating the electromagnetic energy equation.

First, we take the scalar product of the electric field and Ampère's law and the scalar product of the magnetic field and the Faraday's law; i.e.,

$$E \cdot \left[-\nabla \times H + \varepsilon_0 \frac{\partial E}{\partial t} + J + \frac{\partial P}{\partial l} \right] = 0 \tag{3.39}$$

and

$$H \cdot \left[\nabla \times E + \mu_0 \frac{\partial H}{\partial t} + \mu_0 \frac{\partial M}{\partial t} \right] = 0 . \tag{3.40}$$

By summing these two equations and using the following vector identity

$$\nabla \cdot (a \times b) = b \cdot (\nabla \times a) - a \cdot (\nabla \times b) , \tag{3.41}$$

---

[1] John Henry Poynting is one of Maxwell's former students. He derived the equation to describe how much electromagnetic energy passes through an area per second.

we arrive at the local form of the EM energy equation in vacuum, which is known as Poynting's theorem,

$$\boldsymbol{\nabla} \cdot (\boldsymbol{E} \times \boldsymbol{H}) + \frac{1}{2}\frac{\partial}{\partial t}\left[\varepsilon_0 E^2 + \mu_0 H^2\right] + \boldsymbol{E} \cdot \left(\boldsymbol{J} + \frac{\partial \boldsymbol{P}}{\partial t}\right) + \mu_0 \boldsymbol{H} \cdot \frac{\partial \boldsymbol{M}}{\partial t} = 0$$

or

$$\boldsymbol{\nabla} \cdot \boldsymbol{S} + \frac{\partial W}{\partial t} = -\boldsymbol{E} \cdot \left(\boldsymbol{J} + \frac{\partial \boldsymbol{P}}{\partial t}\right) - \mu_0 \boldsymbol{H} \cdot \frac{\partial \boldsymbol{M}}{\partial t} = 0 \tag{3.42}$$

with

$$\boldsymbol{S} = \boldsymbol{E} \times \boldsymbol{H} \quad \text{and} \quad W = \frac{1}{2}\left[\varepsilon_0 E^2 + \mu_0 H^2\right] , \tag{3.43}$$

where W is the energy density of the EM field and $\boldsymbol{S}$ is Poynting's vector (the area density of electromagnetic power flow). Poynting's vector is energy transferred per unit time per unit cross-sectional area (or power unit area). So in vacuum (i.e., $\boldsymbol{J} = \boldsymbol{0}$, $\boldsymbol{P} = \boldsymbol{0}$, and $\boldsymbol{M} = \boldsymbol{0}$) electric fields and magnetic fields carry half of the total energy.

Equation (3.42) is Poynting's theorem. It does not immediately tell us much about energy in an electromagnetic field; but the physical interpretation becomes clearer if we convert it from differential form into integral form. Consider a given volume D. We integrate the equation of Poynting's theorem in (3.42) by using the divergence theorem to obtain the global form of the EM energy equation; i.e.,

$$\int_{\partial D} \boldsymbol{n} \cdot (\boldsymbol{E} \times \boldsymbol{H})\, dS(\boldsymbol{x}) = \frac{\partial}{\partial t}\left[\int_D \left\{\left(\frac{1}{2}\varepsilon_0 E^2\right) + \left(\frac{1}{2}\mu_0 H^2\right)\right\} dV(\boldsymbol{x})\right]$$
$$+ \int_D \boldsymbol{E} \cdot \left(\boldsymbol{J} + \frac{\partial \boldsymbol{P}}{\partial t}\right) dV(\boldsymbol{x})$$
$$+ \mu_0 \int_D \boldsymbol{H} \cdot \frac{\partial \boldsymbol{M}}{\partial t} dV(\boldsymbol{x}) . \tag{3.44}$$

Notice that we define $\boldsymbol{n}$ here to be an inward-pointing vector so that we can switch the order of $\boldsymbol{E}$ and $\boldsymbol{H}$. The term in the bracket in the right-hand side of this equation is the energy contained in volume D. The first term of this bracket is the energy stored in the electric field and the second term is the energy stored in the magnetic field. In the vacuum, the energy density of the EM field in the vacuum is

$$W = \frac{1}{2}\left[\varepsilon_0 E^2 + \mu_0 H^2\right] = \varepsilon_0 E^2 \tag{3.45}$$

because $B = \mu_0 H = E/c_0$ in the free space. Let us consider the particular case of a sinusoidal traveling wave of the form

$$\boldsymbol{E}(\boldsymbol{x}, t) = \boldsymbol{i}_1 E_0 \cos(kz - \omega t) \quad \text{and} \quad \boldsymbol{B}(\boldsymbol{x}, t) = \boldsymbol{i}_2 \frac{E_0}{c_0}\cos(kz - \omega t) . \tag{3.46}$$

The energy density and Poynting's vector of an electromagnetic field are

$$W = \frac{1}{2}\left[\varepsilon_0 E^2 + \mu_0 H^2\right] = \varepsilon_0 E^2 = \varepsilon_0 E_0^2 \cos^2(kz - \omega t) \tag{3.47}$$

$$S = E \times H = c_0 \varepsilon_0 E^2 k = c_0 W k = c_0 \varepsilon_0 E_0^2 \cos^2(kz - \omega t) k \qquad (3.48)$$

with $k = i_1 \times i_2 = i_3$. Since the average value of $\cos^2(kz - \omega t)$ over one period is $1/2$, the average values of the energy density and Pointing's vector are

$$W_{\text{avg}} = \frac{1}{2} \varepsilon_0 E_0^2 \quad \text{and} \quad |S|_{\text{avg}} = \frac{1}{2} c_0 \varepsilon_0 E_0^2 . \qquad (3.49)$$

$|S|_{\text{avg}}$ is known as the wave intensity. It is the average amount of energy per unit area per unit time that passes through (or hits) a surface.

Exercise: A source in the subsurface emits a cosine wave with an average power of 5×10^{-3} watts. (i) Determine the cosine wave intensity. (ii) Determine E_0 and B_0. We assume that the source radiates equally in all directions in a 31.623 m radius.

It is worthwhile to emphasize that ions and/or electrons themselves do not really travel all that much. Instead, what really travels is the Poynting vector which captures the interaction between ions and/or electrons. Imagine a lattice in a vacuum at rest made of magnets all located in such a way to repel each other. The interaction between them is such that they don't move. Now if we displace one of the magnets, you'll see the displacement spread. However, magnets themselves barely move, but the interaction spreads at light speed. That is the way Poynting vector flows through conductors, allowing electricity to find its way into our homes and buildings.

3.1.4. Linear Constitutive Equations

We have six unknown fields (ρ_E, P, M, J, H, and E), including five vectorial quantities, but only four equations (Ampere's law, Faraday's law, and two Gauss' laws). Therefore, the macroscopic Maxwell's equation system is underdetermined. From a physical point of view, the electromagnetic field must be uniquely determined once the sources that generate the fields and the distribution of matter are given. Maxwell's equations for matter remain incomplete without another set of relations. These relations are called constitutive relations. Our next task is to describe these constitutive relations, which describe material behavior (i.e., matter distribution). They relate EM field intensities (primary fields) and EM displacement fields (secondary fields), as illustrated in Figure 3.4.

A second reason these relationships are between macroscopic quantities is that they are derived from experimental observations of material properties. In other words, they can be established without knowing the relations between local microscopic fields.

In constitutive equations, the primary fields, E and H, can be thought of as inputs to the secondary fields, D, B, and J (or P, M, and J) as outputs or predicted responses. In most general cases, constitutive equations are arbitrary relationships between the primary and secondary fields; i.e.,

$$J = J(E, H) , \quad P = P(E, H) , \quad M = M(E, H) \qquad (3.50)$$

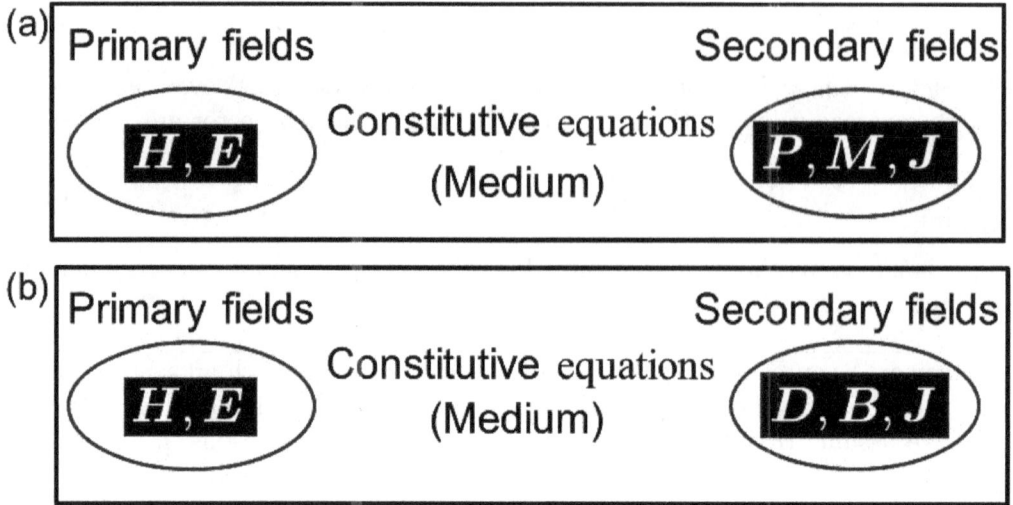

Figure 3.4. (a) The constitutive relationships are the relationships between the primary fields, E and H, and the secondary fields, P, M, and J. (b) Alternatively, the constitutive relationships are the relationships between the primary fields, E and H, and the secondary fields, D, B, and J. These relationships depend on the matter (medium) in which the fields exist. There are no constitutive equations for the microscopic scale.

or

$$J = J(E, H) , \quad D = D(E, H) , \quad B = B(E, H) . \tag{3.51}$$

If these relationships are linear, the medium is considered linear in its electromagnetic behavior. Otherwise, the medium is considered nonlinear in its electromagnetic behavior. Nonlinear materials are desirable in some applications such as light phase modulation. In other applications, however, they are undesirable. If the medium parameters that describe the relationships between the primary and secondary fields are invariant with orientation (that is, they are scalar), at a given point in space, the medium is considered isotropic at this point. Otherwise, the medium is anisotropic. In anisotropic materials, properties depend on the x-, y-, z-direction and constitutive relations may be written component-wise in matrix (or tensor) form. A wide class of materials have linear time-independence (i.e. properties of materials do not change with time), and isotropic behavior in the presence of weak EM fields (E and H fields are sufficiently small). Such materials include crystalline minerals. Crystalline minerals are made up of randomly oriented microscopic crystals and behave as isotropic media on a macroscopic scale. We will now focus on time-independent linear isotropic media.

Electrical conductivity. Table 3.4 shows parameters introduced by the discoveries of the linear relationships between primary and secondary fields. Equation $J(x, t) = \sigma(x)E(x, t)$, which is known as Ohm's law, relates the current density, J, to the electric field, E, by the electrical conductivity, σ, which is a material's ability to pass electric current. The unit of conductivity is Siemens per meter (S/m).

Electrical conductivity varies between materials by more than 27 orders of magnitude, the greatest variation in any physical property. The reciprocal of electrical

Table 3.4. Parameters introduced by the discoveries of the linear relationships between primary and secondary fields in the case of time-independent linear isotropic media. In isotropic media, all these parameters are scalar (i.e., tensors of rank zero).

Primary	Secondary	Parameter
$\boldsymbol{E}(\boldsymbol{x},t)$	$\boldsymbol{J}(\boldsymbol{x},t) = \sigma(\boldsymbol{x})\boldsymbol{E}(\boldsymbol{x},t)$	σ: conductivity
$\boldsymbol{E}(\boldsymbol{x},t)$	$\boldsymbol{P}(\boldsymbol{x},t) = \varepsilon_0\chi^{(e)}(\boldsymbol{x})\boldsymbol{E}(\boldsymbol{x},t)$	$\chi^{(e)}$: Electric susceptibility
$\boldsymbol{E}(\boldsymbol{x},t)$	$\boldsymbol{D}(\boldsymbol{x},t) = \varepsilon(\boldsymbol{x})\boldsymbol{E}(\boldsymbol{x},t)$	ε: Permittivity
	$= \varepsilon_0\varepsilon_r(\boldsymbol{x})\boldsymbol{E}(\boldsymbol{x},t)$	ε_r: Relative permittivity
$\boldsymbol{H}(\boldsymbol{x},t)$	$\boldsymbol{M}(\boldsymbol{x},t) = \chi^{(m)}(\boldsymbol{x})\boldsymbol{H}(\boldsymbol{x},t)$	$\chi^{(m)}$: Magnetic susceptibility
$\boldsymbol{H}(\boldsymbol{x},t)$	$\boldsymbol{B}(\boldsymbol{x},t) = \mu(\boldsymbol{x})\boldsymbol{H}(\boldsymbol{x},t)$	μ: Magnetic permeability
	$= \mu_0\mu_r(\boldsymbol{x})\boldsymbol{H}(\boldsymbol{x},t)$	μ_r: Relative magnetic permeability

conductivity is electric resistivity; i.e.,

$$\varrho(\boldsymbol{x}) = \frac{1}{\sigma(\boldsymbol{x})} \; . \tag{3.52}$$

It is a measure of how strongly a given material opposes the flow of an electric current. The unit of electrical resistivity is the ohm-meter, i.e.,

$$\frac{\text{Siemens}}{\text{meter}} = \frac{1}{\text{ohm} \times \text{meter}} \; . \tag{3.53}$$

As shown in Figure 3.5, using conductivity values we can classify materials into electric conductors, electric insulators, and semiconductors. A conductor is an object or material that allows the movement of electric charges freely. An insulator is an object or type of material that strongly resists the flow of electric charges, meaning it does not conduct charges. A semiconductor is an object or type of material whose electrical conductivity is between that of a conductor and an insulator. Most metals are good electrical conductors. The most utilized materials in the construction of electrical equipment are copper and aluminum because of their great conductivity. Also, alloys of copper with other elements are used. The most utilized materials as insulators in the construction of electrical equipment are rubber, glass, plastic, and wood because of their very poor conductivity. Semiconductors form the backbone of our technological society. What makes semiconductors so important is their unique electrical behavior. They exhibit extreme sensitivity (exponential dependence) to the presence of minute quantities of solutes, which are known as dopants, thus giving us the possibility of controlling the electrical conductivity of a semiconductor by doping. All microchip processors are based on applications of semiconductors such as silicon (Si) and germanium (Ge). In addition, many devices used as sensors are also based on these semiconductor materials.

The electrical conductivity of rocks mostly depends on the porosity, temperature, and salinity of the water and water-rock interaction. Remember, porosity is the ratio between the pore volume and the total volume of a material. The saturated fluid is the dominant electrical conductor in the rock, and the degree of saturation

Table 3.5. Conductivities of some rocks, metals, and fluids, their relative permittivities at 1 MHz (Keller and Frischknecht, 1966, and Daniels and Alberty, 1966), and their magnetic susceptibilities. We added the conductivity, relative permittivity, and susceptibility of vacuum as reference values. The electric conductivity of air is almost zero and zero in a vacuum (i.e., free space). Salinity (i.e., the amount of salt dissolved in the water) is proportional to conductivity. Groundwater here is assumed to be fresh water.

Material	σ (S/m)	$\varepsilon/\varepsilon_0$	$\chi^{(m)}$
Vacuum	0	1	0
Igneous and metamorphic			
Basalt	$10^{-6} - 10^{-3}$	12	$300 - 182000$
Granite	$10^{-6} - 2 \times 10^{-4}$	$4 - 7$	$10 - 50000$
Gabbro	$10^{-7} - 10^{-3}$		$800 - 76000$
Peridotite	$10^{-5} - 10^{-2}$		$95500 - 196000$
Marble	$4 \times 10^{-9} - 10^{-2}$	$1 - 8$	$30 - 25000$
Slate	$2.5 \times 10^{-8} - 1.7 \times 10^{-3}$		$0 - 38000$
Gneiss	$10^{-6} - 10^{-2}$		$250 - 25000$
Sedimentary			
Dolomite			$-12.5 - 20000$
Limestone	$2.5 \times 10^{-3} - 0.02$	$4 - 8$	$10 - 25000$
Sandstone	$2.5 \times 10^{-4} - 0.125$	$2 - 10$	$0 - 21000$
Shale	$5 \times 10^{-4} - 0.05$	$5 - 15$	$60 - 18600$
Soils and fluids			
Oil	$3 \times 10^{-12} - 13 \times 10^{-12}$	2	
Coal	$0.5 \times 10^{-3} - 2 \times 10^{-3}$	$3.5 - 8$	$0 - 1000$
Salt	$1 \times 10^{-5} - 1 \times 10^{-3}$	$3 - 15$	-10
Clays	$0.01 - 1$	$5 - 40$	$10 - 500$
Groundwater	$0.01 - 1$	80	0
Sea water	4	0.2	0
Air	$3 \times 10^{-15} - 8 \times 10^{-15}$	1	0
Plasma	$10^3 - 10^6$		
Metals at 20° C			
Silver	6.78×10^7		-2.6×10^{-5}
Cooper	5.96×10^7		-0.98×10^{-5}
Gold	4.50×10^7		0
Aluminum	3.80×10^7		2.2×10^{-5}
Sodium	2.16×10^7		0.72×10^{-5}
Tungsten	2.00×10^7		6.8×10^{-5}
Nickel	1.43×10^7		99

determines the bulk of the conductivity. All dried constituents of reservoir rocks have a high electrical resistivity (they are electrical insulators). Table 3.5 gives us conductivity values of common rocks, fluids, and metals (Keller and Frischknecht 1966, and Daniels and Alberty 1966). Igneous and metamorphic rocks typically have low conductivity values. Sedimentary rocks, which are usually more porous and have higher water content, normally have higher conductivity values than igneous and metamorphic rocks. Hydrocarbons, dry rock, and fresh water are highly resistive (that is, they are nonconductive).

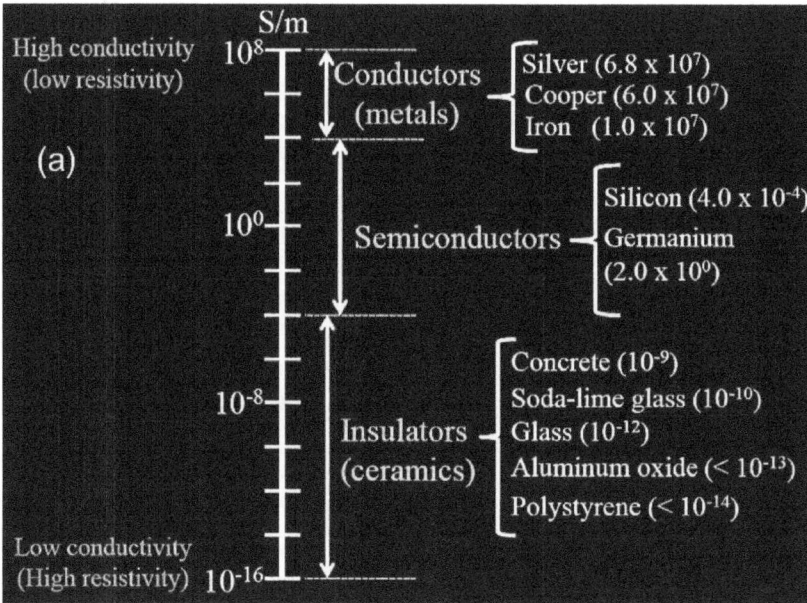

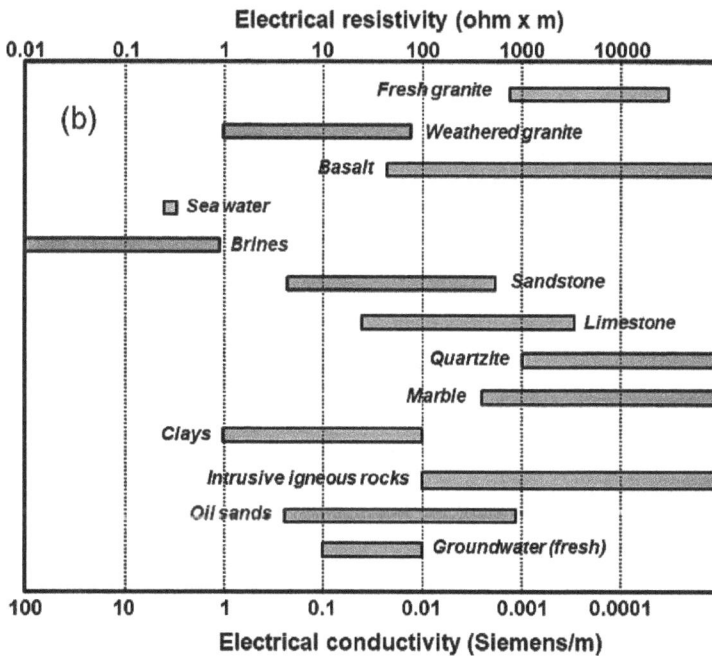

Figure 3.5. (a) An illustration of variations in electrical conductivity. It varies between materials by more than 27 orders of magnitude, the greatest variation in any physical property. The values of the conductivity allow us to group materials into electric conductors, electric insulators, and semiconductors. (b) Resistivities of some common rocks and minerals. These values are taken from Keller and Frischknecht (1966) and Daniels and Alberty (1966). Notice that the reciprocal of resistivity is conductivity.

Salt water (brine) is very conductive. Thus porous formation has electrical conductivity, which depends upon the nature of the fluids filling the pore space, because the rock matrix is non-conducting, and the usual saturating fluid is conductive brine. Therefore, contrasts of conductivity are produced when the brine is replaced with nonconductive hydrocarbons. Note that in a vacuum (that is, free space), the conductivity sigma is zero. Actually, one of the key differences between matter and a vacuum is that matter has nonzero electrical conductivity.

As we described earlier, electricity flow is mostly about the interaction between ions and/or electrons. Pure water does not conduct electricity because it does not have free electrons available and therefore there is nothing to interact with. What makes tap water conductors are the dissolved salts in it. Salts do not create free electrons, but ions. Sodium is an effective electricity conductor because it has a single electron in its valence shell. This results in an unstable atomic structure, giving away electrons. In more general terms, metal conductivity depends on the number of free electrons available.

Permittivity. The other parameters in Tables 3.1 and 3.4 are the electric and magnetic susceptibilities, permittivity, and magnetic permeability. The electric susceptibility, $\chi^{(e)}$, relates to the electric field, E, with the electric polarization, P, and the magnetic susceptibility, $\chi^{(m)}$, relates to the magnetic field, H, with the magnetization, M. Both susceptibilities are dimensionless.

Electric susceptibility is a measure of how easily a material polarizes in response to an electric field. Permittivity or absolute permittivity, ε, is the measure of resistance that is encountered when forming an electric field in a medium. Thus, permittivity relates to a material's ability to resist an electric field. The use of the word *permittivity*, in this case, is unfortunate because the word stems from *permit* which suggests the inverse quantity. So, the convention is permittivity. In Table 3.4, we also introduced the relative permittivity of an electric material, ε_r, which is the ratio of the permittivity of the electric material over the permittivity of free space (vacuum).

Permittivity is directly related to electric susceptibility, as we can see from the following derivations:

$$D_i(\boldsymbol{x}, t) = \varepsilon_0 E_i(\boldsymbol{x}, t) + P_i(\boldsymbol{x}, t)$$

$$= \varepsilon_0 \left[1 + \chi^{(e)}(\boldsymbol{x}) \right] E_i(\boldsymbol{x}, t)$$

$$= \varepsilon(\boldsymbol{x}) E_i(\boldsymbol{x}, t) = \varepsilon_0 \varepsilon_r(\boldsymbol{x}) E_i(\boldsymbol{x}, t) . \tag{3.54}$$

So we have

$$\varepsilon(\boldsymbol{x}) = \varepsilon_0 \left[1 + \chi^{(e)}(\boldsymbol{x}) \right] \quad \text{or} \quad \varepsilon_r(\boldsymbol{x}) = \frac{\varepsilon(\boldsymbol{x})}{\varepsilon_0} = 1 + \chi^{(e)}(\boldsymbol{x}) . \tag{3.55}$$

In other words, permittivity eliminates the need for electric susceptibility, or vice versa. Table 3.5 shows the relative permittivity of some common materials.

Magnetic materials. Magnetic permeability, μ, is the measure of the ability of a material to support the formation of a magnetic field within itself. Hence, it is the degree of magnetization that a material obtains in response to an applied magnetic field. The reciprocal of magnetic permeability is magnetic reluctivity. In

Table 3.4, we also introduced the relative permeability of a magnetic material, which is the ratio of the permeability of the magnetic material over the permeability of free space (a vacuum). A material with a relative permeability of about 1 (i.e., $\mu_r \approx 1$) is considered a non-magnetic material. Non-magnetic materials include air, aluminum, gold, and copper.

Magnetic permeability is directly related to magnetic susceptibility, as we can see from the following derivations:

$$\boldsymbol{B}(\boldsymbol{x},t) = \mu_0 \left[\boldsymbol{H}(\boldsymbol{x},t) + \boldsymbol{M}(\boldsymbol{x},t) \right]$$

$$= \mu_0 \left[1 + \chi^{(m)}(\boldsymbol{x}) \right] \boldsymbol{H}(\boldsymbol{x},t)$$

$$= \mu(\boldsymbol{x}) \boldsymbol{H}(\boldsymbol{x},t) = \mu_0 \mu_r(\boldsymbol{x}) \boldsymbol{H}(\boldsymbol{x},t) \ . \tag{3.56}$$

So we have

$$\mu(\boldsymbol{x}) = \mu_0 \left[1 + \chi^{(m)}(\boldsymbol{x}) \right] \quad \text{or} \quad \mu_r(\boldsymbol{x}) = \frac{\mu(\boldsymbol{x})}{\mu_0} = 1 + \chi^{(m)}(\boldsymbol{x}) \ . \tag{3.57}$$

In other words, permeability eliminates the need for magnetic susceptibility. Table 3.5 shows the magnetic susceptibility of rocks and common materials. Rocks owe their magnetic character to the generally small proportion of magnetic minerals, namely, magnetite (Fe_3O_4) and hematite (Fe_2O_3). Igneous rocks are usually highly magnetic due to their relatively high magnetite content. The proportion of magnetite in igneous rocks tends to decrease with increasing acidity.

In summary, another great success of Maxwell's equations lies in their simple characterization of materials at a macroscopic scale in terms of conductivity, permittivity, and permeability. Knowing these quantities tells us all we need to know about the materials under consideration at the macroscopic scale. Moreover, the values of these properties allow us to group material in electric conductors, electric insulators, and semiconductors, and as we see later, also in diamagnetic materials, paramagnetic materials, and ferromagnetic materials. In vacuum (that is, free space) the conductivity is zero ($\sigma = 0$ S/m), the permittivity is $\varepsilon = \varepsilon_0 = 8.8542 \times 10^{-12}$ F/m, and the permeability is $\mu = \mu_0 = 4\pi \times 10^{-7}$ H/m.

Quiz: List two examples of conductors, two examples of insulators, and two examples of semiconductors.

Magnetic properties of minerals and rocks. All materials can also be classified into three categories based on their bulk magnetic susceptibility as diamagnetic, paramagnetic, or ferromagnetic (e.g., Blakely, 1995), as shown in Table 3.6.

- Diamagnetic materials have small and negative susceptibility (i.e., $\chi^{(m)}$ is very small and negative). Magnetic susceptibility for most known diamagnetic materials is between -10^{-7} and -10^{-5}. Diamagnetic materials include many metals (e.g., aluminum, bismuth, copper, lead, germanium, silver, gold, diamond, and mercury), non-metallic elements (e.g., boron, silicon, phosphorus, and sulfur), graphite, quartz, feldspar, marble, salt (NaCl), water (H_2O), and gas (H_2).

Table 3.6. Magnetic permeability of diamagnetic, paramagnetic, and ferromagnetic materials.

Materials	Magnetic susceptibily	Magnetic permeability
Diamagnetic	$\left[-10^{-5}, -10^{-7}\right]$	$\mu < \mu_0$ or $\mu_r < 1$
Paramagnetic	$\left[10^{-5}, 10^{-3}\right]$	$\mu > \mu_0$ or $\mu_r > 1$
Ferromagnetic	$\chi^{(m)} \gg 1$	$\mu \gg \mu_0$ or $\mu_r \gg 1$

- Paramagnetic materials have small and positive susceptibility (i.e., $\chi^{(m)}$ is very small and positive). Paramagnetic materials generally have magnetic susceptibility, of the order of 10^{-5} for aluminum, barium, calcium, magnesium, platinum, sodium, titanium, tungsten, uranium, liquid (O_2), and gas (O_2).

- Ferromagnetic materials have large and positive susceptibility (i.e., $\chi^{(m)}$ is large and positive). Ferromagnetic materials include iron, nickel, and cobalt, and compounds containing these elements. Note that the ferromagnetic materials here include ferrimagnetic materials, which also have large and positive susceptibility and their susceptibilities are smaller than those of ferromagnetic materials. Ferrites are the most useful ferrimagnetic materials.

This grouping is important for the interpretation of anomalies of geomagnetic fields associated with geophysical properties in the Earth's crust and for the study of changes of the geomagnetic field over time, as we will discuss in the next chapter.

Note that ferromagnetic materials have a permanent magnetization (also known as remanent magnetization). In fact, ferromagnetism is the strongest form of magnetism. It is responsible for most of the magnetic behavior encountered in everyday life, and is the basis for all permanent magnets, including the metals that are attracted to them. They are used to enhance the magnetic flux density ($\boldsymbol{B}$) produced when an electric current is passed through the material. Applications of ferromagnetism include cores for electromagnets, electric motors, transformers, generators, data storage, recording media (magnetic micro-actuator, magnetic sensors, dynamic microphone, ribbon microphone, etc.), and magnetic levitation (maglev trains and magnetic suspension). Ferromagnetic forces are used to counteract the effects of the gravitational acceleration and any other acceleration.

It is important to keep in mind the distinction between induced magnetization and remanent magnetization. Induced magnetization, $\boldsymbol{M}_I$, exists only in the presence of an external magnetic field. It is very important in ore exploration. Remanent magnetization, $\boldsymbol{M}_R$, is the part of initial magnetization that remains after the external field disappears or changes in character. Remanent magnetization is frozen within the rock, and the rock remains magnetized in a field-free area. The remanent magnetization captures the record of the past magnetic field and Figure 3.6 is central to the paleomagnetism, which we will discuss in Ikelle (2024).

The ratio $|\boldsymbol{M}_R|/|\boldsymbol{M}_I|$, which is known as the Königsberger ratio (Clark, 1997, and Mussett and Khan, 2000), is used to characterize rocks. Rocks with a high Königsberger ratio are good recorders of the ancient geomagnetic field. They are mafic rocks, such as basalt and gabbro, and also granite. Limestone, for instance, typically has a very small Königsberger ratio. Diamagnetism and paramagnetism are examples of induced magnetization and for this reason the diamagnetic and

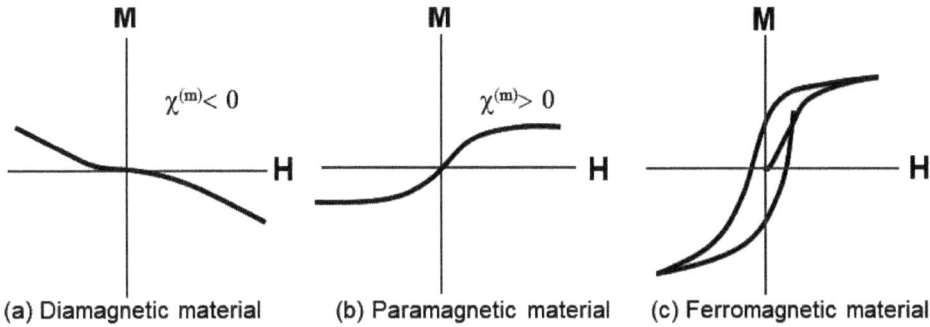

Figure 3.6. All materials can be classified in terms of their magnetic behavior falling into one of five categories depending on their bulk magnetic susceptibility. The three most common types of magnetism are diamagnetism, paramagnetism, and ferromagnetiism. (a) A diamagnetic material has a small and negative susceptibility. (b) A paramagnetic material has a small and postive susceptibility. (c) A ferromagnetic material has a large and positive susceptibility.

paramagnetic materials are usually referred to as non-magnetic, whereas ferromagnetic materials are referred to as magnetic because part of their initial magnetization remains after the external field disappears—that is, ferromagnetic materials have a permanent magnetization. It is the studies of remanently magnetized rocks that have allowed us to determine that the magnetic north and south poles have reversed through geologic time.

The most important rock-forming minerals with remanent magnetic properties are magnetite (Fe_3O_4), hematite (Fe_2O_3), and ilmenite ($FeTiO_3$). First discovered magnets were pieces of iron-bearing rock called loadstone (magnetite). The magnetite accounts for only about 1.5 percent of crustal minerals and is by far the most important magnetic mineral in geophysical mapping (Clark and Emerson, 1991). Sedimentary rocks are usually non-magnetic. Metamorphic rocks probably make up the largest part of the Earth's crust and have a wide range of magnetic susceptibilities. Igneous rocks (e.g., basalt, gabbro, diorite, and granite) also show a wide range of magnetic properties.

3.1.5.   Conductors, Semiconductors, Insulators, and Dielectrics

Energy gap. Again, we can also group materials as an insulator, semiconductor, and a conductor based on an electron energy band. Again, an isolated atom consists of a dense central nucleus surrounded by a cloud of electrons (see Chapter 2 of Ikelle, 2023a). This cloud of electrons is organized into a set of discrete energy bands, as illustrated in Figure 3.7a. These bands describe the range of energies that an electron within the solid may have and the range of energies that it may not have. Electrons in an individual atom are restricted to a well-defined energy band and their energy changes within the atom only take place between one band and another. A conduction band is formed by grouping the range of energy levels of the free electrons. Generally, the conduction band is empty, but when external energy is applied, the electrons in the valence band jump into the conduction band and become free electrons. Electrons in a conduction band have higher energy

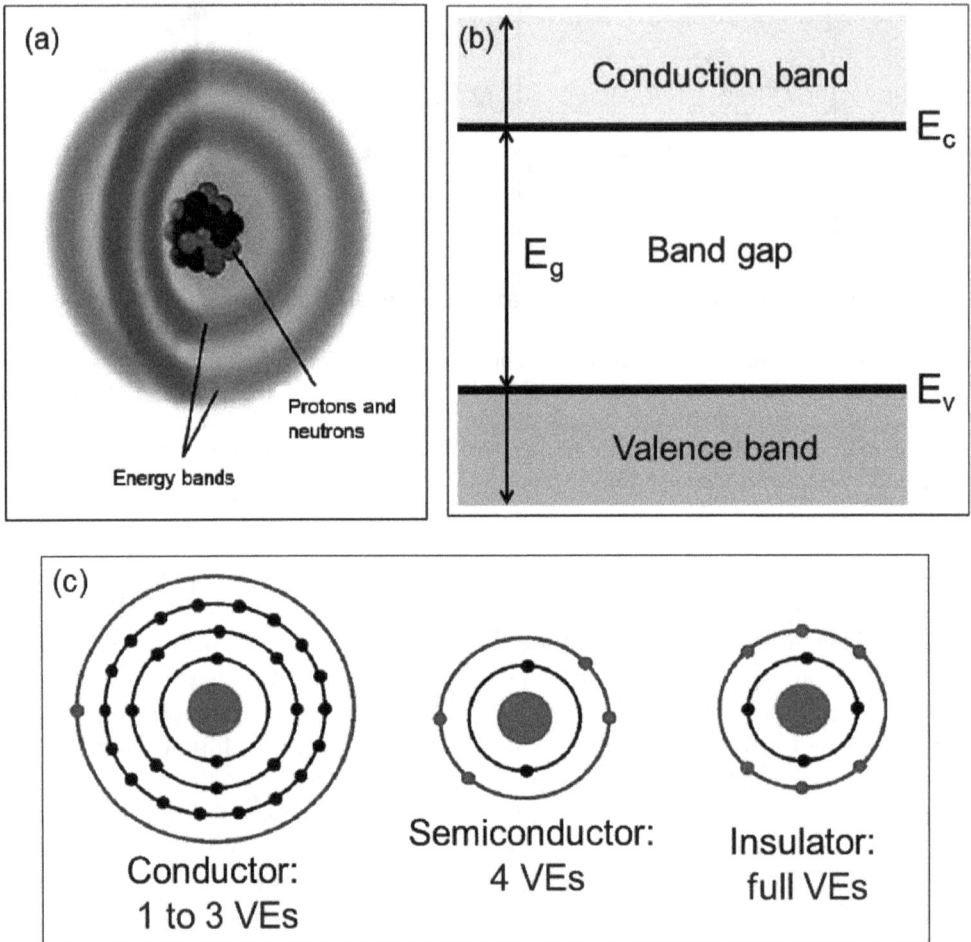

Figure 3.7. (a) Another illustration of an atom's electrons. The cloud of electrons surrounding the nucleus is organized into a set of discrete energy bands. (b) An illustration of the three very critical energy bands in solids, namely, the valence band, the conduction band, and the forbidden band (also known as the energy gap and band gap). Conduction bands are formed by grouping free electron energy levels. (c) Atomic structures of conductors, insulators, and semiconductors. Conductors have between one and three valence electrons (VEs) in their outer orbit shell. Semiconductors have four valence electrons in their outer orbit shell. Insulators have between five and eight valence electrons in their outer orbit shell.

than electrons in a valence band. Conduction-band electrons are not bound to the atom's nucleus. The forbidden band is between the valence and conduction band. The forbidden band is the major factor in determining the electrical conductivity of a solid; therefore it can be used to classify materials as insulators, conductors, and semiconductors.

Figures 3.7b and 3.8 and Table 3.7 show typical energy-band diagrams of semiconductors and insulators. They show the top edge of the valence band, denoted by E_v, and the bottom edge of the conduction band, denoted by E_c. The difference between E_c and E_v is the bandgap energy, or energy gap, E_g. The electrons in

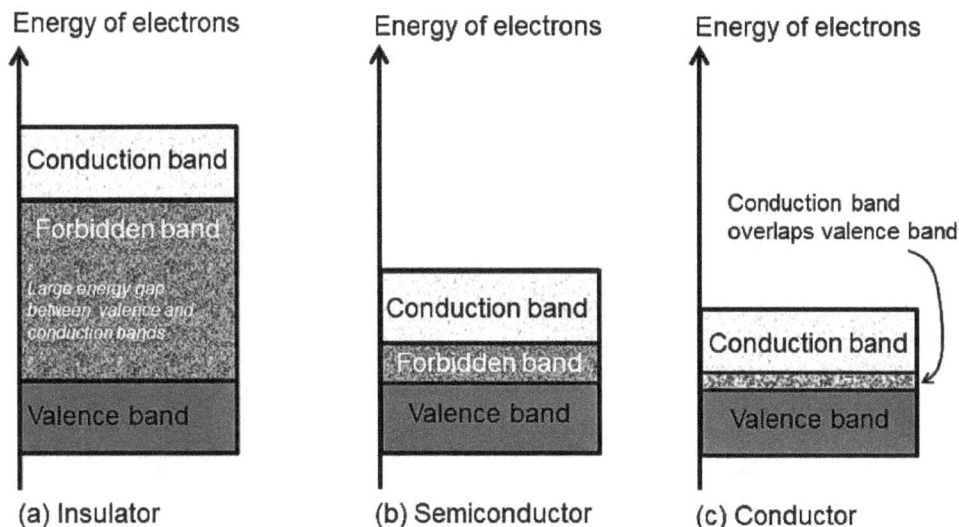

Figure 3.8. (a) An illustration of insulators' energy bands. The large forbidden band ensures that at normal temperatures no electron in the valence band can reach the conductive band. (b) An illustration of semiconductor energy bands. The small forbidden band ensures that thermal energy allows a small fraction of electrons in the valence band to bridge the forbidden band. This allows them to reach the conductive band. (c) An illustration of conductor energy bands. There is no forbidden band in these cases because the valence band overlaps the conductive band.

insulators' valence band are separated by a very large forbidden band from the empty conduction band (e.g., $E_g = 7$ eV for diamond). It would take a lot of energy to make an electron jump through the gap and cause the insulator to break down. The electrons in semiconductor valence bands are separated by a small forbidden band from the empty conduction band of about 0.15 eV $< E_g < 3.5$ eV

Table 3.7. Intervals for conduction and energy gaps of conductors, semiconductors, and insulators. The energy gap (forbidden band) for conductors is zero or very small. For semiconductors, it is more than in conductors, but less than in insulators. For insulators, it is very large, σ in S/m.

Properties	Conductors	Semiconductors	Insulators
Conductivity	Very high $\sigma > 10^4$	Between those of conductors and insulators $10^{-4} < \sigma < 10^4$	Negligible $\sigma < 10^{-4}$
Forbidden band (Energy gap)	Zero or very small	More than in conductors but less than in insulators ($E_g = 0.7$ eV for Ge).	Very large (e.g., $E_g = 9$ eV for SiO_2).
Examples	Silver, copper, and gold.	Silicon, carbon, and germanium.	Mica, phosphorus, glass, and rubber.

(e.g., $E_g = 1.1$ eV for silicon). The forbidden band is narrow enough that thermal excitations or other excitations can bridge the gap. In conductors such as metals, the valence band overlaps the conduction band. In summary, the energy gap (forbidden band) of conductors is zero or very small. It is higher in semiconductors than in conductors, but lower in insulators. For insulators, it is very large.

Quiz: If a semiconductor is transparent to light with a wavelength longer than 0.87 μm, what is its band-gap energy?

The conductivity of a semiconductor is described by an Arrhenius equation of the form

$$\sigma = \sigma_0 \exp\left[-\frac{E_g}{2k_B T}\right] , \qquad (3.58)$$

where E_g is the energy gap, k_B is Boltzmann's constant, T is the absolute temperature, and σ_0 is a constant. If we know conductivity at one temperature, we can determine σ_0 by substituting this temperature value into the Arrhenius equation. What happens when a semiconductor is heated up? As more and more electrons acquire sufficient energy to overcome the band gap, the semiconductor becomes a better conductor of electricity. This is one key distinction between semiconductors and metals. As metals heat up, their electrical conductivity decreases.

P-type and n-type doping of semiconductors. Let us now discuss how semiconductors become sometimes very effective conductors, thanks to doping.

Semiconductors do not conduct and behave like insulators. There is an energy gap of 1.1 eV for silicon and 0.7 eV for germanium between the valence band and the conduction band. The valence band is full but the conduction band is empty, especially at very low temperatures. However, the energy gap between the two bands is so small that electrons can jump across it by adding thermal energy alone or even light energy of a suitable wavelength. Therefore, heating a semiconductor specimen or shining a light on it may cause its electrical conductivity. For example, at 20 degrees Celsius (or 293 degrees Kelvin), silicon has 0.0009 S/m conduction electrons per cubic centimeter; at 150 degrees Celsius (or 423 degrees Kelvin), silicon's conduction is thus more than 550 times higher. Therefore a semiconductor starts conducting only when electrons gain enough energy and move to the conduction band. This type of semiconductor is called an intrinsic semiconductor (pure semiconductor) because it conducts on its own.

There is another type of a semiconductor called an extrinsic semiconductor (also a doped semiconductor or impure semiconductor). An extrinsic semiconductor is basically an intrinsic semiconductor to which a very small amount of impurity has been added to increase the conductivity. The process of adding the impurity to the semiconductor is called doping, and the semiconductor is thus called a doped semiconductor. Doping with an impurity can have a quite marked effect on the electrical properties of the material. The addition of a 10^{-8} impurity atom in germanium can increase the conductivity of germanium by 12 times at 300 degrees Kelvin.

There are two types of extrinsic semiconductors: p-type and n-type. In p-type semiconductors, trivalent (three valent electrons) impurities such as boron are

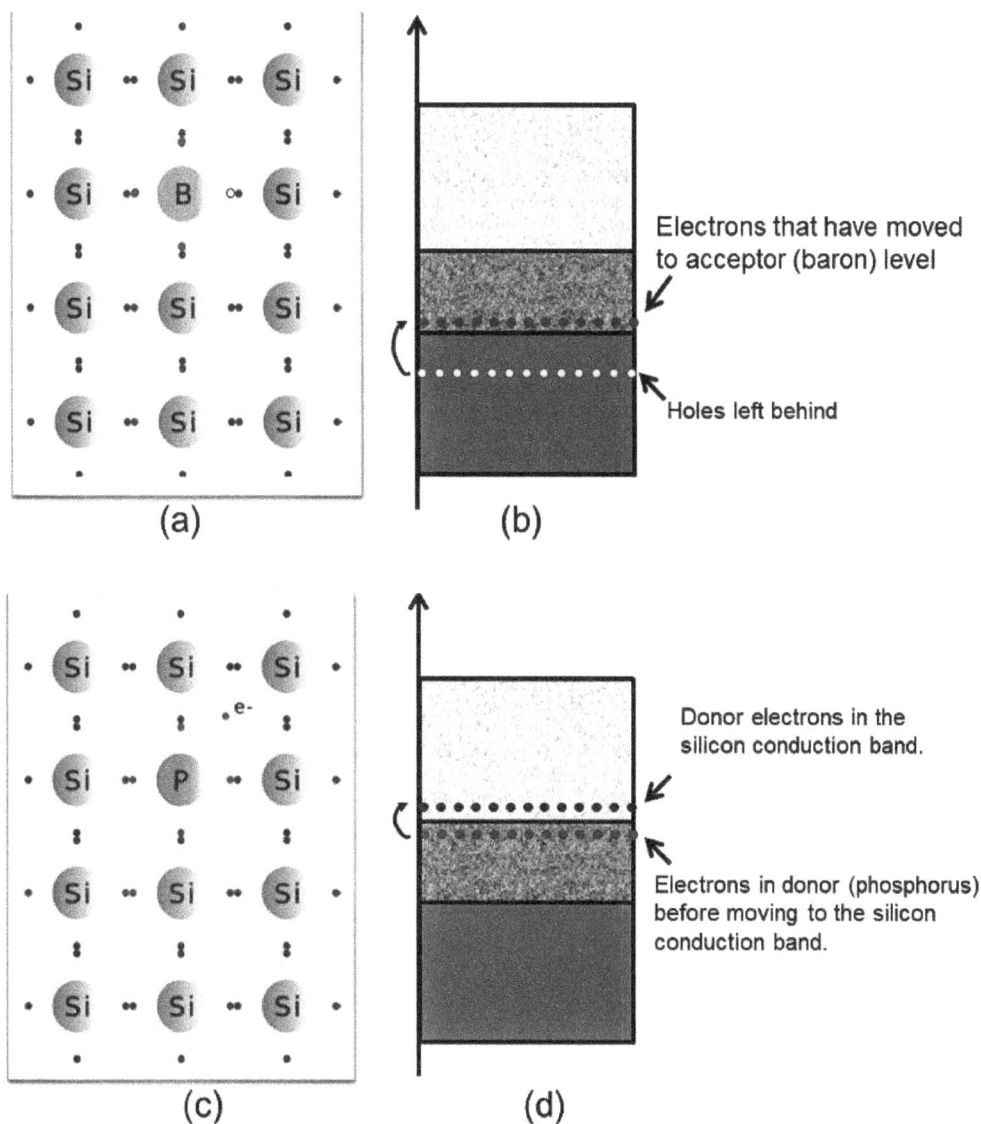

Figure 3.9. (a) An illustration of what p-type semiconductors look like. Trivalent impurities such as boron are added to intrinsic silicon or germanium. (b) The trivalent-atom levels fall just above the full silicon valence band. When an electron jumps up to the boron level, it leaves behind a hole in the valence band. It is the movement of holes within the valence band that causes the greatest conduction in a p-type semiconductor. (c) An illustration of n-type semiconductor doping. Pentavalent impurities such as phosphorous are added to intrinsic silicon or germanium. These impurities add extra electrons to the semiconductor. (d) The pentavalent energy levels fall just below the semiconductor's empty conduction band. Conduction is mainly due to electron movement.

added to intrinsic silicon or germanium semiconductors, as illustrated in Figure 3.9a. The particular impurities are chosen to reduce the gap that electrons have to jump to reach the conduction band. Trivalent impurities are the most appropriate choices. For example, silicon electrons in the valence band just have to make a very small jump to reach the boron levels (acceptor levels) that fall just above the full valence band of silicon, as illustrated in Figure 3.9a. The process of adding trivalent impurities automatically creates holes because the impurity has only three valence electrons to offer to the four valence electron semiconductors, i.e., when an electron jumps up to the boron levels, it leaves behind a hole in the valence band. Hence, p-type doping creates many holes in the semiconductor. It is the movement of holes within the valence band that causes the greatest conduction in a p-type semiconductor. That is, the holes are majority charge carriers in p-type semiconductors and electrons are minority carriers; p stands for positive charge as holes act as positive charges.

In n-type semiconductors, pentavalent (five valent electrons) impurities such as phosphorous are added to intrinsic silicon or germanium semiconductors. Impurities (dopants) have five valence electrons and four electrons are used for covalent bonds with surrounding silicon atoms, as illustrated in Figure 3.9b. Thus one electron left over is only loosely bound. Pentavalent (five valent electrons) impurities are chosen to reduce the gap that electrons have to jump to reach the conduction band. Phosphorus is chosen because its energy levels fall just below the empty conduction band of silicon. The extra electrons from phosphorus can very easily jump from the phosphorus energy levels into the conduction band with very small energy. This will help with conduction. As an example, the doping fraction of a nano amount of phosphorus in silicon yields about 5×10^{16} conduction electrons per cubic centimeter at room temperature. This is a gain of 5 million over intrinsic silicon.

In summary, a doped semiconductor is characterized as a p-type when holes dominate its electrical conductivity and as an n-type when electrons dominate its electrical conductivity. P-type doping consists of adding trivalent (three valent electrons) impurity atoms such as boron, gallium, and indium to intrinsic silicon or germanium semiconductors. The N-type doping consists of adding pentavalent (five valent electrons) impurity atoms such as phosphorus, arsenic, and antimony to intrinsic silicon or germanium semiconductors. Conduction can occur either by negative electrons moving within the conduction band (n-type semiconductors) or by positive holes moving within the valence band (n-type semiconductors).

3.1.6. Nonlinear Constitutive Equations

Atomic vapors and non-metallic solids are generally well described by linear responses. However, in reality, everything in nature is nonlinear, it is just a matter of degree. This nonlinearity can be modeled by constitutive equations. In Ikelle (2023c) we describe how linear constitutive equations associated with the equations of motion (i.e., strain and strain relations) can be extended to be nonlinear. Here, we are considering nonlinear constitutive equations associated with Maxwell's equations. In other words, $\boldsymbol{P}$ and $\boldsymbol{E}$ are related in a manner that is analogous to strain and stress in elastic media—that is, $\boldsymbol{P} = \boldsymbol{P}(\boldsymbol{E})$. As discussed in the previous subsection, for small enough values of $|\boldsymbol{E}|$ this relationship is considered linear—that is, $\boldsymbol{P} = \varepsilon_0 \chi^{(e;1)} \boldsymbol{E}$, where $\chi^{(e;1)} = \chi^{(e)}$. Similarly, one can introduce the second constitutive equation, $\boldsymbol{H} = \boldsymbol{H}(\boldsymbol{B})$, which reduces $\boldsymbol{H} = \mu \boldsymbol{B}$ give for linear magnetic materials. Experience suggests that most materials are linearly magnetic. The nonlinearity is primarily determined by the polarization and electric

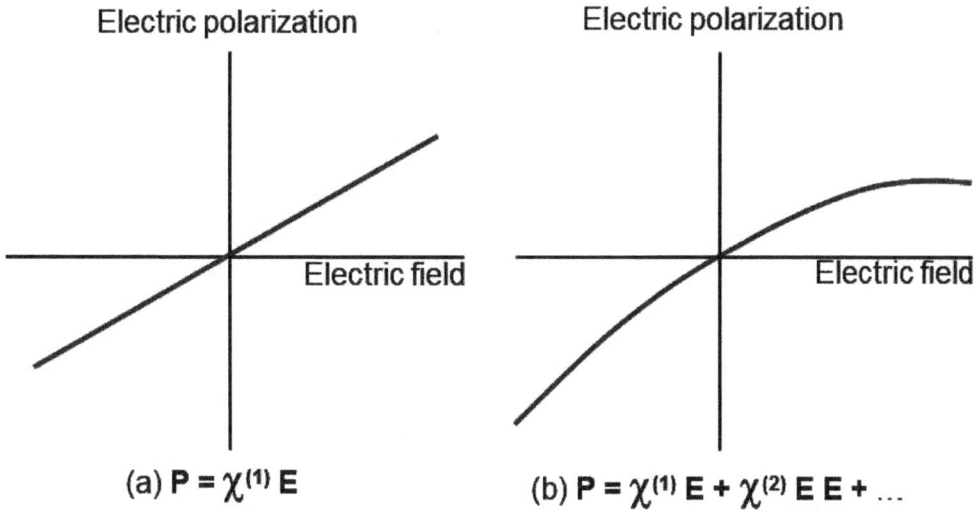

Figure 3.10. The polarization P induced in a medium when electric field E is applied. The response can be (a) linear or (b) nonlinear.

field intensity relation (see Figure 3.10). Therefore, the extension to nonlinear electromagnetism here will focus on the relation between electric polarization and electric field intensity.

Nonlinear susceptibilities. Following the approach adopted in Ikelle (2023c) in the derivations of some constitutive equations, we start from the energy density of interaction between the polarization and the electric field. This energy is

$$\mathcal{U} \;=\; \varepsilon_0 \left[P_i^{(0)} E_i + \frac{\chi_{ij}^{(e;1)} E_i E_j}{2} + \frac{\chi_{ijk}^{(e;2)} E_i E_j E_k}{3} + \frac{\chi_{ijkl}^{(e;3)} E_i E_j E_k E_l}{4} + ... \right] \quad (3.59)$$

and the polarization is related to the electric field intensity, in Cartesian coordinates, as follows:

$$P_i \;=\; \frac{\partial \mathcal{U}}{\partial E_i} = \varepsilon_0 P_i^{(0)} + \underbrace{\varepsilon_0 \sum_{j=1}^{3} \chi_{ij}^{(e;1)} E_j}_{\text{linear}}$$

$$+ \underbrace{\sum_{k=1}^{3}\sum_{l=1}^{3} \chi_{ikl}^{(e;2)} E_k E_l + \sum_{m=1}^{3}\sum_{n=1}^{3}\sum_{p=1}^{3} \chi_{imnp}^{(e;3)} E_m E_m E_p + ...}_{\text{nonlinear}} \quad (3.60)$$

or

$$\boldsymbol{P} \;=\; \varepsilon_0 \left(\underbrace{\boldsymbol{P}^{(0} + \chi^{(e;1)} \boldsymbol{E}}_{\text{linear}} + \underbrace{\chi^{(e;2)} \boldsymbol{E}^2 + \chi^{(e;3)} \boldsymbol{E}^3 + ...}_{\text{nonlinear}} \right) \quad (3.61)$$

with

$$\chi_{ijk\ldots}^{(e;n)} = \frac{1}{n_1!n_2!n_3!} \left.\frac{\partial^n P_i}{\partial E_j \partial E_k \ldots}\right|_{\boldsymbol{E=0}} , \qquad (3.62)$$

where n_i represents the number of fields polarized along the i-th cartesian direction, $\boldsymbol{\chi}^{(e;n)}$ is called the $(n+1)$-th order electric susceptibility tensor; $\boldsymbol{\chi}^{(e;1)}$ has nine constants, $\boldsymbol{\chi}^{(e;2)}$ has 27 constants, and $\boldsymbol{\chi}^{(e;3)}$ has 81 constants. The first two terms in the right-hand side of (3.61) constitute linear electromagnetism and the subsequent terms constitute nonlinear electromagnetism. The zero-th order term, $\boldsymbol{P}^{(0)}$, represents a polarization of the medium that exists in the absence of an applied field (i.e., static polarization). This zero-th order term plays a significant role in vacuum polarization, especially when dealing with the Casimir effect that we will describe later. Most nonlinear materials are usually characterized by their $\boldsymbol{\chi}^{(e;2)}$- and $\boldsymbol{\chi}^{(e;3)}$-responses. $\boldsymbol{\chi}^{(e;2)}$-materials have the greatest nonlinear response, but they must lack inversion symmetry, otherwise their response will be almost zero. A centrosymmetric material is one for which the even powers of the susceptibility expansion are zero.

For simplicity, let us consider the case of isotropic nonlinear media. Equation (3.60) reduces to

$$P_i = \varepsilon_0 \left(\chi^{(e;1)} E_i + \chi^{(e;2)} E_i^2 + \chi^{(e;3)} E_i^3 + \ldots \right) . \qquad (3.63)$$

We can get a rough estimate of the nonlinear susceptibilities by analyzing the displacements of the electrons from the nuclei. Assuming that linear susceptibility is of the order of unity (i.e., $\chi^{(e;1)} \sim 1$) and that the electrons are displaced a distance as large as the atomic size, which is roughly of the order of the Bohr radius, $a_0 = \varepsilon\hbar^2/(me^2) \approx 5 \times 10^{-9}$ cm., it follows that the second-order susceptibility can be estimated as $\chi^{(e;2)} \approx 1/E_{\mathrm{at}} \approx 1.94 \times 10^{-12}$ m/V, where $E_{\mathrm{at}} = e/(4\pi\varepsilon_0 a_0^2) \approx 1.94 \times 10^{-12}$ m/V is the field binding electrons to a nucleus (Boyd, 1999). Similarly, one can estimate as $\chi^{(e;3)} = 1/E_{\mathrm{at}}^2 \approx 3.78 \times 10^{-24}$ m^2/V^2. A typical value of the third-order susceptibility is between 10^{-21} and 10^{-24} m^2/V^2. In more general, the total energy required to liberate an electron from its lattice site is roughly one electronvolt and the separation between lattice sites is about 10^{-10} meters, the characteristic electric field for strong nonlinearities is about 10^{10} V/m (i.e., 0.01 volt per picometer). Correspondingly, since $\chi_{ijk}^{(e;2)}$ has dimensions $1/(\text{electric field})$ and $\chi_{ijkl}^{(e;3)}$ has dimensions $1/(\text{electric field})^2$, rough upper limits on their Cartesian components are: $\chi_{ijk}^{(e;2)} \leq 100$ pm/V and $\chi_{ijkl} \leq 100$ pm/V^2, where pm is picometer (10^{-12} meter).

Table 3.7 shows some crystals[2] with significant second-order susceptibilities. Potassium dihidrogen phosphate (KH_2PO_4), also known as KDP, is one of the most widely used nonlinear crystals as of 2022, not because of its nonlinear susceptibilities (which are quite modest) but because it can sustain large electric fields without suffering damage and it is highly birefringent (different light speeds in different

---

[2]Crystals are repetitive arrangements of molecules in 3-dimensional space. Common examples include diamond crystals, and one easy way of forming crystals on your own, is to dissolve common salt in water to saturation, and leave it unattended overnight.

directions and for different polarizations). Material with a significant $\chi_{ijk}^{(e;2)}$ is known to be noncentrosymmetric, which means the material lacks inversion symmetry; that is, if a fixed point is taken as the origin, then every other point x becomes $-x$. An inversion has a single fixed point. A centrosymmetric material, on the other hand, has an inversion center as its symmetry. KDP crystals are noncentrosymmetric and fused silicon is centrosymmetric.

Sum- and difference-frequency generation. Let us consider the circumstance in which the field incident upon a second-order nonlinear medium consists of two distinct frequency components, which we represent in the form

$$E(t) = E_1 \exp\left(-i\omega_1 t\right) + E_2 \exp\left(-i\omega_2 t\right) + \text{c.c.} . \tag{3.64}$$

where c.c. is complex conjugate. Then the second-order response to polarization in our nonlinear medium is

$$P^{(2)}(t) = \varepsilon_0 \chi^{(e;2)} \left[E(t)\right]^2$$

$$= \varepsilon_0 \chi^{(e;2)} \left\{ \underbrace{E_1^2 \exp\left(-2i\omega_1 t\right) + E_2^2 \exp\left(-2i\omega_2 t\right)}_{SHG} \right.$$

$$+ \underbrace{2E_1 E_2 \exp\left[-i(\omega_1 + \omega_2)t\right]}_{SFG} + \underbrace{2E_1 E_1^* \exp\left[-i(\omega_1 - \omega_2)t\right]}_{DFG} \left. \right\}$$

$$+ \varepsilon_0 \chi^{(e;2)} \text{ c.c.} + \underbrace{2\varepsilon_0 \chi^{(e;2)} \left[E_1 E_1^* + E_2 E_2^*\right]}_{OR} . \tag{3.65}$$

So if we have an input field of two frequencies (ω_1 and ω_2), the polarization will have components of twice each frequency ($2\omega_1$ and $2\omega_2$) that we have labeled second-harmonic generation (SHG), sum and difference frequencies ($\omega_1 + \omega_2$, $\omega_1 - \omega_2$) that we have labeled sum-frequency generation (SFG) and difference-frequency generation (DFG), respectively, and zero frequency that we have labeled optical rectification (OR). In other words, the nonlinear effects of materials create *new* spectral components by shifting spectral energy to new frequencies. The last term here represents a DC term in the output, which amounts to a spatially uniform time-invariant electric field. In the case of ultra-fast pulses passing through a $\chi^{(e;2)}$-nonlinear medium, the rapid building up of a DC field, followed by its rapid decay will lead to the emission of electromagnetic waves. We will come across this effect again in the context of generating terahertz (THz) waves. Notice that in the SFG process two input photons annihilate giving rise to one photon at the sum frequency, $\omega_3 = \omega_1 + \omega_2$. In the DFG process, however, annihilation generates a difference frequency photon $\omega_3 = \omega_1 - \omega_2$. As an illustration, if we have two fields with frequencies 200 THz and 300 THz, then the polarization will have components of twice each frequency (400 and 600 THz), the sum and difference frequencies (500 and 100 THz), and zero.

Figure 3.11 illustrates the nonlinear effect of electromagnetism. For linear media, no new frequencies are created so there is no distortion of the incident waveform.

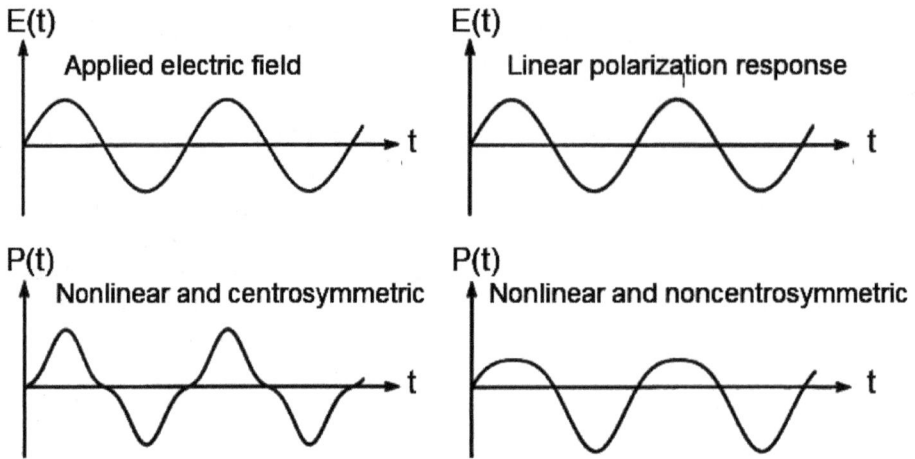

Figure 3.11. Waveforms associated with the atomic response. Note, for the centrosymmetric media, the time-averaged response is zero, whereas for the noncentrosymmetric media the time-average response is nonzero, because the medium responds differently to an electric field pointing, say, in the upward direction than to one pointing downward.

However, nonlinear media that exhibit significant waveform distortions exhibit induced polarization. Also we can distinguish nonlinear media with noncentrosymmetry from those with centrosymmetry. The time-average response for centrosymmetric nonlinear media is zero despite the distortions. Only odd harmonics of the fundamental frequency are present in the response to centrosymmetric nonlinear media. The responses to centrosymmetric nonlinear media contain both even and odd harmonics. Their time-averaged responses are nonzero because they respond differently to an electric field pointing, say, in the upward direction than to one pointing downward.

Let us now consider three incoming fields of frequencies ω_1, ω_2, and ω_3;

$$E(t) = E_1 \exp(-i\omega_1 t) + E_2 \exp(-i\omega_2 t) + E_3 \exp(-i\omega_2 t) + \text{c.c.} . \qquad (3.66)$$

The field with three distinct frequency components is the most general possibility for a third-order interaction. The field $P^{(3)} = \chi^{(e;3)}[E(t)]^3$ generate the 22 different frequency components in Table 3.8. In other words, the algebra becomes very lengthy if all three terms are written explicitly. Following is the case of a medium with non-instantaneous responses to perturbations:

$$
\begin{aligned}
P(\boldsymbol{x}, t) &= \varepsilon_0 \int_{-\infty}^{+\infty} \chi^{(e;1)}(\tau) E(\boldsymbol{x}, t - \tau) d\tau \\
&+ \varepsilon_0 \int \int_{-\infty}^{+\infty} \chi^{(e;2)}(\tau_1, \tau_2) E(\boldsymbol{x}, t - \tau_1) E(\boldsymbol{x}, t - \tau_2) d\tau_1 d\tau_2 \\
&+ \varepsilon_0 \int \int \int_{-\infty}^{+\infty} \chi^{(e;3)}(\tau_1, \tau_2, \tau_3) E(\boldsymbol{x}, t - \tau_1) E(\boldsymbol{x}, t - \tau_2) \\
&\quad E(\boldsymbol{x}, t - \tau_3) d\tau_1 d\tau_2 d\tau_3 + \dots \qquad (3.67)
\end{aligned}
$$

Table 3.8. Some secpnd- and third-order nonlinear susceptibility of nonlinear crystals Materials used in 4-wave mixing. Picometer per volt (pm/V) $= 10^{-12}$ m/V. Adapted from Yariv and Yeh (2007).

Nonlinear crystals	$\chi^{(e;2)}$ (pm/V)	$\chi^{(e;3)}$ (pm/V)^2
Tellurium (Te)	$\chi^{(e;2)}_{333} = 1300$	
CdGeAs$_2$	$\chi^{(e;2)}_{321} = 900$	
Selenium (Se)	$\chi^{(e;2)}_{333} = 320$	
KH$_2$PO$_4$ (KDP)	$\chi^{(e;2)}_{321} = 0.88$	
Beta barium borate (BBO)	$\chi^{(e;2)}_{222} = 4.4$	
Fused Silica (SiO$_2$)		$\chi^{(e;3)}_{1111} = 56.4$
SF$_6$ glass		$\chi^{(e;3)}_{1111} = 587$
CS$_2$ liquid		$\chi^{(e;3)}_{1111} = 6,400$
Polymeric crystal		$\chi^{(e;3)}_{1111} = 6.53 \times 10^5$
GaAs	$\chi^{(e;2)}_{321} = 740$	

In the Fourier domain, this expression becomes

$$P(\boldsymbol{x},t) = \varepsilon_0 \chi^{(e;1)}(\omega)E(\boldsymbol{x},\omega)$$

$$+ \quad \varepsilon_0 \int_{-\infty}^{+\infty} \chi^{(e;2)}(\omega - \omega_2, \omega_2)E(\boldsymbol{x}, \omega - \omega_2)E(\boldsymbol{x}, \omega_2)d\omega_2$$

$$+ \quad \varepsilon_0 \int\int_{-\infty}^{+\infty} \chi^{(e;3)}(\omega - \omega_2 - \omega_3, \omega_2, \omega_3)E(\boldsymbol{x}, \omega - \omega_2 - \omega_3)$$

$$E(\boldsymbol{x}, \omega_2)E(\boldsymbol{x}, \omega_3)d\omega_2 d\omega_3 + ... \tag{3.68}$$

The frequency of the nonlinear response is always the sum of the frequencies of the input fields.

An example of nonlinear wave equation. We here start from Maxwell's equations in Table 3.1. Taking the curl of Faraday's law leads to

$$\boldsymbol{\nabla} \times \boldsymbol{\nabla} \times \boldsymbol{E} = -\frac{1}{c_0}\frac{\partial}{\partial t}\left(\boldsymbol{\nabla} \times \boldsymbol{B}\right) , \tag{3.69}$$

where $c_0^2 = 1/(\varepsilon_0 \mu_0)$. Inserting Ampere's law into this equation yields,

$$\boldsymbol{\nabla} \times \boldsymbol{\nabla} \times \boldsymbol{E} + \frac{1}{c_0^2}\frac{\partial^2 \boldsymbol{E}}{\partial t^2} = -\mu_0 \frac{\partial^2 \boldsymbol{P}}{\partial t^2} , \tag{3.70}$$

or after breaking $\boldsymbol{P}$ as linear and nonlinear electric polarizations, $\boldsymbol{P} = \boldsymbol{P}_\mathrm{L} + \boldsymbol{P}_\mathrm{NL}$, we arrive at

$$\boldsymbol{\nabla} \times \boldsymbol{\nabla} \times \boldsymbol{E} + \frac{\varepsilon^{(1)}}{c_0^2}\frac{\partial^2 \boldsymbol{E}}{\partial t^2} = -\mu_0 \frac{\partial^2 \boldsymbol{P}_\mathrm{NL}}{\partial t^2} , \tag{3.71}$$

where $\varepsilon^{(1)} = 1 + \chi^{(e;1)}$ and

$$P_{\text{NL}} = \chi^{(e;2)} E^2 + \chi^{(e;3)} E^3 + ... \tag{3.72}$$

Using the relation $\boldsymbol{\nabla} \times \boldsymbol{\nabla} \times \boldsymbol{E} = -\boldsymbol{\nabla}^2 \boldsymbol{E} + \boldsymbol{\nabla}(\boldsymbol{\nabla} \cdot \boldsymbol{E})$ and assuming that there are no free charges or currents (i.e., assuming (i.e., $\boldsymbol{\nabla} \cdot \boldsymbol{E} = 0$), we obtain

$$\boldsymbol{\nabla}^2 \boldsymbol{E} - \frac{\varepsilon^{(1)}}{c_0^2} \frac{\partial^2 \boldsymbol{E}}{\partial t^2} = \mu_0 \frac{\partial^2 \boldsymbol{P}_{\text{NL}}}{\partial t^2} . \tag{3.73}$$

Noncentrosymmetric and centrosymmetric media. Here is the 1D equation of motion of the electron position, x, when an electric field, E:

$$m \frac{d^2 x}{dt^2} = \underbrace{- 2m\gamma \frac{dx}{dt}}_{\text{damping}} \underbrace{- m\omega_0^2 x - max^2 + mbx^3}_{\text{restoring force}} - eE \tag{3.74}$$

and by using $P = -Nex$ as

$$\frac{d^2 P}{dt^2} = -2\gamma \frac{dP}{dt} - \omega_0^2 P + \frac{a}{Ne} P^2 + \frac{b}{N^2 e^2} P^3 + \frac{Ne^2}{m} E , \tag{3.75}$$

where a and b are parameters that characterize the strength of the nonlinearity, $-e$ is the charge of the electron, γ is the damping parameter, and N is the number of dipoles per unit volume. We arrived at this equation by assuming that the restoring force is a nonlinear function of the displacement of the electron from its equilibrium position. We also retained the linear and nonlinear terms in the Taylor series expansion of the restoring force in the displacement x. The potential energy function corresponding to this form of the restoring force is

$$U(x) = - \int \tilde{F}_{\text{restoring}} dx = \frac{1}{2} m\omega_0^2 x^2 + \frac{1}{3} max^3 - \frac{1}{4} mbx^4 . \tag{3.76}$$

This potential function is illustrated in the Figure 3.12. When $a = b = 0$, the potential that the atomic electron feels is perfectly parabolic. When $a \neq= 0$ and $b = 0$, the electron is no longer in a perfect parabola. This model describes noncentrosymmetric crystals. When $a = 0$ and $b \neq 0$, the potential function is symmetric (i.e., $U(x) = U(-x)$) and the potential energy function contains even powers of x. This model illustrates centrosymmetric. When $a \neq 0$ and $b \neq 0$, (3.74) describes noncentrosymmetric media. Let us now derive $\chi^{(e;1)}$, $\chi^{(e;2)}$, and $\chi^{(e;3)}$.

Linear susceptibility. This case corresponds to $a = b = 0$ in (3.74). By taking the Fourier transform of this equation, we obtain

$$\left[\left(\omega_0^2 - \omega^2 \right) x + 2i\omega\gamma \right] x = -\frac{e}{m} E \quad \text{or} \quad x(\omega) = \frac{-eE}{mD(\omega)} , \tag{3.77}$$

where

$$D(\omega) = \omega_0^2 - \omega^2 + 2i\omega\gamma . \tag{3.78}$$

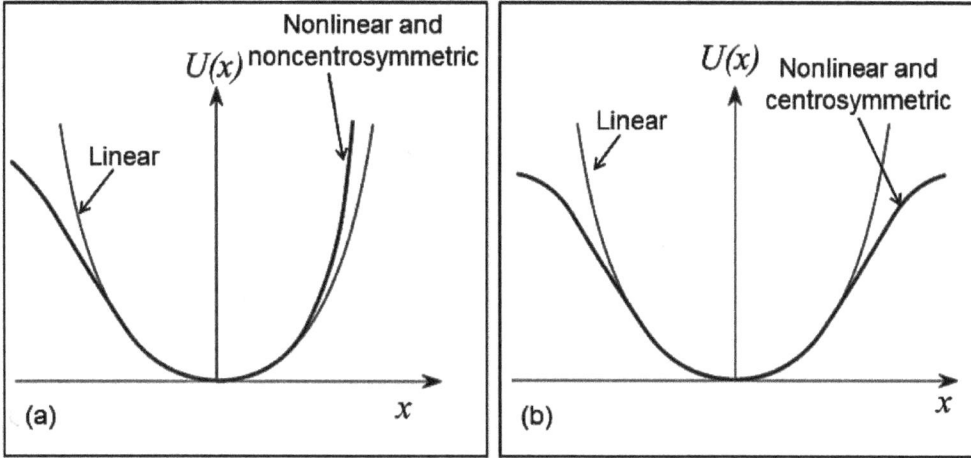

Figure 3.12. Potential energy function for (a) a noncentrosymmetric medium and (b) a centrosymmetric medium. forms associated with the atomic response.

Using the fact that $P(\omega) = -Nex(\omega)$ and also $P(\omega) = \varepsilon_0 \chi^{(e;1)}(\omega)E$, we can then obtain $\chi^{(e;1)}$ as follows:

$$\frac{Ne^2 E}{mD(\omega)} = \varepsilon_0 \chi^{(e;1)}(\omega)E \implies \chi^{(e;1)}(\omega) = \frac{Ne^2}{m\varepsilon_0 D(\omega)} \ . \tag{3.79}$$

Notice that $\chi^{(e;1)}$ is a complex quantity. Its real part is related to the index of refraction of the medium while its imaginary part is related to the absorption coefficient.

Second-order nonlinear susceptibility. This case corresponds to $a \neq 0$ and $b = 0$ in (3.74). Let us seek a solution in the form of a power series expansion $x = \lambda x^{(1)} + \lambda^2 x^{(2)} + \dots$. In order for this power series to be a solution to (3.74) for any value of the coupling strength λ, we require that the terms in (3.74) proportional to λ, λ^2, etc., each satisfy the equation separately. These terms lead

$$\frac{d^2 x^{(1)}}{dt^2} + 2\gamma \frac{dx^{(1)}}{dt} + \omega_0^2 x = \frac{-eE(t)}{m} \tag{3.80}$$

$$\frac{d^2 x^{(2)}}{dt^2} + 2\gamma \frac{dx^{(2)}}{dt} + \omega_0^2 x^{(2)} + a \left[x^{(1)}\right]^2 = 0 \ . \tag{3.81}$$

By taking the Fourier transform of (3.80) and using a general form for the electric field [i.e., $E = \sum_n E(\omega_n) \exp(-i\omega t)$], we arrive at

$$x^{(1)} = -\frac{e}{m} \frac{\sum_n E(\omega_n) \exp(-i\omega_n t)}{D(\omega_n)} \ . \tag{3.82}$$

By taking the Fourier transform of (3.80) and using a general form for the electric field [i.e., $E = \sum_n E(\omega_n) \exp(-i\omega t)$], we arrive at

$$x^{(2)} = -\frac{ea}{m^2} \frac{\sum_n \sum_m E(\omega_n) E(\omega_m) \exp[-i(\omega_m + \omega_n)t]}{D(\omega_m)D(\omega_n)D(\omega_m + \omega_n)} \ . \tag{3.83}$$

From the following two expressions

$$P^{(2)} = -Nex^{(2)} = \frac{Ne^2 a}{m^2} \frac{\sum_n \sum_m E(\omega_n)E(\omega_m)\exp[-i(\omega_m + \omega_n)t]}{D(\omega_m)D(\omega_n)D(\omega_m + \omega_n)}$$

and

$$P^{(2)} = \frac{ea}{m^2} \sum_n \sum_m \chi^{(e;2)}(\omega_m, \omega_n)E(\omega_n)E(\omega_m)\exp[-i(\omega_m + \omega_n)t]$$

we obtain

$$\chi^{(e;2)}(\omega_1, \omega_2) = \frac{Ne^3 a}{\varepsilon_0 m^2} \frac{1}{D(\omega_1 + \omega_2)D(\omega_1)D(\omega_2)} . \tag{3.84}$$

Miller (1964) noted that the following ratio

$$\alpha = \frac{\chi^{(e;2)}(\omega_1, \omega_2)}{\chi^{(e;1)}(\omega_1 + \omega_2)\chi^{(e;1)}(\omega_1)\chi^{(e;1)}(\omega_2)} \tag{3.85}$$

is nearly constant for all noncentrosymmetric crystals.

Third-order nonlinear susceptibility. This case corresponding to $a = 0$ and $b \neq 0$ in (3.74). We now seek a solution in the form of a power series expansion $x = \lambda x^{(1)} + \lambda^2 x^{(2)} + \lambda^3 x^{(3)}\ldots$. By using the above derivations for second-order nonlinear susceptibility, we arrive at

$$\chi^{(e;3)}(\omega_1, \omega_2, \omega_3) = \frac{Ne^4 b}{\varepsilon_0 m^3} \frac{1}{D(\omega_1 + \omega_2 + \omega_3)D(\omega_1)D(\omega_2)D(\omega_3)} .$$

3.1.7.   Electric Dipole Moments

Eltrostatics and magneto-statics. Before we introduce dipoles and dipole moments, let us recall the fundamentals of electrostatics described in Ikelle (2017b, 2020, and 2023b). Again, electrostatics is the branch of electromagnetics that deals with the effects of electric charges at rest (i.e., the fields, including the forces, do not depend on the velocity of the particles). Magneto-statics is another branch of electromagnetics dealing instead with the effects of electric charges in steady motion (that is steady current or DC). These branches of electromagnetics are developed under the assumption of static behavior of the electromagnetic field, i.e.,

$$\frac{\partial \boldsymbol{E}}{\partial t} = \frac{\partial \boldsymbol{H}}{\partial t} = \frac{\partial \boldsymbol{D}}{\partial t} = \frac{\partial \boldsymbol{B}}{\partial t} = 0 . \tag{3.86}$$

That is, the solutions of Maxwell's equations approach a time-independent limit as time tends toward infinity when, at a certain instant $t = t_0$, given sources of a constant magnitude in time are switched on. Under static assumptions, the macroscopic Maxwell's equations, in Table 3.1, reduce to

$$\nabla \times \boldsymbol{H} = \boldsymbol{J} \tag{3.87}$$

Table 3.9. Frequencies generated during third-order interactions.

RHG	SFG & DFG	SFG & DFG	RHG	SFG & DFG	SFG & DFG
ω_1	$2\omega_1 \pm \omega_2$	$\omega_1 + \omega_2 + \omega_3$	ω_2	$2\omega_1 \pm \omega_3$	$\omega_1 + \omega_2 - \omega_3$
ω_3	$2\omega_2 \pm \omega_1$	$\omega_1 - \omega_2 + \omega_3$	$3\omega_1$	$2\omega_2 \pm \omega_3$	$-\omega_1 + \omega_2 + \omega_3$
$3\omega_2$	$2\omega_3 \pm \omega_1$		$3\omega_3$	$2\omega_3 \pm \omega_2$	

$$\nabla \times \boldsymbol{E} = 0 \tag{3.88}$$

$$\nabla \cdot \boldsymbol{D} = \rho_E \tag{3.89}$$

$$\nabla \cdot \boldsymbol{B} = 0 . \tag{3.90}$$

Notice that the equations of the electric and magnetic fields are no longer mutually coupled. So in the static limit, Maxwell's equations split into decoupled electrostatic and magneto-static equations.

Faraday's law and Gauss's law for the electric field [that is, equations (3.88) and (3.89)] are electrostatic equations. The constitutive equation of electrostatics is only between electric displacement $\boldsymbol{D}$ and the electric field intensity $\boldsymbol{E}$, as shown in Table 3.9. Ampère's law and Gauss's law for the magnetic field [that is, equations (3.87) and (3.90)] are magneto-static equations. One of the constitutive equations of magneto-statics is between electric displacement $\boldsymbol{B}$ and the magnetic field intensity $\boldsymbol{H}$, as shown in Table 3.9.

Another important EM equation that we described in the previous chapter is the *continuity equation*. It describes the fact that charge can neither be created nor destroyed. As we impose that the time derivatives are zeros, thus

$$\nabla \cdot \boldsymbol{J} = 0 . \tag{3.91}$$

The currents described by this equation are called steady currents. A steady current refers to a continuous flow that has been going on forever, without change and

Table 3.10. Maxwell's equations of electrostatics and magnetostatics.

Electrostatics	Law name
$\nabla \times \boldsymbol{E} = 0$	Faraday
$\nabla \cdot \boldsymbol{D} = \rho_E$	Gauss (electric)
$\boldsymbol{D} = \varepsilon \boldsymbol{E}$	Constitutive equation

Magnetostatics	Law name
$\nabla \times \boldsymbol{H} = \boldsymbol{J}$	Ampère
$\nabla \cdot \boldsymbol{B} = 0$	Gauss (magnetic)
$\boldsymbol{B} = \mu \boldsymbol{H}$	Constitutive equation
$\nabla \cdot \boldsymbol{J} = 0$	Ohm

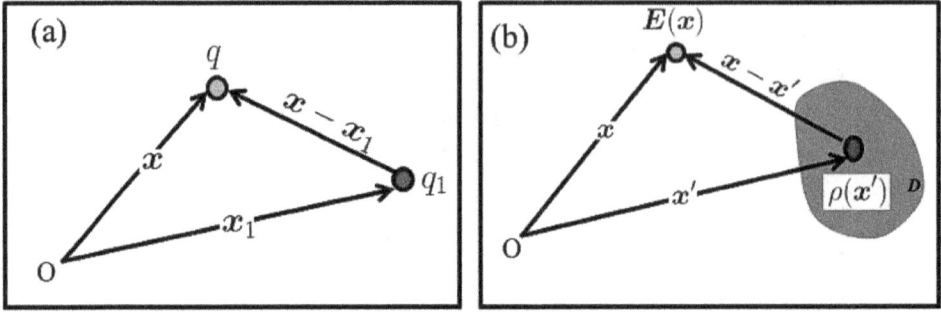

Figure 3.13. (a) Geometry of Coulomb interaction between two charges. Coulomb's law describes how a static electric charge, q, located at a point, $\boldsymbol{x}$, relative to the origin, O, experiences an electrostatic force from a static electric charge, q_1, located at $\boldsymbol{x}_0$. Like charges repel and unlike charges attract. (b) An illustration of the electric field intensity due to each of the charge distributions $\rho(\boldsymbol{x}')$. The Coulomb's law, due to $\rho(\boldsymbol{x}')$, is a superposition of the fields due to the numerous point charges.

without charge filling up anywhere. Thus steady currents produce magnetic fields that are constant in time. We are interested in electrostatics and magnetostatics because, in most part, electromagnetic effects on molecules, including on greenhouse gases, can be described by electrostatics, some well-logging studies for subsurface exploration can be described by magnetostatics, and the geomagnetic field that we will discuss in Ikelle (2024), also changes slowly over time and thus can be described by magnetostatics.

We will start our study of electrostatics by investigating the two fundamental laws governing electrostatic fields: Coulomb's law and Gauss's law. Both of these laws are based on experimental studies. Although Coulomb's law is applicable to finding the electric field due to any charge configuration, it is easier to use Gauss's law when the charge distribution is symmetrical. Special problems can be solved with much effort using Coulomb's law, but will be solved with ease by applying Gauss's law.

The electrostatic force law and the electric field. We begin with the electrostatic force between two point charges (see Figure 3.13a). This force controls the separation of electric charges which is associated with electrostatic effects. Given two such charges, q and q_1, at positions $\boldsymbol{x}$ and $\boldsymbol{x}_1$, respectively, and force, F_1, acting on an electrically charged particle with charge, q, located at $\boldsymbol{x}$, due to the presence of a charge q_1 located at $\boldsymbol{x}_1$ is given by the expression (*Coulomb's law*)

$$\boldsymbol{F}_1(\boldsymbol{x}) = \frac{1}{4\pi\varepsilon_0} \frac{qq_1}{|\boldsymbol{x} - \boldsymbol{x}_1|^3} (\boldsymbol{x} - \boldsymbol{x}_1) . \tag{3.92}$$

where ε_0 is the permittivity of free space (vacuum). Note that this formula only makes sense if $\boldsymbol{x}$ and $\boldsymbol{x}_1$ are in different positions. The force on q due to q_1 is equal in magnitude but opposite in direction to the force on q_1 due to q (like charges repel and unlike charges attract). This formula assumes that the electric charges are in a vacuum or in free space. When the electric charges are placed in a dielectric material, we simply have to replace the permittivity of free space, ε_0, with the permittivity of the dielectric material, ε. Note also that this force is the same if the particles are stationary or moving.

Without loss of generality, we might as well put the second charge at the origin $\boldsymbol{x}_1 = 0$, and the force, $\boldsymbol{F}_1$, acting on an electrically charged particle with charge q located at $\boldsymbol{x}$, due to the presence of a charge q_1 located at $\boldsymbol{x}_1 = 0$ given by the expression

$$\boldsymbol{F}_1(\boldsymbol{x}) = \frac{1}{4\pi\varepsilon_0} \frac{qq_1}{|\boldsymbol{x}|^3} \boldsymbol{x} \ . \tag{3.93}$$

This force is known as Coulomb's law of electrostatics, and is an experimental fact. It does not depend on the velocity of the particles and is the same if the particles are stationary or moving. Note that the magnitude of this force is inversely proportional to the square of the distance of separation of the two charges, $1/|\boldsymbol{x}|^2$, and why? For light and sound intensity, the inverse-square law is a result of the energy being equally distributed on a spherical surface. Since the area of a sphere is $4\pi|\boldsymbol{x}|^2$, the sound or light intensity decreases as $1/|\boldsymbol{x}|^2$. For electrostatic forces there is nothing that is spread out evenly over a spherical surface. Their dependence on the $1/|\boldsymbol{x}|^2$ law is a result of the geometry of space.

If we have N charges q_i at positions $\boldsymbol{x}_i$, $i = 1, \ldots, N$, then an additional charge q at position $\boldsymbol{x}$ experiences this force

$$\boldsymbol{F}(\boldsymbol{x}) = \frac{q}{4\pi\varepsilon_0} \sum_{i=1}^{N} \frac{q_i}{|\boldsymbol{x} - \boldsymbol{x}_i|^3} (\boldsymbol{x} - \boldsymbol{x}_i) \ . \tag{3.94}$$

So, to get the total force on charge q due to all the other charges, we simply add up (superpose) the Coulomb force from each charge q_i. In other words, electrostatic forces obey the principle of superposition. We define the electric static intensity field, $\boldsymbol{E}$, to be the force on a unit test charge (i.e., $q = 1$) at position $\boldsymbol{x}$:

$$\boldsymbol{F}(\boldsymbol{x}) = q\boldsymbol{E}(\boldsymbol{x}) \iff \boldsymbol{E}(\boldsymbol{x}) = \frac{1}{4\pi\varepsilon_0} \sum_{i=1}^{N} q_i \frac{(\boldsymbol{x} - \boldsymbol{x}_i)}{|\boldsymbol{x} - \boldsymbol{x}_i|^3} \ . \tag{3.95}$$

Thus the electric field, which is a fundamental quantity in electromagnetic theory, as described in the previous section, is just a mathematically convenient way of describing the force of a unit test charge. Note that, in contrast to the previous section, the electric field here is analytically defined and time-independent. Generally, it depends on the position and the charges. That is a result of electrostatic approximation.

So far, we have only considered forces and electric fields due to point charges, which are essentially charges occupying a very small physical space, that is, at the microscopic scale. Again, at macroscopic scale we use the concept of charge density introduced in the previous section. Let $\rho_E(\boldsymbol{x})$ be the charge density per unit volume. The electric field intensity due to each of the charge distributions $\rho_E(\boldsymbol{x})$ may be regarded as the summation of the field contributed by the numerous point charges making up the charge distribution. We simply replace in (3.95) a point charge q_i at position $\boldsymbol{x}_i$ by $\rho_E(\boldsymbol{x}')dV(\boldsymbol{x}')$, and the sum by a volume integral to obtain the electric field intensity, as shown in Figure 3.13b, as follows:

$$\boldsymbol{E} = \frac{1}{4\pi\varepsilon_0} \int_{D_\epsilon} \frac{\rho_E(\boldsymbol{x}')}{|\boldsymbol{x} - \boldsymbol{x}'|^3} (\boldsymbol{x} - \boldsymbol{x}')\, dV(\boldsymbol{x}') \ . \tag{3.96}$$

Actually, under the assumption of linear superposition, the electric field associated with charge density in (3.95) is valid for an arbitrary distribution of electric charges, including discrete charges. Consider the following discrete distribution of charges: $\rho_E(\boldsymbol{x}') = \sum_i^N q_i\delta(\boldsymbol{x}' - \boldsymbol{x}_i)$. We can move from the electric field due to charge density to the electric field due to point charges:

$$\begin{aligned}
\boldsymbol{E} &= \frac{1}{4\pi\varepsilon_0} \int_{D_\epsilon} \frac{\rho_E(\boldsymbol{x}')}{|\boldsymbol{x} - \boldsymbol{x}'|^3} (\boldsymbol{x} - \boldsymbol{x}')\, dV(\boldsymbol{x}') \\
&= \frac{1}{4\pi\varepsilon_0} \int_{D_\epsilon} \frac{\sum_i^N q_i\delta(\boldsymbol{x}' - \boldsymbol{x}_i)}{|\boldsymbol{x} - \boldsymbol{x}'|^3} (\boldsymbol{x} - \boldsymbol{x}')\, dV(\boldsymbol{x}') \\
&= \frac{1}{4\pi\varepsilon_0} \sum_i^N \frac{q_i}{|\boldsymbol{x} - \boldsymbol{x}_i|^3} (\boldsymbol{x} - \boldsymbol{x}_i) \ .
\end{aligned} \tag{3.97}$$

Returning to point charges, we can verify that

$$\boldsymbol{E}(\boldsymbol{x}) = \frac{q}{4\pi\varepsilon_0} \frac{\boldsymbol{x}}{|\boldsymbol{x}|^3} = -\boldsymbol{\nabla}\phi_E(\boldsymbol{x}) \ , \tag{3.98}$$

with

$$\phi_E(\boldsymbol{x}) = \frac{1}{4\pi\varepsilon_0} \frac{q}{|\boldsymbol{x}|} \ , \tag{3.99}$$

by using the identity

$$\boldsymbol{\nabla}\left(\frac{1}{|\boldsymbol{x} - \boldsymbol{x}'|}\right) = -\frac{(\boldsymbol{x} - \boldsymbol{x}')}{|\boldsymbol{x} - \boldsymbol{x}'|^3} \ . \tag{3.100}$$

The quantity ϕ_E is known as the electrostatic potential for point charge, q. It is a scalar. Because electric fields add like vectors (the superposition principle), we can obtain the potential of N point charges as follows

$$\phi_E(\boldsymbol{x}) = \frac{1}{4\pi\varepsilon_0} \sum_{i=1}^N \frac{q_i}{|\boldsymbol{x} - \boldsymbol{x}_i|} \ . \tag{3.101}$$

For charge density, the electrostatic field and potential can be described as follows:

$$\boldsymbol{E}(\boldsymbol{x}) = \frac{1}{4\pi\varepsilon_0} \int_{D_\epsilon} \frac{\rho_E(\boldsymbol{x}')}{|\boldsymbol{x} - \boldsymbol{x}'|^3} (\boldsymbol{x} - \boldsymbol{x}')\, d\boldsymbol{x}' = -\boldsymbol{\nabla}\phi_E(\boldsymbol{x}) \tag{3.102}$$

with

$$\phi_E(\boldsymbol{x}) = \frac{1}{4\pi\varepsilon_0} \int_{D_\epsilon} \frac{\rho_E(\boldsymbol{x}')}{|\boldsymbol{x} - \boldsymbol{x}'|}\, dV(\boldsymbol{x}') \ . \tag{3.103}$$

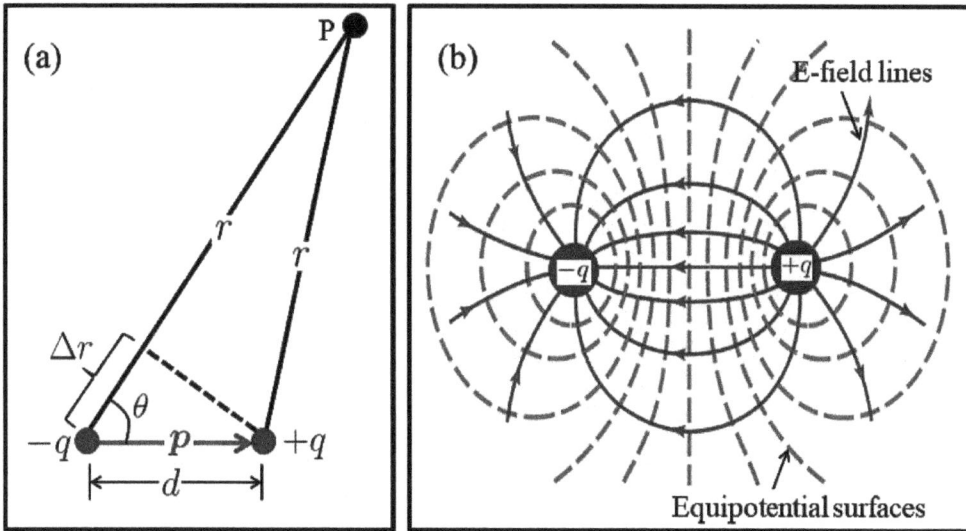

Figure 3.14. (a) An electric charge dipole consists of a pair of equal and opposite point charges separated by a small distance (i.e., much smaller than the distance at which we observe the resulting field). The vector, $\boldsymbol{p}$, is the dipole moment vector. (b) An illustration of the electric field lines and scalar electric potential lines (i.e., equipotential surfaces) for a dipole. Equipotential surfaces are perpendicular to the E-field lines.

As $\boldsymbol{E}$ is a gradient its curl is zero ($\boldsymbol{\nabla} \times \boldsymbol{E} = 0$; Faraday's law), and we thus deduce that Coulomb's electric field also satisfies Faraday's law in addition to Gauss's law (i.e., the two Maxwell's equations of electrostatics are satisfied). Note that the electrostatic potential falls as $1/|\boldsymbol{x}|$ while the electric field intensity decreases as $1/|\boldsymbol{x}|^2$. Note also that the electrostatic potential is zero "at infinity."

From Gauss's law and electrical potential, we have

$$\boldsymbol{\nabla} \cdot \boldsymbol{E}(\boldsymbol{x}) = \frac{\rho_E(\boldsymbol{x})}{\varepsilon_0} \iff \boldsymbol{\nabla}^2 \phi_E(\boldsymbol{x}) = \frac{\rho_E(\boldsymbol{x})}{\varepsilon_0} , \tag{3.104}$$

which is called Poisson's equation. It reduces to Laplace's equation when $\rho_E = 0$; i.e.,

$$\boldsymbol{\nabla}^2 \phi_E(\boldsymbol{x}) = 0 . \tag{3.105}$$

We described the solution to this equation in Chapter 1.

Electric potential and electric field of a dipole. Let us compute the electric potential and electric field of a dipole (see Figure 3.14a). Remember that an electric charge dipole consists of a pair of equal and opposite point charges separated by a short distance (i.e., much shorter than the distance at which we observe the resulting field), as illustrated in Figure 3.14a. The potential due to an electric dipole is the sum of the potentials due to each charge:

$$\phi_E(\boldsymbol{x}) = \frac{1}{4\pi\varepsilon_0} \left[\frac{q}{r} - \frac{q}{r + \Delta r} \right] = \frac{q}{4\pi\varepsilon_0} \left[\frac{\Delta r}{r^2(1 + \Delta r/r)} \right] \tag{3.106}$$

Table 3.11. A Matlab code for computing electric field line of electric points charges.

```
clear all; clear figure
n = 5; x=rand(n,1)-0.5; y=rand(n,1)-0.5; % n. of charges and locations
q = rand(n,1); q = q - mean(q); % charge
ke = 8.9875e9;xi = linspace(-1,1,33); yi = linspace(-1,1,33);
[XI YI] = meshgrid(xi,yi); zi = complex(XI,YI); z = complex(x,y);
[ZI Z]=ndgrid(zi(:),z(:)); dZ=ZI-Z; Zn=abs(dZ); E = (dZ./Zn.^3)*(q(:)*ke);
E = reshape(E,size(XI)); Ex = real(E); Ey = imag(E);
figure; quiver(XI, YI, Ex./E, Ey./E); hold on
plot(x, y, 'or'); axis equal
```

with $r = |\boldsymbol{x}|$. Using the fact that $(1+x)^n \approx 1 + nx$ or $\Delta r(1 + \Delta r/r)^{-1} \approx \Delta r$ in the far-field approximation (i.e., $\Delta r \ll r$), we have $\Delta r = d\cos\theta$ and therefore

$$\phi_E(\boldsymbol{x}) \approx \frac{qd\cos\theta}{4\pi\varepsilon_0 r^2} = \frac{p\cos\theta}{4\pi\varepsilon_0 r^2} \ , \tag{3.107}$$

where $p = qd$ is the dipole moment. We can then deduce the electric field from the potential as follows:

$$\boldsymbol{E}(\boldsymbol{x}) = -\boldsymbol{\nabla}\phi_E(\boldsymbol{x}) = -\left[\frac{\partial\phi_E}{\partial r}\mathbf{i}_r + \frac{1}{r}\frac{\partial\phi_E}{\partial\theta}\mathbf{i}_\theta\right] \approx \frac{p}{4\pi\varepsilon_0 r^2}\left[\cos\theta\mathbf{i}_r + \sin\theta\mathbf{i}_\theta\right] \ . \tag{3.108}$$

We can see that using potentials instead of fields can make solving problems much easier. Figure 3.14b shows electric field lines for an electric dipole and equipotential lines of the electric dipole. Equipotential lines are surfaces over which ϕ_E is a constant. Notice that the direction of field lines is radially outward for a positive charge and radially inward for a negative charge. Because the electric field is the negative of the gradient of the electric scalar potential, the electric field lines are everywhere normal to the equipotential surfaces and point in the direction of decreasing potential.

Computational exercise: (1) Table 3.11 provides a Matlab code to compute electric field lines caused by electric point charges. Use this code to reproduce Figure 3.15. Vary n (the number of charges) to gain more intuition about the representation of electric field lines. (2) Table 3.12 provides a Matlab code for computing the electric field due to a dipole in a 2-D plane by using Coulomb's law. Use this code to reproduce the results shown in Figure 3.16.

Electric dipole moments of molecules. The size and orientation of a dipole is often measured by its dipole moment; i.e.,

$$\boldsymbol{p} = \sum_i q_i\boldsymbol{x}_i \ , \tag{3.109}$$

where $\boldsymbol{p}$ is the dipole moment vector, q_i is the magnitude of the ith charge, and $\boldsymbol{x}_i$ is the vector representing the position of ith charge. The unit of $|\boldsymbol{p}|$ is generally Debye

Table 3.12. A Matlab code for computing the electric field due to a dipole in a 2-D plane by using the Coulomb's law.

```
clear all; const = 9; Nx = 101; Ny = 101; EF = zeros(Nx,Ny); ex = EF; ey = EF;
r = EF; rr2 = EF; Q = [1,-1]; X = [5,-5]; Y = [0,0]; % Array of charges
for k = 1:2
    q = Q(k); % Compute the unit vectors
    for i=1:Nx
        for j=1:Ny
            rr2(i,j) = (i-51-X(k))∧2+(j-51-Y(k))∧2;
            r(i,j) = 0.01+sqrt(rr2(i,j));ex(i,j) = ex(i,j)+(i-51-X(k))./r(i,j);
            ey(i,j) = ey(i,j)+(j-51-Y(k))./r(i,j);
        end
    end
    rr2 = 0.01+rr2; EF = EF + q.*const./rr2;
end
xrange = (1:Nx)-51; yrange = (1:Ny)-51; contourrange = -8:0.02:8;
contour(xrange,yrange,EF',contourrange,'linewidth',1.2);
axis([-25 25 -15 15]);colorbar('location','eastoutside','fontsize',12);
xlabel('x ','fontsize',14);ylabel('y ','fontsize',14);
title('Electric field distribution, E (x,y) in V/m','fontsize',14);
```

(D) with 1 Debye $= 3.336 \times 10^{-30}$ C m. It is named after Peter Debye (1884-1966), who pioneered the study of dipole moments and of electrical interactions between particles; he won the Nobel Prize for Chemistry in 1936. For a two-charge separated

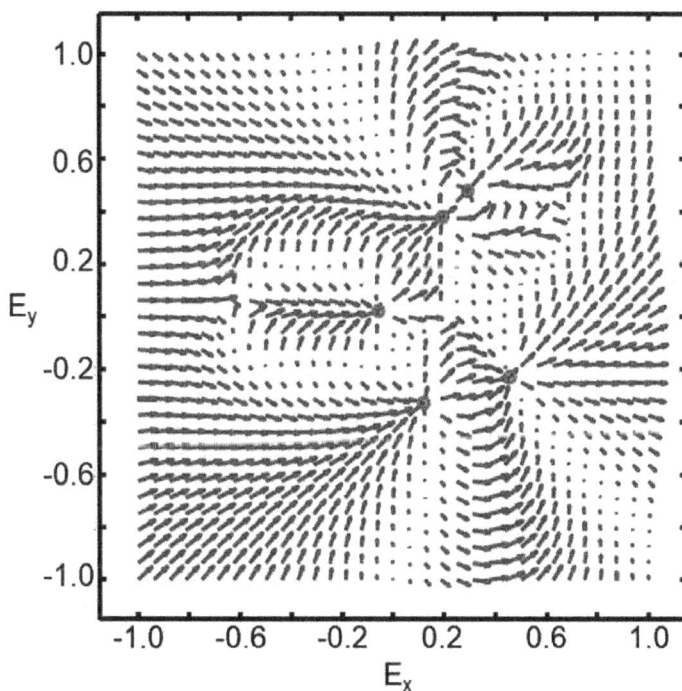

Figure 3.15. An illustration of electric field lines due to electric point charges (circles).

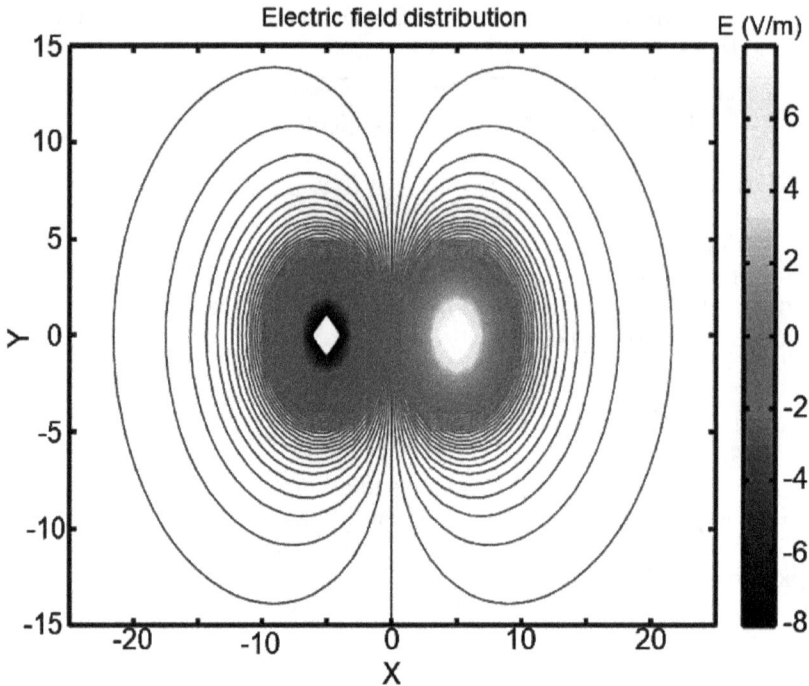

Figure 3.16. An illustration of the electric field due to a dipole in a 2-D plane. We used Coulomb's law to compute the electric field.

dipole system ($q_1 = -q$ and $q_2 = +q$; or the more commonly used terms δ^+ and δ^- using the chemical nomenclature) like diatomic molecules or when considering a bond dipole within a molecule, equation (3.109) can be simplified to

$$\boldsymbol{p} = q(\boldsymbol{x}_2 - \boldsymbol{x}_1) = q|\boldsymbol{x}_2 - \boldsymbol{x}_1|\mathbf{i}_r \ , \tag{3.110}$$

where $\boldsymbol{x}_1$ and $\boldsymbol{x}_2$ are the vectors that define the position of the two charges in space, $\mathbf{i}_r$ is the direction of line of the two charges (or the orientation of the dipole is along the axis of the bond with working diatomic molecules). Note that when $|\boldsymbol{x}_1| = |\boldsymbol{x}_2|$, the dipole moment in (3.110) is zero. In other words, when the center of gravity of the positive charge coincides with that of the negative charge, the dipole moment in (3.110) is zero. For continuous charge distributions, the sum is replaced by an integral over space and we use charge density instead of discrete charges. We have

$$\boldsymbol{p} = \int_V \rho_E(\boldsymbol{x}') \, \boldsymbol{x}' \, dV(\boldsymbol{x}') \ , \tag{3.111}$$

where $\rho_E(\boldsymbol{x}')$ is again the charge density. Consider, for example, a simple system of a single electron and a proton separated by 100 pm. When a proton and an electron are close together, the dipole moment (degree of polarity) decreases. In this case, the dipole moment is calculated as follows:

$$|\boldsymbol{p}| = q|\boldsymbol{x}_2 - \boldsymbol{x}_1| \;\; = \;\; \left(1.60 \times 10^{-19}\,\text{C}\right)\left(1.0 \times 10^{-10}\,\text{m}\right)$$

$$= \;\; \left(1.60 \times 10^{-29}\,\text{C}\cdot\text{m}\right)\left(\frac{1\,\text{D}}{3.336 \times 10^{-30}\,\text{C}\cdot\text{m}}\right)$$

$$= \;\; 4.80\,\text{D} \,. \tag{3.112}$$

As a proton and an electron get farther apart, the dipole moment increases.

Our next task is to determine whether molecules are polar or nonpolar. Before that, let us recall some basic chemical nomenclature of molecules, namely homonuclear, heteronuclear, linear, and nonlinear molecules. Homonuclear molecules are molecules composed of various numbers of atoms with only one chemical element. H_2, O_2, N_2, and ozone (O_3) are homonuclear molecules. A heteronuclear molecule is a molecule made up of atoms of more than one chemical element. For example, the water molecule (H_2O) is heteronuclear because it has atoms of two different elements, hydrogen (H) and oxygen (O), and HF is also a heteronuclear molecule. A linear molecular geometry is composed of a central atom connected to two other atoms oriented at a bond angle of 180 degrees. O=C=O is a linear molecule (see Figure 3.17). Nonlinear molecules, on the other hand, have irregular geometries. In other words, nonlinear molecules have atoms not arranged straight. The chemical structure of these molecules has a zig-zag or cross-linked structure.

It is important to examine the notion of electronegativity reviewed in Ikelle (2023a) to determine whether a molecule is nonpolar or polar. Let us also recall that the electronegativity (EN) of an element is a measure of its ability to attract shared electrons in a covalent bond with another element. Electronegativity is measured on the Pauling scale, which ranges from 0 to 4 (e.g., Ikelle, 2023a). F, O, N, and Cl are the most electronegative elements and Cs and Fr with electronegative at 0.7 are the least electronegative elements. Consider a generic diatomic molecule, XY, containing elements X and Y. When the absolute value of the difference in electronegativity of the two bonded atoms is smaller than 0.5 [i.e., $|\text{EN}(X) - \text{EN}(Y)| < 0.5$], XY is considered nonpolar. In these cases, both atoms exert almost equal attraction to the bonding electrons. Hence, diatomic molecules containing elements of identical electronegativity, such as H_2, O_2, Cl_2, and N_2, are nonpolar [i.e., $|\text{EN}(X) - \text{EN}(Y)| = 0 < 0.5$]. When the absolute value of the difference in electronegativity of the two bonded atoms is equal or greater than 0.5 and smaller than 1.7 [i.e., $0.5 \leq |\text{EN}(X) - \text{EN}(Y)| < 1.7$], XY is covalently polar; that is, we have a bond with an uneven electron density distribution. The oxygen atom (EN=3.5) is more electronegative than the hydrogen atom (EN=2.1) and attracts electrons more strongly. O–H bond is strongly polarized because their difference in electronegativity is 1.4. We have a bond with an uneven distribution of electron density because there is one pair of unshared electrons (i.e., one lone pair of electrons). C–H, with a difference of 0.4, is nonpolar; C–N bond, with a difference of -0.5, is relatively nonpolar; and C–O, with a difference of -1.0, is polar. The polar molecules include H_2O, HCl, CH_3Cl, NH_3, etc. (e.g., $|\boldsymbol{p}|_{HCl} = 1.03$ Debye; $|\boldsymbol{p}|_{H_2O} = 1.84$ Debye), as described in Figure 3.17. The arrows placed parallel to the atom lines indicate the dipole moments. Homonuclear diatomic molecules like

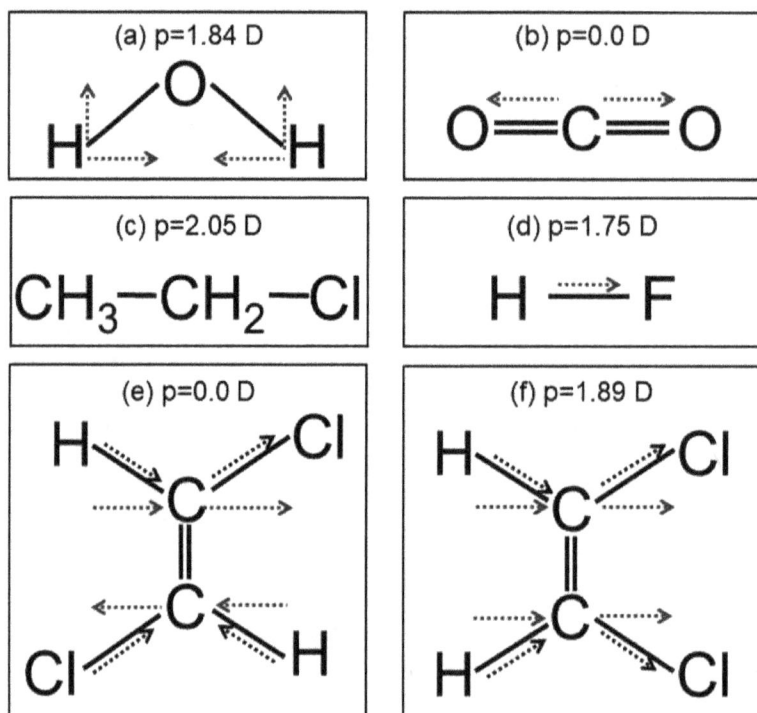

Figure 3.17. Examples of dipole moments of molecules. The dipole moments of bonds are determined by using electronegativities (e.g., Keller, 2022a). The dipole moments of molecules are vector sums of dipole moments. Dipole moment of (a) H_2O, (b) CO_2, (c) C_2H_5Cl, (d) HF, and (e) and (f) two isomers of $C_2H_2Cl_2$, one is polar and the other one is non polar. Dipole moment of (a) H_2O, (b) CO_2, (c) C_2H_5Cl, (d) HF, and (e) and (f) two isomers of $C_2H_2Cl_2$, one is polar and the other one is non polar.

nitrogen (N_2), hydrogen (H_2), oxygen (O_2), and chlorine (Cl_2) have zero dipole moments due to their symmetrical charge distributions and similar electronegativity. These molecules are perfectly nonpolar molecules because their electronegativity differences are zero. Other nonpolar molecules include CH_4, C_2H_6, C_6H_6, CCl_4, and CBr_4.

Often, we can recognize if a molecule is polar or nonpolar without calculating its properties. When the center of gravity of the positive charge of a given molecule does not coincide with the center of gravity of the negative charge, polarity arises in the molecules and the dipole moment is nonzero. Based on Figure 3.18, benzene has the same center of gravity for both positive and negative charges, so its dipole moment (C_6H_6) is zero. A molecule with a V" shape is polar because its total dipole moment is nonzero. Figure 3.17 shows that the water molecule has a V" shape and is polar. Note that alkenes (C_nH_{2n}) are hydrocarbons containing carbon-carbon and carbon-hydrogen bonds. They are nonpolar because carbon-carbon bonds have zero dipole moment and the electronegativity difference of carbon-hydrogen bonds is less than 0.5. Alkanes (C_nH_{2n+2}) are also nonpolar molecules, since they contain only nonpolar carbon-carbon and carbon-hydrogen bonds.

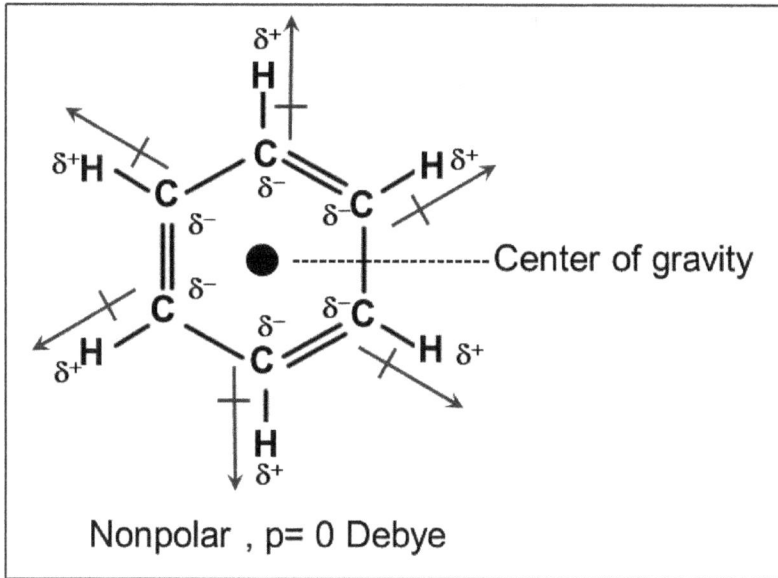

Figure 3.18. Dipole moments of benzene molecule (C_6H_6).

Electric multipole expansion. To further explain the notion of the dipole moment, it useful to expand the electric potential defined above; i.e.,

$$\phi_E(\boldsymbol{x}) = k_\varepsilon \sum_{i=1}^{N} \frac{q_i}{|\boldsymbol{x} - \boldsymbol{x}_i|} \approx \underbrace{\frac{k_\varepsilon}{|\boldsymbol{x}|} \left[\sum_i q_i \right]}_{\text{Monopole}} + \underbrace{\frac{k_\varepsilon}{|\boldsymbol{x}|^2} \left[\sum_i q_i \boldsymbol{x}_i \right]}_{\text{Dipole}}$$

$$+ \underbrace{\frac{k_\varepsilon}{2|\boldsymbol{x}|^3} \left[\sum_i q_i \left(3x_i^{(j)} x_i^{(k)} - \delta_{jk} |\boldsymbol{x}_i|^2 \right) \right] \mathbf{i}_j \mathbf{i}_k}_{\text{Quadrupole}} + ..., \qquad (3.113)$$

where $k_\varepsilon = 1/4\pi\varepsilon_0$ and $x_i^{(k)}$ denotes the kth coordinate of the position of the ith particle (see Figure 3.19). This expansion is known as *multipole expansion*. The first bracketed expression is just the total charge, also called the monopole moment. Because this term depends $1/|\boldsymbol{x}|$, the source looks like a point whose charge is the total charge of the source when we are far away from a source. The second bracketed expression is the dipole moment defined earlier. The whole second term dipole term contributes to the potential too, but it falls off as $1/|\boldsymbol{x}|^2$, so will in general be negligible far away from the source unless the monopole moment vanishes. If the monopole moment vanishes, the dipole term will be the leading-order term. The next bracketed expression

$$\left[\sum_i q_i \left(3x_i^{(j)} x_i^{(k)} - \delta_{jk} |\boldsymbol{x}_i|^2 \right) \right] \mathbf{i}_j \mathbf{i}_k \qquad (3.114)$$

is the traceless quadrupole moment. A quadrupole moment is constructed from two dipoles. The third term of a potential is only significant when both the monopole

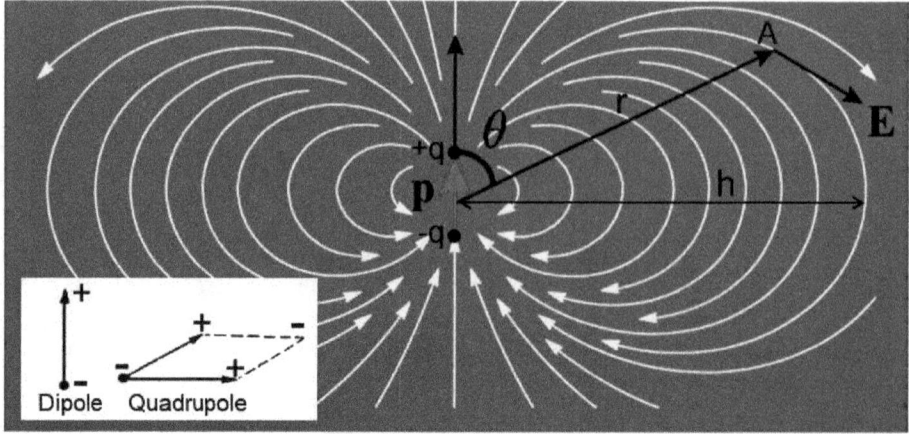

Figure 3.19. Electric dipole (of moment p) field configuration with electric field vectors E in the $r = h\sin^2\theta$ and θ directions to an arbitrary observation A; h is the equatorial field line distance. We also append here the charge distribution for electric dipole and quadrupole configurations.

moment and the dipole moment are zero. We can also write this expansion in integral form for a continuous matter distribution:

$$
\phi_E(\boldsymbol{x}) = k_\varepsilon \int_V \frac{\rho_E(\boldsymbol{x}')}{|\boldsymbol{x} - \boldsymbol{x}'|} \, dV(\boldsymbol{x}')
$$

$$
\approx \underbrace{\frac{k_\varepsilon}{|\boldsymbol{x}|} \left[\int_V \rho_E(\boldsymbol{x}') \, dV(\boldsymbol{x}') \right]}_{\text{Monopole}} + \underbrace{\frac{k_\varepsilon}{|\boldsymbol{x}|^2} \left[\int_V \rho_E(\boldsymbol{x}') \, x^{(j)'} \, dV(\boldsymbol{x}') \right] \mathbf{i}_j}_{\text{Dipole}}
$$

$$
+ \underbrace{\frac{k_\varepsilon}{2|\boldsymbol{x}|^3} \left[\int_V \rho_E(\boldsymbol{x}') \left(3x_i^{(j)'} x_i^{(k)'} - \delta_{jk} |\boldsymbol{x}_i'|^2 \right) dV(\boldsymbol{x}') \right] \mathbf{i}_j \mathbf{i}_k + \dots .}_{\text{Quadrupole}}
$$

For an electric field of pure dipole (i.e., the monopole moment is zero) the dipole term becomes the leading-order term; i.e.,

$$
Q = \int_V \rho_E(\boldsymbol{x}) \, dV(\boldsymbol{x}) = 0 \implies \phi_E \approx \frac{k}{|\boldsymbol{x}|^2} \boldsymbol{p} \cdot \mathbf{i}_r . \tag{3.115}
$$

For example, if the dipole is located at the origin and oriented along the z-axis (i.e., $\boldsymbol{p} = p_m \mathbf{i}_z$), (3.115) reduces to

$$
\phi_E = k_\varepsilon \frac{p_m \mathbf{i}_z \cdot \mathbf{i}_r}{|\boldsymbol{x}|^2} = k_\varepsilon \frac{p_m \cos\theta}{|\boldsymbol{x}|^2} . \tag{3.116}
$$

As described in Ikelle (2023b), the gravitational field can also expand in the same way.

Dipole nature of dielectric materials. Dielectric materials are a particular form of electric insulator. Dielectric materials have electric dipoles, which are atoms or

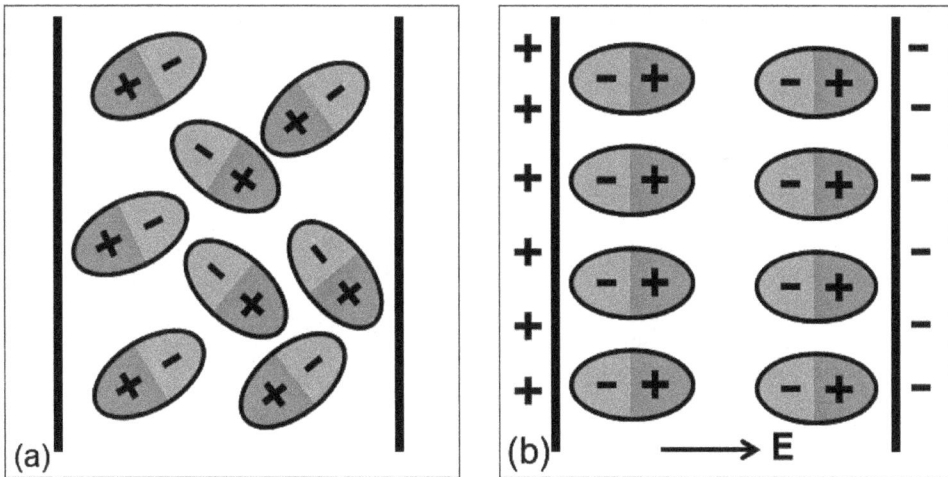

Figure 3.20. (a) Polar molecules in a dielectric without an electric field. (b) Example of a state of polar molecules in a dielectric after the application of an electric field.

molecules with positive and negative charges at opposite ends. In the presence of an applied electric field, it can be polarized by acquiring a significant electric polarization (i.e., $|\boldsymbol{P}|$ significantly increases). Positive charges are displaced in the direction of the electric field and negative charges shift in the opposite direction. Note that electric charges do not flow through the material as they do in an electric conductor. Instead, they only slightly shift from their average equilibrium positions causing dielectric polarization. The number and behavior of these dipoles determine materials' dielectric properties. An increase in electric polarization strength implies an increase in relative permittivity, as described in (3.55).

Most dielectric materials are solids. Examples include porcelain (ceramic), mica, glass, plastics, and metal oxides. Some liquids and gases can serve as efficient dielectric materials. Distilled water is a fair dielectric. Vacuum is an exceptionally efficient dielectric. Dry air is an excellent dielectric and is used in variable capacitors and transmission lines. One of the key reasons for introducing dielectrics is that they help us make better capacitors (see Chapter 5). Dielectrics are also critical for explaining various phenomena in electronics, optics, solid-state physics, and cell biophysics.

As we increase the magnitude of the applied electric field, the magnitude of the opposing internal electric field increases simultaneously. At some point the dielectric material breaks down under the external field. For example, air breaks down and charges pass through it to produce lightning. The maximum electric field you can apply without destroying the material is called the dielectric strength of the material.

3.1.8. Magnetic Dipole Moments

Magneto-statics: Vector magnetic potential. The fundamental equations of magneto-statics, including the continuity equation, are given in Table 3.10, along with the electrostatic equations. They require steady currents. A steady current refers to a continuous flow that has been carried on forever, without change and without charge filling up anywhere (i.e., $\nabla J = 0$). Thus steady currents produce magnetic fields that are constant over time.

The analogy between electrostatics and magneto-statics. Again, if a charge, q, at position $\boldsymbol{x}$ is moving with velocity $\boldsymbol{v} = d\boldsymbol{x}/dt$, it experiences a Lorentz force:

$$\boldsymbol{F} = \boldsymbol{F}_e + \boldsymbol{F}_m = q\boldsymbol{E} + q\boldsymbol{v} \times \boldsymbol{B} \ . \tag{3.117}$$

The magnetic force is $\boldsymbol{F}_m = q\boldsymbol{v} \times \boldsymbol{B}$ and the electric force is $\boldsymbol{F}_e = q\boldsymbol{E}$. By replacing q by $q\boldsymbol{v}$, q' by $q'\boldsymbol{v}'$, and the simple multiplication operation by the cross-product in Coulomb's law, we can go from the electric force, in (3.92), to the magnetic force; i.e.,

$$\boldsymbol{F}_m(\boldsymbol{x}) = \frac{qq'}{4\pi\varepsilon_0} \frac{\boldsymbol{v}}{c_0} \times \left[\frac{\boldsymbol{v}'}{c_0} \times \frac{(\boldsymbol{x} - \boldsymbol{x}')}{|\boldsymbol{x} - \boldsymbol{x}'|^3} \right] = \frac{\boldsymbol{v}}{c_0} \times \left[\frac{\boldsymbol{v}'}{c_0} \times \boldsymbol{F}_e(\boldsymbol{x}) \right] \ , \tag{3.118}$$

with $c_0^2 = 1/(\mu_0\varepsilon_0)$. A similar analogy exists between the magnetic field and the electric field. By analogy to Coulomb's formula, in (3.96), the magnetic field for charge, q', at $\boldsymbol{x}'$ moving with velocity, $\boldsymbol{v}'$,

$$\boldsymbol{F}_m = q\boldsymbol{v} \times \boldsymbol{B} \Longrightarrow \boldsymbol{B}(\boldsymbol{x}) = \frac{\mu_0 q'}{4\pi} \frac{\boldsymbol{v}' \times (\boldsymbol{x} - \boldsymbol{x}')}{|\boldsymbol{x} - \boldsymbol{x}'|^3} = \frac{\boldsymbol{v}'}{c_0} \times \frac{\boldsymbol{E}(\boldsymbol{x})}{c_0}$$

or

$$\boldsymbol{B}(\boldsymbol{x}) = \nabla \times \left(\frac{\mu_0}{4\pi} \frac{q'\boldsymbol{v}'}{|\boldsymbol{x} - \boldsymbol{x}'|} \right) = \nabla \times \boldsymbol{A}(\boldsymbol{x}) \ , \tag{3.119}$$

where

$$\boldsymbol{A}(\boldsymbol{x}) = \frac{\mu_0}{4\pi} \frac{q'\boldsymbol{v}'}{|\boldsymbol{x} - \boldsymbol{x}'|} \tag{3.120}$$

is the magnetic vector potential. We can verify that the divergence of $\boldsymbol{B}$ is zero (i.e., $\nabla \cdot \boldsymbol{B} = 0$; Gauss's law).

The integral form of Ampère's law. At the macroscopic scale, we have to replace $q\boldsymbol{v}$, which represents an individual charge in motion, by $\boldsymbol{J}(\boldsymbol{x}) = \rho_E(\boldsymbol{x})\boldsymbol{v}(\boldsymbol{x})$, which represents the flow of charges. The macroscopic differential form of Ampère's law, in Table 3.10, captures this point via current density. The integral form of Ampère's law captures the same point via the electric current instead of current density; i.e.,

$$\nabla \times \boldsymbol{H}(\boldsymbol{x}) = \boldsymbol{J}(\boldsymbol{x}) \Longleftrightarrow \iint_S \{[\nabla \times \boldsymbol{H}(\boldsymbol{x})] \cdot \boldsymbol{n}(\boldsymbol{x})\} \ dS(\boldsymbol{x})$$

or

$$\iint_S \{[\nabla \times \boldsymbol{H}(\boldsymbol{x})] \cdot \boldsymbol{n}(\boldsymbol{x})\} \ dS(\boldsymbol{x}) = \iint_S [\boldsymbol{J}(\boldsymbol{x}) \cdot \boldsymbol{n}(\boldsymbol{x})] \ dS(\boldsymbol{x}) \tag{3.121}$$

where S is any surface that is not necessarily closed and $\boldsymbol{n}$ is the unit normal vector of surface S pointing outward, away from the volume enclosed by the surface. By

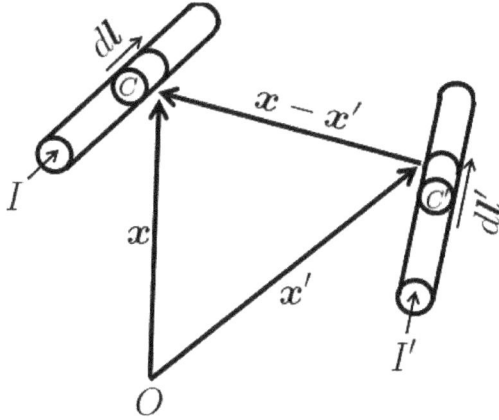

Figure 3.21. Ampère's law of force. A small loop C, carrying a static electric current I through its tangential line element dl located at x, experiences a magnetostatic force from a small loop C', carrying a static electric current I' through the tangential line element dl' located at x'.

applying Stokes' theorem (see Ikelle 2020) to the left hand side of the equation in (3.121), we obtain the integral form of Ampère's law, i.e.,

$$\oint_C [\boldsymbol{B}(\boldsymbol{x}) \cdot dl(\boldsymbol{x})] = \mu_0 I , \tag{3.122}$$

where

$$I = \iint_S [\boldsymbol{J}(\boldsymbol{x}) \cdot \boldsymbol{n}(\boldsymbol{x})] \, dS(\boldsymbol{x}) . \tag{3.123}$$

The current, I, is the flux associated with the current density vector, $\boldsymbol{J}$. The line integral of the magnetic field carried out along the boundary C of a surface S. C is always a closed curve. Sometimes magnetic fields are difficult to determine with Biot-Savart law which we will describe next. For magnetic fields with high symmetry, Ampère's law in (3.122) can alternatively be used because it can simplify derivations.

The Biot-Savart law and vector potential. Experiments on the interaction between two small loops of electric current, (C, dl, I) and (C', dl', I'), show that they interact through a mechanical force,

$$\boldsymbol{F}(\boldsymbol{x}) = \frac{\mu_0 I I'}{4\pi} \oint_C dl \times \oint_{C'} dl' \times \frac{(\boldsymbol{x} - \boldsymbol{x}')}{|\boldsymbol{x} - \boldsymbol{x}'|^3} = I \oint_C dl \times \boldsymbol{B}(\boldsymbol{x}) ,$$

according to Ampère's law force with

$$\boldsymbol{B}(\boldsymbol{x}) = \frac{\mu_0 I'}{4\pi} \oint_{C'} dl' \times \frac{(\boldsymbol{x} - \boldsymbol{x}')}{|\boldsymbol{x} - \boldsymbol{x}'|^3} = -\frac{\mu_0 I'}{4\pi} \oint_{C'} dl' \times \nabla \left(\frac{1}{|\boldsymbol{x} - \boldsymbol{x}'|} \right) , \tag{3.124}$$

much the same way that electric charges interact. This force at $\boldsymbol{x}$ acts on (C, dl, I) due to (C', dl', I') at $\boldsymbol{x}'$, as illustrated in Figure 3.21. We use the identity

$$\nabla \left(\frac{1}{|\boldsymbol{x} - \boldsymbol{x}'|} \right) = -\frac{(\boldsymbol{x} - \boldsymbol{x}')}{|\boldsymbol{x} - \boldsymbol{x}'|^3} , \tag{3.125}$$

in the derivation (3.124). Equation (3.124) is a fundamental law of magnetostatics known as the Biot-Savart law. It confirms the $\boldsymbol{B}$-field arising at a specified point $\boldsymbol{x}$ from a given current distribution. It is named
after Jean Biot and Felix Savart who discovered equation (3.124). Notice that equation (3.124) can also be written as

$$\boldsymbol{B}(\boldsymbol{x}) = \boldsymbol{\nabla} \times \boldsymbol{A}(\boldsymbol{x}) \tag{3.126}$$

where $\boldsymbol{A}$ is the vector magnetic potential for a line current and is given by

$$\boldsymbol{A}(\boldsymbol{x}) = \frac{\mu_0 I}{4\pi} \oint_C d\boldsymbol{l}' \frac{1}{|\boldsymbol{x} - \boldsymbol{x}'|} \ . \tag{3.127}$$

This vector potential can be transformed as $\boldsymbol{A}(\boldsymbol{x}) \longrightarrow \boldsymbol{A}'(\boldsymbol{x}) = \boldsymbol{A}(\boldsymbol{x}) + \boldsymbol{\nabla}\Phi(\boldsymbol{x})$, where $\boldsymbol{a}(\boldsymbol{x}) = \boldsymbol{\nabla}\Phi(\boldsymbol{x})$ is an arbitrary vector field, without affecting the magnetic field vector $\boldsymbol{B}(\boldsymbol{x})$ because the curl of $\boldsymbol{a}(\boldsymbol{x})$ is zero (i.e., $\boldsymbol{\nabla} \times \boldsymbol{\nabla}\Phi(\boldsymbol{x}) = 0$). The transformation from $\boldsymbol{A}(\boldsymbol{x})$ to $\boldsymbol{A}'(\boldsymbol{x})$ is referred to as gauge transformation for the magnetostatic vector potential. An analogous gauge transformation for the scalar potential of the electrostatic field did not arise because it is just a trivial constant potential. In some ways, the vector magnetic potential $\boldsymbol{A}(\boldsymbol{x})$ is analogous to the scalar electric potential $\phi_E(\boldsymbol{x})$. In classical physics, the vector magnetic potential is viewed as an auxiliary function with no physical meaning. However, there are phenomena in quantum physics (atomic and sub-atomic physics) that suggest that the vector magnetic potential is a real (i.e., measurable) field.

Magnetic multipole. Notice we can expand the magnetic potential in a multiple as we did for the electrical potential. By using the spherical coordinates defined in Ikelle (2023b), we can also write magnetic potential in terms of spherical harmonics just like scalar gravitational and electrical potentials. We start by writing the position differences between the observation point and the position on C, as follows:

$$|\boldsymbol{x} - \boldsymbol{x}'| = \left[r^2 - 2rr' \cos\gamma + (r')^2\right] \tag{3.128}$$

where

$$\cos\gamma = \cos\theta \cos\theta' + \sin\theta \sin\theta' + \cos(\lambda - \lambda') \tag{3.129}$$

represents the angle subtended between $\boldsymbol{x} = (r, \theta, \lambda)$ and $\boldsymbol{x}' = (r', \theta', \lambda')$. According to the definition of spherical harmonics in Ikelle (2020), we can write

$$\frac{1}{|\boldsymbol{x} - \boldsymbol{x}'|} = \frac{1}{r} \sum_{n=0}^{\infty} \left(\frac{r'}{r}\right)^n P_n^0(\cos(\gamma)) \tag{3.130}$$

for $r' < r$, where $P_n^0(x)$ are associated Legendre functions (see Ikelle, 2020). By using (3.130), the magnetic vector potential can be written as

$$
\begin{aligned}
\boldsymbol{A}(\boldsymbol{x}) &= \frac{\mu_0 I}{4\pi} \oint_C dl' \frac{1}{|\boldsymbol{x} - \boldsymbol{x}'|} \\
&= \frac{\mu_0 I}{4\pi} \left\{ \sum_{n=0}^{\infty} \left[\frac{1}{r^{n+1}} \oint_C (r')^n P_n^0(\cos(\cos\gamma)) dl \right] \right\} \\
&= \frac{\mu_0 I}{4\pi} \left\{ \frac{1}{r} \oint_C dl' + \frac{1}{r^2} \oint_C r' \cos\gamma \, dl' \right. \\
&\quad + \left. \frac{1}{r^3} \oint_C (r')^2 \left(\frac{3}{2} \cos^2\gamma - \frac{1}{2} \right) dl' + ... \right\} .
\end{aligned}
$$

The last equality in (3.131), which is expressed as a multipole expansion, is obtained by assuming that distance $|\boldsymbol{x} - \boldsymbol{x}'|$ is large. The first term, called the *monopole* term, is zero because the line integral of dl' is zero for any closed loop. The second term, called the *dipole* term, is usually the dominant term in the multipole expansion. The third term is called the *quadrupole* term. The vector potential generated by the dipole term is

$$
\boldsymbol{A}_{\text{dipole}}(\boldsymbol{x}) = \frac{\mu_0}{4\pi} \frac{\boldsymbol{m} \times \boldsymbol{x}}{|\boldsymbol{x}|^2} \implies \boldsymbol{B}_{\text{dipole}}(\boldsymbol{x}) = \boldsymbol{\nabla} \times \boldsymbol{A}_{\text{dipole}}(\boldsymbol{x}) , \tag{3.131}
$$

where $\boldsymbol{m}$ is called the magnetic dipole moment of the current loop. It is defined as

$$
\boldsymbol{m} = \frac{1}{2} I \oint_C (\boldsymbol{x}' \times d\boldsymbol{l}') . \tag{3.132}
$$

For points far away from the loop compared to its size, the dipole is a good approximation. So an important difference between the magnetic vector potential and the scalar gravitational and electrical potentials is that the magnetic vector potential does not include the monopole term. That is another way of confirming that at the surface of the Earth we are still not able to locate a magnetic monopole.

For completeness, we can also obtain the magnetic multipole using volume integral instead of line integral. By using the current density $\boldsymbol{J}$ defined in Tables 3.1 and 3.10, we can alternatively write the magnetic potential as a volume integral of the current density $\boldsymbol{J}$ and expand the result to obtain the magnetic multipole; i.e.,

$$
\begin{aligned}
A_i(\boldsymbol{x}) &= k_M \int \frac{J_i(\boldsymbol{x})}{|\boldsymbol{x} - \boldsymbol{x}'|} dV(\boldsymbol{x}') \\
&\approx \underbrace{\frac{k_M}{|\boldsymbol{x}|} \left[\int_V J_i(\boldsymbol{x}') dV(\boldsymbol{x}') \right]}_{\text{Monopole}} + \underbrace{\frac{k_M}{|\boldsymbol{x}|^2} \left[\int_V J_i(\boldsymbol{x}') x^{(j)'} dV(\boldsymbol{x}') \right] \mathbf{i}_j}_{\text{Dipole}} \\
&\quad + \underbrace{\frac{k_M}{2|\boldsymbol{x}|^3} \left[\int_V J_i(\boldsymbol{x}') \left(3x_i^{(j)'} x_i^{(k)'} - \delta_{jk} |\boldsymbol{x}_i'|^2 \right) dV(\boldsymbol{x}') \right] \mathbf{i}_j \mathbf{i}_k}_{\text{Quadrupole}} + ...
\end{aligned}
$$

where $k_M = \mu_0/(4\pi)$ and A_i is the i-component of the vector $\boldsymbol{A}$. So we can see that the leading-order contribution to the magnetic vector potential is given by

$$A_i \approx k_M \int_V J_i(\boldsymbol{x}')dV(\boldsymbol{x}') = k_M \int_V \left[\boldsymbol{\nabla} \cdot (x_i'\boldsymbol{J}) - x_i'\boldsymbol{\nabla} \cdot \boldsymbol{J} \right] dV(\boldsymbol{x}') = 0 \ .$$

We use the continuity equation (i.e., $\boldsymbol{\nabla} \cdot \boldsymbol{J} = 0$) in these derivations. This is the monopole term; the monopole moment is just the total current. The total current always vanishes for a static current distribution, because otherwise there would be a net movement of charge over time. Thus, not only are there no magnetic monopoles, but there is also no way to set up a static current distribution that produces a $1/|\boldsymbol{x}|$ magnetic vector potential. The magnetic vector potential always falls off as $1/|\boldsymbol{x}|^2$ or faster for a static current distribution. The dipole term is

$$A_i \approx \frac{k_M}{|\boldsymbol{x}|^2} \left[\int_V J_i(\boldsymbol{x}')x^{(j)'}dV(\boldsymbol{x}') \right] \boldsymbol{i}_j \ . \tag{3.133}$$

The bracketed object is a second-rank tensor; it has two free indices: i and j. This second-rank tensor is the magnetic dipole moment. Likewise, the magnetic quadrupole moment is a third-rank tensor; it has two free indices: i, j, and k.

Example: The magnetic field due to a circular current loop. Consider a circular loop of radius a in the $x - y$ plane carries a steady current I, as shown in Figure 3.22b. By using the definitions in Figure 3.22b, we have

$$\boldsymbol{l}' = a(\cos\theta\,\boldsymbol{i}_1 + \sin\theta\,\boldsymbol{i}_2) \Longrightarrow d\boldsymbol{l}' = a\,d\theta(-\sin\theta\,\boldsymbol{i}_1 + \cos\theta\boldsymbol{i}_2) \tag{3.134}$$

and

$$\frac{(\boldsymbol{x} - \boldsymbol{x}')}{|\boldsymbol{x} - \boldsymbol{x}'|^3} = \frac{-a\cos\theta\boldsymbol{i}_1 - a\sin\theta\boldsymbol{i}_2 + z\boldsymbol{i}_3}{[a^2 + z^2]^{3/2}} \ , \tag{3.135}$$

with $\boldsymbol{x} = [0, 0, z]^T$ and $\boldsymbol{x}' = a[\cos\theta, \sin\theta, 0]^T$. From (3.124) we can deduce the magnetic field at a distance z along the axis of a circular loop with radius a and current I as follows:

$$\begin{aligned}
\boldsymbol{B} &= \frac{\mu_0 I}{4\pi} \int_0^{2\pi} d\theta\, \frac{(-\sin\theta\boldsymbol{i}_1 + \cos\theta\boldsymbol{i}_2) \times (-\cos\theta\boldsymbol{i}_1 - \sin\theta\boldsymbol{i}_2 + z\boldsymbol{i}_3)}{a[1 + (z/a)^2]^{3/2}} \\
&= \frac{\mu_0 a I}{4\pi} \int_0^{2\pi} d\theta\, \frac{z\cos\theta\boldsymbol{i}_1 + z\sin\theta\boldsymbol{i}_2 + a\boldsymbol{i}_3}{a[1 + (z/a)^2]^{3/2}} \\
&= \boldsymbol{i}_3\, \frac{\mu_0 I}{2a[1 + (z/a)^2]^{3/2}} \ .
\end{aligned} \tag{3.136}$$

Remember that $\boldsymbol{i}_1 \times \boldsymbol{i}_3 = -\boldsymbol{i}_2$, $\boldsymbol{i}_2 \times \boldsymbol{i}_3 = \boldsymbol{i}_1$, and $\boldsymbol{i}_2 \times \boldsymbol{i}_1 = -\boldsymbol{i}_3$. The x and y components of $\boldsymbol{B}$ are zero. Thus, we see that B_z is the only non-vanishing component of the magnetic field. This conclusion can also be reached by using the symmetry arguments.

Example: The magnetic field due to a solenoid (helical coil). Consider a solenoid of length l with N turns (loops). The solenoid can be considered as stacked up

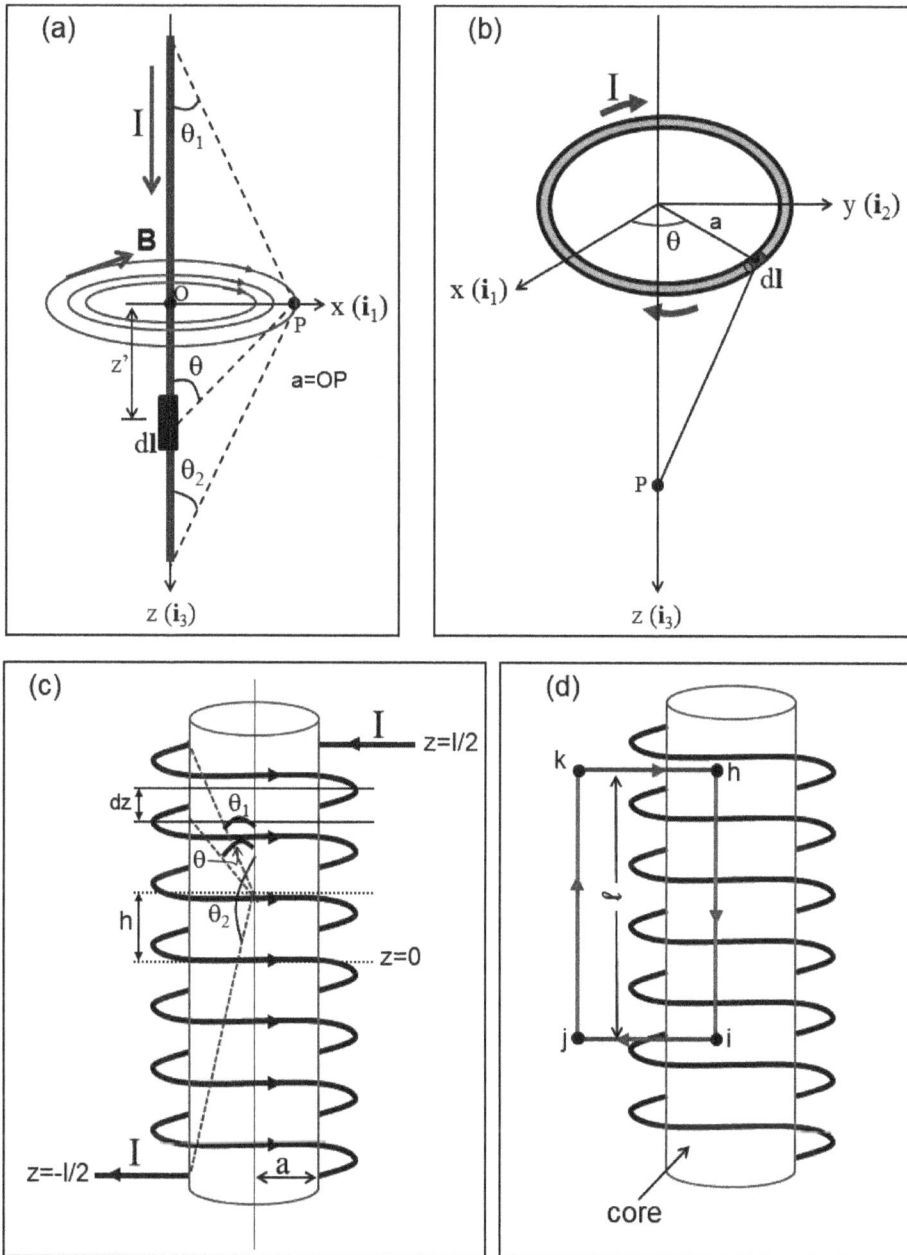

Figure 3.22. (a) Magnetic field of a single wire. The shape of the magnetic field around a long, straight wire carrying electric current. (b) Looping the wire turns the magnetic field into a toroidal (donut) shape. (c) A coil wrapped around a solenoid. (d) An illustration of the rectangular path used to calculate the magnetic field strength inside an ideal infinite solenoid based on Ampere's law in (3.122).

circular coils here along the z-axis, as illustrated in Figure 3.22c. The number of turns within z and $z + dz$ is $N dz/l = n dz$, where n is the number of turns per unit length wrapped around a cylindrical tube of radius a. The current carried by these turns is $nI dz$. Based on the results in the previous example, the field at P due to the turns between z and $z + dz$ is

$$dB = i_3 \frac{\mu_0 a^2}{2[a^2 + (z-h)^2]^{3/2}} (nI dz) . \tag{3.137}$$

By using the definitions in Figure 3.22c, we have

$$z - h = \frac{a}{\tan\theta} \implies dz = -\frac{a}{\sin^2\theta} d\theta \tag{3.138}$$

and

$$\frac{a^2}{[a^2 + (z-h)^2]^{3/2}} = \frac{\sin^3\theta}{a} \tag{3.139}$$

and integrating (3.137) over the entire length of the solenoid, we obtain

$$\begin{aligned} \boldsymbol{B} &= i_3 \frac{\mu_0\, n\, I}{2} \int_{-l/2}^{l/2} \frac{a^2}{[a^2 + (z-h)^2]^{3/2}} dz \\ &= -i_3 \frac{\mu_0\, n\, I}{2} \int_{\theta_1}^{\theta_2} \sin\theta\, d\theta = i_3 \frac{\mu_0\, n\, I}{2} [\cos\theta_1 - \cos\theta_2] . \end{aligned} \tag{3.140}$$

For an infinite solenoid, $\theta_1 = 0$ and $\theta_2 = \pi$, we obtain

$$\boldsymbol{B} = i_3\, \mu_0\, n\, I = i_3 \frac{\mu_0 NI}{l} . \tag{3.141}$$

These results are valid for points inside the solenoid but $|\boldsymbol{B}| = 0$ outside the solenoid.

We can also use Ampere's law in (3.122) to calculate the magnetic field strength inside an ideal infinite solenoid. We choose a rectangular path of length l and width w, as illustrated in Figure 3.22d, with one part of the path inside the solenoid and one part of the path outside. The line integral of $\boldsymbol{B}$ along this loop is

$$\oint_C \boldsymbol{B}(\boldsymbol{x}) \cdot d\boldsymbol{l}(\boldsymbol{x}) = \int_h^i \boldsymbol{B}(\boldsymbol{x}) \cdot d\boldsymbol{l}(\boldsymbol{x}) = Bl , \tag{3.142}$$

assuming that the $\boldsymbol{B}$-field is constant. The contributions along paths ij and kh are zero because $\boldsymbol{B}$ is perpendicular to $d\boldsymbol{l}$ along these paths. We assume that the path jk on the outside is sufficiently far from the solenoid, compared to the diameter of the solenoid, that the field is approximately equal to zero for an ideal solenoid, which is infinitely long with turns tightly packed. We can deduce that

$$\oint_C [\boldsymbol{B}(\boldsymbol{x}) \cdot d\boldsymbol{l}(\boldsymbol{x})] = \mu_0 NI \implies B = \frac{\mu_0 NI}{l} . \tag{3.143}$$

We used the fact that the total current passing through the area defined by the integration path is NI where N is the number of turns and I is the current in the loop. So the right-hand side of Ampere's law is $\mu_0 NI$.

As illustrated in Figure 3.23b, a loop creates a dipole-shaped magnetic field and many loops in a solenoid shape create a dipole-shaped magnetic field (see Figure 3.23b) similar to that of a bar magnet described in Ikelle (2023b). So the motion of electrical charges (current) is the only source of magnetism and a changing magnetic

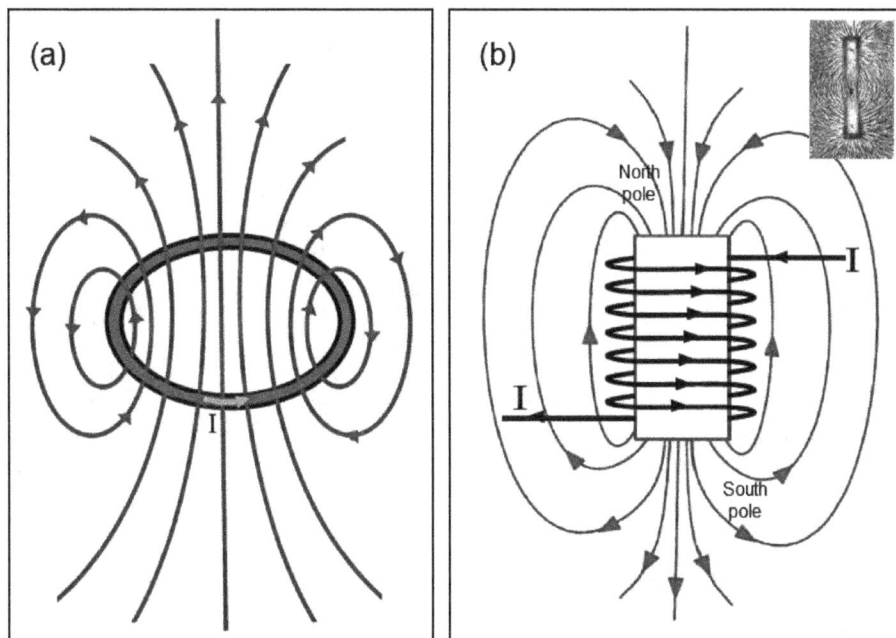

Figure 3.23. (a) Magnetic-field lines due to a current loop. (b) Magnetic-field lines due to a solenoid carrying current.

field in the vicinity of a wire or coil will induce a voltage (current) in the wire or coil.

Dipole nature of magnetic materials. Just like dielectric materials, magnetic materials also have dipoles, but magnetic dipoles (i.e., small magnets). Again magnetic dipoles is a bar magnet (Figure 3.24) with a north pole and a south pole. In other words, on the atomic (microscopic) scale, some atoms of magnetic materials

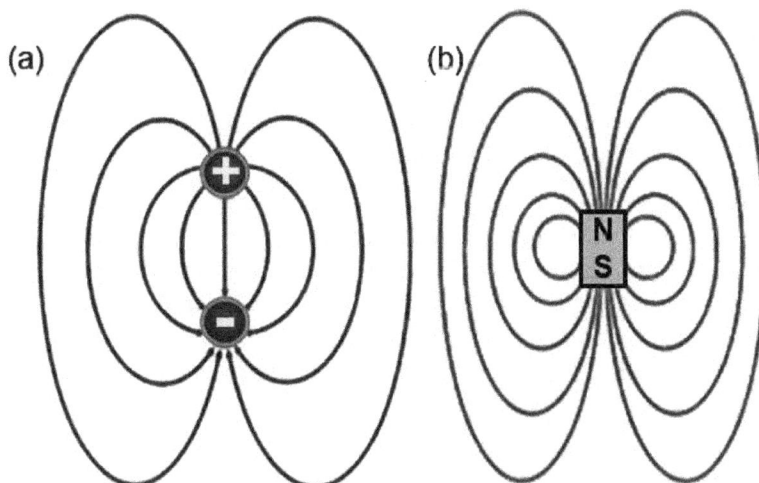

Figure 3.24. (a) Electric dipole and (b) Magnetic dipole. Electric dipoles can be isolated to point charges. A point source of magnetic north or south has never been isolated or discovered on Earth.

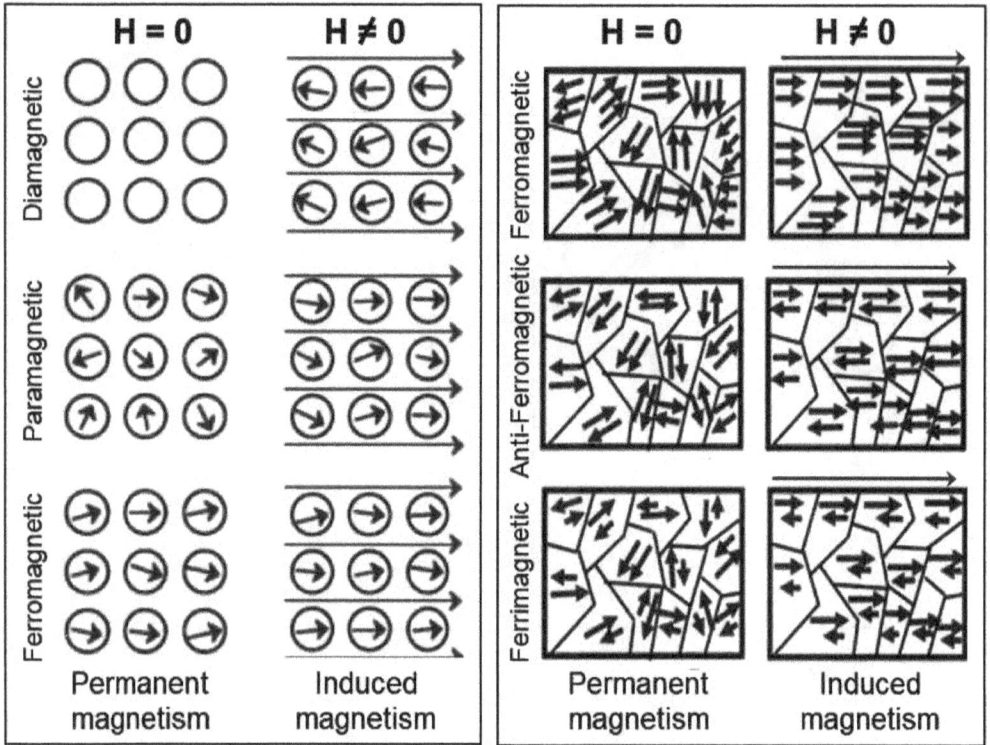

Figure 3.25. (left) Magnetization of diamagnetic, paramagnetic, and ferromagnetic materials. Diamagnetic materials do not have permanent magnetic atoms. The induced atoms orient themselves opposite the applied field. Paramagnetic materials have randomly permanent magnetic atoms. The induced atoms are parallel to the applied field direction and orient themselves along the direction of the applied field. Ferromagnetic materials have permanent aligned magnetic atoms that produce large magnetic moments. The induced atoms are parallel to the applied field direction and align themselves in the direction of the applied field. (right) Magnetization of ferromagnetic, antiferromagnetic, and ferrimagnetic materials.

behave as magnetic dipoles; that is to say, they are magnetic. Other atoms can be non-magnetic. These dipoles interact with an applied magnetic field to give a characteristic (macroscopic) magnetic flux density (magnetic displacement) in the material. Magnetic materials are classified as diamagnetic, paramagnetic, and ferromagnetic materials.

In the absence of an applied magnetic field, each atom of a diamagnetic material has a net zero magnetic dipole moment. Figure 3.25a illustrates this. Basically, in diamagnetic materials, for every electron with a certain magnetic moment, there is one with an opposite magnetic moment. This results in zero net magnetic moment for the whole atom. That is for every electron circling the nucleus clockwise, there is one circling counter-clockwise. Diamagnetic materials have no net permanent magnetic moments and magnetization. In the presence of an external magnetic field, the induced magnetic dipole moments align themselves opposite to the applied field (i.e., magnetization develops in the direction opposite to the magnetic field in accordance with Figure 3.25). In other words, diamagnetic materials exhibit no permanent magnetism, and the induced magnetic moment disappears when the applied field is withdrawn.

Paramagnetic materials have permanent dipole magnetic moments whereas diamagnetic substances have zero dipole moments. In paramagnetic materials, the atoms carry a net magnetic moment due to a few more electrons moving clockwise than counter-clockwise or vice versa. However, magnetic moments on different atoms point randomly. That is, without an external magnetic field, each atom of a paramagnetic material has a net non-zero (but weak) magnetic dipole moment. These magnetic dipole moments are randomly oriented, as illustrated in Figure 3.25a, so that the macroscopic magnetization is zero. In the presence of an external magnetic field, the magnetic dipoles align themselves with the external field.

In ferromagnetic materials, the atoms carry a net magnetic moment. However, in contrast to paramagnetic materials the magnetic moments are not random, but aligned. This alignment is associated with the domain. Domains are regions of many atoms with aligned dipoles, as depicted in Figure 3.25b. In the absence of an applied magnetic field, the domains are randomly oriented so that the net macroscopic magnetic field is zero. When a specimen of ferromagnetic material is placed in a magnetic field, the magnetic moments of its domains tend to rotate in alignment with the direction of the applied field.

The border class of ferromagnetic materials also includes two classes of magnetic materials. They are antiferromagnetic and ferrimagnetic materials. In antiferromagnetic materials, the magnetic moments of individual atoms are strong, but adjacent atoms align in opposite directions as illustrated in Figure 3.25b. In other words, atoms have anti-parallel magnetic moments. The macroscopic magnetization of the material is negligible even in the presence of an applied field, as illustrated in Figure 3.25b. Anti-ferromagnetic materials include transition metals, chromium and manganese, and many of their compounds such as manganese oxide, cobalt oxide, nickel oxide, chromium oxide, manganese sulfide, manganese selenide, and cupric chloride.

Ferrimagnetic materials exhibit behavior between ferromagnetism and antiferromagnetism. In the absence of an applied magnetic field, the domains are randomly oriented so that the net macroscopic magnetization is zero. This is illustrated in Figure 3.25b. In other words, atoms have mixed parallel and anti-parallel aligned magnetic moments. In the presence of an applied magnetic field, the domains align themselves with the external magnetic field. Ferrites are the most useful ferrimagnetic materials. Ferrites are a ceramic material containing compounds of iron.

In ferromagnetic materials, the relationship between magnetic displacement and magnetic field intensity is very non-linear (i.e., the relationship between B and H is nonlinear) because permeability depends on the previous history of the material (a hysteresis loop between B and H). In other words, ferromagnetic materials have memory. Starting with an unmagnetised ferromagnetic sample. Both B and H are zero. We are at point O on the magnetisation curve in Figure 3.26. If H is increased in a positive direction then B increases from O to saturation point a. Now if H is reduced to zero, B also decreases but does not reach zero. B will move from a to b due to residual magnetism within the sample. To reduce B to zero we have to apply a force called a coercive force. This coercive force reverses the magnetic field rearranging the atomic magnets until the sample becomes unmagnetised (i.e., at c). An increase in this reverse current causes the sample to magnetise in the opposite direction. Increasing H further will cause the sample to reach its saturation point but in the opposite direction (i.e., point d). This point is symmetrical to point b.

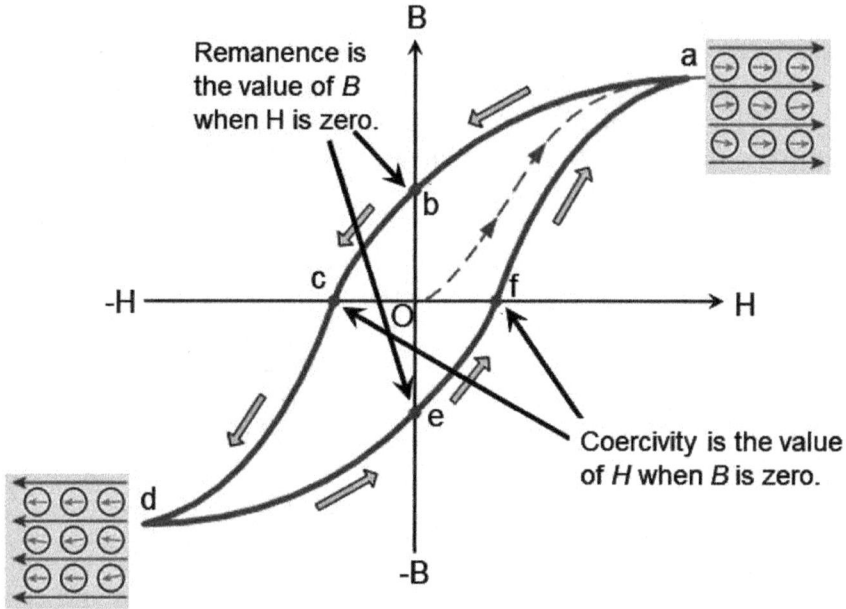

Figure 3.26. Ferromagnetic hysteresis loop.

If H is reduced again to zero the residual magnetism present in the sample will be equal to the previous value but in reverse at point e. Again increasing H in a positive direction will result in B reaching zero at point f. As before increasing H further in a positive direction will cause the sample to reach saturation at point a. The effect of magnetic hysteresis shows that ferromagnetic materials have memory because they remain magnetised after the external magnetic field has been removed. Note that soft ferromagnetic materials such as iron or silicon steel have very narrow magnetic hysteresis loops resulting in very small amounts of residual magnetism.

3.1.9. Ground Penetration Radar

Surface electromagnetic surveys (i.e., without drilling a borehole) used to reconstruct electric resistivity (or conductivity) models of the subsurface include magnetotelluric (MT) surveys and ground penetrating radar (GPR). In the MT surveys, the EM fields are generated naturally by telluric currents caused by solar wind interactions with the Earth's magnetosphere and ionosphere and thunderstorms, lightning (Figure 3.27). The period of telluric currents ranges between periods of 0.1 s and 10^5 s (or between frequencies of 10 Hz and 10^{-5} Hz). Telluric currents with frequencies higher than 5 Hz are caused by world-wide thunderstorms and those with frequencies less than 0.5 Hz are generated from solar wind interactions with the Earth's magnetosphere and ionosphere. We have a dead-band between 0.5 Hz and 5 Hz in which natural EM fluctuations are very low in intensity and thus negligible. So tens of kilometers of depth of resistivity can be reconstructed from MT data because the deep penetration of low frequencies is large.

GPR uses radar pulses (electromagnetic radiation in the microwave band of the radio spectrum) to emit electromagnetic energy into the ground and uses the reflected signals to detect subsurface structures. These pulses are usually in the range 10 MHz to 2.6 GHz. Because of frequency-dependent attenuation mecha-

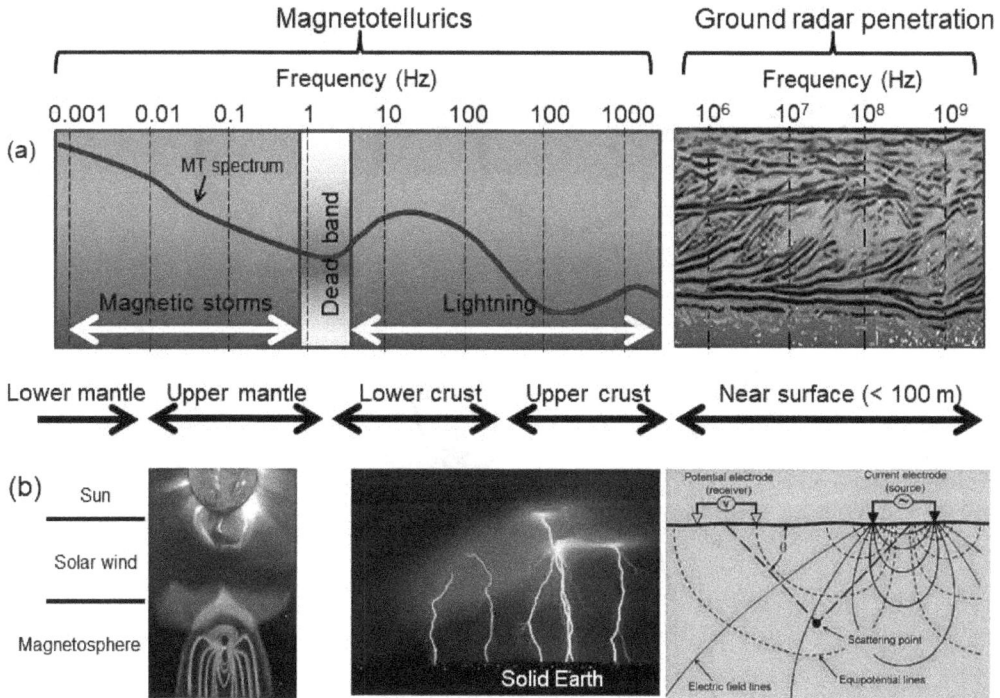

Figure 3.27. (a) Spectrum of MT current sources and ground-penetration-radar (GPR) sources. The MT sources produce signals between $10 * [-5]$ Hz and 10 Hz as a result of thunderstorms (usually near the equator) and the interaction of solar wind with the Earth's magnetic field. Central Africa (e.g., Cameroon, Northern Congo, and Uganda), northern South America (e.g., Colombia, Venezuela, and Northern Brazil), all of Central America, and Indonesia are the major lightning storm centers. Ground-penetrating radars emit electromagnetic pulses into the ground. These pulses are usually in the range 10 MHz to 2.6 GHz. We have a dead-band between 0.5 Hz and 5 Hz where natural electromagnetic fluctuations are very low.

nisms, higher frequencies (e.g., GPR) do not penetrate as far as lower frequencies (e.g., MT). However, higher frequencies can provide a higher resolution description of resistivity than lower frequencies. The principles involved in GPR are similar to seismology (see Chapter 1 of Ikelle (2023a)), except GPR methods use electromagnetic energy rather than acoustic/elastic energy, and energy may be reflected at boundaries where subsurface electrical properties change rather than subsurface mechanical properties, as is the case with seismic energy.

Almost all modern spacecraft landing on Mars are equipped with GPR systems in order to learn about the underground of Mars and to search for possible living things, as such lives may only be likely underground. Possible evidence of humanoid lives includes finding underground caves and pipes and buried cables, determining the thickness and structure of layers and ice glaciers, and detecting buried containerized hazardous waste. The question is how deep we can realistically expect to see with GPRs? We here provide some answers to this question by discussing EM wave propagation in a linear homogeneous and isotropic conductor—that is, by assuming that there is no crystalline structure, no inhomogeneities, no voids, and no defects in the conductor.

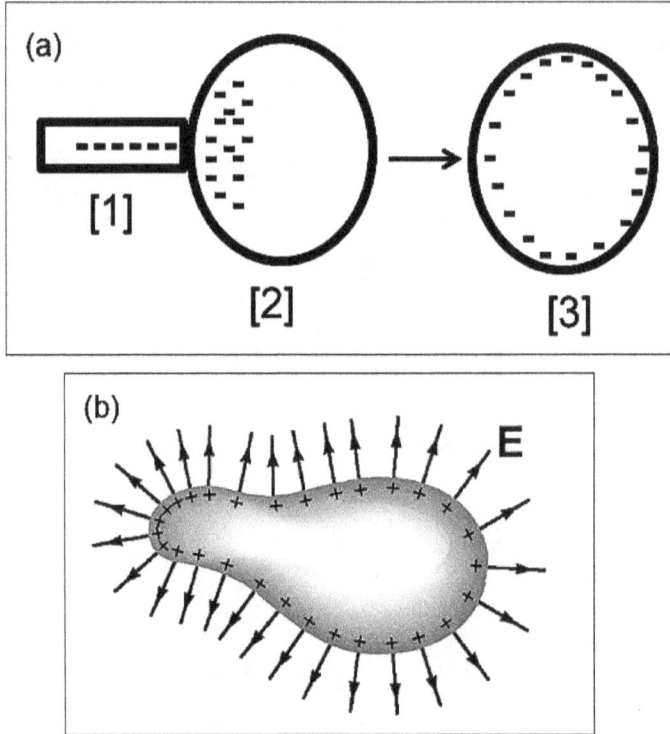

Figure 3.28. An illustration of how electrons are free to move around within a *conductor*. Remember that electrons in insulators are closely and tightly joined to atoms and this is the main reason that no current flows through *insulators*. (a) [1] When charges are transferred to a conductor, [2] they repel each other and start to move apart. [3] They end up distributed over the surface of the conductor. In other words, the electrons quickly rearrange and move to a configuration where the electric field inside the conductor vanishes. Thus, all the charges on a conductor are at its surface. (b) Another illustration of the fact that the electric field inside a conductor is equal to zero. The electric field is perpendicular to the surface of the conductor, just outside the conductor.

The differential equation of the free-charge density. Before we write Maxwell's equations for a conductor, it is useful to rederive the differential equations for the free-charge density because it allows us to make the assumption that we are dealing with an uncharged conductor.

Substituting the Ohm's law, $\boldsymbol{J} = \sigma \boldsymbol{E}$, into the continuity equation (charge conservation), in (3.13), and using Gauss's law of electricity, $\boldsymbol{\nabla} \cdot \boldsymbol{D} = \rho_E$, to eliminate the electrical fields from the continuity equation, we obtain

$$\frac{1}{\rho_E} \frac{\partial \rho_E}{\partial t} + \frac{\sigma}{\varepsilon} = 0 \Longrightarrow \rho_E(\boldsymbol{x}, t) = \rho_E(\boldsymbol{x}, 0) \exp\left[-\frac{\sigma}{\varepsilon} t\right] , \qquad (3.144)$$

with $\rho_E(\boldsymbol{x}, 0)$ as the initial charge density. This relation is actually another expression of the continuity equation. It shows that if we put some charge on a conductor, it will flow out to the edges in a characteristic time $\tau_c = \varepsilon/\sigma$ (see Figures 3.28, 3.29, and 3.30). Consider, for example, a conductor made of pure copper with a conductivity $\sigma = 5.95 \times 10^7$ S/m. If we assume $\varepsilon = 3\varepsilon_0 = 3 \times 8.85 \times 10^{-8}$ F/m, then

$$\tau_c = \frac{\varepsilon}{\sigma} = 4.5 \times 10^{-19} \text{ second .} \qquad (3.145)$$

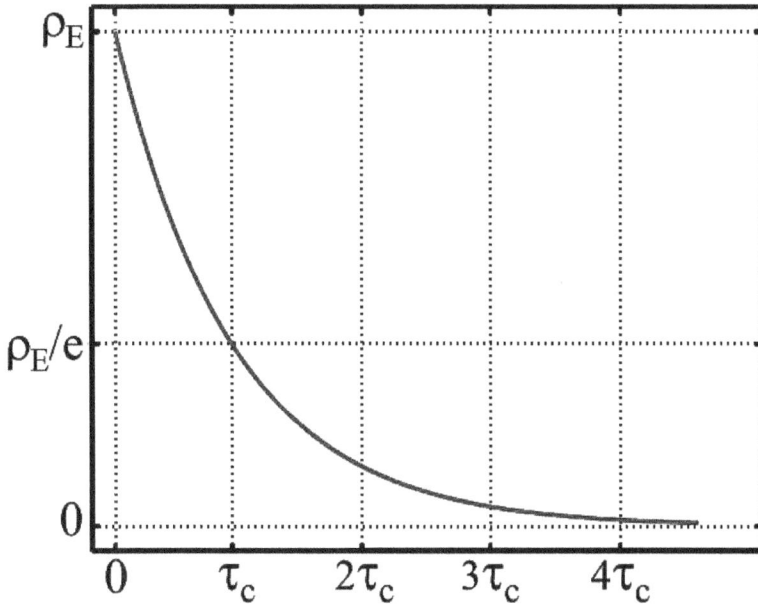

Figure 3.29. An illustration of the general solution of the differential equation for charge density, which is a damped exponential. Characteristic damping time is $\tau_c = \varepsilon/\sigma$, where ε is the permittivity and σ is the conductivity. This solution tells us that any charge density initially present at time zero is exponentially damped and dissipated in a characteristic time, τ_c.

However, the characteristic collision time of free electrons in pure copper is $\tau_{\text{collision}} = 3.2 \times 10^{-13}$ second; hence $\tau_c \ll \tau_{\text{collision}}$. We can see that the characteristic time in pure copper is very small compared to the collision time in pure copper. So, if we are willing to wait a short time (e.g., a pico of a second) then, any initial free-charge density accumulated inside a good conductor at time zero will have dissipated away/damped out, and from that time onward, free-charge density can safely be assumed to be zero in a conductor. That is the assumption we are making here for Maxwell's equations in a conductor.

The EM wave equation. For a good conductor with $\rho_E(\boldsymbol{x}, t > \Delta t) = 0$ (e.g., $\Delta t = 20\tau_c$), we can write Maxwell's equations for $t \geq \Delta t$ (an uncharged ohmic conductor) as follows:

$$-\boldsymbol{\nabla} \times \boldsymbol{H} + \sigma \boldsymbol{E} + \varepsilon \frac{\partial \boldsymbol{E}}{\partial t} = 0 \tag{3.146}$$

$$\boldsymbol{\nabla} \times \boldsymbol{E} + \mu \frac{\partial \boldsymbol{H}}{\partial t} = 0 \tag{3.147}$$

$$\boldsymbol{\nabla} \cdot \boldsymbol{E} = 0 \tag{3.148}$$

$$\boldsymbol{\nabla} \cdot \boldsymbol{H} = 0 \, . \tag{3.149}$$

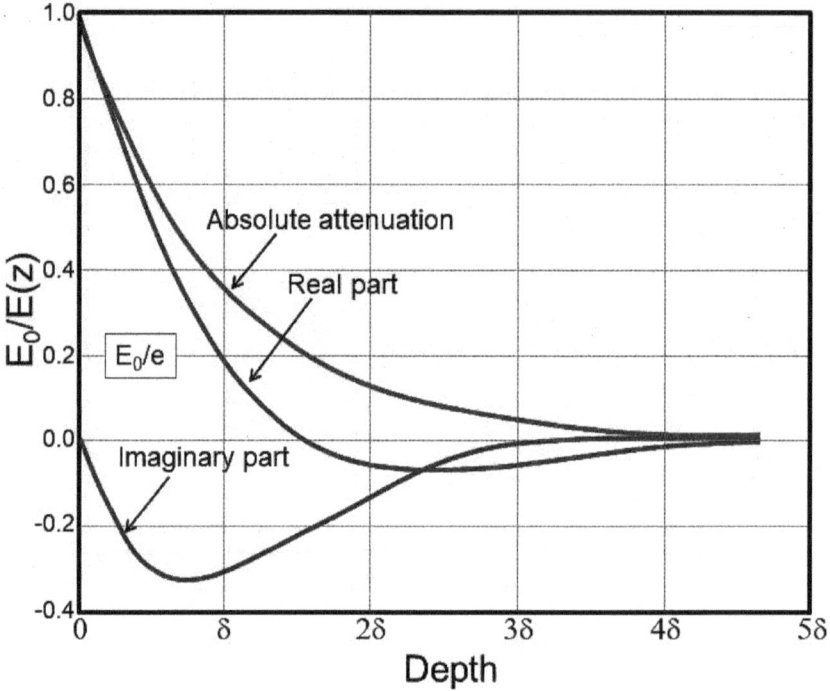

Figure 3.30. The depth-dependent behavior of the E-field. For the absolute value, the E-field attenuates to $1/e$ strength of the surface value at the skin-depth δ.

Taking the curl of Faraday's law in (3.147), then making use of Gauss's law for the electric field in (3.148) and Ampère's law, in (3.146), we obtain the following electromagnetic wave equation for the field E in a conductor

$$\nabla^2 E = \mu\sigma\frac{\partial E}{\partial t} + \mu\varepsilon\frac{\partial^2 E}{\partial t^2} \ . \tag{3.150}$$

Similarly by taking the curl of Ampère's law, then making use of Gauss's law for the magnetic field and Faraday's law, we obtain the following wave equation for the magnetic field H for a conductor

$$\nabla^2 H = \mu\sigma\frac{\partial H}{\partial t} + \mu\varepsilon\frac{\partial^2 H}{\partial t^2} \ . \tag{3.151}$$

Note that the wave equations for electric and magnetic fields in a conductor have additional single-time derivative terms compared to similar equations in vacuum (Ikelle, 2023b) and for acoustic media (Ikelle, 2023c). These terms are zero for poor conductors (that is, when σ is zero or very small). They are also referred to as dissipative terms because they allow current to flow through the medium.

Let us focus on the wave equation of the electric field. The general form of the solution to the wave equation in (3.150) is

$$E(x, t) = E_0(x, t) \exp\left[-i\omega t\right] \ . \tag{3.152}$$

Substitution of this field in the wave equation gives

$$\nabla^2 E = \mu\sigma\frac{\partial E}{\partial t} + \mu\varepsilon\frac{\partial^2 E}{\partial t^2} \Longleftrightarrow \nabla^2 E = -i\omega\mu\sigma E_0 - \omega^2\mu\varepsilon E_0 \ , \tag{3.153}$$

where $-i\omega\mu\sigma\boldsymbol{E}_0$ represents conduction current and $\omega^2\mu\varepsilon\boldsymbol{E}_0$ represents displacement current. The ratio of displacement current to conduction current, $R_0 = \omega\varepsilon/\sigma$, determines the relevant term in a given application. For a low-frequency signal at about 1 Hz and conductivity of about 0.01 S/m (i.e., magnetotellurics and assuming a free space value of dielectric permittivity, we have $R_0 = 5.56 \times 10^{-9}$. The conduction current is dominant in this case, i.e.,

$$\boldsymbol{\nabla}^2\boldsymbol{E} = \mu\sigma\frac{\partial\boldsymbol{E}}{\partial t} + \mu\varepsilon\frac{\partial^2\boldsymbol{E}}{\partial t^2} \implies \boldsymbol{\nabla}^2\boldsymbol{E} = \mu\sigma\frac{\partial\boldsymbol{E}}{\partial t} \; ; \text{(diffusion eq.)} \qquad (3.154)$$

That is, if we ignore the second-time derivative term of the wave equation, we end up with a diffusion-type equation. For a high-frequency signal at about 1 GHz and conductivity of about 10^{-5} S/m (i.e., ground-penetrating radar), and assuming a free space value of dielectric permittivity, we have $R_0 = 5561$. The displacement current is dominant and the signal travels as an electromagnetic wave, i.e.,

$$\boldsymbol{\nabla}^2\boldsymbol{E} = \mu\sigma\frac{\partial\boldsymbol{E}}{\partial t} + \mu\varepsilon\frac{\partial^2\boldsymbol{E}}{\partial t^2} \implies \boldsymbol{\nabla}^2\boldsymbol{E} = \mu\varepsilon\frac{\partial^2\boldsymbol{E}}{\partial t^2} \; ; \text{(wave eq.)} \qquad (3.155)$$

So for high-frequency methods such as ground-penetrating radar (GPR) dielectric permittivity is the most important property.

The 1D-EM wave equation and skin depth. Let us consider the one-dimensional (along z-axis) version of the wave equation, shown in (3.150), in order to gain more intuition on magnetic energy lost with depth. We can assume the following complex plane-wave solution to this wave equation

$$\boldsymbol{E}(z,t) = \boldsymbol{E}_0 \exp\left[i(kz - \omega t)\right] \; , \qquad (3.156)$$

which is a decaying exponential in the direction of the propagation of the EM wave and $\boldsymbol{E}_0$ is the electric intensity at $z = 0$ (e.g., the Earth's surface). Substituting the above solution, in (3.150), we get the following dispersion relation

$$k^2 = \mu\varepsilon\omega^2 + i\mu\sigma\omega = [k_r + ik_i]^2 \; , \qquad (3.157)$$

with

$$k_r = \omega\sqrt{\frac{\mu\varepsilon}{2}\left[1 + \sqrt{1 + \left(\frac{\sigma}{\omega\varepsilon}\right)^2}\right]} \; , \qquad (3.158)$$

$$k_i = \frac{\omega\mu\sigma}{2k_r} = \omega\sqrt{\frac{\mu\varepsilon}{2}\left[-1 + \sqrt{1 + \left(\frac{\sigma}{\omega\varepsilon}\right)^2}\right]} \; . \qquad (3.159)$$

This relation tells us that the wave vector, $\boldsymbol{k} = k\mathbf{i}_3 = (k_r + ik_i)\mathbf{i}_3$, which is a vector in the z-direction, is a complex quantity. Notice that when the conductivity is zero (that is when we are dealing with dielectric media), the wave vector is real. This result physically means that EM wave propagation in a conductor is dispersive (that is, EM wave propagation is frequency dependent).

The real part of the wave vector in a conductor, k_r, is related to the wavelength of the wave ($\lambda = 2\pi/k_r$ and $v = \omega/k_r$) while the imaginary part, k_i, is related to its amplitude. As far as frequencies much smaller than $\varepsilon\mu$ [see equation (3.157)], the distance, δ, that a wave travels before its amplitude falls below $\exp[-1] = 0.368$ of the original value at $z = 0$ is

$$\delta = \frac{1}{|k_i|} = \left(\frac{2}{\omega\mu\sigma}\right)^{\frac{1}{2}} = \left(\frac{1}{\pi\mu}\right)^{\frac{1}{2}} \left(\frac{T}{\sigma}\right)^{\frac{1}{2}} \approx 503 \left(\frac{T}{\sigma}\right)^{\frac{1}{2}}, \qquad (3.160)$$

where $T = 2\pi/\omega$ is the period in seconds. In the last approximation of (3.160), we assume that $\mu = \mu_0 = 4\pi \times 10^{-7}$ H/m. This distance is known as the skin depth. It describes the penetration depth of EM waves as a function of the conductivity of the material and the frequencies used. At the skin depth, field amplitudes are reduced by about 37 percent. For example, in the case of copper, $\delta \approx 6$ cm at 1 Hz and $\delta \approx 2$ mm at 1 kHz, the high-frequency waves are found to penetrate less into the conductor. Table 3.6 shows skin depths for some layers of the subsurface as a function of conductivity and frequency only. These depths tell how quick an electromagnetic wave decays in a medium. For a perfectly conducting medium (i.e., $\sigma \to \infty$) then skin depth is zero (no penetration).

GPR simulation of Antarctic subglacial lakes. Along with seismic exploration, ground-penetrating radar (GPR) is a popular geophysical method for high-resolution imaging of the shallow subsurface over a large area. GPR is an attractive tool for Antarctica exploration due to its portability and non-destructive nature, as well as its rapid data acquisition and excellent signal penetration through the ice sheet. Figure 2.61 shows a section of Antarctica's shallow subsurface containing Lake Vostok. This is one of the 70 or so lakes beneath the Antarctic ice sheet. The discovery and/or confirmation of these lakes has been made possible by GPR data-acquisition surveys and GPR data imaging, especially at the time when liquid water from none of the lakes was not sampled directly yet because the lakes are subject to high pressure (about 355 bars), low temperatures (about $-3°$degrees Celsius), and permanent darkness.

GPR systems are typically composed of a source (transmitter) that transmits electromagnetic (EM) signals into the solid Earth and receiver antennas that capture the reflected, refracted, and scattered signals. The shapes of collected signals are used to identify lakes and bedrock beneath ice sheets. Sometimes we limit GPR data acquisition to a straight line and assume the medium is 2D. For example, the medium can vary in x and z, and is invariant in y and acquisition is carried out along the x-axis for a fixed z. In other words, the Earth model is assumed to be invariant along the crosslines. This is, of course, not true. Because of the structural features of lakes and bedrock beneath ice sheets, reflected, refracted, and scattered signals outside the vertical plane of acquisition will be recorded in the 2D data. When the source is fired, electromagnetic energy propagates outward in an expanding wavefront. In the absence of any variation in the geologic structure outside the vertical plane of the 2D acquisition line, only reflections returning from within the vertical plane would be recorded. The presence of reflections from outside the plane is one of the inherent problems with 2D GPR data. It is usually difficult to distinguish reflection events within the vertical plane of the 2D seismic line from

Transverse Electric (TE) $[H_y, E_x, E_z]$	Transverse Magnetic (TM) $[H_x, H_z, E_y]$
$$\mu\frac{\partial H_y}{\partial t} = -\frac{\partial E_x}{\partial z} + \frac{\partial E_z}{\partial x}$$	$$\mu\frac{\partial H_x}{\partial t} = \frac{\partial E_y}{\partial z}$$
$$\varepsilon\frac{\partial E_x}{\partial t} + \sigma E_x = -\frac{\partial H_y}{\partial z}$$	$$\mu\frac{\partial H_z}{\partial t} = -\frac{\partial E_y}{\partial x}$$
$$\varepsilon\frac{\partial E_z}{\partial t} + \sigma E_z = \frac{\partial H_y}{\partial x}$$	$$\varepsilon\frac{\partial E_y}{\partial t} + \sigma E_y = \frac{\partial H_x}{\partial z} - \frac{\partial H_z}{\partial x}$$

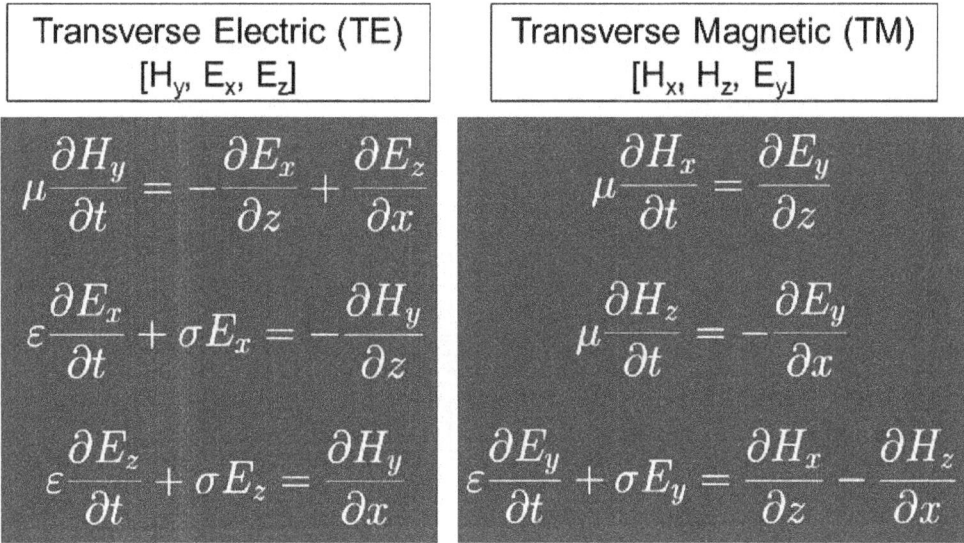

Figure 3.31. Field patterns propagating independently in 2D media according to Maxwell's equations. In the transverse electric (TE) mode, we have $E_z = 0$ and $H_z \neq 0$. In Transverse magnetic (TM) mode, we have $E_z \neq 0$ and $H_z = 0$.

out-of-plane reflections. If these reflections are ignored in our imaging schemes, we can get inaccurate subsurface representations. Any reflection of the ice sheets will indicate the presence of bedstrocks and lakes if the ice sheets remain homogeneous. Thus 2D acquisition suffices to detect bedrocks and lakes. Because the contracts of ice/liquid water, ice/bedrock, and liquid water/bedrock are significant different strengths as shown in Figure 2.61, we can even identify lakes and bedrock.

When we are limited to two dimensions (2D), we have to assume that both electric and magnetic fields have no dependence on one of the three classical Cartesian dimensions x, y, and z—for example, y; i.e.,

$$\frac{\partial \boldsymbol{E}}{\partial y} = \frac{\partial \boldsymbol{H}}{\partial y} = \boldsymbol{0} \ . \tag{3.161}$$

The fact that $\boldsymbol{E}$ is perpendicular to $\boldsymbol{H}$ means that we have two possible versions of Maxwell's equations known as transverse electric (TE) mode [$\boldsymbol{H} = [0, H_y, 0]^T$, $\boldsymbol{E} = [E_x, 0, E_y]^T$] and transverse magnetic (TM) mode [$\boldsymbol{H} = [H_x, 0, H_z]^T$, $\boldsymbol{E} = [0, E_y, 0]^T$]. So in TE mode, all the electric field components are transverse to the direction of propagation. In TM mode there is only an electric field along the propagation direction. Figure 3.31 shows Maxwell's equations corresponding to these two modes.

Figure 3.32 show examples of electromagnetic wave propagation through the 2D Antarctic subglacial lake model in Figure 2.61. In Figure 3.32, we used TM mode and displayed E_y. Due to limited space, only four of these snapshots of wave propagation through this model are shown in this figure. The source used was

$$s(t) = 0.353 - 0.488\cos\left(2.28\pi\, t\, f_c\right)$$

$$+ \ 0.145\cos\left(4.562\pi\, t\, f_c\right) - 0.010\cos\left(6.844\pi\, t\, f_c\right) \tag{3.162}$$

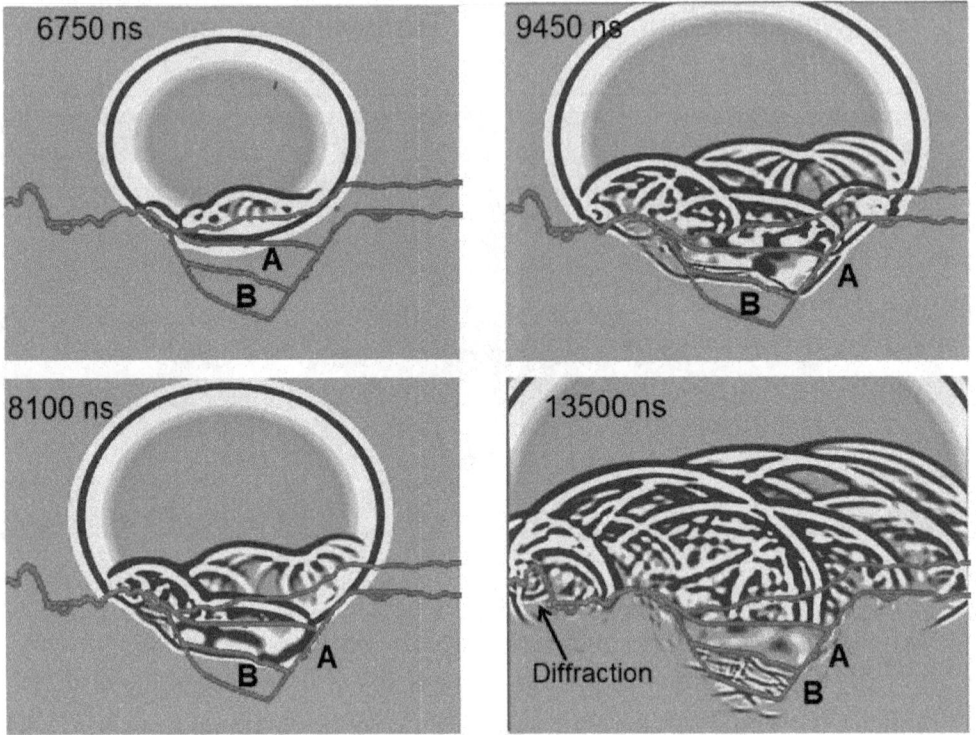

Figure 3.32. Snapshots of electromagnetic wave propagation using transverse magnetic (TM) mode. The quantity shown here is E_y.

where $f_c = 2.83$ MHz. We also used the finite-difference modeling formulation and code described in Appendix A to obtain the results in Figure 3.32. The model in Figure 2.61 was discretized in 1200×1200 with a grid space $dx = dz = 3.2$ m. This grid is quite small but enough to provide a clear insight of key feature electromagnetic wave in Antartica in the presence of a liquid water lake. Actually, Lake Vostok is at least 240 km long and 50 km wide, and lies between 3,750 m (over the south of the lake) and 4,150 m (over the north) beneath the central east Antarctic ice sheet.

Let us return to Figure 3.32. We can recognize some of the reflections and transmissions at the various interfaces. Pictures of snapshots in Figure 3.32 are very complex, especially for $t \geq 8100$ ns. The multiple reflections, refractions, and diffraction are caused the large differences in permittivity, electromagnetic permeability, and electric conductivity between the lake's liquid water and ice and bedrock and the ice (see Figure 2.61). These differences are so large that areas almost no signal pass through the bedrock. In the area denoted "A" of lake the wavelength of the signal has significant increase due to the large contrast mostly in permittivity between ice and liquid water and ice and bedrock. with no reflections. As discussed in Chapter 1 of Ikelle (2023c), diffraction waves radiates downward as well as upwards. The downward diffracted waves in the bedrock latterally entered in area B of where they are trapped in a layer with two boundary with very contracts in electromagnetic properties. Unfortunately the signal cannot move upward thus it cannot be recorded by sensors located at or above ice sheet surface. So the detection of the target below the lake and even inside the bedrock is difficult because we have areas with no reflections. Those interested in exploration inside the bedrock like oil and gas industry most overcome this challenge. The seismic survey, which two wave types (P- and S-waves) with different wave types can be used to overcome as it currently done to explore reservoirs below salt structures.

The snapshots provide us an insight into the electromagnetic wave interaction with the Antarctica subsurface. However, the output of GPR experiments are data recorded at the surface of the ice sheet or above it as illustrated in Figure 3.33a. The GPR device (e.g., airplanes) moves longitudinally through the predefined line at or above the ice emitting and receiving electromagnetic signals. The returning waves from the subsurface due to the changes permittivity, the permeability, and electrical conductivity are coded in the reflected GPR data. The GPR data recorded in this process are then imaged, based on arrival time and strengths of reflections to obtain a model of the subsurface. The time it takes for the wave to travel from the source to the receivers is recorded in the GPR data. From these traveltimes we can reconstruct the depth of the reflector at which the recorded energy has been reflected. Furthermore, the magnitude of the reflected wave allows us to determine the contrast in physical properties that caused the reflection. Thus we reconstruct the locations of the various discontinuities of our geological model and the contrasts of physical properties which characterize these discontinuities. Examples of such reconstructions are discussed in a later section.

In Figure 3.33b, we have assumed that our data acquisition is carried out in the ice sheet to reduce the propagation in the ice sheet because the ice sheet is considered homogeneous. That is, for the propose of demonstration, sources and receivers are located inside the ice are close to the bedrock and lake which cause the reflection, scattering, and refraction. Actually, sources and receivers in most GPR are located at the air as depicted in Figure 3.33a.

GPR data are generally described using the concept of an event. An event is a coherent seismic energy corresponding to the wave that have traveled from the source to a receiver via some path through the subsurface. In Figure 3.34, we have indicated examples of GPR events. Events of interest for exploration of the subsurface are events which reflect or diffract in the subsurface. Free-surface-reflection (air/ice) events are not included in Figure 3.34. We treated this inference as an absorbing boundary (see Appendix A).

Note that there are some unverifiable/confirmed reports about a lake 1.5 kilometers below the Martian south pole. If such water were discovered it automatically means extraterrestrial life in at least microbial form. Whenever there is water on Earth there is almost always some biological life form small and large.

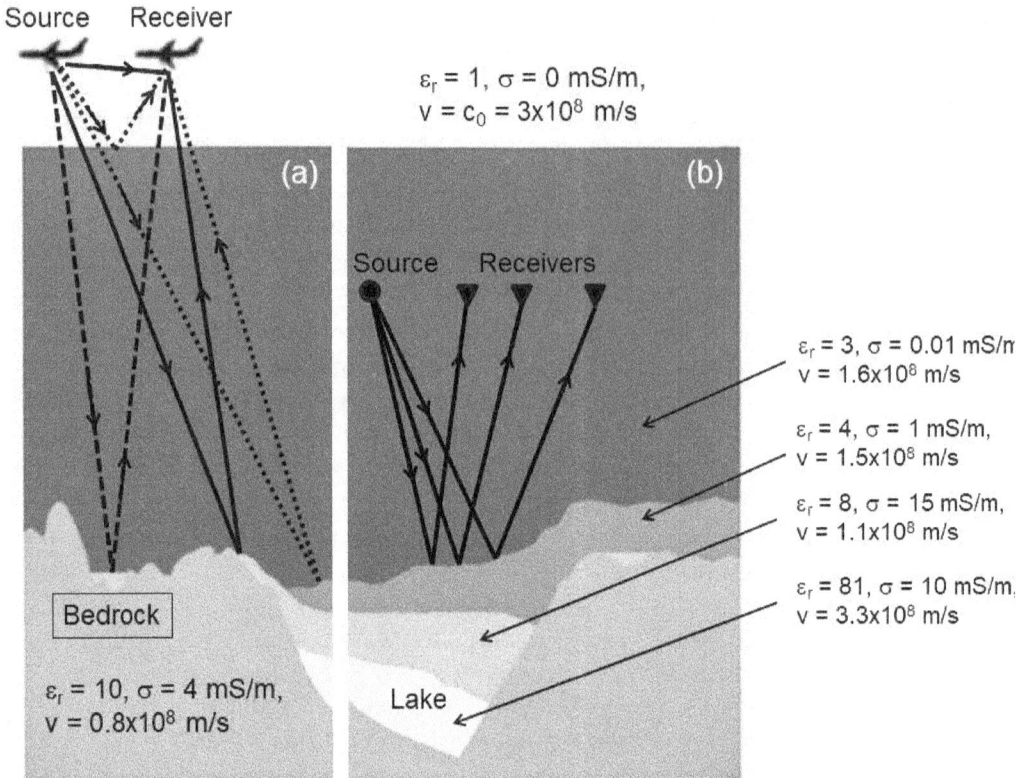

Figure 3.33. (a) An example of source and receiver distributions in GPR experiments. (b) The source and receiver distributions used for Figure 3.34.

Elasticity Versus Electromagnetism. Elastic waves (including acoustic waves, which are limited to wave propagation in liquids and gases) and electromagnetic waves are two important ways through which energy is transported in the world around us. These two types of waves are central to modern science. Elastic waves travel through solids, liquids, and gases, as described in Ikelle (2023b). They are primarily used to enhance the understanding of the solid Earth, including the hazards associated with it, to explore Earth's energy resources and minerals, and to produce Earth's energy.

Electromagnetic waves differ from elastic waves in that they do not require a medium in order to propagate. Electromagnetic waves can travel not only through gases, liquids, and solid materials but also through the vacuum of space. This ability is possible because electromagnetic waves consist of an electric field and a magnetic field, both of which are oscillating. As previously mentioned, a change in an electric field causes a magnetic field to change, and vice versa. Through this relationship, electromagnetic waves are able to propagate through space by themselves. Table 3.13 summarizes some of the key differences between electromagnetism and elasticity.

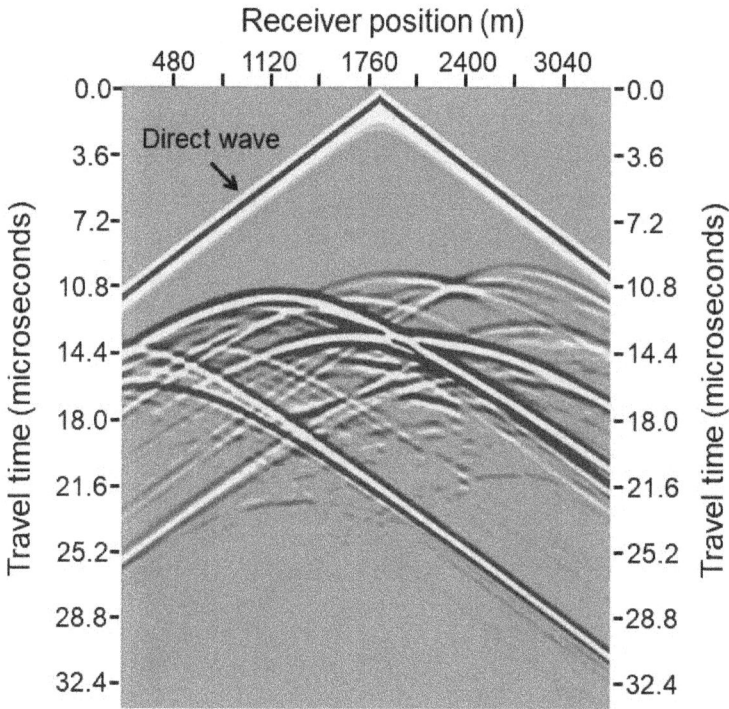

Figure 3.34. This is a simulated shot gather for the model in Figure 2.61. Note the direct wave (D) describes the wave propagation in the ice sheet. The other events are due to reflection and scattering by the bedrock and the lake. The quantity shown here is E_y.

Light and sound are two manifestations of electromagnetism and elasticity, respectively, which play a vital role in human life. Light triggers the sensation of seeing (a difference in color) and sound stimulates hearing (a difference in pitch). Apart from being a wave, light exhibits the properties of particles, as described in Ikelle (2023b). Light can be emitted and absorbed as tiny energy packets known as photons. Intensity, frequency or wavelength, direction and polarization are some

Table 3.13. Skin depths for some of the layers of solid Earth in meters. We are using the averages of the conductivity in these layers. These skin depths are indicators of the penetration of the magnetic field in solid Earth. A 1-Hz (i.e., $T = 1$ s) magnetic field can only penetrate about the first 5 kilometers of the crust. The outer core is almost a perfect conductor into which a 1-Hz magnetic field cannot penetrate. Even the longest-period ($T = 1$ year; the last column) magnetic field cannot penetrate the outer core.

Layer	σ (S/m)	T (s) 0.001	T (s) 1	T (s) 3600	T (s) 3.15×10^7
Outer core	3×10^5	0.029	0.92	55.1	5154
Mantle	10	5.03	159	9543.8	893×10^3
Sea water	4	7.95	251.5	15.1×10^3	1410×10^3
Crust	0.01	159	5030	302×10^3	2820×10^3

Table 3.14. Key differences between electromagnetism and elasticity.

Elasticity	Electromagnetism
(i) Elastic waves do not transmit energy through a vacuum.	(i) EM waves transmit energy through solids, liquids, and gases and in vacuum.
(ii) Elastic waves require a medium (solids, liquids, gases) to transport their energy from one location to another.	(ii) EM waves are produced by the vibration of charged particles.
(iii) Elastic waves are caused by a sudden deformation or a movement of a part of the medium.	(iii) EM waves that are produced by the Sun subsequently travel to Earth through the vacuum of outer space. In general terms, they are created by blackbody radiation.
(iv) Elastic waves are P- and S-waves.	(iv) All light waves are examples of electromagnetic waves.
(v) In water, P-wave velocity is 1500 m/s and S-wave velocity is zero.	(v) In a vacuum, the speed of light is 3×10^8 m/s. Light travels more slowly in air than in a vacuum, and even more slowly in water.
(vi) Sounds are elastic waves produce by vibrations. *Hearing* is based on sound waves.	(vi) *Seeing* is based on EM waves. *Smelling* is also based on EM waves.
(vii) Sound waves cannot excite atoms.	(vii) EM waves can excite atoms.

of the primary properties of light. Light travels much faster than sound. Both light and sound undergo refraction, diffraction and interference. While propagating through a medium, both light and sound suffer loss of speed, change in direction or get absorbed. Both are used for medical diagnostic and therapeutic as discussed in through the three volumes of this book. For example, diagnostic ultrasound is a noninvasive diagnostic technique used to image inside the body. Therapeutic ultrasound uses sound waves to interact with tissues in order to destroy diseased or abnormal tissues such as tumors, dissolve blood clots, or deliver drugs to specific locations in the body. In most cases, ultrasound therapies are noninvasive. No incisions or cuts need to be made to the skin, leaving no wounds or scars. Similarly, infrared therapy is nowadays widely used in dentistry, and in autoimmune diseases, to name a few. It is used to treat muscle pain, joint stiffness, and arthritis, to name a few.

Quiz: Why we are sometimes are able to hear sounds, but we are unable see emitting source?

3.2.   Application: Infrared Spectroscopy Technique

In Ikelle (2023a), we described the compositions of compounds, molecules, minerals, and rocks. We also described some functional groups, which are specific groupings of atoms within molecules that have their own characteristic properties, regardless of the other atoms present in a molecule. Some of the common examples of functional groups include alcohol (containing a hydroxyl (OH) group), alkanes (contains C–C), alkenes (hydrocarbons containing C=C), alkynes (hydrocarbons containing C≡C), and aromatic compounds (containing benzene) (see also, for example, Soderberg (2016) for more details). However, we have not yet addressed a large number of questions regarding determining the nuclear and electronic configurations of molecules and their behavior. These questions include: What is the structure of a given molecule? What are its bond lengths? How strong or stiff are its bonds? What are its symmetries? Where is its electron density? How much do the nuclei move (vibration/rotation)? These questions can be addressed using spectroscopy techniques. Spectroscopy is the study of spectral information. A physical stimulus is applied to a molecule, and the response is detected and interpreted. For example, upon irradiation with infrared light (physical stimulus) on a molecule, certain bonds respond by vibrating faster. This response can be detected and translated into a visual representation (i.e., a spectrum). From the spectrum, we can determine patterns. In addition to physical parameters, we can interpret these patterns in terms of atomic and molecular structures, even electronic structures. So, spectroscopy techniques allow us to determine the structure of a molecule, the bond lengths and stiffnesses of a molecule, the symmetries of a molecule, the vibrations and rotations of atoms of a molecule, and many more features of molecules. Our task in the remaining sections of this chapter is to describe some of the key spectroscopy techniques and some of their applications.

Basically, a spectrometer consists of an electromagnetic radiation source that emits the desired light (ultraviolet, visible, infrared, gamma rays, X-rays, etc.) and a detector that includes a signal amplifier. Some prominent spectroscopy techniques are infrared spectroscopy (also known as vibrational spectroscopy), Raman spectroscopy, nuclear magnetic resonance spectroscopy, X-ray spectroscopy, and Mossbauer spectroscopy. We here describe four of these techniques along with their applications in determining the chemical composition of matter, including determining the composition of living-thing molecules and organs; determining the composition of stars, minerals, and rocks; finding extrasolar elements in space; and measuring the temperature and density of gases. These four techniques are infrared spectroscopy, Raman spectroscopy, terahertz spectroscopy, and nuclear magnetic resonance (NMR). The first three trigger molecular vibrations through irradiation with infrared and terahertz light. They provide mostly information about the presence or absence of certain functional groups. NMR spectroscopy is based on the excitation of the nucleus of atoms through radiofrequency irradiation. It provides extensive information about molecular structure and atom connectivity.

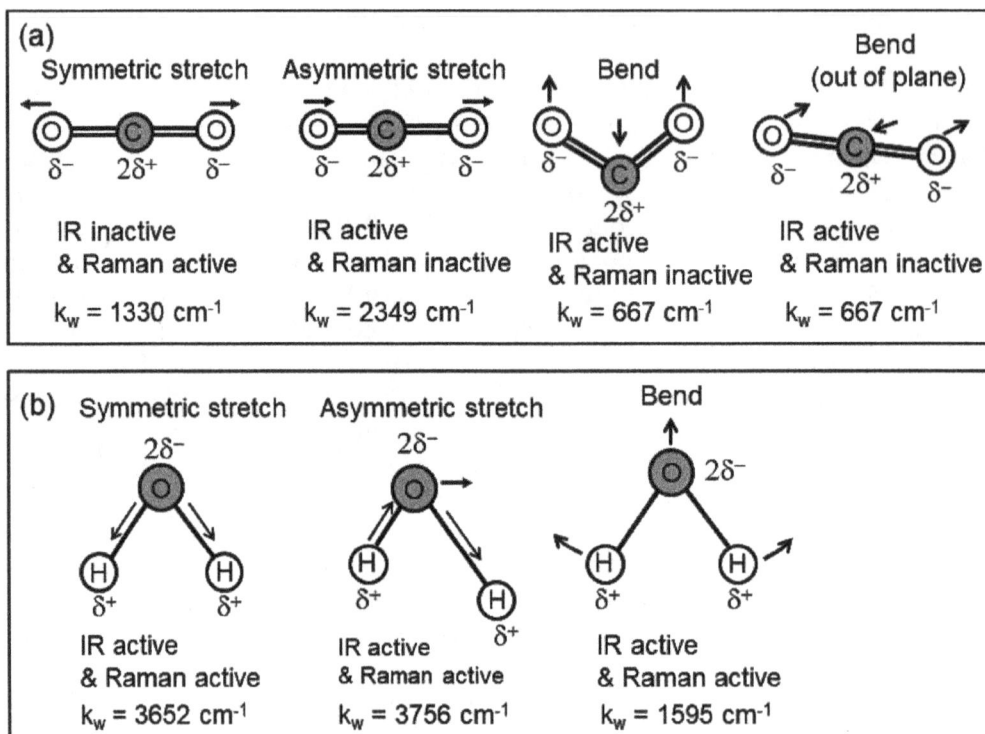

Figure 3.35. (a) Spring and ball model of the vibration modes of CO_2 with their wavenumbers. (b) Spring and ball model of the vibration modes of H_2O with their wavenumbers. Atoms are here the ball, and bonds are the ticks.

3.2.1. Molecular vibrations

We start with a brief background on molecular vibrations. See also Ikelle (2023a and 2023b) for more details on molecular vibrations. Every molecule containing n atoms has $3n$ degrees of freedom (vibrational degrees of freedom + translational degrees of freedom + rotational degrees of freedom). Nonlinear molecules can have three translations, three rotations, and $3n - 6$ vibrational degrees of freedom (also known as the number of vibrational fundamentals). Linear molecules can have three translations, two rotations, and $3n-5$ vibrations. Let us consider the CO_2 molecule, for example. It is a linear polyatomic molecule. Carbon dioxide (i.e., $n = 3$) can have $3n - 5 = 4$ vibrations: two CO bond stretches and two OCO bond angles (see Figure 3.35a). The triatomic molecule H_2O also has three modes of vibration (i.e., $3n - 6 = 3(3) - 6 = 3$). These modes are a symmetrical stretch, a bending or deformation mode, and an asymmetrical stretch as shown in Figure 3.35b. Notice that the vibrations of water (H_2O) and carbon dioxide (CO_2) are very different despite the fact that the two molecules have the same number of atoms. Thus, vibration modes can be used to differentiate molecules.

For a diatomic molecule, like HCl and O_2 (i.e., $n = 2$), we have $3(2)-5 = 1$ vibration mode (i.e., there is only one vibration), which is a simple stretch. This vibration changes the polarizability of the molecule but it does not induce any dipole change since there is no dipole in the molecule (see also Figure 3.17). In the case of O–O, the vibration is symmetric about the center. For nonlinear molecules like C_6H_6, we have $3n - 6 = 3(12) - 6 = 30$ vibration modes.

A molecular vibration can be thought of as a harmonic oscillator. The energy level of any molecular vibration is

$$\mathcal{U}_v = \frac{h}{2\pi}\left(v + \frac{1}{2}\right)\sqrt{\frac{\kappa_s}{m_R}} = \left(v + \frac{1}{2}\right)hf = 2\pi c_0 h\left(v + \frac{1}{2}\right)k_w \ ,$$

with

$$m_R = \frac{m_1 \times m_2}{m_1 + m_2} \quad \text{and} \quad k_w = \frac{1}{2\pi c_0}\sqrt{\frac{\kappa}{\mu}} \ ,$$

where v is vibrational quantum number (positive integer values, including zero, only), $h = 6.626 \times 10^{-34}$ Js is the Planck's constant, κ_s is the force constant of the bond, m_1 and m_2 are masses of the individual atoms involved in the vibration, m_R is reduced mass, and k_w is wavenumber (cm^{-1}; 1 cm^{-1} $\leftrightarrow$ 29,979 MHz $\leftrightarrow$ 1.2398×10^{-4} eV), and c_0 is the speed of light. The force constants of C–C, C=C, C≡C, C–O, and N–O are 5×10^5 dynes/cm, 10×10^5 dynes/cm, 15×10^5 dynes/cm, 19×10^5 dynes/cm, and 16×10^5 dynes/cm, respectively—that is, as bond strength increases, wavenumber increases (e.g., C≡C > C=C > C–C).

Also, as the mass decreases, the wavenumber increases (e.g., C–H > C–C > C–Cl > C–Fe). The vibrational energy, and therefore the frequency of vibration, is directly proportional to the strength of the bond and inversely proportional to the mass of the molecular system. Thus, different chemical functional groups vibrate at different frequencies. A vibrating molecular functional group can absorb radiant energy to move from the lowest ($v = 0$) vibrational state to the first excited ($v = 1$) state, and the frequency of radiation that will cause this absorption to occur must be identical to the initial frequency of vibration of the bond. The fundamental absorption frequency is referred to as this. Molecules also can absorb radiation to move to a higher excited state ($v = 2$ or 3). The so-called zero point energy occurs when $v = 0$ where $\mathcal{U}_v = (1/2)hf$ and this vibrational energy cannot be removed from the molecule.

3.2.2. Formulation of the Infrared Spectroscopy Technique

As described in Ikelle (2023b) and shown in Figure 3.36, the EM spectrum includes, from lowest energy/longest wavelength to highest energy/shortest wavelength, radio waves, microwaves, infrared, visible, ultraviolet, X-ray, and gamma-ray (see Figure 3.36); the energy of a wave is proportional to its frequency, as described in the previous subsection. Different energies affect molecules in different ways. Radio waves can cause nuclear spin transitions; microwaves can cause rotational motions; infrared can cause molecular vibrations; visible and near ultraviolet can cause electronic transitions; and far ultraviolet, X-rays, and gamma-rays can cause ionization[3].

---

[3]Ionization of an atom occurs when the excitation energy is high enough so the electron leaves the atom. This can be triggered by a high-energy photon (photo ionization), by heating (thermal ionization) or a collision with external electrons or ions.

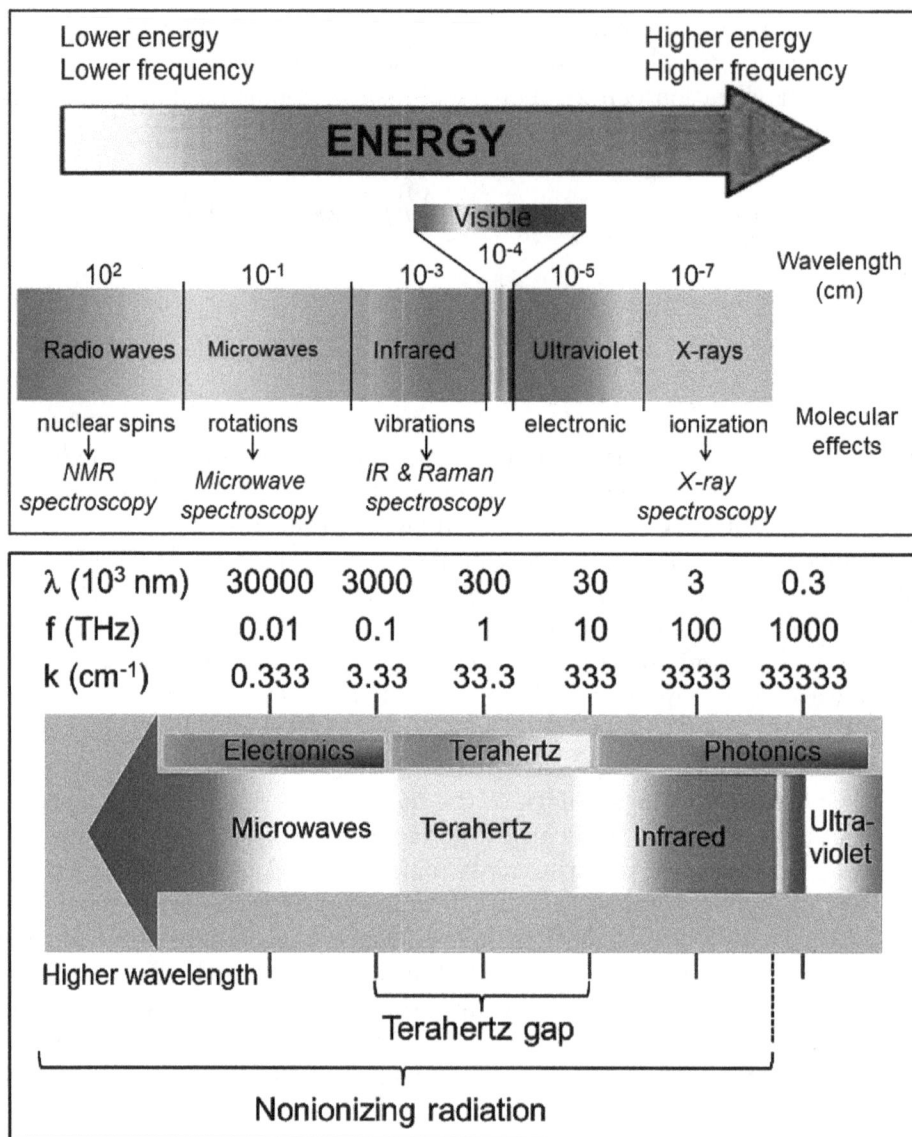

Figure 3.36. (Top) EM spectrum and their effects on molecules. When we shine IR light on a compound, the energy only causes the molecular bonds to vibrate. (Bottom) Teraherz (THz) wave position in the EM spectrum; λ is the wavelength, f is the frequency; and k is the wavenumber. 1 nm $= 10^{-9}$ meter $= 10$ Å, 1 μm $= 1{,}000$ nm.

Quiz: What is a zero-point energy?

Infrared radiation is not the highest energy radiation but it has just the right energy to make chemical bonds stretch or bend. As described in Ikelle (2023a and 2023b), chemical bonds are not rigid or immovable sticks; rather, they are flexible, and are capable of both stretching and bending (see also Figure 3.35). In fact, they are always in motion, they vibrate, and they can absorb light when the light energy is comparable to the bond-vibration energy—that is, when bonds are irradiated with polychromatic light that includes photons of energy that match the energy difference between two vibrational energy levels (e.g., $\mathcal{U}_1 - \mathcal{U}_0$). Because these energy differences of chemical bonds are on the order of 0.5 and 0.005 eV, which correspond to infrared light (IR), they are sufficient to induce molecular vibrations. Thus, vibrational spectroscopy can be based on infrared absorption. Such vibrational spectroscopy is known as *infrared* spectroscopy and is commonly referred to as IR spectroscopy. It is routinely used to identify functional groups such as alcohols, amines, and carbonyl groups, in unknown compounds and molecules by shining infrared light through compounds or molecules we want to identify.

The IR spectrum is obtained by plotting the intensity (absorbance or transmittance) versus the wavenumber, which is proportional to the energy difference between the ground and the excited vibrational states, as depicted in Figure 3.37. Absorbance is generally calculated as $A = \log_{10}(\mathcal{U}_{inc}/\mathcal{U}_{abso})$, where $\mathcal{U}_{inc}$ is the incident energy and $\mathcal{U}_{abso}$ is the absorbed energy. It states how much of the light the sample absorbs. Transmittance is the opposite of absorbance, $T = \mathcal{U}_{abso}/\mathcal{U}_{inc}$. So, decreases in transmittance mean energy is being absorbed by a specific bond, and, in turn, a specific bond is being excited. The relations between regions where changes in transmittance occur and functional groups are well known. These relationships form the basis of the identification of the functional groups present in the molecule for a given IR spectrum.

The effect of molecular vibration is visible in IR spectra only when IR photon energy transferred to the molecule via absorption produces a change in the dipole moment of the molecule. More precisely, IR absorption is only possible when the change in the dipole moment, $\boldsymbol{p}$, with respect to a change in the vibrational amplitude, Q, is nonzero; i.e., Here, the electric field is considered to be uniform over the whole molecule since λ is much larger than the size of most molecules. In terms of quantum mechanics, IR absorption is an electric dipole operator-mediated transition where the change in the dipole moment, $\boldsymbol{p}$, with respect to a change in the vibrational amplitude, Q, is greater than zero; i.e.,

$$\left| \left(\frac{\partial \boldsymbol{p}}{\partial Q} \right)_0 \right| \neq 0 \, , \tag{3.163}$$

where $\boldsymbol{p}$ is the magnitude of the dipole moment as defined in (3.109, 3.110, and 3.111). So, in a molecule with a center of symmetry, the vibrations that are symmetric about the center of symmetry are IR-inactive (e.g., all the symmetrical diatomic molecules such as H_2, N_2, and Cl_2 are IR-inactive). The vibrations that are not symmetrical about the center of symmetry are IR-active (e.g., symmetric stretching in CO_2 is IR-inactive, whereas asymmetric stretching is IR-active).

Once an infrared spectrum has been recorded, the next stage of this experimental technique is interpretation. The X–H stretching region, where X can be O, C, and N are generally located between 4000 and 2500 cm^{-1}. The triple-bond region is between 2500 and 2000 cm^{-1} and the double-bond region is between 2000 and 1500 cm^{-1}. Using the infrared spectrum of a compound alone is not always sufficient to identify a compound. It is normal to use infrared spectroscopy in conjunction with other techniques, such as Raman spectroscopy and NMR spectroscopy, which we will describe later. Advances in computerized inverse-problem solutions, including artificial intelligence systems, have extended the range of information available from IR instruments (see Ikelle, 2020d). IR bands can be classified as strong, medium,

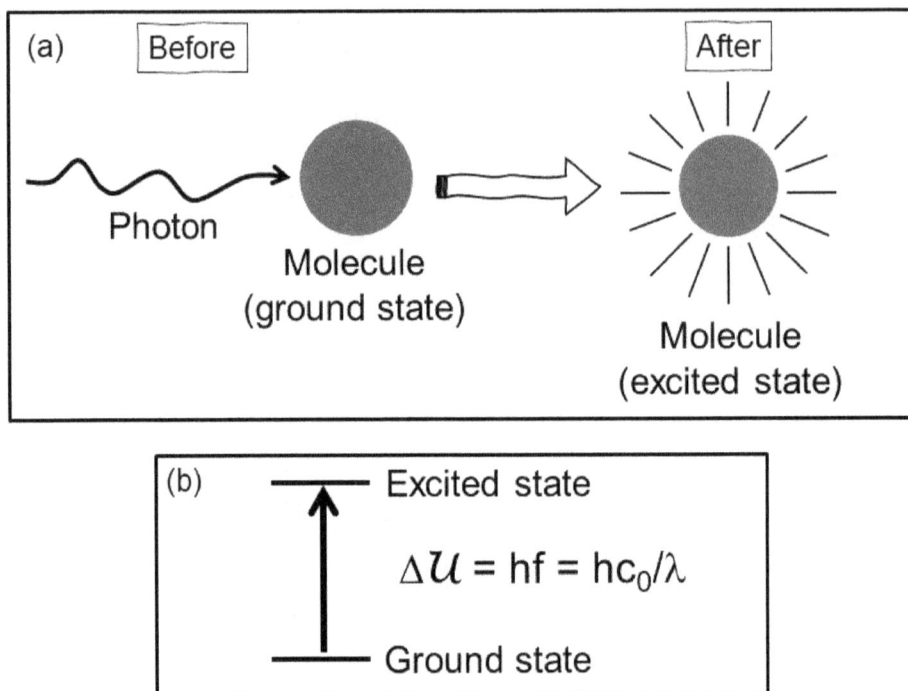

Figure 3.37. (a) What happens when a sample is irradiated? When a molecule absorbs a photon, it takes up the photon's energy to reach an *excited state* of some sort. (b) The photon energy has to be equal to the difference $\Delta \mathcal{U}$ between some pair of energy levels of the molecule in order for absorption to occur.

or weak, depending on their relative intensities in the infrared spectrum. A strong band covers most of the transmittance-axis like at 950 cm^{-1} in Figure 3.38a, at 1250 cm^{-1} in Figure 3.38b, and at 675 cm^{-1} in Figure 3.38d. A medium band falls to about half of the transmittance-axis like at 3000 cm^{-1} in Figure 3.38b and at 775 cm^{-1} in Figure 3.38d. A weak band falls to about one third or less of the transmittance-axis like at 1800 cm^{-1} in Figure 3.38a and 3500 cm^{-1} in Figure 3.38b.

Research exercise: (1) Explain what element 115 is and how it is made. Name one of its stable isotopes along with its half-life. Also describe its current industrial application, if any. (2) Physicist Robert Feynman predicted the periodic table would end at element 137. Did we reach this limit?

3.2.3. Infrared spectroscopy of organic molecules

Figure 3.38 shows typical IR spectra. One hundred percent transmittance means that every wavelength passes through the sample. However, a lower transmittance level means that some of the energy is absorbed by the compound, resulting in downward spikes. The horizontal axis indicates the position of an absorption band. Wavenumbers (in cm^{-1}), instead of frequency, are used as a conventional way to represent IR spectra. The wavenumbers of infrared radiation are normally in the range of 4000 to 500 cm^{-1}. In Figure 3.38, the spikes represent absorption bands

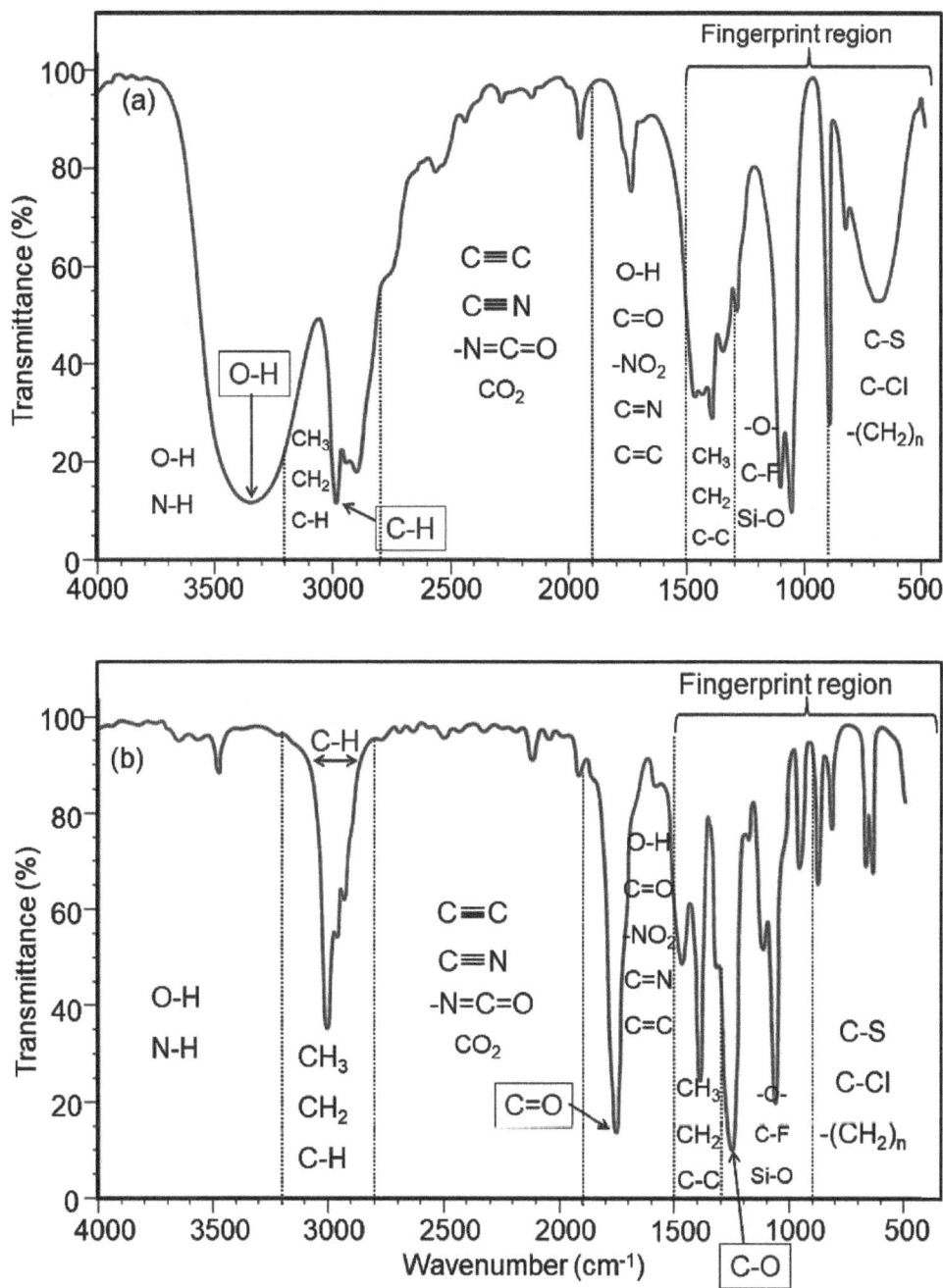

Figure 3.38. *(Continued.)*

in an IR spectrum. Molecular molecules have a variety of covalent bonds, and each bond has a different vibration mode, so the absorption spectrum of a compound usually has several bands. Figure 3.38a shows the spectrum of ethanol (C_2H_5OH). As with most alcohols, it has significant O–H stretching vibrations as well as C–O stretching vibrations. The O–H stretch of alcohols appear in the region 3500-3200 cm^{-1}. It is a very intense and broad band. The C–O stretch is in the region

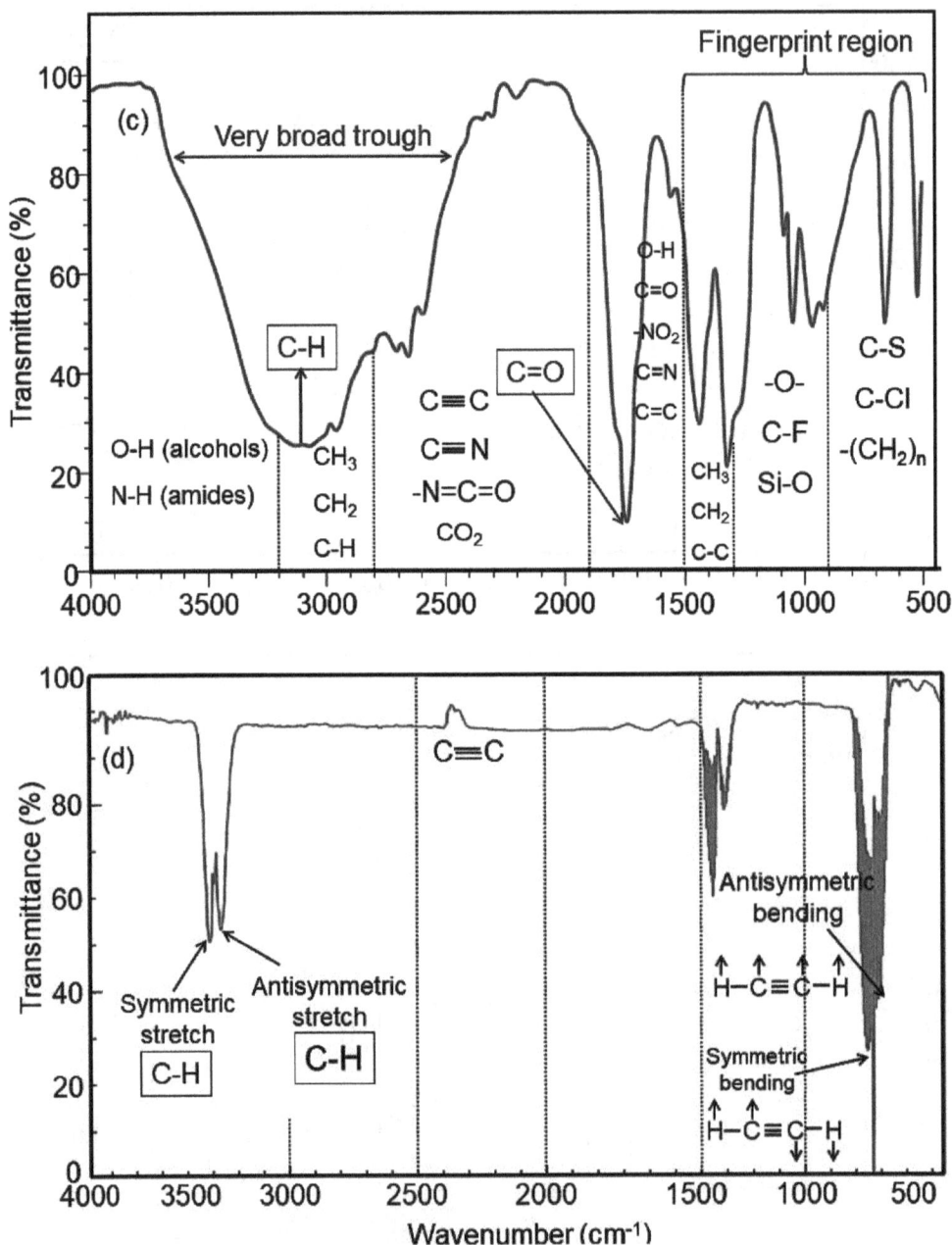

Figure 3.38. IR spectra in wavenumbers (horizontal axis) and intensity (vertical axis) of (a) ethanol (C_2H_5OH), (b) ethyl ethanoate ($C_4H_8O_2$), (c) ethanoic acid (CH_3COOH), and (d) acetylene (C_2H_2). The region between 1500 and 400 cm^{-1} is known as the fingerprint region. It is rarely used to identify particular functional groups. The dotted lines indicate typical regions of vibrational spectra. Not all covalent bonds display bands in the IR spectrum. Only polar bonds do so. A typical example of a broad band is that displayed by O–H bonds found in alcohols. Data from NIST.

Table 3.15. IR spectroscopy is useful in organic chemistry because it enables you to identify different functional groups. Each functional group contains certain bonds, and these bonds always show up at same places in the IR spectrum.

Functional group	Bond	Absorption range [cm^{-1}]	Intensity
Alcohol	O–H stretch	3,600–3,200	strong and broad
	O–H stretch	3,700–3,500	strong and sharp
Alkane	C–H stretch	3,100–2,850	strong
Alkene	=C–H stretch	3,100–3,000	medium
Alkyne	C–H stretch	3300	strong
	C≡C stretch	2,300–2,100	variable
Amine	N–H stretch	3,500–3,300	medium
Aromatic	C–H stretch	3,100–3,000	medium and weak
Nitrile	C≡N	2,300–2,200	

of 1260-1050 cm^{-1}. Figure 3.38b shows ethyl ethanoate ($C_4H_8O_2$), which is also known as ethyl acetate. The most characteristic absorption band in the spectrum of ethyl ethanoate is that from the stretching vibration of the carbonyl double bond C=O, at 1720 cm^{-1}. The C–H bond stretching of all hydrocarbons occur in the range of 2800-3200 cm^{-1}, and the exact location can be used to distinguish between alkane, alkene and alkyne (Table 3.15).

Figure 3.38c shows ethanoic acid (CH_3COOH). The broad absorption bound around 3000 cm^{-1} is characteristic of ethanoic acids. Notice that in the three examples, there are many bands in the long-wavelength 400-1400 cm^{-1} region. This part of the spectrum is called the *fingerprint* region. While it is usually very difficult to pick out any specific functional group from this region and rarely used for identification of particular functional groups, the fingerprint region nevertheless contains valuable information. As we can see in Figure 3.38, the pattern of absorbance bands in the fingerprint region is unique to every molecule. Table 3.15 lists the locations and intensities of absorption produced by typical functional groups. In Ikelle (2024), we describe how artificial intelligence can be used to determine an unknown molecule using infrared spectroscopy.

Figure 3.38d shows the IR spectrum of acetylene (C_2H_2), a polyatomic linear molecule with 7 normal vibrational modes because $n = 4$ (i.e., $3 \times 4 - 5 = 7$). This figure explicitly shows the five distinct wavenumbers that result from two of these modes being doubly degenerate. Assuming that the molecule that generates this IR spectrum is known, we can identify it without even using the fingerprint region. Unfortunately, this example is an exception. Determining unknown molecules from infrared spectroscopy requires advanced computation methods (see Ikelle, 2024).

Quiz: Determine the molecule in Figure 3.39 using infrared spectroscopy?

For acetylene, two fundamentals appear in the IR spectrum at 3281.9 and 730.3 cm^{-1}. Besides the fundamental stretch and bending modes that appear in the IR spectrum, other transitions can also appear such as combination bands (two

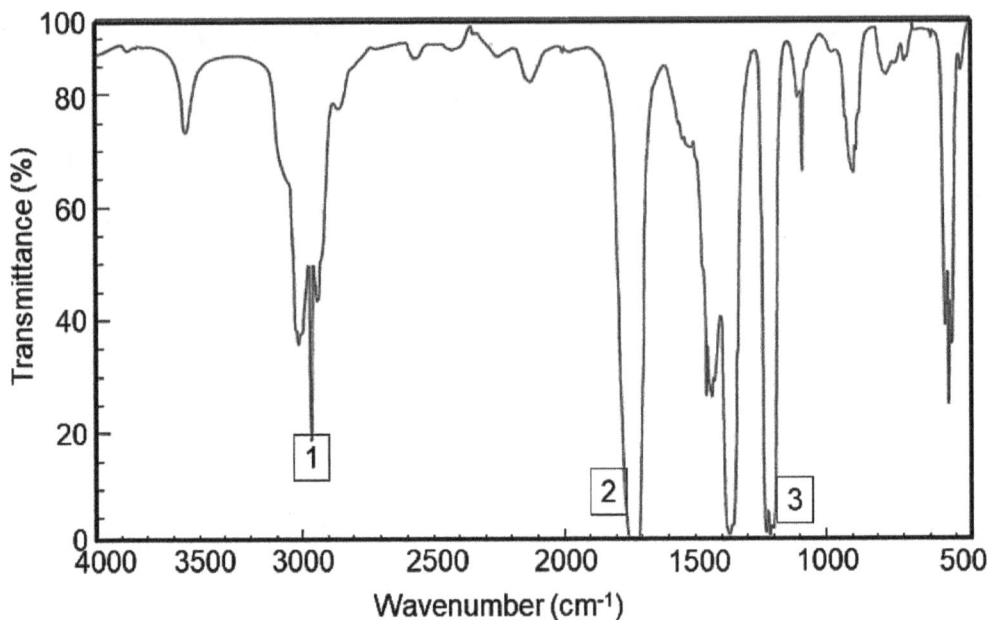

Figure 3.39. An example of an IR spectrum.

fundamental modes added together). In this experiment, you will be analyzing the combination band of the two (asymmetric and symmetric) bending modes at 1328.18 and 1048.49 cm^{-1} for C_2H_2 and C_2D_2, respectively. A table showing the atomic displacement patterns of the normal modes of acetylene is given below. Only five modes are shown because the bending modes come in degenerate pairs, i.e., there are two equivalent modes with the same frequency. The partners to the bending modes depicted involve bending out of the plane of the page (as represented by the + and - signs).

3.2.4. Infrared Spectroscopy: Inorganic Molecules

Except for water, the discussion of IR spectra has focused on organic matter. Actually, as with organic compounds, inorganic compounds and by extension minerals and rocks can produce infrared spectra. As described in Ikelle (2023a), a rock is a naturally occurring cohesive aggregate of grains of one or more minerals, and a mineral is a naturally occurring inorganic, crystalline solid with a definite composition (or range of compositions). Kaolinite ($Al_2(Si_2O_5)(OH)_2$), muscovite $KAl_3(Si_3O_{10})(OH)_2$, calcite ($CaCO_3$), plagioclase feldspar ($CaAl_2Si_2O_8$ or $NaAlSi_3O_8$), alkali feldspar ($KAlSi_3O_8$), olivine (Fe_2SiO_4 or Mg_2SiO_4), and quartz (SiO_2) are examples of minerals. A rock can also include non-minerals, such as organic matter within a coal bed. Note that liquids and non-crystalline materials such as glass are not minerals. There are about fifty minerals in the crust and 98.5 percent of the crust mass of these minerals is from eight chemical elements: 46.6 percent from oxygen (O), 27.7 percent from silicon (Si), 8.1 percent from aluminum (Al), 5 percent from iron (Fe), 3.6 percent from Calcium (Ca), 2.8 percent from sodium (Na), 2.6 percent from potassium (K), 2.1 percent from magnesium (Mg),

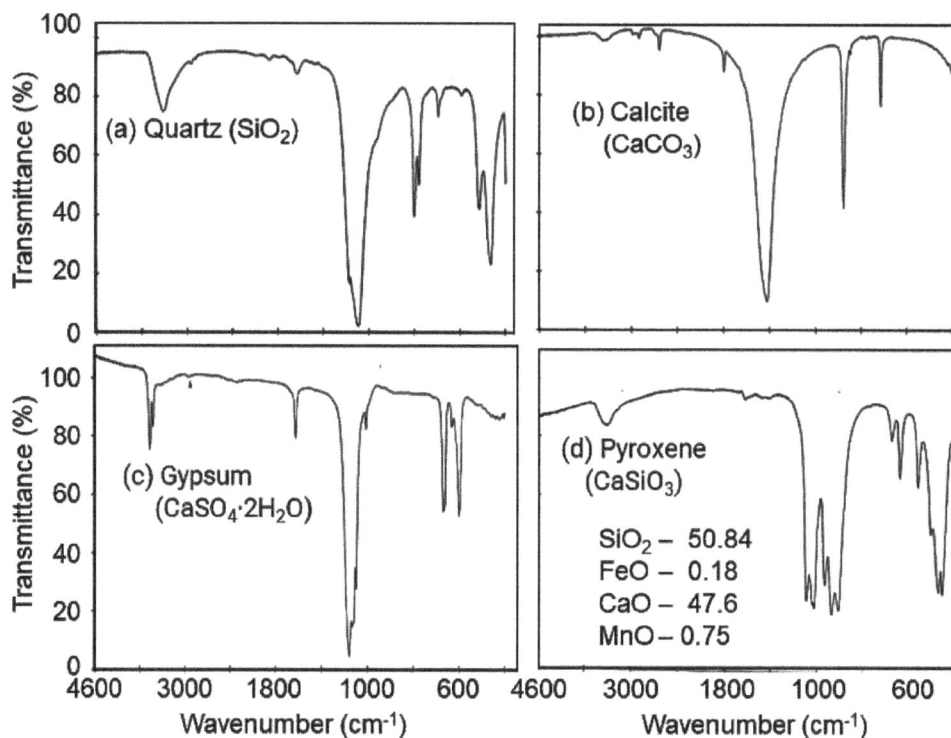

Figure 3.40. IR spectra of silicate minerals (a) and (d); and non-silicate minerals (b) and (c).

and 1.5 percent from all others. They can be divided into two broad classes, silicate and non-silicate minerals. In the crust, silicate minerals make up 92 percent, while non-silicate minerals make up 8 percent. All silicate minerals contain silicon and oxygen in varying proportions, ranging from SiO_2 to SiO_4.

Quartz, feldspar, amphibole, pyroxene, and kaolinite are all silicate minerals. Non-silicate minerals include calcite, gypsum, pyrite, and halite. Quartz is three times more hardy than gypsum, twice as hard as calcite, and just slightly more hardy than alkali feldspar. Corundum (Al_2O_3) is the hardest mineral. The crystal lattice (the lattice atoms are held in place by atomic bonds) controls the shape and many properties of minerals. There are also certain minerals called oxides. These minerals are used to describe the chemical composition of rocks. Examples of oxides includes silica (SiO_2), titanium dioxide (TiO_2), aluminum oxide (Al_2O_3), iron oxide (FeO), manganese oxide (MnO), magnesium oxide (MgO), calcium oxide (CaO), and potassium oxide (K_2O).

As we can see in Figures 3.40, 3.41, and 3.42, the IR spectra of the minerals have broad infrared bands. These bands are fewer in number compared to those of organic molecules, and appear predominantly at lower wavenumbers. The fewer number of infrared bands indicates a small number of components in the minerals that interact with infrared light. For example, simple inorganic compounds, such as NaCl, KBr, and ZnSe, do not produce any vibrations in the mid-infrared (4000-400 cm^{-1}) region; that is why they are used in the construction of infrared windows.

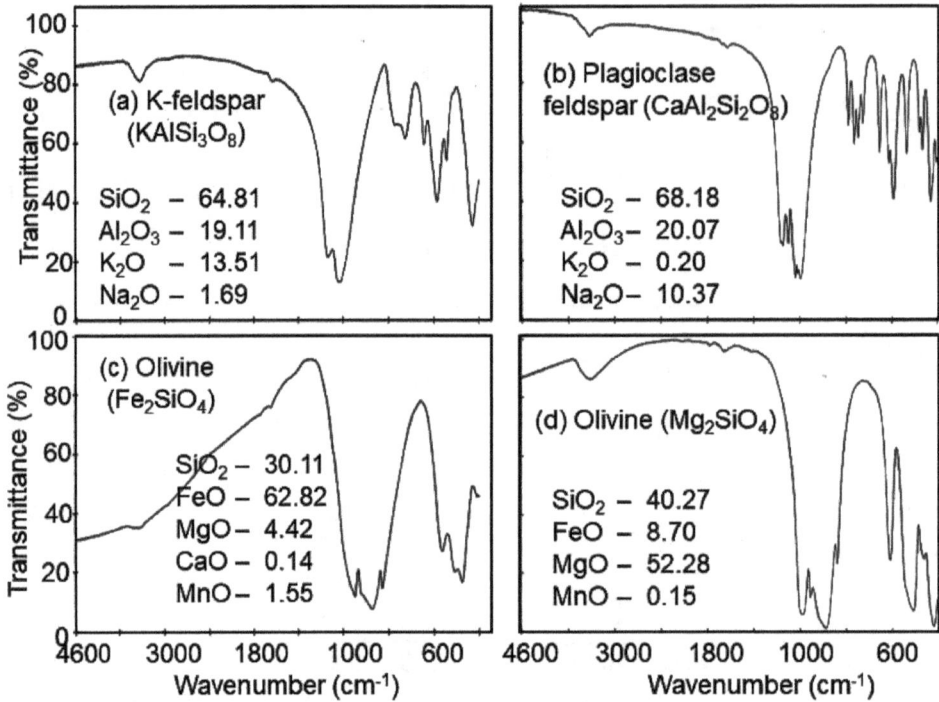

Figure 3.41. IR spectra of silicate minerals.

A key aspect of studying the spectroscopy of rocks and minerals is the need to understand planetary compositions, including the rocky planets of our solar system. Using such studies, we can infer the main stages of the geological evolution of planets and exoplanets. In Figures 3.40, 3.41, and 3.42 we demonstrate how spectroscopy of rocks can be inferred from minerals because minerals are the building blocks of rocks. Figure 3.40 demonstrates that we can distinguish non-silicate minerals from silicate minerals. Quartz and pyroxene are silicate minerals and calcite (a carbonate mineral) and gypsum (a sulfate mineral) are non-silicate minerals. Figure 3.40a shows the typical major absorption features (peaks) of quartz: 1100 to 950 cm^{-1} (asymmetric stretching vibration of the Si$-$O groups with a peak maximum at about 1080 cm^{-1}); 800 and 780 cm^{-1} (symmetric stretch vibrations); 700 cm^{-1}, 520 cm^{-1}, and 450 cm^{-1} (symmetric and asymmetric Si$-$O bending modes). Figure 3.40b shows the spectrum of the carbonate mineral calcite. The fundamental vibrations arise from the carbonate CO_3^{-2} ion at 1400 cm^{-1} (asymmetric stretch), at 875 cm^{-1} (out-of plane bending vibration), and at 727 cm^{-1} (in-plane bending vibration). Figure 3.40c illustrates the fundamental vibrations are at 3550 cm^{-1} (symmetric stretch vibration), at 1700 cm^{-1} (O-H bending vibration mode), at 1130-1080 cm^{-1} (anti-symmetric vibration mode), at 680 (stretching mode of sulphate group), and at about 600 (bending mode of sulphate group). These peaks confirm the identity of the sample. These observations are in excellent agreement with X-ray diffractometry (XRD) analysis, also called X-ray crystallography, which identifies crystal structures and orients atomic planes in materials.

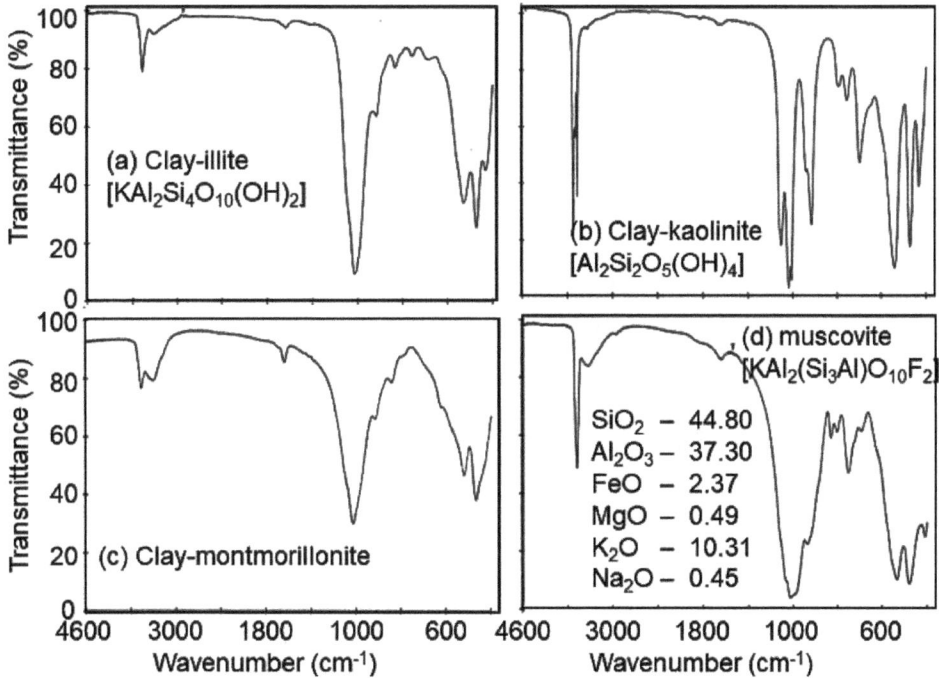

Figure 3.42. IR spectra of clay minerals.

Pyroxene in Figure 3.40a displays absorption features ranging from 1100 to 980 cm^{-1}, 800 to 700 cm^{-1}, and 650 to 375 cm^{-1}. Pyroxenes are useful indicators of igneous processes. The four minerals have distinct behavior from 2000 to 400 cm^{-1}. Further experiments show that we can successfully discriminate silicate minerals from non-silicate minerals by using IR spectroscopy. If we focus on only the first dominant peak of each mineral, we can see that these peaks are in front of the 1100 cm^{-1} wavenumber for non-silicate minerals and after this wavenumber for silicate minerals. Figure 3.41, which describes only silicate minerals, confirms this observation. Variations in the amount of silica in a given mineral or rock affect the spectra trend before the first peak and the wavenumber of the first peak. The wavenumber of this first peak decreases with the amount of silica. The variation after this peak varies with other oxides in the mineral or rock.

The spectra of clay minerals are shown in Figure 3.42. To highlight the illite sample is actually a silicate. Its variations are highly dependent on its quartz content, and therefore, several signals are overlapping or may be assigned to quartz vibrations. The peaks located at 3630 cm^{-1} corresponds to the stretching vibrations of the -OH groups. The spectra of montmorillonite are quite similar to illite with only subtle differences in the spectral features. Kaolinite and muscovite can be distinguished easily from other clay minerals by their clearly defined absorption regions at around 3700 cm^{-1}. This region includes in-phase and out-of-phase motion modes, and the stretching vibration of inner-surface -OH groups.

In summary, IR spectroscopy refers to measurements of the absorption of infrared radiation by organic and inorganic matter, including solids, liquids, and gases. Different functional groups in organic matter absorb infrared radiation at different frequencies, allowing us to identify these groups. All symmetric diatomic molecules such as H_2, N_2, and Cl_2, etc., are IR-inactive. The symmetrical stretching of the $C=C$ bond in ethylene (center of symmetry) is IR-inactive. The symmetrical stretching in CO_2 is IR-inactive, whereas asymmetric stretching is IR-active. The mid-infrared spectrum (4000-400 cm^{-1}) can roughly be divided into four regions and the nature of a group frequency may generally be determined by the region in which it is located. The regions are generalized as follows: the X–H (e.g., O–H, C–H and N–H) stretching region (4000-2500 cm^{-1}), the triple-bond region (2500-2000 cm^{-1}), the double-bond region (2000-1500 cm^{-1}) and the fingerprint region (1500-600 cm^{-1}). O–H stretching produces a broad band that occurs in the range 3700-3600 cm^{-1}. By comparison, N–H stretching is usually observed between 3400 and 3300 cm^{-1}. This absorption is generally much sharper than O–H stretching and may, therefore, be differentiated. Multivariate data analysis for determining rock compositions from the IR spectra of oxides is described in Ikelle (2024).

Box 3.1: Functional Groups in Organic Molecules

Organic molecules are constructed by functional groups, which demonstrates the importance of functional groups. Moreover, the properties and reaction chemistry of a particular functional group can be remarkably independent of the environment. Therefore, it is only necessary to know about the chemistry of a few generic functions to predict the chemical behaviour of thousands of natural chemicals. The number of functional groups is unknown. However, the common functional groups are: Alcohol ($-OH$), Aldehyde ($-CHO$), Alkene ($-C=C-$), Alkyne ($-C\equiv C-$), Amide ($-CONH_2$), Amine ($-NH_2$), Anhydride ($-COO-CO-$), Carboxylic acid ($-COOH$), Ester ($-COO-$), Ketone ($-CO-$), Nitrile ($-CN$), and Thiol ($-SH$).

3.3. Application: Raman Spectroscopy Technique

3.3.1. Formulation of Raman Spectroscopy

Raman spectroscopy is another vibrational spectroscopic technique that is actually complementary to IR measurements. Raman spectroscopy is complementary to IR spectroscopy in that some vibrations are only Raman active, some are only IR active, and some are both (see Figure 3.43). In contrast to IR spectroscopy, Raman spectroscopy relies on light scattering from matter instead of light absorption. As described in Ikelle (2023b) and illustrated in Figure 3.43, such scattering can either be elastic (also known as Rayleigh scattering), where kinetic energy is conserved, or inelastic (also known as Raman scattering), where energy is lost.

Raman spectra are based on the inelastic scattering of photons. In this scattering, the frequency of the scattered light is shifted from the frequency of the incident beam. Therefore, the energy of the scattered photons differs from that of the incident photons. The energy differences lie in the same range as the transitions probed by the direct absorption of mid-IR quanta, although photons of UV, visible, or near-infrared light are used to induce scattering. The inelastic scattering of photons in the context of spectroscopy was first discovered by the Indian scientist Chandrasekhra Venkata Raman (1888-1970) in 1928 and is thus denoted as the Raman effect. His discovery of inelastic scattering won him the Nobel Prize in physics in 1930.

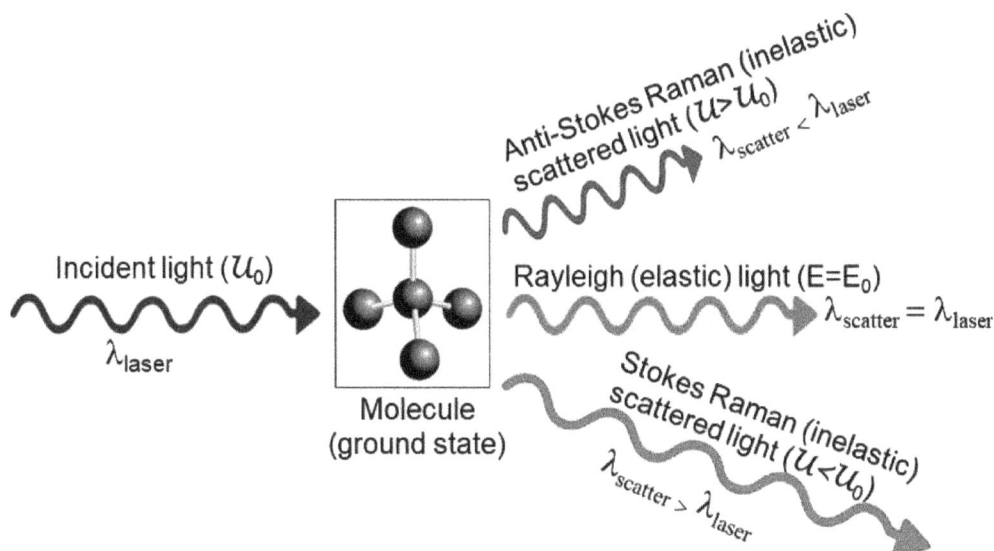

Figure 3.43. Schematic representation of the Raman effect. A laser is used to irradiate the sample with monochromatic radiation. Laser sources are available for excitation in the UV, visible, and near-IR regions (785 and 1064 nm). Scattering consists of Rayleigh and Raman scattering. There is no energy lost in elastically scattered Rayleigh light. Raman scattered photons lose some energy relative to the incident energy.

To be more precise about the shift of the frequency of the scattered light from the frequency of the incident beam, let us consider a molecule's interaction with an electromagnetic wave, $\boldsymbol{E} = \boldsymbol{E}_0 \cos(2\pi f_0 t)$, where f_0 is the frequency of the incident beam. The oscillating electric field can induce the following electric dipole moment:

$$\boldsymbol{p}^{(\text{ind})} = \tilde{\boldsymbol{\alpha}}(f)\boldsymbol{E} = \tilde{\boldsymbol{\alpha}}(f)\boldsymbol{E}_0 \cos(2\pi f_0 t)\,, \tag{3.164}$$

where $\tilde{\boldsymbol{\alpha}}(f)$ is the polarisability tensor—$\boldsymbol{p}$, $\boldsymbol{E}$, and $\boldsymbol{E}_0$ are vectors and $\tilde{\boldsymbol{\alpha}}(f)$ is a symmetric third-order tensor. The polarisability varies with time as it describes the response of the electron distribution to the movements of the nuclei that oscillate at the normal mode frequency f_k. Thus, we can express $\tilde{\boldsymbol{\alpha}}(f)$ as follows:

$$\tilde{\boldsymbol{\alpha}}(f) = \tilde{\boldsymbol{\alpha}}_0(f_0) + \left(\frac{\partial \tilde{\boldsymbol{\alpha}}}{\partial Q_k}\right)_0 (2\pi f_k t)\ . \tag{3.165}$$

By substituting (3.165) into (3.164), we obtain

$$
\begin{aligned}
\boldsymbol{p}^{(\text{ind})} &= \boldsymbol{E}_0 \left[\tilde{\boldsymbol{\alpha}}_0(f_0) + \left(\frac{\partial \tilde{\boldsymbol{\alpha}}}{\partial Q_k}\right)_0 (2\pi f_k t)\right] \cos(2\pi f_0 t) \\[2mm]
&= \underbrace{\boldsymbol{E}_0 \tilde{\boldsymbol{\alpha}}_0 \cos(2\pi f_0 t)}_{\text{Rayleigh scattering}} + \underbrace{\boldsymbol{E}_0 \left(\frac{\partial \tilde{\boldsymbol{\alpha}}}{\partial Q_k}\right)_0 Q_k \cos\left[2\pi(f_0 + f_k)t\right]}_{\text{Anti-Stokes Raman scattering}}
\end{aligned}
$$

$$
+ \underbrace{\boldsymbol{E}_0 \left(\frac{\partial \tilde{\boldsymbol{\alpha}}}{\partial Q_k}\right)_0 Q_k \cos\left[2\pi(f_0 - f_k)t\right]}_{\text{Stokes Raman scattering}}\ . \tag{3.166}
$$

Table 3.16. Characteristics of infrared and Raman bands (cm^{-1}) of organic and inorganic molecules.

Molecule	k_{w1}	IR	Ram	k_{w2}	IR	Ram	k_{w3}	IR	Ram
NH_3	3337	yes	yes	950	yes	yes	3444	yes	yes
CO_2	1388	no	yes	667	yes	no	2349	yes	no
$CaCO_3$				879	yes	yes	1492	yes	yes
KNO_3				828	yes	yes	1370	yes	yes
SO_3^{2-}	967	yes	yes	620	yes	yes	933	yes	yes
HCN	3310	yes	yes	715	yes	yes	2097	yes	yes
H_2O	3652	yes	yes	1595	yes	yes	3756	yes	yes
O_3	1103	yes	yes	701	yes	yes	1042	yes	yes
CO	2169	yes	yes						
N_2	2358	no	yes						

3.3.2. Raman versus IR

We have three scattering terms illustrated in Figure 3.43—that is, Rayleigh scattering and the two Raman scatterings—which depend on different frequencies. Rayleigh scattering leaves the frequency of incident light unchanged (i.e., no energy is transferred between photon and molecule), while Raman scattering has frequencies $(f_0 - f_k)$ and $(f_0 + f_k)$, which are shifted from the frequency of incident light. Scattering with frequency $(f_0 + f_k)$ is known as anti-Stokes Raman scattering, while that with frequency $(f_0 - f_k)$ is known as Stokes Raman scattering. In Stokes Raman scattering, energy of photon decreases and vibrational energy of molecule increases whereas in anti-Stokes Raman scattering, energy of photon increases and vibrational energy of molecule decreases. Also, from equation (3.166), we can see that a vibration is Raman active only if the polarizability of molecule changes during vibration; i.e.,

$$\left| \left(\frac{\partial \tilde{\alpha}}{\partial Q_k} \right)_0 \right| \neq 0 . \tag{3.167}$$

So if the induced dipole is not constant, inelastic scattering is allowed (i.e., Raman active). In other words, for Raman scattering to occur, a molecule must undergo a change in polarizability, but does not need to undergo a change in dipole moment. The difference in frequency between incident and scattered light is equal to the frequency of vibration of the molecule. Symmetrical vibrations, which cannot be observed by IR spectroscopy, can be observed by Raman spectroscopy. For example, a diatomic molecule such as oxygen gas with a simple stretch of the O−O bond, infrared absorption will change its polarizability but will not induce any dipole change since there is no dipole in the molecule and the vibration is symmetric about the center. Thus, oxygen gas will give a band in the Raman spectrum and no band in the infrared spectrum. A molecule such as NO will appear in both the infrared and Raman spectrums since there are both dipole and polarizability changes. More examples are shown in Table 3.16.

Figure 3.44 shows some comparison of Raman and IR spectra for the same molecules. Notice how the Raman intensities can be different from those of the

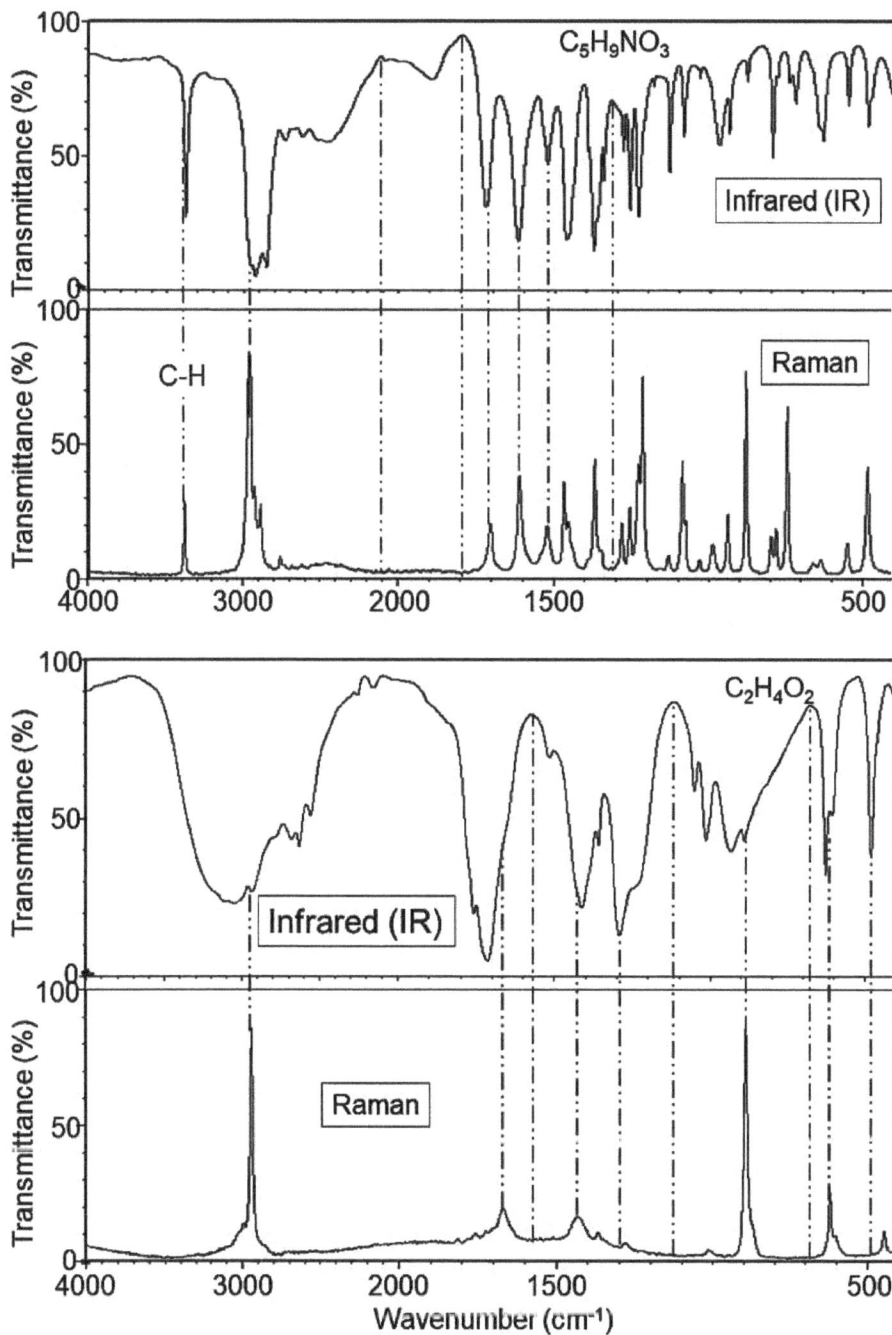

Figure 3.44. (a) IR and Raman spectra of butanoic acid ($C_5H_9NO_3$). (b) IR and Raman spectra of acetic acid ($C_2H_4O_2$). The peaks provide information about composition, concentration and more. Peaks can occur because different molecules interact with light differently in the IR and Raman methods. Data from NIST.

IR spectra. They are often in opposition to those in the infrared spectrum. Some entire IR bands can be missing from Raman spectra of the same molecule, and vice versa.

In fact, *fingerprint spectral* is actually a term related to forensic studies of chemical components present in our fingerprints and related to the identification of authors. Raman spectroscopy is used to analyze a wide variety of forensic evidence, including fingerprints, at a crime scene. Fingerprints are primarily made up of the eccrine and sebaceous glands. Eccrine sweat consists predominantly of water (98 percent), and a mixture of organic (e.g., amino acids, proteins and lactate) and inorganic material (e.g., Na^+, K^+, Cl^-, and metal ions). Sebaceous secretions are predominantly composed of fatty acids, glycerides, cholesterol, squalene and a variety of lipid esters.

Raman spectroscopy has proven to be a powerful tool in the analysis of fingerprints due to its high sensitivity to samples of complex chemical composition. Figure 3.44 shows an example of fingerprint Raman spectra in the fingerprint region 800-1500 cm^{-1} with multiple peaks related to the chemical composition of fingerprints. Further analysis of such data is carried out in Ikelle (2024).

3.4. Application: Terahertz Spectroscopy and Unconscious Awareness

A detailed division of the electromagnetic spectrum includes terahertz (THz) radiation (i.e., frequencies from about 0.1 THz to 10 THz; wavelengths of radiation from 3×10^6 nm to 3×10^4 nm; wavenumbers from 3.33 cm^{-1} to 333 cm^{-1}; or equivalently the energy from 0.42 to 42 MeV), as depicted in Figure 3.36. Terahertz radiation is also known as terahertz gap, terahertz waves, T-waves, T-rays, terahertz light, T-light, T-lux, submillimetre radiation, and simply THz. The main motivation for discussing terahertz spectroscopy in this chapter is that a significant number of molecules, including molecules associated with anesthetic drugs, have vibrational activities that expand to encompass terahertz (THz) radiation. Some of the vibrations in the low-frequency region of the THz spectrum are intermolecular (see Table 3.17). That is one of the differences between infrared spectroscopy and terahertz spectroscopy.

At its core, THz spectroscopy is conceptually similar to infrared spectroscopy (IR) technique. The interpretations of the resulting spectra, however, differ significantly. The interpretation of IR absorptions is based on the vibrations associated with individual bonds (e.g., $O-H$ stretching at about 3300 cm^{-1}) describing specific functional groups. Terahertz frequencies are much lower in energy and are generated by inter-molecular motions which does not lend them to interpretation as

Table 3.17. Molecular modes and activity in the terahertz region of the electromagnetic spectrum.

Wavenumber (cm^{-1})	Interaction type
3.33–4000	Molecular rotations (gases)
3.33–300	Crystalline phonon vibrations
300–800	Intramolecular bonding vibrations
80–800	Low frequency skeletal and torsional deformations
50–1000	Intermolecular bonding vibrations

functional group-specific absorptions. THz spectra can be used for describing crystalline systems. These systems are not only determined by the individual molecular structure, but also by the arrangement of atoms and molecules within the solid. They can also be used for identifying materials that are chemically identical but have different crystalline forms, commonly known as polymorphs, for learning about amorphous systems including DNA and proteins, and even predicting amorphous stability.

THz radiation is non-ionizing (i.e., does not damage bonds), non-invasive, and non-destructive. In addition, they are phase-sensitive to polar substances, and have high penetration capabilities. THz radiation can pass through most opaque materials to visible light, including plastics, papers, clothes with hidden weapons, and propagates for tens of meters. Since many nonpolar and nonmetallic packaging materials are transparent at terahertz frequencies, they allow THz radiation to penetrate through concealed barriers. This allows THz radiation to identify crystalline and in particular potentially dangerous materials inside sealed environments. Furthermore, the THz spectral range contains significant portions of the spectra of a large number of crystalline materials. These materials include explosives, illicit drugs, as well as most other chemicals in powder form. The lowest frequencies in this region are associated with intermolecular motion and the highest frequencies are associated with intramolecular motion. So THz spectroscopy can be used to detect harmful gases, pesticides, toxic chemical compounds, and drugs and to investigate nanomaterials.

Some may wonder why terahertz spectroscopy is not in all textbooks. Essentially, the upper atmosphere of the Earth contains gases that almost entirely absorb terahertz light from our Sun before they reach the Earth's surface. We are therefore left with the daunting task of artificially generating terahertz radiation on Earth and even strong-field THz pulses for some applications. This challenge has limited human ability to benefit from terahertz applications in our lives so far. In fact, the reference to the terahertz region as the "terahertz gap" stems from difficulties in producing and detecting terahertz light. Increasingly, optics and microwave theories are leading to technology that generates and detects terahertz radiation. In this section, we describe some of the features and applications of terahertz spectroscopy.

3.4.1. Anesthetic Drugs and Unconsciousness

Anesthesia is one of the world's greatest medical discoveries, which occurred by chance ("a gift from aliens") for mankind. It selectively and reversibly blocks consciousness while sparing non-conscious brain (e.g., cerebellum) activities, enabling invasive surgery. Consciousness is defined here as the loss of sensation and the loss of memory for a short period of time. Therefore, we use consciousness here in very narrow and concrete terms whether someone is awake or not. After strokes or traumatic brain injuries, some people may never fully wake up; that is, they will never recover from unconscious awareness. Consciousness includes feelings, pain, love, prayer, meditation, and free will. Although highly significant, these characteristics of consciousness are outside the realm of modern science including current quantum mechanics and quantum computing. This is because of the challenges of proving existing theories. It looks more and more like modern science must be expanded to provide a framework on which consciousness characteristics can be studied.

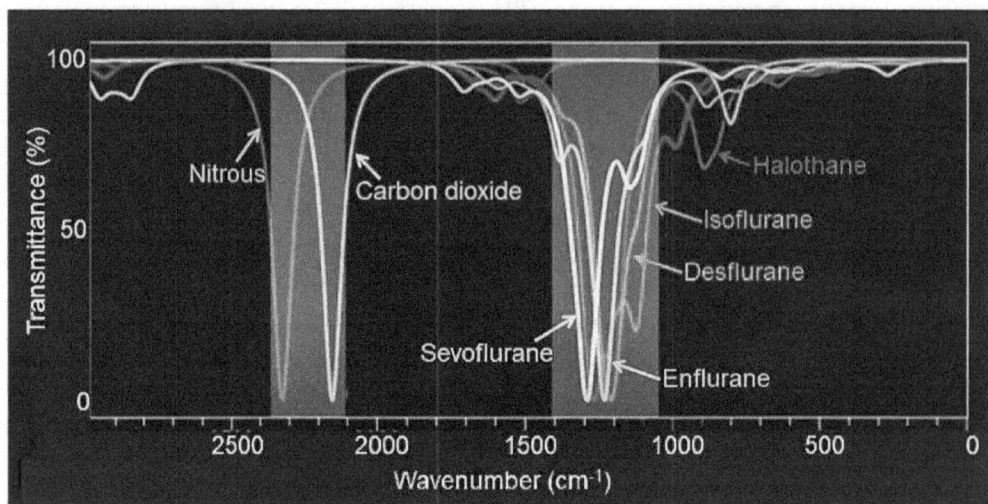

Figure 3.45. Spectroscopy of five anesthetic molecules along with carbon dioxide and nitrous oxide. These five anesthetic molecules play a central role in operating rooms around the world. Spectroscopy of five anesthetic molecules along with carbon dioxide and nitrous oxide.

Let's continue with our focus, conscious awareness. Nowadays, the gas analyzer on the anesthesia machine, which includes IR technology, is a technological marvel found in most operating rooms in all countries of the world. Despite more than 170 years of the use of anesthetic drugs to induce temporary unconsciousness, including loss of sensation, in clinical settings, our understanding of the mechanisms of these drugs is still incomplete. Also, the mechanism by which anesthesia acts and how the brain produces conscious experience remains unknown. Their understanding may help explain consciousness.

Most inhalational anesthetics in clinical use today are halogenated ethers (e.g., Urban and Barann, 2002), such as isoflurane ($C_3H_2ClF_5O$), desflurane ($C_3H_2F_6O$), enflurane ($C_3H_2ClF_5O$), halothane ($C_2HBrClF_3$), and sevoflurane ($C_4H_3F_7O$). The spectra of these anesthetics are shown in Figure 3.45, including enflurane and halothane whose usages are decreasing these days. We added nitrous oxide (N_2O) and carbon dioxide (CO_2) to emphasize how anesthetic spectra differ from some molecules containing chemical elements of significant importance in human bodies and human cells. In addition, N_2O and CO_2, along with O_2, are used to monitor the continued functions of the non-conscious brain activities such as breathing and sleeping during the anesthetic period.[4] We can see, with no exceptions, that all anesthetics have very small transmittance (large absorbance) between 1050 cm^{-1} and 1400 cm^{-1} and almost total transmittance (no absorbance) outside this window— we will call this window *conscious window*. Moreover, it is rare to find molecules whose responses to terahertz or infrared light have the shape of band-limited filters. When a biological or chemical sample is exposed to anesthetics, their interactions

---

[4]As discussed in the previous subsection, infrared spectroscopy cannot measure O_2 but Raman spectroscopy can.

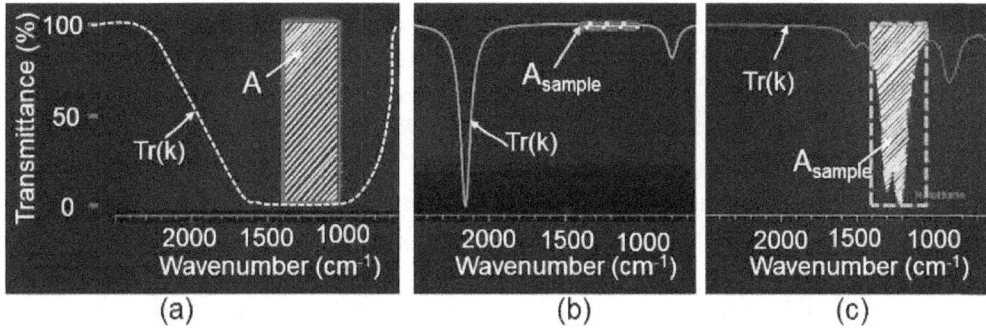

Figure 3.46. (a) An illustration of spectroscopy of a totally unconscious sample. (b) An illustration of spectroscopy of an always unconscious sample such as CO_2. (c) Spectroscopy of a partially unconcisous sample. $Tr(k)$ is the transmittance curve, and A and A_{sample} are areas between the transmittance curves and the 100 percent transmittance line.

are limited to a single narrow window when represented in the wavenumber domain. Anesthetic drugs do not affect chemical samples outside the conscious window. In other words, anesthetic gases spare non-aware brain activities. They affect only conscious awareness and short-term memory.

Is the interaction between anesthetics and chemical samples linear or nonlinear? The terahertz spectrum operates on a microscopic scale, defined as wavelengths smaller than 1/10th of a millimeter. Therefore the superposition of electromagnetic radiation (or light) works in this region and we can consider anesthetic drugs' interactions with human body molecules as linear. So, as illustrated in Figure 3.46, when anesthetic drugs and chemical samples are represented in the wavenumber domain, we see that the action of anesthetic drugs on chemical samples can be described as a classical band-limited filtering operation on chemical samples. The product of these two spectra is a convolution in the time domain.

3.4.2. A Quantification of Consciousness; Conscious Awareness

The key takeaway point so far is that we can shut down our consciousness and fully recover as long as our action is limited to the conscious window. Because strokes or traumatic brain injuries affect non-conscious brain activities, people cannot fully recover. In the 2D space of transmittance and wavenumber, the conscious window can be defined as $[0, 100] \times [1050, 1400]$. Thus, any chemical sample with 100 percent transmittance in the conscious window is, by definition, totally unconscious. Figure 3.42 shows some clay minerals that are almost "unconscious." However the organic molecules in Figure 3.38 are.

Let us denote by A the area $[0, 100] \times [1050, 1400]$ of the conscious window. Let also by A_{sample} the area $[0, Tr(k)] \times [1050, 1400]$, where $Tr(k)$ is the transmission curve of the sample under consideration. We can define the level of consciousness of any chemical sample exposed to any anesthetic drug as follows:

$$l_{\text{consc}} = \frac{A_{\text{sample}}}{A} . \tag{3.168}$$

The level of consciousness of the subject varies between 0 and 1, where $l_{consc} = 0$ indicates that the subject subjected to an anesthetic drug is totally unconscious and $l_{consc} = 1$ indicates that the drug has no effect on the subject. It remained fully conscious; no drugs were absorbed. From this definition, we can see that in Figure 3.42 the unconsciousness level is not zero but quite small.

It is surprising to note that minerals have "consciousness." In other words, some rocks have memory as described in Ikelle (2023c), but they are not conscious aware. Organic and inorganic molecules presented in this chapter have memory but they are not necessarily conscious entities. So it is likely that some matter in our universe is consciousness and therefore the entire universe may be consciousness. Figure 3.47 shows that caffeine, whiskey, and cocaine have high consciousness levels, but not at anesthetic levels. Caffeine, whiskey, and cocaine affect non-conscious brain actions such as sleeping and respiration.

Alcohol causes blurred vision, slurred speech, unsteady walk, and more. These usually disappear once you become sober again. However, if you drink often for long periods of time, alcohol can affect your brain permanently. This effect does not reverse once you become sober again. Long-term effects include memory issues and reduced cognitive function.

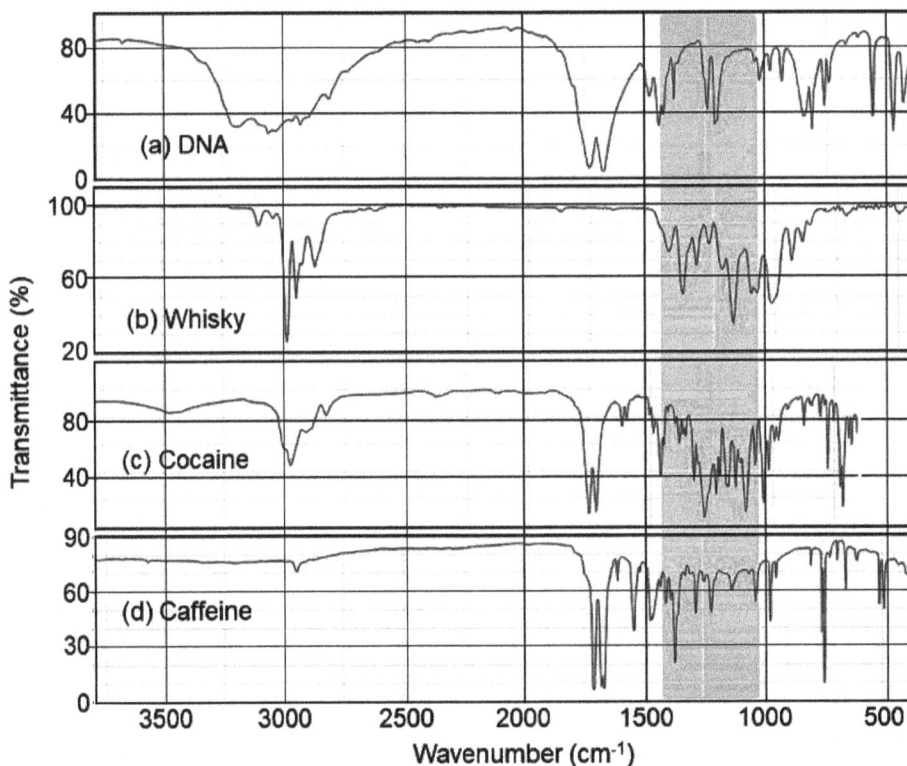

Figure 3.47. Examples of spectroscopy, including their conscious window, of (a) a DNA molecule known as thymine ($C_5H_6N_2O_2$, (b) a whiskey molecule known as cyclohexyl propionate ($C_9H_{16}O_2$), (c) a cocaine molecule $C_{17}H_{21}NO_4$, and (d) caffeine 8-chloro ($C_8H_9ClN_4O_2$). Data from NIST Data Standard Reference.

3.4.3. Terahertz Frequencies: the Transition from Quantum to Classical Mechanisms

Terahertz radiation corresponds to frequencies between the microwave and infrared regions of the electromagnetic spectrum (wavelengths of 1mm to 1μm). With the infrared, we operate on a quantum scale. We can view terahertz (THz) as a transition from quantum to classical mechanics scales or from micro to macro scales. Strangely enough, consciousness falls into this transition zone. While quantum mechanics processes and experiments are probabilistic, those of classical mechanics are not. During the transition, the outcomes of processes and experiments are highly likely. As such, unconsciousness and consciousness are not totally quantum-mechanics-based or classical-mechanics-based processes. This conclusion also allows us to sidestep the fact that unconsciousness and consciousness occur at room temperature in contradiction with some current quantum mechanics formulations.

3.4.4. Loss Memory, Enhanced Memory, and Missing Time

When general anesthesia is performed on a patient, he/she misses time during the active period of the anesthesia drugs. Our memory captures nothing during this period. However, our memory recalls events before and after anesthesia. In other words, anesthesia drugs only affect temporary memories. So the idea put forward by a number of studies that there are short- and long-term memories (e.g., Mangal, 2022) makes sense. Short-term and/or working memories are the gateway to long-term memory. Long-term memory is commonly called memory. It stores our learned facts, concepts, theories, routine behaviors, tasks we have mastered, etc. In addition, the fact that we can restore our memory in cases of Alzheimer's, Parkinson's, and stroke reinforces the concept that there are, at least, two types of memory systems. So the anesthesia drugs primarily affect short-term and/or working memories and only affect long-term memory in the sense that it will be altered and reconstructed in response to changes in our moods or fleeting states of mind during the active period of the anesthesia drugs. By adding some impurities to the anesthesia we can control their resistance and conductivity in a given subject and the duration of their effectiveness.

We have used three key concepts in this subsection so far: wavenumbers, light, and drugs. Light and drugs exist as physical objects to observe. Although no data is lost or added when we pass from the time domain to the wavenumber representation, this representation provides a framework for the determination, or reconstruction of some obscure or hidden aspects of signal data. It can even create desirable effects using the wavenumber representation.

In theory, we can design light to produce a wavenumber response similar to the wavenumber response of drugs and therefore use light instead of drugs. It is a very challenging task to use for medical and clinical purposes. This is because after shining terahertz light on a brain, controlling the reverberations of light inside the brain is near impossible with our present technology. The reverberations change the desired wavenumber response by introducing notches in the wavenumber response. The notches narrow the wavenumber response and duration reverberations can have undesirable effects. These reverberations may explain the December 26-28, 1980 Rendlesham Forest incident in the UK (BBC News, December 26, 2020, and

Weaponization of an Unidentified Aerial Phenomenon by by John Burroughs and James Worrow, December 15, 2020). During December 26-28, 1980, USAF security personnel stationed at RAF Woodbridge in the UK near Rendlesham Forest were engulfed in mysterious lights for three nights. Exposure to this radiation turned out to be debilitating for the health of these individuals. We can design desirable terahertz spectrums and terahertz light to search for plants and mixtures of plants that emit these wavenumbers.

3.4.5.   Can A Machine Be Conscious?

Despite the enormous advances in computing technology, there is not even a hint that computers/AIs will one day simulate consciousness in the large sense. Computation is a limited implementation of classical and quantum physics. As these sciences do not explain consciousness, computers cannot make up the missing information. The day a science develops a framework to simulate consciousness computers may one day be able to do so. We further examine this issue in Chapters 4 and 5.

3.5.   Application: NMR Spectroscopy

NMR (nuclear magnetic resonance) spectroscopy, discovered by Nobel Prize Physics winners, Hensen Black and Torrey Parcell, is based on light's interaction with matter, just like the three forms of spectroscopy described above. Light, however, is made up of radio waves of various frequencies. No electronic, vibrational, or rotational transitions are created in this process because the energy at radio-wave frequencies is too small. However, it is significant enough to cause a change in the spin of nucleons (protons and neutrons)—that is, the absorption of a radio-frequency photon promotes a nuclear spin from its ground state to its excited state. The critical notion of *spin* of particles is critical in this section.

3.5.1.   Quantum Spin

In the above discussion and in the previous chapters, we described various properties of matter, atoms, and even sub-atomic particles, including electrons and photons. Like its charge, the spin of particles such as electrons and photons is another fundamental observable quantity of a particle, which means that a particle has a magnetic dipole moment. This property is an electromagnetic orientation possessed by all particles with spin. It plays a crucial role in quantum physics. Here, spin does not imply movement, but simply a weird choice of terminology. A particle's spin is the quantized intrinsic angular momentum and magnetic moment; meaning that it takes only discrete values. This angular momentum is

$$L = \hbar\sqrt{s(s + 1)}\,,$$

where $s = \{0, 1/2, 1, 3/2, 2, ...\}$ is the spin, and $\hbar = 1.055 \times 10^{-34}$ J s is the reduced Planck constant—the intrinsic angular momentum here that is different from the classical angular momentum defined in Ikelle (2023b and 2023c). Note that s is restricted to non-negative integers and half integers with $s = 1$ for photons and

$s = 1/2$ for matter particles such as electrons and protons. Larger composite particles such as heavy nuclei or many electron atoms can have many different possible s values. Zero and integer spin particles are called bosons. Half-integer $(1/2, 3/2, 5/2, ...)$ spin particles are called fermions.

Regardless of the physics involved, a particle spin is an experimental fact. The first experiment was performed by Gerlach and Stern (1922). They consider atoms with electrons horizontally through a magnetic field oriented up-down. One would expect the magnetic field to deflect atoms based on the electron's spin. Regardless of the magnetic field orientation, *electrons always point in the direction of the magnetic field or in the opposite direction of the magnetic field.* Note that, in classical physics, different orientations of a magnetic field would cause a continuous range of deflections in the apparatus. However, only two separate deflections are observed in quantum mechanics. Chapter 4 illustrates this weirdness of the quantum world. No matter how you make an electron, including when it does not align with the magnetic field, when you place it in the magnetic field one of two things can happen when you turn on the magnetic field. It is pointing up—then no photons are emitted. If it is pointing downwards, then the electron emits a photon of a frequency corresponding to the energy of jumping up and down. The spin of an electron in an atom determines the direction of the magnetic field produced by it.

In many atoms (such as carbon-12) the spins are paired against each other, such that the nucleus of the atom has no overall spin. However, in some atoms (such as hydrogen-1 and carbon-13) the nucleus does possess an overall spin. The rules for determining the net spin of a nucleus are as follows:

- If the number of neutrons and the number of protons are both even, then the nucleus has no spin (i.e., $s = 0$); for example, carbon-12, oxygen-16, and sulfur-32.

- If the number of neutrons plus the number of protons is odd (i.e., number of protons is odd and number of neutrons is even or number of protons is even and number of neutrons is odd), then the nucleus has a half-integer spin; for example, hydrogen-1 with $s = 1/2$, carbon-13 with $s = 1/2$, fluorine-19 with $s = 1/2$, nitrogen-15 with $s = 1/2$, phosphorus-31 with $s = 1/2$, sodium-23 with $s = 3/2$, aluminum-27 with $s = 5/2$, oxygen-17 with $s = 5/2$, and cobalt-59 with $s = 7/2$.

- If the number of neutrons and the number of protons are both odd, then the nucleus has an integer spin (e.g., 1, 2, 3); for example, hygrogen-2 with $s = 1$, and nitrogen-14 with $s = 1$.

Quantum mechanics tells us that a nucleus with spin s will have $2s + 1$ possible orientations; e.g., a nucleus with spin 1/2 will have 2 possible orientations, just like an electron. To summarize, spin interacts with magnetic fields, both inside and externally. Spin can affect bulk properties.

3.5.2. Nuclear Magnetic Resonance (NMR)

Spectroscopy is generally used to identify compounds. It is based on the principle that the major nuclei of atoms possess magnetic moments and angular momentum.

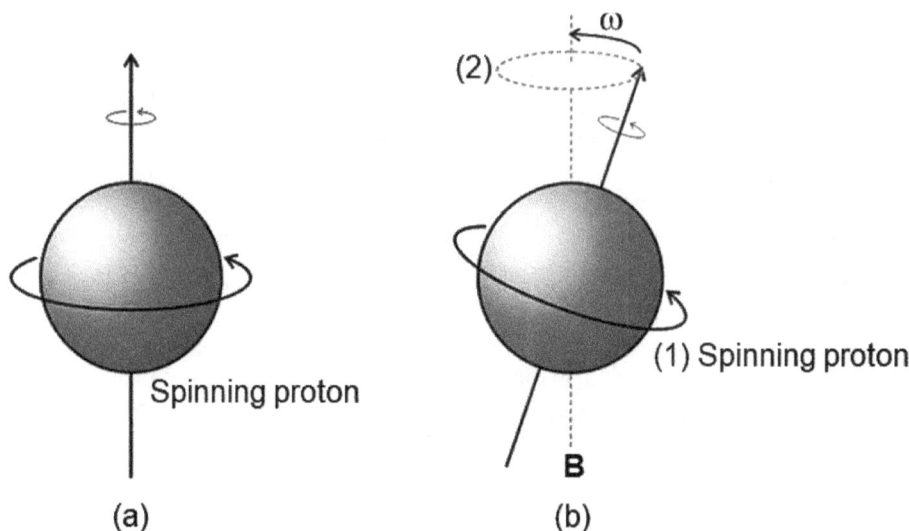

Figure 3.48. (a) Major nuclei of atoms possess magnetic moments and angular momentum. Spinning charges on a proton generate a magnetic dipole. (b) In the event of an external magnetic field being applied to a nucleus, we have rotation of the nuclear momentum about its own axis (1) and about the magnetic field axis (2).

In the event of an external magnetic field being applied to a nucleus, its rotational axis will change. It will go through a precession around its rotation axis before an external magnetic field is applied (see Figure 3.48). The frequency of rotation of the nucleus around the previous axis is known as precession frequency or Larmor frequency (e.g., for a proton in a magnetic field with $B_0 = 1.41$ Telsa, the precision frequency is around 60 MHz). This frequency is proportional to the magnetic field magnitude. In other words, the stronger the external magnetic field intensity, the higher the precession frequency is. Due to the fact that protons are charged particles, the nucleus also produces an electric field that oscillates at radio frequency. When the radio frequency of this electric field matches the precession frequency, we have nuclear magnetic resonance and energy is transferred from the radio waves to the nucleus. This transfer of energy causes the nucleus to move from a lower to a higher state, giving rise to absorption.

3.5.3. Chemical Shift

Another key feature of nuclear magnetic resonance to be considered is the effects of electrons surrounding the nucleus on the net magnetic field of the nucleus. When an external magnetic field is applied, electrons surrounding the nucleus circulate. This creates a small secondary magnetic field in the opposite direction of the external magnetic field. So that the magnitude of the net magnetic field in the nucleus is

$$B_{\text{loc}} = (1-\sigma)B_0 , \tag{3.169}$$

where B_0 is the magnitude of the applied external magnetic field, B_{loc} is the actual magnetic field felt by the nucleus, σB_0 the effect of electron motion, and σ is known as the shielding constant or screening constant. The constant σ is very small—typically 10^{-5} for protons and less than 10^{-3} for other nuclei according to Becker (1969). The difference between the resonance frequency associated with B_0 (nucleus without electrons) and the resonance frequency associated with B_{loc} (nucleus with electrons) is known as the chemical shift. The magnitude of the shift depends upon the type of nucleus and the details of electron motion in nearby atoms and molecules.

Since σ is so small, it is difficult to determine resonance frequencies and differences. Actually no practical attempt is made to measure them. Instead, a reference compound is added to the solution of the substance to measure the resonance frequency of the nucleus relative to the resonance frequency of the reference nucleus, as follows:

$$\delta = \frac{f_0(\text{Hz}) - f_{ref}(\text{Hz})}{f_{ref}(\text{Hz})} \times 10^6 = \frac{f_0(\text{Hz}) - f_{ref}(\text{Hz})}{f_{ref}(\text{MHz})} \text{ ppm} , \qquad (3.170)$$

where f_0 is the resonance frequency of the nucleus under consideration, f_{ref} is the resonance frequency of the reference nucleus, ppm stands for parts per million, and δ is the practical measurement of *chemical shift*. It is customary to adopt the resonance frequency of tetramethylsilane ($Si(CH_3)_4$), also known as TMS, as the reference frequency. There are a few reasons why TMS is used as a reference. It is a rare example of carbon connected to a less electronegative element, silicon. This shields it and therefore causes it to appear at lower ppm where other protons do not emit signals. Therefore, the peak at 0 ppm can be ignored when analyzing the NMR spectrum. Second, it has a low boiling point which makes purifying the sample easier if needed. The precision of NMR spectroscopy allows this chemical shift to be measured. The study of chemical shifts has produced a large store of information about chemical bonds and the structure of molecules.

So NMR detects the quantum transition of nuclear spins under a strong external magnetic field. To be detected, the nuclei must have at least two different spin states. This is equivalent to a non-zero spin (i.e., $s \neq 0$). Therefore, isotopes with $s = 0$ (e.g., carbon-12, oxygen-16, and silicon-28) are invisible in NMR. They are called NMR inactives. Active NMR isotopes (i.e., visible by NMR) include carbon-13, oxygen-17, and silicon-29, which have integer or half integer spins. Here are two examples. In proton NMR of benzyl alcohol in a 60 MHz instrument, the resonance frequency of OH is 144 Hz higher than the TMS frequency. The chemical shift is then

$$\delta = \frac{f_0 - f_{ref}}{f_{ref}} = \frac{144 \text{ Hz}}{60 \text{ MHz}} = 2.4 \text{ ppm} . \qquad (3.171)$$

In a proton NMR of benzyl alcohol in a 300 MHz NMR instrument, the measured difference in frequency between CH_2 and TMS is $\Delta f = f_0 - f_{ref} = 1380$ Hz. We can deduce that the chemical shift is 4.6 ppm; i.e.,

$$\delta = \frac{\Delta f}{f_{ref}} = \frac{1380}{300} = 4.6 \text{ ppm} . \qquad (3.172)$$

On a δ-scale, proton NMR chemical shifts range from -10 to 30 ppm. Most metal hydrides have negative chemical shifts. The majority of organic compounds have

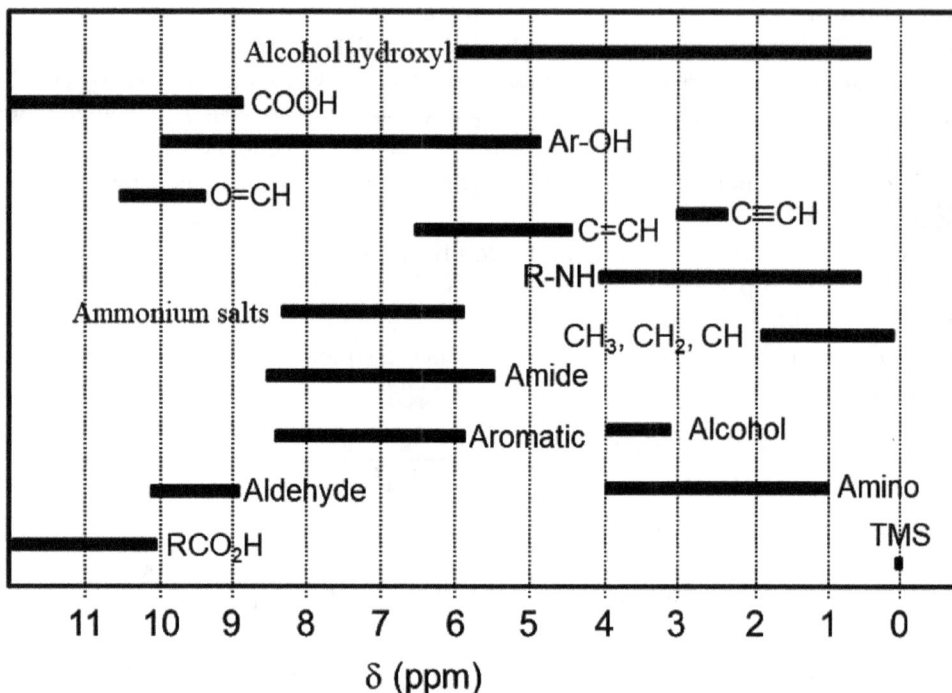

Figure 3.49. The chemical shifts for various 1H resonances. Note magnetic field increases with decreasing chemical shift.

positive chemical shifts between 0 and 13 ppm. A simple illustration is provided by the NMR spectrum is shown in Figure 3.49. One of the key factors influencing chemical shift is the electonegativity of nuclei introduced in Ikelle (2023a) and in the first section of this chapter. For example, the chemical shifts of CH_3H, CH_3Cl, and CH_3F are 0.23, 3.05, and 4.26. Notice that three molecules have the common CH_3 group. But they are attached to H (with 2.2 electonegativity), Cl (with 3.16 electonegativity), and F (with 3.98 electonegativity). There is no doubt that chemical shifts are affected by the electonegativity of the chemical elements in a molecule. Other factors are hydrogen bonds of nuclei and intramolecular and intermolecular forces.

Carbon plays an instrumental role in our lives and is widespread on our planet (see Ikelle, 2023a), so alternative NMR spectroscopy methods have been developed, such as carbon-13 NMR spectroscopy. Carbon-13 NMR spectroscopy determines the types and number of carbon atoms present in a molecule or substance. It yields structural information much more directly than a proton spectrum such as quaternary carbons and functional groups (CN, C=O, and C=NR). Its chemical shift varies from -20 to 220 ppm.

In summary, NMR data contains information about individual atoms in molecules. This includes a specific signal for every nucleus, the relative molar concentration of every magnetically unique nucleus that is in a magnetically unique location. It also includes the bond angle between spin-coupled nuclei, and the relative motion of nuclei in the molecule(s). These data can likewise be used to determine whether or not groups of nuclei are in a chemical bond network. They can also be used to determine the distance between nuclei groups.

Nuclear magnetic resonance (NMR) is mostly used in scientific research to determine the effects of a magnetic field on the nucleus of atoms and more broadly the physical, chemical and biological properties of matter, including molecular structure changes (or is invariant) with temperature, molecular concentration, pH, and ionic strength. The popular medical imaging technique known as magnetic resonance imaging (MRI) that we will discuss in Ikelle (2024) is based on the same principle as NMR. It is an advanced medical imaging technique that uses computer-generated radio waves and a magnetic field to create detailed images of internal organs. The procedure creates complex pictures of water-containing body parts. Just like NMR, MRI relies on radiation data to identify unknown compounds. While NMR uses radiation frequencies to generate information, MRI produces information based on radiation intensity. NMR spectroscopy determines matter's chemical structure. In MRI imaging, the goal is to obtain detailed images of the body. It is possible to differentiate between different types of tissue with MRI.

3.6. Application: Spectroscopy of Human Eyes

The three main components of the eye are the pupil, lens, and retina, as illustrated in Figure 3.50. The pupil allows visible light to enter the eye. The lens focuses on visible light to create an image. Retina then detects the light and converts it into electrochemical signals which are then sent to the brain. In the human vision process, we assume that the light source is sufficiently distant from the lens so the incident rays can be described as parallel light rays.

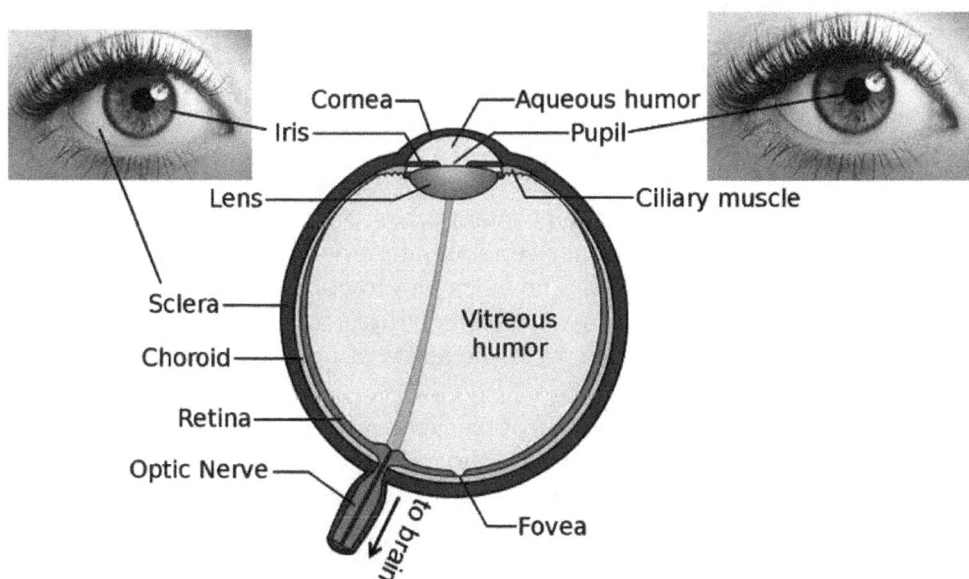

Figure 3.50. The human eye is roughly spherical, about 2.4 cm in diameter. In the eye, the cornea and the lens are the eye's refractive elements. The eye is filled with clear, jelly-like fluids called the aqueous humor and the vitreous humor.

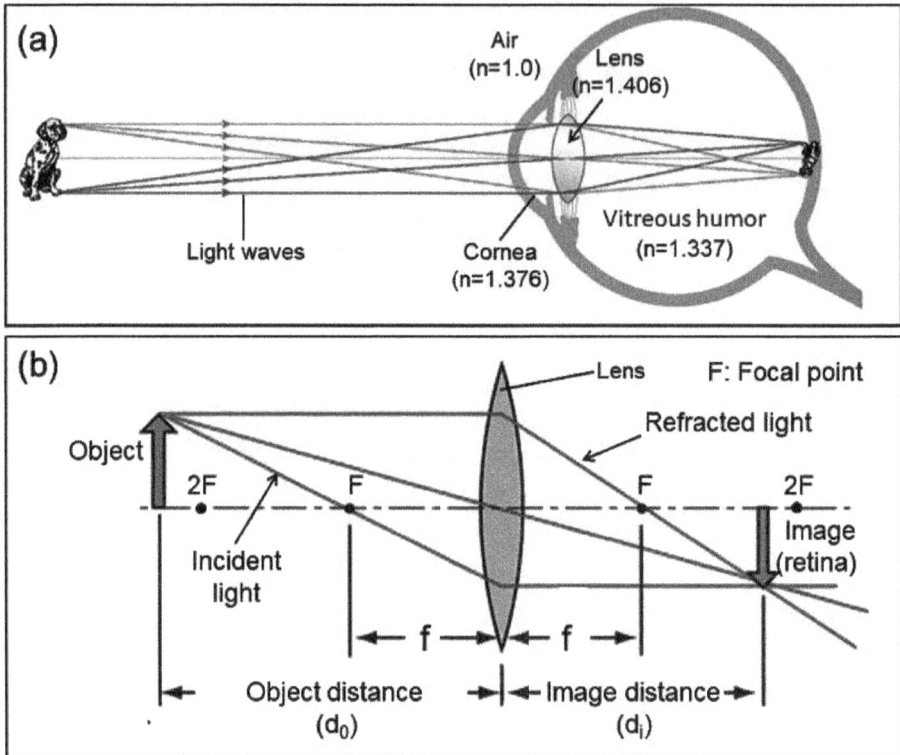

Figure 3.51. (a) An image is formed on the retina with light rays converging most at the cornea and upon entering and exiting the lens. Rays from the top and bottom of the object are traced and produce an inverted real image on the retina. The distance to the object is drawn smaller than scale. (b) Refraction occurs at the anterior surface of the lens. The image of the object on the retina is much smaller than the original object.

3.6.1.   Cornea, Pupil, and Lens

As discussed in Ikelle (2023b), light can be considered as waves or particles. Here, we refer to it as waves. We can use geometric rays to describe these waves. Light rays are bent, or refracted, when they cross an interface between two materials that have different indices of refraction. The index of refraction of a material is the ratio of the speed of light in a vacuum to the speed of light in the medium—that is, the refractive index is $n = c_0/v$, where c_0 is the speed of light and v is the velocity of the material.

The lens system of our eyes consists of two lenses: the cornea on the front surface of the eye, and the crystalline lens inside the eye. The space between the lenses is filled with a clear gel-like fluid called the aqueous humor. Additionally, the inner cavity of the eye is not completely empty. As a result, it is filled with another transparent, gel-like fluid called vitreous humor. Note that the cornea and lens do not contain any blood vessels. The absence of blood (a viscous liquid) ensures transparency of the cornea and lens. Between the lenses is the iris, an opaque, colored membrane. This lens system uses refraction to bend parallel light rays (see Figure 3.51). Light enters the lens at the front surface, and it leaves the lens at the back surface.

For completeness, let us describe the optical geometry of lenses in general, including lenses in our eyeglasses and cameras. The focal length of a thin lens, f, is related to the object distance, d_o, and the image distance, d_i, as depicted in Figure 3.51; i.e.,

$$\frac{1}{f} = \frac{1}{d_i} + \frac{1}{d_0} \ .$$

(3.173)

If the object is very far from the lens, the object distance is considered to be infinity. In this case, the rays from the object are parallel, $1/d_0$ equals zero, and the image distance equals the focal length. This leads to the definition of the focal point as the place where a lens focuses incoming parallel rays from a distant object. The size of an image can be different from the size of the object. The magnification, M, of the image is defined by:

$$M = -\frac{d_i}{d_0} \ .$$

(3.174)

If $|M|$ is greater than 1, the image is larger than the object; if $|M|$ is less than 1, the image is smaller. The magnification, M, which can be positive or negative, represents both the size and orientation of the image. If M is positive, the image is upright, or in the same orientation as the object. If M is negative, the image is inverted, or in the opposite orientation to the object. If the object is upright, then the inverted image appears upside-down.

Let us return to the functions of our eye-lens systems. Our crystalline lens prevents ultraviolet light from reaching our retina while transmitting all visible light. They contain protective pigments that absorb harmful ultraviolet radiation and safely dissipate its energy. Ultraviolet photons have enough energy to break apart proteins in the retina, so any leakage of ultraviolet light is bad for our retina. Aqueous humor and vitreous humor selectively absorb light of shorter than 380 nanometres wavelength, because the photons of such light, with energy greater than 3 electron volts, would destroy molecules that are essential to the functioning of our eyes. Infrared light is also absorbed by these liquids. In other words, our lens systems and their liquids limit our vision to visible light. The cornea and sclera will be damaged by strong exposure, but the damage is generally quickly self-repaired. Also, our skin becomes darker when exposed to ultraviolet light. Moreover ultraviolet light can harm DNA. Only animals with very short life spans can see ultraviolet light because their eyes would be damaged over a longer life span.

3.6.2.   Retina

The retina reacts only to visible light. The cells of retina, called cones, absorb the incoming photons and then emit electrical signals to the brain, which interprets the resulting electrochemical signals in all of the colors we see. There are three types of cones, each with a range of one of the basic three colors (red, yellow, and blue). The combination of electrochemical impulse concentration of each type of cone along with their location gives us our vision. None of the cones are sensitive to ultraviolet radiation.

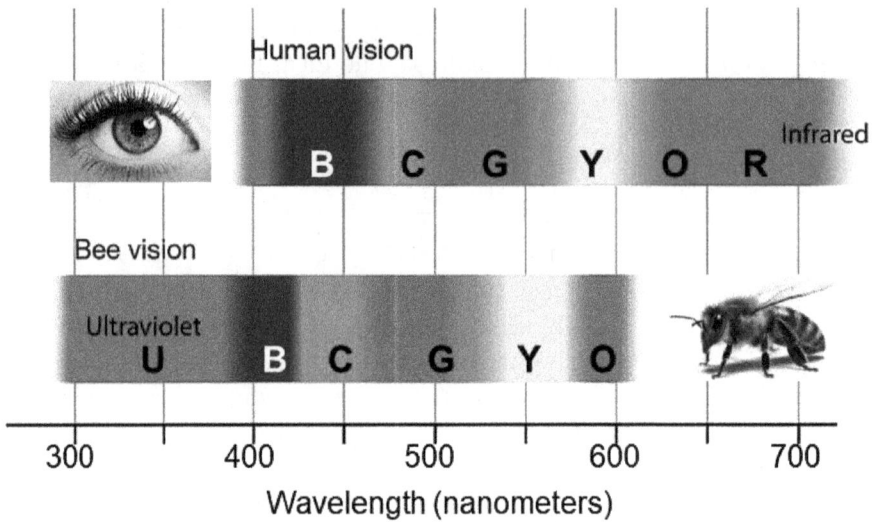

Figure 3.52. Bee vision versus human vision. Bees can see color and they are actually trichromatic. Humans base their color combinations on red, blue and green, while bees base their colors on ultraviolet light, blue and green. U=ultraviolet (less than 440 nm); B = blue (440-485 nm); C= cyan (485-510 nm); G=green (510-565 nm); Y= yellow (565-590 nm); O= orange (590-625 nm); and R= red (greater 625 nm).

3.6.3.   Sunlight and Earth's Atmosphere Filtering

As described earlier, the electromagnetic spectrum includes radio waves, microwaves, infrared, visible, ultraviolet, X-ray, and gamma-ray. All these lights are emitted by stars—that is, why stars are blackbodies (see Chapter 2). As illustrated in Figure 3.52, the amount of solar radiation reaching us is not identical for all parts of the electromagnetic spectrum. The visible band is the dominant band of solar radiation that reaches us. All other light is attenuated by our magnetosphere and atmosphere. Radio waves, microwaves, infrared, ultraviolet, X-rays, and gamma rays used on Earth are mostly artificially generated or are the result of refractions, reflections, or nonlinear interactions. Basically we can nonlinearly mix visible frequencies to produce frequencies outside the visible band, as described in the previous section and in Ikelle (2023b).

Our eyes have evolved to detect only visible light, which is the dominant light band of sunlight reaching Earth. The wavelengths of visible light vary from 380 to 700 nanometers. For a wavelength range from 10^6 to 10^{-12} meters, visible light occupies, on this logarithmic scale, about 1/60 of the entire electromagnetic spectrum. In other words, our eyes only see a tiny range of the electromagnetic spectrum— that is, there is a lot we do not observe. While there are no other 3-D inhabitants on Earth's surface that we cannot observe, we are missing some things that happen there. We have developed sensors that can sense wavelengths we cannot see.

Visible range is large enough for humans to see clean water, food, and evade danger because most physical dangers to our lives are in the visible spectrum. But why have we evolved to see only a thin slice of the electromagnetic spectrum? In order to detect light outside the visible range, our eyes must contain molecules that can absorb this light and undergo a chemical reaction in response. Our bodies are predominantly composed of carbon-based compounds in water solution, just like all other living things. Some parts of the electromagnetic spectrum, such as ultraviolet light, are energetic enough to break the carbon-carbon bonds of organic molecules as well as break up water molecules themselves. Our human vision system and its habitat on Earth evolved to avoid exposure to the energetic parts of the electromagnetic spectrum. In other words, chemical reactions do not allow an organic molecule to dissolve in water solution. On the long wavelength end of the spectrum, including visible light, photons do not carry enough energy to trigger chemical reactions with organic molecules in water solution—that is, it is not possible to create organic molecules in water solution that can detect that wavelength. Thus chemical reactions associated with short wavelength lights forbid human eyes to see higher than the visible band.

Even if we were able to modify our eyes in order to see beyond the visible spectrum, the computation requirements and the amount of data may overwhelm our current brain capacity. We may need to physically increase our brain size. Other creatures on Earth detect light outside the visible range. Bees and butterflies see well into the ultraviolet spectrum for the limited purpose of identifying nectar in flowers (see Figure 3.52b). However, the bee's eye lacks receptors that are sensitive to the portion of the electromagnetic spectrum that we humans see as red. Pit vipers detect light very deeply into the infrared. Nocturnal animals like snakes, for example, also use part of the infrared spectrum in order to move around at night. In fact, after the meteorite impact that wiped out dinosaurs from Earth about 60 million years ago, the only surviving mammals were nocturnal. Their evolution had allowed them to lose most of their color vision.

Box 3.2: Ray Approximation

A snapshot of wave propagation at a specific time represents a wavefront. Rays are defined as lines normal to the wavefront (i.e., they point in the direction of propagation). Both of these definitions refer to a homogeneous medium and to a medium with slowly varying physical properties. Notice that the rays are straight lines when the medium is homogeneous, and they can take any arbitrary curved lines when the medium is heterogeneous (see Ikelle, 2023c). So when we talk about rays instead of waves, we are making the assumption that the waves are traveling in a series of layers in which the interfaces are continuous and the physical properties in each layer are either homogeneous or smooth. Rays of light are assumed to approximate electromagnetic waves and rays to approximate elastic waves.

When the medium contains discontinuous interfaces, non-homogenous layers and/or non-smooth layers—the wavelength of the source of waves is not too large compared to the scale of heterogeneities—rays cannot be used because we can no longer form lines normal to the wavefronts. Rays, for example, cannot approximate elastic wave propagation within a subsurface containing faults, pinchouts, and unconformities. In such cases, the laws of reflection and refraction are no longer adequate because the energy is diffracted rather than reflected or refracted (see Figure 3.53).

Ray optics treats light as either traveling in a straight line, refracting at a surface or reflecting at a surface and then continuing in a straight line to another surface (see Figure 3.51). Ray optics is very useful for designing ordinary lenses. However, it is unable to predict the diffraction pattern of finite-sized apertures. It also fails to predict a spot. And in laser resonators where the path is long compared to the diameter of the beam, ray optics pretty much fail altogether.

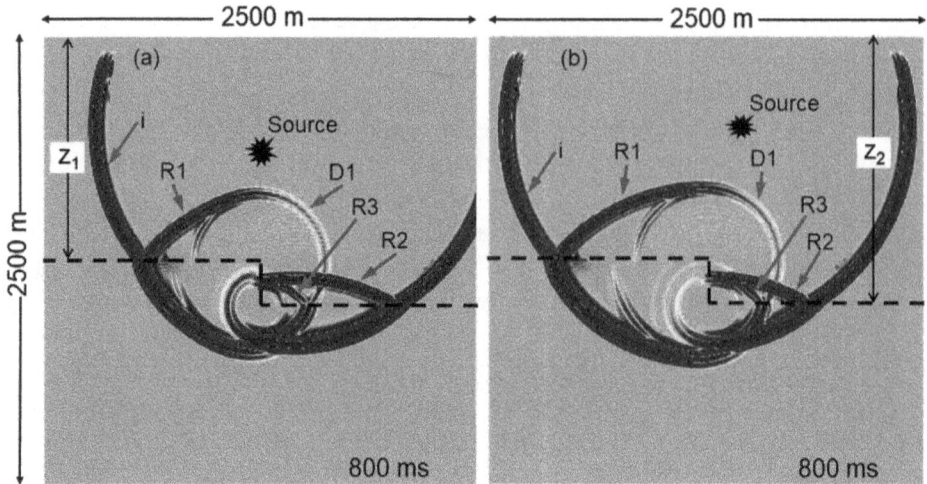

Figure 3.53. Diffraction and reflection patterns over the vertical step fault. In (a) the source is directly over the step and (b) the source is displaced 200 m from the diffracting edge ($z_1 = 500$ m and $z_2 = 750$ m). The properties of the top medium here are [$V_P = 1500$ m/s, $V_S = 0.0$ m/s, and $\rho = 1.0$ g/cc]. Those of the bottom medium are [$V_P = 2000$ m/s, $V_S = 0.0$ m/s, and $\rho = 2.0$ g/cc]. R1: reflection from the interface at z_1; D1: first diffraction from the top corner; R2: reflection from the interface at z_2; R3: reflection from the interface at z_2 followed by a reflection from the step.

To summarize, almost our entire anatomy and physiology, including our eyes, evolved to adapt to our surroundings. But this evolution has not included molecules that can withstand the higher energies necessary for full human vision of the electromagnetic spectrum. If we are exposed to high intensity of light beyond the ultraviolet spectrum, such as bright sunlight or when surrounded by snow, we need sunglasses to prevent irreversible damage to our retinas. Note also that, when the sun sets down and the sky turns dark, human eyes cannot see objects. With infrared telescopes or infrared cameras, we can see a very busy night sky. There are several objects that become active at night and make the night sky busier than it would otherwise be.

Quiz: Why does the cornea have the strongest refraction?

3.7.   Application: Spectroscopy of Stars

3.7.1.   Optical Telescopes

Modern technology allows us, through detectors like optical telescopes, to see what we cannot see with our naked eyes. Optical telescopes are optical instruments designed to make distant objects appear nearer. It is an arrangement of lenses or curved mirrors, by which EM radiation from 300 GHz to 10^{-21} Hz (this includes the infrared, visible spectrum, ultraviolet, X-ray and gamma ray spectra) is collected, focused on a small area, and the resulting image magnified, as illustrated in

Figures 3.54 and 3.55. In other words, optical telescopes are EM radiation collectors that they then turn into images. Again, lenses are materials that let light pass through themselves. It can be crystal, glass, or plastic. A mirror is essentially a reflective surface. The more polished or shiny the mirror is, the more light bounces off its surface. Optical telescopes observe objects at very large distances. Therefore, electromagnetic radiation can be assumed to originate from infinity and be represented by parallel rays. Optical telescopes measure the intensity of extraterrestrial electromagnetic fields.

As illustrated in Figures 3.54 and 3.55, there are two main types of optical telescopes: (1) refracting telescopes that use lenses to focus incoming EM radiation by bending EM radiation (i.e., by changing the direction of EM radiation) to a focus point, just like our eyes, and (2) reflecting telescopes that use curved mirrors to focus EM radiation by refracting light rays to the focus point. The amount of bending in refracting telescopes depends on the type of lens's material and the wavelength of EM radiation. In reflecting telescopes, the angle of incidence of EM radiation upon a reflective surface is equal to the angle of reflection from that surface. Most of today's telescopes are multi-lens systems. For example, most of today's refracting telescopes use the so-called Cassegrain design. Light enters from the top. Bounces off the primary mirror. Bounces off the secondary mirror. Light passes through a hole to focus, as illustrated in Figure 3.54. Gallileo invented the first two-lens refracting telescope and Isaac Newton invented the basic design of the reflecting telescope.

The size of optical telescope is usually described by diameter, d, of its primary lens or mirror. The collecting area of a lens or mirror is $A = \pi(d/2)^2$. Light-gathering power area is proportional $\pi(d/2)^2$. For $d = 10$ meters, area is $P = 7.85 \times 10^5$ cm^2. Light gathering power is 1.96 million times that of the eye, assuming $d = 7.1$ mm for the eye. Larger telescope are capable of taking images with greater detail and also better angular resolution. Angular resolution is inversely proportional to the diameter of the mirror.

By far the most professional telescopes today are reflector telescopes. Problems with refractors include deviations from the actual focus point known as aberrations, glass defects, and distortion caused by weight. Reflecting telescopes have fewer problems with aberrations, glass defects, support, and distortion. Moreover, reflecting telescope diameters can reach 10 meters while refracting telescope diameters only reach 1 meter. Therefore large reflector telescopes have high resolution.

Optical telescopes can see celestial objects. From very hot to cold, they include hot interstellar gas, black holes, supernovas, gas jets, pulsars, solar flares, hot young stars, galaxies, and interstellar dust clouds. With the gamma rays and x-rays of the optical spectrum we can see interstellar gas, black holes, supernovas, gas jets, pulsars, and solar flares. With the ultraviolet part of the optical spectrum, we can also observe solar flares in addition to hot young stars. The visible part of the optical spectrum allows us to view stars, including the sun, and galaxies. Using the infrared part of the optical spectrum, we can observe interstellar dust clouds.

Optical telescopes are negatively affected by bad weather (clouds are problematic) and stray radiation (such as the Sun, the Moon, and light pollution from cities). So the world's largest and most professional telescopes are located in Hawaii, Chile, Arizona, the Canary Islands, and Mexico. These telescopes are far from light

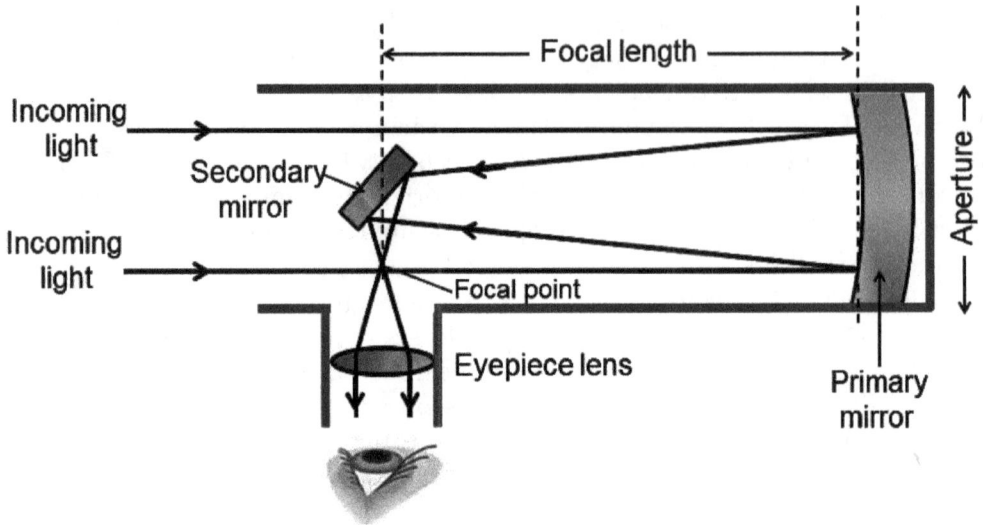

Figure 3.54. A schematic diagram of reflecting telescope. Reflecting telescopes use a large parabolic mirror to direct light onto a flat secondary mirror, that in turn directs the image at a right angle to focus at your eye.

pollution, at high altitudes with a steady atmosphere. The largest refracting telescope in the world is at Yerkes Observatory, in Williams Bay, Wisconsin, United States. It has a 1-meter objective. Objective is the term used for the lens at the top of the optical tube. The size of the objective relates directly to how dim an object can be discerned. The Hubble Space Telescope is a reflector telescope. Most common-man telescopes are called Dobsonians, after John Dobson whose simple optical telescope design popularized telescopes starting in the late 60s. The 2.5-meter Hooker Newtonian reflector telescope was used by Edwin Hubble to discover that the Universe is expanding (Ikelle, 2024). The largest reflecting telescope with a 39-meter diameter primary mirror (actually several mirrors joined together), known as the European Extra-Large Telescope, is expected to start collecting light in 2025.

> **Quiz:** Suppose that you are living on a planet which is 150 million light years away from Earth. If you focus your telescope on Earth today, what species will you see?

3.7.2.   The Birth of the Famous Hubble Telescope

In the 1960s, NASA and USAF (US Air Force) developed a program known as MOL[5] (Manned Orbiting Laboratory) to send crews on 30-day missions in space, and return to Earth (*Reflections on national reconnaissance and the manned orbiting laboratory* by Courtney Homer; https://bookstore.gpo.gov). MOL contained one of the first CCD (charge coupled device) cameras capable of transmitting images from

---

[5] Astronauts of MOL may have landed on the Moon well before Armstrong and Burden of Apollo 11. Tentacles of the US security state and many other so-called developed countries are deeply involved in science editing, which nowadays includes so-called equity. The US has always had two or more space programs, including the public/civilian one known as NASA.

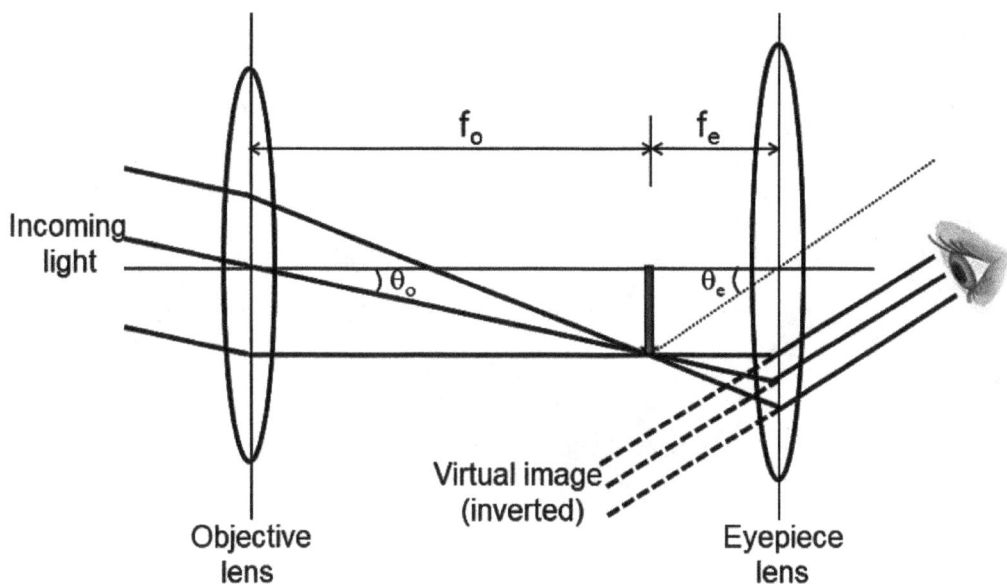

Figure 3.55. A schematic diagram of reflecting telescope. Refracting telescopes use two or more lenses to bend light to a focus; here two two lenses are used. Objective is the classical term for the lense at the top of the optical tube.

an orbiting satellite down to Earth. These sophisticated lenses had keyholes, like a door keyhole. All the keyholes were looking down at Earth from orbit. Someone had the bright idea to turn one keyhole around to look out from Earth out into space and that is how the famous Hubble telescope was born. The electronic camera system of the keyhole-12 was so effective that a satellite pointing at Earth could read the writing on a baseball on the ground. In the 1980s, the Hubble telescope with its keyholes pointed at the stars, toward the moon, and toward Mars to study their surfaces in minute detail. Searching for alien bases and alien activities was part of this process.

On April 24, 1990, Hubble was carried to space aboard the space shuttle Discovery. It was deployed into Earth orbit a day later. Hubble was the first major optical telescope in space, the ultimate mountaintop. Above the distortion of the atmosphere, far above rain clouds and light pollution, Hubble has an unobstructed view of the universe. Despite its small mirror (only 2.5 meters in diameter) compared to behemoth Earth-bound telescopes, it can see more clearly than Earth-bound telescopes. It is named after American astronomer Edwin P. Hubble (1889–1953) who confirmed an expanding universe and provided the foundation for the big-bang theory.

The Hubble telescope's top accomplishments include measuring the expansion and acceleration rate of the universe; finding that black holes are common among galaxies; providing key new insights into the lives of stars from birth to death; uncovering the first evidence that planet formation accompanies star birth; characterizing the atmospheres of planets around other stars; and monitoring weather changes on planets across our solar system.

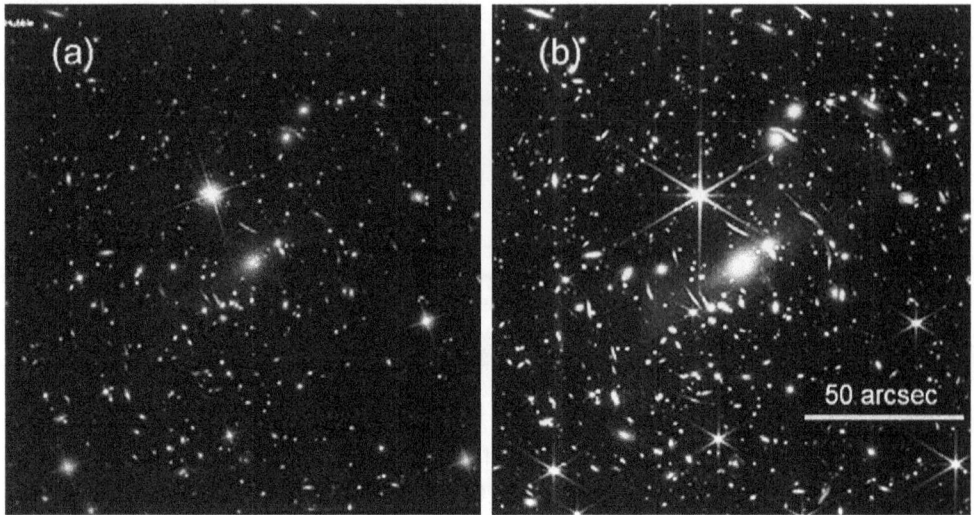

Figure 3.56. A comparison of (a) a view of the Hubble's image taken in 2017 and (b) the JWST's image image taken in 2022 of deepest space. Webb primarily look at the Universe in the infrared, while Hubble studies it primarily at ultraviolet wavelengths. The bar at the bottom corresponds to 50 arcsec, which is approximately the maximum size of Jupiter observed from Earth. Adapted from NASA/STScI; NASA/ESA/CSA/STScI.

3.7.3. James Webb Space Telescope and Search of Exoplanets

The James Webb Space Telescope (JWST) is another NASA optical space telescope that was launched on December 25, 2021. Compared with current telescopes in 2023, including Hubble which has orbited Earth since 1990, it is generally more powerful. It is now at the L_2 Lagrangian point in space (about 1.6 million kilometers from Earth) at -223 degrees Celsius. JWST is now sending us images of galaxies very far away as they existed soon after the big bang. This is described in the next subsection. It will eventually examine the light coming from some of the closest exoplanets and possibly detect signs of life.

The mirror is one of the most essential parts of a telescope. It determines the wavelength of light a telescope can collect. Hubble collects light with a 2.4 meter primary mirror while JWST has a 6.5 meter primary mirror. This is because JWST can collect more light than any other telescope, including infrared light. It can also detect faint objects that other telescopes cannot detect. Capturing infrared, however, is critical as JWST can view through a lot of stuff out there, particularly dust clouds that exist throughout the galaxy (as shown in Figure 3.56). JWST is actually looking in infrared, a longer wavelength than the light we observe with our eyes, a light that Hubble is not tuned to sense. Thus it can see through things like dust and gas that would normally block our view. Infrared's brilliance allows it to see the furthest objects in the universe. It can look into galaxies' cores, including our own, where dust and gas block visible light from reaching us. It can also peer into molecular clouds where stars form. Figure 3.56b represents the first public JWST image released by NASA in July 2022. It represents an eight-hour exposure of deep space with a bright star at the center of the field. Circular events are galaxies. We have a cluster of galaxies that gravitate toward one another. The distorted events describe Einstein's gravitational lensing phenomenon. A gravitational lens occurs when a huge amount of matter, like a cluster of galaxies, creates a gravitational field. This distorts and magnifies the light from distant galaxies behind it but in the same line of sight. These distorted objects are further away than the galaxies.

The European Space Agency is a NASA project partner of JWST. NASA spends $8.8 billion on Webb design and construction, with an additional $860 million to support five years of missionary work That's a total of $9.7 billion in the United States. The European Space Agency has invested around $760 million in this project, contributing to two of Webb's four devices and providing the Ariane launch The Canadian Space Agency (CSA) also provided approximately $200 million. With all its accessories, JWST is so big that it must fold to fit in the launcher and remotely unfolded at its arrival in space. The Webb has a fuel supply designed to remain stable for ten years in orbit. It is not excluded that the robot may refill the fuel tank in the future, but the outlook is currently unplanned.

In addition to mirrors, JWST also has a near infrared camera (NIRCam), a near infrared spectrometer (NIRSpec), four Mid-Infrared Instruments (MIRI), a fine Guidance Sensor (FGS), and a near Infrared Imager and Slitless spectrometer (NIRISS). These different instruments help JWST explore the universe. These instruments take pictures, study object light, and keep the telescope steady and on track. With this package, JWST can see almost everything in the sky, especially large celestial objects like galaxies and stars.

The two most important goals of JWST are (1) to collect images and data for the study of provide the formation of early galaxies and stars, the formation (birth) of new stars and exoplanets and (2) to collect spectroscopy of exoplanet atmospheres for determining their composition and inferring potential life. The JWST telescope can detect gases such as nitrogen, oxygen, argon, carbon dioxide, methane, water vapor, neon, ozone, nitrous oxide, methyl chloride, bromide, phoshine, and sulfur gases like methyl mercaptan which is produced by bacteria in wastelands, inland soils, and coastal ecosystems. Because we are still not certain about the definition of extraterrestrial life inferring the existence of past or existing lives from these biosignatures will become one of the most challenging scientific challenges of our time with a number of false positives. JWST has already discovered TRAPPIST-1b, the first exoplanet in the TRAPPIST-1 system, and Proxima b in the Proxima system. TRAPPIST-1 b, the innermost of seven known planets in the TRAPPIST-1 system, orbits its star at a distance of 0.011 AU. It completes one circuit in 1.51 Earth days. TRAPPIST-1 b is slightly larger than Earth, but has around the same density, which indicates a rocky composition Webb's measurement of the mid-infrared light given off by TRAPPIST-1 b suggests that the planet does not have a substantial atmosphere. More interesting results regarding extraterrestrial life are expected for TRAPPIST-1 e (see Chapter 6).

Proxima b's JWST data are highly anticipated. We can expect that it will provide more information to confirm and explain the discovery of artificial lights on Proxima b (e.g., Tabor and Loeb, 2021). The obvious question is: Do we have another civilization in our interstellar neighborhood? There is a great deal of anticipation about the answer to this question. It is a necessity to mention that JWST is not able to directly image exoplanets. Instead, the WFIRST telescope may do it before the end of this decade. The Nancy Grace Roman Space Telescope (shortened to Roman or the Roman Space Telescope, or the Wide-Field Infrared Survey Telescope or WFIRST) is another NASA infrared space telescope in development and scheduled to launch by May 2027 primarily for extra-solar planets and life.

As a result of JWST's current survey of hundreds of thousands of distant galaxies across the sky, we may learn more about dark matter's potential role in galaxies' history and future. Note also that the JWST is not the first infrared telescope. There are Earth-based infrared observatories. The Hubble Space Telescope has near-infrared capability, and the Spitzer Space Telescope focuses on infrared observation, for example. Because they are located close to Earth or at the Earth's surface they are affected by the strong infrared absorption of the Earth's atmosphere.

3.7.4.   Seeing into the Past and at What Scale?

The only way we can see objects around us is when they reflect and/or emit light and light strikes our eyes. The same holds for distant celestial objects because light travels at a finite and very large speed of 300,000 kilometers per second. So every time we see the Moon, we go back in time by

$$t = \frac{d_{\text{earth-object}}}{c_0} = \frac{384,400 \text{ km}}{300,000 \text{ km/s}} = 1.281 \text{ s} \qquad (3.175)$$

where c_0 is the speed of light and $d_{\text{earth-object}}$ is the distance between the Moon The light entering our eyes left the Moon 1.281 seconds ago (not very far back— but it is history). If the object is the Sun, $d_{\text{earth-object}}$ is 150 million kilometers. When we see the Sun, we see it as it was 8.3 minutes ago. Solar flares headed toward Earth and will reach us in 8.3 minutes. If we do not have warnings from satellite monitoring stations around it and/or solar monitoring stations on Earth, the damages will be unprecedented. Located in the constellation Centaurus in the southern hemisphere, Proxima Centauri (Figure 3.57) is a red dwarf star in the triple star system (two stars orbiting a third). It is the closest star outside our Solar System, at $d_{\text{earth-object}} = 4.2$ light-years[6] away. When we look at Proxima Centauri on Earth in April 2023, what we see is Proxima Centauri in February 2019. When you look at the Andromeda Galaxy, which is $d_{\text{earth-object}} = 2.537$ million light years from us, we see what it was 2.537 million years ago. In other words, if the objects are very far, the light takes some time to reach us. Some celestial objects are so far away that light travels for years to reach Earth. Hypothetically, if we could instantly travel to a star 66 million light-years away, we could see how it looks today and how Earth was 66 million years ago, in dinosaur's time. With a very powerful telescope we could even see dinosaurs roaming the Earth. Unfortunately current telescopes are not powerful enough for such resolution. Currently, we do not possess telescopes capable of capturing spacecraft landing on the Moon, let alone seeing an animal or a Neanderthal millions of light-years from Earth.

The James Webb telescope (JWST), currently located at the Lagragian L_2 point, can look back many light years back but not all the way to the Big Bang. The earliest light we saw was 13.799 billion years ago. That is almost at the Big Bang, but not quite: it came 378,000 years after the Big Bang. Prior to that, the universe was not

---

[6]a light-year is defined as the distance of light traveling for one year.

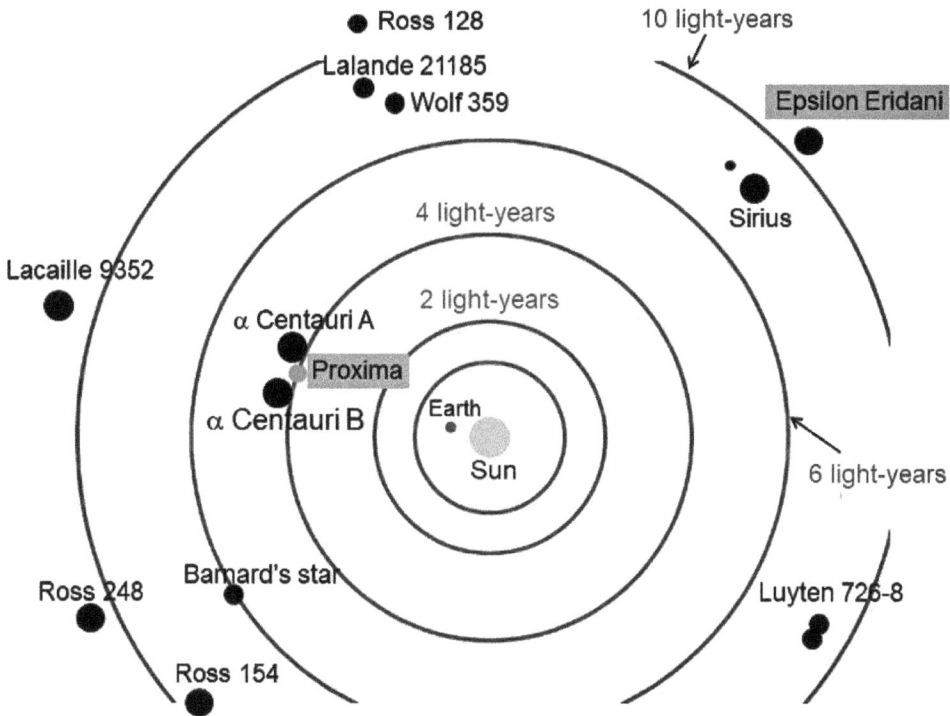

Figure 3.57. This map covers all the stars in our 11 light-year radius. Proxima Centauri is about 4.22 light years away. Only two stars on this map have exoplanets. Proxima Centauri has one exoplanet and Epsilon Eridani has two exoplanets.

transparent to electromagnetic radiation. The reason is that the early universe was hot and dense, and filled with plasma. Plasma is a mix of ions and free electrons (see Chapter 6). It is not like gas, with atoms consisting of nuclei and orbiting electrons. Despite this, the electrons were so energetic that they couldn't stay in the atoms. So they zoomed freely about. As a result, the plasma was opaque. 378,000 years after the Bang, the universe had expanded and cooled, meaning electrons could stay in their atomic nuclei. So the universe became filled with gas instead of plasma, and cleared up. The epoch during which charged electrons and protons first became bound to form electrically neutral hydrogen atoms, about 378,000 years ago, is called the *Recombination Era*. The earliest light is actually the after glow of that plasma prior to the Recombination Era. That is the earliest telescope working on the electromagnetic spectrum (including radio waves, IR, visible light, UV, X-rays, and gamma rays) can see. JWST has returned images of the first galaxies to emerge in the early universe. These are about 46 billion light years away. It is worth repeating that the first galaxies formed after the Big Bang may have been very bright. However, the wavelengths of this light have shifted into the infrared spectrum, thus the importance of JWST infrared instruments that collect light.

You may be wondering how JWST can see objects 46 billion light years away that may even have died by now. JWST can collect one photon from a given light source, wait however long it takes for another to hit it, and add them together over time. Eventually, a picture forms of the object emitting the light. The longer you

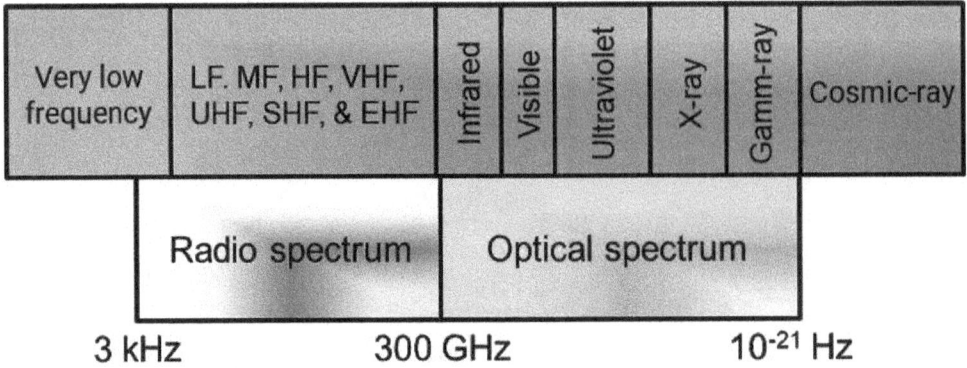

Figure 3.58. Definitions of radio and optical spectra.

collect photons from the light source, the better your picture. Our brains are not capable of storing individual photon strikes for later assembly, so we miss out on viewing much of the ancient universe directly.

The other interesting feature of infrared is that when a telescope looks at stars billions of kilometers far from us, when the light from those stars reaches the telescope, we can recognize the universe is expanding because the wavelengths of the light are stretched (or elongated). This phenomenon is known as the red shift. It is defined as follows (de Sitter, 1934; Hubble, 1929; and Parijskij, 2001)

$$\text{Redshift} = \frac{d_{\text{now}}}{d_{\text{emit}}} - 1 = \frac{\lambda_{\text{now}}}{\lambda_{\text{emit}}} - 1 \,, \tag{3.176}$$

where d_{now} and λ_{now} distance between object and the telescope and wavelength, respectively, at the observation time and d_{emit} and λ_{emit} distance between object and the telescope and wavelength, respectively, at the emission time.

3.7.5. Radio Telescopes

Radio astronomy telescopes record information from the 3 KHz band all the way to 300 GHz (wavelengths of light from about 10 meters to less than about 3 millimeters—this is, long wavelengths or low photon energies) of radio waves originating from outside the Earth, while optical astronomy telescopes record waves from 300 GHz to 10^{-21} Hz, as mentioned earlier (see Figure 3.58). An optical telescope, also known as visual telescopes, is a light gathering/focusing device that creates a magnified image for viewing or photography while a radio telescope is a focusable radio wave[7] receiver, used to search for radio transmissions from space. In other words, optical telescopes allow us to observe space and radio telescopes allow us to hear from space. One day, seeing space and hearing from space will be merged to create actual narrated reports from space.

Radio telescopes are large parabolic or spherical dishes used to catch radio waves from the sky (see Figure 3.59). The dish contains a primary reflecting surface that collects light and focuses it onto a central receiving antenna, often called a

---

[7]Radio waves are electromagnetic waves, not sound waves. We do not listen to radio EM data.

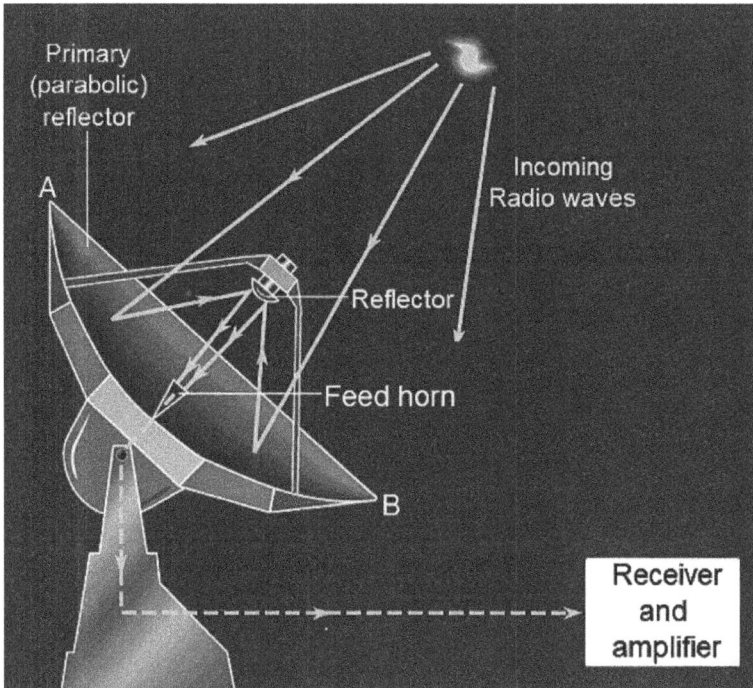

Figure 3.59. An illustration of the most familiar type of radio telescope. The distance AB is the diameter of the dish, approximately 100 m for the Green Bank Telescope in West Virginia, USA. Antennas convert electromagnetic radiation into electrical currents in conductors. The output signals are extremely weak and need to be amplified considerably to have a chance to detect them.

dipole. Radio telescopes observe electromagnetic fields emitted from extraterrestrial sources. Because radio waves have wavelengths much longer than those of optical light and the radio signals from space are very weak, we need a wide receiving area to capture radio waves. The wider the telescope, the more energy it can trap and the fainter the signal it can pick up, and the more detail it can see in that signal. That is why most radio telescopes are generally very large, especially compared to optical telescopes. Telescope resolution $\theta = 1.02\lambda/D$, λ is the smallest separation you can measure and D is the diameter of a single dish. In the case of an array of dishes, D is the longest baseline for an array.

Because the atmosphere absorbs most of the electromagnetic spectrum, only the optical and radio bands reach the ground. Therefore, they are the only ones accessible to ground-based instruments, such as optical and radio telescopes. Rayleigh scattering by interstellar dust and the earth's atmosphere prevents daytime optical telescope recordings, but radio telescope recordings are not affected. Radio opacity from water vapor, hydrosols, and molecular oxygen can limit radio telescope recordings. Also note that the radio spectrum offers 100 times more observable wavelengths than the optical spectrum.

The sources of radio emission are blackbodies (e.g., planets and stars) and non-blackbodies such as supernovae, pulsars, and synchrotron emission. These physical processes have signatures primarily at radio wavelengths which form the basis of radio telescope activities. Those who are blackbodies are thermal emitters, as described in Chapter 2, while those who aren't blackbodies aren't thermal, as described in Ikelle (2024). The highest contribution to radio signals is free electrons. Any acceleration of a charged particle, like an electron, makes photons. These photons dominate sky brightness in the radio regime. Here is a recap of the emission mechanisms that give rise to bright radio waves:

- Synchrotron emission (non-thermal radiation) occurs when relativistic electrons are incident on a magnetic field. An electron in a magnetic field begins to spiral, and synchrotron emission is the emission that comes out as a result of that spiraling motion. Synchrotron emission typically peaks at low radio frequencies. Common sources of synchrotron emissions are (1) electrons jetted relativistically from a supermassive black hole, (2) charged particles ramming into an ambient plasma after shock waves as in supernova remnants, (3) charged electrons from distant sources that are travelling through the diffuse magnetic field in our galaxy.

- Thermal Bremsstrahlung arises from an ionized cloud—atoms in the cloud become ionized when their electrons become stripped from the atom—and occurs when a free electron passes by an ion, and changes direction due to that encounter. That change in direction is an acceleration, and causes emission that is peaked at radio wavelengths. The most prominent source of thermal bremsstrahlung is HII[8] regions that reside in star-forming regions around young stars.

- Spectral line emission (thermal radiation) involves the transition of electrons in atoms from a higher energy state to lower energy state. When this happens, a photon is emitted with the same energy as the energy difference between the two levels. The emission of this photon at a certain discrete energy shows up as a distinct *line* or wavelength in the EM spectrum. Many of these occur at radio wavelengths, and most prominently the HI line.

- Pulsars have emission that is brighter at low radio frequencies, and whose physical emission process is not fully understood.

- Blackbody emission (thermal radiation) occurs in radio waves. Most prominently it comes from the cosmic microwave background, and from the Sun. Most other blackbody sources are too faint or difficult to see compared to the other above-listed mechanisms.

Radio astronomy, that is, the detection of long wavelength radiation from extraterrestrial sources, was discovered by Karl Jansky (1905-1950) and Grote Reber (1911-2002) in 1932 and 1937, respectively. Jansky was investigating the sources of static that might interfere with radio voice transmissions at short wavelengths. After recording signals from all directions for several months, Jansky identified three types: (1) nearby thunderstorms, (2) distant thunderstorms, and (3) a faint steady hiss of unknown origin. Jansky finally figured out that the type of radiation was

---

[8]HII to refer to ionised hydrogen and HI for neutral hydrogen.

coming from the Milky Way and was strongest in the direction of the center of our Milky Way galaxy, in the constellation of Sagittarius.

Grote Reber learned about Karl Jansky's discovery of radio waves from the Milky Way Galaxy and wanted to follow up his discovery. Reber built a parabolic dish reflector in his backyard. He spent hours every night scanning the skies with his radio telescope. He had to do the work at night because there was too much interference from sparks in automobile engines during the daytime. In the end, he was able to detect radio emissions from the Milky Way with a receiver operating at 160 MHz (1.9 meters wavelength). In the years 1938 to 1943, Reber made the first surveys of radio waves from the sky and published his results both in engineering and astronomy journals. This ensured radio astronomy's future.

Again, There is a lot to see beyond what we see with our eyes. This is because visible light is just a very small portion of the electromagnetic spectrum and I'm understating it. The spectrum consists of different radiation with varying levels energy. The longer wavelengths(radio, infrared, microwave) being the lowest energy and the shorter wavelengths (UV, X-ray and Gamma) being the highest energy. All the objects in space emit different kinds of radiation and in different amounts.

China's FAST telescope is a radio telescope on the earth's surface. Its detectable wavelength ranges from 10 cm to 4.3m. It is the biggest of its kind with a 500 meter diameter with a 196,000 square meters collecting area. Previous champion was the Arecibo Observatory. Apparently FAST sees three times further into space than the Arecibo Observatory and is able to survey the skies 10 times faster than the Arecibo Observatory. Its main purpose is to survey neutral hydrogen in the Milky Way and other galaxies. It also detects new pulsars (both galactic and extragalactic) and the search for extraterrestrial life (this probably being of the most interest to them). Comparison of FAST and JWST in the public press is misleading. Again, the James Webb Space Telescope is mainly an infrared telescope that orbits the Earth in space.

3.7.6. Composition of Stars

The combined use of spectroscopy[9] and large telescopes has allowed us to determine motions of stars, galactic streams, supernovae, exoplanets, solar flare activities. and chemical compositions of stars. Most of the time large telescopes collect light coming from the star and other space objects and this light is automatically separated into single-color images by filters placed in front of a camera. The filters allow only certain colors through to the detector. In other words, filters are designed to separate the different wavelengths of light before they hit the detector. Single-color images are then superimposed to form true color images—that is how images of space objects are obtained. Common color used in the star spectroscopy and their representative wavelengths include cherry red ($\lambda = 660$ nm), red ($\lambda = 630$ nm), orange-red ($\lambda = 610$ nm), orange ($\lambda = 5950$ nm), amber ($\lambda = 590$ nm), yellow ($\lambda = 580$ nm), chartreuse ($\lambda = 565$ nm), grass green ($\lambda = 550$ nm), emerald green ($\lambda = 520$ nm), teal ($\lambda = 500$ nm), aqua ($\lambda = 490$ nm), azure ($\lambda = 480$ nm), blue ($\lambda = 465$ nm), indigo ($\lambda = 450$ nm), violet ($\lambda = 420$ nm), and deep violet ($\lambda = 390$ nm). The spread-out light spectrum can be seen in Figure 3.60. The dark lines are

---

[9]Spectroscopy is the real key to current picture of the universe. Without it we still be in dark about our univere.

Continuous spectrum

Hydrogen emission spectrum

Hydrogen absorption spectrum

Hydrogen absorption spectrum

400 500 600 700

Wavelength (nanometers)

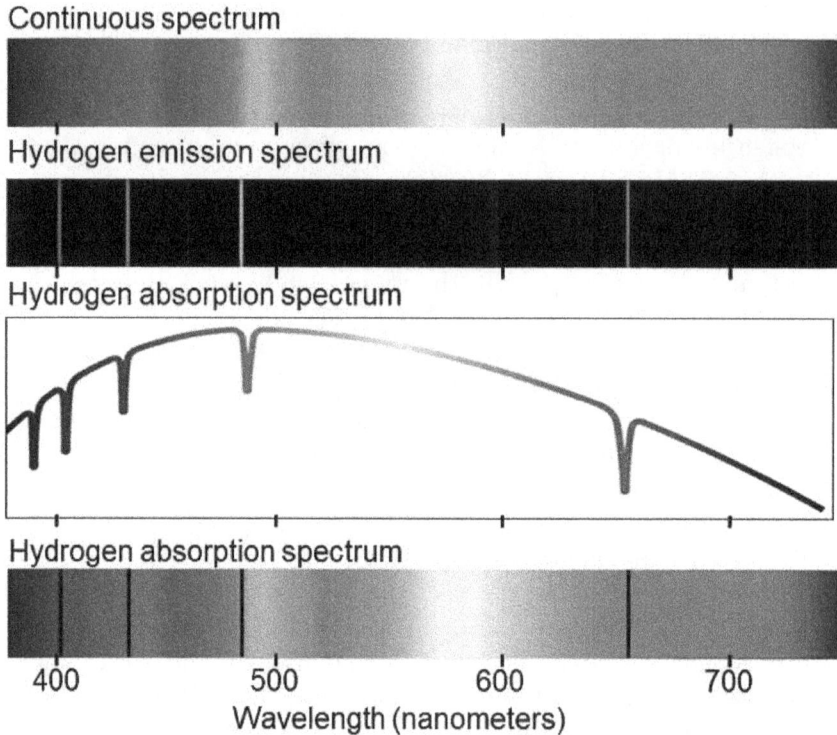

Figure 3.60. A comparison of continuous, emssion, and absorption spectrum of hydrogen along emission- and absorption-spectrum lines. The absorption spectrum (bottom) contains black lines at exactly the same position as the lines in the emission spectrum appear. The spectrum that originates from stars is usually an absorption spectrum.

hydrogen lines at 410, 434, 486 and 656 nm. They are characteristic of hydrogen. Hubble can collect light in a wavelength range from 200 nm in the ultraviolet to 1700 nm in the infrared (the range of human vision is approximately 380-700 nm). We can use more than 76 different filters to record specific wavelengths or wavelength ranges.

Because the energy levels of electrons in atoms and molecules are quantized and the absorption and emission of electromagnetic radiation only occurs at specific wavelengths, that is why spectra are not smooth but punctuated by "lines" of absorption or emission with varying thickness, as depicted in Figure 3.60. The precise positions (wavelength) of lines indicate emission and/or absorption locations. Each chemical element has specific and unique wavelengths at which it can absorb or emit. The absorption spectra of stars and other space objects can be described as superpositions of the absorption patterns of their chemical constituents. Thus we can decode the light coming from stars and other objects in space to determine their chemical composition. By collecting spectra of a star over a long period of time, it allows us to determine whether a star shows variation (e.g., it gets brighter and fainter with time, in a repeatable pattern). This continuous recording can occasionally capture unique moments in a star's life, such as the death of a star and a supernova explosion. The absorption lines can also help us estimate the temperature. Both the presence and strength of lines of different elements indicate hotter or cooler stars. For example, hot stars have strong hydrogen lines in their spectra. While cool stars have strong lines of calcium and sodium. We can use Wien's law, described in Chapter 2, to deduce temperatures from wavelengths:

$$T(\text{Kelvin}) = \frac{3 \times 10^7}{\lambda(\text{Angstroms})} \,. \tag{3.177}$$

Any object heated above 500 degrees Celsius starts to glow red. As it is heated further (so long as it does not burst into flames or vaporize), it will change color in a sequence: orange, yellow, white. Each color has its own temperature. Spectra can also tell us how fast and in what direction a star is traveling. All the absorption lines in a moving star's spectrum will appear shifted by the same amount with respect to the spectrum of a non-moving star. A shift toward the lower wavenumbers (blue-shift) indicates that the star is moving toward us and a shift toward the higher wavenumbers (red-shift) indicates that a star is moving away from us. In a similar way we can analyze a star's rotation. Anyway, once we determine the temperature of a star, we can deduce its luminosity and its age, as described in Ikelle (2023b).

Figure 3.61 shows the spectra of 98 chemical elements in the stars. We can code and decode star's and other space object's chemical makeup as a linear composition of these 98 spectra. The more elements an object contains, the more complicated its spectrum can be (see Figure 3.62). Other factors, such as motion, can affect the positions of spectral lines, though not the spacing between the lines from a given element. Fortunately, computer modeling allows us to tell many different elements and compounds apart even in a crowded spectrum, and to identify lines that appear shifted due to motion. Note that what we see and characterize is not the spectra of the star's interiors but the spectra of star's atmospheres and sometimes even the atmospheres of their exoplanets. Note also that some telescopes are located above the Earth's atmosphere to capture infrared light and other bands of light which are blocked by our Earth's atmosphere.

There are cases of more than one star within a system. Binary-star systems (e.g., Procyon A&B and Sirus A&B) are very common and even more common than single-star systems, and can have planets (see Ikelle, 2024). Binary stars are gravitationally connected to another star. One of the stars is most likely much smaller and not visible to the naked eye. The two stars orbit a common center and rotate in opposite directions. The brightest star in the northern night sky is known as Sirius A. It is a binary star. Its companion is now a white dwarf known as Sirius B. We must emphasize that the distance between two stars does not determine whether they form a binary system or not. A binary system indicates that two stars are gravitationally bound to each other—that is, they orbit each other regardless of their distances from one another. Binary stars can consist of two stars touching each other, but this is only a hypothetical possibility. The widest separation between stars in a binary system is still unknown. Separations of over 0.03 light-years have been reported.

> **Quiz:** The average temperature of the Sun is 6000 degrees Kelvin, humans is 310 degrees Kelvin, molecular clouds is 20 degrees Kelvin, and cosmic background is 2.7 degrees Kelvin. Determine the light of the electromagnetic spectrum associated with the sun, humans, molecular clouds, and cosmic background.

A triple-star system (e.g., Alpha Centauri, EZ Aquarii, and Epsilon Indi) consists of a double star with a third star much farther away. Alpha Centauri is a

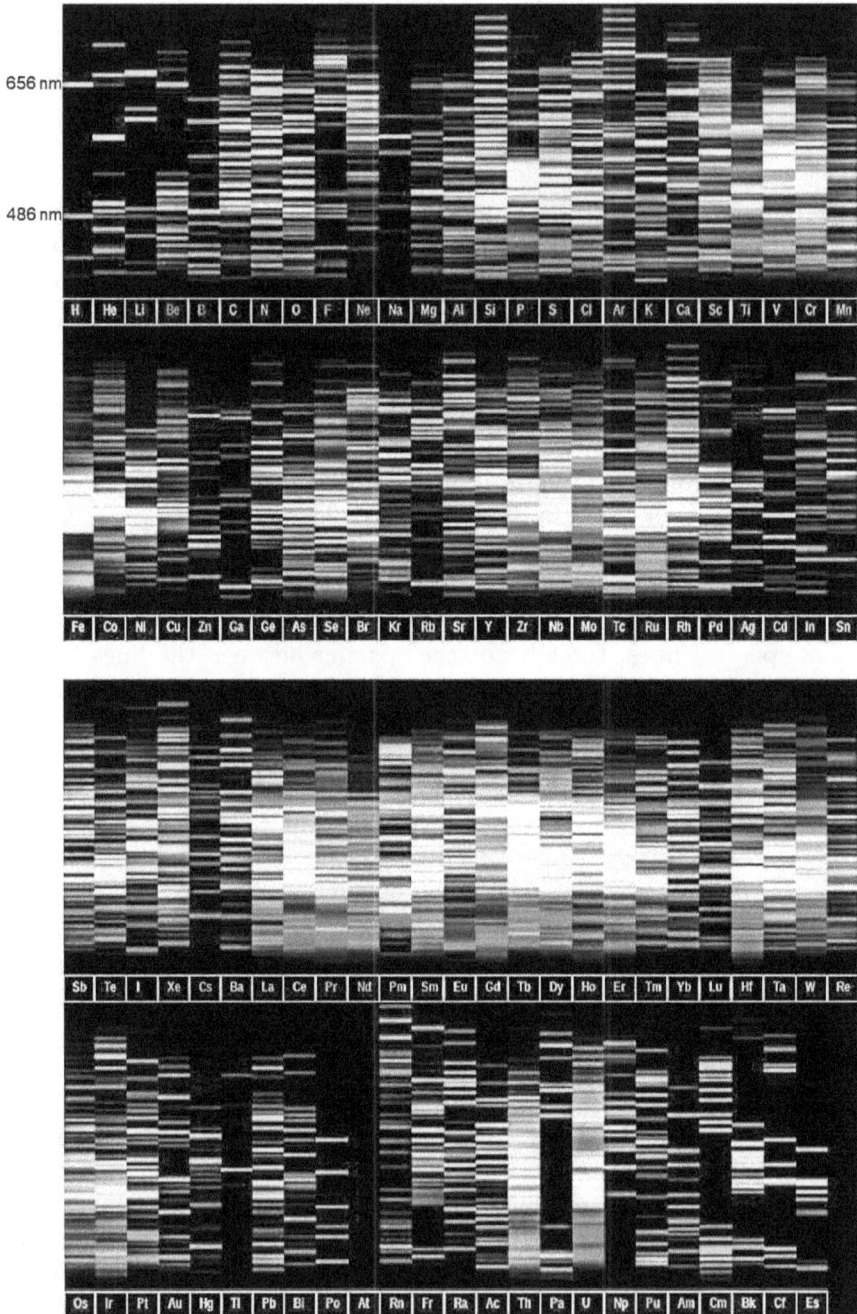

Figure 3.61. Each element of the periodic table has unique sets of colors and different strengths of those colors. Here is the periodic table for determining the star chemical composition. Source: http://umop.net/spectra/

good example of a triple-star system—Alpha Centauri A and B orbit about as far apart as the Sun and Uranus, with Proxima 0.21 light-years away. If you stood on a planet orbiting Alpha Centauri A, then Alpha Centauri B would be a mini-sun about 100 times dimmer than A, and Proxima would look like just another star. A four-star system is two doubles a considerable distance apart from each other.

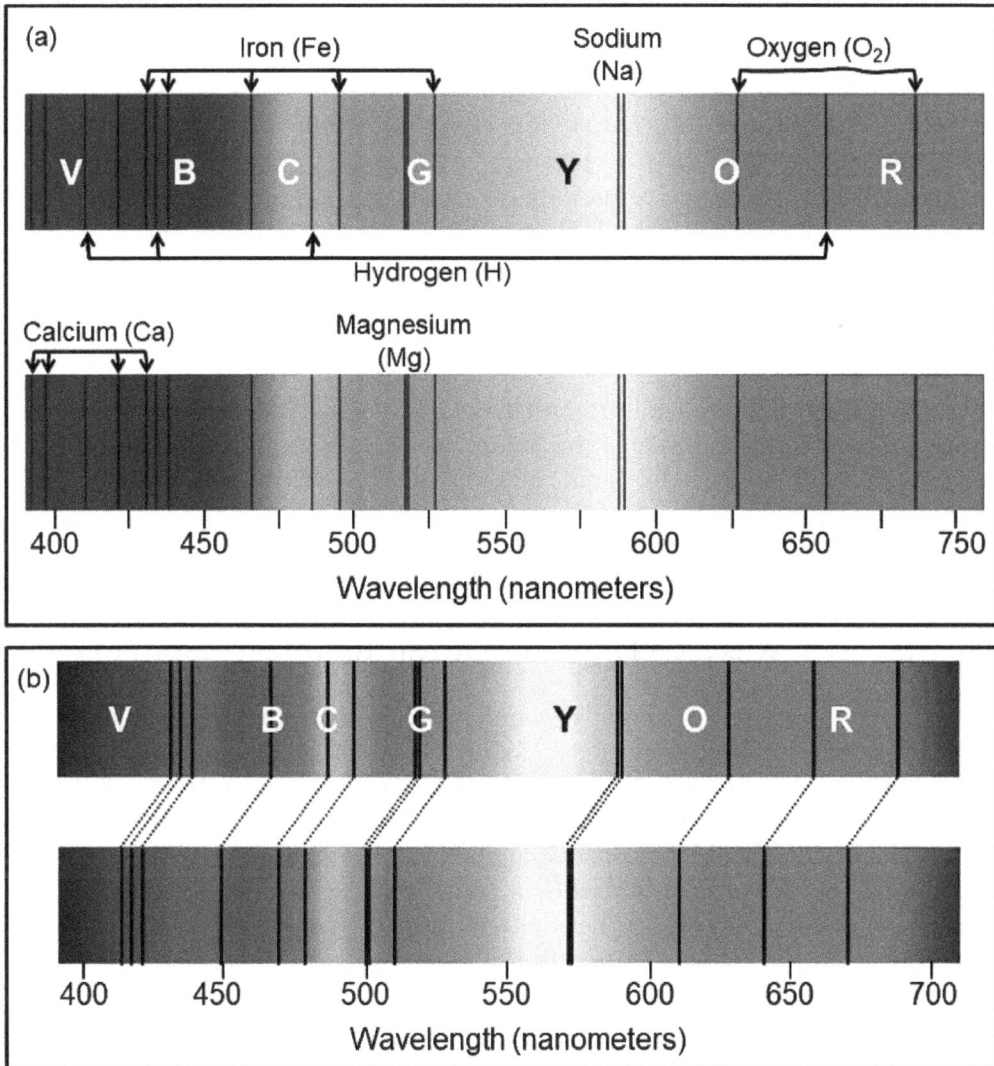

Figure 3.62. (a) The Sun spectral lines. The absorption spectrum of our Sun and its chemical composition. (b) Absorption spectra can also reveal another aspect of a distant star, its velocity. The lines can be shifted in the red or blue direction of the spectrum. A shift in the red direction is known as redshift. This indicates that the star is moving away from us. A shift in the blue direction, as shown here, is known as blueshift. It indicates that the star is moving toward us. V=violet (380-440 nm); B = blue (440-485 nm); C= cyan (485-510 nm); G=green (510-565 nm); Y= yellow (565-590 nm); O= orange (590-625 nm); and R= red (greater 625 nm).

Systems up to six stars have been found, but they are very rare, and more than six is very unlikely.

Notice that the search of exoplanets (new planets orbiting other stars than the Sun) and the observations of distant galaxies and the expansion of our universe are predominantly based spectroscopy of the light absorption, emission, and scattering of stars and other space objects collected by telescopes. Unfortunately, the light emitted by some matter in our universe, called *dark matter*, is not invisible to any telescope we have today, as discussed in Chapter 1. So James Webb telescope now in space cannot see dark matter and nobody can, as we do not know what this substance really is. But these telescope have no problems seeing through or pass

dark matter because dark matter does not interact with electromagnetic radiation—
that is, dark matter does not make any other thing invisible in our universe. Objects
in space appear to be obscured by dark matter, which scatters, reflects, and refracts
any light. The light from distant stars is slightly distorted in ways that allow the
dark matter distribution to be mapped.

Although Isaac Newton's apparatus can be considered a crude spectroscope, it
is generally recognized that the spectroscope was invented by Joseph Fraunhofer
(1787-1826), Gustav Robert Kirchhoff (1824-1887), and Robert Wilhelm Bunsen
(1811-1899) around 1860. Joseph Fraunhofer (1787-1826) showed in 1818 that by
measuring the exact wavelengths of the colors, each element emitted its own spectral
pattern.

In 1859, Bunsen and Gustav Robert Kirchhoff (1824-1887) developed the modern version of this instrument called a flame spectroscope. This allowed them to
precisely identify elements by their emission spectra—even new elements within
mixtures and compounds.

Box 3.3: MeerKAT: South African Radio Observatory

Breakthrough Listen is a scientific research program that analyzes radio signals in
search of extraterrestrial intelligence (SETI) beyond Earth. The basic idea is to
detect and record the signatures of technologies used for communication, air travel,
space travel, propulsion, or other purposes by aliens just like natural phenomena
such as pulsars, jets and lobes of active galactic nuclei, and the cosmic microwave
background have signatures primarily at radio frequencies. These signatures may
reveal advanced life on worlds other than our own. There are a significant number
of radio and optical telescopes around the globe that can record such signatures,
even at interstellar distances. Breakthrough Listen employs the Automated Planet
Finderin California for optical search. For radio search, the program employs the
Robert C. Byrd Green Bank Telescope in West Virginia, USA, and the CSIRO
Parkes Radio Telescope in New South Wales, Australia, the VERITAS Cherenkov
Telescope in Arizona, USA, and the MeerKAT telescope in South Africa. These
telescopes are sensitive enough to hear common aircraft radar transmitted to us
from any of the 1000 nearest stars. During the 10 years of Breakthrough Listen,
100 million dollars will be committed. MeerKAT telescope, the Southern Hemisphere's largest radio telescope, is now participating in the Breakthrough Listen
program. It is made up of 64 satellite dishes (Figure 3.59). Each stands 20 meters
tall. The dishes are connected over five miles of dry land, 630 kilometers northeast
of Cape Town. Moreover, there is no need to physically move its antennas. The
64 dishes can monitor a large area of the sky and scan 64 objects at the same
time. MeerKat's targets include Alpha Centauri A, Alpha Centauri B, and Proxima Centauri. In 2020, Breakthrough Listen detected an odd radio signal coming
from Proxima Centauri, the closest star to our own Sun and a member of Alpha
Centauri. MeerKat is expected to search over one million stars in our interstellar
neighborhood in the next two years. It is possible that all alien communications
are telepathic and the machines, including engines, are silent. Therefore, the signals we are recording may be the result of blowing winds, explosions of cosmic
objects, and collisions between cosmic objects. Anyway, we must identify alternative measurements to constrain Breakthrough Listen measurements.

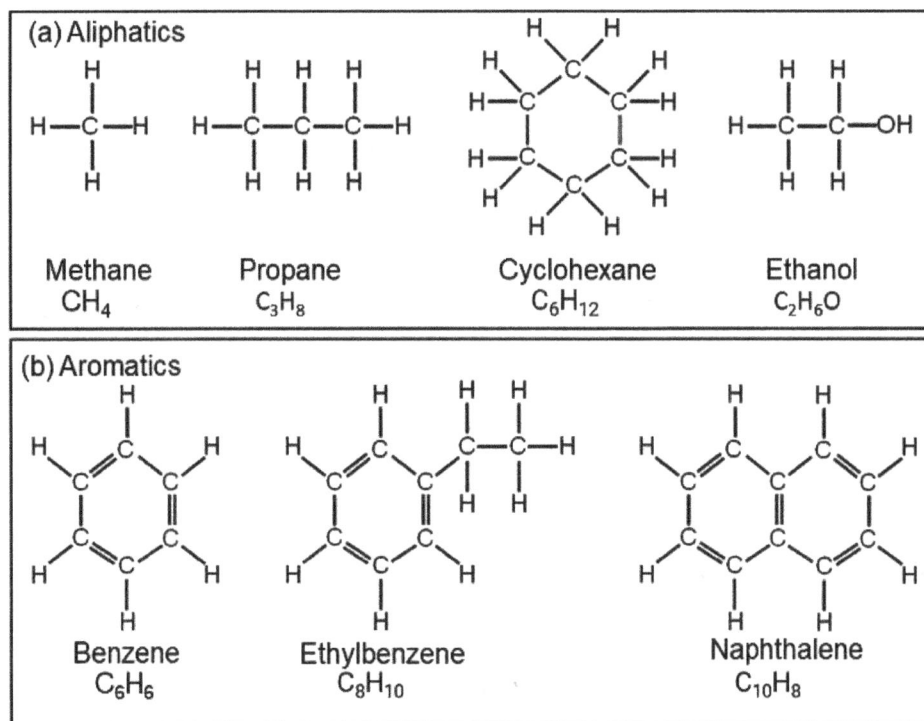

Figure 3.63. (a) Examples of aliphatic hydrocarbons. This group of hydrocarbons forms the most abundant fraction in crude oil. (b) Examples of aromatic hydrocarbons. These hydrocarbons are cyclic, planar compounds that resemble benzene. Benzene is the simplest aromatic hydrocarbon.

3.8. Essay: Spectroscopy of Cancer-causing Chemicals: Polycyclic Aromatic Hydrocarbons

Hydrocarbons are divided into two major categories: aliphatic (non-aromatic) and aromatic (arenes), as illustrated in Figure 3.63. The chemical structure of aromatic compounds cannot be properly represented in the same way as aliphatic compounds. We here focus on polycyclic aromatic hydrocarbons, also known as PAHs.

PAHs are a class of organic compounds that contain only carbon and hydrogen. They are usually made up of two to ten aromatic rings in a linear, cluster, or angular arrangement (e.g., Sahoo et al., 2020). Figure 3.64 shows reactions associated with the formation of the first aromatic ring. These reactions are based on experimental and computational studies, involving several different reactants and reaction mechanisms (e.g., Reizer et al., 2022). Rings are drawn on two vertical sides (when possible), with as many rings as possible in a line and the rest of the structure sitting in the top right quadrant. The simplest representatives of such members are naphthalene ($C_{10}H_8$) with two aromatic rings, anthracene ($C_{14}H_{10}$) with three aromatic rings, acenaphthylene ($C_{12}H_8$) with two aromatic rings and quadrant, pyrene ($C_{16}H_{10}$), fluorene ($C_{13}H_{10}$), and fluoranthene ($C_{16}H_{10}$) (see Figure 3.65). PAHs are solid at ambient temperature and the least volatile hydrocarbons. As described by Achten and Andersson (2015), the boiling points of the PAHs [200-300 degrees

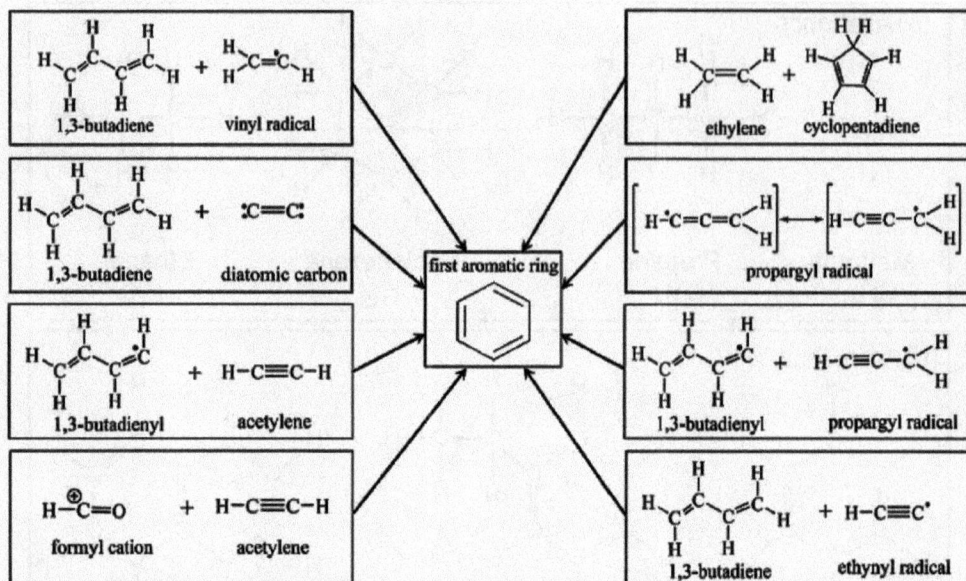

Figure 3.64. The most well-known reactant pairs involved in the formation of the first aromatic ring of PAHs. Adapted from Reizer et al., 2022.

Celsius (2-ring PAHs), 300-400 degrees Celsius (3-ring PAHs), 300-450 degrees Celsius (4-ring PAHs), 350-500 degrees Celsius (5-ring PAHs), and 500-600 degrees Celsius (6-10-ring PAHs)] are markedly higher than those of the n-alkanes of the same carbon number (e.g., 174 degrees Celsius for 10 carbons, 235 degrees Celsius for 13 carbons, and 287 degrees Celsius for 16 carbons). Nonetheless, light PAHs (2-3 rings) are present in the environment mainly in the vapor form, while heavier PAHs (more than five rings) are predominantly adsorbed on organic particulate matter, usually on small particles (< 2.5 μm); four-ring compounds have an intermediate behavior. PAHs are uncharged, non-polar, lipophilic (they dissolve much easier in lipids (a class of oily organic compounds) than in water), and planar. Larger PAHs are generally insoluble in water, while some PAHs are soluble and known contaminants in drinking water. They are usually colorless. PAHs with five or six rings are the most common.

PAHs are found in coal and oil deposits. They are also produced by the incomplete combustion or pyrolysis of organic matter—for example, in engines and incinerators, coke production, waste burning, and when biomass burns in forest fires (Kozak et al., 2017; Li et al., 2019; Lin et al., 2019; and Sahoo et al., 2020). Industrial activity that produces and distributes PAHs includes aluminum, iron, steel, road paving, and asphalt manufacturing; coal gasification, tar distillation, shale oil and gas production; and the production of coke. Over 100 PAHs have been identified in Earth's atmospheric particulate matter and about 200 in tobacco smoke.

Polycyclic aromatic hydrocarbons (PAHs) are carcinogenic organic compounds. Cancers, including skin, lung, bladder, liver, and stomach cancers, represent a primary human health risk of exposure to PAHs. People who smoke tobacco products,

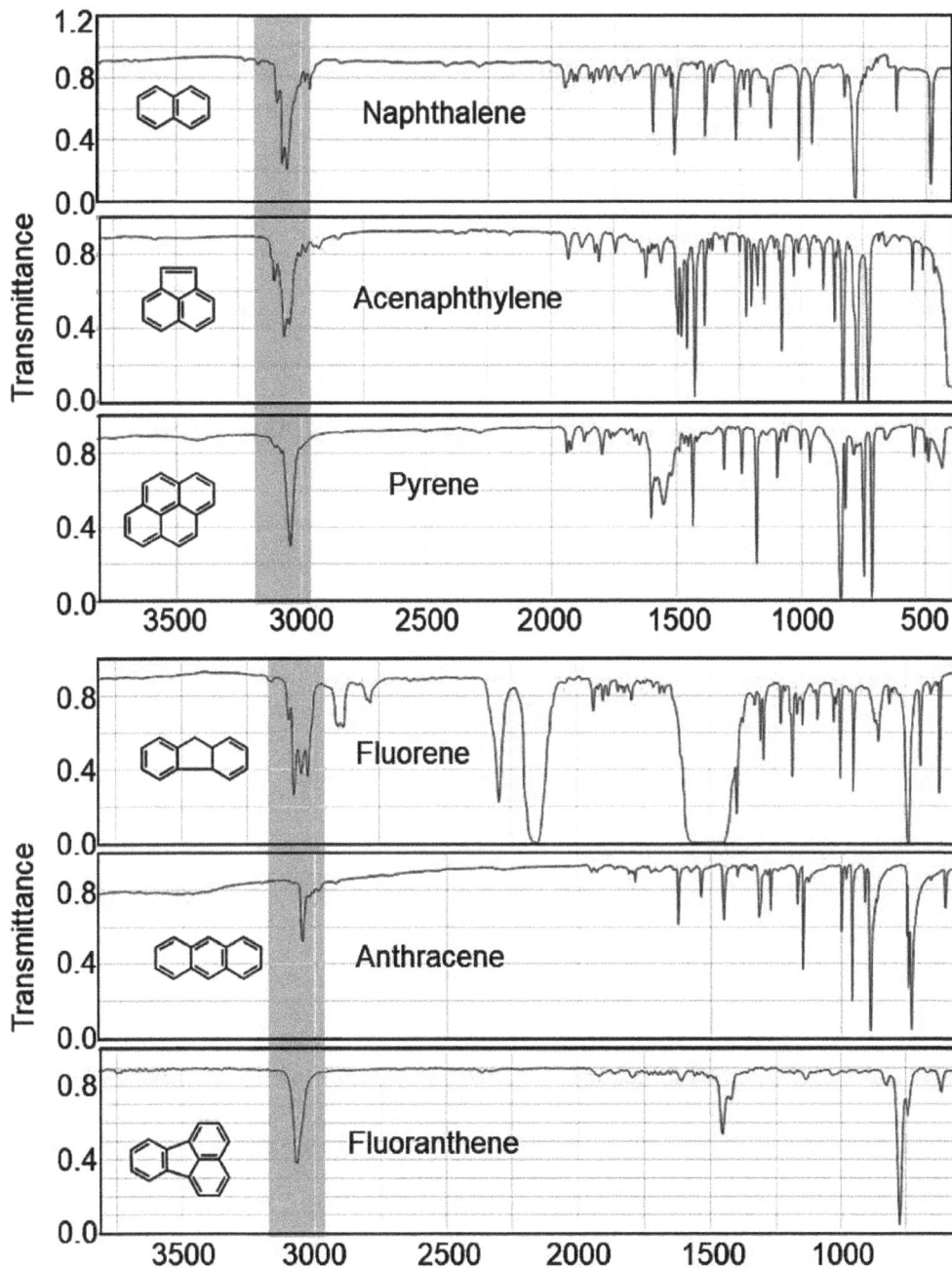

Figure 3.65. *(Continued.)*

or who are constantly exposed to second-hand smoke, are among the most highly exposed groups; tobacco smoke contributes to 90 percent of indoor PAH levels in smokers' homes. In fact PAHs comprise the largest class of cancer-causing chemicals and are ranked ninth among chemical compounds threatening humans. Exposure to PAHs has also been linked to cardiovascular disease and poor fetal development.

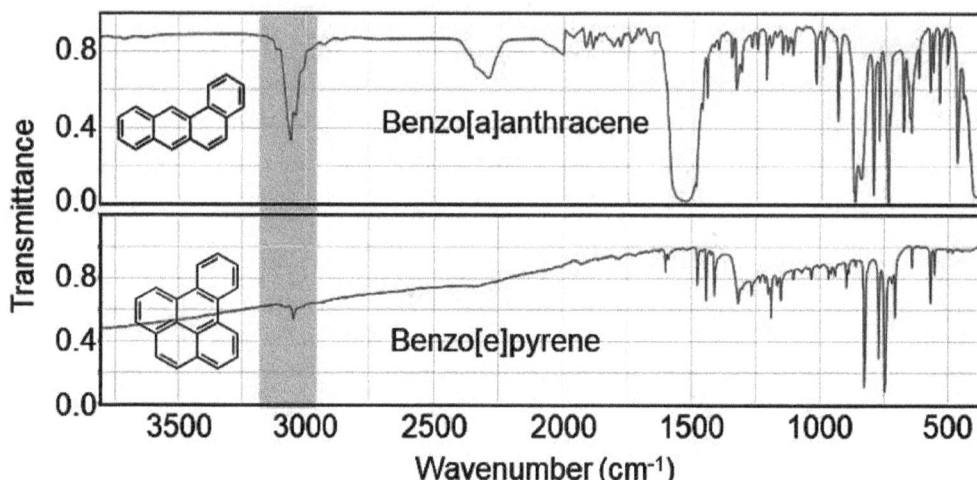

Figure 3.65. IR spectra of some PAHs using NIST (National Institute of Standards and Technology) data.

Benzo[a]pyrene ($C_{20}H_{12}$) can trigger lung cancer in smokers, and methylcholan-threne ($C_{21}H_{16}$) can trigger colon cancer when it is fermented from bile acids by Salmonella in the gut (Tarashi et al., 2019). PAHs comprise the largest class of cancer-causing chemicals and are ranked ninth among chemical compounds threatening humans. Many PAHs are also genotoxic, mutagenic, teratogenic, and in addition to being carcinogenic. They bioaccumulate in living organisms' soft tissues. Interestingly, many are not directly carcinogenic, but act like synergists. Yet PAHs are considered starting materials for abiotic syntheses of materials required by the early forms of life.

As discussed in Chapter 2, PAHs have been identified in Titan's atmosphere. In other words, PAHs can also be found at extremely low temperatures in the absence of oxygen in the atmosphere. Remember that for billions of years, the Earth was cold and oxygen was absent. The presence of PAHs in the atmosphere reinforces the idea that PAHs may have been present on early Earth and contributed to life on Earth. Moreover, some of the PAHs on Earth today may have originated from a period when Earth was cold and oxygen was absent. Titan may be on an evolutionary path that may not be so different from the one that Earth has been on.

PAHs are carcinogenic for reasons not well established. PAH carcinogenicity is related to their ability to bind to DNA, triggering a series of disruptive effects that can result in tumor initiation. The fact that PAHs are planar molecules just like DNA bases, namely guanosine, adenosine, cytosine and thymine is considered one of the reasons PAHs are carcinogenic because PAHs can intercalate between nucleic acid base pairs causing frameshift mutations (Kathuria, 2015). When mutations occur, cancer can result if they are not repaired. Also PAH molecules can be metabolized by enzymes into molecules with rings opened up, which then react directly with DNA. The metabolism opens up the ring, leading to highly reactive radicals that attack DNA. PAHs with four or more rings have shapes that allow them to interact with these enzymes in order to produce carcinogenic metabolites. Smaller PAHs with 2-4 rings are thought to promote cancer by blocking gaps between cells, interfering with signaling mechanisms between cells, increasing cell division chances, and presumably giving an advantage to cells with cancerous changes. Benzene itself causes leukemia due to its metabolism in the liver. So PAHs on their own are not dangerous. Thare are so mostly when ingested into the digestive system.

Polycyclic aromatic hydrocarbons undergo biotransformation. As a result of this process, highly reactive metabolites are generated, which interact with cellular DNA and form PAH-DNA adducts. Measurement of these adducts serves an important purpose in human biomonitoring for exposure to PAH carcinogenesis and they are also used as tools in molecular epidemiology studies (Poirier et al., 1998; Perera, 2000, 2011; Farmer et al., 2003). PAH-DNA adducts have been used as biomarkers to assess human exposure to toxicants mixtures in Poland (Perera, 2000), Bulgaria, the Czech Republic, and Slovakia (Taioli et al., 2007). Bulgaria had the highest average personal exposure to carcinogenic PAHs, followed by Slovakia and the Czech Republic according to these studies. Peripheral blood samples collected from residents of high-exposure regions revealed an association between environmental pollution and significant increases in levels of PAH-DNA adduct, sister chromatid exchanges, chromosomal aberrations, and carcinogenic PAHs oncogene expression.

Figure 3.65 shows the infrared spectroscopy of a number of PAHs. We can see that their cancerous effects are between 2950 cm^{-1} and 3150 cm^{-1}. The strongest of the C$-$H stretching typically falls in the 3080-3040 cm^{-1} (3.25-3.29 μm) range is included in this region. Notice that for PAHs with three to four rings such as anthracene ($C_{14}H_{10}$), fluoranthene ($C_{16}H_{10}$), and benzo[a]anthracene ($C_{18}H_{12}$), the cancerous region narrows considerably with a single peak. As we move to five and six ring PAHs like Benzo[a]pyrene ($C_{20}H_{12}$) and Dibenzo[a,h]pyrene ($C_{24}H_{14}$), the narrow cancerous region is replaced by a steep dip window which spreads over the entire non-fingerprint window of the IR spectrum. In other words, cancer spreads widely in the body when it is related to PAHs with five or more rings.

Treatment of cancer comes down to removing the absorption strip of wavenumbers through light configurations and molecules designs. Current treatment attempts are generally carried out at high temperatures.

Titan's atmosphere teaches us that PAHs can be cultured in a pristine very-low temperature environment and even cancer treatment can be carried out in this pristine environment. Remember that PAHs are at Titan's atmosphere temperatures of 90 degrees Kelvin (-183.15 degrees Celcius). In signal processing, correcting spectra from a notch like Figure 3.65 is known as deghosting (see Ikelle and Amundsen, 2018). It can be done here by combining emitting and absorbing molecules/isotopes, through drugs and/or through the manipulation of light.

3.9. Essay: Some Biosignature Gases and Light Signatures

3.9.1. Some Biosignature Gases

The exoplanet study is currently being transformed by the James Webb Space Telescope (JWST). We will shortly gain our first insight into conditions on rocky, potentially Earth-like worlds beyond our solar system such as the TRAPPIST-1d, TRAPPIST-1e, and TRAPPIST-1f exoplanets. JWST will detect starlight-absorbing gases that compose an exoplanet's atmosphere using spectroscopy. These gases are called biosignatures. The biosignatures include hydrogen (H_2), oxygen (O_2), ozone (O_3), water (H_2O), methane (CH_4), nitrous oxide (N_2O), hydrogen sulfide (H_2S), methane (CH_4), carbon dioxide (CO_2), phosphine (PH_3), and ammonia (NH_3).

Most of these gases are present in the atmospheres of our solar system planets. For example, phosphine occurs naturally in hydrogen-dominated atmospheres like gas giants Jupiter and Saturn. It has been detected in Venus' atmosphere along with ammonia. On Earth, microorganisms produce phosphine, such as animals' intestinal tracts. If no life was present, we would not expect phosphine to occur in large quantities. The recurring idea that life exists on Venus is mainly related to phosphine in Venus' atmosphere. Additionally, this comment showed that, even if we discover and accurately measure an exoplanet's biosignature gas, we cannot be certain whether that exoplanet has present, past, or future life. Such a discovery might narrow the search. Additional data will then be sought, including the percentage of each gas, before a conclusive answer can be reached.

The chemistry of beings is generally similar to that of their planet or exoplanet atmosphere. As we can see in Figure 3.66, ammonia, another potential biosignature gas, exhibits resonance at millimeter wavelengths. When projected onto humans or any porous objects, such wavelengths can cause reverberations in small cavities such as those found in the human brain. This can have severe effects on us or even death. Thus beings originating from exoplanets or planets with an atmosphere containing sufficient ammonia will pose a danger and hazardous to humans.

Ammonia is an inorganic colorless gas of nitrogen and hydrogen with a distinct pungent smell. Biologically, it is a common nitrogenous waste, particularly among aquatic organisms. It contributes significantly to the nutritional needs of terrestrial organisms by serving as a precursor to 45 percent of the world's food. Thus an atmosphere with sufficient ammonia may indicate life.

NASA's James Webb Space Telescope has captured the distinct signature of water, along with evidence for clouds and haze, in the atmosphere surrounding a hot, puffy gas giant planet, WASP-96 b, orbiting a distant Sun-like star using an instrument on JWST known as the Near Infrared Imager and Slitless Spectrograph (FGS-NIRISS). NIRISS observes light from the wavelengths of 0.8 to 5.0 microns. NIRISS provides near-infrared imaging and spectroscopic capabilities. As the only instrument capable of aperture mask interferometry, NIRISS can capture images of bright objects at a resolution broader than other imagers. The spectroscopic mode is capable of observing exoplanets through spectroscopy. WASP-96 b is located roughly 1,150 light-years away from us and is a gas giant without a direct analog in our solar system. With a mass 0.48 of Jupiter's mass and a radius 1.2 times Jupiter's radius, WASP-96 b is much puffier than any planet orbiting our Sun. And with a temperature of more than 538 degrees Celsius, it is significantly hotter. WASP-96 b orbits extremely close to its Sun-like star, about 0.0453 AU with an orbital period of 3.5 days. The Near-Infrared Imager and Slitless Spectrograph (NIRISS) is one of Webb's four scientific instruments. NIRISS provides near-infrared imaging and spectroscopic capabilities. As the only instrument capable of aperture mask interferometry, NIRISS has the unique ability to capture images of bright objects at a resolution greater than other imagers.

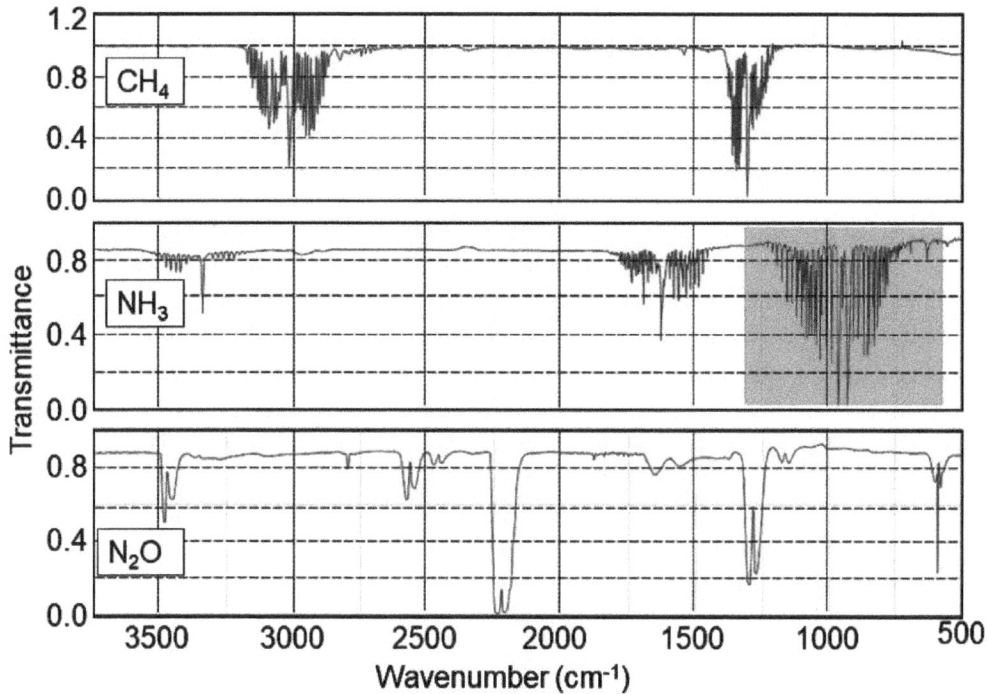

Figure 3.66. IR spectra of three possible biosignatures using NIST (National Institute of Standards and Technology) data. The region at which ammonia exhibits resonance at millimeter wavelengths has been highlighted.

When the gas composition of the atmosphere of an exoplanet has been measured, we can use the following formula (see Chapter 1 of Ikelle, 2023a)

$$O_2/N_2 \approx -10.07\,CO_2 + 3462 \,, \tag{3.178}$$

to determine the exoplanet's suitability for biologic life and oxygenic photosynthesis. Here O_2/N_2 is measured in per meg (One per meg is 0.0001 percent) and CO_2 in ppm, where ppm stands for parts per million (i.e., 1 ppm = 0.0001 percent) and 4.8 per meg is equivalent to 1 ppm (i.e., 1μmole per mole of dry air). The use of ppm and even ppb (parts per billion) is another way of expressing how small some gases like O_2 are in the Earth's atmosphere today.

3.9.2. Light Signatures

We may be able to observe glowing and stars at light-year distances as JWST shows these days (e.g., Figure 3.56). Also at light-year distances observing directly light generated by a planet, moon, exoplanet, or exomoon with an atmosphere is another way of discovering potential life or even civilization. Detectable lights include auroras, nuclear radiation and light related to uncontrolled nuclear fission or fusion explosions. They also include abrupt light and darkness related to falling large meteorites, lights from VE-8 and VE-9 type volcanic eruptions, huge flares from stars, and lights from exoplanets containing their own stars, as described in Chapter 6. Such lights can be detected by telescopes, especially space telescopes, or probes like NASA's Voyager 1 and Voyager 2. The probes have exited our solar system and are in the interstellar medium. As of November 2023, Voyager 1 was 19 billion kilometers away.

Auroras are produced by planets and exoplanets with significant atmospheres and large global magnetic fields. The occurrence of these lights at the north and south poles of a newly formed exoplanet can be considered a monumental event because these lights tell us that an advanced civilization will eventually emerge from that exoplanet. On the other hand, large flares from a star can totally overwhelm exoplanet magnetospheres, almost shutting them down leading to mass extinctions.

A celestial body's uncontrolled nuclear fission (nuclear chain reaction) or fusion (thermonuclear) explosion will radiate much light. As an example, the highly powerful Tsar Bomba from the Soviet Union will cause temperatures to rise to millions of degrees Celsius abruptly, which is 1,500 times more powerful than Hiroshima and Nagasaki combined. At such a detonation temperature, nitrogen atoms in the atmosphere fuse, releasing energy. Atmospheric nitrogen ignition might fuse hydrogen in the oceans. The nuclear explosion could turn the whole planet or exoplanet into a fusion bomb, vaporizing it. A planet or exoplanet might even transform into a glowing star. The reaction stops when we run out of fissile material. To understand the effects of these explosions, consider the following analogy. Suppose you live in a neighborhood where neighbors do not mind doing business outside their houses. If one house in your neighborhood caught fire, everyone near and afield from the burning house would become attentive and willing to help in one form or another to avoid the fire spreading to other houses. That is what happens between civilizations spread on other planets light-years away from us. As a result of the Second World War when weapons illuminated Earth's sky and after the Hiroshima and Nagasaki explosions, there is a space folklore story that mysterious visitors from other civilizations visited Earth. In fact, some of them are keeping a close watch on nuclear weapons to prevent their potential release into space. Therefore, this folklore story is partially supported by the above analogy. This is why extraterrestrial encounters are more likely to occur in places with advanced nuclear weapons than randomly around the world. In other words, lighting or darkening a planet's atmosphere suddenly is one way to attract and identify other civilizations nearby. If Mars' atmosphere were vaporized by two nuclear explosions as suggested by Brandenberg (2021), the study of Mars geology and planetary evolution becomes an even more critical test case. The many bizarre geological structures revealed by the six probes that have landed on Mars so far make the discoveries even more significant. One will also expect evidence of Mars' explosions on Earth to be discovered. Anyway, by looking at the stars any advanced civilization has mastered the phenomena of fission and fusion nuclear reactions. It is very likely that artificial fission and/or fusion reactions are the most evident tools and signs of identifying and locating other advanced civilizations in our galaxy and elsewhere in our universe.

The early solar system especially in the first billion years or so of the earth's history was frequently hit by large impacts of one hundred kilometers in diameter or more. Many of these impacts will exceed our current nuclear bombs. The Chicxulub impact was about 100 million megatons, in the energy it released, or two million times more powerful than the Tsar Bomba. It is linked to the mass extinction of dinosaurs; the impact may have instantaneously wiped out the dinosaur civilization.

Any nearby civilization will have noticed and felt the light and radiation generated by such an impact.

The quantity, known as the volcanic explosiveness index (VEI), measures volcanic eruption explosiveness. Newhall and Self devised it (1982). Volume of ejecta (i.e., volcanic materials thrown out of solid Earth into the atmosphere), eruption cloud height, and eruption duration are the determinants of VEI. The VEI treats lava, lava bombs, and ignimbrite alike. VEI-8 volcanoes can produce more than 1.0×10^{12} (a trillion) cubic meters m^3 of ejecta and have cloud-column heights of over 40 km. As a result of such an eruption, advanced civilizations light-years away will be able to see it, and any civilization nearby would also be attracted to it.

As discussed in Chapters 1 and 6, some exoplanets have their own stars (i.e., their gravitational pull has surrounded them with stars). Space telescopes or similar systems can observe the lights from these exoplanets at very large distances. Until now, such lights have not been well interpreted scientifically.

Was the light emanating from Proxima b picked by JWST? Most likely not, not yet, as of November 2023. Scientists and the public at large are waiting for the definitive answer to this question. As mentioned earlier, Proxima is the nearest exoplanet to us. As described in Figure 3.57 potential civilizations on Proxima and Epsilon Eridani exoplanets must be our primary concerns. Yet it is 4.5 light years away from us. With our current space-travel technology, it would take 70,000 or so years or 750 homo-sapiens-sapiens (modern humans) generations to reach Proxima b. Current data about Proxima b are hard to interpret. It has a large mass density of 6 g/cm^3, larger than Earth. According to the mass densities of oxides (see chapter 3 of Ikelle, 2023b), such planets are expected to have an enormous massive metallic core and heat-producing radioactive elements, as well as a substantial magnetosphere. Yet its close proximity to its star (closer than Mercury is to the Sun and TRAPPIST-1b is to TRAPPIST-1) suggests it does not have an atmosphere. This means its large mass density is essential for Proxima b to avoid being broken apart by its star's gravitational pull. Moreover the gravitational roles of Alpha Centauri A and B in the four-body configuration (see Figure 3.57) are still an open question. In fact Proxima Centauri is also known as Alpha Centauri C, and the planet Proxima b is also known as Alpha Centauri cb because of the three-star system. Essentially, James Webb will determine whether Proxima-b has an atmosphere, its gas composition, but will not confirm life on the planet. It is highly likely that Proxima contains two additional planets, Proxima c and Proxima d. Alpha Centauri A and Alpha Centauri B, the other two stars in the Alpha Centauri star system, also likely containtwo planets, Alpha Centauri Ab and Alpha Centauri Bc. The confirmation of these planets is pending, as of November 2023.

3.10.   Essay: Decoding of Molecule, Mineral, and Star Spectroscopy

The star spectra can be interpreted as functions of elements of the periodic table and the spectra of these elements. These spectra are presented in Figure 3.61. We denote these elements and their spectrum as U_{jk}, where k are these elements and j represents spectra associated with these elements. Similar, we can describe mineral spectra as functions of oxides and the spectra of these oxides. So we can also refer to these oxides and their spectrum as U_{jk}, where k represents these oxides and j denotes the spectra associated with these oxides. In Ikelle (2024) we introduced novel elementary molecules and their spectra. Spectra of organic molecules can be described as functions of these elementary molecules and their spectra. Again we can denote these elementary molecules and their spectra as U_{jk}, where k are these elements and j denotes the spectra associated with these elements. So we can formulate the decoding of spectra (i.e., determining the chemical composition of a given star, molecular, and mineral spectra) as follows:

$$X_i = \sum_j \sum_k A_{ijk} U_{jk} \; , \qquad (3.179)$$

where X_i the spectra we wish to decode. Both X_i and U_{jk} are known. The decoding process comes down to reconstructing A_{ijk}. As described in Ikelle (2010, 2017, and 2024) we can determine these weighted terms irrespective of the values of the indexes i, j, and k, including undetermined cases in which $i < k$ and/or $i < j$. Examples of such estimations are given in Ikelle (2024).

Chapter 4

A View of our Universe from Quantum Computing

As discussed in Ikelle (2023a), nature is made of quantum mechanics with counterintuitive and non-classical concepts at its heart. These concepts include wave-particle duality, the absence of exact predictability, the discrete spectrum of energy levels, and the inseparability of the universe. We have already discussed wave-particle duality, the discrete spectrum of energy levels, and the absence of exact predictability in Ikelle (2023a, 2023b, and 2023c). Our focus here is on the universe's inseparability. Various parts of the universe are interconnected, so that no matter how far apart they are, they affect each other. At the sub-atomic level, this inseparability is known as *quantum entanglement.*

Quantum entanglement happens when two same-type quantum particles, such as photons or electrons, become linked such that an effect on one automatically and instantaneously affects the other, even if the particles are trillions of kilometers apart (for example, one particle is on Earth and the other particle is 40.2 trillion kilometers from us in Proxima Centauri). For example, changes in the spin, momentum, and/or polarization of one entangled particle automatically and instantaneously affect its twin, irrespective of the distance between them. We are dealing here with short- and long-distance correlations in which light speed is not involved. Thus quantum entanglement does not violate Einstein's special theory of relativity. This phenomenon is expected to manifest at the macroscopic level and lead to profound technological advancements and societal changes. Because entanglement can occur among many quantum particles, it occurs among atoms, molecules, living species, and organic and inorganic matter in some form. In this chapter, we introduce the key principles of quantum entanglement and some applications. Some of them are still untouchable, and yet they are already here.

One of the major applications of quantum entanglement is in the so-called *quantum computing.* Quantum computing has two schools of thought. One school aims to address current bottleneck applications in classical physics. These include weather predictions, climate modeling, optimizing radiotherapy treatments, energy exploration, seismic survey optimization, reservoir optimization, and reserve and spot trading optimization. Compared to traditional digital computing, quantum computing dramatically reduces computation time and energy consumption. Perhaps,

DOI: 10.1201/9781032620619-4

one day, this school will be extended to solve elastic wave equations with nonlinear stress-strain relations, Navier-Stokes equations with nonlinear stress-strain relations, and Maxwell's equations with nonlinear constitutive relations described in the previous section as a way of simulating nature. Because our nonlinear constitutive relations will remain speculative or incomplete for the foreseeable future, it is very unlikely this approach will ever reveal to us how nature does things.

Another school of thought posits that nature is based on quantum mechanics. Since nature is based on quantum mechanics, this approach to quantum computing can help us further understand how nature works. At the end of the day, we have to finally figure out if the complexities and the large number of anomalies reported by modern sciences are real or just the effects of our modern mathematical monstrosities and our weird theories. Like our daily lives, we suspect that the universe is a series of ups and downs, a series of cycles in time, fractal patterns all evolving in cycles. Anyway, this approach will finally teach humans how nature works and how to live in harmony with our universe. It represents the ultimate test of quantum mechanics and a quantification of quantum mechanics' limits. We elaborate on these two quantum computing approaches.

Key sentences: *Ultimately we will have to understand and computationally simulate the universe to control some natural occurrences.*

4.1. Schrödinger Equation

4.1.1. Wave Functions

Schrödinger equation is one of the most famous and important equations of physics. It is central to our modern understanding of the universe. He advances the idea of particle duality first formulated by Louis de Broglie (1892-1987). Again, particle duality means that light waves can also behave like particles and particles like electrons can also behave like waves. A particle of mass m moving at speed v can be associated with a plane wave of wavelength, $\lambda = h/p = h/mv$ where $p = mv$ is the particle's momentum and $h = 6.63 \times 10^{-34}$ Js $= 4.14 \times 10^{-15}$ eV s is the Planck constant. In addition, if the particle has energy U, its frequency is $f = U/h$ ($\omega = U/\hbar$). So you can describe particles with $[U, p]$ and waves with $[f, \lambda]$.

In 1926, Erwin Rudolf Josef Alexander Schrödinger (1887-1961) predicted that quantum particles would obey the following quantum wave equation:

$$i\hbar \underbrace{\frac{\partial \Psi(\boldsymbol{x}, t)}{\partial t}}_{\text{Total energy}} = \underbrace{-\frac{\hbar^2}{2m} \boldsymbol{\nabla}^2 \Psi(\boldsymbol{x}, t)}_{\text{Kinetic energy}} + \underbrace{V(\boldsymbol{x}, t) \Psi(\boldsymbol{x}, t)}_{\text{Potential energy}} , \qquad (4.1)$$

where $\psi(\boldsymbol{x}, t)$ is the so-called wave function, $V(\boldsymbol{x}, t)$ is the potential energy, $\hbar = h/2\pi = 1.055 \times 10^{-34}$ Js $= 6.59 \times 10^{-16}$ eV sis the reduced Planck constant, and m is the particle mass. This equation was postulated—it was not derived—yet the predictions based on it have been experimentally verified. When asked how Schrödinger came up with this equation, Richard Feynmann said: "It is not possible to derive it from anything you know. It came out of Schrödinger's mind." The key parameter is the potential energy of the particle under consideration. In response to forces applied to a particle, which is related to its potential energy as

$$F_k(\boldsymbol{x}, t) = -\frac{V(\boldsymbol{x}, t)}{\partial x_k} \, , \tag{4.2}$$

where $\boldsymbol{F}$ are the forces applied to the particle, we have a wave function, $\psi(\boldsymbol{x}, t)$. The wave function is a mathematical description of the quantum state of a particle. Its square amplitude $|\Psi(\boldsymbol{x}, t)|^2 = \Psi^*(\boldsymbol{x}, t)\Psi(\boldsymbol{x}, t)$ is the probability of finding a particle in a certain time and space. That is, at locations where $|\Psi(\boldsymbol{x}, t)|^2$ is big there is a large probability to detect the particle, where $|\Psi(\boldsymbol{x}, t)|^2$ is small, there is a smaller probability to locate the particle, and where it vanishes, there is no chance of finding the particle. This interpretation was made by Max Born (1882-1970). It also leads to the normalization of Ψ to ensure that the sum of the probabilities must be unity. Note that Schrödinger equation obeys the superposition principle; that is If $\Psi_1(\boldsymbol{x}, t)$, $\Psi_2(\boldsymbol{x}, t)$, etc., are different solutions of Schrödinger equation for a given potential $V(\boldsymbol{x}, t)$ then any arbitrary linear combination

$$\Psi(\boldsymbol{x}, t) = \sum_{i=1}^{N} c_k \Psi_k(\boldsymbol{x}, t) \, , \tag{4.3}$$

where c_k are constants, is also a solution of Schrödinger equation.

Let us contrast this equation with wave propagation described in Ikelle (2023c) for classical particles. The Schrödinger equation is not derived from Newton's equation of motion: force equals the product of mass and acceleration, as we did with the elastic and acoustic wave equations. In a homogeneous nonviscous fluid, the wave equation is

$$\rho \frac{\partial^2 \chi(\mathbf{x}, t)}{\partial t^2} = K \frac{\partial^2 \chi(\mathbf{x}, t)}{\partial x_k^2} \quad \text{or} \quad \frac{\partial^2 \chi(\mathbf{x}, t)}{\partial t^2} = \frac{1}{c^2} \frac{\partial^2 \chi(\mathbf{x}, t)}{\partial x_k^2} \, , \tag{4.4}$$

with $k = 1, 2, 3$ or $k = x, y, z$, where χ is the fluid pressure, ρ is the mass density, K is the bulk modulus, and $c = \sqrt{K/\rho}$ is the wave speed. Consider the case where $\boldsymbol{x} = [x, 0, 0]^T$. Notice that

$$\chi(\boldsymbol{x}, t) = A \cos(kx - \omega t) \tag{4.5}$$

is a solution of (4.4) with $c = \omega/k$ while

$$\psi(\boldsymbol{x}, t) - A \cos(kx - \omega t) \tag{4.6}$$

is not a solution of (4.1) even if $V(\boldsymbol{x}, t) = V_0 = $ constant. Similar, $\psi(\boldsymbol{x}, t) = A \cos(kx - \omega t) + iB \sin(kx - \omega t)$ is a solution of (4.1) for $V(\boldsymbol{x}, t) = V_0 = $ constant and $\chi(\boldsymbol{x}, t) = A \cos(kx - \omega t) + iB \sin(kx - \omega t)$ is not a solution of (4.4). In notches, you can see the difference between these two fundamental physics equations. This is despite their similarity in terms of time and space derivations and sinusoidal behavior in some of their solutions. These differences also capture the difference between quantum and classical mechanisms. In classical mechanics we deal with a real-value field solution and a complex-value field solution in quantum mechanisms.

As described below the probabilist nature of quantum mechanics solutions arises from the complex-value representations of its solutions. Notice that, for a constant potential, the Schrödinger equation has sinusoidal traveling wave solutions with constant wavelength and frequency. This result follows from the fact that if the potential is constant the force acting on the particle is zero (see equation (4.2)).

Let us recall that the total energy for a particle is the sum of kinematic energy and potential energy; i.e.,

$$U = \underbrace{\frac{1}{2}mv^2}_{\text{kinematic}} + \underbrace{V(\boldsymbol{x},t)}_{\text{potential}} = \frac{1}{2}\frac{p^2}{m} + V(\boldsymbol{x},t) \ . \tag{4.7}$$

Potential energy provides information about the forces acting on the particle. It is the term $V(\boldsymbol{x},t)$ in (4.1).

Computational exercise: We use the derivations in Appendix A to develop the code in Table 4.1 for solving Schrödinger's equation in time-space domain using the finite difference technique. Use these commands to reproduce the results in Figure 4.1. These results describes the propagation of a wave packet that represents a free particle (electron). The wave packet a sinusoidal function and a Gaussian envelope.

In the most interesting cases in physics, we actually have $V(\boldsymbol{x},t) = V(\boldsymbol{x})$; that is, potential energy is time independent. Wave functions are found in the form

Table 4.1. A Matlab code for simulating wave functions in space-time domain.

```
clear all; close all;
nx = 501; nt = 4000; L = 5e-9; me=9.10938291e-31; hbar=1.054571726e-34;
x=linspace(0,L,nx); dx=x(2)-x(1); dt=0.1*2*me*dx^2/hbar; t=0:dt:nt*dt;
psiR = zeros(nt,nx); psiI = zeros(nt,nx); xx = x.*1e9;
nx1 = round(nx/6); s = L/25; wL = 1.6e-10; AN= 5.3113e+04;
yR = AN*exp(-0.5.*((x-x(nx1))./s).^2).*cos((2*pi).*(x-x(nx1))./wL);
yI = AN*exp(-0.5.*((x-x(nx1))./s).^2).*sin((2*pi).*(x-x(nx1))./wL);
psiR(1,:) = yR; psiI(1,:) = yI;
for it = 1:nt-1
    yR=yR-0.1*([0,yI(1:end-1)]-2*yI+[yI(2:end),0]); psiR(it+1,:)=yR;
    yI=yI+0.1*([0,yR(1:end-1)]-2*yR+[yR(2:end),0]); psiI(it+1,:)=yI;
end
for it = 2:nt
    psiI(it,:) = 0.5*(psiI(it,:)+psiI(it-1,:));
end
yP = psiR(nt,:);plot(xx,yP,'b','linewidth',2); hold on
yP = psiR(1,:); plot(xx,yP,'b','linewidth',1); hold off
grid on
% yP = psiI(nt,:); plot(xx,yP,'r','linewidth',2); hold on
% yP = psiI(1,:); plot(xx,yP,'r','linewidth',1); hold off
% yP = psiR(nt,:).^2+psiI(nt,:).^2; plot(xx,yP,'k','linewidth',2); hold on
% yP = psiR(1,:).^2 +psiI(1,:).^2; plot(xx,yP,'k','linewidth',1); hold off
```

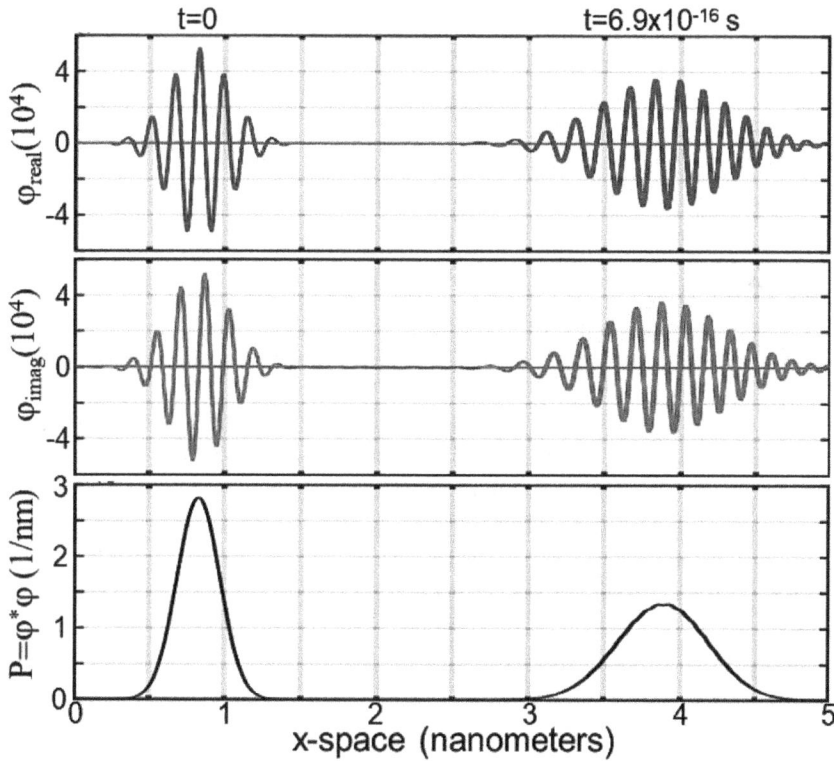

Figure 4.1. Using two instants of free propagation of a one-dimensional wave packet, we illustrate the time evolution of a Gaussian wavepacket. (Top) The real values of the wave function. (Middle) The imaginary values of the wave function (Bottom) The probability density function associated with the wave function: $P = |\Psi|^2$. As time passes the packet moves to the right and spreads.

$\Psi(x, t) = \psi(x)\phi(t)$ in these cases. We will limit ourselves to $x = [x, 0, 0]^T$. By substituting this expression for (4.1), we arrive at

$$\frac{1}{\psi(x)}\left[-\frac{\hbar}{2m}\frac{d^2\psi(x)}{dx^2} + V(x)\psi(x)\right] = i\hbar\frac{1}{\phi(t)}\frac{d\phi(t)}{dt} = C \tag{4.8}$$

where C is constant because the time-independent part of this equation is equal to the x-independent part of the independent part. Let us start by solving the time-dependent part of this equation; i.e.,

$$i\hbar\frac{1}{\phi(t)}\frac{d\phi(t)}{dt} = C \implies \phi(t) = \exp\left[-i\frac{C}{\hbar}t\right] = \exp\left[-2\pi i f t\right] = \exp\left[-i\frac{U}{\hbar}t\right] , \tag{4.9}$$

where U is the total particle energy. Since $f = C/h$ and $f = U/h$, C has been replaced by U. The time-independent part of (4.8) can be written as an eigenfunction equation as follows:

$$-\frac{\hbar}{2m}\frac{d^2\psi(x)}{dx^2} + V(x)\psi(x) = U\psi(x) . \tag{4.10}$$

This equation is called the time-independent Schrödinger equation because the time variable t does not enter the equation. The solution to this equation is an eigenfunction. So when eigenfunctions are determined, we obtain the solution of Schrödinger equation as

$$\Psi(\boldsymbol{x}, t) = \psi(\boldsymbol{x}) \exp\left[-i\frac{U}{\hbar}t\right] . \tag{4.11}$$

Quiz: Show that $\Psi(x,t) = A\sin(kx - \omega t + \phi)$ is not a solution to Schrödinger equation, where k is the wavenumber, ω is the angular frequency and ϕ is a phase constant. The wave is moving in the positive x-direction.

Examples of potential energy. One of the key components of the Schrödinger equation is potential energy. Let us briefly look at some known examples. We start with the Gamow model of α radioactivity, introduced in 1928. Briefly speaking, Gamow assumed that the α particles (Helium-4) were already bound to the nucleus. The potential they feel is the sum of two terms: The first is due to the averaged nuclear force, and can be described as a Woods-Saxon potential $V_{\mathrm{VW}}(r)$. The second one is simply the Coulomb potential $V_C(r)$ due to the electromagnetic interaction between the α particle and the other nucleons:

$$V_{\mathrm{VW}}(r) = -V_0\left[1 - \exp\left(\frac{r-R}{a}\right)\right]^{-1} \quad \text{and} \quad V_C(r) = \frac{V_0}{r} , \tag{4.12}$$

where $V_{\mathrm{VW}}(r)$ is related to the averaged nuclear force and is known as the Woods-Saxon potential $V_{\mathrm{VW}}(r)$ and $V_C(r)$ is simply the Coulomb potential $V_C(r)$ due to the electromagnetic interaction between the α particle and the other nucleons. We can write the net potential felt by the α particle as $V(r) = \kappa V_{\mathrm{VW}}(r) + (-\kappa)V_C(r)$ where κ can be constant or a function of r, e.g. a linear function. As discussed in Ikelle (2023b), alpha emission is a radioactive process involving two nuclei X and Y, which can be described in general as follows:

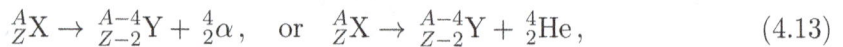

$$^A_Z X \to {}^{A-4}_{Z-2}Y + {}^4_2\alpha, \quad \text{or} \quad {}^A_Z X \to {}^{A-4}_{Z-2}Y + {}^4_2\mathrm{He}, \tag{4.13}$$

where $^A_Z X$ is the generic unstable nucleus, $^{A-4}_{Z-2}Y$, the helium-4 nucleus being known as an alpha particle. We will use alpha particles as a quantum tunneling example in the next section.

The Morse potential can be a better approximation of a potential energy surface when working with diatomic molecules. It is given by (Morse, 1929)

$$V(r) = D_e\left[1 - \exp(a(r-r_0))\right]^2 , \tag{4.14}$$

where $r-r_0$ is the distance from a potential well at r_0, a and D_e are parameters that relate to the "width" and "depth" of the potential. These parameters are molecular specific.

The potential energy can be a time-dependent potential. Here is an example, where the time dependent part models system's interaction with a photon. In this case with a laser pulse

$$V(r,t) = V(r) + A\cos\left(\omega_p t\right) , \tag{4.15}$$

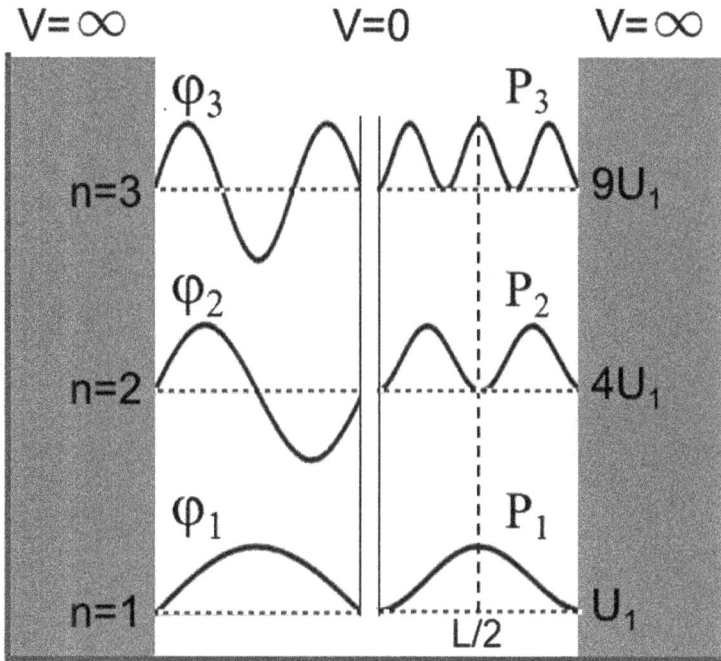

Figure 4.2. An illustration of potential energy with associated wave functions and particle position probabilities. The barriers outside a one-dimensional box have an infinitely large potential, while the interior has a constant, zero potential. Wave functions for the first three states are Ψ_1, Ψ_2, and Ψ_3. The probability density functions associated with these states are $P_1 = |\Psi_1|^2$, $P_2 = |\Psi_2|^2$, and $P_3 = |\Psi_3|^2$.

where the time dependent part model can be used to model a system's interaction with a photon. Here A is a constant, and ω_p is the frequency of the time-dependent perturbation.

Example: Energy quantization. Consider a particle of mass m confined to a 1D potential well, as illustrated in Figure 4.2. The particle is free to move around inside the box but cannot escape. This example is often called the infinite square wall. The potential energy can be described as

$$\begin{cases} V(x) = 0 & \text{for } 0 \leq x \leq L \\ V(x) = \infty & \text{for other } x \end{cases} \tag{4.16}$$

The wave function is zero (i.e., $\Psi(x) - 0$) outside the box because the probability of finding particles outside the box is zero by definition. The time-independent Schrödinger equation in (4.10) reduces to

$$-\frac{\hbar^2}{2m}\frac{d^2\psi}{dx^2} = U\psi \implies \frac{d^2\psi}{dx^2} = -k^2\psi , \tag{4.17}$$

where $k = \sqrt{2mU}/\hbar$. We search for a solution in the form $\psi(x) = C_1 \sin(kx) + C_2 \cos(kx)$, C_1 and C_2 are arbitrary constants. The boundary conditions are that

$$\psi(0) = 0 \implies C_2 = 0 \quad \text{and} \quad \psi(L) = 0 \implies C_1 \sin(kL) = 0 . \tag{4.18}$$

Since $\sin(kL) = 0$ implies that $kL = n\pi$, we have *energy quantization*; i.e.,

$$k = \frac{\sqrt{2mU_n}}{\hbar} = \frac{n\pi}{L} \iff U_n = \frac{n^2\pi^2\hbar^2}{2mL^2} = \frac{n^2h^2}{8mL^2} = n^2U_1 \ , \qquad (4.19)$$

where $n = 1, 2, 3, \ldots$. The particle energy has become quantized and can only take certain discrete values depending on the value of n with each value of n corresponding to a different energy and in turn this corresponds to a different eigenfunction, as illustrated in Figure 4.2. We can also notice that the lowest possible energy state is not zero as would be the case in classical physics. The lowest available energy state is U_1 corresponding to $n = 1$. This is called the ground state and U_n with $n > 1$ are called excited states. Though this effect is only noticeable in quantum situations, the same rules hold in the macroscopic world. In general we see that the energy of the confined particle is inversely proportional to both the mass m and the square of the confinement scale L. Taking the case of an electron we can write

$$U_1 = \frac{\pi^2(\hbar c_0)^2}{2(m_e c_0^2)L^2} = \frac{\pi^2 \left(197.3 \text{ eV nm}\right)^2}{2\left(511,000 \text{ eV}\right)L^2} \approx \frac{0.3759}{L^2 \, (\text{nm})^2} \ .$$

Therefore, if we confine an electron in a one dimensional box of length 0.1 nm we obtain a ground state energy $U_1 \approx 37.59$ eV, a reasonable approximation to atomic electron energies. In order to cause the electron to change from the ground state to the first excited state the photon energy must be

$$U_{\text{photon}} = \frac{hc_0}{\lambda} = U_2 - U_1 = 3U_1 \implies \lambda = \frac{hc_0}{3U_1} = \frac{1240 \text{ eV nm}}{3 \times 37.59 \text{ nm}} \approx 10.99 \text{ nm} \ .$$

Energy quantization is a general feature of quantum mechanical systems and can be understood in terms of Heisenberg uncertainty principle. This roughly speaking implies that whenever you confine a particle to a region of space there is inherent uncertainty in the particle's momentum and therefore its energy. It is this quantum uncertainty that gives rise to non-zero ground state energy when a particle is restricted to a region of space. So the wave function of a particle confined to an infinite square wall is

$$
\begin{aligned}
\Psi_n(x,t) &= \psi_n(x)\phi_n(t) \\[2mm]
&= \psi_n(x)\exp\left[\frac{-iU_n}{\hbar}i\right] \\[2mm]
&= C_1\sin\left(\frac{n\pi x}{L}\right)\exp\left[\frac{-iU_n}{\hbar}i\right] \\[2mm]
&= \sqrt{\frac{2}{L}}\sin\left(\frac{n\pi x}{L}\right)\exp\left[\frac{-iU_n}{\hbar}i\right] \ . \qquad (4.20)
\end{aligned}
$$

We used the fact that the probability of finding the particle located in the region $0 \le x \le L$ is by definition 1 to determine C_1; i.e.,

$$\int_0^L |\Psi(x,t)|^2 dx = 1 \implies C_1^2 \int_0^L \sin^2\left(\frac{n\pi x}{L}\right) dx = 1 \implies C_1 = \sqrt{2/L} \ .$$

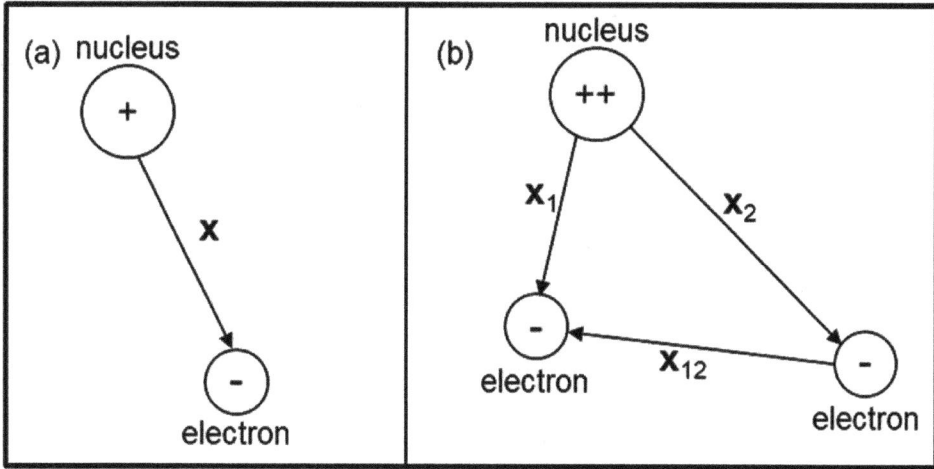

Figure 4.3. a) The nucleus (+) and electrons (e-) of the hydrogen atom. b) The nucleus (++) and electrons (e-) of the helium atom.

The process of determining C_1 is known as *normalization of wave functions*. Note that the probability density function in this example is time-independent because $\Psi_n^*(x,t)\Psi_n(x,t)$ is time-independent.

As illustrated in Figure 4.2, particles vary between states. We are most likely to find our particle in the center of the box in the ground state. In the $(n = 2)$-state, there is zero probability of finding the particle in the center of the box. As n increases the quantum mechanical probability density oscillates more and more. As n nears infinity, the quantum mechanical probability becomes constant and the particle has equal probability to be found at any location within the box. In this case, quantum mechanics predictions approach classical mechanics predictions.

Example: Schrödinger's wave equation for hydrogen atom. The hydrogen atom, consisting of an electron and a proton, is a two-particle system (see Figure 4.3a). As described in Chapters 3 and 5, the potential energy of atoms can be calculated by using Coulomb's equation for two charges; i.e.,

$$V(r) = \frac{kq_1q_2}{r} = \frac{Ze(-e)}{4\pi\varepsilon_0 r} \underset{Z=1}{\Longrightarrow} V(r) = -\frac{e^2}{4\pi\varepsilon_0 r} \,, \tag{4.21}$$

where $r = |x|$ is the distance between electron and the nucleus, $q_1 = Ze$, is the charge of the nucleus (i.e., the proton in the case of hydrogen), the charge of electron is $q_2 = -e$, and ε_0 is the permeability of free space. For hydrogen we have $Z = 1$. The potential energy is three-dimensional and spherically symmetric. We will work with spherical coordinates (r, θ, ϕ). Let us start by recalling what the Laplacian in spherical coordinates is (e.g., Ikelle and Amundsen, 2018; Chapter 17)

$$\nabla^2 = \frac{1}{r^2}\frac{\partial}{\partial r}\left(r^2\frac{\partial}{\partial r}\right) + \frac{1}{r^2\sin\theta}\frac{\partial}{\partial\theta}\left(\sin\theta\frac{\partial}{\partial\theta}\right) + \frac{1}{r^2\sin^2\theta}\frac{\partial^2}{\partial\phi^2} \,. \tag{4.22}$$

In the three-dimensional case of a particle in a spherically symmetric potential wave function, the solutions derived earlier can be applied almost immediately. This symmetry implies $\partial/\partial\theta = \partial/\partial\phi = 0$ and $\psi(r,\theta,\phi) = \psi(r)$. The time-independent Schröndinger equation can be written as follows:

$$-\frac{\hbar^2}{2m_e}\boldsymbol{\nabla}^2\psi(r) + V(r)\psi(r) = 0 \implies -\frac{\hbar^2}{2m}\frac{1}{r^2}\frac{d}{dr}\left(r^2\frac{d}{dr}\right)\psi(r) + V(r)\psi(r) = 0 \ .$$

It is easy to check that if we write the wave function $\psi(r) = u(r)/r$, the function $u(r)$ obeys the following one-dimensional equation

$$-\frac{\hbar^2}{2m_e}\frac{d^2u(r)}{dr^2} + V(r)u(r) = 0 \quad \text{with } r > 0 \ . \tag{4.23}$$

This equation is exactly like a particle in one dimension, except that here r is only positive, and $u(r)$ must go to zero at the origin.

Let us consider the full three-dimensional case. We now use $\psi(r,\theta,\phi)$ with all its dependencies to solve

$$-\frac{h}{2m_e}\boldsymbol{\nabla}^2\psi(r,\theta,\phi)\psi - \frac{e^2}{4\pi\varepsilon_0 r}\psi = U\psi \ , \tag{4.24}$$

where m_e is the electron mass, assuming the following form

$$\psi(r,\theta,\phi) = R(r)\Theta(\theta)\Phi(\phi) = R(r)Y(\theta,\phi) \ , \tag{4.25}$$

with $Y(\theta,\phi) = \Theta(\theta)\Phi(\phi)$. In other words, the variations of the radius, the polar angle, and the azimuthal angle of the solution to Schröndinger equation are separated. For atoms and molecules, the wave function is actually a function of the coordinates of all the electrons and nuclei. We have here used the Born-Oppenheimer approximation by ignoring the nucleus coordinates. This approximation is based on the fact that nuclei have much larger masses than electrons; i.e.,

$$\frac{\hbar^2}{2m_{\text{nucleus}}} \ll \frac{\hbar^2}{2m_e}$$

for these terms of the Schrödinger equation, where m_{nucleus} is the nucleus mass and m_e is the electron mass. To a reasonable approximation, one can solve the Schrödinger equation only for electrons and assume the nuclei are frozen.

By substituting (4.25) and (4.22) into (4.24), we arrive at

$$\frac{1}{\Phi}\frac{d^2\Phi}{d\phi^2} = -\frac{\sin^2\theta}{R}\frac{d}{dr}\left(r^2\frac{dR}{dr}\right) - \frac{\sin\theta}{\theta}\frac{d}{d\theta}\left(\sin\theta\frac{d\Theta}{d\theta}\right)$$

$$- \frac{2m}{\hbar^2}r^2\sin^2\theta\left[U + \frac{e^2}{4\pi\varepsilon_0 r}\right] = -m^2 \ , \tag{4.26}$$

where m^2 is an integer constant (not to be confused with the mass). We used the fact that the first term of these equalities is independent of r and θ and the second term is independent of ϕ to determine that these equalities must be constant. We can deduce the first term of this first fundamental Schröndinger equation for hydrogen

$$\frac{d^2\Phi}{d\phi^2} + m^2\Phi = 0 \implies \Phi = C_\phi \exp\left[im\phi\right] \ , \tag{4.27}$$

where C_ϕ is a constant of integration. We can divide the second equality by $\sin^2 \theta$ to arrive at

$$\frac{1}{R}\frac{d}{dr}\left(r^2\frac{dR}{dr}\right) + \frac{2m}{\hbar^2}r^2\left[U + \frac{e^2}{4\pi\varepsilon_0 r}\right] = -\frac{1}{\theta\sin\theta}\frac{d}{d\theta}\left(\sin\theta\frac{d\Theta}{d\theta}\right)$$

$$-\frac{m^2}{\sin^2\theta} = l(l+1) . \qquad (4.28)$$

The first term only depends on the radius r, and the second term depends on the angle θ, ϕ. The separating constant is $l(l+1)$. We can deduce the second and third Schröndinger equations for hydrogen

$$\frac{d}{dr}\left(r^2\frac{dR}{dr}\right) + \frac{2r^2 m_e}{\hbar^2}\left[U + \frac{e^2}{4\pi\varepsilon_0 r}\right]R - l(l+1)R = 0 \qquad (4.29)$$

and

$$\sin\theta\frac{d}{d\theta}\left(\sin\theta\frac{d\Theta}{d\theta}\right) + \left[l(l-1)\sin^2\theta - m^2\right]\Theta = 0 \Longrightarrow \Theta(\theta) = P_l^m(\cos\theta) , \quad (4.30)$$

where again l is a non-negative integer and m is an integer. Equation (4.30) is the well known to have Legendre polynomials with argument $\cos\theta$ which is generally denoted $P_l^m(\cos\theta)$. The product of $Y(\theta,\phi) = \Theta(\theta)\Phi(\phi)$ actually represents spherical harmonics; i.e.,

$$Y(\theta,\phi) = \Theta(\theta)\Phi(\phi) = \sqrt{\frac{2l+1}{4\pi}\frac{(l-m)!}{(l+m)!}}P_l^m(\cos\theta)\exp\left[im\phi\right] . \qquad (4.31)$$

Derivations of the solution of the radial equation in (4.29) are quite lengthy and do not provide any significant educational benefits. So we borrow the results from Liboff (2003) and Sullivan (2012). The radial equation solution can be written as

$$R_{nl}(r) = \left(\frac{2Z}{na}\right)^{3/2}\sqrt{\frac{(n-1-l)!}{2n[(n+l)!]}}\left(\frac{2Zr}{na}\right)^l L_{n-l-1}^{2l+1}\left(\frac{2Zr}{na}\right)\exp\left(-\frac{Zr}{na}\right) , \quad (4.32)$$

where

$$a = \frac{4\pi\varepsilon_0\hbar^2}{m_e e^2} \quad \text{and} \quad L_n^k = \frac{1}{n!}\exp(x)x^{-k}\frac{d^n}{dx^n}\left[\exp(-x)x^{n+k}\right] \qquad (4.33)$$

are Bohr radius and associated Laguerre functions, respectively. The full normalized hydrogen wave function is

$$\psi_{nlm}(r,\theta,\phi) = R_{nl}(r)\Theta(\theta)\Phi(\phi) = R_{nl}(r)Y_l^m(\theta,\phi) . \qquad (4.34)$$

So when you solve the Schröndinger equation of hydrogen, the key outcomes are quantum numbers n, l, and m; $n = 1, 2, 3, ...$ is known as the principal quantum number (energy level), $l = 0, 1, 2, 3, ..., n-1$ as the angular momentum quantum number, and $m = -l, -l+1, ..., 0, ..., l-1, l$ as the magnetic quantum number.

The ground state wave function is ψ_{100} also known as the 1s orbital. When $l = 0$ we have the s orbital, $l = 1$ is the p orbital, and $l = 2$ is the d orbital.

The Schrödinger equation for helium. The Helium atom has two electrons and a single nucleus. The nucleus carries a $Z = +2e$ charge. Figure 4.3b shows a schematic representation of a helium atom with two electrons whose coordinates are given by the vectors $\boldsymbol{x}_1$ and $\boldsymbol{x}_2$. The electrons are separated by a distance $\boldsymbol{x}_{12} = |\boldsymbol{x}_1 - \boldsymbol{x}_2|$. There is a fixed origin in the nucleus of the coordinate system. As with the hydrogen atom, helium nuclei are so much heavier than an electron that the nucleus is assumed to be the center of mass. Fixing the origin of the coordinate system at the nucleus allows us to exclude translational motion of the center of mass from our quantum mechanical treatment. The Schrödinger equation for helium can be written as follows:

$$\left[-\frac{\hbar}{2m_e} \left(\sum_{i=1}^{2} \boldsymbol{\nabla}_i^2 \right) + \sum_{i=1}^{2} V(\boldsymbol{x}_i) + \sum_{i \neq j} V(\boldsymbol{x}_{ij}) \right] \psi(\boldsymbol{x}_1, \boldsymbol{x}_2) = U\psi(\boldsymbol{x}_1, \boldsymbol{x}_2), \quad (4.35)$$

where

$$V(\boldsymbol{x}_i) = -\frac{Ze^2}{4\pi\varepsilon|\boldsymbol{x}_i|} \quad \text{and} \quad V(\boldsymbol{x}_{ij}) = \frac{e^2}{4\pi\varepsilon_0|\boldsymbol{x}_i - \boldsymbol{x}_j|}. \quad (4.36)$$

In this equation, the nucleus position is the origin, $\boldsymbol{\nabla}_i$ are the Laplacian operators with respect to the coordinates of ith electron. The term $V(\boldsymbol{x}_{ij})$ represents the electron-electron repulsion taken as a Coulomb interaction based on the absolute value of the electron-electron separation. If this term was not present, then Schrödinger's equation of helium would be the sum of two Schrödinger's equations for hydrogen-like atoms, in which case the wave function would be simply the product of two hydrogen atomic orbitals, one for each electron. So if we neglect electron-electron repulsion in the Helium atom problem, we can solve the Schröndinger equation of helium as a product of the individual electron wave functions (here approximated as hydrogen-like wave functions):

$$\psi(\boldsymbol{x}_1, \boldsymbol{x}_2) = \psi^{(H)}(\boldsymbol{x}_1)\psi^{(H)}(\boldsymbol{x}_2) \implies \psi_{n_1 l_1 m_1 n_2 l_2 m_2} = \psi^{(H)}_{n_1 l_1 m_1} \psi^{(H)}_{n_2 l_2 m_2},$$

where $\psi^{(H)}$ is the hydrogen-like wave function. Unfortunately, even the simplified Schrödinger equation in (4.35) cannot be solved exactly. This is not a surprise. The helium atom is a three-body problem: two electrons plus a helium nucleus. Three-body problems, even in classical physics, cannot be solved analytically. The culprit is the term $V(\boldsymbol{x}_{ij})$.

For completeness and numerical approximations and solutions, let us now include the nucleus in the Schrödinger's equation of helium atom; i.e.,

$$\left[-\frac{\hbar}{2m_n} \boldsymbol{\nabla}_{\boldsymbol{x}'}^2 - \frac{\hbar}{2m_e} \left(\sum_{i}^{N} \boldsymbol{\nabla}_i^2 \right) + \sum_{i}^{N} V(\boldsymbol{x}'_{ij}) + \sum_{i \neq j} V(\boldsymbol{x}_{ij}) \right] \psi(\boldsymbol{x}_1, \boldsymbol{x}_2, \boldsymbol{x}')$$

$$= U\psi(\boldsymbol{x}_1, \boldsymbol{x}_2, \boldsymbol{x}'), \quad (4.37)$$

where

$$V(\boldsymbol{x}'_{ij}) = -\frac{Ze^2}{4\pi\varepsilon|\boldsymbol{x}' - \boldsymbol{x}_i|} \quad \text{and} \quad V(\boldsymbol{x}_{ij}) = \frac{e^2}{4\pi\varepsilon_0|\boldsymbol{x}_i - \boldsymbol{x}_j|} \,, \tag{4.38}$$

with $N = 2$ for helium atom, $\boldsymbol{x}'$ is the nucleus's position, and m_n is the mass of nucleus.

The Schrödinger equation for multi-electron atoms and molecules. Let us consider a system consisting of N electrons and K nuclei. The time-independent Schröndinger equation becomes

$$\left[-\frac{\hbar}{2m_n} \left(\sum_{i=1}^{K} \boldsymbol{\nabla}^2_{\boldsymbol{x}'_i} \right) - \frac{\hbar}{2m_e} \left(\sum_{i=1}^{N} \boldsymbol{\nabla}^2_i \right) + \sum_{j=1}^{K}\sum_{i=1}^{N} V(\boldsymbol{x}'_{ij}) + \sum_{i=1}^{K}\sum_{j=i+1}^{K} V(\boldsymbol{x}''_{ij}) \right.$$

$$\left. + \sum_{i}^{N}\sum_{j=i+1}^{N} V(\boldsymbol{x}_{ij}) \right] \psi(\boldsymbol{x}_1, ..., \boldsymbol{x}_N, \boldsymbol{x}'_1, ..., \boldsymbol{x}'_K)$$

$$= U\psi(\boldsymbol{x}_1, ..., \boldsymbol{x}_N, \boldsymbol{x}'_1, ..., \boldsymbol{x}'_K) \ .$$

where

$$V(\boldsymbol{x}'_{ij}) = -\frac{Z_je^2}{4\pi\varepsilon|\boldsymbol{x}'_j - \boldsymbol{x}_i|} \,, \ V(\boldsymbol{x}_{ij}) = \frac{e^2}{4\pi\varepsilon_0|\boldsymbol{x}_i - \boldsymbol{x}_j|} \,, \ V(\boldsymbol{x}''_{ij}) = \frac{Z_jZ_ie^2}{4\pi\varepsilon_0|\boldsymbol{x}'_i - \boldsymbol{x}'_j|} \ .$$

The first term is the kinetic energy of nuclei, the second term is the kinetic energy of electrons, $V(\boldsymbol{x}'_{ij})$ represents the electron attractions to nuclei, $V(\boldsymbol{x}''_{ij})$ represents nucleus-nucleus repulsions, and $V(\boldsymbol{x}_{ij})$ represents the electron-electron repulsions. Solving the Schrödinger equation for N-electron atoms and K nuclei has turned out, so far, to be analytically and exactly impossible and numerically prohibitive expensive in classical-mechanics-based computers because the large numbers of variable; for an atom like sodium, $N = 11$ and $K = 1$ which corresponds $3N = 34$ variables including the time; and water contains 10 electrons and 3 nuclei with $Z_O = 8$, $Z_{H1} = Z_{H2} - 1$ for oxygen and hydrogen nuclei. The most basic approximations to the exact solutions involve writing a multi-electron wave function as a simple product of single-electron wave functions, and obtaining the energy of the atom in the state described by that wave function as the sum of the energies of the one-electron components. The wave functions are written:

$$\psi(\boldsymbol{x}_i, ..., \boldsymbol{x}_N) \approx \psi_1(\boldsymbol{x}_1)\psi_2(\boldsymbol{x}_2)\psi_3(\boldsymbol{x}_3)...\psi_N(\boldsymbol{x}_N) \ . \tag{4.39}$$

By writing the multi-electron wave function as a product of single-electron functions, we conceptually transform a multi-electron atom into a collection of individual electrons located in individual orbitals whose spatial characteristics and energies can be separately identified. So each function, $\psi_i(\boldsymbol{x}_i)$, is associated with an orbital energy, U_n. For atoms these single-electron wave functions are called atomic orbitals. It is important to note that this form does not imply fully independent electrons, since each electron's dynamics (and hence wave function) is governed by the effective field/potential of all the other electrons in the atom. The one-electron orbitals are well-approximated by hydrogen-atom-like wave functions.

Hamiltonian (quantum mechanics). The Hamiltonian is named after William Rowan Hamilton, who developed a revolutionary reformulation of Newtonian mechanics, known as Hamiltonian mechanics, which was historically important to the development of quantum physics. Similar to vector notation, it is typically denoted by $\hat{H}$, where the hat indicates that it is an operator. It can also be written as H or $\breve{H}$. In quantum mechanics, the Hamiltonian of a system is an operator corresponding to the total energy of that system, including both kinetic energy and potential energy of all particles associated with the system. For a one-particle system, the Hamiltonian is

$$\hat{H} = \hat{T} + \hat{V} = -\frac{\hbar^2}{2m_p}\boldsymbol{\nabla}^2 + V(\boldsymbol{x}, t) \tag{4.40}$$

where

$$\hat{V} = V = V(\boldsymbol{x}, t) \quad \text{and} \quad \hat{T} = -\frac{\hbar^2}{2m_e}\boldsymbol{\nabla}^2 \tag{4.41}$$

and m_p is the particle's mass. Although this is not the technical definition of the Hamiltonian in classical mechanics, it is the form it most commonly takes in quantum mechanics. Combining these yields the form used in the Schrödinger equation:

$$\hat{H}\Psi(\boldsymbol{x}, t) = i\hbar\frac{\partial\Psi(\boldsymbol{x}, t)}{\partial t} \,, \tag{4.42}$$

as described in (4.27). In the time-independent Schrödinger equation, the operation may produce specific values for the energy called energy eigenvalues. This situation can be shown in the form

$$\hat{H}\psi_i(\boldsymbol{x}) = U_i\psi_i(\boldsymbol{x}) \,, \tag{4.43}$$

where the specific values of energy, U_i, are called energy eigenvalues and the functions ψ_i are called eigenfunctions (wave functions), as discussed earlier.

Computational exercise: We use the derivations in Appendix A to develop the code in Table 4.2 for solving the eigenvalues U and eigenvectors of ψ_i using the function **eig**. Use these commands to reproduce the results in Figure 4.4.

For N-particle systems like atoms, molecules, Earth, and Universes, the Hamiltonian has the following form

$$\hat{H} = -\frac{\hbar^2}{2}\sum_{i=1}^{N}\frac{1}{m_j}\boldsymbol{\nabla}_j^2 + V(\boldsymbol{x}_1, \boldsymbol{x}_2, ..., \boldsymbol{x}_N, t) \tag{4.44}$$

where

$$\boldsymbol{\nabla}_j^2 = \frac{\partial^2}{\partial x_j^2 + \partial y_j^2 + \partial z_j^2} \tag{4.45}$$

Table 4.2. A Matlab code for simulating wave functions.

```
clear all; close all;
L = 2*pi; N = 200; x = linspace(0, L, N)'; dx = x(2) - x(1);
hbar = 1; m = 1;
H0 = (diag(ones((N-1),1),-1)-2*diag(ones(N,1),0) ...
+ diag(ones((N-1),1),1))/(dx∧2);
H = -(1/2)*(hbar∧2/m)*H0;
[ psi, U] = eig(H);
plot(x,psi(:,3), 'k'); hold; plot(x,psi(:,4), 'b');
plot(x,psi(:,5), 'r–'); xlabel('x (m)');
ylabel('unnormalized wave function (10/m)');
ax = gca; ax.XLim = [0 2*pi];
```

Combining these yields the form used in the Schrödinger equation:

$$\hat{H}\Psi(\boldsymbol{x}_1, \boldsymbol{x}_2, ..., \boldsymbol{x}_N, t) = i\hbar\frac{\partial \Psi(\boldsymbol{x}_1, \boldsymbol{x}_2, ..., \boldsymbol{x}_N, t)}{\partial t} \ , \tag{4.46}$$

where m_j is the mass of the j particle.

4.1.2. Quantum Tunneling

Mount Everest's peak is the highest altitude above mean sea level at 8.848 kilometers, a major barrier to mankind and other mammals. Good luck climbing it if you are not among the 0.001 percent of the fittest athletes in the world. Most of us can

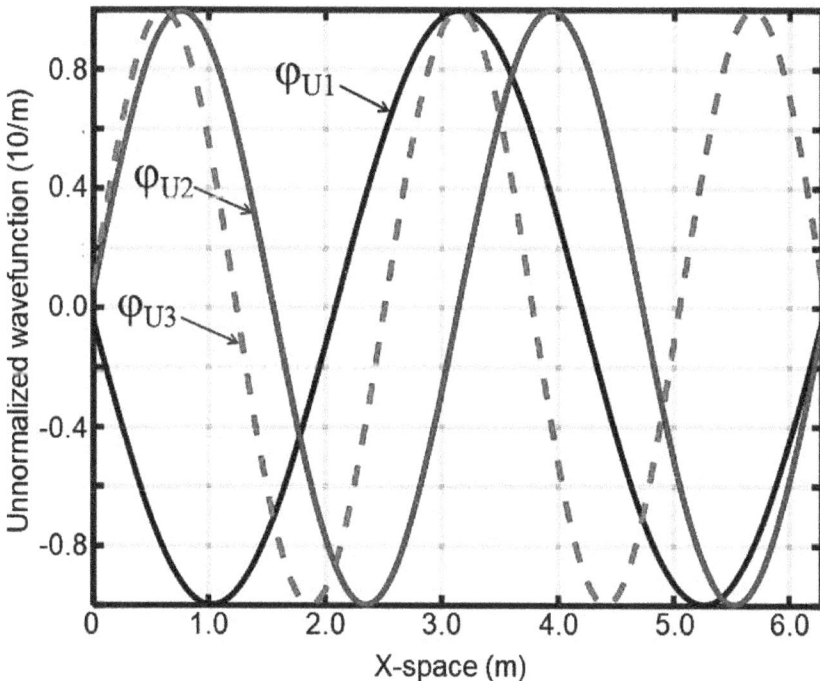

Figure 4.4. Combined plot of the lowest three definite-energy wave functions.

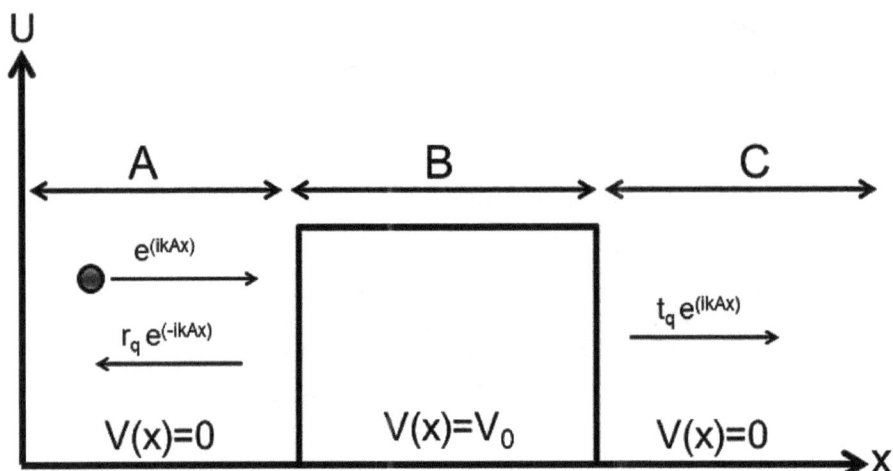

Figure 4.5. Schematic diagram indicating a potential barrier. The kinetic energy (U) is higher than the potential (V) in regions (A) and (C). For region (B), the wave vector becomes imaginary (as $V > U$).

only wait for the announced construction of a rail tunnel under Mount Everest to start disappearing from one side of the mount and appearing on the other side of Mount Everest without hiking up the mount. In other words, even trains and cars do not cross hills without climbing. You can search or look for a tunnel, and then go through that tunnel to reach the other side of the hill without hiking it even if your fitness is not among the highest. Similar phenomena exist at the quantum level. Tunneling through an energy barrier allows a quantum particle to disappear at one location and appear at another location immediately. This subsection describes spontaneous tunneling at the quantum level.

As illustrated in Figure 4.5, a particle traveling in region A, where the potential energy is zero ($V = 0$), encounters a barrier, which has a potential energy of $V(x) = V_0$. If this particle were a classical particle, it would bounce back. That is the end of matter. A quantum particle can sometimes pass even if its potential energy is less than its barrier energy. The configuration can be described as follows:

$$\begin{cases} \text{Region A } [-\infty, 0], & V(x) = 0, & \psi_{xx}(x) + k_A^2 \psi = 0, & k_A = \sqrt{2mU}/\hbar \\ \text{Region B } [0, L], & V(x) = V_0, & \psi_{xx}(x) + k_B^2 \psi = 0, & k_B = \sqrt{2m(V_0 - U)}/\hbar \\ \text{Region C } [L, \infty], & V(x) = 0, & \psi_{xx}(x) + k_A^2 \psi = 0, & k_A = \sqrt{2mU}/\hbar \end{cases}$$

Here ψ_{xx} denote twice the derivatives of ψ with respect to x. In each of these regions the Schrödinger equation will be solved, then the solutions will be stitched together in such a way as to make it smooth at any boundary. Let us now turn to the solutions of the time-independent Schrödinger equation for each region of this configuration. By using the results in the previous subsection we have

$$\begin{cases} \text{Region A } [-\infty\ 0], & \phi_A = A_1 \exp(ik_A x) + A_2 \exp(-ik_A x) \\ \text{Region B } [0\ L], & \phi_B = B_1 \exp(k_B x) + B_2 \exp(-k_B x) \\ \text{Region C } [L\ +\infty], & \phi_C = C_1 \exp(ik_A x) + C_2 \exp(-ik_A x) \end{cases} \qquad (4.47)$$

Note that in region B, we use real exponentials instead of complex exponentials because we assume $U < V_0$. For the less interesting cases of $U > V_0$, the solution

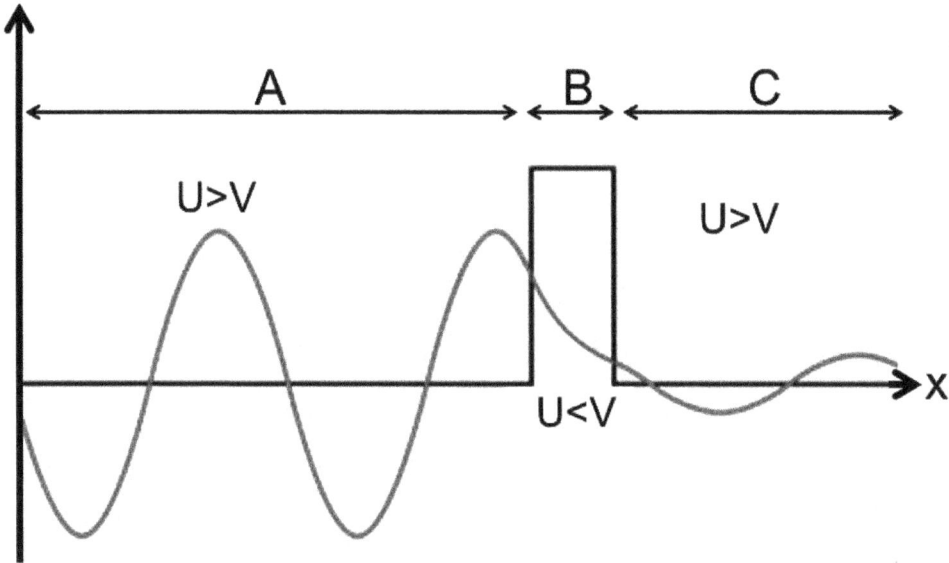

Figure 4.6. Sketch the wave function $\psi(x)$ corresponding to a particle with energy U. The wave function oscillates in regions A and C because $U > V(x)$, and decays exponentially in region B because $U < V(x)$. That is, waves do not end abruptly at a wall or barrier, but taper off quickly. If the barrier is thin enough, the probability function may extend into region C, through the barrier.

to the time-independent Schrödinger equation in region B will also involve complex exponentials. Real exponential means that the wave function decays rapidly; there is no propagation here (Figure 4.6). The furst term of ψ_A is the incident wave and its second term corresponds to a reflection at the barrier. Inside the barrier there is no propagation. In other words the particle is localized. There is no wave from region C reflecting on the barrier. So the coefficient C_2 of ϕ_C is zero ($C_2 = 0$).

The objective is to determine constants in (4.47) and thus deduce solutions to the Schrödinger equation. Actually what we are interested in are transmission and reflection coefficients which allow us to evaluate quantum tunneling. We can define $r_q = A_2/A_1$ as the reflection coefficient and $t_q = C_1/A_1$ as the transmission coefficient. As the wave function and its derivative with respect to x must be continuous for $x = 0$ and $x = L$, we have the following boundary conditions:

$$\psi_A(0) = \psi_B(0), \ \ \psi_B(L) = \psi_C(L), \ \ \frac{d\psi_A}{dx}(0) = \frac{d\psi_B}{dx}(0), \ \text{and} \ \frac{d\psi_B}{dx}(L) = \frac{d\psi_C}{dx}(L) \, .$$

We then obtain

$$\begin{cases} 1 + r_q = (B_1/A_1) + (B_2/A_1) \\ ik_A - ik_A r_q = k_B(B_1/A_1) - k_B(B_2/A_1) \\ (B_1/A_1)\exp(k_B L) + (B_2/A_1)\exp(-k_B L) = t_q \exp(ik_A L) \\ k_B(B_1/A_1)\exp(k_B L) - k_B(B_2/A_1)\exp(k_B L) = ik_A t_q \exp(ik_A L) \\ r_q r_q^* + t_q t_q^* = 1 \end{cases} \quad (4.48)$$

By solving this system of equations we arrive at

$$T_q = t_q t_q^* = \left[1 + \frac{\sinh^2(k_B L)}{4\eta(1 - \eta)} \right]^{-1}, \quad (4.49)$$

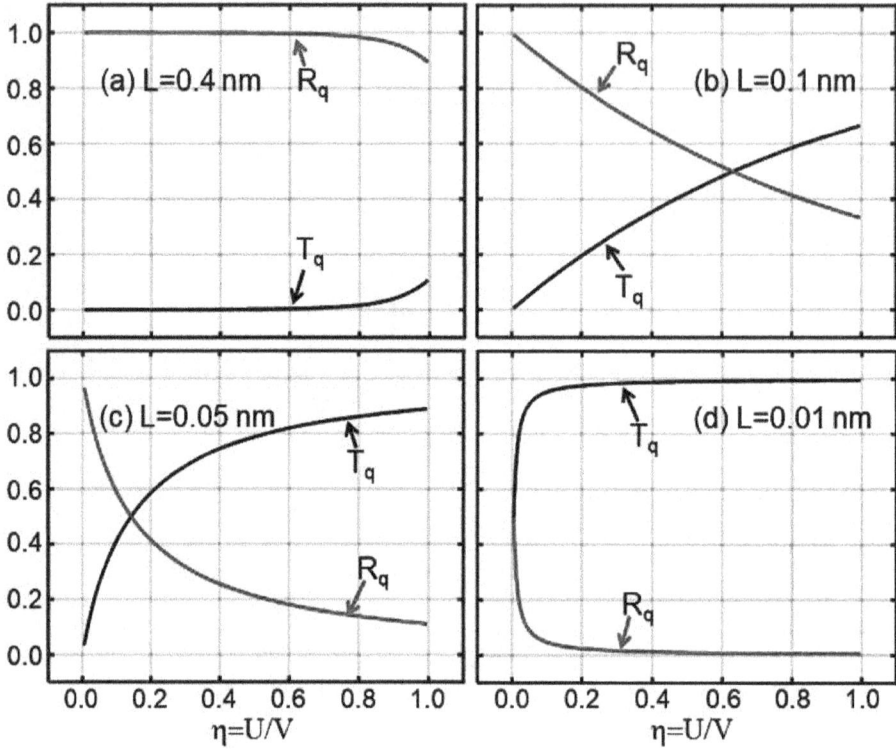

Figure 4.7. Transmission and reflection coefficients of quantum particles for hitting a rectangular potential barrier of various length with $U < V_0$.

where $\eta = U/V_0$. We can use $R_q = 1 - T_q$ to deduce R_q. Figure 4.7 shows transmission and reflections for various energy barrier lengths. The smaller the barrier length, the higher the number of particles that can pass through the energy barrier despite their potential energy being smaller than the barrier energy, according to these results. That is quantum tunneling.

> **Computational exercise:** We use the code in Table 4.3 for computing reflection and transmission coefficients to reproduce the results in Figure 4.7.

Let us consider for completeness, the case in which $U > V_0$; that is, the particle energy is much higher than the barrier height. Since the particle can propagate

Table 4.3. A Matlab code for simulating reflection and transmission coefficients.

```
clear all; close all;
me=0.0017; he=4.135*10^(-15); V0=1*10^(-6);
eta=0.005:0.001:0.995; [n1,n2]=size(eta); Lect = 0.01*10^(-9);
for iax=1:n2
    kB00 = sqrt(2*me*V0*(1.0-eta(iax))); kB = kB00/he;
    TI00 = sinh(kB*Lect)*sinh(kB*Lect); TI11 = 4*eta(iax)*(1-eta(iax));
    TINV = 1 + (TI00/TI11); TT(iax) = TINV^(-1);
end
```

freely in region B as well, there is no real exponential in region B. In other words, all exponentials are now complex-valued. Using the above boundary conditions, we arrive at the following system of equations

$$
\begin{cases}
1 + r_q = (B_1/A_1) + (B_2/A_1) \\
ik_A - ik_A r_q = k'_B(B_1/A_1) - k'_B(B_2/A_1) \\
(B_1/A_1)\exp(ik'_B L) + (B_2/A_1)\exp(-ik'_B L) = t_q \exp(ik_A L) \\
k'_B(B_1/A_1)\exp(ik'_B L) - k'_B(B_2/A_1)\exp(ik'_B L) = ik_A t_q \exp(ik_A L) \\
r_q r_q^* + t_q t_q^* = 1
\end{cases} \qquad (4.50)
$$

where

$$
k'_B = \frac{\sqrt{2m(U - V_0)}}{\hbar} = \frac{\sqrt{2mV_0(\eta - 1)}}{\hbar} . \qquad (4.51)
$$

Notice that we now work with k'_B instead of k_B to ensure that the wavenumber is the region B are positive. The solution to the system in (4.47) for T_q is

$$
T_q = t_q t_q^* = \left[1 + \frac{\sin^2(k'_B L)}{4\eta(\eta - 1)} \right]^{-1} , \qquad (4.52)
$$

where $\eta = U/V_0$. We deduce $R_q = 1 - T_q$. Figure 4.8 shows transmission and reflections for various energy barrier lengths. Notice also that η now varies from 1 to 2 because $U > V_0$. The large variations in Figure 4.8a are due to the reverberations between the barriers walls. As the barrier length reduces the reverberations decrease because the incoming wavelength increases and becomes larger the barrier length. At certain points, there is almost 100 percent chance of particles being transmitted.

Alpha decay: quantum tunneling effect. The nuclear equation for Polonium-210 alpha decay can be written as follows:

$$
{}^{210}_{84}\text{Po} \rightarrow {}^{206}_{82}\text{Pb} + {}^{4}_{2}\text{He} .
$$

Polonium-210 has 84 protons and 126 neutrons in its nucleus. During this reaction, an α-particle, which is a Helium-4 nucleus, is emitted by the nucleus. The alpha particle forms within the atom's nucleus. However, classical mechanics says it should never escape since it lacks energy. But it does so by quantum tunneling through the energy barrier.

Quantum tunneling described above can explain the alpha decay introduced in Ikelle (2023b). Unstable nuclei do not spontaneously decay to form more stable daughter nuclei. For instance, polonium-210 with a half-life of 138 days emits alpha particles, which carry high amounts of energy that can damage or destroy genetic material in cells inside the body. The fact that some unstable nuclei take a long time to decay means there must be some type of barrier in place that the alpha particle must overcome. This barrier is created by weak and strong nuclear forces from other nucleons (protons and neutrons) and the Coulomb repulsive forces. These forces produce an energy potential $V(x)$ as defined in (4.2). Alpha particles have a kinetic energy of 4 to 9 MeV, less than the Coulomb barrier, around 26-30 MeV.

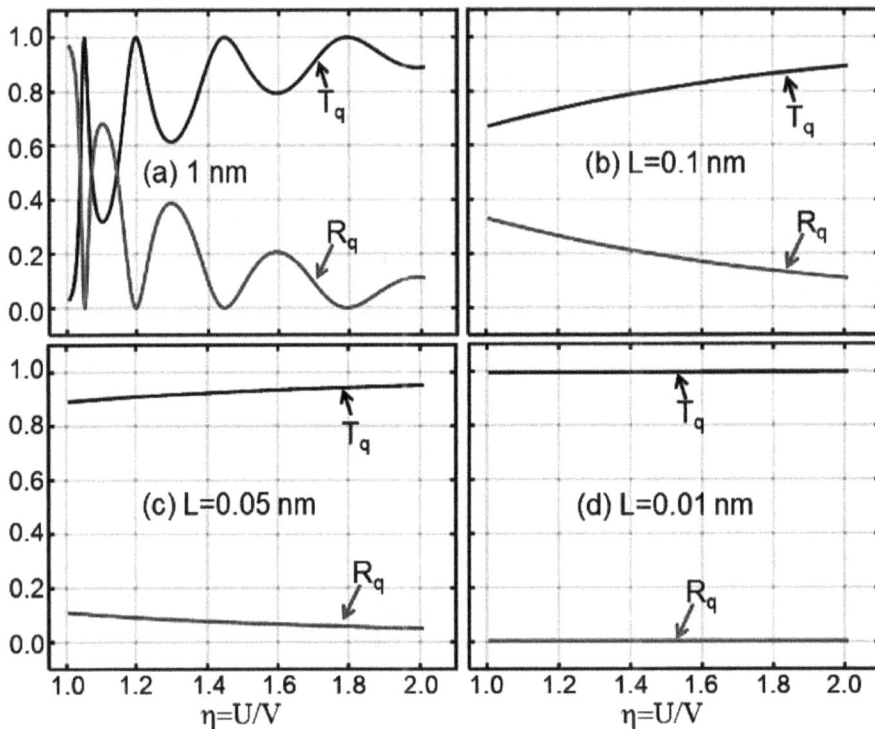

Figure 4.8. Transmission and reflection coefficients of quantum particles for hitting a rectangular potential barrier of various length with $U > V_0$.

For instance, Polonium α-decay is 5.403 MeV. So quantum particles, including alpha particles, have a small, nonzero probability of tunneling through a finitely high potential barrier, even if they do not have enough kinetic energy to do so as required by classical mechanics.

For an alpha particle, the strong but short ranged nuclear force creates a spherical finite depth well having a steep wall more or less coinciding with the nucleus surface. However, we must also include the electrostatic repulsion between the α-particle and the rest of the nucleus. So the potential energy is

$$V = k\frac{q_1 q_2}{r} = \frac{k(Z-2)e \times 2e}{r} \tag{4.53}$$

outside the nucleus (see Figure 4.9). The α-particles are emitted with spherical symmetry, so the wave function can be written $\psi(r) = u(r)/r$, as discussed above, and Schrödinger's equation is given in (4.23). Alpha particles with higher energy will penetrate through a narrower repulsion barrier since the barrier falls by $1/r$ outside the nuclear range, increasing the probability of finding an alpha particle outside the barrier. As we can see in Figure 4.9, the wave function will exponentially decay with an increase in penetration distance until it emerges from the other side of the barrier with a small but nonzero amplitude. The probability of an alpha particle emerging from the other side of the barrier exponentially decreases with an increase in the length of the barrier, potential barrier height, and particle mass.

So radioactive decay is the product of protons or neutrons escaping the nucleus by tunneling. As discussed in Ikelle (2023b), the radioactivity of minerals in the

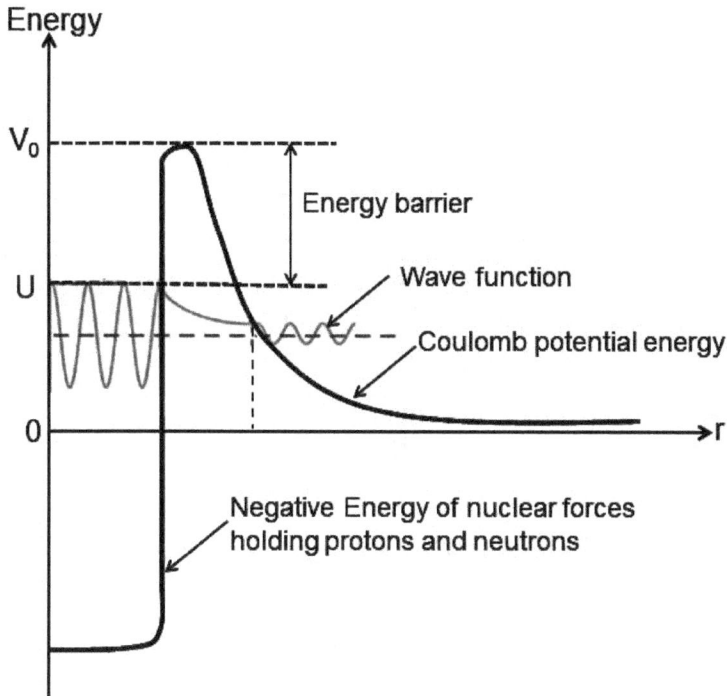

Figure 4.9. Potential energy against distance r from the nucleus center. The thick curve represents the combined effect of nuclear forces and Coulomb repulsive forces. Nuclear attractive forces make potential negative while the Coulomb forces make positive potential energy. The thin curve is wave function with an exponential decay of the quantum wave function of the alpha particle inside the barrier. However, it does not become zero.

core of the Earth provides the heat needed to keep a comfortable balance between gaining heat from the Sun and losing heat into outer space. Without quantum tunneling, the Earth would be a very cold.

Hydrogen fusion: Quantum tunneling effect. In the Sun's core, four hydrogen nuclei at very high temperatures and pressure are burned into one helium nucleus, which contains two protons and two neutrons (see Figure 4.10). This nuclear reaction is categorized as fusion because it combines four hydrogen nuclei with low masses to form one helium nucleus of a larger mass. It produces a lot of photon energy. Because of Einstein's equation $U = mc^2$, a small amount of mass gets multiplied by the square of the speed of light (a very large number), resulting in a huge amount of energy being released from a very small amount of mass (similar to a nuclear bomb). These fusion reactions are sustained over times because there are enough particles (density) that are energetic enough (temperature) and collide often enough. The energy released then begins to move by photon radiation outward toward the outer layers of the Sun. It travels extremely slowly through the radiative zone—the radiative zone-named for the primary mode of transporting energy across it (see Chapter 2)—to the convection zone.

A photon can take as long as 50 million years to travel through the radiative zone. The energy in the form of heat then moves by convection to the surface—convection is the flow of heat through a fluid (see Chapter 2), which is plasma in

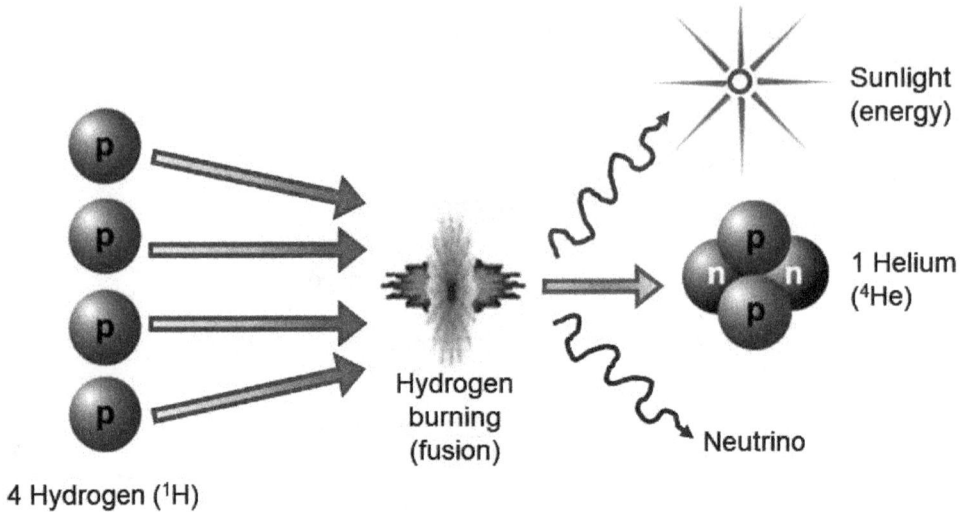

Figure 4.10. Fusion is a process whereby four hydrogen nuclei at very high temperatures and pressures are burned into one helium nuclei plus some other particles and a lot of energy. This process happens in the Sun's core. The release of energy in the form of sunlight from fusion takes 8 minutes and 20 seconds to reach Earth. Sunlight travels through space in the form of electromagnetic wave radiation.

this case. The plasma at the bottom of the convective zone is extremely hot, and it bubbles to the surface where it loses its heat to space. Once the plasma cools, it sinks back to the bottom of the convective zone. Anyway, once the energy reaches the Sun's surface it is mainly transmitted by charged particles (mainly protons, which are hydrogen ions, H^+, and electrons) to the rest of the solar system and in all directions at an average of speed of about 400 km/s (almost 1.5 million kilometers per hour), but it can range from 200 km/s to as high as 1000 km/s. These streams of charged particles are called the *solar wind* or plasma (Kivelson and Russel, 1995). The solar wind is invisible to the naked eye.

So the stars (including our Sun) works by squishing hydrogen nuclei together and turns it into helium as illustrated in Figure 4.10. Each second, our Sun manages to squish 600 million tons of hydrogen into helium and producing energy in the process. Before thermonuclear fusion process begins, the hydrogen nuclei have to overcome the Coulomb's barrier. The barrier is created by the electrostatic effect and the hydrogen nuclei have to breach this barrier to get close enough for the fission to happen. Temperature and pressure at stars are totally capable of making these nuclei fuse together in the traditional way. Quantum tunneling allows them to go right through Coulomb's energy barriers and fuse anyway. The chance of this happening is extremely low, however, since a star is composed of an uncountable number of hydrogen nuclei, this is an extremely common thing, as we are all alive and warm.

Deep space chemistry. Interstellar space (deep space) is cold (under 50 degrees Kelvin), bleak, and fairly empty. Chemical reactions slow as temperatures decrease because there is less energy to cross the energy-reaction barrier. Particles are almost completely stationary and there are very few particles thus chemical reactions are rare. So deep space is supposed to be theoretically an inhospitable environment for chemical reactions and therefore unsuitable for molecules production based on our experience on Earth. The conditions are very different from Earth. This is because there is no medium, no liquids, so traditional wet chemistry in a test tube is not relevant in deep space. And yet even in this frigid cold and low-density environment, abundant chemical reactions occur. Not only do they occur, but they generate amazingly complex organic molecules. For example, deoxyribose—a 19-atom molecule with five carbons, ten hydrogen and four oxygen atoms that is a key component of DNA—is found in interstellar space (Furukawa et al., 2019; Nuevo et al., 2023). Knowing that such complex molecules exist in deep space one has to wonder how they are constructed at deep-space temperatures where there is insufficient energy for collisions to overcome any activation barrier to reaction. The only gas-phase chemical reactions that can proceed at such low temperatures are radical-radical reactions and ion-molecule reactions, both barrier-less. Reactions that do not operate via these mechanisms use quantum tunneling. In other words, the interaction of OH with methanol can emerge on the product side of the reaction without having to go over the top of the reaction barrier.

Chemical reactions in space can occur in the gas phase (top), as part of gas–grain interactions (lower left), through exposure to ionizing radiation (energetic photons and/or particles, lower middle), and via thermal cycling (lower right). These different processes occur in different space environments and yield different populations of products.

Quantum tunneling can result in the production of complex organic molecules such as amino acids (the building blocks of proteins), sugars (the building blocks of carbohydrates), and nucleobases (the building blocks of RNA and DNA), as described in Ikelle (2023a). Quantum tunneling may prove vital molecules not only for life, but even for its origin as quantum tunneling is critical to the construction of many organic molecules.

Scanning tunneling microscopy (STM) is an extremely sensitive method for imaging surfaces at the atomic level (i.e., nanometers). As described in Chapter 3 of Ikelle (2023c), it is primarily used to study features in much smaller dimensions. Quantum tunneling is the principle behind STM. STM development in 1981 earned its inventors, Gerd Binnig and Heinrich Rohrer, the Nobel Prize in Physics in 1986 (Binnig and Rohrer, 1983).

Figure 4.11a illustrates STM with a metallic surface and a small nanometer-size tip that comes very close to the sample. The tip is generally made of high conductive metal. Different types of samples include metals and organic molecules. STM is used today in biology, geology, chemistry, and physics to describe microscopic surface arrangements at the atom level. Electron tunneling is through a potential barrier formed by a gap between the outermost atoms on the metal tip and the sample. When an electric potential difference is applied between the tip and the sample, the electrons from the sample tunnel to the tip. Basically, the electrons from the occupied states in the valence band of the sample travel through the barrier to the unoccupied states at the tip. The thick arrow in Figure 4.11b indicates electrons tunneling from filled states in the sample into empty states at the tip according to:

$$I = I_0(V) exp\left(-2z\frac{\sqrt{2m_e(V_0 - U)}}{\hbar}\right) \qquad (4.54)$$

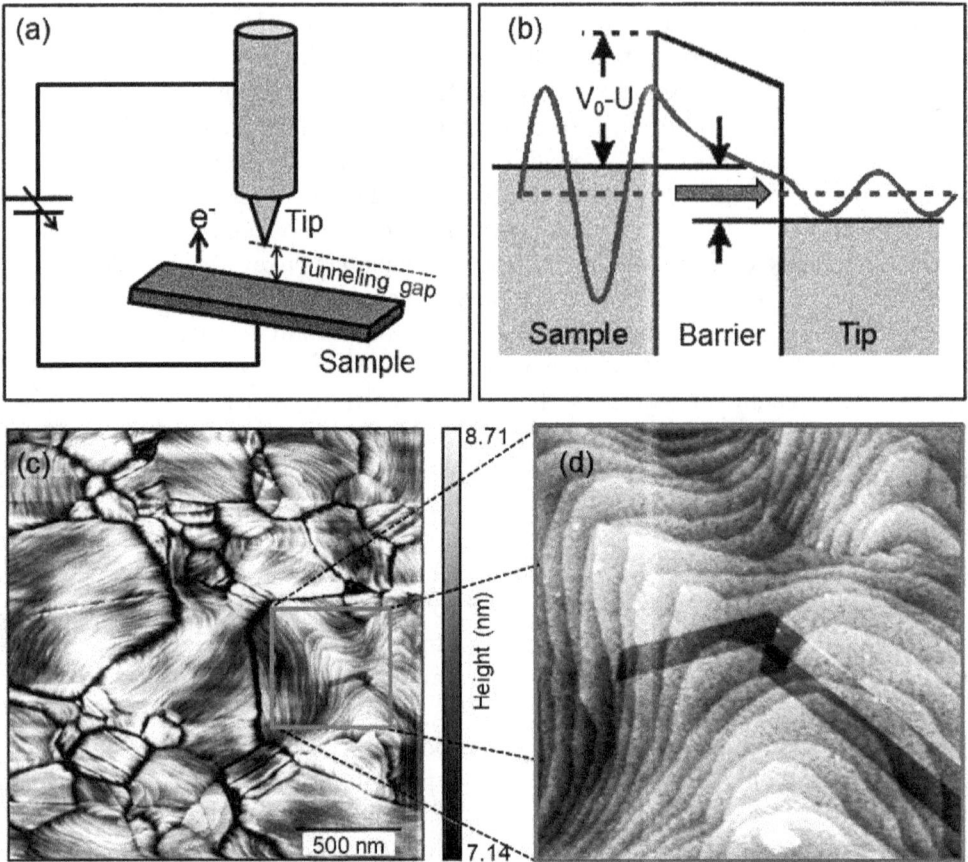

Figure 4.11. (a) A scanning tunneling microscope (STM) setup (STM). (b) An illustration of the tunnelling process. (c) High-resolution STM image of ultra-flat gold surface. (d) Zoom-in view of the image reveals terraces of gold. (c) and (d) are adapted from https://www.platypustech.com/

where z is the tunneling gap width, U is electron energy, V_0 is barrier energy, and V is external gap voltage. The small electric current produced by tunneled electrons is amplified and sent to a computer that records the tunneling current. Depending on the distance between the tip and the sample, the magnitude of the tunneling current is used to determine the surface of the specimen. The tunneling current decreases as the distance between the tip and the sample increases. It allows imaging of the surface at the atomic level and also provides a three-dimensional profile of the surface. The rapid decay of the current with distance z is due to the very small probability of tunneling across the vacuum barrier at large distances, with characteristic decay length relationship proportional to $k = \sqrt{2m_e(U - V_0)}/\hbar$, hence, the tunnel current depends exponentially on tip-surface separation. The rapid decay of the current with separation z, leads to a high resolution in the z direction. Both the tip and the sample are generally conductive or semi-conductive materials. However modern STMs allow for the analysis of a broad range of systems, including non-conducting systems as well as samples immersed in a liquid environment.

By attaching the tip to a rotator, it can move from one place to another to produce the desired scans of a given sample. To detect the specimen's surface structure, we can either have the tip at a constant height or the current is constant. Because the surface has mountains and valleys, the tip height is variable to ensure a constant current. In this way the surface image is based on the tip height. Alternatively, we can keep the tip height constant, and then use the current flow variation to form the sample image.

Figures 4.11c and 4.11d show gold surfaces imaged using a STM. The images reveal that each grain is composed of several stacked gold terraces. By close inspection of the images, we observe that the terrace-to-terrace height difference is about 0.2 nanometers.

In summary the important features of the STM technique are tunneling barrier shape and size, the applied voltage to sample and tip, and the extremely sharp conducting tip that can distinguish features smaller than 0.1 nanometer with a 0.01 nanometer depth resolution. This means that individual atoms can routinely be imaged and manipulated.

Deep space travel. In Ikelle (2024), important topics are discussed in great detail, including the motivations for interstellar travels. We have just discussed the potential for energy barriers in space and how we can deal with them. First, let us recall that the nearest star to us is Proxima Centauri, also known as Alpha Centauri because it is a member of the three star system including Alpha Centauri A and B. Proxima Centauri is a red dwarf that we can only see using telescopes. It is about 4.2 light years from Earth. Actually, we are primarily interested in visiting exoplanets associated with planetary systems rather than the stars themselves. As of 2023, Proxima Centauri has one exoplanet, Proxima Centauri b. Among the first destinations for interstellar travel will be this exoplanet. For reference the nearest galaxy to us is Andromeda, about 2.5 million light-years away.

Our journey to a potential exoplanet is underway, although we do not know which one it is. In 1977, NASA launched the Voyager I and II probes. These probes were not intended for interstellar spacecraft. They were not designed to last 46 years and counting. They are now about 0.0025 light years away. They have already escaped Sun gravity. They are not pointed toward Proxima Centauri b at this time. If they were redirected in the Proxima Centauri b direction, they would reach Proxima Centauri in 75,000 Earth years (or 75 millennia) from today. By then, our civilization may have gone extinct due to simultaneous earthquakes, volcanic eruptions, total solar flare, or nuclear Armageddon. More realistic scenarios are under investigation as discussed in Ikelle (2024).

One day, perhaps in this millennium, deep space vehicles for long distance travel to other stars will occur. The first major challenge will then be finding paths to avoid cosmic obstacles high radiation, zones newborn star, and other obstacles that are near impermeable on the basis of classical physics. One such barrier has been identified in our neighborhood, 11,500 km above Earth, by Baker et al. (2014). They discovered a nearly impenetrable barrier that can fry satellites and degrade space systems during intense solar storms. The barrier is located in the Van Allen radiation belts—a collection of charged particles gathered in place by Earth's magnetic field. The Van Allen belts themselves were detected in 1958 by Explorer 1. In the decades since, researchers have learned that their size can change or merge, or separate into three belts occasionally. But generally the inner belt stretches from 650 to 9,650 km above Earth's surface and the outer belt stretches from 13,500 to 58,000 km above the surface. The most recent data from the Van Allen probes show that the inner edge of the outer belt (at roughly 11,500 km at altitude) is, in fact, highly pronounced. For the fastest, highest-energy electrons, this edge is a sharp boundary that, under normal circumstances, electrons cannot penetrate. It's almost as if these electrons run into a glass wall in space. So in the first step, we need to develop methods of identifying potential energy barriers for spacecraft and humans. A kind of quantum barrier and tunnelling at slightly higher scales. After all, at the galactic scale, humans and aircraft can be seen as subparticles. Anyway, the laws of physics are what they are regardless of the scale and quantum physics has nothing to do with size.

4.1.3. Heisenberg's Uncertainty Principle

One of the fundamental results in quantum mechanics is Heisenberg uncertainty principle. It tells us that it is impossible to simultaneously know where a particle is located and how fast it is moving. Louis de Broglie says particles like electrons behave like waves. The wave function described in the previous subsections provides meaning for particle-like waves and Heisenberg uncertainty principle.

Let us start with a plane $\Psi(x,t) = A\cos(kx - \omega t)$. As described earlier, this wave function solves the classical particle wave equation, but not the quantum particle wave equation. First we can notice that the probability associated with $\Psi(x,t)$ is

$$P(x,t) = \int_{-\infty}^{\infty} \Psi^*(x,t)\Psi(x,t)dx = A^2 \int_{-\infty}^{\infty} \cos^2(kx - \omega t)dx = A^2 \times \infty . \quad (4.55)$$

Basically we have a non-normalizable wave function because $P(x,t)$ is not equal to one by definition of a probability density function. We can also verify that the wave velocity and the velocity of the corresponding particle are different; i.e.,

$$c_{\text{wave}} = \frac{\omega}{k} = f\lambda = \left(\frac{U}{h}\right)\left(\frac{h}{p}\right) = \frac{U}{p} = \frac{\frac{1}{2}mc^2_{\text{particle}}}{mc_{\text{particle}}} = \frac{1}{2}c_{\text{particle}} . \quad (4.56)$$

We expected $c_{\text{wave}} = c_{\text{particle}}$. This is not the case. So we cannot associate this wave function with a quantum particle because we have incorrect velocity and a non-normalizable wave function.

Consider adding two wave functions as follows:

$$\begin{aligned}
\Psi(x,t) &= \cos(kx - \omega t) + \cos\left[(k + \Delta k)x - (\omega + \Delta\omega)t\right] \\
&= 2\cos\left[\frac{\Delta k}{2}x - \frac{\Delta\omega}{2}t\right]\cos\left[\frac{(2k + \Delta k)}{2}x - \frac{\Delta\omega}{2}t\right] \\
&\approx 2\underbrace{\cos\left[\frac{\Delta k}{2}x - \frac{\Delta\omega}{2}t\right]}_{\text{modulator}}\cos(kx - \omega t) \quad (4.57)
\end{aligned}$$

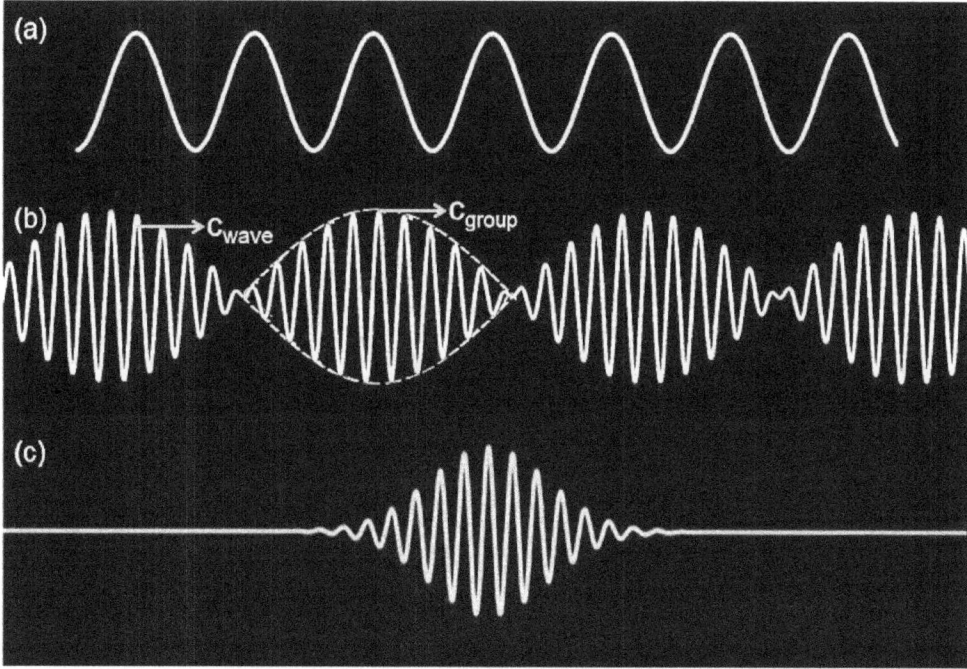

Figure 4.12. (a) This wave does not represent a quantum particle because it is not localized. (b) This wave does not represent a quantum particle because it is not localized. (c) This wave represents a quantum particle because it is localized.

In these derivations we assume that $k \gg \Delta k$ and $\omega \gg \Delta \omega$. In Figure 4.12b, we can see that the combination of these two planes produces a series of wave packets that move at different speeds from their individual components. The velocity of individual waves, associated with the second cosine and given in (4.56), is obviously different from the velocity of the whole group; i.e.,

$$c_{\text{group}} = \frac{d\omega}{dk} = \frac{d}{dk}\left(\frac{\hbar k^2}{2m}\right) = \frac{p}{m} = \frac{mc_{\text{particle}}^2}{m} = c_{\text{particle}}. \qquad (4.58)$$

So it is actually the group wave velocity that represents the particle velocity.

We still have to form a single group to ensure wave function normalization. The solution is to select a modulator in (4.57) that is nonperiodic and local so that we can obtain a single isolated wave packet. Let us consider a zero-mean Gaussian probability density as the modulator; i.e.,

$$\Psi(x) = \underbrace{\left(\frac{2\Delta x^2}{\pi}\right)^{1/4} \exp\left(-\frac{x^2}{4\Delta x^2}\right)}_{\text{modulator}} \exp\left[i\left(k_0 x - \omega t\right)\right] \qquad (4.59)$$

where Δx is the standard deviation and thus $\Delta k = 1/(2\Delta x)$ which implies that $\Delta x \Delta k = 1/2$. As we can see in Figure 4.12c, we now have a localized wave packet with Δx being the width of the wave packet. As the width of the modulator function increases Δk gets smaller. We now have a well-defined wave packet that describes particle-like particles. The probability associated with this wave function is

$$|\Psi|^2 = \Psi^*\Psi = \frac{1}{\sqrt{2\pi a}} exp\left(-\frac{x^2}{2\alpha}\right) , \tag{4.60}$$

which is the probability density function of the Gaussian distribution. Since the standard deviation provides information about the spread or width of the corresponding probability distribution. Furthermore we see that the changes in Δx and Δk are such that the product of these two always remains

$$\Delta x \Delta k = \frac{1}{2} \implies \Delta x \Delta p = \frac{\hbar}{2} . \tag{4.61}$$

This last expression tells us that there is a fundamental limit to the precision with which a particle's position and momentum can be known. As the uncertainty in the particle's position decreases the uncertainty in the particle's momentum increases. Likewise if the uncertainty in the particle's momentum decreases, the uncertainty in the particle's position increases. It is now clear that this relationship comes from the wave packets associated with quantum particles and the principle of superposition.

It is worth emphasizing that we arrived at this result using a Gaussian modulator. Since the Gaussian distribution is the maximum entropy continuous distribution when the variance is known, it comes up when we look at many particles in accordance with the central limit theorem. So the Gaussian modulator is actually a limit. For a Gaussian modulator, $\Delta x \Delta p = \hbar/2$ is the absolute minimum total uncertainty. If we repeat this analysis for other non-Gaussian probability density functions (see Chapter 2) as the modular function we will find the following inequality

$$\Delta x \Delta p \geq \frac{\hbar}{2} , \tag{4.62}$$

which is known as Heisenberg uncertainty principle. It is said that precise knowledge of position is incompatible with precise knowledge of momentum since the product of their probability-distribution uncertainties $\Delta x \Delta p$ must be greater than or equal to $\hbar/2$.

Example. Let us look at a specific application of Heisenberg uncertainty principle. We first need to recall the notion of mean and standard deviation. Suppose that we make N identical measurements of the physical parameter X which can be the particle position x or its momentum p. The mean value of an observable quantity X is

$$\langle X \rangle (t) = \int_{-\infty}^{+\infty} \Psi^*(x,t)\hat{X}\Psi(x,t)dx , \tag{4.63}$$

where $\hat{X}$ is the quantum-mechanical operator corresponding to the observable quantity X and $\Psi(x,t)$ is the quantum wave function of the particle associated with the observable quantity X. So the mean values of position and momentum are

$$\langle x \rangle (t) = \int_{-\infty}^{+\infty} \Psi^*(x,t)\, x\, \Psi(x,t)dx$$

and

$$\langle p \rangle (t) = -i\hbar \int_{-\infty}^{+\infty} \Psi^*(x,t) \frac{\partial \Psi(x,t)}{\partial x} dx ,$$ (4.64)

respectively, where $\hat{X} = x$ for position and $\hat{X} = -i\hbar\partial/\partial x$ for momentum. Determining the uncertainty of X comes down to estimating its standard deviation; i.e.,

$$\Delta X(t) = \left[\langle X^2 \rangle (t) - \langle X \rangle^2 (t) \right]^{1/2}$$ (4.65)

where

$$\langle X^2 \rangle (t) = \int_{-\infty}^{+\infty} \Psi^*(x,t) \hat{X} \hat{X} \Psi(x,t) dx .$$ (4.66)

Therefore the uncertainties in x and p are

$$(\Delta x)^2 (t) = \int [x - \langle x \rangle (t)]^2 |\Psi(x,t)|^2 \, dx$$

$$(\Delta p)^2 (t) = (i\hbar)^2 \int_{-\infty}^{+\infty} \Psi^*(x,t) \left\{ \frac{\partial^2 \Psi(x,t)}{\partial x^2} - \left[\frac{\partial \Psi(x,t)}{\partial x} \right]^2 \right\} dx .$$ (4.67)

Let us illustrate these results with the time evolution of the wave packet described in Figure 4.1. Table 4.4 shows the calculations of the formulas above. Figure 4.13 plots the results as $\Delta x \Delta p$ variation with time. We can see that $\Delta x \Delta p$ is greater than $\frac{\hbar}{2}$, irrespective of time. That is, at the time of evolution of the wave packet, the Heisenberg uncertainty principle is always satisfied. We can also see how wave packets spread as a result of the increasing uncertainty. The uncertainty in position increases with time but the uncertainty in momentum does not increase since there is no force acting on the electron to change its momentum. You can decrease the wave packet width and run a simulation. You will notice that decreasing the spatial width of the initial wave packet increases the spread of momenta. This increases the rate at which the wave packet spreads. Again, the *uncertainty principle* tells us that it is not possible to find a state in which a particle can have definite values in both position and momentum. Hence, the classical view of a particle following a well-defined trajectory does not hold in quantum mechanics.

In summary, the Heisenberg uncertainty principle tells us that particle position and momentum cannot be simultaneously and precisely determined. The uncertainty principle applies only to subatomic particles like electrons, positrons, photons, etc. For example, it does not forbid nanotechnology that deals with the position and momentum of large particles as atoms and molecules. This is because atoms and molecules are large. Quantum mechanical calculations by the Heisenberg uncertainty principle place almost no limit on how well atoms and molecules can be held in place.

Table 4.4. A Matlab code for computing $\Delta x \Delta p$ associated with the Heisenberg Uncertainty Principle.

```
clear all; close all; nx = 501; nt = 4000; L = 5e-9;
me = 9.10938291e-31; hbar = 1.054571726e-34;
x = linspace(0,L,nx); dx = x(2)-x(1); dt = 0.1*2*me*dx∧2/hbar;
psiR = zeros(nt,nx); psiI = zeros(nt,nx);
ix1 = round(nx/6); s = L/25; wL = 1.6e-10; AN= 5.3113e+04;
yR = AN*exp(-0.5.*((x-x(ix1))./s).∧2).*cos((2*pi).*(x-x(ix1))./wL);
yI = AN*exp(-0.5.*((x-x(ix1))./s).∧2).*sin((2*pi).*(x-x(ix1))./wL);
for it = 1:nt-1
    for ix = 2:nx-1
        yR(ix)=yR(ix)-0.1*(yI(ix+1)-2*yI(ix)+yI(ix-1));
    end
    psiR(it+1,:) = yR;
    for ix = 2:nx-1
        yI(ix) = yI(ix) + 0.1*(yR(ix+1)-2*yR(ix)+yR(ix-1));
    end
    psiI(it+1,:) = yI;
end
for it = 2:nt
    psiI(it,:) = 0.5*(psiI(it,:) + psiI(it-1,:));
end
sc=2*ones(nx,1); sc(2:2:nx-1)=4; sc(1)=1; sc(nx)=1; sc=((L/(nx-1))/3)*sc;
for icx=1:nt
    y = psiR(icx,:) + 1i.*psiI(icx,:); fn = conj(y).*x.*y; X1 = fn*sc;% <x>
    fn = conj(y).*x.*x.*y; X2 = fn*sc;% <x**2>
    fn = gradient(y,dx); fn = conj(y).*fn; P1=(-1i*hbar)*fn*sc; % <p>
    fn = 4*del2(y,dx); fn = conj(y).*fn; P2=hbar∧2*fn*sc; % <p**2>
    DelX = abs(sqrt(X2 - X1∧2)); DelP = abs(sqrt(P2 - P1∧2));
    DelXP0 = DelX*DelP; bnu(icx) = DelXP0; bna(icx) = 0.53e-34;
end
tta = 0:dt:(nt-1)*dt; plot(tta, bnu,'k'); hold; plot(tta, bna,'r')
grid; axis([0 (nt-1)*dt -0.53e-34 (1+0.02)*DelXP0])
```

4.2. Quantum Entanglement: Nature Inseparability

In this section, we introduce the key principles of quantum entanglement. Quantum entanglement, first recognized by Einstein, Podolsky, and Roson (1935) and Schrödinger (1935), is a physical phenomenon in which the quantum states of a many-particle system cannot mathematically be factorized into a Kronecker product of single-particle wave functions, even when the particles are separated by large distances. Entangled states have been produced in laboratories (e.g., Wu and Shaknov, 1950; Freedman and Clauser, 1972; and Aspect et al., 1983) and exploited to test the contradiction between classical local hidden variable theory and quantum mechanics by using Bell's inequality (Bell, 1964). To begin, we introduce the Dirac notation, which is used in most mathematical descriptions of quantum mechanics, including quantum entanglement.

Quantum measurements. Before we discuss quantum measurements, let us reintroduce the notions of system and state in the context of quantum physics. Again, the state of a physical system is one of the basic notions of physics. It refers to those

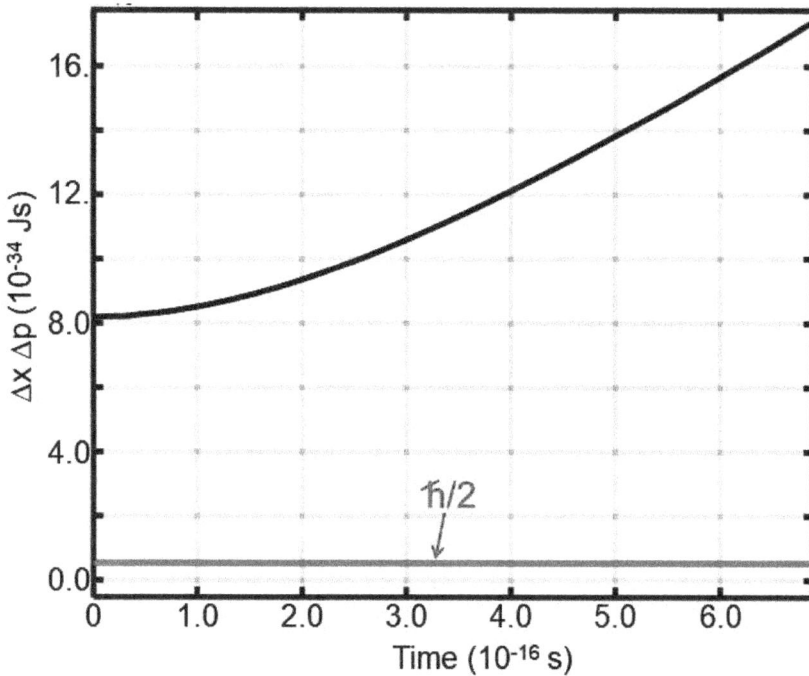

Figure 4.13. The Heisenberg Uncertainty Principle applies to wavepacks as they expand and spread over time.

aspects of a system that distinguish it from other similar systems. We humans, for instance, are made of similar particles (e.g., electrons, photons, and atoms), yet no two persons are exactly the same. In other words, a system of elementary particles can be configured to give various personalities or states.

As discussed in Ikelle (2023a), quantum-mechanics measurements only provide probabilities for the possible outcomes of a given experiment. They do not provides the actual result. Probabilities are also associated with possibilities in classical physics. However, these probabilities are just indications of insufficient information/data needed to make a definite determination. When the necessary information becomes available for some reason, the final result can be determined. In contrast, the quantum-mechanics measurements are probabilistic, irrespective "missing information" or no "missing information."

As illustrated in Figures 4.14 and 4.15, a quantum measurement consists of (1) a microscopic system whose properties are being measured, (2) the measuring apparatus which interacts with the system being measured, and (3) the environment surrounding the apparatus. The apparatus is a classical system. If the microscopic system is a single particle, the apparatus could be designed to measure its energy, or its position, or its momentum or its spin, or some other property. As in classical physics, these properties are known as observables. The key point here is that the microscopic system can be modified after quantum measurements have been taken. Let us consider the example of counting photons in a single mode cavity field by photodetectors as described by Ikelle (2023b). The mode of operation of photodetectors, which represent the classical apparatus, is to absorb a photon and create a pulse of current. So we may well be able to count the number of photons

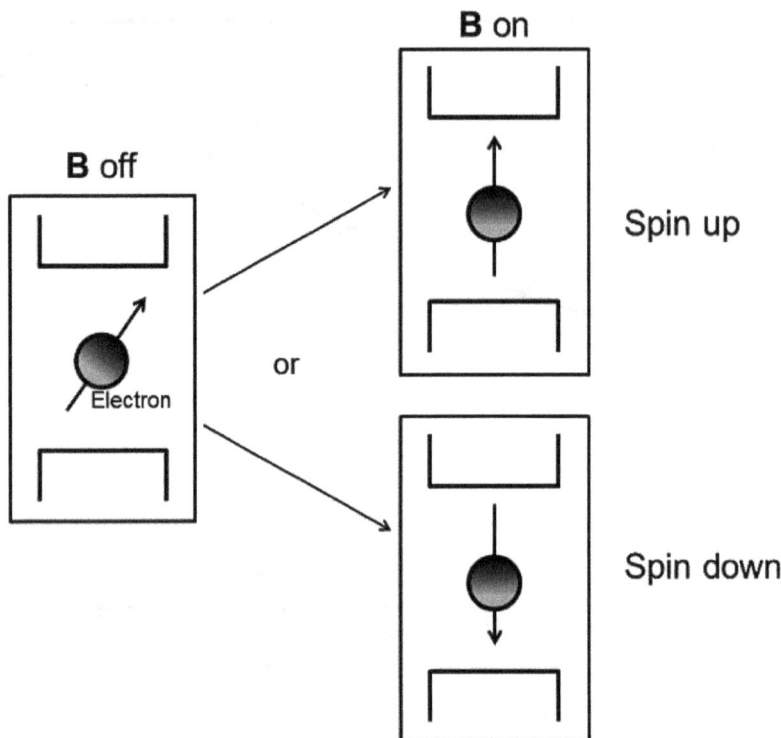

Figure 4.14. Assume that an electron is in a vacuum and not oriented by a magnetic field. When the magnetic field is active, one of two things can happen. After circling around the orientation of the magnetic field for a period of time, the electron will release its precession energy. It then points straight up or straight down.

in the field. By doing so, however, no photons are left behind after counting is done. In other words, we have performed the quantum measurement but we have left the cavity field in the vacuum state after the measurement is completed. Note, however, that we fiddle around with the process by which we put some photons back in the cavity after measurements. Let us consider a second example in which a classical apparatus measures the energy of a quantum system. The apparatus will measure the energy as if a classical system were substituted for the quantum system. After the measurement has been performed, the microscopic system survives the measurement, and even if we were to immediately remeasure the same quantity, we would get the same result.

To recap, an observable is any property of a system that can be measured; e.g., spin, polarization, position, momentum, energy, and angular momentum, magnetic moment, and so on. In classical mechanics, all observables are functions of generalized coordinates and momenta. So a knowledge of the physical state implies a knowledge of all possible observables of the system. In quantum mechanics, for every observable there exists a corresponding hermitian operator. In the next section, we will define the rule according to which knowing a physical state at some point t implies knowing the expectation value of every observable. These operators along x-, y-, and z-axes are

$$\sigma_1 = \begin{pmatrix} 0 & 1 \\ 1 & 0 \end{pmatrix}, \quad \sigma_2 = \begin{pmatrix} 0 & -i \\ i & 0 \end{pmatrix}, \quad \sigma_3 = \begin{pmatrix} 1 & 0 \\ 0 & -1 \end{pmatrix}, \tag{4.68}$$

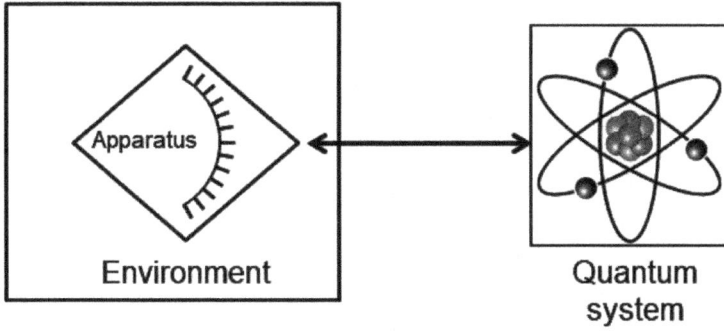

Figure 4.15. A quantum system interacts with a classical physics measuring device in the presence of its surrounding environment. The outcome of the measurement is recorded by the measuring apparatus.

where $\boldsymbol{\sigma}_1$, $\boldsymbol{\sigma}_2$, and $\boldsymbol{\sigma}_3$ are the observables of spin particles along x-, y-, and z-axes, respectively. We can verify the eigenvectors of $\boldsymbol{\sigma}_1$ are: $[1/\sqrt{2}, 1/\sqrt{2}]^T$ with eigenvalue $\lambda = 1$ and $[1/\sqrt{2}, -1/\sqrt{2}]^T$ with eigenvalue $\lambda = -1$. We can also verify that the eigenvectors of $\boldsymbol{\sigma}_2$ are: $[1/\sqrt{2}, i/\sqrt{2}]^T$ with eigenvalue $\lambda = -1$ and $[1/\sqrt{2}, -i/\sqrt{2}]^T$ with eigenvalue $\lambda = +1$; the eigenvectors of $\boldsymbol{\sigma}_3$ are: $[1, 0]^T$ with eigenvalue $\lambda = 1$ and $[0, 1]^T$ with eigenvalue $\lambda = -1$. We also have $\boldsymbol{\sigma}_1^2 = \boldsymbol{\sigma}_2^2 = \boldsymbol{\sigma}_3^2 = \boldsymbol{I}$, where $\boldsymbol{I}$ is the identity matrix. Moreover, the three matrices share the same eigenvalues and the square of each eigenvalue is 1. Because eigenvalue is a measurement and the fact that the square of each eigenvalue is 1 means that we can only measure +1 and -1. There is no eigenvectors in common between $\boldsymbol{\sigma}_1$, $\boldsymbol{\sigma}_2$, and $\boldsymbol{\sigma}_3$ which basically means that we cannot simultaneously make two measurements along two directions. In other words, the condition to measure two quantities simultaneously is that observables share the same eigenvectors. The eigenvalues are Note that for any given arbitrary vector $\boldsymbol{a} = [a_1, a_2]^T$ with $|a_1|^2 + |a_2|^2 = 1$, there exist a normal vector $\boldsymbol{n} = [n_1, n_2, n_3]^T$ (i.e., a direction) with $n_1^2 + n_2^2 + n_3^2 = 1$, such that

$$(\boldsymbol{\sigma} \cdot \boldsymbol{n}) \begin{pmatrix} a_1 \\ a_2 \end{pmatrix} = \begin{pmatrix} a_1 \\ a_2 \end{pmatrix}, \tag{4.69}$$

or

$$(n_1 \boldsymbol{\sigma}_1 + n_2 \boldsymbol{\sigma}_2 + n_3 \boldsymbol{\sigma}_3) \begin{pmatrix} a_1 \\ a_2 \end{pmatrix} = \begin{pmatrix} a_1 \\ a_2 \end{pmatrix}, \tag{4.70}$$

and the eigenvalue is 1. Consequently, the spin will only go up or down in the direction indicated by $\boldsymbol{n}$ if the above condition is met. Note also that, for $\boldsymbol{n} = [\cos\theta, 0, \sin\theta]^T$, we have

$$\boldsymbol{\sigma} \cdot \boldsymbol{n} = \begin{pmatrix} \cos\theta & \sin\theta \\ \sin\theta & -\cos\theta \end{pmatrix}. \tag{4.71}$$

You get an answer when you measure a certain quantity. That answer corresponds to some eigenvalue. You leave the system with that eigenvalue and eigenvector. So if you find the spin along the n-direction to be up you maintain the system in the state with a spin along the n-direction that is up. To recap, an observable is any property of a system that can be measured; e.g., spin.

A quantum particle has no exact location. Its location is a probability function until you measure it. Then you sort of force it to make up its mind. That is an anthropomorphic explanation, but close enough for our purposes. So, if you put a supply of quanta on one side of a barrier, they cannot cross. But because their position is uncertain, each quantum has a small, but non-zero probability of being on the wrong side of the barrier when "forced to make up its mind."

Quantum measurement methodology and understanding always involve classical-physics measurements through classical physics apparatus, including classical computers. As we will discuss in greater detail later, current designs and proposals do not envisage stand-alone quantum computers. They require peripheral classical computers. This requirement may be another sign of incomplete quantum mechanics formulation.

Box 4.1: Kronecker Product

The Kronecker product, denoted by $\otimes$, is an operation on two matrices of arbitrary size, resulting in a block matrix. It is a special case of a tensor product. If A is an $M \times N$ matrix and B is a $P \times Q$ matrix, then $A \otimes B$ is the $MP \times NQ$ matrix; i.e.,

$$A \otimes B = \begin{pmatrix} a_{11}B & \dots & a_{1n}B \\ \vdots & \ddots & \vdots \\ a_{m1}B & \dots & a_{mn}B \end{pmatrix}$$

$$= \begin{pmatrix} a_{11}b_{11} & \dots & a_{11}b_{1q} & \dots & \dots & a_{1n}b_{11} & \dots & a_{1n}b_{1q} \\ a_{11}b_{21} & \dots & a_{11}b_{2q} & \dots & \dots & a_{1n}b_{21} & \dots & a_{1n}b_{2q} \\ \vdots & \ddots & \vdots & & & \vdots & \ddots & \vdots \\ a_{11}b_{p1} & \dots & a_{11}b_{pq} & \dots & \dots & a_{1n}b_{p1} & \dots & a_{1n}b_{pq} \\ \vdots & & \vdots & \ddots & & \vdots & & \vdots \\ \vdots & & \vdots & & \ddots & \vdots & & \vdots \\ a_{m1}b_{11} & \dots & a_{m1}b_{1q} & \dots & \dots & a_{mn}b_{11} & \dots & a_{mn}b_{1q} \\ a_{m1}b_{21} & \dots & a_{m1}b_{2q} & \dots & \dots & a_{mn}b_{21} & \dots & a_{mn}b_{2q} \\ \vdots & \ddots & \vdots & & & \vdots & \ddots & \vdots \\ a_{m1}b_{p1} & \dots & a_{m1}b_{pq} & \dots & \dots & a_{mn}b_{p1} & \dots & a_{mn}b_{pq} \end{pmatrix}.$$

Let now consider two vectors: $x = [x_1, ..., x_M]^T$ and $y = [y_1, ..., y_N]^T$. Their product is

$$x \otimes y = [x_1 y^T, ..., x_M y^T]^T = [x_1 y_1, x_1 y_2, ..., x_1 y_N, x_2 y_1, ..., x_M y_N]^T \;;$$

e.g., $[1, 0]^T \otimes [0, 1]^T = [0, 1, 0, 0]^T$. The Matlab function for performing this product is kron.

Table 4.5. Dirac notations.

Vector	$\lvert a \rangle = [a_1, a_2, ..., a_n]^T$	a is a column vector
Scaled vector	$\alpha \lvert a \rangle = [\alpha a_1, \alpha a_2, ..., \alpha a_n]^T$	α is a complex value
Matrix on vector	$M \lvert a \rangle = m_{ij} a_j$	M is a matrix
Dual vector	$\langle b \rvert = [b_1^*, b_2^*, ..., b_n^*]$	b is a row vector
Inner product	$\langle b \lvert a \rangle = b_i^* a_i = \langle a \lvert b \rangle^*$	$\langle b \lvert a \rangle = \langle a \lvert b \rangle^*$
Inner product	$\langle b \lvert M \lvert a \rangle = \langle a \lvert M \lvert b \rangle^*$	M is Hermitian
Expectation of M	$\langle a \lvert M \lvert a \rangle = \langle a \lvert M \lvert a \rangle^*$	
Eigen equation	$M \lvert a \rangle = \lambda_a \lvert a \rangle$	λ_a is eigen-value
Eigenvalue	$\lambda_a = \langle a \lvert M \lvert a \rangle / \langle a \lvert a \rangle$	λ_a is real value
Outer product	$\lvert a \rangle \lvert b \rangle = \lvert a \rangle \otimes \lvert b \rangle$	Tensor product
Outer product	$\lvert a \rangle \langle b \rvert = a_i b_j^*$	Matrix

4.2.1. Dirac Notations

Dirac's bra & ket notation is a very convenient notation that enhances the clarity of mathematical operations. As illustrated in Table 4.5, Dirac notation allows us to easily distinguish between scalars, vectors, and operators and to easily construct inner products, outer products, and tensor products. The name bra & ket notation comes from the word bracket (simply because it is like splitting a bracket into two halves < | and | >). The colon vector, $\boldsymbol{a}$, is denoted as $\lvert a \rangle$, the row vector, $\boldsymbol{b}$, is denoted as $\langle b \rvert$, the inner product of a and b is denoted $\langle b \lvert a \rangle$. Let $\boldsymbol{M}$ be a matrix, $c = M \lvert a \rangle$ is vector and $\langle b \lvert M \lvert a \rangle = \langle b \lvert c \rangle$ is also an inner product. When M is a hermitian matrix (i.e., $m_{ij} = m_{ji}^*$, where m_{ij} are the components of M and the diagonal of M is, by definition, made of real values only), then we have $\langle b \lvert M \lvert a \rangle = \langle a \lvert M \lvert b \rangle^*$. The eigenvalues of an Hermitian operator are real since the inner product of any vector with itself is a real number. Notice that, for an Hermitan matrix M with two distant eigenvalues—that is, $M \lvert a \rangle = \lambda_a \lvert a \rangle$ and $M \lvert b \rangle = \lambda_b \lvert b \rangle$—if $\lambda_a \neq \lambda_b$ then $\boldsymbol{a}$ and $\boldsymbol{b}$ are orthogonal. Suppose that a and b are eigenstates of an Hermitian operator M, and $\lambda_a \neq \lambda_b$. We can verify that

$$M \lvert a \rangle = \lambda_a \lvert a \rangle \implies \langle b \lvert M \lvert a \rangle = \lambda_a \langle b \lvert a \rangle \quad \text{and} \quad M \lvert b \rangle = \lambda_b \lvert b \rangle$$

or

$$\langle b \lvert M \lvert a \rangle = \langle M b \lvert a \rangle \iff \langle b \lvert \lambda_a \lvert a \rangle = \langle \lambda_b b \lvert a \rangle \iff (\lambda_a - \lambda_b)\langle b \lvert a \rangle = 0 .$$

As $\lambda_a \neq \lambda_b$, $\boldsymbol{a}$ and $\boldsymbol{b}$ are orthogonal. Also if two matrices, M and N, share a common set of eigenvectors, they commute; i.e., $[M, N] = MN - NM = 0$ then $MN = NM$. If two eigenstates of an Hermitian operator correspond to different eigenvalues, then the two eigenstates are orthogonal. We can verify that

$$M \lvert a_i \rangle = \lambda_i^{(a)} \lvert a_i \rangle \iff NM \lvert a_i \rangle = \lambda_i^{(a)} N \lvert a_i \rangle = \lambda_i^{(a)} \lambda_i^{(b)} \lvert a_i \rangle$$

while

$$N \lvert a_i \rangle = \lambda_i^{(b)} \lvert a_i \rangle \iff MN \lvert a_i \rangle = \lambda_i^{(b)} M \lvert a_i \rangle = \lambda_i^{(b)} \lambda_i^{(a)} \lvert a_i \rangle$$

therefore

$$[M, N] \lvert a_i \rangle = (MN - NM) \lvert a_i \rangle = \left(\lambda_i^{(b)} \lambda_i^{(a)} - \lambda_i^{(b)} \lambda_i^{(a)} \right) \lvert a_i \rangle = 0 .$$

In quantum mechanics, for every observable M there exists a corresponding hermitian operator. Knowledge of the physical state a implies a knowledge of the expectation value of every observable, according to the rule

$$< M >=< a|M|a > \, ,$$

$< M >$ is the average or expected value of an observable for a system in state $|a\rangle$. Although the result of a single measurement is probabilistic, we are usually interested in the average outcome, which gives us more information about the system and the observable.

Remember that the value of the measurement of an observable (e.g., matrices in (4.68)) is one of the observable eigenvalues. The probability of obtaining one particular eigenvalue is given by the modulus square of the inner product of the state vector of the system with the corresponding eigenvector. The state of the system immediately after the measurement is the normalized projection of the state prior to the measurement onto the eigenvector subspace. If B is an observable with eigenvalues $\lambda_k^{(b)}$ and eigenvectors $|b_k\rangle$ [i.e., $B|b_k\rangle = \lambda_k^{(b)}$], given a system in the state $|\psi\rangle$, the probability of obtaining $\lambda_k^{(a)}$ as the outcome of the measurement of B is

$$p(b_k) = |\langle b_k|\psi\rangle|^2 = \langle \psi|P_k|\psi\rangle \, , \qquad (4.72)$$

where $P_k = |b_k\rangle\langle b_k|$.

So, in quantum mechanics, we mostly deal with vectors and linear operators on vectors. The vectors are represented by bras and kets. Hence the attraction of vectors in mathematical descriptions of quantum mechanics, which are probably the most convenient and efficient notations for vectors and operators on vectors. So, in contrast to the classical obscure view, at least from a mathematical point of view, quantum mechanism is not so obscure. It essentially uses a linear complex-valued vector space with scalar products, which is by definition a complex Hilbert space and encompasses linear transformations through matrices acting on vectors. Note that we use complex numbers in quantum mechanics to effectively deal with time and reversibility, especially the evolution of states with time.

In quantum mechanics pure states are unit vectors in a Hilbert space—that is, each direction corresponds to a single unique physical state. For example, after a spin measurement along the z-direction we know with certainty the state of the spin of the system along this direction. Such states are referred to as pure states.

Example: qubit system. Consider two vectors $|u\rangle = [1,0]^T$ and $|d\rangle = [0,1]^T$ representing the quantum bit, also known as the qubit. Vector $|u\rangle$ can be interpreted as a spin of electron pointing up and vector $|d\rangle$ as a spin of an electron pointing down. We can definite an electron *state* as $|a\rangle = a_u|u\rangle + a_d|d\rangle$, where $|\langle u|a\rangle|^2 = a_u a_u^*$ is the probability of finding the electron pointing up and $|\langle d|a\rangle|^2 = a_d a_d^*$ is the probability of finding the electron pointing down. Because the sum of probability must be 1, we have the following condition $\langle a|a\rangle = a_u a_u^* + a_d a_d^* = 1$. We can verify that $\langle u|u\rangle = \langle d|d\rangle = 1$ and $\langle u|d\rangle = \langle d|u\rangle = 0$. So vectors u and d are orthogonal, as one may have excepted. Notice that $a_u = a_d = 1/\sqrt{2}$ is a valid choice–the electron points horizontally in this case—but $a_u = a_d = 1$ is not because the sum of probabilities is not 1. Notice also that knowing the probabilities is not enough to determine the coefficients a_u and a_d; e.g.,

$$|a\rangle = \cos(\theta_1/2)\,|u\rangle + \exp(i\phi_1)\sin(\theta_1/2)\,|d\rangle \tag{4.73}$$

with $a_u = \cos(\theta_1/2)$, $a_d = \exp(i\phi_1)\sin(\theta_1/2)$, $|a_u|^2 = \cos^2(\theta_1/2)$, and $|a_d|^2 = \sin^2(\theta_1/2)$. This choice makes sense because the sum of probabilities is 1. However we cannot recover a_d from $|a_d|^2$.

Box 4.2: The Schröodinger Equation in Dirac Notation

Our objective is to rewrite the Schröodinger equation described in the previous section in Dirac notation. We can do so by transforming the differential equation into a matrix equation. If we divide the x into a grid of N equally spaced points $[x_1, x_2, ..., x_N]$, we can express the wave function as $\Psi(x,t) = [\Psi(x_1,t), \Psi(x_2,t), ..., \Psi(x_N,t)]^T$. The Schrödinger equation gives the time-evolution of some wave function due to some Hamiltonian:

$$i\hbar\frac{\partial}{\partial t}\,|\psi\rangle = \hat{H}\,|\Psi\rangle \tag{4.74}$$

where $\hat{H}$ is the Hamiltonian (i.e., the sum of the kinetic and potential energies) defined in the previous section. If we separate the time-dependent part of the wave function from the spatial part of the wave function, we can analyze the eigenvalues of the spatial wave function using the time-independent Schrödinger equation:

$$\hat{H}\,|\psi\rangle = U\,|\psi\rangle \tag{4.75}$$

where $\psi(x) = [\psi(x_1), \psi(x_2), ..., \psi(x_N)]^T$. We can find exact solutions to the time-independent Schrödinger equation for simple potentials (square wells, square barriers, harmonic oscillators, etc.). However, we usually cannot find an exact solution for more complicated potential distributions. In these cases, we have to turn to numerical methods as described in Appendix A.

A qubit is the quantum analog of the classical bit, which is generally represented as being 0 and 1 or 1 and -1 in ordinary computers. In contrast, a qubit is the simplest interesting quantum system and a fundamental unit of information used in quantum computers (Martin-Lopez et al., 2012). Notice also that superposition refers to the fact that any vector is a linear combination of two quantum states. Linear combinations of qubits are also valid quantum states once normalized.

The outer product of two states is an important operation that outputs a matrix given two states. The outer product of the two states (see also Table 4.5) is generally denoted as $|a\rangle\langle b|$; e.g.,

$$|a\rangle\langle b| = [a_1, a_2]^T [b_1^*, b_2^*] = \begin{pmatrix} a_1 b_1^* & a_1 b_2^* \\ a_2 b_1^* & a_2 b_2^* \end{pmatrix}.$$

In this notation, any matrix can be written as a linear combination of outer products between bit-string states. For a 2×2 matrix, we have

$$\boldsymbol{A} = \begin{pmatrix} a_{00} & a_{01} \\ a_{10} & a_{11} \end{pmatrix} = a_{00}\,|u\rangle\langle u| + a_{01}\,|u\rangle\langle d| + a_{10}\,|d\rangle\langle u| + a_{11}\,|d\rangle\langle d|,$$

and the observables in (4.68), also known as the Pauli operators, can be expressed in terms of qubits as follows:

$$\sigma_1 = |u\rangle \langle d| + |d\rangle \langle u| \,, \quad \sigma_2 = -i\,|u\rangle \langle d| + i\,|d\rangle \langle u| \,, \quad \text{and} \quad \sigma_3 = |u\rangle \langle u| - |d\rangle \langle d| \,.$$

Together with the identity operator, the Pauli operators form a basis for two qubit Hermitian operators, i.e., any single qubit operator can be written as a linear combination of σ_1, σ_2, σ_3, and I.

The classic terminology of quantum states are (i) *basis states* or *classical states*, which are individual states in some computational basis, such as $|u\rangle$ and $|d\rangle$, (ii) *pure states* which are superpositions of basis states, such as $|a\rangle = a_u |u\rangle + a_d |d\rangle$, and (iii) *mixed states* which are classical probability distributions over pure states $\rho = \sum_i p_i |a\rangle \langle b|$. We further describe and discussed mixed states in the next subsection.

Example: Multi-qubit systems. The mathematical structure of a qubit generalizes to higher dimensional quantum systems for applications such as quantum computers which contain many numbers of qubits. So it is necessary to know how to construct the combined state of a system of qubits given the states of the individual qubits. The joint state of a system of qubits is described using the tensor product, $\otimes$. Mathematically, taking the tensor product of two states is the same as taking the Kronecker product of their corresponding vectors. Say we have two single qubit states $|a\rangle = [a_1, a_2]^T$ and $|a'\rangle = [a'_1, a'_2]^T$. Then the full state of a system composed of two independent qubits is given by,

$$|aa'\rangle = |a\rangle \otimes |a'\rangle = [a_1 a'_1, a_1 b'_2, b_2 a'_1, b_2 b'_2]^T \,.$$

Sometimes the $\otimes$ symbol is dropped. For example, $|a\rangle \otimes |a'\rangle$ is shortened to $|aa'\rangle$, and $|u\rangle \otimes |u\rangle \otimes |u\rangle$ is shortened to $|uuu\rangle$.

Consider the tensor product of two qubits, $|a\rangle_1$ and $|a\rangle_2$ residing in the Hilbert spaces $\mathcal{H}_1$ and $\mathcal{H}_2$, respectively. The basis vectors of the composite system, $\mathcal{H} = \mathcal{H}_1 \otimes \mathcal{H}_2$, are

$$|uu\rangle = |u\rangle \otimes |u\rangle = [1,0,0,0]^T \,, \quad |ud\rangle = |u\rangle \otimes |d\rangle = [0,1,0,0]^T \,,$$

$$|du\rangle = |d\rangle \otimes |u\rangle = [0,0,1,0]^T \,, \quad |dd\rangle = |d\rangle \otimes |d\rangle = [0,0,0,1]^T \,,$$

and

$$|ddu\rangle = |d\rangle \otimes |d\rangle \otimes |u\rangle = [0,0,0,0,0,0,1,0]^T \,,$$

the first vector of, say $|du\rangle$, represents electron 1 and the second vector indicates electron 2. Using these basis states we can represent a general two-qubit state as

$$|a_{12}\rangle = \alpha_0 |uu\rangle + \alpha_1 |ud\rangle + \alpha_2 |du\rangle + \alpha_3 |dd\rangle \tag{4.76}$$

where

$$|\alpha_0|^2 + |\alpha_1|^2 + |\alpha_2|^2 + |\alpha_3|^2 = 1 \,, \tag{4.77}$$

Table 4.6. Transforms of qubits. The pair, say, (up, up) corresponding spins of (electron 1, electron 2). These τ-matrices are analog to *sigma*-matrices. σ_1, σ_2, and σ_3 act on electron 1 only and τ_1, τ_2 and τ_3 act on electron 2.

(up, up)	(up, down)	(down, up)	(down, down)
$\sigma_3\|uu\rangle = \|uu\rangle$	$\sigma_3\|ud\rangle = \|ud\rangle$	$\sigma_3\|du\rangle = -\|du\rangle$	$\sigma_3\|dd\rangle = -\|dd\rangle$
$\sigma_2\|uu\rangle = i\|du\rangle$	$\sigma_2\|ud\rangle = i\|dd\rangle$	$\sigma_2\|du\rangle = -i\|uu\rangle$	$\sigma_2\|dd\rangle = -i\|dd\rangle$
$\sigma_1\|uu\rangle = \|du\rangle$	$\sigma_1\|ud\rangle = \|dd\rangle$	$\sigma_1\|du\rangle = \|uu\rangle$	$\sigma_1\|dd\rangle = \|dd\rangle$
$\tau_3\|uu\rangle = \|uu\rangle$	$\tau_3\|ud\rangle = -\|ud\rangle$	$\tau_3\|du\rangle = \|du\rangle$	$\tau_3\|dd\rangle = -\|dd\rangle$
$\tau_2\|uu\rangle = i\|ud\rangle$	$\tau_2\|ud\rangle = -i\|uu\rangle$	$\tau_2\|du\rangle = i\|dd\rangle$	$\tau_2\|dd\rangle = -i\|du\rangle$
$\tau_1\|uu\rangle = \|ud\rangle$	$\tau_1\|ud\rangle = \|uu\rangle$	$\tau_1\|du\rangle = \|dd\rangle$	$\tau_1\|dd\rangle = \|du\rangle$

and α_i are complex numbers. Here is another superposition (i.e., state):

$$|a'\rangle = \frac{1}{\sqrt{2}}|u\rangle|u\rangle + \frac{1}{\sqrt{2}}|d\rangle|d\rangle \qquad (4.78)$$

This state means that, if we measure the spin of the 2-particle system, we have a probability of $1/2$ to get $|u\rangle|u\rangle$ (e.g., both particles have spin up) and a probability of $1/2$ to get $|d\rangle|d\rangle$ (e.g., both particles have spin down).

Let us label the spin operators for electron 1 as $\boldsymbol{\sigma}_1$, $\boldsymbol{\sigma}_2$, and $\boldsymbol{\sigma}_3$, as described in (4.68) and electron 2 as $\boldsymbol{\tau}_1$, $\boldsymbol{\tau}_2$, and $\boldsymbol{\tau}_3$. The actions of these operators are shown in Table 4.6.

In general, the dimension of a composite qubit system consisting of N qubits will be 2^N, i.e., the dimension of product spaces grows exponentially. Likewise, a state consisting of N qubits can be written as follows:

$$|a_N\rangle = \alpha_0 |uu\ldots uuu\rangle + \alpha_1 |uu\ldots uud\rangle + \ldots + \alpha_{N-1} |dd\ldots ddd\rangle \quad .$$

Also note that the superposition principle is still applicable in these cases. In summary, we used the Kronecker product to construct multiqubits. The growth of vectors and matrices as we move from one qubit to two and from two to three is an indication of the potential of quantum computers to have capacities that surpass classical computers in storage and solving a number of problems.

> **Quiz:** An operator that measures spin along the direction $\boldsymbol{n} = [\cos\theta, 0, \sin\theta]^T$ is given in (4.71). Show that it has eigenvectors $|\theta_+\rangle = \cos(\theta/2)|u\rangle + \sin(\theta/2)|d\rangle$ and $|\theta_-\rangle = -\sin(\theta/2)|u\rangle + \cos(\theta/2)|d\rangle$.

Quantum measurement. As described in the previous subsection, a quantum corresponds to transforming the quantum information (stored in a quantum system) into classical information. Measurement of a qubit, $|a\rangle$, corresponds typically to reading out a classical bit, which is 0 or 1. One of the key features of quantum mechanics is that measurement outcomes are probabilistic. The probability of obtaining $|u\rangle$ after measurement is $|\langle u|a\rangle|^2$ and the probability of obtaining $|d\rangle$ after measurement is $|\langle d|a\rangle|^2$. Now consider a three qubit, $|a_3\rangle$ state, but we only measure the first qubit and leave the other two qubits undisturbed. The probability of obtaining $|u\rangle$ in the first qubit is

$$|\langle uuu|a_3\rangle|^2 + |\langle uud|a_3\rangle|^2 + |\langle udu|a_3\rangle|^2 + |\langle udd|a_3\rangle|^2 .$$

The state of the system after this measurement will be obtained by normalizing the state,

$$\langle uuu|a_3\rangle\,|uuu\rangle + \langle uud|a_3\rangle\,|uud\rangle + \langle udu|\phi\rangle\,|udu\rangle + \langle udd|a_3\rangle\,|udd\rangle .$$

If $a_3 = (1/\sqrt{2})(|uuu\rangle + |ddd\rangle)$, we see that the probability of getting $|u\rangle$ in the first qubit will be 0.5, and if this result is obtained, the final state of the system would change to $|uuu\rangle$. On the other hand, if we were to measure $|d\rangle$ in the first qubit we would end up with a state $|ddd\rangle$. Note that, if as a consequence of a measurement, the state of the system undergoes a discontinuous change of state, the instantaneous change in state is known as *the collapse of the state vector*.

Quantum Gates. Classical computers use gates to compute bits. There are also gates for quantum computers for working with qubits. In quantum information theory, including gate-based quantum computing, we often refer to small unitary transformations as *gates*. Gates can transform qubits at each computing step. We use some of these gates so often that they have special names and symbols associated with them. For example, the matrix

$$\text{NOT} = X = \begin{pmatrix} 0 & 1 \\ 1 & 0 \end{pmatrix}$$

is called the NOT gate. It flips bits. We can verify that $\text{NOT}\,|u\rangle = |d\rangle$. The $\sqrt{\text{NOT}}$ gate,

$$\sqrt{\text{NOT}} = \frac{1}{2}\begin{pmatrix} 1+i & 1-i \\ 1-i & 1+i \end{pmatrix} ,$$

satisfies the property that

$$\sqrt{\text{NOT}}\sqrt{\text{NOT}} = \text{NOT} . \tag{4.79}$$

The Hadamard gate is

$$H = \frac{1}{\sqrt{2}}\begin{pmatrix} 1 & 1 \\ 1 & -1 \end{pmatrix} . \tag{4.80}$$

The Hadamard gate is useful because it maps the $\{|u\rangle, |d\rangle\}$ basis to the $\{|+\rangle, |-\rangle\}$ basis, and vice versa; i.e.,

$$H\,|u\rangle = |+\rangle = \frac{(|u\rangle + |d\rangle)}{\sqrt{2}} , \quad H\,|d\rangle = |-\rangle = \frac{(|u\rangle - |d\rangle)}{\sqrt{2}} , \tag{4.81}$$

$H\,|+\rangle = |u\rangle$ and $H\,|-\rangle = |d\rangle$. We can that the Hadamard gate create superposition. Another example of a basis changing gate is

$$\frac{1}{\sqrt{2}}\begin{pmatrix} 1 & 1 \\ i & -i \end{pmatrix} \tag{4.82}$$

This gate switches the $\{|u\rangle, |d\rangle\}$ basis to the $\{|i\rangle, |-i\rangle\}$ basis.

The Hadamard gate is used in quantum computation to create superpositions. Equation (4.81) shows two such examples. The minus sign in the second expression distinguishes between classical and quantum computing. Let us apply the Hadamard gate to something that is really a superposition, say, $(|u\rangle + |d\rangle)/\sqrt{2}$, we arrive at

$$H\frac{1}{\sqrt{2}}\left(|u\rangle + |d\rangle\right) = \frac{1}{\sqrt{2}}H\,|u\rangle + \frac{1}{\sqrt{2}}H\,|d\rangle \;. \tag{4.83}$$

This is a small example of the interference effect. Quantum interference is when subatomic particles interact with and influence themselves and other particles while in a probabilistic superposition state. Quantum states can be influenced by it when measured. Quantum interference enables certain algorithms to speed up over classical computing methods. Quantum interference arises from the wave-like nature of quantum particles, such as electrons or photons, and it is a phenomenon where quantum states can combine constructively or destructively, leading to unique computational advantages.

One of the most common two-qubit operations we will encounter in quantum computing is the CNOT gate. This flips the second bit if and only if the first bit is 1. The matrix corresponding to this operation is

$$\mathrm{CNOT} = \begin{pmatrix} 1 & 0 & 0 & 0 \\ 0 & 1 & 0 & 0 \\ 0 & 0 & 0 & 1 \\ 0 & 0 & 1 & 0 \end{pmatrix}. \tag{4.84}$$

We can verify that

$$\mathrm{CNOT}\,|du\rangle = \mathrm{CNOT}\left[\begin{pmatrix} 0 \\ 1 \end{pmatrix} \otimes \begin{pmatrix} 1 \\ 0 \end{pmatrix}\right] = \begin{pmatrix} 1 & 0 & 0 & 0 \\ 0 & 1 & 0 & 0 \\ 0 & 0 & 0 & 1 \\ 0 & 0 & 1 & 0 \end{pmatrix}\begin{pmatrix} 0 \\ 0 \\ 1 \\ 0 \end{pmatrix}$$

$$= \begin{pmatrix} 0 \\ 0 \\ 0 \\ 1 \end{pmatrix} = \begin{pmatrix} 0 \\ 1 \end{pmatrix} \otimes \begin{pmatrix} 0 \\ 1 \end{pmatrix} = |dd\rangle$$

$$\mathrm{CNOT}\,|dd\rangle = \mathrm{CNOT}\left[\begin{pmatrix} 0 \\ 1 \end{pmatrix} \otimes \begin{pmatrix} 0 \\ 1 \end{pmatrix}\right] = \begin{pmatrix} 1 & 0 & 0 & 0 \\ 0 & 1 & 0 & 0 \\ 0 & 0 & 0 & 1 \\ 0 & 0 & 1 & 0 \end{pmatrix}\begin{pmatrix} 0 \\ 0 \\ 0 \\ 1 \end{pmatrix}$$

$$= \begin{pmatrix} 0 \\ 0 \\ 1 \\ 0 \end{pmatrix} = \begin{pmatrix} 0 \\ 1 \end{pmatrix} \otimes \begin{pmatrix} 1 \\ 0 \end{pmatrix} = |du\rangle \;.$$

Note that the gate matrices introduced above are all reversible. Actually, all the gate matrices used today in quantum computing are reversible. Quantum computing does not have an AND gate because such gate is not reversible. Advanced

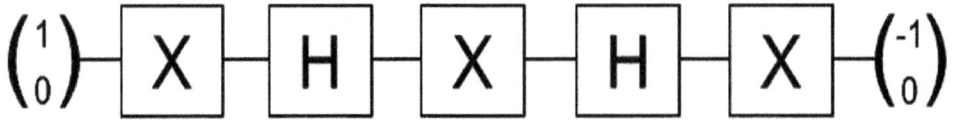

Figure 4.16. An example of quantum computing where X is the NOT gate and H is the Hadamard gate.

computation is based in the construction chains of logic gates, as illustrated in Figure 4.16.

Quiz: (1) Determine $H \otimes H \ket{ud}$ as a function of $\ket{-}$ and $\ket{-}$.
(2) We consider a system of n qubits. A number of quantum-computing algorithms use the n-bit Hadamard transformation which is the tensor product of a Hadamard transformation for each bit:

$$W_n = \underbrace{H \otimes H \otimes H \otimes \dots \otimes H}_{n \text{ times}} ,$$

where W_n is the Walsh-Hadamard transformation. Determine $W_3 \ket{uuu}$ function of the Walsh-Hadamard transformation.

4.2.2.   A Formulation of Quantum Entanglement

Quantum entanglement is one of the key components of modern quantum information theory, with endless potential applications. How can we define and quantify that the state of our quantum system is entangled? The goal of this subsection is to answer these question, starting with the definition of entangled state and Schmidt decomposition. This decomposition provides one of the simplest methods for determining whether a pure state is separable or entangled. Gühne and Tóth (2009) and Horodecki et al. (2009) are two of the key references used in developing this subsection.

Mathematical definition of quantum entanglement Let us consider two electrons with the following states: $\ket{a_1} = \alpha_1 \ket{u} + \beta_1 \ket{d}$ and $\ket{a_2} = \beta_1 \ket{u} + \beta_2 \ket{d}$ with $\alpha_1^2 + \alpha_2^2 = 1$ and $\beta_1^2 + \beta_2^2 = 1$. Let us emphasize that each electron has its own state. In mathematical terms, $\ket{a_1}$ is a two-dimensional vector of $\mathcal{H}_1$ and $\ket{a_2}$ is a two-dimensional vector of $\mathcal{H}_2$. The two electrons can be written as the product of states:

$$\ket{a_1 a_2} = \alpha_1 \alpha_2 \ket{uu} + \alpha_1 \beta_2 \ket{ud} + \beta_1 \alpha_1 \ket{du} + \beta_1 \beta_2 \ket{dd} . \tag{4.85}$$

We have *a state of two separate states*, which is also known as a product state; entangled particles cannot be factored into completely separate local terms without reference to the other. In quantum physics world, it is common to call electron 1 Alice's electron and electron 2 Bob's electron. Then for a product state Alice's and Bob's electrons are independent of one another. (In terms of their states, that is. The electrons will still interact electrically and magnetically with one another.) The fact that they are independent means that Alice can do whatever she want with

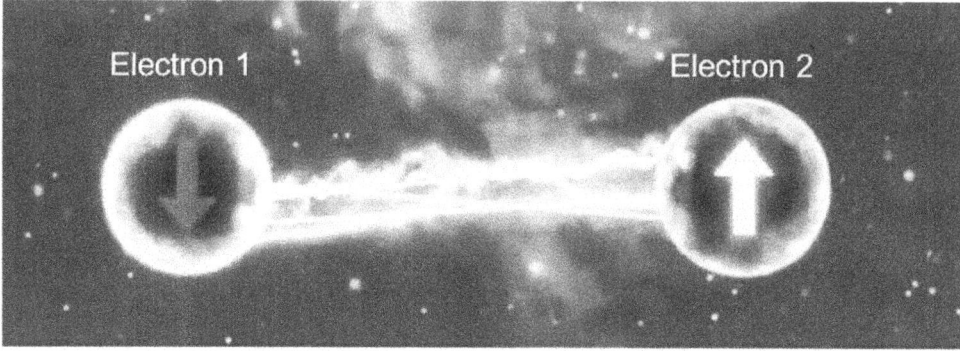

Figure 4.17. An illustration of a phenomenon in the quantum world known as entanglement. The entanglement of particles describes a relationship between their fundamental properties (e.g., spins, like here, momentum, position, or polarization).

her electron and it won't effect the state of Bob's electron and vice-versa (see also Figure 4.17).

Let us now construct another

$$|v\rangle = \frac{1}{\sqrt{2}} \left[|u\rangle |d\rangle + |u\rangle |d\rangle \right] = \frac{1}{\sqrt{2}} \left[|u\rangle_1 |d\rangle_2 + |u\rangle_1 |d\rangle_2 \right] . \tag{4.86}$$

We added the subscripts in the second equality to explicitly differentiate between the orientations of the spins of electron 1 and electron 2. This state has the following properties: if you measure electron 2 and found it down then electron 1 must be up. If you find electron 2 up then electron 1 is down. So you can determine the property of the second electron 2 by only measuring the properties of electron 1. It does not matter what the distance between them is. In other words, each electron no longer has its own state in (4.86). The electrons are correlated. Notice that we cannot arrive at $|v\rangle$ by selecting particular values of α_1, α_2, β_1, and β_2 because $|v\rangle$ is *entangled state*. It is easy to see that these are not product states for if they were it would imply the contradiction: $\alpha_1 \alpha_2 = \beta_1 \beta_2 = 0$, $\alpha_1 \beta_2 = \beta_1 \alpha_2 = 1/\sqrt{2}$—it is clear that there is no solution here. By using Tables 4.5 and 4.6, we can also verify that the averages of $|v\rangle$ are zero; i.e.,

$$\langle v|\sigma_1|v\rangle = \langle v|\sigma_2|v\rangle = \langle v|\sigma_3|v\rangle = 0 \tag{4.87}$$

and

$$\langle v|\tau_1|v\rangle - \langle v|\tau_2|v\rangle = \langle v|\tau_3|v\rangle = 0 . \tag{4.88}$$

These results also hold for $\boldsymbol{\sigma} \cdot \boldsymbol{n}$ and $\boldsymbol{\tau} \cdot \boldsymbol{n}$, where $\boldsymbol{n}$ normal unit vector—that is, there is no direction of definite spin for either Alice's or Bob's electron. The fact that these averages are zero means that there is a $1/2$ probability for the electron to be up along the $\boldsymbol{n}$ direction and a $1/2$ probability for the electron to be down along the $\boldsymbol{n}$ direction. Whatever the spin direction of one electron is the other is in the opposite.

State $|v\rangle$ is generally known as the triplet of state. It can happen naturally if we place two electron close enough in proximity because each electron has its own magnetic field and each electron responds to the other electron in the opposite direction. Eventually, the electrons form the state of (4.86). Also when two electrons collide they naturally create this state.

Quiz: Is the following quantum state

$$\left[\frac{1}{\sqrt{2}}, 0, 0, \frac{1}{\sqrt{2}}\right]^T \tag{4.89}$$

entangled?

So far, we have mostly used electrons and electron spins to illustrate entanglement. Again, quantum entanglement means that the two particles have some correlation. These two particles can electrons. For example electrons can be spin entangled. Photons can be polarization entangled. They do not need to have the same frequency, nor do they have to be formed at the same time. A pair of photons or electrons get entangled, they remain connected even when separated at longer distances. The phenomenon is mostly a pair based phenomenon where in either two photons or two electrons can undergo quantum entanglement. So, collectively, n number of photons can undergo entanglement, but would happen in a pair. For example electrons can be spin entangled. Photons can be polarization entangled. Entanglement applies to all quantum systems. The possibility of that an electron and a photon can be entangled is still open question. If such entanglement it must be based on properties that photons and electrons, that is spin.

Note that entanglement between three electrons is possible. Here are the two typical examples of entangled three-particle states:

$$|GHZ\rangle = \frac{1}{\sqrt{2}}\left(|uuu\rangle \pm |ddd\rangle\right) \tag{4.90}$$

named after Greenberger, Horne, and Zeilinger (1989). Again, we can see that when one particle is observed, it takes a specific value and its entangled pair takes the opposite value at the same instant. Suppose, for example, that two of the measurements are σ_1^A and σ_1^B and the third measurement is σ_2^C. We can verify

$$\begin{aligned}
\sigma_1^A \otimes \sigma_1^B \otimes \sigma_2^C \,|GHZ\rangle &= \sigma_1^A \otimes \sigma_2^B \otimes \sigma_1^C \,|GHZ\rangle \\
&= \sigma_2^A \otimes \sigma_1^B \otimes \sigma_1^C \,|GHZ\rangle \\
&= +\,|GHZ\rangle
\end{aligned} \tag{4.91}$$

In other words, the product of the three measurements always equals +1. $|GHZ\rangle$ is also an eigenstate of $\sigma_2^A \otimes \sigma_2^B \otimes \sigma_2^C$; i.e.,

$$\sigma_2^A \otimes \sigma_2^B \otimes \sigma_2^C \,|GHZ\rangle = -\,|GHZ\rangle \tag{4.92}$$

In other words, the product of these three measurements must give -1.

Another common famous 3-qubit states are the example is (Dür et al., 2000)

$$|W\rangle = \frac{1}{\sqrt{3}}\left(|uud\rangle + |udu\rangle + |duu\rangle\right) \tag{4.93}$$

In the multipartite setting, separability and entanglement are defined in a similar way, using product tensors.

The Schmidt decomposition. Let us assume that $\mathcal{H} = \mathcal{H}_1 \otimes \mathcal{H}_2$ is a Hilbert space of a $m \times n$-vectors made of complex values, $\mathcal{H}_1$ a Hilbert space of a m-vectors made of complex values, and a $\mathcal{H}_2$ Hilbert space of a n-vectors made of complex values. If we have orthonormal sets of bases $\{|a_i\rangle\}_{i=1}^r$ of $\mathcal{H}_1$ and $\{|b_i\rangle\}_{i=1}^r$ of $\mathcal{H}_2$, a unit vector $|v\rangle$ of $\mathcal{H}$ can be written v as follows:

$$|v\rangle = \sum_{i=1}^r \gamma_i |a_i\rangle \otimes |b_i\rangle , \qquad (4.94)$$

where γ_i are strictly positive real scalars and $\sum_{i=1}^r \gamma_i^2 = 1$. A pure state $|v\rangle$ is called separable (or sometimes a product state) if it can be reduced to

$$|v\rangle = |a\rangle \otimes |b\rangle , \qquad (4.95)$$

where $|a\rangle$ pertains to $\mathcal{H}_1$ and $|b\rangle$ pertains to $\mathcal{H}_2$. Otherwise, $|v\rangle$ is called entangled. Another words, state $|v\rangle$ is separable iff its Schmidt coefficients, $\{\gamma_1, \gamma_2, \gamma_3, ...\}$, are $\{1, 0, 0, ...\}$.

Let us look at some specific examples. For a quantum systems with $n = m = 2$, the following state

$$|v\rangle = \frac{1}{\sqrt{2}} (|u\rangle_1 |d\rangle_2 - |u\rangle_1 |d\rangle_2) \qquad (4.96)$$

is entangled. This state is the most famous of all entangled states and is usually known as the singlet state or an EPR pair, after Einstein, Podolsky, and Rosen (1935). Note that this state is a sum over states and cannot be written in a simpler form; this is what makes it entangled. The following two states

$$|v\rangle = \frac{1}{\sqrt{2}} (|ud\rangle + |du\rangle) \quad \text{and} \quad |w\rangle = \frac{1}{\sqrt{2}} (|uu\rangle + |dd\rangle) \qquad (4.97)$$

are also entangled. For example, in $|w\rangle$, if one electron is measured to have spin up, then the other electron will also have spin up simply because they were entangled in such a way that their spins are correlated. However, just because a state is written as a sum of terms of the bases does not necessarily mean that it is entangled. Consider, for example,

$$|v\rangle = \frac{1}{\sqrt{2}} (|u\rangle_1 |d\rangle_2 + |d\rangle_1 |d\rangle_2) = \frac{1}{\sqrt{2}} (|u\rangle_1 + |d\rangle_1) |d\rangle_2 = |h\rangle_1 |d\rangle_2$$

where $|h\rangle_1 = |u\rangle_1 + |d\rangle_1$. This state is not entangled. In more general terms, a state is said to be entangled if it cannot be written in the form in (4.95). For example, $|v\rangle = |u\rangle_1 \otimes |d\rangle_2$ is not entangled.

Let us now describe a way to check whether or not a state is entangled later. Consider a linear map between $\mathcal{H} = \mathcal{H}_1 \otimes \mathcal{H}_2$ and a matrix space of $m \times n$ matrices, $\mathcal{M}$. This mapping can be defined as

$$\Gamma(|v\rangle) = \sum_{i=1}^r \gamma_i |a_i\rangle (|b_i\rangle)^* = UDV , \qquad (4.98)$$

where $|v\rangle$ is a state of $\mathcal{H}$. As for most matrices, we also used the singular value decomposition (SVD) to rewrite $\Gamma = UDV$, where U is a $n \times n$ matrix, V is a $m \times m$ matrix, D is a $n \times m$ diagonal matrix D, r is the rank of Γ, γ_i is the i-th nonzero diagonal entry of D, $|a_i\rangle$ is the i-th column of U, and $|b_i\rangle^*$ is the i-th row of V. The matrices U and V are both unitary ($UU^\dagger = U^\dagger U = I$ and $VV^\dagger = V^\dagger V = I$, where where $U^\dagger$ and $U^\dagger$ are the transposed, complex conjugate of U and V, respectively). State $|v\rangle$ if and only if the matrix $\Gamma(|v\rangle)$ has rank 1. For example, for $|v\rangle = |uu\rangle$ of $\mathcal{H}$ (which is clearly separable) we have

$$\Gamma(|v\rangle) = \begin{pmatrix} 1 & 0 \\ 0 & 0 \end{pmatrix} \qquad (4.99)$$

which has rank 1. On the other hand, for $|v\rangle = (1/\sqrt{2})(|uu\rangle + |dd\rangle)$, which is known as the Bell state, we have

$$\Gamma(|v\rangle) = \begin{pmatrix} 1 & 0 \\ 0 & 1 \end{pmatrix} , \qquad (4.100)$$

which has rank 2. So $|v\rangle$ is entangled.

Bell's inequality (1964) is a simple inequality of college-entry-level set theory. Think of a set of particles that can be classified into three categories based on three properties that we denote A, B, and C (Figure 4.18). Let $\mathcal{N}(X \text{ and } Y)$ be the number of elements with the properties X and not Y, where X and Y represent the properties A, B, and C. Bell's inequality states that:

$$\mathcal{N}(A, \text{ and not } B) \leq \mathcal{N}(A, \text{ and not } C) + \mathcal{N}(C, \text{ and not } B)$$

and

$$n_1 + n_4 \leq n_1 + n_2 + n_7 + n_4 \quad \text{or} \quad \cancel{n_1} + \cancel{n_4} \leq \cancel{n_1} + n_2 + n_7 + \cancel{n_4} ,$$

n_i, (with $i = 1, 2, ..., 7$) represents the number of particles in the various regions of Figure 4.18. This inequality is evidently correct based on classical logic, as we can see in Figure 4.18. In other words, classical physics always obeys Bell's inequality but theory and experiments demonstrate that the quantum world we live in does not in certain situations, namely when the set can include particles that are quantum entangled, that classical logic can consider them as hidden from us—some particles are uncounted for. Another way to interpret this result is to think of it in terms of the so-called *locality principle*, which is the idea that physical process at one location does not affect physical process at another location. In Einstein's words, there is no spooky action at a distance. Classical physics always obeys the inequality but quantum physics does not in certain situations when the objects at the two locations are quantum entangled. In classical theories, states are represented by points in a phase space (real-value space). However, in quantum mechanics states are complex-value vectors in a complex-value vector space. These two spaces are completely different things and a side effect of that is that Bell's inequality is violated.

Let us consider a set of atoms and electrons, and their properties include spin, mass, velocity, and position. In quantum mechanics, there are pairs of values that

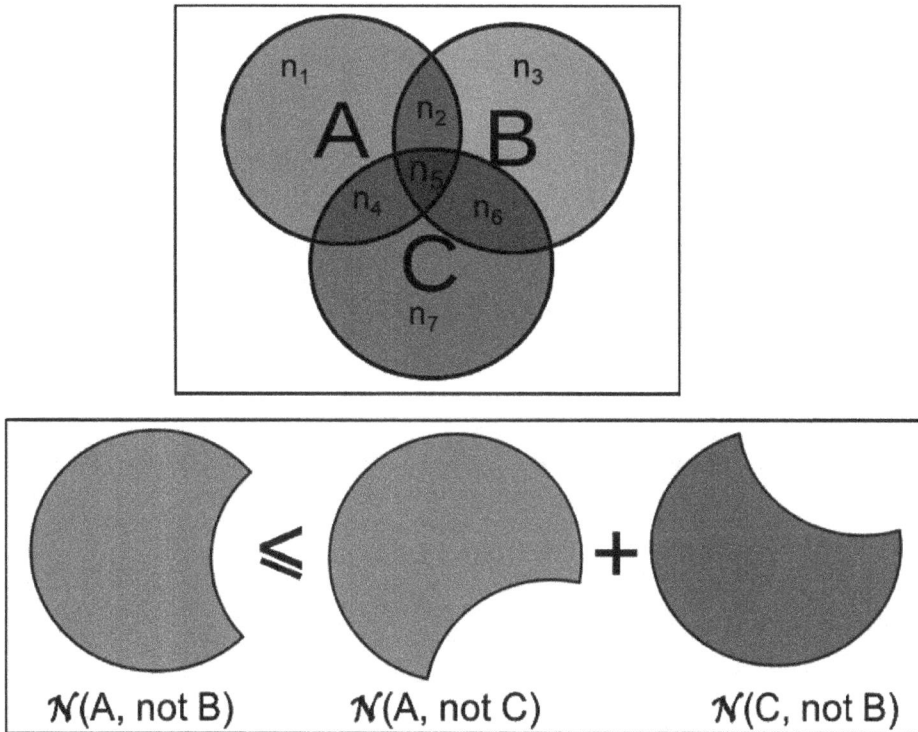

Figure 4.18. A description of Bell's theorem. (Top) A simple Venn diagram of three sets labeled A, B, and C that can overlap one another. We also divided these sets into regions. The numbers of particles in these regions are denoted as n_1, n_2, ..., n_7. Adapted from Leonard Susskind's lectures. (Bottom) Bell's inequality in the form of a Venn diagram.

cannot be simultaneously measured to any desired precision. For example, the velocity and location of these particles can be simultaneously measured. At any given time, the velocity, location, or both of a particle are uncertain. Maybe there are hidden variables but they were there and if you did know their values you could completely predict the state of the system. That is, the universe is not really inherently uncertain. Alternatively, quantum mechanics itself is incomplete. The Newton equations of motion and Maxwell's equations are correct, almost at every scale. However, the constitutive relations appended to these equations are incomplete and imprecise, as discussed in Ikelle (2023c) and the previous chapter. The Newton equations of motion and Maxwell's equations are correct, almost at every scale. For example, in Newtonian mechanics, the motions of two non-interacting particles can be the results of the equations of motion of each particle separately and the motion of the system is simply the combination of those two separate solutions. Not so for quantum mechanically entangled particles. Motion of one depends on the other, even if there is no physical interaction between them after they have been created.

4.2.3. Density Matrix and Mixed States

So far we have dealt with pure states. They happens all the time in nature at the subatomic level. Preserving the entanglement whilst interacting with the environment is much harder. Thus in practice, we cannot guarantee a perfect production of pure states due the apparatus involved in such productions and simply the information totally or incomplete description of states. So we must assign probabilities to the production of states—that is, density operator, or density matrix provide a way to to these state with pure states as particular cases of density matrix. In other density matrix is the most general way of representing quantum systems. Our objective in this subsection is to introduce the idea of density matrix and non-pure states, which are known as mixed states.

Suppose we have an apparatus which prepares quantum systems in certain states. For instance, this apparatus could be an oven producing electrons, or a quantum optics setup producing photons. But suppose that this apparatus is imperfect, it produces a state $|\psi_i\rangle$ with a certain probability p_i, with i varying from 1 to n. Let M be an observable. The expectation value of M is

$$< M > = \sum_i p_i \langle \psi_i | M | \psi_i \rangle = \sum_i p_i \mathrm{Tr}\left(M |\psi_i\rangle \langle \psi_i|\right) = \mathrm{Tr}\left(M\rho\right) \qquad (4.101)$$

with

$$\rho = \sum_i p_i |\psi_i\rangle \langle \psi_i| , \qquad (4.102)$$

where $0 \leq p_i \leq 1$ for all i, $\sum_i p_i = 1$, and ρ is density matrix. Then we have $< M > = \mathrm{Tr}\left(A\rho\right)$. For a pure state, $|\phi\rangle$, we have $n = 1$ and $p_1 = 1$, and the matrix density reduces to $\rho = |\psi\rangle \langle \psi|$. So in this pure state it is not necessary to use ρ at all. One may simply continue to use $|\psi\rangle$. A state which is not pure is usually called a mixed state. In this case ket notation is not suitable, we must use ρ.

In summary, a density matrix is a matrix that describes a quantum system in a mixed state, a statistical ensemble of several quantum states. A mixed states is a distribution over quantum states, $\{p_i, \psi_i\}$, meaning that a state with probability p_i the state is $|\psi_i\rangle$. Thus, we can think of a pure state as a degenerate case of a mixed state where all the probabilities are 0 or 1. Unfortunately quantum entanglement is a very fragile quantity. It is easily destroyed by way of quantum decoherence as a quantum particle exchanges information with the environment.

Example. Suppose a machine trying to produce qubits in the state $|u\rangle$, end up producing $|u\rangle$ with probability p and a state $|\psi\rangle = \cos(\theta/2) |u\rangle + \sin(\theta/2) |d\rangle$, with probability $1 - p$, where θ is small angle enough that $|\psi\rangle$ can be characterized as a small deviation from $|u\rangle$. The density matrix for this system is

$$\rho = p |u\rangle \langle u| + (1 - p) |\psi\rangle \langle \psi|$$

$$= \begin{pmatrix} p + (1 - p) \cos^2 \frac{\theta}{2} & (1 - p) \cos \frac{\theta}{2} \sin \frac{\theta}{2} \\ (1 - p) \sin \frac{\theta}{2} \cos \frac{\theta}{2} & (1 - p) \sin^2 \frac{\theta}{2} \end{pmatrix} . \qquad (4.103)$$

We can generalize ρ to $n \times n$ by constructing $|\psi_i\rangle$ with k from 1 to $n - 1$. For $\theta = pi/4$ and $p = 1/2$ (i.e., the machine 50-50 percent chances of producing $|u\rangle$ or $|\phi\rangle$, ρ reduces to

$$\rho = \frac{1}{2} \begin{pmatrix} 1 & 0 \\ 0 & 1 \end{pmatrix} , \qquad (4.104)$$

Hence, we have complete random information. In more general, if the probability are equally weigted of the mixed states are the probability are equally weigted, we have complete random information. This case is known as *ambiguity of mixtures*.

Properties of the density matrix. The density matrix satisfies a number of properties. First, the density matrix is a Hermitian operator: $\rho^\dagger = \rho$ are the result of the fact that the probabilities p_i are real. The consequence of hermiticity of ρ means that all of the eigenvalues of density matrix are real.

Second, $\text{Tr}(\rho) = 1$, which implies that some probability adds up to one. Actually $\text{Tr}(\rho)$ is the sum of eigenvalues of ρ. These eigenvalues themselves can be taught as probabilities for different states. The eigenvalues go with the eigenstates (i.e., eigenvectors). Because ρ is hermitian matrix, its eigenstates are all mutually orthogonal. As mentioned above example, when $\lambda_1 = \lambda_2 = ... = \lambda_n = 1/n$, we have a complete ignorance and a maximal knowledge when we are dealing with a pure state (i.e., $n = 1$ and $p = 1$ hence $\rho = |\psi\rangle\,\bar\psi$). In addition, a mixed state can be defined as a density matrix that has more than one nonzero eigenvalue.

Third, the density matrix is positive semi-definite; i.e., if $|\phi\rangle$ is an arbitrary quantum state then

$$\langle\phi|\rho|\phi\rangle = \sum_i p_i |\langle\phi|\psi_i\rangle|^2 \geq 0 \ . \tag{4.105}$$

An equivalent definition of a positive semi-definite operator is one whose eigenvalues are always non-negative.

Entropy of a quantum system. In Chapter 2, we introduce the entropy of classical systems. It is given by:

$$S = -\sum_i^n p_i \log p_i \ . \tag{4.106}$$

If only one p_i (i.e., $n = 1$ and $p_1 = 1$) is non-zero, we know exactly the state that the system is in and the entropy (lack of information) is zero; $S = 1 \log 1 = 0$. If there is equal probability $p_1 = p_2 = ... = p_n = 1/n$ then:

$$S = -\sum_{i=1}^n \frac{1}{n} \log(1/n) = \log n = S_{\text{max}} \ , \tag{4.107}$$

where $S_{\text{max}} = \log n$ is the maximum entropy of this system. For a quantum system, entropy is

$$S = -\text{Tr}\,(\rho \log \rho) \tag{4.108}$$

As for the classical case, if the system is definitely in one state, then $S = 1 \log 1 = 0$. If there is equal probability $1/n$ for the system to be in any of n states and the probability to be in any state other than these n states is zero, then: $S = \log n$. As an example, in the two dimensional case of electron spin, the entropy is:

$$S = -\text{Tr}\left(\rho \log \rho\right)$$

$$= \text{Tr}\begin{pmatrix} 1/2 & 0 \\ 0 & 1/2 \end{pmatrix}\begin{pmatrix} \log(1/2) & 0 \\ 0 & \log(1/2) \end{pmatrix}$$

$$= \log 2 \tag{4.109}$$

We use the density matrix in (4.104).

Quantum entanglement is very sensitive to the natural environment. For example, the natural tendency of a qubit in a quantum computer is to entangle with the environment as quantum particles exchange information with the environment. This unwanted entanglement represents information loss, which is known as *decoherence*. For this reason, it is extremely difficult to construct a maximally entangled pair of quantum particles in nature. So most examples of maximally entangled quantum particles are created in laboratories where humans can minimize quantum particles' exchange of information with the environment. Various quantum technologies aim at isolating entangled states in a controlled environment and controlling their evolution in an actual environment by not allowing them to act on their own. Also note that two maximally entangled quantum particles cannot be separated into separate particles without destroying their entanglement.

Box 4.3: Micius Satellite Tests Quantum Entanglement

Yin et al. (2017) used the Micius satellite, which was launched in 2016 into low Earth orbit and is equipped with an assemblage of crystals and lasers, to successfully demonstrate the satellite-based entanglement distribution to receiver stations separated by more than 1200 km. Micius's assemblage of crystals and lasers generates entangled photon pairs, then splits and transmits them on separate beams to ground stations in its line of sight on Earth (see Figure 4.19). Yin et al. (2017) have entangled particles such as electrons and photons, as well as larger objects such as superconducting electric circuits.

The Micius satellite, named after an ancient Chinese philosopher, was launched in August 2016. The satellite is the foundation of the $100 million Quantum Experiments at Space Scale (QUESS) program, one of a number of Chinese space science projects that are on par with the United States and Europe.

The satellite-based distribution of entangled photon pairs to two locations was separated by 1203 kilometers on Earth, through two satellite-to-ground downlinks with a summed length varying from 1600 to 2400 kilometers. They observed a survival of two-photon entanglement and a large violation of Bell inequality.

4.2.4. Source of Quantum Entanglements

In general terms, the creation of an entangled particle pair requires some nonlinear interaction and conservation of energy and linear and angular momentum by the

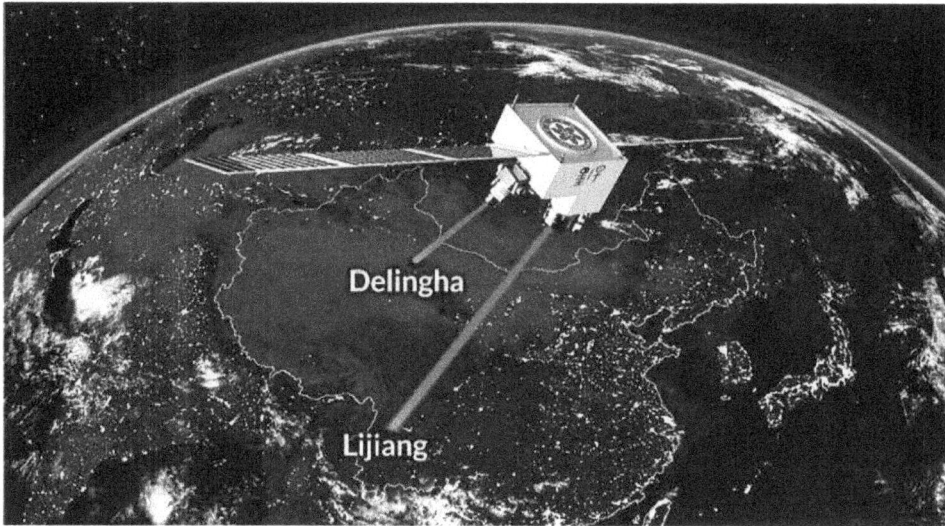

Figure 4.19. Using the quantum-communications satellite Micius successfully sent entangled photons to two cities in China. Micius beamed photons to Earth, separating them by more than 1200 kilometers. Adapted from https://www.sciencenews.org/article/quantum-satellite-shatters-entanglement-record.

quantum entanglement process. Let us emphasize that quantum entanglement is the property that the state of one particle determines the state of the other due to the requirement that the conservation laws, a pillar of modern physics and quantum mechanics in particular, must be satisfied. The conservation laws are really the origin of entanglement. The seemingly weird properties come from quantum mechanics. Here we discuss the entanglements of electrons and photons.

Photon entanglements. One can create a maximally entangled pair of photons by a process known as down conversion. In this process a single photon of frequency f passing through a suitable nonlinear crystal which creates two maximally entangled ower energy photons, each of frequency $f/2$, at the same time. The process must preserve energy, momentum, and angular momentum (polarization). So these three properties are entangled in the two photons. If the process is energy degenerate, then the two emitted photons, called the signal and idler, have the same energy and the entangled properties are momentum and angular momentum (polarization). Possible choices of nonlinear crystals are noncentrysymmetric nonlinear $\chi^{(2)}$-crystals in Table 3.6, including usually beta barium borate (BBO), described in the previous chapter 3—a $\chi^{(2)}$ medium is a nonlinear crystal in which the second term in a general nonlinear expansion of the polarisability is the dominant nonlinear term. Scattering events on such crystals satisfy the conservation of momentum and energy. Without this type of nonlinear crystals, a single photon will not automatically split into two. Anyway, all photon pairs emitted from the crystal are entangled.

Electron entanglements can occur in many ways, although only a few are really practical. For example, the electrons in a specific electron orbital (shell) are entangled with one another in orbitals on a single atom because the electrons in a specific electron orbital on an atom have strong correlations with each other due to their mutual Coulomb interaction. Electrons in an atom more general form spin singlet states (4.96), which ensure they are in entangled states. A similar argument will also hold for the spin triplet configuration (4.86), in which case the electrons occupy different orbital states. These entanglements are not very useful because they are local, while in most practical cases we are interested spatially separable electron pairs.

Another way to entangle electrons is to collide them. The challenge is that electrons do not physically interact, as they cannot get nearer than the Planck length (the Planck length is 10^{-35} meters and the Planck time is 10^{-43} seconds). Remember that each electron has an electromagnetic field. When two electrons get close to each other, each electron begins to respond to the changes in the electromagnetic field due to the other electron. The closer they are, the stronger the repulsive force between the electrons becomes. As a result, the two electrons never really bounce off each other but they can be near enough to ensure that they entangle.

Let us now consider a practical example. If two separate electrons absorb maximally entangled photons, the electrons become spin-entangled because the polarization of light carries spin information that can be transferred to the electrons. Entanglement occurs in optical cavities. Basically, an optical cavity is made of two very smooth partially reflecting surfaces placed very close to each other and the space between them is a vacuum. The whole apparatus is then cooled to near absolute zero temperature. Two neutral atoms trapped in an optical cavity can then be entangled with each other using an entangled pair of photons. One photon would be absorbed by each atom and the two atoms would then become entangled with each other. Notice we do not use electrons directly, rather we use atoms.

Atom entanglements.   Entanglement is not limited to subatomic particles. Atoms, ions, and other massive particles can also become entangled. Such entanglements are generally carried out by an experiment involving an optical cavity—two opposing imperfect mirrors—containing two or more atoms, as illustrated in Figure 4.20. Light is shone into one side of the cavity and allowed to bounce back and forth between the mirrors. Some of the light will eventually escape through the opposite side of the cavity, where it is captured by a detector. The leakage is due to the fact that mirrors here are imperfect—that is, not all the light rays that strike them are reflected back. Atoms will spontaneously emit photons out of the cavity due to the coupling of the atom to the modes of the electromagnetic field of free space on either side, as illustrated in Figure 4.20. One can then monitor changes in the cavity transmission resulting from coupling to atoms falling through the cavity. In optical cavity experiments, spherical mirrors are often used as shown in Figure 4.20. This setup is known as a Fabry-Perot configuration. A cavity is designed to confine light (electromagnetic field) between two reflectors. Atoms or other particles placed between two reflectors interact with the electromagnetic field. Thus an optical cavity is an arrangement of two highly reflective mirrors (either dielectric or metallic) placed parallel to each other at a small distance. This arrangement is widely used as a platform to amplify light-matter interactions. Trapping single atoms in this arrangement with high-quality factor mirrors opens up endless possibilities for the

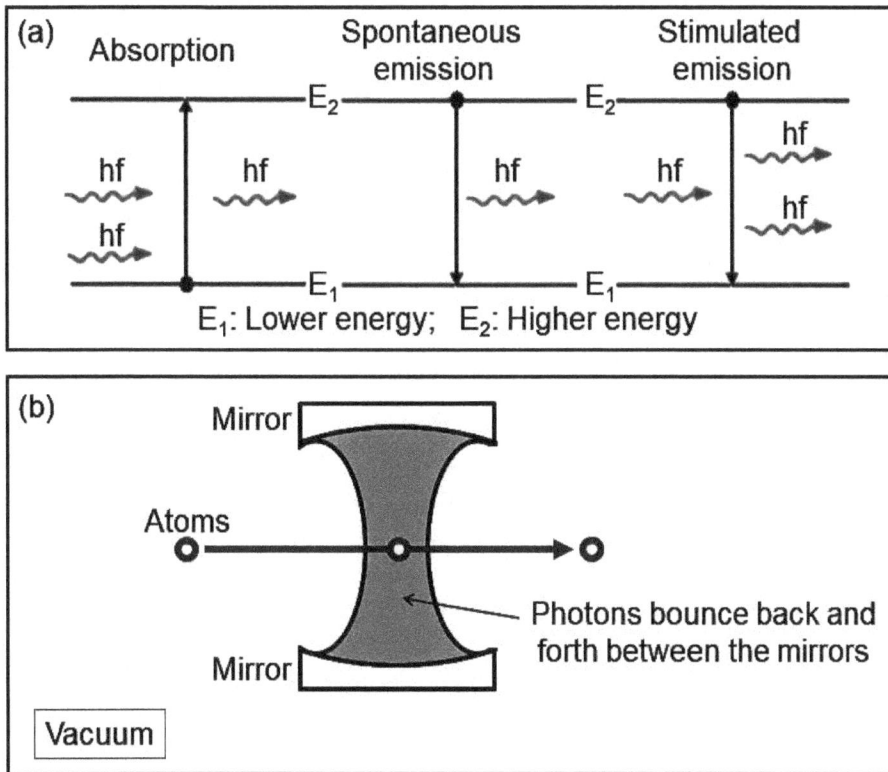

Figure 4.20. (a) An atom interacts with light in free space in three ways. Absorption and stimulated emission are time reversals of each other. (b) An illustration of an optical cavity in vacuum for atom-photon entanglement, for example. The photons bounce back and forth inside a small cavity between the two mirrors. The mirrors are so reflective that a single photon can travel long enough between the mirrors before it is lost. The trapped photon can be used for many quantum manipulations during its lifetime. The process starts with a laser field which drives the atom from the ground $|g\rangle$ state to excited $|e\rangle$ in the cavity to from $|e\rangle$ to $|f\rangle$. This scheme will work for one atom as well as multiple atoms.

observation and manipulation of the dynamics of single particles and for controlling their interactions with single-mode photons.

While in vacuum without cavity a single photon[1] sent onto an atom would interact once and then depart, in an optical cavity the photon bounces back and forth many times, increasing the chance to interact with the matter. As a photon is emitted or absorbed by an atom a transfer of momentum occurs modifying its initial motion state. Due to the fact that light can influence atoms' motion, lasers make it possible to manipulate their state. One can also monitor the spontaneous emission of the atoms into transverse modes not confined by the cavity. In this case, an

---

[1] As discussed in Chapter 2, spontaneous emission occurs for matter and radiation when they reach thermal equilibrium. Nowadays, spontaneous emission, absorption, and stimulated emission are considered to be inherent properties of matter. Absorption and stimulated emission are time reversals of each other as part of classical physics. Quantum physics includes spontaneous emission.

excited atom in an initially empty cavity will emit one (and only one) photon, which can then be trapped and reabsorbed again, a phenomenon known as vacuum Rabi oscillations. The presence of the cavity has made the spontaneous emission from the atom, usually an irreversible process, into a coherent and reversible oscillation. In both the optical and the microwave domains, strong coupling of single atoms and photons has been achieved by using electromagnetic resonators of small mode volume with quality factor mirrors.

Consider atoms with *two* stable ground states $|g\rangle$ and $|f\rangle$. Suppose that both atoms are initially prepared in an equal superposition of the two ground states $(|g\rangle + |f\rangle)/\sqrt{2}$. By performing a non-destructive measurement that measures N, the number of atoms in state $|f\rangle$, the state vector is projected into the maximally entangled state $|\phi\rangle = (|gf\rangle + |fg\rangle)/\sqrt{2}$ provided we get the outcome $N = 1$. If the outcomes $N = 0$ or $N = 2$ are achieved, the resulting state is $|gg\rangle$ or $|ff\rangle$, respectively, so that the initial state can be recreated, and we can repeat the measurement until the desired entangled quantum state is obtained. On average two measurements suffice to produce the state (Sorensen and Molme). With $M = 3$ atoms in the cavity, we can produce $W = (|ffg\rangle + |fgf\rangle + |gff\rangle)/\sqrt{3}$ by performing the detection $4/3$ times on average. (Sorensen and Molme) Also, one could entangle a small subset of the atoms by leaving all other atoms in $|g\rangle$ so that they do not contribute to N. One can combine multiples, especially for quantum computing (see Ikelle, 2024). A component of an entangled state for a set of atoms at one site can be transferred to an atom in another set at a remote location. By simple repetition any component of the original state may be transferred in this way to create nonlocal entanglements.

Through an experiment involving an optical cavity, as described above, containing about 3100 rubidium-87 atoms that are cooled to a temperature of near absolute zero, McConnell et al. (2015) had managed to entangle about 2910 atoms within a much larger ensemble of 3100 atoms. To do this they fired an extremely weak polarized laser pulse into the cavity. Occasionally, just one photon in the pulse will bounce back and forth in the cavity and interact with nearly all of the atomic spins. This succession of interactions is what entangles the atoms. This photon can then leave the cavity and be detected. Such entangling photons are identifiable because their polarizations have been rotated by 90 degrees by atomic interactions. So, whenever such a "herald" photon was detected, the physicists immediately measured the direction of the total atomic spin. They repeated this process many times over to determine the Wigner function of the entangled atoms. Instead of a Gaussian disc, the distribution resembled a ring of positive probability surrounding an inner region of negative probability. This hole of negative probability is the hallmark of entanglement. Furthermore, the researchers were able to calculate that the entanglement involved about 2910 of the 3100 atoms. In this experiment, the atomic spins were entangled in two states that lie on opposing sides of the hole of negative probability (see Ikelle, 2024). This process obeys the laws of energy, momentum, and angular momentum (polarization) conservation.

Atoms can be entangled remotely using light through a method called measurement-induced entanglement or entanglement swapping (Halder et al., 2007). In this method, entanglement is transferred (teleported) between two photons that were created independently and had never interacted before. The atoms are first entangled with photons. These photons are only measured in a specific way such that they only give information about the atom pair and not about individual atoms. The atom/photon entanglement is local; the swapping is remote, at the midpoint between the two atoms. There is no superluminal action at a distance occurring. Everything is causal. Atoms are not the only entities that can swap entanglements. In fact, it was first used to entangle remote photons. Entanglement can be transferred, or swapped, between two particles that originated from different sources and were formerly completely independent. Halder et al. (2007) entangled two autonomous photons from continuous sources—two independent pairs of entangled photons, say, A1-A2 and B1-B2, were emitted by autonomous sources. By taking a joint measurement of one photon in each pair (A1 and B1), these photons fall into an entangled state (later verified using detectors), one of the four so-called Bell states (i.e., $|v\rangle = (1/\sqrt{2})(|uu\rangle + |dd\rangle)$).

There are several other ways to entangle atoms. Laser cooling is one example of this. A group of N helium-4 atoms can be cooled by a laser to near zero Kelvin (-273.15 degrees Celsius) and produce a Bose-Einstein condensate. The condensate's N-atoms are entangled since they are all identical and whatever happens to one atom (like spin state) must happen to the rest of them.

4.3. Application: Quantum Computing

4.3.1. Classical Mechanics-Based Computers

Present computers works with bits—that is, all information/data are stored, processed, and exchanged as a series of 1's and 0's. These series are able to represent and store any number, letter, or symbol. A computer with N bits can then be in 2^N possible states; e.g.,

$$\text{Lucas} = \underbrace{0100\,1100}_{\text{L}}\ \underbrace{0111\,0101}_{\text{u}}\ \underbrace{0110\,0011}_{\text{c}}\ \underbrace{0110\,0001}_{\text{a}}\ \underbrace{0111\,0011}_{\text{s}}$$

$$49.27\text{N} = \underbrace{0011\,0100}_{4}\ \underbrace{0010\,1001}_{9}\ \underbrace{0010\,1110}_{.}\ \underbrace{0011\,0010}_{2}\ \underbrace{0011\,0111}_{7}\ \underbrace{0100\,1110}_{\text{N}}\ ,$$

for $N = 8$. The number of bits in a binary word determines the maximum decimal value that can be represented by that word. A 8 bit processor means the processor is capable of working with 8 bit binary numbers at once. In order to perform calculations and store numbers in memory, computers use tiny switches called *transistors*. The transistors are the smallest and the fundamental unit of a processor. They allow electrons (electricity) to flow through or block the flow of electrons. The flow of electrons is represented by 1 and the blocking of that flow by 0. The actual calculations are carried out by using circuits known as logic gates, which themselves are made up of a number of transistors connected together. When logic gates are chained together, you can begin to do advanced calculations. So conventional computers store numbers in memory and uses those stored numbers to perform simple calculations such as addition and subtraction. When computers perform a bunch of these operations in a sequence, it becomes an algorithm and that is when computers start to become useful. The more transistors there are, the faster the processing speed and the greater its efficiency.

Quiz: Propose five physical ways of constructing qubits.

4.3.2. Quantum Mechanics-Based Computers

Engineers and scientists have crammed several millions of transistors onto silicon pieces the size of a human thumbnail. The result is unprecedented computation power growth in four decades. However, we have reached the limit because our current transistors are almost atom-sized. They cannot get smaller in classical physics. If we further reduce the size of a transistor then things will start getting weird responses because we are now in quantum physics—for example, atomic and subatomic particles can exist in more than one state at a time. Examples include the electron that can be taken as spin up and spin down; or the polarization of a single photon in which the two states can be taken as vertical polarization and horizontal polarization. In a classical system, a bit must be in one state or the other. Thus we have to reconfigure our computers to operate in quantum physics to continue computing power growth—that is, we have to make a significant paradigm shift.

The key changes are that we now work with qubits and multiqubits instead of bits along with quantum gates. We have to use photons, atoms' nuclei, or electrons in place of transistors to store, process, and exchange data. For example, instead of two bits as an input and two bits as an output, we now deal with a two-dimensional vector $|u\rangle = |0\rangle = [1, 0]^T$ and its orthogonal companion $|d\rangle = |1\rangle = [0, 1]^T$. Moreover, we can make superpositions of these two values—that is, $|a\rangle = a_0 |0\rangle + a_1 |1\rangle$ where a_0 and a_1 are complex values with $|a_0|^2 + |a_1|^2 = 1$. We measure the state $|0\rangle$ with probability $|a_0|^2$ and the state $|0\rangle$ with probability $|a_1|^2$. The main advantage of quantum computers is the possibility of solving some problems that in classical computing would expend an excessive amount of time, such as factorization and information retrieval in a database.

Superposition is a fundamental concept in quantum mechanics, and it plays a crucial role in quantum computing. Superposition can lead to constructive interference, thus enhancing the right answer. It can also lead to destructive interference, which facilitates the identification of wrong answers as wrong answers generally destructively interfere. We need more than one qubit to make long sequences or arrays of qubits, and therefore quantum computers, attractive. For example, instead of two bits as an input and two bits as an output, we now deal with four input and output states, $|00\rangle$, $|01\rangle$, $|10\rangle$, and $|11\rangle$ along with their superpositions. As a result, N qubits can store 2^N numbers (commensurate with the many superpositions possible) where N bits can contain N numbers at any specific point in time. In addition, only one operation is needed to manipulate the state of N qubits (or 2^N numbers) where 2^N operations would be needed in a classical computer.

As with standard computers, we have also developed algorithms for quantum computers which come down to qubit operations. These operations are carried out by simple logic gates that operate on your bits. Classical Boolean circuits consist of AND, OR, and NOT gates on an N-bit register. The NOT gate just slips a bit so 0 goes to 1 and 1 becomes 0. The AND gate takes two input bits and outputs a bit of

1 if and only if both input bits are one. These three gates are sufficient to implement any classical operation you want on a string of N bits. Quantum computing has analogous gates that act one qubit at a time. As we defined in the previous section, we have described a NOT-gate that swithch $|0\rangle$ to $|1\rangle$ and vice versa. The Hadamard gate allows superposition. For example, if we apply the Hadamard-gate on $|0\rangle$ we end up $(|0\rangle + |1\rangle)/\sqrt{2}$. Hadamard-gate also includes interference. There are also gates that work on controlled-NOT gates which are kinds of if-then statements. If the first bit is 1, you flip the second bit. Interaction between two quibits via controlled-NOT, we have $|0a\rangle \longrightarrow |0a\rangle$ and $|1a\rangle \longrightarrow |1(1-a)\rangle$. These three gates are enough for most quantum computing. Quantum computing does not have an AND gate because it is not reversible.

Finally, we need to bring quantum computing outcomes to the classical world through strings of 0's and 1's that we can analyze/interpret. From the programmer's point of view the challenge is that the output may even come out as a superposition. For any quantum problem, any linear combination of solutions is also a solution. In contrast, classical problems generally end with a single solution or multiple independent solutions. In other words, quantum computer algorithms do not provide the correct/desired answer. The answers must be checked with a conventional computer. In some cases, we even need artificial intelligence and machine learning tools (see Ikelle, 2024) to determine the desired/correct answer. Sometimes, the quantum program needs to be run several times before the correct output can be obtained. Anyway, it is worthwhile emphasizing that current designs and proposals do not envisage stand-alone quantum computers. Unlike classical computing, they require peripheral computing. This requirement may be a sign of an incomplete quantum mechanics formulation.

> **Quiz:** A three-qubit quantum computer can store $2^3 = 8$ possible combinations of $|0\rangle$ and $|1\rangle$ simultaneously. As a result, a three-qubit quantum computer does calculations 8 times faster compared to a 3-bit conventional computer; a n-qubit computer has $N = 2^n$ possible states. Compare a 64-qubit quantum computer with a 64-bit classical one.

To summarize, a quantum computer is an assemblage of gates and qubits. Thus, when switching from classical computers to quantum computers, bits become qubits, the gates become unitary[2] quantum gates, a classical circuit evolves into a quantum circuit, and reading the circuit output turns into measuring the final state. Technology-wise, all qubits must first be constructed. Run a circuit that produces the appropriate interference: computational paths resulting in the correct output should be constructively interfering with others destructively. In addition to superposition, qubits can also exhibit entanglement, where two or more qubits are correlated. This property allows for even more computational power, as operations on one qubit can affect another qubit instantaneously, regardless of distance.

Quantum computing still faces enormous scientific and technological challenges. As mentioned earlier, any reasonable two-level physical system can be used to implement a qubit: electron spin, photon polarization, and ground state versus excited

---

[2]The unitary matrix is a complex square matrix whose columns (and rows) are orthonormal. It has the remarkable property that its inverse is equal to its conjugate transpose; mathematically, $\boldsymbol{F}\boldsymbol{F}^\dagger = \boldsymbol{F}^\dagger\boldsymbol{F} = \boldsymbol{I}$, where $\boldsymbol{F}^\dagger$ is the Hermitian adjoint of $\boldsymbol{F}$.

state are three such examples. For example $|0\rangle$ can be spin up and $|1\rangle$ can be spin up; an ion trapped somewhere can be identified by its ground state with $|0\rangle$ and an excited state with $|1\rangle$. One of the main differences between current quantum-computer prototypes is the technological challenges of constructing qubits. Current quantum computers must operate at temperatures below zero Celsius to facilitate sub-atomic operations as such operations are carried out in nature in a pristine environment, away from noise and heat, where quantum decoherence is minimal or non-existent. Quantum computers, which operate in a pristine environment, must overcome quantum decoherence and noise. In addition, they must be able to be entanglement between any other qubits, and have a significant coherence time for each qubit. Coherence affects the amount of operations a qubit can execute.

Quiz: Suppose that you have a function $f(x)$. There exist y with $y \neq x$ and $|x - y| = kP$, where k is an integer and P is an arbitrary number, such that $f(x) = f(y)$. Characterize this function.

4.3.3. Shor's Quantum Factorization Algorithm

Quantum algorithms are instructions for a quantum computer. However, unlike algorithms for classical computers their results cannot be guaranteed to be correct. A typical quantum algorithm is a series of quantum state transformations using quantum gates followed by a measurement. As described earlier, these gates are unitary (reversible). Quantum algorithms have already shown their superiority over classical algorithms. One of them is Shor's algorithm (Shor, 1994 and 1997). Take this simple equation:

$$M = pq \, , \tag{4.110}$$

where M is a known positive integer and p and q are unknown prime numbers. For a given p and q we can easily deduce M. The reverse is almost impossible in large numbers. This negative result is used today for digital security, including banking information security, computer security, internet security, and telecommunication security. Solving this inverse will have far reaching ramifications in our current societal set up because it allows some to bypass digital security. The Shor quantum algorithm in a quantum computer allows us to recover p and q. It makes it possible to factorize an integer in polynomial time whereas the most efficient classical algorithms carry out the factorization in exponential time.[3] Because of Shor's algorithm, modern popular culture claims that quantum will end secrecy soon and everything will be screened from rooftops, as described by the Bible.[4] Anyway, the quantum Fourier transform (QFT), which is the quantum implementation of the discrete Fourier transform, is a significant component of Shor's algorithm. So we will start by describing QFT before deducing the Shor factorization algorithm.

---

[3] In 2022, there is no computer in the world that can execute Shor's algorithm. Therefore, we shouldn't expect precise estimates of running time, as that will depend upon the details of the computer that the algorithm is running on.

[4] What you have said in the dark will be heard in the daylight, and what you have whispered in the ear in the inner rooms will be proclaimed from the roofs (Luke 12:3).

The quantum Fourier transform (QFT) works in the same way as the classical Fourier transform. However it has a very elegant implementation that is exponentially faster than its classical counterpart. The classical Fourier transform maps the vector $\boldsymbol{x} = [x_1, x_2, ..., x_N]^T$ into $\boldsymbol{y} = [z_1, z_2, ..., z_N]^T$ as follows:

$$z_k = \frac{1}{\sqrt{N}} \sum_{j=0}^{N-1} x_j \exp\left(\frac{2\pi j k}{N}\right), \tag{4.111}$$

where N is the number of evenly spaced samples of $\boldsymbol{x}$ with $x_j = j\Delta x$; x_j are complex numbers, with $j = 0, ..., N-1$ and y_k are complex numbers, with $k = 0, ..., N-1$. The QFT does pretty much the same thing. However it acts on quantum states. In other words, QFT works directly on qubits. We assume that we have a n-qubit computer with $N = 2^n$. Thus a quantum basis vector

$$\begin{aligned}
|x\rangle &= \left|2^{n-1}x_1 + 2^{n-2}x_2 + ... + 2^0 x_n\right\rangle \\
&= \underbrace{|x_1 x_2 x_3 ... x_n\rangle}_{n-\text{qubits}} = |x_1\rangle \otimes |x_2\rangle \otimes |x_3\rangle \otimes ... \otimes |x_n\rangle, \tag{4.112}
\end{aligned}$$

where $x_k \in \{0, 1\}$. In addition to the multi-qubit notations introduced in the previous section, we have added the decimal representation through the first equality. This equality allows us to switch from binary representations to decimal representations and vice versa. So we have, in a 3-qubit computer for example, $|5\rangle = \left|2^2 \times 1 + 2^1 \times 0 + 2^0 \times 1\right\rangle = |101\rangle = |1\rangle \otimes |0\rangle \otimes |1\rangle$. From the definition of multi-qubits in the previous section, we can map $|x\rangle$ into $|z\rangle$, as follows:

$$\begin{aligned}
|z\rangle &= \text{QFT}\,|x\rangle = \frac{1}{\sqrt{N}} \sum_{y=0}^{N-1} \exp\left(\frac{2\pi i x y}{N}\right) |y\rangle \\
&= \frac{1}{\sqrt{N}} \sum_{y=0}^{N-1} \exp\left(2\pi i x \sum_{k=1}^{n} \frac{y^k}{2^k}\right) |y_1 y_2 y_3 ... y_n\rangle \\
&= \frac{1}{\sqrt{N}} \sum_{y=0}^{N-1} \prod_{k=1}^{n} \exp\left(2\pi i \frac{x y^k}{2^k}\right) |y_1 y_2 y_3 ... y_n\rangle \\
&= \frac{1}{\sqrt{N}} \left(|0\rangle + \exp\left(2\pi i \frac{x}{2^1}\right)|1\rangle\right) \otimes \left(|0\rangle + \exp\left(2\pi i \frac{x}{2^2}\right)|1\rangle\right) \\
&\quad \otimes \left(|0\rangle + \exp\left(2\pi i \frac{x}{2^3}\right)|1\rangle\right) \otimes ... \otimes \left(|0\rangle + \exp\left(2\pi i \frac{x}{2^n}\right)|1\rangle\right) \\
&= \frac{1}{\sqrt{N}} \bigotimes_{j=1}^{n} \left(|0\rangle + \exp\left(2\pi i \frac{x}{2^j}\right)|1\rangle\right). \tag{4.113}
\end{aligned}$$

Here, the larger tensor product symbol functions as an indexed product of the Kronecker tensor product. In this form, QFT can be described as multiple tensor

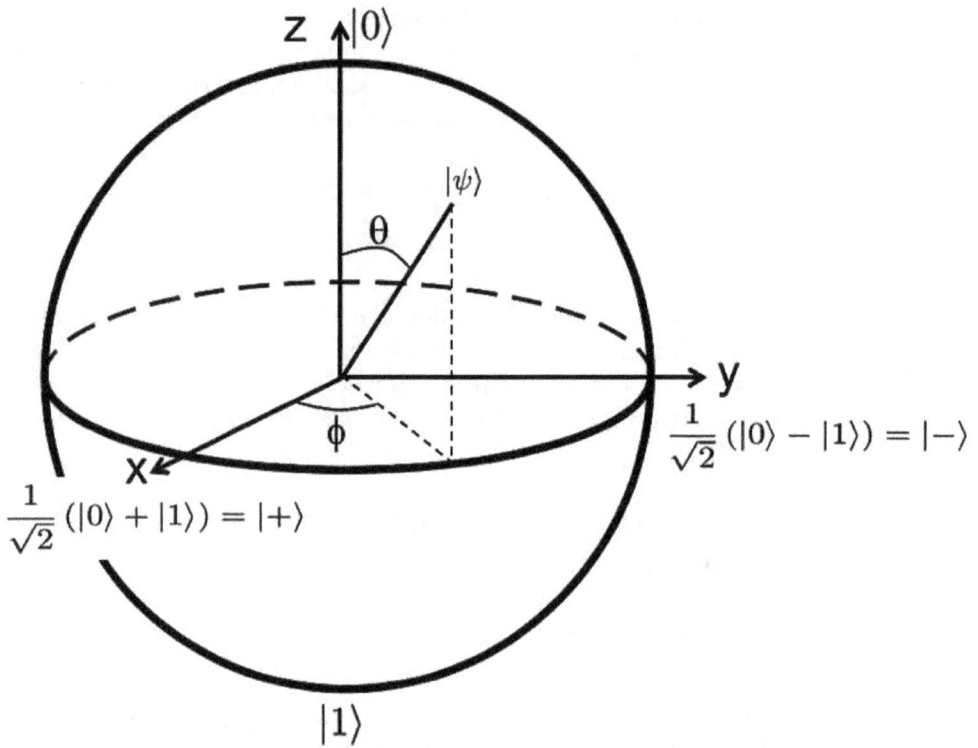

Figure 4.21. A one-qubit QFT. We move from basic qubits, $|0\rangle$ and $|1\rangle$ to $|+\rangle$ and $|-\rangle$.

products of quantum states. For one-qubit system ($N = 2^n = 2^1 = 2$), we can see that QFT is

$$\text{QFT} \, |0\rangle = \frac{1}{\sqrt{2}} \left(|0\rangle + |1\rangle \right) = |+\rangle = \text{H} \, |0\rangle$$

and

$$\text{QFT} \, |1\rangle = \frac{1}{\sqrt{2}} \left(|0\rangle - |1\rangle \right) = |-\rangle = \text{H} \, |1\rangle \ . \tag{4.114}$$

As illustrated in Figure 4.21, a one-qubit QFT is like a 90 degree rotation of one basis to another. For multiqubit systems, the tensor products also include rotation gates (also known as controlled phase shift gates) in addition to the Hadamard gate. The rotation gates can be written as follows:

$$\text{R}_k = \begin{bmatrix} 1 & 0 \\ 0 & \exp\left(\frac{2\pi i}{2^k}\right) \end{bmatrix} \tag{4.115}$$

or

$$\text{R}_k \, |x_j\rangle = \exp\left(\frac{2\pi i x_j}{2^k}\right) |x_j\rangle \ . \tag{4.116}$$

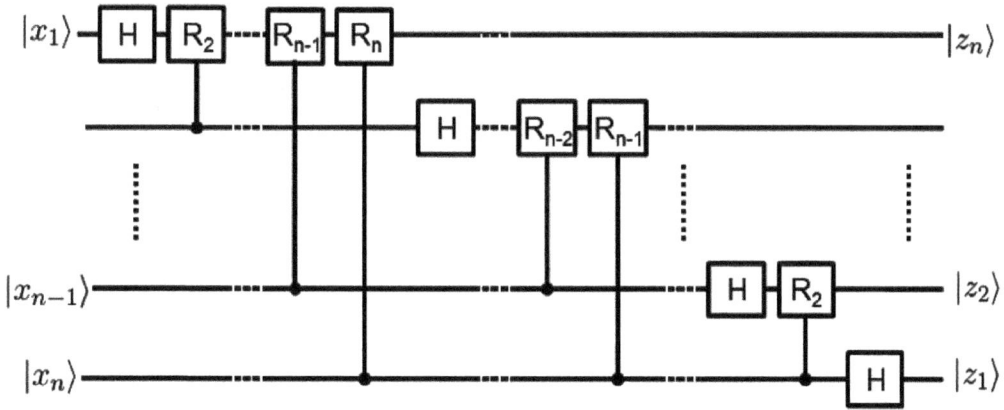

Figure 4.22. QFT circuit is composed of H gates and the controlled version of R_k.

Note that R_k is a unitary gate like the other quantum gates introduced earlier. QFT represents a time evolution operator composed of n^2 gates in parallel, some Hadamard gates, and some controlled phase shift gates.

Figure 4.22 shows a QFT quantum circuit associated with (4.113). Let us digress a little to define a quantum circuit. So a quantum circuit means a sequence of quantum gates acting on qubits. It can also explicitly include measurements and initializations of qubits. The horizontal axis is time, starting on the left and ending on the right. Horizontal lines are qubits, double lines represent classical bits. The items connected by these lines are operations performed on the qubits. These lines define the sequence of events, and are usually not physical cables. Current quantum circuits are generally written for ideal quantum computers because real quantum hardware does not exist yet. When they eventually become available, further vendor-specific limitations like lower decoherence time (or shorter execution time because of its interaction with the physical environment) may further constrain the current quantum circuit.

The QFT circuit is composed of H gates and the controlled version of R_k. The QFT algorithm starts with a Hadamard operation and $n-1$ conditional rotations at the first qubit $|x_1\rangle$, totalizing n operations. It is followed by the application of another Hadamard operation, and $n-2$ rotations, to the second qubit, $|x_2\rangle$, totalizing $(n-1)$ operations and so on. Coppersmith (1994) discovered the quantum Fourier transform.

Considering that $N = 2^n$, the QFT circuit has a computational cost of $O((\log_2 N)^2)$, that is exponentially faster than the FFT algorithm, which has a computational cost of $O(N \log_2 N)$. Table 4.7 compares the costs of FFT (Fast Fourier Transform) and QFT (Quantum Fourier Transform) algorithms. Notice the huge advantage of QFT when the bit amount increases. So the main differences between the quantum Fourier transform (QFT) and the classical fast Fourier transform (FFT) are that (1) the QFT performs a Fourier transform on a superposition of states all at once, whereas the FFT operates on one data sample at a time in sequence; (2) the QFT requires fault-tolerant quantum computers with high qubit counts while the FFT is implemented widely on conventional digital hardware.

Table 4.7. FFT and QFT algorithms costs comparison.

$N = 2^n$	FFT $N \log_2 N$	QFT $(\log_2 N)^2$	FFT/QFT
4	8	4	2.0
8	24	9	2.67
16	64	16	4
32	160	25	6.40
64	384	36	10.67
128	896	49	18,29
256	2.048	64	32.00
512	4.608	81	56.89
1024	10.24	100	102,40
2048	22.528	121	186,18
4.096	49.152	144	341,33
8.192	106.496	169	630,15

Quantum Factorization. "The problem of distinguishing prime numbers from composite numbers and of resolving the latter into their prime factors is known to be one of the most important and useful in arithmetic. It has engaged the industry and wisdom of ancient and modern geometers to such an extent that it would be superfluous to discuss the problem at length... Further, the dignity of the science itself seems to require that every possible means be explored for the solution of a problem so elegant and so celebrated." (Carl Friedrich Gauss, Disquisitiones Arithmeticae, 1801). Hence, it was remarkable when, that Peter Shor (1994) showed that quantum computers could efficiently factor numbers; that is, for a given positive integer M, we can find prime integers p and q such that $M = pq$.

Modular arithmetic. The basic idea of factorizing M based on the Shor's algorithm comes down to finding the period of M. So we first have to an integer M can be described as periodic. We need to recall two key tools of modular arithmetic to do so, namely, the notion of integer modulo and function is greatest common divisor (gcd) modulo. The Matlab function of gcd is gcd. Here are illustrations of these two notions using for $M = 3$ and $M = 21$.

$a \equiv b \pmod M$	$\gcd(a, M)$	$a^r \equiv b \pmod M$	$\gcd(a^r, M)$
$1 \equiv 1 \pmod 3$	$\gcd(1,3) = 1$	$2^0 \equiv 1 \pmod{21}$	$\gcd(1,21) = 1$
$2 \equiv 2 \pmod 3$	$\gcd(2,3) = 1$	$2^1 \equiv 2 \pmod{21}$	$\gcd(2,21) = 1$
$3 \equiv 0 \pmod 3$	$\gcd(3,3) = 3$	$2^2 \equiv 4 \pmod{21}$	$\gcd(4,21) = 1$
$4 \equiv 1 \pmod 3$	$\gcd(4,3) = 1$	$2^3 \equiv 8 \pmod{21}$	$\gcd(8,21) = 1$
$5 \equiv 2 \pmod 3$	$\gcd(5,3) = 1$	$2^4 \equiv 16 \pmod{21}$	$\gcd(16,21) = 1$
$6 \equiv 0 \pmod 3$	$\gcd(6,3) = 3$	$2^5 \equiv 11 \pmod{21}$	$\gcd(32,21) = 1$
$7 \equiv 1 \pmod 3$	$\gcd(7,3) = 1$	$2^6 \equiv 1 \pmod{21}$	$\gcd(64,21) = 1$
$8 \equiv 2 \pmod 3$	$\gcd(8,3) = 1$	$2^7 \equiv 2 \pmod{21}$	$\gcd(128,21) = 1$
$9 \equiv 0 \pmod 3$	$\gcd(9,3) = 3$	$2^8 \equiv 4 \pmod{21}$	$\gcd(256,21) = 1$
$12 \equiv 0 \pmod 3$	$\gcd(12,3) = 3$	$3 \equiv 3 \pmod{21}$	$\gcd(3,21) = 3$
$15 \equiv 0 \pmod 3$	$\gcd(15,3) = 3$	$7 \equiv 7 \pmod{21}$	$\gcd(7,21) = 7$

So we can write $a \equiv b \pmod{M}$ means where $b = a - qM$, q is some quotient M; $b \in \{0, \ldots M-1\}$ is the remainder; e.g., consider the ratio 5/3, the quotient is 1 and the remainder is 2. The formal way to write this statement is $5 \equiv 2 \pmod{3}$. Notice that modular arithmetic of M is periodic. Notice also that if the $\gcd(a, M) > 1$, then we have factorized M by pure luck. Let us look two more examples. Suppose that that the factorization is a solution to the equation $a^2 \equiv 1 \pmod{M}$, where M is an odd integer and r=2, then

$$a^2 - 1 \equiv 0 \pmod{M} \iff (a+1)(a-1) = 0 \pmod{M}$$

Thus the factor of M is either $\gcd(a+1, M)$ or $\gcd(a-1, M)$. For $M = 21$ and $a = 8$, we have

$$8^2 = 64 \equiv 1 \pmod{21} \iff 8^2 - 1^2 = (8-1) \times (8+1) \equiv 1 \pmod{21}$$

We cab deduce $\gcd(9, 21) = 3$ and $\gcd(7, 21) = 7$ as the prime factors of 21.

Let us consider a larger integer, $M = 314191$, and a random guess $a = 127$ with $\gcd(a, m) = \gcd(127, 314191) = 1$. So a does not factorized M. Now we see r such that $a^2 = 127^r \equiv 1 \pmod{314191}$. By classical search, $r = 17388$. This implies that

$$127^{17388} - 1 = 314191 \iff \left(127^{17388/2}\right) - 1\right) \left(127^{17388/2}\right) + 1\right) = 314191$$

We can deduce $\gcd(127^{8694} + 1, 314191) = 829$ and $\gcd(127^{8694} - 1, 314191) = 379$ are the two factors of $M = 314319$. Finding r for large integers, where r occurs as periodic bumps of interval r, is prohibitively expensive (or near impossible) by using classical computers. Shor's algorithm is designed to finding this period by using quantum computer since quantum computers run multiple steps simultaneously due to their ability to store qubits in superposition. Table 4.8 describes the key steps of the Shor's algorithm. It shows how the factorization is turn to the problem of period finding.

Period finding problem. Suppose we have a function $f(r)$ that is periodic with period r. Its inverse Fourier transform $\hat{f}$ is also periodic with period M/r. That is the key property on which finding the period for Shor's algorithm is based. The

Table 4.8. The keys of Shor's algorithm. We turn the factorization of M to finding the period of a^r, where r is the period.

(1) Select $a < M$; e.g., $a = 13$ and $M = 15$.
If $\gcd(a, M) = 1$ (i.e., a and M share no common divider) goto step #2.
If $\gcd(a, M) > 1$ then there is no need for a quantum computer.
(2) Find the smallest r such that $a^r \equiv 1 \pmod{M}$; e.g., $r = 4$.
$[13^0 \equiv 1, 13^1 \equiv 13, 13^2 \equiv 4, 13^3 \equiv 7, 13^4 \equiv 1, 13^5 \equiv 13, 13^6 \equiv 4, \ldots] \pmod{15}$
(3) If r is even then $x \equiv a^{r/2} \pmod{M}$; e.g., $x = 13^2 \equiv 4 \pmod{15}$
if $x + 1 \not\equiv 0 \pmod{M}$, p and/or $q \in \{\gcd(x+1, M), \gcd(x-1, M)\}$.
e.g., $4 + 1 \not\equiv 0 \pmod{15}$, p and/or $q \in \{\gcd(3, 15) = 3, \gcd(5, 15) = 5\}$.
(4) Else (i.e., r is odd) Select another a and to back to step #2.

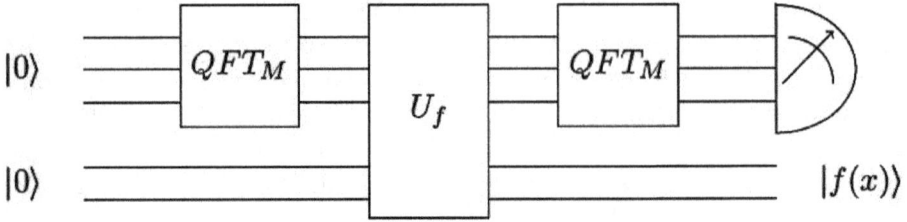

Figure 4.23. Circuit for period finding. The interference that occurs in the final step is one reason quantum computers are so well equipped for period finding. We call it interference because it is additions in the phase that cause cancellations.

basic approach consists of three key steps: (1) prepare the periodic superposition which includes r in its phase, (2) then take its Fourier transform to get rid of the linear shifts and produce $y = kM/r$, and (3) repeat these two steps until there are enough such y's so that we can compute their greatest common divisor and solve for r.

We start by constructing this unitary transform

$$U_f |y\rangle = \begin{cases} |xy\rangle & \text{if } 0 \leq y \leq M-1 \\ |y\rangle & \text{otherwise} \end{cases} .$$

We then performed these sequences of operations as illustrated in Figure 4.23:

$$|0\rangle |0\rangle \underset{QFT}{\Longrightarrow} \frac{1}{\sqrt{M}} \sum_{x=0}^{N-1} |x\rangle |0\rangle \underset{U_f}{\Longrightarrow} \frac{1}{\sqrt{M}} \sum_{x=0}^{M-1} |x\rangle |f(x)\rangle$$

$$\underset{\text{meas}}{\Longrightarrow} \sqrt{\frac{r}{M}} \sum_{k=0}^{M/r-1} |kr + x_0\rangle |f(x_0)\rangle \underset{QFT}{\Longrightarrow} \frac{1}{M} \sqrt{r} \sum_{k=0}^{r-1} \alpha_y |y\rangle$$

with

$$\alpha_y = \sum_{k=0}^{M/r-1} \exp\left[2\pi i(x_0 + kn)y\right] , \tag{4.117}$$

where $f(x) = a^x \pmod{M}$, with $a^1 = a^r = 1$, all powers in between are distinct and then it repeats; r is the period; meas stands for measure. The registers are initially in the state $|0\rangle \otimes |0\rangle$. Next we apply the Fourier Transform to register 1. Next we measure the second register. If y is a multiple of M/r, then $\alpha_y = 1/\sqrt{r}$. This should be viewed as constructive interference due to the final QFTM. If y is not a multiple of M/r, α_y is 0. This can be viewed as destructive interference due to the final QFT.

Let us look at an example in which $f(x) = x \pmod{2}$ with $r = 2$. We will use a 3-qubit system so that $N = 8$. We have

$$|0\rangle |0\rangle \underset{QFT}{\Longrightarrow} \frac{1}{\sqrt{8}} \sum_{x=0}^{7} |x\rangle |0\rangle \underset{U_f}{\Longrightarrow} \frac{1}{\sqrt{8}} \sum_{x=0}^{7} |x\rangle |x \pmod{2}\rangle$$

Table 4.9. Comparison between classical and quantum factorization algorithms' computational costs. The constant c here is 3 and m is the number of bits needed for p and q.

m	Classics $\exp\left[c\, m^{1/3} \left(\log(m)\right)^{2/3}\right]$	Quantum $c\, m^2 \log(m) \log\left(\log(m)\right)$	Classics/Quantum
16	5.1982e+03	74.6005	69.6808
32	2.7078e+05	821.0971	329.7754
64	5.3615e+07	5.6986e+03	9.4085e+03
128	6.1987e+10	3.3528e+04	1.8488e+06
256	7.3016e+14	1.8073e+05	4.0401e+09
512	1.8050e+20	9.2226e+05	1.9571e+14
1024	2.3914e+27	4.5322e+06	5.2764e+20

$$\underset{\text{meas}}{\Longrightarrow} \sqrt{\frac{1}{2}}\left(|1\rangle + |3\rangle + |5\rangle + |7\rangle\right) \otimes |1\rangle \underset{\text{QFT}}{\Longrightarrow} \sqrt{2}\left(|0\rangle - |4\rangle\right)$$

with

$$\alpha_y = \sum_{k=0}^{M/r-1} \exp\left[2\pi i (x_0 + kn)y\right] .$$

Finally, if we take a few measurements we will measure both $|0\rangle$ and $|4\rangle$. Therefore $N/r = 4$, and since $N = 8$, it is clear that $r = 2$. Table 4.9 shows a comparison of Shor's algorithms for classical and quantum computers.

4.3.4. Will Quantum Computing Work?

The real interest in quantum computers for scientists isn't in Shor's algorithm or Grover's search or any of that (Grover, 1996 and 1997). However, they are very significant milestones in the field. Quantum computation must be developed in order to provide us with some insights into the way the universe works. This includes how light appeared 378 million years after the big bang and the occurrence of life, including conscious beings. Quantum computers represent the only hope of simulating humans from beginning to end. Classical computers cannot and will not do it. Furthermore, building a real quantum computer would be a direct test of quantum mechanics (putting away any doubt that there is at least something like an exponential vector space in which the state of a system lives) and may even provide ways of improving quantum computation.

In fact, there are at least three areas where scientific developments are still needed before seriously considering quantum computing. There is (i) qubit development, (ii) the link between quantum mechanics and the world we experience every second and minute of our lives on Earth, and (iii) where to build quantum computers. These problems have not yet been overcome in Shor's algorithm and Grover's search algorithm, so it is difficult to believe that technology is just around the corner. It is estimated that qubit coherence times today are measured in microseconds, that simultaneity does not yet exist in our daily experiments, and that subatomic particles operate in a world that is easily accessible on earth. There are also exoplanets where nuclear fusion is more natural than on Earth. There are other exoplanets in plasma states where the sub-atomic world is more naturally accessible. In fact, even on Earth, not everything is possible everywhere. There are variations in temperature, atmosphere conditions, soils, and underground contents that determine what is possible and what is not.

The linkage between quantum mechanics and our world is treated today as appending a classical computer to a potential quantum computer. The brain-and-mind model, with the mind potentially being in a sub-atomic world (most photons) and the brain being in an atomic and molecular world, tells us how this linkage might be. Quantum computing may be more effective by mimicking this model. And decoherence must be treated as a component of quantum computing rather than a nuisance. It provides the link and explanation of why physics on a large scale appears classical. Moreover, consciousness in its multiple forms may naturally emerge from such a computing model.

If one turns to the simulation of the world around us, as described by quantum mechanics, even the number of particles needed in constructing the energy potential of systems to solve Schrödinger equation is just imaginable. A human weighing about 70 kg contains about 1.5×10^{28} electrons. The definition of energy potential and the state of a system for so many particles as $\Psi(x_1, x_2, x_3, ..., x_n, t)$, where $n = 1.5 \times 10^{28}$ is imaginable. In addition, one has to consider the inter-connectivity and interactions of these particles. Such interconnections are central to our lives and vary considerably with time. Also no quantum wave function or a set of quantum wave functions exists in isolation from the rest of the universe. Any way, anyone who solves the above problems as formulated by quantum mechanics in their entirety and builds an adequate quantum computer, will control the world. They will become a menace to mankind.

4.3.5. The Future Is Not Fixed

In the previous chapter, we discussed the possibility that beings at very long distances might be able to see into the past at other far away locations because of the time it takes light to travel from one location to another. Similar observations are impossible at any location about the future of any system. Actually, the derivations in this chapter tell us that we cannot view the future irrespective of location and time because the outcome of schrödinger equation are probabilistic. The wave function is a probabilistic quantity. Predicting the future means predicting the wave function of a given system. Because systems are not static in time and space, the energy potential is always changing in time and space.

Most wave functions are in general approximately periodic or cyclic, consistent with the famous anecdote that "history keeps repeating itself." The cycles are the basis of future predictions such as whether there will be a VEI = 4 or higher volcanic eruption in the Tokyo region in the next ten years. Even the so-called public seers, such as Nostradamus (December 1503- July 1566) are very vague about the time and location of their prophecies. Note that wave functions are never discontinuous as are all responses to wave equations. Thus the popular use of the notion of anomalies in the sciences must be revisited. Those that claim to predict the future are actually seeing events associated with the high probability of the wave function if they can access quantum wave functions. To put it another way, we monitor the wave functions and see how the likelihood of certain events increases over time.

Generally, the concept of searching for other types of life forms and civilizations involves finding Earth-like planets, exoplanets, moons, and exomoons. We often overlook the issue of finding celestial bodies on the same evolutionary path as Earth. If such a body is found and it is behind us on this path we can significantly improve our understanding of our own early planet and our universe. If such a body is ahead of us on this evolutionary path, it could also help us predict our own future and plan for it.

One day, perhaps in this century, humans will be able to establish instantaneous communications with other civilizations light years away from us. This is without light speed. Such interactions will allow other civilizations to see our past and instantly transmit images and streams. We will finally know who constructed the pyramids and for what purpose. We will also know how humans appeared/arrived on Earth, what happened 278 million years ago, etc.

Chapter 5

Spontaneous Technology

One of our main aims in this chapter is to describe the motion of electrically charged particles and fluids in human bodies, the atmosphere, underground, and vacuum. The motion of charged particles in human bodies drives communication between cells, on the one hand, and between cells and the environment, on the other. They are central to the functioning of our five main senses (smell, touch, sight, taste, and hearing) and the communication of our brains and minds with the rest of our bodies. Without the motion of these charged particles in our bodies, you could not read this chapter or stand straight. In addition, these motions are also central to self-healing.

There is also the motion of charged particles and fluids underground, in the atmosphere, and even in the vacuum. These motions can also spontaneously produce significant amounts of energy essential for the exploration of the solid Earth's crust. This can even produce electricity for households and industry daily.

In more general terms, spontaneous systems, also known as self-organized systems, are very different from our current complex-arrange systems because they are very robust with respect to natural and anthropogenic hazards. Moreover they extend to other areas such as architecture, nuclear fission, nuclear fusion, and mechanical engines. There are tools developed and perfected by nature or higher intelligences that we do not need to reinvent. We really disadvantage future generations if these systems and the technologies associated with them are not integrated into our current education systems. This is necessary even when modern science and technology cannot explain some of them. In the case of natural disasters like total Sun flares or simultaneous multiple earthquakes, our modern technology, including records of our civilization, will be lost. Relying on spontaneous technology can be one of the keys to our survival in such cases.

Key sentences: *One of the most significant scientific challenges of the twenty-first century is understanding the human mind and brain, including the bioelectric signaling system that coordinates all body functions. Thanks to these bioelectric signals, one day our minds, brains, and physical machines, including computers, will form a seamless entity.*

5.1. Electricity in the Human Body

In this section we will discuss the spontaneous generation of electricity in our bodies (i.e., electricity production that occurs continuously in living individuals), the

DOI: 10.1201/9781032620619-5

electric excitability of tissues, and the behavior of electric and magnetic fields in and around our bodies. Bioelectromagnetism is electrical and magnetic activities. We begin by recalling the anatomy and physiology of the human nervous system on which bioelectromagnetism is based.

5.1.1.   Brain and Mind Interaction

The human brain is one of humans' organs—like eyes, nose, hands etc.—and probably the most complex living physical structure known to mankind in our solar system. Along with our minds, it shapes our thoughts, hopes, dreams, and imaginations, and coordinates all body functions. The brain influences the immune system's response to disease and determines how well people respond to medical treatments. It allows us to find novel ways of feeding the world, to improve the length and quality of life on Earth, to facilitate short- and long-distance communication and transportation, and to predict, adapt to, and mitigate natural hazards. In short, the mind and brain make us humans.

Note that plants, which are also living things like humans, have no brains. Other living things also possess conscious awareness, which is a property of the brain. Microorganisms react to the environment. Houseflies are more consciously aware than microorganisms, they have senses, they can find food, avoid enemies, find mates, and sexually reproduce. But they do not think. Some birds have considerable mental acuity. They can count, learn to use tools, and communicate in various ways. But they do not write. Chimpanzees are close to us, bright as three-year-olds. They have complex social lives, learn things, teach them to their offspring, and communicate in various ways. Their brains are like ours in shape. But they do not write or develop advanced transportation technologies.

Human brain and mind interaction is not analogue to software and hardware in classical computer. The human brain is spongy and has around 1.4 kilograms of mass (Carter, 2019) of fatty tissue inside a skull that protects it. It occupies space, metabolizes sugar, and performs trillions of operations per second using electrochemical signals, as we will discuss later. The brain is responsible for a wide range of functions, including perception, movement, emotion, and cognition. In contrast to the mind, the brain is a tangible organ, as illustrated in Figure 5.1. The mind is not an intangible organ to the naked human eyes and even to our entire electromagnetic spectrum as we understand it today. In any case, we have come to accept that it exists and captures our feelings, thoughts, sensations, language, memory, perception, and conscious (being awake) experience. The mind appears more linked to subjective experiences arising from brain activity.

The question of how mind and brain interact, whether brain events cause mind events or vice-versa, remains a formidable scientific challenge because the activities of the mind have yet to be physically observed, especially within our understanding of the electromagnetic spectrum. Because computer terminology such as memory, calculations, scale, encoding, decoding, transmitters, and neuronetwork are common in brain studies, it is sometimes suggested that the brain and mind work simultaneously, just like a computer's hardware and software, in which the mind is analogous to inputting data and running code under quantum mechanics and the tangible brain is analogous to computer hardware based on classical mechanics and physics. Many have even suggested that the interaction between the brain

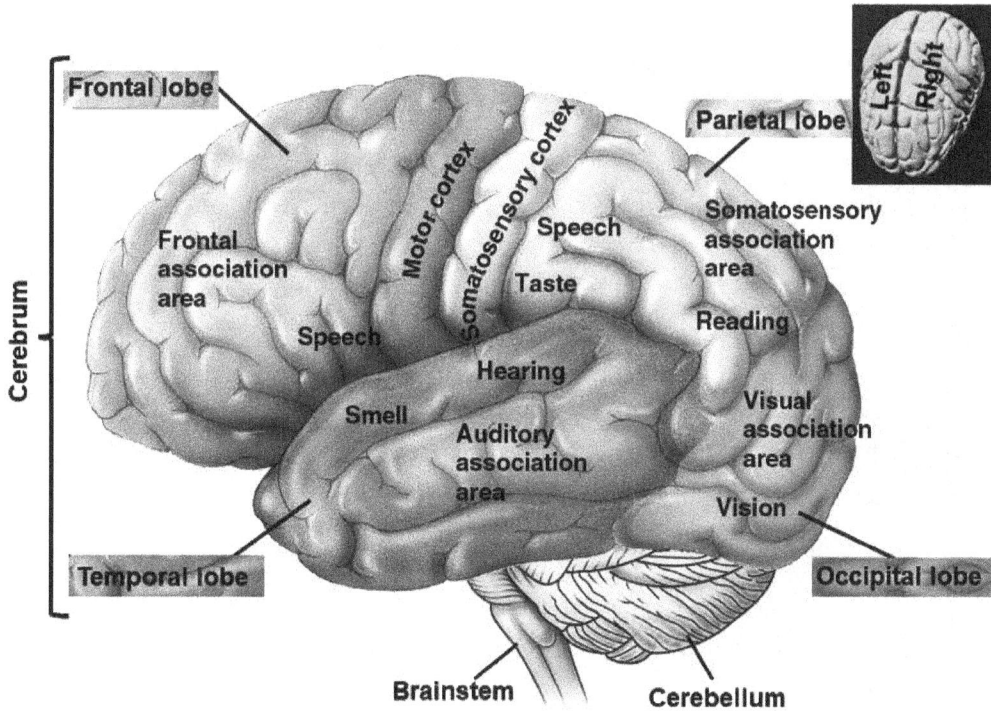

Figure 5.1. The brain anatomy. Some of the brain's main structures and functions. The structures share some functions and are highly connected. We have attached here a full view of the brain with the hemispheres.

(tangible part) and mind (intangible part) is comparable to that between an artificial intelligence system and its originator. These metaphors are wrong and must be retired. Remember that metaphors allow us to relate qualities from one thing to another. We know almost nothing about the mind, including its qualities. Moreover human memory storage appears to be directly linked to our emotions; a process without an analog in computers. Also, we do not know how human memory is biologically put together, let alone compare it to computer memory. Thus it makes no sense at this point to consider any metaphor between a spontaneous system and passive devices like current-era computers.

An hypothetical model of our mind. One thing we know is that the mind interfaces with our body through the brain, the actual nerve center with more than 86 billion[1] nerve cells (Herculano-Houzel, 2009; Azevedo et al., 2009) and trillions of specialized connections called synapses. As a result of our genome, development, learning, and experience, these connections govern how the brain works. Neurons that are very closely interconnected via axons and dendrites (see Figure 5.2).

---

[1]Total neuron quantity in the human brain was not directly measured. The brains of four recently deceased men were liquefied. Samples from the soups were measured, the number of neurons in each sample, and then scaled up to find the neuronal content of each brain bisqué (i.e., a smooth, creamy, highly seasoned soup of French origin). The idea that the human brain has 86 ± 5 billion neurons is a precise tally of individual brain cells. It is average.

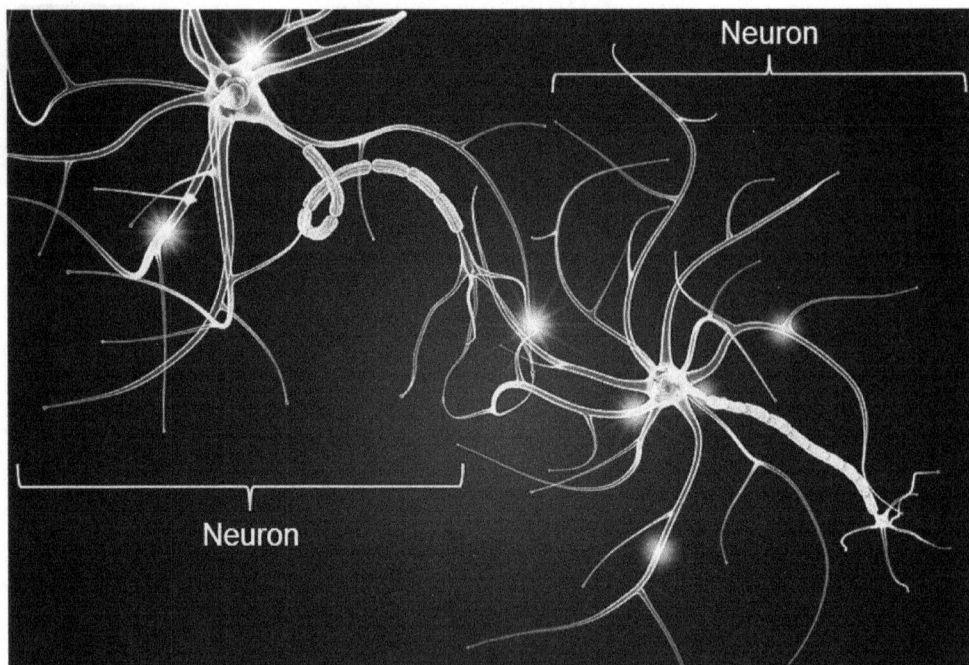

Figure 5.2. Illustration of two neurons connected by spontaneous electricity. The nervous system, including the brain, is made up of billions of connected neurons. Human behavior, thoughts, and feelings are driven by this interconnected web. A single neuron can be nearly 1.2 meters long.

Given that our brain is dominated by interactions between charged particles, it is very likely that our brain and mind interact through photons (as waves rather than particles; remember that photons are simultaneously waves and particles as shown in Ikelle, 2023b) and that our mind may well be a complex encoding of photons, just like light. A system of this type is intangible, untouchable, colorless, weightless, produces no sound in the classical time-space domain, and has no representation in 4D (i.e., in time and 3D space) since photons have no representation in 4D. Remember that photons are zero mass elementary particles and therefore occupy no space and their energy cannot be described by $\mathcal{U} = mc^2$. They are characterized only by their energy (i.e., $\mathcal{U} = hf$, where h is Planck's constant and f is the photon frequency) and spin. In more general terms, a particle's energy is $\mathcal{U}^2 = m^2 c_0^4 + p^2 c_0^2$, where p is momentum. So when $m = 0$, $\mathcal{U} = pc_0$. The momentum of photons is $p = hf/c_0$.

As we discussed in Ikelle (2023a and 2023b), our bodies die through molecular changes caused by aging, sicknesses, and/or accidents. Minds that operate at photon levels are not affected and are likely to survive. A mind, as a physical quantum entity, can only die by actions at sub-atomic levels such as nuclear explosions or any other action that affects atomics, large doses of terahertz radiation (millimeter waves), or any other sub-atomic levels and associated frequencies. Remember that electrons create and destroy photons. An electron moving in a strong magnetic field generates photons from its acceleration. Similarly, when a photon of the right wavelength strikes an atom, it disappears and imparts all its energy to kick the electron to an entirely different energy level. A new photon is created and emitted when the electron falls back into its original position. Nuclear explosions create and destroy photons in all these ways.

Brain diseases. It is critical to reiterate that the societal benefits of the studies of biological intelligence are not limited to stimulating the construction of advanced computers and artificial intelligence systems; they include potentially better treatments for diseases of the brain, such as Alzheimer's disease, Parkinson's disease, schizophrenia, depression, post-traumatic stress disorder, autism spectrum disorders, and so on. General brain injury symptoms may include: headache, nausea or vomiting, feeling confused or disoriented, dizziness, feeling tired or drowsy, speech problems, such as slurring, sleeping more or less than usual, dilation of one or both pupils, and fluid draining from your nose or ears, sensory problems, such as blurry vision or a ringing in your ears, trouble remembering things or difficulty concentrating, mood swings or unusual behavior. In addition, a consequence of a deeper understanding of the human mind is an understanding of the human decision-making process. In the latter part of this section, we further describe the key cells of the human brain and some of their functions. We also discuss the connection between human brain diseases.

It is remarkable that none of the diseases mentioned above are mind diseases. In other words, the mind does not fatigue. Also, emerging technologies for treating brain diseases seem to include millimeter-sized implant devices and are not affected by changes in the mind in nature.

5.1.2.   The Brain: An Electrochemical Network

The brain processes information and controls bodily functions. It is made up of billions of neurons that communicate through electrical and chemical signals. The brain is responsible for a wide range of functions, including perception, movement, emotion, and cognition.

Brain anatomy and functions. The brain has three main parts: the cerebrum, cerebellum and brainstem (see Figure 5.1). The cerebrum is the largest part of the brain and is composed of right and left hemispheres separated by a groove called the interhemispheric fissure (and also known as the longitudinal fissure). Each hemisphere can be divided into four lobes: frontal, parietal, temporal and occipital. In other words, each brain lobe has left and right sides. Each lobe can be divided further. For example, the temporal lobe can be subdivided into the amygdala, thalamus, hypothalamus (helps maintain homeostasis), etc (see Figure 5.3). There is no clear level at which subdivision becomes impossible. However, our brain does grow new neurons, not through division but through neurogenesis (the growth and development of nervous tissue) from neuronal stem cells. However, the neurogenesis does not keep up with attrition, and for the most part it is confined to certain areas of the brain, such as the hippocampus. The motor cortex in the frontal lobe gives the brain its characteristic wrinkled appearance. Although we now know that most brain functions rely on many different regions across the entire brain working in conjunction, it is still true that each lobe carries out the bulk of certain functions.

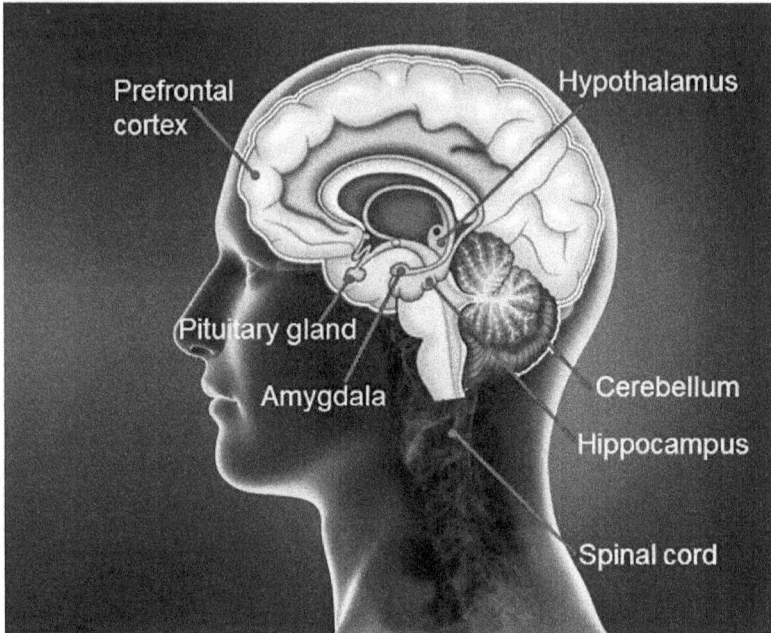

Figure 5.3. The limbic system includes the hypothalamus, the hippocampus, the amygdala, and several other nearby areas. It may be responsible for our emotional life, and influence the formation of memories.

The frontal lobe manages thinking, emotions, personality, judgment, self-control, muscle control and movement, memory storage and more. The temporal lobe of your brain manages our emotions, smell, processing information from our senses, storing and retrieving memories, and understanding language. The area of temporal lobe known as the amygdala manages emotions like fear and anxiety. It also contributes to how you feel when you get a reward and learning-related emotions. The hippocampus is another key area of the temporal lobe. It stores declarative memories that we can access, remember, and describe. Declarative memories include memories of events or memorized facts and information. Our hippocampus also helps with recognition memory, which is our ability to recognize something—such as objects, sounds or faces—based on stored memories. The parietal lobe is a key part of our understanding of the world around us. It processes our touch sense and assembles input from our other senses into a form you can use. Our parietal lobe also helps us understand where we are in relation to other things around us. Our occipital lobe processes visual signals and works cooperatively with many other brain areas. It plays a crucial role in language and reading, storing memories, recognizing familiar places and faces, and more. Common conditions and disorders affecting the brain include Alzheimer's disease (i.e., a decline in memory, thinking, learning and organizing skills over time), headaches, memory loss or forgetfulness, brain tumors (including *brain cancer*), seizures (e.g., epilepsy), and strokes. All four lobes are generally affected by these diseases.

Our cerebellum is part of your brain that coordinates and regulates a wide range of functions and processes in both our brain and body. These functions include balance, learning, emotion regulation, coordinating movement, and attention. It holds more than half of the neurons (cells that make up your nervous system) in our whole body.

The brainstem connects the brain to the spinal cord. It consists of three major parts: the midbrain, pons, and the medulla oblongata. The midbrain controls eye movement and processes visual and auditory information. The pons is a group of nerves that connect different parts of the brain. The pons also contain the start of some cranial nerves. These nerves control facial movements and transmit sensory information. The medulla oblongata is the lower part of the brain. It acts as the control center for the heart and lungs. It regulates many important functions, including breathing, heart rate, blood flow throughout the body, alertness, sleep patterns, sneezing, and swallowing.

Humans are animals! This means that human brains have a lot in common with many other animal brains. Almost all animal brains have the same basic parts: parts to help us move, think, and sense the world around us. Although the basic parts of the brain are the same among most animals, every animal's brain does something a little bit different and special. For example, cats have very good eyesight, and have more brainpower for their sense of sight. Similarly, dogs have a very good sense of smell, so the part of their brain that can identify different scents is very powerful compared to other animals.

Neurons (i.e., nerve cells in the brain). Neurons are information messengers. Basically, they are sensors and wires that transmit data inside the brain and around the body. One thing that makes neurons so special is that, unlike other cells, neurons do not reproduce or regenerate, as mentioned earlier. They are not replaced once they die. When a neuron is damaged along the way to another neuron, axons face a hostile environment full of molecular *stop signs* that signal *no trespassing* to axons. Some stop signs are part of the myelin sheath (see Figure 5.4). Some build a protective wall around the injury neuron to keep damage from spreading. The latter are made by brain cells called *astrocytes*.

It is through the process of *neurogenesis* that neurons form in the brain. Neurogenesis is crucial when an embryo develops, but also continues in certain brain regions after birth and throughout our lifespan. The new neurons are produced in the hippocampus regions of our brains. Again, the newly formed neurons are not replacements; they form new connections to occasionally distant parts of the cortex to aid in memory consolidation. Current estimates show no more than 40 new Hippocampus neurons each day. This amounts to about a million new neurons over a lifetime, but that is not much compared to the 16 billion neurons in the cortex.

The most common brain cells are neurons (about 86 billions) and glials (about 84 billions roughly the same number of neurons as there are stars in the Milky Way, around 100 billion). Although neurons are the most famous brain cells, both neurons and glial cells (or glia) are necessary for proper brain functioning. There are actually many more glial cells than neutrons. Neurons send and receive electrochemical signals to and from the brain and nervous system, which consists of the central and peripheral nervous systems. The nervous system includes nerve cells, muscle, and the spinal cord, or gland cells. A typical neuron has 10,000 connections thus the brain itself contains perhaps 860 trillion connections ($86 \times 10^9 \times 10^4$) altogether or

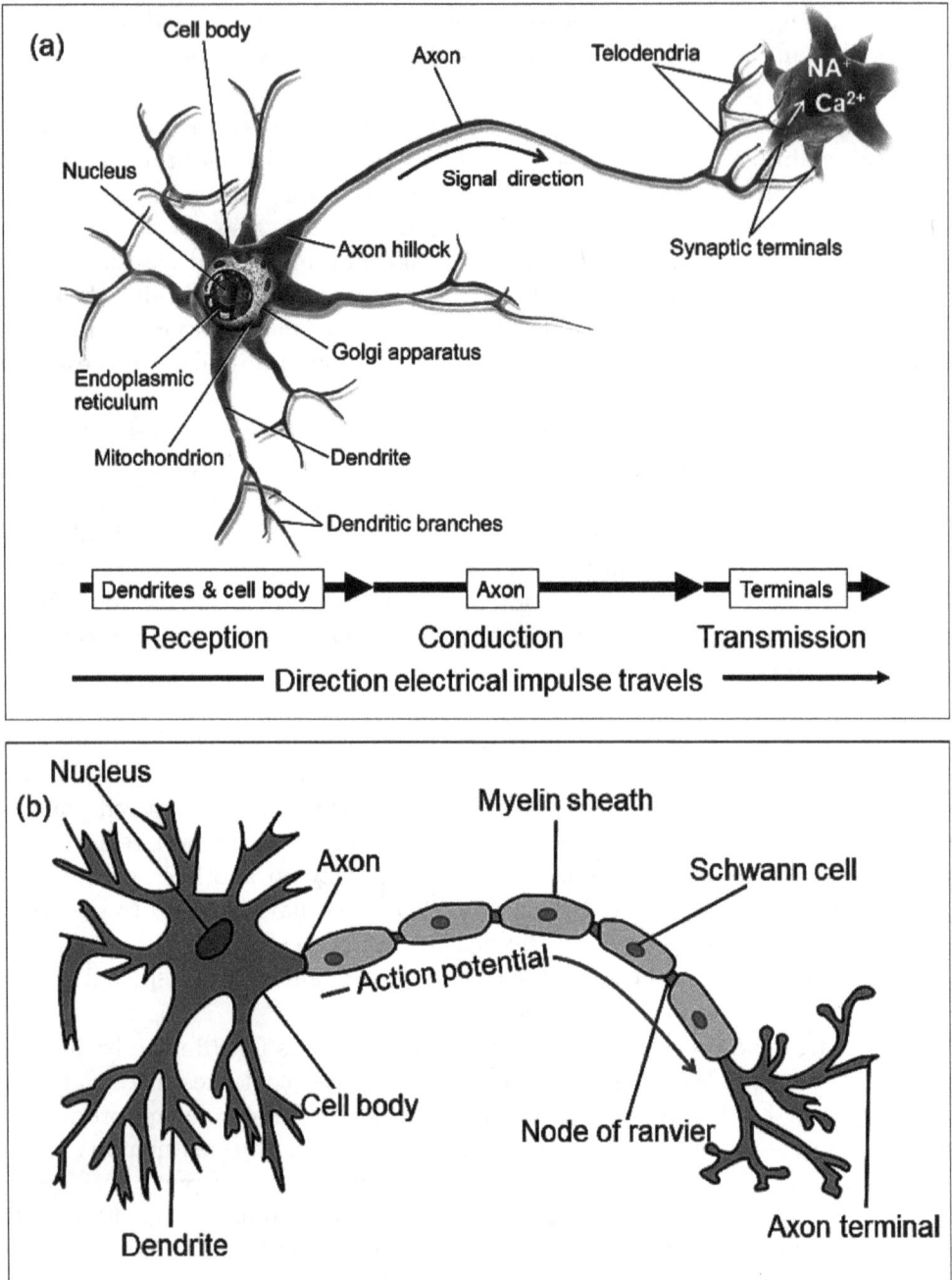

Figure 5.4. (a) A simplified rendering of a neuron. All neurons have three parts: dendrites, cell body, and axon. A single axon transmits signals away from the cell body, while multiple dendrites do the same. The arrows indicate the direction in which signals are conveyed. The dendrite receive information, the cell body processes and integrate that information, the axon carries from one part of the neutron to another, and the axon terminal transmits the information to the next cell. Neuron varies in shape and size depending upon their function and location. A bundle of axons traveling together is called a nerve. (b) A description of an axon; an axon is a tube-like structure that carries electrical impulses from the cell body to the axon terminals, which transmit the impulse to another neuron. The signal propagates in an axon as an unattenuated nerve impulse. Starting with receiving a chemical signal from another neuron, the sequence of transmission of an electrical impulse through a neuron is: dendrites $\longrightarrow$ cell body $\longrightarrow$ axon $\longrightarrow$ axon terminal.

about 86,000 km of interconnections. If we very roughly approximate the brain by a computer operating at about 100 Hz (frequency of neuron's vibration), we end up with 86×10^{15} (or $86 \times 10^9 \times 10^4 \times 10^2$) operations per second per brain.

The glia, also known as glial cells or neuroglia, hold neurons together like glue. They also participate in brain signaling, and are necessary for neurons' healthy function. However, glial cells do not send nor receive electrochemical signals. They maintain homeostasis (the state of steady internal chemical and physical conditions maintained by living systems; homeostasis is any self-regulating process by which biological systems tend to retain stability while adjusting to conditions that are optimal for survival), form myelin in the peripheral nervous system, and provide support and protection for neurons. As illustrated in Figure 5.4, myelin is an insulating layer, or sheath that forms around nerves, including those in the brain and spinal cord. It is made up of proteins and fatty substances.

Neurons are made of the same materials as other cells in the body. They share the same organelles other cells have (i.e. mitochondria, ribosomes etc.) and have the exact same DNA. Neurons are molecular masses of organized molecules such as phospholipids and proteins. At the atomic level they are made of different atoms and ions.

A neuron has three basic parts: the cell body (also known as the soma), dendrites that receive signals from other neurons using their many branches, and the axon that sends signals out to surrounding neurons through the axon terminal, and carries some types of information back to the soma, as shown in Figure 5.2. The cell body contains the nucleus. The nucleus contains genetic material in the form of chromosomes, as described in Ikelle (2023a). The axon extends tens, hundreds, or even tens of thousands of times the soma diameter in length. Many neurons have only one axon, but this axon may—and usually will—undergo extensive branching, enabling communication with many target cells. The part of the axon where it emerges from the soma is called the axon hillock. The axon terminal is found at the end of the axon, at the farthest tip of the axon's branches, from the soma and contains synapses. You can think of them as signal transmitters. Although not included in some neuron diagrams, some long axons are covered with myelin sheaths. Dendrites also extend a few hundred micrometers from the soma. Thus, a neuron receives multiple electro-chemical signals (as we will see in the next subsection) through the dendrites and soma and sends out signals down the axon; the signals cross from the axon of one neuron to a dendrite of another. Neutrons receive and transmit information that allows our bodies to respond. Also neurons are highly specialized in cellular signal processing and transmission. Given their diversity of functions performed in different parts of the nervous system, there is a wide variety in their shape, size, and electrochemical properties.

Neurons are classified into three types based on their location. (1) Sensory neurons respond to stimuli such as (but are not limited to) touching, hearing, seeing, tasting, and smelling. These neurons are activated by sensory input from the environment. They react by producing internal stimuli. For example, the skin contains millions of sensory receptors that gather information related to touch, pressure, temperature, and pain. These receptors send it to sensory neurons (i.e., part of the brain) for processing and reactions. The tongue contains small groups of sensory cells called taste buds that respond to chemicals in foods. Taste buds react to sweet, sour, salty, bitter, and savory flavors. The taste buds send messages to sensory neurons for processing taste. (2) Motor (efferent) neurons receive signals from the brain and spinal cord to control everything from muscle contractions to glandular output. (3) Interneurons connect neurons to other neurons within the same region of the brain or spinal cord.

In addition to the five main senses (smell, touch, sight, taste, and hearing) that we humans possess, there is one more powerful sense called the vestibular sense that enables our body to recognize movements and use them to maintain balance. This system is situated in the ears, as shown in Figure 5.3. It is composed of otolith organs and semicircular canals. We can sense the direction and speed of linear acceleration (speed changes without change in direction) due to the otolith organs while the semicircular canals enable us to sense the direction and speed of angular acceleration (speed changes along with change in direction). The vestibular sense works with Earth's gravity and provides sensory information about motion, equilibrium, and spatial orientation. Take gravity away and we can no longer figure out where the ground and ceiling are because we lose spatial perception. It is our vestibular system that allows us to walk a tightrope or twist while diving. It actively keeps track of the position of our arms and legs and allows us to perform these tricks without losing balance.

The brain also makes new cells, as do most organs in our bodies. They are known as stem cells (a type of empty cell). These cells are unspecialized. They are like blank sheets of paper. Similar to blank sheets of paper, they can become different types of specialized cells. When stem cells remain alive, they can divide into other stem cells and specialized cells. There are many stem cells in embryo brains, which means that as the brain grows, all kinds of specialized cells grow. In other words, brain cells are born as embryos. Stem cells generate neurons during brain development and childhood. Interestingly, stem cell production persists in the brain into adulthood.

To summarize, the human brain has 170 billion cells. The cerebral cortex has 16 billion neurons and 61 billion glia. The cerebellum has 69 billion neurons and only 16 billion glial cells. The other parts of the brain have, together, 7 billion glial cells and 700 million neurons.

Emotion and depression. Emotion triggers brain memory. Because the brain encodes a memory, it signals when it should or should not encode something. Because our memories define who we are, our emotions are very critical and we do not need to work too hard to suppress them. Some of our emotions are externally triggered, collectively triggered, or internally triggered. Nearly all Cameroonian soccer fans watch their country's world cup matches in expectation of victories. These fans feel emotions both individually and collectively. Victims of robbery experience externally generated emotions, but not together. Birth generates internal emotions in a woman. All these three examples will be stored in our memories because of the emotional energy associated with them. Without emotion energy we would unnecessarily encode information we might not need to and that is a huge cost to the brain.

From the electromagnetic point of view, emotional energy images light up neurons in our brain. These neurons produce electric energy that travels to the visual cortex at the back of your brain. The cortex interprets the sensory world. It interprets the visual, auditory, and tactile worlds. So all the senses come to the brain's surface. The signal then travels to the Hippocampus, which creates memory. Directly adjacent to it is the amygdala (Figure 5.3), which is the emotional center. The Hippocampus and Amygdala are so densely interconnected that their neurons, dendrites, and connections between them are almost inextricable. As a result, images of soccer games and the robber, for example, immediately trigger emotion in the amygdala. So we have evolved from visual perception to memory and emotional centers. This translates into experiences in the body, including fear, through the signals that travel to the Hypothalamus, which is just above the surface of the brain. The Hypothalamus is the place from which the commands for the autonomic nervous system in the body emanate and thus range of experiences that the autonomic nervous system mediates in our body.

So if you feel threatened, your person's brain circuitry made up of the Hippocampus, Amygdala, and Hypothalamus activates another system in the body called the endocrine stress system. This system includes stress hormones. The signal enters the bloodstream and carries a hormone from the hypothalamus to the adrenal glands. In the adrenal glands the hormone cortisol is released. A sustained release of these hormones can occur in depressed people (i.e., people under extreme emotional and cognitive stress for a long period of time). You damage your memory with sustained depression. Your Hippocampus shrinks and your Amygdata increases. A bias toward strong emotional reactions developed at the cost of memory as the emotional center expanded.

Figure 5.5 shows the functional topography of the brain of a depressed patient. It turns out that there is not just one spot. Depression symptoms include mood

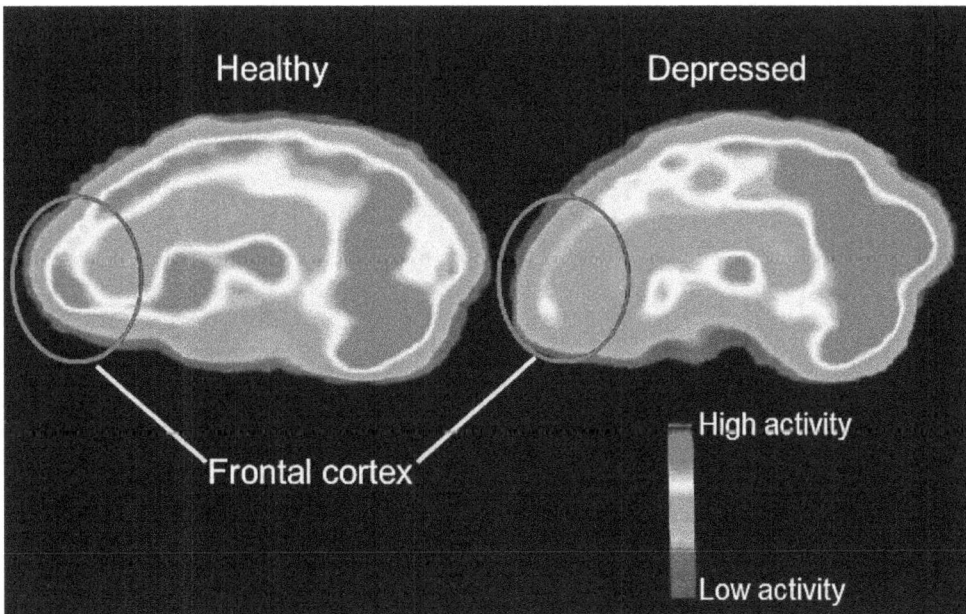

Figure 5.5. Left: A scan of a healthy brain, including healthy blood flow to the frontal lobe. Right: A scan of a depressed brain, including reduced flow to the frontal and cingulate cortex.

swings and lack of drive, thought, and action. Thus it is not surprising that there are multiple brain areas acting alone or not acting together that are abnormal. As we don't know exactly what initiates depression and what stage depression is, the patterns in Figure 5.5 are very dynamic and non-repeating.

Intelligence, learning, and memory. Intelligence is still a vague concept. When we see intelligence forms, we may not even recognize them. Anyway, we here assume that intelligence is a psychological construct that we invented to describe and explain some source of variance in human behavior. Like all behavior, intelligence manifests itself from neural activity and other biological signals. These patterns of neural activity include neuron firing. Which neurons? Which brain areas? What causes some people to have higher intelligence than others? What can you do to improve your intelligence?

One of the most fascinating and puzzling aspects of the human brain is its natural biological growth from a tiny brain to an enormous brain capable of understanding and manipulating the world at large. This growth is very complex and still largely not understood. As we grow and learn, messages travel from one neuron to another over and over, creating connections, or pathways, in the brain. As we get older, the brain has to work harder to form new neural pathways. This makes it difficult to master complex tasks or change set behavior patterns. That is why it is important to keep challenging the brain to learn new things and make fresh connections; it helps keep the brain active over the course of a lifetime. This observation also suggests that environments and conditional environments are factors in the growth of human intelligence. Capturing the growth of the human brain is one of the major challenges of artificial intelligence systems.

Memory is another complex brain function. Things we have done, learned, and seen are first processed in the cortex. Then, if we sense that this information is significant enough to remember permanently, it is passed inward to other regions of the brain (such as the hippocampus and amygdala) for long-term storage and retrieval. As these messages travel through the brain, they create pathways that serve as the basis of memory. Reading, for example, which is a central component of our intelligence in modern society, requires vast knowledge and therefore vast storage capacities.

Human declarative knowledge appears to be stored holographically, in connection patterns. For a long time, neurobiologists thought information must be localized, with memory stored in one cell or a group of cells. Now we know the information that makes up each memory is stored across networks of connections between many cells. A particular cell or set of cells may store parts of many memories. As more memories are stored, less strong memories are not deleted or just get fuzzy and fuzzy. The details blur until eventually they're no longer re-callable at all.

Sleep remains one of modern science's most intriguing mysteries. Without enough sleep, people have trouble focusing and responding quickly. In fact, sleep loss can have as big an effect on productivity as drinking too much alcohol. It is also critical for our emotional health. Some suggest that a lack of sleep increases the risk of a variety of health problems, including diabetes, cardiovascular disease and heart attacks, strokes, depression, high blood pressure, obesity, and infections. The current understanding based on electroencephalography (EEG, see a later subsection) experiments is that the brain is highly active during sleep. Heart rate, blood pressure, and body temperature all fall during sleep.

During sleep, some memories are erased and some are recovered. During sleep, glia erase memories you don't need, allowing your brain to make new, stronger connections in the future. One can also use a part of the terahertz-frequency spectrum at a certain intensity to clear some memories from the brain although side effects are still to be determined (see Chapter 3). Another approach is to implant devices (such as deep brain stimulation; DBS) in the brain to disrupt connectivity in neural circuits. Alternatively the implanted devices can be used to reestablish the functionality of neural circuits disrupted by depression, obsessive-compulsive disorder, brain injury, infection, and other psychiatric disorders (Benabid, 2007; Hampson et al., 2013; and Lozano and Lipsman, 2013). One of the fundamental unresolved questions about sleep is: Is it what our minds are doing during this period of recharge of body energy?

Attention. Cognitive aspects include attention. How do we prioritize incoming sensory information, filter and format it in a way appropriate for both short-term and long-term action planning? Suppose, for example, that we need to understand what one person is saying when others speak at the same time (the cocktail-party problem)? This problem is known as the cocktail-party problem, which was formulated by Colin Cherry and his coworkers more than sixty years ago. Figure 5.6 provides an illustration of this problem, which involves several people speaking simultaneously in a room containing two microphones that represent human ears [(Cherry, 1953, 1957, and 1961); (Cherry and Taylor, 1954); (Cherry and Sayers, 1956 and 1959); and (Sayers and Cherry, 1957)]. If we have I persons speaking, the sound signal produced is a mixture of I sound signals (Figure 5.6). Our brain focuses on one output and ignores the other signals up to some point. How does our brain filter out undesirable signals? We are not born with attention to focus on one person. Infants capture the entire mixture. As we grow, our dictionary forms multiple words, expressions, and sentences that we can access almost instantaneously. Additionally, our dictionary contains significant redundancy over time, allowing us to extrapolate, even interpolate, from partial sounds. That is a number of features of our civilization are acquired through time, and therefore there are alternative ways to develop them and even render them more effective.

We have not built a decoding machine for the cocktail party problem so far despite multiple efforts. This is because our criteria for constructing the dictionary fail to include redundancy, extrapolation, and interpretation. A more effective dictionary must mimic the human brain's focus and attention if we want effective AI systems.

Microtubules (e.g., Hameroff and Penrose, 1996) are tiny tube-like elements of the cell cytoskeleton[2] . Figure 5.7 shows a neuron with the nucleus and microtubes—

---

[2]The cytoskeleton is a microscopic network of protein filaments and tubules in the cytoplasm of many living cells, giving them shape and coherence. It extends from the cell nucleus to the cell membrane and is composed of similar proteins. It helps cells maintain their shape and internal organization. It also provides mechanical support that enables cells to perform essential functions like division and movement.

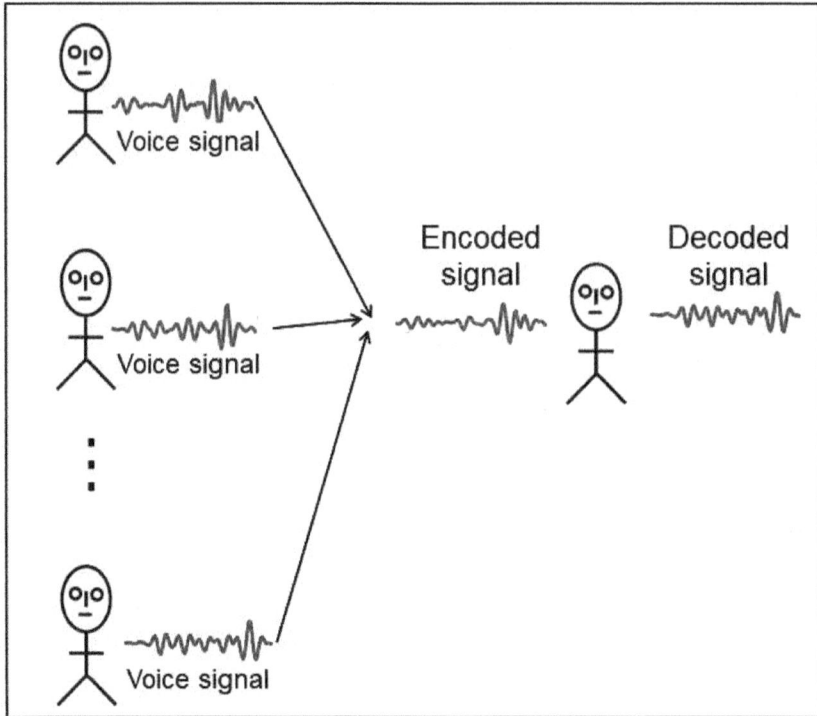

Figure 5.6. Cocktail-party problem. If I people speak at the same time in a room containing two microphones, then the output of each microphone is a mixture of two voice signals. Given these two signal mixtures, a decoding process aims at recovering the original I voice signals.

remember that a cell has only one nucleus. They are straight, hollow, fibrous shafts, with a diameter of about 25 nanometers (i.e., down to the atom level which is between 0.1 and 0.5 nanometers) that contribute to the form of eukaryotic cells (prokaryotes do not have them). Microtubules in a cell vary between 200 nanometers and 25 micrometers. This depends on the task of a particular microtubule and the state of the cell's life cycle. They are actually the largest cytoskeletal filaments in cells and are found throughout the cytoplasm. They are the skeleton on which neurons are built and give structure and shape to a cell. They form a network in the brain that serves as a transportation route for vesicles, granules, organelles like mitochondria, and chromosomes via special attachment proteins. They are becoming essential for understanding brain functions, including memory loss. So a neuron is not a water bag. It is highly structured internally with a cytoskeleton and microtubes.

Microtubules are biopolymers that are composed of subunits made from an abundant globular cytoplasmic protein known as tubulin, as illustrated in Figure 5.7. Each subunit of the microtubule consists of two slightly different but closely related simple units called α-tubulin and β-tubulin that are bound very tightly together to form heterodimers, as the tubulin is made up of two different subunits creating a dimer. In a microtubule, the subunits are organized to form 13

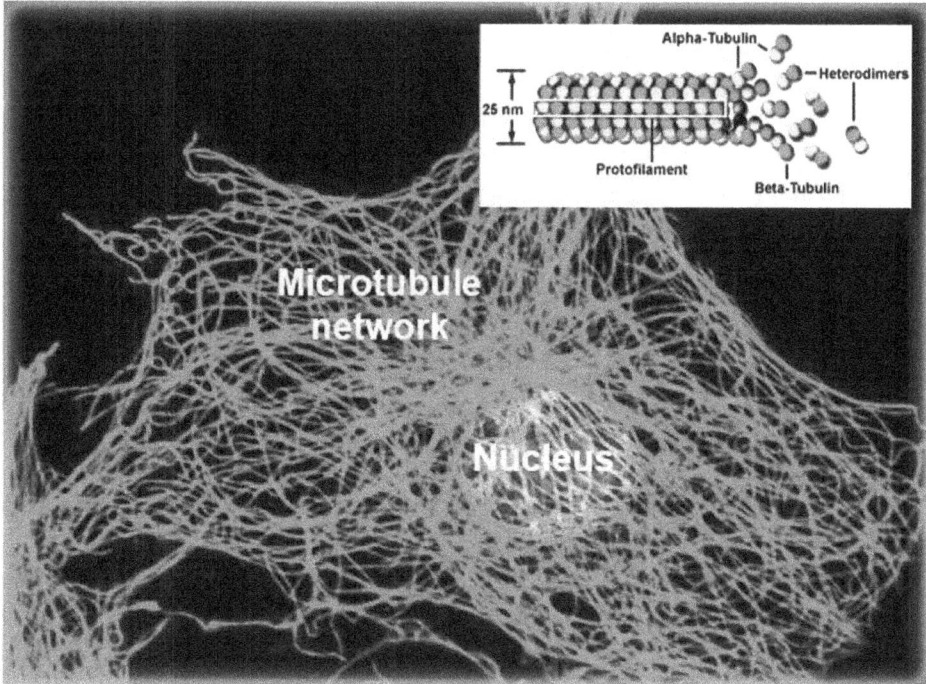

Figure 5.7. A microtubule network as seen through a fluorescence optical microscope. These microtubules are straight, hollow cylinders found throughout the cytoplasm of all eukaryotic cells (prokaryotes do not have them). They are about 25 nanometers in diameter and their lengths vary between 200 nanometers and 25 micrometers. We appended here an organization of tubulin dimers inside one protofilament. Note that α-tubulin is oriented toward the minus end, whereas β-tubulin is toward the plus end. Adapted from https://www.olympus-lifescience.com/en/microscope-resource/galleries/confocal/cells/gelu/gelusb1/

parallel protofilaments. This organization gives the structure polarity, with only the α-tubulin proproteins exposed at one end and only β-tubulin proteins at the other.

By adding or removing globular tubulin proteins, polymeric microtubule length can be increased or decreased. Because the two ends of a microtubule are not the same, the rate at which growth or depolymerization occurs at each pole is different. The end of a polarized filament that grows and shrinks the fastest is known as the plus end and the opposing end is called the minus end. For all microtubules, the minus end is the one with exposed α-tubulins. In an animal cell, it is this end that is located at the centriole-containing centrosome found near the nucleus, while the plus end, comprised of exposed β-units, is projected out toward the cell's surface. Microtubules are continuously assembled and disassembled so that tubulin monomers can be transported elsewhere to build microtubules when needed. It is important to note that the microtubules in dendrites are discontinuous, with some polarities mixed—one pointing one way, the other next to it pointing the other way—while those in axons, for example, and all other non-neuronal cells are continuous and unbroken.

If we very roughly approximate the brain as a computer with microtubules as bits (or qubits) operating at 10 million Hertz (frequency of microtubule's vibration), with a billion tubulins per neuron, we will end with 10^16 operations per second per neuron 86×10^{25} (or $86 \times 10^9 \times 10^9 \times 10^7$) operations per second per brain. For the foreseen future, artificial processes and systems will never match the density of computation and efficiency of the human brain.

Quiz: How are neuron cells different from other cells in our bodies?

5.1.3.   Limits of Humanoid's Brains

Life expectancy of 120 years is a challenge and a potential limit. The human brain grows three times in size in the first year of life along with the body while maintaining a brain-to-body ratio of between 1:40 and 1:50.[3] The body and head continue to grow until we are about 18 years old. The human brain begins to lose some memory abilities as well as some cognitive skills in our late 20s. As we age, the brain shrinks, as does the body. The brain-to-body ratio stays at about 1:40.

Human brain evolution requires metabolic trade-offs. When this trade-off holds, our brain, which is only about two percent of our body mass, uses up to 20 percent of the blood in our body, 20 percent of our oxygen intake, and about 20 percent of all the energy in our body. Neurons in our brain are the main consumers of this disproportional amount of oxygen, blood, and energy. Moreover neurons need oxygen and energy to survive. As energy production decreases, the brain receives less energy and shrinks. Consequently some neurons died. If the wrong neurons died off, so would we. Also, neuronal death creates a ton of garbage in our brains that kill us.

Aging is characterized by a progressive loss of brain volume at an estimated rate of 5 percent per decade after age 40. This rate is still faster after age 70. The cerebral gyri (plural of gyrus) become narrower, the sulci (plural of sulcus) become wider, the cortex gets thinner, the ventricles are noticeably enlarged, and there's more space between the brain and skull leading to neurons dying. Our cerebral cortex, the wrinkled outer layer of the brain, thins as you age. Aging is also noticeable in the frontal lobe, which processes memory, emotions, impulse control, problem-solving, social interaction, and motor function. As an example of an old-age brain disorder, Alzheimer's disease slowly destroys thinking and memory, thus destroying neurons and the connections between them (Figure 5.8). It ultimately interferes with a person's ability to perform their regular daily activities. Currently, the disease cannot be cured.

By 100 years old, we have lost more than one quarter of the brain, including a quarter of its neurons. Almost half of the brain is lost at 150 years old. Current life span is between 70 and 80 years, which corresponds to 15 and 20 percent brain loss. Because the brain is still working, all components of the brain including neurons, neuroglia, fat, and water have decreased by near the same percent in terms of their number and volume. Thus our neurons decline from 86 billions on average to 68.6 billions when we are 80 years old. At this age, not only the brain shrinks and our weight significantly decreases, but more importantly, we lose neuron connections

[3]Brain–body mass ratio, also known as the brain–body weight ratio, is the ratio of brain mass to body mass.—the ratio of brain weight to the entire body weight (e.g., small ants 1:7; small birds 1:12; mouse 1:50; cat 1:100; dog 1:125; elephant 1:560; and hippopotamus 1:2789).

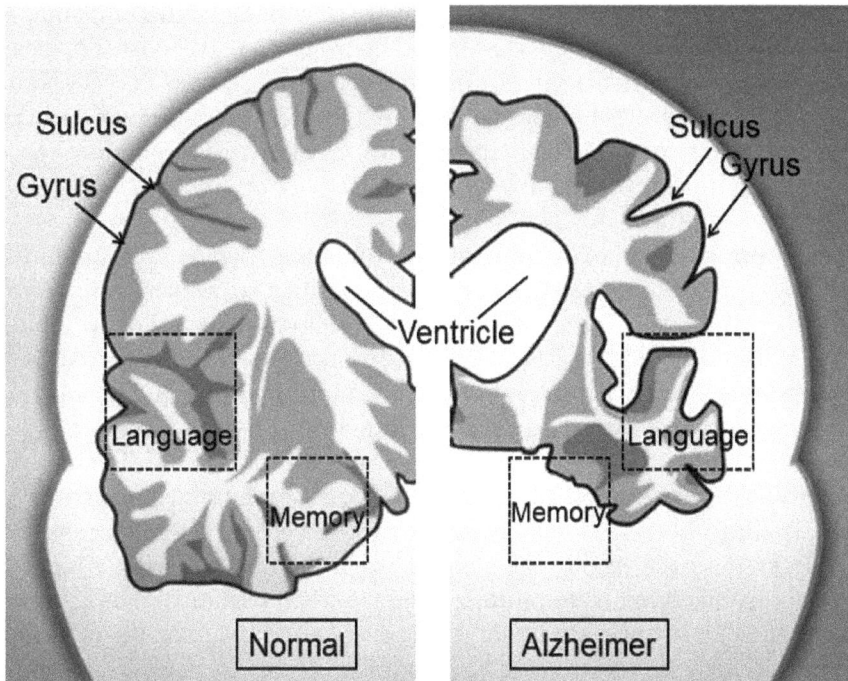

Figure 5.8. These images represent a cross-section of the brain seen from the front. The cross-section on the left shows a normal brain and the one on the right represents a brain with Alzheimer's disease.

that determine our memory and therefore who we are. Even in the 1950s, many people lived into their 70s and 80s. Thus it was easily sampled from this group to establish that hygiene, life style adoption, and reducing risky activities were sufficient to help a large number of people to reach 70 to 80 years. There is no obvious path to increasing life expectancy beyond 80 years. Living in good health has not stopped multitude of old-age diseases that can damage the brain thus limit our life expectation.

Cells in our bodies, other than the brain, are constantly renewed and repaired. In contrast, the brain cells capacity to repair themselves is extremely limited and is only effective in certain forms of injury where the axonal anatomy that allows transport and reconstruction of components is preserved in some way. Only a few regions, such as the hippocampus, generate new neurons in the brain as we age. The vast majority of neurons are laid down during early development and are with you for life. In other words, the brain continues to add neurons even when we reach our 70s in limited regions. However, the losses far outnumber the gains—which means significant net losses as we age. The good news is that neurons in human brains can live for 120 years. To extend a healthy life expectancy to 100 years or more, we must figure out how 90 percent or more of our neurons can stay healthy after 100 years. Ikelle (2022d) discusses various methods of doing so, including stimulation and noninvasive surgery/treatments. It is possible, however, that human life expectancy could be limited to 120 years based on neurons' life expectancy. Extending beyond this limit means that passing the limit may require massive genetic manipulation so that most neurons can live past 120 years. By the way, creating more neutrons is only attractive attractive at very young age otherwise new neurons create new paths and therefore new memory while losing the old memory. New neurons can change who we are. Moreover, they can change the brain-to-body ratio undesirably.

Brain–body mass ratio of humanoids. Humanoid beings are nonhuman creatures or beings with characteristics such as the ability to walk upright resembling humans, with two arms and two legs. Figure 5.9 shows a sketch of two humanoids with different brain-to-body ratios. As the brain-to-body ratio increases beyond that of humans, the brain becomes larger with more neurons and more connectivity between neurons compared to those of humans. There is no doubt that their bodies are different from ours to account for the environment. Mathematically, the brain-to-body ratio can increases to infinity where the humanoid is entirely brain made. The brain-to-body ratio can also decrease to zero as the body mass becomes infinitely larger than the brain. So, at least, some humanoids with a higher brain-to-body ratio are likely to have more mental capability than humans. In artificial neuron networks, it is suggested that neuron connectivity plays a critical role in determining intelligence. Moreover, the complexity of the neural circuitry of the cerebral cortex is considered to be correlated to the brain's coherence and predictive power. This is, therefore, a measure of some intelligence. The anatomy of the humanoid body are very different from between two significant brain-to-body ratios, including significant difference in genome and proteome.[4]

Figure 5.9 shows one possible skull form featuring a large brain. One can also consider a vertically and/or horizontally elongated skull. In some environments, the brain may contain less fat and water than ours. In those cases, the skull may not need to be larger than ours to contain a larger amount of neurons than ours. The increase in humanoids' brain-to-body ratio with their increasing brain size is monotone but not necessarily linear. Also the chemistry of body parts is expected to change drastically as the brain-to-body ratio of humanoids increases. For example, bones can be hollow with special struts inside to strengthen them.

Gravity and temperature on a planet or exoplanet can also influence the brain-to-body ratio. As the brain-to-body ratio increases, the gravity likely decreases from the one on Earth the body mass decreases while the gravity increases from the one on Earth the body mass of humanoids increases. Because bones and muscles represent a significant amount of body mass and are subjected to rapid losses as gravity decreases, as discussed in Chapter 1, it is very likely humanoids at low gravity have less bones and muscles or have an entirely different body anatomy. As the brain-to-body ratio increases, the temperature increases from

---

[4]A genome refers to the complete set of genetic material (DNA) of an organism, including all genes and non-coding regions. A human's genome contains all the information necessary for it to develop, function, and reproduce. It is the genetic makeup of an individual. Proteome, on the other hand, combines the two words protein and genome. It refers to the complete set of proteins expressed by an organism's genome, tissue, or cell at a particular time. Proteins are the functional products of genes and carry out a wide range of biological processes within cells, including structural support, transport, signaling, and enzymatic catalysis. Proteomes are dynamic and change in response to developmental, environmental, or disease-related factors.

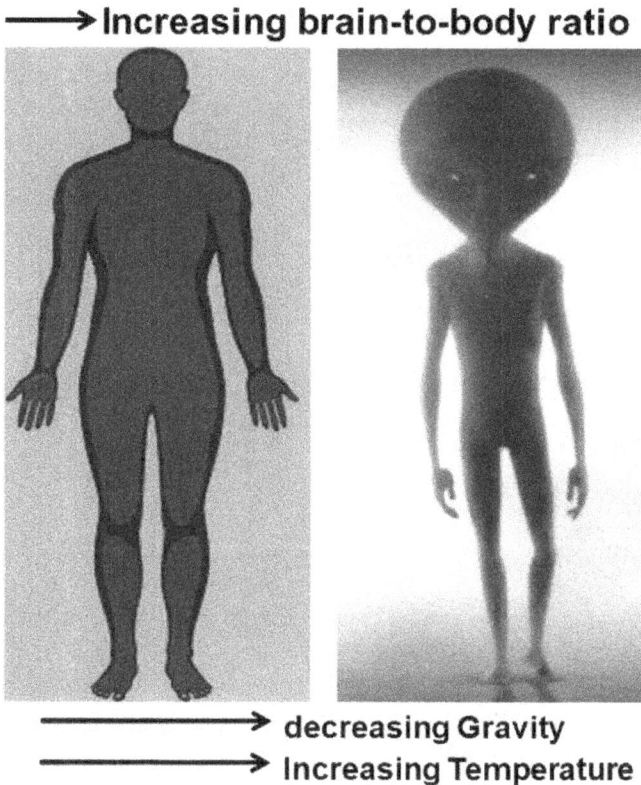

Figure 5.9. Sketch of two humanoids with different brain-to-body ratios.

Earth. But when the temperature decreases from Earth, humanoids' body mass increases. In general, humanoids have a large body mass at low temperatures for extra protection, as discussed in Chapters 1 and 2.

From apes to humans. During our period of evolution from ape to human, not only the number of brain's neurons increased drastically but also our brain-to-body ratio also increase from 1:180 to 1:40. Accordingly, the human brain substantially grows in size, and the rest of our body mass decreases from that of apes. How did we evolve from an ape with 6.4 billion neurons to humans with 86 billion neurons? Possible answers include spontaneous genetic manipulation and/or artificial genetic manipulation. There will have been more than one binary pair of apes that have made it into human binary pairs as a result of spontaneous genetic manipulation, which means multiple binary pairs of apes made it into human binary pairs. The fact that only one binary pair of ape made to a human binary pair, according to archeology data that date an early human in an Ethiopia 233,000 years, suggests that this evolution likely occurs through an artificial genetic manipulation. The fact that apes never have tails may be critical in this selection to be turn apes to walk upright primates.

Human brains have much in common with animal brains (see Figure 5.10). For example, chimpanzees have a cortex that uses information coming in from our senses to assist them understand the world, and sends signals out to other parts of the

Animal:	Ape	Human	Elephant
Brain's picture:	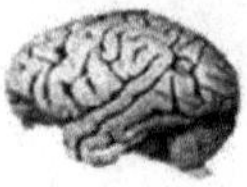	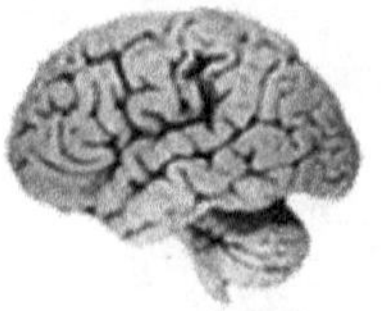	
Brain's weight:	0.083 kg	1.36 kg	2.85 kg
Brain's neurons:	22×10^9 neurons	86×10^9 neurons	251×10^9 neurons
Cerebral cortex's neurons:	6×10^9 neurons	16×10^9 neurons	6×10^9 neurons
Brain-to-body ratio:	1:180	1:40	1:560

Figure 5.10. Size comparisons between the brains of apes, humans, and elephants.

brain and body to help it move and communicate just like humans. The cortex is also responsible for learning, thinking, and decision-making. If you compare the human brain to that of other animals in terms of neurons, we can see in figure 5.10 that human brains do not have more neurons than elephant brains, for example. However the human cerebral cortex has nearly three times as many neurons as the large cerebral cortex of an elephant. Despite the size of the African elephant cerebral cortex, the 5.6 billion neurons in it pale in comparison to the average 16 billion neurons concentrated in the much smaller human cerebral cortex. Unless we were ready to concede that the elephant, with three times more neurons in its cerebellum (and, therefore, in its brain), must be more cognitively capable than we humans, we could rule out the hypothesis that the total number of neurons in the cerebellum was in any way limiting or sufficient to determine the cognitive capabilities of a brain.

5.1.4. Membrane Potential: Signaling within the Body and with the Environment

Our objective in this subsection is to describe how electric currents drive communication between cells, on the one hand, and between cells and the environment external to the cell, on the other. These currents are electrochemical because they are generated by the movement of ions instead of electrons. For example, remember that an ion is an atom or molecule with a net electric charge due to the loss of one or more electrons (anions) or the gain of one or more electrons (cations). The ions that reside in human cells include sodium (Na^+), chloride (Cl^-), potassium (K^+), calcium (Ca^{++}), and proteins (A^-). These ions move across the cell membranes (also known as the plasma membrane or cytoplasmic membrane) that separate the interior of all cells from the outside environment (also known as the extracellular space).

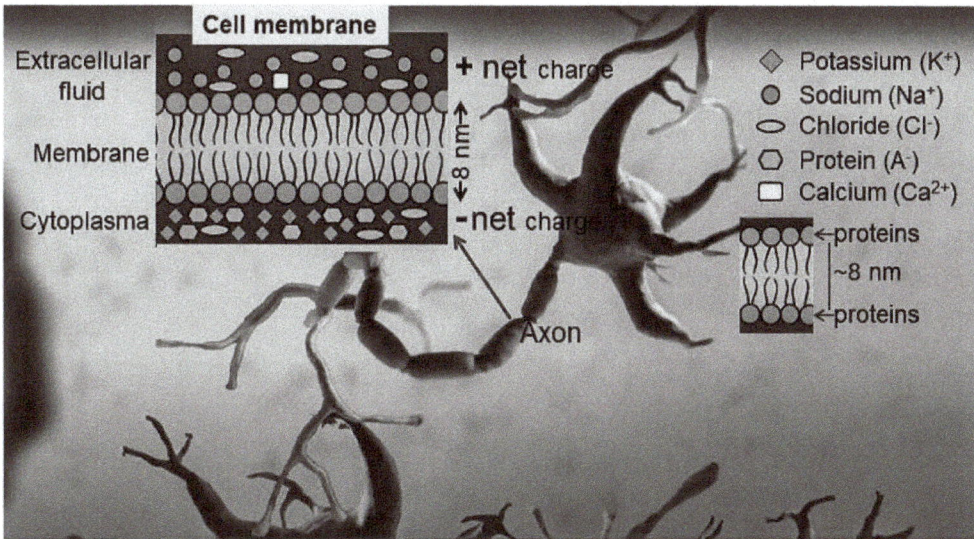

Figure 5.11. Illustration of the transmission of signal in our bodies through the cell membranes. For a typical neuron at rest, sodium, chloride, and calcium are concentrated outside the cell, whereas potassium and other anions are concentrated inside. This ion distribution leads to a negative resting membrane potential.

As described in Ikelle (2023a), a cell membrane consists of a lipid bilayer with embedded proteins. It separates the neuron from the extracellular fluid surrounding it (see Figure 5.11). It is a series of lipid (fat) molecules oriented with the heads facing outwards, and the tails facing inwards. Within the neurons themselves, there is a mass of proteins used for every function imaginable. Some are kinases and phosphatases. Some regulate metabolism and energy consumption. Others are involved in the production and degradation of neurotransmitters after use. Proteins are ubiquitous in the cell. Neurons have their own proteins. On the phospholipid bilayer, protein receptors span the membrane, from the intracellular fluid to the extracellular fluid. This allows communication via neurotransmitters. Charged atoms, called ions, are an integral part of the neuron. Intracellular fluid contains ions such as sodium, potassium and chloride. Without these ions, and the receptors to facilitate their movement, neurons could not send messages.

Electrical signaling, electrical gradient, and resting membrane potential. As described above, various ions, including sodium, potassium, and chloride, are unequally distributed between the inside and the outside of the cell. The presence and movement of these ions are not only critical when a neuron is fired but also when the neuron is at rest. As we can see in Figure 5.12, intracellular environment is rich in K^+ while extracellular environment has abundance of Na^+ and Cl^-—that is, sodium is always more concentrated outside the cell and potassium is always more concentrated inside the cell. Extracellular environment resembles sea water (i.e., high Na^+, low K^+, and high Cl^-) while intracellular environment reverses these concentrations (i.e., low Na^+, high K^+, and low Cl^-).

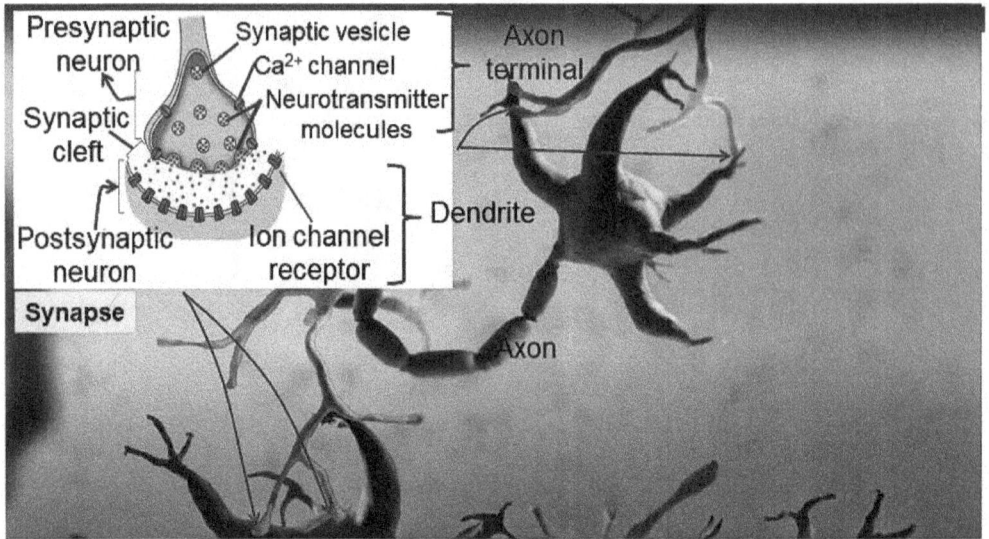

Figure 5.12. Illustration of a chemical synapse. The presynaptic neuron releases neurotransmitter molecules into the synaptic cleft. The molecules bind to receptors on the postsynaptic cell and make it more or less likely to fire.

The main functions of a cell membrane are (1) to protect the cell from the environment, (2) to control the movements of ions in and out of cells' organelles, (3) to serve as cell signaling and ion conductivity, and (4) to serve as the attachment surface for several extracellular structures, including the cell wall, the carbohydrate layer called the glycocalyx, and the intracellular network of protein fibers (i.e., the cytoskeleton).

The cell membrane thickness is between 7.5 and 10.0 nanometers. From the bioelectric viewpoint, the ionic channels (see Figure 5.11) constitute the most significant part of the cell membrane. They are macromolecular pores through which sodium, potassium, and chloride ions pass through the membrane. About 1 million ions flow through an open channel per second. Electric current within our bodies is caused by these ions flows. The resulting electric impulses propagate along a nerve fiber without attenuation (Beck, 1888). In other words, signal transmission down neuron axons is an all-or-nothing process. When the cell body is stimulated above the threshold, the axon transmits the same action potential at the same speed and in the same direction, regardless of the extent above the threshold or duration of the input. It differs from signal propagation through a physical electrical circuit. As we discussed in the previous chapter, in a copper wire, electrons drift along the signal path, but the signal itself moves as a compression wave rather than a transverse wave as in the biological system.

Consider the case where only our neuron system contains potassium and sodium ions. A neuron is at rest when it is not sending a signal. In a typical neutron in its resting state, the concentration of sodium ions is higher outside the cell than inside. Potassium concentration is the opposite, with more potassium ions inside the cell than outside. This ion separation occurs right at the cell membrane and creates a chemical gradient across the membrane. We also have an electrical gradient, because, at rest, there are more positively charged particles outside the cell relative to those inside. The difference in net charge inside and outside the cell is called membrane potential. So a resting membrane potential means no current flows across the membrane. However, potassium concentration is normally between 30 and 50 times greater in the intracellular space compared to the extracellular space. So the resting membrane potential is generally negative (i.e., there is excess negative charge inside compared to outside) and varies from cell to cell, from about -20 mV (millivolts) to -100 mV. Actually, all human cells have a negative resting membrane potential, including neurons. For example, in a typical neuron, the resting membrane potential is -70 mV (i.e., at rest, the potential inside a neuron is 70 mV less than outside the neuron). In a typical skeletal muscle cell, its value is -90 mV, and in most other mammalian cells, the membrane potential is around -50 mV. This potential difference can be maintained because the lipid bilayer of the cell membrane acts as a barrier to ions diffusion.

In more general terms, we have two environments with different potentials. The difference in electric potential between the interior and the exterior of a biological cell is called the *membrane potential* (also transmembrane potential or membrane voltage). Electrical signals generated at the neuronal membrane travel from dendrites to synapses. They can be measured with electrodes, amplified using instrumentation amplifiers, and used to monitor the functioning of various organs like the heart, brain, eye, muscles, etc., as we will discuss later. Note that intracellular A^- help keep the cytosol negatively charged, and Cl^- in the extracellular fluid can participate in electrical signaling. Note also that, except under very unusual circumstances, the overall electrical balance of our body is neutral: for every negative charge there is a positive one.

Synapse. The junction (boundary) between the axon of one neuron and the dendrite of another, through which the two neurons communicate, is known as a synapse. In other words, a synapse is the interface between two neurons. There are two types of synapses: electrical synapses and chemical synapses. Electrical synapses are less common in the human nervous system. Electrical synapses are simple pores between two cells that allow ions to pass through. They allow the passage of that electric signal through to a neighboring cell. Most neurons are connected together by a much more complicated structure called a chemical synapse. A chemical is released in the very small space between the two neurons. This space is known as the synaptic cleft, as shown in Figure 5.12. This chemical is taken up by the downstream neuron, on the other side of the cleft.

Action potential: depolarization, hyperpolarization, and repolarization.
So in many cells, the resting membrane potential never changes; the cells can do their work without affecting their electrical balance. But in other cells, especially nerve and muscle cells, changes in the resting potential occur when ions move in or out of cells. Ions cannot move across the membrane at will. They cross the membrane through ion channels. These channels are selectively permeable, meaning that they only allow one, or a small subset of ions, to pass through (see Figure 5.13). As ions move through a channel and cross it, they cause the membrane potential to change from its resting potential to its action potential.

Box 5.1: Ion Concentration and Diffusion

There are several ways of defining or specifying a solution's concentration. Number density and concentration describe electrical processes on the atomic scale. Number density (n) is

$$n = \frac{\text{number of particles in solution}}{\text{volume of solution}} \; .$$

The amount of substance per volume is called concentration. It is defined as

$$C = \frac{\text{amount of ion species}}{\text{volume of solution}} \; .$$

The unit of concentration is mole per cubic meter. A more practical unit is the mole per liter. A solution with a concentration of 1 mole per liter is often called a molar solution. A concentration of 10^{-3} mole per liter is called millimolar (mM). For concentrations in this chapter, we will mostly use the mM unit.

Consider the distribution of charge in the presence of a voltage (battery) in Figure 5.14. The energy difference between charges on the top and bottom plate is:

$$\Delta \mathcal{U} = \mathcal{U}(x = d) - \mathcal{U}(x = 0) = q\left[V(d) - V(0)\right] = q\Delta V \; ,$$

where $\mathcal{U}$ is the energy, d is the distance between the two parallel plates, ΔV is the voltage as illustrated in Figure 5.14. From Boltzmann random thermal motion of the ions, we have (see Chapter 2)

$$C(d) = C(0)\exp\left(-\frac{\Delta \mathcal{U}}{k_0 T}\right) = C(0)\exp\left(-q\frac{\Delta V}{k_0 T}\right) \iff \Delta V = -\frac{k_0 T}{q}\ln\left[\frac{C(d)}{C(0)}\right],$$

where k_0 is the Boltzmann constant and T is the absolute temperature. The last equation here is known as the Nernst relation. It tells us that the potential is indeed a result of the concentration difference between the two solutions. Ion diffusion refers to the tendency of ions to move from regions of high concentration to regions of lower concentration.

When the outside stimulation is large enough to bring the membrane in the neuron body up from say -70 mV (resting stage) to the threshold of -55 mV or higher, this triggers an action potential at the axon hillock, which then travels down the axon. Once the cell membrane reaches the threshold voltage, the sodium channel opens and Na^+ rushes into the cell because of the electrochemical gradient. The membrane potential moves positive. This is called *depolarization*. The voltage gradient can go up to +40 mV. As the membrane potential becomes positive, the sodium channel shuts down and the potassium channels because of the potassium-electrochemical gradient. Potassium ions flow out of the cell, making it less positive and eventually negative. This process is called *repolarization*. For a brief period,

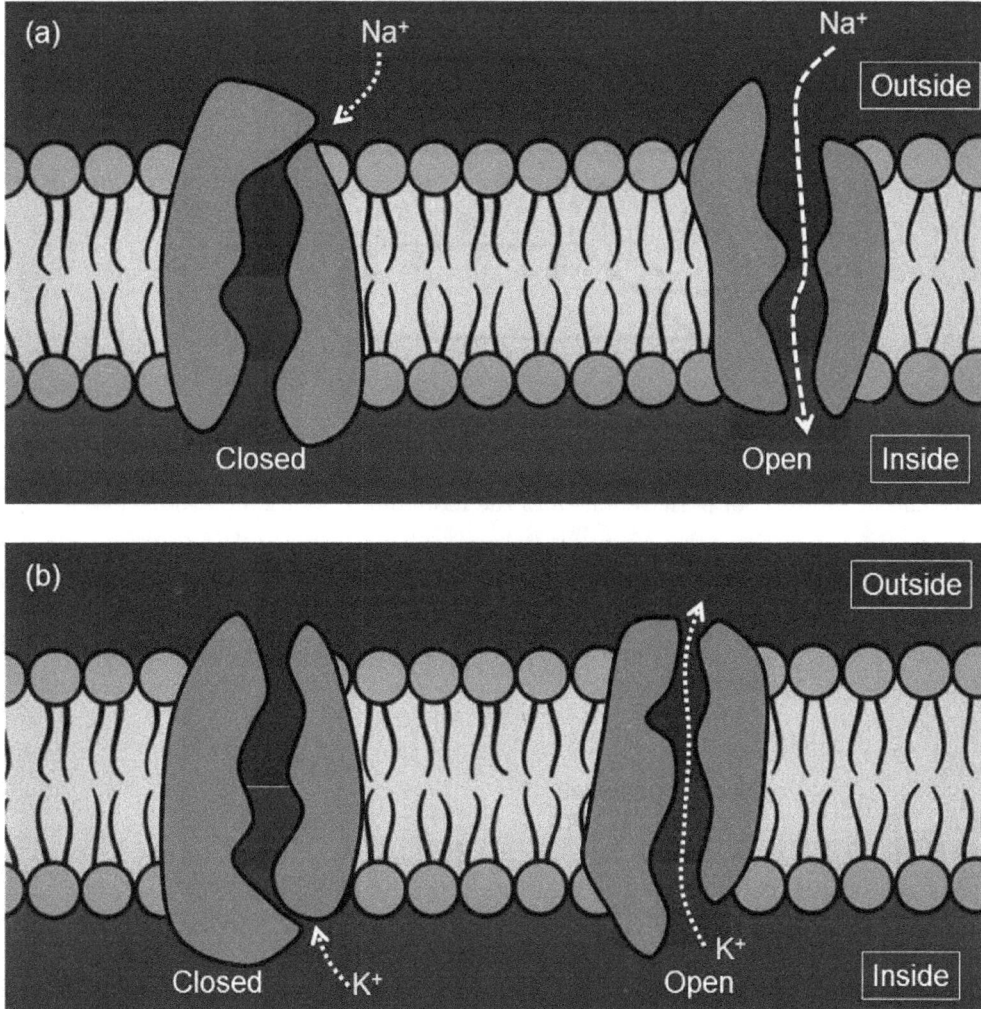

Figure 5.13. Transport of sodium and potassium through voltage-gated channels. The cytoplasm of the cell is also referred to as the inside of the cell. The sodium-potassium exchange pump mechanism. (a) Sodium ions are moved from outside the cell to inside and (b) potassium ions are moved from inside the cell to outside.

the membrane potential is *hyperpolarization*—that is, it is more negative than the resting potential. Throughout the sodium-potassium pump, chemical gradients are restored by moving more sodium ions out than potassium ions in. This process returns the membrane potential back to its resting potential.

For clarity, let us add some specifics to the notions of depolarization, hyperpolarization, and repolarization (see Figure 5.15). For example, a change of the electrical charge from -70 mV to -40 mV is partial depolarization; and a change from -70 mV to 0 mV is complete depolarization. A change from -70 mV to -80 mV is hyperpolarization. In *repolarization*, a change in the electrical gradient returns a cell to its original resting membrane potential. Again, a cell's resting state is not electrical neutrality (0 mV) but a negative cell interior of about -70 mV for neurons.

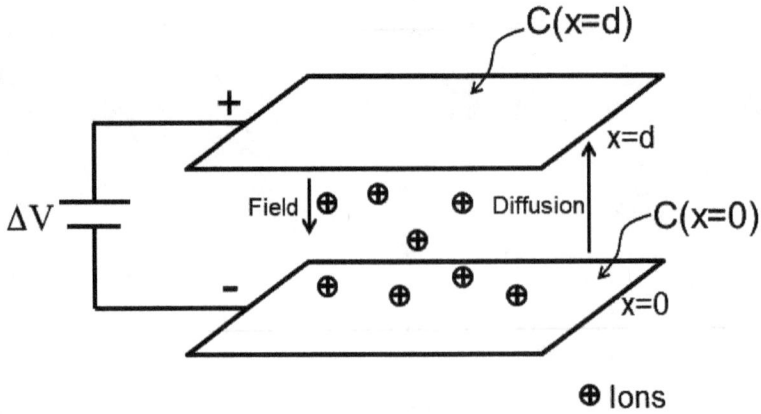

Figure 5.14. The distribution of charge in the presence of a voltage (battery). Here the membrane is permeable to the positive ions only. The electric field is set up in the membrane by diffusion of positive ions into the more dilute solution on the left.

Also, in general terms, the membrane is highly permeable to K^+—that is, potassium can get through (in and out) the membrane with ease because there are many open potassium channels in the membrane. These potassium channels do not allow other ions to go through the membrane (i.e., membranes are semi-permeable). For

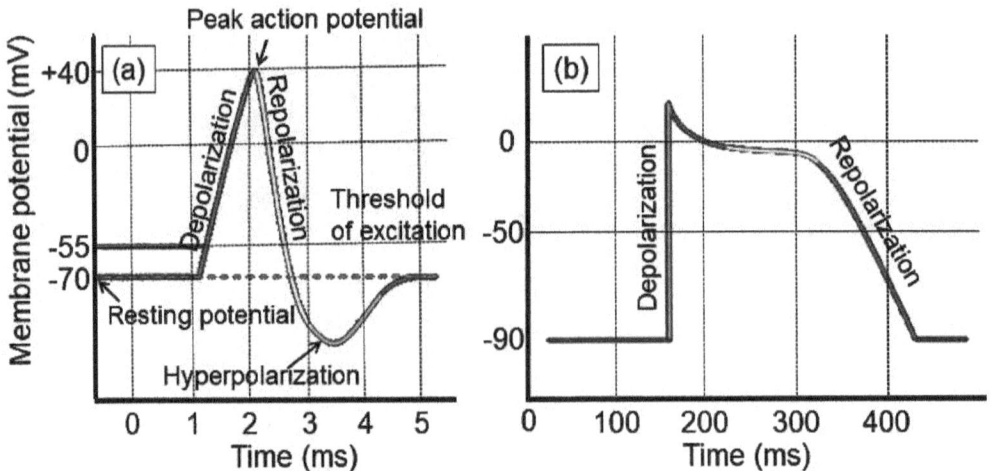

Figure 5.15. (a) Neuronal action potentials. Action potentials can be triggered by a small depolarization of the resting membrane potential (around -70 mV). When depolarization reaches a threshold, this triggers the opening of voltage-gated Na^+ channels (-55 mV). This opening causes a rapid influx of Na^+ ions which depolarize the membrane toward +40 mV. The Na^+ channels inactivate (a form of closure) within a millisecond or so, and the voltage-gated K^+ channels open. This efflux of K^+ ions repolarizes the membrane back to -70 mV or beyond to more negative potentials, causing an after hyperpolarizing potential. (b) Cardiac action potential. Rapid Na^+ influx through voltage-gated Na^+ channels depolarizes the cell membrane. Voltage-gated K^+ channels open, beginning repolarization. Voltage-gated Ca^{++} channels open and the influx of Ca^{++} positive charges is balanced by the efflux of K^+ positive charges, producing a plateau phase. Ca^{++} channels close and K+ channels remain open, repolarizing the membrane potential to -90 mV.

example, the majority of the negative charges inside cells, which are proteins, cannot get out—that is, the membrane is not permeable to proteins because they are too big (proteins are large molecules). Cations Na^+ are more concentrated outside than inside and are driven into the cell by the electrical potential difference across the membrane through Na^+ channels. Again, the resting membrane potential corresponds to Na^+ and Cl^- more concentrated outside the cell, and K^+ and organic anions (organic acids and proteins) more concentrated inside. The combination of concentration gradients and a semi-permeable membrane creates an electrical potential.

Membrane potential between the interior and the exterior of a cell is particularly important to neurons, because rapid changes in the membrane resting potential of neurons produce about the nervous impulse, which is the basis of neuronal signaling. In muscle cells, changes in the membrane potential cause contraction. In endocrine cells, changes in the membrane potential result in the release of hormones. In non-excitable cells, such as glia, epithelial cells, and adipose cells (and others), the resting membrane potential does not change appreciably over time.

The Nernst equation defines the electrochemical equilibrium between a concentration gradient and an electrical potential (i.e., the total energy inside the cell is equal to the total energy outside), for one particular ion. The electrical potential that exactly balances a concentration gradient is called the Nernst potential (also called the reversal potential). The Nernst equation reveals how the reversal potential of an ion depends on the inside and outside concentrations of this ion. It is a simple algebraic formula to calculate membrane potential as follows (Nernst, 1888 and 1989):

$$\underbrace{zFV_{in}}_{\text{electrical energy}} + \underbrace{RT\ln[\text{ion}]_{in}}_{\text{chemical energy}} = \underbrace{zFV_{out}}_{\text{electrical energy}} + \underbrace{RT\ln[\text{ion}]_{out}}_{\text{chemical energy}}$$

or

$$\Delta V = V_{in} - V_{out} = \frac{RT}{zF}\ln\left[\frac{(\text{ion})_{out}}{(\text{ion})_{in}}\right] \approx 27\frac{1}{z}\ln\left[\frac{(\text{ion})_{out}}{(\text{ion})_{in}}\right] \text{ mV}, \qquad (5.1)$$

where R is the gas constant, T is the absolute temperature (in Kelvin), F is the Faraday's constant [$9.649 \times 10^4 C/mol$], and z is the valence of the ion ($z = 1$ for cations and $z = -1$ for anions), ion_{out} and ion_{in} are chemical concentrations of outside (extracellular) and inside (intercellular) of the cell, respectively. We used $T = 310$ degrees Kelvin ≈ 37 degrees Celsius for the approximation in (5.1). Note that, in bioelectromagnetism, the quantity of ions is usually expressed in moles (one mole equals the molecular weight in grams is 6.0225×10^{23}, Avogadro's number of molecules) and Faraday's constant converts quantity of moles to quantity of charge for a univalent ion. So zF actually represents the electrical charge. Table 5.1 shows four ions' potentials. We can see that Na^+ tends to depolarize the cell (make the inside more positive), Ca^{++} tends to depolarize the cell (make the inside more positive), K^+ tends to hyperpolarize the cell (make the inside more negative), and Cl^- tends to hyperpolarize the cell (make the inside more negative).

Table 5.1. Ion concentrations and the Nernst potential for ions within the cell (mM = 10^{-6} Mole/mm^3 and 1 Mole contains about $\sim 6.023 \times 10^{23}$ ions).

Material	z z	Extracellular (mM)	Intracellular (mM)	ΔV (mV)
K$^+$	1	20	400	-80
Na$^+$	1	400	50	+59
Cl$^-$	-1	450	40	-65
Ca^{++}	2	2	2×10^{-4}	+124

As a general rule, when ΔV is determined by two or more species of ions, the influence of each species is determined not only by the concentrations of the ion inside and outside the cell but also by the ease with which the ion crosses the membrane, as described by the Goldman-Hodgkin-Katz equation. This equation calculates the electrical equilibrium potential across the cell's membrane in the presence of more than one ion, taking into account the membrane's permeability. It is derived from the Nernst equation, which is applicable to one ion species. In terms of electrical current flow, the membrane's conductance (1/resistance) provides a convenient measure of how readily the ion crosses the membrane. Another convenient measure is the permeability (P) of the membrane to that ion in velocity units, cm/s. This measure is similar to that of a diffusion constant, which measures the rate of solute movement in solution (see Table 5.2). The dependence of membrane potential on ionic permeability and concentration is given quantitatively by the Goldman-Hodgkin-Katz equation (Goldman, 1943):

$$\Delta V = \frac{RT}{zF} \ln \left(\frac{P_{K^+} \left(K^+\right)_{out} + P_{Na^+} \left(Na^+\right)_{out} + P_{Cl^-} \left(Cl^-\right)_{out}}{P_{K^+} \left(K^+\right)_{in} + P_{Na^+} \left(Na^+\right)_{in} + P_{Cl^-} \left(Cl^-\right)_{in}} \right) \ \text{mV} \,,$$

where P_X is the relative membrane permeability for X ion species. Note that we are here using the natural logarithm instead of $\log_1 0$. The Nernst potential for an ion does not depend on membrane permeability to that ion.

Table 5.2. Diffusion coefficient of ions at 0° C and 25° C. Diffusion coefficient, D, is the proportionality factor of the mass of a substance diffusing in time through the surface normal to the diffusion direction. At a concentration gradient of unity, the diffusion coefficient is calculated as the mass of the substance diffuses through a unit surface in a unit amount of time. The unit of D here is 10^{-6} cm^2/s. Adapted from Li and Gregory, 1974.

Ion	237.15 K	298.15 K	Ion	237.15 K	298.15 K
K$^+$	9.86	19.6	HCO$_3^-$		11.8
Na$^+$	6.27	13.3	OH$^-$	25.6	52.7
Cl$^-$	10.1	20.3	F$^-$		14.6
Ca^{++}	3.73	7.93	HS$^-$	9.75	17.3
H$^+$	56.1	93.1	HSO$_4^-$		13.3
Ba^{++}	4.04	8.43	H$_2$PO$_4^-$		8.46

The resting potential is reached when the total membrane current is zero; i.e.,

$$I_{Na^+} + I_{K^+} + I_{Cl^-} = 0 \qquad (5.2)$$

or

$$g_{K^+}[\Delta V_{res} - \Delta V_{K^+}] \quad + \quad g_{Na^+}[\Delta V_{res} - \Delta V_{Na^+}]$$

$$+ \quad g_{Cl^-}[\Delta V_{res} - \Delta V_{Cl^-}] = 0 , \qquad (5.3)$$

where

$$g_{Na^+} = \frac{I_{Na^+}}{\Delta V_{res} - \Delta V_{Na^+}} , \quad g_{K^+} = \frac{I_{K^+}}{\Delta V_{res} - \Delta V_{K^+}} ,$$

and

$$g_{Cl^-} = \frac{I_{Cl^-}}{\Delta V_{res} - \Delta V_{Cl^-}} , \qquad (5.4)$$

where ΔV_{res} is membrane voltage [mV], ΔV_{Na^+}, ΔV_{K^+}, ΔV_{Cl^-} are Nernst voltage [mV] for sodium, potassium, and chlorine, respectively, I_{Na^+}, I_{K^+}, I_{Cl^-} are the electric current carried by sodium, potassium and chlorine (leakage current) per unit area mA/m^2, respectively, g_{Na^+}, g_{K^+}, g_{Cl^-} are membrane conductance (1/resistance) per unit area for sodium, potassium, and chlorine, respectively—also referred to as the leakage conductance S/m^2. We can alternatively express the resting membrane potential in (5.3) as follows:

$$\Delta V_{res} = \frac{g_{K^+}\Delta V_{K^+} + g_{Na^+}\Delta V_{Na^+} + g_{Cl^-}\Delta V_{Cl^-}}{g_{K^+} + g_{Na^+} + g_{Cl^-}} . \qquad (5.5)$$

Thus, the membrane's resting potential is dominated by the most permeable ions. In these derivations, we only used the three ions that influence bioelectrical phenomena: K^+, Na^+, and Cl^-. Note that Ca^{++} also contributes to bioelectricity to a few tissues, including the heart.

Quiz: (1) What is the reversal potential of Ca^{++} at body temperature?
(2) Determine the resting membrane potential of a cell with $g_{Na^+}/g_{K^+} = 0.03$ and $g_{Cl^-}/g_{K^+} = 0.1$.

5.1.5.   Electrophysiological Activity

Nowadays, medicine routinely measures electric or magnetic signals generated by (1) the spontaneous activity of living tissues inside the body and (2) electronic devices outside biological tissues that can stimulate electric and/or magnetic responses. These electric and magnetic fields are used for medical diagnostics and therapy. The magnetic signals are included here through the link between electric and magnetic fields discussed in the previous subsection. We here describe some of the classical bioelectromagnetic methods used in clinics. We start by some electromagnetic background on calculation of electrical potential and magnetic fields around an inhomogeneous body like the brain.

Bioelectric sources and conductors. The brain radius is about 0.25 m whereas the wavelength of electrochemical signals is about 500 m—that is more than 2000 times the brain radius. Therefore the use of Maxwell's quasi-static equations in Chapter 3 is justified. Also, there are two current densities involved in this problem. The resulting current arises from the bioelectric activity of nerve and muscle cells due to the conversion of energy from chemical to electric form. This current is nonconservative and we denote it $\boldsymbol{J}^{(a)} = \boldsymbol{J}^{(a)}(\boldsymbol{x}, t)$, where $\boldsymbol{x}$ is the point at which the field is evaluated. It is responsible for the electric field, $\boldsymbol{E} = \boldsymbol{E}(\boldsymbol{x}, t)$, and the second current, $\sigma\boldsymbol{E}$, where σ is the electric conductivity of the brain, is the result of an applied field in a resistive material, such as the brain. So the total current is [see Chapter 5 of Ikelle (2023a) and Chapter 12 of Ikelle (2020)]

$$\boldsymbol{J} = \boldsymbol{J}^{(a)} + \sigma\boldsymbol{E} \underset{\boldsymbol{E}=-\boldsymbol{\nabla}\phi}{\Longleftrightarrow} \boldsymbol{J} = \boldsymbol{J}^{(a)} - \sigma\boldsymbol{\nabla}\phi \,, \tag{5.6}$$

where $\phi = \phi(\boldsymbol{x}, t)$ is the scalar potential. We also used the fact that $\boldsymbol{E} = -\boldsymbol{\nabla}\phi$. Using $\boldsymbol{\nabla} \cdot \boldsymbol{J} = 0$ in Chapter 3, we can deduce that

$$\boldsymbol{\nabla} \cdot \boldsymbol{J}^{(a)} = \sigma\boldsymbol{\nabla}^2\phi \Longrightarrow \phi = -\frac{1}{4\pi\sigma}\int_V \left(\frac{1}{|\boldsymbol{x}|}\right)\boldsymbol{\nabla} \cdot \boldsymbol{J}^{(a)}dV$$

$$= \frac{1}{4\pi\sigma}\int_V \boldsymbol{J}^{(p)}\boldsymbol{\nabla} \cdot \left(\frac{1}{|\boldsymbol{x}|}\right)dV \,. \tag{5.7}$$

The first equation here allows us to compute the electric potential under the assumption that brain conductivity is smooth, but it is not because of the various regions in the brain, as illustrated in Figure 5.1. To account the regions, we divide the brain into a finite number of homogeneous regions. We denote the boundaries of these regions as S_m. At these boundaries both the electric potential ϕ and the normal component of the current density must be continuous:

$$\phi^{(-)}(S_m) = \phi^{(+)}(S_m)$$

and

$$\sigma_m^{(-)}\boldsymbol{\nabla}\phi^{(-)}(S_m) \cdot \boldsymbol{n}_m = \sigma_m^{(+)}\boldsymbol{\nabla}\phi^{(+)}(S_m) \cdot \boldsymbol{n}_m \,, \tag{5.8}$$

where the superscripts (-) and (+) represent the opposite sides of the boundary and $\boldsymbol{n}_m$ is the normal vector to S_m directed from the (-) region to the (+) one. If we define $\psi = 1/|\boldsymbol{x}|$, according to Green's theorem (Smyth, 1968), we have

$$\int_{S_m} \left[\sigma_m^{(-)}\left(\psi^{(-)}\boldsymbol{\nabla}\phi^{(-)} - \psi^{(-)}\boldsymbol{\nabla}\phi^{(-)}\right)\right.$$

$$\left. - \sigma_m^{(+)}\left(\psi^{(+)}\boldsymbol{\nabla}\phi^{(+)} - \psi^{(+)}\boldsymbol{\nabla}\phi^{(+)}\right)\right] \cdot \boldsymbol{n}_m dS_m$$

$$= \int_{V_m} \left[\psi\boldsymbol{\nabla} \cdot (\sigma_j\boldsymbol{\nabla}\phi) - \phi\boldsymbol{\nabla} \cdot (\sigma_j\boldsymbol{\nabla}\psi)\right]dV_m \,. \tag{5.9}$$

By substituting the boundary conditions in (5.8) and the first in (5.7) into (5.9), we obtain (Geselowitz, 1967 and 1970):

$$\phi \;=\; \frac{1}{4\pi\sigma}\int_V J^{(a)}\nabla\left(\frac{1}{|\boldsymbol{x}|}\right)dV + \sum_m \int_{S_m}\left(\sigma_m^{(+)}-\sigma_m^{(-)}\right)\phi\nabla\left(\frac{1}{|\boldsymbol{x}|}\right)\cdot\boldsymbol{n}_m dS_m. \quad (5.10)$$

This equation evaluates the electric potential anywhere within the brain.

The current density throughout a volume conductor produces a magnetic field given by the following relationship (Stratton, 1941; Jackson, 1968):

$$\begin{aligned}
4\pi\boldsymbol{H} \;&=\; \int_V \boldsymbol{J}\times\nabla\left(\frac{1}{|\boldsymbol{x}|}\right)dV \\
&=\; \int_V \boldsymbol{J}^{(a)}\times\nabla\left(\frac{1}{|\boldsymbol{x}|}\right)dV - \int_V \sigma\nabla\phi\times\nabla\left(\frac{1}{|\boldsymbol{x}|}\right)dV \\
&=\; \int_V \boldsymbol{J}^{(a)}\times\nabla\left(\frac{1}{|\boldsymbol{x}|}\right)dV - \sum_m \int_{V_m}\sigma_m\nabla\phi\times\nabla\left(\frac{1}{|\boldsymbol{x}|}\right)dV \\
&=\; \int_V \boldsymbol{J}^{(a)}\times\nabla\left(\frac{1}{|\boldsymbol{x}|}\right)dV - \sum_m \int_{V_m}\sigma_m\nabla\times\left[\phi\nabla\left(\frac{1}{|\boldsymbol{x}|}\right)\right]dV \;.
\end{aligned}$$

In the last equality here, we use the vector identity $\nabla\times(\phi\boldsymbol{a}) = \phi\nabla\times\boldsymbol{a} + \nabla\phi\times\boldsymbol{a}$, with $\boldsymbol{a} = \nabla(1/|\boldsymbol{x}|)$, and the fact $\nabla\times\nabla\Psi = 0$, where Ψ is an arbitrary function of $\boldsymbol{x}$; i.e.,

$$\nabla\times\left[\phi\nabla\left(\frac{1}{|\boldsymbol{x}|}\right)\right] = \phi\nabla\times\nabla\left(\frac{1}{|\boldsymbol{x}|}\right) + \nabla\phi\times\nabla\left(\frac{1}{|\boldsymbol{x}|}\right)$$

or

$$\nabla\times\left[\phi\nabla\left(\frac{1}{|\boldsymbol{x}|}\right)\right] = \nabla\phi\times\nabla\left(\frac{1}{|\boldsymbol{x}|}\right)\;. \quad (5.11)$$

Finally, applying this divergence theorem (see Appendix A of Ikelle, 2020), we arrive at

$$\begin{aligned}
4\pi\boldsymbol{H} \;=\;& \int_V \boldsymbol{J}^{(a)}\times\nabla\left(\frac{1}{|\boldsymbol{x}|}\right)dV \\
&+ \sum_m \int_{S_m}\left(\upsilon_m^{(+)}-\upsilon_m^{(-)}\right)\phi\nabla\left(\frac{1}{|\boldsymbol{x}|}\right)\times\boldsymbol{n}_m dS_m\;. \quad (5.12)
\end{aligned}$$

This equation describes the magnetic field outside a finite volume conductor (brain) containing internal (electric) volume sources $\boldsymbol{J}^{(a)}$ and inhomogeneities $\left(\sigma_m^{(+)}-\sigma_m^{(-)}\right)$. Equations (5.10) and (5.12) are the basis of electrophysiological activity and measurements. These equations were first derived by Geselowitz (1970).

Electroencephalogram (EEG) measures the brain's spontaneous electrical activity on the surface of the brain for 20 to 30 minutes at a time. Spontaneous

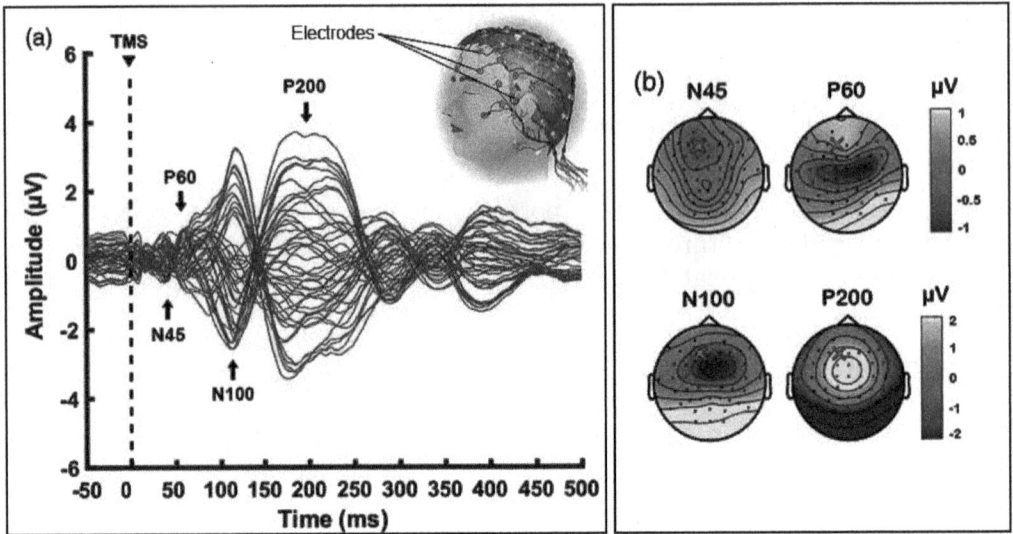

Figure 5.16. (a) A plot of the resulting time series of the averaged potentials as well as (b) a topography plot (waveforms) of the potential measured at the electrodes at the peak of the P200, P60, N45, and N100 components at various time points. Note that the EEG waveforms depend on measurements taken at various locations.

activity means that it occurs continuously in living individuals. Here we focus on brain activity. Note that brain activity can also occur in response to a stimulus, whether electric, auditory, visual, etc. In general, such stimuli produce a relatively small response EEG is a set of plots of voltage, as defined in (5.10), versus time of the brain's electrical activity using multiple electrodes placed on the scalp (see appended picture in Figure 5.16a). These plots are called electroencephalograms (Figure 5.16a). The electrodes consist of small metal discs with thin wires. The EEG amplitude is about 100 μV when measured on the scalp, and about 1-2 mV when measured on the brain surface. Voltage fluctuations are due to ionic current within brain neurons, as described earlier. The bandwidth of this signal is from 0.1 Hz to 50 Hz. EEG is used to diagnose sleep disorders, strokes, depth of anesthesia, coma, encephalopathies, brain tumors, brain damage from head injury, brain dysfunction, and epilepsy. Epilepsy is one of the world's most common neurological diseases. It affects more than 40 million people worldwide.

An EEG signal is generally analyzed using so-called brain waves—alpha, beta, delta, theta, and gamma waves—as first proposed by German psychiatrist Hans Berger. These classifications provide indications of brain activity sources. Figure 5.17 illustrates these brain waves. Delta waves have a frequency range of 0.5-4 Hz and amplitudes ranging between 20 and 400 μV. They are detectable in infants and deeply sleeping adults. The theta waves have a frequency range of 4-8 Hz and their amplitudes range between 5 and 100 μV. In adults, theta waves occur under stress. High levels of theta waves are considered abnormal. Alpha waves have a frequency spectrum of 8-13 Hz and amplitudes of 2-10 muV. When the eyes are closed, alpha waves dominate the EEG signal. Alpha waves are about rest, meditation, and sleep. Beta waves have frequencies of 13-30 Hz and amplitudes of

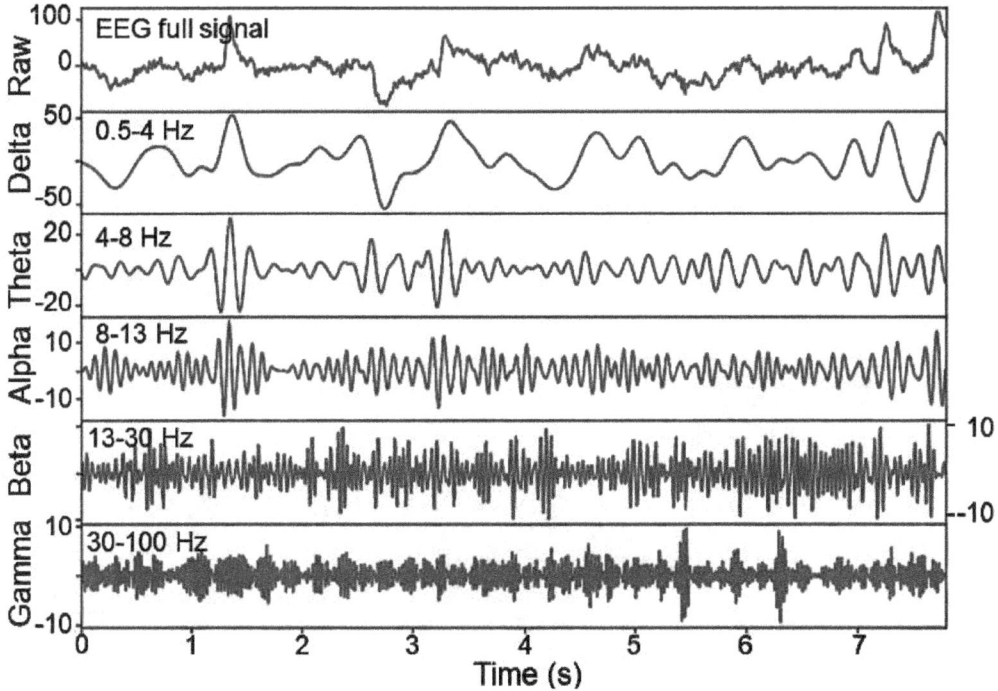

Figure 5.17. The signal channel contains raw EEG signals and corresponding frequency bands: Delta (0.1 to 4 Hz), theta (4 to 8 Hz), alpha (8 to 14 Hz), beta (14 to 30 Hz), and gamma (30 to 100 Hz).

1-5 MHz. If you're resisting or suppressing motion or solving a math problem, beta waves typically have small amplitudes and high frequencies. A gamma wave has a frequency spectrum between 30-100 Hz and amplitudes less than 2 μV. Gamma waves are seen in REM (rapid eye movement), learning moments, and extreme happiness.

The first recording of the human brain's electric field was made by German psychiatrist Hans Berger in 1924 at University Hospital Jena in Jena (Germany). He gave this recording the name electroencephalogram (EEG) (Berger, 1929). From 1929 to 1938 he published 20 scientific papers on the EEG under the same title "Über das Elektroenkephalogram des Menschen."

To summarize, EEG signals are today an indispensable tool for studying the brain and diseases associated with old age. We can also simulate these data by using equation (5.10). Performing inversion to identify sources of EEG signals is still an open issue. Yet, we hope EEG technology will improve significantly to monitor brain neuron health. As discussed before, neurons are essential for extending the life expectancy on Earth to 100 years.

Magnetoencephalogram (MEG) also maps brain activity like EEG but by measuring magnetic fields produced by neurons' electrical activity instead of electric potential. In other words, MEG measures magnetic fields induced by electrical currents occurring naturally in the brain, as defined in (5.10), using very sensitive magnetometers.

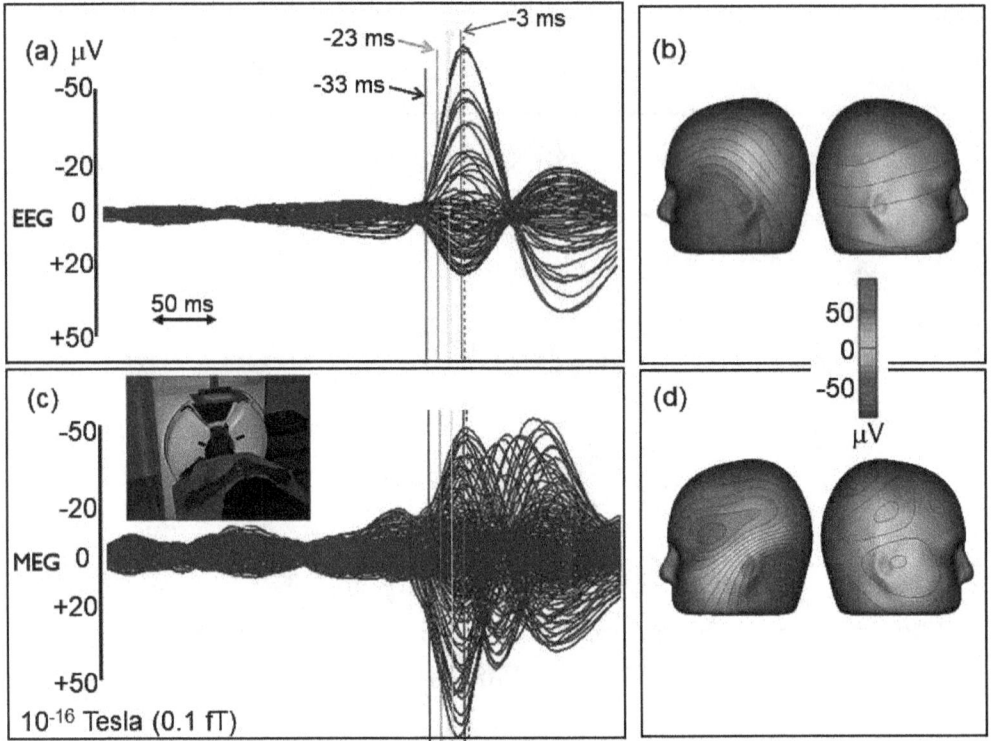

Figure 5.18. (a) Some magnetoencephalography signals (MEG). (b) Contour plot shows the direction and amplitude of the response at 100 ms on a scale of 20 femtoTesla (fT)/step. (c) Other magnetoencephalography signals (MEG). We also included a sketch of a patient with a large helmet full of magnetic sensors is placed on the patient's head. Each detector outputs a time-varying magnetic field on the scalp. (d) The contour plot shows the direction and amplitude of the response at 100 ms.

The patient typically sits under or lies down inside the MEG scanner, which resembles a whole-head hair dryer, but contains an array of magnetic sensors (see Figure 5.18c). Arrays of SQUIDs (superconducting quantum interference devices) are currently the most common used magnetometers (Hämäläinen et al., 1993; and Boto et al., 2018). While the magnitude of fields associated with an individual neuron is negligible, the effect of multiple neurons excited together in a specific area generates a measurable magnetic field outside the head. These neuromagnetic signals generated by the brain are extremely small—a billionth of the earth's magnetic field strength. Therefore, MEG scanners require superconducting sensors such as SQUID sensors to amplify the signals. SQUID sensors are sensitive to neural activity induced at femoTelsa (i.e., 10^{-15} Tesla), about 100 million times smaller than the Earth's magnetic field.

MEG includes perceptual and cognitive brain processes, localizing regions affected by pathology before surgical removal, and determining the function of various parts of the brain. MEG can be applied in a clinical setting to find evidence of abnormalities (Carlson, 2010; and Braeutigam, 2013).

To summarize, the source of MEG and EEG signals is the brain's electric activity. But magnetic measurements differ in sensitivity distribution. If the

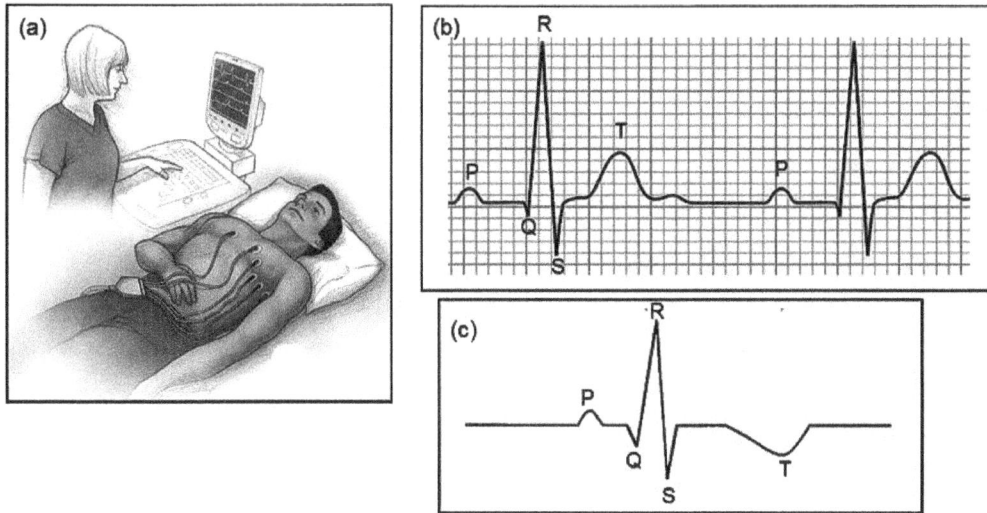

Figure 5.19. (a) A typical ECG test with electrodes attached to your arms and chest. (b) A typical electrocardiogram with a nomenclature of events. (c) An example of an abnormal ECG with an inverted T-event. It indicates ischemia.

electrodes of an electric lead are placed on a spherical volume conductor (the head) and lie on the axis of symmetry of a magnetic lead, the electric and magnetic lead fields are normal to each other everywhere in the volume conductor. The MEG, therefore, allows detection of source components not sensed by the EEG.

Electrocardiogram (ECG) measures the electrical activity of the heartbeat, including the rate and rhythm of heartbeats, the size and position of the heart chambers, the presence of any damage to the heart's muscle cells or conduction system, the effects of heart drugs, and the function of implanted pacemakers (Braunwald, 1997). It is a plot of voltage, as defined in (5.10), versus time of the heart's electrical activity using electrodes placed on the skin (see Figure 5.19), over the chest, or thorax and heartbeat. In other words, the records of heart potentials on the skin are electrocardiograms (ECG) because the rhythmical action of the heart is controlled by an electrical signal initiated by spontaneous stimulation of special muscle cells located in the right atrium. These cells make up the senatorial (SA) node or the pacemaker (Figure 5.20). Remember that the heart is composed of a cardiac muscle, called the myocardium. It consists of four compartments: the right and left atria and ventricles. Its main function is to pump blood to the systemic and pulmonary circulation (see Ikelle, 2023c). More deeply, the heart generates electrical current through contracting its muscle cells. These cells can self-stimulate, generating cardiac rhythm, a regular sequence of heart beats.

ECG recordings are similar to EEG recordings (see Figure 5.17), especially when interpreted by visual inspection. ECGs are more patient-friendly than EEG because it does not require electrode placement on the scalp, but it is substantially more costly than EGG. The key advantage of ECG over EEG is that magnetic signals are less distorted by differences in conductivity between the brain, skull, and scalp than electric signals.

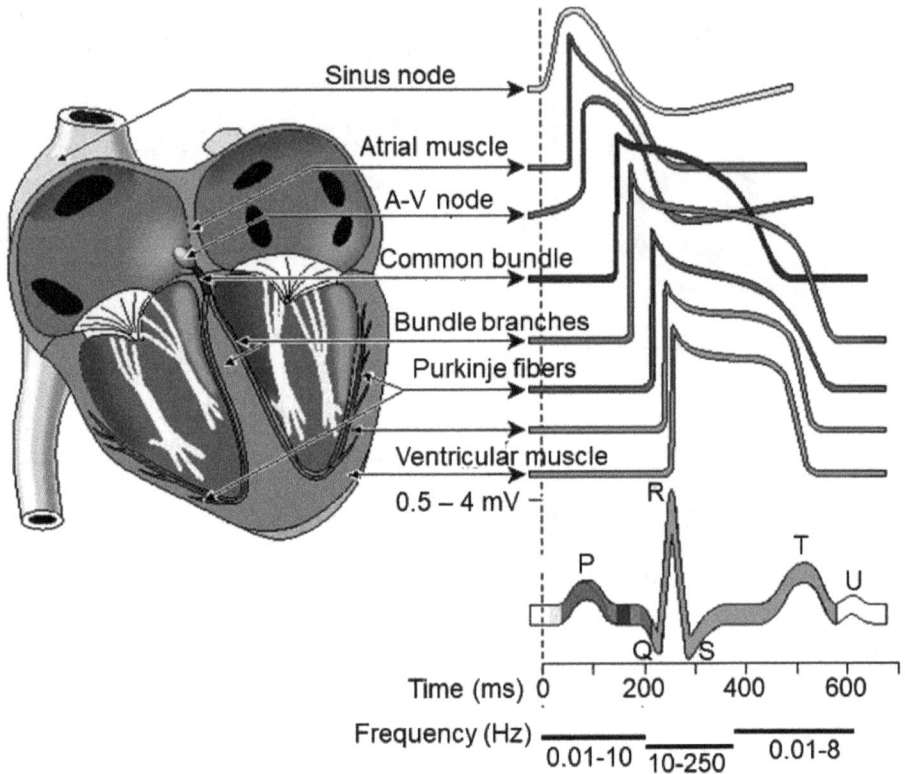

Figure 5.20. A schematic representation of the heart and normal electrical activity, including the waveforms of specialized cells in the heart and corresponding ECG signals. The different waveforms for each heart specialized cell are shown. The latency shown approximates that found in healthy hearts. (Adapted from Malmivuo and Plonsey, 1995).

In heart muscle cells, electric activation is similar to nerve cells—that is, from sodium ions and potassium ions flowing across the cell membrane. The electrodes detect the small electrical changes caused by cardiac muscle depolarization followed by repolarization during each cardiovascular cycle (heartbeat). Figure 5.20 shows the waveforms of action impulse observed in different specialized cardiac tissues. A healthy heart has an orderly progression of depolarization that starts with pacemaker cells in the sinoatrial node (SA node), spreads throughout the atrium, and passes through the atrioventricular node (AV node) down into the bundle of His (named after German physician Wilhelm His, Jr., 1863-1934) and into the Purkinje fibers (named after Jan Evangelista Purkinje, 1787-1869), spreading down and to the left throughout the ventricles. Again the SA nodal cells are self-excitatory pacemaker cells. They generate an action potential at 70 per minute. Between them, the internodal atrial connects the SA and AV nodes. It regulates the passage of the cardiac impulse from the atria to the ventricles. From the inner side of the ventricular wall, the many activation sites form a wavefront. This wavefront propagates through the ventricular mass toward the outer wall. This process results from cell-to-cell activation. After each ventricular muscle region depolarizes, repolarization occurs. The termination of activity appears to propagate from the epicardium (the outer side of the cardiac muscle) toward the endocardium (the inner side of the cardiac muscle). By comparing EEG (Figures 5.16 and 5.17) and ECG (Figures 5.19 and 5.20), we can see that an ECG has a pattern of spikes (events) whereas an EEG shows multiple waves.

The sign of ECG waves depends upon the direction of the electric dipole vector, the polarity, and the position of the electrodes of the measuring instrument. In a clinical examination, six transverse plane ECGs are usually made in addition to the six frontal plane ECGs.

A normal ECG exhibits a specific pattern of five recognizable events in the cardiac cycle. These events are P-event, QRS-event, T-event, P-R interval, and S-T segment. P-event shows activation of the right atrium, which suggests atrial depolarization (atrial systole). Its duration is about 80 ms. An enlarged P-event indicates a larger atrium. The QRS event shows rapid depolarization (systole) between the right and left ventricles. Its duration is between 80 and 100 ms. Just after the QRS event begins, the ventricles contract. An enlarged QRS event suggests a heart attack. Longer P-Q interval implies more time for impulses to travel through the atria and reach the ventricles. It happens in coronary artery disease and rheumatic fever when scar tissue forms in the heart. It happens in coronary artery disease and rheumatic fever when scar tissue forms in the heart. T-event shows ventricles repolarization (diastole). Its duration is about 160 ms. ST-event demonstrates repolarization of the interventricular septum. Elevated S-T segment above the base line represents acute myocardial infraction and a depressed S-T segment implies that heart muscles receive insufficient oxygen. Flat T-event indicates insufficient oxygen supply to heart muscle as it occurs in coronary artery disease. Elevated T-event may indicate increased levels of potassium ions in the blood as in hyperkalemia. P-R interval represents the time required for an impulse to travel through the atria to reach the ventricles.

To summarize, ECG indicates the rate and rhythm or pattern of heart contraction. It gives a clue about the condition of heart muscle. This includes whether the heart is normal, enlarged, if certain regions of the heart are damaged, and irregularities in the heart's rhythm known as *arrhythmia*. ECG also helps to determine the location and amount of injury caused by a heart attack and later assess the extent of recovery. ECG is used to diagnose blood clots, hypotension, dizziness, high blood pressure, congestive heart failure, etc. The typical ECG waveform is characterized by P-QRS-T-U complexes (see Table 5.3). Each complex has a particular spectral content: P wave, with a duration of approximately 150 ms, and spectral content up

Table 5.3. The major electrical events of the normal heart cycle. Notice the atrial repolarization is rarely seen and is unlabeled.

Intervals	Standar duration (s)	Event in the heart during interval
PR	0.12-0.21	Atrial depolarization
QRS	0.08-0.11	Ventricular depolarization
QT	0.35-0.42	Ventricular depolarization
ST	0.27-0.33	Ventricular repolarization

to 10 Hz; the QRS complex has a relatively higher amplitude compared with the other waves, with a duration of approximately 100 ms in a normal heartbeat.

Quiz: (1) Determine the potential level across the membrane that will exactly prevent net diffusion of Na^+, K^+, Cl^-, Ca^{++}, and HCO_3^-.
(2) Resting membrane potential: There are different kinds and concentrations of ions inside and outside cell membranes. Define them.
(3) Propose a method for diagnosing and identifying brain tumors.
(4) The SQUID magnetometer can also record the brain's magnetic field. During the alpha rhythm ,the magnetic field from the brain is about 10^{-11} Tesla. How does this magnetic field compare to that of the Earth's magnetic field?

Magnetocardiogram (MCG) measures the magnetic fields produced by electrical currents in the heart using extremely sensitive devices such as SQUID sensors (see Figure 5.21). It can be viewed as the magnetic version of ECG just like MEG is viewed as the magnetic version of EEG. The first MCG measurements were made by Baule and McFee (1963) using two large coils placed over the chest, connected in opposition to cancel out the relatively large magnetic background. The use of SQUID sensors marked the beginning of magnetocardiography (Zimmerman et al., 1970; and Cohen et al., 1970). As we can see in Figure 5.21b, magnetocardiograms are similar to electrocardiograms (MCG), with more local variations. In patients with left ventricular hypertrophy, MCG showed T-event inversion more frequently than ECG. MCG provides different information from ECG. MCG in addition to ECG can improve diagnostics and treatment. MCG is now considered one of the ways of monitoring the fetal heart (see Figures 5.21c and 5.21d).

5.1.6. Analogy between Cell Membranes and Electric Capacitors

As discussed in the previous subsections of this section, electrical signals are fundamental to understanding nervous system functions. In particular the electrical properties of cells which determine how electrical signals spread along the plasma membrane. Analogy between cell membranes and electric capacitors provide us an alternative way to develop intuition of electrical properties of cells. In this subsection we explore the electrical characteristics of cell membranes as electrical conductors and insulators. These passive electrical properties arise from the physical and chemical properties of the membrane material and from the ion channels in the membrane. We will start by recalling electric capacitors.

Capacitors. A capacitor is an electrical device for storing electrical potential energy. It consists of two conductors (metal plates) separated by free space or another dielectric medium (see Figure 5.22). The two conductors are close but not touching and carry equal and opposite charges when charged (i.e., when voltage[5] is applied to them for a finite charge time). The two conductors can be of any shape, but the parallel-plate type is the most common. In electric circuit diagrams, a capacitor is represented by two equal parallel lines. The net effect of charging a

---

[5]Again, current is the actual flow of charged carriers, while the difference in potential is the force that causes that flow. This difference is known as voltage.

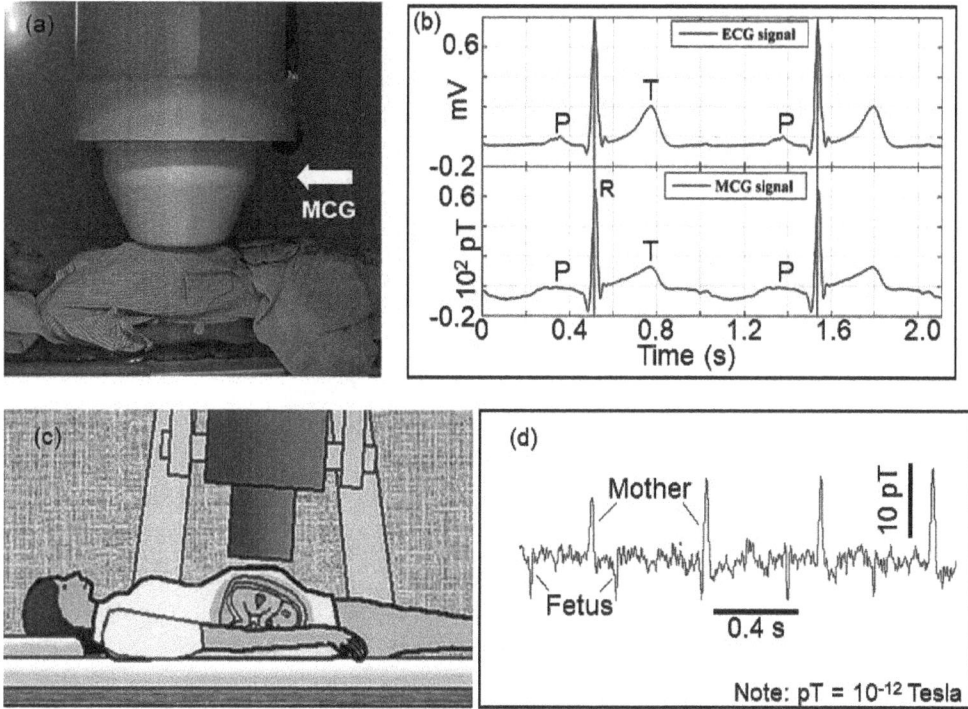

Figure 5.21. A monitoring device for magnetocardiography (MCG) and its signals. MCG systems measure the induced magnetic field at the torso level with an array of super-conductive quantum interference devices (SQUID) magnetometers. (a) A patient receives MCG. (b) A magnetocardiogram signal (c) Fetal magnetocardiography measures the magnetic activity of the fetal heart caused by electrical activity within the fetal heart. As the magnetic fields over the maternal abdomen are tiny, superconducting devices are used to record these fields. These measurements are conducted in a magnetically shielded room, because fetal cardiac signals have only a very small magnitude. The peak magnetic field of the fetal heart is about 3 picoTesla (or 3×10^{-12} Tesla). (d) Fetal MCG signal (mother and child).

capacitor is to remove charge from one plate and add it to the other plate; that is what a battery or any other source of electricity does when connected to a capacitor.

Let us mathematically illustrate the process of charging and discharging a capacitor with conductors separated by free space.

Step #1. By using Gauss's law (Chapter 3), we can verify that the potential obeys Laplace's equation; i.e.,

$$\boldsymbol{\nabla} \cdot (\varepsilon \boldsymbol{E}) = \rho \Longleftrightarrow -\boldsymbol{\nabla} \cdot (\varepsilon \boldsymbol{\nabla} \phi) = \boldsymbol{\nabla} \rho \Longleftrightarrow \boldsymbol{\nabla}^2 \phi(\boldsymbol{x}) = 0 \,, \qquad (5.13)$$

where ϕ is the energy potential. We assume that ε and ρ are constant and use the fact $\boldsymbol{E} = -\boldsymbol{\nabla}\phi$. We determine the potential between the plates by solving Laplace's equation with the two plates representing the boundary conditions; i.e.,

$$\boldsymbol{\nabla}^2 \phi(\boldsymbol{x}) = 0 \Longleftrightarrow \frac{d^2 \phi(x)}{dx^2} = 0 \,, \qquad (5.14)$$

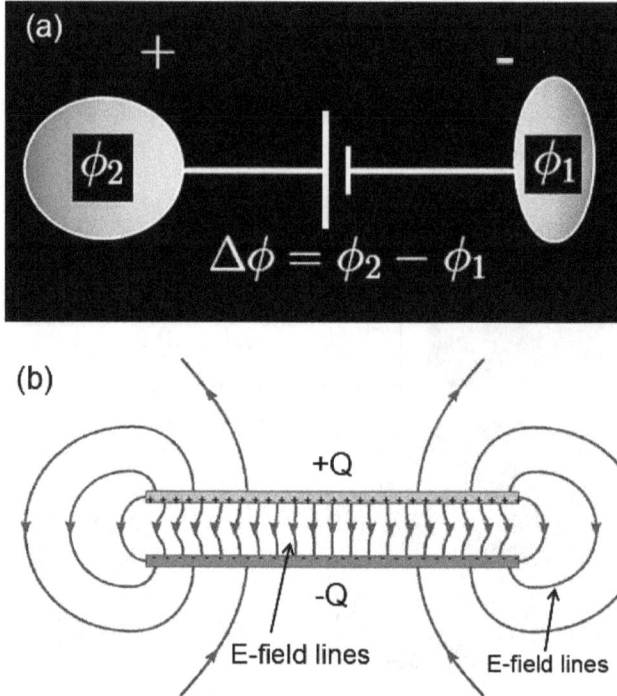

Figure 5.22. (a) Electrical capacitors are charge-storing devices consisting of two conducting plates separated by an insulating barrier such as free space or a dielectric material like ceramic, polymer films, or aluminum oxide. If two conductors are charged (i.e., when a voltage is applied to them for a finite period of time), they carry equal and opposite charges. (b) Here are the electric-field lines (E-field lines) of two parallel conductor planes. Note: The E-field line inside is pretty uniform. E-field lines outside are negligible. Charges accumulate at plate inner edges.

with $\phi(x = 0) = 0$ and $\phi(d) = \phi_{12}$. The solution is $\phi(x) = (\phi_{12}/d)x$, where d is the distance between the two plates (electrodes) and also the thickness of the dielectric medium.

Step #2. Evaluate the electric field between the plates as a negative potential gradient; i.e.,

$$\boldsymbol{E} = -\boldsymbol{\nabla}\phi(\boldsymbol{x}) = -\boldsymbol{i}_1 \frac{d\phi(x)}{dx} = -\boldsymbol{i}_1 \frac{\phi_{12}}{d} \ . \tag{5.15}$$

Step #3: Evaluate the surface charge on conductor 2:

$$q = \varepsilon \boldsymbol{i}_n \cdot \boldsymbol{E} = \varepsilon(-\boldsymbol{i}_1) \cdot \left(-\boldsymbol{i}_1 \frac{\phi_{12}}{d}\right) = \frac{\varepsilon\phi_{12}}{d} \ . \tag{5.16}$$

Step #4: Evaluate the total charge on conductor 2:

$$Q = qA = \frac{\varepsilon A \phi_{12}}{d} \ , \tag{5.17}$$

where A is the surface area of the plate (electrode).

Step #5: Evaluate the capacitance of a parallel-plate capacitor:

$$C = \frac{Q}{\phi_{12}} = \frac{\varepsilon A}{d} .$$

(5.18)

Step #6: Evaluate the energy stored (see Chapter 3) in the capacitor:

$$\mathcal{U}_E = \text{Energy density} \times \text{Volume}$$

$$= \left(\frac{1}{2}\varepsilon E^2\right) \times (Ad) = \frac{1}{2}\frac{\varepsilon A \phi_{12}}{d}\phi_{12} = \frac{1}{2}\frac{Q^2}{C} .$$

(5.19)

So the quantity $C = Q/V$, where $V = \Delta\phi$ is the voltage or the difference in potential between the two plates, is called the capacitance [unit is the farad (F)]. It is a measure of a conductor configuration's ability to store charge. It is related to the energy stored in a capacitor as follows $\mathcal{U}_E = \frac{1}{2}Q^2/C$. Conventional capacitors yield capacitance in the range of 0.1 to 1 μF with a voltage range of 50 to 400 V. Specific resistance is the reciprocal of conductance (equivalent to 1/conductance).

Neurons and axons are essentially minute capacitors. The cell membrane consists of a double lipid layer (i.e., meaning that it consists of two layers of lipid (fat) molecules) tails facing inward and their hydrophilic (water-loving) heads facing outward. The two layers of phospholipids create a fluid, semipermeable membrane that controls the movement of ions and molecules in and out of the cell. So, to answer your question, the cell membrane is composed of two layers of lipid molecules, resulting in a total of two layers.) that separates ions in the extracellular space from ions and charged proteins in the cytoplasm. These lipid membranes are excellent electrical insulators. Thus the cell membrane is an electrical insulator separating opposing charges inside and outside the cell, the plasma membrane behaves as a capacitor (see Figure 5.23). That is, the cell membrane has capacitance. From the

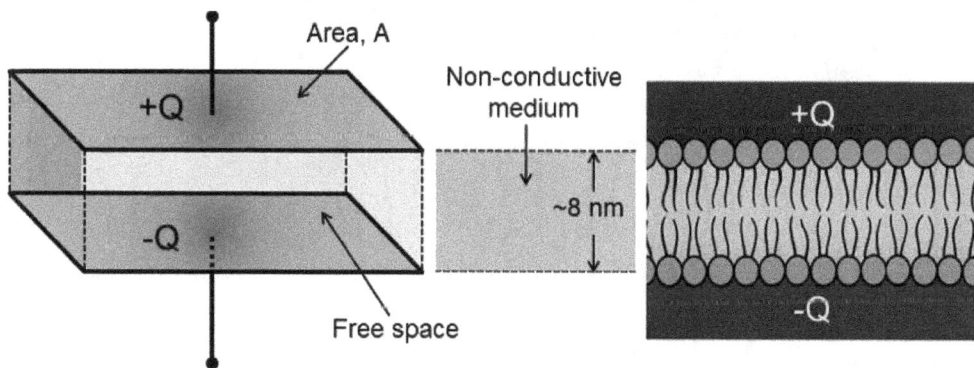

Figure 5.23. A representation of cell membrane as capacitor. (left) An idealized capacitor consists of two parallel conductors separated by free space or a dielectric material like ceramic, polymer films, or aluminum oxide. (middle) A cross section of an idealized capacitor. (right) An illustration of a cell membrane without ion channels as a capacitor that is formed by a piece of lipid membrane. The two plates are, in fact, the electrolyte solutions on either side of the membrane.

above discussion, membrane capacitance is related to the thickness of the insulating lipid portion of the membrane as follows:

$$C = \frac{\varepsilon_0 \kappa}{d} \approx \frac{8.85 \times 10^{-14} \times 5}{8 \times 10^{-9}} = 5.53 \times 10^{-5} \text{ F/cm}^2 = 55.3 \ \mu\text{F/cm}^2 \ , \qquad (5.20)$$

where C is the capacitance of the plasma membrane, d is the distance between the extracellular space from ions and charged proteins in the cytoplasm, ε_0 is the permittivity constant ($\varepsilon_0 = 8.85 \times 10^{-8} \ \mu\text{F/cm}$), and κ is the dielectric constant of the insulating material separating the two conducting plates ($\kappa = 5$ for membrane lipid). The estimate of d by electron microscopy reported earlier is 8 nanometers. This membrane capacitance is just an indication because $V = Q/C$, as defined earlier, means that a change in membrane voltage is followed by a change in membrane capacitance. A typical nerve cell membrane capacitance value is about 1 $\mu\text{F/cm}^2$.

So, neurons and axons can be considered miniature capacitors. However, there are some substantial differences between the two. Conventional capacitors are designed to store electrical energy, whereas neurons and axons use polarization to generate electrical impulses (action potentials). Polarization occurs when positive and negative charges are imbalanced inside the cell. Neurons and Axons use polarization to generate electrical impulses (action potentials). In short, while neurons and axons share some similarities with capacitors, they are not exactly the same thing.

Electrical response of the cell membrane to injected current. As illustrated in Figure 5.24, if a constant current, I, is injected into the cell, then charge, Q, is added to the membrane capacitor at a constant rate ($I = dQ/dt$). Ion channels provide another path for the injected charge to move across the membrane. This

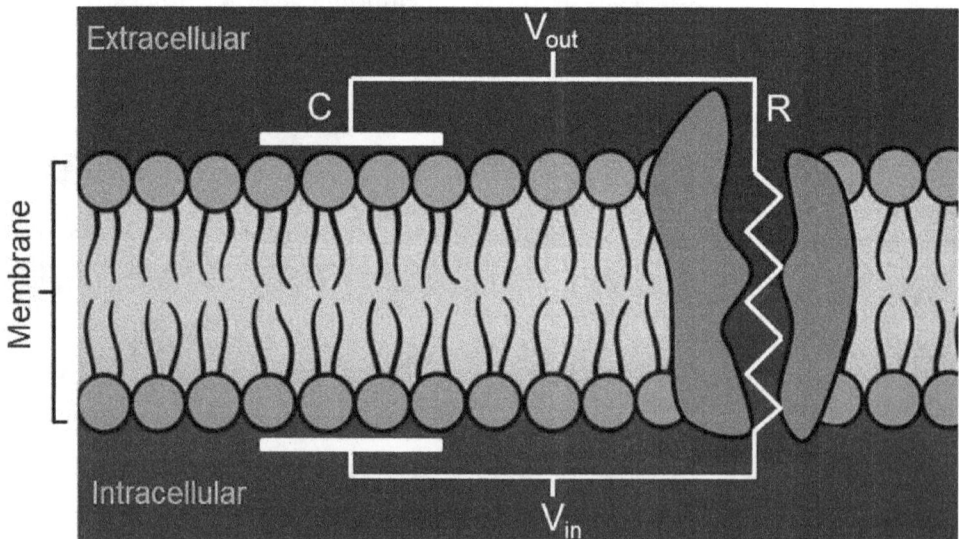

Figure 5.24. Electrical analogue of the cell membrane as a combination of a resistor and capacitor. The lipid bilayer is represented by a capacitor and the resistor represents the permeability of the ionic channel.

is instead of being added to the membrane capacitor charge. So when current is injected into the cell it can either charge the capacitor, or flow through the conductor and out of the cell, or a combination of the two. Initially, all current will flow into the capacitor. This is because the capacitor starts off with the resting potential voltage across it, and holds the resistor at that same voltage (because they are in parallel). So the overall circuit of the membrane consists of parallel resistances and combined parallel capacitance (see Figure 5.24) and the response is

$$I = I_R + I_C = \underbrace{\frac{V}{R}}_{\text{Resistor}} + \underbrace{C\frac{dV}{dt}}_{\text{Capacitor}} \implies V(t) = IR\left[1 - \exp\left(\frac{t}{RC}\right)\right], \qquad (5.21)$$

V is the change in the membrane voltage [mV], I is the stimulus current per unit area [μA/cm^2], R is membrane resistance times unit area [kΩcm^2], t is stimulus time [ms], $\tau = RC$ is membrane time constant [ms], and C is membrane capacitance per unit surface μF/cm^2. Thus, the voltage rises exponentially during injection of a constant current, I. The product, RC, is the exponential time constant of the voltage rise, abbreviated $\tau = RC$ (stimulus time). The asymptotic value of the voltage is $V_\infty = IR$, which is the voltage expected when all current is flowing through the membrane resistance. Initially, all injected charge flows onto the membrane capacitor. However, as the charge accumulates, more and more charge flows instead through the resistor. This is until finally all of the current flows through the resistive path. When the current injection terminates, the accumulated charge on the capacitor discharges through the parallel resistance. This voltage decay is also exponential, with the same time constant, τ.

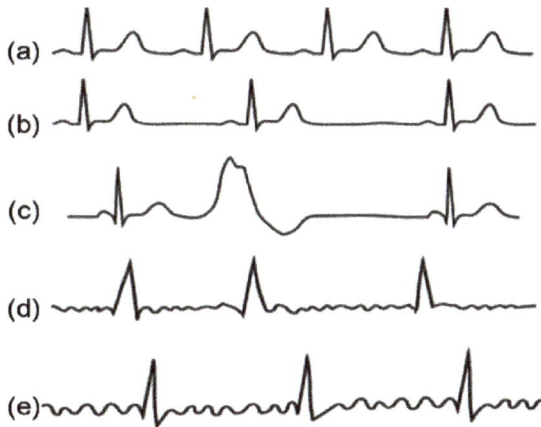

Figure 5.25. Electrocardiograms (ECGs). The ECG does not depict the heart's physical state or its function, but rather electrical activity.

Quiz: Figure 5.25 shows electrocardiograms (ECGs or EKGs). Determine the ones corresponding to healthy patients.

Computational exercise: (1) Use the Matlab code in Table 5.4 to reproduce the EEG signal in Figure 5.26 and its beta, alpha, theta, and delta waves. (2) Modify the code to reproduce to reproduce also the gamma waves.

Table 5.4. A Matlab code for simulating beta, alpha, theta, and delta waves.

```
Zc=0.057501127785; Zri=0.0001;Zlc=.001; time=0:1:128;
AA= [2195 1998 1932 2014 2015 1965 1900 1883 1883 1966 1916 1932 1916 ...
1883 1916 1769 1752 1769 1785 2015 2080 2080 2162 2129 2097 2014 ...
1949 1982 1998 1818 1801 1851 1736 1506 1457 1687 1850 1966 2047 ...
1982 1966 2146 2195 2506 2637 2637 2604 2342 2179 1867 1818 1785 ...
1753 1720 1769 2031 2031 2162 2162 2145 2227 2424 2440 2424 2326 ...
1900 2178 2178 2146 2194 1998 2015 1982 1785 1752 1523 1523 1752 ...
1736 1867 1851 1867 2047 2097 2129 2064 2048 2031 2031 2244 2277 ...
2276 2260 2113 2080 2162 2129 2129 2244 2309 2391 2473 2473 2572 ...
2604 2719 2670 2358 2359 2244 2146 2145 2080 1998 2113 2015 2064 ...
2179 2227 2621 2686 2441 2309 2129 2080 1998 1932 1818 1];
ffx=AA';
%% Orginal
plot(time,ffx,'k'); grid; axis([-10 138 -100 3000]);
fs=500; Afft=fft(ffx,1024); Lfft=length(Afft); L=(0:Lfft-1)*(fs/Lfft);
plot(L,20*log((abs(Afft)))); grid; axis([-20 520 80 260]);
%%delta wave
[nn,fo,ao,ww] = firpmord([0.3, 4.0]/(fs/2), [1 0], [Zc, Zri]);
qq = firpm(nn,fo,ao,ww); xdel=filter(dfilt.dffir(qq),ffx);
plot(time,xdel,'k'); grid; axis([-10 138 -20 450]);
Afft=fft(xdel,1024);
plot(L,20*log((abs(Afft)))); grid; axis([-20 520 80 200]);
%%theta wave
[nn,fo,ao,ww] = firpmord([3.75 4 7.75 8]/(fs/2), [0 1 0], [Zle, Zc,Zri]);
qq = firpm(nn,fo,ao,ww); xth=filter(dfilt.dffir(qq),ffx);
plot(time,xth,'k'); grid; axis([-10 138 -8 10]);
Afft=fft(xth,1024);
plot(L,20*log((abs(Afft)))); grid; axis([-20 520 10 120]);
%%alpha wave
[nn,fo,ao,ww] = firpmord([7.75 8 13.75 14]/(fs/2), [0 1 0], [Zle, Zc,Zri]);
qq = firpm(nn,fo,ao,ww); xalp=filter(dfilt.dffir(qq),ffx);
plot(time,xalp,'k'); grid; axis([-10 138 -10 15]);
Afft=fft(xalp,1024);
plot(L,20*log((abs(Afft)))); grid; axis([-20 520 20 130]);
%%beta wave
[nn,fo,ao,ww] = firpmord([13.5 14 29.5 30]/(fs/2), [0 1 0], [Zle, Zc,Zri]);
qq = firpm(nn,fo,ao,ww); xbet=filter(dfilt.dffir(qq),ffx);
plot(time,xbet,'k'); grid; axis([-10 138 -15 20]);
Afft=fft(xbet,1024);
plot(L,20*log((abs(Afft)))); grid; axis([-20 520 -50 130]);
```

To summarize, electrophysiology is the study of the electromagnetic properties of biological cells and tissues. It is done through the measurement and interpretation of electromagnetic responses of the body, particularly those of the heart and nervous system. As an example, an electrophysiology study of the heart might involve placing electrodes inside the heart to measure its electrical activity. This would enable us to detect abnormalities. No detectable abnormalities means the patient's electrophysiology is normal. This can be a valuable diagnostic tool, as it can help rule out certain conditions or diseases that might cause symptoms. Abnormalities may indicate an underlying condition or disease, such as an arrhythmia or neurological disorder. They are used to diagnose and treat a variety of conditions such as epilepsy, heart rhythm disorders, and arrhythmias. Treatments include medication

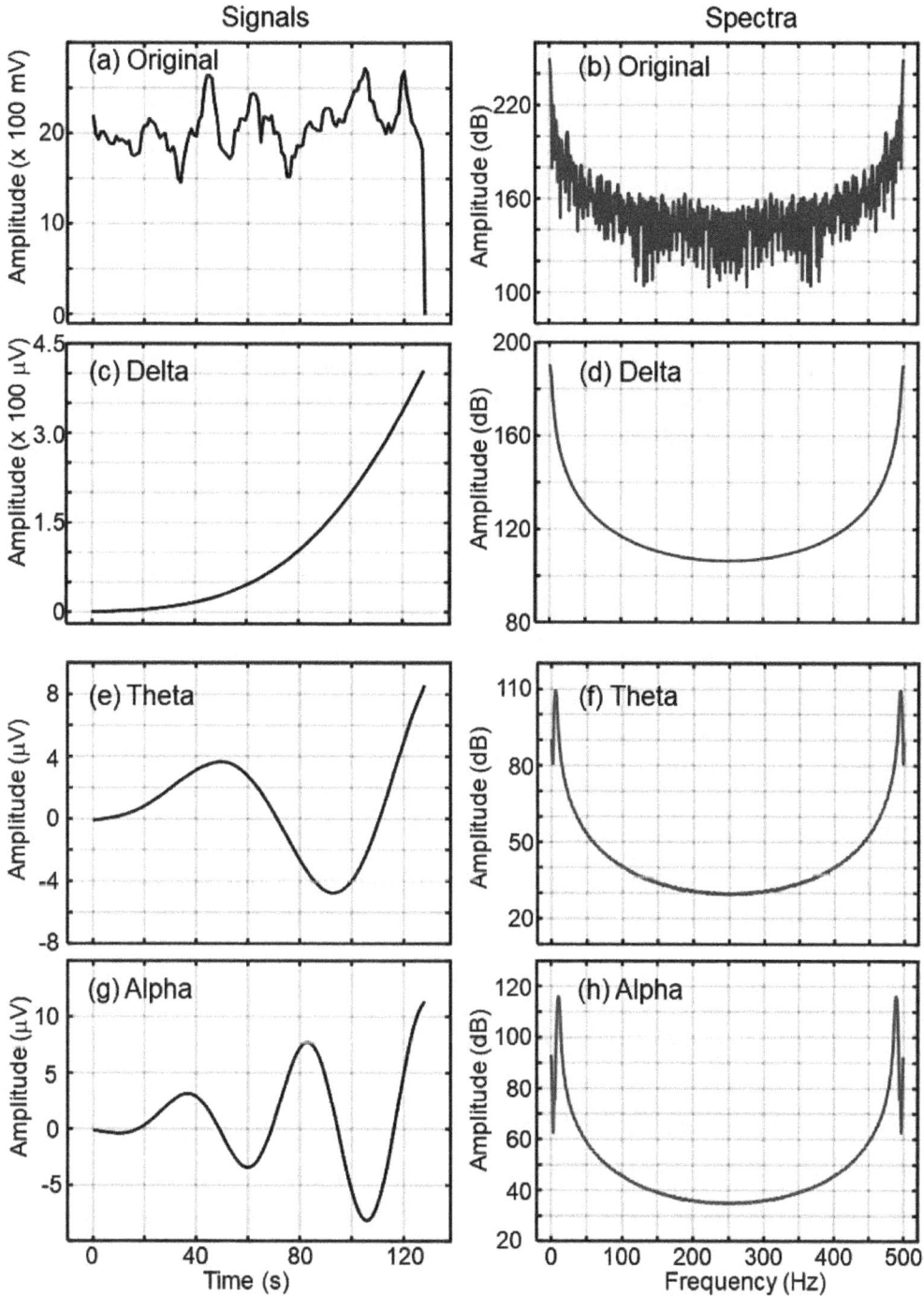

Figure 5.26. *(Continued.)*

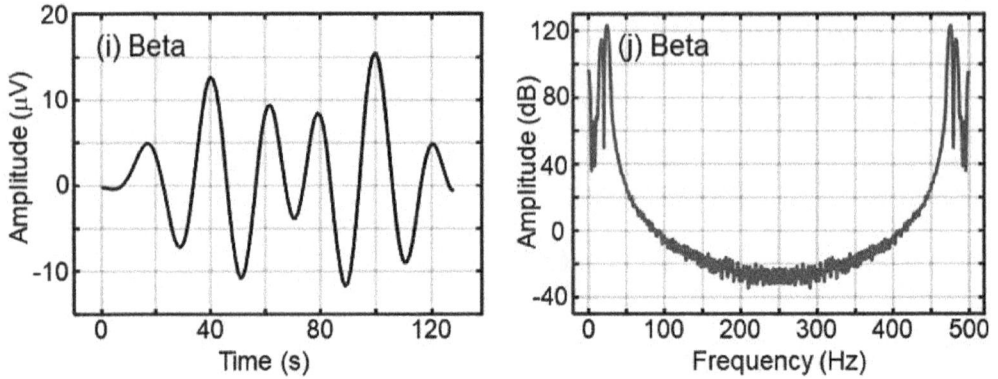

Figure 5.26. (a and b) EEG wave samples and its (c and d) delta wave, (e and f) theta wave, (g and h) alpha wave, and (i and j) beta wave.

to keep the rhythm normal, or to speed it up or slow it down, putting a pacemaker in the heart to keep the heartbeat fast enough or to electrically resynchronize several portions of the heart, shocking the heart out of serious/fatal rhythms, removing a small portion of the conduction system to fix a malfunctioning area and re-routing the electricity. In more general terms, electrophysiology has improved our understanding of the electromagnetic properties of the human body and how they relate to health and disease.

5.2.   Electricity from the Ground

Our objective in this section is to describe some possible ways electricity can be spontaneously generated from the ground and vacuum. We will start by introducing the integral form of Maxwell's equations and Faraday's induction. These equations are fundamental to understanding electricity generation.

5.2.1.   Electromagnetic Induction

Integral form of Maxwell's equations. The four Maxwell's equations derived in Chapter 3 can alternatively be written in the integral form, as shown in Table 5.5. The integral forms of Gauss's laws for electric and magnetic fields are obtained by using the divergence theorem and Ampère's and Faraday's laws are obtained by applying the Stokes's theorem. Let us look at a specific example. It is easy to make the transition from Maxwell's differential equations to integral equations with the help of these two theorems. For example, by integrating both sides over a closed volume D surrounded by the closed surface ∂D of Gauss's laws for electric field and using the divergence theorem, we obtain the integrated form; i.e.,

$$\oiint_{\partial D} [\boldsymbol{D}(\boldsymbol{x}, t) \cdot \boldsymbol{n}(\boldsymbol{x})] \, dS(\boldsymbol{x}) = Q_E(t) \tag{5.22}$$

with

$$Q_E(t) = \iiint_D \rho(\boldsymbol{x}, t) \, dV(\boldsymbol{x}) , \tag{5.23}$$

Table 5.5. Differential and integral forms of Maxwell's equations. These two forms are related through Stokes's and divergence theorems (e.g., Ikelle, 2020). Note that $dS = ndS$ and S is any surface, not necessarily closed, whereas ∂D is a closed surface (the boundary of the volume D). The line integrals of the electric and magnetic fields are closed paths along the boundary C of the surface S; C is always a closed curve. Note that Gauss' laws are integrals over closed surfaces whereas Ampére's laws are integrals over closed paths.

Differential form	Integral form	
$\nabla \times \boldsymbol{H} = \boldsymbol{J} + \dfrac{\partial \boldsymbol{D}}{\partial t}$	$\oint_C \boldsymbol{H} \cdot dl = \iint_S \left(\boldsymbol{J} + \dfrac{\partial \boldsymbol{D}}{\partial t} \right) \cdot d\boldsymbol{S}$	Ampere
$\nabla \times \boldsymbol{E} = -\dfrac{\partial \boldsymbol{B}}{\partial t}$	$\oint_C \boldsymbol{E} \cdot dl = -\iint_S \dfrac{\partial \boldsymbol{B}}{\partial t} \cdot d\boldsymbol{S}$	Faraday
$\nabla \cdot \boldsymbol{D} = \rho$	$\oiint_{\partial D} \boldsymbol{D} \cdot d\boldsymbol{S} = \iiint_D \rho \, dV$	Gauss (electricity)
$\nabla \cdot \boldsymbol{B} = 0$	$\oiint_{\partial D} \boldsymbol{B} \cdot d\boldsymbol{S} = 0$	Gauss (magnetic)

where $\boldsymbol{n}$ is normal unit vector normal to surface ∂D and pointing outward, Q_E is the total charge within volume D. Similarly, we can obtain the integral form of Gauss's laws for magnetic fields by using the divergence theorem. We can obtain the integral form of the charge conservation law in the same way; i.e.,

$$\oiint_{\partial D} [\boldsymbol{J}(\boldsymbol{x}, t) \cdot \boldsymbol{n}(\boldsymbol{x})] \, dS(\boldsymbol{x}) = -\frac{d}{dt} \left[\iiint_D \rho(\boldsymbol{x}, t) \, dV(\boldsymbol{x}) \right] . \qquad (5.24)$$

The left-hand side represents the total amount of charge flowing outwards through the surface ∂D per unit time. The right-hand side represents the amount by which charge decreases inside the volume V per unit time. In other words, charge does not disappear into (or is created out of) nothingness—it decreases in a region of space only because it flows into other regions.

Let us integrate the differential form of Faraday's law over the surface S bounded by the contour C. We arrive at

$$\iint_S \{ (\nabla \times \boldsymbol{E}(\boldsymbol{x}, t)) \cdot \boldsymbol{n}(\boldsymbol{x}) \} \, dS(\boldsymbol{x}) = -\frac{d}{dt} \iint_S [\boldsymbol{B}(\boldsymbol{x}, t) \cdot \boldsymbol{n}(\boldsymbol{x})] \, dS(\boldsymbol{x}) . \qquad (5.25)$$

Note S is any surface and not necessarily closed. By applying Stokes' theorem to the left hand-side of (5.25), we obtain the integral form of Faraday's law; i.e.,

$$\oint_C [\boldsymbol{E} \cdot dl(\boldsymbol{x})] = -\frac{d}{dt} \iint_S [\boldsymbol{B}(\boldsymbol{x}, t) \cdot \boldsymbol{n}(\boldsymbol{x})] \, dS(\boldsymbol{x}) . \qquad (5.26)$$

The line integral of the electric field carried out along the boundary C of a surface S. C is always a closed curve. A similar application of Stokes' theorem to the differential form of Ampere's law yields the integral form of Ampère's law in Table 5.5. We can deduce the integral form of Maxwell's equations for electrostatic and magneto-statics by assuming that the time derivative of the fields in Table 5.5 is zero.

Faraday's law of induction. The integral form of Faraday's law in (5.26) can also be written as

$$\mathcal{E}(t) = -\frac{d\Phi_B(t)}{dt} , \qquad (5.27)$$

where $\mathcal{E}(t)$, which is known as the electromotive force (emf) or induced voltage, is defined as follows:

$$\mathcal{E}(t) = \oint_C \boldsymbol{E}(\boldsymbol{x}, t) \cdot d\boldsymbol{l}(\boldsymbol{x}) , \qquad (5.28)$$

and $\Phi_B(t)$ is the magnetic flux and defined as follows:

$$\Phi_B(t) = \iint_S [\boldsymbol{B}(\boldsymbol{x}, t) \cdot \boldsymbol{n}(\boldsymbol{x})] \, dS(\boldsymbol{x}) . \qquad (5.29)$$

The closed line integral of the electric field carried out along the boundary C of the surface S. As illustrated in Figure 5.27, C is always a closed curve.

Let us expand on the definition of electromotive force and magnetic flux. Consider a region of an electric field and the motion of a charge q from A to B along a defined path. As we discussed in Chapter 3, at each and every point along the AB path, the electric field exerts a force on this charge and, hence, does a certain amount of work in moving the charge to another point along this path. The total

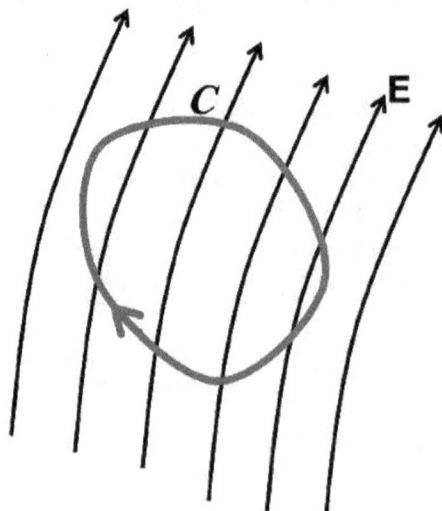

Figure 5.27. Closed path C in a region of electric field.

amount of work done from A to B is the line integral, in ((5.28), from A to B, in which we use force $\boldsymbol{F} = q\boldsymbol{E}$, instead of $\boldsymbol{E}$, without the assumption of a closed path. Suppose now that we are dealing with a unit of charge and that the path AB is a closed curve, the field in moving the charge from A to B becomes exactly $\mathcal{E}(t)$, as defined in (5.28). In other words, the line integral of $\boldsymbol{E}$ around a closed path is the work per unit charge done by the field in moving a charge around the closed path. It is the voltage around the closed path and is known as electromotive force.

Let us now turn to the magnetic flux. For the particular case of a uniform magnetic field, the magnetic flux, in (5.29), becomes the scalar product of the surface area and the magnetic field; i.e.,

$$\Phi_B = [\boldsymbol{B} \cdot \boldsymbol{n}]\, S = B\, S \cos\theta \ . \tag{5.30}$$

The magnetic flux is maximized when the magnetic field vector is perpendicular to the surface [i.e., $\theta = 0$]; that is, the surface is oriented normal to the magnetic field lines, as illustrated in Figure 5.28c.

If, however, the surface is oriented parallel to the magnetic field lines, as shown in Figure 5.28a, there are no magnetic field lines crossing the surface; the magnetic flux is zero. If the normal to the surface makes an angle θ with the magnetic field lines, as shown in Figure 5.28b, then the amount of magnetic flux flowing along the surface S is $B\, S \cos\theta$ where $B\cos\theta$ is the component of $\boldsymbol{B}$ normal to the surface and the component tangential to the surface is $B\sin\theta$. So magnetic flux characterizes field lines crossing an open surface S. It depends on the size of S, the magnitude of $\boldsymbol{B}$, and the angle between the normal to the surface and the magnetic field.

Let us now consider a non-uniform magnetic field. The magnetic flux crossing S can be found by dividing the surface S into a number of infinitesimal surfaces $dS(\boldsymbol{x})$ on which the magnetic field is uniform. For each infinitesimal surface, $dS(\boldsymbol{x})$, represented by $\boldsymbol{x}$, the magnetic flux is $\boldsymbol{B}(\boldsymbol{x}) \cdot \boldsymbol{n}(\boldsymbol{x})dS(\boldsymbol{x})$. The total magnetic flux in (5.29) is the sum of the contributions of these infinitesimal surfaces.

Let us return to Faraday's induction law, in (5.27). The minus sign indicates that the direction of the electromotive force is opposite to the change in the magnetic flux that produces the voltage. This equation was discovered experimentally in 1831 by Michael Faraday and can be stated as follows. Electric current (or voltage) can be induced by changing the magnetic field. The induced emf effectively depends on the rate of change of magnetic flux. For the special case of a coil of wire (an inductor), composed of N loops (turns) with the same area, as illustrated in Figure 3.22, the Faraday induction law, in (5.27), becomes

$$\mathcal{E}(t) = -N\frac{d\Phi_B(t)}{dt} \ . \tag{5.31}$$

We can see that the amount of voltage produced by the generator can be optimized by changing the number N of turns of wire in the coil. Because the magnetic flux is proportional to the current $I(t)$, we can also express (5.31) as

$$\mathcal{E}(t) = -L\frac{dI(t)}{dt} \ , \tag{5.32}$$

Where L is called the *inductance*. It is a measure of an inductor's resistivity to current change. The unit of L is Henry (H). The two expressions, in (5.31) and

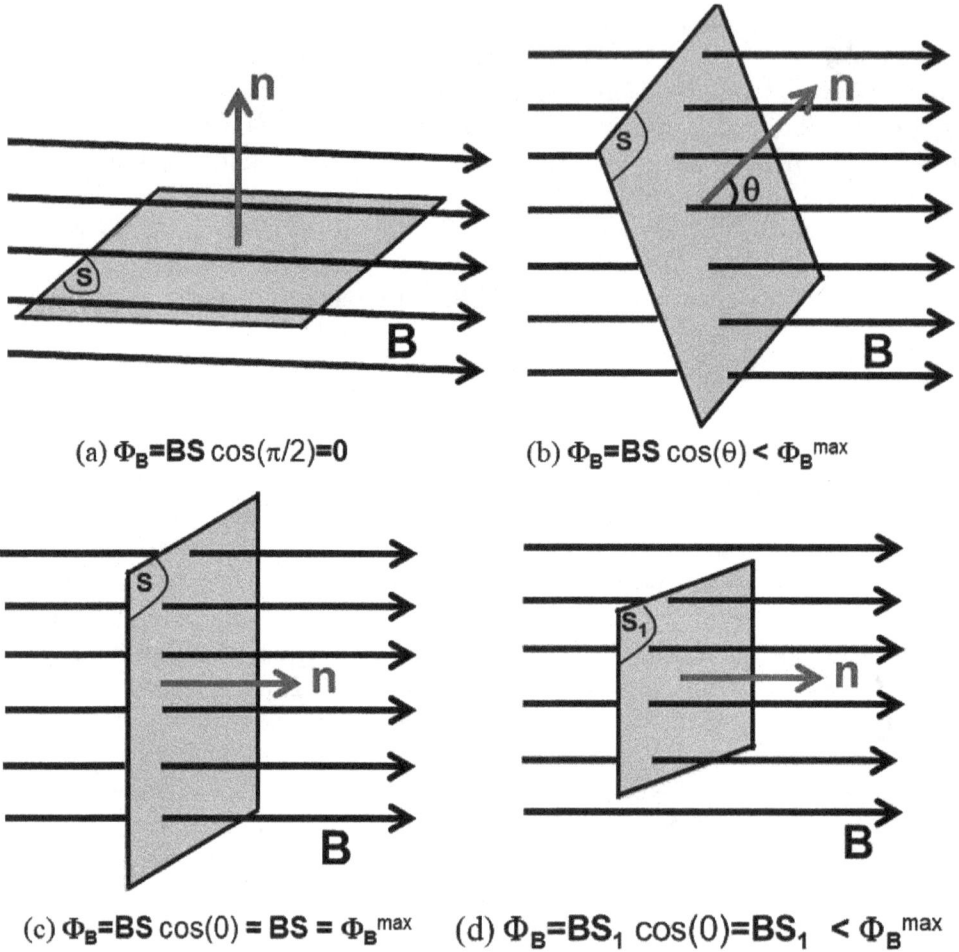

Figure 5.28. The magnetic flux depends on the angle between B and n, the magnitude of B, and the size of the area, S. (a) The flux is zero when B and n are perpendicular. (b) As the angle between B and n decreases, the flux increases. (c) The flux has its maximum value when B and n are parallel. (d) As the size of the area, S, decreases, the flux also decreases.

(5.32), can be combined to yield

$$L = N \frac{\Phi_B(t)}{I(t)} \ . \tag{5.33}$$

If $dI(t)/dt > 0$ (i.e., the current is increasing), the external source is doing positive work to transfer energy to the inductor. Thus, we charge the inductor and the internal energy $\mathcal{U}_B$ of the inductor is increased. The total work done by the external source to increase the current from 0 to I can be obtained as follows:

$$\frac{d\mathcal{U}_B}{dt} = LI\frac{dI}{dt} = \frac{d}{dt}\left(\frac{1}{2}LI^2\right) \Longrightarrow \mathcal{U}_B = \frac{1}{2}LI^2 \ , \tag{5.34}$$

where $\mathcal{U}_B$ is the magnetic energy stored in the inductor. The above expression is analogous to the electric energy stored in a capacitor, $\mathcal{U}_E = \frac{1}{2}Q^2/C$. On the other hand, if the current is decreasing with $dI(t)/dt < 0$, the external source takes energy away from the inductor, causing its internal energy to decrease, and we discharge the inductor.

Let us compute the inductance of a solenoid with N turns, length l, and radius a with a current I flowing through each turn, as shown in Figures 3.22 and 3.23. As described in Chapter 3, the magnetic field inside a solenoid is

$$\boldsymbol{B} = \frac{\mu_0\,N\,I}{l}\boldsymbol{i}_3 = \mu_0 n I \boldsymbol{i}_3 \ , \tag{5.35}$$

where $n = N/l$ is the number of turns per unit length. The magnetic flux through the solenoid is

$$\Phi_B = BS = \mu_0 n I \times (\pi a^2) \Longrightarrow L = N\frac{\Phi_B}{I} = \mu_0 n^2 \pi a^2 l = \frac{\mu_0 N^2 S}{l} \ . \tag{5.36}$$

We see that L depends only on the geometrical factors (n, a, and l) and is independent of the current I. We can evaluate the magnetic energy stored in the solenoid as follows

$$\mathcal{U}_B = \frac{1}{2}LI^2 = \frac{1}{2\mu_0}\left(\frac{\mu_0\,N\,I}{l}\right)^2 lS = \frac{1}{2}\frac{B^2}{\mu_0}lS = W_B V \ , \tag{5.37}$$

where W_B is the magnetic energy density stored in the solenoid, as defined in Chapter 3, and $V = S \times l$ is the solenoid volume.

Note that the electromotive force (emf) in superconductors' interior is zero because their electrical resistivity is zero in superconductors. Hence (5.31) reduces to

$$\frac{d\Phi_B(t)}{dt} = 0 \Longrightarrow B = \frac{\Phi_B}{S} = \text{constant} \ . \tag{5.38}$$

That is, the magnetic field inside a superconductor cannot change over time. Therefore, the magnetic field remains constant inside the superconductor after the external magnetic field is turned off. Table 5.6 summarizes the induction equations.

Table 5.6. Integral forms of Maxwell's equations and induction equations.

Integral form	Induction equation
$\oint_C \boldsymbol{H} \cdot d\boldsymbol{l} = \iint_S \left(\boldsymbol{J} + \dfrac{\partial \boldsymbol{D}}{\partial t}\right) \cdot d\boldsymbol{S}$	$\mathcal{H} = \varepsilon_0 \dfrac{\partial \Phi_E}{\partial t} + I_E$
$\oint_C \boldsymbol{E} \cdot d\boldsymbol{l} = -\iint_S \dfrac{\partial \boldsymbol{B}}{\partial t} \cdot d\boldsymbol{S}$	$\mathcal{E} = -\dfrac{\partial \Phi_B}{\partial t}$

Electric induction. In the previous paragraphs, the electric fields and magnetic fields were considered to have been produced by stationary charges and moving charges (currents), respectively. We have seen that by varying the magnetic field with time, an electric field could be generated through electromagnetic induction. Also imposing an electric field on a conductor gives rise to a current which in turn generates a magnetic field; i.e.,

$$\mathcal{H} = \varepsilon_0 \frac{\partial \Phi_E}{\partial t} + I_E \tag{5.39}$$

where

$$\mathcal{H} = \oint \boldsymbol{H} \cdot d\boldsymbol{l} \tag{5.40}$$

$$\Phi_E = \int\int \boldsymbol{D} \cdot d\boldsymbol{S}$$

and

$$I_E = \iint_S (\boldsymbol{J} \cdot \boldsymbol{n})\, dS \; . \tag{5.41}$$

The current, I_E, is the flux associated with the current density vector $\boldsymbol{J}$.

Logging tool. Well logging involves measuring the physical properties of rocks surrounding a borehole by using sensors located in a borehole. Well logs provide in situ measurements of porosity, lithology, hydrocarbon presence, and other rock properties of interest. Remember that the tools used to take well-logging measurements must cope with extremely tough conditions downhole. These conditions include high temperatures and pressures, inhospitable chemical conditions, and the physical constraints imposed by the physics of the measurements and the borehole geometry. It should also be remembered that we are interested in the properties of rocks in undisturbed conditions, and the act of drilling the borehole is the single-most disturbing thing we can do to a formation. Therefore, it is worthwhile to understand the borehole environment.

Drilling mud (also known as lubricating mud) is a key ingredient in the drilling process and has a significant influence on log measurements and interpretation. The drilling process uses lubricating mud pumped down through the drill pipe (see Figure 5.29a). This lubricating mud keeps the drill bit cool and lubricated. It prevents formation fluids from entering the wellbore (the mud column has higher pressure than the pore pressure of the formation) by carrying the cuttings, or chips, away from the drilled formation to prevent the well from clogging. The excess borehole pressure over the formation pressure prevents blowouts. Drilling mud is usually a mixture of bentonite clay and oil or water, plus barite ($BaSO_4$), a high specific gravity nonmetallic mineral, to regulate density.

When the drilling mud encounters porous and permeable layers (loose formations), the drilling mud will flow into the formation because of the difference in fluid pressure. This flow is called invasion. The zone where much of the original fluid

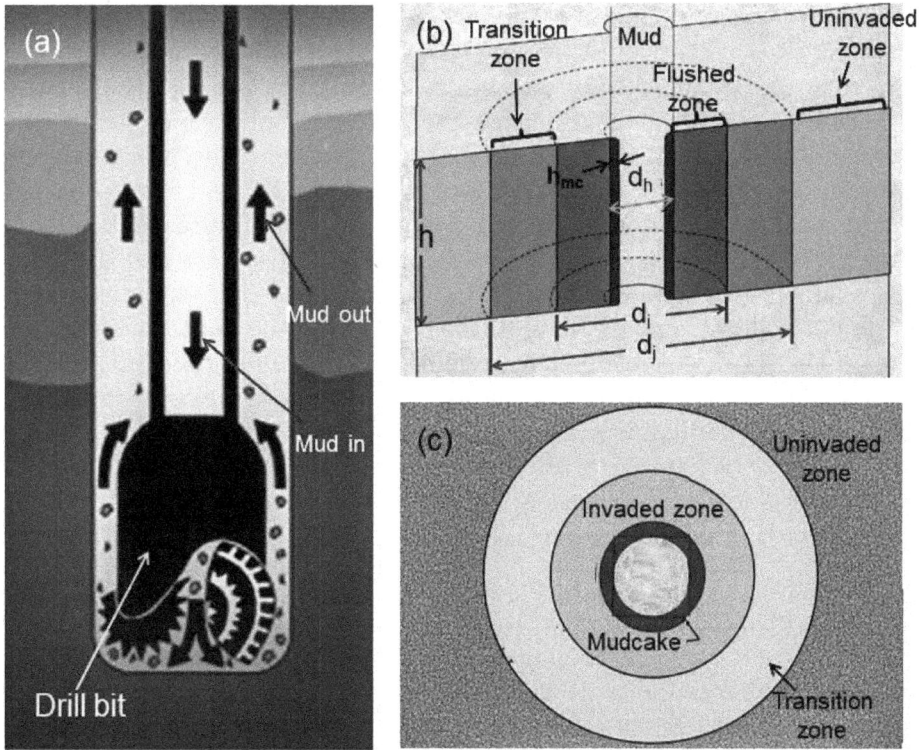

Figure 5.29. (a) An illustration of the lubricating mud (drilling mud) pumped down through the drill pipe. The drill bit is kept cool and lubricated by the drilling mud. Drilling muds are also used to remove drilled materials away from the drill-bit and transport them to the surface. (b) A vertical slide of the borehole showing the mudcake, flushed zone (the zone where the mud filtrate replaced the reservoir fluids), the transition zone, and the uninvaded zone further into the formation. (c) A horizontal slice of the borehole.

is replaced by mud filtrate is called *the invaded zone*. The depth of *mud-filtrate invasion* into the invaded zone is called the diameter of invasion, denoted here by d_j (see Figure 5.29b). The invasion diameter is generally expressed as a ratio of d_j over d_h (where d_h represents the borehole diameter). During the invasion, the solid particles (i.e., clay minerals from the drilling mud) in the drilling mud will be left at the borehole wall, with the rock acting as an efficient filter. The buildup of mud particles on the inner wall of the borehole is often referred to as *mudcake* (h_{mc}). Mudcake is an indicator of porous and permeable formations. The amount of invasion depends upon the permeability of the mudcake and not on the porosity of the rock. In general, an equal volume of mud filtrate can invade low-porosity and high-porosity rocks if the drilling muds have the same amounts of solid particles. The solid particles in drilling mud coalesce and form an impermeable mudcake. The mud cake then acts as a barrier to further invasion. Because an equal volume of fluid can be invaded before an impermeable mudcake barrier forms, the invasion diameter is greater in low-porosity rocks. This occurs because low-porosity rocks have less storage capacity or pore volume to fill with the invading fluid, and, as a result, pores throughout a wider volume of rock are affected. The remaining liquid part of the drilling mud enters the porous and permeable formation, pushing back reservoir fluids. This part of drilling mud is called the *mud filtrate*.

The zone where mud filtrate replaces reservoir fluids is called the *flushed zone* (i.e., d_i in Figure 5.29b). The flushed zone extends only a few inches from the wellbore and is part of the invaded zone. If the invasion is deep or moderate, the flushed zone is completely cleared of its formation water by mud filtrate. When oil is present in the flushed zone, the degree of flushing by mud filtrate can be determined from the difference between water saturation in the flushed zone and the uninvaded zone. Usually, about 70-95 percent of the oil is flushed out—the remaining oil is called residual oil. A zone further into the rock where the replacement of reservoir fluids with mud filtrate is incomplete is called *The transition zone* ($d_j - d_i$ in Figure 5.29b). A zone further into the formation where the original pore content is not contaminated by mud filtrate is called the *uninvaded (virgin) zone*. Efforts are generally made to minimize the invaded zone by (i) ensuring the mud is sufficiently saline to avoid wash-outs increasing the borehole diameter, and (ii) by setting the mud weight such that the mud pressure is only slightly greater than the formation pressure.

5.2.2. Resistivity Well-Logging

Again, the electric resistivity of a substance is a measure of its opposition to the passage of electrical current. Igneous, metamorphic, and dry sedimentary rocks are poor conductors of electricity—that is, they have very high resistivities. Most sedimentary rocks, which form the bulk of oil and gas reservoirs, contain pores filled with water containing dissolved salts. While the matrices of sedimentary rocks are not electrically conductors, their salted-water (brine) filled pores allow electricity to pass through them. Thus, the resistivities of these rocks are directly related to their water saturations. Table 5.7 illustrates the resistivity of some typical subsurface materials. Notice the range of resistivity variation for salt water, which depends on the concentration of NaCl. Rock materials are insulators. Reservoir rocks have detectable conductivity due to electrolytic conductors in the pore space. The conductivity of clay minerals, for example, is greatly increased by the presence of an electrolyte. In some cases, rock resistivity may result from the presence of

Table 5.7. Resistivity of some Earth formations. Oil and gas are good insulators therefore have high resistivities. The concentrations of dissolved salts are measures of salinity which is expressed by part per million (ppm).

Material	R (ohm-m)	Material	R (ohm-m)
Marble	$5 \times 10^7 - 10^9$	Clay/shale	$2 - 10$
Quartz	$10^{12} - 3 \times 10^{14}$	Saltwater sand	$0.5 - 10$
Petroleum	2×10^{14}	Oil sand	$5 - 10^3$
Distilled water	2×10^{14}	"Tight" limestone	10^3
Saltwater ($15°$ C; 2 kppm)	3.4	Oil	High resistivity
Saltwater ($15°$ C; 10)	0.72	Gas	High resistivity
Saltwater ($15°$ C; 20)	0.38	Water-based-mud	Low resistivity
Saltwater ($15°$ C; 100)	0.09	Oil-based-mud	High resistivity
Saltwater ($15°$ C; 200)	0.06	Rock's matrices	High resistivity

metal, graphite, or metal sulfides. Note that oil and gas are good insulators and therefore have high resistivities. Brine is a good conductor therefore it has low resistivity.

Remember that fluid saturation is the ratio of the fluid volume to the pore volume in a given sample; i.e.,

$$S_w = \frac{V_w}{V_p} \quad , \quad S_o = \frac{V_o}{V_p} \quad , \quad \text{and} \quad S_g = \frac{V_g}{V_p} \quad \text{with} \quad S_w + S_o + S_g = 1 \ ,$$

where V_w is the volume of water, V_o is the volume of oil, and V_g is the volume of gas, V_p is the volume of pores, S_w is water saturation, S_o is oil saturation, and S_g is gas saturation. In oil and gas exploration and production, it is generally assumed that all pore space not filled with water is occupied by hydrocarbons when rocks are not dry.

In the context of well-logging, we deal with quasi-time-independent electromagnetic fields. Electrostatics and magneto-statics are the branches of electromagnetics corresponding to this case (see Chapter 3). For a charge, q, in a sphere with radius $r = |\boldsymbol{x}|$, the electric current can be written as

$$I = \int_S J dS = \int_S (\sigma E) dS = \frac{\sigma q}{4\pi\varepsilon_0 r^2} 4\pi r^2 = \frac{\sigma q}{\varepsilon_0} = \frac{q}{R_t \varepsilon_0} \ , \tag{5.42}$$

where $J = |\boldsymbol{J}|$, the electric field is

$$\boldsymbol{E}(\boldsymbol{x}) = \frac{1}{4\pi\varepsilon} q \frac{\boldsymbol{x}}{|\boldsymbol{x}|^3} = -\boldsymbol{\nabla}\phi(\boldsymbol{x}) \ , \tag{5.43}$$

$E(\boldsymbol{x}) = |\boldsymbol{E}(\boldsymbol{x})|$, and the electric potential is

$$\phi(\boldsymbol{x}) = V(\boldsymbol{x}) = \frac{1}{4\pi\varepsilon_0} \frac{q}{|\boldsymbol{x}|} \ . \tag{5.44}$$

By substituting (5.44) into (5.42), we arrive at

$$V(\boldsymbol{x}) = \frac{1}{4\pi|\boldsymbol{x}|} R_t I \quad \text{or} \quad R_t = 4\pi|\boldsymbol{x}|\frac{V}{I} = K\frac{V}{I} \ , \tag{5.45}$$

where $K = 4\pi|\boldsymbol{x}|$ is constant for most device.

Resistivity logging consists of electrodes emitting current of intensity I (active source) and electrodes measuring voltage between two points (receiver), V, as illustrated in Figure 5.30. We deduce the resistivity R_t for various depths, z, as follows:

$$R_t(z) = K\frac{V(z)}{I} \ , \tag{5.46}$$

where K is the tool constant related to electrode spacing, the subscript t stands for "true," and R_t is for true resistivity or resistivity in the uninvaded zone of the formation. We added the adjective *uninvaded* in the properties of the rocks in undisturbed drilling conditions because the act of drilling the borehole is the single most disturbing thing that can happen to a formation, as illustrated in

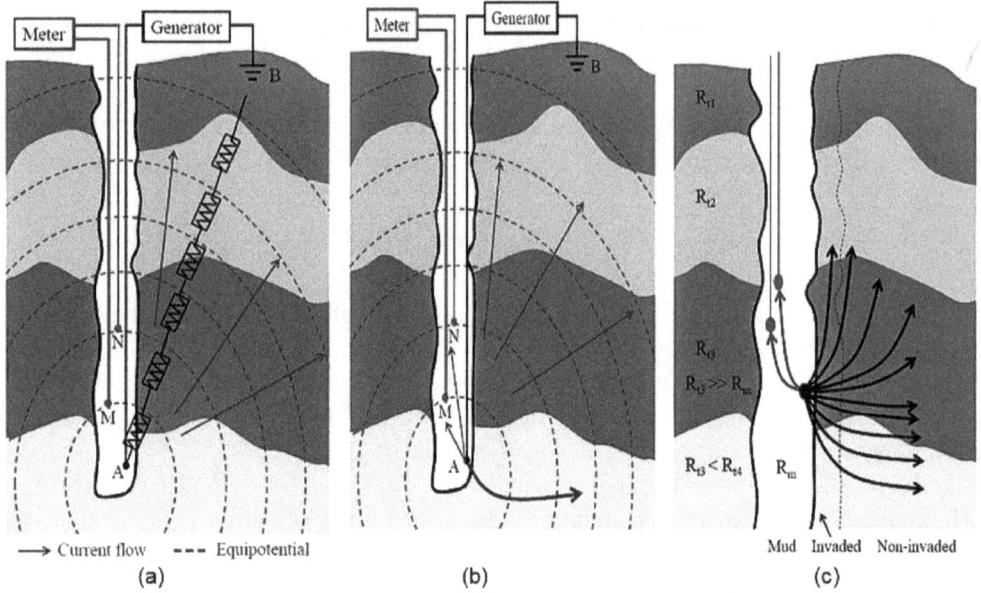

Figure 5.30. (a) Resistivity logging consists of electrodes emitting current and electrodes measuring voltage between two points as illustrated here. B is a surface electrode and A, M, N electrodes are mounted on the sonde. Electric current flows into the formation through the electrode pair A-B, and voltage is measured on the electrode pair M-N. (b) An illustration of the fact that current flows in the mud rather than the formation. In this case, the measured resistivity does not reflect the formation's resistivity accurately. (c) The media surrounding the logging tool include borehole mud, the invaded zone, and the uncontaminated zone (uninvaded zone). Only virgin formation resistivity (uninvaded zone) is desired in resistivity logging.

Figure 5.29. Mud in the borehole is very conductive and the current tends to flow in the mud rather than the formation and we end up with erroneous measurements. Moreover, some drilling mud will flow into the formation some point, depending on how porous and permeable the formation is. This is because of the difference in fluid pressure between the over-pressured wellbore and the formation pressure. We can overcome this issue by generating electric current by induction or by focusing electrical current laterally in the formation through the use of electrode arrays, as we will next discuss. These tools can cope with extremely highly resistive muds (oil-based muds or gas as a borehole fluid). The induction log measures conductivity, and hence is sometimes called the conductivity log.

Focusing borehole tools. A resistivity well-logging system consists of one current emitter electrode and a voltage measurement electrode (see Figure 5.30). In a borehole filled with very conductive mud, current flows through the mud rather than through the formation. In this case, the resistivity from the injected current results in a voltage that will not reflect the formation's resistivity. The second difficulty with this technique is that the injected current can end up in adjacent shoulder beds of much lower resistivity than the actual formation directly opposite the current electrode. In this case, the apparent resistivity will again be representative not

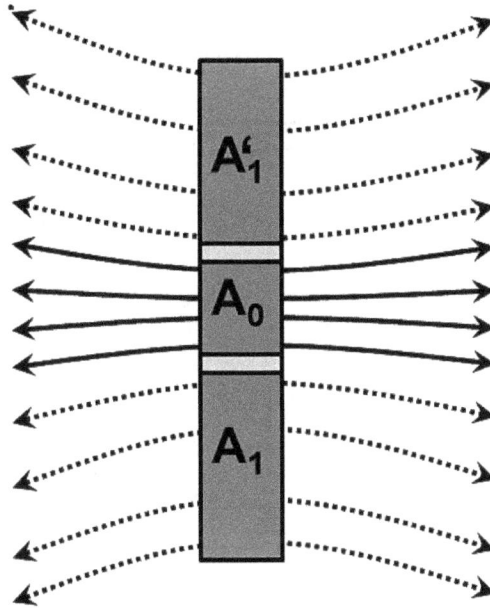

Figure 5.31. An illustration of three current emitting electrodes, A_0, A_1, and A_1' in resistivity logging. The potential of electrodes A_1 and A_1' is held constant and at the same potential as the middle electrode A_0. As current flows only when a potential difference exists, there is no current flow (or very little current flow) in the vertical direction. The current is mostly focused laterally and originates from A_0.

of the resistive bed, but, more likely, of the less resistive shoulder bed. Focusing borehole tools forces the current to flow horizontally in the formation. This prevents measurements from being affected by drilling mud or invaded parts. Consider, for example, the case of three emitting electrodes of the same elongations A0, A1, and A1', as illustrated in Figure 5.31—the potential of electrodes A1 and A1' is held constant and at the same potential as the middle electrode A0. Since current flows only if a potential difference exists, there should, in principle, be no current flowing in the vertical direction. This is because the potential drop is zero along this vertical direction, the current will be focused on the formation. The current bubble therefore emanates horizontally from the central measurement electrode.

Three electrodes, as illustrated in Figure 5.31, explain the focusing method. However, in practice, seven or more electrodes are used because three-electrode tools show some effects of bed boundaries on resistivity measurements. The electrodes A1 and A1' are no longer elongated: instead, additional monitoring electrodes are introduced to impede the flow of current parallel to the sonde through the borehole mud. This is achieved by varying the bucking current of the guard electrodes so that the potential drop between the pairs of monitor electrodes is zero. Since the potential drop is zero along this vertical direction, the current will focus on the formation.

In the induction case, the logging tools have several transmitter-receiver coils, but the basic principles of the idea can simply be described by considering single transmitter and receiver coils, as shown in Figure 5.32. A high-frequency current

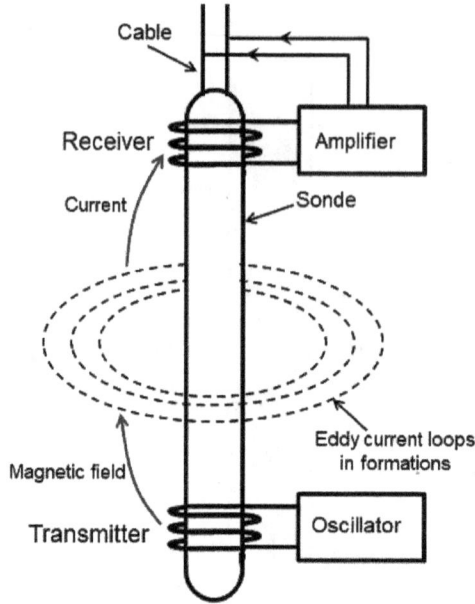

Figure 5.32. Induction log schematic. A sonde is a nonconductive tool made from non-conductive material. The sonde consists of 2 wire coils, a transmitter and a receiver. High frequency alternating current (20 kHz) of constant amplitude is applied to the transmitter coil. This induces an alternating magnetic field around the sonde that induces secondary currents in the formation. These currents flow in coaxial loops around the sonde. They create their own alternating magnetic field, which induces currents in the sonde receiver coil. The received signal is measured, and its size is proportional to the formation's conductivity.

(typically 20 to 40 kHz) is sent through the transmitter coil, which induces an alternating electrical current in the formation, creating a magnetic field in the formation that induces an alternating current in the receiver coil. Induction tools provide good resistivity measurements regardless of the borehole fluid.

In summary, there are two types of resistivity well-logging tools: induction tools and electrode tools. In the latter, electric current is sent into the formations by an array of electrodes. These tools work with brine-based muds. In the former, tools generate a magnetic field that induces an electric current in the formation. These tools are generally used with oil-, air-, or fresh-water-based muds.

Saturation estimation. Resistivity is determined from log data (see Figure 5.33c). Archie's water saturation equation determines:

$$S_w(z)^n = \frac{F R_w}{R_t(z)} \quad \text{or} \quad \log(R_t(z)) = \log(F R_w) - n \log(S_w(z))$$

where R_w is water resistivity in the formation, R_t is true formation resistivity, and F is the formation resistivity factor. By working in small portions of the formation in which F is constant, we can solve/reconstruct by water saturation of the formation using the statistics methods described in Ikelle (2024).

Nowadays, resistivity is measured not only vertically but also at different azimuths around the borehole. In other words, the above equation can alternatively

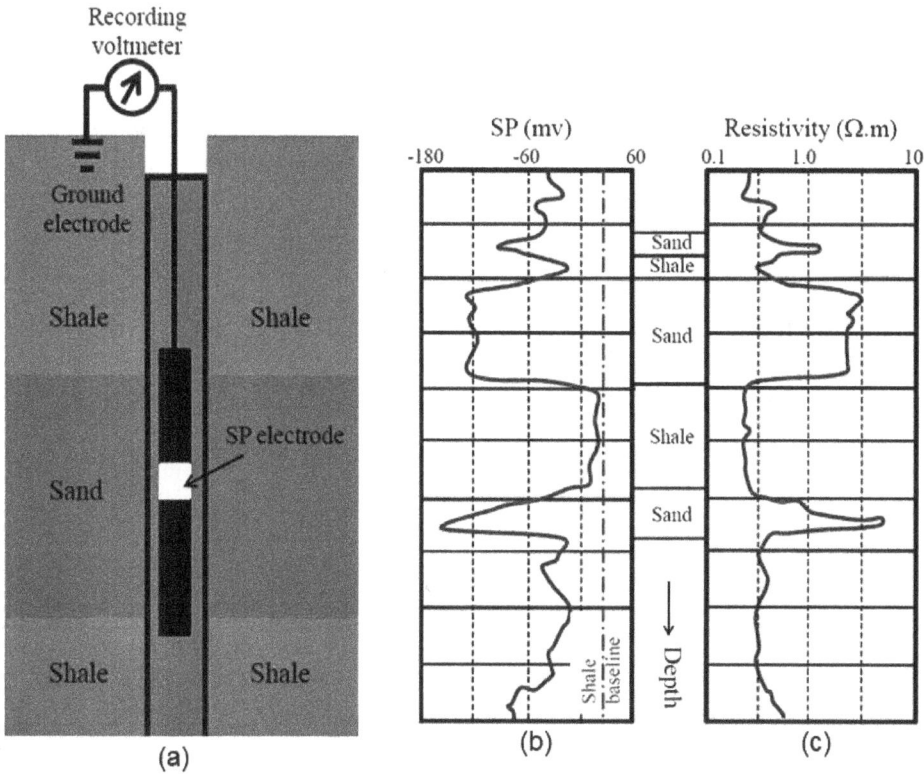

Figure 5.33. (a) A schematic diagram of a typical SP measurement tool. It consists of an electrode suspended in the borehole. Another electrode fixed to the surface. A voltmeter that measures the potential difference between the borehole electrode and the ground electrode. SP can occur when a permeable and porous formation (e.g., sand) is sandwiched between two impermeable formations (e.g., shale). The SP measurement tool is considered one of the cheapest geophysical methods in terms of equipment requirements and simplicity to use. (b) SP log. (c) Resistivity log.

be written as

$$[S_w(z,\theta)]^n = \frac{F'(\theta)R_w}{R_t(z)} \text{ or } \log[R_t(z)] = \log[F(\theta)R_w] - n\log[S_w(z,\theta)]$$

where θ is the azimuthal angle. Again, by working in small portions of the formation in which F is independent of z, we can solve/reconstruct the water saturation by using the statistics methods described in Ikelle (2024).

5.2.3.   Spontaneous Electricity from the Well-logging

Historically, the first well-logged measurements were recordings of the electric resistivity of rock formations as a function of depth by sending electric current into the formation, as described in the previous subsection. Unexpectedly, *noise* was repeatedly observed in recordings. This noise was finally determined to be caused by a second electric current source. This second source was naturally or spontaneously in the ground.[6] In other words, there were two electric potential sources in the ground: the man-made electric potential source and the spontaneously occurring electric potential source. These two sources have been turned into two types of electric well-logging: resistivity logging described earlier and spontaneous potential (SP) logging. We focus on logging SP wells in this subsection. This method is also intended to recover the saturation and resistivity of rock formations. From these parameters one can also identify impermeable zones such as shale, and permeable zones such as sand.

SP Logging Experiment. Figure 5.33a shows the typical SP tool for measuring potential or DC voltage difference between two points. One point is the electrode on the surface. Second point is the electrode taken downhole by wireline. SP logging is passive; SP experiments do not require electric currents to be sent to the ground. That is why SPs are also referred to as *self potentials* —SP actually means self potential. However, the SP experiment requires conductive drilling mud such as salted-water-based mud in the borehole to operate. It does not produce the desired measurements when the borehole is filled with non-conductive drilling mud such as oil-based drilling mud because such mud does not provide electrical continuity between SP electrodes and rock formation.

A single moving electrode suspended in the borehole allows us to vary measurements of potential differences with depth. This is relative to a reference electrode fixed at the Earth's surface, in a mud pit, or other suitable locations. Remember that an electrode is a conductor through which electricity enters or leaves an object or region. Typically, recorded potential differences are as small as 1 mVolt. Note that SP logs are difficult to run offshore. Sea waves on the rig leg generally induce wavy patterns in the recorded SP log that are difficult to filter.

So, there are three requirements for the existence of an SP current (i.e., for the creation of flow ions):

- A conductive borehole fluid (i.e., salted-water-based mud). SP does not work with non-conductive drilling mud (e.g., oil-based). Therefore, it is worthwhile to recall that hydrocarbons, dry rock, and fresh water are highly resistive (that is, they are nonconductive). Saltwater (brine) is very conductive.

- A formation must consist of a semi-permeable-permeable-semi-permeable configuration. In other words, a porous and permeable formation is sandwiched between two semi-permeable formations such as shale. Figure 5.34b illustrates this. The *semi-permeable* rocks, such as shale, allow some ions to pass through and block other ions just like cell membranes in human bodies. Na^+ ions are more concentrated in the permeable zone than in shale and therefore tend to flow into the shale. They are driven into the shale by the electrical potential difference across the interface. Also, the shale is impervious to Cl^- anions. In more general terms, the semi-permeable formation must be ion-selective to allow preferential ions' transportation. Ion movement induces current flow.

---

[6]Whereas in modern civilization electricity mostly comes from generators that produce electrical energy by electromagnetic induction, electric current is spontaneously generated in the ground by the motion of ions. Maybe civilizations that lived or may have lived underground may have effectively harnessed this natural electricity source.

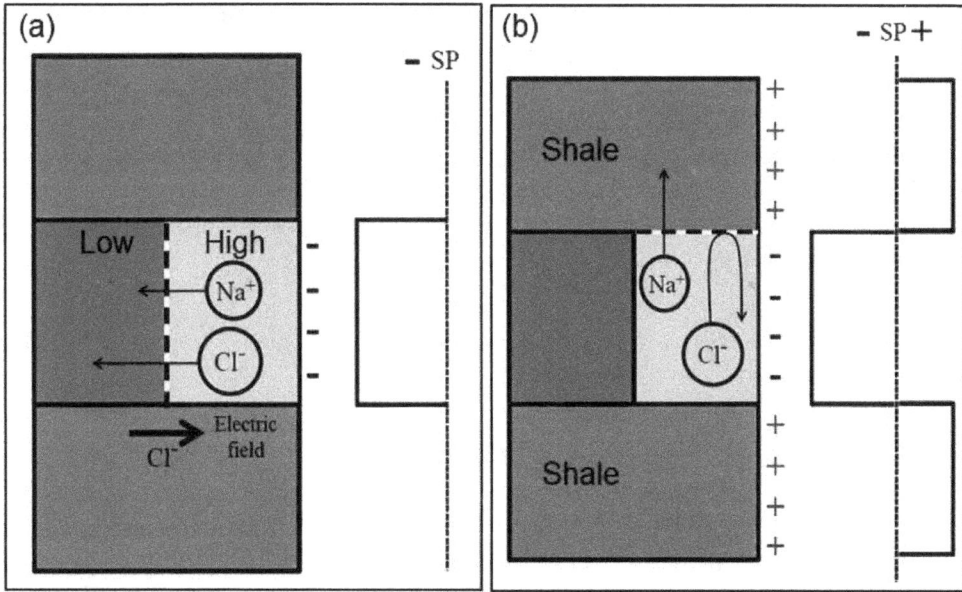

Figure 5.34. (a) An illustration of the diffusion potential. It is the result of the salinity difference between the mud filtrate and the formation fluid. Chloride ions are more mobile and diffuse into the invaded zone than sodium ions. The net result is a flow of negative charge into the invaded zone, which creates a charge imbalance (potential difference) called diffusion potential. The diffusion potential causes a current to flow (from negative to positive) from the invaded zone into the non-invaded zone. (b) An illustration of membrane potential. Consider a permeable formation with thick shale beds above and below. Shales are permeable to the Na^+ ions but impervious to the Cl^- ions. That is Na^+ ions can move through the shale from the more concentrated to the less concentrated NaCl solution.

- Differences in salinity concentrations and/or in fluid pressure salinity (the measure of all the salts dissolved in water) between the borehole fluid and the formation fluid are the main sources of SP.

Voltage spontaneous source. In the ground, there are many possible ways to generate voltage spontaneously. The SP can arise from the movement of electrically charged ions in the fluid relative to the fixed rock and movements of ions in a fixed porous rock and from electrical interactions between the various chemical constituents of rocks and fluids. The two most effective sources are diffusion potential (also known as liquid-junction potential) and membrane potential. In one case, we have the difference in salinity (sodium chloride, NaCl) between the two salt-water-bearing formations, for example, and in the other case we have semi-permeable rocks, such as shale, that allow some ions to pass through and block the passage of other ions.

Diffusion Potential (also known as liquid-junction potential). In this case, spontaneous potentials due to the difference in salinity between the two salt-water-bearing formations (see Figure 5.34a). This source of SP is known as a diffusion potential. It exists at the junction between the invaded and non-invaded zones, and is the direct result of the difference in salinity between the mud filtrate and the formation fluid.

Let us now confine ourselves to the case of sodium chloride, NaCl, a common salt found in formation waters including mud filtrate and formation water. So the two solutions are mud filtrate and formation water. The salt has two species of ions: Cl^- and Na^+. The current flowing across the junction between solutions of different salinities is produced by an electromagnetic force (emf) called liquid-junction potential, ΔV (see Box 5.1):

$$C_w = C_{mf} \exp\left[-\frac{q}{k_0 T}\Delta V\right] \Longleftrightarrow \Delta V = -\frac{k_0 T}{q}\log\left(\frac{C_w}{C_{mf}}\right) , \qquad (5.47)$$

where C_w and C_{mf} are the sodium chloride concentrations (i.e., concentration is the amount of ion species over volume of solution) of the two solutions at formation temperature, T, k_0 is Boltzmann constant; q is the charge of ion species with $q = e$ for Na^+ and $q = -e$ for Cl^-. Basically, the potential difference between the two solutions very rapidly reaches an equilibrium value proportional to the logarithm of the concentration ratio. This equation is known as the Nernst relation. We can see that the potential difference is caused by the difference in mobility of anions (Cl^-) and cations (Na^+) because Na^+ and Cl^- ions can diffuse (move) from each solution to the other. The diffusion of ions here refers to the tendency of ions to move from regions of high NaCl concentration to regions of lower NaCl concentration; that is, the self-diffusion of the dissolved ions in the water. This diffusion is a direct result of the random thermal motion of the ions.

The rate at which a particular ion species diffuses depends inversely on its mass. Since Na^+ is considerably smaller than Cl^-, Cl^- ions have a greater mobility than Na^+ ions (see Table 5.2),[7] the net result of this ion diffusion is a flow of negative charges (Cl^- ions) from the more concentrated to the less concentrated solution. In the well-logging experiments the diffusion potential can occur at the interfaces between invaded and noninvaded zones, defined earlier and illustrated in Figure 5.29. Remember that when the drilling mud encounters porous and permeable layers, the mud will flow into (or invade) the formation. The uninvaded zone is the zone further into the formation in which the original pore content is not contaminated by mud filtrate. The difference in salinity between the fluid invaded zone and the formation fluid in the uninvaded zone is the main source of spontaneous diffusion potential in well logging. Assume that the water in the uninvaded zone is more saline than the water in the invaded zone. When the two liquids come into contact at the interface between the invaded and noninvaded zones, diffusion will occur. Ions from the high salinity formation water will diffuse into the invaded zone to balance the salinities out. Chloride ions are more mobile and diffuse into the invaded zone than sodium ions. The net result is a flow of negative charge into the invaded zone, which results in a charge imbalance (potential difference) called the diffusion potential. The diffusion potential causes a current to flow (from negative to positive) from the invaded zone into the non-invaded zone.

Because the salt content in water is roughly proportional to its conductivity, as we can see in Table 5.7, we can rewrite (5.47) in terms of the resistivity of

---

[7]The mass of Cl^- ion is 5.9×10^{-26} kg and that of Na^+ ion is 3.8×10^{-26} kg.

the formation water, the mud filtrate, and the formation temperature. T is the formation temperature in degrees Celsius as follows:

$$\Delta V = (65 + 0.24 \times T) \log \left(\frac{R_{mf}}{R_w} \right) , \tag{5.48}$$

with

$R_{mf} < R_w \Longrightarrow \Delta V > 0$; i.e., SP > 0,

$R_{mf} > R_w \Longrightarrow \Delta V < 0$; i.e., SP < 0,

$R_{mf} \approx R_w \Longrightarrow \Delta V \approx 0$; i.e., no SP,

and where T is the formation temperature in degrees Celsius, R_{mf} is the resistivity of the mud filtrate, and R_w is the formation water resistivity. The most popular SP relationship version is the equation (5.48). Actually, this is the only equation you need for spontaneous potential logging. Note that if the total SP deflection can be found from the log and the formation temperature is known and the resistivity of mud filtrate can be measured, the desired quantity of formation water resistivity, R_w, can be found. When the mud filtrate salinity is lower than formation water, that is, the resistivity of mud filtrate is greater than the resistivity of formation water, spontaneous potential deflection will be a negative response as shown in Figure 5.35. Spontaneous potential deflects to the right when mud filtrate salinities exceed formation water, that is, when mud filtrate resistivity is less than formation water resistivity. When mud filtrate resistivity equals formation water resistivity, there is no SP response. The SP response in shale is relatively constant and is called the shale baseline.

In permeable beds the SP log will do the following relative to the shale baseline: (1) negative deflection to the left of the shale baseline where resistivity of mud filtrate is greater than resistivity of formation water so spontaneous potential deflection will be a negative response; (2) positive deflection to the right of the shale baseline where resistivity of mud filtrate is less than resistivity of formation water; (3) no deflection where resistivity of mud filtrate equals to resistivity of formation water. The SP curve can be suppressed by thin beds, shaliness, and gas.

The presence of hydrocarbons (e.g., oil, natural gas, condensate) will reduce the response on an SP log because the interstitial water contact with the well bore fluid is reduced. This phenomenon is called hydrocarbon suppression and can be used to identify rocks with commercial potential. The SP curve is usually 'flat' opposite shale formations because there is no ion exchange due to the low permeability, low porosity properties (tight) thus creating a baseline. Tight rocks other than shale (e.g. tight sandstones, tight carbonates) will also result in poor or no response to the SP curve because of no ion exchange.

Membrane potential (sometimes called the shale potential) exits at the junction between the non-invaded zone and the shale (or other semi-permeable rock) sandwiching the permeable bed (see Figure 5.34b).Similarly to the previous section, membrane potential is composed of an ion-selective membrane which is a layer of shale from the top and bottom of the permeable formation.. Due to the presence

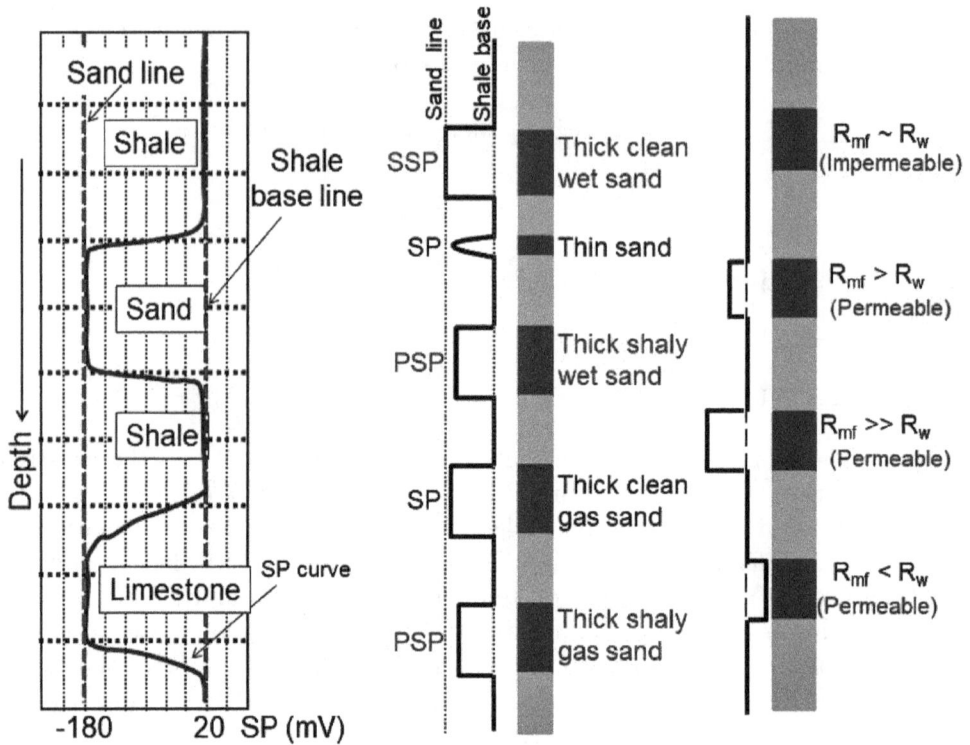

Figure 5.35. (left) An illustration of an SP log. SP logging readings in shale are fairly constant (shale base line). SP may deflect to the right (negative) or right (positive) of the shale base line, depending on the relative salinities of formation water and mud filtrate; these occur only in porous, permeable beds. Limestone is low in permeability unless porous or fractured. Sandstone deflects left. If the sonde encounters a fluid that is a better conductor than drilling mud (such as salt water), the SP will deflect to the left; if the fluid is a poor conductor (such as fresh water or oil), it will deflect to the right. (middle) PSP = SP log read in a thick homogeneous shaly sand zone, SSP = SP log read in a thick clean sand zone. (right) SP log is an indicator of bed permeability but not quantitative.

of electric double layer, shale allows passing of Na^+ ions from saline formation water, in non-invaded zone, to borehole. Chlorine ions cannot therefore pass through shale. That is because shale are cation exchangers (i.e., negatively charged) and repel anions such as chlorine ions. As sodium ions migrate from non-invaded zone to borehole, potential is induced. The current then flows through shale from the non-invaded zone of the permeable formation to the borehole. Such case is true for more saline formation water. The magnitude of diffusion potential is usually 20 percent of membrane potential.

The genesis of membrane potential can be found in the cell biology described in the previous section. Again cells (such as neurons, muscle cells, some endocrine cells, and some other cells in the body) are the basic building blocks of all living things. The human body is composed of trillions of cells. They provide structure for the body, take in nutrients from food, convert them into energy, and carry out specialized functions. Cells are separated from the outside environment by the cell membrane. In geophysics, the cell membrane is replaced by shale or any

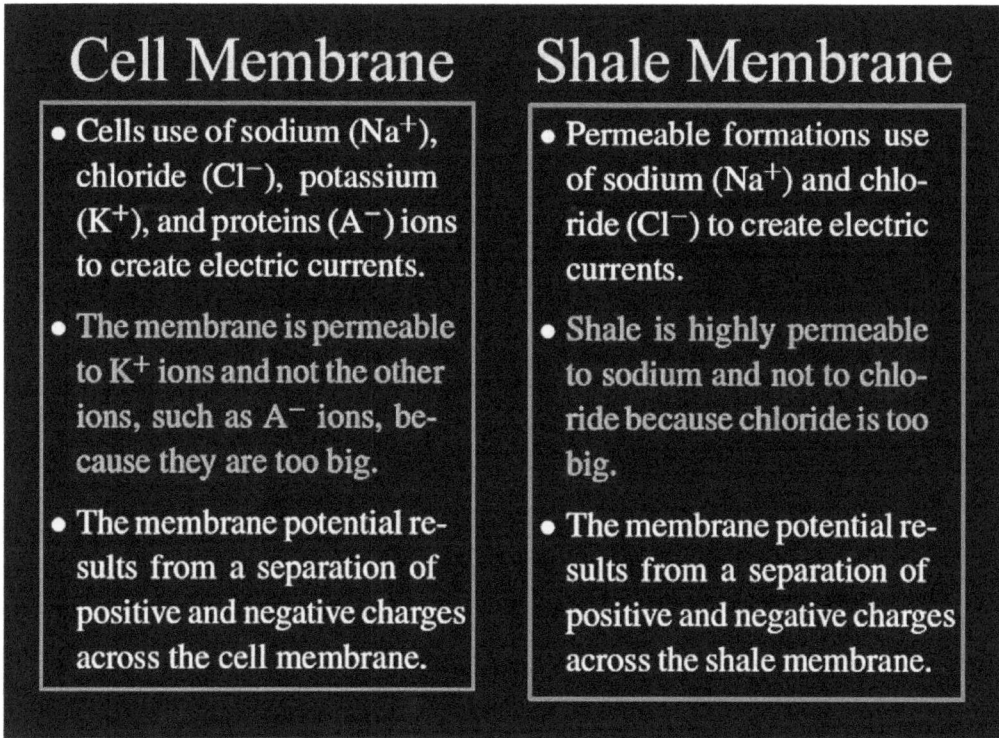

Figure 5.36. An illustration of the parallel of membrane potential in cell biology and SP well-logging.

impermeable rocks and the cells are replaced by permeable and porous formations. We have parallel membrane potentials between cell biology and SP well-logging. (see also Figure 5.36).

SP logs. SP log measurements range from 1 millivolt to 1 volt. Figure 5.33b shows a typical SP log example. The readings of SP logged in shale are usually fairly constant and follow a straight line on the log, called the shale base line. In permeable formations, the SP curves show excursions from the shale base line either to the left (negative) or to the right (positive) of the shale base line, depending on the relative salinities of the formation water and of the mud filtrate. If the formation fluid is a better conductor than drilling mud (such as salt water), the curve will deflect to the left. If the fluid is a poor conductor (such as fresh water or oil), it will deflect to the right. Even small SP deflections indicate bed permeability. Thus, the SP log detects permeable beds and their boundaries. However, the present SP logs cannot quantify permeability or porosity. In other words, large permeabilities are not necessarily associated with large deflections, and vice versa. Note that some permeable beds might give no deflection when there is no difference in salinity between formation fluids and mud filtrates. Figure 5.35 illustrates the typical SP response in shale, sand, and limestone.

Notice in Figure 5.35 that the sand line is drawn through the maximum (usually negative) excursions of the SP curve adjacent to the thickest permeable beds. The separation of the sand line from the shale baseline, measured in mV, is known as static spontaneous potential (SSP). So SSP is the maximal deflection of the spontaneous potential curve in a thick, clean, porous and permeable bed. We can empirically define the shale volume in a water-shale sandstone formation as

$$V_{\text{shale}} = 1 - \frac{\text{PSP}}{\text{SSP}}, \tag{5.49}$$

where PSP is the deflection of the spontaneous potential curve in a thick homogeneous shaly sand zone; i.e., PSP = SP log read in a thick homogeneous shaly sand zone, SSP = SP log read in the thick clean sand zone. The volume of shale in the sand is a rough indicator of the amount of shale in shaly sand reservoirs.

Quiz: Propose a method for determining formation water resistivity using SP logs.

5.2.4. Application: Artificial Geothermal Energy from Dry Oil Reservoirs

Geothermal energy is created by heat that escapes as steam from a hot spring, as illustrated in Figure 5.37a. Geothermal systems are generally categorized into (1) steam-dominated systems, (2) liquid-dominated systems, and (3) enhanced geothermal systems (EGS). In the steam-dominated and liquid-dominated systems, steam or hot water is extracted from naturally occurring fractures within rock in the subsurface, and cold fluid is then injected into the ground to replenish the depleted fluid. EGS consist of impermeable rocks containing hot water. Wells are used to pump cold fluid into hot rocks to gather heat, which is then extracted by pumping the fluid to the surface. When an EGS-type reservoir lacks sufficient connectivity through natural fractures, the reservoir is fractured by using high-pressure fluid injections to increase permeability and facilitate fluid movement through rock, just as is done with hydraulic fracturing of tight oil-and-gas reservoirs.

The heat from the solid earth shows that 99 percent of the earth's volume has temperatures greater than 1000 degrees Celsius (see Figure 5.37b). The total heat content is estimated to be about 10^{31} Joules, so the geothermal resource base is sufficiently large to contribute to global electricity demand. However, geothermal systems have little capacity because there is a limited number of suitable sites, and the heat can be depleted if used too rapidly.

Some depleted oil and gas reservoirs are utilized today to storage capture CO_2, the so-called CO_2 sequestration (e.g., Ikelle and Amundsen, 2005 and 2018; Ikelle, 2017 and 2020). depleted oil and gas reservoirs can also be utilized for electricity production in the same configuration as standard geothermal energy production. We inject fluid into the ground to heat up and produce vapor, which is then used to drive a turbine that generates electricity. Essentially, we are using technology that generates power from coal, making the steam turbine a fundamental component of electricity generation. Typical input is hot water, or any suitable fluid, or steam, and the output is electricity or heat. The existence of wells and surface equipment in depleted oil and gas fields can minimize the investment costs for heat and power production from these wells. This makes geothermal energy extraction from

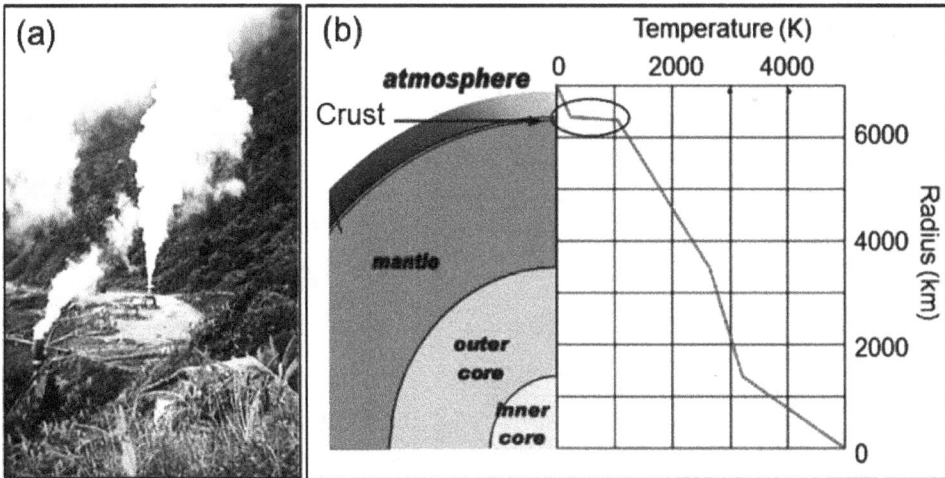

Figure 5.37. (a) An illustration of how the earth's heat—called geothermal energy—escapes as steam at a hot spring in Nevada. (b) The Earth's temperature gradient.

depleted wells profitable. Furthermore, geothermal production will leave no additional footprint on the environment. The reservoirs that were taped for oil and/or gas production have hundreds and even thousands of wells, like supergiant oilfields such as the Ghawar oilfield. These wells are used for injecting fluids in reservoirs to increase the pressure and/or fracking rocks and for production oil and gas.

There are more than 20 million abandoned oil and gas wells with a depth exceeding 3 kilometers. Geothermal resources that are currently used are primarily extracted from wells of less than 2 kilometers depth. Because abandoned oil and gas wells are deeper than geothermal reservoirs, abandoned wells can produce more heat on average based on the solid Earth temperature gradient. The key parameter here is temperature. Wells at temperatures lower than 75 degrees Celsius cannot be used for electricity generations. The bottom hole temperatures of abandoned oil or gas wells more than 3 kilometers deep are generally more than 250 degrees Celsius. Therefore, abandoned oil wells produce more heat than standard geothermal wells.

Global geothermal power generation capacity is estimated at 16,127 MW (mega watts) at 2022 with about 3,200 active wells. The 20 million or so-deep wells have the potential to increase geothermal power several folds. Moreover, as standard geothermal energy, abandoned well-based geothermal energy can be replenished with new water or any other suitable fluid over a short period—say months. Therefore making them renewable. Oil and gas take billions of years to replenish. Note that the amount of electricity potential produced by self-potential in the ground is too small to be commercially viable, even with abandoned wells. The cost of the mud production further make the SP solution for electricity produced and used at the Earth not attractive.

5.2.5. Application: Spontaneous Nuclear Fission Reactors, the Gabon Case.

The subsurface is characterized by spontaneous nuclear fission reactions, as described by Ikelle (2020 and 2023b). What we are interested in here is a spontaneous nuclear fission reactor near the Earth's surface. Therefore, natural tectonic forces, water cycles, and rock cycles must combine to gather and uplift enough fissionable materials (i.e., nuclei like uranium-235 that naturally split when they absorb sub-atomic particles like neutrons) near the surface of the planet at a critical mass so that nuclear fissions can occur. Nuclear fission chain reactions (nuclear fission chain reactions) must be controlled. When radioactive materials are depleted the system must safely shut down and radioactive waste must be safely buried in the ground.

Here is an example of a uranium-235 fission reaction:

$$^{235}_{92}\mathrm{U} + {}^{1}_{0}\mathrm{n} \rightarrow \left({}^{236}_{92}\mathrm{U}\right)^{*} \rightarrow {}^{141}_{56}\mathrm{Ba} + {}^{92}_{36}\mathrm{Kr} + 3\,{}^{1}_{0}\mathrm{n} + 166\ \mathrm{MeV}\ ,$$

which were discovered in 1938 by Otto Hahn and Fritz Strassman. The fission products (also called fission fragments), three neutrons, and about 166 MeV (more than 10 million times the energy per reaction of a chemical reaction such as coal burning) are released. As described in Ikelle (2023b), these specific fission reactions are governed by statistical probability, including the one above. Figure 5.38 shows other examples. Interestingly, the fact is that energy is produced and two to three neutrons are generated. These neutrons could hit other uranium-235 atoms and trigger a chain reaction. Nuclear bombs and nuclear power plants release enormous amounts of energy through fission chain reactions. Billions of reactions occur each second in a fission chain reaction. The products of this reaction chain are enormous.

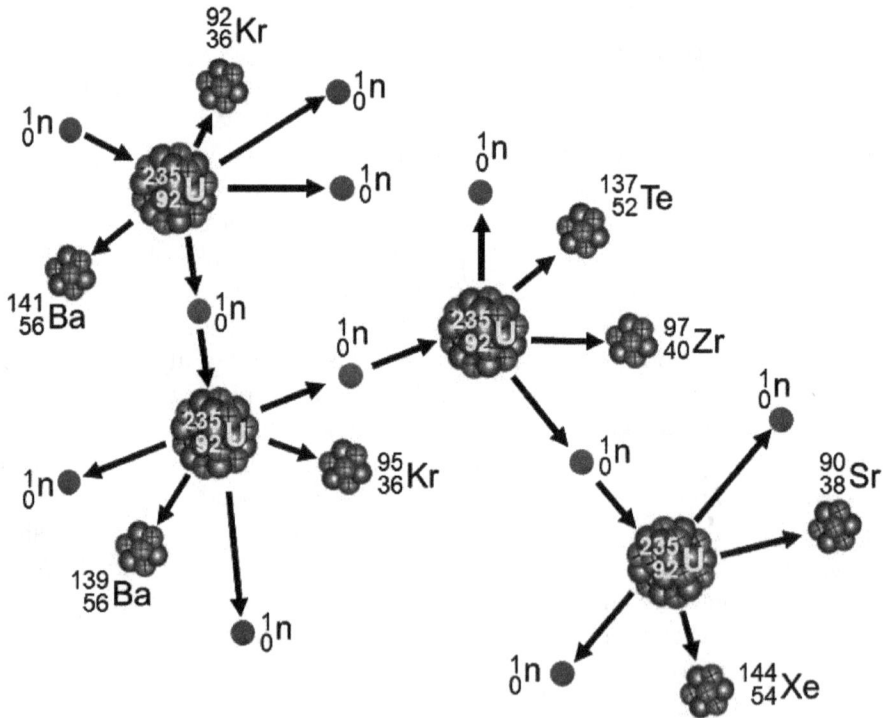

Figure 5.38. A neutron colliding with a uranium-235 nucleus splits it into two nuclei (typically one heavier and one lighter nucleus) and 2 or 3 free neutrons. When neutrons strike the next uranium nucleus, they may cause next fissions, and the next, and the next, and the next, whilst building up a chain reaction.

As described in Ikelle (2023b), they include ceasium-133, caesium-134, iodine-135, xenon-135, zirconium-93, molybdenum-99, Technetium-99, Strontium-90, yttrium-90, zirconium-90, neodymium-142, neodymium-143, and ruthenium-99. Some of these isotopes are stable. They characterize fission reactions and can be used to trace the origin of fission nuclear reactions billions of years later near the Earth's surface.

The reactions mentioned above are processed in a controlled manner in man-made nuclear reactors. The chain reaction is stopped or controlled by neutron poison (also called a neutron absorber, a nuclear poison, or a rod) and neutron moderators. Neutron poison $_{54}^{135}$Xe is the most common use because it has a large neutron absorption cross-section, about 3.5×10^6 barnes (see Ikelle, 2023b). Neutron moderators, which contain water [light water (H_2O) and heavy water (D_2O)] or graphite (C), allow us to change the portion of neutrons that cause more fissions. They slow down neutrons from their initial speed of 20,000 km/s to 2 km/s. The controlled (nuclear fission) chain reaction produces heat, which boils water, produces steam, which drives a turbine, and generates electricity. So the main components of a nuclear reactor are the supply of fissionable materials such as uranium-235 in sufficient quantities, the boiler, where fission reactions take place, the neutron moderator, and disposal of radioactive waste.

Spontaneous uranium-enrichment. Today approximately 0.72 percent of all natural mined uranium consists of uranium-235, 99.27 percent uranium-238, and 0.0055 percent uranium-234. The uranium-235-uranium-238 ratio is 0.00720. Even uranium samples brought back from the moon by Apollo astronauts contained 0.72 percent uranium-235. Yet the vast majority of nuclear power reactors use uranium-235 as fuel. However, 0.72 percent uranium-235 is not enough to power these reactors. Most uranium-235-based civil and commercial nuclear reactors require about 3-5 percent uranium-235. Therefore, naturally mined uranium must be improved through a process known as uranium enrichment. It consists of increasing the amount of uranium-235 in natural uranium relative to the other isotopes. State-of-the-art enrichment technology involves a gas centrifuge (Khan, 1986 and 1987; Bogovalov and Borman, 2016). Enrichment increases the proportion of the fissile isotope U-235 about five- to sevenfold from the 0.72 percent of uranium-235 found in natural uranium for civil nuclear reactors. Note that nuclear bomb construction requires enrichment level of 90 percent. Actually the enrichment of uranium and the control of fission chain reactions are central differences between civil nuclear power generation and military nuclear weapons.

But 1.7 billion years ago, which is more than two full half-lives of uranium-235, the amount of uranium-235 was about 3.58 percent of all uranium because 1.7 billion years is about 2.42 half-lives of uranium-235 (uranium-235 has a long half-life of 700 million years). There was approximately 3.58 percent uranium-235 in the earth's uranium at the time Gabon spontaneous nuclear reactors operated about 1.7 billion years ago. In other words, there was enough uranium-235 to trigger a self-sustaining nuclear reactor under the right conditions 1.7 billion years ago. Man-made uranium enrichment was necessary. Today, we know of only one self-sustaining nuclear reactor from 1.7 billion years ago. Most of the large uranium accumulations are found in Australia (about 45 percent), Kazarkhistan (about

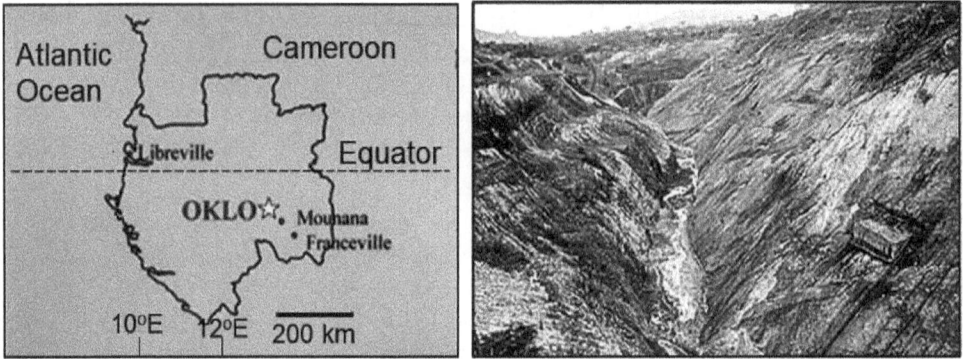

Figure 5.39. (a) The Oklo reactor location. (b) A View of the Oklo uranium mine in Gabon.

15 percent), Canada (about 14 percent), and southern Africa (about 10 percent). Because uranium is well distributed around the Earth's crust, it is possible erosion, weathering, and/or metamorphism may have made the identification of other past self-sufficient nuclear reactors difficult to determine today.

Formation and operation of the Oklo fission reactor. About 1.7 billion years ago, at the Oklo mine in Gabon (Figure 5.39), uranium was gathered together by mineralizing volcanic fluids that flowed over ancient sandstones rich in organic carbon and uranium veins. Tectonic forces uplifted the sandstones toward the surface along with their uranium mineral veins toward the granite bedrock that makes up most of Earth's crust.[8] As they were uplifted toward the surface groundwater would penetrate the rocks and surround the uranium. Weathering of magmatic rocks and bacterial activity concentrated the uranium enough to start a nuclear chain reaction. The reactor zones themselves were centimeter to meter thick layers of highly enriched uranium, buried within the uranium ore layer (Figure 5.40). There were sixteen of them. Each reactor operated intermittently for a few years to hundreds of thousands of years. The total time period over which the reactors operated is thought to be about a million years.

At Oklo the relative abundance of uranium-235 was 3.58 percent of all uranium 1.7 billion years ago. This is one of the major reasons nuclear fission started. In addition to sufficient uranium-235, a natural gas reactor also requires neutron absorbers and moderator concentration. The absorber and moderator in this case was groundwater. Water that infiltrated the formation along faults slowed down the emitted neutrons to a sufficient degree, absorbing some of the emitted neutrons. This helped sustain stable nuclear fission. Without water to slow neutrons, controlled fission would not have been possible. The atoms would not have split. The water not only slows down radioactive decay neutron products, but also triggers a chain reaction within the densely concentrated uranium deposits in the first place. Eventually the heat released by ongoing decay would boil the mediated water. This would create powerful plumes of radioactive steam that

---

[8]Large uranium accumulations are associated with granite, sedimentary rocks, and groundwater.

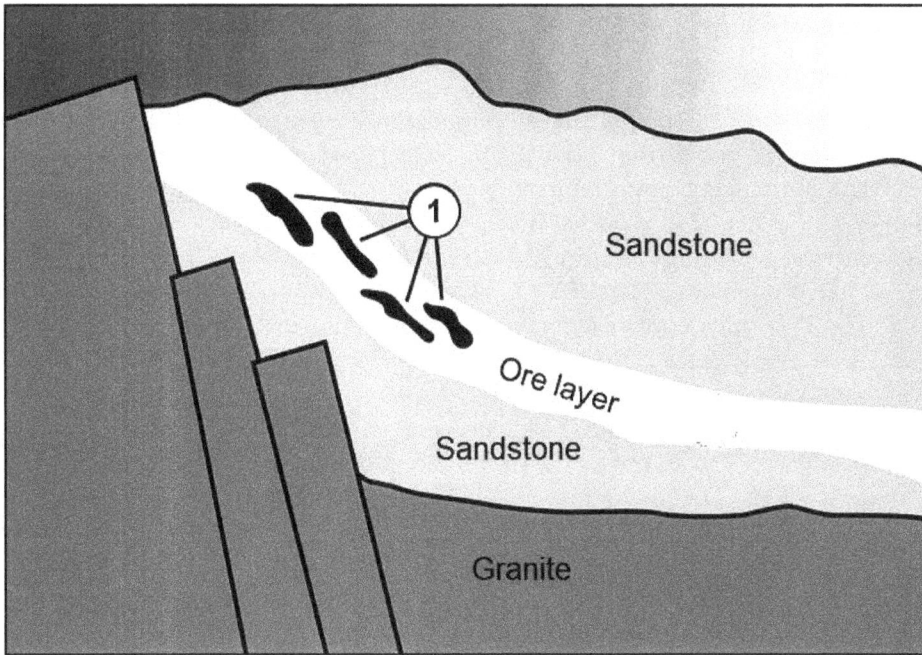

Figure 5.40. Simplified geology of the Oklo natural nuclear reactors. The reactors are indicated by a circle around the number 1. They are located in porous, water-soaked sandstone. Oklo consists of 16 sites at which self-sustaining nuclear fission reactions are thought to have taken place approximately 1.7 billion years ago, and ran for hundreds of thousands of years. Adapted from Wikipédia.

burst out of the solid Earth like toxic geysers. After hundreds of millennia, Oklo's uranium, once so rich, was depleted to below-average levels for fission reactors, and the spontaneous reactor fell silent. Over time the Oklo reactor produced large quantities of toxic plutonium and cesium-isotopes, which have since decayed into stable nuclei. During this process, no harmful radioactivity leaked into the environment. That is how a self-sustaining nuclear reaction occurred naturally 1.7 billion years ago in Franceville, Gabon and lasted for about 1 million years. It is estimated to have averaged under 100 kW of thermal power during that time (Meshik, 2005; Mervine, 2011; and Gauthier-Lafaye et al, 1996).

Spontaneous radioactive waste containment. Just like man-made nuclear reactors, Oklo fission reactors also produce radioactive waste. How did the spontaneous reactor dispose of its waste? The waste stayed in place and moved less than 3 meters in 1.7 billion years (Gauthier-Lafaye et al., 1996). This is despite the fact that the waste was not packaged in fuel bundles, not encapsulated, and was subjected to violent temperature swings for several millions of years. It was washed through water for hundreds of thousands of years. The chief finding was that long-lived waste—the transuraniums like plutonium and americium and other such Actinides—binds chemically to rock in a reducing environment (little or no free oxygen) and remains entirely immobile. In order to be sustainable, geological repositories of nuclear waste must meet this criteria. Based on the findings of about

Oklo nuclear reactors, we are building underground radioactive waste repositories today.

KBS-3 (an abbreviation of kärnbränslesäkerhet, nuclear fuel safety) is a technology for disposal of high-level radioactive waste developed in Sweden by Svensk Kärnbränslehantering AB (SKB). It is based on the findings of the Oklo reactor that a reducing environment is present. Additional barriers include the following to reinforce the hypothesis of a reducing environment. The fuel remains in the fuel rods, which are clad in Zirconium alloy. Fuel rods are held in cast iron holders. Cast iron ensures that the environment will remain reducing even if water enters the 50 mm thick copper capsule that encapsulates the fuel bundles and their holders. A layer of water-absorbing Bentonite clay is added around the capsules. The clay acts as soft padding to keep the capsule from being subjected to bedrock movements. It is also meant to be wet, because when it wets it swells to a pressure of 50 atmospheres, and is pressed into all the cracks and fissures around the bore hole, made 500 meters down into geologically stable bedrock, with a reducing environment and only small water movement.

Proof of Oklo fission reactors. Acording to Bodu et al. (1972), Neuilly et al. (1972), Raffenach et al. (1976), Naudet (1991), Gauthier-Lafaye et al. (1996), and Davis et al (2014), uranium samples from Oklo contain only 0.6 percent uranium-235 while all natural uranium today has 0.72 percent uranium-235—a significant difference that indicates the occurrence of large fissions. Large quantities of ancient (no longer radioactive) fission product waste are embedded in the natural uranium ore, confirming that spontaneous nuclear fission reactions occurred at Oklo some 1.7 billion years ago. The typical neodymium contains 27.1 percent neodymium-142 while in Oklo it consists of only 6 percent.

The typical neodymium contains 27.1 percent neodymium-142 while in Oklo it consists of only 6 percent. In addition neodymium in Oklo contains neodymium-143 which is consistent with nuclear fission reactions.[9] There is also a large concentration of ruthenium-99 between 27 and 30 compared to the typical 12.7 percent. Ruthenium is one of the fission products in nuclear reactors. The excess is the result of technetium-99 (fission products) decaying into ruthenium-99 by emitting an electron; i.e.,

$$_{43}^{99}\text{Tc} \rightarrow \ _{44}^{99}\text{Ru} + \ _{-1}^{0}\beta \ .$$

Technetium-99 is a long-lived radioactive technetium isotope with a half life of 211,000 years. It is a uranium-235 product.

Thorium nuclear reactors. Nature gave us two nuclides capable of nuclear reactions without significant input: thorium and uranium. Unlike uranium, thorium does not fission (i.e., thorium does not split) even at high energy neutrons. In other words, thorium is not a nuclear fuel itself. Other heavy fissile nuclei are uranium-233, plutonium-239 and plutonium-241. Each of these is produced artificially in a

---

[9]Neodymium stable isotopes include neodymium-142 (27.1 percent), neodymium-146 (17.2 percent), neodymium-143 (12.2 percent), neodymium-145 (8.3 percent), and neodymium-148 (5.76 percent). Neodymium radioactive ones include neodymium-144 (23.80 percent) and neodymium-150 (5.64 percent).

nuclear reactor, from the fertile nuclei of thorium-232, uranium-238, and plutonium-240 respectively. So uranium-235 is the only naturally occurring thermally fissile isotope.

Thorium-232 cannot undergo fission by thermal neutrons, but can be converted into uranium-233, which is a fissile material—that is, it is a nuclear fuel. Then nuclear fusion can start. So far this process has not proven to be commercially viable. Uranium-238 is also a naturally occurring fertile isotope.

5.3. Essay: Casimir Effect and Ultracapacitors

5.3.1. Ultracapacitors: Energy Storage

The conventional capacitors, as described in the first section, yield capacitance in the range of 0.1 to 1 μF with a voltage range of 50 to 400 V. To increase the capacitance, as per (5.18), one has to increase the permittivity, ε, and/or surface area, A, and decrease the thickness, d. The current solutions increase permittivity and decrease thickness through electrode surfaces (plates) and electrode material. These solutions are known as supercapacitors or ultracapacitors with capacitance values in the order of 1 to 100 farads; that is, ultracapacitors have capacitance values more than one million times greater than conventional capacitors. They typically store only 10 to 100 times more energy per unit volume than conventional capacitors. This is because ultracapacitors have a smaller voltage than conventional capacitors. The classical example is electrostatic double-layer capacitors (EDLCs).

EDLCs share a similar mechanism to conventional capacitors. However, instead of storing charges in the dielectric layer, EDLCs utilize the interfaces between the electrode and the electrolyte for energy storage (e.g., Béguin et al., 2009). As shown in Ikelle (2023b), an EDLC has two double layers separated by a thin interface (between 3×10^{-10} and 8×10^{-10} meters). Each double layer typically has a carbon electrode and an electrolyte solution (e.g., water). The two phases come into contact with each other at the interface (separatator) where positive and negative charges accumulate. The thin interface with charges stored at it acts as EDLC's dielectric. The fact that the thickness of the effective dielectric is exceedingly thin translates to a very high capacitance. In addition to the very small separation thickness, a larger area of electrode/electrolyte interface than that in conventional capacitors can further increase the capacitance of EDLC. Moreover, they are capable of extremely rapid charge and discharge rates with the ability to be cycled millions of times. The charge time of EDLCs is about 1 second, and their discharge time is approximately 1 second. Lithium-ion batteries charge between 3 and 5 minutes, while EDLCs discharge between 3 and 5 minutes. Note also that batteries store their energy through chemical reactions, as described in Ikelle (2022b), while in ultracapacitors, energy is stored within the electrical double layer, which is a physical storage mechanism (see for example, Ikelle, 2020, Chapter 16). Since charge is stored physically, there is no direct degradation mechanism that limits ultracapacitors' cycle life. Only side effects prevent the infinite cycle life of ultracapacitors. The cycle life of batteries is much shorter and heavily dependent on how much energy is cycled into and out of the battery on a given cycle. This is also known as the discharge depth. In contrast, ultracapacitors have no limitations on discharge depth.

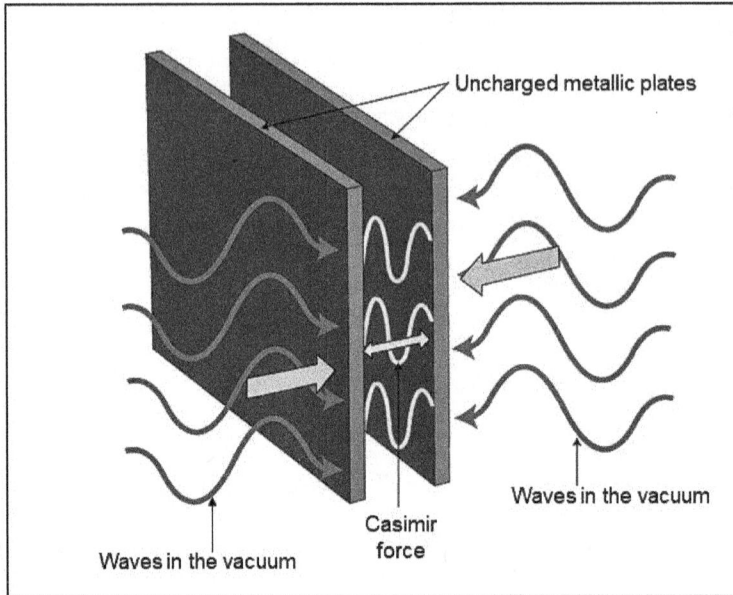

Figure 5.41. An illustration of the Casimir effect. The interior region of parallel conducting uncharged plates. The region experiences a reduced quantum vacuum energy density owing to the boundary conditions the plates impose on the fields. This energy generates a measurable attractive force that pushes the plates together.

5.3.2. Casimir Effect: Generating Energy from Vacuum

The Casimir effect describes a force arising from arising from between two uncharged conductive plates (neutral plates) in a vacuum, placed a few nanometers apart, without any external electromagnetic field. In other words, the Casimir effect also deals with plates like capacitors. However, unlike capacitors, the Casimir effect plates are neutral.

In a classical description, the lack of an external field also means that there is no field between the plates, and no force would be measured between them. Quantum mechanics completely changed our notion of vacuum. All fields—in particular electromagnetic fields—have fluctuations. In other words at any given moment their actual value varies around a constant mean value. Even a perfect vacuum at absolute zero has fluctuating fields known as "vacuum fluctuations," the mean energy of which corresponds to half the energy of a photon. The Casimir effect tells us that, on a very small scale, uncharged conductive plates generate a force which is (Casimir, 1948)

$$\frac{F_c}{A} = -\frac{d\mathcal{U}}{da} = -\frac{\hbar c_0 \pi^2}{240 a^4} \Longrightarrow \mathcal{U} = -\frac{\pi^2 \hbar c_0 \, A}{720 a^3} \tag{5.50}$$

with F_c is Casimir force, $\mathcal{U}$ is Casimir energy, A is area of the plates, $\hbar = 1.055 \times 10^{-34}$ J.s is reduced Planck constant, c_0 = speed of light, and a = distance between the plates with $A \gg a^2$. The force is negative, indicating that the force is attractive: by moving the two plates closer together, the energy is lowered. Although it is quantum in nature, it manifests itself macroscopically.

For example, for two parallel plates of area $A = 1$ cm^2 separated by a distance of $d = 1$ μm the force of attraction is 1.3×10^7 Newtons. This force is certainly within laboratory force-measuring techniques.

The Casimir effect is named after Dutch physicist Hendrik Casimir, who predicted the effect on electromagnetic systems in 1948. It was not until 1997 that a direct experiment by S. Lamoreaux quantitatively measured the Casimir force to within 5 percent of the value predicted by the theory (Lamoreaux, 1977). This is a profound result in the sense that the origin of this force cannot be traced back 10 one of the four fundamental forces of nature (gravity, electromagnetism, and the two nuclear forces), but is a force that is entirely due to a modification of the quantum vacuum.

Standing electromagnetic waves and Casimir effect. As described in Ikelle (2017b, 2020, and 2022b), Maxwell's equations have wave-like solutions for electric and magnetic fields in free space. In the one-dimensional configuration, the magnetic and electric fields satisfy the following one-dimensional wave equation:

$$\left(\frac{\partial^2}{\partial x^2} - \mu_0 \varepsilon_0 \frac{\partial^2}{\partial t^2} \right) E_y(x,t) = 0$$

and

$$\left(\frac{\partial^2}{\partial x^2} - \mu_0 \varepsilon_0 \frac{\partial^2}{\partial t^2} \right) B_y(x,t) = 0 \ , \tag{5.51}$$

where k is the wavenumber and ω is angular frequency. Let us examine the situation where there are two sinusoidal plane electromagnetic waves, one traveling in the $+x$-direction, with

$$E_y^{(1)}(x,t) = E_0 \cos(kx - \omega t) \quad \text{and} \quad B_z^{(1)}(x,t) = B_0 \cos(kx - \omega t)$$

and the other traveling in the negative x-direction, with

$$E_y^{(2)}(x,t) = E_0 \cos(kx + \omega t) \quad \text{and} \quad B_z^{(2)}(x,t) = B_0 \cos(kx + \omega t) \ .$$

For simplicity, we assume that these electromagnetic waves have the same amplitudes and wavenumbers. Using the superposition principle and the identities $\cos(\alpha \pm \beta) = \cos\alpha \cos\beta \mp \sin\alpha \sin\beta$, we arrive at

$$E_y(x,t) = E_y^{(1)}(x,t) + E_y^{(2)}(x,t) = 2E_0 \sin(kx) \sin(\omega t)$$

and

$$B_z(x,t) = B_z^{(1)}(x,t) + B_z^{(2)}(x,t) = 2B_0 \cos(kx) \cos(\omega t) \ .$$

One may verify that the total fields E_y and B_z still satisfy the wave equation stated in equation (5.51), even though they no longer have the form of functions of $kx \pm \omega t$. The waves described by the equations in (5.51) are known as *standing waves*, which do not propagate but simply oscillate in space and time. That is the case for waves between plates in the Casimir effect. In other words, the Casimir effect destroys all propagating wave modes by destructive (out of phase) superposition, except standing waves. Also, as illustrated in Figure 5.38, these standing waves have a limited number of frequencies, whereas all waves and all frequencies are present in the region outside the two plates. Moreover, the outside of the interior of the two plates is symmetrical with respect to the interior of the two plates.

The fifth force: As we mentioned above, the Casimir force is not due to any of the four forces—that is, gravity, electromagnetism, and the weak and strong nuclear forces—responsible for every event, action, and reaction that occurs in our ordinary matter universe. Hence, it is sometimes speculated that the Casimir force is the fifth force and that it is related to a dimension or dimensions outside our 4D (time-space) world (see Ikelle, 2024).

Free space vacuum energy has been estimated to be about 4 GeV per cubic meter. In other words, vacuum energy is greater than nuclear energy (each deuterium-tritium fusion event releases 17.6 MeV, a U-235 fission releases about 200 MeV, and each deuterium-deuterium fusion event releases about 4 MeV) by several orders of magnitude. Hence the rush to harness it for consumption and stellar travel, and to weaponize it. By trying to do this in the same way we do with wind energy, we are likely to make mistakes. The effect of wind energy on destroying living-thing species will end up being several times more destructive than burning hydrocarbons may have caused the environment. Anyway, the correct way to harness any available energy resource is to copy or mimic nature's use. So far, it has been difficult to identify how nature uses vacuum energies (see also Ikelle, 2024).

Chapter 6

Plasma, Magnetohydrodynamics, Accretion, and Exoplanets

Solids, liquids, and gases are made of "neutral" atoms (i.e., atoms with a balance between negative and positive charges). Plasma, on the other hand, does not contain neutral atoms; that is, it has no atomic structure like in solids, liquids, and gases. It consists of electrons, protons, and ions. Ions are "charged" atoms (i.e., atoms that have lost electrons orbiting their nuclei) such as H^-, H_2^+, H_3^+, O^-, O_2^+, and C^-. The process of ripping atoms of their electrons is known as *ionization*. It occurs at very high temperatures, say 10,000 degrees Kelvin or above, like in stars, including our sun, and galaxies. In addition, these media sustain such high temperatures (i.e., ionized matter), which is why trillions of stars, billions of galaxies, and the spaces between them are all made of plasma. Everything in outer space is plasma, including auroras, solar wind, solar corona and sunspots, the magnetospheres of Earth and Jovian planets, comet tails, gaseous nebulae, stellar interiors and atmospheres, galactic arms, quasars, pulsars, novas, and black holes. Actually, plasma comprises over 99 percent of the visible universe (i.e., 99 percent of the non-dark matter in the universe).

As humans, we find ionization very comforting because it creates the plasma state from our common neutral-atomic matter. This process is familiar and intuitive to us. Actually, the most significant part of Figure 6.1 is the idea of *deionization*, which encompasses nuclear fusion. It is the process behind everything on Earth, living and nonliving alike. We can turn all humans into liquid, then into gas, and finally into plasma. One day, perhaps in this millennium, we will also understand how to go from plasma to humans. That is, humans, including their minds (souls), are actually biological forms of plasma. Genesis 3:19—"... For dust you are, and to dust you shall return."

The charged particles from the Sun do not reach the Earth because they are blocked by the Earth's magnetic field, as we will discuss later. However, the Earth's magnetic field allows light from the Sun to pass due to the fact that light is composed of charge-balanced (neutral) atoms.

DOI: 10.1201/9781032620619-6

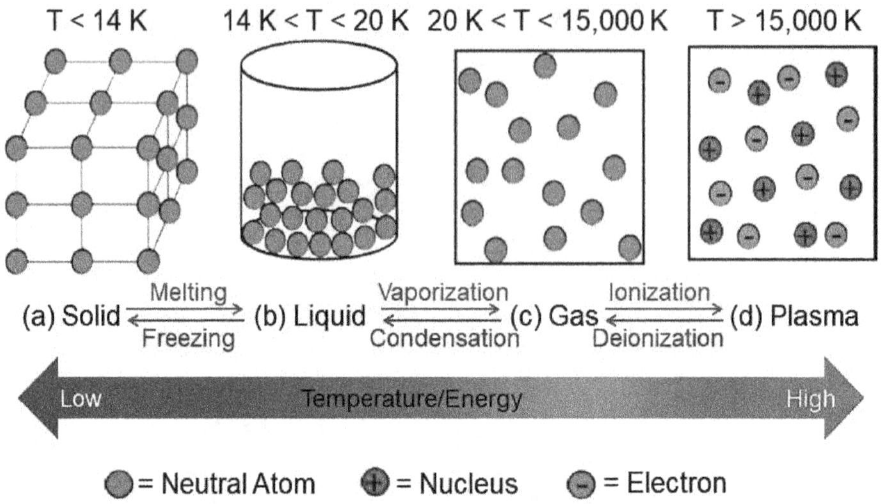

Figure 6.1. A solid has a specific shape and volume. A liquid does not have a specific shape, but has a specific volume. Its shape is determined by its container. A gas is without shape and diffuses indefinitely unless contained on all sides. Plasma is a fluid without defined shape or volume. Unlike a gas, it has unbound charged particles. It is hotter and less dense than gases. In addition, it is subject to electromagnetic forces to a much broader extent than gases, which are generally marginally affected by them if at all. The temperatures here are those of hydrogen at 1 atmospheric pressure.

So, with the exception of rocky planets, rocky exoplanet, rocky moons, and asteroids, all objects in our universe are made of plasma. For example, stars are spheres of plasma in hydrostatic equilibrium (i.e., gravitational force is balanced by pressure gradients), Jovian planets are large spheres of gas, albeit with rocky/metallic cores; planet atmospheres are stratified, gaseous fluids retained by the planet's gravity; white dwarfs and neutron stars are fluids; proto-planetary disks are dense disks of gas (mostly H and He, and some molecules: H_{20}, CO, CO_2, CH_4, and NH_3) and dust (The dust is made up of microscopic mineral grains made of metals, graphites, and silicates, silicon, magnesium, iron, as well as carbon in various forms) surrounding newly formed stars out of which planetary systems form; inter-stellar medium is the plasma in between the stars in a galaxy; intergalactic medium is the plasma in between galaxies; and intra-cluster medium is the hot plasma in clusters of galaxies. Among these fluids are neutral fluids, charged fluids (also known as plasmas), collisionless fluids (also known as deep-space plasma), and radiative processes. Radiative processes, in the form of electromagnetic radiation (photons), allow us to receive information on Earth from objects on the other side of our universe. So the dynamics of fluids described in Ikelle (2023c) primarily for terrestrial fluid types (e.g., water, oils, and gases) must be extended to include the dynamics of neutral fluids, electrically charged fluids, collisionless fluids, and radiative processes in order to fully understand the formations of stars and planets, stellar and galactic environments, stellar interiors and their evolution, lightning and storms in the sky, time-dependent turbulent convection and rotation, even the parts of the interiors of rocky planets and asteroids, and many other phenomena of our universe. Our task in this chapter is to attempt to extend fluid dynamics to encompass some of these phenomena.

We will describe neutral fluids by their density, $\rho(\boldsymbol{x})$, velocity field, $\boldsymbol{v}(\boldsymbol{x}, t)$, pressure, $\boldsymbol{p}(\boldsymbol{x}, t)$, and internal energy density, $\mathcal{U}(\boldsymbol{x}, t)$ (or equivalently, the temperature, $T(\boldsymbol{x}, t)$), as we did in Ikelle (2023b and 2023c). Their time evolutions will be described by the continuity equation, the Navier-Stokes equation, and the energy equation. For plasmas, we will add the Maxwell equations. The system of equations made of the continuity equation, the Navier-Stokes equations, the energy equation, and Maxwell equations is known as the hydrodynamic equations. When the induction equation is used to describe the time-evolution of the magnetic field and the electrical field is ignored, this system of equations is known as magnetohydrodynamics or MHD for short. We begin by discussing one of the most critical plasmas of life.

Key sentence: *Wisdom and provenance have always been sought from the sky and stars by our ancestors. It makes no sense to attend and leave college, irrespective of the majoring discipline, knowing less about the sky and the heavens than the Donga tribe of Mali knew 5,000 years ago.*

6.1. Solar Structure and Wind

6.1.1. Sun Structure

As discussed in Ikelle (2023a), the sun is an enormous ball with a mean radius of about 110 times the radius of the Earth (i.e., $r_{\text{sun}} \approx 110\, r_{\text{earth}}$ with $r_{\text{earth}} = 6371$ km) and mass of 1.99×10^{30} kg, about 99 percent of the mass of the solar system. The sun is in the plasma state, which is the fourth state of matter because it is so heated that no matter can survive as a liquid or a solid. In fact, the sun is so hot that many of the atoms in it are ionized, that is, stripped of one or more of their electrons. This removal of electrons from their atoms means that there is a large quantity of free electrons and positively charged ions in the sun, making it an electrically charged environment, quite different from the neutral environment on the Earth's surface. In a later section, we will discuss how the earth's electromagnetic field and its atmosphere block most of the charged particles raining on Earth from the Sun. Only light is let through.

Despite the fact that the sun is not solid, it does have a definite internal structure, as illustrated in Figures 6.2a and 6.3a. The layers of the sun's radius ($r_{\text{sun}}^{\text{core}} \approx 0.3 r_{\text{sun}} = 33\, r_{\text{earth}}$) with an average temperature of approximately 12 million degrees Kelvin, where nuclear fusion occurs, (2) the *radiative zone* (approximately 280,000 kilometers thick) with an average temperature of about 5 million degrees Kelvin, where heat is transferred by radiation, (3) the *convective zone* (about 210,000 kilometers thick), with an average temperature of about 1 million degrees Kelvin, where heat is transferred by convection, and (4) the *photosphere*, which is the visible surface, with an average temperature of about 5000 degrees Kelvin with a 500 kilometers thick. The *chromosphere* is a layer in the sun's atmosphere between about 400 km and 2100 km above the photosphere. The temperature in the chromosphere varies between about 4000 K at the bottom and 8000 K at the top. It gets hotter if you move further away from the photosphere, unlike in the lower layers, where it gets hotter if you move closer to the center of the sun. Above the chromosphere, we have the *Corona*, with an average temperature of nearly a

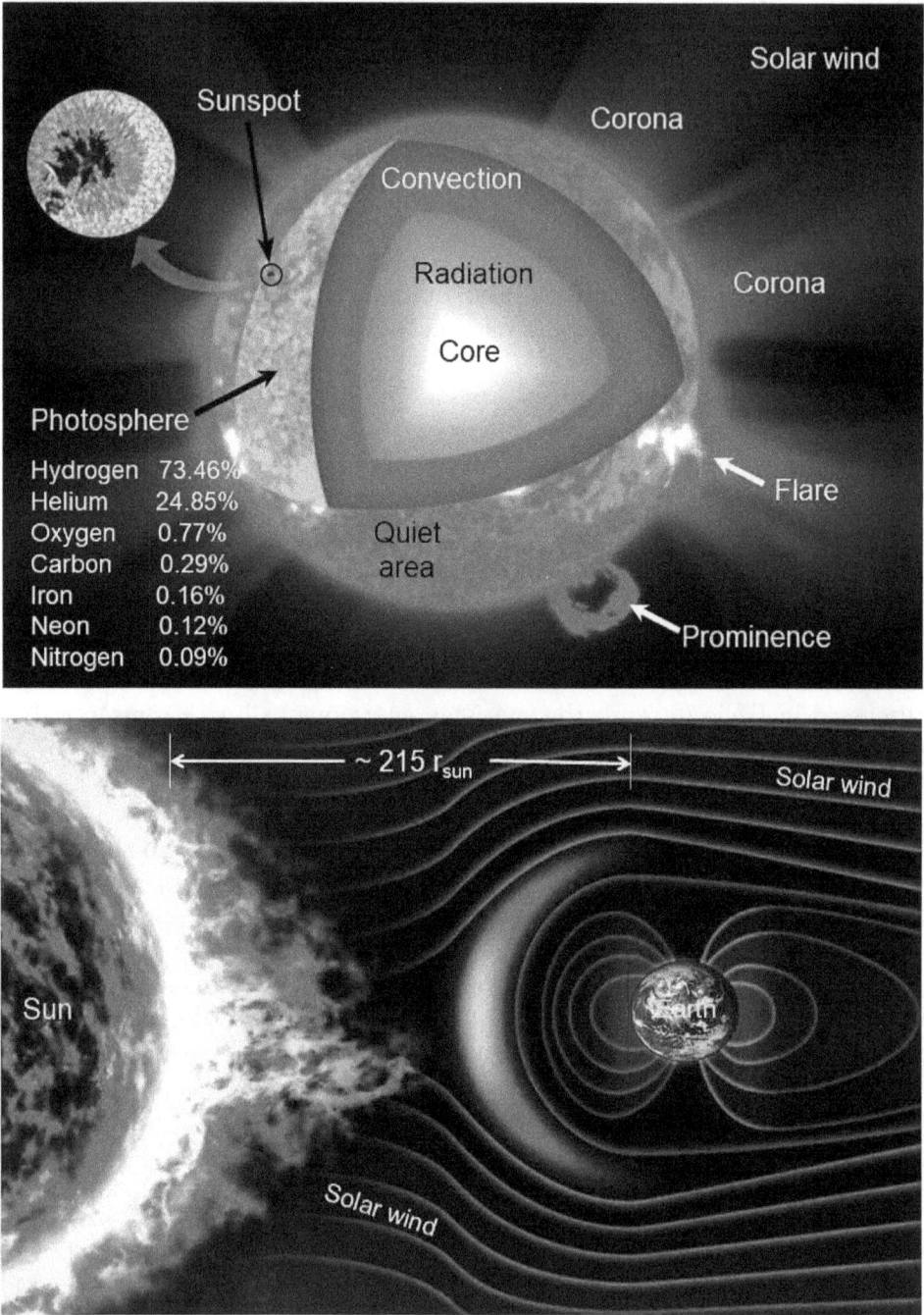

Figure 6.2. (top) An illustration of the structure of the Sun. (bottom) The Sun releases a constant stream of particles and magnetic fields called the solar wind. Solar wind streams reach planetary surfaces unless thwarted by atmospheres, magnetic fields, or both such as Earth.

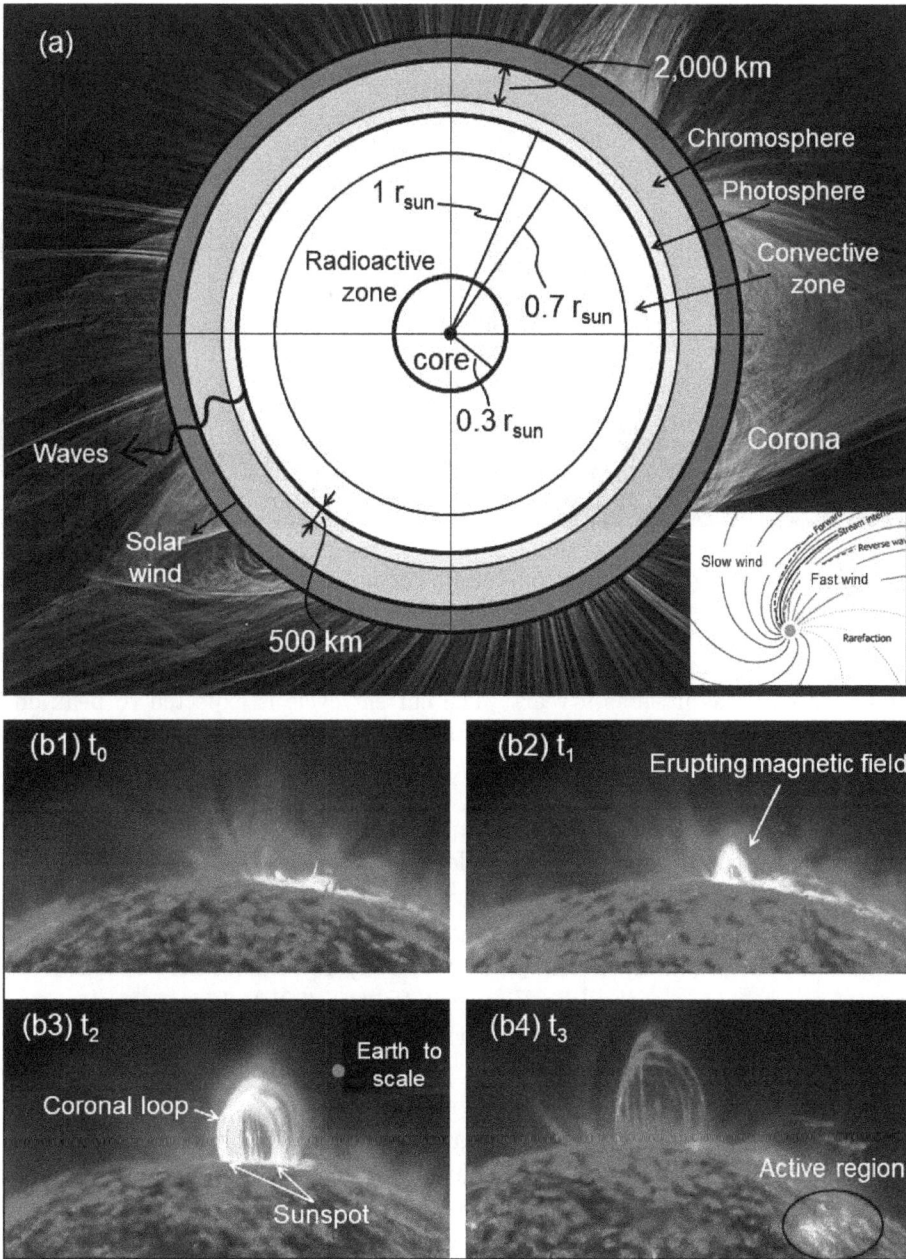

Figure 6.3. (a) The overall structure of the Sun. (b) The restless Sun with $t_3 > t_2 > t_1 > t_0$. Snapshots of an eruption on the sun that produced a magnetic display known as coronal rain. Magnetic fields are invisible, but charged plasma is forced to move along the lines. This shows up brightly at the extreme ultraviolet wavelength of 30 nanometers, and outlines the fields as it slowly falls back to the solar surface. (b1) The pre-eruption state of the active region. (b2) and (b3) formation of the coronal loop and the expanding plasma structure. (b4) After eruption. Source: NASA images.

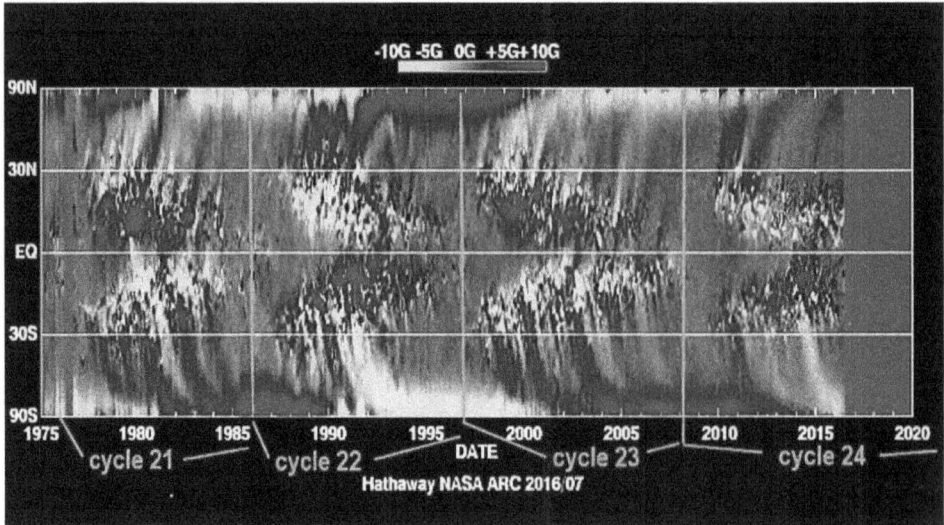

Figure 6.4. The magnetograms (time series of the solar magnetic field as function of time) over the spans of the last 45 years. The Sun reverses all of its magnetic fields to their opposite polarization in a 11-year cycle, and back again in the next cycle, for a full cycle time of 22 years. The 11-year cycle is an average. Since 1700 some cycles have lasted as long as 15 years and as little as 9 years. The current cycle is expected to peak on 2024 declining to 2031. Adapted from an image of NASA MSF Center.

million degrees Kelvin. The region of rapid temperature rise occurs between the chromosphere and the Corona and is called the *transition region*. It is only a few tens of kilometers thick.

The sun's atmosphere (i.e., photosphere, chromosphere, and corona) also has some hot and cool regions intermixed in these layers and numerous features with lifetimes of seconds to months (see Figures 6.2b and 6.3b). For example, *sunspots*, which are the coolest parts of the solar surface (the temperature inside the sunspots is typically between 3,000 and 4500 degrees Kelvin, lower than the surrounding plasma which is about 5800 degrees Kelvin), are the most visible solar structures. The temperature contrast with the surrounding photosphere renders sunspots as dark spots in visible light emission. They are dark blotches on the sun's surface and are caused by magnetic field loops that form in pairs and shift around with the field. They are strong magnetic concentrations and are relatively cool because the magnetic field suppresses convection motion. Sunspots can measure up to 80,000 km in diameter. A sunspot evolves over a lifetime of about a few days or weeks. It is subject to expansion, contraction, rotation and differential motion relative to the surrounding plasma. Given their strong concentration of a single magnetic polarity, sunspots are hosts of coronal loops, prominences, reconnection events, solar flares, and coronal mass ejections, etc.

The sunspot cycle follows a roughly 11-year cycle. Magnetic field changes over 11 years and then flips over (Figure 6.4). Some sunspot cycles are more intense than others. Also there are some periods with almost no sunspots. These periods are known as the Maunder minimum. The last Maunder minimum occurred from 1645 to 1715. During this 70-year or so period fewer than 100 sunspots were observed.

This contrasts with the typical 80,000-100,000 sunspots seen in modern times over a similar time span (Maunder, 1894; and Beckman and Mahoney, 1998). Sunspot numbers are effective indicators of the sun's dynamics and the variation of solar magnetic fields. Note that, in 2017 the sun was totally blank, there were no sunspots 25 percent of the time, and the solar cycle was historically weak. Solar cycle 24 starting in the spring of 2019 and all the way through solar cycle 26 expected to start around 2030 the Sun's solar output could be even weaker taking the Sun down to a level of weak solar output not seen since the 1600s.

Furthermore, the sun's magnetic field can also be disturbed. These disturbances generate solar flares and coronal mass ejections, which are actually two types of immense explosions associated with the release of magnetic energy stored in the solar corona (see Figures 6.2 and 6.3). A solar flare is a sudden release of huge amounts of energy (up to 6×10^{25} Joules over minutes/hours) that can be observed at almost all frequencies of the EM spectrum (Benz 2008), from radio waves to X-rays so we can only see the visible part of it which is 350 nm to 750 nm. The mechanisms of flare emission will be discussed in Chapter 4. Solar irradiation flux is now continuously monitored by spacecraft. If a flare is directed toward the Earth, it takes 2 to 3 days for accelerated particles to reach the Earth. The particle flux may damage satellites, distort the ionosphere and affect electricity grid distribution and telecommunications. Predicting the occurrence rate of solar flares is a key part of space weather forecasting (see Chapter 4).

A sun's Corona mass ejection (CME) is also a large burst of plasma out of the solar atmosphere, induced by the rearrangement of the magnetic field in the corona. It can eject billions of tons of coronal material. Actually, the main difference between CMEs and solar flares is that solar flares affect all layers of the solar atmosphere, while most CMEs originate from the corona only. CMEs are often associated with other large-scale transient activities, e.g., solar flares, erupting prominences, and radio bursts. CMEs travel outward from the Sun at speeds ranging from slower than 250 km/s to as fast as near 3000 km/s. The fastest Earth-directed CMEs can reach our planet in 15 to 18 hours. Slower CMEs can take several days to reach our planet. They expand in size as they propagate away from the Sun. Larger CMEs can cover a size comprising nearly a quarter of the space between Earth and the Sun by the time they arrive on our planet. The Sun produces several CMEs per day near solar maxima, whereas near solar minima, it produces one every few days. CMEs oriented toward the Earth may damage satellites and affect communication systems. It is a vital part of space weather prediction to estimate the probability of CME occurrence and its arrival time at the Earth. The physics of CMEs will be explored in Chapter 4. Note that *coronal holes* are dark areas in the sun's Corona because they are cooler, less dense regions than the surrounding plasma. They were discovered at the beginning of the X-ray telescope era. They can develop at any time and location on the Sun, but are most prevalent and stable at the sun's north and south poles; but these polar holes can grow and expand to lower sun latitudes. Persistent coronal holes are long-lasting sources of high-speed solar wind streams. Also, note that *coronal loops* are associated with closed magnetic field lines that connect magnetic regions on the solar surface (see Figure 6.3b). Some coronal loops last for days or weeks but most change quite rapidly.

There are also the sun's prominences and filaments, which are structures formed from plasma suspended by magnetic fields in the sun's atmosphere. Prominences are cool dense plasma structures suspended over the sun's surface, with temperatures much lower than the surrounding Corona. Their typical lifespan is on the order of several days or even months. Some prominences break apart and induce CMEs. Prominences observed on the sun's disk are known as sun filaments.

Space weather can affect space activities, including human satellites and space-crafts (current artificial satellites originated from Earth orbit Earth in the thermosphere and the exosphere), communications of all kinds, from telephone to television, navigational and geographical positioning systems, and even the operation and integrity of electric power grids. It also affects the safety of space visitors and immigrants. Moreover, our use of space technology on Earth continues to grow at a rapid pace. GPS (Global Positioning System) receivers are now in nearly every cell phone and in many automobiles, and trucks, and we become more exposed to the risks of space weather and therefore it is becoming more and more important to be able to predict when space weather may get worse, and in what way. For example, a solar flare observed by the GOES-13 spacecraft on December 5, 2006, was so intense that it damaged the camera taking the picture and disrupted satellite navigation systems on Earth. On March 13, 1989 a *600 nT magnetic storm interrupted the entire power grid* in Quebec, Canada. On September 1-2, 1859, a large and intense solar storm, a kind of global aurora, engulfed the whole globe for 18 hours. This type of storm is known as the Carrington Event. The 1859 Carrington Event disrupted all telecommunication services on Earth. There will be a similar event in this decade (that is before 2031).

6.1.2. From Solar Plasma to Earth Atmosphere

A photon can take as long as 50 million years to travel through the radiative zone. The energy in the form of heat moves by convection to the surface—convection is the flow of heat through a fluid (see Chapter 2), which is plasma in this case. The plasma at the bottom of the convective zone is extremely hot, and it bubbles to the surface where it loses its heat to space. Once the plasma cools, it sinks back to the bottom of the convective zone. Once the energy reaches the Sun's surface it is mainly transmitted by charged particles (mainly protons, which are hydrogen ions, H^+, and electrons) in all directions of the rest of the solar system and at an average of speed of about 400 km/s (almost 1.5 million kilometers per hour), but it can range from 200 km/s to as high as 1000 km/s. These streams of charged particles, primarily electrons and protons, are called the *solar wind* or plasma (Kivelson and Russell, 1995)—a stream can amount to one million tons of matter per second. The solar wind is invisible to the naked eye. The effects of the sun's activity are transmitted to Earth and other planets via the solar wind. Therefore, the solar wind is a critical element of the coupled sun-terrestrial system. It carries the Sun's magnetic field (known as the interplanetary magnetic field (IMF)). Note that there is slow solar wind with a speed of about 300 km/s and a temperature of 1.4×10^6 between 1.4×10^6 degrees Kelvin and a fast solar wind with speed at about 750 km/s, a temperature of 1.8×10^5, as measured by Ulysses mission and we will discuss these figures in the third section of this chapter along with the sources of

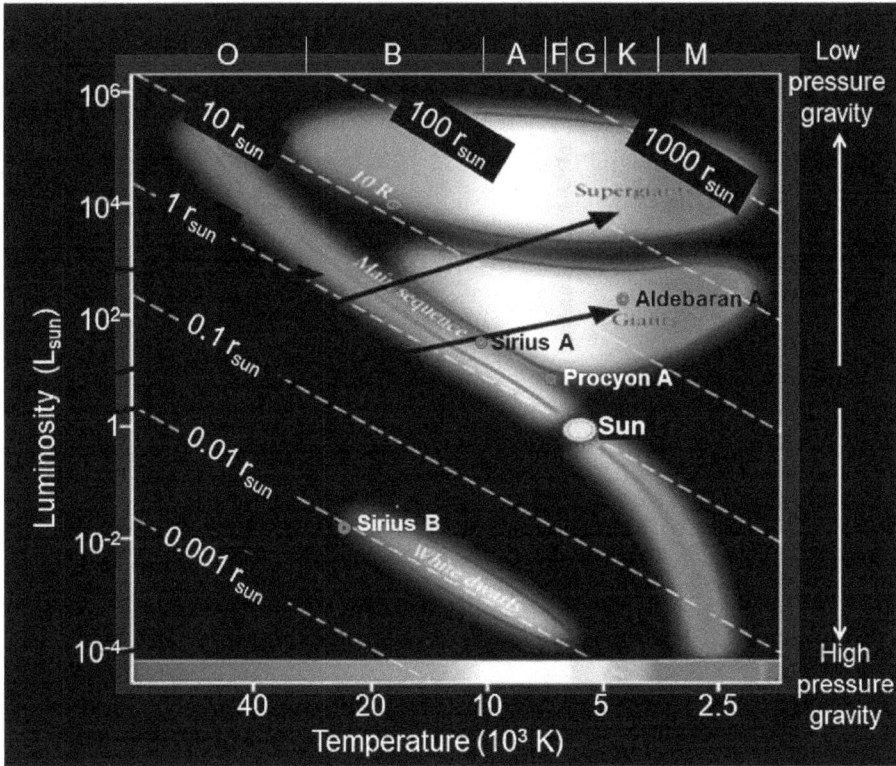

Figure 6.5. About 90 percent of stars lie on the main sequence (e.g., the Sun). In some cases a relatively cool star can be quite luminous if it has a radius greater than $10r_{sun}$ (e.g., red giants and red supergiants). A relatively hot star can have very low luminosity if its radius is very small, about 0.01 r_{sun} (e.g., white dwarfs). (O, B, A, F, G, K, M) is the spectral class of stars with subdivision from 0 and 9. The sun is G2 star. The greater the surface gravity of the star, the greater the pressure and the higher the density of the gas at any given level in the atmosphere.

slow and fast solar wind. These sources are obviously different. The region of space that is dominated by the solar wind is referred to as the heliosphere.

Solar flares may have destroyed Venus' atmosphere. A planet with either an atmosphere and/or a magnetic field, such as Earth and Jupiter, can absorb or deflect the huge amount of charged particles (e.g., protons, electrons, helium nuclei, and tiny amounts of silicon, sulfur, calcium, chromium, nickel, neon, and argon ions) released by the Sun. Without a magnetic field to deflect these particles, the solar wind will eventually knock away a planet's atmosphere, through collision by atomic collision. This is probably what happened to Venus billions of years ago. We will later study in some detail the plasma physics processes of the solar wind in later sections. Note that, as we expand our human horizon to locate and characterize exoplanets, we will also include the plasma discussion of other stars in this chapter. For that reason, it is useful to note that stellar winds can relate stars to their exoplanets.

Let us recall that stars can be classified based on luminosity versus temperature, as illustrated in Figures 6.5 and 6.6. These figures are based on the work by Ejnar

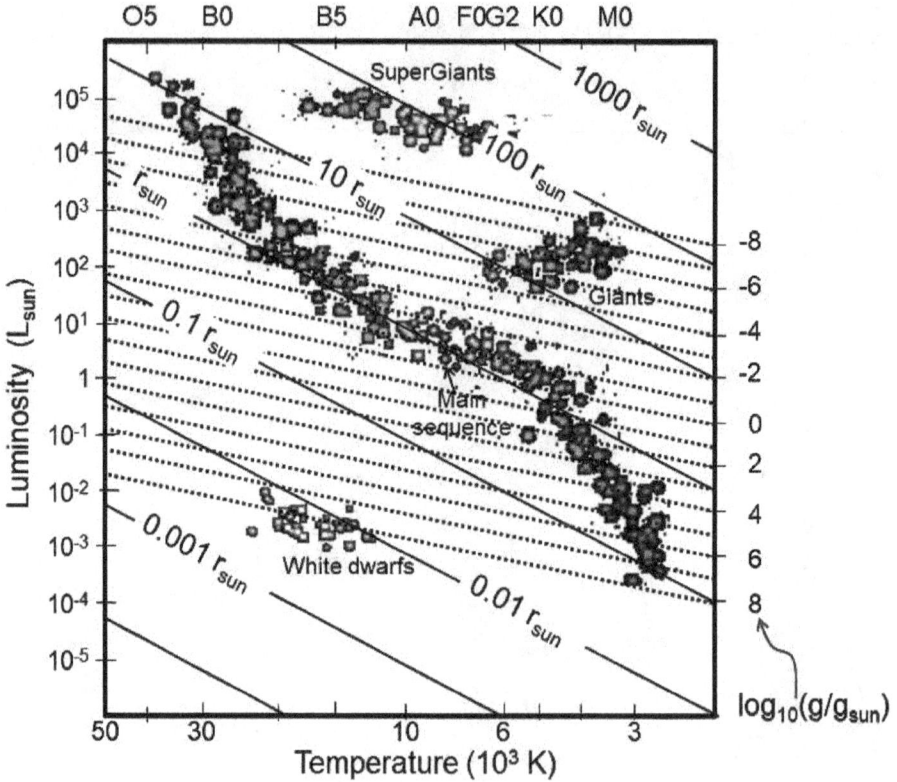

Figure 6.6. Another illustration of Hertzsprung-Russell diagram which includes the surface gravity versus luminosity with $L_{\text{sun}} = 3.86 \times 10^{26}$ W.

Hertzsprung and Henry Norris Russell. It is probably the most fundamental and powerful illustration tool in astrophysics. We can also add to it the seven classes of the spectral sequence: O, B, A, F, G, K, M (see Table 6.1). These classes range from hot to cool with O being the hottest and M the coolest. Each spectral type is further divided into 10 subdivisions, e.g., G0, G1, G2, ..., G9, where G2 corresponds to the sun. We can also use gravity to classify stars (see Figures 6.5 and 6.6). By using luminosity versus temperature (surface); i.e.,

$$L = 4\pi r^2 \sigma T^4 \implies \frac{L}{L_{\text{sun}}} = \left(\frac{r}{r_{\text{sun}}}\right)^2 \left(\frac{T}{T_{\text{sun}}}\right)^4 \tag{6.1}$$

or

$$r = \sqrt{\frac{L}{4\pi\sigma T^2}} \implies \log_{10}\left(\frac{r}{r_{\text{sun}}}\right) = \frac{1}{2}\log_{10}\left(\frac{L}{L_{\text{sun}}}\right) - 2\log_{10}\left(\frac{T}{T_{\text{sun}}}\right) \tag{6.2}$$

and the fact that surface gravity is

$$g = \frac{G_0 m}{r^2} = \frac{4\pi G_0 m \sigma T^4}{L} \, , \tag{6.3}$$

Table 6.1. Spectral class summary: Upsilon Orionis is a O-star; Pleiades, Eta Centauri, and Vega, and Sirus A are B-stars; Diadem and Porrima are A-stars; Tabby star and Procyon A are F-stars; Alpha Centuri B, Tau Ceti, and Sun are G-stars, Alpha Centuri C, Epsilon Eridani, Epsilon Indi are K-stars; and TRAPPIST-1, Alpha Centuri A and EZ aquarii A, B, and C are M-stars. Tabby star is in the constellation Cygnus approximately 1,470 light-years (450 parsecs) from Earth with $L = 4.68 L_{\text{sun}}$. This star goes through a series dimming and brightening that are still not fully explained.

	T_{eff} (10^3 K)	Color	m/m_{sun}	r/r_{sun}	L/L_{sun}
O	28-50	Blue	20-60	9-15	90,000-800,000
B	10-28	Blue-white	3-18	3-8.4	95-52,000
A	7.5-10	White	2-3	1.7-2.7	8-55
F	6-7.5	White-yellow	1.1-1.6	1.2-1.6	2.0-6.5
G	4.9-6.0	Yellow	0.85-1.1	0.85-1.1	0.66-1.5
K	3.5-4.9	Orange	0.65-0.85	0.65-0.85	0.10-0.42
M	2.0-3.5	Red	0.08-0.05	0.17-0.63	0.001-0.08

we can deduce how the gravity of stars is related to their luminosity and temperature; i.e.,

$$\log_{10}\left(\frac{g}{g_{\text{sun}}}\right) = \log_{10}\left(\frac{m}{m_{\text{sun}}}\right) - \log_{10}\left(\frac{L}{L_{\text{sun}}}\right) + 4\log_{10}\left(\frac{T}{T_{\text{sun}}}\right). \qquad (6.4)$$

Remember that $\log_{10}(g_{\text{sun}}) = 4.4$ and $\log_{10}(T_{\text{sun}}) = 3.76$ for $T = 5770$ degrees Kelvin. Note that gravity, temperature, and luminosity are all tied together, as illustrated in Figure 6.6. Surface gravity is a function of mass and radius—but mass and radius are very difficult to determine, so we often just stick to the surface gravity of stars. Our forms of life are influenced by the gravity of stars and exoplanets. The K-star and M-star are at temperatures where a significant number of neutral atoms are expected to exist and therefore matter in gas, liquid, or even solid form can exist.

The sun's brightness, also known as solar irradiance, mathematically redefined above is the amount of solar energy that hits the earth at a given location or point. In most cases, the units used are watts per meter. The sun's brightness on the earth varies from one specific point to the next. It is probably the most well-known feature of the Sun to the public. As discussed in Ikelle (2020) and Chapter 2, all weather and climates on the planets of our solar system, including Earth, from the surface of the planets out into space are influenced by the dynamics of the Sun. As a black body, it emits electromagnetic energy across the entire electromagnetic spectrum. Most of the energy from the Sun is emitted at visible wavelengths (approximately from 350 to 750 nanometers). The output from the sun at these wavelengths is nearly constant. At ultraviolet wavelengths (approximately from 120 to 350 nanometers), the sun's brightness variability is larger over the course of the solar cycle, with changes of up to 15 percent. At shorter wavelengths, the Sun changes between 30 and 300 percent over a few minutes. These wavelengths are absorbed in the upper atmosphere so they have minimal impact on the climate of Earth. At the other end of the light spectrum, at Infrared wavelengths (approximately from 750 to 10,000

nanometers), the Sun is very stable and only changes by one percent or less over
the solar cycle.

Quiz: (1) Find the relation between gravity, temperature, and luminosity of stars
in the main sequence. (2) Find the radius of Sirius B from its surface temperature
$T = 25,000$ degrees Kelvin and its luminosity $L = 3.84 \times 10^{26}$ Watts.

In Chapter 4, we will discuss the dimming of the sun up until 2070 (about 50
years), which will cause the earth to cool as a result. By 2030, we expect to see
less sunspot production and less ultraviolet radiation reaching Earth. We will see
a disruption of the sun's 11-year cycle of variable sunspot activity. This type of
disruption of the sun's 11-year cycle is known as the *Maunder Minimum*. The last
Maunder Minimum occurred between 1645 and 1715, a span of time when parts of
the world became extremely cold.

6.1.3.   The Sun's Magnetic Field: A Dynamo

Solar atmospheric activities are actually solar atmospheric magnetic field activities.
As a result, the magnetic field of the sun is a foundation for studying its solar
atmosphere. As described in Ikelle (2023b) and Chapter 3, magnetic fields can be
generated by electric currents and changing electric fields through the Ampère's law
as follows:

$$\nabla \times \boldsymbol{B} = \mu_0 \boldsymbol{J} + \mu_0 \varepsilon_0 \frac{\partial \boldsymbol{E}}{\partial t}$$

with

$$\boldsymbol{J}(\boldsymbol{x}, t) = \sum_i q_i \int \boldsymbol{v}\, f_i(\boldsymbol{x}, \boldsymbol{v}, t)\, d^3\boldsymbol{v}$$

under the non-monopole condition that $\nabla \cdot \boldsymbol{B} = 0$, where $\boldsymbol{B}$ is the magnetic field,
where $\boldsymbol{E}$ is the electric field, $\boldsymbol{J}$ is the current density, μ_0 is the magnetic permeability
of the free space, ε_0 is the permittivity of the free space, q_i is the particle charge
for the i-species, v is particle velocity, and f_i is the distribution function of the
i-species. So a magnetic field can be generated by electric currents. Therefore, we
have a two-step process, first generating electricity and then inducing a magnetic
field. Actually that is how the magnetic field is created in the Sun. As discussed
earlier, our sun is a huge plasma with charge particles of flowing, convecting, moving
charge particles. These motions produce electrical currents, which in turn create the
solar magnetic field. Why do we study electric current systems and their relation
with the magnetic field was pioneered by Gauss (1839) who was first pointed out
the possibility of electric currents in space altering the magnetic field observed on
the ground, Carrington (1860) who found a relation between auroral displays and
magnetometer perturbations during the superstorm of that year, Stewart (1882)
who first noticed that solar radiation ionizes the upper atmosphere to allow for
electric currents to flow in this region and Birkeland (1908, 1913) who discovered
that field-aligned currents connect the solar wind to the Earth's ionosphere, leading
to the aurora.

The Sun actually acts as a dynamo, a device that converts mechanical energy to electromagnetic energy or vice versa. Bicycles and flywheels are examples of dynamo. Magnets are attached to the axles of bicycles. To lighten the bicycle, an electric field and current are induced by the rotating magnet. In other words, spinning pedals convert kinetic energy into electrical energy and light energy. We can also store electrical energy by spinning wheels. Basically, a flywheel is a disk with a certain amount of mass that rotates at very high speeds, storing energy as kinetic and/or rotational energy. Modern high-tech flywheels are built with the disk attached to a rotor in an upright position to reduce gravity's influence (see Ikelle, 2020). They are charged by a simple electric motor that simultaneously acts as a generator in the process of discharging. When energy is extracted, the flywheel's rotational speed declines as a consequence of conservation of energy. Added energy causes an increase in the speed of the flywheel. Most flywheel systems use electricity to accelerate and decelerate flywheels. The spinning speed of a modern single flywheel reaches up to 16,000 rpm (revolution per minute) and offers a capacity up to 25 kilowatt hours (kWh), which can be absorbed and injected almost instantly.

The fact that the Sun's magnetic field changes dramatically over the course of just a few years, and the fact that it changes in a cyclical manner indicates that the magnetic field continues to be generated within the Sun. So the Sun creates its electromagnetic fields and the atmosphere of the sun is affected by them. This gives rise to a wide variety of phenomena for which there is no parallel in the other three states of atomic matter, including the emission of high-energy radiation and the formation of supersonic ionised winds. These facts are consistent with the idea of sun dynamo. Many celestial bodies including the earth, Jovian planets, and stars can be powered by dynamo theory as the mechanism by which a rotating, convecting, and electrically conducting fluid creates and maintains a magnetic field. Note that solar eruptions, such as major solar flares and coronal mass ejections, are the result of the rapid release of magnetic energy into thermal and kinetic energy. Observations of the solar wind taken by satellites such as the Advanced Composition Explorer, and the SDO (Dynamics Observatory) and Ulysses missions have greatly improved our knowledge and understanding of solar wind properties, as we will discuss in the next sections.

6.2. Motion of Charged Particles in Collisionless Plasmas

Charged particles and their motions are fundamental to understanding electric current systems in the atmosphere (lightening, storms, etc.) and the magnetic fields around us and in our universe. They allow us to understand the interactions between the Sun's radiation and the Earth's radiation, as discussed in the third section of this chapter. So it is important to familiarize ourselves with the motion of charged particles in prescribed electric and/or magnetic fields, which is our objective in this section.

When an electromagnetic wave propagates through plasma[1] deep in the Earth's atmosphere, billions and billions of charges are interacting with the wave and with

---

[1]About 99 percent of matter in our universe is in the plasma state. Yet we rarely encounter naturally occurring plasma in our everyday lives—almost exclusively in the form of lightning bolts.

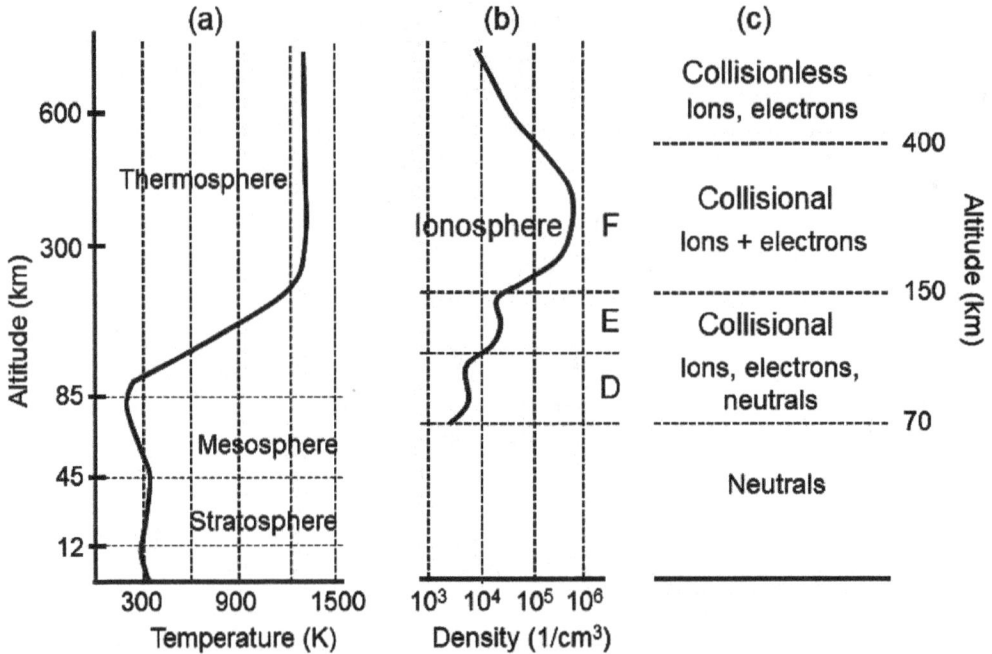

Figure 6.7. An illustration of three distinct plasma regimes ranging from highly ionized collisionless plasmas to weakly ionized collisional particles. (a) Earth's atmosphere layers. (b) Earth's ionosphere layers. (c) Electrons, ions, and neutral particles and their collisions in the magnetosphere, ionosphere, and thermosphere. At high altitudes, greater than 400 km, charged particles can be treated as collisionless. In the F-region ionosphere, collisions between electrons and ions are not negligible. Collisions between neutral and charged particles are significant in the E-region ionosphere.

each other. We will come to such problems later in this chapter. But for now we just want to discuss the much simpler problem of the motion of a single charge in a given field. We can then disregard all other charges except, of course, those charges and currents that exist somewhere to produce the fields we will assume to exist in the region under consideration. In other words, we consider a model in which the motion of each particle is treated individually and collisions between particles are infrequent. This model is generally valid in a plasmatic atmosphere, very far from the Earth's surface, because deep-space plasmas are generally collisionless such as galaxies and dark matter halos. More importantly, the dark matter that we will discuss in the next chapter may also obey this model. Also plasma in stars in a galaxy and grains in Saturn's rings may also obey this model.

What are specific examples of places in space where the collisionless model of charged particles holds? As described in Ikelle (2023a and 2023b) and in Figure 6.7, the ionosphere is a layer of plasma formed by stripping electrons from atomic oxygen and nitrogen by highly energetic ultraviolet and X-ray solar radiation. Again, the process of stripping electrons from atoms is known as ionization. The ionosphere ranges from 70- to about 600-km altitude (from the middle of the mesosphere up to the magnetosphere), and the atmospheric layers below the ionosphere remain mostly neutral. At altitudes of 400 kilometers and above, collisions between particles are rare. This part of the ionosphere is an example of a location in space where the collisionless model of charged particles holds.

Solar Extreme Ultraviolet (EUV) is solar radiation that covers the wavelengths 10-120 nm of the electromagnetic spectrum. It is highly energetic and it is absorbed in the upper atmosphere, which not only heats the upper atmosphere but also ionizes it, creating the ionosphere. Solar EUV radiation changes by a factor of ten over the course of a typical solar cycle. This variability produces similar variations in the ionosphere and upper atmosphere. Solar EUV variations are one of the three primary drivers of ionospheric variability. Solar Extreme-Ultraviolet (EUV) radiation originates in the corona and chromosphere of the Sun's atmosphere. The solar EUV spectrum, between 1 and 120 nm, is dominated by spectral lines from hydrogen (H), helium (He), oxygen (O), sodium (Na), magnesium (Mg), silicon (Si), and iron (Fe). The EUV photons reach Earth and are completely absorbed in the upper atmosphere above 80 km. The thermosphere of the earth, 80 to 600 km in altitude, is heated predominantly by solar EUV radiation. The EUV photons also ionize the atmosphere creating electrons, which form the ionosphere. Solar EUV irradiance varies by as much as an order of magnitude on time scales of minutes to hours (solar flares), days to months (solar rotation), and years to decades (solar cycle). The highly varying EUV radiation causes the thermosphere and ionosphere to vary over similar magnitudes and time scales. Because solar EUV radiation is absorbed by the upper atmosphere it is impossible to measure from the ground. Thus, measurements must be made from rockets and satellites. It is difficult to build and maintain sensors that can measure the solar EUV radiation so for many years people relied on proxies for solar EUV such as the Sunspot Number or the F10.7 cm radio flux.

The motion of a single charged particle subject to arbitrary electric ($\boldsymbol{E} = \boldsymbol{E}(\boldsymbol{x},t)$) and magnetic ($\boldsymbol{B} = \boldsymbol{B}(\boldsymbol{x},t)$) fields in the absence of collisions is described by the basic equation of motion—that is, Lorentz's force acting on this single particle equals the product of the particle mass and particle acceleration (second Newton's law of motion):

$$m\frac{d\boldsymbol{v}(\boldsymbol{x},t)}{dt} = q\boldsymbol{E}(\boldsymbol{x},t) + q\boldsymbol{v}(\boldsymbol{x},t) \times \boldsymbol{B}(\boldsymbol{x},t) , \tag{6.5}$$

where m is the particle mass, $\boldsymbol{v} = \boldsymbol{v}(\boldsymbol{x},t)$ is the particle velocity, q is its charge, and t is time. If we assume that $\boldsymbol{E} = \boldsymbol{0}$ and we take the dot product of both sides of (6.5) with $\boldsymbol{v}$, we obtain

$$\boldsymbol{v} \cdot \left[m\frac{d\boldsymbol{v}}{dt} \right] = q\left(\boldsymbol{v} \cdot [\boldsymbol{v} \times \boldsymbol{B}]\right) \implies \frac{d}{dt}\left[\frac{1}{2}mv^2 \right] = 0 , \tag{6.6}$$

with $v = |\boldsymbol{v}|$. We used the fact that $\boldsymbol{v} \cdot (\boldsymbol{v} \times \boldsymbol{B}) = 0$ (see Ikelle, 2020). We can see that the kinetic energy, $\frac{1}{2}mv^2$, is constant and therefore v is a constant. Because the kinetic energy is equal to the work done upon the charged particle by Lorentz's force, we can also observe that the magnetic field does not do any work even when it is non-uniform.

The differential equation, in (6.5), along with initial conditions, completely determines the motion of a charged particle in terms of m/q. The solutions to this equation tell us that charged particles can undergo different types of motions, including gyro, drift, and bounce motions. We here describe these three motions—that is, how charged particles and fluids move in prescribed $\boldsymbol{E}$- and $\boldsymbol{B}$-fields.

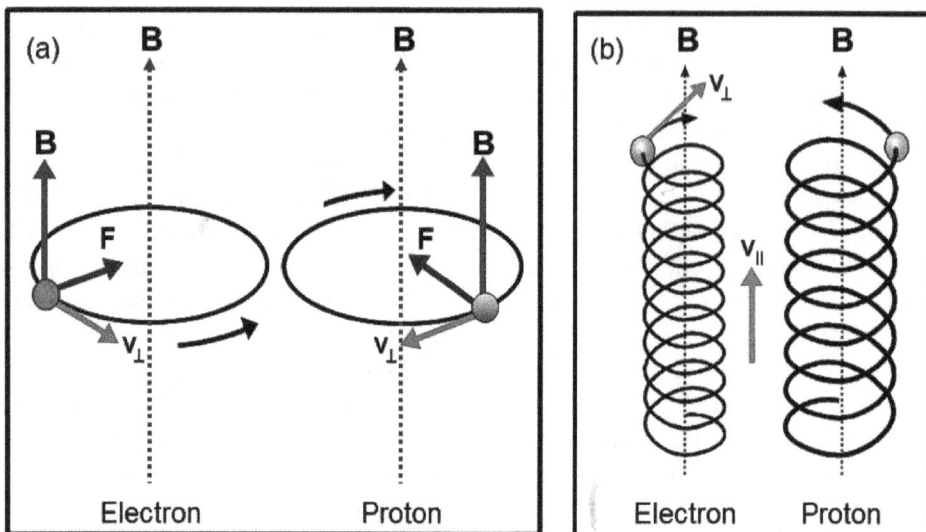

Figure 6.8. (a) Circular orbit in a uniform magnetic field. Protons (q positive) rotate anticlockwise (left-handed) . Electrons ($q = -e$ negative) rotate clockwise about the magnetic field. $v_\perp$ gives circular component and $v_\parallel$ gives constant motion along a uniform field. (b) When $v_\perp$ is nonzero, the total motion is along a helix in a steady, uniform magnetic field. When $v_\perp = 0$, charged particles move freely in a constant magnetic field.

6.2.1. Gyromotion (Larmor Motion)

Let us start with the case of a charged particle moving perpendicular to a magnetic field with no electric field (i.e., $\boldsymbol{E} = \boldsymbol{0}$), that is, the Lorentz's force acting on the single particle will be perpendicular to $\boldsymbol{v}$ and $\boldsymbol{B}$, as depicted by Figure 6.8a. We assume that the magnetic field is uniform and steady along the vertical direction (i.e., $\boldsymbol{B} = B_z \boldsymbol{i}_z = [0, 0, B_z]^T$). From (6.5) and using the definition of a vector cross product in Ikelle (2020), we obtain

$$\frac{dv_z}{dt} = 0 , \quad \frac{dv_x}{dt} = \frac{qB_z}{m}v_y , \quad \text{and} \quad \frac{dv_y}{dt} = -\frac{qB_z}{m}v_x . \tag{6.7}$$

We can see that in a steady uniform magnetic field, particle motion is distinct in directions parallel (z-axis) and perpendicular (x- and y-axes) to the magnetic field. If the charged particle is parallel to the magnetic field, which is here the $\boldsymbol{i}_z$-direction, the magnetic field exerts no force[2] on the particle and the velocity remains constant because $dv_z/dt = 0$. By taking the second derivatives of the second and third equations in (6.7), we arrive at

$$\frac{d^2 v_x}{dt^2} = \frac{qB_z}{m}\frac{dv_y}{dt} = -\left(\frac{qB_z}{m}\right)^2 v_x$$

and

$$\frac{d^2 v_y}{dt^2} = -\frac{qB_z}{m}\frac{dv_x}{dt} = -\left(\frac{qB_z}{m}\right)^2 v_y . \tag{6.8}$$

---

[2]Newton's first law states that if an object experiences no net force, then its velocity is constant.

Table 6.2. Gyrofrequencies and gyroradii at three layers of the atmosphere. B_z is the magnitude of the magnetic field and U is the kinetic energy. The first row represents the solar wind above the magnetosphere, the second row also represents the magnetosphere, and the last row represents the ionosphere. Protons have a much larger gyroradius than electrons.

B_z (nT)	U (nT)	Electron Ω_g (rad/s)	Electron r_g (km)	Proton Ω_g (rad/s)	Proton r_g (km)
5	10	850	2.09	0.463	89.4
100	1000	17000	1.05	9.3	44.7
50000	0.1	8.5×10^6	2.09×10^{-5}	4630	8.94×10^{-4}

These equations are similar to those of a simple harmonic oscillator. We can verify that the general solutions to the equations, in (6.8), are

$$v_x(t) = v_\perp \cos{(\Omega_g t + \phi)} \Longrightarrow x(t) = x_0 + \frac{v_\perp}{\Omega_g} \sin{(\Omega_g t + \phi)} \ , \tag{6.9}$$

$$v_y(t) = \mp v_\perp \sin{(\Omega_g t + \phi)} \Longrightarrow y(t) = y_0 \pm \frac{v_\perp}{\Omega_g} \cos{(\Omega_g t + \phi)} \ , \tag{6.10}$$

$$v_z(t) = v_{z0} = \text{constant} \Longrightarrow z(t) = z_0 + v_{z0}t \ , \tag{6.11}$$

where $\Omega_g = (|q|B_z)/m$, $v_\perp$ is constrained by the initial velocity perpendicular to $\boldsymbol{B}$, and $\mp$ accounts for positive or negative charge, q. We can also verify that

$$v_\perp^2 = v_x^2 + v_y^2 \ , \qquad v_\parallel = v_z \ , \tag{6.12}$$

and

$$[x(t) - x_0]^2 + [y(t) - y_0]^2 = r_g^2 \ , \tag{6.13}$$

where $r_g = v_\perp / \Omega_g = (mv_\perp)/(|q|B_z)$, $v_\perp$ is the velocity associated with the circular motion of the charged particles around the magnetic field lines, and $v_\parallel$ is the velocity associated with the motion of the charged particle along the magnetic field lines as depicted in Figure 6.8b. Setting $t = 0$ in (6.9) and (6.10), we obtain $\tan\phi = -v_y(0)/v_x(0)$. That is, the phase, ϕ, is used to match the initial velocity. So a charged particle in a uniform magnetic field rotates in a circular path perpendicular to $\boldsymbol{B}$-field lines around the point (x_0, y_0, z). Figure 6.8b illustrates particle motion. The radius of the orbiting circle, r_g, is known as a gyroradius. All particles of the same species rotate at the same angular frequency (the angular gyrofrequency), Ω_g. For example, protons and electrons rotate in opposite directions.

Let us look at some numerical examples of gyroradius (also known as Larmor radius) and gyrofrequency (also known as cyclotron frequency). By using the above definition and the properties of electrons and protons in Table 6.2, we can write the gyroradius of an electron and a proton as follows:

$$r_g^{(e)} = \frac{\sqrt{2m_e}}{e} \frac{\sqrt{U}}{B_z} \approx 3.3 \times 10^{-6} \frac{\sqrt{U}}{B_z}$$

and

$$r_g^{(p)} = \frac{\sqrt{2m_p}}{e} \frac{\sqrt{U}}{B_z} \approx 1.41 \times 10^{-4} \frac{\sqrt{U}}{B_z} \; , \qquad (6.14)$$

where $U = \frac{1}{2}mv^2 = \frac{1}{2}m_e v_e^2 = \frac{1}{2}m_p v_p^2$ is kinetic energy in electron volt (eV), B_z is in Tesla, $r_g^{(e)}$ and $r_g^{(p)}$ are in meters, and $r_g^{(p)}/r_g^{(e)} = \sqrt{m_p/m_e} \approx 42.86$. For $U = 10$ eV and $B_z = 5 \times 10^{-9}$ Telsa, we have $r_g^{(e)} = 2087$ m and $r_g^{(p)} = 89400$ m. The corresponding values of angular gyrofrequencies are

$$\Omega_g^{(e)} = \frac{e}{m_e} B_z \approx 1.7 \times 10^{11} B_z = 850 \; (\text{rad/s}) \qquad (6.15)$$

$$\Omega_g^{(p)} = \frac{e}{m_p} B_z \approx 9.26 \times 10^{7} B_z = 0.463 \; (\text{rad/s}) \; ; \qquad (6.16)$$

for solar corona, $\Omega_g^{(p)} = 1$ MHz, $r_g^{(p)} = 0.1$ m, $\Omega_g^{(e)} = 1.8$ GHz, $r_g^{(e)} = 73$ μm. Again, the electron volt (eV) is a unit of energy commonly used in atomic (microscopic scale) and nuclear physics. It is equal to the kinetic energy acquired by an electron accelerated through an electric potential of 1 volt. Table 6.2 shows additional specific values of angular gyrofrequency and gyroradius for some of Earth's atmosphere layers.

Let us point out that the above result, which states that a charged particle in a uniform magnetic field rotates in a circular path perpendicular to $\boldsymbol{B}$-field lines, is valid for any uniform magnetic field, regardless of the direction of $\boldsymbol{B}$. Basically, we can solve equations (6.5), with $\boldsymbol{E} = 0$, using vector analysis. By using the definition $\boldsymbol{v} = \boldsymbol{v}_\perp + \boldsymbol{v}_\parallel$, where $\boldsymbol{v}_\perp$ is perpendicular, and $\boldsymbol{v}_\parallel$ is parallel to $\boldsymbol{B}$, the equation of motion in (6.5) can be split into two equations as follows:

$$\frac{d\boldsymbol{v}_\parallel}{dt} = \frac{q}{m} \left(\boldsymbol{v}_\parallel \times \boldsymbol{B} \right) = 0 \;\; \text{and} \;\; \frac{d\boldsymbol{v}_\perp}{dt} = \frac{q}{m} \left(\boldsymbol{v}_\perp \times \boldsymbol{B} \right) \; . \qquad (6.17)$$

The first equation tells us that $\boldsymbol{v}_\parallel$ is constant. The second equation tells us that $d\boldsymbol{v}_\perp/dt$ is always perpendicular to $\boldsymbol{v}_\perp$ (see Ikelle, 2020). Thus $\boldsymbol{v}_\perp$, and consequently $\boldsymbol{v}_\perp \times \boldsymbol{B}$, are constant. That is, acceleration is constant in magnitude and always perpendicular to $\boldsymbol{B}$ and $\boldsymbol{v}_\perp$, and therefore the second equation in (6.17) describes circular motion.

Let us digress a little bit from plasma physics to note that the case in which the electric field is considered alone in the study of the trajectory of particles (i.e., $\boldsymbol{B} = 0$) is the basis of generating accelerated charged particle beams in a wide range of applications, including X-ray generation from electron beams in medical and industrial applications (see Ikelle, 2023c), electron microscopes (see Ikelle, 2023a), neutron generation (see Ikelle, 2023b), and television and (CRT) monitors. The equation of particle motion in (6.5) reduces to

$$m\frac{d\boldsymbol{v}}{dt} = q\boldsymbol{E} = -q\boldsymbol{\nabla}\phi \; , \qquad (6.18)$$

where ϕ is the electrostatic potential, as introduced in Chapter 3. Energy conservation becomes

$$\left[m\frac{d\boldsymbol{v}}{dt}\right] \cdot \boldsymbol{v} = (-q\boldsymbol{\nabla}\phi) \cdot \boldsymbol{v} = -q\frac{d\phi}{dt} \implies \frac{d}{dt}\left(\frac{1}{2}mv^2 + q\phi\right) = 0 , \tag{6.19}$$

thus $\frac{1}{2}mv^2 + q\phi = $ constant. That is, a particle gains kinetic energy $q\phi$ when it falls through a potential drop-ϕ. For an accelerator system with the electric potential of the particle source, ϕ_s, and two parallel conductors that carry equal and opposite potential, ϕ_a, we have

$$\frac{1}{2}mv_x^2 + q\phi_s \implies v_x = \sqrt{\frac{-2q\phi_s}{m}} \tag{6.20}$$

and

$$m\frac{dv_z}{dt} = qE_a \implies v_z = \frac{q}{m}E_a t = \frac{q}{m}E_a \frac{x}{v_x} . \tag{6.21}$$

where $\boldsymbol{E} = E_a \boldsymbol{i}_z$ is the electrostatic field of the accelerator. So we have

$$\frac{dz}{dt} = v_z$$

or

$$z = \int_0^t v_z dt' = \frac{q}{m}E_a \int_0^t t' dt' = \frac{q}{m}E_a \frac{t^2}{2} = \frac{q}{m}E_a \frac{x^2}{2v_x^2} . \tag{6.22}$$

Hence the deviation of the particle at the output of the accelerator is

$$z_0 = \frac{q}{m}E_a \frac{L^2}{2v_x^2} = \frac{1}{4}\frac{\phi_a}{\phi_s}\frac{L^2}{d} , \tag{6.23}$$

where d is the distance separating the two conductors in the z direction and L is the conductor's length in the x-direction. We used the fact that $E_a = -\phi_a/d$ (see Chapter 5).

To summarize, the motion of a charged particle in a uniform magnetic field is the sum of uniform motion, at constant velocity $\boldsymbol{v}_\parallel$ in a direction parallel to $\boldsymbol{B}$, and uniform circular motion, with velocity $\boldsymbol{v}_\perp$ in a plane orthogonal to $\boldsymbol{B}$. The specific frequency of revolution of circular motion, known as gyrofrequency (or cyclotron frequency), and the specific radius, known as gyroradius (or Larmor radius), depend on the charge and mass of particles and the magnitude of the magnetic field. In other words, charged particles move freely along a constant magnetic field, but any velocity perpendicular to the field causes them to orbit around the magnetic-field lines. This particle motion is often referred to as gyromotion because its path is a helix. The center of gyromotion is known as the *guiding center* or *Larmor center*. Without the magnetic field, the charged particles randomly move, as illustrated in Figure 6.9.

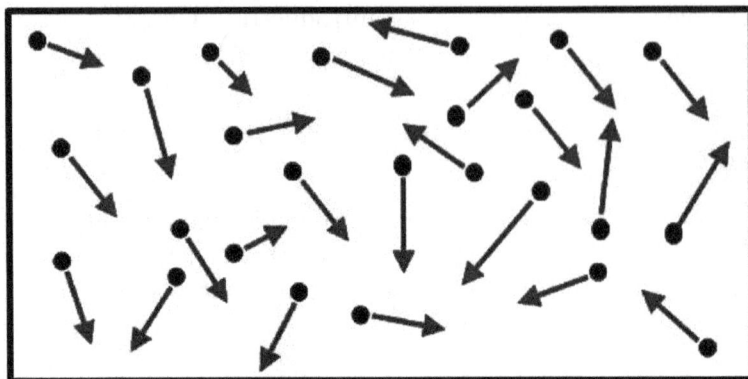

Figure 6.9. Charged particles in the absence of magnetic fields move almost randomly—they do not follow magnetic field lines.

Note that for the relativistic case, we drop the assumption of non-relativistic charged particles (i.e., $|\boldsymbol{v}| \ll c_0$). By replacing v by v/γ in the definition of the Larmor frequency and radius, we obtain

$$\Omega'_g = \frac{qB_z}{m\gamma} \quad \text{and} \quad r'_g = \frac{\gamma m v}{qB_z} , \tag{6.24}$$

where

$$\gamma = \left(1 - \frac{v^2}{c_0^2}\right)^{-1/2} , \tag{6.25}$$

and Ω'_g and r'_g are known as gyration frequency and gyration radius, respectively. When $\gamma = 1$, we return to the earlier definitions of the Larmor frequency and radius. Note also that, as the velocity of matter, v, approaches the speed of light, c_0, γ asymptotically approaches zero, $\gamma \to 0$. So the Larmor-frequency increase ensures that it is impossible to reach the speed limit in the space-time domain.

6.2.2. Drift Motion

Uniform electric and magnetic fields. Let us consider particle motion in uniform magnetic field $\boldsymbol{B}$ and electric field $\boldsymbol{E}$ which are orthogonal. We can write the velocity as follows:

$$\boldsymbol{v} = \boldsymbol{w} + \frac{\boldsymbol{E} \times \boldsymbol{B}}{B^2} = \boldsymbol{w} + \boldsymbol{v}_E \tag{6.26}$$

with

$$\boldsymbol{v}_E = \frac{\boldsymbol{E} \times \boldsymbol{B}}{B^2} , \tag{6.27}$$

where $B = |\boldsymbol{B}|$ and $\boldsymbol{v}_E$ is, by definition, a constant but depends on the field directions and magnitudes. Inserting the expression in (6.26) into (6.5), we obtain

$$m\frac{d\boldsymbol{w}}{dt} + \underbrace{m\frac{d}{dt}\left(\frac{\boldsymbol{E} \times \boldsymbol{B}}{B^2}\right)}_{=0} = q\left[\boldsymbol{E} + \left(\frac{\boldsymbol{w} + \boldsymbol{E} \times \boldsymbol{B}}{B^2}\right) \times \boldsymbol{B}\right] . \tag{6.28}$$

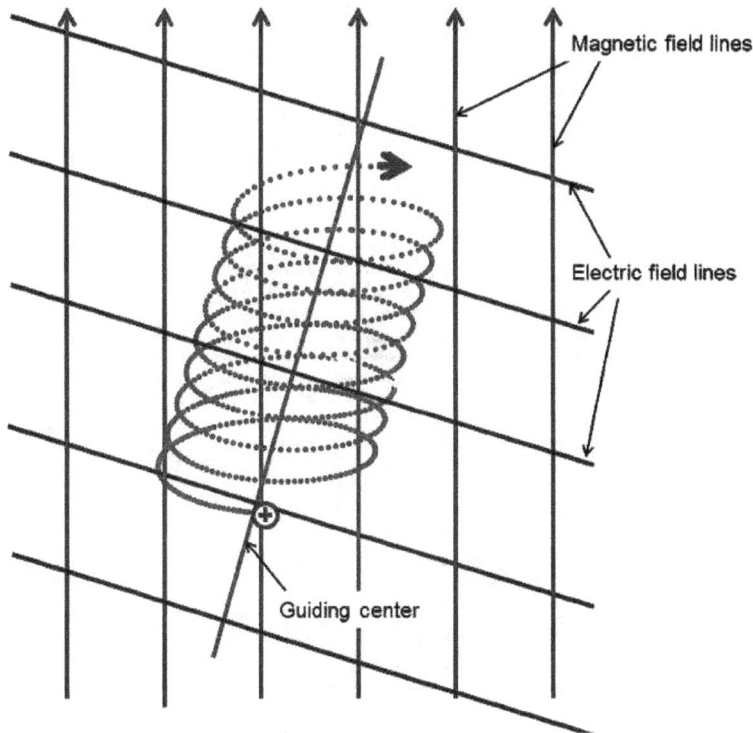

Figure 6.10. An illustration of $E \times B$ drift of a gyromotion for $E = [E_x, 0, 0]^T$ and $B = [0, 0, B_z]^T$. We show the spatial location of the particles and their trajectories. The result is that the guiding center moves perpendicular to both E and B. The particle drifts across the field. The $E \times B$ drift is independent of charge and mass. Both electrons and ions move together.

The second term on the lefthand side of (6.28) is zero because the electric and magnetic fields are both time invariant. Also because magnetic field B and electric field E orthogonal, we have the following identity

$$(E \times B) \times B = \underbrace{B (B \cdot E)}_{=0} - E (B \cdot B) = -B^2 E \ . \tag{6.29}$$

By using this identity, we deduce an equation of a particle similar to that of gyromotion in a uniform magnetic field; i.e.,

$$m \frac{dw}{dt} = q w \times B \ . \tag{6.30}$$

Since $v = w + v_E$, the resulting motion is a sum of gyromotion and a drift with velocity v_E orthogonal to both the electric and magnetic fields (see Figure 6.10). In other words, by including an electric field (perpendicular to B), the charged particle starts to drift perpendicular to E and B. All particles drift in the same direction at v_E, regardless of charge or mass (i.e., q and m), as illustrated in Figures 6.11a and 6.12a.

Note that the expression of the drift velocity, v_E, can be generalized, by replacing the electric force qE with any non-EM force F acting on the particle; i.e., the

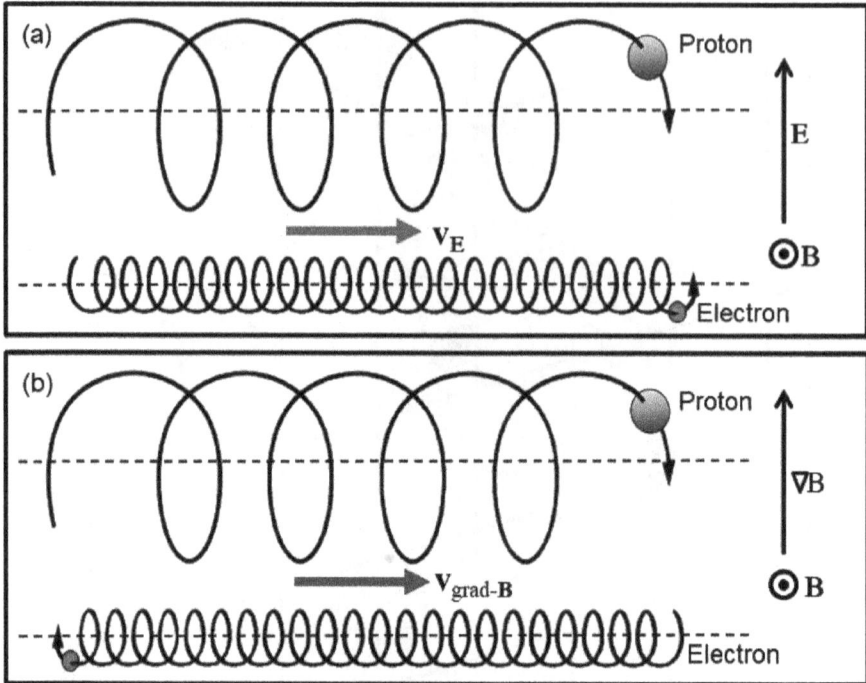

Figure 6.11. (a) $\boldsymbol{E} \times \boldsymbol{B}$ drift. Both electrons and ions move together. Notice that the $\boldsymbol{E} \times \boldsymbol{B}$ drift is independent of charge and mass. Therefore, no currents are generated by this drift motion. (b) Particle drifts due to a magnetic field gradient. Grad-B drift is a drift associated with a spatially non-uniform magnetic field. $\boldsymbol{B}$ is approximately uniform with lines of forces being straight, but their density increases in a particular direction, say, y-direction and this density is described by ∇B. Drift, in this case, is perpendicular to both ∇B and B and ions and electrons drift in opposite directions. Currents are generated by this drift motion.

associated equation of motion becomes

$$
m\frac{d\boldsymbol{v}}{dt} = q\left(\boldsymbol{v} \times \boldsymbol{B}\right) + \boldsymbol{F}
$$

$$
= q\left(\frac{\boldsymbol{F}}{q} + \boldsymbol{v} \times \boldsymbol{B}\right), \tag{6.31}
$$

and we replace $\boldsymbol{v}_E$ with

$$
\boldsymbol{v}_F = \frac{\boldsymbol{F} \times \boldsymbol{B}}{qB^2}. \tag{6.32}
$$

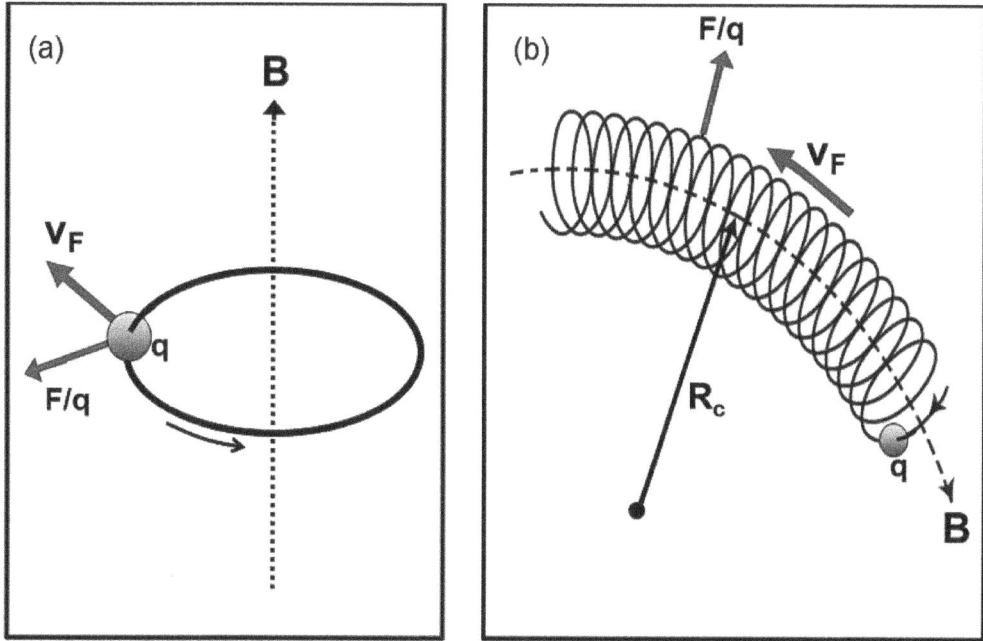

Figure 6.12. (a) $\boldsymbol{F} \times \boldsymbol{B}$ drift. We assume that force $\boldsymbol{F}$ on the charge is perpendicular to the $\boldsymbol{B}$-field and that $\boldsymbol{F}$ is charge-independent (i.e., gravity). (b) A particle motion around a curved magnetic field produces a curvature drift, which is associated with a spatially non-uniform magnetic field. The drift is caused by the outward centrifugal force, $\boldsymbol{F}$. As charged particles move along curved magnetic field lines, they experience centrifugal forces perpendicular to the magnetic field. Electrons and ions go in opposite directions (unlike $\boldsymbol{E} \times \boldsymbol{B}$) and currents are generated.

Exercise: Consider a charged particle in a uniform, constant electric field with no magnetic field. The equation of particle motion in (6.5) reduces to $m(d\boldsymbol{v}/dt) = q\boldsymbol{E}$. Suppose that $\boldsymbol{E} = E_x \boldsymbol{i}_x$ and $\boldsymbol{v}(t = 0) = \boldsymbol{0}$. Hence, the (x, y)-plane contains both the electric field and the initial velocity. Since the acceleration normal to this plane is zero, and if the initial velocity normal to it is zero, the motion remains in this plane. Verify that the subsequent motion is $x(t) = (qE_x/2m)t^2$ and $v_x(t) = (qE_x/m)t$. Verify also that the kinetic energy of the particle is $U_E = mv_x^2/2 = [(qE_x)^2/(2m)]t^2$. The particle trajectories are, in this case, parabolic.

In the case of a gravitational field, for example, we have

$$\boldsymbol{v}_F = \frac{m}{q} \frac{\boldsymbol{g} \times \boldsymbol{B}}{B^2} , \qquad (6.33)$$

and $\boldsymbol{v}_F$ is called gravitational drift (Goldston and Rutherford, 1995). So exerting an external force on charged particles can also produce a drift of the charged particles. Because gravitational drift depends on the charge of the particles, protons and electrons drift in opposite directions.

Computational exercise: (1) Use the Matlab code, in Table 6.3, to reproduce the gyromotion in Figure 6.8b. (2) Use the Matlab code, in Table 6.3, to reproduce the gyromotion with the $\boldsymbol{E} \times \boldsymbol{B}$ drift in Figure 6.10.

Quiz: Let us consider that $\boldsymbol{B} = [0, 0, B_z)]^T$ and $\boldsymbol{E} = [E_x, 0, 0]^T$. Verify that the drift velocity of the particle with charge q is $\boldsymbol{v}_E = [0, -E_x/B_z, 0]^T$, which means that this drift is perpendicular to both $\boldsymbol{E}$ and $\boldsymbol{B}$. Verify also that

$$v_y = \mp \sin\left(\frac{|q|B_z}{m} + \phi_0\right) - \frac{E_x}{B_z} . \tag{6.34}$$

The particle guiding center drifts in the direction of $-\boldsymbol{i}_2$ (i.e., -y direction). So add an electric field (perpendicular to $\boldsymbol{B}$) and the plasma starts to drift perpendicular to $\boldsymbol{E}$ and $\boldsymbol{B}$ ($\boldsymbol{E} \times \boldsymbol{B}$ drift).

Nonuniform magnetic fields have different effects on the dynamics of charged particles. For instance, a charged particle in a magnetic field with a nonzero gradient undergoes the so-called *grad-B drift* (Baumjohann and Treumann, 1997). Charged particles moving along curved magnetic field lines experience a centrifugal force perpendicular to the magnetic field. Hence charged particles experience drift in their motion. This kind of drift is known as curvature drift (Baumjohann and Treumann, 1997). There are many other drifts associated with the motion of charged particles in nonuniform magnetic fields. We will concentrate on grad-B drift and curvature drift.

Table 6.3. (Top) A code for simulating a moving charged particle in a uniform magnetic field environment; $\boldsymbol{x}(t) = [r_g \sin(\Omega_g t + \theta), r_g \cos(\Omega_g t + \theta), v_{\parallel} t]^T$. (Bottom) A code for simulating $\boldsymbol{E} \times \boldsymbol{B}$ drift of a gyromotion for $\boldsymbol{E} = [E_x, 0, 0]^T$ and $\boldsymbol{B} = [0, 0, B_z]^T$.

```
clear all; clear figure; Bz=-5; m=5; vpara=[0 0 0.5]; vper=[3 4 0]; q=-1;
rg=m*(norm(vper))/(q*Bz); theta=atan(vper(2)/vper(1))+pi/2;
xc=rg*cos(theta); yc=rg*sin(theta); wg=norm(vper)/rg; figure
plot3(-15:0.1:15,0,0); hold on; plot3(0,-15:0.1:15,0); plot3(0,0,-15:0.1:15)
xlim([-15 15]); ylim([-15 15]); t=0; tic
for n=1:500
    dt = toc; tic; x=xc+rg*cos(wg.*t+theta); y=yc+rg*sin(wg.*t+theta);
    z=vpara(3)*t; plot3(x,y,z,'-.'); hold on ; t=t+dt; pause(0.00000000215)
end
q=+1; rg = m*(norm(vper))/(q*Bz);
xc=rg*cos(theta); yc=rg*sin(theta); wg= norm(vper)/rg; t=0; tic
for n=1:400
    dt = toc; tic; x=xc+rg*cos(wg.*t+theta); y=yc+rg*sin(wg.*t+theta);
    z=vpara(3)*t; plot3(x,y,z,'-.r'); hold on; t=t+dt; pause(0.0000000025)
end
```

```
Bz=5; Ex = 2900; m=5; vpara=[0 0 0.5]; vper=[3 4 0];
q=-1; rg=m*(norm(vper))/(q*Bz); theta=atan(vper(2)/vper(1));
xc=rg*cos(theta); yc=rg*sin(theta); z0=-10; wg=norm(vper)/rg; figure
plot3(-15:0.1:15,0,0); hold on; plot3(0,-15:0.1:15,0);
plot3(0,0,-15:0.1:15); xlim([-15 15]); ylim([-15 15]); t=0; tic
for n=1:700
    dt = toc; tic; x=xc+rg*cos(wg.*t+theta);
    y=yc+rg*sin(wg.*t+theta)-(Ex/Bz)*t*0.0001;
    z=z0+(vpara(3)*t); plot3(x,y,z,'-.'); hold on; t=t+dt; pause(0.0000000025)
end
```

Although we cannot analytically solve (6.5) for the most general nonuniform magnetic fields, it can still be solved for some classes of nonuniform magnetic fields. We here describe the motion of charged particles in nonuniform magnetic fields for the case in which the magnetic field can be expressed as follows (Baumjohann and Treumann, 1997):

$$B(x,t) \approx B_0 + B_1(x,t) = B_0 + (x \cdot \nabla) B(x,t) , \qquad (6.35)$$

with

$$B_1(x,t) = (x \cdot \nabla)B(x,t) , \qquad (6.36)$$

where B_0 is a uniform magnetic field and $|B_1| = |(x \cdot \nabla)B|$ is very small compared to $|B_0|$. This approximation means that the magnetic field strength can be expanded in a Taylor expansion for distances $|x - x_0| < |x_0|$, where $r_0 = |x_0|$ is the gyroradius. We have retained only the first two terms of the Taylor series. We also decompose the particle velocity and position as follows:

$$v = v_0 + \delta v \quad \text{and} \quad x = x_0 + \delta x , \qquad (6.37)$$

with

$$v_0 = \frac{dx_0}{dt} \quad \text{and} \quad \delta v = \frac{d(\delta x)}{dt} , \qquad (6.38)$$

and with $|\delta v| \ll |v_0|$ and $|\delta x| \ll |x_0|$. By substituting these decompositions in (6.5) along with the assumption that $E = 0$, we obtain

$$
\begin{aligned}
m\frac{dv(x,t)}{dt} &= qv(x,t) \times B(x,t) \\
&= q\left[v \times B_0 + (v_0 + \delta v) \times (x \cdot \nabla) B\right] \\
&\approx q(v \times B_0) + F_{\text{grad}}(x_0, v_0) \qquad (6.39)
\end{aligned}
$$

with

$$F_{\text{grad}}(x_0, v_0) = v_0 \times (x_0 \cdot \nabla) B . \qquad (6.40)$$

All the second-order correction terms have been discarded because we are dealing with small changes.

Now we have to deal with the fact that v_0 and x_0 are periodic functions as described in the previous subsection by average $F_{\text{grad}}(x_0, v_0)$ over time and phase. Remember that, for $B = B_z i_z$, we have

$$x_0 = [x_0 + r_0 \sin(\Omega_0 t + \phi), y_0 \pm r_0 \cos(\Omega_0 t + \phi), z_0 + v_{z0} t]^T \qquad (6.41)$$

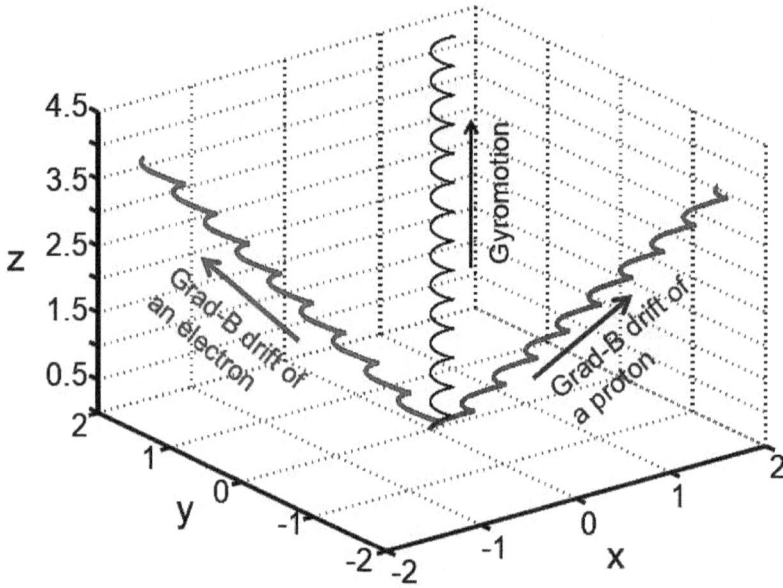

Figure 6.13. Another illustration of the grad-B drift for electron and proton. Electrons and protons drift in opposite directions. Gyromotion (black) is due to a uniform magnetic field. The nonuniform magnetic field used here is $\boldsymbol{B} = [0, 0, B_z]^T$, where $B_z = B_{z0} + x dB_x/dx + y dB_y/dy$ and $B_{z0} = 30$ nT. The uniform field is $\boldsymbol{B} = [0, 0, B_{z0}]^T$ with $B_{z0} = 30$ nT.

$$\boldsymbol{v}_0 = [v_\perp \cos(\Omega_0 t + \phi), \mp v_\perp \sin(\Omega_0 t + \phi), v_{z0}]^T \tag{6.42}$$

with $r_0 = v_\perp/\Omega_0 = (mv_\perp)/(|q|B_z)$. To reduce unnecessary algebra length, we assume that the gyromotion is centered at the origin of the frame $(0,0,0)$ and initial velocity $v_{z0} = 0$. Next we average $\boldsymbol{F}_{\text{grad}}$ as follows:

$$
\begin{aligned}
\boldsymbol{F}_{\text{grad,averg}} &= \frac{1}{2\pi t_0} \int_{-\pi}^{\pi} \int_{-t_0/2}^{t_0/2} \boldsymbol{F}[\boldsymbol{x}_0(t,\phi), \boldsymbol{v}_0(t,\phi)] \, dt \, d\phi \\
&= -\frac{U_\perp}{B} \left[\frac{\partial B_z}{\partial x}, \frac{\partial B_z}{\partial y}, \frac{\partial B_z}{\partial z} \right]^T = -\frac{U_\perp}{B} \boldsymbol{\nabla} B_z ,
\end{aligned}
\tag{6.43}
$$

where $t_0 = 2\pi/\Omega_0$ and $U_\perp = (mv_\perp^2)/2$. Note that we have used the fact that $\boldsymbol{\nabla} B = 0$ to replace $\partial B_x/\partial x + \partial B_y/\partial y$ with $-\partial B_z/\partial z$ in this integration. Note also the equivalence of (6.39) and (6.31). Based on this equivalence, we can insert $\boldsymbol{F}_{\text{grad,averg}}$ into (6.32) to obtain the drift velocity.

$$\boldsymbol{v}_{\text{grad-B}} = \frac{\boldsymbol{F}_{\text{grad,averg}} \times \boldsymbol{B}}{qB^2} = \frac{U_\perp}{qB^3} [\boldsymbol{B} \times \boldsymbol{\nabla} B_z] . \tag{6.44}$$

The effect of this velocity on charged particles is called the *grad-B drift* (Baumjohann and Treumann, 1997). Because $\boldsymbol{v}_{\text{grad-B}}$ depends on q, the grad-B drift is in opposite directions for electrons and protons, as illustrated in Figure 6.11b and Figure 6.13. These figures show particle drifts due to a magnetic-field gradient.

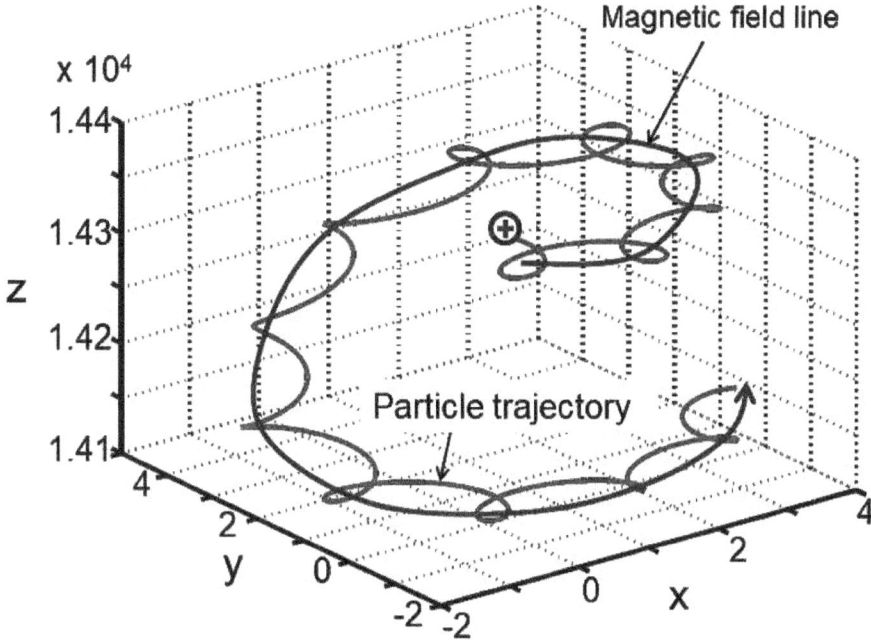

Figure 6.14. Another illustration of the curvature drift (i.e., B-field lines are curved) shows the guiding center trajectory of a charged particle in the presence of a nonuniform magnetic field. The nonuniform magnetic field used here is $\boldsymbol{B} = [B_r \cos(\Omega_g t), B_r \sin(\Omega_g t), 0]^T$, where $B_r \propto \sqrt{x^2 + y^2}$, $\Omega_g = mq/B_{z0}$, and $B_{z0} = 30$ nT.

In our above derivations of $\boldsymbol{v}_{\text{grad-B}}$ we assume that $\boldsymbol{B} = B_z \boldsymbol{i}_z$. Actually the drift velocity in (6.44) is valid for any nonuniform magnetic field with a small nonzero gradient, regardless of the direction of $\boldsymbol{B}$. We simply have to replace B_z with B; i.e.,

$$\boldsymbol{v}_{\text{grad-B}} = \frac{U_\perp}{qB^3}[\boldsymbol{B} \times \boldsymbol{\nabla}B] = \frac{\boldsymbol{F} \times \boldsymbol{B}}{qB^2}, \qquad (6.45)$$

where $\boldsymbol{F} = -\mu_{\text{mag}}\boldsymbol{\nabla}B$ and $\mu_{\text{mag}} = U_\perp/B$.

Curvature drift. We have seen that the grad-B drift causes charged particles to drift transverse to the B-field in opposite directions when they have opposite charges. The field lines of this nonuniform magnetic field are assumed to be approximately straight lines but their density increases in a particular direction. These variations in density are described by ∇B. In addition to the grad-B drift there is the *curvature drift* which is a consequence of curved magnetic field lines. In other words, when the $\boldsymbol{B}$-field lines are curved and the particle has $v_\parallel$ along the field lines that is nonzero, another drift occurs.

Charged particles moving along curved magnetic-field lines experience centrifugal forces (Figure 6.12b). Assume the radius of curvature, R_c, is large compared to the gyroradius, r_0 (i.e., $r_0 \ll R_c$). As described in Ikelle (2023a), the outward centrifugal force is

$$\boldsymbol{F}_{\text{cent}} = \frac{mv_\parallel^2}{R_c}\frac{\boldsymbol{R}_c}{R_c}, \qquad (6.46)$$

where $R_c = |\boldsymbol{R}_c|$ and the equation of charged particles in (6.31) becomes

$$m\frac{d\boldsymbol{v}(\boldsymbol{x},t)}{dt} = q\left(\frac{\boldsymbol{F}_{\text{cent}}}{q} + \boldsymbol{v}\times\boldsymbol{B}\right). \tag{6.47}$$

This force can be directly inserted into (6.32) to obtain the centrifugal drift velocity as follows:

$$\boldsymbol{v}_{\text{drift}} = \frac{\boldsymbol{F}_{\text{cent}}\times\boldsymbol{B}}{qB^2} = \frac{mv_\parallel^2}{qR_c^2}\left(\frac{\boldsymbol{R}_c\times\boldsymbol{B}}{B^2}\right) = \frac{2U_\parallel}{qR_c^2}\left(\frac{\boldsymbol{R}_c\times\boldsymbol{B}}{B^2}\right), \tag{6.48}$$

where $U_\parallel = (mv_\parallel^2)/2$. Drift is therefore into or out of the page depending on the sign of q. Figure 6.14 shows an example of the guiding center trajectory of a charged particle in a non-homogeneous magnetic field with a toroidal geometry. The particles drift around the curvature of the magnetic field lines.

Because B is proportional to $1/R_c$ as we can see in (6.48), we can verify that $\boldsymbol{\nabla}B/B = \boldsymbol{R}_c/R_c^2$ [i.e., $\boldsymbol{\nabla}(R_c^{-1}) = -\boldsymbol{R}_c/R_c^3$]. Thus, we can alternatively write the drift velocity as follows:

$$\boldsymbol{v}_{\text{drift}} = \frac{2U_\parallel}{q}\left(\frac{\boldsymbol{B}\times\boldsymbol{\nabla}B}{B^3}\right). \tag{6.49}$$

In this form, we can directly compare grad-B and curvature drifts. To summarize, gradient and curvature drifts are in the same direction and are proportional to particle energies. Finally we can write the total guiding center drift as

$$\boldsymbol{v}_D = \underbrace{\frac{\boldsymbol{E}\times\boldsymbol{B}}{B^2}}_{E\times B\,\text{drift}} + \underbrace{\frac{\boldsymbol{F}\times\boldsymbol{B}}{B^2}}_{\text{Ext}-\text{force drift}} + \underbrace{\frac{U_\perp}{qB^3}[\boldsymbol{B}\times\boldsymbol{\nabla}B]}_{\text{Grad}-\text{B drift}} + \underbrace{\frac{2U_\parallel}{qB^3}[\boldsymbol{B}\times\boldsymbol{\nabla}B]}_{\text{Curvature drift}}.$$

These are just examples of particle drift, which is the motion of the particle's guiding center. There are a variety of known as well as unknown drifts caused by a combination of magnetic fields and external forces. The last exercise of this chapter along with the codes in Tables 6.4 and 6.6 provide readers with opportunities to experiment with the drifts described here and to explore possible new drifts. Note that particle drifts can be used to explain and describe many phenomena, such as planetary fields, coronal loops, and tokamaks.

6.2.3.   Magnetic Mirrors

Let us stay with the case of charged particle motion in a nonuniform magnetic field with zero electric field. We can see in Figure 6.15 that as particles move in regions of strong magnetic fields, they can reflect because these regions act like mirrors. The bounce motion of a charged particle is the motion up and down along (parallel to) the magnetic field between mirrors. Our objective here is to explain this mirror effect behind bounce motion. We will start by introducing the second constant of charged particle motion, known as the magnetic moment. The other constant introduced in (6.6) is the conservation of the kinematic energy of charged particles.

Table 6.4. A Matlab code for simulating single particle motion in the presence of a magnetic field using finite-differences to compute field derivatives with respect to time and gyromotion (i.e., circular motion about a magnetic field line).

```
gam=5.5;q=1.602e-19; m=1.67e-27;Bx0=0; By0=0;B0 = 0.0000011
if(icase =3)
    niter = 49000; dt = 1e-4;x=-0.0+zeros(niter,1); y=zeros(niter,1); z=y;
    vx=0.2+zeros(niter,1); vy=vx; vz=vx; Bz0=0.00000003;
    rL = 0.2*(m/(q*Bz0)); wc= (q*Bz0)/m;
end
if(icase==3)
    niter = 919009; dt = 1e-3; x=-0.00001+zeros(niter,1);
    y=0.0001+zeros(niter,1); z=zeros(niter,1); vx=0.2+zeros(niter,1);vy=vx;
    vz=vx; Bz0=0.0000003; rL = 0.2*(m/(q*Bz0)); wc= (q*Bz0)/m;
end
for it=1:niter
    x(it+1) = x(it) + vx(it)*dt; y(it+1) = y(it) + vy(it)*dt;
    z(it+1) = z(it) + vz(it)*dt;
    if(icase==1) % uniform B
        Bx = 0.0; By = Bz0*0.0; Bz = Bz0;
    end
    if(icase==2) % gradB drift
        Bx = 0; Bz = Bz0+(rL*cos(wc*(it-1)*dt)*0.007*Bz0/(rL*0.01));
        By = 0; Bz = Bz-(rL*sin(wc*(it-1)*dt)*0.007*Bz0/(rL*0.01));
    end
    if(icase==3) % Curvature drift
        xx = x(it+1)*x(it+1)+y(it+1)*y(it+1); Br = rL*Bz0*gam*sqrt(xx);
        Bx = Br*cos(wc*(it-1)*dt);By = Br*sin(wc*(it-1)*dt); Bz = 0.0;
    end
    if(icase==4) ;% mirror
        Br=B0; Bx = Br*cos(wc*(it-1)*dt);By = Br*sin(wc*(it-1)*dt);
        Bz = Bz0+(rL*cos(wc*(it-1)*dt)*0.007*Bz0/(rL*0.01));
        Bz = Bz-(rL*sin(wc*(it-1)*dt)*0.007*Bz0/(rL*0.01));
    end
    fx = (vy(it)*Bz - vz(it)*By)*dt*q/m; fy = (vz(it)*Bx - vx(it)*Bz)*dt*q/m;
    fz = (vx(it)*By - vy(it)*Bx)*dt*q/m;
    vx(it+1) = vx(it) + fx; vy(it+1) = vy(it) + fy; vz(it+1) = vz(it) + fz;
end
if(icase==3)
    ni = niter; xx = x(ni-2600:ni); yy = y(ni-2600:ni); zz = z(ni-2600:ni);
    plot3(xx, yy,zz,'b');
end
if(icase =3) plot3(x,y,z,'r'); end
```

The magnetic moment. Because we are dealing with particle motion in a nonuniform magnetic field the component of particle velocity along the magnetic field, $\boldsymbol{v}_\parallel$, is not necessarily zero. We can describe particle motion parallel to the magnetic field lines as follows:

$$m\frac{d\boldsymbol{v}_\parallel}{dt} = -\mu_{\mathrm{mag}}\frac{dB}{dz} \; , \tag{6.50}$$

Table 6.5. Some properties of various plasmas in nature. L is the macroscopic length scales of interest or system size, B_{eq} is the magnetic field, and t_0 is the observation time.

	L (km)	B_{eq} (T)	t_0 (s)
Tokamak	0.020	3	3×10^{-6}
Magnetosphere Earth	4×10^4	3×10^{-5}	6
Solar coronal loop	10^5	3×10^{-2}	15
Magnetosphere neutron star	10^3	10^{11}	10^{-2}
Accretion disc YSO	1.5×10^6	10^{-4}	7×10^5
Accretion disc AGN	4×10^{15}	10^{-4}	2×10^{12}
galactic plasma	10^{18}	10^{-8}	10^{15}

Table 6.6. A Matlab code for simulating single particle motion in the presence of a magnetic dipole (Earth-like magnetic dipole). The outcome is the trajectory and dynamics of charged particles (here protons).

```
close all; clear all;
e = 1.602176565e-19; mp = 1.672621777e-27; me = 9.10938291e-31;
c = 299792458; Re = 6378137 ; U = 15e6; % kinetic energy in eV
m = mp; vabs = c/sqrt(1+(m*c^2)/(U*e)); pitcha = 30.0; vx0 = 0.0;
vy0 = vabs*sin(pitcha*pi/180); vz0 = vabs*cos(pitcha*pi/180);
tf = 100; time = 0:0.01:tf;
[t,xx] = ode45(@Lorenz,time,[4*Re ;0; 0; vx0; vy0; vz0]); % solve eqn
plot3(xx(:,1)/Re,xx(:,2)/Re,xx(:,3)/Re,'r'); % trajectory in 3D
xlabel('x[Re]'); ylabel('y[Re]'); zlabel('z[Re]')
%plot(xx(:,1)/Re,xx(:,2)/Re,'r'); % trajectory in 2D
hold on; sphere(30); hold off;grid on
```

```
function xxf = Lorenz(t,xxin)
xxf = zeros(6,1); Beq = 3.07e-5; Re = 6378137; e = 1.602176565e-19;
mp = 1.672621777e-27; me = 9.10938291e-31; c = 299792458;
qem = e/mp; % qom=e/mp for proton & qom=-e/me for electron
usB00 = -Beq*Re^3/(xxin(1)^2 + xxin(2)^2 + xxin(3)^2)^2.5;
Bx = 3*xxin(1)*xxin(3)*usB00; By = 3*xxin(2)*xxin(3)*usB00;
Bz = (2*xxin(3)^2 -xxin(1)^2- xxin(2)^2)*usB00;
xxf(1) = xxin(4); xxf(2) = xxin(5); xxf(3) = xxin(6); % (dx/dt,dy/dt,...)
xxf(4) = qem*(xxin(5)*Bz - xxin(6)*By); % dvx/dt = qem*(vv x B)-x
xxf(5) = qem*(xxin(6)*Bx - xxin(4)*Bz); % dvy/dt = qem*(vv x B)-y
xxf(6) = qem*(xxin(4)*By - xxin(5)*Bx); % dvz/dt = qem*(vv x B)-z
end
```

where

$$\mu_{\mathrm{mag}} = \frac{mv_\perp^2}{2B} = \frac{|q|v_\perp^2}{2\Omega_0} = \frac{U_\perp}{B} = \frac{\text{perpendicular kinetic energy}}{\text{magnetic field strength}} . \tag{6.51}$$

We are using the component of particle velocity along the magnetic field created by the grad-B drift. Quantity μ_{mag} is known as the magnetic moment, and because it is constant, it is conserved during particle motion. Let us prove that μ_{mag} is indeed

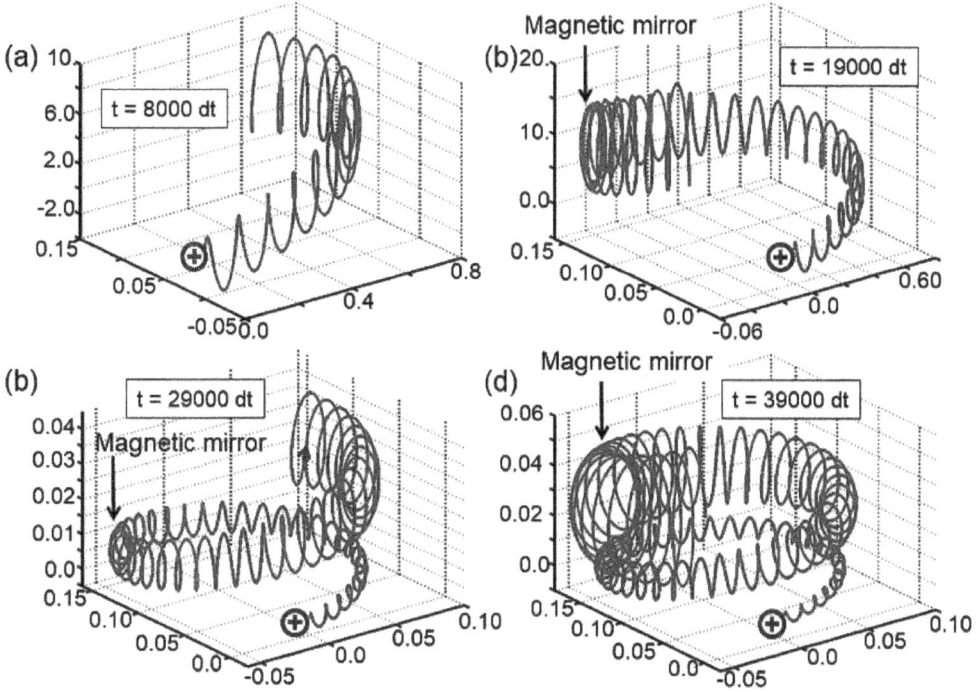

Figure 6.15. An illustration of a magnetic mirror with a snapshot of a charged particle motion at $t = 8000\,dt$, $t = 19000\,dt$, $t = 29000\,dt$, and $t = 39000\,dt$. The nonuniform magnetic field used here is $\boldsymbol{B} = B_{r0}\boldsymbol{i}_r + B_z\boldsymbol{i}_z = [B_{r0}\cos(\Omega_g t), B_r\sin(\Omega_g t), B_z]^T$, where $B_z = B_{z0} + x\,dB_x/dx + y\,dB_y/dy$, $\Omega_g = mq/B_{z0}$, $B_{r0} = 1100$ nT, and $B_{z0} = 30$ nT. More details can be found in Table 6.4. The gyroradius is larger where the field is weaker and smaller where the field is stronger. Notice that particles can reflect (bounce) on the magnetic mirror.

a constant during particle motion. Note that one can alternatively introduce the magnetic moment by using current flowing along a closed loop that encloses an area A. The current loop with area A and current I has magnetic moment $\mu_{\mathrm{mag}} = IA/c_0$, where c_0 is the speed of light. A charged particle moving along its Larmor radius is such a loop with $A = \pi r_g^2$ and $I = q(\Omega/2\pi)$ with $\Omega = v_\perp/r_g$, we obtain the result in (6.51). Note also that the magnetic moment is an adiabatic invariant, which means that it is conserved under slow changes in an external variable. Our next task is to prove this result.

If we take the dot product of both sides of (6.50) with $\boldsymbol{v}_\parallel$, we obtain

$$\frac{d}{dt}\left(\frac{1}{2}mv_\parallel^2\right) = -\mu_{\mathrm{mag}}v_\parallel\frac{dB}{dz} = -\mu_{\mathrm{mag}}\frac{dz}{dt}\frac{dB}{dz} = -\mu_{\mathrm{mag}}\frac{dB}{dt}\,. \tag{6.52}$$

By using the conversation of kinematic energy, we can construct a second independent equation with μ_{mag} as follows:

$$\frac{d}{dt}\left(\frac{1}{2}mv_\parallel^2 + \frac{1}{2}mv_\perp^2\right) = \frac{d}{dt}\left(\frac{1}{2}mv_\parallel^2 + \mu_{\mathrm{mag}}B\right) = 0\,. \tag{6.53}$$

We used the definition of the magnetic moment, $\frac{1}{2}mv_\perp^2 = \mu_{\mathrm{mag}}B$. By equating (6.52) and (6.53), we arrive at

$$\frac{d(\mu_{\mathrm{mag}}B)}{dt} - \mu_{\mathrm{mag}}\frac{dB}{dt} = 0 \quad \Longleftrightarrow \quad \frac{d\mu_{\mathrm{mag}}}{dt} = 0 \, , \tag{6.54}$$

which is satisfied for constant values of μ_{mag} with respect to time. We used, in the above derivations, the fact that the magnetic field slowly varies in time.

Conservation of the magnetic moment is basically conservation of angular momentum about the guiding center. As we defined in Ikelle (2023b), the angular momentum is $\boldsymbol{L} = m\boldsymbol{x}_0 \times \boldsymbol{v}$ and its magnitude is

$$|\boldsymbol{L}| = m|\boldsymbol{x}_0|v\sin\theta = mr_0 v_\perp = \frac{2m}{|q|}\mu_{\mathrm{mag}} \, , \tag{6.55}$$

where θ is the angle between $\boldsymbol{x}_0$ and $\boldsymbol{v}$, as defined in Ikelle (2020), and $v_\perp = v\sin\theta$. As $\boldsymbol{L}$ is constant, μ_{mag} is constant as well.

The mirror force. Now consider the magnetic field in Figure 6.16. When the field is uniform, field lines run parallel, the particle moves along the ellipse, and the Lorentz force is perpendicular to both $\boldsymbol{B}$ and $\boldsymbol{v}_\perp$ and points toward the guiding center. When the field is nonuniform (Figure 6.16) field lines converge toward the right in this case, the magnetic field now makes an angle with respect to the magnetic field line, and thus the Lorentz force now has a non-zero component in the z-direction. Hence, the particle will be accelerated away from the direction in which the field strength increases. The particle will be reflected back, and the region of increasing magnetic field thus acts as a reflector, known as a magnetic mirror. To make this result quantitative, let us compute the Lorentz force in the z-direction in a cylindrical coordinate system:

$$F_z = \frac{q}{c_0}\left(\boldsymbol{v} \times \boldsymbol{B}\right)_z = \frac{q}{c_0}v_\perp B_r \, , \tag{6.56}$$

where $\boldsymbol{B} = B_r\boldsymbol{r} + B_z\boldsymbol{z}$, B_r is the magnetic field component in the cylindrical r-direction, in which the z-direction is as indicated in Figure 6.16. We assume that our magnetic field is axisymmetric (i.e., $\partial\boldsymbol{B}/\partial\theta = 0$). Using the Maxwell equation $\boldsymbol{\nabla} \cdot \boldsymbol{B} = 0$ and, again, working in cylindrical polar coordinates, we have

$$\frac{1}{r}\frac{\partial(rB_r)}{\partial r} + \frac{\partial B_z}{\partial z} = 0 \quad \Longrightarrow \quad \frac{\partial(rB_r)}{\partial r} = -r\frac{\partial B_z}{\partial z} \, . \tag{6.57}$$

If $\partial B_z/\partial z$ is given at $r = 0$ and does not vary much with r, then

$$rB_r = -\int_0^r \frac{\partial B_z}{\partial z}dr \approx -\frac{1}{2}r^2\left[\frac{\partial B_z}{\partial z}\right]_{r=0} \Longrightarrow B_r = -\frac{1}{2}r\left[\frac{\partial B_z}{\partial z}\right]_{r=0} \, .$$

We obtain the vertical component of the Lorentz force

$$\boldsymbol{\nabla} \cdot \boldsymbol{B} = \frac{1}{r}\frac{\partial\left(rB_r\right)}{\partial r} + \frac{\partial B_z}{\partial z} = 0$$

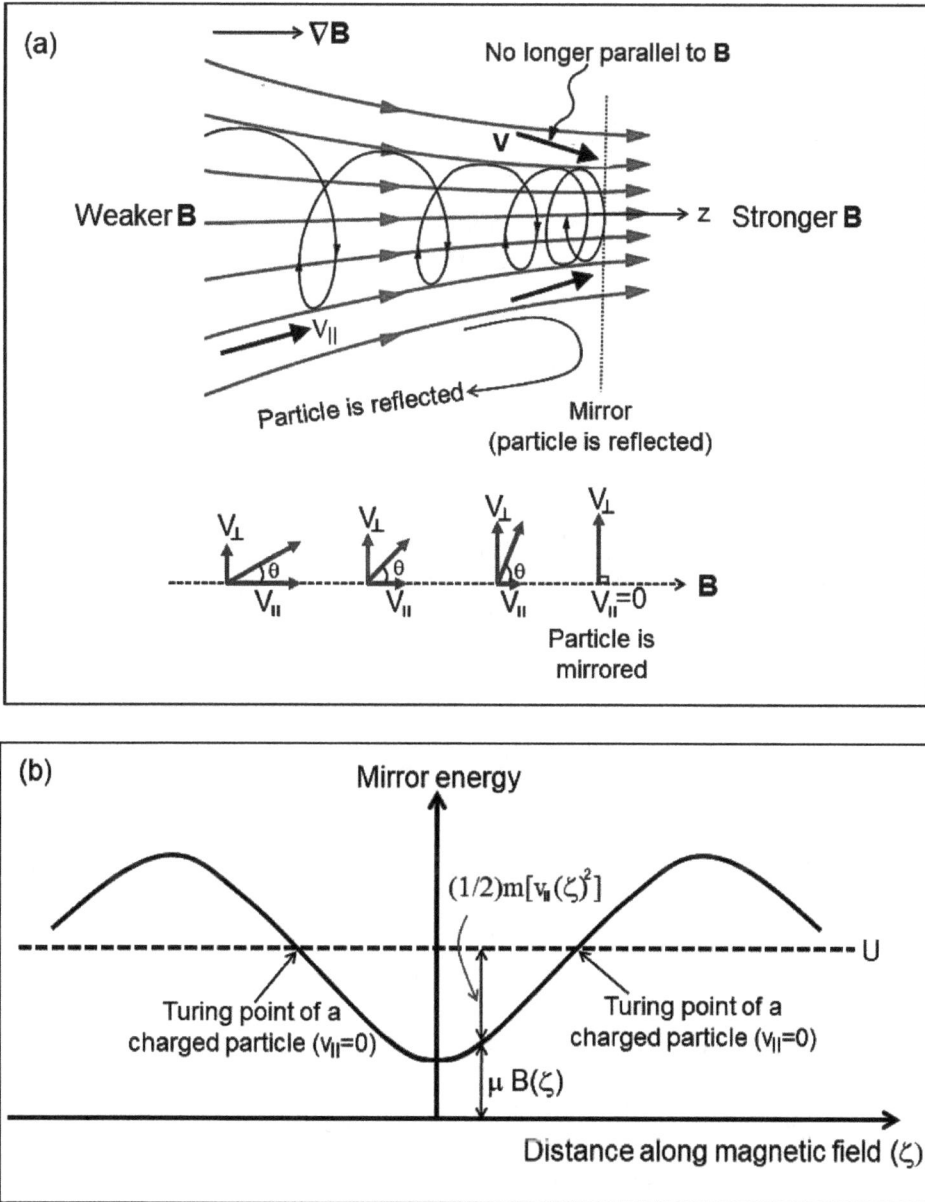

Figure 6.16. (a) An illustration of a magnetic mirror (i.e., magnetic reflector) and pitch angles between v and B. (b) An interpretation of the energy of a magnetic mirror. As $B(\zeta)$ increases, the conservation of energy implies that $v_\parallel(\zeta)$ decreases. At some turning point $v_\parallel(\zeta) = 0$, and the particle turns around.

or

$$B_r = -\frac{r}{2}\frac{\partial B_z}{\partial z} \implies F_z = -\mu_{\text{mag}}\frac{\partial B_z}{\partial z} \,. \tag{6.58}$$

In these derivations, we also assume that $\partial B_z/\partial z$ is independent of r for the first implication. We then obtain B_r by integrating the first equation with respect to r. We can see that the Lorentz force has a non-zero component, proportional to the magnetic moment of the charged particle, in the direction opposite to that in which the magnetic field strength increases; it accelerates the particle along the field line in the direction of decreasing field magnitude. Note that by analogy to the electrostatic forces on a charge, $\boldsymbol{F} = -q\boldsymbol{\nabla}\phi$, the magnetic field in (6.58) behaves like a potential field.

Figure 6.16b shows an interpretation of the energy of a magnetic mirror. This interpretation is based on constant kinematic energy and constant magnetic moment; i.e.,

$$U = \frac{1}{2}mv_\parallel^2 + \frac{1}{2}mv_\perp^2 = \text{cst} \quad \text{and} \quad \mu_{\text{mag}} = \frac{mv_\perp^2}{2B} = \text{cst}$$

or

$$U = \frac{1}{2}m\left[v_\parallel(\zeta)\right]^2 + \mu_{\text{mag}}B(\zeta) = \text{cst} , \tag{6.59}$$

where ζ is the distance along the magnetic field and cst stands for constant. As $B(\zeta)$ increases, the conservation of energy implies that $v_\parallel(\zeta)$ decreases. At some turning point $v_\parallel(\zeta) = 0$, and the particle turns around.

Loss cone. We are now ready to address the notion of *loss cone*, as depicted in Figure 6.17. The magnetic field is stronger close to the coils than in between. Suppose the particle starts out with kinetic energy $U = \frac{1}{2}m\left(v_\perp^2 + v_\parallel^2\right)$ and magnetic moment μ_{mag}, which are both conserved as the charged particle moves. As the particle moves in the direction along which the strength of $\boldsymbol{B}$ increases (e.g., toward one of the coils in Figure 6.17), $v_\perp$ must increase in order to guarantee conservation of the magnetic moment. The particle will eventually reach a region of sufficiently strong B, where it is not possible for the particle to penetrate further into regions without violating either magnetic moment conservation, kinetic energy, or both: the particle will be reflected back. In other words, the region of increasing magnetic field acts as a reflector, known as a *magnetic mirror*.

Let us denote the velocity between the coils by $v_{\perp,1}$ and $v_{\parallel,1}$ and velocity at the coils by $v_{\perp,2}$ and $v_{\parallel,2}$. At the reflection point, we have $v_{\parallel,2} = 0$. Because the magnetic moment in (6.51) is constant, we move from a weaker $\boldsymbol{B}_{\text{min}}$ to a stronger $\boldsymbol{B}_{\text{max}}$, as the gyromotion speeds up from $\boldsymbol{v}_{\perp,1}$ to $\boldsymbol{v}_{\perp,2}$; i.e.,

$$\frac{mv_{\perp,1}^2}{2B_{\text{min}}} = \frac{mv_{\perp,2}^2}{2B_{\text{max}}} \iff \frac{mv_1^2\sin\theta_1^2}{2B_{\text{min}}} = \frac{mv_1^2\sin\theta_2^2}{2B_{\text{max}}}$$

or

$$\frac{\sin\theta_2^2}{\sin\theta_1^2} = \frac{B_{\text{max}}}{B_{\text{min}}} . \tag{6.60}$$

So, as B increases, the perpendicular component of the particle velocity increases: the particle moves more and more perpendicular to $\boldsymbol{B}$. We will eventually reach a critical point where $\theta_2 = \pi/2 = 90$ degrees. We then have

$$\sin\theta_c = \sqrt{\frac{B_{\text{max}}}{B_{\text{min}}}} , \tag{6.61}$$

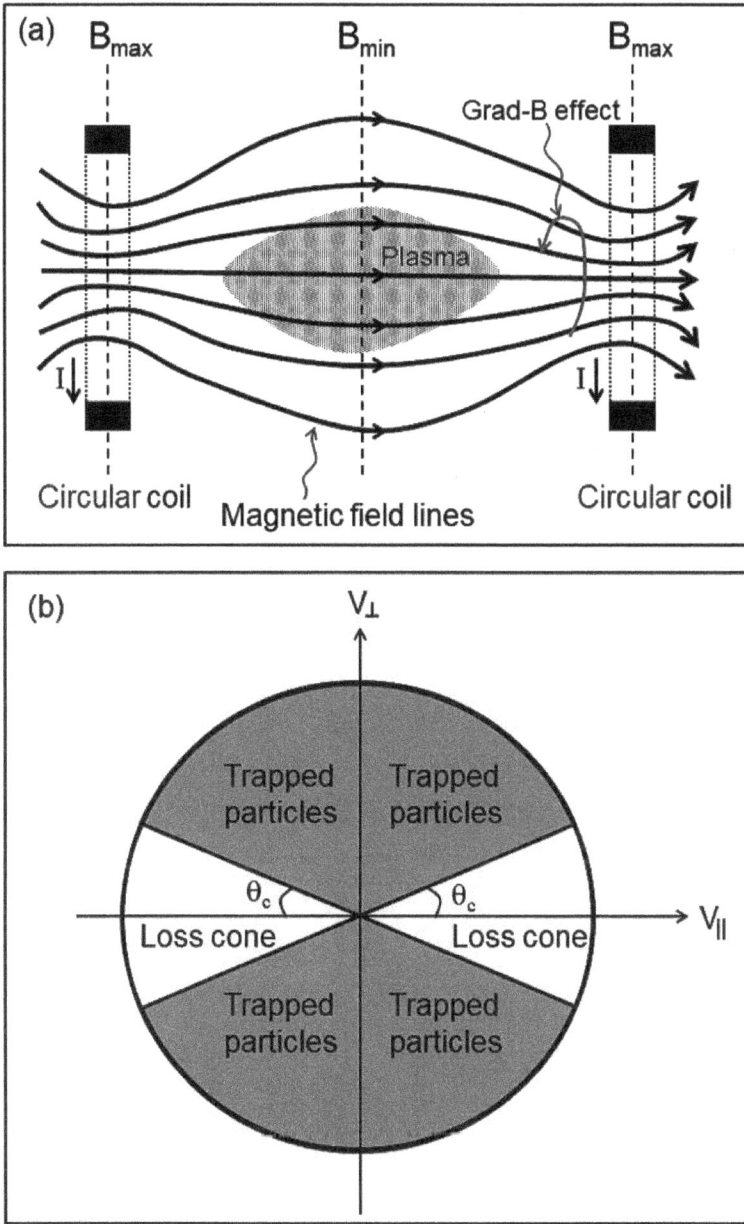

Figure 6.17. (a) An illustration of a magnetic bottle (i.e., mirror trapping). We can get a magnetic bottle by combining two magnetic mirrors, which are here two coils. A magnetic bottle can confine charged particles by both spiraling and reflection off the mirrors. The curved lines depict magnetic field lines. The magnetic field strength is larger at the coils than in the middle, creating magnetic mirrors in between charged particles can be trapped. (b) The critical angle, $\theta_c = \sin^{-1}[(B_{max}/B_{min})^{1/2}]$, divides velocity space into a loss cone and a magnetic trapping region. The charged particles are trapped in magnetic trapping regions if $\theta > \theta_c$ and escape in loss cones if $\theta < \theta_c$.

where B_{max} is the magnitude of the magnetic field at $\theta_2 = 90$ degrees. Particles are confined to a mirror if they have $\theta > \theta_c$. Otherwise they escape the mirror (i.e., $\theta < \theta_c$; particles with a value of θ smaller than θ_c cannot be reflected). That portion of velocity space occupied by escaping particles is called the *loss-cone*. The opening angle of the loss cone is not dependent on mass or charge. Electrons and protons are lost equally if the plasma is collisionless.

Box 6.1: Some Terminology

- **White dwarfs and neutron stars** are stellar remnants.

- **Proto-planetary disks** are the dense disks of gas and dust (metals, graphites, silicates)surrounding newly formed stars out of which planetary systems form.

- **Inter-Stellar Medium (ISM)** is the gas in between the stars in a galaxy. The ISM has roughly a three-phase structure: it consists of a dense, cold (about 10 degrees Kelvin) molecular phase, a warm (about 104K) phase, and a dilute, hot (about 106K) phase. Stars form out of the dense molecular phase, while the hot phase is caused by supernova explosions.

- **Accretion**[a] **disks** are gaseous, viscous disks in which the viscosity (enhanced due to turbulence) causes a net rate of radial infall toward the center of the disk, while angular momentum is being transported outwards (accretion).

- In celestial astronomy, a **constellation** is an area on the celestial sphere in which a group of visible stars forms a perceived pattern or outline, typically representing an animal. Nowadays the sky is divided into eighty-eight constellations with defined boundaries. The constellation Orion contains two of the ten brightest stars in the sky—Rigel (Beta Orionis) and Betelgeuse (Alpha Orionis)—and a number of famous nebulae—the Orion Nebula, De Mairan's Nebula and the Horsehead Nebula.

[a]Accretion refers to the growth in mass of any celestial heavy object due to its gravitational attraction

---

One can design a machine with many coils (i.e., rings of electric current), as magnetic mirrors, surrounding the middle of the magnetic field to confine the paths of charged particles and even optimize the topology of the magnetic field by adjusting the number of coils and their orientation. The objective of optimization is to reduce as much as possible ion/electron leakages. A magnetic bottle is a particular case of such a machine with two magnetic mirrors placed close together. For example, two parallel coils separated by a small distance, carrying the same current in the same direction will produce a magnetic bottle between them (see Figure 6.17). Particles near each end of the bottle experience a magnetic force toward the center of the region; particles with appropriate speeds spiral repeatedly from one end of the region to the other and back—they will spiral toward the other end of the bottle, and when they reach the other coil, they get reflected back and so on. So a magnetic bottle is similar to a glass bottle, but made using magnetic fields, with both sides having a neck.

> **Computational exercise:** We now use the differential equation solvers in Table 6.4 to predict particle motions in the presence of nonuniform magnetic fields. (1) Validate this code by reproducing the results in Figure 6.8b. (2) These codes contain four examples of nonuniform magnetic fields. Modify this code as necessary to reproduce the results in Figures 6.13, 6.14, and 6.15.

Magnetic bottles can be used to temporarily trap charged particles. It is easier to trap electrons than ions, because electrons are so much lighter (they have a lower mass). This technique is used to confine the high energy of plasma in fusion experiments, as we will discuss later. Magnetic confinement is actually the general principle of magnetic confinement fusion. However, magnetic bottles are leaky. Particles that make up a loss cone will leak out of the magnetic bottle.

6.2.4. Charged Particles in Earth-like Magnetic Fields and Van Allen Radiation Belts

Our objective in this subsection is to study the trajectory and dynamics of charged particles (electron, proton, alpha particles, and ionospheric oxygen ions) in the presence of the earth-like's magnetic field. We approximate the Earth's magnetic field as a giant dipole field. In other words, we consider an approximate model of charged particle motion in a strong dipole magnetic field (see also Table 6.5). In the Cartesian coordinate system, this dipole can be expressed as follows (Öztürk, 2012):

$$\boldsymbol{B} = \frac{\mu_0}{4\pi|\boldsymbol{x}|^3} \left[3\left(\boldsymbol{M}\cdot\frac{\boldsymbol{x}}{|\boldsymbol{x}|}\right)\frac{\boldsymbol{x}}{|\boldsymbol{x}|} - \boldsymbol{M} \right] = -\frac{B_{eq}R_{earth}^3}{|\boldsymbol{x}|^5}\left[3xz,\ 3yz,\ 2z^2 - x^2 - y^2\right]^T , \quad (6.62)$$

with

$$\boldsymbol{M} = -M_0\boldsymbol{i}_z \quad \text{and} \quad M_0 = (4\pi/\mu_0)B_{eq}R_{earth}^3 , \quad (6.63)$$

where $B_{eq} = 3.07 \times 10^{-5}$ Tesla is the magnetic field strength at the Earth's surface measured at magnetic equator (i.e., at $x = R_{earth}$, $y = 0$, and $z = 0$) and R_{earth} is the Earth's radius, $|\boldsymbol{x}|^2 = x^2 + y^2 + z^2$, and $\boldsymbol{M}$ is the dipole moment. For Earth, we conventionally take $\boldsymbol{M}$ to be anti-parallel to the z-axis because the magnetic north pole is near the geographic south pole, as we will discuss later in this chapter. The trajectory and dynamics of charged particles (electron, proton, alpha particles) in this magnetic field configuration can be obtained by solving this system of six ordinary differential equations:

$$\frac{d\boldsymbol{x}}{dt} = \boldsymbol{v} \quad \text{and} \quad \frac{d\boldsymbol{v}}{dt} = \frac{q}{m}(\boldsymbol{v}\times\boldsymbol{B}) , \quad (6.64)$$

where $m = \gamma m_0$, where m_0 is the rest mass and $\gamma = (1 - v^2/c_0^2)^{1/2}$ is the relativistic factor. Notice that $\gamma = 1$ for non relativistic motions. The velocity vector is $\boldsymbol{v} = [v_x, v_y, v_z]^T$ and its magnitude is estimated from the kinetic energy, $\mathcal{U}_k$ of particle using the following equation:

$$v = c_0\left[1 - \left(\frac{m_0c_0^2}{m_0c_0^2 + \mathcal{U}_k}\right)^2\right]^{1/2} . \quad (6.65)$$

The effect of the Earth's gravitational field on the motion of these charged particles is also neglected here. We numerically solve the system of ordinary differential equations in (6.62) by using the ODE45 Matlab solver (the fourth-order Runge-Kutta method with fifth-order corrections), as shown in Table 6.6. The results are v_x, v_y, v_z, x, y, and z. We first compute the velocity components $[v_x, v_y, v_z]^T$ and then deduce $[x, y, z]^T$ using the estimates of velocity components. At the initial time, $t = 0$, the velocity components are $[v_{x0} = v \sin \theta \cos \psi, v_{y0} = v \sin \theta \sin \psi, v_{z0} = v \cos \theta]^T$. The magnitude of velocity, v for ions and electrons of the same kinetic energy will be different due to their mass. Here, r_L is gyro-radius and ψ is gyrophase.

Figure 6.18 shows the trajectories of protons with 5 MeV and 15 MeV kinetic energy, a typical energy range for radiation belts. These protons start at $[4R_{\text{earth}}, 0, 0]^T$ with an equatorial pitch angle (angle between the velocity and field vectors) $\theta = 30°$ and $\psi = 90°$ so that $v_{y0} = v \sin \theta$, $v_{z0} = v \cos \theta$, and $v_{x0} = 0$. We followed them for 100 seconds. As we can see in Figure 6.18, charged particles are trapped in the magnetic mirrors formed by a dipolar planetary magnetic field, starting at $4R_{\text{earth}}$. Notice that the protons do not track single lines of the magnetic field; instead, they spiral toward one of the poles while drifting laterally. The drift velocity of the particle is not constant, since it depends on its radial position as described earlier. The particle with 5 MeV kinematic energy never reaches the poles; it bounces back and forth in the neighborhood of the equator. So the motions of these protons are the superposition of (1) $E \times B$ drift, (2) polarization drift, (3) gradient drift, and (4) curvature drift. These motions were described in the previous subsections. Due to the charge dependency of $\nabla B \times B$ drift, the proton moves westward.

Another illustration of highly energetic ionized particles (protons and electrons) trapped in the magnetic mirrors formed by the Earth's magnetic field is shown in Figure 6.19. The trapped particles move back and forth between the North and South poles of the Earth's magnetic field, and they experience curvature drift (in opposite directions for the electrons and protons). In other words, the Earth's magnetic field creates its own magnetic bottle. Most of these particles come from the solar wind and cosmic rays. Protons, electrons, helium ions, and ionospheric oxygen ions are commonly seen species in the trapped regions of the Earth's magnetic field. By trapping the solar wind, the Earth's magnetic field deflects those energetic particles and protects the atmosphere from destruction. These confinements of charged particles are known as *Van Allen radiation belts*.

These belts are named after James Van Allen, who discovered them in 1958 using data obtained by instruments aboard the Explorer 1 satellite. These belts are regions that contain very energetic particles, especially energetic electrons and protons, that are trapped in the Earth's magnetic field, and around the Earth in orbits. Belts are highly dynamic, increasing and decreasing on minute-to-year time scales. The high levels of radiation caused by energetic electrons and protons make Van Allen radiation belts very harsh regions for satellites, spacecraft, and space travelers. So, the Van Allen radiation belts are composed of electrons with kinematic energies up to several MeV and protons with energies up to several hundred MeV. Earth has two such belts, and sometimes others may be temporarily created. Electrons with typical energies above 0.1 million electron volts (MeV) are found in both an inner belt (from about 1.5 to 3 R_{earth} above Earth's center in the equatorial plane), and an outer belt (from about 3 to 10 R_{earth}). High-energy protons, with

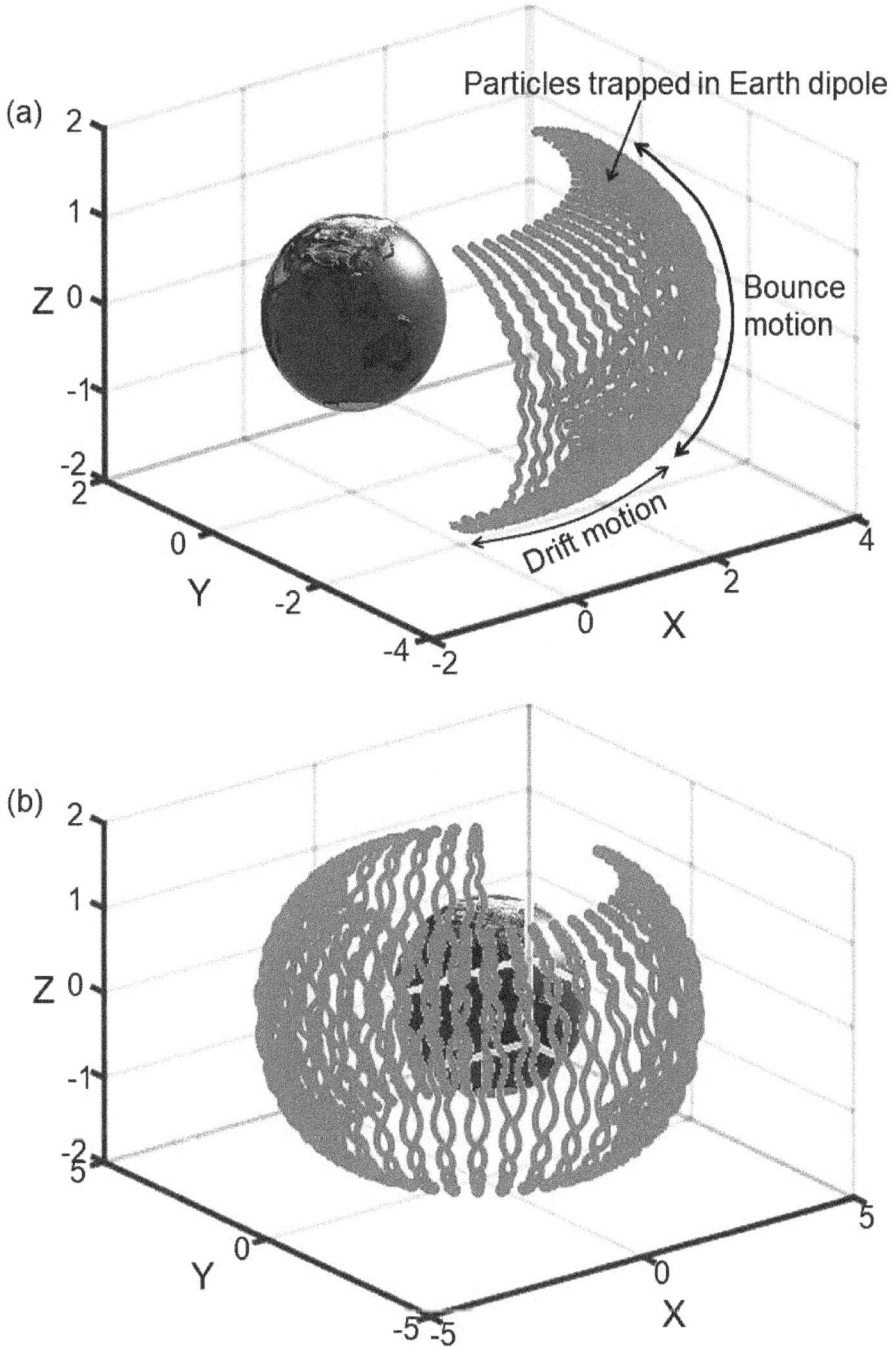

Figure 6.18. (a) Simulated three-dimensional trajectories of a 5 MeV kinetic energy proton in the Earth's dipole field with $\theta = 30^{\circ}$ starting at $[4R_{\mathrm{earth}}, 0, 0]^{T}$ for 50 s and $\psi = 90^{\circ}$. (b) The same as (a) for a 15 MeV kinetic energy proton.

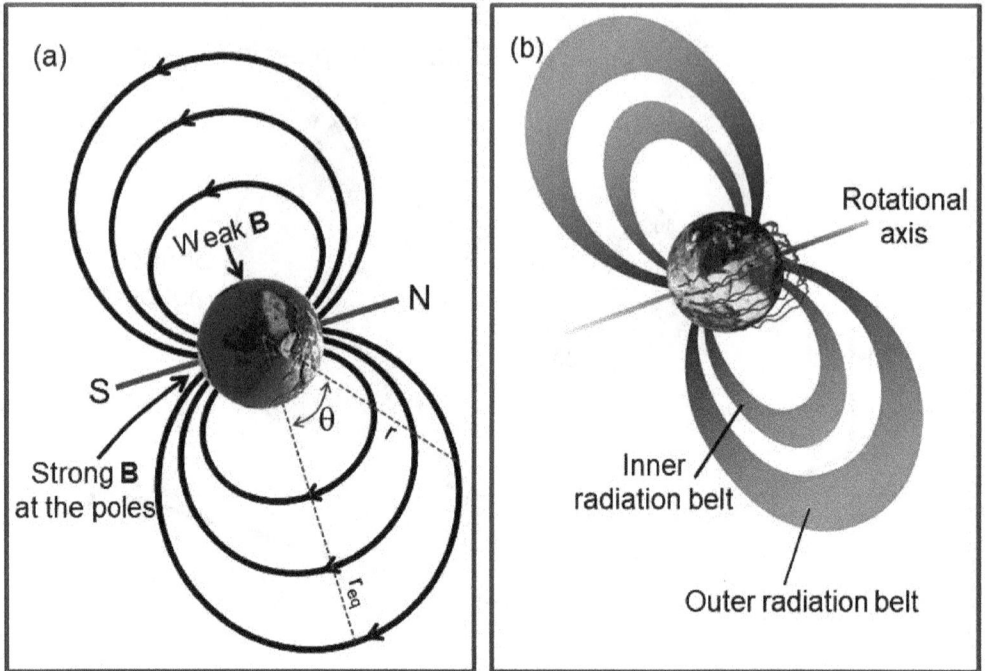

Figure 6.19. (a) Particles (electrons, protons, and ions) are trapped in a dipole field along magnetic field lines. (b) The two main Earth's Van Allen belts, which are actually magnetic bottles. They consist of highly energetic ionized particles trapped in the Earth's magnetic fields. They can be found from 640 km above the ground up to a distance of about 9 R_{earth}.

typical energies exceeding more than 10 MeV, form a belt that extends from about 1.1 to 3 R_{earth}. In other words, the inner Van Allen belt is made up of high-energy protons and electrons. The outer Van Allen belt is made up of lower-energy electrons. The two main belts are concentric, donut-shaped regions of charged particles held in place around the Earth by the planet's magnetic field. Both belts are dynamic and can expand and contract in response to solar storms and other changes in the space environment.

Like most magnetic bottles, the Van Allen radiation belts have some leaks—that is, some charged particles escape from the Van Allen radiation belt of Earth by not turning back at the poles, and instead they enter the atmosphere and collide with air molecules, resulting in auroras (see Figure 1.20). During the Northern Lights, known as aurora Borealis in the Northern Hemisphere (Arctic region), and the Southern (Antarctica region) Lights, known as aurora Australis in the Southern Hemisphere, occur in magnetic poles between 100 kilometers and 1000 kilometers above the Earth's surface (in the ionosphere)—people living in the south and north poles experience aurora regularly. The most suitable time and place to see Earth's aurorae is during long winter nights at high latitudes, which is why they are also known as the polar, northern or southern lights. A big aurora illuminates the dark sky with colorful light rays. The Van Allen belts were formed by the Earth's magnetic field, which deflects charged particles from the solar wind and traps them in the Earth's magnetosphere. These particles are then accelerated by the magnetic field and become trapped in the Van Allen belts.

Figure 6.20. (a) The Van Allen radiation belts are composed of electrons with energies up to several MeV and proton with energies up to several hundred MeV. (b) Aurora as seen here at ground level. (c) Global view of the development of the aurora observed by a spacecraft's camera. (d) A bright, shimmering aurora illuminates a dark sky with bright rays of light. (Credit to the University of Alaska Geophysical Institute.)

Seen from space (Figure 6.20c), the aurora is an oval of glowing light centered on the Earth's magnetic poles, which follow the magnetic sphere. The only part lit up is where solar particles enter the atmosphere. They create a chemical reaction that sheds energy, which we see in the form of light. The most common color is green due to excited oxygen transition. However, the aurora can appear in a variety of colors. For example red and pink auroras are also observed and measurements from space show that the aurora emits X-rays and ultraviolet light too (Figure 6.20b). In other words, the Sun is ultimately responsible for auroras. Their formation is the result of the collisions between billions of charged particles from the solar wind and the atoms and molecules in the planet's atmosphere. In other words, the Sun ultimately causes auroras. They are the result of collisions between billions of charged particles from the solar wind and the atmosphere of the planet's atoms and molecules. These collisions give the sky a colorful glow. Collisions with energized oxygen glow the sky red and green. Collisions with energized nitrogen can produce a blue or crimson color or both at once making the aurora seem purple. Notice also that most of the sky remains dark during an Aurora event. The only part lit up is where solar particles enter the atmosphere. They create a chemical reaction that sheds energy, which we see in the form of light.

At this point in this book series, you probably thought you knew everything about space travel hazards. In fact, there is one hazard we still have to discuss. This is missions passing through the Van Allen belts, which are seas of deadly radiation, on their way to the Moon and/or Mars. In addition to protection materials, the common strategy involves minimizing travel time through the belts, avoiding regions of high-energy and high-density charged particles like the inner belt, and selecting the travel date as much as possible based on the weather. For example, the travel path selected by the Apollo program in the 1970s in Figure 6.20 shows that astronauts avoided the inner belt and the high-intensity parts of the outer belts–the region between the inner and outer Van Allen belts lies at 2 to 4 Earth radii and is considered the safe zone. Apollo astronauts were in and out of Van Allen belts in one to two hours. The total radiation received by the astronauts varied from mission-to-mission but was measured to be between 1.6 and 11.4 milligrays, according to NASA. This is much less than the standard of 50 milligrays per year set by the United States Atomic Energy Commission for people who work with radioactivity. As of yet, these travel paths and doses have not been independently verified. Radiation doses appear to be at least 10 times smaller than one might expect (see Ikelle, 2023b). Anyway, the radial location and intensity of the electron radiation belts are extremely variable, and predicting their variability is one of the major challenges in space travel.

Our current quantitative knowledge of the intrinsic magnetic fields of the other seven major planets of our solar system is that they can likewise be approximated as dipoles with different magnetic field strengths and orientations as we will discuss later. In contrast to Mars ($B_{eq} < 0.5$ nanoTesla) , Venus ($B_{eq} < 2$ nanoTesla), and Mercury $B_{eq} = 330$ nanoTesla), the four most distant planets to the sun, Jupiter ($B_{eq} = 428,000$ nanoTesla), Saturn ($B_{eq} = 21,000$ nanoTesla), Uranus ($B_{eq} = 23,000$ nanoTesla), and Neptune ($B_{eq} = 14,000$ nanoTesla) have significant magnetic field strength, with nanoTesla $= 10^{-9}$ Tesla $= 10^{-5}$ Gauss. Our moon can also be approximated as a dipole with $B_{eq} < 0.2$ nanoTesla. Because Jovian planets have strong magnetic fields, and strong near the magnetic poles, charged particles can also be trapped on field lines that extend from one magnetic pole to the other since they bounce off the high field region near one pole and then bounce off the high field region near the other. High-energy ions and electrons trapped on magnetic field lines in this way comprise the Van Allen belts at Jupiter and Saturn. These radiation belts have been discovered at Jupiter and Saturn by NASA Pioneers 10 (1973) and 11 (1974) and at Uranus by NASA Voyager 2 (1979).

Computational exercise: We use the differential equation solvers in Table 6.6 to predict particle motions in the presence of a dipole magnetic Earth field. Reproduce Figure 6.18.

6.2.5. Auroras on Jupiter

Jupiter's magnetic field, for example, stretches over 4 million miles from its surface. It performs the same function as Earth's magnetic field. Jupiter is shielded from the high-energy charged particles raining from the sun, by its magnetic field, except at its poles where they can enter the atmosphere and light it up just like on Earth.

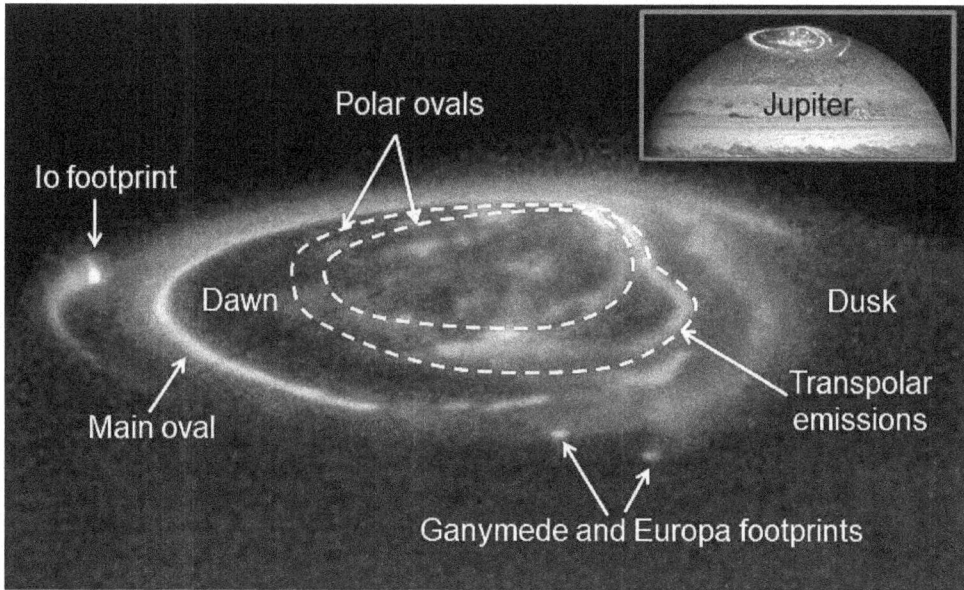

Figure 6.21. UV image of Jupiter's aurora taken by Hubble Space Telescope. It shows the three different emission regions. Main oval co-rotates with Jupiter and it is not Sun-aligned.

In other words, Jupiter also has auroras like Earth. However, they are invisible to the eye because they lie in the UV and X-ray portions of the electromagnetic spectrum. Figure 6.21 is an aurora image of Jupiter taken by NASA/ESA Hubble Space Telescope. Jupiter's aurora cover an area larger than Earth. Moreover, Jupiter's auroras are also hundreds of times more energetic than auroras on Earth and, unlike those on Earth, they never cease.

Notice that Jupiter's auroras are different from those of Earth because Jupiter's auroras have more than one source. Jupiter's strong magnetic field at the poles grabs charged particles from its surroundings. This includes not only the charged particles within the incoming solar wind but also the particles thrown into space by its orbiting moons, especially Io. There are numerous and large volcanoes on this moon, which we will discuss in the next sections.

While Earth and Jupiter display auroral ovals, the circular regions of auroral emission around their magnetic poles, only Jupiter has auroral emission located inside the auroral ovals, commonly known as polar auroras. Figure 6.21 illustrates the three primary types of aurora at Jupiter as noted by Bhardwaj and Gladstone, 2000. These are central emissions (also known as the main auroral oval), two polar emissions near each pole, and outer emissions. The main aurora is oval and steady, co-rotates with the planet, and corresponds to a few hundred kilometers horizontally in Jupiter's atmosphere. Auroral emissions are also observed at the feet of the flux tubes of Io, Europa, and Ganymede. They are made up of moon footprints. Figure 6.21 shows bright spots caused by electric current from Jupiter's moons. While the magnetosphere interaction with Callisto is thought to be much weaker than the other satellites, any Callisto aurora would be difficult to separate from the main aurora. The Io-related aurora includes a "wake" signature that extends around Jupiter. The third type of Jupiter aurora is the highly variable polar aurora that occurs at higher latitudes than the main aurora. When exposed to solar wind, any planet with a significant magnetic field produces auroras.

Quiz: Perpendicular forces lead to drift of the particles; i.e.,

$$
\boldsymbol{v} = \underbrace{v_{\parallel}\boldsymbol{b}}_{\#1} + \underbrace{\boldsymbol{v}_g}_{\#2} + \underbrace{\frac{\boldsymbol{E}\times\boldsymbol{B}}{B^2}}_{\#3} + \underbrace{\frac{m}{qB^2}\frac{d\boldsymbol{E}_{\perp}}{dt}}_{\#4} + \underbrace{\frac{mv_{\parallel}^2 + 0.5mv_{\perp}^2}{qB}\frac{\boldsymbol{B}\times\boldsymbol{\nabla}B}{B^2}}_{\#5},
$$

where the subscripts $\parallel$ and $\perp$ refer to the parallel and perpendicular components with respect to the magnetic field, respectively, $\boldsymbol{b} = \boldsymbol{B}/B$ is the unit vector along the magnetic field, m is the particle mass, q is the particle charge, B is the magnetic field strength, and $\boldsymbol{v}_g$ is the rotating perpendicular velocity.
Define the drift terms #1, #2, #3, #4, and #5.

6.2.6. Fusion Reactors by Magnetic Confinement

As we discussed in the previous two books of this book series, nuclear fusion is the most basic energy production mechanism in the universe. A starry night displays of countless fusion reactors, burning their hydrogen nuclei and converting them into helium, releasing energy in the process. The energy released by our own star, the sun, is what makes our Earth habitable and our existence possible. Out of all possible fusion reactions the one with the largest cross-section is that of the reaction between deuterium ($_1^2$D) and tritium ($_1^3$T); i.e.,

$$
_1^2\text{D} + {}_1^3\text{T} \rightarrow {}_2^4\text{He}\,(3.5\,\text{Mev}) + {}_0^1\text{n}\,(14.1\,\text{Mev})\,, \tag{6.66}
$$

where $_1^2$D $= {}_1^2$H and $_1^3$T $= {}_1^3$H (two isotopes of hydrogen). Basically, the combination of these isotopes of hydrogen produces helium atoms which carry 3.5 MeV energy and free neutrons having energies of 14.1 MeV. The classical approach to achieving this fusion reaction is to confine ions in environments that permit highly dense plasma at temperatures of the order of 100 million degrees Celsius. There are three ways to confine plasma: gravitational confinement described in Ikelle (2023b) and Chapter 1, inertial confinement that we will describe in the next chapter, and magnetic confinement, which is our focus in this subsection. Most fusion tests on Earth have been limited to magnetic confinement.

As described earlier, in magnetic confinement, charged particles are confined by a magnetic field of appropriate geometry and strength. Such confinement is indifferent to heat action and does not contain impurities (i.e., almost no leakage and no impurities so that the fusion reaction can occur). Constructing a magnetic confinement for fusion reactors is challenging because it must fulfill these two conditions. We cannot confine particles in a straight magnetic field since particles can move freely along them even if not across them. The only choice left is to bend the magnetic field lines so that they have no ends. One of the basic geometries of a magnetic field that has developed to confine plasma for fusion is *tokamak* (see

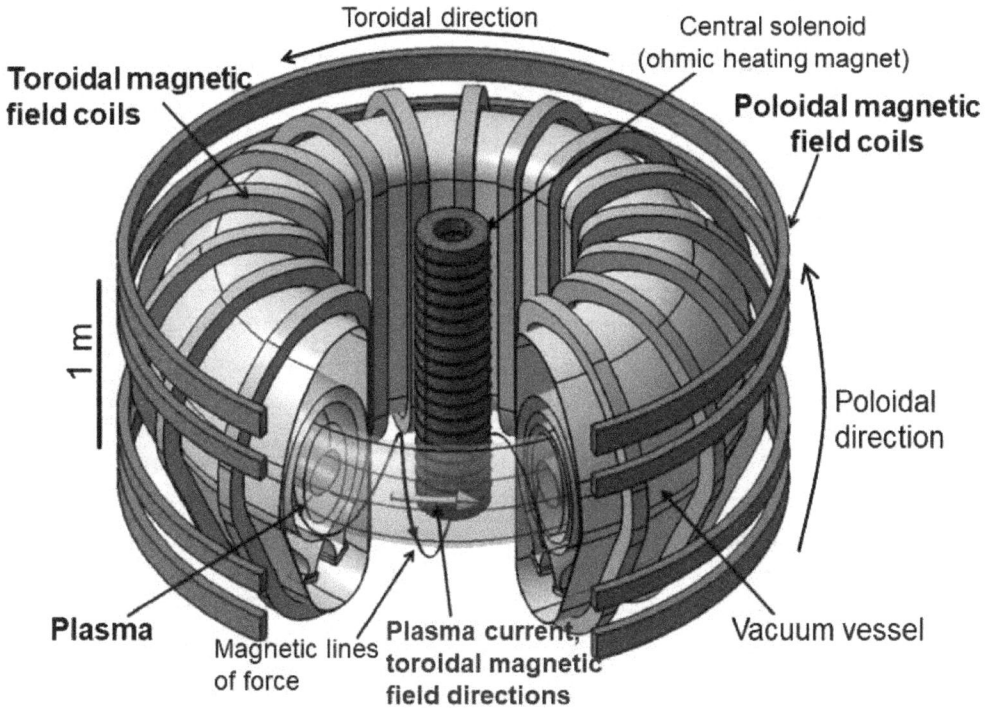

Figure 6.22. Schematic of a tokamak fusion reactor along with its poloidal and toroidal field coils, ohmic coils, vacuum vessel and nested plasma flux surfaces with magnetic field lines. There are two types of magnetic fields: toroidal fields generated by coils, and poloidal fields generated by plasma current. Poloidal field coils control plasma position and shape. Adapted from Kikuchi et al. (2020).

Figure 6.22). A tokamak stands for "magnetic toroidal chamber" in Russian. Tokamaks were initially conceptualized in the 1950s by Soviet physicists Igor Tamm and Andrei Sakharov. The first working tokamak was attributed to the work of Natan Yavlinsky—that is, he created an interface between a 100 million-degree plasma and a room temperature wall. In a tokamak the magnetic field lines bend into circles so that they have no ends; that is the idea behind tokamaks.

As illustrated in Figure 6.22, tokamaks are torus(donut)-shaped fusion facilities where the charged particles are confined using very strong magnetic fields in a vacuum with the magnetic coils arranged in a closed ring thus avoiding end-losses. They contain powerful electromagnets that generate a very strong magnetic field. This main field component in a tokamak is the field in the toroidal direction (around the torus) which is generated by a set of identical toroidal field coils (arranged in the poloidal plane, i.e. the plane perpendicular to the toroidal direction). Usually 16 to 32 toroidal coils are employed. They have rotational symmetry around the vertical axis (which is here chosen to be the z-axis). Alone, the toroidal field cannot totally confine a plasma because the charged particles drift due to the magnetic field gradients and curvature which cause opposite vertical drifts for the differently charged species—ions drift up and electrons down. Electric fields result from the separation of charges and the loss of plasma. Before we discuss remedying this problem, let's get a better understanding of these drifts.

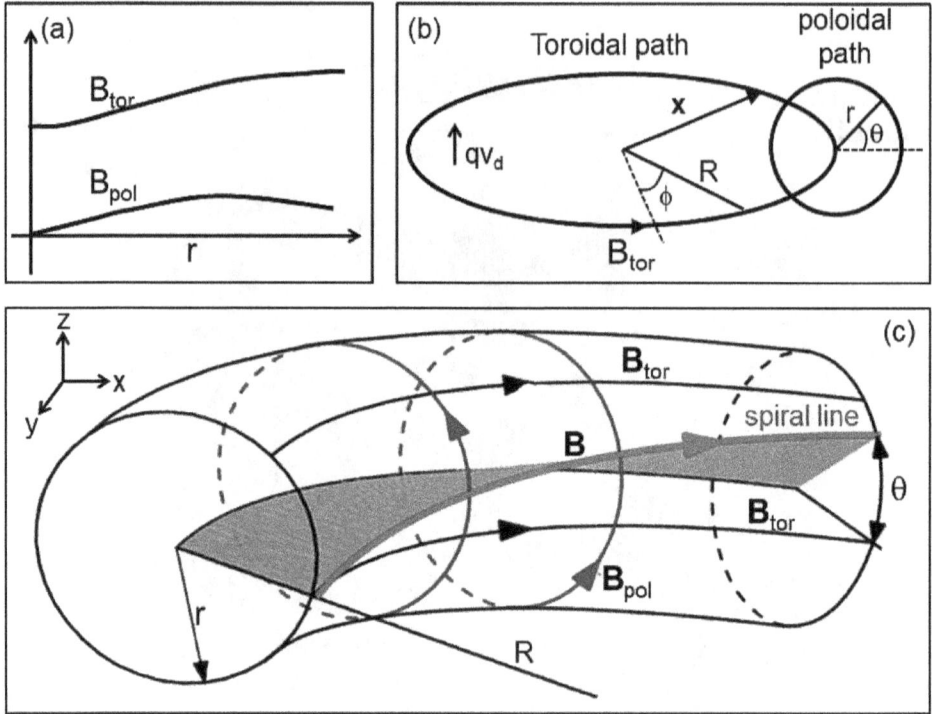

Figure 6.23. (a) Toroidal and poloidal magnetic field. (b) Toroidal and poloidal field lines. (c) Coordinate system, (r, θ, ϕ), for plasma quantities and tokamak field lines. r is the radius, θ is the poloidal angle, ϕ is the toroidal angle, $x = (R + r \cos \theta) \cos \phi$, $y = (R + r \cos \theta) \sin \phi$, and $z = r \sin \theta$.

Basically, the toroidal magnetic field is curved and its strength is not homogeneous. The vertically projected view of the tokamak can be described as a circular curve in the z-plane with the center of the circle at $R = 0$ (see Figure 6.23). By integrating the Maxwell equation (Ampère's law) over the enclosed surface (S), we arrive at

$$\boldsymbol{\nabla} \times \boldsymbol{B}_{\text{tor}} = \mu_0 \boldsymbol{J} \Longrightarrow \int_S \boldsymbol{n} \cdot (\boldsymbol{\nabla} \times \boldsymbol{B}_{\text{tor}}) \, dA = \mu_0 \int_S \boldsymbol{n} \cdot \boldsymbol{J} \, dA$$

or

$$2\pi R B_{\text{tor}} = \mu_0 I \ , \tag{6.67}$$

where $\boldsymbol{n}$ is the unit vector perpendicular to the surface and $\boldsymbol{B} = B_{\text{tor}} \boldsymbol{i}_\phi$, the Stokes theorem has been applied to transform the surface integral into an integral along the curve, B_{tor} is the toroidal component of the magnetic field (i.e., the component in the direction of the unit vector $\boldsymbol{i}_r$), and I is the total current flowing through the surface. This current is due to the current in the toroidal field coils that cross the surface S. We assume that I is constant so that $B_{\text{tor}} = C/R$, where $C = \mu_0 I/(2\pi)$ is constant, thus the magnetic field varies as $1/R$. The gradient of the toroidal magnetic field is

$$\boldsymbol{\nabla} B_{\text{tor}} = \boldsymbol{\nabla} \left(\frac{C}{R} \right) = -\frac{C}{R^2} \boldsymbol{i}_r = -\frac{B_{\text{tor}}}{R} \boldsymbol{i}_r \Longrightarrow \frac{\boldsymbol{B}_{\text{tor}} \times \boldsymbol{\nabla} B_{\text{tor}}}{B_{\text{tor}}^2} = \frac{1}{R} \boldsymbol{i}_r \ .$$

Thus the toroidal-magnetic field inhomogeneity causes the following drifts; i.e.,

$$
\begin{aligned}
\boldsymbol{v}_d &= \frac{mv_\parallel^2 + 0.5\,mv_\perp^2}{qB_{\text{tor}}}\frac{\boldsymbol{B}_{\text{tor}}\times\boldsymbol{\nabla}B_{\text{tor}}}{B_{\text{tor}}^2} \\
&= \frac{mv_\parallel^2 + 0.5\,mv_\perp^2}{qB_{\text{tor}}R}[\boldsymbol{i}_\phi\times(-\boldsymbol{i}_r)] = \frac{mv_\parallel^2 + 0.5\,mv_\perp^2}{qB_{\text{tor}}R}\boldsymbol{i}_z .
\end{aligned}
\tag{6.68}
$$

This drift leads to charge separation. Because the sign of drift depends on the sign of the charge q, ions and electrons drift parallel to the symmetry axis, in opposite directions, thus leaving their magnetic field line and to an additional vertical electric field. A second drift appears now in the crossed-field arrangement of vertical electric and toroidal magnetic fields. In other words, an $\boldsymbol{E}\times\boldsymbol{B}_{\text{tor}}$ drift appears perpendicular to both $\boldsymbol{E}_{\text{tor}} = -E_{\text{tor}}\boldsymbol{i}_z$ and $\boldsymbol{B}_{\text{tor}} = B_{\text{tor}}\boldsymbol{i}_\phi$ directions. This drift can be expressed as follows:

$$
\boldsymbol{v}_E = \frac{\boldsymbol{E}_{\text{tor}}\times\boldsymbol{B}_{\text{tor}}}{B_{\text{tor}}^2} = -\frac{E_{\text{tor}}}{B_{\text{tor}}}(\boldsymbol{i}_z\times\boldsymbol{i}_\phi) = -\frac{E_{\text{tor}}}{B_{\text{tor}}}\boldsymbol{i}_r .
\tag{6.69}
$$

It is directed outward and thus causes the plasma torus to expand radially. So the drifts in (6.68), associated with the gradient of $\boldsymbol{B}_{\text{tor}}$, set up an electric field and this electric field results in $\boldsymbol{E}\times\boldsymbol{B}_{\text{tor}}$ drifts, which drive plasma (electrons and ions) to the chamber wall, destroying the confinement.

Let us now turn our attention to the solution to the toroidal confinement problem. We adopt the idea of preventing the charge separation resulting from gradient-B drifts and thereby avoid the $\boldsymbol{E}_{\text{tor}}\times\boldsymbol{B}_{\text{tor}}$ drift. The gradient-B drifts and resulting charge separation can be canceled out by twisting the toroidal field lines to form helices—that is, the twisting of the field lines in Figure 6.23. The basic idea is that the tokamak device includes a poloidal magnetic field along the toroidal magnetic field. The net magnetic field is $\boldsymbol{B} = \boldsymbol{B}_{\text{tor}}+\boldsymbol{B}_{\text{pol}}$, where $\boldsymbol{B}_{\text{pol}} = B_{\text{pol}}\boldsymbol{i}_\theta$ is the poloidal magnetic field. Let us verify that we indeed have that we have spiral magnetic field lines by introducing $\boldsymbol{B}_{\text{pol}}$. The equation of motion of a particle generated by a poloidal magnetic field is:

$$
r\frac{d\theta}{dt} = \frac{B_{\text{pol}}}{B_{\text{tor}}}v_\phi = \frac{B_{\text{pol}}}{B_{\text{tor}}}\frac{B_{\text{tor}}}{B}v_\parallel = \frac{B_{\text{pol}}}{B_{\text{pol}}}v_\parallel
\tag{6.70}
$$

and $r = $ constant, where $B^2 = B_{\text{pol}}^2 + B_{\text{tor}}^2$. Now add on to this motion the cross field drift in the $\boldsymbol{i}_z$ direction; i.e.,

$$
r\frac{d\theta}{dt} = \frac{B_{\text{pol}}}{B}v_\parallel + v_d\cos\theta \quad\text{and}\quad \frac{dr}{dt} = v_d\sin\theta .
\tag{6.71}
$$

By eliminating time from these two equations we arrive at

$$
\frac{dr}{dt}\frac{dt}{d\theta} = \frac{dr}{d\theta} \implies \frac{dr}{r} = \frac{v_d\sin\theta}{(B_{\text{pol}}/B)v_\parallel + v_d\cos\theta}d\theta .
$$

Let us assume B_θ, B, $v_\parallel$, v_d are constants. By integrating the above orbit equation, we obtain

$$
\ln r = -\ln\left|\frac{B_{\text{pol}}}{B}v_\parallel + v_d\cos\theta\right| + C .
\tag{6.72}
$$

where C is the constant of integration. By setting $r = r_0$ when $\theta = \pi/2$, we arrive at

$$r = r_0 \left[1 + \frac{B v_d}{B_{\mathrm{pol}} v_\parallel} \cos\theta \right]^{-1} \approx r_0 - \Delta_t \cos\theta \, , \qquad (6.73)$$

where $\Delta_t = (B v_d / B_{\mathrm{pol}}) v_\parallel r_0$. The approximation here is that $\Delta_t / r_0 \ll 1$. Provided Δ / r_0 is small, particles will be confined.

To summarize, the development of a tokamak comes down to a construct of toroids with field lines, which encircle the tori toroidally and poloidally and with a field line density, which is higher on the inside than on the outside (torus effect) so that charged particles (electrons and ions) that form the plasma spiral along magnetic field lines. As a result of trapped particles in magnetic confinement, more nuclear collisions occur, increasing fusion efficiency. The primary difficulty in magnetic confinement construction is reducing plasma leakages to insignificant levels.

Note that, as described in Ikelle (2023b), helium-3 can be fused with deuterium to form helium-4 and an energetic proton that can be used to produce electrical power. The challenge is obtaining helium-3 which is rare on Earth but plentiful in the gas giants, Jupiter, Saturn, Uranus, and Neptune and in Earth's Moon.

Box 6.2: Some Nuclear Fusion Reactions

As discussed in Ikelle (2023b) and above, whenever the heavier element has a lower potential energy compared to the sum of potential energies of two separate nuclei, the fusion reaction is plausible. According to the experiment, iron with atomic number 26 has the lowest level of potential energy, so it would be the most stable nucleus. This shows that the fusion of lighter elements than iron always generates energy. But the released energy depends on the reaction cross section as well as the energy obtained from every individual reaction:

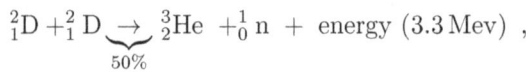

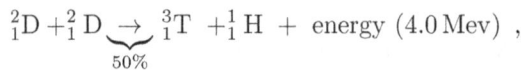

$$^2_1\mathrm{D} + {}^3_1\mathrm{T} \rightarrow {}^4_2\mathrm{He} + {}^1_0\mathrm{n} \ (14.1 + \text{ energy } (17.6\,\mathrm{Mev}) \, ,$$

$$^2_1\mathrm{D} + {}^2_1\mathrm{D} \underset{50\%}{\rightarrow} {}^3_2\mathrm{He} + {}^1_0\mathrm{n} + \text{ energy } (3.3\,\mathrm{Mev}) \, ,$$

$$^2_1\mathrm{D} + {}^2_1\mathrm{D} \underset{50\%}{\rightarrow} {}^3_1\mathrm{T} + {}^1_1\mathrm{H} + \text{ energy } (4.0\,\mathrm{Mev}) \, ,$$

$$^2_1\mathrm{D} + {}^3_2\mathrm{He} \rightarrow {}^1_1\mathrm{H} + {}^4_2\mathrm{He} + \text{ energy } (18.4\,\mathrm{Mev}) \, ,$$

$$^3_2\mathrm{He} + {}^3_2\mathrm{He} \rightarrow 2{}^1_1\mathrm{H} + {}^4_2\mathrm{He} + \text{ energy } (12.9\,\mathrm{Mev}) \, ,$$

$$^2_1\mathrm{D} + {}^6_3\mathrm{Li} \rightarrow 2{}^4_2\mathrm{He} + \text{ energy } (22.4\,\mathrm{Mev}) \, ,$$

$$^1_1\mathrm{H} + {}^{11}_5\mathrm{B} \rightarrow 3{}^4_2\mathrm{He} + \text{ energy } (8.7\,\mathrm{Mev}) \, .$$

6.3. Magnetized Fluids: Magnetohydrodynamics

As described in the previous section, charged particles and their motions are fundamental to understanding electric systems in deep space and the electric and magnetic fields around us. In the previous section, we adopted a more kinetic view of
planetary ion acceleration by treating particles in deep space as single particles in
a collisionless medium. In a collisionless plasma, particle motion takes place on a
much smaller scale than the typical distance traveled by a particle before its direction of travel is significantly changed due to particle collisions (i.e., a mean-free
path). Plasma in dark matter and intracluster medium of galaxy clusters can be
considered collisionless. In a collisional plasma, particle motion takes place on a
scale much larger than the mean-free path. All fluids considered in this section are
collisional.

So, here, we consider the charged particles and their motions as electrically
conducting fluids in motion. There are so many charged particles, in the order of
billions in deep space, that they behave on a macroscopic scale as an electrically
conducting plasma fluid that is driven by magnetic fields. Moreover, this plasma
model assumes that collisions between particles are frequent (see Figure 6.23). The
word *magnetohydrodynamics* (MHD) is used for phenomena where the magnetic
field induces the motion, including the velocity field, of an electrically conducting
and non-magnetic fluid. An electrically conducting plasma fluid in the presence of
a magnetic field is one such phenomenon. So, we here treat these charged particles
and their motions as electrically conducting fluid in motion. There are so many
charged particles, in the order of billions in deep space, that they behave on a
macroscopic scale as electrically conducting plasma fluid that is driven by magnetic
fields. Moreover, this plasma model assumes that collisions between particles are
frequent (see Figure 6.24).

Besides deep-space exploration, there are many applications for MHD on the
Earth's surface. In the metallurgical industry, magnetic fields are used to stir,
pump, levitate, and heat liquid metals because of the high conductivity of metals.
There is no need for standard engines. Some modern vessels and submarines benefit
from the MHD concept by replacing oil/gas-based engines with magnetic fields and
using conducting fluids such as hydrogen fluids. This combination is also the basis
of some future aircraft and missile systems.

The MHD is based on the basic physics laws of fluid motion described in Ikelle
(2023c). It establishes a coupling between the laws of fluid motion and Maxwell's
equations. We here describe this coupling, along with the MHD main result, which
is that magnetic fields can induce electrical currents in a conductive fluid. These
currents apply forces to the fluid and change the original magnetic field. We do
not assume that the plasma is collisionless. Actually, the fluid model of a plasma is
most appropriate when a plasma is collisional.

> **Quiz:** What is the difference between a gas and a plasma?

6.3.1. MHD Equations

Various plasmas and their length and time scales. Before we review the
fundamental equations of MHDs, let us describe the various plasmas and their time

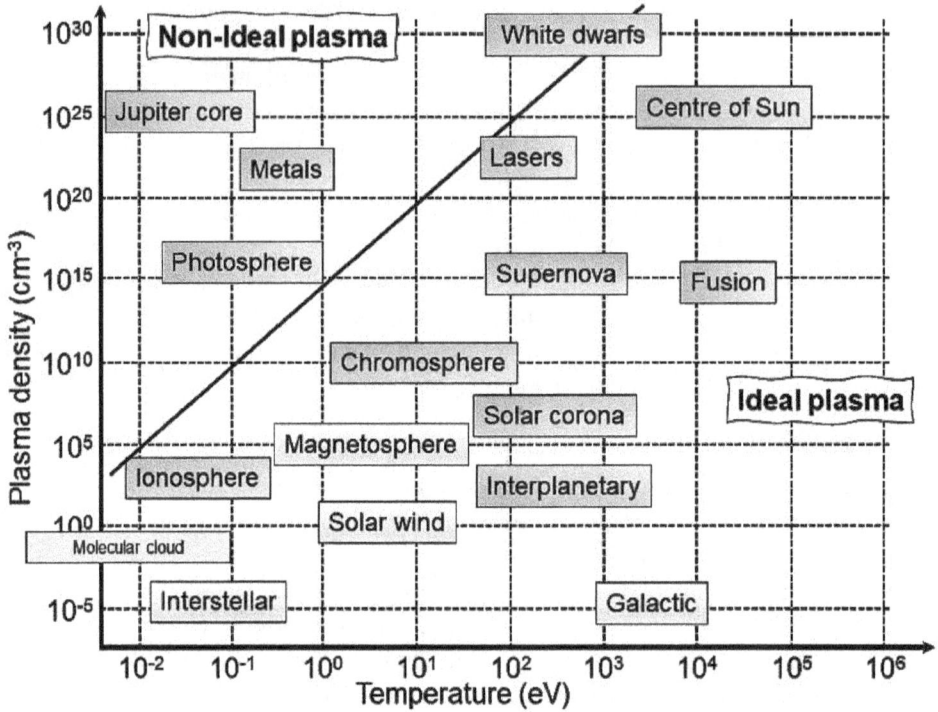

Figure 6.24. Several plasmas are shown relative to number-density and temperature. An ideal plasma is one in which Coulomb collisions are negligible, otherwise the plasma is non-ideal. [Adapted from Myers (1978) and Saigo (2000)].

and length scales. Again, a plasma is an ionized fluid consisting of positive ions and free electrons in proportions resulting in more or less no overall electric charge, typically at low pressures (as in the upper atmosphere and in fluorescent lamps), as described in https://home.cern/science. As illustrated in Figure 6.24, the plasmas in our galaxy vary greatly. Cold interstellar plasma forms molecular clouds ($T \sim 10^{-3}$ electronVolt or 11.6 degrees Kelvin) and diffuse clouds ($T \sim 10^{-2}$ electronVolt or 116 degrees Kelvin). Interstellar plasma and more generally the interstellar medium is the space between stars.

The time scales of MHD plasmas are longer than the inverses of the plasma frequencies and cyclotron frequencies for both ions and electrons: $1/\tau \gg \omega_{pe}, \omega_{pi}, \Omega_{ce}, \Omega_{ci}$, where ω_{pe} and ω_{pi} are plasma frequencies for electrons and ions, respectively, and Ω_{ce} and Ω_{ci} are cyclotron (gyro- or Larmor) frequencies for electrons and ions, respectively. Again the plasma and cyclotron frequencies are

$$\omega_{p\alpha} = \left(\frac{4\pi n_\alpha q_\alpha^2}{m_\alpha}\right)^{1/2} \quad \text{and} \quad \Omega_\alpha = \frac{q_\alpha B}{m_\alpha}, \tag{6.74}$$

respectively, where α represents the various species with $q_\alpha = e$ for electrons and $q_\alpha = Ze$ for ions. The total plasma frequency for a two-component plasma is defined as $\omega_p^2 = \omega_{pe}^2 + \omega_{pi}^2$. Since for most plasmas in nature $\omega_{pe} \gg \omega_{pi}$, we have $\omega_p \approx \omega_e$.

The Debye length tells us how far the electric field of a charged particle can be felt by other particles. A positively charged ion, for example, tends to attract electrons. MHD is valid on length scales longer than the Debye length and electron/ion gyroradii (Larmor radii): $L \gg \lambda_D, r_{Le}, r_{Li}$, where λ_D is Debye length, and r_{Le} and r_{Li} are gyroradii of electrons and ions, respectively. The Debye length and gyroradii are defined as

$$\lambda_D = \left(\frac{T}{4\pi e^2 n_e}\right)^{1/2} = \left(\frac{\lambda_e^2 \lambda_i^2}{\lambda_e^2 + \lambda_i^2}\right)^{1/2} , \quad \lambda_\alpha = \left(\frac{T_\alpha}{4\pi e^2 n_e}\right)^{1/2}$$

and

$$n_\alpha = \frac{\text{number of } \alpha-\text{particles}}{\text{volume (cm}^3}) \quad \text{and} \quad r_{L\alpha} = \frac{m_\alpha v_\alpha}{q_\alpha B} ,$$

where α represents the various species, n_α is the number density in cm^{-3}, and T is the temperature in degrees Kelvin.

In Figure 6.24, we have distinguished ideal and non-ideal plasmas. In the non-ideal plasma model, the plasma can be viscous and/or electrically resistive. In the ideal plasma model, we assume that these two effects are negligible—that is, the plasma is a nonviscous and perfectly conducting medium.

Continuity equation: conservation of mass. Let us now turn to the MHD equations. We start by recalling the mass conservation equation described in Ikelle (2023c). It is:

$$\frac{\partial \rho(\boldsymbol{x}, t)}{\partial t} + \boldsymbol{\nabla} \cdot (\rho(\boldsymbol{x}, t)\boldsymbol{v}(\boldsymbol{x}, t)) = 0 , \tag{6.75}$$

where ρ is the mass density and $\boldsymbol{v}$ is the fluid velocity vector.

Momentum equation (now includes the Lorentz force). Consider plasma made of electrons with mass m_e and charge $q_e = -e$, and ions with mass m_i and charge $q_i = e$. As described in the first section of this chapter, we can define the charge and current density of this plasma as follows:

$$\rho_E = e(Z_i n_i - n_e) \quad \text{and} \quad \boldsymbol{J} = e(Z_i n_i \boldsymbol{v}_i - n_e \boldsymbol{v}_e) , \tag{6.76}$$

where n_e is the electron particle density, n_i is the ion particle density, Z_i is the ion atomic number, v_e is the electron velocity, and v_e is the ion velocity. One can treat the plasma as two fluids with one fluid describing the flow of ions and the other fluid describing the flow of electrons. In deep-space plasmas, it is appropriate to consider the entire plasma as a single fluid because the two flows are slow and move and act in unison. Electrons and ions are forced to act in unison, because of frequent collisions and/or by the action of a strong external magnetic field.

We here assume that quasineutrality applies, i.e., $n = n_e = Z_i n_i$. This assumption is realistic for deep space plasma and it implies that charges vanish, $\rho_E = 0$, but the plasma carries a finite current. We also describe the entire plasma as a single fluid. The mean mass of particles and velocity of the single fluid are defined as follows:

$$m = m_i + m_e \quad \text{and} \quad \boldsymbol{v} = \frac{m_i \boldsymbol{v}_i + m_e \boldsymbol{v}_e}{m_i + m_e} . \tag{6.77}$$

The plasma motion is governed by Newton's second law as follows:

$$\left(\frac{\partial}{\partial t} + \boldsymbol{v}(\boldsymbol{x},t) \cdot \boldsymbol{\nabla}\right) \rho(\boldsymbol{x},t)\boldsymbol{v}(\boldsymbol{x},t) = \boldsymbol{F}(\boldsymbol{x},t) , \tag{6.78}$$

where $\rho = \rho(\boldsymbol{x},t) = m\,n(\boldsymbol{x},t)$ is the plasma mass density, $\boldsymbol{v} = \boldsymbol{v}(\boldsymbol{x},t)$ is the plasma velocity, and $\boldsymbol{F} = \boldsymbol{F}(\boldsymbol{x},t)$ is the force per unit volume acting on the plasma element. Examples of $\boldsymbol{F}$ include the Lorentz force, gravitational force, and pressure gradient force. The MHD motion equation is

$$\underbrace{\left[\frac{\partial}{\partial t} + \boldsymbol{v} \cdot \boldsymbol{\nabla}\right](\rho\boldsymbol{v})}_{\text{momentum density}} = -\underbrace{\boldsymbol{\nabla}p}_{\text{pressure}} - \underbrace{\rho\boldsymbol{\nabla}\Psi}_{\text{gravity}} + \underbrace{\rho_E \boldsymbol{E} + \boldsymbol{J} \times \boldsymbol{B}}_{\text{Lorentz force}} , \tag{6.79}$$

where ρ_E is the electrical charge density, $\boldsymbol{J}$ is the plasma electric current-density vector, $\boldsymbol{B}$ is the magnetic field, $\boldsymbol{E}$ is the electric field, Ψ is gravitational potential (with $\boldsymbol{\nabla}^2\Psi = 4\pi G_0\rho$), and p is the thermodynamic pressure (also called *gas pressure* or *kinetic pressure*, and here as *plasma pressure*), which is the sum of the separate pressures of the ions and electrons, i.e.,

$$p = p_e + p_i ,$$

rather than on each species separately, p_i are pressures of the ions, and p_e are pressures of the electrons. These pressures push plasma from regions of high-plasma pressure to low-plasma pressure. Fluids in which the gravitational potential Ψ is nonzero are characterized as *self-gravitating*. For example, stars and galaxies are self-gravitational. By using Ampère's law without displacement currents (that is, we dropped the displacement current, $\partial\boldsymbol{E}/\partial t$, because its effects are small), i.e.,

$$\boldsymbol{\nabla} \times \boldsymbol{B} = \mu_0 \boldsymbol{J} \Longrightarrow \boldsymbol{J} \times \boldsymbol{B} = -\frac{1}{\mu_0}\boldsymbol{B} \times (\boldsymbol{\nabla} \times \boldsymbol{B})$$

$$= \underbrace{\frac{(\boldsymbol{B} \cdot \boldsymbol{\nabla})\boldsymbol{B}}{\mu_0}}_{\text{tension}} - \boldsymbol{\nabla}\underbrace{\left[\frac{B^2}{2\mu_0}\right]}_{\text{pressure}} , \tag{6.80}$$

where $B^2 = \boldsymbol{B} \cdot \boldsymbol{B}$, $B^2/2\mu_0$ is magnetic energy density, which is generally thought of as magnetic pressure, and $(\boldsymbol{B} \cdot \boldsymbol{\nabla})\boldsymbol{B}/\mu_0$ is the magnetic curvature force, which is generally thought of as a tension that tries to straighten curved field lines. Magnetic tension force acts toward the center of curvature of field lines, as illustrated in Figure 6.25. So the Lorentz force equals magnetic pressure plus magnetic tension. Thus, the parameter β, defined as

$$\beta = \frac{\text{plasma pressure}}{\text{magnetic pressure}} = \frac{\text{thermal pressure}}{\text{magnetic pressure}} = \frac{p}{B^2/(2\mu_0)} , \tag{6.81}$$

can be employed to determine the importance of these terms in the Lorentz force and total pressure = magnetic pressure + thermal pressure = $p + B^2/2\mu_0$. If $\beta \ll 1$, the

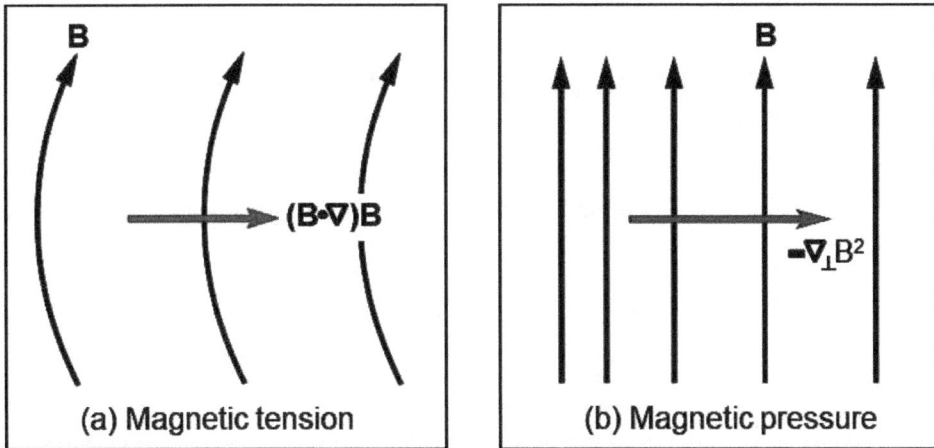

Figure 6.25. The decomposition of the Lorentz force into (a) magnetic tensor (i.e., $-B^2\boldsymbol{x}/(4\pi|\boldsymbol{x}|^2)$) and (b) magnetic pressure sounds (i.e., $\boldsymbol{\nabla}_\perp(B^2/(8\pi))$), where B is the magnitude of the magnetic field vector.

magnetic field dominates—that is the case for solar corona and tokamaks ($\beta < 0.1$). If $\beta \gg 1$, plasma pressure forces dominate—that is the case in the stellar interior. If $\beta \sim 1$, pressure/magnetic forces are both important—that is the case for the solar chromosphere, and parts of the solar wind and interstellar medium. Most modern reconstructions of the coronal magnetic field assume that the magnetic pressure force and magnetic tension force are balanced (i.e., the Lorentz force vanishes in the Corona).

Quiz: (a) Show that for solar wind, the ratio of the gas pressure gradient term and the Lorentz force known as the plasma-β is

$$\beta = \frac{3.5 \times 10^{-21}nT}{B^2}$$

where n is in m^{-3}, T in degrees Kelvin and B in Gauss. Determine β in the solar Corona, photospheric magnetic flux tubes, and the solar wind near Earth's orbit. (b) In a configuration with a magnetic field of 5 Tesla, what is the pressure exerted on it? (c) Determine the magnetic pressure and magnetic tension of $\boldsymbol{B}(x) = [0, \exp(-x^2), 0]^T$. (d) Determine the magnetic pressure and magnetic tension of $\boldsymbol{B}(y) = [2y, 1, 0]^T$

In (6.80), we used the identity

$$(\boldsymbol{\nabla} \times \boldsymbol{B}) \times \boldsymbol{B} = -\boldsymbol{\nabla}\left(B^2\right) + (\boldsymbol{B} \cdot \boldsymbol{\nabla})\boldsymbol{B} . \tag{6.82}$$

The assumption that the MHD plasma is quasineutral (i.e., $\rho_E = 0$) allows us to reduce equation (6.79) to

$$\left[\frac{\partial}{\partial t} + \boldsymbol{v} \cdot \boldsymbol{\nabla}\right](\rho\boldsymbol{v}) = -\boldsymbol{\nabla}p - \rho\boldsymbol{\nabla}\Psi - \frac{1}{\mu_0}\boldsymbol{B} \times (\boldsymbol{\nabla} \times \boldsymbol{B}) \tag{6.83}$$

or

$$\left[\frac{\partial}{\partial t} + \boldsymbol{v} \cdot \boldsymbol{\nabla} \right] (\rho \boldsymbol{v}) = -\boldsymbol{\nabla} p - \rho \boldsymbol{\nabla} \Psi + \frac{(\boldsymbol{B} \cdot \boldsymbol{\nabla}) \boldsymbol{B}}{\mu_0} - \boldsymbol{\nabla} \left(\frac{\boldsymbol{B}^2}{2\mu_0} \right) . \qquad (6.84)$$

By considering $\boldsymbol{B}$ as the source, we can generate plasma motion, which describe the forward time evolution of the plasma quantities ρ, $\boldsymbol{v}$, Ψ, and p.

The induction equation shows that the magnetic field also evolves under the influence of the plasma velocity $\boldsymbol{v}$. Let us derive this evolution equation. Ohm's law, $\boldsymbol{J} = \sigma \boldsymbol{E}$, is actually only valid when the electric field is experienced by the plasma element in its rest frame. For moving plasma at velocity $\boldsymbol{v}$, Ohm's law is (Spitzer, 1962)

$$\boldsymbol{J} = \sigma \left(\boldsymbol{E} + \boldsymbol{v} \times \boldsymbol{B} \right) . \qquad (6.85)$$

We will revisit this law in the section on generalized Ohm's law. By combining this equation with Faraday's law, we arrive at the induction equation for resistive MHD

$$
\begin{aligned}
\frac{\partial \boldsymbol{B}}{\partial t} &= -\boldsymbol{\nabla} \times \boldsymbol{E} \\[2mm]
&= \boldsymbol{\nabla} \times (\boldsymbol{v} \times \boldsymbol{B}) - \boldsymbol{\nabla} \times \left(\frac{1}{\sigma} \boldsymbol{J} \right) \\[2mm]
&= \underbrace{\boldsymbol{\nabla} \times (\boldsymbol{v} \times \boldsymbol{B})}_{\text{induction (generation)}} - \underbrace{\boldsymbol{\nabla} \times \left(\frac{1}{\mu_0 \sigma} \boldsymbol{\nabla} \times \boldsymbol{B} \right)}_{\text{diffusion (dissipation)}} \\[2mm]
&= \boldsymbol{\nabla} \times (\boldsymbol{v} \times \boldsymbol{B}) - \frac{1}{\mu_0 \sigma} \boldsymbol{\nabla}^2 \boldsymbol{B} ,
\end{aligned}
\qquad (6.86)
$$

where we allow σ to vary in space. We used the vector identity

$$\boldsymbol{\nabla} \times \boldsymbol{\nabla} \times \boldsymbol{B} = \boldsymbol{\nabla} (\boldsymbol{\nabla} \cdot \boldsymbol{B}) - \boldsymbol{\nabla}^2 \boldsymbol{B} . \qquad (6.87)$$

The first equality, in (6.86), is Faraday's law, the second equation is obtained by substituting (6.85) in Faraday's law, and the third equality is obtained by using Ampère's law without the displacement current, which is small, as mentioned earlier. Now, if $\boldsymbol{B}$ is a solution to the induction MHD equations, so is $-\boldsymbol{B}$. Regardless of the flow field, initial conditions, or boundary conditions, there is no preference for *normal* versus *reversed* field polarity. If σ is a constant and $\sigma \to \infty$ (i.e., plasma is considered a perfect conductor and therefore collisionless), (6.86) becomes

$$\frac{\partial \boldsymbol{B}}{\partial t} = \boldsymbol{\nabla} \times (\boldsymbol{v} \times \boldsymbol{B}) . \qquad (6.88)$$

So velocity $\boldsymbol{v}$ changes, in (6.83), by the action of the magnetic field. Now the magnetic field also changes by the action of plasma velocity (i.e., plasmas can generate their own magnetic fields) , as described in (6.88). In other words, in a perfectly conducting fluid like in the Earth's outer core and space plasma, the

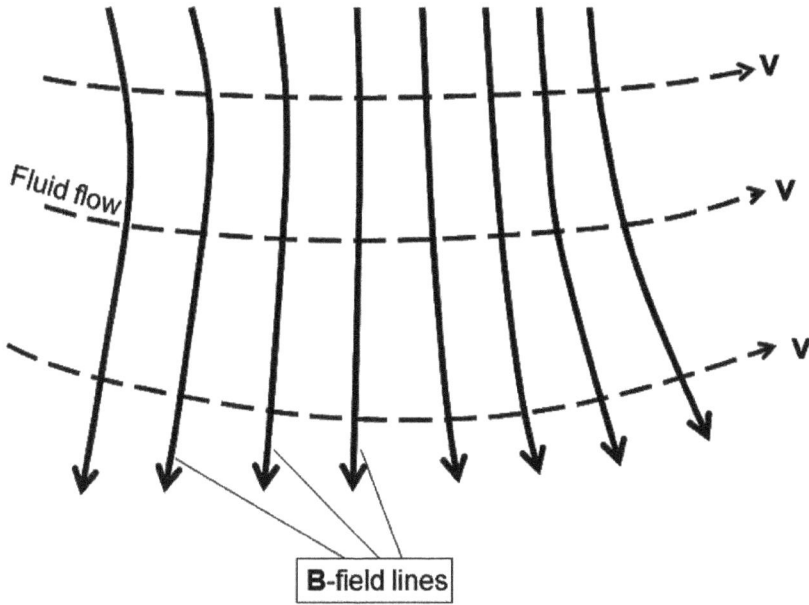

Figure 6.26. In ideal MHD, magnetic field lines move with the fluid: Magnetic field is frozen to the plasma. The topology/connectivity of the magnetic field is invariant.

field lines are stuck to the material and move as the material moves, i.e., "the field lines are frozen into the plasma" (Alfvén, 1943), as illustrated in Figure 6.26. The "frozen-in-flow" approximation implies that the magnetic field remains frozen to the perfectly conducting plasma or outer core. This approximation is also known as collisionless MHD or ideal MHD. When σ is finite, the field can move through the material, causing electric currents to be induced.

The fact that the magnetic flux is conserved in the plasmas is the explicit proof of the frozen-in-flow condition—that is, the rate of change of the flux is zero; i.e.,

$$
\begin{aligned}
\frac{D\Phi}{Dt} &= \frac{D}{Dt} \int_S \boldsymbol{B} \cdot d\boldsymbol{S} \\
&= \int_S \frac{\partial \boldsymbol{B}}{\partial t} \cdot d\boldsymbol{S} + \oint_C \boldsymbol{B} \cdot (\boldsymbol{v} \times d\boldsymbol{l}) \\
&= -c \int_0 \boldsymbol{\nabla} \times \boldsymbol{E} d\boldsymbol{S} - \oint_C (\boldsymbol{v} \times \boldsymbol{B}) \cdot d\boldsymbol{l} \\
&= -c \oint_C \boldsymbol{E} \cdot d\boldsymbol{l} + c \oint_C \boldsymbol{E} \cdot d\boldsymbol{l} = 0 \,,
\end{aligned} \tag{6.89}
$$

where (see next section)

$$
\frac{D}{Dt} = \frac{\partial}{\partial t} + \boldsymbol{v} \cdot \boldsymbol{\nabla} \tag{6.90}
$$

and where S is a surface bounded by a close curve $\partial S = C$, Φ is the magnetic flux through S. We used (6.88) which corresponds to assuming an ideal MHD.

This result implies that the $\boldsymbol{B}$-flux through a co-moving fluid loop is constant and is known as Alfvén's theorem. In other words, in ideal MHD (collisionless plasmas), the plasma cannot move across B field lines. If two plasma elements are initially connected by a field line, they will remain connected all the time; the magnetic topology is preserved in ideal MHD. Also this approximation implies that no current is generated by the motion of the magnetic field because $\boldsymbol{v}_i = \boldsymbol{v}_e$ (see the fourth section). The frozen-in condition can break in non-ideal circumstances $[\eta = 1/(\mu_0\sigma) \neq 0$; magnetic diffusivity is nonzero]. In particular, magnetic reconnection (e.g., solar corona with massive sources of energy and Earth's magnetosphere interaction with solar wind) is a process that breaks magnetic field topology, as we will see later.

Thermal energy equation: conservation of energy. Energy conservation can be written as

$$\frac{D}{Dt}\left(\frac{\rho v^2}{2}\right) = \rho\boldsymbol{v}\cdot\frac{D\boldsymbol{v}}{Dt} = -\rho\boldsymbol{v}\cdot\boldsymbol{\nabla}\Phi - \boldsymbol{v}\cdot\boldsymbol{\nabla}p + \frac{1}{4\pi}\boldsymbol{v}\cdot[(\boldsymbol{\nabla}\times\boldsymbol{B})\times\boldsymbol{B}] \ . \quad (6.91)$$

which is a statement that the kinetic energy of a fluid element changes as work is done by the forces. Using these two identities

$$\boldsymbol{\nabla}\cdot(\boldsymbol{A}\times\boldsymbol{B}) = (\boldsymbol{\nabla}\times\boldsymbol{A})\cdot\boldsymbol{B} - \boldsymbol{A}\cdot(\boldsymbol{\nabla}\times\boldsymbol{B}) \quad (6.92)$$

and

$$\frac{\partial}{\partial t}\left(\frac{B^2}{2}\right) = \boldsymbol{B}\cdot\frac{\partial\boldsymbol{B}}{\partial t} = c_0\boldsymbol{B}\cdot(\boldsymbol{\nabla}\times\boldsymbol{E}) \ , \quad (6.93)$$

where $\boldsymbol{E} = -(\boldsymbol{v}/c_0)\times\boldsymbol{B}$, the equation of energy conservation becomes:

$$\frac{\partial}{\partial t}\left(\mathcal{U} + \frac{B^2}{2\mu_0}\right) + \boldsymbol{\nabla}\cdot\left(\mathcal{U}\boldsymbol{v} + p\boldsymbol{v} + \frac{\boldsymbol{E}\times\boldsymbol{B}}{\mu_0}\right) = 0$$

or

$$\frac{\partial e}{\partial t} + \boldsymbol{\nabla}\cdot(e\boldsymbol{v} + \boldsymbol{\Pi}\cdot\boldsymbol{v}) = 0 \ , \quad (6.94)$$

where

$$\mathcal{U} = \frac{1}{2}\rho v^2 + \rho\epsilon + \rho\Psi \ , \quad e = \mathcal{U} + \frac{B^2}{2\mu_0} \ ,$$

and

$$\boldsymbol{\Pi} = \left(p + \frac{B^2}{2\mu_0}\right)\boldsymbol{I} - \frac{\boldsymbol{B}\boldsymbol{B}}{\mu_0} \ , \quad (6.95)$$

and where ϵ is the thermal energy density, $B^2/2\mu_0$ is the energy density of the magnetic field, and $(\boldsymbol{E}\times\boldsymbol{B})/\mu_0$ is Poynting flux. Note that $\boldsymbol{\Pi}$ is a second-rank stress tensor. In this case of a polytropic ideal gas, we have $\epsilon = p/(\gamma-1)\rho$, where $\gamma = \text{constant}$.

$$\frac{\partial \rho}{\partial t} + \boldsymbol{\nabla} \cdot (\rho \boldsymbol{v}) = 0$$

$$\left[\frac{\partial}{\partial t} + \boldsymbol{v} \cdot \boldsymbol{\nabla}\right](\rho \boldsymbol{v}) = -\boldsymbol{\nabla} p - \rho \boldsymbol{\nabla} \Psi - \frac{1}{\mu_0} \boldsymbol{B} \times (\boldsymbol{\nabla} \times \boldsymbol{B})$$

$$\frac{\partial \boldsymbol{B}}{\partial t} = \boldsymbol{\nabla} \times (\boldsymbol{v} \times \boldsymbol{B})$$

$$\frac{\partial}{\partial t}\left(\mathcal{U} + \frac{B^2}{2\mu_0}\right) + \boldsymbol{\nabla} \cdot \left(\mathcal{U}\boldsymbol{v} + p\boldsymbol{v} + \frac{\boldsymbol{E} \times \boldsymbol{B}}{\mu_0}\right) = 0$$

$$\boldsymbol{\nabla} \cdot \boldsymbol{B} = 0$$

Figure 6.27. MHD equations of ideal plasma. A closed set of 8 equations for 8 unknowns, ρ, $\boldsymbol{v}$, $\boldsymbol{B}$, and p.

If $p = p(\rho)$ (i.e., the equation of state is barotropic), then the energy equation is not needed to close the set of equations. There are two barotropic equations of state that are used in space and planetary physics: the isothermal equation of state ($p = c_{is}^2 \rho$ where c_{is} is the isothermal sound speed and is constant), which describes a fluid for which cooling and heating always balance each other to maintain a constant temperature, and the adiabatic equation of state ($p = c_s^2 \rho^\gamma$, where c_s is the adiabatic sound speed and is constant), in which there is no net heating or cooling (other than adiabatic heating or cooling due to the compression or expansion of volume). We replace (6.94) by $p = c_s^2 \rho$ or $p = c_s^2 \rho^\gamma$, where γ is the ratio of specific heats. The adiabatic energy equation can be written as

$$\frac{D}{Dt}\left(\frac{p}{\rho^\gamma}\right) = 0 \iff \frac{\partial p}{\partial t} = -p\boldsymbol{\nabla} \cdot \boldsymbol{v} - \gamma p \boldsymbol{\nabla} \cdot \boldsymbol{v} . \tag{6.96}$$

For strongly collisional conditions, this equation is suitable. Note that no equation of state exists for a collisionless fluid.

Summary: equations of MHD of ideal plasma are summarizes in Figure 6.27. They consist of equations of conservation of mass (6.75), conservation of momentum (6.83), induction (6.88), and conservation of energy and (6.94). There is a constraint that $\boldsymbol{\nabla} \cdot \boldsymbol{B} = 0$. These equations are for ideal plasma because all dissipative processes (finite viscosity, electricity resistivity, and thermal conductivity) were neglected. Notice that these magnetohydrodynamic equations have embedded within them the usual equations of acoustics (see Chapter 1 of Volume 1), which follow by taking

$$\frac{\partial \rho}{\partial t} + \nabla \cdot (\rho v) = 0$$

$$\left[\frac{\partial}{\partial t} + v \cdot \nabla \right] (\rho v) = -\nabla p - \rho \nabla \Psi - \frac{1}{\mu_0} B \times (\nabla \times B)$$

$$+ \nu \rho \left[\nabla^2 v + \frac{1}{3} \nabla (\nabla \cdot v) \right]$$

$$\frac{\partial B}{\partial t} = \nabla \times (v \times B) - \nabla \times \left(\frac{1}{\mu_0 \sigma} \nabla \times B \right)$$

$$\frac{\partial}{\partial t} \left(\mathcal{U} + \frac{B^2}{2\mu_0} \right) + \nabla \cdot \left(\mathcal{U} v + p v + \frac{E \times B}{\mu_0} \right) = -\frac{B \times (\nabla \times B)}{\mu_0 \sigma}$$

$$\nabla \cdot B = 0$$

Figure 6.28. MHD equations of non-ideal plasma.

$B = 0$. Consequently, the magnetohydrodynamic equations contain the familiar sound speed $c_s = \sqrt{\gamma p / \rho}$, where ρ and p refer to the fluid density and pressure in the unperturbed state of the medium.

Quiz: Rewrite MHD equations for static equilibria (i.e., $\partial / \partial t = 0$ and $v = 0$).

Summary: Equations of MHD of non-ideal plasma are summarizes in Figure 6.28. They consist of equations (6.75), (6.83), (6.86), and (6.94). They contain dissipative effects through the parameter η is the magnetic diffusivity, connected with the electrical conductivity σ, $\eta = 1/(\mu\sigma)$ and the external viscous force acting on a unit of volume of the plasma which can be written as follows:

$$F_{\text{ex}} = \nu \rho \left[\nabla^2 v + \frac{1}{3} \nabla (\nabla \cdot v) \right] , \tag{6.97}$$

ν is the coefficient of kinematic viscosity. One can also consider ambipolar diffusion, Ohmic diffusion, and the Hall effect which we will discuss later. Ambipolar diffusion refers to the interaction between neutral and charged particles. It can be seen as a friction term. It enables the neutral field to respond to magnetic forces, via collisions with charged particles. The Ohmic diffusion results from the collision of electrons with neutrals. Last, the Hall effect is due to the drift between positively and negatively charged species.

Quiz: A wave is said to be stationary if $\partial/\partial t = 0$ for all quantities entered into the wave equations. Let us assume that these quantities also only vary along the x-axis. Show that the following are solutions to MHD equations:

$$\frac{\partial(\rho v_x)}{\partial x} = 0 \Longleftrightarrow \rho v_x = \text{const} \tag{6.98}$$

$$\rho v_x \frac{\partial v_x}{\partial x} = -\frac{\partial}{\partial x}\left(p + \frac{B^2}{\mu_0}\right) \Longleftrightarrow \rho v_x^2 + p + \frac{B^2}{\mu_0} = \text{const} \tag{6.99}$$

$$\rho v_x \frac{\partial \boldsymbol{v}_\perp}{\partial x} = \frac{B_x}{\mu_0} \frac{\partial \boldsymbol{B}_\perp}{\partial x} \Longleftrightarrow \rho v_x \boldsymbol{v}_\perp - \frac{B_x}{\mu_0} \boldsymbol{B}_\perp = \text{const} \tag{6.100}$$

$$\frac{\partial}{\partial x}\left(B_x \boldsymbol{v}_\perp - v_x \boldsymbol{B}_\perp\right) = 0 \Longleftrightarrow B_x \boldsymbol{v}_\perp - v_x \boldsymbol{B}_\perp = \text{const} \tag{6.101}$$

where $\perp$ stands for y and z-components and B_x is constant because divergence of $\boldsymbol{B}$ is zero.

Again, MHD do not deal with individual particles but regard plasma as a continuous medium under the assumption that the length-scales of interest are larger than the mean-free path for collisions between particles, which is about 3 cm in the chromosphere, 30 km in the Corona, and about the average distance between Sun and Earth in the solar wind. MHD often works well in a collisionless plasmas, where this assumption is not satisfied: for instance, in the solar wind, the mean-free path is about an astronomical unit. In a collisionless plasma, there are two significant modifications to MHD normal equations. First, the pressure is not isotropic, and so one needs a pressure tensor; second, Ohm's law generalizes with new terms representing electron inertia, a Hall term and electron stress as discussed in the next chapter.

6.3.2. MHD Equations Are Scale-Independent

Solutions to MHD equations are applicable from nanometer (10^{-9} m) scales to galactic (10^{21} m) scales. They provide us with a powerful tool to investigate the interaction of a plasma (i.e., an electrically-conducting fluid) with magnetic fields. For example, these solutions to MHD systems of equations will be used in this chapter and in later chapters to attempt to understand star formation, the interior of stars and their atmospheres, planetary magnetospheres, and Earth's core.

MHD equations are scale independent just like Navier-Stokes equations described in Chapter 3 of Ikelle (2023b). In other words, MHD equations can be made dimensionless by choosing units of length, mass, and time, based on typical magnitudes: l_0 for length scale, ρ_0 for plasma density, and B_0 for magnetic field at some representative position. The unit of time then follows by exploiting the Alfvén

Table 6.7. Scales of actual plasmas. YSO stands for young stellar object and AGN (active galactic nucleus) for active galactic nucleus.

	l_0 (m)	B_0 (T)	t_0 (s)
Ionosphere	$10^4 - 10^5$	5×10^{-5}	100
Exosphere	10^7	5×10^{-5}-5×10^{-8}	10^5 s
Interplanetary space	10^{11}	10^{-8}	2×10^6
Interstellar space	3.2×10^{19}	10^{-10}-10^{-9}	10^{16}
Solar chromosphere	10^8	10^{-1}-10^{-4}	10^3
Solar corona	$10^8 - 10^9$	10^{-2}-10^{-5}	10^3

speed,

$$v_0 = \frac{B_0}{\sqrt{\mu_0 \rho_0}} \quad \Longrightarrow \quad t_0 = \frac{l_0}{v_0} \; . \tag{6.102}$$

We will discuss the Alfvén speed later in this section. By means of this basic triplet l_0, B_0, t_0 (and derived quantities ρ_0 and v_0), we create dimensionless independent variables and associated differential operators:

$$\bar{l} = \frac{l}{l_0} \; , \quad \bar{t} = \frac{t}{t_0} \quad \Longrightarrow \quad \bar{\boldsymbol{\nabla}} = l_0 \boldsymbol{\nabla} \; , \quad \frac{\partial}{\partial \bar{t}} = t_0 \frac{\partial}{\partial t} \; , \tag{6.103}$$

and dimensionless dependent variables:

$$\bar{\rho} = \frac{\rho}{\rho_0} \; , \quad \bar{\boldsymbol{v}} = \frac{\boldsymbol{v}}{v_0} \; , \quad \bar{p} = \frac{p}{\rho_0 v_0^2} \; , \quad \bar{\boldsymbol{B}} = \frac{\boldsymbol{B}}{B_0} \; , \quad \bar{\boldsymbol{g}} = \frac{l_0}{v_0^2} \boldsymbol{g} \; . \tag{6.104}$$

We can verify that barred MHD equations are identical to unbarred ones (except that μ_0 is eliminated); that is, barred ideal MHD equations, for example, are independent of the size of the plasma (l_0), magnitude of the magnetic field (B_0), and density (ρ_0), and thus of time scale (t_0).

   This property is important because it allows us to use the same MHD equations across a large spectrum of physics processes—from model plasmas in magnetically-confined laboratory experiments to the Sun and large-scale phenomena, from accretion disks of the order of hundreds of astronomical units (typical of protoplanetary disks) to hundreds or thousands of kiloparsec in active galactic nuclei (AGN), as we can see in Tables 6.5 and 6.7.

6.3.3.   Application: Parker's Solar Wind Model

Un-magnetized wind. Parker (1958) assumes that the solar wind can be described as isothermal (i.e., $\boldsymbol{B} = 0$ and $T = T_0$) and steady (i.e., $\partial/\partial t = 0$) expanding plasma. The ideal MHD equations in this case reduce to:

$$\boldsymbol{\nabla} \cdot (\rho \boldsymbol{v}) = 0 \; , \quad \rho \left(\boldsymbol{v} \cdot \boldsymbol{\nabla} \right) \boldsymbol{v} = -\boldsymbol{\nabla} p - \rho \boldsymbol{\nabla} \Psi \; , \quad \text{and} \quad p = \rho R T$$

with $T = T_0$. To solve the MHD equations in this case, let us focus on the spherically symmetric solution. Under this assumption, we take the velocity $\boldsymbol{v}$ is taken as purely radial, $\boldsymbol{v} = v(r)\boldsymbol{i}_r$, $\rho = \rho(r)$, and the gravitational acceleration $-\boldsymbol{\nabla}\Psi = \boldsymbol{g} = g\boldsymbol{i}_r$ obeys the inverse square law, $g = G_0 m_{\text{sun}}/r^2$. Assuming that the spherical coordinates are (r, θ, ϕ), the components of $\boldsymbol{v}$ along θ and ϕ are zero. The temperature is constant and, consequently, the sound speed is $c_s^2 = p/\rho$ as $\gamma = 1$. The mass-continuity and momentum-conversation equations become

$$\frac{d\left(r^2\rho v\right)}{dr} = 0 \ \text{(or } r^2\rho v = \text{cst)} \iff \frac{1}{\rho}\frac{d\rho}{dr} = -\frac{1}{r^2 v}\frac{d(r^2 v)}{dr} \tag{6.105}$$

$$\rho v\frac{dv}{dr} = -\frac{dp}{dr} - \frac{G_0 m_{\text{sun}}\rho}{r^2} \underset{c_s = p/\rho}{\Longrightarrow} v\frac{dv}{dr} = -c_s^2\frac{1}{\rho}\frac{d\rho}{dr} - \frac{G_0 m_{\text{sun}}}{r^2} . \tag{6.106}$$

In these derivations we used differential operators in spherical coordinates; i.e.,

$$\boldsymbol{\nabla}a = \frac{da}{dr} \quad \text{and} \quad \boldsymbol{\nabla}\cdot\boldsymbol{A} = \frac{1}{r^2}\frac{d(r^2 A_r)}{dr} , \tag{6.107}$$

where a is an arbitrary scalar function of r and $\boldsymbol{A}$ is an arbitrary vector with only a radial component that depends of r. By substituting (6.105) into (6.106), we arrive at

$$v\frac{dv}{dr} = \frac{c_s^2}{r^2 v}\frac{d\left(r^2 v\right)}{dr} - \frac{G_0 m_{\text{sun}}}{r^2} \iff \left(v - \frac{c_s^2}{v}\right)dv = 2\frac{c_s^2}{r^2}\left(r - r_c\right)dr ,$$

where $r_c = G_0 m_{\text{sun}}/(2c_s^2)$ is the critical radius showing the position where the wind speed reaches the sound speed, $v = c_s$. This last equation can readily be integrated to obtain the Parker equation; i.e.,

$$\left(\frac{v}{c_s}\right)^2 - \log\left(\frac{v}{c_s}\right)^2 = 4\log\left(\frac{r}{r_c}\right) + 4\frac{r_c}{r} + C . \tag{6.108}$$

where C is the constant of integration that can be determined from boundary conditions. There are a number of mathematically admissible classes of solutions to this equation that depend on C. These solutions are illustrated in Figure 6.29.

We can see that if the flow makes a transition from subsonic to supersonic or vice-versa (transonic waves), it must pass at $r = r_c$ in order for the velocity gradient to remain finite. We basically have six classes of solutions with different behavior at small r values and at $r \to \infty$. There are some that are unphysical, such as classes 5 and 6, which predict two wind speeds at a given radius. Also no wind speed over significant ranges of r/r_c, including $r = r_c$. Solutions of class 4 suggest the wind speed is supersonic, irrespective of r. The idea that at the base of the Corona, particles move faster than the speed of sound is consistent with measurements. Although solutions of class 3 are transonic, they also suffer from the fact that particles streaming from the Corona move faster than the speed of sound. This is consistent with the measurements. Solutions of class 1 suggest that the wind speed is subsonic, irrespective of r. They are called the solar breeze solutions. Before the advent of direct measurements, this solution was considered seriously.

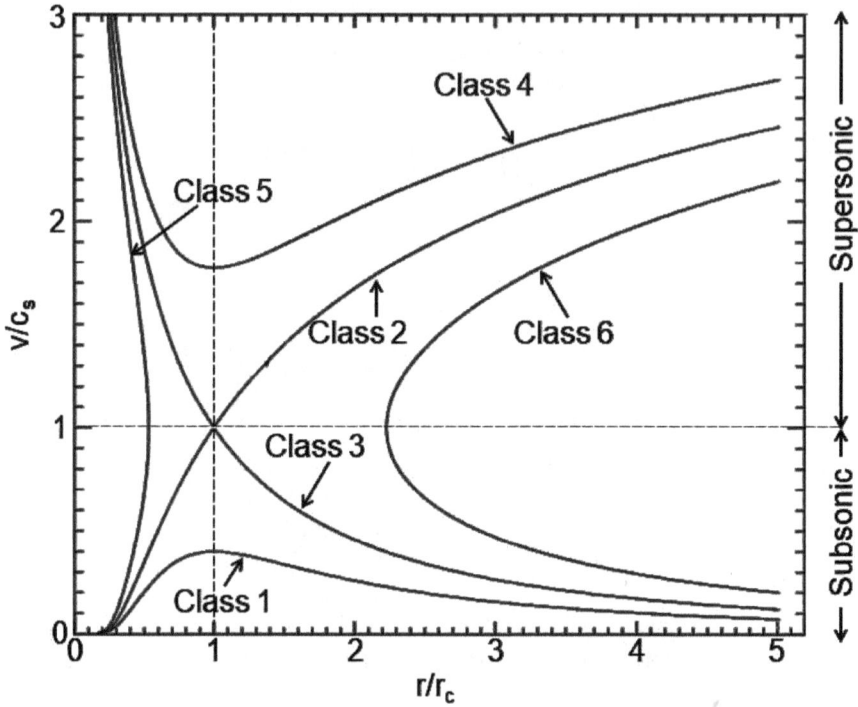

Figure 6.29. The six mathematical classes of solutions to the Parker solar wind model.

Parker (1958) predicted that the physical solutions would be those of class 2. For solar wind, he predicted that the smallest speed of solar wind is at $r = r_{\text{sun}}$, which is the base of the Corona layer (see Figures 6.2 and 6.3). He also found that the wind speed becomes supersonic as we move away from the Corona. His conclusions were met with skepticism until the first sporadic measurements of solar wind speed by Soviet Lunik 3 and Venus 1 in 1959 and later continuous measurements by Mariner 2 in 1966 between 0.3 AU and 1 AU from the sun center.

The class 2 solutions correspond to $C = -3$ in equation (6.108), as can be verified by substituting $v = c_s$ and $r = r_s$. Parker's solution passes through the critical point (the model places the critical point at about 11.5 r_{sun} for a million degree Corona) and the wind is supersonic for $r > r_c$. For example, assuming a typical Coronal temperature is $T = 10^6$ degrees Kelvin (one million degrees Kelvin, the sound speed is

$$c_s = \sqrt{RT} = \left(8.3 \times 10^7 \times 10^6\right)^{1/2} \approx 10^7 cm/s = 100 \text{ km/s} ,$$

and the critical radius is

$$r_c = \frac{G_0 m_{\text{sun}}}{2c_s^2} = 7 \times 10^9 \text{ m} \approx 10 \, r_{\text{sun}} .$$

To put this into context, the radius of Earth's orbit is 214 r_{sun}. Thus, the solar wind is highly super-sonic by the time it reaches Earth. To calculate the actual wind speed from Parker's model we set $r = 214\,r_{\text{sun}}$ and solve for v. Hence,

$$\left(\frac{v}{c_s}\right)^2 - \log\left(\frac{v}{c_s}\right)^2 = 4\log\left(\frac{214}{10}\right) + 4\frac{10}{214} - 3 = 9.44$$

By using the Newton-Raphson method (e.g., Matlab), we obtain $v = 3.45c_s = 310$ km/s. Observations at 1 AU give the quiet solar wind as $v = 320$ km/s. Thus, Parker's solar wind model gives quite an accurate estimation of the velocity. If $T = 2 \times 10^6$ degrees Kelvin, the wind speed is about 600 km/s. As we will discuss later, in the context of wave propagation in plasma, the fact that Parker (1958) ignored the magnetic field in his derivation of solar wind speed does not change his results in any significant way.

Figure 6.30 shows the solar wind speeds measured by the joint ESA/NASA Ulysses mission. Ulysses was launched on October 6, 1990. Its probe was equipped with scientific instruments capable of detecting and measuring solar wind ions and electrons, magnetic fields, radio and plasma waves, dust and gas, X-rays, and gamma rays. In addition to determining that particles of solar wind usually travel at speeds of 300-700 km/s, this mission helped determine that the Sun's magnetic field reverses direction every 11 or so years. High-speed solar wind originates from coronal holes in the sun's polar regions, while slow-speed solar wind originates from streamers. The north and south poles of the Sun have large, persistent coronal holes, so high latitudes are filled with fast solar wind (generally greater than 500 km/s). In the equatorial plane, where the Earth and the other planets orbit, the most common state of the solar wind is the slow speed wind, with speeds less than 400 km/s. This portion of the solar wind forms the equatorial current sheet. In other words, the solar wind at the poles is typically faster than in the ecliptic plane.

As one might expect, the solar wind is always directed away from the Sun, and it carries with it magnetic clouds. The heliospheric magnetic field (HMF), also known as the interplanetary magnetic field (IMF), is a part of the Sun's magnetic field that is carried into interplanetary space by the solar wind. The IMF is considered to be "frozen in" solar wind plasma as described earlier; that is, the IMF is embedded in the plasma and flows outward with the solar wind. Because of the Sun's rotation, the IMF, like the solar wind, travels outward in a spiral pattern. The low-speed wind is denser than the high-speed wind ($n \approx 10^7$ cm^{-3} as opposed to $n \approx 10^6$ cm^{-3}). Thus, the Sun constantly loses mass, through particles carried away from the Sun by the solar wind, at a rate of about 2.5×10^{-14} solar mass per year. Note that the solar wind does not come from the whole of the sun's surface but only from the regions where the magnetic field is open—that is, where field lines emerge. This is only about 20 percent of the whole surface. About 80 percent of the magnetic field is closed, effectively holding in the hot Corona.

The fast wind from the polar coronal holes carries magnetic fields of opposite polarity into the heliosphere, which are then separated by the heliospheric current sheet (HCS) embedded in the slow wind. Measurements over a range of latitudes far from the Sun show that this boundary is not symmetric around the Sun's equator, but is on average displaced southward. This offset must reflect an asymmetry on the Sun; but since there cannot be a mismatch between the inward and outward magnetic flux on the Sun, its origin is unclear. In situ, the HCS is warped and

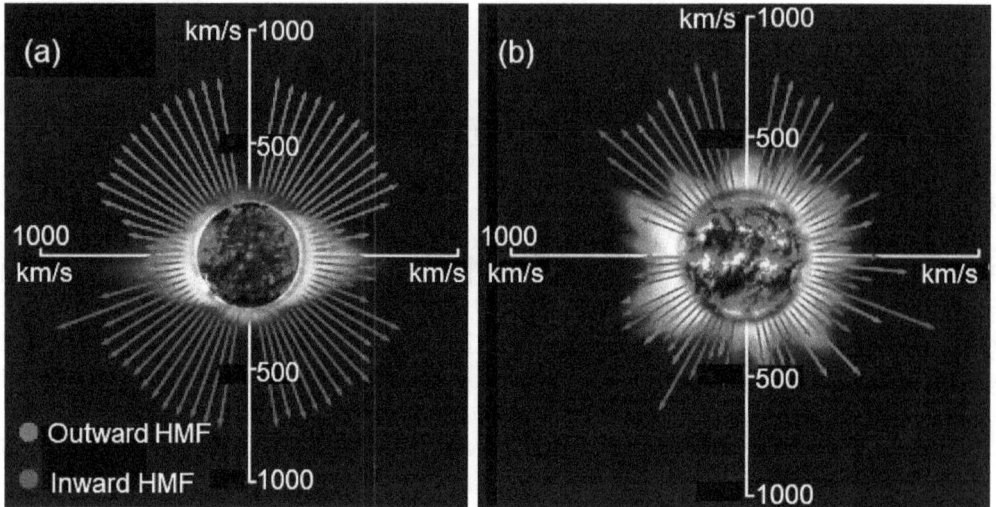

Figure 6.30. Two pictures of the speed of the solar wind measured by the Ulysses spacecraft. The arrows show how fast the solar wind was blowing with longer arrows showing higher speeds. (a) shows the Sun at solar minimum when there are few sunspots and (b) the Sun at solar maximum when the Sun is very active and there are usually lots of sunspots. The heliospheric magnetic field (HMF) is the component of the sun's magnetic field that is dragged out of the solar corona by the solar wind flow to fill the Solar System. Adapted in part from the image of ESA.

deformed by the combined effects of solar rotation and the inclination of the Sun's magnetic axis, effects that are even more prominent at solar maximum.

The energy that heats the Corona and drives the wind comes from the mechanical energy of convective photospheric motions, which is converted into magnetic and/or wave energy. In particular, both turbulence and magnetic reconnection are implicated theoretically and observationally in coronal heating and acceleration. However, existing observations cannot adequately constrain these theories, and the identity of the mechanisms that heat the Corona and accelerate the solar wind remains one of the unsolved mysteries of solar and heliospheric physics. How the coronal plasma is generated, energized, and the way in which it breaks loose from the confining coronal magnetic field are fundamental physical questions with crucial implications for predicting our own space environment, as well as for the understanding of the natural plasma physics of other astrophysical objects, from other stars, to accretion disks and their coronas, to energetic phenomena such as jets, X- and gamma-ray bursts, and cosmic-ray acceleration.

One may be wondering why Corona's temperatures are much higher than significant parts of the Sun's interior. Figure 6.31 shows the temperature, pressure, and mass density profiles in the Sun's atmosphere and confirms these observations. We can see that the plasma in the Sun's atmosphere is highly heterogeneous. The temperature increases in the Corona region from about 0.1 million degrees Kelvin to more than one million degrees Kelvin as we move away from the Sun. In other words, the Corona is heating up as the distance from the core of the Sun increases. This basically means there are other sources of Coronal heat than the Sun's interior

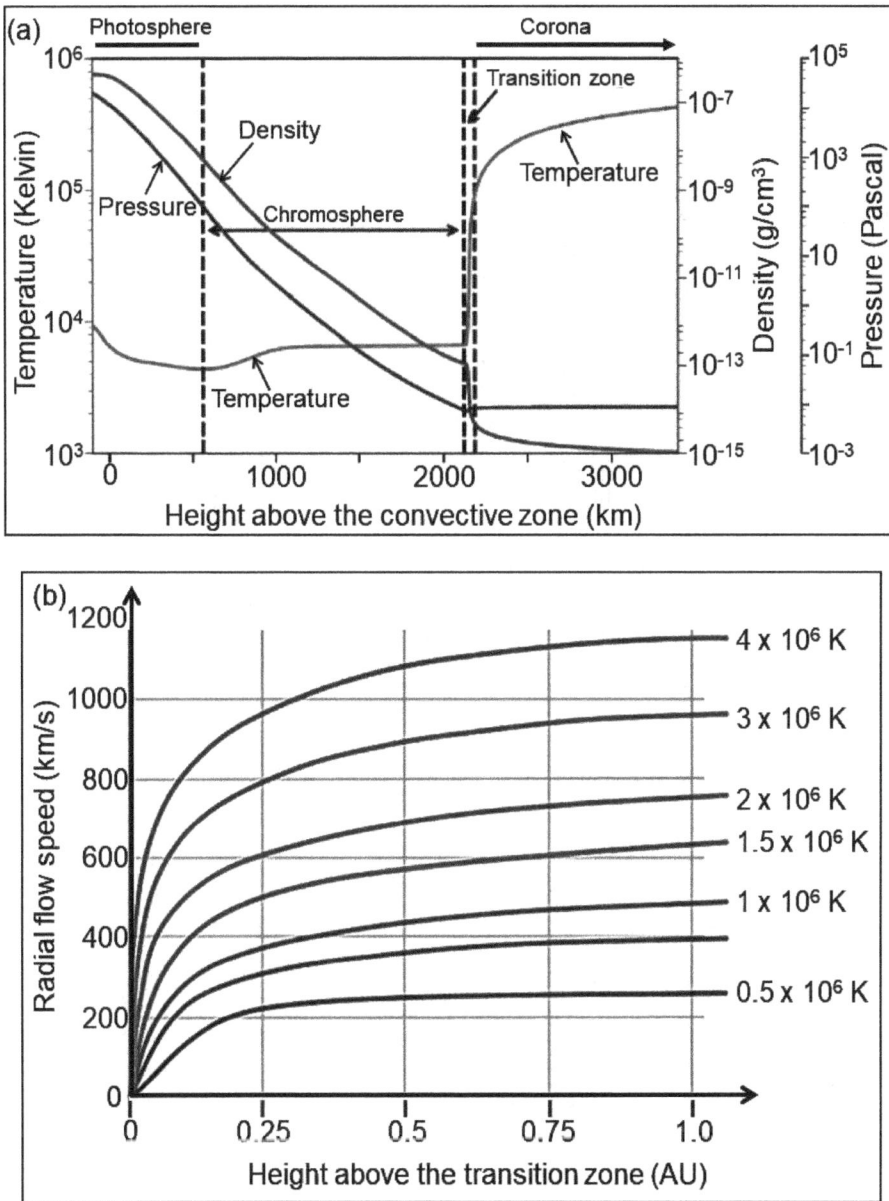

Figure 6.31. (a) The VAL model (Vernazza et al., 1981) of the solar atmosphere. (b) Solutions to the Parker solar-wind equation for some coronal temperatures and heliocentric distances, in astronomical units (AU). Adapted from Parker (1958).

only. The interplanetary medium, including its magnetized cosmic objects such as our solar system's Jovian planets, especially Jupiter and Saturn that radiate more energy than they receive from the Sun, is one source of heat for Coronal plasma because the strength of magnetism directly relates to temperature according to Ikelle (2020). Secondly, there are many cosmic objects in our solar system and near our Sun that have dynamo structures. These objects generate electricity, through Friday's law, which induces magnetic fields, through Ampère law, outside our Sun.

Lastly, the third source is related to the fact that the Corona plasma and solar wind are heterogeneous, meaning that there are electromagnetic mirrors all over the place that create nuclear confinement-type environments that allow charged particles to bounce back and generate large amounts of heating, as described in the previous section.

> **Quiz:** What are the differences between the Navier-Stokes and MHD equations?

Magnetized solar wind. In the above derivations, we assume that the Sun is not rotating and hence has no magnetic field. This assumption is incorrect. However it still allows us to introduce key concepts of solar wind. Let us now discuss how the above solution changes in the case of a magnetized Sun with radial outflow. By working in the spherical coordinates $[i_r, i_\theta, i_\phi]$, we describe the solar wind velocity as $v = [v_r, 0, v_\phi = \Omega_{\text{sun}} r]^T$, where $\Omega_{\text{sun}} = 2.7 \times 10^{-6}$ rad/s is the angular velocity of the Sun's rotation, r is the distance from the Sun's center, and the magnetic field as $B = [B_r, 0, B_\phi]^T$, with v and B depending on r only. So for simplicity, the fluid velocity and magnetic field only have radial and azimuthal components in spherical coordinates. This assumption restricts our description to equatorial magnetized winds with magnetic field lines flowing in and out of the Sun with respect to the equator. This assumption also ensures that the magnetic field is not a global monopole; i.e.,

$$\boldsymbol{\nabla} \cdot \boldsymbol{B} = 0 \Longrightarrow B_r = \frac{B_0 R_{\text{sun}}^2}{r^2} , \tag{6.109}$$

where B_0 is the magnetic field strength at the Sun's surface.

Let us recall that the Sun's outflow has a negligible effect on the Sun's mass and energy. The Sun's mass is expected to decrease by $10^{-5} m_{\text{sun}}$ in a billion years. A significant amount of angular momentum is captured by solar wind, however. At birth, the Sun rotated 30 times more than today. It has lost that angular momentum to the solar wind. Thus, the evolution of the Sun's outflow is dominated by angular momentum, with $L = r v_\phi = r^2 \Omega_{\text{sun}} =$ constant being extracted from the Sun's momentum. These losses can only be explained by magnetized solar wind as described below.

Let us determine the properties of a magnetized solar wind, starting with the induction equation. The induction equation for a static magnetized wind (i.e., $\partial \boldsymbol{B}/\partial t = \boldsymbol{0}$) is

$$\boldsymbol{\nabla} \times (\boldsymbol{v} \times \boldsymbol{B}) = \boldsymbol{0} \Longrightarrow \begin{cases} \frac{\partial}{\partial r} [r (v_\phi B_r - v_r B_\phi)] = 0 \\ \frac{\partial}{\partial \phi} [r (v_\phi B_r - v_r B_\phi)] = 0 \end{cases} . \tag{6.110}$$

By integrating (6.110), we obtain

$$v_r B_\phi - v_\phi B_r = \frac{C}{r} , \tag{6.111}$$

where C is the integration constant. At $r = R_{\text{sun}}$, we have $B_\phi = 0$, $v_\phi = \Omega_{\text{sun}} R_{\text{sun}}$, $C = -\Omega_{\text{sun}} R_{\text{sun}}^2 B_0$. So by integrating (6.110) with the boundary condition at

$r = R_{\text{sun}}$ and using (6.109), we obtain

$$\frac{B_\phi}{B_r} = \frac{v_\phi - r\Omega_{\text{sun}}}{v_r} \ .$$

(6.112)

The conservation of mass and momentum equations must be added to the above equation of frozen-in magnetic field. Assuming that the rate at which the Sun loses mass is constant in a steady-state model, the mass and momentum conservations in spherical coordinates can be expressed as follows:

$$\frac{dm_{\text{sun}}}{dt} = 4\pi\rho r^2 v_r \ \text{ and } \ \rho v_r \frac{d(rv_\phi)}{dr} = \frac{B_r}{\mu_0} \frac{d}{dr}(rB_\phi) \ ,$$

(6.113)

respectively. By multiplying the momentum conservation equation r^2 to take advantage of the constant mass loss, we can integrate the momentum conservation equation to obtain

$$r\left(v_\phi - \frac{B_r B_\phi}{\mu_0 \rho v_r}\right) = L \ \text{ or } \ \underbrace{rv_\phi}_{\text{AM gas}} - \underbrace{\frac{rB_r B_\phi}{\mu_0 \rho v_r}}_{\text{magnetic AM}} = L \ ,$$

(6.114)

where L is the integration constant and AM stands for angular momentum per unit mass. So we can see L is actually the angular momentum of outflow per unit mass of the outflow, which includes the angular momentum of gas and the angular momentum of the magnetic field. Now equations (6.114) and (6.111) can be rewritten in the form

$$v_\phi = r\Omega_{\text{sun}} \frac{v_r^2[L/(r^2\Omega_{\text{sun}})] - v_A^2}{v_r^2 - v_A^2} \ ,$$

(6.115)

where

$$v_A^2 = \frac{B_r^2}{\mu_0\rho}$$

(6.116)

is the Alfvén speed that we will discuss in more detail in the next subsection. Notice that only the radial component of the magnetic field, B_r, affects the azimuthal velocity of solar wind speed. The Alfvénic critical point, which is at a distance r_A from the center of the Sun, is defined as the point at which the radial flow is equal to the Alfvén speed, that is, $v_A = v_r$. Close to the Sun, the solar wind is too weak to modify structure of magnetic field, thus

$$\frac{v^2}{2\rho} \ll \frac{B^2}{\mu_0} \ .$$

(6.117)

Solar magnetic field therefore forces the solar wind to co-rotate with the Sun. When the solar wind becomes super-Alfvénic,

$$\frac{v^2}{2\rho} \gg \frac{B^2}{\mu_0} \ .$$

(6.118)

Transition between the two regimes occurs at the Alfvén radius (r_A), where

$$\frac{v^2}{2\rho} = \frac{B^2}{\mu_0} .$$
(6.119)

Because the Sun's field is well approximated as a dipole [see equations 6.62 and 6.63], $B \approx M_0/R_{\text{sun}}^3$, the Alfvén radius is then

$$r_A \approx \left(\frac{M_0^2}{\mu_0 \rho_{sw} v_r}\right)^{1/6} .$$
(6.120)

At this point, the denominator is zero and therefore the numerator must be zero to use the l'Hospital's rule. The vanishing of the numerator in (6.115) at this critical point occurs when we have $L = r_A^2 \Omega_{\text{sun}}$ and thus determines the angular momentum per unit mass at $r = r_A$. In this case we have $L = r_A^2 \Omega_{\text{sun}}$. So the time of the Sun or any star to that matter to lose angular momentum is

$$\tau_{\text{loss}} = \frac{\text{Angular momentum of the Sun}}{\text{Angular momentum loss rate of the Sun}}$$

$$= \frac{m_{\text{sun}} \Omega_{\text{sun}} R_{\text{sun}}^2}{(dm_{\text{sun}}/dt)\Omega_{\text{sun}} r_A^2}$$

$$= \frac{m_{\text{sun}} R_{\text{sun}}^2}{(dm_{\text{sun}}/dt) r_A^2} .$$
(6.121)

Because $r_A^2/R_{\text{sun}}^2 \sim 10^2 - 10^4$ and $(dm_{\text{sun}}/dt)/m_{\text{sun}} \sim 10^{-4} - 10^{-5}$ in a billion years, the Sun's loss of its angular momentum is essentially due to its magnetized wind. Nearly all the angular momentum of the solar system is in the planetary motions, mostly driven by Jupiter's contribution.

6.3.4. Application: Magnetohydrodynamic Waves

Waves are at the heart of solid earth and fluid dynamics. They are also at the heart of plasma physics. There is nothing that plays a more critical role in plasma life than waves. This subsection introduces the wave properties of plasmas within the MHD approximation. As in solids and fluids, waves in plasmas can also be described as small perturbations that propagate in the medium from one place to another at a later time. Larger perturbations are described in the next subsection.

We start with plasma at the equilibrium state, where nothing depends on time and there are no flows; that is, $\partial/\partial t = 0$ and $v = 0$, $\rho = \rho_0 = $ constant, $p = p_0 = $ constant, and $B = B_0 = $ constant. We then assume that all quantities are slightly perturbed as follows:

$$\rho = \rho_0 + \rho_1(x,t) , \quad v = v^{(1)}(x,t) , \quad B = B_0 + B_1(x,t)$$

and

$$p = p_0 + p_1(x,t) ,$$
(6.122)

where $\boldsymbol{v}^{(0)} = 0$ because the reference MHD is static. We assume no gravity. Subscript and superscript 1 are used for perturbed quantities. When substituting (6.122) into the MHD equations, the equilibrium terms vanish naturally, and those terms in which a perturbation appears more than once will be relatively small. Hence, we are left with only terms in which a perturbation quantity appears once. In essence the MHD equations have been linearized. We then decompose all perturbed quantities into Fourier modes in the form of $\exp[i(\omega t - \boldsymbol{k} \cdot \boldsymbol{x})]$ (e.g., $\boldsymbol{v}^{(1)} = \hat{\boldsymbol{v}}^{(1)} \exp[i(\omega t - \boldsymbol{k} \cdot \boldsymbol{x})]$) to arrive at

$$\frac{\partial \rho_1}{\partial t} + \rho_0 \boldsymbol{\nabla} \cdot \boldsymbol{v}^{(1)} = 0 \implies -\omega \hat{\rho}_1 + \rho_0 \boldsymbol{k} \cdot \hat{\boldsymbol{v}}^{(1)} = 0 \tag{6.123}$$

$$\rho_0 \frac{\partial \boldsymbol{v}^{(1)}}{\partial t} = -\boldsymbol{\nabla} p_1 + \frac{(\boldsymbol{\nabla} \times \boldsymbol{B}_1) \times \boldsymbol{B}_0}{\mu_0}$$

or

$$-\omega \rho_0 \hat{\boldsymbol{v}}^{(1)} = -\hat{p}_1 \boldsymbol{k} + \frac{\left(\boldsymbol{k} \times \hat{\boldsymbol{B}}_1\right) \times \boldsymbol{B}_0}{\mu_0} \tag{6.124}$$

$$\frac{\partial \boldsymbol{B}_1}{\partial t} = \boldsymbol{\nabla} \times (\boldsymbol{v}^{(1)} \times \boldsymbol{B}_0) \implies -\omega \hat{\boldsymbol{B}}_1 = \boldsymbol{k} \times \left(\hat{\boldsymbol{v}}^{(1)} \times \boldsymbol{B}_0\right) \tag{6.125}$$

$$\frac{\partial p_1}{\partial t} + \gamma p_0 \boldsymbol{\nabla} \cdot \boldsymbol{v}^{(1)} = 0 \implies -\omega \hat{p}_1 + \gamma p_0 \boldsymbol{k} \cdot \hat{\boldsymbol{v}}^{(1)} = 0 , \tag{6.126}$$

and $\boldsymbol{\nabla} \cdot \boldsymbol{B}_1 = 0$, where $\boldsymbol{k}$ is the wavevector (it is perpendicular to wavefronts). Taking another derivative in time of the momentum equation yields, substituting (6.125) and (6.126) then in the new momentum equation, and assuming that $\boldsymbol{B}_0 = B_0 \boldsymbol{i}_z$, we arrive at:

$$\frac{\partial^2 \boldsymbol{v}^{(1)}}{\partial t^2} - -\frac{1}{\rho_0} \boldsymbol{\nabla} \left(\frac{\partial p_1}{\partial t}\right) + \frac{1}{\rho_0 \mu_0} \left[\boldsymbol{\nabla} \times \left(\frac{\partial \boldsymbol{B}_1}{\partial t}\right) \times \boldsymbol{B}_0\right]$$
$$= c_s^2 \boldsymbol{\nabla} \left(\boldsymbol{\nabla} \cdot \boldsymbol{v}^{(1)}\right) + c_A^2 \left[\boldsymbol{\nabla} \times \left(\boldsymbol{\nabla} \times (\boldsymbol{v}^{(1)} \times \boldsymbol{i}_z)\right) \times \boldsymbol{i}_z\right] ,$$

or after taking the time-Fourier transform at

$$\omega^2 \hat{\boldsymbol{v}}^{(1)} = c_s^2 \left(\boldsymbol{k} \cdot \hat{\boldsymbol{v}}^{(1)}\right) \boldsymbol{k} + c_A^2 \left\{\boldsymbol{k} \times \left[\boldsymbol{k} \times \left(\hat{\boldsymbol{v}}^{(1)} \times \boldsymbol{i}_z\right)\right] \times \boldsymbol{i}_z\right\} , \tag{6.127}$$

where $c_s = \sqrt{\gamma p_0 / \rho}$ is the adiabatic sound speed and $c_A = \sqrt{B_0^2 / (\mu_0 \rho)}$ is known as the Alfvén speed, in honor of Hannes Alfvén, Swedish scientist who first observed MHD waves. Sound speed and Alfvén speed are the basis of all wave phenomena described by magnetohydrodynamic equations. Other speeds also play an instrumental role, but they can be described in terms of c_s and c_A as we will discuss later.

Remember that the solar atmosphere, which is defined as the part of the Sun from which photons can escape directly into space, is generally divided into the photosphere, chromosphere, transition region and Corona. The regions form approximate concentric spherical shells around the Sun. In the solar Corona, the plasma is mainly made up of hydrogen and helium. By assuming that the mean particle mass of the plasma is $m = 0.6\,m_H$ (Aschwanden, 2004), where $m_H = 1.673 \times 10{-27}$ kg is the mass of hydrogen (proton), and the plasma pressure is $p_0 = (k_B/m)\rho T_0$, where $k_B = 1.38 \times 10^{-23}$ J/K is the Boltzmann constant, and the adiabatic index is $\gamma = 5/3$, the sound speed can alternatively be written as

$$c_s = \left(\frac{\gamma p_0}{rho}\right)^{1/2} = \left(\frac{\gamma k_B T_0}{m}\right)^{1/2} \approx 1.51\, T_0^{1/2} \text{ m/s} \; . \tag{6.128}$$

A coronal temperature of, say, $T_0 = 10^6$ K then yields a sound speed of $c_s = 151$ km/s (or $c_s = 151,000$ m/s. The high temperature of the coronal plasma thus leads to a correspondingly high sound speed, far in excess of the 350 m/s sound speed in the Earth's atmosphere or the 1500 m/s sound speed in water. In the photosphere, the density is about $\rho_0 = 3 \times 10^{-7}$ g/cm^3 and pressure of $p_0 = 2 \times 10^5$ dynes/cm^2, and $\gamma = 5/3$. Thus the sound speed is $c_s = 10.5$ km/s.

Let us now turn to the Alfvén speed. Assuming that $\rho_0 = nm$, where n is the total number density and m is the mean mass, we have

$$c_A = \sqrt{B_0^2/(\mu_0 \rho_0)} = 2.816 \times 10^{12} \times \frac{B_0}{\sqrt{n}} \text{ m/s} \tag{6.129}$$

with B_0 in gauss and $\mu_0 = 4\pi \times 10^{-7}$ henry/m. For example, with a coronal field strength of $B_0 = 10$ gauss, we obtain an Alfvén speed of some $c_A = 890$ km/s. In an active region the field is stronger; for example, with $B_0 = 10^2$ gauss in a medium with an electron number density $n_e = 10^{10}$ particles per cm^3 and a total number density n $= 2 \times 10^6$ particles per cm^3, we obtain $c_A \approx 2000$ km/s. Thus, despite the Corona's high temperature and correspondingly high sound speed, the tenuous nature of the coronal plasma acts to produce a yet higher Alfvén speed. In the photosphere, it is usual to quote values of fluid density ρ_0 directly, basing these values on model computations of the convection zone and the atmosphere above. A density of $\rho_0 = 3 \times 10^{-7}$ g/cm^3 is typical of the surface layers of the Sun (see, for example, Parker 1979a, p. 149). In a magnetic field of $B_0 = 1500$ gauss the corresponding Alfvén speed is $c_A = 0.077 \times 10^5$ m/s. Thus in regions of strong photospheric magnetic fields, the Alfvén speed is about 7.7 km/s and sound and Alfvén speeds are typically comparable. Note that, by using the definition of magnetic tension in (6.80), the Alfvén wave speed can be interpreted as

$$c_A = \left(\frac{B_0^2/\mu_0}{\rho}\right)^{1/2} = \left(\frac{\text{applied tension force}}{\text{mass density}}\right)^{1/2} \; . \tag{6.130}$$

By using the decomposition of the Lorentz force in (6.80) into magnetic pressure and tension, we can verify that Alfvén wave is a transverse wave caused by magnetic tension—that is, Alfvén waves propagate along magnetic field lines and the restoring forces are magnetic tensions. We can also see that Alfvén waves carry information about changes in the magnetic field. Let us now write the $\boldsymbol{k}$ vector in the $x - z$ plane as follows:

$$\boldsymbol{k} = k_\perp \boldsymbol{i}_x + k_\| \boldsymbol{i}_z = k_x \boldsymbol{i}_x + k_z \boldsymbol{i}_z = (k \sin\theta)\boldsymbol{i}_x + (k\cos\theta)\boldsymbol{i}_z \ . \tag{6.131}$$

This decomposition leads to a dispersion relation in the form $\boldsymbol{W}\hat{\boldsymbol{v}}^{(1)} = \boldsymbol{0}$; i.e.,

$$\begin{bmatrix} \omega^2 - c_A^2 k_\|^2 - \left(c_s^2 + c_A^2\right)k_\perp^2 & 0 & -c_s^2 k_\| k_\perp \\ 0 & \omega^2 - c_A^2 k_\|^2 & 0 \\ -c_s^2 k_\| k_\perp & 0 & \omega^2 - c_s^2 k^2 \end{bmatrix} \begin{bmatrix} \hat{v}_x^{(1)} \\ \hat{v}_y^{(1)} \\ \hat{v}_z^{(1)} \end{bmatrix} = \begin{bmatrix} 0 \\ 0 \\ 0 \end{bmatrix} , \tag{6.132}$$

where $k^2 = k_\perp^2 + k_\|^2$. The dispersion along the x-axis (i.e., the first column) involves sound and Alfvén waves. The Alfvén wave is due to the magnetic pressure of the Lorentz force. The dispersion along the y-axis (i.e., the second column; the transverse direction) consists of Alfvén waves only. The Alfvén wave is due to magnetic tension of the Lorentz force. The dispersion along the z-axis (i.e., the third column) involves sound waves only. The nontrivial solution to this system requires that $\det \boldsymbol{W} = 0$, which is equivalent to

$$\left(\omega^2 - c_A^2 k_\|^2\right)\left[\omega^4 - \left(c_s^2 + c_A^2\right)k^2\omega^2 + c_A^2 c_s^2 k_\|^2 k^2\right] = 0 \tag{6.133}$$

or

$$\omega^2 = c_A^2 k_\|^2 = c_A^2 k_z^2 = \frac{[\boldsymbol{k} \cdot \boldsymbol{B}_0]^2}{\mu_0 \rho_0} \ , \quad \text{Alfven waves} \ . \tag{6.134}$$

and

$$\frac{\omega^2}{k^2} = \frac{1}{2}c_+^2 \left\{ 1 \pm \left[1 - 4c_-^2 \frac{k_\|^2}{k^2} \right]^{1/2} \right\} \ , \quad \text{magnetoacoustic waves} \tag{6.135}$$

where

$$c_+^2 = c_s^2 + c_A^2 \iff c_+ = \left(c_s^2 + c_A^2\right)^{1/2} \ , \tag{6.136}$$

and

$$\frac{1}{c_-^2} = \frac{1}{c_s^2} + \frac{1}{c_A^2} \iff c_- = \frac{c_s c_A}{\sqrt{c_s^2 + c_A^2}} \ . \tag{6.137}$$

Altogether, we now have four speeds, c_s, c_A, c_+ and c_-, that characterize the propagation of disturbances in a magnetic atmosphere, and these speeds are ordered thus:

$$c_- \leq \min(c_s, c_A) \leq \max(c_s, c_A) \leq c_+ \ . \tag{6.138}$$

Notice again that without the magnetic field, $\boldsymbol{B} = \boldsymbol{0}$ (i.e., c_A), we have the classical sound wave problem. Equation (6.133) reduces to

$$\omega^4 \left(\omega^2 - k^2 c_s^2 \right) = 0 \tag{6.139}$$

or

$$\text{phase velocity} = \frac{\omega}{k} = c_s \ \text{ and } \ \text{group velocity} = \frac{\partial \omega}{\partial k} = c_s \ , \tag{6.140}$$

with a single velocity associated with one acoustic wave only.

Shear-Alfvén wave. Equation (6.133) has three independent roots in ω^2 corresponding to three different types of waves that can propagate through magnetized fluids, including solar wind, as illustrated in Figure 6.32. One of the roots corresponds to the Alfvén wave (also known as the shear-Alfvén wave) is given (6.134). The corresponding eigenvector from equation (6.132) is $\boldsymbol{v}^{(1)} = [0, v_y^{(1)}, 0]^T$. Thus the velocity perturbation must lie in the y-direction and therefore $\boldsymbol{k} \cdot \boldsymbol{v}^{(1)} = \boldsymbol{v}^{(1)} \cdot \boldsymbol{B}_0 = 0$—that is, they are transverse waves with $\boldsymbol{v}^{(1)} \perp \boldsymbol{k}$ (perpendicular to $\boldsymbol{B}$). By examining equation (6.134), we can observe Alfvén waves are independent of perturbations in density and pressure. They are therefore incompressible. Note also that their group velocity is always parallel to the magnetic field while their phase velocity can be oblique to the magnetic field. Their absolute value of group velocity equals c_A.

Fast and slow magnetosonic (magnetoacoustic) waves. The other two roots are given in (6.135). The two solutions are always real and correspond to fast magnetosonic (+) and slow magnetosonic (-) waves. The corresponding eigenvectors, from equation (6.132), lie in the $x-z$ plane: $\boldsymbol{v}^{(1)} = [v_x^{(1)}, 0, v_z^{(1)}]^T$. As a consequence, we have that $\boldsymbol{k} \cdot \boldsymbol{v}^{(1)} \neq 0$ and, similarly $\boldsymbol{v}^{(1)} \cdot \boldsymbol{B}_0 \neq 0$: these waves are compressive in nature and involve plasma motion in both the parallel and the perpendicular field direction. Notice that, in general $(\omega/k)_{\text{slow}} \leq c_s$ while $(\omega/k)_{\text{fast}} \geq c_s$, as illustrated in Figure 6.32. The properties of these waves depend on the ratio between c_s^2 and c_A^2 and on the relative orientation between $\boldsymbol{k}$ and $\boldsymbol{B}_0$. For parallel propagation (i.e., $\theta = 0$; $\boldsymbol{k}$ parallel to $\boldsymbol{B}_0$) the matrix $\boldsymbol{W}$ in (6.132) is diagonal and the solutions given in (6.135) reduce to

$$\left(\frac{\omega}{k}\right)^2_{\text{fast}} = \begin{cases} c_s^2 & \text{for } c_s > c_A \\ c_A^2 & \text{for } c_s < c_A \end{cases} \ \text{ and } \ \left(\frac{\omega}{k}\right)^2_{\text{slow}} = \begin{cases} c_A^2 & \text{for } c_s > c_A \\ c_s^2 & \text{for } c_s < c_A \end{cases} \ .$$

In a weakly magnetized medium ($c_s > c_A$) the fast magnetosonic wave becomes an acoustic (or sound) wave whereas the slow mode propagates at the Alfvén speed. Conversely, in a strongly magnetized medium where $c_A > c_s$, the fast and slow modes propagate at the Alfvén and sound speeds, respectively. For perpendicular propagation (i.e., $\theta = \pm\pi/2$; $\boldsymbol{k}$ perpendicular to $\boldsymbol{B}_0$) the matrix $\boldsymbol{W}$ in (6.132) is again diagonal and the solutions of the dispersion relation reduce to

$$\left(\frac{\omega}{k}\right)^2_{\text{fast}} = c_A^2 + c_s^2 \ \text{ and } \ \left(\frac{\omega}{k}\right)^2_{\text{slow}} = 0 \ .$$

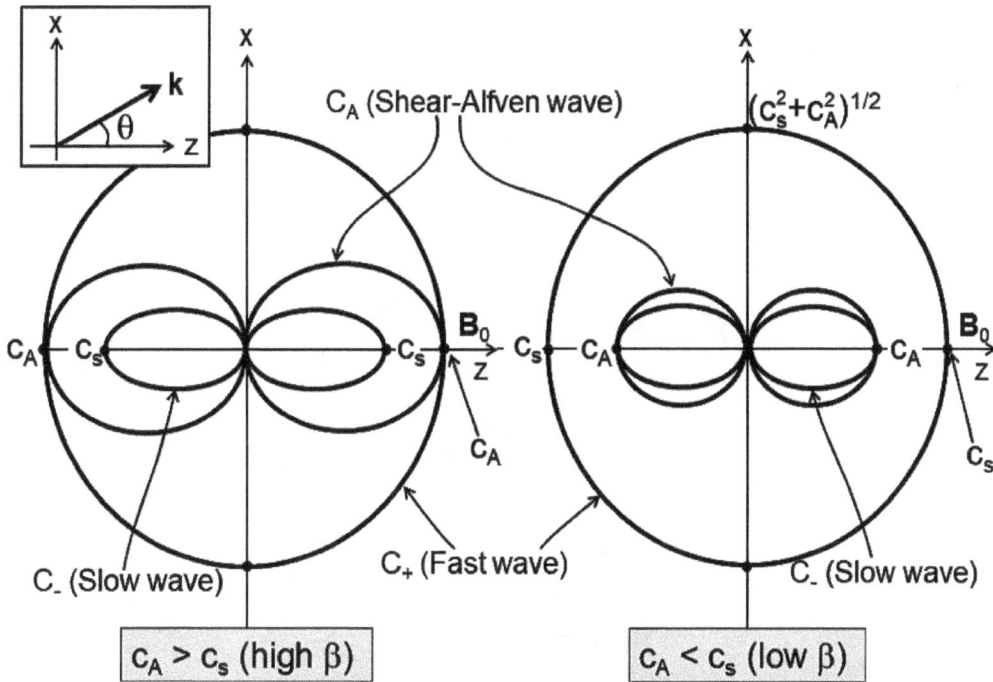

Figure 6.32. Phase velocities of MHD modes. Wave diagram for fast, slow and pure Alfvén waves in the $x - z$ plane. The magnetic field, $\boldsymbol{B}$, is in the z-direction. This diagram is known as Friedrichs diagram. The angle θ describes the direction of propagation with respect to $\boldsymbol{B}_0$. The parameter β is the ratio of plasma pressure to magnetic pressure.

So these magnetosonic waves with the dispersion relations in (6.134) are longitudinal and always perturb the density of the plasma. They become degenerate when $\theta = 0$.

Since both sound and Alfvén speeds are involved, it is obvious that the key parameter demarcating different physical regimes will be their ratio, or, conventionally, the ratio of the thermal to magnetic energies in the MHD medium, known as plasma beta:

$$\beta = \frac{p_0}{B_0^2/\mu_0} = \frac{c_s^2}{c_A^2} \, . \tag{6.141}$$

The magnetosonic waves can conveniently be summarized by the so-called Friedricks diagram in Figure 6.32.

> **Lab:** *Rayleigh-Taylor instability* can be produced by injecting hot colored water into a glass of cold water as illustrated in Figure 6.34; we are dealing with two immiscible fluids of different density. Initially, this configuration is stable but as the colored water cools down and becomes denser than the water below, it becomes unstable. Rayleigh-Taylor instability produces the "mushrooms" falling. Reproduce the results in Figure 6.34.

Seismology of plasma. Wave motion occurs in solids, liquids, gases, and as we have described above in plasmas. Wave type can be determined by looking at the changes in shape and volume in a medium when it is subjected to sudden deformation. A change in shape in a medium produces shear-wave (also known

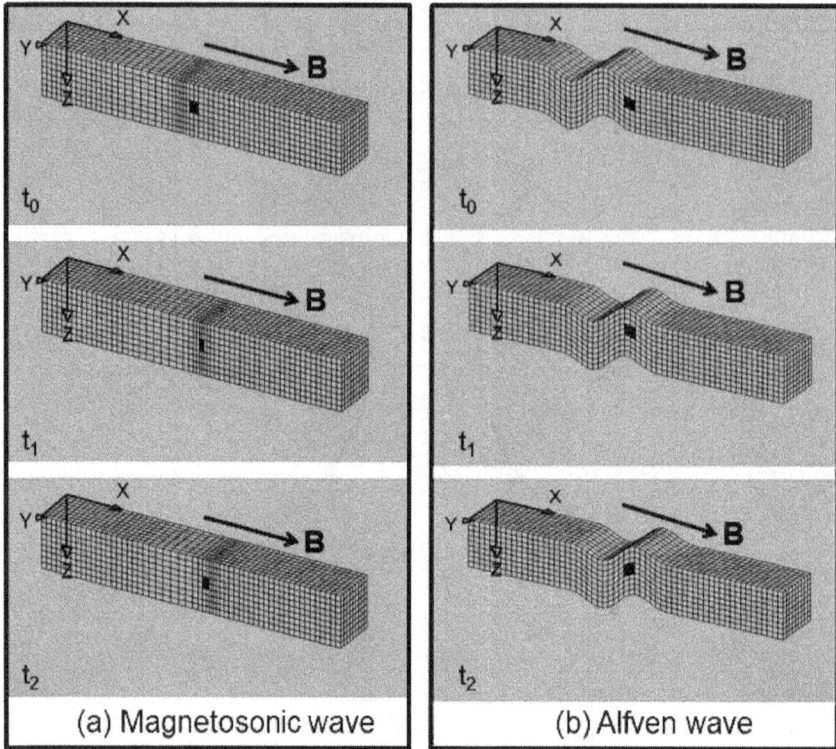

Figure 6.33. Magnetohydrodynamic (MHD) waves include (a) magnetoacoustic waves and (b) Alfvén waves with $t_2 > t_1 > t_0$. There are two magnetoacoustic waves. Only the slow wave that propagates in the direction parallel to the magnetic field, $\boldsymbol{B}$, is shown.

as S-wave) propagation and a change in volume produces compressional wave (P-wave or acoustic wave) propagation. As summarized in Table 6.8 and Figure 6.33, non-viscous fluids support only compressional waves since they cannot be shaped. These waves characterize the compressibility of the medium. Solids support both compressional and shear waves. In fact, there are two shear waves (SH- and SV-waves) since S-waves are polarized in two directions. These three waves characterize the compressibility and rigidity (i.e., elasticity) of solids. Plasma also supports both compressional and shear waves. Also, it has three waves: one shear-wave (Alfvén wave), which is transversely polarized, and two acoustic waves. One of the acoustic waves propagates along the magnetic field lines and the others propagate perpendicularly to the magnetic field lines. So these waves characterize the magnetic field and the particle density of the plasma: slow/fast magnetosonic (compressible) waves, and the Alfvén wave (incompressible).

Table 6.8. Wave-types in solids, liquids, gases, and plasmas.

Solids	liquids and gases	Plasma
•One acoustic wave (P-wave)	•One acoustic wave (P-wave)	•Two acoustic waves (Fast and slow)
•Two shear wave (SV- and SH-waves)	•No shear wave	•One shear wave (Alfvén wave)

P- and S-wave propagation through Earth is the principal tool that scientists have used over the last 100 years or so to develop our current knowledge of Earth's interior, including its main layers, its oil, gas, and water reserves. Similarly, Alfvén and magnetoacoustic waves represent nowadays one of the principal tools in understanding and transporting energy around our solar system, characterizing the solar atmosphere, and predicting space weather, which has become central to communication and the functioning of electrical grids on Earth. In other words, the existence

of MHD waves allows us to understand the elastic and compressible nature of the plasma in the presence of a frozen-in external magnetic field. The forces driving these waves are the gradients of the magnetic and gas pressures, and the magnetic tension.

Lab: *Kelvin-Helmholtz instability.* Acquire a tank as illustrated in Figure 6.35. Fill the tank half full with fresh water ($\rho = 0.998$ g/cm^3) dyed red. Then fill the other half with salt water ($\rho = 1.075$ g/cm^3) dyed blue. There is no mixing between the two types of water. Fresh water stays on top because it has a lower density than salt water. Seal the tank and make sure that there are no air bubbles inside. Then tilt the tank to ensure that two fluids move at different speeds to obtain the results in Figure 6.35.

6.4. MHD Instabilities and Accretion

A plasma, in a steady state (i.e., $\partial/\partial t = 0$), is by definition in a state of equilibrium. If a small perturbation is applied to the plasma and the small perturbation decays with time or just oscillates around the equilibrium configuration we will still consider our configuration stable. In contrast, if this configuration grows with time, the system will become unstable with respect to perturbations. Important phenomena in our universe are caused by unstable configurations, including wave clouds in the Earth's atmosphere described in Chapter 2, mantle convection as described in Chapter 2, sand dunes, particle accelerations, the nature of Jupiter and Saturn's atmospheres, the red spots on Jupiter, solar/stellar eruptions, the interaction of the solar wind with the Earth's magnetosphere, star and galaxy formations, including their spiral arms, accretion disks that we will discuss later and spiral arms, the multiphase nature of the interstellar medium, and supernova explosions in the interstellar medium. In other words, there are several reasons for studying plasmatic instabilities. Actually fluid and plasma instabilities are all around us at all scales. A large portion of nature is made up of fluid and plasma instability, including the process of accretion in the universe which is analogous to Earth's assembly processes. Figures 6.34 and 6.35 illustrate how instability works in certain situations. Our objective in this subsection is to establish the criteria for five such instabilities, namely, convective instability, Jeans instability, magnetorotational (accretion) instability, Rayleigh-Taylor instability, and Kelvin-Helmholtz instability.

Convective instability is concerned with the stability of hydrostatic equilibrium, as described in Chapter 2. Jeans instability is concerned with the stability of a self-gravitational fluid against gravitational collapse. Magnetorotational instability is concerned with the stability of a plasma in orbit around a central object.

(a) t_0 (b) $t_1 > t_0$ (c) $t_2 > t_1$

Figure 6.34. Rayleigh-Taylor (or heavy over light) instability with $t_2 > t_1 > t_0$; a heavy fluid over a light fluid (heavy/light). These two fluids are immiscible. The density difference drives the Rayleigh-Taylor instability.

Rayleigh-Taylor instability and Kelvin-Helmholtz instability are concerned with the stability of an interface with tangential velocity and/or density that continuously changes to zero with respect to the mass flux. We start by introducing the Lagrangian and Eulerian formalisms of small perturbations. These formalisms are essential for the derivation of criteria for these instabilities.

6.4.1.   Eulerian versus Lagrangian Formalisms

The Eulerian formalism uses fixed positions that do not change over time, as we did in the previous subsection. The Lagrangian formalism considers medium and position to be co-moving. Specifically, we are interested in establishing the differences between mathematical Eulerian and Lagrangian differentiation and between the Lagrangian and Eulerian formulations of small perturbation models. We will limit ourselves to non-, See for example, Fiedler and Burton, 2021 and Burton and Noble, 2014, relativistic formulation.

Eulerian and Lagrangian time derivatives. Consider, for example, the plasma density, $\rho(\boldsymbol{x}, t)$ at position $\boldsymbol{x}$ and time t. At time $t + \delta t$ the element will be at position $\boldsymbol{x} + \delta\boldsymbol{x}$. The change in quantity ρ of the plasma element is

$$\frac{\mathrm{D}\rho}{\mathrm{D}t} = \lim_{\delta t \to 0} \left[\frac{\rho(\boldsymbol{x} + \delta\boldsymbol{x}, t + \delta t) - \rho(\boldsymbol{x}, t)}{\delta t} \right] = \lim_{\delta t \to 0} \left[\frac{\partial\rho}{\partial t} + \frac{\delta\boldsymbol{x}}{\partial t} \cdot \boldsymbol{\nabla}\rho + \mathcal{O}(\delta t, |\delta\boldsymbol{x}|) \right], \quad (6.142)$$

where

$$\rho(\boldsymbol{x} + \delta\boldsymbol{x}, t + \delta t) = \rho(\boldsymbol{x}, t) + \frac{\partial\rho}{\partial t}\delta t + \partial\boldsymbol{x} \cdot \boldsymbol{\nabla}\rho + \mathcal{O}(\delta t^2, |\delta\boldsymbol{x}|^2, \delta t|\delta\boldsymbol{x}|),$$

and where $\mathrm{D}/\mathrm{D}t$ is the Lagrangian derivative and $\partial/\partial t$ is the Eulerian derivative, respectively, of the density. So we have the following relation between the Lagrangian and Eulerian derivatives:

$$\underbrace{\frac{\mathrm{D}\rho}{\mathrm{D}t}}_{\text{Lagrangian}} \approx \underbrace{\frac{\partial\rho}{\partial t}}_{\text{Eulerian}} + \underbrace{\frac{\partial\boldsymbol{x}}{\partial t} \cdot \frac{\partial\rho}{\partial\boldsymbol{x}}}_{\text{convective}} = \underbrace{\frac{\partial\rho}{\partial t}}_{\text{Eulerian}} + \underbrace{\boldsymbol{v} \cdot \boldsymbol{\nabla}\rho}_{\text{convective}}. \qquad (6.143)$$

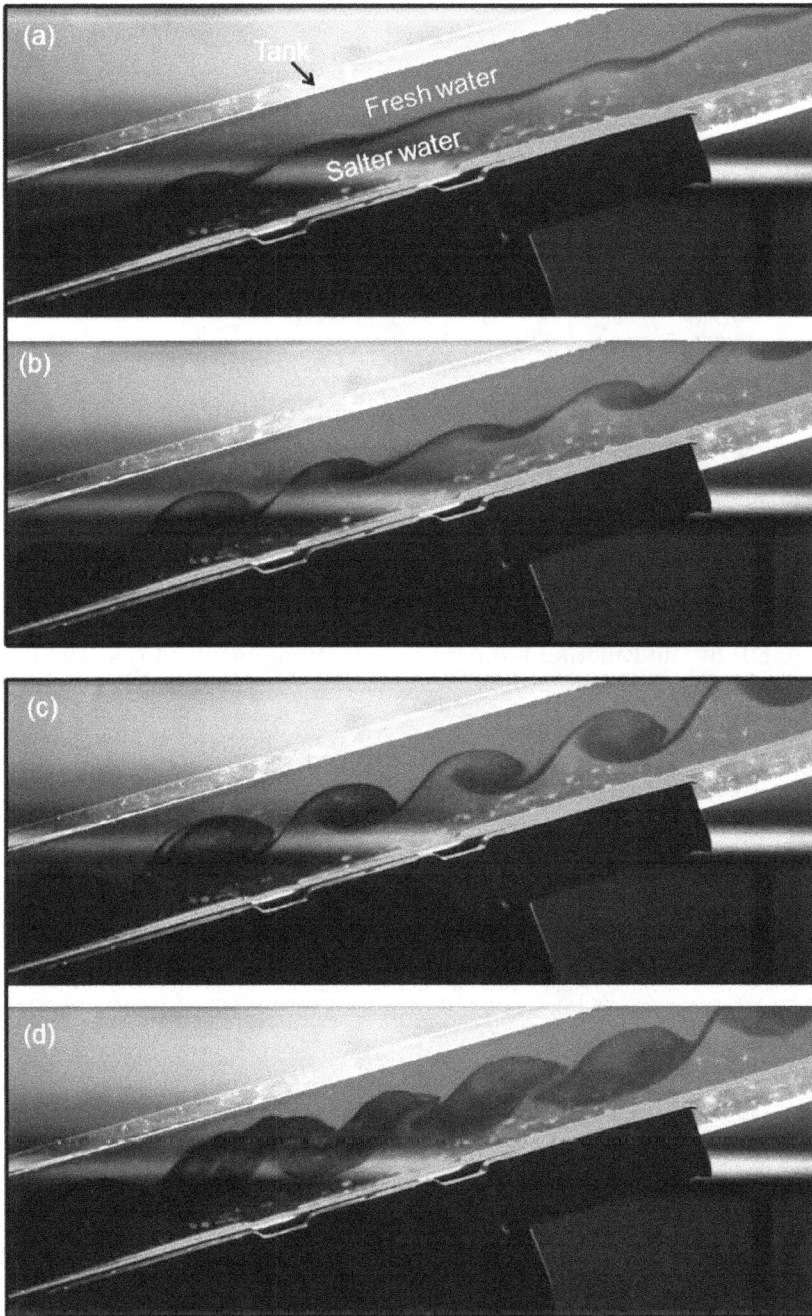

Figure 6.35. (a) Two fluids moving at two different velocities are separated by a flat boundary. The velocity difference drives the Kelvin-Helmholtz instability. In (b), (c), and (d), the interface curls up, as first seen in the famous photograph by Thorpe. Extracted from the experiments of Dr. Barbara Turnbull from Nottingham University in the UK.

This relationship holds for any other quantity. By using these definitions, we can write the continuity and momentum equations in (6.75) and (6.80) with Eulerian time derivatives, in the Lagrangian form as follows:

$$\frac{\mathrm{D}\rho}{\mathrm{D}t} + \rho \boldsymbol{\nabla} \cdot \boldsymbol{v} = 0 \tag{6.144}$$

$$\frac{\mathrm{D}\boldsymbol{v}}{\mathrm{D}t} = -\frac{1}{\rho}\boldsymbol{\nabla}p - \boldsymbol{\nabla}\Psi - \frac{1}{\mu_0}\boldsymbol{B} \times (\boldsymbol{\nabla} \times \boldsymbol{B}) \tag{6.145}$$

or

$$\frac{\mathrm{D}\boldsymbol{v}}{\mathrm{d}t} = -\frac{1}{\rho}\boldsymbol{\nabla}\left(p + \frac{B^2}{2\mu_0}\right) + \frac{1}{\rho}\frac{\boldsymbol{B}\cdot\boldsymbol{\nabla}\boldsymbol{B}}{\mu_0} - \boldsymbol{\nabla}\Psi \ ,$$

where $\boldsymbol{g} = -\boldsymbol{\nabla}\Psi$ is the gravity acceleration and $p + B^2/2\mu_0$ is the total pressure (magnetic plus gas pressure). We can refer specifically to (6.75) as the Eulerian continuity equation and (6.144) as the Lagrangian continuity equation. For incompressible fluids, we have

$$\frac{\mathrm{D}\rho}{\mathrm{D}t} = 0 \Longrightarrow \boldsymbol{\nabla} \cdot \boldsymbol{v} = 0 \ . \tag{6.146}$$

Similarly (6.80) as the Eulerian momentum equation and (6.145) as the Lagrangian momentum continuity equation.

Formulations of Eulerian versus Lagrangian perturbation theories. Just as there is an Eulerian time derivative and a Lagrangian time derivative, there is Eulerian perturbation theory and Lagrangian perturbation theory. Eulerian perturbations are denoted here as 'δ', which measures the change in a quantity. For example, if the equilibrium density at $\boldsymbol{x}$, $\rho_0(\boldsymbol{x})$, is changed at time t by some disturbance to become $\rho(t, \boldsymbol{x})$, then we denote the Eulerian perturbation of the density by

$$\delta\rho = \rho(t, \boldsymbol{x}) - \rho_0(\boldsymbol{x}) \ll \rho_0(\boldsymbol{x}) \ .$$

Again, these perturbations are taken at a fixed position. Lagrangian perturbation theory concerns the evolution of small perturbations about a background state within a particular fluid element as it undergoes a displacement $d\boldsymbol{\xi}$. For example, if the density of a fluid element is displaced from its equilibrium position $\boldsymbol{x}$ to position $\boldsymbol{x} + d\boldsymbol{\xi}$, then the density of that fluid element changes by an amount

$$\begin{aligned} \Delta\rho &= \rho(t, \boldsymbol{x} + \boldsymbol{\xi}) - \rho_0(\boldsymbol{x}) \\[2mm] &\approx \rho(t, \boldsymbol{x}) + \boldsymbol{\xi} \cdot \boldsymbol{\nabla}\rho_0(\boldsymbol{x}) - \rho_0(\boldsymbol{x}) \\[2mm] &\approx \delta\rho(t, \boldsymbol{x}) + \boldsymbol{\xi}(t, \boldsymbol{x}) \cdot \boldsymbol{\nabla}\rho_0(\boldsymbol{x}) \ . \end{aligned} \tag{6.147}$$

This change is a Lagrangian perturbation. By using (6.143), we obtain the Lagrangian velocity perturbation $\Delta\boldsymbol{v}$ as

$$\Delta\boldsymbol{v} = \frac{\mathrm{D}\boldsymbol{\xi}}{\mathrm{D}t} = \left(\frac{\partial}{\partial t} + \boldsymbol{v}_0 \cdot \boldsymbol{\nabla}\right)\boldsymbol{\xi} \ , \tag{6.148}$$

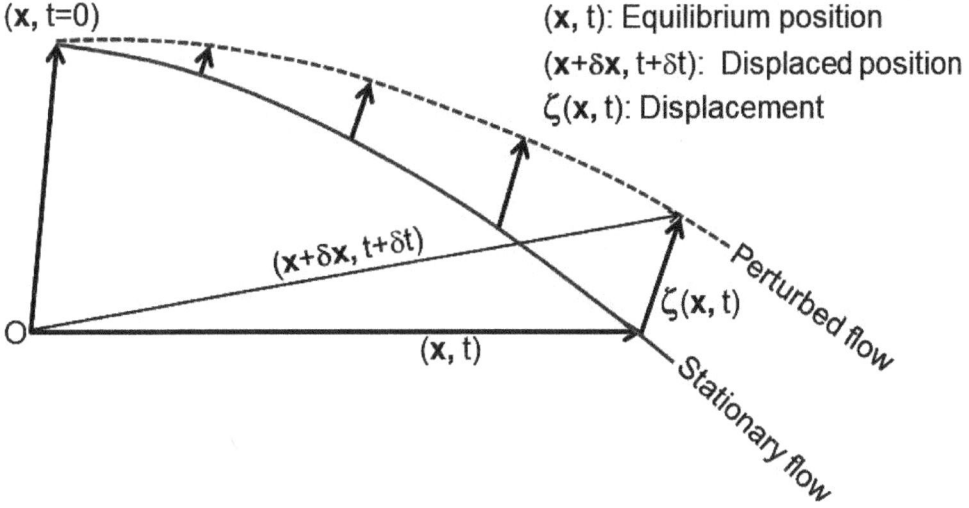

Figure 6.36. Displacement vector field for plasma with stationary background flow. (Adapted from Frieman and Rotenberg, 1960).

where v_0 is the background velocity. Because $\Delta v = \delta v + \xi \cdot \nabla v_0$, we can use (6.148) to obtain

$$\delta v = \frac{\partial \xi}{\partial t} + v_0 \cdot \nabla \xi - \xi \cdot \nabla v_0 \ . \tag{6.149}$$

Note the additional $\xi \cdot \nabla v_0$ term (Figure 6.36), representing a measurement of the background fluid.

Formulations of Eulerian MHD perturbation. Let us seek solutions to ideal MHD equations in the form

$$\rho = \rho_0 + \delta\rho \ , \quad p = p_0 + \delta p \ , \quad \Psi = \Psi_0 + \delta\Psi \ , \quad \text{and} \quad B = B^{(0)} + \delta B \ ,$$

and $v = 0 + \delta v$, where ρ_0 is the equilibrium mass density, v_0 is the flow velocity, p_0 is the equilibrium pressure and B_0 is the equilibrium magnetic field. By inserting these quantities into the ideal MHD equation and assuming that the amplitudes of perturbations are considerably smaller, the Eulerian perturbations $\delta\rho$, δv, δp, δB can be written as follows: equations:

$$\frac{\partial \delta\rho}{\partial t} + \nabla \cdot (\rho_0 \delta v) = 0 \tag{6.150}$$

$$\frac{\partial \delta v}{\partial t} \quad + \quad \rho_0 \nabla (\delta\Psi) + \delta\rho \nabla \Psi_0 + \nabla \delta p$$

$$+ \quad B_0 \times (\nabla \times \delta B) - (\nabla \times B_0) \times \delta B = 0 \tag{6.151}$$

$$\frac{\partial \delta p}{\partial t} + \delta v \cdot \nabla p_0 + \gamma p_0 \nabla \cdot \delta v = 0 \tag{6.152}$$

$$\frac{\partial \delta B}{\partial t} - B_0 \cdot \nabla \delta v + B_0 \nabla \cdot \delta v + v_1 \cdot \nabla B_0 = 0 \tag{6.153}$$

and $\boldsymbol{\nabla} \cdot \delta \boldsymbol{B} = 0$, where ρ_0, $\boldsymbol{v}_0$, p_0, $\boldsymbol{B}_0$ should satisfy the equilibrium conditions of ideal MHD equations.

Box 6.3: Equations of States (EoS)

Equation state relates ρ, T and p as $p = p(\rho, T)$. Most astrophysical fluids can be roughly approximated as ideal gases. The corresponding EoS is

$$p = n k_B T = \frac{k_B}{\mu m_{\mathrm{H}}} \rho T = \frac{\mathcal{R}}{\mu} \rho T \,, \tag{6.154}$$

where μ is the mean particle mass in units of the proton mass m_{H}. Fluids for which p is only a function of ρ are called barotropic fluids. In the isothermal case, T is constant so that $p \propto \rho$. In the adiabatic case, ideal gas undergoes reversible thermodynamic changes such that $p = K \rho^{\gamma}$, where K and γ are constants. In many circumstances, stars behave as polytropes, e.g. fully convective stars with p-ρ relation close to the adiabatic relation. In such a star, assuming monatomic gas with $\gamma = 5/3$, we have $p = K \rho^{5/3}$.

Isothermal case   **Adiabatic case**

$p = \dfrac{\mathcal{R}}{\mu} \rho T$ $p = K \rho^{\gamma}$ EoS

$c_{is}^2 = \left. \dfrac{dp}{d\rho} \right|_T = \dfrac{\mathcal{R}T}{\mu}$ $c_{As}^2 = \left. \dfrac{dp}{d\rho} \right|_S = \dfrac{\gamma p}{\rho} = \dfrac{\gamma \mathcal{R} T}{\mu}$ Phase velocity

We see that c_{is} and c_{As} differ by only $\sqrt{\gamma}$. In ISM, we have $\mu \approx 1$ and $\mu \approx 28$ in the Earth's atmosphere. A fluid system is in hydrostatic equilibrium if $\boldsymbol{v} = \boldsymbol{0}$ and $\partial / \partial t = 0$. By combining continuity equation and momentum equation, we obtain the equation of hydrostatic equilibrium:

$$\boldsymbol{\nabla} p = -\rho \boldsymbol{\nabla} \Psi \,.$$

Assuming a barotropic equation of the state $p = p(\rho)$, this system of equations can be solved.

Frieman–Rotenberg formalism (1960). Let us also seek solutions to ideal MHD equations in terms of $\boldsymbol{\xi}$ (see Figure 6.36). We focus on these perturbations

$$\rho = \rho_0 + \delta \rho \,, \quad p = p_0 + \delta p, \text{ and } \boldsymbol{B} = \boldsymbol{B}^{(0)} + \delta \boldsymbol{B} \,,$$

where ρ_0 is the equilibrium mass density, $\boldsymbol{v}_0$ is the flow velocity, p_0 is the equilibrium pressure and $\boldsymbol{B}_0$ is the equilibrium magnetic field. By inserting these quantities into the ideal MHD equation, we can write them in terms of $\boldsymbol{\xi}$ alone as follows:

$$\Delta \rho \approx \delta \rho + \boldsymbol{\xi} \cdot \boldsymbol{\nabla} \rho_0 \text{ with } \delta \rho = -\boldsymbol{\nabla} \cdot (\rho_0 \boldsymbol{\xi}) \,, \tag{6.155}$$

$$\Delta p \approx \delta p + \boldsymbol{\xi} \cdot \boldsymbol{\nabla} p_0 \quad \text{with} \quad \delta p = -\gamma p_0 \boldsymbol{\nabla} \cdot \boldsymbol{\xi} - \boldsymbol{\xi} \cdot \boldsymbol{\nabla} p_0 , \tag{6.156}$$

$$\Delta \boldsymbol{v} \approx \delta \boldsymbol{v} + \boldsymbol{\xi} \cdot \boldsymbol{\nabla} \boldsymbol{v}_0 \quad \text{with} \quad \delta \boldsymbol{v} = \frac{\partial \boldsymbol{\xi}}{\partial t} + (\boldsymbol{v}_0 \cdot \boldsymbol{\nabla}) \boldsymbol{\xi} - (\boldsymbol{\xi} \cdot \boldsymbol{\nabla}) \boldsymbol{v}_0 , \tag{6.157}$$

$$\Delta \boldsymbol{B} \approx \delta \boldsymbol{B} + \boldsymbol{\xi} \cdot \boldsymbol{\nabla} \boldsymbol{B}_0 \quad \text{with} \quad \delta \boldsymbol{B} = \boldsymbol{\nabla} \times (\boldsymbol{\xi} \times \boldsymbol{B}_0) . \tag{6.158}$$

Equation (6.155) is based on mass conservation, equation (6.156) is based on the equation of state, equation (6.157) is based on the momentum equation, and equation (6.158) is based on the equation of induction. By inserting these quantities into the Lagrangian momentum equation (6.145), we obtain a linearized equation of motion in $\boldsymbol{\xi}$ (Frieman and Rotenberg, 1960):

$$\rho_0 \frac{\partial^2 \boldsymbol{\xi}}{\partial t^2} + 2\rho_0 \boldsymbol{v}_0 \cdot \boldsymbol{\nabla} \left(\frac{\partial \boldsymbol{\xi}}{\partial t} \right) - \boldsymbol{G}(\boldsymbol{\xi}) = 0 , \tag{6.159}$$

where

$$\boldsymbol{G}(\boldsymbol{\xi}) = \boldsymbol{F}(\boldsymbol{\xi}) + \boldsymbol{\nabla} \cdot (\boldsymbol{\xi} \rho_0 \boldsymbol{v}_0 \cdot \boldsymbol{\nabla} \boldsymbol{v}_0 - \rho_0 \boldsymbol{v}_0 \boldsymbol{v}_0 \cdot \boldsymbol{\nabla} \boldsymbol{\xi}) , \tag{6.160}$$

$$\boldsymbol{F}(\boldsymbol{\xi}) = -\boldsymbol{\nabla}(\delta p) - \boldsymbol{B}_0 \times (\boldsymbol{\nabla} \times \delta \boldsymbol{B})$$

$$+ (\boldsymbol{\nabla} \times \boldsymbol{B}_0) \times \delta \boldsymbol{B} + (\boldsymbol{\nabla}\Psi)\boldsymbol{\nabla} \cdot (\rho_0 \boldsymbol{\xi}) , \tag{6.161}$$

where $\boldsymbol{G}$ is the generalized perturbation force acting on the plasma. It contains particular flow terms. In addition, $2\rho_0 \boldsymbol{v}_0 \cdot \boldsymbol{\nabla}$ is an antisymmetric operator known as the *gyroscopic* operator. If $\boldsymbol{\xi} \cdot \boldsymbol{F}(\boldsymbol{\xi}) < 0$, the displacement and force are in opposite directions, the force opposes displacements, and the system will oscillate around the equilibrium. The configuration is stable. The displacement and force are opposite in direction if $\boldsymbol{\xi} \cdot \boldsymbol{F}(\boldsymbol{\xi}) > 0$, and the force encourages displacement, causing the perturbation to grow. The configuration is unstable (see Figure 6.37). If $\boldsymbol{\xi} \cdot \boldsymbol{F}(\boldsymbol{\xi}) = 0$, the perturbation is neutrally stable. Note that the force-like term $\boldsymbol{G}(\boldsymbol{\xi})$ can also be written as the total pressure as follows:

$$\boldsymbol{G} = -\boldsymbol{\nabla}\Pi + \frac{\boldsymbol{B}_0 \cdot \boldsymbol{\nabla}\delta \boldsymbol{B} + \delta \boldsymbol{B} \cdot \boldsymbol{\nabla} \boldsymbol{B}_0}{\mu_0} - g\boldsymbol{\nabla} \cdot (\rho_0 \boldsymbol{\xi})$$

$$+ \boldsymbol{\nabla} \cdot \left(\rho_0 \boldsymbol{\xi} \frac{d\boldsymbol{v}_0}{dt} - \rho \boldsymbol{v}_0 \boldsymbol{v}_0 \cdot \boldsymbol{\nabla} \boldsymbol{\xi} \right) - \frac{\partial(\rho_0 \boldsymbol{v}_0)}{\partial t} \cdot \boldsymbol{\nabla} \boldsymbol{\xi} , \tag{6.162}$$

where

$$\Pi = \delta p + \frac{\boldsymbol{B}_0 \cdot \delta \boldsymbol{B}}{2\mu_0} = -\gamma p_0 \boldsymbol{\nabla} \cdot \boldsymbol{\xi} - \boldsymbol{\xi} \cdot \boldsymbol{\nabla} p_0 + \frac{\boldsymbol{B}_0 \cdot \delta \boldsymbol{B}}{2\mu_0} , \tag{6.163}$$

is the total pressure.

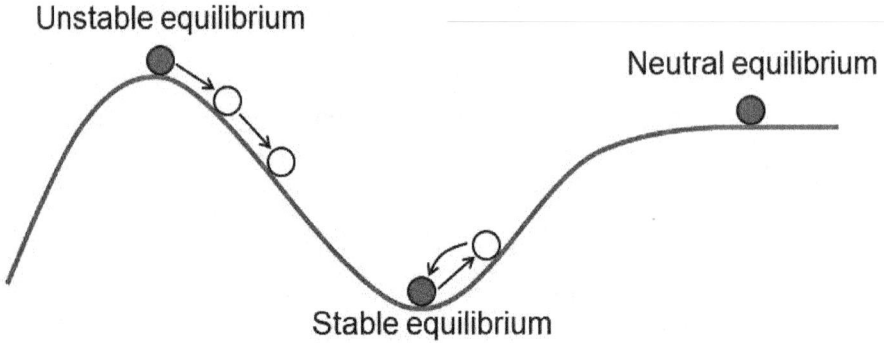

Figure 6.37. If you perturb the stable equilibrium, the ball will oscillate around the equilibrium position. If you perturb the unstable equilibrium, the ball will roll away.

For normal modes, $\boldsymbol{\xi}(\boldsymbol{x}, t) = \hat{\boldsymbol{\xi}}(\boldsymbol{x}) \exp(-i\omega t)$, a quadratic eigenvalue problem is obtained

$$\boldsymbol{G}(\hat{\boldsymbol{\xi}}) + 2i\rho_0\omega\boldsymbol{v}_0 \cdot \boldsymbol{\nabla}\hat{\boldsymbol{\xi}} + \rho_0\omega^2\hat{\boldsymbol{\xi}} = 0 \ . \tag{6.164}$$

When $\boldsymbol{v}_0 = \boldsymbol{0}$ in the background, equation (6.164) reduces to

$$\boldsymbol{F}(\hat{\boldsymbol{\xi}}) = -\rho_0\omega^2\hat{\boldsymbol{\xi}} \ . \tag{6.165}$$

For $\omega^2 > 0$ for all solutions, the equilibrium is stable (oscillatory) and for $\omega^2 < 0$ for all solutions, the equilibrium is unstable (exponential growth). The formulation of stability and instability in terms of dispersion is mathematically more straightforward than that in Figure 6.37. I will therefore use that formulation. So when $\omega^2 > 0$ the system is stable and when $\omega^2 < 0$, the system is unstable or when ω is real-valued frequency the system is stable and when ω is complex-valued frequency, the system is unstable.

Example: MHD of stratified static plasmas in Cartesian coordinates.
Consider the case in which $\rho_0 = \rho_0(z)$, $\boldsymbol{v}_0(z) = [v_{x0}(z), v_{y0}(z), 0]^T = \text{const.}$, $\boldsymbol{B}_0(z) = [B_{x0}(z), B_{y0}(z), 0]^T$, and $p_0 = p_0(z)$; that is, we choose an equilibrium magnetic field in the $x - y$ plane that varies in the z direction and the gravitational acceleration $\boldsymbol{g} = -g\boldsymbol{i}_z$. So we choose z as the stratification direction. Plasmas are considered incompressible here. We can verify that $\boldsymbol{v}_0 \cdot \boldsymbol{\nabla}\boldsymbol{v}_0 = 0$. By using the fact that the background density is independent of time, the mass conservation equation is reduced to $\boldsymbol{\nabla}(\rho\boldsymbol{v}) = 0$. So we can also verify that

$$\boldsymbol{\nabla} \cdot (\rho_0\boldsymbol{v}_0\boldsymbol{v}_0 \cdot \boldsymbol{\nabla}\hat{\boldsymbol{\xi}}) = \rho_0(\boldsymbol{v}_0 \cdot \boldsymbol{\nabla})^2\hat{\boldsymbol{\xi}}$$

and equation (6.164) reduces to

$$\boldsymbol{F}(\hat{\boldsymbol{\xi}}) = -\rho_0\tilde{\omega}^2\hat{\boldsymbol{\xi}} \quad \text{with} \quad \tilde{\omega} = \omega + i\boldsymbol{v} \cdot \boldsymbol{\nabla} = \omega - \Omega_0(z) \ , \tag{6.166}$$

where $\Omega_0(z) = \boldsymbol{k}_0 \cdot \boldsymbol{v}(z)$ represents the local Doppler shift and $\tilde{\omega}(z)$ as the local Doppler shifted frequency observed in a local frame co-moving with the plasma layer at the vertical position z. For static plasmas, $\Omega_0(z)$ is obvious zero.

We can rotate the coordinate system so that $k_x = 0$ so that we choose the displacement of this form

$$\boldsymbol{\xi}(\boldsymbol{x}, t) = \left[\hat{\xi}_x(z; k_x, k_y), 0, \hat{\xi}_z(z; k_x, k_y)\right]^T \exp(ik_x x + ik_y y - i\omega t) , \quad (6.167)$$

where the quantities with hat are Fourier amplitudes. Because we assume incompressibility, the two components are related as follows:

$$\boldsymbol{\nabla} \cdot \boldsymbol{\xi} = \frac{d\hat{\xi}_z}{dz} + ik_x \hat{\xi}_x = 0 . \quad (6.168)$$

In other words, only one component is needed because the other can be deduced from incompressibility. Let us focus on vertical displacement which we will denote $\hat{\xi} = \hat{\xi}_z$. Let us start with the case $\boldsymbol{B} = \boldsymbol{0}$. After a lengthy algebra, we can verify that (6.164) becomes (e.g., Goedbloed and Poedts, 2004; and Goedbloed et al., 2019)

$$\frac{d}{dz}\left[\frac{\gamma p_0 \, \rho_0 \omega^2}{\rho \omega^2 - k_0^2 \gamma p_0} \frac{d\hat{\xi}}{dz}\right] + \left[\rho_0 \omega^2 - \frac{k_0^2 \rho_0^2 \, g^2}{\rho_0 \omega^2 - k_0^2 \gamma p_0} - \frac{d}{dz}\left(\frac{k_0^2 \gamma p_0 \, \rho_0 g}{\rho \omega^2 - k_0^2 \gamma p_0}\right)\right]\hat{\xi} = 0 . (6.169)$$

For incompressible plasmas (i.e., $\gamma \longrightarrow \infty$), we can assume that $\rho|\omega^2| \ll k_0^2 \gamma p_0$ in addition to the assumption we have already made that $B_0^2 = BB_0 \cdot \boldsymbol{B}_0 = 0$. We also assume that gravity is constant. We can verify that (6.169) reduces to

$$\frac{d}{dz}\left[\rho_0 \omega^2 \frac{d\hat{\xi}}{dz}\right] - k_0^2 \rho \left(\omega^2 - N_B^2\right) \hat{\xi} = 0 , \quad (6.170)$$

where

$$\begin{aligned}
N_B^2 &= -\frac{1}{\rho_0}\left(\frac{\rho_0^2 g^2}{\gamma p_0} + g\frac{d\rho_0}{dz}\right) \\
&= -g\left(\frac{1}{\rho_0}\frac{d\rho_0}{dz} - \frac{1}{\gamma p_0}\frac{dp_0}{dz}\right) \\
&= -g\left(\frac{d\ln(\rho_0)}{dz} - \frac{1}{\gamma}\frac{d\ln(p_0)}{dz}\right) \\
&= \frac{g}{\gamma}\frac{d\ln\left(p_0 \rho_0^{-\gamma}\right)}{dz}
\end{aligned} \quad (6.171)$$

is the so-called Brunt-Väisälä frequency. In (6.171), we use the equilibrium equation, $dp_0/dz = -g$. For an isothermal gas $\gamma = 1$ and the temperature gradient is zero. Thus, in an isothermal celestial body the Brunt-Väisälä frequency is zero. An atmosphere with a steep temperature gradient is likely to be convective (with negative N_B^2 and not satisfying the Schwarzschild criterion for stability, as described in Chapter 2) whereas one with a shallow temperature gradient should be stable (and with positive N_B^2). We will return to this frequency in our discussion of convective instability later on. When the magnetic field is nonzero (6.169) becomes

$$\frac{d}{dz}\left[\frac{(\gamma p_0 + B_0^2)\rho_0\left(\omega^2 - \omega_A^2\right)\left(\omega^2 - \frac{\gamma p_0}{\gamma p_0 + B^2}\omega_A^2\right)}{\rho_0\omega^4 - k_0^2(\gamma p_0 + B_0^2)\left(\omega^2 - \frac{\gamma p_0}{\gamma p_0 + B^2}\omega_A^2\right)}\frac{d\hat{\xi}}{dz}\right]$$

$$+\left[\rho_0\left(\omega^2 - \omega_A^2\right) - \frac{k_0^2\rho_0^2 g^2\left(\omega^2 - \omega_A^2\right)}{\rho_0\omega^4 - k_0^2(\gamma p_0 + B_0^2)\left(\omega^2 - \frac{\gamma p_0}{\gamma p_0 + B^2}\omega_A^2\right)}\right.$$

$$+\left.\frac{d}{dz}\left\{\rho_0 g - \frac{\rho_0^2 g\omega^2\left(\omega^2 - \omega_A^2\right)}{\rho_0\omega^4 - k_0^2(\gamma p_0 + B_0^2)\left(\omega^2 - \frac{\gamma p_0}{\gamma p_0 + B^2}\omega_A^2\right)}\right\}\right]\hat{\xi} = 0, \quad (6.172)$$

where $\omega_A = \boldsymbol{k}_0 \cdot \boldsymbol{B}_0/\sqrt{\rho_0\mu_0}$ is the Alfvén angular frequency. We can verify that by setting $B_0 = 0$, which implies that $\omega_A = 0$, we will return to (6.172). For incompressible plasmas (i.e., $\gamma \longrightarrow \infty$), we can assume that $\rho_0|\omega^2| \ll k_0^2(\gamma p_0 + B_0^2)$. We can verify that (6.172) reduces to

$$\frac{d}{dz}\left[\rho_0\left(\omega^2 - \omega_A^2\right)\frac{d\hat{\xi}}{dz}\right] - k_0^2\left[\rho_0\left(\omega^2 - \omega_A^2\right) - \frac{\rho_0^2 g^2}{\gamma p_0 + B_0^2} - \frac{d\rho_0}{dz}g\right]\hat{\xi} = 0 , \quad (6.173)$$

or

$$\frac{d}{dz}\left[\rho_0\left(\omega^2 - \omega_A^2\right)\frac{d\hat{\xi}}{dz}\right] - k_0^2\rho_0\left[\left(\omega^2 - \omega_A^2\right) - N_M^2\right]\hat{\xi} = 0 , \quad (6.174)$$

where

$$N_M^2 = -\frac{1}{\rho_0}\left(\frac{\rho_0^2 g^2}{\gamma p_0 + B_0^2} + \frac{d\rho_0}{dz}g\right) \quad (6.175)$$

is the magnetically modified Brunt-Väisälä frequency because it now contains magnetic contribution.

Example: MHD of stratified stationary plasmas in Cartesian coordinates. Since the equilibrium conditions are the same as for static and stationary plasmas, this implies that the equations in (6.169) through (6.176) are valid for the stationary plasmas by replacing ω with $\tilde{\omega}(z)$. For incompressible magnetized plasmas (i.e., $\gamma \longrightarrow \infty$), for example, (6.176) becomes

$$\frac{d}{dz}\left[\rho_0\left(\tilde{\omega}^2 - \omega_A^2\right)\frac{d\hat{\xi}}{dz}\right] - k_0^2\left[\rho_0\left(\tilde{\omega}^2 - \omega_A^2\right) + \frac{\rho_0^2 g^2}{\gamma p_0 + B_0^2} + \frac{d\rho_0}{dz}g\right]\hat{\xi} = 0 , \quad (6.176)$$

with

$$\Pi = \frac{\rho_0}{k_0^2}\left(\tilde{\omega}^2 - \omega_A^2\right)\frac{d\hat{\xi}}{dz} , \quad (6.177)$$

where $\tilde{\omega}(z) = \omega - \Omega_0(z)$, $\Omega_0(z) = \boldsymbol{k}_0 \cdot \boldsymbol{v}(z)$, $\boldsymbol{k}_0 = [k_x, k_y, 0]^T$ is the horizontal wave vector, and Alfvén angular frequency is $\omega_A = \boldsymbol{k}_0 \cdot \boldsymbol{B}_0/\sqrt{\rho_0\mu_0}$. opens up the possibility of new flow-driven (Kelvin–Helmholtz) instabilities as we will discuss later. When the density is constant and $\rho_0^2 g_0^2 \ll k_0^2(\gamma p_0 + B_0^2)$, (6.176) further reduces

$$\frac{d^2\hat{\xi}}{dz^2} - k_0^2\hat{\xi} = 0 . \qquad (6.178)$$

For the case of two half spaces, with the upper half-space representing the interval $[0, +\infty]$ and the lower half-space representing the interval $[0, -\infty]$, there are two boundary conditions describing the continuity the normal velocity and the pressure balance at $z = 0$; i.e.,

$$\left[\boldsymbol{n} \cdot \hat{\boldsymbol{\xi}}\right]_1^2 = 0 \quad \text{and} \quad \left[\Pi + \boldsymbol{n} \cdot \boldsymbol{\xi}\boldsymbol{n} \cdot \boldsymbol{\nabla}\left(p + \frac{B^2}{2\mu_0}\right)\right]_1^2 = 0 , \qquad (6.179)$$

where $\boldsymbol{n}$ represents the normal vector to the interface of the half spaces. Note that celestial bodies are generally in hydrostatic equilibrium which implies that the profiles: $p(z)$, $B_x(z)$, and $B_y(z)$ are related as follows:

$$\frac{d}{dz}\left[p(z) + \frac{B^2(z)}{2\mu_0}\right] = -\rho g . \qquad (6.180)$$

Example: stratified MHD in cylindrical coordinates. Let us consider the Frieman–Rotenberg formalism in cylindrical coordinates $\boldsymbol{r} = [r, \theta, z]^T$ in a one-dimensional cylindrical equilibrium, described by the density $\rho(r)$, the pressure $p(r)$, the magnetic field components $B_\theta(r)$ and $B_z(r)$ or $\boldsymbol{B} = [0, B_\theta(r), B_z(r)]^T$ and the velocity components $v_\theta(r)$ and $v_z(r)$ or $\boldsymbol{v} = [0, v_\theta(r), v_z(r)]^T$. The equilibrium is restrained by just a single differential equation in the radial coordinate r. The gravitational potential $\Psi(r)$ is considered to be due to a compact object centered on the axis $r = 0$. Because of the cylindrical symmetry, we consider Fourier components of the normal modes of the form

$$\boldsymbol{\xi}(r, \theta, z) = [\xi_r(r), \xi_\theta(r), \xi_z(r)]^T \exp\left[i\left(m\theta + kz - \omega t\right)\right] . \qquad (6.181)$$

The spectral equation for cylindrical plasmas is (e.g., Bondeson et al., 1987)

$$\boldsymbol{F}(\boldsymbol{\xi}) - \underbrace{\left[\left(\frac{\rho_0 v_\theta^2}{r}\right)\boldsymbol{\nabla}\cdot\boldsymbol{\xi} + r\frac{d}{dr}\left(\frac{\rho_0 v_\theta^2}{r^2}\right)\xi_r\right]\boldsymbol{i}_r}_{\sim \text{ centrifugal force}} + 2i\rho\tilde{\omega}\left(\frac{v_\theta}{r}\right)\boldsymbol{i}_z \times \boldsymbol{\xi} + \rho\tilde{\omega}^2\boldsymbol{\xi} = \boldsymbol{0}, \qquad (6.182)$$

where

$$\tilde{\omega} = \omega - \Omega_0 = \omega - \underbrace{\frac{mv_\theta}{r}}_{\text{Coriolis}} - kv_z . \qquad (6.183)$$

Note that in cylindrical coordinates, the spectral equation becomes much more involved since $\boldsymbol{v} \cdot \boldsymbol{\nabla} \boldsymbol{v}$ in (6.160) now yields centrifugal acceleration and the operator $\boldsymbol{v} \cdot \boldsymbol{\nabla}$ gives an additional Coriolis contributions to the frequency in (6.166). Note also that in the absence of rotation, $v_\theta = 0$, the simple "quasi-static" form $\boldsymbol{F}(\boldsymbol{\xi}) = -\rho \tilde{\omega}^2 \boldsymbol{\xi}$ is again obtained.

Let us assume that plasma is incompressible, the $\boldsymbol{\nabla} \cdot \boldsymbol{\xi} = \boldsymbol{0}$, the displacement is limited to the disk, $v_z = 0$ (i.e., a purely azimuthal velocity field), and the vertical magnetic field is zero, $B_\theta = 0$. Furthermore, we restrict the analysis to vertical wave numbers k only; that is, $m = 0$. These assumptions hold for disk accretion which we will discuss later. The spectral problem then simplifies to the solution of

$$r \frac{d}{dr}\left(\frac{1}{r}\frac{d\chi}{dr}\right) - k^2 \left[1 - \frac{\kappa^2(r)}{\omega^2 - \omega_A^2} - \frac{4\omega_A^2 \Omega^2(r)}{\left(\omega^2 - \omega_A^2\right)^2}\right] \chi = 0 \qquad (6.184)$$

where

$$\chi = r\xi , \quad \Omega = \frac{v_\theta}{r} , \quad \text{and} \quad \kappa^2 = \frac{1}{r^3}\frac{d\left(r^4 \Omega^2\right)}{dr} = 2r\Omega\frac{d\Omega}{dr} + 4\Omega^2 , \qquad (6.185)$$

subject to the boundary $\xi(r_1) = \xi(r_2) = 0$. Note that celestial bodies are generally in hydrostatic equilibrium. The cylindrical coordinates equilibrium implies that the profiles: $p(r)$, B_θ, and $B_z(r)$ are related as follows:

$$\frac{d}{dr}\left[p(r) + \frac{B^2(r)}{2\mu_0}\right] + \frac{B_\theta^2(r)}{r} = \rho\frac{1}{r}v_\theta^2(r) - \rho g . \qquad (6.186)$$

Hence, $\rho(r)$, $p(r)$, $v_\theta(r)$, B_θ, and $B_z(r)$ are related. Only the component $v_z(r)$ can be chosen arbitrarily. Notice that linearized MHD equation for stratified plasma can be written as ordinary differential equation (ODEs) in the 1D case, irrespective of the coordinate system, in the following form:

$$\frac{d}{d\alpha}\left[P(\alpha; \omega^2)\frac{d\xi}{d\alpha}\right] + Q(\alpha; \omega^2)\xi = 0 , \qquad (6.187)$$

where α can be z, y, x, and r. Very efficient numerical routines exist to integrate a system of n nonlinear first order differential equations. Therefore, it is a useful exercise to learn to transform the above second order ODE into a system of two first order (linear) ODEs.

6.4.2. Gravitational Collapse

Galaxies contain giant molecular clouds (GMCs). The stability of these clouds is ensured by the balance between self-gravity which pulls matter in and internal cloud pressure which pushes it back out. When cloud pressure exceeds gravity forces, oscillations occur (sound waves). When gravity overcomes cloud pressure, the cloud collapses. For example, small perturbations in a cloud can cause them to grow exponentially and become unstable. This leads to the collapse of the cloud which results in star formation. Jeans instability describes gravitational instability in self-gravitating clouds. Our objective here is to derive this instability. Other forces are also involved in these processes, including magnetic and rotational forces, as well as external pressure. The influence of these forces on Jeans instability will also be discussed.

Box 6.4: Molecular Clouds

A molecular cloud is a vast area of the interstellar medium—the region between stars inside a galaxy—with 99 percent hydrogen gas by number of atoms along with helium. It also contains a trace amount of heavier elements produced by earlier stellar evolution such as CO. A molecular cloud is denser than its surroundings. Also molecular clouds are very inhomogeneous and have much higher-density regions called *clumps* and *cores*. Their clumps can be shaped, scattered, or compressed by various stellar forces, including stellar winds, photon bombardment, and supernova shock waves. As more matter accumulates in a certain area, density, pressure, and temperature rise. As the gas cloud density increases, it loses energy from photon emission, cooling it in the process. As it cools, the monoatomic hydrogen gas forms into diatomic hydrogen molecules (H_2); traces of H_2O, N_2 will also form if heavier elements are present. We now have a molecular cloud. These clouds are extremely cold, typically around 10 K, mainly because some of their molecules are very efficient at radiating away energy. Their masses are generally between 10 m_{sun} to 10^6 m_{sun}, their average density is 100 cm^{-3}, and their sizes are between 10 and 100 parsecs (32.6 light years to 326 light years).

Gravity collapses the cloud. This is where protostars (young stars)[a] form. When a temperature of 15 million degrees Celsius is reached, nuclear fusion occurs, and a star is born. By the time the Sun formed completely, it had used up almost all the mass in its vicinity. It was surrounded by a spinning, swirling disc of leftover material called a solar nebula. Inside this flattened disk, the planets, moons, asteroids and comets that orbit the sun are created.

Clouds are mostly detected by the emission from the CO molecules within them. CO molecules can rotate (see Ikelle, 2023a; and Chapter 2). Molecules that rotate with 1 quantum of angular momentum can spontaneously emit a photon and stop rotating (giving their angular momentum to the photon). These photons have energies of $U = 4.8 \times 10^{-4}$ eV, so the corresponding frequency is $f = U/h = 115$ GHz. This is in the radio part of the spectrum. These photons can penetrate the Earth's atmosphere and be detected by radio telescopes. Radio waves are largely unaffected by dust, allowing us to detect HI clouds—clouds consisting mostly of neutral hydrogen atoms (one proton and one electron). Most of the matter between the stars in the Milky Way Galaxy, as well as in other spiral galaxies, occurs in the form of relatively cold HI gas.

---

[a]In Ikelle (2023b), we discussed the structure of hydrogen-burning stars, which constitute the main sequence in Figures 6.5 and 6.6. This is the phase in which stars spend the majority of their lives. Young stars are those that have not yet reached the main sequence.

Unmagnetized Jeans instability using perturbations. Consider a uniform molecular cloud with density ρ_0 and pressure p_0 without motion $\boldsymbol{v}_0 = \boldsymbol{0}$ in the gravitational potential, Ψ_0. In this uniform gas distribution, we assume small perturbations caused, for example, by a shock wave due to a nearby supernova or a passing spiral arm of the galaxy. As a result, density, pressure, and velocity distributions are perturbed from their original states. We have

$$\rho = \rho_0 + \delta\rho \,, \quad p = p_0 + \delta p \,, \quad \Psi = \Psi_0 + \delta\Psi \,, \quad \text{and} \quad \boldsymbol{v} = \delta\boldsymbol{v},$$

where the amplitudes of perturbations are assumed to be smaller, that is, $|\delta\rho|/\rho_0 \ll 1$, $|\delta p|/p_0 \ll 1$ and $|\delta\boldsymbol{v}|/c_s \ll 1$. Variables change only in the z direction. Using (6.150)-(6.153), we have

$$\frac{\partial\delta\rho}{\partial t} + \rho_0\frac{\partial\delta v_z}{\partial z} = 0 \,, \quad \rho_0\left(\frac{\partial\delta v_z}{\partial t}\right) = -\frac{\partial\delta p}{\partial z} - \rho_0\frac{\partial\delta\Psi}{\partial z} \,,$$

and

$$p_0 = c_{is}^2\rho_0 = \frac{k_B}{\mu m_{\mathrm{H}}}\rho_0 T \quad \text{and} \quad \delta p = c_{is}^2\delta\rho \implies c_{is}^2 = \frac{k_B}{\mu m_{\mathrm{H}}}T \,, \tag{6.188}$$

where μ is the mean molecular weight (with $\mu = 1$ for the gas fully neutral), m_{H} is the mass of hydrogen, and T is the mean cloud's temperature. In addition, we have assumed the perturbation itself to be isothermal so that c_{is} remains unchanged. By substituting the continuity equation and the state equation in the momentum equation, we arrive at

$$\frac{\partial\delta v_z}{\partial t} = c_{is}^2\frac{\partial\delta\rho}{\partial z} - \rho_0\frac{\partial\delta\Psi}{\partial z} \,. \tag{6.189}$$

By taking the derivative with respect to z of (6.189) and substituting the continuity equation and the Poisson equation ($\boldsymbol{\nabla}^2\delta\Psi = 4\pi G_0\delta\rho$) in the resulting equation, we obtain the growth of density perturbation due to the self-gravity; i.e.,

$$\frac{\partial^2\delta v_z}{\partial z\partial t} = -c_{is}^2\frac{\partial^2\delta\rho}{\partial z^2} - 4\pi G_0\rho_0\delta\rho \underbrace{\implies}_{\text{continuity}} \frac{\partial^2\delta\rho}{\partial t^2} = c_{is}^2\frac{\partial^2\delta\rho}{\partial z^2} + 4\pi G_0\rho_0\delta\rho \,. \tag{6.190}$$

If $G_0 = 0$, this equation expresses the propagation of sound waves in a homogeneous medium. In other words, this equation shows how sound waves propagate in a self-gravitational medium. By decomposing the perturbations in Fourier modes of the form $\delta\rho(z,t) \propto \exp[i(kz - \omega t)]$, we obtain the following dispersion relation

$$\omega^2 = c_{is}^2 k^2 - 4\pi G_0\rho_0 \,. \tag{6.191}$$

We can distinguish two cases: $\omega^2 > 0$ the wave is an ordinary oscillatory wave and instability occurs when $\omega^2 < 0$. The critical wavenumber (i.e., the border between these two regimes; $\omega = 0$) and critical wavelength are

$$k_J = \sqrt{\frac{4\pi G_0\rho_0}{c_{is}^2}} \quad \text{and} \quad \lambda_J = \frac{2\pi}{k_J} = \sqrt{\frac{\pi c_{is}^2}{G_0\rho_0}} \,, \tag{6.192}$$

respectively, where λ_J is known as the Jeans wavelength, and ω^2 becomes negative for $k < k_J$. The mass contained in a sphere with a radius $r_J = \lambda_J/2$ is often called the Jeans mass, which gives a typical mass scale above which the cloud collapses. The typical Jeans mass is as follows:

$$
\begin{aligned}
m_J &= \frac{4\pi}{3}\rho_0 r_J^3 = \frac{4\pi}{3}\rho_0 \left(\frac{\lambda_J}{2}\right)^3 \\
&= \frac{\pi^{5/2}}{6} \frac{c_{is}^3}{\left(G_0^3\rho_0\right)^{1/2}} = \frac{\pi^{5/2}}{6}\left(\frac{k_B}{G_0\mu m_{\mathrm{H}}}\right)^{3/2} T^{3/2}\rho_0^{-1/2} ,
\end{aligned} \qquad (6.193)
$$

where T is the mean temperature. We used (6.188) in the derivation of (6.193). So Jeans' mass is small if the cloud temperature is low and initial cloud density is high. If $m > m_J$, the cloud collapses leading to protostar formation. This condition is known as *the Jeans criterion* for gravitational instability. If $m < m_J$, the cloud re-expands after end of external compression. The Jeans instability occurs on smaller and smaller scales as the cloud contracts. This results in an increase in density, leading to fragmentation into many small pieces. In other words, as the cloud disintegrates, a dense, hot core forms and collects dust and gas. Not all of this material becomes a star; the remaining dust may form planets, asteroids, or comets, or it may remain as dust.

By rewriting equation (6.189) of velocity, Jeans wavelength (6.192), and Jeans mass, (6.193), as follows,

$$
c_{is} = 287.53 \, \mu^{-1/2} \left(\frac{T}{10 \text{ K}}\right)^{1/2} \text{ m/s}
$$

$$
\lambda_J = 6.39 \times 10^{-14} \left(\frac{1}{\mu < m >}\right)^{1/2} \left(\frac{T}{10 \text{ K}}\right)^{1/2} \left(\frac{n}{10^9 \text{ m}^{-3}}\right)^{-1/2} \text{ parsec}
$$

and

$$
m_J = 2.02 \times 10^{-12} \, \mu^{-3/2} < m >^{-1/2} \left(\frac{T}{10 \text{ K}}\right)^{3/2} \left(\frac{n}{10^9 \text{ m}^{-3}}\right)^{1/2} m_{\mathrm{sun}}
$$

where $< m >$ is the average mass per particle. In the totally neutral cloud of hydrogen in the form of H_2, $\mu = 2.33$. We can clearly see that the key parameters driving cloud collapse are the cloud's temperature, T, and particle number density, n. When μ and $< m >$ are fixed $\log_{10} c_{is}$ is linearly related to the cloud's temperature, T. Similarly, when μ, $< m >$, and T are fixed $\log_{10} \lambda_J$ is linearly related to cloud's particle number density, n, and $\log_{10} m_J$ is also linearly related to cloud's particle number density.

As we can see in Table 6.9, typical values in molecular clouds, such as $c_{is} = 200$ m/s, $\rho_0 = n < m > = 1.67 \times 10^{-21}$ kg/m^3 and $n = 10^6$ m^{-3} , give us a Jeans wavelength of $\lambda_J = 1.064 \times 10^{18}$ m = 34.5 pc. So a gas cloud with isothermal sound speed c_{is} and density ρ_0 and a mass larger than the Jeans mass $m_J \simeq 529.5 \, m_{\mathrm{sun}}$ will eventually collapse due to its own gravitation. Clouds with a much higher particle

Table 6.9. Physical Properties of Molecular Clouds. The constants used in the derivations of these properties include $G_0 = 6.67 \times 10^{-11}$ m^3/(kgs^2), $k_B = 1.380649 \times 10^{-23}$ m^{2}kgs^{-2}K^{-1}, and $m_H = 1.67 \times 10^{-27}$ kg. The conversion includes 1 pc = 206,265 AU = 3.09×10^{16}. Myrs stands for million years and pc stands parsec. Giant molecular clouds (GMCs) are large clouds with $10^4 m_{sun} < 10^7$ m_{sun} sizes in the range 10-100 pc.

μ	$<m>$ (10^{-27} kg)	c_{is} (m/s)	T (K)	n (m^{-3})	λ_J (pc)	m_J (m_{sun})	t_{ff} (Myrs)	Examples Cloud
2.33	3.34	84.24	2	5×10^7	1.45	3.9	5.15	Cold
2.33	3.34	133.2	5	1×10^{12}	0.0072	0.05	0.0163	Cold
1.00	1.67	287.5	10	10^6	49.5	1565	51.5	
1.00	1.67	287.5	10	10^8	4.95	156.5	5.15	Cold
1.00	1.67	287.5	10	10^{10}	0.495	15.66	0.51	Core/B68
2.00	1.67	203.3	10	10^6	32.42	440.21	51.5	B335
1.00	1.67	642.9	50	10^6	110.6	17,505	51.5	Diffuse
1.00	1.67	642.9	50	10^{10}	1.11	175.05	0.51	
1.00	1.67	909.3	100	10^6	156.5	49,511	51.5	Dark
1.00	1.67	909.3	100	10^{10}	1.565	495.11	0.51	
2.33	1.67	5327.8	8000	10^6	916.8	10^7	51.5	Warm
2.33	1.67	53278	800000	10^4	91,687	10^{11}	515	Hot

density, $n = 10^8$ m^{-3} has $m_J = 52.95 m_{sun}$. These clouds are typically between 10^2 and 10^7 m_{sun} in mass. However, they occupy a tiny volume of the galaxy, because they are denser than the gas around them. With number densities of $n = 10^6$ m^{-3} and temperatures of $T = 8,000$ K, we can get Jean mass as high as $m = 5 \times 10^6$ m_{sun}—in other words, huge clouds can be held up by pressure. Notice that the cloud can be thousands of times the Sun's mass, as shown in Table 6.9. Note also that stars have sizes of only about 10^{-7} pc, and densities of about 1 g/cm^3. Hence, during star formation, densities can increase by more than 20 orders of magnitude.

As described in Chapter 2, a cloud of gases subjected to gravity collapses at

$$t_{ff} = \left(\frac{3\pi}{32 G_0 \rho_0} \right)^{1/2} = 6.65 \times 10^{-8} \ <m>^{-1/2} \ \left(\frac{n}{10^9 \text{ m}^{-3}} \right)^{-1/2} \text{ million years } , \quad (6.194)$$

where time t_{ff} is the so-called free-fall time t_{ff} of the cloud. That is, the time it takes until the cloud contracts to a point at $r = 0$. For interstellar densities $\rho_0 = \times 10^{-15}$ g/cm^{-3} = 10^{-21} kg/m^{-3}, the free-fall time is $t_{ff} \approx 51.7$ million years. This time is very fast by astronomical standards. We should therefore be very lucky to catch a star in formation. In other words, without the opposite forces, especially gas pressure, this cloud would collapse very quickly! Figure 6.38 illustrates how the Jeans instability causes the collapse of interstellar gas clouds and subsequent protostars and star formation. This is when the internal gas pressure is not strong enough to prevent the gravitational collapse of a region filled with matter. Note that self-gravity and associated instability are essential to galaxies and clusters. The results in (6.189), (6.192), (6.193) and (6.194) ignore both angular momentum (rotation) and magnetic fields, which will counteract collapse, and therefore lengthen the true collapse time. The free-fall collapse time given above is a lower limit to a more realistic collapse time calculation.

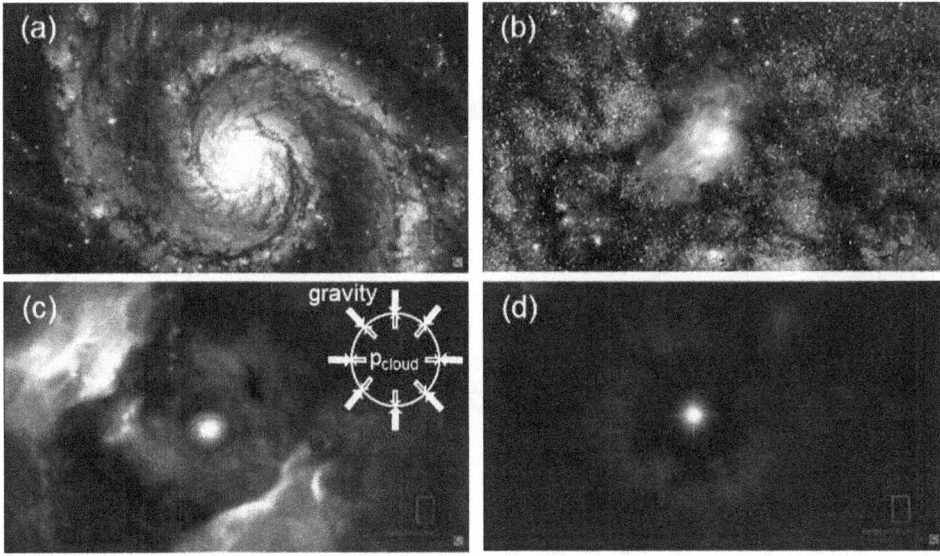

Figure 6.38. (a) Nebula houses baby stars in every spiral arm of the galaxy. (b) A zoom of a baby star. (c) The cloud wants to dissipate due to internal cloud pressure (here p_{cloud}) and gravity wants to pull it together. Collapse occurs if the inward directed gravitational force is bigger than the outward directed pressure force. Material crunches down into a disk which is the beginning of an eventual star. As gravity pulls more gas toward the center of the disk it gets denser and stiffer until it reaches about 15 or so million degrees Kelvin. (d) Hydrogen atoms fuse together to form helium and the star shines.

Unmagnetized Jeans instability using the virial theorem. There are several ways to derive Jeans instability. Probably, the simplest way is through the virial theorem described in Chapter 1. For a spherically-symmetric cloud in hydrostatic equilibrium, assuming constant density, ignoring magnetic fields, and neglecting rotation, the virial theorem relates the cloud's internal kinetic energy and its gravitational potential energy as follows:

$$2K + U = 0 \, , \tag{6.195}$$

where K is the internal kinetic energy and U is the gravitational potential energy. Let us again consider the case of a system perturbed from equilibrium. If gas pressure dominates gravity, $2K > |U|$. If gravity dominates gas pressure, we have gravitational collapse which can be expressed as $2K < |U|$. The cloud's internal kinetic energy is

$$K = \frac{3}{2}\mathcal{N}k_BT = \frac{3}{2}\frac{m}{\mu m_{\text{H}}}k_BT \, , \tag{6.196}$$

where

$$\mathcal{N} = \frac{m}{\mu m_{\text{H}}} \tag{6.197}$$

is the total number of particles. Cloud's gravitational potential energy is

$$U = -\alpha\frac{G_0m^2}{r} = -\alpha G_0 m^{5/3}\left(\frac{3}{4\pi\rho_0}\right)^{-1/3} \tag{6.198}$$

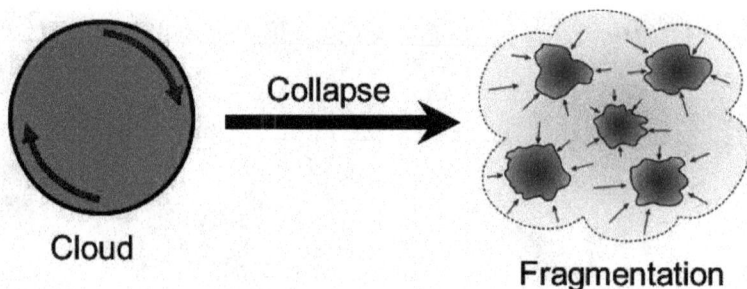

Figure 6.39. An illustration of the cloud collapse when the inwards directed gravitational force is bigger than the outwards directed pressure force. During collapse, cloud sub-regions emerge. These sub-regions start collapsing on their own. This leads to fragmentation.

where α is a small and positive fudge factor of order unity that depends on the internal density structure; for a diatomic molecule like H_2, $\alpha = 5/2$ and for monatomic gas, $\alpha = 3/2$. We use the fact that $m = (4/3)\pi r^3 \rho_0$ in the definition of a cloud's gravitational potential energy. We can deduce the condition for gravitational collapse as follows:

$$2K < U \implies \frac{3mk_BT}{\mu m_H} < \alpha G_0 m^{5/3}\left(\frac{3}{4\pi\rho_0}\right)^{-1/3} \iff m > m_J \,, \qquad (6.199)$$

where

$$m_J = \frac{9}{2\alpha^{3/2}\sqrt{\pi}}\left(\frac{k_B}{G_0\mu m_H}\right)^{3/2} T^{3/2}\rho_0^{-1/2} \,. \qquad (6.200)$$

We return to the Jeans mass in (6.193) by selecting $\alpha = (3/\pi)^2$.

Magnetized Jeans instability: magnatohydrostatic clouds. For a more realistic instability criterion, we should consider a rotating and magnetized plasma in addition to self-gravitating. We will work in cylindrical coordinates (r, θ, z) or (i_r, i_θ, i_z), where i_r is the direction of cloud expansion and contraction, i_θ describes the rotation of the cloud, and i_z describes the direction of cloud collapse. We assume that $\delta v = [\delta v_r, \delta v_\theta, 0]^T$, and $\delta B = [0, 0, \delta B_z]^T$ as shown in Figure 6.39. By taking the Fourier transforms of (6.150)-(6.153) with respect to t and r, we arrive at the following perturbation equations of the mass conservation, gravitational potential, equation of state, momentum conservation, and magnetic flux conservation of the ideal MHD laws

$$\omega\frac{\delta\rho}{\rho_0} - k_r\delta v_r = 0 \,, \quad k^2\Psi + 4\pi G_0\delta\rho = 0 \,, \quad \frac{\delta p}{\delta\rho} = \frac{p_0}{\rho_0} = c_{is}^2 \,, \qquad (6.201)$$

$$-i\omega\delta v_r - 2\Omega_0\delta v_\theta + ik_r\frac{\delta p}{p_0} + k_r\delta\Psi + ik_r c_A^2\frac{\delta B_z}{B_0} = 0 \qquad (6.202)$$

$$\kappa^2\frac{\delta v_r}{2\Omega_0} - i\omega\delta v_\theta = 0 \,, \quad \omega\frac{\delta B_z}{B_0} + k_r\delta v_r = 0 \,, \qquad (6.203)$$

Table 6.10. Physical properties of magnetized molecular clouds. The constants used in the derivation of these properties include $G_0 = 6.67 \times 10^{-11}$ m^3/(kgs^2), $k_B = 1.380649 \times 10^{-23}$ m^{2}kgs^{-2}K^{-1}, and $m_H = 1.67 \times 10^{-27}$ kg. The conversion includes 1 pc $= 3.09 \times 10^{16}$. The rotation is described by $\kappa_0 = \kappa/(4\pi G_0 \rho_0)$ and pc stands parsec.

ρ (kg/m^3)	c_{is} (m/s)	c_A/c_{is}	κ_0	λ_J (pc)	m_J (m_{sun})	$\hat{\lambda}_J$ (pc)	$\hat{m}_J$ (m_{sun})	Examples Cloud
1.67×10^{-19}	84.24	0.2	0	1.45	3.9	1.48	4.18	Cold
1.67×10^{-19}	84.24	0.2	0.2	1.45	3.9	1.65	5.84	Cold
1.67×10^{-19}	84.24	0.4	0.2	1.45	3.9	1.75	6.88	Cold
1.67×10^{-19}	84.24	0.4	0.4	1.45	3.9	2.02	10.58	Cold
1.67×10^{-17}	287.5	0.2	0.0	0.495	15.66	0.504	16.61	Core/B68
1.67×10^{-17}	287.5	0.2	0.2	0.495	15.66	0.564	23.21	Core/B68
1.67×10^{-17}	287.5	0.4	0.4	0.495	15.66	0.688	42.09	Core/B68
1.67×10^{-21}	642.9	0.2	0.0	110.6	17,505	112.8	18,565	Diffuse
1.67×10^{-21}	642.9	0.2	0.2	110.6	17,505	126.2	25,946	Diffuse
1.67×10^{-21}	642.9	0.4	0.4	110.6	17,505	153.8	47,0562	Diffuse

with $\boldsymbol{k} = [k_r, 0, 0]^T$, $\kappa = 4\Omega_0^2 + \partial(\ln\Omega_0^2)/\partial r$, and $\Omega_0 = \frac{v_\theta}{r}$. By substituting equations in (6.201) and (6.203) in (6.202), we arrive at

$$\left[\omega^2 - \kappa^2 - \omega_A^2 - \left(c_{is}^2 k_r^2 - 4\pi G_0 \rho_0\right)\right] \delta\rho(\omega, k_r) = 0 . \tag{6.204}$$

Notice that at the limits $\kappa = 0$ and $c_A = B_0^2/(\mu_0 \rho_0) = 0$, we recover the standard Jeans dispersion relation in (6.191). The instability condition $\omega^2 < 0$ implies that

$$\hat{k}_r^2 < \hat{k}_J^2 = \frac{4\pi G_0 \rho_0 - \kappa^2}{c_{is}^2 + c_A^2}$$

or

$$\hat{\lambda}_J = \frac{2\pi}{\hat{k}_J} = 2\pi \left[\frac{c_{is}^2 + c_A^2}{4\pi G_0 \rho_0 - \kappa^2}\right]^{1/2} \quad \text{and} \quad \hat{m}_J = \frac{4\pi}{3}\rho_0 \left(\frac{\hat{\lambda}_J}{2}\right)^3 . \tag{6.205}$$

This solution is physically meaningful only if $\kappa^2 < 4\pi G_0 \rho_0$. We can see in (6.205) and in Table 6.10 that the Jeans mass increases with magnetic strength through c_A and also with increasing rotation term, κ. In other words, magnetic fields and rotations further reduce the chance of gravitational collapse because $\hat{m}_J$ is greater than m_J. For a given cloud mass m, the condition of gravitational collapse is now $m > \hat{m}_J \geq m_J$. A rotating, self-gravitating, and magnetized plasma stabilizes the cloud. Stars form in the densest and coldest parts of the cloud.

Summary. Stars form in molecular clouds. Examples of molecular clouds include Andromeda, Ophiuchus, Lupus, Serpens, and Orion. Their masses vary between 1 m_{sun} and 10^7 m_{sun} and their sizes vary between 1 and 10^7 pc. Gravitational force is the only force that causes molecular clouds to collapse. All the other known forces oppose collapse—that is, they can prevent collapse (Figure 6.40). They include cloud internal pressure, cloud magnetic field, cloud rotation and external forces related to phenomena such as supernovae, winds, spiral density waves. Thus modern star formation theory is based on the complex interplay between all these processes. Star formation is truly a multi-scale and multi-physics problem, where it is difficult to single out individual processes.

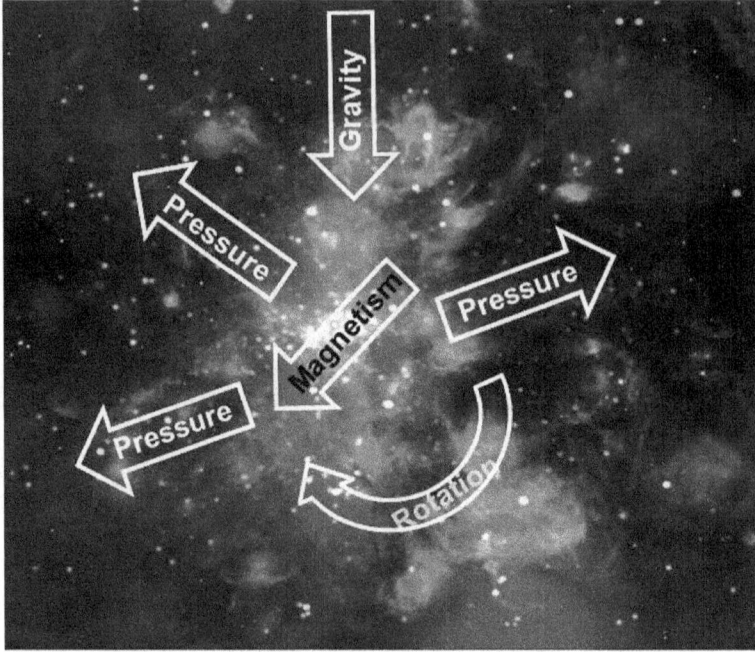

Figure 6.40. Giant molecular clouds are held up by magnetic fields, thermal gas pressure, and rotation. If the clouds become massive enough, they can collapse due to gravity.

Box 6.5: Mean (Average) Molecular Mass

For an ideal gas composed of different atomic species, the equation of state is $p = nkT$, where p is the gas pressure, T is the temperature, k_B is Boltzman's constant, and n is the particle number density, i.e. the number of particles (of all species combined) per unit volume. If $< m >$ is the average mass per particle, we can rewrite the equation of state as

$$p = \frac{n < m >}{< m >} k_B T = \frac{\rho}{< m >} k_B T \quad \text{or} \quad p = \frac{n < m >}{m_H} \frac{m_H}{< m >} k_B T = \frac{\rho}{\mu m_H} k_B T \quad (6.206)$$

where

$$\rho = n < m > \quad \text{and} \quad \mu = \frac{< m >}{m_H}, \quad (6.207)$$

with ρ the total density of the gas mixture, m_H is the proton mass, and μ is *average molecular mass*. It is customary to express the mean mass per particle in terms of the atomic mass unit (AMU), which, for our purposes, we declare to be the proton mass, m_p. For example, suppose that the gas mixture consists of hydrogen only, and all hydrogen atoms are neutral. Then the mean mass per particle in amu would be 1, making $\mu = 1$. If all the hydrogen is fully ionized, the mass of one hydrogen atom (1 amu) would be shared by two particles (the H nucleus and the free electron), and the mean mass per particle would now be 1 amu per 2 particles, or $\mu = 0.5$. Similarly, for a pure He gas, with all He atoms fully ionized, the mass of one He atom (4 amu) would be shared by three particles (the He nucleus and two free electrons), or $\mu = 4/3 = 1.33$.

Quiz: Jeans instability with rotation. To derive a more realistic instability criterion it is necessary to consider rotation. In a rotation frame of reference it is necessary to add the Coriolis and the centrifugal accelerations to the hydrodynamic equations in (6.150)-(6.153). We can combine the centrifugal force and the gravitavional force describe Ψ as the perturbation of the gravitational potential caused by rotation through the centrifugal force. Coriolis acceleration can be added to the momentum equation as an additional source term. Assuming that perturbations vary only along the z-axis, rewrite equations in (6.150)-(6.153) for rotating clouds and in the absence of magnetic effects in the time-space and frequency-wavenumber domains. (2) Deduce the Jeans wavelength for the case in which the angle between the direction of propagation of the perturbation and the rotation axis is 90 degrees.

6.4.3.   Kelvin-Helmholtz and Rayleigh-Taylor Instabilities

Rayleigh-Taylor (RT) and Kelvin-Helmholtz (KH) instabilities are canonical fluid and plasma dynamic phenomena observed in nature, including clouds, planet atmospheres, and supernovae. They occur at the interfaces separating two different fluids or plasmas of different properties. These two instabilities usually appear simultaneously. Moreover, the same mathematical derivations can be used for both. We here derive conditions for these interface instability.

Consider two incompressible homogeneous plane plasma layers with embedded constant magnetic fields $\boldsymbol{B}_1$ and $\boldsymbol{B}_2$, constant velocity fields $\boldsymbol{v}_1$ and $\boldsymbol{v}_2$, and constant densities ρ_1 and ρ_2 as shown in Figure 6.41. We keep the vertical gravity field $g = -g\boldsymbol{i}_z$. The entire configuration is then made up of two regions as follows:

$$\rho_0 = \begin{cases} \rho_1, & z < 0 \\ \rho_2, & z > 0 \end{cases} \quad , \quad \boldsymbol{v} = \begin{cases} \boldsymbol{v}_1, & z < 0 \\ \boldsymbol{v}_2, & z > 0 \end{cases} \quad , \quad \boldsymbol{B}_0 = \begin{cases} \boldsymbol{B}_1, & z < 0 \\ \boldsymbol{B}_2, & z > 0 \end{cases} \quad . \quad (6.208)$$

The magnetic fields in both regions are assumed to be parallel to the x-y plane. We are dealing with stationary plasmas and not static plasmas as shown in Figure 6.41. It is the jumps in the velocity, magnetic field, and density at the interface of the two plasmas that cause the plasmas to become unstable under certain conditions.

So the two plasmas have different velocities, density, and subjected to different magnetic fields as follows (see also Figure 6.41):

- Upper layer $(0 < z \le a)$: $\rho_1 = \text{const.}$, $\boldsymbol{v}_1 = [v_{x1}, v_{y1}, 0]^T = \text{const.}$, $\boldsymbol{B}_1 = [B_{x1}, B_{y1}, 0]^T = \text{const.}$, and $p_1 = p_0 - \rho_1 g z$ with $p_0 > \rho g a$. According to (6.177), we have

  $$\frac{d^2\xi^{(1)}}{dz^2} - k_0^2 \xi^{(1)} = 0 \ , \ \text{with}\, \xi^{(1)}(a) = 0 \Longrightarrow \xi^{(1)} = C_1 \frac{\sinh\left[k_0(a - z)\right]}{\sinh(k_0 a)} \ ,$$

  where C_1 is constant of integration and $\xi^{(1)}$ is the vertical component of the Lagrangian-displacement vector, $\boldsymbol{\xi}$. Again, $\boldsymbol{\xi}$ describes the describes the manner in which the instability displaces fluid element between initial flows $(\boldsymbol{v} = \boldsymbol{v}_0 = \boldsymbol{0})$ and perturbed flows $(\boldsymbol{v} = \boldsymbol{v}_0 + \delta \boldsymbol{v} = \delta \boldsymbol{v} = \partial \boldsymbol{\xi}/\partial t)$.

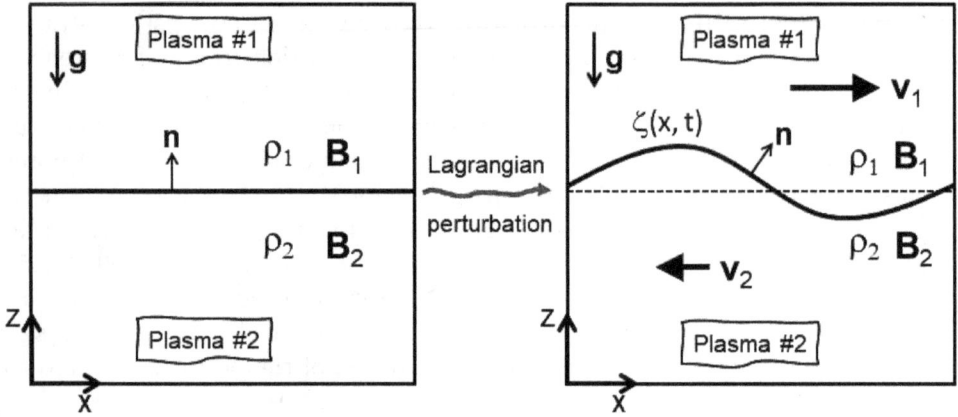

Figure 6.41. Consider two plasmas. The upper ($z > 0$) and lower ($z < 0$) fluids have different densities ρ_1 and ρ_2. There are two different magnetic fields applied to the plasmas. The disturbance at the interface of the plasma is the vertical displacement $z = \xi(x, y, t)$; that is, the perturbed position of the interface at $\xi(x, y, t)$.

- Lower layer ($-b < z \leq 0$): $\rho = $ const., $v_2 = [v_{x2}, v_{y2}, 0]^T = $ const., $B_2 = [B_{x2}, B_{y2}, 0]^T = $ const., and $p_2 = p_0 - \rho_2 g z$. According to (6.177), we have

$$\frac{d^2\xi^{(2)}}{dz^2} - k_0^2\xi^{(2)} = 0 \ , \ \text{with} \ \xi^{(2)}(-b) = 0 \Longrightarrow \xi^{(2)} = C_2\frac{\sinh\left[k_0(z + b)\right]}{\sinh(k_0 b)} \ ,$$

where C_2 is constant of integration.

- The first boundary condition is the kinematic boundary condition that expresses the continuity of the normal component of plasma displacement (or normal velocity) at the interface; i.e.,

$$\xi^{(1)}(0) = \xi^{(2)}(0) \Longrightarrow C_1 = C_2 \ .$$

- The second boundary condition is the dynamic boundary condition expressing the continuity of the total pressure (pressure balance) at the interface, $z = 0$; i.e.,

$$[\Pi - \rho g\xi]_1^2 = 0 \Longrightarrow \left[\frac{\rho}{k_0^2}\left(\tilde{\omega}^2 - \omega_A^2\right)\frac{1}{\xi}\frac{d\xi}{dz} - \rho g\right]_1^2 = 0 \ , \tag{6.209}$$

where Π is total pressure (kinetic plus magnetic) defined in (6.177) for depth-dependent incompressible plasmas (i.e., $\gamma \longrightarrow \infty$), which implies that $\nabla \cdot \xi = 0$.

The standard method for analyzing an instability is to find its dispersion relation. Inserting solutions for ξ_1 and ξ_2 (6.209) yields the dispersion equation:

$$-\rho_1\left[(\omega - \Omega_{01})^2 - \omega_{A1}^2\right]\coth(k_0 a) - k_0\rho_1 g$$

$$= -\rho_2\left[(\omega - \Omega_{02})^2 - \omega_{A2}^2\right]\coth(k_0 b) - k_0\rho_2 g. \tag{6.210}$$

For short wavelengths (i.e., walls effectively at $a \longrightarrow \infty$ and $b \longrightarrow -\infty$), we have $k_0 x \gg 1$ and therefore $\coth(k_0 x) \approx 1$. The above solution reduces to

$$
\omega = \omega^{(r)} + i\omega^{(i)} = \frac{\rho_1 \Omega_{01} + \rho_2 \Omega_{02}}{\rho_1 + \rho_2} \pm \left[-\frac{\rho_1 \rho_2 (\Omega_{01} - \Omega_{02})^2}{(\rho_1 + \rho_2)^2} + \frac{\rho_1 \omega_{A1}^2 + \rho_2 \omega_{A2}^2}{\rho_1 + \rho_2} \right.
$$
$$
\left. - \frac{k_0 (\rho_1 - \rho_2) g}{\rho_1 + \rho_2} \right]^{1/2} , \qquad (6.211)
$$

where $\Omega_{01} = \boldsymbol{k}_0 \cdot \boldsymbol{v}_1$, $\Omega_{02} = \boldsymbol{k}_0 \cdot \boldsymbol{v}_2$, $\omega_{A1} = \boldsymbol{k}_0 \cdot \boldsymbol{B}_1 / \sqrt{\mu_0 \rho_1}$, $\omega_{A2} = \boldsymbol{k}_0 \cdot \boldsymbol{B}_2 / \sqrt{\mu_0 \rho_2}$, $\boldsymbol{k}_0 = [k_x, k_y]^T$, and $k_0^2 = \boldsymbol{k}_0 \cdot \boldsymbol{k}_0$. The wave-equation solutions, ξ, represent either two waves when ω is a real number (or $\omega^2 > 0$) and two modes with complex frequency (an instability)—that is, they depend on the sign of the expression under the square root in (6.211). We can see that ω is real if $\omega^{(i)} = 0$ and more generally when that expression is negative; i.e.,

$$
\underbrace{\frac{(\boldsymbol{k}_0 \cdot \boldsymbol{B}_1)^2 + (\boldsymbol{k}_0 \cdot \boldsymbol{B}_2)^2}{\mu_0}}_{\text{magnetic shear}} < \underbrace{\frac{\rho_1 \rho_2}{\rho_1 + \rho_2} [\boldsymbol{k}_0 \cdot (\boldsymbol{v}_{01} - \boldsymbol{v}_{02})]^2}_{\text{KH drive}} + \underbrace{k_0 (\rho_1 - \rho_2) g}_{\text{RT drive}} , \quad (6.212)
$$

we have an unstable system. The terms KH and RT drives characterize two different instabilities that we describe in the next paragraphs. We have genuine competition between the three terms so that stability depends on the precise choice of all those parameters. So the key reasons of this interface instability are when $\Omega_{01} \neq \Omega_{02}$, $\omega_{A1} \neq 0$, $\omega_{A2} \neq 0$, and/or $\rho_1 < \rho_2$. Note that our above derivations were limited to the case in which $\boldsymbol{k}_0 \parallel \boldsymbol{B}_0$. They actually hold even when $\boldsymbol{k}_0 \perp \boldsymbol{B}_0$ which equivent to setting $\omega_{A2} = \boldsymbol{k}_0 \cdot \boldsymbol{B}_2 / \sqrt{\mu_0 \rho_2} = 0$ and $\omega_{A1} = \boldsymbol{k}_0 \cdot \boldsymbol{B}_1 / \sqrt{\mu_0 \rho_1} = 0$. In other words, $\boldsymbol{k}_0 \perp \boldsymbol{B}_0$ removes the influence of the magnetic field in the interface instability, resulting the hydrodynamic instability.

Kelvin-Helmholtz (KH) instability is an interface instability that arises when two fluids with different densities have a velocity difference across their interface. Let assume that there is no gravity; that is, (6.211) becomes

$$
\omega_{\text{KH}} = \frac{\rho_1 \Omega_{01} + \rho_2 \Omega_{02}}{\rho_1 + \rho_2} \pm \left[-\frac{\rho_1 \rho_2 (\Omega_{01} - \Omega_{02})^2}{(\rho_1 + \rho_2)^2} + \frac{\rho_1 \omega_{A1}^2 + \rho_2 \omega_{A2}^2}{\rho_1 + \rho_2} \right]^{1/2} \quad (6.213)
$$

and ω_{KH} becomes a complex-valued frequency under this condition

$$
\frac{(\boldsymbol{k}_0 \cdot \boldsymbol{B}_1)^2 + (\boldsymbol{k}_0 \cdot \boldsymbol{B}_2)^2}{\mu_0} < \frac{\rho_1 \rho_2}{\rho_1 + \rho_2} [\boldsymbol{k}_0 \cdot (\boldsymbol{v}_{01} - \boldsymbol{v}_{02})]^2 . \quad (6.214)
$$

This instability was proposed by Lord Kelvin (of the temperature unit fame) in 1871 and Hermann von Helmholtz (of the Helmholtz equation fame) in 1868 and is nowadays known as the Kelvin-Helmholtz instability (KHI). Consider the particular case in which $\boldsymbol{v}_1 = [v_{1x}, 0, 0]^T$, $\boldsymbol{v}_2 = [0, 0, 0]^T$ (i.e., stationary), and $\boldsymbol{B}_1 = \boldsymbol{B}_2 = \boldsymbol{B}_0$, and and $\boldsymbol{k}_0 = [k_x, k_y]^T$. The instability criterion in (6.213) reduces to

$$
\underbrace{\frac{2(\boldsymbol{k}_0 \cdot \boldsymbol{B}_0)^2}{\mu_0 \bar{\rho}}}_{\text{magnetic tension}} < \underbrace{\boldsymbol{k}_0 \cdot \boldsymbol{v}_{01}}_{\text{velocity shear}} . \quad (6.215)
$$

where

$$\bar{\rho} = \frac{\rho_1 \rho_2}{\rho_1 + \rho_2} \ . \tag{6.216}$$

Note that when we have $\boldsymbol{k}_0 \perp \boldsymbol{B}_0$, the magnetic field has no effect on Kelvin-Helmholtz instability.

For the hydrodynamic case (i.e., no magnetic field is considered), (6.213) reduces

$$\omega_{\text{KH;hydro}} = \frac{\rho_1 \Omega_{01} + \rho_2 \Omega_{02}}{\rho_{01} + \rho_{02}} \pm \left[-\frac{\rho_1 \rho_2 \left(\Omega_{01} - \Omega_{02}\right)^2}{\left(\rho_{01} + \rho_{02}\right)^2} \right]^{1/2}$$

and the instability condition is

$$\frac{\rho_1 \rho_2}{\rho_{01} + \rho_{02}} \left[\boldsymbol{k}_0 \cdot \left(\boldsymbol{v}_{01} - \boldsymbol{v}_{02}\right)\right]^2 > 0 \ . \tag{6.217}$$

At all wavelengths we have hydrodynamic HK instabilities as long as $\boldsymbol{v}_{01} \neq \boldsymbol{v}_{02}$.

KHI is a ubiquitous physical phenomenon observed in various media involving two fluids flowing at different speeds. Figure 6.35 shows hydrodynamic KH instability only when tilting the tank, because the two fluids travel at different speeds. At their interface, fluid speed differences cause some cloud parts to rise up and others to go down. This steady interference pattern forms rolling waves that grow with increasing velocity differences between the two fluids. So Kelvin-Helmholtz instability waves can arise when two fluids travel at different speeds. The fluids can be air layers at varying temperatures and densities with the upper layer moving faster. They can be layers of varying densities inside the oceans. They can be the sea surface and the windy air above. They can be swiftly moving belts in Jupiter's atmosphere. In fact, weather satellites use Kelvin-Helmholtz instability to measure wind speeds over large bodies of water. Waves are generated by the wind which shears the water at the air/water interface as a result of the Kelvin-Helmholtz instability between the wind and the ocean. The computers on board the satellites determine the ocean roughness by measuring the rolling-wave height.

In Figure 6.42a, we have clouds which are a product of mainly hydrodynamic KH instabilities because we have two fluids subject to distinct winds (i.e., wind shear). In Chapter 2, we discuss in detail the cloud formation of the two fluids—namely water vapor above and dry air below. Remember that air currents vary in speed and direction with height. Barriers such as mountains or hills can split a fluid into two fluids of different velocities. The part of fluid closer to the barrier slows down while the part above is almost not affected by the barrier.

In Chapter 2, we also discussed cloud formation in Titan's atmosphere. Titan is Saturn's largest moon. It has a thick and dense atmosphere, composed mainly of 98,5 percent nitrogen and the remainder is methane and hydrogen. Atmospheric pressure at the surface is higher than Earth's, about 1.45 atm (i.e., Titan's atmosphere has 1.45 times the pressure of Earth's) with a surface gravity of only 0.14 g_{earth} (1.35 m/s^2). The temperature is -180 degrees Celsius. On this webside   http://esamultimedia.esa.int/images/huygens_alien_winds_descent.mp3   you can hear Titan's winds. In other words, Titan's atmosphere has wind and this wind varies with height. This wind and its variations are responsible for cloud formation in Titan's atmosphere.

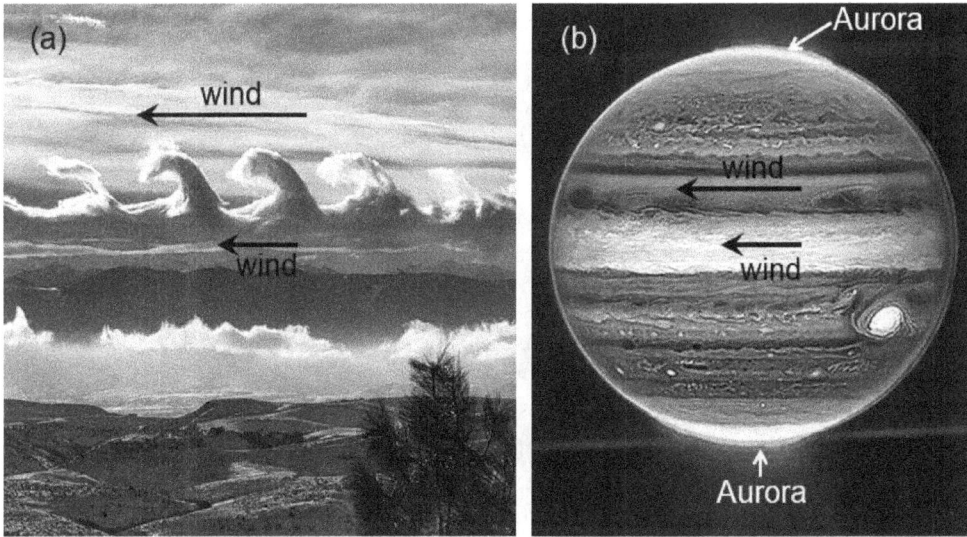

Figure 6.42. (a) An illustration of the Kelvin-Helmholtz instability (KHI) that develops in the clouds in the Earth's atmosphere when velocity shear occurs. Notice disturbances of vorticity along a velocity discontinuity. (b) KHI is observed in Jupiter's atmosphere. In Earth's atmosphere, nitrogen makes up 78 percent of the composition, while oxygen makes up 21 percent, while helium makes up a small portion of the composition on Jupiter's planet. The differential rotation bands help prove that the near Jupiter surface is not a solid body. Jupiter's northern and southern parts of this image from the James Webb Space Telescope are shining Jupiter's auroras.

Quiz: Let us assume the magnetic fields are zeros, $\boldsymbol{v}_1 = -\boldsymbol{v}_2 = [v_{1x}, 0, 0]^T$, $\rho_1 \neq \rho_2$, and $g = 0$. Describe the stability criteria in this case.

Rayleigh-Taylor instability is most familiarly seen on Earth when a heavier fluid is placed on top of a lighter fluid; gravity destabilizes the interface separating the fluids resulting in the growth of imposed perturbations and ultimately culminating in vigorous turbulent mixing. It is the special case for which $\boldsymbol{v}_1 = \boldsymbol{v}_2 = \boldsymbol{0}$ and $g \neq 0$. The dispersion relation in (6.211) becomes

$$\omega_{\mathrm{RT}}^2 = \frac{\rho_1 \omega_{A1}^2 + \rho_2 \omega_{A2}^2}{\rho_1 + \rho_2} - \frac{k_0 \left(\rho_1 - \rho_2 \right) g}{\rho_1 + \rho_2} = \frac{\rho_1 \omega_{A1}^2 + \rho_2 \omega_{A2}^2}{\rho_1 + \rho_2} - k_0 g \, \mathcal{A} \,, \qquad (6.218)$$

where

$$\mathcal{A} = \frac{\left(\rho_1 - \rho_2 \right)}{\left(\rho_1 + \rho_2 \right)} \qquad (6.219)$$

is known as the Atwood number of the system and ω_{RT} becomes a complex-valued frequency under the condition that

$$\frac{\left(\boldsymbol{k}_0 \cdot \boldsymbol{B}_1 \right)^2 + \left(\boldsymbol{k}_0 \cdot \boldsymbol{B}_2 \right)^2}{\mu_0} < k_0 \left(\rho_1 - \rho_2 \right) g \,. \qquad (6.220)$$

Figure 6.43. An illustration of the remnants of a supernova's explosion being driven by a blast wave. This causes instability and leads to intensive interfacial mixing of progenitor star materials. Silicon, sulfur, calcium, iron, and high-energy X-rays contribute to this mixture. (Adapted from an image obtained by NASA's Chandra X-ray Observatory, a space telescope launched by the Space Shuttle Columbia on July 23, 1999).

This instability is named after Lord Rayleigh and G. I. Taylor. For hydrodynamic fluids (i.e., no magnetic field is considered), (6.220) reduces to

$$k_0 \left(\rho_1 - \rho_2 \right) g > 0 , \qquad (6.221)$$

where ρ_2 and ρ_1 are the "bottom" and "top" fluid densities, respectively. Thus, we find that if $\rho_2 > \rho_1$ (the denser fluid is on the bottom) then ω is real and the system is stable, while if $\rho_1 > \rho_2$ (the denser fluid is on top) the frequency is imaginary and the velocity grows exponentially in time, $\xi(t) = \xi(0) \exp(wt)$. Figure 6.34 illustrates the hydrodynamic Rayleigh-Taylor instability between the lighter fluid and pushing the heavier fluid. We can see the fingering of the lighter (lower-density) fluid into the heavier (higher-density) fluid.

Another example of RT instability is illustrated in Figure 6.43. It is a product of supernova explosions. As described in Ikelle (2023b), supernova explosions are violent, disruptive explosions of stars. The lighter debris ejected from a supernova pushes on the heavier interstellar medium, forming a supernova remnant. The blast wave causes Rayleigh-Taylor instabilities which provide conditions for the synthesis of heavy mass elements in addition to light mass elements synthesized by fusion in the star before its explosion. Figure 6.43 shows the mixture of debris ejected from a supernova and the interstellar medium. The result is a supernova remnant. The various brightnesses of Figure 6.43 gases indicate chemical composition differences.

Jupiter's atmosphere. On this website, https://www.planetary.org/space-images/24-jupiter-days-worth-of, we can see that Jupiter's atmosphere has more than twelve parallel bands of ammonia and methane gases. There are different winds in each band between 100 and 1000 km/hour (faster than earth's most violent hurricanes) in opposite directions. Because Jupiter is a fast-spinning planet with a large magnetic field and gravity, none of the parameters in (6.212) can be considered a prior zero. We deal simultaneously with magnetic KHI and RTI, as described (6.212). It is this combination that makes the complex mixtures of vortexes described in Ikelle (2024). By using the Jupiter model in Figure 6.44, he numerically simulates vortexes and even the Great Red Spot. It is not possible based solely on fluid and wind properties. As discussed in Chapter 2, circular and localized clouds reflect the topography of the boundary between the atmosphere and the solid interior of a given planet or moon. To produce red spots, Jupiter's atmosphere must contain a large hill/mountain at the boundary. The red spot is expected to be just above the hill. Based on microwave radiometer and gravity data, Jupiter has an atmosphere 3,000 kilometers thick and the Great Red Spot has roots 300 kilometers deep. Ikelle (2024) determines through modeling that the hill must be at least 2500 km above Jupiter's boundary between its atmosphere and its solids. Notice that the fact that this hill does not show up clearly in Nettelman (2017) gravity spherical harmonics tells us that there are other smaller mountains at this boundary. These mountains alter the weighted averages associated with spher-

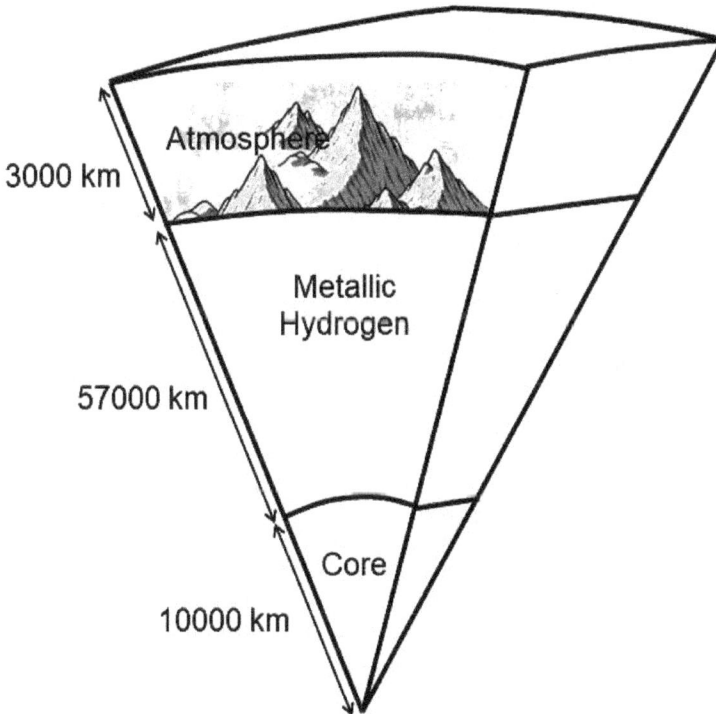

Figure 6.44. Jupiter's interior with mountains below its atmosphere. The mountains have been modeled in Ikelle (2024).

ical harmonic calculations. Note also that Jupiter's total radius is almost 70,000 kilometers.

Rather than being broken up, clouds persist for thousands and even millions of years in Jupiter's atmosphere. This is because Jupiter's atmosphere is very cold, about -110 degrees Celsius. At that temperature gas molecules do not move much and atoms move very slowly. In addition Jupiter has a large gravity (i.e., $g_{\text{jupiter}} \approx 2.5 g_{\text{earth}}$) that wants to pull the atmosphere together. Titan is able to maintain a substantial atmosphere because of its atmosphere is very cold, -180 degrees Celsius. Titan does so despite its low gravity. It is obvious that the shield of Jupiter's atmosphere from the solar wind by its magnetosphere and its long distance from the Sun (Jupiter's distance from the Sun is 9 times the distance from the Sun to Earth) are important in maintaining clouds in Jupiter's atmosphere for thousands of years. A solar wind that blows Jupiter's clouds is much weaker at that distance besides its magnetic shields.

6.4.4.   Convective Instability and Energy Transport in the Stars

As described in Chapter 2, we often encounter fluids heated from below and thus colder further up (e.g., the Earth's core heats the Earth's mantle base; Earth's atmosphere is heated from the Earth's surface; the Sun's radiation zone heats the base of the Sun's convective zone base, etc.). This process results in a temperature gradient: lighter/hotter at the base. Like accretion we will discuss in the next subsection, convection is a fundamental process in celestial objects. For example, convection is central to energy transport inside the stars. The question here is under what conditions this adverse temperature gradient becomes unstable, developing *overturning* motions known as thermal convection. Note that the RTI is an example of convective instability (Kruskal and Schwarzschild, 1954); here we generalize the RTI calculations for a continuously stratified MHD fluid. That is

$$\rho = \rho_0(z) + \delta\rho \,, \quad p = p_0(z) + \delta p \,, \boldsymbol{B} = \boldsymbol{B}_0(z) + \delta\boldsymbol{B}\,, \quad \text{and} \quad \boldsymbol{v} = \delta\boldsymbol{v} \qquad (6.222)$$

and we still have the equilibrium plasma pressure, which is defined from (6.79) by this equation

$$-\frac{1}{\rho_0}\left(\frac{dp_0}{dz} + \frac{dB_0^2}{dz}\right) = g \quad \underset{B_0=0}{\Longrightarrow} \quad -\frac{1}{\rho_0}\frac{dp_0}{dz} = g\,, \qquad (6.223)$$

where g is the gravity acceleration and ρ_0 is the equilibrium plasma density. Under the ideal MHD assumption, we also have

$$\frac{\partial}{\partial t}\left(\frac{p}{\rho^\gamma}\right) = 0 \quad \Longrightarrow \quad \rho_0\left(\alpha^2 + k_0^2\right)\frac{1}{p_0}\frac{\partial\delta p}{\partial t} = \frac{\gamma}{\rho_0}\frac{\partial\delta\rho}{\partial t} \quad \Longrightarrow \quad \frac{\delta p}{p_0} = \gamma\frac{\delta\rho}{\rho_0}\,, \qquad (6.224)$$

as our equilibrium state, but allow perturbations throughout the fluid.

WKB approximation. These perturbations produce a Lagrangian vertical displacement, ξ, described in (6.184). As we did in the previous derivations of instabilities, we would like to determine dispersion relation of this equation by eliminating the derivatives of $\xi = \xi_z$ by taking the Fourier transform of (6.184) with respect z and used the fact that the resulting dispersion only depends on ω^2, which must positive otherwise we have instability (i.e., $\omega^2 < 0$). Unfortunately, such calculations are not readily possible here because the background plasma is depth-dependent, thus the product of background quantities and vertical displacement in (6.184) produce undesired convolution along k_z wavenumber. We can solve (6.184) numerically. Alternatively, by assuming that ξ varies at a very small scale compared to the scale of the background, we can use the WKB (Wentzel–Kramers–Brillouin) approximation (Dingle, 1973; Bleistein, 1984; Ikelle et al., 1988; and Ikelle, 1995; and Ikelle and Amundsen, 2005) to solve this equation; i.e.,

$$\frac{d}{dz}\left[f'(z,\omega^2)\frac{d\xi}{dz}\right] - g'(z,\omega^2)\xi \approx \left[\alpha^2 f'(z,\omega^2) + g'(z,\omega^2)\right]\xi \,, \tag{6.225}$$

under the assumption that ξ has the following form

$$\xi(z,\omega) = \beta(z,\omega)\exp\left[-i\int_{z_0}^{z}\alpha(s)ds\right] \,, \tag{6.226}$$

where $f'(z,\omega^2)$ and $g'(z,\omega^2)$ are two arbitrary functions, α represents a local "wavenumber" in the z-direction and β represents the amplitude of ξ. The implication in (6.225) leads to a local dispersion equation. This approximation is valid here because the equilibrium state allows us to assume that the background pressure and density variations are very smooth. We will first study them in HD and then in MHD.

HD Schwarzschild criteria for convective stability. The local dispersion equation associated with (6.170), for example, reduces to

$$\alpha^2 = -\frac{g'}{f'} = \frac{-k_0^2 \rho_0\left(\omega^2 - N_B^2\right)}{\rho_0 \omega^2} \tag{6.227}$$

which implies that

$$\omega^2 = \frac{k_0^2}{k_0^2 + \alpha^2}N_B^2 \,, \tag{6.228}$$

where N_B is the Brunt-Väisälä frequency in (6.171). So instability just appears to depend on the square of the Brunt-Väisälä frequency: $N_B^2 \geq 0$. It is assumed that the magnetic field is zero and that fluid is incompressible and static in order to reach this result.

A valid method of verifying the results in (6.234) is by comparing them to similar results in Chapter 2 based on thermodynamic arguments. Ideal gases obey $p = (\mathcal{R}/\mu)\rho T$. If we move the gas parcel upward a small distance dz, then the change in pressure is

$$\frac{dp_0}{dz} = \frac{p_0}{T}\frac{dT}{dz} + \frac{p_0}{\rho_0}\frac{d\rho_0}{dz} \quad \Longleftrightarrow \quad \frac{1}{\rho_0}\frac{d\rho_0}{dz} = \frac{1}{p_0}\frac{dp_0}{dz} - \frac{1}{T}\frac{dT}{dz} \,. \tag{6.229}$$

By substituting (6.229) in (6.171) and assuming that μ is constant, we obtain

$$
\begin{aligned}
N_B^2 &= -g\left(\frac{1}{\rho_0}\frac{d\rho_0}{dz} - \frac{1}{\gamma p_0}\frac{dp_0}{dz}\right) \\
&= -g\left(\frac{1}{p_0}\frac{dp_0}{dz} - \frac{1}{T}\frac{dT}{dz} - \frac{1}{\gamma p_0}\frac{dp_0}{dz}\right) \\
&= \frac{g}{T}\left[\frac{dT}{dz} - \left(1 - \frac{1}{\gamma}\right)\frac{T}{p_0}\frac{dp_0}{dz}\right].
\end{aligned}
\tag{6.230}
$$

The last equality here allows us to express the Brunt-Väisälä frequency in terms of the equilibrium temperature gradient. We can see that

$$
N_B^2 \geq 0 \implies \underbrace{\frac{dT}{dz}}_{star} \geq \underbrace{\left(1 - \frac{1}{\gamma}\right)\frac{T}{p_0}\frac{dp_0}{dz}}_{adaibatic} \quad \text{or} \quad \frac{1}{T}\frac{dT}{dz} \geq \left(1 - \frac{1}{\gamma}\right)\frac{1}{p_0}\frac{dp_0}{dz}. \tag{6.231}
$$

As described in Chapter 2, we obtain the Schwarzschild criterion for convective stability. It compares the thermal gradient of the celestial object (e.g., a star) and the adiabatic gradient. If this inequality holds, there is no convection in the region of the celestial body under consideration. *Convection begins if this inequality is violated*; $N_B^2 < 0$. In other words, $N_B^2 \geq 0$ is actually the Schwarzschild criterion for convective stability. For example, in the outer layers of the Sun, cooling is so strong that the absolute value of temperature gradient exceeds the adiabatic temperature gradient. This is the threshold given by the Schwarzschild criteria for convective stability. So the condition for convective instability is that

$$
\frac{d\ln p_0}{d\ln T} < \frac{\gamma}{\gamma - 1}. \tag{6.232}
$$

For a monoatomic gas, $\gamma = 5/3$ and then $d\ln p_0/d\ln T < 2.5$.

MHD Schwarzschild criteria for convective stability of a static plasma.
We now include magnetic fields to seek a modified version of the Schwarzschild stability criterion by using (6.174) instead of (6.170). The local dispersion equation associated with (6.174) is

$$
\alpha^2 = -\frac{g'}{f'} = \frac{-k_0^2\rho_0\left(\omega^2 - \omega_A^2 - N_M^2\right)}{\rho_0\left(\omega^2 - \omega_A^2\right)} \tag{6.233}
$$

which implies that

$$
\omega^2 = \omega_A^2 + \frac{k_0^2}{k_0^2 + \alpha^2}N_M^2, \tag{6.234}
$$

where N_M is the magnetically modified Brunt-Väisälä frequency in (6.172) because it now contains magnetic contribution. Two conditions must be fulfilled to reach convection instability:

$$
N_M^2 < 0 \quad \text{and} \quad |N_M^2| > \omega_A^2 \tag{6.235}
$$

is a cross-over of two branches of the local dispersion equation with the solutions.

MHD Schwarzschild criteria for convective stability of a stationary plasma. We now consider the motion of the celestial object in addition to magnetic fields. We seek a modified version of the Schwarzschild stability criterion by using (6.176). The local dispersion equation associated with (6.176) is

$$\alpha^2 = -\frac{g'}{f'} = \frac{-k_0^2 \rho_0 \left(\tilde{\omega}^2 - \omega_A^2 - N_M^2\right)}{\rho_0 \left(\tilde{\omega}^2 - \omega_A^2\right)} \tag{6.236}$$

which implies that

$$\tilde{\omega}^2 = \omega_A^2 + \frac{k_0^2}{k^2 + \alpha^2} N_M^2 \tag{6.237}$$

or

$$\omega = k_0 v \pm \sqrt{k_0^2 v^2 + \omega_A^2 + \frac{k_0^2}{k^2 + \alpha^2} N_M^2} \,, \tag{6.238}$$

where N_M is the magnetically modified Brunt-Väisälä frequency in (6.172) because it now contains magnetic contribution. Two conditions must be fulfilled to reach convection instability:

$$N_M^2 < 0 \quad \text{and} \quad |N_M^2| > k_0^2 v^2 + \omega_A^2 \tag{6.239}$$

is a cross-over of two branches of the local dispersion equation with the solutions.

6.4.5. Accretion Disks and Magnetorotational Instability (MRI)

As described in Box 6.1, an accretion disk around a compact heavy central gravitating object (e.g., star, planet, neutron star, and black hole) consists of a region of space in which the rapidly rotating material orbiting in the gravitational field of the rotating matter loses energy and angular momentum to the benefit of the central object as it slowly spirals inward (see Figure 6.45). These rotating materials include gas, dust, and other stellar debris that has come close to the central object but not fallen into it. Around a central object, infalling matter forms a flattened band. So, gas moving toward a massive central object tends to circularize, form a disk, and spread inward and outward. This configuration is known as an *accretion disk* (or disc). Examples of disks in astrophysics include planetary rings (e.g., Saturn's ring), protoplanetary disks (such as HL Tau), and spiral disks (e.g., galactic disks). The central objects in accretion disks can be young prostars (with $m_{\text{star}} \approx m_{\text{sun}}$), neutron stars (with $m_{\text{star}} \approx 1.5 m_{\text{sun}}$), and black holes (with $m_{\text{star}} \approx 10^9 \times m_{\text{sun}}$)—it is incredible that such a range of objects can be based on the same formula as we will see later. There are several significant differences between these systems, but they all involve accretion. Fusion reactors share this accretion feature. The macroscopic dynamics of plasmas in both laboratory fusion devices (tokamaks, stellarators, etc.) and celestial objects (stellar coronas, accretion disks, spiral arms of galaxies, etc.) can be described by the same magnetohydrodynamics (MHD) equations, which are scale-independent. Although scale independence of the MHD equations permits analysis of global plasma dynamics in laboratory and astrophysical plasmas by the

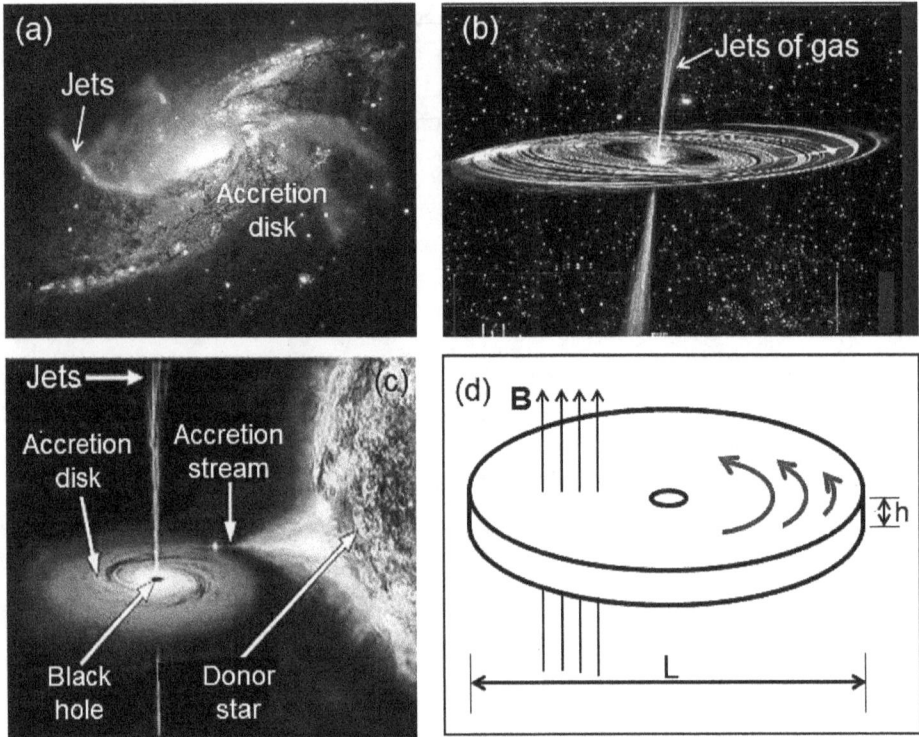

Figure 6.45. (a) Nebula houses baby stars in every spiral arm of the galaxy. (b) A zoom of a baby star. (c) The cloud wants to dissipate due to internal cloud pressure (here p_{cloud}) and gravity wants to pull it together. Collapse occurs if the inward directed gravitational force is bigger than the outward directed pressure force. Material crunches down into a disk which is the beginning of an eventual star. As gravity pulls more gas toward the center of the disk it gets denser and stiffer until it reaches about 15 or so million degrees Kelvin. (d) Hydrogen atoms fuse together to form helium and the star shines.

same techniques, the key differences in the parameters that govern overall force balance should not be lost sight of.

For example, the parameter $\beta = 2\mu_0 p/B^2$ is small for tokamak plasmas and usually large for astrophysical plasmas. So plasma dynamics in tokamaks is always dominated by magnetic fields whereas this may not be the case for astrophysical plasmas.

The properties and observables of accretion disks. Infalling matter powers the central gravitating object with its gravitational energy. This process includes the gradual loss of angular momentum by infalling matter. This allows matter to progressively move inwards, toward the central gravitating object. Therefore, infalling matter on the disk follows a round path until it loses angular momentum before falling into the central object. As material falls into the central object, gravitational potential energy is released. This energy release heats the plasma. The hot plasma radiates away the energy, proving an observational signature of accretion. Note that black holes themselves do not shine, but accretion disks do.

The gravitational force at r is

$$F_{\text{grav}} = \frac{Gm_{\text{object}}m_H}{r^2} \, , \tag{6.240}$$

and the radiation at r is

$$F_{\text{rad}} = \kappa m_H p_{\text{rad}} = \frac{L}{c_0}\frac{\kappa m_H}{4\pi r^2} = \frac{L}{c_0}\frac{\sigma_T}{4\pi r^2} \, , \tag{6.241}$$

where κ is the opacity, p_{rad} is pressure, L is the luminosity, G is the gravitational constant, c_0 is the speed of light, m_{object} is the mass of the gravitating body, m_H is the hydrogen mass. Opacity is the cross-sectional area per unit mass for radiation scattering. It is approximated as $\kappa \approx \sigma_T/m_H$, σ_T is the Thompson cross-section. As photons carry momentum and thus can exert pressure there is a maximum possible luminosity at which gravity can balance the outward pressure of radiation (i.e., $F_{\text{grav}} = F_{\text{rad}}$). So the limit for a steady, spherically symmetric accretion flow is given by the following luminosity:

$$L_{\text{Edd}} = \frac{4\pi Gm_{\text{object}}m_H c_0}{\sigma_T} \approx 1.26 \times 10^{31}\left(\frac{m_{\text{object}}}{m_{\text{sun}}}\right) \text{Watts} \, . \tag{6.242}$$

This is known as the Eddington limit. The approximation uses the known values of m_H, c_0, G, and $\sigma_T = 6.7 \times 10^{-25}$ cm^2. The Eddington limit quantifies object luminosity. It is the maximum luminosity beyond which radiation pressure overcomes gravity. Material outside the object will be forced away rather than falling inwards. Note that this approximation for κ is not valid in all situations. In low mass stars the opacity follows Kramer's Law, $\kappa \propto \rho/T^{7/2}$. Notice that it is also possible for an object to have a luminosity significantly greater than the Eddington luminosity when spherical symmetry and/or accretion is not steady (e.g., supernova explosions). L values can reach 10^{35} Watts.

One of the consequences of the Eddington limit is that we can also limit the rate at which such accretion can occur. Suppose a compact object accretes mass from its surroundings at a rate of dm/dt. Next assume that some fraction of the GPE can be radiated away. If we express this as a proportion of rest-mass energy, ϵ, the luminosity radiated away becomes

$$L = \epsilon\frac{dm}{dt}c_0^2 \implies \frac{dm_{\text{Edd}}}{dt} = \frac{4\pi Gm_{\text{object}}m_H}{\epsilon c_0 \sigma_T} \tag{6.243}$$

or

$$\frac{dm_{\text{Edd}}}{dt} \approx 2 \times 10^{-8}\left(\frac{m_{\text{object}}}{m_{\text{sun}}}\right)\left(\frac{0.1}{\epsilon}\right) m_{\text{sun}}/\text{year} \, . \tag{6.244}$$

We do not have the accretion rate limit because ϵ is unknown. However, for a supermassive central object $m_{\text{object}} = 7 \times 10^8\, m_{\text{sun}}$ and $\epsilon = 0.1$, Eddington accretion rate is $dm_{\text{Edd}}/dt \approx 3\, m_{\text{sun}}/\text{year}$ and the accretion time is

$$t_{\text{acc}} = \frac{m_{\text{object}}}{dm_{\text{Edd}}/dt} = \frac{\epsilon c_0 \sigma_T}{4\pi Gm_H} \approx 2.33 \times 10^8 \text{ years} \tag{6.245}$$

(i.e., 0.233 billion years) is known as the Salpeter timescale.

Table 6.11. A short summary of the accretion disk properties. PPS stands for proto-planetary systems, WD for white dwarfs, BH for black holes, and NS for neutron stars.

Central object	Disk size (km)	T (K)	Luminosity	Disk thickness	Example
• 1-3 m_{sun}	$10^4 - 10^{10}$	$10 - 10^5$	$L \ll L_{\text{Edd}}$	Thin	PPS WD
• 3-20 m_{sun}	$10 - 10^6$	$10^3 - 10^7$	$L < L_{\text{Edd}}$	Thin	BH, NS Binary stars
• $10^6 - 10^9$ m_{sun}	$10 - 10^6$	$10^3 - 10^7$	$L > L_{\text{Edd}}$ $L \gg L_{\text{Edd}}$	Thick	BH, AGN neutron star

Table 6.11 summarizes the accretion disk properties. During star formation, the central part of a dense molecular cloud collapses into a proto-star with a gaseous envelope that eventually settles on a rotating proto-planetary accretion disc. Accretion in such disks triggers planet and planetary system formation. Proto-stars are heavily embedded in gas and dust. They are visible only in the infrared, millimetre or sub-millimetre wavelength bands. Most galaxies have supermassive (millions to billions of solar masses) black holes at their centres (nuclei). In AGN, black hole accretion generates radiative power that usually outshines its host galaxy. The accretion disk is surrounded by hot gas clouds. King (2015) shows that there might be a limit of about 50 billion times the mass of the Sun to how massive the biggest supermassive black holes in the Universe can get. At some point in their size, their accretion disk might get so immense that the friction and other forces between infalling matter might create instability. This might lead to the whole accretion disk collapsing into a multitude of stars.

> **Quiz:** Determine the amount of energy released by 1 kg of infalling matter onto (1) White dwarf with $m_{wd} = 0.85 m_{\text{sun}}$ and $r_{wd} = 6.6 \times 10^6\,\text{m} = 0.0095\,r_{\text{sun}}$ and (2) neutron star with $m_{ns} = 1.5 m_{\text{sun}}$ and $r_{ns} = 10\,\text{km}$.

Magnetorotational Instability (MRI) Criterion. The three fundamental phenomena behind accretion disks are gravity, angular momentum conservation, and dissipation. Dissipation is a crucial part of this process because, without it, a disk would not conserve angular momentum and, hence, rotate forever without change. However the idea of rotating material losing its angular momentum is the most challenging. Remember that the specific angular momentum (i.e. the angular momentum per unit mass) is an increasing function of orbital radius, r; i.e.,

$$\hat{L}(r) = \frac{L}{\rho} = \sqrt{Gmr}\ . \tag{6.246}$$

If angular momentum is conserved with $r = $ constant, rotating material would orbit forever without accretion. To get closer to the central object, the rotating material must lose its angular momentum. The current solution to the accretion disk problem with angular momentum conservation is the MHD instability known as *magnetorotational instability*, where the accretion disk is described as a spinning disk with a central object in the presence of a magnetic field (Velikhov, 1959; Chandrasekhar, 1960; Balbus and Hawley, 1991; Hawley and Balbus, 1991; and Latter et al., 2015). Velikhov (1959) and Chandrasekhar (1961) demonstrated the existence of instability in magnetized rotating flows, but the importance of this instability for accretion disks was not recognized. Magnetorotational instability was rediscovered and applied to accretion disks in breakthrough work by Balbus and Hawley (1991). In equations (6.181) to (6.185), we describe the fundamental mathematics of magnetorotational instability. They are based on simple perturbations in cylindrical coordinates. We will focus only here on extracting from these derivations the dispersion equation that describes magnetorotational instability.

To describe a particle orbiting in a disk, cylindrical coordinates are the natural choice to take advantage of the cylindrical symmetry of this accretion disk configuration. By using the WKB approximation mentioned above the local dispersion equation associated with (6.185) is

$$\left(\alpha^2 + k_0^2\right)\left(\omega^2 - \omega_A^2\right)^2 - k_0^2\kappa^2\left(\omega^2 - \omega_A^2\right) - 4k_0^2\omega_A^2\Omega^2 = 0 \;, \tag{6.247}$$

where κ^2 is the epicyclic frequency and $\Omega(r)$ is the angular frequency [not confused with $\Omega_0(r)$ introduced in KHI]. Basically, the epicyclic frequency indicates how much specific angular momentum (i.e., how much angular momentum per unit mass) will be measured, where κ^2 is the epicyclic frequency and $\Omega(r)$ is the angular frequency [not confused with $\Omega_0(r)$ introduced in KHI]. Epicyclic frequency is a measure of how much specific angular momentum (i.e., angular momentum per unit mass)

$$\hat{L}(r) = \frac{L}{\rho} = r^2\Omega \implies \frac{\partial\hat{L}(r)}{\partial t} = 0 \;, \text{if } \kappa^2 = 0 \;. \tag{6.248}$$

The corresponding two solutions to are

$$\omega^2 = \omega_A^2 + \frac{k_0^2}{2\left(k_0^2 + \alpha^2\right)}\kappa^2 \pm \sqrt{\left(\frac{k_0^2}{2\left(k_0^2 + \alpha^2\right)}\kappa^2\right)^2 + 4\frac{k_0^2}{k_0^2 + \alpha^2}\omega_A^2\Omega^2} \;. \tag{6.249}$$

The expressions for the most global modes ($\alpha^2 \ll k_0^2$) yield the instability criteria,

$$\omega^2 \approx \omega_A^2 + \frac{\kappa^2}{2} \pm \sqrt{\kappa^4 + 16\omega_A^2\Omega^2} \underset{\omega^2 < 0}{\implies} \begin{cases} \kappa^2 < 0 & \text{for } \omega_A = 0 \\ \kappa^2 - 4\Omega^2 < -\omega_A^2 & \text{for } \omega_A \neq 0 \end{cases} \;. \tag{6.250}$$

Equation (6.185) of κ and the above equations allow us to differentiate between the stability of a hydrodynamic disk ($\omega_A^2 = 0$) and the stability of a magnetohydrodynamic disk ($\omega_A^2 \neq 0$).

Keplerian disk configuration. Let us start by recalling that if most of the mass of the system is concentrated at the center, as in the solar system, the speed of any orbiting body, such as a planet, is inversely proportional to the square root of its distance from the center. The fact that spiral galaxies do not have a rotation curve like this indicates that their overall mass distribution is different from the distribution of matter that can be seen. This is a significant piece of evidence for the existence of large amounts of dark matter in the Universe as discussed in Chapter 1.

So in Keplerian disk configuration the rotation speed versus distance from the center of an astronomical system obeys Kepler's laws $\Omega(r) \propto r^{-3/2}$ which implies that $d\Omega/dr < 0$. At the hydrodynamic limit, we can see that fluids are stable when $\kappa^2 \geq 0$. This criterion is satisfied for Keplerian rotation since

$$\Omega^2 = \Omega_K^2 = \frac{Gm}{r^3} \implies \kappa_K^2 = \frac{Gm}{r^3} > 0$$

where G is the gravitational constant, the subscript K identifies this as characteristic of Keplerian orbits. To explain the turbulent increase in dissipation processes in accretion disks, interest shifted to magnetohydrodynamic instabilities. However, MRI says nothing about turbulence and angular momentum transport in accretion disks. The links were established through simulations (Brandenburg et al., 1995; Hawley et al., 1995; Armitage, 1998; and Simon et al., 2012), including the fact that MRI can act as a dynamo, sustaining turbulence and magnetic fields against dissipation in a domain with no external currents. Angular momentum is transported primarily due to MHD stresses. Stronger turbulence and transport occur if the disk is threaded by a net vertical magnetic field.

Let us consider a case in which the disk, centered at r_0, co-rotates with the fluid at an angular velocity Ω_0. We assume that the angular velocity can be approximated as a power-law, i.e.,

$$\Omega = \Omega_0 \left(\frac{r}{r_0} \right)^{-\alpha}, \tag{6.251}$$

where α describes different rotation laws with $\alpha = 3/2$ in the Keplerian case. The MRI dispersion relation in (6.250) is plotted for various values of α in Figure 6.46. We can see the MRI is unstable for a insignificant range of ω_A/Ω_0 (i.e., $\omega^2 < 0$) which includes Keplerian disks and all the positive values of α. The instability disappears for $\alpha < 0$. Notice that for large values of ω_A (i.e., large values of the magnetic field, B) the system is stable irrespective of α values. As a result, Keplerian disks with a small magnetic field are always unstable. So magnetorotational instability causes turbulence, which drives angular momentum transport and accretion in sufficiently ionized disks Even with the success of modeling turbulent processes in accretion disks by means of magnetorotational instability, it is still an open question whether nature actually uses this process.

In summary, accretion disks require that (i) $\partial \Omega^2/\partial r < 0$, (ii) weak magnetic fields are present in the disk, and (iii) the ionization fraction of the disk is high enough to couple the magnetic field to the fluid. These prerequisites are satisfied by most astronomical disks. Colliding particles transfer energy and angular momentum between themselves. Whoever ends up with to little of those quantities drop in their orbits and get swallowed by the central object such as the black hole.

6.4.6.   Star Formation: Gravitational Collapse and Accretion Disk

One of the goals of magnetohydrodynamics is to help us understand our origin (i.e., how and when life occurred on Earth). As we further discuss in Ikelle (2024), our

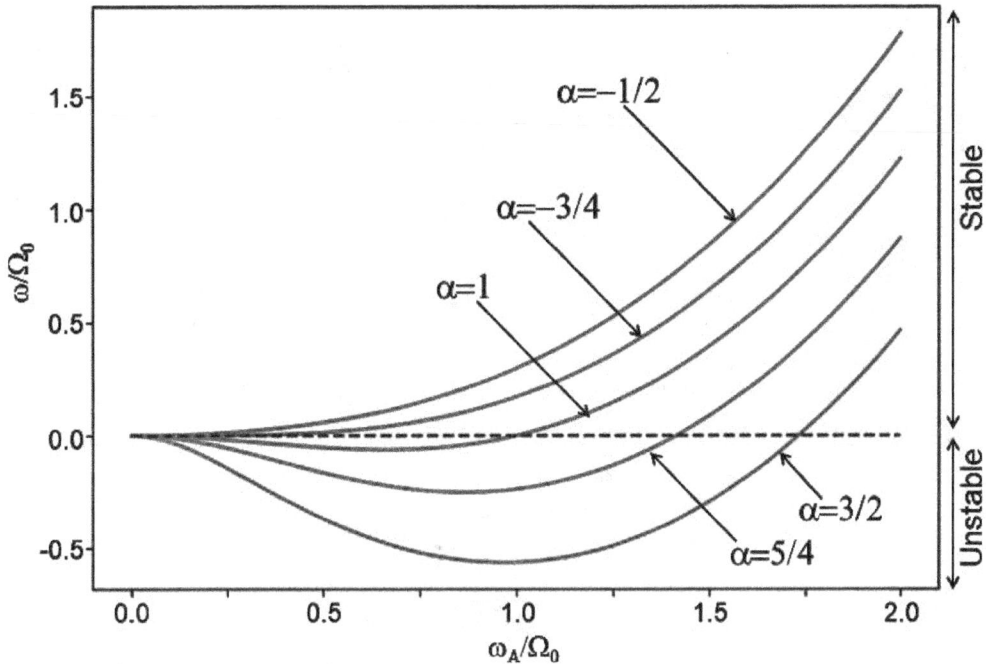

Figure 6.46. The MRI dispersion relation in ideal MHD is plotted for different rotation laws. The negative branch of the dispersion relation solution corresponds to an unstable accretion disk with respect to MRI.

origin is intertwined with the formation of the planets, our solar system, the interstellar medium, and nearby stars. It is therefore extremely critical to understand the processes leading to star and planet formation. In this subsection, we will focus on star formation. As far as we know, star formation is an essential prerequisite for planet formation. So we will discuss planet formation in the next subsection. Spitzer, Meatel, Field, Shu, and Monscharies are some of the pioneers in studying star formation. We apologize for not including all references.

Thanks to powerful telescopes like Hubble (launched 1990) and Kepler (launched 2009), we can survey, catalogue, and deduce the characteristics of thousands of star systems at every stage of their life cycle from their births to their violent deaths (Ikelle, 2023b). There are multiple protostar systems currently being studied with actively accreting protoplanetary disks. An example is the extremely young HL Tauri system, Figure 6.45c. The grooves in the disk are carved out by the planets. More earlier data on star formation were used by Shu et al. (1987) to provide us with an elegant description of all the key steps of star formation, as illustrated in Figure 6.47. It turns out that this description is essentially based on the gravitational collapse and disk accretion described earlier. We here use Shu's model of main Sequence star formation.

Gravitational collapse: The construct of the central object. All stars form in cold molecular clouds which are vast high-density regions in the interstellar medium (ISM) mostly made of hydrogen and helium. Often the clouds contain small

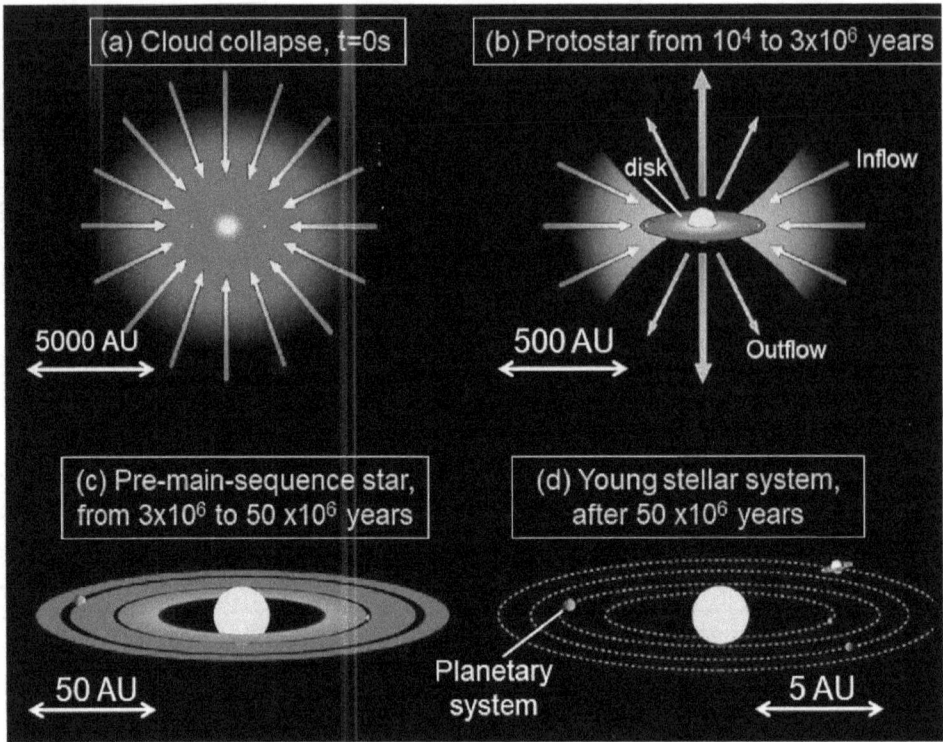

Figure 6.47. An illustration of star formation. (a) Dark cloud cores gradually contract until magnetic support is overcome and inside-out collapse begins at t=0. (b) A phase of both high accretion and supersonic outflow occurs in deeply embedded protostars. (c) Gradual clearing by the outflow leaving only the young T Tauri star and a residual protoplanetary accretion disk, which, on time scales of 1-10 million years, (d) leads to the formation of a mature planetary system. (Adapted from Mark McCaughrean at the Astrophysikalisches Institut Potsdam, and M.R. Hogerheijde).

amounts of more organic molecules, which are central to life, hydrocarbons, silica, carbon monoxide, water, and methane. These molecules move within the cloud, creating pressure. These molecules also have mass, so they create gravity in the cloud. The unevenly distributed mass within the cloud further multiplies molecular interactions within the cloud. The clouds also rotate. The average cloud radius is usually between 6×10^6 AU and 6.5×10^6 AU. Their masses are between 10 m_{sun} and 10^6 m_{sun} and temperatures are between 10 and 50 degrees Kelvin. Supernovae can send shockwaves through interstellar space and other phenomena in space, causing a cloud to become gravitationally unstable. Gravity then overcomes all the other forces in the cloud. Eventually, the cloud fragments into protostars, or areas of high particle density. The protostars disperse and separate from the overall cloud. Even when protostars are detached from the overall cloud, gas and dust continue to rain down. These protostars range in size from 2000 AU to 20000 AU and lead to star formation. Clouds produce multiple protostars. Because each protostar can become a star, clouds rarely form a single star. A cloud produces a cluster of stars. Figure 6.47 illustrates the star-formation process from the protostar stage.

The Milky Way galaxy, for example, contains a large number of molecular clouds, ranging from giant systems with masses in the range of 10^5–10^6 m_{sun} and sizes in the order of 50-100 pc (pc = parsec = 3.26,156 light-years= 206, 265 AU), temperatures between 10 and 50 degrees Kelvin, and densities ranging from $1,000$ particles per cm^3 in molecular clouds to $100,000$ particles per cm^3 in dense cores.

Protostars are also experimental facts because Hubble Space Telescope found protostars in the Orion Nebula. They are opaque to visible light and are called dark nebula. In order to identify them, IR and radio telescopes are used. In other words, we need long wavelengths to peer into dark clouds.

Accretion: The protostellar phase. A protostar has its own unique gravity. In protostars, clouds can also experience a disequilibrium between gravity, which can cause acceleration of the cloud collapse, and plasma pressure, turbulence, magnetic fields, and radiation pressure, which oppose gravity. The cloud within the protostar eventually collapses because it continues to gain mass, especially at its center from infalling gas, and gravity has overcome all the other forces acting on the protostar cloud. The infalling gas also releases kinetic energy in the form of heat and the temperature and pressure in the center of the protostar rise up. In addition to the protostar cloud collapsing, the cloud continues to rotate, pulling protostar contents into a vortex. This creates a disc around a sphere. Additionally, the increase in matter contraction causes the protostar's temperature to increase. The speed at which the cloud rotates throws gas into the central disk, compressing matter around the center and concentrating mass. The process continues for about a hundred thousand years until a central stable sphere of hot gas that no longer contracts forms at the center of the disk. This is where the star will arise. At some point, the pressure stops the infall of more gas into the core and the object becomes stable as a protostar. But the star's envelope continues to grow as infalling material is accreted.

Accretion: T Tauri phase and thermonuclear fusion. In-falling matter collects around the hot sphere which now acts as a gaseous oven by raising the mass compression density and ultimately the temperature. Eventually the mass and pressure are sufficient that the protostar ignites with the first stage of the star's great furnace. At this point the star is not hot or massive enough to begin nuclear fusion because its energy and heat source only comes from the gravitational compression of the envelope which compresses hot swelling gas around the star increasing its mass and density often accompanied by characteristic twin polar jets of outflow gas. Before long a substantial amount of the proto-stellar nebula's matter is concentrated on the protostar. The envelope around the star depletes and the infant star becomes visible for the first time. This is first seen in the far infrared spectrum and later in the visible spectrum. It is a young stellar object. Around this time, provided that the star's mass is at least 0.08 m_{sun}, the protostar's internal temperature reaches a value high enough (at least three million kelvins) for thermonuclear fusion to begin. Over tens of millions of years, it will slowly transition into their main sequence burning phase. In the meantime the young star generates some kind of outward pressure and erupts jets from its photosphere. This blow away the last part of the gaseous envelope. This envelope takes about a million years to become transparent and clear completely by a strong stellar wind is produced which stops the infall of new mass—that is, stellar wind breaks out along the rotational axis of the system, reversing infall and sweeping material into two outwardly expanding shells of gas and dust; infall still occurs in equatorial regions (see Figure 6.47b). The young star is now fully visible and classified as a T Tauri Star (see Ikelle, 2023b, for more details). Thermonuclear fusion begins in its core, its mass is fixed, and its future evolution is set. A star shines because of thermonuclear reactions in its core, which release enormous amounts of energy by fusing hydrogen into helium.

A star in the T-Tauri phase can lose up to 50 percent of its mass before settling down as a main sequence star, thus we call them pre-main sequence stars. As we can see in Figure 6.47, the arrows indicate how the T-Tauri stars will evolve into the main sequence. They begin their lives as cool stars, then heat up and become bluer and slightly fainter, depending on their initial mass. Very massive young stars are born so quickly that they only appear in the main sequence with such a short T-Tauri phase that they are never observed.

Outflow phase. Once a protostar becomes a hydrogen-burning star, a strong stellar wind forms, usually along the rotation axis. Thus, many young stars have a bipolar outflow, a flow of gas out of the poles of the star. This is, the outflow angle opens up with time, terminating infall and revealing a newly formed star optically with a circumstellar disk (Figure 6.47c). Radio telescopes easily see this feature. One consequence of this collapse is that young T Tauri stars are usually surrounded by massive, opaque, circumstellar disks. These disks gradually accrete onto the stellar surface, and radiate energy both from the disk (infrared wavelengths), and from the position where material falls onto the star at optical and ultraviolet wavelengths. Somehow a fraction of the material accreted onto the star is ejected perpendicular to the disk plane in a highly collimated stellar jet. The circumstellar disk eventually dissipates, when planets form.

Brown dwarf. If a protostar forms with a mass less than $0.08\ m_{\text{sun}}$, its internal temperature will never reach a value high enough for thermonuclear fusion to begin. It becomes a failed star known as a brown dwarf, halfway between a planet and a star. A brown dwarf is heavier than a gas-giant (Jovian) planet but not large enough to be a star. It was only in the mid-1990s that brown dwarfs were identified. Now scientists are tackling a host of intriguing questions about brown dwarfs: How many brown dwarfs are there? What is their mass range? Is there a continuum of objects down to Jupiter's mass? And did they all originate the same?

Anyway, brown dwarfs still emit energy, mostly in the IR, due to their collapse potential converted into kinetic energy. There is enough energy from the collapse to cause the brown dwarf to shine for over 15 million years. They eventually fade and cool to become black dwarfs. Brown dwarfs show that accretion disks do not necessarily lead to celestial objects powered by fusion. The amount of available matter is key to accretion outcome. Thus, the idea that accretion disks do not apply to planet formation may be wrong.

Quiz: Is Jupiter a brown dwarf star? 0.0000000

All stars have the same life cycle. The sun is an average star so it would expand its outer layers and end up as a black dwarf like the smaller stars. A giant blue star, on the other hand, would blow out in a massive explosion and end up as a neutron star (or black hole for the largest stars). Figure 6.48 shows the Sun's cycle.

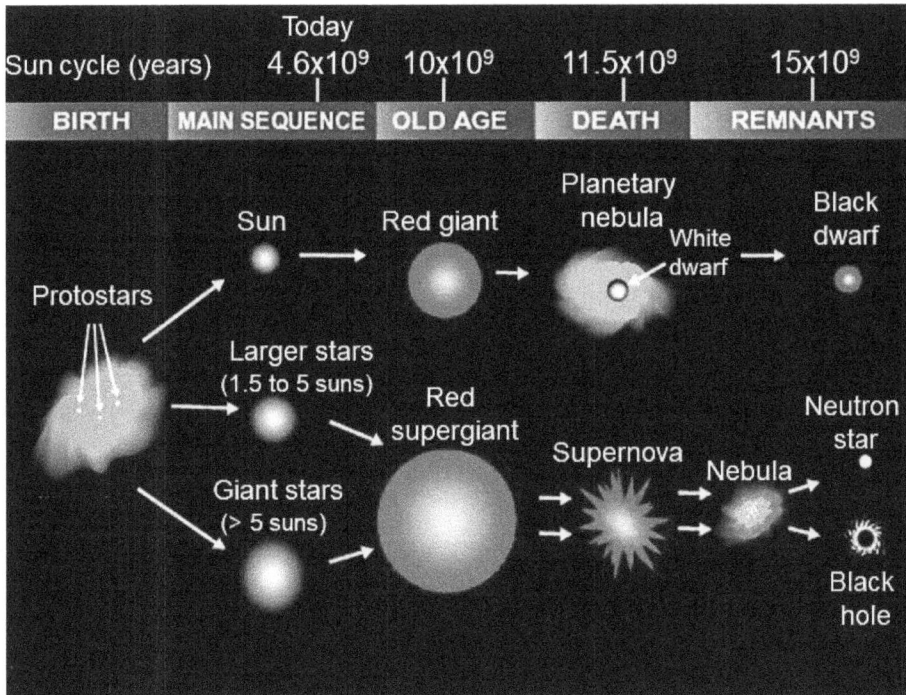

Figure 6.48. Illustration of the life cycle of the stars including the Sun. Our Sun was created 4.6 million years ago from the gravitational collapse of a giant molecular cloud. It eventually became hot enough to begin the nuclear fusion of hydrogen. So far the Sun has used 37 percent of this fuel and will run out of hydrogen in 5 billion years. At this point the Sun will enlarge and become a red giant, engulfing the inner solar system. When all the fuel is used the Sun will collapse and become a white dwarf.

Predictive model of star formation. In order to fully understand star formation from a scientific perspective, we need laws of physics that allow us to go from molecular cloud to star formation. Such laws must integrate gravitational collapse, accretion disks, and fusion, as well as the possibility that one process can fail to occur. MIID can be used to describe each process separately. The challenge is to automate the cascade process. Ikelle (2023a) describes a sequence using AI to learn how this cascade is constructed.

6.4.7. Accretion of Planets and Oumuamua

Planet formation. Planets are the end products of cosmic evolution that started with galaxies. It appears that some planets and exoplanets contain life and even intelligent life. They may be the final destinations for matter and energy in the universe. Accordingly, planets are formed mainly from remnants of the star's formation and infalling matter that occurs after the star's formation. There is a possibility that we will get another main sequence star, a brown star, or planets. This depends on the amount and composition of the remaining materials. Furthermore, planets are not expected to be necessary in the same state of matter. Some planets have one state of matter, some may have two states of matter, some may have three states of matter like Earth, or all four states depending on the location of the accretion disk of planets in the interstellar medium. The temperature and chemical composition of the accretion disk are key parameters.

Planet formation modeling is still a formidable problem in science. There is still a struggle to find a model that fits all known observations (see Ikelle, 2024). For example, it is believed that the rocky planets of our solar system have been subjected to multiple violent collisions in the past. They may even have been the result of these collisions. Even so, the rocky planets have a smaller eccentricity than the Jovian planets: Earth has 0.0167086 eccentricity, Venus 0.006772, Mercury 0.2656, Mars 0.0934, Jupiter 0.048498, Saturn 0.0555, Uranus 0.046381, and Neptune 0.009456 eccentricity. Our view is that planets and exoplanets were formed by accretion, just like stars and galaxies. Contrary to stars, planet formation does not involve nuclear fusion. If the remaining hydrogen and helium were large enough after Sun formation, Jupiter may have been a star.

The Sun's formation included an accretion disk of gas and dust, as described above. The chemical content of this disk included hydrogen, helium, water, carbon monoxide, carbon dioxide, methane, ammonia, hydrogen sulfide, metals, graphite, and silicates. The various planets in our solar system were accreted by a disk cloud of gas and dust. Metals, graphite, and silicates are part of dust. Because hydrogen and helium were the dominant components in the disk, Jupiter and Saturn, which are rich in helium and hydrogen, were formed first. Neptune and Uranus were the second planets. Last but not least, rocky planets with large amounts of silicate form. We know that Jupiter and Saturn were created because unlike Uranus, Neptune and the rocky planets, Jupiter and Saturn are rich in helium and hydrogen. They are also the largest planets because they have the largest materials stores to draw from. By the time the other rocky planets were formed, much of the hydrogen and helium in the primordial cloud in the solar system had disappeared from the accretion disk. Jovian planets end up in the region beyond the Frost line, between 2.7 and 3.4 AU from the forming Sun, where the conditions are conducive to the formation of gas giants because the lighter, volatile gases such as hydrogen and helium would have been at a low enough temperature to have been readily captured by the forming outer planets. They may follow the same path as the stars without reaching the fusion temperature.

The rocky planets are formed from the remaining material in the accretion disk after the formation of Jovian planets. They began as dust grains orbiting the central protostar. Through direct contact and self-organization, these grains formed into clumps up to 200 meters in diameter, which in turn collided to form larger bodies (planetesimals) of $\sim$ 10 kilometres (km) in size (see Figure 6.49). These gradually increased through further collisions, growing at the rate of centimetres per year over the course of the next few million years. Planetesimals are central objects in the context of an accretion disk. Planetesimals which over a 100 million year period, merged to eventually produce the inner worlds: Mercury, Venus, Earth and Mars. Moreover, in the warmer parts of the disk, closer to the star, only rocky planetesimals could survive. This is because ice and other volatile materials evaporate or sublimate. It is clear that a study of the variations of the magnetorotational instability criterion must be made to link the chemical composition of accretion disk with the type of planets and exoplanets that are ultimately formed.

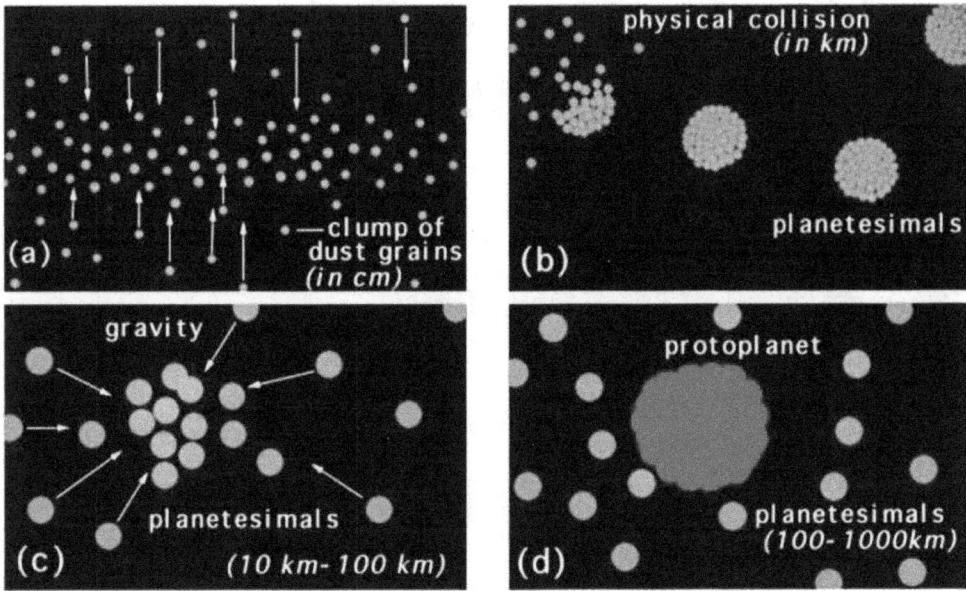

Figure 6.49. Planet formation. (a) The formation of centimeter-sized particles. (b) Physical collision on the kilometer scale. (c) Gravitational accretion on the 10-100 km scale. (d) Protoplanets from accretion heat. For each protoplanet that forms, there are thousands of left over planetisimal objects. These planetisimals accrete to form a planet. (Adapted from http://homework.uoregon.edu/pub/class/123/fss.html).

Weathering and possible plate tectonics on the Oumuamua parent exoplanet. Oumuamua is the first interstellar object detected passing through the Solar System. Robert Weryk discovered it using the Pan-STARRS telescope at Haleakala Observatory, Hawaii, on October 19, 2017. This was approximately 40 days after Oumuamua passed its closest point to the Sun on 9 September. When it was first observed, it was about 33 million km (0.22 AU) from Earth (about 85 times as far away as the Moon) and already heading away from the Sun.

Figure 6.50 shows that Oumuamua is a very unusual object. Its shape is different from those of planets, moons, asteroids, and comets orbiting the Sun for billions of years. Such extreme elongation suggests that Oumuamua may be the result of the weathering process described in Chapter 3 of Ikelle (2023a) and Chapter 1 of Ikelle (2023b) rather than disk accretion. Oumuamua's parent exoplanet might also have active tectonics, just like the Earth, due to such internal weathering processes.

Accretion on earth: aerospace science. When large amounts of materials come together, they can spin and flatten out around a compact central object. Accretion disks are powerful engines in the universe, as described above. Moreover, accretion disks are widespread in the Universe. They can convert gravitational energy into radiation with high efficiencies, reaching 40 percent. This process can be used on Earth to make giant airplanes and spacecraft, including our own version of Noah's Ark. Such design can also include learning about the accretion disk which may have

Figure 6.50. Artist's concept of Oumuamua. It is a dark red object and highly-elongated metallic or rocky, about 1,000 feet (400 meters) long. Adapted from Image of ESO/M. Kornmesser

huge empty space in its interior like Our Moon which has a significant empty space (or a homogeneous space) as described in Ikelle (2023c and 2024) and Chapter 1. So how disk accretion process can accomodate a hole in the middle of the central object. If you have ever tried to dig a big enough hole and struggled with the dirt, soil, gravel or whatever trying to fill the hole back in, you would already know the answer. Gravity pulls in matter to fill holes. It's just that simple.It can have a significant impact on interstellar travel (Ikelle, 2024). Nature always chooses the path that minimizes energy consumption given a fixed time. Because accretion disks are widespread, we can say that nature has picked them out as the optimal approach. Modern science and technology aim to learn from nature and duplicate how nature does things.

6.5. Generalized Ohm's Law and Magnetic Reconnection

When introducing equation (6.85), we alluded to differences in Ohm's law of a fluid in motion and that of a fluid at rest. Actually, that is the difference between the generalized Ohm's law and the standard Ohm's law in Chapter 3. This is the key difference between ideal MHD (figure 6.27) and non-ideal MHD (figure 6.28). This section derives generalized Ohm's law. By using generalized Ohm's law, we also discuss magnetic reconnection and magnetic monopoles. We will start by re-deriving the standard Ohm's law from the MHD perspective to facilitate the comparison between the two laws.

6.5.1. Microscopic Interpretation of Electrical Conductivity

Let us write the motion equation, in (6.79), for electrons and ions separately:

$$m_e \left[\frac{\partial}{\partial t} + \boldsymbol{v}_e \cdot \boldsymbol{\nabla} \right] (n_e \boldsymbol{v}_e) = -\boldsymbol{\nabla} p_e - n_e e \left(\boldsymbol{E} + \boldsymbol{v}_e \times \boldsymbol{B} \right) + m_e n_e \nu_{ei} \left(\boldsymbol{v}_i - \boldsymbol{v}_e \right), (6.252)$$

$$m_i \left[\frac{\partial}{\partial t} + \boldsymbol{v}_i \cdot \boldsymbol{\nabla} \right] (n_i \boldsymbol{v}_i) = -\boldsymbol{\nabla} p_i + n_i e \left(\boldsymbol{E} + \boldsymbol{v}_i \times \boldsymbol{B} \right) - m_e n_i \nu_{ei} \left(\boldsymbol{v}_i - \boldsymbol{v}_e \right) , (6.253)$$

where subscript e is used for electrons and i for ions. We used the fact that $\rho_e = m_e n_e$, $\rho_i = m_i n_i$, $\rho_{E,e} = -n_e e$, and $\rho_{E,i} = Z_i n_i e$, and again assumed that $Z_i = 1$ (i.e., the ions are protons). The last term in (6.252) describes the rate of momentum transfer during collisions between ions and electrons. It represents the mean momentum transferred between ions and electrons and ν_{ei} is electron-ion collision frequency. It is equal, with opposite signs, to the last term in (6.253) because momentum conservation requires the two terms to be of opposite signs. We assume no gravity. We did not include the effects of ion-to-ion and electron-to-electron collisions in (6.252) and (6.253) because the like particle collisions do not change the total momentum (which is averaged over all particles of that species) whereas collisions between unlike particles do exchange momentum between the species. Let us now assume that the plasma is at rest or in uniform motion (particle acceleration is zero) and the pressure and magnetic field are zero. Equation of motion in (6.252) reduces to

$$0 = -n_e e \boldsymbol{E} - m_e n_e \nu_{ei} \boldsymbol{v}_e \iff \boldsymbol{J} = \sigma \boldsymbol{E} , \qquad (6.254)$$

where

$$\boldsymbol{J} = -n_e e \boldsymbol{v}_e = \rho_{E,e} \boldsymbol{v}_e \quad \text{and} \quad \sigma = \frac{e^2 n_e}{m_e \nu_{ei}} \qquad (6.255)$$

and σ is the electric conductivity. This formula for electric conductivity allows us to link the macroscopic quantity, σ, defined in Chapter 3, to the microscopic scale. By increasing the electron particle density, n_e, in (6.255), we increase plasma conductivity, and by increasing the ion-electron collision frequency, ν_{ei}, we increase plasma resistivity. It is clear that the law, $\boldsymbol{J} = \sigma \boldsymbol{E}$, does not work when $\boldsymbol{B} \neq \boldsymbol{0}$, $\boldsymbol{v}$ is not a constant vector, and/or p is not constant.

Let us turn to the derivation of generalized Ohm's law. We again assume electrical neutrality: $n_i = n_e = n$. By multiplying equation (6.252) by $-e/m_e$ and adding it to equation (6.253) multiplied by c/m_i, and neglecting the inertia terms, which are $(\boldsymbol{v}_e \cdot \boldsymbol{\nabla}) \boldsymbol{v}_e$ and $(\boldsymbol{v}_i \cdot \boldsymbol{\nabla}) \boldsymbol{v}_i$, we obtain

$$\frac{\partial}{\partial t} \left[en \left(\boldsymbol{v}_i - \boldsymbol{v}_e \right) \right] = -\frac{e}{m_i} \boldsymbol{\nabla} p_i + \frac{e}{m_e} \boldsymbol{\nabla} p_e + e^2 n \left(\frac{1}{m_e} + \frac{1}{m_i} \right) \boldsymbol{E}$$

$$+ e^2 n \left[\left(\frac{\boldsymbol{v}_e}{m_e} \times \boldsymbol{B} \right) + \left(\frac{\boldsymbol{v}_i}{m_i} \times \boldsymbol{B} \right) \right]$$

$$- en \nu_{ei} \left[\left(\boldsymbol{v}_i - \boldsymbol{v}_e \right) - \frac{m_e}{m_i} \left(\boldsymbol{v}_i - \boldsymbol{v}_e \right) \right] . \qquad (6.256)$$

Since $m_i \gg m_e$ then $e/m_e \gg e/m_i$ and $n_e e^2/m_e \gg n_i e^2/m_i$, we neglect the small terms of the order of the ratio m_e/m_i. Also in thermal equilibrium, electron kinetic pressures are similar to ion pressures (i.e., $p_e \approx p_i$). By omitting these terms, we obtain the generalized Ohm's law (Rossi and Olbert, 1970; and Greene, 1973), i.e.,

$$\frac{\partial \boldsymbol{J}}{\partial t} = \frac{e^2 n}{m_e} \left[\boldsymbol{E} + (\boldsymbol{v} \times \boldsymbol{B}) \right] - \frac{e}{m_e} (\boldsymbol{J} \times \boldsymbol{B}) - \nu_{ei} \boldsymbol{J} + \frac{e}{m_e} \boldsymbol{\nabla} p_e , \qquad (6.257)$$

$$\boldsymbol{J} = \sigma \left[\boldsymbol{E} + (\boldsymbol{v} \times \boldsymbol{B}) \right] - \frac{\Omega_g^{(e)}}{\nu_{ei}} \frac{\boldsymbol{J} \times \boldsymbol{B}}{B} - \frac{1}{\nu_{ei}} \frac{\partial \boldsymbol{J}}{\partial t} + \frac{e}{m_e} \boldsymbol{\nabla} p_e , \qquad (6.258)$$

or

$$\boldsymbol{E} = -\boldsymbol{v} \times \boldsymbol{B} + \frac{1}{\sigma} \boldsymbol{J} + \frac{\boldsymbol{J} \times \boldsymbol{B}}{e n_e} - \frac{1}{n_e e} \boldsymbol{\nabla} p_e + \frac{m_e}{n e^2} \frac{\partial \boldsymbol{J}}{\partial t} , \qquad (6.259)$$

where $\Omega_g^{(e)} = eB/m_e$ is the electron gyrofrequency of the electron gyromotion, introduced earlier. Generalized Ohm's law, or more generally magnetohydrodynamics, explains that changing flows, pressure gradients, or inertial stresses produce electric currents.

By assuming that $\boldsymbol{B} = \boldsymbol{0}$, a slowly varying current density, and p_e is constant, the generalized Ohm's law reduces to the standard Ohm's law, $\boldsymbol{J} = \sigma \boldsymbol{E}$. If $\boldsymbol{B} \neq \boldsymbol{0}$, but a slowly varying current density, and p_e is constant, the generalized Ohm's law reduces to (6.85). In ideal MHD, the plasma is assumed to conduct ideally, there is no electron pressure gradient, and the current density is assumed to change only slowly. The generalized Ohm's law reduces to $\boldsymbol{E} = -\boldsymbol{v} \times \boldsymbol{B}$.

Assuming pressure p_e is constant and current density $\boldsymbol{J}$ is smooth, equation (6.258) becomes

$$\boldsymbol{J} = \sigma \left[\boldsymbol{E} + (\boldsymbol{v} \times \boldsymbol{B}) \right] - \frac{\Omega_g^{(e)}}{\nu_{ei}} \frac{\boldsymbol{J} \times \boldsymbol{B}}{B} . \qquad (6.260)$$

One can verify that the solution to (6.260) is

$$\boldsymbol{J} = \sigma \boldsymbol{E}_{\parallel}^{(\text{eff})} + \sigma_{\perp} \boldsymbol{E}_{\perp}^{(\text{eff})} + \sigma_H \boldsymbol{b} \times \boldsymbol{E}_{\perp}^{(\text{eff})} \qquad (6.261)$$

where

$$\sigma_{\perp} = \sigma \left[1 + \left(\frac{\Omega_g^{(e)}}{\nu_{ei}} \right)^2 \right]^{-1} \quad \text{and} \quad \sigma_H = \sigma \frac{\Omega_g^{(e)}}{\nu_{ei}} \left[1 + \left(\frac{\Omega_g^{(e)}}{\nu_{ei}} \right)^2 \right]^{-1} ,$$

and where $\boldsymbol{E}^{(\text{eff})} = \boldsymbol{E} + \boldsymbol{v} \times \boldsymbol{B}$, $\boldsymbol{b} = \boldsymbol{B}/B$ is the unit vector along the magnetic field, $\boldsymbol{E}_{\parallel}^{(\text{eff})}$ is the component of $\boldsymbol{E}^{(\text{eff})}$ parallel to $\boldsymbol{b}$, and $\boldsymbol{E}_{\parallel}^{(\text{eff})}$ is the component of $\boldsymbol{E}^{(\text{eff})}$ perpendicular to $\boldsymbol{b}$. We can see that the presence of a magnetic field in plasma not only changes the conductivity magnitude. It also causes plasma electric conductivity to vary with direction when $\sigma \neq \sigma_{\perp}$. The third term here is the Hall current.

6.5.2. Three-fluid Ohm's Law

The three-fluid model considers electrons, ions, and neutral particles as three fluids—that is, a plasma does not have to consist only of electrons and ions. Neutral particles can also be present in plasmas, especially in partially ionized plasmas. It also takes into account their collisions, that is, electron-ion, electron-neutral, and ion-neutral collisions. This model is particularly useful in describing the interaction between the magnetosphere, ionosphere, and thermosphere (see Figure 6.7). The corresponding equations of motion are

$$m_e \left[\frac{\partial}{\partial t} + \boldsymbol{v}_e \cdot \boldsymbol{\nabla} \right] (n_e \boldsymbol{v}_e) = -\boldsymbol{\nabla} p_e - n_e e \left(\boldsymbol{E} + \boldsymbol{v}_e \times \boldsymbol{B} \right)$$

$$+ \quad m_e n_e \nu_{ei} \left(\boldsymbol{v}_i - \boldsymbol{v}_e \right)$$

$$+ \quad m_e n_e \nu_{en} \left(\boldsymbol{v}_e - \boldsymbol{v}_n \right) , \qquad (6.262)$$

$$m_i \left[\frac{\partial}{\partial t} + \boldsymbol{v}_i \cdot \boldsymbol{\nabla} \right] (n_i \boldsymbol{v}_i) = -\boldsymbol{\nabla} p_i + n_i e \left(\boldsymbol{E} + \boldsymbol{v}_i \times \boldsymbol{B} \right)$$

$$- \quad m_e n_i \nu_{ei} \left(\boldsymbol{v}_i - \boldsymbol{v}_e \right)$$

$$+ \quad m_i n_e \nu_{in} \left(\boldsymbol{v}_i - \boldsymbol{v}_n \right) , \qquad (6.263)$$

$$m_n \left[\frac{\partial}{\partial t} + \boldsymbol{v}_n \cdot \boldsymbol{\nabla} \right] (n_n \boldsymbol{v}_n) = -\boldsymbol{\nabla} p_n + m_n n_n \nu_{ni} \left(\boldsymbol{v}_n - \boldsymbol{v}_i \right)$$

$$+ \quad m_n n_n \nu_{ne} \left(\boldsymbol{v}_n - \boldsymbol{v}_e \right) , \qquad (6.264)$$

where the subscripts e, i, and n are associated with electrons, ions, and neutrals, respectively, ν_{ei} is the electron-ion collision frequency, ν_{ni} is the neutral-ion collision frequency, and ν_{ne} is the neutral-electron collision frequency. Compared to (6.252) and (6.253), we now have added the effects of neutrals on charged particles to the collision terms, in (6.262) and (6.263). Notice that equation (6.264) can be ignored in the derivations of the generalized Ohm's law because it does not include the electric current density or electromagnetic fields. The derivations are the same for two fluids. We multiply equation (6.262) by $-(e\nu_{in})/m_e$ and add it to equation (6.263) multiplied by $(e\nu_{en})/m_i$, and neglect the inertia terms. We eliminate $\boldsymbol{v}_n$ in this process. We then arrive at the generalized Ohm's law by using the fact that $m_i \gg m_e$ neglects the small terms of the order of the ratio m_e/m_i. However, the coefficients associated with the electric current-density and electromagnetic-field vectors, in (6.258), for example, include ion-neutral and electron collision frequencies. The physical benefits of the three fluid models are gained from the examination of these different coefficients. For the particular case of partially ionized plasmas (i.e., electron-ion collisions are rare), the generalized Ohm's law pertaining to these three fluid models is

$$\boldsymbol{J} = \sigma_P \left[\boldsymbol{E} + (\boldsymbol{v} \times \boldsymbol{B}) \right] - \alpha_H \frac{\boldsymbol{J} \times \boldsymbol{B}}{B} , \qquad (6.265)$$

with

$$\sigma_P = \frac{e^2 n_e}{m_e} \frac{m_i \nu_{in} + m_e \nu_{en}}{m_i \nu_{in} \nu_{en}} \quad \text{and} \quad \alpha_H = \frac{e}{m_e} \frac{m_i \nu_{in} - \frac{m_e}{m_i} m_e \nu_{en}}{m_i \nu_i \nu_{en}} \,,$$

assuming pressure p_e is constant and current density, J, is smooth, $\nu_{in} \neq 0$, and $\nu_{en} \neq 0$. One can verify that the solution to (6.265) is

$$\boldsymbol{J} = \sigma_P \boldsymbol{E}_{\parallel}^{(\text{eff})} + \sigma_{\perp} \boldsymbol{E}_{\perp}^{(\text{eff})} + \sigma_H \boldsymbol{b} \times \boldsymbol{E}_{\perp}^{(\text{eff})} \qquad (6.266)$$

where

$$\sigma_{\perp} = \sigma_P \left(1 + \alpha_H^2\right)^{-1} \quad \text{and} \quad \sigma_H = \sigma \alpha_H \left(1 + \alpha_H^2\right)^{-1} \,.$$

6.5.3.   Magnetic Reconnection: A Departure from the Ideal MHD

As described in Biskamp (2000), Yamada et al. (2010), and Ikelle (2020), magnetic reconnection is about two magnetic fields that point in opposite directions. As they move closer to each other, they break apart such that the field lines are rearranged into a pair of two U shapes. This is illustrated in Figure 6.51. At their core, magnetic reconnections are explosions that affect magnetic fields and plasmas in the vicinity of explosions. These explosions release large amounts of energy in a very small space. Magnetic field topology changes suddenly. Field lines break and reconnect.

Magnetic reconnection is a key phenomenon in astrophysical, space, and fusion plasmas. This process is observed in a variety of space and astrophysical situations, including solar flares (Giovanelli, 1946), coronal mass ejections and the aurora borealis (see Ikelle, 2024), the Earth's magnetopause and magnetotail (Dungey, 1961), etc. Already in the 1950s, observations of solar flare reconnection were well established. In less than 20 minutes, they would flake and then die off. People could see them from a distance. It is also frequent in laboratory plasmas, such as tokamaks, spheromaks, and reversed-field pinches, although they are undesirable

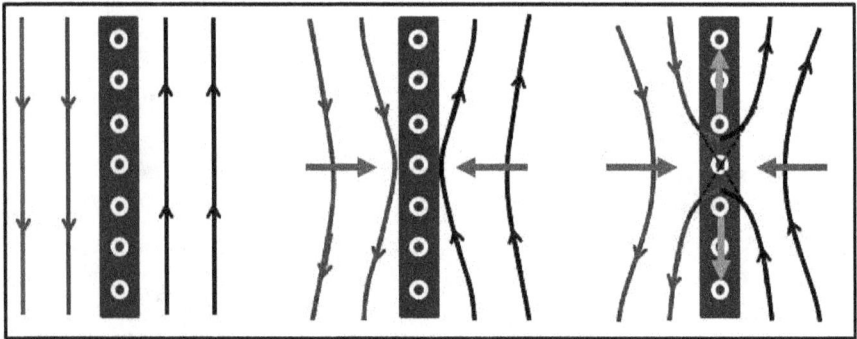

Figure 6.51.   Illustration of the phenomenon of reconnection in which two magnetic field lines break and connect. Field lines' directions are critical to reconnection. The two magnetic field lines must be opposites. As they get close to each other (from left to right), they break and reconnect. Two U-shaped field lines are reconnected to one another.

in these cases. It cannot happen according to the ideal MHD. We need to add additional terms to Ohm's law to allow reconnection. In the generalized Ohm's law, there are factors such as resistivity, Hall current, electron inertia and pressure, and ambipolar diffusion that are related to weakly collisional or collisionless plasma. Since the generalized Ohm is one of the equations of the non-ideal MDH model, the non-ideal MHD can be used to model magnetic reconnection in space. Note that we return to this phenomenon in Ikelle (2024) because it plays a crucial role in the understanding of the Sun and magnetosphere interaction. Magnetic reconnection is the key mechanism for energy release in solar Corona and solar flares (see Figure 6.3).

Another review of the differences between ideal and non-ideal MHD. The difference between ideal and non-ideal MHD resides in the magnetic induction equation. In more general terms, this equation can be written as follows:

$$\boldsymbol{E} = -\boldsymbol{v} \times \boldsymbol{B} + \boldsymbol{R} , \tag{6.267}$$

where $\boldsymbol{R}$ is the operator that differentiates the two MHD formulations. For ideal MHD, $\boldsymbol{R} = \boldsymbol{0}$. For non-ideal MHD, we can have

$$\boldsymbol{R} = \frac{1}{\sigma} \boldsymbol{J} + \frac{\boldsymbol{J} \times \boldsymbol{B}}{e n_e} - \frac{1}{n_e e} \boldsymbol{\nabla} p_e + \frac{m_e}{n e^2} \frac{\partial \boldsymbol{J}}{\partial t} \tag{6.268}$$

as defined in (6.259). Classical electromagnetism generally teaches us that magnetic field lines do not break; they are always continuous. The missing caveat is that this statement is only true when matter obeys the ideal MHD, as in Figure 6.27—that is, $\boldsymbol{E} = -\boldsymbol{v} \times \boldsymbol{B}$. The ideal MHD plasma behaves like a perfectly conducting fluid. All the examples in the third section of this chapter are limited to ideal MDH. To prove this theorem, we consider the following flux:

$$\Phi = \int\!\!\int_{S(t)} \boldsymbol{B} \cdot \boldsymbol{n} dS \tag{6.269}$$

across a moving surface with velocity $\boldsymbol{u}_\perp$. Flux can change due to $\boldsymbol{B}$ changing in time and/or the surface S moving/distorting as the fluid moves. Similar to equation (6.89), we can verify that

$$
\begin{aligned}
\frac{\partial \Phi}{\partial t} &= \int\!\!\int_{S(t)} \frac{\partial \boldsymbol{B}}{\partial t} \cdot \boldsymbol{n} dS - \oint_{\partial S(t)} \boldsymbol{u}_\perp \times \boldsymbol{B} \cdot d\boldsymbol{l} \\
&= \int\!\!\int_{S(t)} \boldsymbol{\nabla} \times (\boldsymbol{v} \times \boldsymbol{B}) \cdot \boldsymbol{n} dS - \oint_{\partial S(t)} \boldsymbol{u}_\perp \times \boldsymbol{B} \cdot d\boldsymbol{l} \\
&= \oint_{\partial S(t)} (\boldsymbol{v} - \boldsymbol{u}_\perp) \cdot d\boldsymbol{l} \\
&= 0 \quad \text{if} \quad \boldsymbol{u}_\perp = \boldsymbol{v} .
\end{aligned}
\tag{6.270}
$$

where $\boldsymbol{S} = S\boldsymbol{n}$, with $\boldsymbol{n}$ a unit vector normal to S, and dS the surface element of S. As a result, the plasma is connected to the field lines. We use in these derivations the fact $\partial \boldsymbol{B}/\partial t = \boldsymbol{\nabla} \times \boldsymbol{E}$ and $\boldsymbol{E} + \boldsymbol{v} \times \boldsymbol{B} = \boldsymbol{0}$ (i.e., $\boldsymbol{R} = \boldsymbol{0}$) implies that the induction equation is $\partial \boldsymbol{B}/\partial t = \boldsymbol{\nabla} \times (\boldsymbol{v} \times \boldsymbol{B})$. We also use Stokes's theorem to move from a surface integral over S to a line integral of δS.

If $\boldsymbol{R} \neq 0$, the induction equation is $\partial \boldsymbol{B}/\partial t = \boldsymbol{\nabla} \times (\boldsymbol{v} \times \boldsymbol{B}) + \boldsymbol{\nabla} \times \boldsymbol{R}$ and the field lines can indeed break. In the event of two opposing field lines approaching each other, the electric current density becomes extremely large. This results in the magnetic field lines breaking and reconnecting.

From a physics point of view, we compare frozen flux (ideal MHD) and unfrozen flux (non-ideal MHD). As we discussed earlier, Hannes Alfvén proved that if $\boldsymbol{R} = \boldsymbol{0}$ plasma particles are frozen (locked) in the magnetic field. That is, with this constraint, we cannot decouple the particles from the field lines; the magnetic field lines are frozen into the plasma motion. Therefore we cannot break the field lines. If two plasma elements are initially connected by a field line, they will remain connected. Magnetic field topology/connectivity is invariant. Based on observations, field lines break up. Particles are no longer frozen in magnetic field lines if one of the terms in (??eq:chpmhdf.01re]) is nonzero. Nature may be more flexible. It allows plasma to not be locked with a magnetic field in some places in space. In other words, the reconnection breaks topological invariance; By reconnection we mean the change of topology of a (part of) a field configuration, when neighboring field lines running in different directions "touch" and exchange the directions of their paths. There are other issues regarding theories of reconnection developed in Chapters 1 and 2 of Ikelle (2024) , including the changes in topology of the magnetic field.

An Example. First of all, let us note that the reconnection, or changes in the topology of magnetic field lines, occurs trivially in magnetic field lines in a vacuum. Let us look at a specific example.

Remember that the magnetic field lines can be defined as

$$\frac{dx}{B_x} = \frac{dy}{B_y} = \frac{dz}{B_z} . \tag{6.271}$$

The solution to this system of differential equations defines a field line. Consider the following vacuum magnetic field in a vacuum,

$$\boldsymbol{B} = B_0 \left[y \cos(\omega t - kz), x \cos(\omega t - kz), 0 \right]^T . \tag{6.272}$$

where B_0 and k are constants. By using (6.271), the magnetic field lines are given by

$$x dx = y dy \implies y^2 - x^2 = \pm a^2 = \text{constant} . \tag{6.273}$$

Therefore, the magnetic field lines are hyperbolas. The solution $y = \sqrt{x^2 \pm a^2}$ represents one set of field lines and $y = -\sqrt{x^2 \pm a^2}$ represents antiparallel field lines for $x^2 \pm a^2 > 0$. By using the following Matlab commands,

```
x = −4 : 0.01 : 4; y2 = x.²; y = sqrt(y2); plot(x, y,′ k′); hold; plot(x, −y,′ k′)
y2 = 1 + x.²; y = sqrt(y2); plot(x, y); plot(x, −y);
y2 = 4 + x.²; y = sqrt(y2); plot(x, y); plot(x, −y);
x = −4 : 0.01 : −2; y2 = −4 + x.²; y = sqrt(y2); plot(x, y,′ r′); plot(x, −y,′ r′);
```

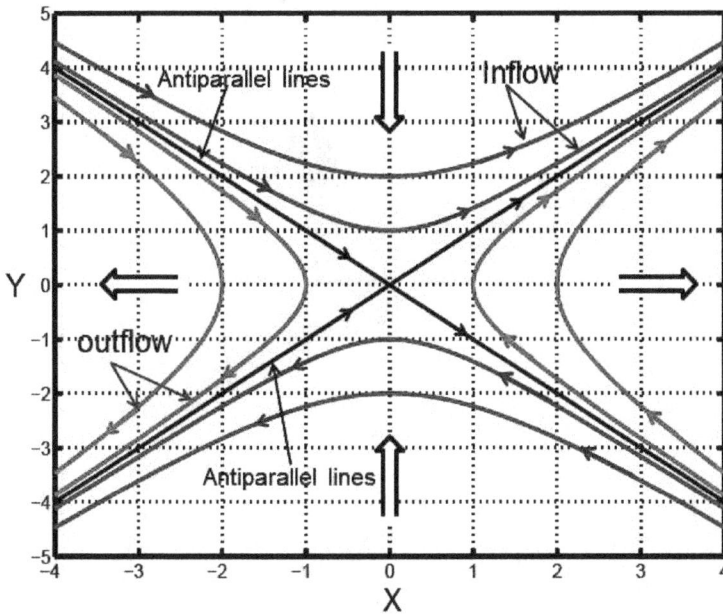

Figure 6.52. Cartoon of reconnection in a vacuum. The reconnection occurs when magnetic field lines merge, rejoin, and snap apart. Antiparallel magnetic field lines are reconfigured by reconnection. Magnetic field lines do not intersect.

$x1 = 2 : 0.01 : 4; y2 = -4 + x1.^2; y = \text{sqrt}(y2); \text{plot}(x1, y,'r'); \text{plot}(x1, -y,'r');$
$x = -4 : 0.01 : -1; y2 = -1 + x.^2; y = \text{sqrt}(y2); \text{plot}(x, y,'r'); \text{plot}(x, -y,'r');$
$x1 = 1 : 0.01 : 4; y2 = -1 + x1.^2; y = \text{sqrt}(y2); \text{plot}(x1, y,'r'); \text{plot}(x1, -y,'r');$

we can verify the reconnection (or merging) mechanism that allows momentum and energy exchange between otherwise segregated regions (see also Figure 6.52). Unfortunately, when dealing with magnetized plasma instead of vacuum, reconnection may occur only for particular magnetic geometries. The charged particles that are considered frozen in the plasma must produce special dissipation processes for this reconnection to occur. These dissipation processes are dependent on magnetic geometries. In other words, the process of reproducing the complex reconnection pictures is more complex than that of reproducing the simple picture in Figure 6.52. We further discuss this issue in Ikelle (2024) in the context of Earth and Jovian magnetosphere.

6.6. Application: Exoplanets

An exoplanet, also known as an extrasolar planet, orbits a star (or multiple stars) other than the Sun. There are also free-floating exoplanets, called rogue planets, that orbit the galactic center and are untethered to any star. Exoplanet exploration is crucial (1) in answering fundamental questions such as: are we the only people in this universe and where do we come from? (2) in providing mankind with

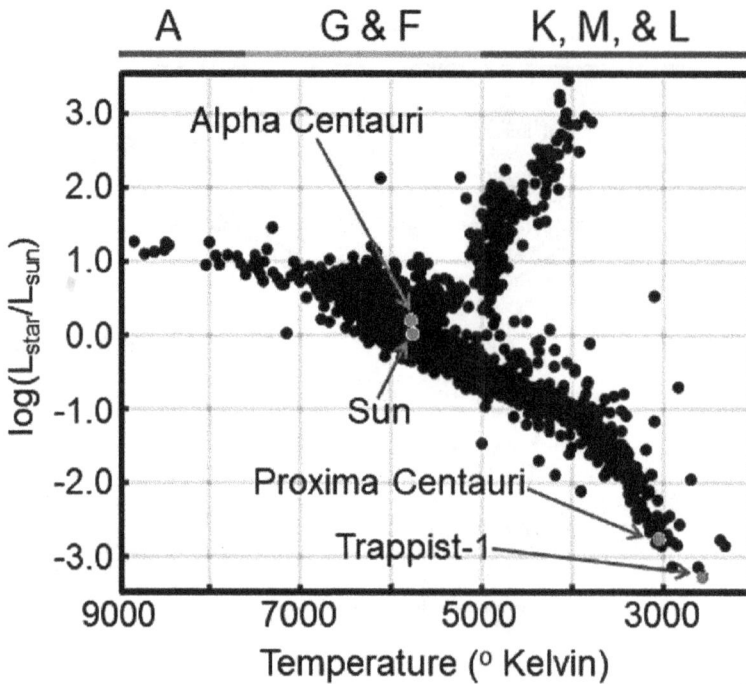

Figure 6.53. Illustration of the luminosity versus temperature of stars associated with currently known exoplanets. Compare to Figure 6.5 we can see that there are no active exoplanets associated with white dwarfs or supergiant stars. Source of data: https://exoplanetarchive.ipac.caltech.edu/.

alternative places to immigrate, and (3) in learning new technology and cures that can improve lives on Earth. For example, identifying lives in places of elevated radiation can help us understand the necessary changes to reduce human vulnerability to radiation effects. In the event of magnetic reversals, as well as nuclear explosions on Earth, this understanding and associated cures and preventatives may prove helpful. This section explores the relationship between stars and exoplanets and exoplanet classification.

As of April 2023, there were 5,322 exoplanets in 4000 planetary systems[3] and 603 multiple planetary systems have been confirmed. In other words, most planetary systems have only one exoplanet. This count of confirmed planets is still too small compared to the number of stars and galaxies in the billions. Our galaxy, the Milky Way, contains at least 100 billion stars, including our Sun. We are already heading into the billions of planets in the galaxy if every ten stars contain more than one planet. However, it is a large enough sample for statistical analysis.

Most of these exoplanets orbit L, M, K, G, F stars (that is white-yellow, yellow, orange, and red stars as described in Figures 6.5 and 6.6 and Table 6.1), and only 15 of the 5,322 exoplanets orbit A stars (white stars). None of them orbit O (blue) and B (blue-white) stars, as illustrated in Figure 6.53. Blue stars and bluish white stars have a high surface temperature above 10,000 degrees Kelvin sometimes exceeding 50,000 degrees Kelvin. Blue stars form from colliding and merging binary stars near

---

[3] A planetary system is a set of gravitationally bound non-stellar objects in or out of orbit around a star or star system. Multiplanetary systems are stars with at least two planets.

Table 6.12. The current confirmed exoplanets nearest to us. Ly stands for light-year. The lowercase letter "b" stands for the exoplanet, in the order it was found. The first exoplanet found is always named b, with the ensuing planets named c, d, e, f and so on.

Planet	Ly	Radius (r_{earth})	Mass (m_{earth})	Orbital period	Star
• Proxima Centauri b	4.24	1.03	1.07	9.9 days	Proxima Centauri (M)
• Epsilon Eridani b	10.5	13.75	209.88	7.3 years	Epsilon Eridani (K)
• Ross 128 b	11	1.11	1.4	9.9 days	Ross 128 (M)
• Tau Ceti e	11.9	1.81	3.93	162.9 days	Tau Ceti (G)
• Tau Ceti f	11.9	1.81	3.93	1.7 years	Tau Ceti (G)
• Tau Ceti g	11.9	1.18	1.75	20 days	Tau Ceti (G)
• Tau Ceti h	11.9	1.19	1.83	49.4 days	Gliese 1061 (M)
• GJ 1061 b	12	1.1	1.37	3.2 days	Gliese 1061 (M)
• GJ 1061 c	12	1.18	1.74	6.7 days	Gliese 1061 (M)
• GJ 1061 d	12	1.16	1.64	13 days	Gliese 1061 (M)
• Epsilon Indi A b	12	12.87	1033.5	45.2 years	Epsilon Indi (K)

galaxies' nuclei. They only last a few million years, so it is not a surprise that no exoplanet orbiting blue stars has been found. If such a giant star has an exoplanet, there is not enough time for life to emerge based on our current knowledge of Earth. Nearly two billion years passed after single-cell organisms appeared before complex life forms evolved. Moreover, life is unlikely in eventual exoplanets of blue stars because of frequent supernova explosions and radiation from the supermassive black hole at the galaxy center. This is related to their proximity to galaxies' nuclei. LBV 1806-20, discovered in January 2004, is a blue star. It is about 28,000 light years away from us, about two million times brighter than the Sun. A blue hypergiant called R136a1 shines nearly nine million times brighter than the Sun. It has an estimated mass of over 250 Suns and a volume large enough to contain 27,000 Suns within it.

A few exoplanets are as far away as 27,000 light-years (i.e., 255.4 quadrillion kilometers) from us. The few nearest exoplanets to us are given in Table 6.12. Note that most of the exoplanets discovered so far are in a relatively small region of our galaxy, the Milky Way, therefore there is no guarantee that these discoveries will extend readily to the whole universe. NASA's Kepler Space Telescope finds more planets than stars in our galaxy.

As you will notice in this section, the convention for naming exoplanets is an extension of the system used for naming multiple-star systems as adopted by the International Astronomical Union (IAU). For exoplanets orbiting a single star, the IAU designation is formed by taking the designated or proper name of its parent star. It is then added a lower case letter. Letters are given in the order of each planet's discovery around the parent star. This is so that the first planet discovered in a system is designated "b" (the parent star is considered "a") and later planets are given subsequent letters. If several planets in the same system are discovered at the same time, the closest one to the star gets the next letter, followed by the other planets in order of orbital size.

So far, only two planetary systems have four stars. In other words, the highest number of possible stars in a planetary system is four. This number may be the maximum even when we discover billions of exoplanets. The number of planetary systems with three stars is 39 and with binary stars there are a total of 562. Thus about 3400 of the 4000 planetary systems have only one star like our planetary system. The maximum of exoplanets in a given planetary system is 8 and all planetary systems with 8 exoplanets have a single star just like our solar system if one ignores Pluto; i.e., maximum number of exoplanets in a single-star planetary system is 8; maximum number of exoplanets in a binary-star planetary system is 5; Maximum exoplanets in a three-star and binary star is 2; and maximum of exoplanets in a four-star star is 1. Actually the number of exoplanets in a given planetary system decreases with the number of stars in the system. No exomoon or extrasolar moon (i.e., a natural satellite) that orbits an exoplanet or other non-stellar extrasolar body has been detected.

In this section, we aim to locate Earth-like exoplanets, exoplanets with potential for biological life, and exoplanets that own stars rather than the other way around. We will start with a brief overview of how exoplanets are discovered.

6.6.1.   Exploration of Exoplanets

Searching for exoplanets has several methods. So far, there is no one method that is significantly better. Each is suited to different circumstances. When possible, we confirm exoplanet discoveries with a method different from that used for the original discovery. Here are the most successful methods:

- *Transit method: measuring star dimming (Rosenblatt, 1971).* By observing periodic dips in brightness, we can infer the presence of an exoplanet and estimate its size and orbit. Almost three quarters (about 73 percent) of confirmed exoplanets have been discovered using the transit method. However, this large percentage of discoveries does not mean this method is superior to others. Instead, it is because Kepler Space Telescope missions, which are based on this method, have been productive. This proportion of discoveries is dropping steadily nowadays and discoveries by direct imaging are growing steadily. Note that Kepler detects at least three transits before flagging it as a candidate exoplanet. Explanet candidates are also verified by direct observations. The additional transits are also used to estimate how long it takes the exoplanet candidate to orbit its star. The exoplanets of TRAPPIST-1 were discovered using the transit method.

- *Radial velocity method: monitoring star's spectra.* This method is based on the wobbles exoplanets experience when orbiting stars. These wobbles can be seen as a change in wavelength in a star's spectra by monitoring its spectral lines. These shifts, known as Doppler shifts, reveal the planet's presence, mass, and orbit. This method works for almost all planetary systems but it is really complicated and requires 4m or 8m class telescopes. Larger wobbles are easier to detect. Consequently, exoplanets that are more massive and closer to their stars are easier to detect because they produce the largest wobbles. About twenty percent of exoplanets are discovered this way. This method gives an estimate of the planet's mass, m_p. In fact, this method sets a lower bound to the planetary mass because it measures $m_p \sin i$ where i is the angle of inclination of the planet's orbit with respect to the plane tangent to the celestial sphere, which is the major source of error in the estimate of planet surface gravity, $g = Gm_p/r_p^2$.

- *Direct imaging.* A massive telescope is put in an occulting disk to block out the star and see the planet. This method is challenging because the planets are often faint and overwhelmed by their host stars' glare. However, it provides valuable information about an exoplanet's characteristics, such as its temperature and atmosphere.

- *Gravitational microlensing: detecting double brightening.* When a massive object, such as a star, passes in front of a distant star, its gravity acts like a lens, bending and magnifying the light from the background star. If an exoplanet orbits the foreground star, it can cause additional perturbations in the observed light. We can infer the exoplanet's presence and properties by studying these fluctuations. This method is still in its infancy in 2021.

Space telescopes have found thousands of planets by observing *transits*, the slight dimming of light from a star when its tiny planet passes between it and our telescopes. This count is expected to rise to hundreds of thousands within a decade with the advent of robotic telescopes lofted into space.

One of the weaknesses of the current methods of exoplanet search is the limited number of M stars associated with exoplanets. M stars make up the bulk of the Milky Way's population, about 75 percent. So they are the most common by far! Sunlike stars are 6 percent of the population, and K dwarfs are 13 percent of the population. Yet Figure 6.53 shows that our current sample has much more exoplanets associated with F and G stars. Most M stars are tiny red dwarfs, with less than 50 percent of the Sun's mass. However, some are actually giants and supergiants, like Betelgeuse. Some familiar M stars include Betelgeuse (red giant), and the red dwarfs Proxima Centauri, Barnard's star, and Gliese 581.

Another weakness of the current methods of exoplanet search is the inability to detect rogue stars. Unlike other exoplanets, rogue planets orbit the galaxy directly, without rotating around a star. If something orbits a star, we can either detect it as it passes between us and its star (transit method), observe the star wobble because of the exoplanet's gravity (radial velocity method), literally photograph it (direct imaging method) or use other methods. Because rogue stars do not have any parent star to pass in between the star or gravitationally affect the star, they remain undetectable. The fact that they do not produce light further complicates things. Gravitational microlensing methods will find rogue worlds floating in interstellar space, when they become available.

Note that rouge exoplanets are part of space folklore, especially Nibiru. It is believed to have been inhabited by ancient aliens known as the Anunnaki. Niburi orbits our Sun for 3600 years. The planet is sometimes very close to the Sun and sometimes far from the Sun. Occasionally, these types of tales are correct and must be investigated further as more data becomes available. Exoplanets with strange orbital periods and large eccentricities can even be found in the current exoplanet database (Table 6.13). For example, exoplanet COCONUTS-2 b takes more 1.1 million years to orbit its star. There are times when the orbital period is so long

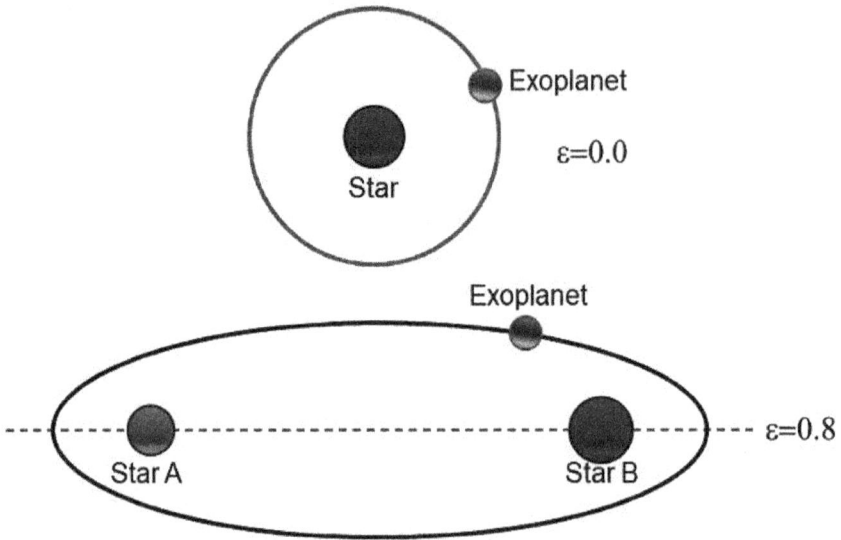

Figure 6.54. Illustration of orbital eccentricity. Eccentricity is the deviation of a planet's orbit from circularity—the higher the eccentricity, the greater the elliptical orbit. Illustration of orbital eccentricity. It is the deviation of a planet's orbit from circularity—the higher the eccentricity, the greater the elliptical orbit. Planets can revolve around two stars which describe two foci of the ellipse. WASP-96 b and KOI-13 b are examples of zero eccentricity exoplanets. Kepler-130 d is an example of a 0.80-eccentricity exoplanet

that the transit method cannot be used. Figure 6.54 shows how a planet or exoplanet revolves around two stars. The planet is sometimes very close to a star and other times very far from the same star.

One additional weakness of current exoplanet search methods is their inability to detect exomons alongside exoplanets. No moon has been confirmed around any of the thousands of exoplanets discovered so far. It is a real possibility that some unconfirmed exoplanets are due to ambiguities, especially for transit, created by complex star–exoplanet–exomoon systems. Such systems can make the star-exoplanet transit curves less repetitive and distorted in a significant way. This weakness may be a significant factor in the current exoplanet sample.

6.6.2.   Why is the Exoplanet Radius Distribution Bimodal?

Bimodal distributions have two peaks. In a continuous probability distribution, modes are peaks in the distribution. Exoplanet's radii follow a bimodal distribution as shown in figure 6.55. The highest apex, known as the major mode, is at $1.3\ r_{\text{earth}}$ and the lower, known as the minor mode, is at $13.6\ r_{\text{earth}}$. The first study with two peaks was made by Fulton et al. (2017). It looks more and more likely that a larger sample of exoplanets will not change this fact. Because exoplanet growth is likely to occur in M stars, the major mode may be larger than Figure 6.55.

First of all, this bimodal distribution persists even when we limit ourselves to exoplanets of single star-type.

- A bimodal distribution is observed for M-star exoplanets with the major mode at $1.25\ r_{\text{earth}}$ and the minor mode at $12.3\ r_{\text{earth}}$. Major mode occurrences are

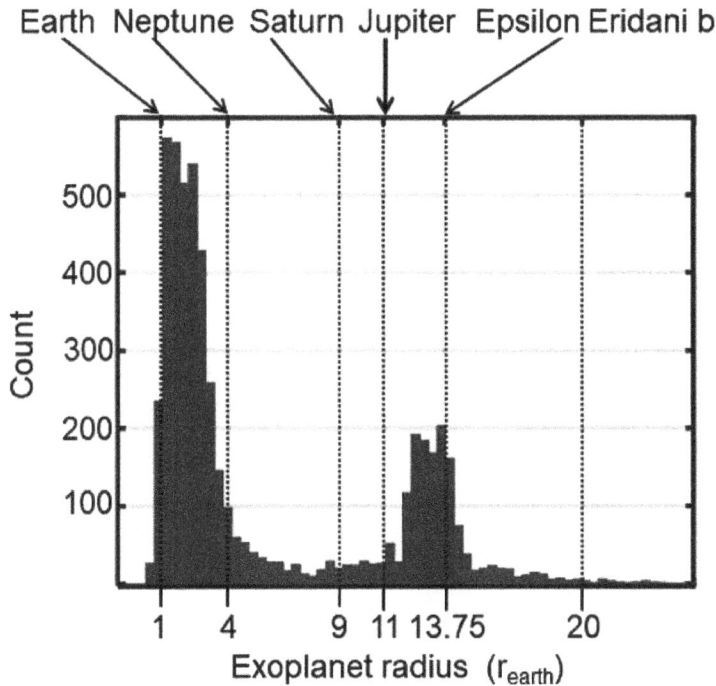

Figure 6.55. Histogram of exoplanet's radius. Proxima Centauri b is the closest exoplanet to Earth, at 4.2 light years. It orbits Proxima Centauri every 11.186 Earth days at a distance of about 0.049 AU. This is over 20 times closer to Proxima Centauri than Earth is to the Sun.

Table 6.13. Some exoplanets exhibit large eccentricities and/or large orbital periods. We have abbreviated some names and parameters. "CO-2 b" stands for COCONUTS-2 b, "CO-2 A" for COCONUTS-2 A, "CFH 98 b" for CFHTWIR-Oph 98 b, "CFH 98 A" for CFHTWIR-Oph 98 A, "VHS b" for VHS J125601.92-125723.9 b, "VHS" for VHS J125601.92-125723.9, # for the number of stars, OS-M for orbit semi-major axis, "sm" for search method, "I" for Direct imaging, and "R" for Radial velocity method. Source of data: https://exoplanetarchive.ipac.caltech.edu/.

Planet	Star	#	Type	sm	Orbital period (Earth days)	OS-M (AU)	Eccen-tricity
CO-2 b	CO-2 A	1	M	I	402000000	7506	
CFH 98 b	CFH 98 A	1	M	I	8040000	200	
Oph 11 b	Oph 11	1	M	I	7300000	243	
VHS b	VHS	2	M	R	5800000	350	0.68
b Cen AB b	b Cen A	2	A	I	1790000	556	0.40
HD 26161 b	HD 26161	1		R	10943	10.1	0.922
HD 80869 b	HD 80869	1	G	R	1753	2.91	0.92
Kepler-1704	b Kepler-1704	1		T	988	2.03	0.92
HD 20782 b	HD 20782	2	G	R	597	1.36	0.95
HD 156846 b	HD 156846	2	G	R	359	1.12	0.85

3.84 times greater than minor mode occurrences. In simple terms, the leftover from M-stars formations produce predominantly of rocky exoplanets. That most materials left after M-star formation are dust.

- Exoplanets of K-stars follow a bimodal distribution with the major mode at 13.5 r_{earth} and the minor mode, is at 2.48 r_{earth}. Notice that the major mode is now a gas giant and the minor mode is a rocky planet. Major mode occurrences are 1.22 times greater than minor mode occurrences. There are only slightly more gas exoplanets than rocky exoplanets.

- The distribution of exoplanets from F-stars follows a bimodal distribution with a major mode of 13.4 r_{earth} and a lower minor mode of 2.5 r_{earth}. Major mode occurrences are 1.3 times greater than minor mode occurrences. There are only slightly more gas exoplanets than rocky exoplanets.

- In the case of G-star exoplanets, there is a bimodal distribution in which the major mode holds for 13.4 r_{earth}, and the minor mode holds for 2.5 r_{earth}. Major mode occurrences are twice as minor mode occurrences. There are more gas exoplanets than rocky exoplanets. There is more hydrogen left over after the formation of G-stars than dust. In other words, gas exoplanets are more common in G-star planetary systems than in F-star, K-star, and M-star planetary systems. In contrast, rocky exoplanets are less common in G-star planetary systems than in F-star, K-star, and M-star planetary systems.

We can see that radii of exoplanets associated with the two modes are almost unvariant with star type.

We basically have two groups of exoplanets. We find most planets within our solar system to be just on the fringes of these two groups, which is strange. The two groups are exoplanets with solid surfaces and gaseous exoplanets. The explanation for the two groups is that the accretion processes at work in planet formation have two types of infalling materials (hydrogen and helium on side and dust on the other) that are different in these two cases. Rocky exoplanets involve significant supernova remnants compared to gaseous exoplanets, which require hydrogen and helium just like star formation. In essence, we have a single mechanism for exoplanet formation with a disproportionate amount of matter and a variety of types of matter. This leads to two groups of exoplanets.

The chemical composition of matter corresponding to the major and minor modes in Figure 6.55 can be defined using only the planets of our solar system. A major pick corresponds to the mixture of liquids and solids at the surface of exoplanets being over 71 percent liquid, which means they have more liquid than the earth, which means that exoplanets have more liquid on their surface than us. The major mode suggests that exoplanets with liquid surfaces are abundant. The minor mode suggests that planets have plasma with high particle densities. This type of plasma can have a mass density several times that of a solid.

With the two main groups of exoplanet populations, we can see that the current 5,000 or so exoplanets do not display remarkable diversity. This is not what one expects. Also shown in Figure 6.56 is the histogram of the exoplanet's temperature, which has different statistics from the exoplanet's radius. We included Mercury, Venus, Earth, Mars, Jupiter, Saturn, Neptune, and Uranus to show the

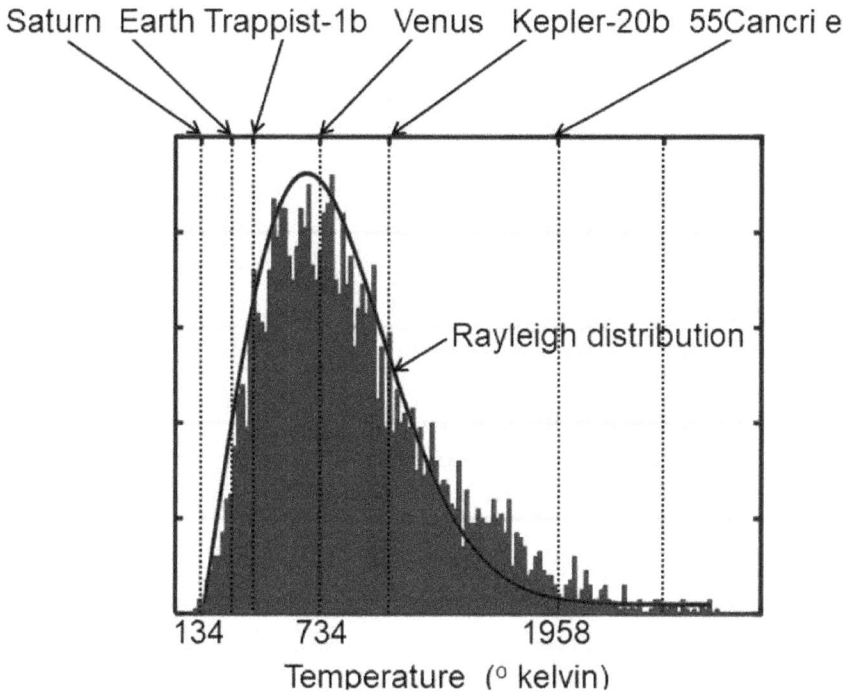

Figure 6.56. Histogram of the exoplanet's temperature. Kepler-20b is an exoplanet orbiting Kepler-20 in a system 922 light years from Earth. 55 Cancri e is an exoplanet in 55 Cancri A's orbit.

differences are significant. Gravity, mass, and density all have distinct statistics. In other words, statistics suggest complex relationships between exoplanet parameters. Note that exoplanet temperatures are generally not directly measured. They are estimated from the effective temperature of the star, T_{eff}, the exoplanet's radius, r_p, along with the semi-major axis of the exoplanet, d, by using the fact that exoplanets are black bodies, as

$$T_p^4 = \frac{1}{4}\left(\frac{r_p}{d}\right)^2 T_{\text{eff}}^4 . \qquad (6.274)$$

The albedo of the exoplanet is ignored in this formula.

6.6.3. Exoplanets with Their Own Stars: Spontaneous Fusion

Since the 1930s, there have been several attempts to create our own stars on Earth. As described earlier, the idea is to replicate the sun's fusion reaction on Earth. The tokamak described in Figure 6.21, for example, was designed to do so. It mixed two types of hydrogen gas, deuterium and tritium, for example, to obtain the desired plasma. The tokamat must reach 150 million degrees Celsius to obtain plasma. Large magnets in the Tokamat walls contain plasma and stop it from touching the walls where it can cool down. The mixture of deuterium and tritium produces helium and spews out neutrons. The neutrons fly out of the plasma and hit the tokamak walls. They carry energy. The hope is that this energy will be greater

Table 6.14. Some exoplanets have their own sun. We have abbreviated some names and parameters. m_p and r_p stand for planet mass and radius, m_s and r_s stand for stellar mass and radius, g_p is the planet's gravity, and g_s is stellar gravity, jm is Jupiter's mass, jr is Jupiter's radius, sm is solar mass, and sr is solar radius. Source of data: https://exoplanetarchive.ipac.caltech.edu/.

Planet	Star	Type	m_p (jm)	r_p (jr)	m_s (sm)	r_s (sr)	g_p m/s^2	g_s m/s^2
ups And c	ups And	F	1.1	13.98	1.6	1.3	299.7	146.4
nu Oph c	nu Oph	K	1.07	24.66	14.6	2.7	558.7	3.47
HD 125390 b	HD 125390	G	1.06	27.20	6.5	1.8	627.9	11.91
Kepler-203 c	Kepler-203		0.22	2.36	1.11	1	1260	222.4
K2-77 b	K2-77		0.21	1.9	0.76	0.8	1170	379.5
K2-416 b	K2-416	M	0.24	5.2	0.55	0.55	2400	498.2
KOI-4777.01	KOI-4777		0.045	0.312	0.4	0.41	3900	702.2
K2-137 b	K2-137	M	0.057	0.5	0.29	0.46	3900	1500

than the energy used for fusion. This energy will use fusion energy to heat up water to turn it into steam to drive a turbine and generate electricity.

Additionally, tritium is almost non-existent on earth. Most of it is produced in fission reactors. With more nuclear fission reactors being shut-down, the global tritium supply is less than 20 kg. A typical fusion reactor consumes tritium reserves in less than three months. The Moon possesses Helium which can also be used for fusion; thus Moon exploration. The ultimate challenge is to sustain the fusion reaction for longer than the current 30 seconds—just like the sun shines more than 30 seconds a day—and to hold on to this very hot plasma for a long time. In non-scientific language, the objective will remain out of reach for at least 100 years.

We will realize sooner rather than later that there is another path. One possible direct method is to steal plasma directly from the Sun's photosphere. Alternative solutions are to learn how some exoplanets ended with their own shining stars. In other words, there are exoplanets with gravity greater than neighboring stars, so they can capture a small star. The current database of exoplanets, as we can see in Figure 6.57 (also Table 6.14), contains about 150 exoplanets with their own planets. Sooner than 100 years we learn to manipulate gravity well enough (see Ikelle, 2024) that we can bring a smaller star at the boundary of our magnetosphere to augment our energy resources well before we can master how creating and sustain our star on Earth.

To sum up, there are two schools for fusion nuclear reactor construction on Earth. Over 100 years ago, Earth materials were used to construct such reactors in the old school. A different approach involves obtaining plasma from stars, including our Sun, and other exoplanets that possess such matter directly. One of the challenges in taking plasma from the extraterrestrial world is the construction of containers of extracted plasma to transport it home. A container such as this could be constructed with iridium as an ingredient. Iridium (Ir) is one (metal) of the elements of the Periodic Table of chemical elements, with an atomic number (i.e. number of protons) of 77, atomic mass (average mass of the atom) of 192.217 amu, melting point of 2446°, and boiling point of 4428°. It has two stable isotopes: iridium-191 (37.3 percent) and iridium-193 (62.7 percent).

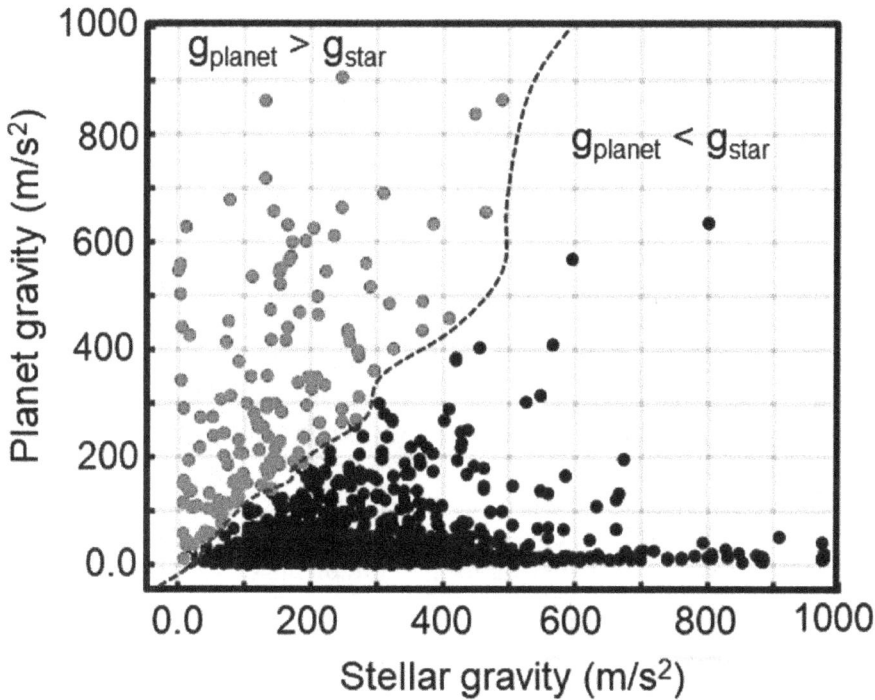

Figure 6.57. Scatterplot of exoplanet gravity versus stellar gravity.

Iridium is created in supernova explosions. On Earth, it is one of the rarest elements. It is also the most corrosion-free element on the Periodic Table of Elements, even at temperatures as high as 2400°. It also has the highest density of all the elements. It has a density of 22.65 g/cm³. By comparison, lead density is 11.34 g/cm³ and iron density is 7.874 g/cm³. Today, iridium is commercially recovered as a byproduct of copper or nickel mining. Iridium ore is found in Brazil, the United States, Myanmar, South Africa, Russia, and Australia. Pure iridium is so rare in the Earth's crust that there are only about 2 parts per billion located in the crust. It is present however in greater quantities inside asteroids or comets since they have not been affected by extreme temperatures and gravitation.

Impacts from asteroids on Earth are called the *iridium layer* because of the ejected material left behind. Such a layer (K-T boundary) separates the Cretaceous and Triassic periods. This layer contains a higher iridium concentration than the Earth's crust. Dinosaur fossils are not found above the K-T boundary which lends support to an asteroid impact of global proportions causing their dinosaur extinction that occurred around 66 million years ago. Chicxulub crater on the Yucatán Peninsula of Mexico was the site of an asteroid impact that caused this mass extinction.

6.6.4. Earth-like Exoplanets with Potential Life

We know that there are relationships between stars' properties as described in (6.1)-(6.4), for example. These relationships make it straightforward to find planetary systems likely to contain exoplanets that are likely to have life. It is desirable to have similar relationships between exoplanet properties. Some of these relationships are emerging as we can see in Figure 6.58. We can see that exoplanets' sizes (radii) and masses (weights) are related. Essentially, we have two main regimes here, the low mass regime where mass and radius are linearly related, and the high mass regime where radius and mass are essentially independent. The first regime may include rocky and gaseous bodies as we know them in our solar system. The second regime may include mostly plasma materials with high particle density as Figure 6.58b suggests. The mass density of such plasma can be 1,000 times that of a solid. Basically, we can see exoplanet compositions can vary from very rocky (like Venus) to liquid to gas and finally to very dense plasma.

Figure 6.58c shows the relationship between the surface gravity versus mass density. There are three regimes to this relationship. The low-density regime corresponds to gaseous and liquid exoplanets. Surface gravity grows linearly with the mass density in the regime denoted 1. Line curve 2 represents exoplanets with rocky surfaces. The line curve marked 3, corresponds to exoplanets with significant plasma matter. Figure 6.58d shows that common exoplanet's temperature is between 200 and 1400 K. Notice that direct measurements of surface temperature are

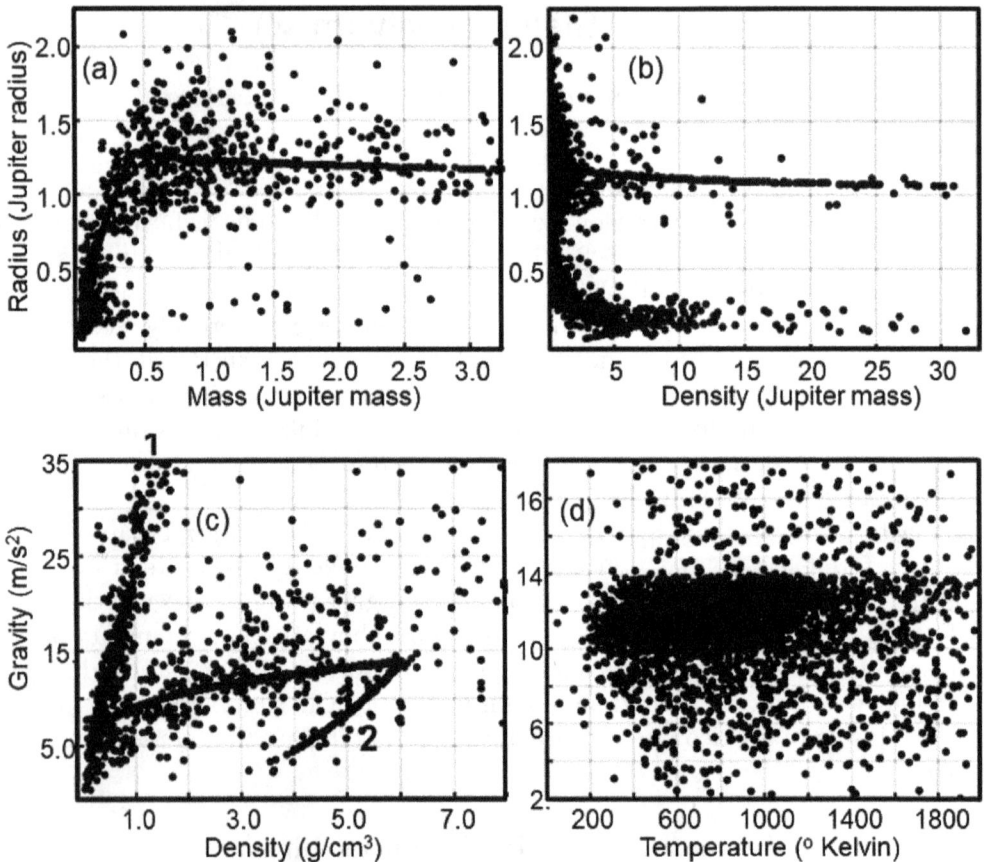

Figure 6.58. (a) Scatterplot of exoplanet radius versus mass. (b) Scatterplot of exoplanet radius versus density. (c) Scatterplot of exoplanet gravity versus density. (d) Scatterplot of exoplanet gravity versus temperature. Notice that some exoplanets have an average temperature over 1,500 K. They likely contain their own light sources.

not available at this time. Many physical processes affect the temperature distribution on the planet's surface. As described in (6.274). However, the dominating influence is an energy balance between stellar radiation input and radiative surface heat loss. With the further assumptions of a uniform planetary surface temperature, no filtering of incoming radiation, and black body emission, the only variables are stellar luminosity and the radial distance of the exoplanet from the star. Thus the effects of the planet's albedo and magnetosphere are ignored. Albedo will decrease Earth's temperature, but magnetosphere will increase it.

In the search for Earth-like exoplanets, the exoplanet's surface temperature, radius, gravity, and density play a crucial role. The surface temperature is related to the exoplanet's magnetic field and therefore can tell us about the possible protection of the exoplanet from incoming charged particles from the star. Our assumption here is that Earth-like exoplanets have temperatures between 270 and 370 degrees Kelvin. Surface gravity is considered to be between 8.6 and 11 m/s^2 because multicellular organisms can move around like us at these gravity range and gravity of about 10.7 m/s^2, the optimal gravity for our biologic type because it causes slow muscle loss for organisms with biology, tells us that one can live a longer healthy life than us. Density is critical to ensure that we deal with a rocky surface at least partially. We set the mass density to be between 2.5 and 6.5 g/c^3. We also assume that the exoplanet's radius is less than 2.0 r_{jupiter}. We did not constrain the exoplanet mass because it is related to radius, gravity, and density. Moreover we reduced our sample from about 5400 exoplanets to about 3800 due to errors associated with uncertainties in planet masses estimations. In examining the current exoplanet catalog, such as https://exoplanetarchive.ipac.caltech.edu/, we only find six exoplanets based on these constraints (see Table 6.15). All six planets are associated with planetary systems with only one star. Four exoplanets are second and third (just like Earth) from their stars. Earth-like exoplanets are rare. As with most rare things, Earth may contain rare things that attract visitors. We add TRAPPIST-1's planets to Table 6.15 for reference. According to Figure 6.58, some exoplanets have temperatures over 1500 K. That is, some exoplanets are nearly as hot as some M stars, so their plasma directly produces light without processing. In other words, some exoplanets can create their own light. Examples of such exoplanets include WASP-121 b, Kepler-76 b, Kepler-78 b, TOI-2260 b, and TOI-561 b.

To summarize, the atmosphere, surface, and interior of a celestial rocky body make it straightforward to discover why and how a planet became magnetized. In addition, it characterizes its inhabitants. One of the most precise ways to investigate the properties of a planet, moon, exoplanet, or exomoon is through the analysis of seismic waves near surface and deep subsurface, as we have done on Earth with the exploration of solid Earth. As discussed in Ikelle and Amundsen (2005 and 2018) and Ikelle (2020 and 2023c), seismic experiments and the resulting seismic data have allowed us to reveal Earth's detailed internal structure. They can also be used to reveal the detailed internal structures of other planets, especially exoplanets and exomoons with rocky or icy surfaces, even gas or plasma as the forst layer. Seismic tools should be extended to deep-space exploration. Perhaps, in this century, seismic experiments will be possible on Mars and Mercury, as well as on the Jovian moons. The prospect of such experiments is far away for exoplanets and exomoons. For now we have to rely on macroscopic data such as density, gravity, radiation level, and magnetic field to characterize the atmosphere, surface, and interior of exoplanets.

Table 6.15. Some Earth-like exoplanets are based on gravity, temperature, density, and radius. We have abbreviated some names and parameters. r_p is for planet radius, g_p is planet gravity, er is Earth's radius, jr is Jupiter's radius, and Dist is the planet-star distance. Source of data: https://exoplanetarchive.ipac.caltech.edu/.

Planet	Tempera-ture (K)	r_p (er)	r_p (jr)	g_p m/s^2	ρ g/cm^3	Stellar Temp. (K)	Dist (pc)
Kepler-1052 c	302	2.3	0.211	10.83	2.66	5818	943.4
Kepler-1126 c	305	1.45	0.15	9.79	4.87	5678	635.7
Kepler-138 d	345	1.51	0.135	9.4	3.6	3841	66.86
Kepler-1649 b	307	1.017	0.091	10.2	5.38	3240	92.19
TOI-1266 c	344	1.56	0.139	9.3	3.18	3600	36.01
TOI-700 e	273	0.953	0.085	9.2	5.19	3459	31.13
TRAPPIST-1c	380	1.097	0.099	10.7	4.89	2550	12.1
TRAPPIST-1g	200	1.13	0.103	8.54	5.04	2550	12.3

There are some minerals called oxides. These minerals describe rocks' chemical compositions. Table 6.16 shows some of these key oxides. Let ρ_i be the i-th oxide with i varying from 1 to N. The density of rocky exoplanets can be described as

$$\rho_{\text{exoplanet}} = \sum_{i=1}^{N} \alpha_i \rho_i \quad \text{with} \quad \sum_{i=1}^{N} \alpha_i = 1 \ .$$

In this equation, α_i represents the ratio of the volume occupied by each oxide to the entire volume of an exoplanet. N represents the number of oxides on that exoplanet. Assuming that $N = 3$ (i.e., crust, mantle, and core), we can determine realistic values of ρ_1, ρ_2, and ρ_3 along with the portion of exoplanet volume they occupy. In addition, we can constrain the final solutions based on the relation between the radius of exoplanets, exoplanet types, and star-type discussed earlier. In Ikelle (2024), we discussed how to construct AI systems for estimating the composition of exoplanets with a number of $N > 10$. When dealing with fluid planets, Table 6.16 must include basic organic molecules.

Table 6.16. Key oxides used to describe chemical compositions of rocks. The density are in g/cm^3,

Oxides	ρ	Oxides	ρ
Silica (SiO$_2$)	2.65	Titanium dioxide (TiO$_2$)	4.23
Aluminum oxide (Al$_2$O$_3$)	3.99	Hematite (Fe$_2$O$_3$)	5.27
Iron oxide (FeO)	5.75	Magnese oxide (MnO)	5.43
Magnesium oxide (MgO)	3.60	Calcium oxide (CaO)	3.34
Sodium oxide (Na$_2$O)	2.27	Potassium oxide (K$_2$O)	2.13
Water (H$_2$O)	1.00	Carbon dioxide (CO$_2$)	1.98
Phosphorus pentoxide (P$_2$O$_5$)	2.39	Uraninite (UO$_2$)	11.0
Thorianite (ThO$_2$)	10.0		

6.6.5. Drake's Equation

Are there aliens out there who are intelligent enough to broadcast a signal to us for a long distance that we can detect? In 1961, Astrophysicist Frank Drake proposed a way of calculating the probability of us finding intelligent alien life in our galaxy. This calculation is known as Drake's equation. It can be written as follows:

$$N = R_* \, f_P \, n_e \, f_1 \, f_i \, f_c \, L \qquad (6.275)$$

where N is the number of intelligence alien civilizations in the Milky Way galaxy that could communicate with humans. More precisely, N gives us an idea of the odds of finding them. The parameters are: R_* is the mean rate of star formation in our galaxy, f_P is the fraction of stars that have planets, n_e is the average number of planets that could support life per star with planets, f_1 is the fraction of life-supporting planets that develop life, f_i is the fraction of planets where life acquires intelligence, f_c is the fraction of intelligent civilizations that develop communication, and L is the average length of time that civilizations can communicate.

In less than 14 billion years, our galaxy has reached 200 billion stars. So $R_* = 15$ year^{-1} (i.e., 15 stars formed per year, on average over the galaxy's life). The current exoplanet database confirms that G, F, K, M, and L stars have plenty of exoplanets and can sustain life. O, B, and even A stars are very likely lifeless. Thus f_P is likely nonzero. There are also observations of "embryo" exoplanets. There are also exoplanets with atmospheres and material compositions suitable for life (see Chapter 3). Thus n_e is likely nonzero. Based on temperature, surface gravity, and density, a small number of exoplanets will support life at some point. So f_1 is likely nonzero. In Ikelle (2023) we described some messages such as crop circle formations, terahertz signals, and even the response to the so-called Arecibo Message that suggests that there is at least one exoplanet where life becomes intelligent. So f_i is likely nonzero. And finally, L is the length of time a civilization exists, a million years. This is the time to star's flare, simultaneously earthquakes, technologically advance enough and reach the level of self destruction in accordance with the second law of thermodynamics. With advances in science and space observations, we will refine these parameters with time. Here are examples of these parameters:

- $R_* = 15$ year^{-1} (15 stars formed per year).

- $f_P = 0.5$ (more than one half of all stars formed will have exoplanets)

- $n_e = 0.01$ (stars with exoplanets will have 1 planets capable of developing life)

- $f_1 = 0.5$ (50 percent of these planets will develop life)

- $f_i = 1$ (100 percent of which will develop intelligent life)

- $f_c = 0.01$ (1 percent of which will be able to communicate)

- $L = $ somewhere between 1,000,000 and 100,000,000 years

Inserting the above values into Drake's equation, we can estimate that there are between 375 and 375,000 alien civilizations out there. These civilizations have reached a level of intelligence able to broadcast a signal out to us for long enough that we can spot them. In order to reach $N = 1$ (that is, there are no other intelligent life forms in our Galaxy with which terrestrial researchers can communicate), we have had to make some of these numbers completely unrealistic, such as $f_c = 2.7 \times 10^{-5}$.

Appendix A

Advanced Numerical Modeling

Modern science includes Maxwell equations, elastic wave equations, Navier–Stokes equations, heat-advection diffusion equations, and Schrödinger equation. They have become the prevailing language of modern science as they provide the foundation for unlocking many mysteries of nature and the universe. These mysteries involve spatial and temporal evolution. These equations explicitly or implicitly are the theoretical underpinning of almost all disciplines such as biology, chemistry, climatology, ecology, economics, engineering, genetics, geology, neuroscience, physics, physiology, sociology, and cosmology. They play a key role in understanding these disciplines. Our objective in this appendix is to describe numerical solutions to these equations using the finite-difference technique.

The problem of simulating heat transfer, electromagnetic wave propagation, and elastic wave propagation corresponds to the problem of solving differential equations under a set of initial, final, and boundary conditions. One of the most successful numerical techniques for solving differential equations is finite-difference modeling (FDM). It consists of discretizing wavefields and the geological model at some space-time gridpoints and of numerical approximations of derivatives of wave equations (e.g., Yee, 1966; Madariaga, 1976; Virieux, 1986; and Ikelle and Amundsen, 2005 and 2018). When adequate discretization in space and time that permits accurate computation of wave equation derivatives is possible, the finite-difference technique is by far the most accurate tool for simulating elastic wave propagation through complex models.

A.1. Difference Operators

Finite-difference formulas are the basis of the numerical differential equation solution methods in this appendix. Throughout the appendix, we have employed the basic notions of forward, backward, and central-difference operators used throughout the appendix. We limit these definitions to second-order operators. Higher-order operators will be introduced as we go along. So to develop approximations of differential elastodynamic equations, we will use the following operators:

$$\text{forward}: \ \frac{\partial g(x)}{\partial x} \approx \frac{1}{\Delta x}\left[g(x + \Delta x) - g(x)\right],$$

DOI: 10.1201/9781032620619-A

$$\text{backward}: \quad \frac{\partial g(x)}{\partial x} \approx \frac{1}{\Delta x}\left[g(x) - g(x - \Delta x)\right],$$

$$\text{central}: \quad \frac{\partial g(x)}{\partial x} \approx \frac{1}{2\Delta x}\left[g\left(x + \frac{1}{2}\Delta x\right) - g\left(x - \frac{1}{2}\Delta x\right)\right],$$

$$\text{average}: \quad g(x) \approx \frac{1}{2}\left[g\left(x + \frac{1}{2}\Delta x\right) + g\left(x - \frac{1}{2}\Delta x\right)\right],$$

where $g(x)$ is an arbitrary continuous function of x. Obviously these formulas are only approximations, but they become more accurate as Δx gets smaller. If the forward-difference operator or the backward operator is applied, the error is approximately halved when the increment is halved. In the case of the central difference, the error is reduced by approximately a factor of 4 when Δx is halved. In other words, as Δx gets smaller, the error of the forward difference approximation also gets smaller in a linear fashion [i.e., $\mathcal{O}(\Delta x)$], whereas the central-difference approximation is accurate up to order $\mathcal{O}\left(\Delta x^2\right)$. So for small values of Δx, the central-difference approximation is better than both the forward and backward approximations.

A.2.   Finite-Difference Modeling of the Heat-Diffusion Equation

Again the finite-difference method allows us to approximate solve partial differential equations and integral equations. There are basically two types of finite-difference methods: the explicit finite-difference method and the implicit finite-difference method. In the next subsections, we will describe the explicit finite-difference method. Here we describe the implicit finite-difference method for the heat equation and contrast it with the explicit finite-difference method. We consider the implicit finite-difference method for the heat equation because of its importance in heat transfer and fluid-flow studies, as discussed in Ikelle (2023b). The pros and cons of explicit and implicit finite-difference methods are outlined in Table A.1.

Table A.1.   A comparison of the explicit and implicit finite-difference methods.

Explicit method	Implicit method
• Easy to set up (simple algorithm)	• Complicated to set up (more involved procedure)
• Constraint on Δt (For a given Δx, stability constraints impose a maximum limit on Δt.)	• No constraint on Δt (Stability is maintained over a larger Δt, and fewer timesteps are required.)
• Less computer time is involved.	• More computer time is required because of the matrix-manipulation algorithms.
	• Potential truncation errors are caused by a large Δt.

A.2.1.   The Heat (or Diffusion) Equation

The heat equation (or diffusion equation) can be written as follows:

$$\rho c_P \frac{\partial T(x,t)}{\partial t} + \rho c_P w \frac{\partial T(x,t)}{\partial x} = \frac{\partial}{\partial x}\left[k(x)\frac{\partial T(x,t)}{\partial x}\right] + q'(T) , \tag{A.1}$$

where k is the thermal conductivity (or diffusion coefficient), w is the flow velocity, t is a time variable, x is a spatial variable, ρ is the density, c_P is the heat capacity, q' is the heat generation, and $T(x,t)$ is the unknown temperature field. The variation of $q'(T)$ with temperature can be used to simulate convection in the context of fluid-solid conduction, blood perfusion in the context of bioheat, and other sources of heat as long as q' is linearly related to temperature. The nonlinear temperature dependence of q' will be addressed in the second section. We assume that x is limited to the interval $[0, L]$. If k is constant, (A.1) reduces to

$$\frac{\partial T(x,t)}{\partial t} = D\frac{\partial^2 T(x,t)}{\partial x^2} - w\frac{\partial T(x,t)}{\partial x} + \frac{q'(T)}{\rho c_P} , \tag{A.2}$$

where $D = k/(\rho c_p)$ is the thermal diffusivity. To find a well-defined solution to this equation, we need initial and boundary conditions. We can define the initial conditions as $T(x,0) = c(x)$ and the boundary conditions as $T(0,t) = a(t)$ and $T(L,t) = b(t)$, where $a(t)$, $b(t)$, and $c(t)$ are arbitrary continuous functions.

A.2.2.   The Explicit Finite-Difference Method

By using forward-difference approximations, as defined earlier, we can approximate the time derivatives in (A.2) as follows:

$$\frac{\partial T(x,t)}{\partial t} \approx \frac{T_{n+1,j} - T_{n,j}}{\Delta t} ; \tag{A.3}$$

and by using central-difference approximations, as defined above, we can approximate the first- and second-order spatial derivatives in (A.2) as follows:

$$\frac{\partial T(x,t)}{\partial x} \approx \frac{T_{n,j+1} - T_{n,j-1}}{2\Delta x}$$

and

$$\frac{\partial^2 T(x,t)}{\partial x^2} \approx \frac{T_{n,j+1} - 2T_{n,j} + T_{n,j-1}}{(\Delta x)^2} , \tag{A.4}$$

where $T_{n,j} = T(j\Delta x, n\Delta t)$. By substituting (A.3) and (A.4) in (A.2.5.), we arrive at the following explicit finite-difference scheme for the heat equation:

$$\frac{T_{n+1,j} - T_{n,j}}{\Delta t} = D\left[\frac{T_{n,j+1} - 2T_{n,j} + T_{n,j-1}}{(\Delta x)^2}\right] - w\frac{T_{n,j+1} - T_{n,j-1}}{2\Delta x} + \frac{q'(T_{n,j})}{\rho c_P} \tag{A.5}$$

or, upon solving for $T_{n+1,j}$,

$$T_{n+1,j} = (\alpha - \varsigma)T_{n,j+1} + (1 - 2\alpha)T_{n,j} + (\alpha + \varsigma)T_{n,j-1} + F(T_{n,j}) , \tag{A.6}$$

where

$$\alpha = \frac{D\,\Delta t}{(\Delta x)^2}\,, \quad \zeta = \frac{w\,\Delta t}{2\Delta x}\,, \quad \text{and} \quad F(T_{n,j}) = \frac{\Delta t}{\rho c_P}q'(T_{n,j})\,. \tag{A.7}$$

The initial condition is $T_{0,j} = c(j\Delta x)$, and the boundary conditions are $T_{n,0} = a(n\Delta t)$ and $T_{n,J} = b(n\Delta t)$ with $L = J\Delta x$. Notice that the explicit finite-difference method in equation (A.26) contains only one unknown, $T_{n+1,j}$ (i.e., the temperature at $t + \Delta t$). This unknown can be obtained directly from the known values of T at t. In other words, since we know $T_{n,j+1}$, $T_{n,j}$, and $T_{n,j-1}$, we can compute $T_{n+1,j}$ by "marching" from one timestep to another. Notice also that equation (A.6) can be written in matrix form as

$$\boldsymbol{u}^{(n+1)} = \boldsymbol{A}\boldsymbol{u}^{(n)} + \boldsymbol{b}^{(n)}\,, \tag{A.8}$$

where

$$\boldsymbol{A} = \begin{pmatrix} 1-2\alpha & \alpha - \zeta & 0 & \cdots & 0 \\ \alpha + \zeta & 1-2\alpha & \alpha - \zeta & \ddots & 0 \\ 0 & \alpha + \zeta & 1-2\alpha & \ddots & \vdots \\ \vdots & \ddots & \ddots & \ddots & \alpha - \zeta \\ 0 & \cdots & 0 & \alpha + \zeta & 1-2\alpha \end{pmatrix} \tag{A.9}$$

$$\boldsymbol{u}^{(n)} = [T_{n,1}, ..., T_{n,J-1}]^T \tag{A.10}$$

$$\boldsymbol{b}^{(n)} = [(\alpha + \zeta)T_{n,0} + F(T_{n,1}), F(T_{n,2}), ..., F(T_{n,J-2}), (\alpha - \zeta)T_{n,J} + F(T_{n,J-1})]^T \tag{A.11}$$

where $\boldsymbol{A}$ is a $(J-1) \times (J-1)$ matrix and $\boldsymbol{u}^{(n)}$ and $\boldsymbol{b}^{(n)}$ are $(J-1)$-component vectors. We will later contrast equation (A.8) with the implicit solution. Let us provide a specific example of $F(T_{n,j})$. In bioheat, blood perfusion can be expressed as $q'(T) = c_b\rho_b\omega_b[T_b(t) - T(x,t)]$. The corresponding descretized $F(T_{n,j})$ is

$$F(T_{n,j}) = \frac{c_b\rho_b}{\rho c_p}\omega_b\left[(T_b)_n - T_{n,j}\right] \tag{A.12}$$

assuming that c_b, ρ_b, and ω_b are constants.

Explicit finite-difference methods are relatively simple to numerically implement. The main drawback is that their solutions are not always stable. Let us elaborate a bit on the stability condition of (A.4). We can arrive at the stability condition of (A.4) by seeking a solution of (A.4) in the form

$$T_{n,j} = \beta_n(k)\exp(-ikj\Delta x)\,, \tag{A.13}$$

with the condition that $|\beta_n(k)| \leq 1$. In the form shown in (A.13), the solution of (A.4) comes down to estimating $\beta_n(k)$, which is generally known as the amplification-factor function. By substituting (A.13) in (A.4) and ignoring the source term for a moment, we obtain

$$\beta_n(k) = 1 - 4D \sin^2 \left(\frac{k\Delta x}{2} \right) , \tag{A.14}$$

and by using condition $|\beta_n(k)| \leq 1$, we obtain the following stability condition

$$\alpha \leq \frac{1}{2} \iff \frac{D\,\Delta t}{(\Delta x)^2} \leq \frac{1}{2} , \tag{A.15}$$

for (A.4). If this condition is not satisfied, the explicit finite-difference solution becomes unstable and starts to wildly oscillate. However, this condition sometimes imposes severe limitations when modeling small-scale heterogeneities because it requires Δt to be very small; hence a very large number of timesteps are needed in such computations.

A.2.3. The Implicit Finite-Difference Method with Backward Differences in Time

Instead of using the forward-difference approximations for the time derivative, the implicit finite-difference scheme uses the backward-difference approximations; i.e.,

$$\frac{\partial T(x,t)}{\partial t} \approx \frac{T_{n,j} - T_{n-1,j}}{\Delta t} . \tag{A.16}$$

One can also use the forward-difference approximation for the time derivative and simply evaluate the spatial derivatives at the new timestep, $(n+1)$. So the implicit finite-difference of equation (A.2.5.) can be written as follows:

$$\frac{T_{n+1,j} - T_{n,j}}{\Delta t} = D \frac{T_{n+1,j+1} - 2T_j^{n+1} + T_{n+1,j-1}}{(\Delta x)^2}$$
$$- w \frac{T_{n+1,j+1} - T_{n+1,j-1}}{2\Delta x} + F(T_{n,j}) , \tag{A.17}$$

or

$$T_{n,j} = -(\alpha + \zeta)T_{n+1,j-1} + (1 + 2\alpha)T_{n+1,j} - (\alpha - \zeta)T_{n+1,j+1} + F(T_{n,j}) . \tag{A.18}$$

These equations are just like equation (A.26), with the exception that instead of at $t = n\Delta t$. However, this difference has significant consequences; we no longer have an explicit relationship for $T_{n+1,j-1}$, $T_{n+1,j}$, and $T_{n+1,j-1}$; in the explicit methods, the temperatures are evaluated at (n) (e.g., $T_{n,j}$) and in the implicit methods, the temperatures are evaluated at $(n+1)$ (e.g., $T_{n+1,j}$). Thus a set of simultaneous linear equations must be solved at each time to obtain T_j^{n+1}. Equation (A.18), along with the boundary conditions introduced earlier, can be written as

$$\boldsymbol{B}\boldsymbol{u}^{(n+1)} = \boldsymbol{u}^{(n)} + \boldsymbol{b}^{(n)} , \tag{A.19}$$

where

$$
\boldsymbol{B} = \begin{pmatrix}
1+2\alpha & -\alpha+\zeta & 0 & \cdots & 0 \\
-\alpha-\zeta & 1+2\alpha & -\alpha+\zeta & \ddots & 0 \\
0 & -\alpha-\zeta & 1+2\alpha & \ddots & \vdots \\
\vdots & \ddots & \ddots & \ddots & -\alpha+\zeta \\
0 & \cdots & 0 & -\alpha-\zeta & 1+2\alpha
\end{pmatrix}, \qquad (A.20)
$$

and where $\boldsymbol{B}$ is a $(J-1)\times(J-1)$ matrix like matrix $\boldsymbol{A}$ in (A.9) of the explicit finite-difference method. However, in (A.19) we are dealing with matrix inversions and not simply a series of multiplication operations as in (A.8). The possible solutions to the linear system in (A.19) include (i) a direct inversion of $\boldsymbol{B}$ (based on the fact that $\boldsymbol{B}$ is a tri-diagonal positive definite Toeplitz matrix), (ii) the use of the LU-decomposition of $\boldsymbol{B}$, and (iii) classical iteration methods for linear equation systems.

Let us now estimate the amplification-factor function associated with (A.18) to determine the stability condition of the implicit finite-difference modeling. By substituting (A.13) in (A.18) and ignoring the source term for a moment, we obtain

$$
\beta_n(k) = \left[1 + 4\alpha \sin^2 \left(\frac{k\Delta x}{2} \right) \right]^{-1}, \qquad (A.21)
$$

which always satisfies the stability condition, $|\beta_n(k)| \leq 1$. Therefore the implicit scheme is unconditionally stable.

A.2.4. The Implicit Finite-Difference Method with Forward Finite Differences in Time

In the explicit scheme, we use a combination of a forward finite difference in time and a central finite difference in space. For the implicit scheme, we use a combination of a backward finite difference in time and a central finite difference in space. Sometimes, it is desirable to combine the stability of an implicit method with the accuracy of a method in a single scheme. The average of the spatial derivatives of the explicit and implicit schemes,

$$
\frac{T_{n+1,j} - T_{n,j}}{\Delta t} = \frac{D}{2(\Delta x)^2} [(T_{n+1,j+1} - 2T_{n+1,j} + T_{n+1,j-1})
$$

$$
+ \ (T_{n,j+1} - 2T_{n,j} + T_{n,j-1})] \qquad (A.22)
$$

or

$$
\alpha T_{n,j-1} + 2(1-\alpha)T_{n,j} + \alpha T_{n,j+1} = -\alpha T_{n+1,j-1} + 2(1+\alpha)T_{n+1,j}
$$

$$
- \ \alpha T_{n+1,j+1} , \qquad (A.23)
$$

provides such a scheme. Now the unknown value $T_{n+1,j}$ is expressed in terms of both known quantities at n and unknown quantities at $(n+1)$. The scheme remains implicit, as the above equation cannot result in a solution at grid point j. The amplification factor function associated with (A.23) is

$$\beta_n(k) = \frac{1 - 2\alpha \sin^2\left(\frac{k\Delta x}{2}\right)}{1 + 2\alpha \sin^2\left(\frac{k\Delta x}{2}\right)} \ . \tag{A.24}$$

We can verify that $|\beta_n(k)| \leq 1$. Therefore the new implicit scheme in (A.23) is also unconditionally stable.

A.2.5.   The Case of Variable Conductivity

Some numerical solutions described can be solved analytically. The numerical method, however, can be used in cases of analytic solution or even when the analytic solution is unknown. The case of variable thermal conductivity described in () is one such example. This equation may also be written as follows:

$$\rho c_P \frac{\partial T(x,t)}{\partial t} + \left[\rho c_P w - \frac{\partial k(x)}{\partial x}\right] \frac{\partial T(x,t)}{\partial x} = k(x)\frac{\partial^2 T(x,t)}{\partial x^2} + q'(T) \ . \tag{A.25}$$

By substituting (A.3) and (A.4) in (A.2.5.), we arrive at the following explicit finite-difference scheme for the heat equation:

$$\frac{T_{n+1,j} - T_{n,j}}{\Delta t} = D_j \left[\frac{T_{n,j+1} - 2T_{n,j} + T_{n,j-1}}{(\Delta x)^2}\right]$$

$$- \left(w - \frac{D_{j+1} - D_{j-1}}{2\Delta x}\right)\frac{T_{n,j+1} - T_{n,j-1}}{2\Delta x} + \frac{q'(T_{n,j})}{\rho c_P}$$

or, upon solving for $T_{n+1,j}$,

$$T_{n+1,j} = \left[\alpha_j + 0.25(\alpha_{j+1} - \alpha_{j-1}) - \zeta\right] T_{n,j+1} + (1 - 2\alpha_j)T_{n,j}$$

$$+ \left[\alpha_j - 0.25(\alpha_{j+1} - \alpha_{j-1}) + \zeta\right] T_{n,j-1} + F(T_{n,j}) \ ,$$

where $\alpha_j = D_j \Delta t/(\Delta x)^2$, $\zeta = w\Delta t/(2\Delta x)$, and $F(T_{n,j}) = q'(T_{n,j})/\rho c_P$.

A.2.6.   Multidimensional Thermal Calculation

The three-dimensional heat equation can be written as follows:

$$\frac{\partial T(\mathbf{x},t)}{\partial t} + \boldsymbol{\nabla} T(\mathbf{x},t) = D\boldsymbol{\nabla}^2 T(\mathbf{x},t) + \frac{q'(\mathbf{x})}{\rho c_p} \ . \tag{A.26}$$

The discretized version of equation (A.26) is given as follows:

$$
T_{n+1,i,j,k} = T_{n,i,j,k} + \frac{D\Delta t}{\Delta x^2}\left[T_{n,i+1,j,k} - 2T_{n,i,j,k} + T_{n,i-1,j,k}\right]
$$

$$
+ \frac{D\Delta t}{\Delta y^2}\left[T_{n,i,j+1,k} - 2T_{n,i,j,k} + T_{n,i,j-1,k}\right]
$$

$$
+ \frac{D\Delta t}{\Delta z^2}\left[T_{n,i,j,k+1} - 2T_{n,i,j,k} + T_{n,i,j,k-1}\right] + \frac{q'(i,j,k)\Delta t}{\rho c_p}\ .
$$

These equations are explicit since the unknown nodal temperatures at time $n+1$ are determined by known temperatures at time n in each time step. Initial conditions must be defined so that the temperature of each node is known at time $t = 0$. Assuming that $\Delta x = \Delta y$, stability criterion for 2-D interior nodes is $\alpha = D\Delta t/(\Delta x)^2 \geq 1/4$.

Implicit method, obtained by evaluating all other temperatures at time $n+1$ instead of n. This gives a backward-difference method, which in two-dimensional form is

$$
T_{n,i,j,k} = [1 - 4\alpha]\,T_{n+1,i,j,k}
$$

$$
- \alpha\,(T_{n+1,i+1,j,k} + T_{n+1,i-1,j,k} + T_{n+1,i,j+1,k}
$$

$$
+ T_{n+1,i,j-1,k} + T_{n+1,i,j,k+1} + T_{n+1,i,j,k-1}) + F(T_{n,i,j})\ ,
$$

where $F(T_{n,i,j}) = q'(i,j,k)\Delta t\rho c_p$. Thus, the new temperature at node i,j,k depends on the new unknown temperatures at the other adjacent nodes. Consequently, a simultaneous solution is required using Gauss-Seidel iteration or matrix inversion.

A.3. Nonlinear Heat-Transfer Modeling

We now consider one-dimensional heat transfer in a media with temperature-dependent thermal conductivity; i.e.,

$$
\rho c_P\left(\frac{\partial T}{\partial t} + u_z \frac{\partial T}{\partial z}\right) - \frac{\partial}{\partial z}\left[k(T)\frac{\partial T}{\partial z}\right] = q'(T) \tag{A.27}
$$

or

$$
\rho c_P\left(\frac{\partial T}{\partial t} + u_z \frac{\partial T}{\partial z}\right) - \frac{\partial k(T)}{\partial T}\left(\frac{\partial T}{\partial z}\right)^2 - k(T)\frac{\partial^2 T}{\partial z^2} = q'(T)\ . \tag{A.28}
$$

The finite-difference scheme for this equation is

$$
\rho c_P\left[\frac{T_{n,j} - T_{n-1,j}}{\Delta t}\right] = -\rho c_P u_z \tilde{T}_{n,j} + \frac{\partial k}{\partial T}(T_{n,j})\,\tilde{T}_{n,j}^2 + k(T_{n,j})\frac{\partial^2 T_{n,j}}{\partial z^2} + q'(T)\ .
$$

$$
\tag{A.29}
$$

Solving (A.29) for the second spatial derivative of the temperature $T_{n,j}$, we get

$$\frac{\partial^2 T_{n,j}}{dz^2} = \frac{T_{n,j+1} - 2T_{n,j} + T_{n,j-1}}{(\Delta z)^2} + f\left(T_{n,j}, \tilde{T}_{n,j}, T_{n-1,j}\right) , \tag{A.30}$$

where

$$f\left(T_{n,j}, \tilde{T}_{n,j}, T_{n-1,j}\right) = \frac{1}{k(T_{n,j})} \left\{ \rho c_P \left[\frac{T_{n,j} - T_{n-1,j}}{\Delta t} \right] + \rho c_P u_z \tilde{T}_{n,j} \right.$$
$$\left. - \frac{\partial k}{dT}(T_{n,j}) \left(\tilde{T}_{n,j}\right)^2 - q'(T_{n,j}) \right\} , \tag{A.31}$$

and

$$\tilde{T}_{n,j} = \frac{T_{n,j+1} - T_{n,j+1}}{2\Delta z} = [\partial T / \partial z](n\Delta t, j\Delta z) , \tag{A.32}$$

$$T_{n,1} = T(t = n\Delta t, z = a) = \alpha(t = n\Delta t) , \tag{A.33}$$

$$T_{n,N} = T(t = n\Delta t, z = b) = \beta(t = n\Delta t) , \tag{A.34}$$

where $\tilde{T}_{n,j}$ is the temperature gradient and f is a nonlinear function. Starting from the initial condition $T_{0,j}$, we can solve successively (A.30) for $n = 1, 2, ..., N$ by using the Newton method. Our implementation of the Newton method requires the partial derivatives of $f\left(T_{n,j}, \tilde{T}_{n,j}, T_{n-1,j}\right)$ with respect to $T_{n,j}$ and $\tilde{T}_{n,j}$. Introducing the notation $f_{n,j} = f\left(T_{n,j}, \tilde{T}_{n,j}, T_{n-1,j}\right)$, we get

$$q_{n,j} = \frac{\partial f}{\partial T}(n\Delta t, j\Delta z)$$
$$= \frac{1}{k(T_{n,j})} \left[-f_{n,j} \frac{\partial k(T_{n,j})}{\partial T} + \frac{\rho c_P}{\Delta t} - \frac{\partial^2 k(T_{n,j})}{dT^2} \left(\tilde{T}_{n,j}\right)^2 - \frac{\partial q'}{\partial T}(T_{n,j}) \right] , \tag{A.35}$$

$$p_{n,j} = \frac{\partial f}{\partial \tilde{T}}(n\Delta t, j\Delta z) = \frac{1}{k(T_{n,j})} \left\{ \rho c_P u_z - 2\frac{\partial k(T)}{\partial T} \tilde{T}_{n,j} \right\} . \tag{A.36}$$

Introducing the column-vector $\boldsymbol{G}_n = [G_{n,1}, G_{n,1}, ..., G_{n,N}]^T$ with components

$$G_{n,1} = T_{n,1} - T_a(t_n) , \quad G_{n,N} = T_{n,1} - T_b(t_n) , \tag{A.37}$$

and

$$G_{n,j} = T_{n,j+1} - 2T_{n,j} + T_{n,j-1} - \Delta z^2 f_{n,j} , \tag{A.38}$$

the system of nonlinear equations (A.30) and the boundary conditions (A.33) and (A.34) can be written as one equation:

$$\boldsymbol{G}_n(\boldsymbol{u}_n) = 0 , \tag{A.39}$$

where

$$u_n = [T_{n,1}, T_{n,2}, ..., T_{n,N}]^T .$$

(A.40)

Starting with some initial guess $u^{(0)}$, the nonlinear system (A.39) can be solved by the Newton iterative method:

$$u_n^{(k+1)} = u_n^{(k)} - \left[L^{(k)}\right]^{-1} G(u_n^{(k)}) ,$$

(A.41)

where

$$L^{(k)} = \frac{\partial G_n}{\partial u_n}\left(u_n^{(k)}\right) .$$

(A.42)

Calculating the elements of the Jacobian, we get

$$L_n^{(k)}(1,1) = 1 ; \quad L_n^{(k)}(N,N) = 1 ; \quad L_n^{(k)}(i,i) = -2 - \Delta z^2 q_{n,i}^{(k)} ;$$

$$L_n^{(k)}(i,i-1) = 1 + \frac{1}{2}\Delta z p_{n,i}^{(k)} ; \quad \text{and} \quad L_n^{(k)}(i,i-1) = 1 - \frac{1}{2}\Delta z p_{n,i}^{(k)}.$$

In practice, the iteration process is usually ended when the difference between the solutions of two successive iterations is smaller than a predefined threshold.

A.4.   Madariaga-Virieux Staggered-grid Scheme

The fields involved in elastic wave equations are as follows: τ, stress tensor in Pa (Pascal = newton/meter^2); v, particle-velocity vector in m/s (meter/second); F, volume-source density of external forces in N/m^3 (newton/meter^3); I, volume-source density of the external stress-source rate in Pa/s. Using these quantities, the equations of the linear elastic wave propagation can be written as

$$\frac{\partial v_i(\boldsymbol{x},t)}{\partial t} - \sigma(\boldsymbol{x})\frac{\partial \tau_{ij}(\boldsymbol{x},t)}{\partial x_j} = F_i(\boldsymbol{x},t) \quad \text{(momentum conservation)} ,$$

(A.43)

$$\frac{\partial \tau_{ij}(\boldsymbol{x},t)}{\partial t} - c_{ijpq}(\boldsymbol{x})\frac{\partial v_q(\boldsymbol{x},t)}{\partial x_p} = I_{ij}(\boldsymbol{x},t) \quad \text{(stress} - \text{strain relations)} ,$$

(A.44)

with $\boldsymbol{x} = [x,y,z]^T = [x_1,x_2,x_3]^T$, $i,j,p,q = 1,2,3$ or $i,j,p,q = x,y,z$, and where c_{ijpq} is the elastic stiffness tensor of the fourth rank (i.e., the reciprocal of the compliance tensor), σ is the specific volume (i.e., the reciprocal of the mass density of the medium), and the symbol T indicates a transpose. The stiffness tensor $c = c(\boldsymbol{x})$ is symmetric at each point $\boldsymbol{x}$; that is, it satisfies $c_{ikpq} = c_{jipq} = c_{jiqp} = c_{ijqp}$ in addition to $c_{ijpq} = c_{pqji}$. These symmetries allow us to alternatively denote the tensorial stiffness, c_{ijkl}, as a 6×6 matrix C_{IJ}, where the subscripts I and J run from 1 to 6, with $ij \to I$, according to $11,22,33,23,31,12 \leftrightarrow 1,2,3,4,5,6$.

To solve these equations, we need to specify the appropriate boundary and initial conditions. The initial conditions are that the stress and particle-velocity fields and their time derivatives are zero before the seismic source is fired; i.e.,

$$\mathbf{v} = \partial_t \mathbf{v} = \mathbf{0} \ , \ t \leq 0 \ , \tag{A.45}$$

$$\boldsymbol{\tau} = \partial_t \boldsymbol{\tau} = \mathbf{0} \ , \ t \leq 0 \ . \tag{A.46}$$

The boundary conditions for modeling seismic-wave propagation are essentially determined by the free-surface boundary: the air-solid interface in land seismics and the air-water interface in marine seismics. Throughout this section, we will assume a planar free-surface boundary. Let the free surface be at a depth of $z = 0$. Then the boundary conditions are

$$\tau_{xz}(x, y, z = 0, t) = \tau_{yz}(x, y, z = 0, t) = \tau_{zz}(x, y, z = 0, t) = 0 \ . \tag{A.47}$$

We consider that the rest of the medium is unbounded.

Our goal in this subsection is to numerically solve equations (A.43)-(A.44) under the boundary and initial conditions in (A.45)-(A.47) by using the Madariaga-Virieux staggered-grid finite-difference techniques that we will describe below. As in most finite-difference modeling, the first step in staggered-grid finite-difference modeling is to discretize the geological model and the quantities that characterize the wavefield—i.e., the particle velocity and stresses in this case. So we discretize both the time and space domains as follows:

$$\begin{aligned} t &= n\Delta t \ , \ n = 0, 1, 2, ..., N \ , \\ x &= i\Delta x \ , \ i = 0, 1, 2, ..., I \ , \\ y &= j\Delta y \ , \ i = 0, 1, 2, ..., J \ , \\ z &= k\Delta z \ , \ k = 0, 1, 2, ..., K \ , \end{aligned} \tag{A.48}$$

where Δt is the time interval (timestep), Δx is the spacing interval (grid size) in the x-direction, Δy is the spacing interval (grid size) in the y-direction, and Δz is the spacing interval (grid size) in the z-direction. The total time for the data length is $N\Delta t$. The size of the subsurface model is $I\Delta x \times J\Delta y \times K\Delta z$. We will call this discretization the reference grid (also as the unstaggered grid or natural grid). This terminology will allow us to distinguish the discretization in (A.48) and other discretization associated with the staggered-grid techniques, which we will discuss later.

The standard finite-difference calculations are carried out on natural grid cells (Figure A.1). In these calculations, each quantity in the differential equations (A.43)-(A.44) is defined as a function of the indices n, i, j, and k, in accordance with the following two examples:

$$\begin{aligned} \sigma(x, y, z) &= \sigma\left[i\Delta x, j\Delta y, k\Delta z)\right] = \sigma_{i,j,k} \\ \tau_{xz}(x, y, z, t) &= \tau_{xz}(i\Delta x, j\Delta y, k\Delta z, n\Delta t) = [\tau_{xz}]_{i,j,k}^n \end{aligned} \tag{A.49}$$

Basically, all quantities in equations (A.43)-(A.44) share the same gridpoints, which are the cell vertices.

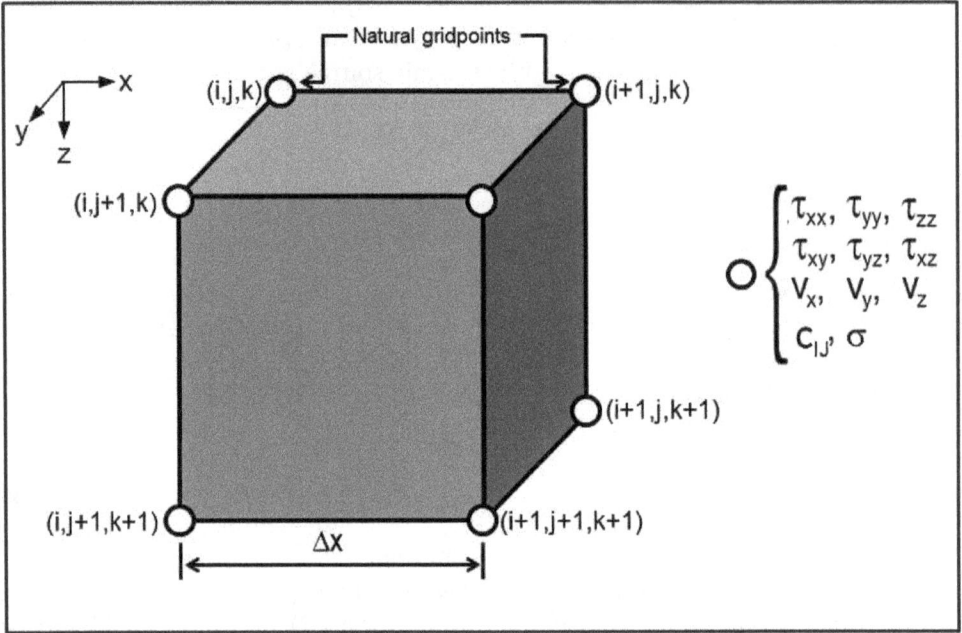

Figure A.1. An illustration of the natural unstaggered grid structures for solving the 3D elastodynamic equations. All wavefield components are defined at gridpoints in their natural locations at the cell vertices (i.e., all wavefield components are defined at each vertex).

In the staggered-grid technique, not all quantities in the differential equations (A.43)-(A.44) are gridded at the points of the reference grid. Some quantities are defined at half a gridpoint off the reference grid, say, $x = \left(i \pm \frac{1}{2}\right)\Delta x$ instead of $x = i\Delta x$. Figure A.2 shows examples of staggered gridding configurations of equations (A.43)-(A.44) for media with only orthorhombic symmetry. The reason for limiting ourselves will be discussed later in this subsection. Notice that the normal stresses, specific volume, and stiffnesses are located at the same points. Two questions come to mind: (i) what is the basis of locating some quantities on the reference grid and others off the reference grid? and (ii) why did we select to work with staggered-grid finite-difference techniques instead of the standard finite-difference technique? Let us start with the second question. Our objective in locating some quantities on the reference grid and others off the reference grid is to end with a finite-difference scheme in which the differential operators act only on the wavefield variables, not on the medium parameters. Remember that the wavefield is by definition continuous in space and time, and therefore suitable for differentiation with respect to space and time variables, whereas medium parameters are generally discontinuous functions with respect to space variables and therefore are less suitable for differentiation with respect to space. It is therefore desirable to avoid the differentiation of the medium parameters whenever possible. The staggered-grid technique allows us to do so.

Let us now turn to the question of the locations of the physical quantities in the staggered-gridding configuration. First of all, there is no unique way of locating physical quantities of elastic wave equations in the staggered-gridding configuration.

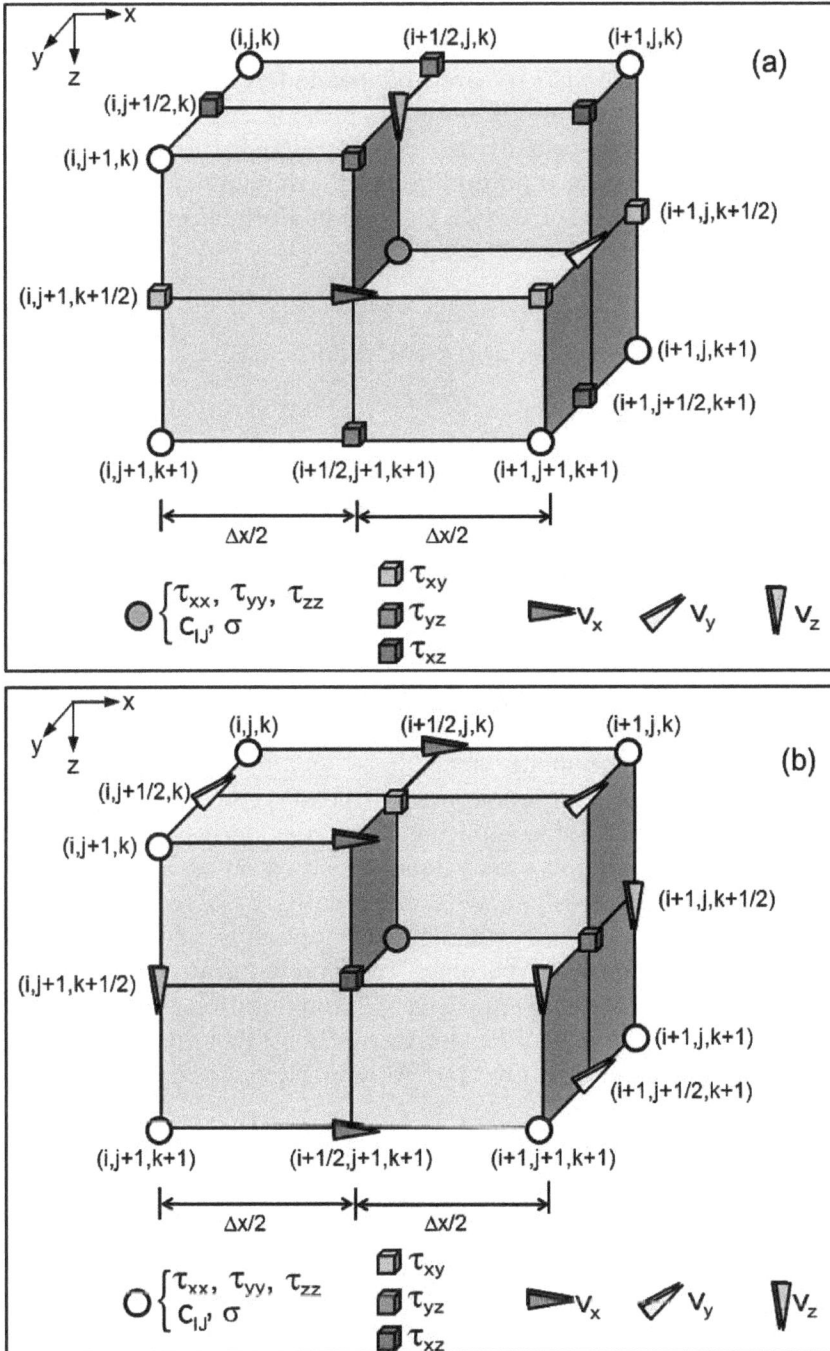

Figure A.2. (a) An example of a staggered gridding of (A.43)-(A.44), under the assumption that the medium has an orthorhombic symmetry. (b) Another example of a staggered gridding of the same equations.

For example, Figure A.2 are two valid solutions. The objective in selecting locations for the physical quantities in the staggered-gridding configuration is that various differential operators must require the wavefield components only at the staggered gridpoints, where these components are defined; the differential operators must not require some interpolation to achieve this objective. When this objective is met, the components of the particle velocity are updated independently from the stresses. This result yields an effective implementation, although its computer code requires some additional thinking (especially in terms of the indexing of various variables) when compared to the standard finite-difference technique.

Using the definitions in Figure A.2a, the discrete form of quantities in equations (A.43)-(A.44) can be written as follows:

$$C_{IJ}(x,y,z) = C_{IJ}\left[(i+\tfrac{1}{2})\Delta x, (j+\tfrac{1}{2})\Delta y, (k+\tfrac{1}{2})\Delta z)\right] = [C_{IJ}]_{i+1/2,j+1/2,k+1/2},$$
$$\sigma(x,y,z) = \sigma\left[(i+\tfrac{1}{2})\Delta x, (j+\tfrac{1}{2})\Delta y, (k+\tfrac{1}{2})\Delta z)\right] = \sigma_{i+1/2,j+1/2,k+1/2},$$

for the medium parameters, and

$$
\begin{aligned}
\tau_{pp}(x,y,z,t) &= \tau_{pp}\left[(i+\tfrac{1}{2})\Delta x, (j+\tfrac{1}{2})\Delta y, (k+\tfrac{1}{2})\Delta z, n\Delta t)\right] = [\tau_{pp}]_{i+1/2,j+1/2,k+1/2}^{n}, \\
\tau_{yz}(x,y,z,t) &= \tau_{yz}\left[(i+\tfrac{1}{2})\Delta x, j\Delta y, k\Delta z, n\Delta t)\right] = [\tau_{yz}]_{i+1/2,j,k}^{n}, \\
\tau_{xz}(x,y,z,t) &= \tau_{xz}\left[i\Delta x, (j+\tfrac{1}{2})\Delta y, k\Delta z, n\Delta t)\right] = [\tau_{xz}]_{i,j+1/2,k}^{n}, \\
\tau_{xy}(x,y,z,t) &= \tau_{xy}\left[i\Delta x, j\Delta y, (k+\tfrac{1}{2})\Delta z, n\Delta t)\right] = [\tau_{xy}]_{i,j,k+1/2}^{n}, \\
v_x(x,y,z,t) &= v_x\left[i\Delta x, (j+\tfrac{1}{2})\Delta y, (k+\tfrac{1}{2})\Delta z, (n+\tfrac{1}{2})\Delta t)\right] = [v_x]_{i,j+1/2,k+1/2}^{n+1/2}, \\
v_y(x,y,z,t) &= v_y\left[(i+\tfrac{1}{2})\Delta x, j\Delta y, (k+\tfrac{1}{2})\Delta z, (n+\tfrac{1}{2})\Delta t)\right] = [v_y]_{i+1/2,j,k+1/2}^{n+1/2}, \\
v_z(x,y,z,t) &= v_z\left[(i+\tfrac{1}{2})\Delta x, (j+\tfrac{1}{2})\Delta y, k\Delta z, (n+\tfrac{1}{2})\Delta t)\right] = [v_z]_{i+1/2,j+1/2,k}^{n+1/2},
\end{aligned}
\tag{A.50}
$$

with $p = x, y, z$ for the normal stresses.

Before we provide the final equations in the staggered-grid technique, let us recall some basic formulae for computing first-order derivatives. The first-order derivatives in the finite-difference technique are based on an approximation of the Taylor series. The derivatives can be approximated by a second-order operator, a fourth-order operator, or even a higher-order operator. The higher the order, the longer the difference operator becomes, the more accurate the actual first-order derivatives will be, and the more expensive the computation time will be.

For an arbitrary α, the first-order derivatives of a function $g(x)$ can be approximated by a three-point formula given by Abramowitz and Stegun (1964):

$$
\begin{aligned}
\frac{\partial g\,(x+\alpha\Delta x)}{\partial x} = & \frac{1}{\Delta x}\left[\left(\alpha - \frac{1}{2}\right)g\,(x-\Delta x) - 2\alpha g(x)\right. \\
& \left. + \left(\alpha + \frac{1}{2}\right)g\,(x+\Delta x)\right] + \dots .
\end{aligned}
\tag{A.51}
$$

The evaluated equation (A.51) at $\alpha = 0$ gives us the derivatives at the reference grid:

$$
\frac{\partial g(x)}{\partial x} \approx \frac{1}{\Delta x}\left[-\frac{1}{2}g\,(x-\Delta x) + \frac{1}{2}g\,(x+\Delta x)\right] .
\tag{A.52}
$$

Similarly, the evaluated equation (A.51) at $\alpha = 1/2$ gives us

$$
\frac{\partial g\,(x+\tfrac{1}{2}\Delta x)}{\partial x} \approx \frac{1}{\Delta x}\left[-g(x) + g\,(x-\Delta x)\right] .
\tag{A.53}
$$

We can also obtain the derivatives in the reference grid by shifting equation (A.53) a half grid-point in the negative direction; i.e.,

$$\frac{\partial g(x)}{\partial x} \approx \frac{1}{\Delta x} \left[-g\left(x - \frac{1}{2}\Delta x\right) + g\left(x + \frac{1}{2}\Delta x\right) \right] . \tag{A.54}$$

Equation (A.54) is the second-order operator for approximating the first-order derivatives. Our time derivatives here are based on this formula.

For spatial derivatives, we will use the fourth- or higher-order approximation. We can obtain the fourth-order approximation by using a five-point formula. For an arbitrary α, the first-order derivatives of a function $g(x)$ can be approximated by a five-point formula, as follows:

$$\begin{aligned}
\frac{\partial g(x + \alpha \Delta x)}{\partial x} = \frac{1}{\Delta x} \Bigg[& \frac{1}{12} \left(2\alpha^3 - 3\alpha^2 - \alpha + 1\right) g\left(x - 2\Delta x\right) \\
& - \frac{1}{6} \left(4\alpha^3 - 3\alpha^2 - 8\alpha + 4\right) g\left(x - \Delta x\right) \\
& + \frac{1}{2} \left(4\alpha^3 - 3\alpha\right) g(x) \\
& + \frac{1}{6} \left(4\alpha^3 - 3\alpha^2 - 8\alpha + 4\right) g\left(x + \Delta x\right) \\
& - \frac{1}{12} \left(2\alpha^3 - 3\alpha^2 - \alpha + 1\right) g\left(x + 2\Delta x\right) \Bigg] + \dots .
\end{aligned} \tag{A.55}$$

The evaluated equation (A.55) at $\alpha = 0$ gives us the derivatives at the reference grid:

$$\begin{aligned}
\frac{\partial g(x)}{\partial x} \approx \frac{1}{\Delta x} \Bigg[& \frac{1}{12} g\left(x - 2\Delta x\right) - \frac{2}{3} g\left(x - \Delta x\right) + \frac{2}{3} g\left(x + \Delta x\right) \\
& - \frac{1}{12} g\left(x + 2\Delta x\right) \Bigg] .
\end{aligned} \tag{A.56}$$

Similarly, evaluated equation (A.55) at $\alpha = 1/2$ gives us

$$\begin{aligned}
\frac{\partial g\left(x + \frac{1}{2}\Delta x\right)}{\partial x} \approx \frac{1}{\Delta x} \Bigg[& \frac{1}{24} g\left(x - \Delta x\right) - \frac{9}{8} g(x) + \frac{9}{8} g\left(x + \Delta x\right) \\
& - \frac{1}{24} g\left(x + 2\Delta x\right) \Bigg] .
\end{aligned} \tag{A.57}$$

We can also obtain the derivatives in the reference grid by shifting equation (A.57) a half-grid-point in the negative direction; i.e.,

$$\begin{aligned}
\frac{\partial g(x)}{\partial x} \approx \frac{1}{\Delta x} \Bigg[& \frac{1}{24} g\left(x - \frac{2}{3}\Delta x\right) - \frac{9}{8} g\left(x - \frac{1}{2}\Delta x\right) \\
& + \frac{9}{8} g\left(x + \frac{1}{2}\Delta x\right) - \frac{1}{24} g\left(x + \frac{3}{2}\Delta x\right) \Bigg] .
\end{aligned} \tag{A.58}$$

Table A.2. The coefficients, c_m^M for the second- to tenth-order, finite-difference operators. This Matlab command can be used to compute the double factorial for a given n: k = cos(pi * n) − 1; df = 2^((−k+2*n)/4) * pi^(k/4) * gamma((n + 2)/2).

		$c_m^{(M)}$			
M	$m = 1$	$m = 2$	$m = 3$	$m = 4$	$m = 5$
1	1				
2	9/8	−1/24			
3	75/64	−25/384	3/640		
4	1225/1024	−245/3072	49/5120	−5/7168	
5	19845/16384	−735/8192	567/40960	−495/229376	35/294912

Equation (A.58) is the fourth-order operator for approximating the first-order derivatives. This formula can be generalized to a $2M$th-order approximation, as follows:

$$\frac{\partial g(x)}{\partial x} \approx \frac{1}{\Delta x} \sum_{m=1}^{M} c_m^{(M)} \left\{ g\left[x + \left(m - \frac{1}{2} \right) \Delta x \right] - g\left[x - \left(m - \frac{1}{2} \right) \Delta x \right] \right\}, \quad \text{(A.59)}$$

where

$$c_m^{(M)} = 2(-1)^{m-1} \left(\frac{1}{2m-1} \right)^2 \frac{[(2M-1)!!]^2}{[2(M-m)]!![2(M+m-1)]!!} \quad \text{(A.60)}$$

and

$$(2M-1)!! = \Pi_{k=1}^{M}(2k-1) = \frac{(2M)!}{2^M M!}. \quad \text{(A.61)}$$

Table A.2 provides the coefficients $c_m^{(M)}$ for the second- to tenth-order finite-difference operators.

By using these formulae, the partial-differential equations in (A.43) can be approximated as

$$[v_x]_{i,j+1/2,k+1/2}^{n+1/2} = [v_x]_{i,j+1/2,k+1/2}^{n-1/2}$$

$$+ \quad [\Delta t b_x \left(D_x \tau_{xx} + D_y \tau_{xy} + D_z \tau_{xz} + F_x \right)]_{i,j+1/2,k+1/2}^{n}$$

$$[v_y]_{i+1/2,j,k+1/2}^{n+1/2} = [v_y]_{i+1/2,j,k+1/2}^{n-1/2}$$

$$+ \quad [\Delta t b_y \left(D_x \tau_{xy} + D_y \tau_{yy} + D_z \tau_{yz} + f_y \right)]_{i+1/2,j,k+1/2}^{n},$$

$$[v_z]_{i+1/2,j+1/2,k}^{n+1/2} = [v_z]_{i+1/2,j+1/2,k}^{n-1/2}$$

$$+ \quad [\Delta t b_z \left(D_x \tau_{xz} + D_y \tau_{yz} + D_z \tau_{zz} + f_y \right)]_{i+1/2,j+1/2,k}^{n},$$

where

$$b_x = \frac{1}{4} \left[\sigma_{i,j,k} + \sigma_{i-1,j,k} \right] ,$$

$$b_y = \frac{1}{2} \left[\sigma_{i,j,k} + \sigma_{i,j-1,k} \right] ,$$

$$b_z = \frac{1}{2} \left[\sigma_{i,j,k} + \sigma_{i,j,k-1} \right] .$$

The operators D_x, D_y, and D_z denote the first-order spatial derivative for x, y, and z, respectively. For example, the spatial-derivative operators along the x-axis, D_x, is evaluated as follows:

$$D_x g_{i,j,k} \approx \frac{1}{\Delta x} \sum_{m=1}^{M} c_m^{(M)} \left(g_{i+m-1/2,j,k} - g_{i-m+1/2,j,k} \right) .$$

Notice that the computation of v_x, for example, requires the specific volume at gridpoint $[i\Delta x, (j + \frac{1}{2})\Delta y, (k + \frac{1}{2})\Delta z]^T$. Unfortunately, the specific volume is not defined at this gridpoint. We took the arithmetic average of the specific volume at gridpoints $[(i - \frac{1}{2})\Delta x, (j + \frac{1}{2})\Delta y, (k + \frac{1}{2})\Delta z]^T$ and $[(i + \frac{1}{2})\Delta x, (j + \frac{1}{2})\Delta y, (k + \frac{1}{2})\Delta z]^T$ to obtain the specific volume at $[i\Delta x, (j + \frac{1}{2})\Delta y, (k + \frac{1}{2})\Delta z]^T$. We have denoted this specific-volume average as b_x. We use the same approach for the computations of v_y and v_z and end by introducing the arithmetic averages b_y and b_z for the specific volumes at $[(i + \frac{1}{2})\Delta x, j\Delta y, (k + \frac{1}{2})\Delta z]^T$ and $[(i + \frac{1}{2})\Delta x, (j + \frac{1}{2})\Delta y, k\Delta z]^T$, respectively. Notice also that the differential operators act only on the wavefield variables, not on the specific volume. Thus the differentiation of the medium parameters is not necessary in this scheme, and therefore the discontinuity of the specific volume is not modified by the differential operators.

Let us turn to the staggered-grid differentiation of the equations in (A.44). The idea that the results of the various differentiation operations in the staggered-grid technique are all naturally centered to coincide with the gridpoints and the quantities associated with the gridpoints does not work for some terms of equations (A.44). To add some concreteness to this point, let us consider the staggered-grid computation of τ_{xx}; i.e.,

$$
\begin{aligned}
[\tau_{xx}]_{i+1/2,j+1/2,k+1/2}^{n+1} &= [\tau_{xx}]_{i+1/2,j+1/2,k+1/2}^{n} \\
&= \Delta t \left[C_{11} D_x v_x + C_{12} D_y v_y + c_{13} D_z v_z \right. \\
&+ C_{14} \left(D_z v_y + D_y v_z \right) + C_{15} \left(D_z v_x + D_x v_z \right) \\
&+ \left. C_{16} \left(D_y v_x + D_x v_y \right) + I_{xx} \right]_{i+1/2,j+1/2,k+1/2}^{n+1/2} . \quad \text{(A.62)}
\end{aligned}
$$

The computations of the terms of τ_{xx} associated with C_{14}, C_{15}, and C_{16} (i.e., $D_z v_y$, $D_y v_z$, $D_z v_x$, $D_x v_z$, $D_y v_x$, and $D_x v_y$) require values of the particle-velocity components at some gridpoints where these components are not defined. For example, we can verify that

$$[D_z v_y]_{i+1/2,j+1/2,k+1/2}^{n+1/2}$$

$$\approx \frac{1}{\Delta x} \sum_{m=1}^{M} c_m^{(M)} \left([v_y]_{i+m,j+1/2,k+1/2}^{n+1/2} - [v_y]_{i-m+1,j+1/2,k+1/2}^{n+1/2} \right) ; \quad (A.63)$$

that is, the computation of $D_z v_y$ requires the values of v_y at gridpoints where v_y is not defined. A similar observation can be made for $D_y v_z$, $D_z v_x$, $D_x v_z$, $D_y v_x$, and $D_x v_y$. In contrast, we can verify that

$$[D_x v_x]_{i+1/2,j+1/2,k+1/2}^{n+1/2}$$

$$\approx \frac{1}{\Delta x} \sum_{m=1}^{M} c_m^{(M)} \left([v_x]_{i+m,j+1/2,k+1/2}^{n+1/2} - [v_x]_{i-m+1,j+1/2,k+1/2}^{n+1/2} \right) , \quad (A.64)$$

which shows that the computation of $D_x v_x$ is naturally centered such that it requires values of v_x only at the gridpoints at which v_x is defined. We can also verify that the computations of $D_y v_y$ and $D_z v_z$ are centered such that only the gridpoints at which v_y and v_z, respectively, are defined are involved in these computations. So some interpolations of the particle velocity are needed for the computations of τ_{xx} and, for that matter, all the stresses because of the computations of $D_z v_y$, $D_y v_z$, $D_z v_x$, $D_x v_z$, $D_y v_x$, and $D_x v_y$. Such interpolations can reduce the overall accuracy of the finite-difference solution. Therefore, it is desirable to avoid them.

Note that a point-source representation of a body-force type of source can be implemented directly in the equation of motion, as described in (A.62)-(A.64). The body-force source-time function is not differentiated with respect to time in these equations; it is simply scaled by the specific volume and Δt.

The moment-tensor source can be implemented through the updating equations of stresses or through the updating equations of the particle velocity. In the particle-velocity implementation, each component of the moment tensor is implemented by using the corresponding couple of the body forces with a discrete arm's length between the forces. For example, in the staggered-grid velocity-stress formulation, the I_{xx} component of the moment tensor is equivalent to a couple of the forces $(I_{xx}/\Delta x)$ acting along the x-axis in the opposite direction. Because these forces are applied at one gridpoint, the appropriate volume is one grid cell (i.e., $\Delta x \Delta y \Delta z$). Therefore, the particle-velocity update for the I_{xx} component of the stress tensor at gridpoint $[i_s \Delta x, j_s \Delta y, k_s \Delta z]^T$ is

$$[v_x]_{i_s+1,j_s+1/2,k_s+1/2}^{n+1/2} = [v_x]_{i_s+1,j_s+1/2,k_s+1/2}^{n+1/2} + \Delta t \, b_x(i_s,j_s,k_s) \frac{I_{xx}(i_s,j_s,k_s)}{\Delta x \Delta y \Delta z}$$

$$(A.65)$$

$$[v_x]_{i_s-1,j_s+1/2,k_s+1/2}^{n+1/2} = [v_x]_{i_s-1,j_s+1/2,k_s+1/2}^{n+1/2} - \Delta t \, b_x(i_s,j_s,k_s) \frac{I_{xx}(i_s,j_s,k_s)}{\Delta x \Delta y \Delta z} .$$

$$(A.66)$$

However, the equivalent body forces for the representation of the I_{xy} component of the stress tensor are not located along the grid line i_s, and they must be averaged from four equivalent body forces ($I_{xy}/2\Delta y$) acting along the x-axis in the opposite direction with a force arm of length $2\Delta y$. Therefore the particle-velocity update for the I_{xy} component of the moment tensor at gridpoint $[i_s\Delta x, j_s\Delta y, k_s\Delta z]^T$ is

$$[v_x]_{i_s\pm1,j_s+1/2,k_s+1/2}^{n+1/2} = [v_x]_{i_s\pm1,j_s+1/2,k_s+1/2}^{n+1/2} + \Delta t\, b_x(i_s,j_s,k_s)\frac{I_{xy}(i_s,j_s,k_s)}{\Delta x\Delta y^2\Delta z}$$

$$\text{(A.67)}$$

$$[v_x]_{i_s\mp1,j_s-1/2,k_s+1/2}^{n+1/2} = [v_x]_{i_s\mp1,j_s-1/2,k_s+1/2}^{n+1/2} - \Delta t\, b_x(i_s,j_s,k_s)\frac{I_{xy}(i_s,j_s,k_s)}{\Delta x\Delta y^2\Delta z}\ .$$

$$\text{(A.68)}$$

Analogously, the remaining components of the moment tensor can be implemented as equivalent body forces centered at grid node $[i_s\Delta x, j_s\Delta y, k_s\Delta z]^T$.

Implementation of the explosive source can be included either by using normal stresses or by using particle velocity. Note that this moment-tensor implementation allows us to model an explosive source by using equal diagonal elements and vanishing non-diagonal elements of the moment tensor.

A.4.1.   Stability and Dispersion Conditions

The dispersion and stability conditions are two important criteria that must be met to successfully simulate wave propagation with a finite-difference algorithm. The dispersion condition allows us to avoid errors resulting from the approximations we make in the computation of spatial derivatives. The stability condition allows us to avoid errors associated with the recursive computation (timestep by timestep) of the wavefield components. As we can see in (A.62)-(A.64), for example, in the staggered-grid finite-difference equations, the quantities characterizing the wave motion are computed recursively, timestep by timestep. For instance, computing the components of the particle velocity at timestep $(n+1/2)$ for the components of the particle velocity and at timestep $(n+1)$ for the stress components requires the previous timestep $(n-1/2)$ of the components of the particle velocity and the timestep (n) of the stress components. However, this recursive computation can be a source of numerical instability. In fact, errors introduced by the numerical solution can propagate and be magnified during the timestepping of the finite-difference scheme, causing significant instabilities during the computation, and artifacts in the resulting data. Such an instability is very unlikely to occur if the instability condition is met. Our objective in this subsection is to describe the dispersion and stability conditions for the staggered-grid finite-difference schemes that we introduced in previous subsections, namely the Madariaga-Virieux and partially staggered-grid schemes. We did not mention the schemes based on staggered grids with multiple-momentum equations because they have the same time and spatial-difference operators as the Madariaga-Virieux schemes. Therefore they have the same stability and dispersion conditions.

To derive the stability and dispersion conditions for the finite-difference algorithms, we start by constructing a second-order partial differential equation for the particle-velocity field under the assumptions (i) that the stiffness tensor is isotropic, (ii) that nonzero components of the stiffness tensor are constant, and (iii) that the source terms are zero. So by substituting the stress-strain equations (A.44) into the velocity-stress equations in (A.43) under these assumptions, we arrive at

$$\frac{\partial^2 v_p(\boldsymbol{x}, t)}{\partial t^2} = \eta_{pqrs} \frac{\partial^2 v_r(\boldsymbol{x}, t)}{\partial x_q \partial x_s} \ , \tag{A.69}$$

with

$$\eta_{pqrs} = \left(V_P^2 - 2V_S^2\right) \delta_{pq}\delta_{rs} + V_S^2 \left(\delta_{pr}\delta_{qs} + \delta_{ps}\delta_{qr}\right) \ , \tag{A.70}$$

where V_P and V_S are P-wave and S-wave velocities, respectively. As is well known, an arbitrary wave can be expanded in terms of a spectrum of plane waves, which can also be viewed as the eigenmodes of the wave equation. So the finite-difference methods must be stable for an arbitrary plane wave. Consequently, our derivations of the stability condition can be based on the following trial plane-wave solution,

$$v_p(\boldsymbol{x}, t) = \zeta_p \exp\left[i\left(\boldsymbol{k} \cdot \boldsymbol{x} - \omega t\right)\right] = \zeta_p \exp\left(-i\omega t\right) \exp\left[i\left(k_1 x_1 + k_2 x_2 + k_3 x_3\right)\right], \tag{A.71}$$

where $\boldsymbol{k} = k_j \boldsymbol{a}_j$ and $\{\boldsymbol{a}_1, \boldsymbol{a}_2, \boldsymbol{a}_3\}$ are three mutually perpendicular base vectors of unit length for each base vector, with $\boldsymbol{a}_3$ pointing vertically downward. If we assume that the time dependence of this solution is such that

$$v_p(\boldsymbol{x}, t + \Delta t) = \alpha v_p(\boldsymbol{x}, t) \ , \tag{A.72}$$

where α is the so-called growth factor, then, for a finite-difference scheme to be stable, we must have a growth factor $|\alpha| \leq 1$ for all the values of k_j. Our next task is to make this condition more user-friendly by explicitly writing it in terms of the P-wave velocity and timestep. By using the fact that any plane wave with a representation given in (A.71) satisfies the divergence-free condition for the propagation of P-waves, which is our concern in the mathematics of dispersion and stability conditions, (A.69) reduces to

$$\frac{1}{V_P^2} \frac{\partial^2 \boldsymbol{v}(\boldsymbol{x}, t)}{\partial t^2} = \langle \boldsymbol{\nabla}, \boldsymbol{\nabla} \rangle \, \boldsymbol{v}(\boldsymbol{x}, t) \ , \tag{A.73}$$

where the gradient operator is

$$\boldsymbol{\nabla} = \frac{\partial}{\partial x_q} \boldsymbol{a}_q \ . \tag{A.74}$$

By using the tensor notation, we can alternatively write (A.73) as

$$\frac{1}{V_P^2} \frac{\partial^2 v_r(\boldsymbol{x}, t)}{\partial t^2} = \langle \boldsymbol{a}_p, \boldsymbol{a}_q \rangle \frac{\partial^2 v_r(\boldsymbol{x}, t)}{\partial x_p \partial x_q} \ . \tag{A.75}$$

By replacing the partial differentials in (A.73) by the finite-difference operators defined in (A.54) and (A.59) and by using (A.72) to express the time finite-difference operators as a function of α, we arrive at

$$\frac{1}{(V_P \Delta t)^2} \left(\frac{\alpha^2 - 2\alpha + 1}{\alpha}\right) v_r(\boldsymbol{k}, t) = -\beta^2 v_r(\boldsymbol{k}, t) \ . \tag{A.76}$$

For the Madariaga-Virieux staggered scheme, β^2 is

$$
\beta^2 = -\sum_{a=1}^{3}\sum_{b=1}^{3} \langle \boldsymbol{a}_a, \boldsymbol{a}_b \rangle \left\{ -\frac{2i}{\Delta x_a} \sum_{m=1}^{M} c_m^{(M)} \sin\left[\left(m - \frac{1}{2}\right) k_a \Delta x_a\right] \right\}
$$

$$
\left\{ -\frac{2i}{\Delta x_b} \sum_{m=1}^{M} c_m^{(M)} \sin\left[\left(m - \frac{1}{2}\right) k_b \Delta x_b\right] \right\}
$$

$$
= 4 \sum_{a=1}^{3} \left[\left\{ \frac{1}{\Delta x_a} \sum_{m=1}^{M} c_m^{(M)} \sin\left[\left(m - \frac{1}{2}\right) k_a \Delta x_a\right] \right\} \right]^2 . \tag{A.77}
$$

In this expression, we have used the fact that the unit vectors are orthonormal. For the partially staggered grid scheme, β^2 is

$$
\beta^2 = -\sum_{a=1}^{3}\sum_{b=1}^{3} \langle \boldsymbol{a}_a, \boldsymbol{a}_b \rangle \left\{ -\frac{i}{2\Delta x_a} \sum_{q=1}^{4} e_{qa} \sum_{m=1}^{M} c_m^{(M)} \sin\left[\left(m - \frac{1}{2}\right) e_{qc} k_c \Delta x_c\right] \right\}
$$

$$
\left\{ -\frac{i}{2\Delta x_a} \sum_{q=1}^{4} e_{qa} \sum_{m=1}^{M} c_m^{(M)} \sin\left[\left(m - \frac{1}{2}\right) e_{qr} k_r \Delta x_r\right] \right\}
$$

$$
= \sum_{a=1}^{3} \left[\left\{ \frac{1}{2\Delta x_a} \sum_{q=1}^{4} e_{qa} \sum_{m=1}^{M} c_m^{(M)} \sin\left[\left(m - \frac{1}{2}\right) e_{qc} k_c \Delta x_c\right] \right\} \right]^2 . \tag{A.78}
$$

We can construct α (A.76) as follows:

$$
\alpha = \left(1 - \frac{1}{2}\beta^2 V_P^2 \Delta t^2\right) \pm \Delta t \sqrt{\beta^2 V_P^2 \Delta t^2 - 4} . \tag{A.79}
$$

As pointed out earlier, for the finite-difference algorithm to be stable, α must be smaller than 1 for all values of k; i.e., we are interested only in the upper bound for the timestep that guarantees stable numerical solutions. From (A.79) we see that this condition is satisfied if and only if

$$
\beta_{\max}^2 V_{P,\max}^2 \Delta t^2 \leq 4 , \tag{A.80}
$$

where $V_{P,\max}$ is the maximum P-wave velocity in the computational volume and $\beta_{\max}$ is the maximum value of β. By replacing V_P by V_S in (A.73), we arrive at a similar expression as in (A.80) for S-waves. Because we are interested only in an upper bound for the timestep that guarantees stable numerical solutions, we can ignore the condition associated with V_S. So our staggered-grid solutions derived earlier may diverge unless the condition in (A.80) is fulfilled. Let us emphasize that, strictly speaking, this condition is valid only under two assumptions made here; i.e., the medium is homogeneous,[1] and the time derivatives are approximated

---

[1] In general, it is not yet possible to derive stability conditions analytically for heterogeneous media. Research is ongoing to develop reliable and easy-to-use formulae for a given heterogeneous medium.

Table A.3. The sum of the absolute values of coefficients c_m^M for the second- to tenth-order of the finite-difference operators and the corresponding stability conditions for the Madariaga-Virieux staggered scheme (MVSG) and the partially staggered grid scheme (PSG) with $\eta = \Delta x / V_{P,\max}$.

M	$\sum_{k=1}^{M} \left| c_m^{(M)} \right|$	MVSG	PSG
1	1	$\Delta t \leq 0.577\eta$	$\Delta t \leq \eta$
2	7/6	$\Delta t \leq 0.495\eta$	$\Delta t \leq 0.857\eta$
3	149/120	$\Delta t \leq 0.465\eta$	$\Delta t \leq 0.805\eta$
4	69157/53760	$\Delta t \leq 0.449\eta$	$\Delta t \leq 0.777\eta$
5	6797417/5160960	$\Delta t \leq 0.438\eta$	$\Delta t \leq 0.759\eta$

by the second-order difference operator in (A.54). However, it has been shown in practice that this condition is sufficient, even for heterogeneous media.

To gain more insight into the condition in (A.80), let us consider the particular case of equal grid spacings (i.e., $\Delta x = \Delta y = \Delta z$). We can verify that the stability condition in (A.80) reduces to

$$\frac{V_{P,\max}\Delta t}{\Delta x} \leq \frac{1}{\sqrt{3}\sum_{k=1}^{M}\left|c_m^{(M)}\right|} \tag{A.81}$$

for the Madariaga-Virieux staggered-gridding configuration. For the partially staggered-gridding configuration, (A.80) reduces to

$$\frac{V_{P,\max}\Delta t}{\Delta x} \leq \frac{1}{\sum_{k=1}^{M}\left|c_m^{(M)}\right|} . \tag{A.82}$$

Notice that for a given grid spacing, Δx, the stability criterion in (A.81) for the Madariaga-Virieux staggered-gridding configuration is more stringent than the stability criterion in (A.82) for the partially staggered gridding configuration. This difference occurs because the spatial-differentiation directions of the partially staggered grid enlarge the spacing between the gridpoints compared to the spacing between gridpoints along one of the coordinate axes. Table A.3 summarizes the stability condition for various values of the order of the difference operator (i.e., M).

Let us now turn to the dispersion condition. From the derivations of the stability condition, we can deduce the dispersion by using the fact that the plane-wave solution in (A.71) includes a time dependence of the form, $\exp(-i\omega t)$. By taking the Fourier transform with respect to the time and space of (A.72) and (A.73), we arrive at

$$\frac{4}{\Delta t^2}\sin^2\left(\frac{\omega\Delta t}{2}\right)v_p(\boldsymbol{k},\omega) = V_P^2\beta^2 v_p(\boldsymbol{k},\omega) . \tag{A.83}$$

From this equation we can deduce that

$$\omega\Delta t = 2\sin^{-1}\left(\frac{\Delta t}{2}V_P\beta\right) \tag{A.84}$$

or

$$\frac{V_{\text{FDM}}}{V_P} = \frac{G}{\pi r} \sin^{-1}\left(\frac{\Delta t}{2} V_P \beta\right) , \tag{A.85}$$

where

$$r = \frac{V_P \Delta t}{\sqrt{3}} \sqrt{\frac{1}{\Delta x^2} + \frac{1}{\Delta y^2} + \frac{1}{\Delta z^2}} \tag{A.86}$$

$$G = \frac{2\pi V_{\text{FDM}}}{\sqrt{3}\,\omega} \sqrt{\frac{1}{\Delta x^2} + \frac{1}{\Delta y^2} + \frac{1}{\Delta z^2}} = \frac{2\pi}{\sqrt{3}\,k} \sqrt{\frac{1}{\Delta x^2} + \frac{1}{\Delta y^2} + \frac{1}{\Delta z^2}} \tag{A.87}$$

$$k^2 = k_x^2 + k_y^2 + k_z^2 . \tag{A.88}$$

Here $V_{\text{FDM}} = \omega/k$ is the velocity with which the wave propagates through numerical grids, G is the number of gridpoints per wavelength, r is the CFL (Courant Friedrich Lewy) number, and G/r is the number of timesteps per cycle. So the input of the dispersion analysis are V_P, Δx, Δy, Δz, $\boldsymbol{k}$, and Δt. So when the ratio V_{FDM}/V_P equals one, there is no dispersion; the velocity with which the wave propagates through numerical grids equals the medium velocity, V_P. If V_{FDM}/V_P is far from 1, dispersion errors will occur.

Let us look at the particular case in which $\Delta x = \Delta y = \Delta z$. By substituting (A.78) into (A.85), we can verify that (A.85) reduces to

$$\frac{V_{\text{FDM}}}{V_P} = \frac{1}{\pi r H} \sin^{-1}\left(r \sqrt{\sum_{a=1}^{3}\left[\sum_{m=1}^{M} c_m^{(M)} \sin\left[2\pi\left(m - \frac{1}{2}\right) H a_a\right]\right]^2}\right) , \tag{A.89}$$

for the Madariaga-Virieux scheme with $H = 1/G$ and

$$\boldsymbol{a} = \frac{\omega}{V_P}\boldsymbol{k} = [\cos(\theta)\sin(\phi), \sin(\theta)\sin(\phi), \cos(\phi)]^T . \tag{A.90}$$

By substituting (A.78) into (A.85) and using the trigometric identity

$$\sin(a)\cos(b)\cos(c) \quad = \quad \frac{1}{4}\,[\sin(a+b+c) + \sin(a+b-c) \\ + \sin(a-b+c) + \sin(a-b-c)] , \tag{A.91}$$

we can verify that (A.85) reduces to

$$\frac{V_{\text{FDM}}}{V_P} = \frac{1}{\pi r H} \sin^{-1}\left(r\left\{\left[\sum_{m=1}^{M} c_m^{(M)} \sin(A_1)\cos(A_2)\cos(A_3)\right]^2\right.\right.$$

$$+ \left[\sum_{m=1}^{M} c_m^{(M)} \sin(A_2)\cos(A_3)\cos(A_1)\right]^2$$

$$+ \left.\left.\left[\sum_{m=1}^{M} c_m^{(M)} \sin(A_3)\cos(A_1)\cos(A_2)\right]^2\right\}^{1/2}\right) , \tag{A.92}$$

with

$$A_a = A_a(m) = 2\pi \left(m - \frac{1}{2} \right) H a_a \,, \tag{A.93}$$

for partially staggered-grid schemes. Notice that we have here used the Virieux parameter H instead of G and made these equations depends on r, H, the order of the difference operator (i.e., M), and direction of propagation θ and ϕ. So the goal is to select the grid spacing such that, for a given velocity and time interval, the ratio V_{FDm}/V_P is close to 1. We can form an idea about the dispersion schemes by plotting the ratio V_{FDM}/V_P as a function of r, H, and the order of the difference operator (i.e., M) for the Madariaga-Virieux staggered-grid schemes, as example. Figure A3 show these plots. Notice that the range of variations of the CFT number, r, in these plots are not the same. In fact, r must be less than a certain constant r_{max} for the scheme to be stable. The maximum value for this number for stability in staggered grids of the Madariaga-Virieux scheme is (Saenger et al., 2000)

$$r_{max} = \sqrt{3} \sum_{m=1}^{M} \left| c_m^{(M)} \right| \,. \tag{A.94}$$

In summary, there are different ways of reducing the grid dispersion for a given finite-difference scheme. Possible solutions include minimizing H, selecting r smaller than r_{max} and using high-order spatial differentiations.

A.4.2. Free-surface Boundary Conditions

The free-surface boundary condition given in (A.47) states that the normal stresses, τ_{zz}, and the shear stresses, τ_{xz} and τ_{yz}, are zero at $z = 0$. Remember that the difference operators in (A.59) for approximating the spatial derivatives in the staggered-grid schemes described in the previous subsections require gridpoints at positions such as $z = -\frac{1}{2}\Delta z$, $z = -\Delta z$, and $z = -\frac{3}{2}\Delta z$ for the evaluation of the z-derivative of the wavefield components at $z = 0$ and even at a couple of gridpoints below $z = 0$. These positions are above the free surface and therefore outside our computational grid. That is why the evaluation of the wavefield at and near the free surface is generally treated separately from the evaluation of the wavefield at gridpoints farther inside the computational grid. We here describe the strategy proposed by Levander (1988) for computing the wavefield components at and near the free surface. This strategy is based on the assumption that the values of τ_{zz}, τ_{xz}, and τ_{yz} above the free surface are antisymmetric of the values of these stresses inside the computational grid with respect to the free surface, as follows:

$$\tau_{pz}(x, y, -z) = -\tau_{pz}(x, y, z) \,, \tag{A.95}$$

with $p = x, y, z$. The antisymmetry ensures that the stresses τ_{pz} are zero at $z = 0$. The practical implementation of this antisymmetric assumption of stresses can be carried out for the Madariaga-Virieux gridding configuration in in Figure A.2a, with

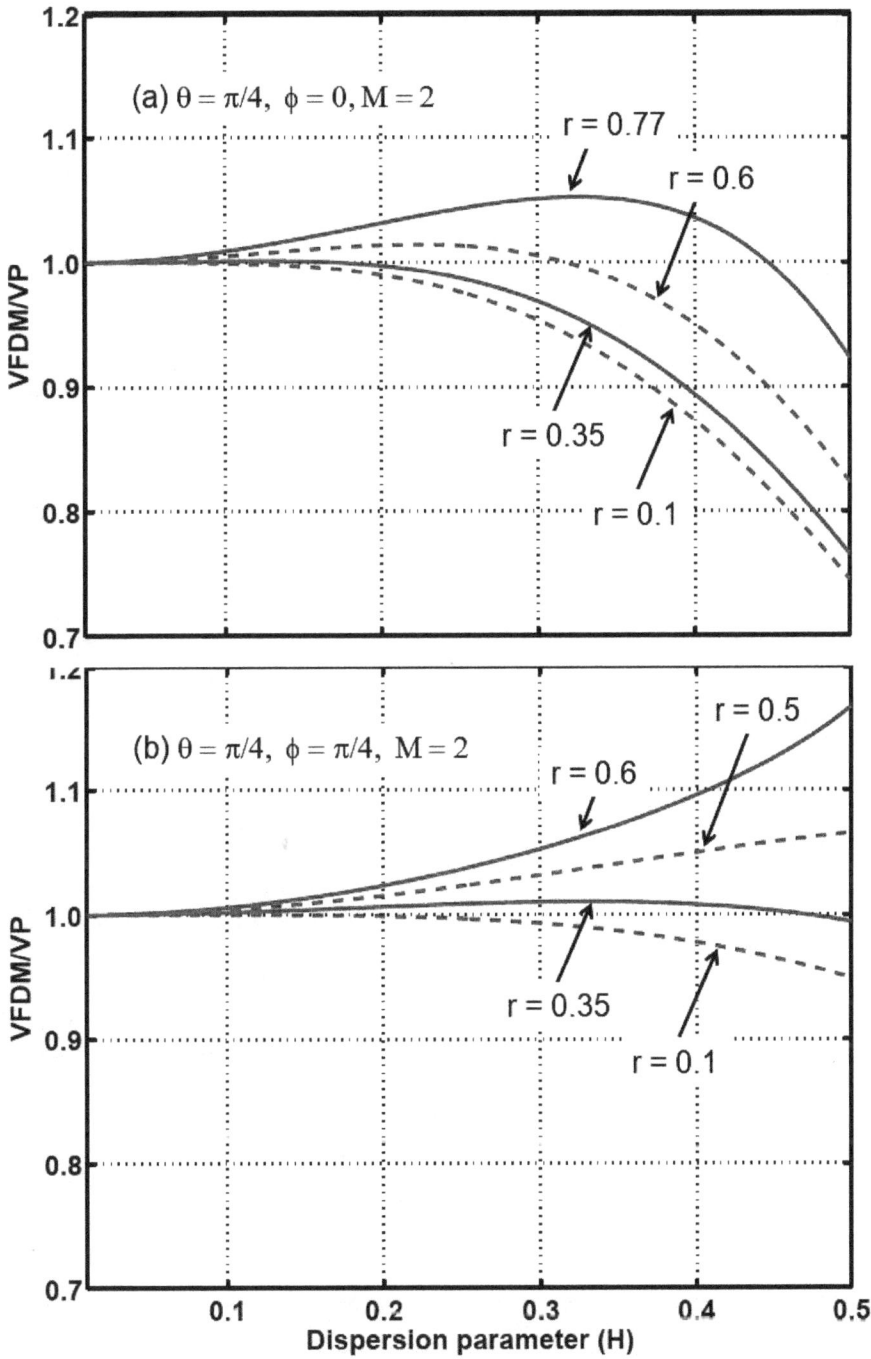

Figure A.3. *(Continued).*

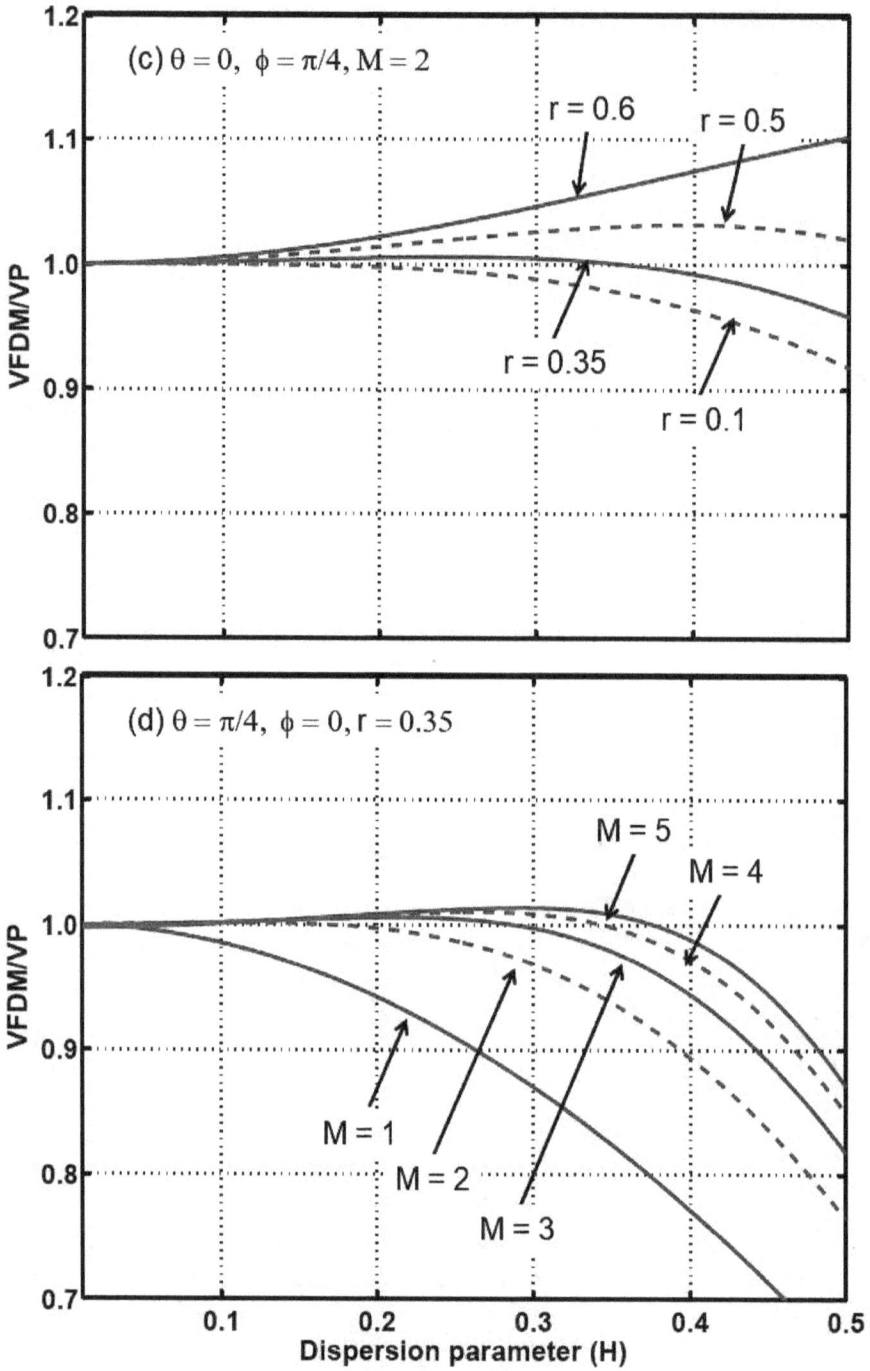

Figure A.3. *(Continued).*

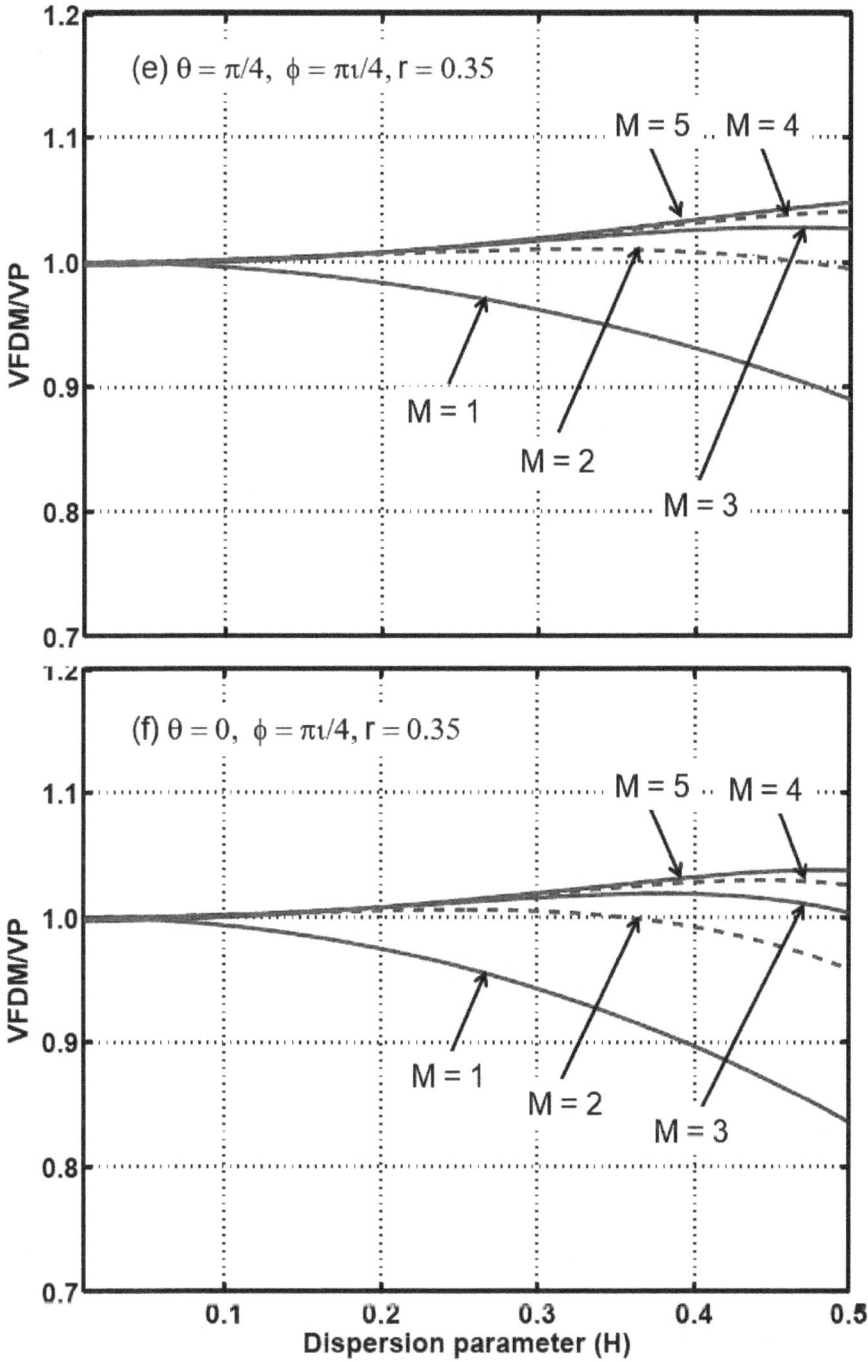

Figure A.3. Curves showing modeling errors caused by numerical dispersion for the Madariaga-Virieux staggered-grid scheme for various directions of propagation, various CFT numbers, and various orders of spatial differentiations.

$2M$th-order difference operators, as follows:

$$\begin{cases} [\tau_{zz}]^n_{i+1/2,j+1/2,-m/2} = - [\tau_{zz}]^n_{i+1/2,j+1/2,m/2} \\ [\tau_{yz}]^n_{i+1/2,j,0} = 0 \\ [\tau_{xz}]^n_{i,j+1/2,0} = 0 \\ [\tau_{yz}]^n_{i+1/2,j,-m} = - [\tau_{yz}]^n_{i+1/2,j,m} \\ [\tau_{xz}]^n_{i,j+1/2,-m} = - [\tau_{xz}]^n_{i,j+1/2,m} \end{cases} , \qquad (A.96)$$

where $m = 1, 2, .., M$. Notice that the normal stress and the shear stresses are treated differently in these equations because in Figure A.2a the shear stresses τ_{xz}, and τ_{yz} are zero at the free surface and τ_{zz} is not. The particle-velocity components with values above the free surface are also needed in updating the wavefield from one timestep to another, based on the updating equations (A.62)-(A.64) associated with the configuration in Figure A.2a. For example, $[v_z]^{n+1/2}_{i+1/2,j+1/2,0}$ can be obtained from (A.62) by using the antisymmetric values of τ_{zz} for the computation of $D_z\tau_{zz}$; $[v_x]^{n+1/2}_{i,j+1/2,-1/2}$ can be obtained from (A.62) with second-order difference operators by using the fact that τ_{xz} is zero at $z = 0$, and Hooke's law for τ_{xz}; $[v_y]^{n+1/2}_{i+1/2,j,-m/2}$ can be obtained from (A.62) with second-order-difference operators by using the fact that τ_{xz} is zero at $z = 0$, and Hooke's law for τ_{xz}, etc.

As we mentioned earlier, the construct of the Madariaga-Virieux gridding configuration in Figure A.2a is not unique. So the practical implementation of the boundary condition in (A.47) will vary with the construct. For example, for the construct in Figure A.2b, the boundary condition in (A.47) can be implemented as

$$\begin{cases} [\tau_{zz}]^n_{i,j,0} = 0 \\ [\tau_{xz}]^n_{i+1/2,j,-m/2} = - [\tau_{xz}]^n_{i+1/2,j,m/2} \\ [\tau_{yz}]^n_{i,j+1/2,-m/2} = - [\tau_{yz}]^n_{i,j+1/2,m/2} \\ [\tau_{zz}]^n_{i,j,-m} = - [\tau_{zz}]^n_{i,j,m} \end{cases} , \qquad (A.97)$$

where $m = 1, 2, .., M$.

The implementation of the boundary conditions in the case of staggered grids with multiple-momentum equations is more straightforward:

$$\begin{cases} [\tau_{pz}]^n_{i+1/2,j,0} = [\tau_{pz}]^n_{i,j+1/2,0} = 0 \\ [\tau_{pz}]^n_{i,j,-m/2} = - [\tau_{pz}]^n_{i,j,m/2} \\ [\tau_{pz}]^n_{i,j+1/2,-m/2} = - [\tau_{pz}]^n_{i,j+1/2,m/2} \\ [\tau_{pz}]^n_{i+1/2,j+1/2,-m/2} = - [\tau_{pz}]^n_{i+1/2,j+1/2,m/2} \end{cases} . \qquad (A.98)$$

Based on these conditions, the updating equations can be used with further modification. Moreover, we can notice that computing all stress components on all cell walls leads to an exact satisfaction of boundary conditions, which is not possible with the usual staggered gridding, as we noted earlier.

A.4.3. Absorbing Boundary Conditions

Another major requirement of finite-difference modeling is the introduction of absorbing boundaries in the finite-difference code to accommodate for the fact that the

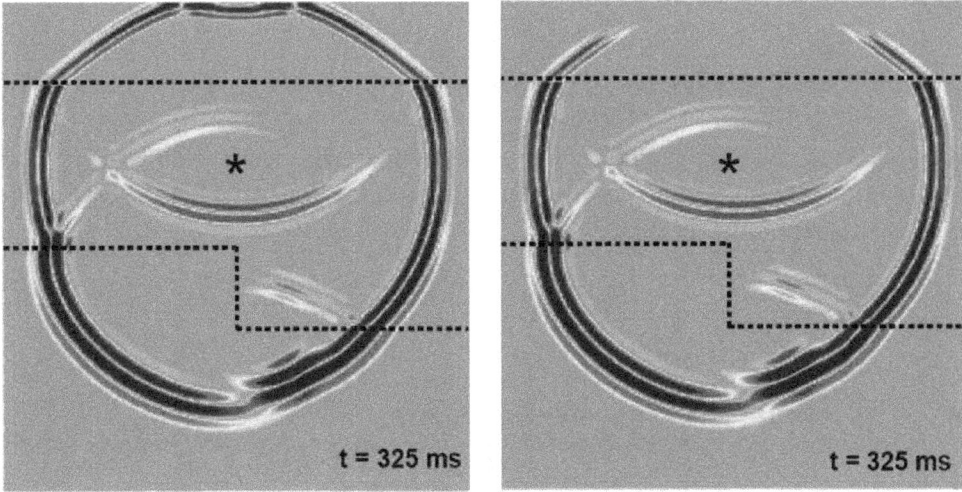

Figure A.4. (left) A snapshot of wave propagation through a medium containing a free surface. We have used the staggered finite-difference modeling technique for these computations. (right) The same snapshot with an absorbing boundary condition instead of a free surface. We have used the Cerjan's absorbing boundary condition. The dotted lines indicate the interfaces.

subsurface is a half-space with infinite lateral boundaries. The absorbing boundaries consist of an additional band of gridpoints around the actual computational domain capable of absorbing the incoming waves. As illustrated in Figure A.4, the absorbing boundary conditions can also be used for generating data without free-surface reflections by replacing the free surface with an absorbing boundary. We here selected to adapt the very effective perfectly matched layer (PML) absorbing conditions for our finite-difference solution.

To get some insight into the idea behind the PML technique, let us start by describing the PML solution for homogeneous media. In a homogeneous and isotropic medium, the solutions to equations (A.43) and (A.44), when the source terms I_{jk} and F_j are zero, is a plane wave of the form

$$v(x, \omega) = A \exp\left[-i\omega\left(t - \frac{k}{\omega}x\right)\right],$$ (A.99)

where A represents the amplitude and polarization of the plane wave, and $k = [k_1, k_2, k_3]^T$ is its wave vector. The objective of the PML method is to modify equations (A.43) and (A.44), such that in the absorbing region along the x-axis, for example, the solution takes the form

$$v(x, \omega) = A \exp\left[-i\omega\left(t - \frac{k}{\omega}x\right) - \gamma_1 x_1\right],$$ (A.100)

with $\gamma_1 > 0$. In other words, we want to have plane waves that decay exponentially in the absorbing regions (Berenger, 1994; Chew and Liu, 1996). We can attain this objective by introducing a new complex-valued variable, $\tilde{x}$, that we define as follows:

$$\tilde{x}_b = x_b + \frac{i}{\omega} \int_0^{x_b} d_b(\eta)d\eta \quad (b = 1, 2, 3) \ , \tag{A.101}$$

where $d_b(x_b)$ is the decay factor. We assume that the decay factor is zero [i.e., $d_b(x_b) = 0$] inside the main computational domain and positive [i.e., $d_b(x_b) = 0$] in the absorbing region. The typical form of the decay factor is

$$d_b(x_b) = \left(\frac{x_a}{L}\right)^2 d_0 \tag{A.102}$$

with

$$d_0 = \frac{3V_{P,\max}\log(R_c)}{2L} \ , \tag{A.103}$$

where L is the thickness of the PML layer and R_c is the target theoretical reflection coefficient, generally chosen as 0.001.

By writing the spatial derivatives of the stresses and the particle-velocity equations (A.43) and (A.44), as follows:

$$\frac{\partial}{\partial \tilde{x}_b} = \frac{1}{\varepsilon_b}\frac{\partial}{\partial x_b} \ , \tag{A.104}$$

with

$$\varepsilon_b(x_b, \omega) = 1 + i\frac{d_b(x_b)}{\omega} \ , \tag{A.105}$$

and by taking the Fourier transform of equations (A.43) and (A.44) with respect to time, we arrive at

$$-i\omega v_a(\boldsymbol{x}, \omega) = \sigma(\boldsymbol{x})\frac{1}{\varepsilon_b(x_b, \omega)}\frac{\partial \tau_{ab}(\boldsymbol{x}, \omega)}{\partial x_b} \tag{A.106}$$

$$-i\omega \tau_{\alpha\beta}(\boldsymbol{x}, \omega) = c_{\alpha\beta\eta\zeta}(\boldsymbol{x})\frac{1}{\varepsilon_\eta(x_\eta, \omega)}\frac{\partial v_\zeta(\boldsymbol{x}, \omega)}{\partial x_\eta} \ . \tag{A.107}$$

Notice that rather than defining a new symbol to express this physical quantityafter it has been Fourier-transformed, we have used the same symbol with different arguments, as the context unambiguously indicates the quantity currently under consideration. For example, $v_a(\boldsymbol{x}, \omega)$ is the Fourier transform of $v_a(\boldsymbol{x}, t)$ with respect to time. We will use this convention for the remaining part of this section. Finally, notice that we have not included the source terms here because they are generally zero in the absorbing regions.

To facilitate the algebra of the PML equations in (A.106) and (A.107), we found it convenient to redefine the stress field so that the system of equations in (A.43) and (A.44) and the ones in (A.106) and (A.107) can have the same mathematical structures. In other words, we can rewrite equations (A.106) and (A.106) as follows:

$$-i\omega v_a(\boldsymbol{x}, \omega) = \frac{1}{\varepsilon_1\varepsilon_2\varepsilon_3}\sigma(\boldsymbol{x})\frac{\partial \tilde{\tau}_{ab}(\boldsymbol{x}, \omega)}{\partial x_b} \tag{A.108}$$

$$-i\omega\tilde{\tau}_{\alpha\beta}(\boldsymbol{x},\omega) = \frac{\varepsilon_1\varepsilon_2\varepsilon_3}{\varepsilon_\beta\varepsilon_\eta}c_{\alpha\beta\zeta\eta}(\boldsymbol{x})\frac{\partial v_\zeta(\boldsymbol{x},\omega)}{\partial x_\eta} \,, \tag{A.109}$$

with

$$\tilde{\tau}_{ab}(\boldsymbol{x},\omega) = \frac{\varepsilon_1\varepsilon_2\varepsilon_3}{\varepsilon_b}\tau_{ab}(\boldsymbol{x},\omega) \,. \tag{A.110}$$

Note that the tensor $\tilde{\boldsymbol{\tau}}$ is nonsymmetric (i.e., $\tilde{\tau}_{ab} \neq \tilde{\tau}_{ba}$). We can also notice that $\tilde{\boldsymbol{\tau}} = \boldsymbol{\tau}$ across the boundary between the non-PML region and the PML region, which means that $\tilde{\boldsymbol{\tau}}$ is just a continuation of $\boldsymbol{\tau}$ in the PML region. Therefore, we can replace $\tilde{\boldsymbol{\tau}}$ by $\boldsymbol{\tau}$ in equations (A.108)-(A.109) and arrive at the following PML equations:

$$-i\omega v_a(\boldsymbol{x},\omega) = \frac{1}{\varepsilon_1\varepsilon_2\varepsilon_3}\sigma(\boldsymbol{x})\frac{\partial \tau_{ab}(\boldsymbol{x},\omega)}{\partial x_b} \tag{A.111}$$

$$-i\omega\tau_{\alpha\beta}(\boldsymbol{x},\omega) = \frac{\varepsilon_1\varepsilon_2\varepsilon_3}{\varepsilon_\beta\varepsilon_\eta}c_{\alpha\beta\zeta\eta}(\boldsymbol{x})\frac{\partial v_\zeta(\boldsymbol{x},\omega)}{\partial x_\eta} \,. \tag{A.112}$$

Because our finite-difference modeling is carried out in the time domain, we have to transform equations (A.111)-(A.112) back to the time domain by taking their inverse Fourier transforms with respect to time. Due to the frequency dependence of the coefficients $\varepsilon_j = \varepsilon_j(x_j,\omega)$, time-convolution operators will appear in the resulting equations. These time-convolution operators must be eliminated or modified to accommodate for the fact that our finite-difference solution here is recursive in time (i.e., the computation of the wavefield values at a given time depends explicitly on the wavefield values at earlier times), but the standard time-convolution operations are not recursive. They are summations over time. One way to eliminate the time-convolution operators from the PML equations is to factorize equations (A.111)-(A.112) into new equations, with only one factor ε_b per equation. To illustrate this approach, let us consider the following expression:

$$w_{ab}(\boldsymbol{x},\omega) = c_b(x_b,\omega)v_a(\boldsymbol{x},\omega) - \left[1 + i\frac{d_b(x_b)}{\omega}\right]v_a(\boldsymbol{x},\omega) \,, \tag{A.113}$$

which is a convolution in the time domain. By multiplying this expression by $-i\omega$, we eliminate the convolution operator; i.e.,

$$-i\omega w_{ab}(\boldsymbol{x},\omega) = -i\omega v_a(\boldsymbol{x},\omega) + d_b(x_b)v_a(\boldsymbol{x},\omega) \,, \tag{A.114}$$

which corresponds to

$$\frac{\partial w_{ab}(\boldsymbol{x},t)}{\partial t} = \frac{\partial v_a(\boldsymbol{x},t)}{\partial t} + d_b(x_b)v_a(\boldsymbol{x},t) \,. \tag{A.115}$$

This idea was proposed by Oskooi and Johnson (2011) for the PML solutions of Maxwell's equations. To extend this idea to the elastic case, we need to factorize equations (A.111)-(A.112). Whereas the factorization of (A.111) is obvious, that of (A.112) requires more attention because of the presence of the ratio of damping factors in this equation. Actually, this ratio is a product of two damping factors; i.e.,

$$\frac{\varepsilon_1 \varepsilon_2 \varepsilon_3}{\varepsilon_\eta} = \varepsilon'_\eta \varepsilon''_\eta = \begin{cases} \varepsilon_2 \varepsilon_3 & \text{if } \eta = 1 \\ \varepsilon_3 \varepsilon_1 & \text{if } \eta = 2 \\ \varepsilon_1 \varepsilon_2 & \text{if } \eta = 3 \end{cases} , \tag{A.116}$$

with $\varepsilon'_\eta \equiv \{\varepsilon_2, \varepsilon_3, \varepsilon_1\}$ and $\varepsilon''_\eta \equiv \{\varepsilon_3, \varepsilon_1, \varepsilon_2\}$. This idea was proposed by Oskooi and Johnson (2011) for the PML solutions of Maxwell's equations. By using equations (A.111)-(A.112), we can factorize equations (A.111) and (A.112) as follows:

$$-i\omega \varepsilon_1 w_{a1}(\boldsymbol{x}, \omega) = \sigma(\boldsymbol{x}) \frac{\partial \tau_{ab}(\boldsymbol{x}, \omega)}{\partial x_b}) , \tag{A.117}$$

$$-i\omega \varepsilon_2 w_{a2}(\boldsymbol{x}, \omega) = -i\omega w_{a1}(\boldsymbol{x}, \omega) , \tag{A.118}$$

$$-i\omega \varepsilon_3 v_a(\boldsymbol{x}, \omega) = -i\omega w_{a2}(\boldsymbol{x}, \omega) , \tag{A.119}$$

$$-i\omega e_{\zeta\eta;1}(\boldsymbol{x}, \omega) = -i\omega \varepsilon'_\eta \frac{\partial v_\zeta(\boldsymbol{x}, \omega)}{\partial x_\eta} , \tag{A.120}$$

$$-i\omega e_{\zeta\eta;2}(\boldsymbol{x}, \omega) = -i\omega \varepsilon''_\eta e_{\zeta\eta;1}(\boldsymbol{x}, \omega) , \tag{A.121}$$

$$-i\omega \varepsilon_\beta \tau_{\alpha\beta}(\boldsymbol{x}, \omega) = c_{\alpha\beta\zeta\eta}(\boldsymbol{x}) e_{\zeta\eta;2}(\boldsymbol{x}, \omega) . \tag{A.122}$$

By taking the Fourier transform of these equations with respect to time, we can deduce the PML equations in the time domain; i.e.,

$$\frac{\partial w_{a1}(\boldsymbol{x}, t)}{\partial t} + d_1(x_1) w_{a1}(\boldsymbol{x}, t) = \sigma(\boldsymbol{x}) \frac{\partial \tau_{ab}(\boldsymbol{x}, \omega)}{\partial x_b}) , \tag{A.123}$$

$$\frac{\partial w_{a2}(\boldsymbol{x}, t)}{\partial t} + d_2(x_2) w_{a2}(\boldsymbol{x}, t) = \frac{\partial w_{a1}(\boldsymbol{x}, t)}{\partial t} , \tag{A.124}$$

$$\frac{\partial v_a(\boldsymbol{x}, t)}{\partial t} + d_3(x_3) v_a(\boldsymbol{x}, t) = \frac{\partial w_{a2}(\boldsymbol{x}, t)}{\partial t} , \tag{A.125}$$

$$\frac{\partial e_{\zeta\eta;1}(\boldsymbol{x}, t)}{\partial t} = \frac{\partial v_\zeta(\boldsymbol{x}, t)}{\partial t \partial x_\eta} + d'_\eta \frac{\partial v_\zeta(\boldsymbol{x}, t)}{\partial x_\eta} , \tag{A.126}$$

$$\frac{\partial e_{\zeta\eta;2}(\boldsymbol{x}, t)}{\partial t} = \frac{\partial e_{\zeta\eta;1}(\boldsymbol{x}, t)}{\partial t} + d''_\eta e_{\zeta\eta;1}(\boldsymbol{x}, t) , \tag{A.127}$$

$$\frac{\partial \tau_{\alpha\beta}(\boldsymbol{x},t)}{\partial t} + d_\beta(x_\beta)\tau_{\alpha\beta}(\boldsymbol{x},t) = c_{\alpha\beta\zeta\eta}(\boldsymbol{x})e_{\zeta\eta;2}(\boldsymbol{x},t) , \qquad (A.128)$$

where $d' = [d_2(x_2), d_3(x_3), d_1(x_1)]^T$ and $d'' = [d_3(x_3), d_1(x_1), d_2(x_2)]^T$. These equations can be solved by using the finite-difference operators in (A.54) and (A.59), just as we did earlier in the non-PML equations in (A.43) and (A.44).

Let us now look at a numerical example of PML boundary conditions. We start by recalling the absorbing boundaries introduced by Cerjan et al. (1985), which was until recently the preferred method. The reason why we are recalling this solution is that the contrast between this Cerjan's solution and the PML solution provides a good indication of the accuracy and effectiveness of the PML solution. The Cerjan's boundary conditions, also known as the damping boundary conditions, are created by surrounding the numerical model with a strip of grids. The stress and particle-velocity fields are multiplied by the factor

$$G(i) = \exp\left\{ -\left[\frac{\alpha}{iabmax}(iabmax - i) \right]^2 \right\} \quad \text{for } 1 \le i \le iabmax , \qquad (A.129)$$

where $iabmax$ is the strip width in gridpoints and α is a constant determined by trial and error for the optimal absorbing-boundary conditions. For $iabmax = 60$ gridpoints, the optimum value of α is 0.3. The results in Figure A.4 are based on these values. The model in this example is 2D with a line source.

Let us now turn to the PML boundary conditions. First of all, we note that in a 2D medium with a line source, the equations (A.123)-(A.128) reduce to

$$\frac{\partial w_1(\boldsymbol{x},t)}{\partial t} + d_1(x_1)w_1(\boldsymbol{x},t) = \sigma(\boldsymbol{x})\left[\frac{\partial \tau_{11}(\boldsymbol{x},t)}{\partial x_1} + \frac{\partial \tau_{13}(\boldsymbol{x},t)}{\partial x_3} \right] \qquad (A.130)$$

$$\frac{\partial w_3(\boldsymbol{x},t)}{\partial t} + d_1(x_1)w_2(\boldsymbol{x},t) = \sigma(\boldsymbol{x})\left[\frac{\partial \tau_{31}(\boldsymbol{x},t)}{\partial x_1} + \frac{\partial \tau_{33}(\boldsymbol{x},t)}{\partial x_3} \right] \qquad (A.131)$$

$$\frac{\partial v_1(\boldsymbol{x},t)}{\partial t} + d_3(x_3)v_1(\boldsymbol{x},t) = \frac{\partial w_1(\boldsymbol{x},t)}{\partial t} \qquad (A.132)$$

$$\frac{\partial v_3(\boldsymbol{x},t)}{\partial t} + d_3(x_3)v_3(\boldsymbol{x},t) = \frac{\partial w_3(\boldsymbol{x},t)}{\partial t} \qquad (A.133)$$

$$\frac{\partial e_{11}(\boldsymbol{x},t)}{\partial t} = \frac{\partial}{\partial t}\left[\frac{\partial v_1(\boldsymbol{x},t)}{\partial x_1} \right] + d_3(x_3)\left[\frac{\partial v_1(\boldsymbol{x},t)}{\partial x_1} \right]$$

$$\frac{\partial e_{13}(\boldsymbol{x},t)}{\partial t} = \frac{\partial}{\partial t}\left[\frac{\partial v_1(\boldsymbol{x},t)}{\partial x_3} \right] + d_1(x_1)\left[\frac{\partial v_1(\boldsymbol{x},t)}{\partial x_3} \right]$$

$$\frac{\partial e_{31}(\boldsymbol{x},t)}{\partial t} = \frac{\partial}{\partial t}\left[\frac{\partial v_3(\boldsymbol{x},t)}{\partial x_1} \right] + d_3(x_3)\left[\frac{\partial v_3(\boldsymbol{x},t)}{\partial x_1} \right] \qquad (A.134)$$

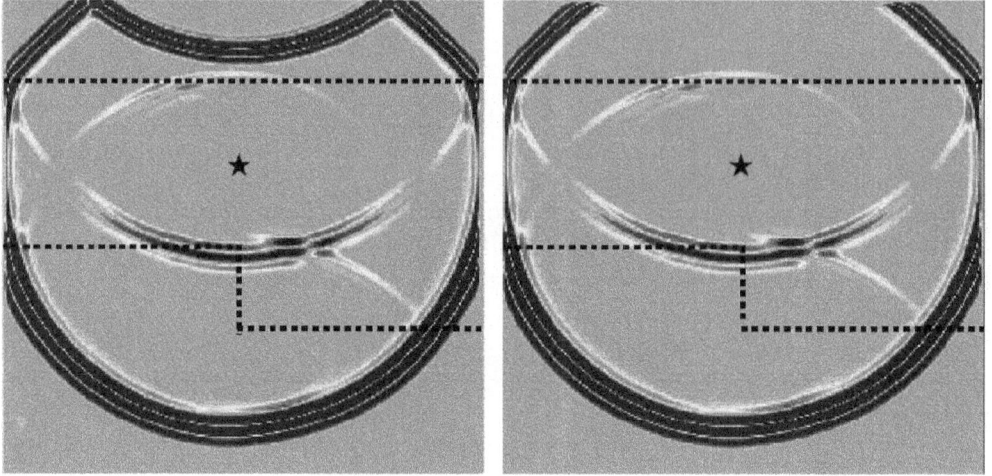

Figure A.5. (left) A snapshot of wave propagation through a medium containing a free surface. We have used the staggered finite-difference modeling technique for these computations. (right) The same snapshot with an absorbing boundary condition instead of a free surface. We have used the PML's absorbing boundary condition. The dotted lines here indicate the interfaces.

$$\frac{\partial e_{33}(\boldsymbol{x},t)}{\partial t} = \frac{\partial}{\partial t}\left[\frac{\partial v_3(\boldsymbol{x},t)}{\partial x_3}\right] + d_1(x_1)\left[\frac{\partial v_3(\boldsymbol{x},t)}{\partial x_3}\right] \tag{A.135}$$

$$\frac{\partial \tau_{jk}(\boldsymbol{x},t)}{\partial t} + d_k(x_k)\tau_{jk}(\boldsymbol{x},t) = c_{jkpq}(\boldsymbol{x})e_{pq}(\boldsymbol{x},t) \ . \tag{A.136}$$

Ikelle and Amundsen (2018) provides the Matlab implementation of these equations in a Madariaga-Virieux scheme. Figure A.5 shows the application of the PML scheme based on this code. The absorbing length is only 10 gridpoints. We can see how accurate the PML boundary conditions are, even for a strip of 10 gridpoints, especially when they are compared to the results of damping boundary conditions in Figure A.4. The key steps in the 3D finite-difference staggered-grid modeling algorithms described in this section are given in Figure A.6. The large size of the 3D models is still one of the impediments in the running of 3D FDM.

A.5.   Modeling of the Schrödinger Equation

A.5.1.   Fimite-Difference Scheme

As described in Chapter 4, the time evolution of a particle of mass m, described by a wave function $\Psi(\boldsymbol{x},t) = \Psi_R(\boldsymbol{x},t) + i\Psi_I(\boldsymbol{x},t)$, is

$$-\frac{\hbar^2}{2m}\boldsymbol{\nabla}^2\Psi(\boldsymbol{x},t) + V(\boldsymbol{x})\Psi(\boldsymbol{x},t) = i\hbar\frac{\partial\Psi(\boldsymbol{x},t)}{\partial t} \ . \tag{A.137}$$

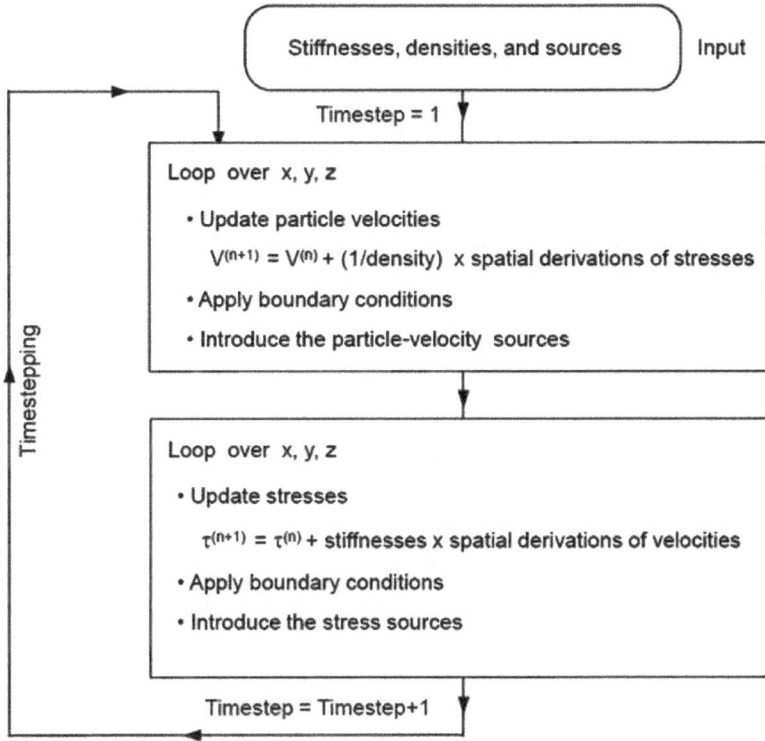

Figure A.6. A flowchart of the main steps of a numerical modeling in the Cartesian coordinates.

where $V(\boldsymbol{x})$ is the potential energy profile. We assume that the potential V does not depend on time. Using the centered differences to discretize time derivatives and Laplacian operators, one obtains

$$
\hbar \frac{[\Psi_R]_{i,j,k}^{n+1} - [\Psi_R]_{i,j,k}^{n}}{\Delta t} = -\frac{\hbar^2}{2m} \left[\frac{[\Psi_I]_{i+1,j,k}^{n+1/2} - 2[\Psi_I]_{i,j,k}^{n+1/2} + [\Psi_I]_{i-1,j,k}^{n+1/2}}{(\Delta x)^2} \right.
$$

$$
+ \frac{[\Psi_I]_{i,j+1,k}^{n+1/2} - 2[\Psi_I]_{i,j,k}^{n+1/2} + [\Psi_I]_{i,j-1,k}^{n+1/2}}{(\Delta y)^2}
$$

$$
+ \left. \frac{[\Psi_I]_{i,j,k+1}^{n+1/2} - 2[\Psi_I]_{i,j,k}^{n+1/2} + [\Psi_I]_{i,j,k-1}^{n+1/2}}{(\Delta z)^2} \right]
$$

$$
+ [V]_{i,j,k} [\Psi_I]_{i,j,k}^{n+1/2} \tag{A.138}
$$

$$\hbar\frac{[\Psi_I]_{i,j,k}^{n+1/2} - [\Psi_I]_{i,j,k}^{n-1/2}}{\Delta t} = \frac{\hbar^2}{2m}\left[\frac{[\Psi_R]_{i+1,j,k}^n - 2[\Psi_R]_{i,j,k}^n + [\Psi_R]_{i-1,j,k}^n}{(\Delta x)^2}\right.$$

$$+ \frac{[\Psi_R]_{i,j+1,k}^{n+1/2} - 2[\Psi_R]_{i,j,k}^n + [\Psi_R]_{i,j-1,k}^n}{(\Delta y)^2}$$

$$+ \left.\frac{[\Psi_R]_{i,j,k+1}^n - 2[\Psi_R]_{i,j,k}^n + [\Psi_R]_{i,j,k-1}^n}{(\Delta z)^2}\right]$$

$$- [V]_{i,j,k}[\Psi_R]_{i,j,k}^n \qquad (A.139)$$

Using the same derivation as the previous section, we ensure that the scheme is stable by selecting the time step as follows:

$$\Delta t < 2\left[\frac{2\hbar}{m}\left(\frac{1}{(\Delta x)^2} + \frac{1}{(\Delta y)^2} + \frac{1}{(\Delta z)^2}\right) + \frac{\max_{i,j,k}[|V|]_{i,j,k}}{\hbar}\right]^{-1}. \qquad (A.140)$$

We can again apply the PML boundary condition.

A.5.2. Time-Space Solution in 1D Case

The Schrödinger equation for a one-dimensional quantum system is given as

$$-\frac{\hbar^2}{2m}\frac{d^2\psi}{dx^2} + V(x)\psi(x) = U\psi(x). \qquad (A.141)$$

For simplicity, we choose our system of units such that the particle mass m is the unit of mass, and $\hbar$ is also the unit of angular momentum. Furthermore, we assume that the potential V does not depend on time. The reduced Schrödinger equation is

$$-\frac{1}{2}\frac{d^2\Psi(x,t)}{dx^2} + V(x,t)\Psi(x,t) = i\frac{\partial\Psi(x,t)}{\partial t}. \qquad (A.142)$$

The Schrödinger equation is effectively a reaction-diffusion equation. Using the central difference for the second order derivative, the Laplace operator can be calculated as

$$\nabla^2\Psi(x,t) \approx \frac{\Psi(x+\Delta x,t) - 2\Psi(x-\Delta x,t) + \Psi(x+t)}{\Delta x^2} \qquad (A.143)$$

we can verify (A.142) reduces to

$$\frac{\partial\Psi(x,t)}{\partial t} = f(x,\Psi), \qquad (A.144)$$

where

$$f(x,t) = i\left[\frac{\Psi(x+\Delta x,t) - 2\Psi(x-\Delta x,t) + \Psi(x+t)}{2\Delta x^2} - V(x)\Psi(x,t)\right] \qquad (A.145)$$

Now let's turn our attention to time-evolution. Then $\Psi(x, t+\Delta t)$ can be computed on each point of the grid using a Runge Kutta of 4th order method; i.e.,

$$\Psi(x, t + \Delta t) = \Psi(x, t) + \frac{\Delta t}{6} \left(k_1 + 2k_2 + 2k_3 + k_4\right) , \qquad (A.146)$$

where

$$k_1 = f(x, \Psi) , \quad k_2 = f\left(\Psi + \frac{\Delta t}{2} k_1\right) ,$$

$$k_3 = f\left(\Psi + \frac{\Delta t}{2} k_2\right) , \quad \text{and} \quad k_4 = f\left(\Psi + \Delta t\, k_3\right) . \qquad (A.147)$$

One can use the Matlab function `ode45` for numerical implementation. For cases where the Hamiltonian is space independent, the algorithm reduces to the Runge Kutta application.

A.5.3. Time-Independent Solution in the 1D Case

As discussed in Chapter 4, in this case Schröndinger equation reduces to an eigenfunction equation

$$\hat{H}\psi(x) = U\psi(x) . \qquad (A.148)$$

After sampling, $\psi(x)$ becomes an eigenvector and U becomes an eigenvalue. More explicitly, we have

$$-\frac{\hbar^2}{2m} \frac{d^2\psi}{dx^2} + V(x)\psi(x) = U\psi(x) . \qquad (A.149)$$

For simplicity, we use units like $m = 1$ and $\hbar = 1$. Hence our equation becomes

$$-\frac{1}{2} \frac{d^2\psi}{dx^2} + V(x)\psi(x) = U\psi(x) . \qquad (A.150)$$

Using the second-order centered derivative formula, we can discretize (A.150) as follows:

$$-\frac{1}{2}\left(\frac{\psi_{j+1} - 2\psi_j + \psi_{j-1}}{\Delta x^2}\right) + V_j\psi_j = U\psi_j \qquad (A.151)$$

or

$$\left[\hat{T}_{ij} + \hat{V}_{ij}\right]\psi_j = U\psi_i , \qquad (A.152)$$

where

$$\hat{T} = -\frac{1}{2\,(\Delta x)^2} \begin{pmatrix} -2 & 1 & 0 & \cdots & \cdots & 0 \\ 1 & -2 & 1 & \cdots & \cdots & 0 \\ 0 & 1 & -2 & \cdots & \cdots & 0 \\ \vdots & \vdots & \vdots & \ddots & & \vdots \\ \vdots & \vdots & \vdots & & \ddots & 1 \\ 0 & 0 & 0 & \cdots & 1 & -2 \end{pmatrix} , \qquad (A.153)$$

$$\hat{V} = \begin{pmatrix} V(x_1) & 0 & \dots & \dots & \dots & 0 \\ 0 & V(x_2) & 0 & \dots & \dots & 0 \\ \vdots & 0 & \ddots & & & \vdots \\ \vdots & \vdots & & \ddots & & \vdots \\ \vdots & \vdots & & & \ddots & 0 \\ 0 & 0 & \dots & \dots & 0 & V(x_N) \end{pmatrix} . \tag{A.154}$$

Appendix B

Answers to Some of the Quizzes and Exercises

Chapter 1:

Suppose it takes a planet 3.09 days to make one orbit (assumed to be circular) around the Sun. The distance of this planet from our Sun is 6.433×10^6 km. (1) How fast is this planet moving? (2) What is the mass of the planet?

(1) The orbital speed of the planet is

$$T = \frac{2\pi r}{v} \implies v = \frac{2\pi r}{T} = \frac{2\pi \times 6.433 \times 10^9 \,\mathrm{m}}{3.09 \times 86400 \,\mathrm{s}} = 151 \,\mathrm{km/s}.$$

(2) Using the fact that

$$v = \sqrt{\frac{Gm_1}{r}} \implies m_1 = \frac{4\pi^2 r^3}{GT^2}$$

$$= \frac{4\pi^2 \times (6.433 \times 10^9 \,\mathrm{m})^3}{(6.67 \times 10^{-11} \,\mathrm{N\,m^2\,kg^{-2}}) \times (3.09 \times 86400 \,\mathrm{s})^2}$$

$$= 2.21 \times 10^{30} \,\mathrm{kg}$$

What is the force of Earth's gravity on a 100-kg man at the to the top of Mount Everest (h = 8850 m)?

$$F = \frac{G \, m_{\mathrm{earth}} \, m_{\mathrm{human}}}{(r_{\mathrm{earth}} + h)^2}$$

$$= \frac{(6.67 \times 10^{-11}) \, (6 \times 10^{24} \,\mathrm{kg}) \, (100 \,\mathrm{kg})}{(6.4 \times 10^6 + 8850)^2 \,\mathrm{m^2}}$$

$$= 974.354 \,\mathrm{N}$$

The distances from the Sun to the planets and the orbital periods of planets with respect to the Sun are the key parameter in the analysis of our solar system. These parameters

DOI: 10.1201/9781032620619-B

increase with the planet distance to the Sun (e.g., orbital periods of Mercury, Venus, Mars, Jupiter, Saturn, Uranus, and Neptune compared with that of the Earth are 0.241, 0.615, 1.88, 11.9, 29.4, 83.80, 164, 248, respectively). However, the gravities, masses, and radii of the planets do not increase or decrease with the planet distance to the Sun (e.g., gravities at surface of Mercury, Venus, Mars, Jupiter, Saturn, Uranus, and Neptune compared with that of the Earth are 0.378, 0.894, 0.379, 2.54, 1.07, 0.8, 1.2, and 0.059, respectively), why?

Because there is no relationship between orbital distance and density; the order of our planets, from densest to lightest is Earth, Mercury, Venus, Mars, Neptune, Jupiter, Uranus, and Saturn. The current planetary positions are not necessarily where they were formed, so there might have been a more orderly sorting in the early solar system. The planetary composition of the gas giants is clearly different from the rocky planets. Their size is also dramatically different for two reasons. First, the original planetary nebula contained more gasses and ices than metals and rocks. There was abundant hydrogen, carbon, oxygen, nitrogen, and less silicon and iron, giving the outer planets more building material. Second, the stronger gravitational pull of these giant planets allowed them to collect large quantities of hydrogen and helium, which could not be collected by weaker gravity of the smaller planets. Jupiter's massive gravity further shaped the solar system and growth of the inner rocky planets. As the nebula started to coalesce into planets, Jupiter's gravity accelerated the movement of nearby materials, generating destructive collisions rather than constructively gluing material together. These collisions created the asteroid belt, located between Mars and Jupiter (see Chapter 1). This asteroid belt is the source of most meteorites that currently impact the Earth. Study of asteroids and meteorites help us to determine the age of Earth and the composition of its core, mantle, and crust. Jupiter's gravity may also explain Mars's smaller mass, with the larger planet consuming material, as it migrated from the inner to outer edge of the solar system. The rarity of the heavy elements partly explains the limited size of the terrestrial planets.

(a) What is the eccentricity of the Earth's orbit? (b) Show that the equation of an ellipse can be also written as

$$r = \frac{a(1 - e^2)}{1 + e\cos\theta} \ .$$

where r and θ are the distance and angle, respectively, as defined in Figure 1.12.

The measured distances semi-major and semi-minor axes give for the eccentricity of the Earth's orbit $e = a^2 - b^2)/a^2 = 0.01674$. Use $x = r\cos\theta$ and $y = r\cos\theta$ to obtain the above equation of an ellipse.

The apparent weight w of an object depends on the latitude θ at which it is measured. In contrast, the true weight is entirely due to the Earth's gravity and its magnitude equals $w = mg_e$ and is the same everywhere (m is the mass of the object). (a) If the tangential speed of a point on the surface of the Earth at the equator is v_θ, then find the apparent weight of the object at the equator. (b) What is the apparent gravitational acceleration at the equator?

(a) Remember that

$$v = \frac{dx}{dt} = \frac{dr}{dt}i_r + r\frac{di_r}{d\theta}\frac{d\theta}{dt} = \frac{dr}{dt}i_r + r\frac{d\theta}{dt}i_\theta$$

At the equator v_θ is the component of velocity along i_θ; i.e.,

$$v_\theta = R\frac{d\theta}{dt}$$

where R is the Earth radius. The weight at the equator is

$$w = mg_e - \frac{mv_\theta^2}{R} = m\left[g_e - R\left(\frac{d\theta}{dt}\right)^2\right]$$

where $\omega = d\theta/dt$ is known as angular speed of the earth. (b) The apparent gravitational acceleration at the equator is

$$g_{eq} = g_e - R\omega^2 = 9.81\,\text{m/s}^2 - \left(\frac{2\pi}{86164\,\text{s}}\right) \times 0.637 \times 10^6\,\text{m} = 9.78\,\text{m/s}^2\ .$$

We use the fact that $\omega = 2\pi/T$ where $T = 86164$ s is the rotation period of the earth. The earth completes one rotation in 86,164 s (about 24 hours).

(1) As we can see in Table 1.2, Mars and Mercury have nearly identical gravitational pull at the surface, despite the fact that Mars is larger and more massive than Mercury. Explain why?
(2) Why is it helpful to have the mass and radius of a planet?
(3) If you weigh 100 kg on Earth, how you weigh in the other planets of our solar system?
Although Mars is more massive than Mercury, Mercury has a higher density (see Table 1.6). This results in the two planets having a nearly identical gravitational pull at the surface.

Black hole. Given a massive star of mass m_1, calculate the radius (called the Schwarzschild Radius) for which not even light emanating from the surface of the star can escape. Determine the Schwarzschild radius of the Sun and the mean mass density it would have if its mass were contained in the corresponding spherical volume (the spherical surface bounding this volume is called the event horizon). Note that the current understanding of our universe is that new black holes will continue to form in the next 1,000 years so.
A black hole is the remains of a star that has collapsed under its own gravitational force. Remember that minimum initial velocity an object to escape the gravitational field of the Earth is

$$v = \sqrt{\frac{Gm_1}{R}}$$

where m_1 and R are mass and radius of solid Earth, respectively. To calculate the Schwarzschild radius we use the above equation for our star and replace the escape speed v by $c = 300 \times 10^6$ km/s, the speed of light,

$$c = \sqrt{\frac{Gm_1}{R_s}}$$

or

$$R_s = \frac{2Gm_1}{c^2} = \frac{2 \times \left(6.67 \times 10^{-11}\,\text{N}\,\text{m}^2\,\text{kg}^{-2}\right) \times \left(1.99 \times 10^{30}\,\text{kg}\right)}{\left(3.0 \times 10^8\,\text{m/s}^2\right)} = 2950\,\text{m}\ .$$

The escape speed is very large due to the concentration of a large mass into a sphere of very small radius. If the escape speed exceeds the speed of light, radiation cannot escape and it appears black. Schwarzschild radius, R_s, is the critical radius at which the escape speed equals $c = 300 \times 10^6$ km/s.

The Schwarzschild radius is evidently much smaller than the radius of the sun, which is 6.96×10^8 m). If the sun were all contained within the event horizon, then its mean mass density would be

$$\rho = \frac{m_1}{(4\pi/3)R^3} = \frac{3 \times \left(1.99 \times 10^{30}\,\text{kg}\right)}{4\pi \times (2950\,\text{m})^2} = 1.84 \times 10^{19}\,\text{kg/m}^3\ .$$

There is evidence that supermassive black holes exist at the centers of galaxies, probably all galaxies.

What is the farthest location on the Earth's surface from the center of the Earth?
Top of the Equator's Chimborazo mount is the highest point on Earth.

(1) What is the variation of range of g on the Earth surface? (2) What are the differences between topographic surface, ellipsoidal surface, and the geoid?
(1) 9.78 - 9.832 m/s^2 (equator-to-pole)
(2) Topographic surface = The actual surface on the Earth
The Geoid = the particular gravity equipotential surface that coincides with the mean sea level. This choice is totally arbitrary. But it is makes sense because the surface of a fluid in equilibrium and it is "easy" to locate. Reference ellipsoid = the ellipsoid that best fits the geoid. This ellipsoid is a theoretical (mathematical) representation of the shape of the Earth's surface.

Despite technological advances in the last 100 years or so, gravity measurements reflect a perceptible displacement of a mass relative to some reference (e.g., test mass). The first gravity data collected in the United States was obtained by G. Putman around 1890. Putnam used a pendulum gravity meter based on the relation between gravity and the period of a pendulum:

$$g = \frac{4\pi^2 l}{T^2}$$

where l is the length of the pendulum and T is the pendulum period. Assuming the period of a pendulum is known to be 1 s, find the pendulum displacement for a gravity anomaly of 10 mGal.

$$l = \frac{T^2 \Delta g}{4\pi^2} = 2.5\mu m \ .$$

You are in a city that is 9184 km from New York City, 6,904 km miles from Paris, and 11800 km from Shanghai. What city are you in?
Look on a map. There is only one city that has those distances, and that is Douala in Cameroon. Knowing three distances uniquely locates the position. GPS works in a similar manner, but instead of measuring distances to cities, it measures distances to satellites. And even though the satellites are moving, their locations when they broadcast their signals are known, so the computer in your GPS receiver can calculate its position.

Satellite systems for gravity measurements can be divided into four categories: (1) displacements of a satellite in orbit relative to tracking stations on the ground; (2) displacements of a satellite and the sea surface with respect to each other; (3) displacements of two satellites with respect to each other; and (4) relative displacements of two masses within a single satellite. Use this grouping to categorize GRACE and GOCE.
Group #1 missions generally rely on accurately measuring a satellite's position and comparing it to its theoretical position given a model of the Earth's (or other planet's) gravity field, assuming no anomalous mass distribution. In early efforts, this involved tracking a single satellite in its orbit and the departure of its true position from an expected position. Group #2 missions measure altitude using satellite-mounted laser altimeters. Variations in altitude measured by laser are a particularly good way to track variation in gravity at sea. Because the altimeter is measuring the ocean height variation, this method essentially maps

Table B.1. Lagrange points.

$m_1 - m_2$	R (10^9 m)	L_1 (10^9 m)	L_2 (10^9 m)	L_3 (10^9 m)
Earth–Moon	0.3844	0.326	0.4489	-0.38168
Sun–Mars	227.94	226.86	229.03	-227.94

gravity variation on the ocean surface rather than at the height of the satellite. Group #3 Modern gravity missions, such as GRACE and GRACE-FO, map variation in the gravity field by placing two satellites in orbit along the same orbital track. Gravity anomalies cause the distance between the two satellites to deviate from their expected distance of separation. The size of the gravity anomaly (or the excess mass that will be detected) depends on the altitude of the satellites and the distance between them. Group #4: GOCE.

(1) Determine the Lagrange points, L_1, L_2, and L_3, for the Earth–Moon-satellite system.
(2) Determine the Lagrange points L_1, L_2, and L_3, for the Sun–Mars-satellite system. L_1, L_2, and L_3 of the Earth–Moon-satellite and Earth–Moon-satellite systems.

Are there always two high tides a day?
No. There are places in the world which experience approximately two high tides a day and there are other places which experience a double-high water or doublelow water. These occurrences are caused by the shape of the coastline and the water depth. In some parts of the world there is just one high and low tide each day (e.g., Karumba, Australia), or even a mixed tide.

The distance between the lunar far side and the Earth is 500,000 km. Define the Earth-Moon L_2 point.
60,000 kilometres beyond the Moon.

Are there tides on Jupiter?
Jupiter is a gas giant without a surface or oceans. Hence there are no tides as we know them on Earth. When the term *tidal effects* refers to the massive gravitational influence that this large body has on its moons, Jupiter still has no moons large enough to cause significant tides. The largest tides will come from Io, since it is the closest large moon (Ganymede is larger, but farther away from Jupiter). Tidal forces from Io on Jupiter will be about 8 percent weaker than from the Moon on Earth. Given how massive Jupiter is, the tides are negligible on Jupiter.

There is no physics that suggests an exoplanet cannot have its own star (see Figure 1.54). A Jupiter-like exoplanet can have its own small star. After all, a fusion reactor is designed to build our own Sun on Earth. It will turn out that nature already does so spontaneously. This is what we need to locate and duplicate. Discuss.
see Chapter 6.

(1) What crustal density (ρ_{cc}) is needed to explain the 5 km high Tibetan Plateau? Assume $\rho_{sc} = 2.8$ g/cc, $\rho_m = 3.1$ g/cc, and $H_2 = 30$ km.
(2) How deep a root is needed to support the Tibetan plateau?
(1) At the compensation depth, pressure is equal at all points (see Figure 3.22). Thus at the compensation depth, the pressure below the mountain must equal the pressure below the Indian plains

$$\rho_{cc} = \left(\frac{H_2}{h_1 + H_2} \right) \rho_{sc} = 2.4 \, \text{g/cc} \, . \tag{B.1}$$

with $h_1 = 5$ km. The Tibetan Plateau is the largest mountain mass on Earth. Most of the Plateau—an area larger than France—stands at 5 km in elevation or higher! (2) The thickness of the crustal root is

$$H_1 = h_1 \frac{\rho_{sc}}{\rho_m - \rho_{sc}} = 47 \, \text{km} \tag{B.2}$$

and the crustal thickness $h_1 + H_2 + H_1 = 82$ km. Seismic imaging results show that the crustal thickness in Southern Tibet is between 75 km and 85 km.

Three points fixed relative to each other are sufficient to define a reference frame in three-dimensional space.
(1) Consider $\boldsymbol{x}_1 = (0,0,0)$, $\boldsymbol{x}_2 = (x_{21},0,0)$, and $\boldsymbol{x}_3 = (x_{31}, x_{32}, 0)$ are the centers of three spheres with radii r_1, r_2, and r_3, respectively. Determine the point of intersection, (x, y, z), of these three spheres.
(2) Use this case to explain why we used four spheres in Figure 1.57 instead of three.
(1) Let (x, y, z) be the point of intersection of three spheres, then we have

$$x^2 + y^2 + z^2 = r_1^2 \, , \quad (x - x_{21})^2 + y^2 + z^2 = r_2^2 \, ,$$

and

$$(x - x_{31})^2 + (y - x_{32})^2 + z^2 = r_3^2 \, .$$

Thus

$$x = \frac{r_1^2 - r_2^2 + x_{21}^2}{2 \, x_{21}} \, , \, y = \frac{r_1^2 - r_3^2 + x_{31}^2 + x_{32}^2 - 2 \, x_{31} x}{2 \, x_{32}}$$

and

$$z = \pm \sqrt{r_1^2 - x^2 - y^2} \, .$$

(2) We have two possible solutions: $(x, y, \sqrt{r_1^2 - x^2 - y^2})$ and $(x, y, -\sqrt{r_1^2 - x^2 - y^2})$. The fourth sphere allows us to select one of them. (3) Imaginary numbers mean that there is no intersection point.

What is the difference between linear momentum and angular momentum? Provide mathematical definitions of these two quantities.
see Ikelle (2023b), Chapter2.

Chapter 2:

Calculate the enthalpy of reaction for the following reactions:

$$\text{(a)} \quad 3 \, Fe_2O_3(s) + 2 \, NH_3(aq) \longrightarrow 6 \, FeO(s) + 3 \, H_2O(l) + N_2(aq)$$

$$\text{(b)} \quad 2 \, N_2O_5 \longrightarrow 4 \, NO_2 + O_2 \, .$$

$$\begin{aligned}
\text{(a)} \quad \Delta H_{\text{rxn}} &= \sum_i n_i \Delta H_{\text{products}}^{(i)} - \sum_k n_k \Delta H_{\text{products}}^{(k)} \\
&= [6\Delta H_{\text{FeO}} + 3\Delta H_{\text{H}_2\text{O}} + \Delta H_{\text{N}_2}] - [3\Delta H_{\text{Fe}_2\text{O}_3} + 2\Delta H_{\text{NH}_3}] \\
&= [6 \times (-266.3) + 3 \times (-285.8) + 0] \, \text{kJ/mol} = 10.9.6 \, \text{kJ/mol} \, .
\end{aligned}$$

Hence, for a general reaction of the form

$$\alpha A + \beta B \longrightarrow \gamma C + \eta D$$

then

$$\Delta H_{\text{rxn}} = [\gamma \Delta H_C + \Delta H_D] - [\alpha \Delta H_A + \beta \Delta H_B]$$

(b) We start with known enthalpies; i.e.,

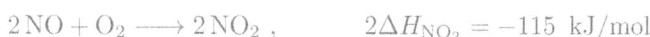

$$4\,NO + 3\,O_2 \longrightarrow 2\,N_2O_5 , \qquad 2\Delta H_{N_2O_5} = -445 \text{ kJ/mol}$$

$$2\,NO + O_2 \longrightarrow 2\,NO_2 , \qquad 2\Delta H_{NO_2} = -115 \text{ kJ/mol}$$

then

$$\Delta H_{\text{rxn}} = [4\Delta H_{NO_2} + \Delta H_{O_2}] - 2\Delta H_{N_2O_5}$$

$$= 2 \times (-115) + 0 - (-445) = 215 \text{ kJ/mol} .$$

(1) The amount of heat needed to change one gram of ice at at 0 degree Celsius to one gram of liquid water at 0 degree Celsius, 3.35×10^7 Joules. What is the change in entropy when 100 grams of ice at 0 degree Celsius melt into 100 grams of liquid water at 0 degree Celsius?
(2) Suppose that two reservoirs are brought into direct contact. An amount of heat, Q, is transferred from the high temperature reservoir (heat source) to the low temperature reservoir (heat sink). Determine the entropy change and show that heat transfer is irreversible.
(3) No heat is supplied in adiabatic expansion, but work is done. Show that, for an ideal gas, this work is $W = c_V(T_B - T_A)$, where T_A is the temperature at the beginning of the adiabatic process and T_B is the temperature at the end of the adiabatic process.
(1) Using converting the temperature of 0 degree Celsius into 273 degrees Kelvin, we find

$$\Delta S = \frac{\Delta Q}{T} = \frac{3.35 \times 10^7 \text{ Joules}}{237 \text{ K}} = 1.23 \times 10^5 \text{ Joules/K}$$

(2) the entropy change is

$$\Delta S = Q \left(\frac{1}{T_{\text{cold}}} - \frac{1}{T_{\text{hot}}} \right) > 0 .$$

Because $\Delta S > 0$, the heat transfer is irreversible. No work, W, is produced by this process. When the two reservoirs are not allowed to come into direct contact and, instead, an engine is used to convert some of the heat, Q_H to work and exhausts the remaining heat Q_C to the low temperature reservoir, we have the basic principle of the heat engine. (3) To find the work, the gas does expanding adiabatically from V_A to V_B for the adiabat $PV^\gamma = P_A V_A^\gamma = P_B V_B^\gamma$,

$$W = \int_{V_A}^{V_B} PdV = P_A V_A^\gamma \int_{V_A}^{V_B} \frac{dV}{V^\gamma} = P_A V_A^\gamma \frac{V_B^{1-\gamma} - V_A^{1-\gamma}}{1-\gamma} = \frac{P_B V_B - P_A V_A}{1-\gamma} .$$

By using the ideal gas law, $PV = RT$, we arrive at

$$W = R \frac{T_B - T_A}{1-\gamma} = c_V (T_B - T_A) .$$

(1) Is a human body a blackbody?
(2) Are stars blacbodies?
(1) A proof of Wien's displacement law. For values of interest to climate science, the exponential term is much larger than unity. By using this assumption, verify that (2.72) can alternatively be written as follows;

$$\log I(\lambda, T) = \log a_1 - 5 \log \lambda - \frac{a_2}{\lambda T} . \qquad \text{(B.3)}$$

Verify also that

$$\frac{\partial \left[\log I(\lambda, T) \right]}{\partial \lambda} = -\frac{5}{\lambda} + \frac{a_2}{\lambda^2 T} . \qquad \text{(B.4)}$$

(2) Estimate the surface temperature of from the sensitivity of the human eye to light at 550 nm. We assume that the atmosphere is for all frequencies in the visible range similarly transparent. Humans, at normal body temperature (around 37 degrees Celsius, or 310 Kelvin), radiate most strongly in the infrared domain but not in all incident electromagnetic radiation. Therefore the human body is not a blackbody. Humans feel the heat radiated by the sun with increased temperature. The radiative energy incident on our skin is absorbed by the molecules of water (remember that our bodies content is about 70 percent water) in our bodies because the internal vibrations of the water molecules occur at the same frequencies as infrared radiation. However our bodies possess thermoregulatory systems that can eliminate the excess of heat from our bodies, maintaining a quasi-stable internal temperature. A star is a near-perfect blackbody. The distribution of energy between different wavelengths (colors) depends strongly on the temperature. Actually, many cosmic objects radiate roughly like true blackbody, including our universe.

(1) A proof of Wien's displacement law. For values of interest to climate science, the exponential term is much larger than unity. By using this assumption, verify that (2.72) can alternatively be written as follows;

$$\log I(\lambda, T) = \log a_1 - 5 \log \lambda - \frac{a_2}{\lambda T} .$$

Verify also that

$$\frac{\partial \left[\log I(\lambda, T) \right]}{\partial \lambda} = -\frac{5}{\lambda} + \frac{a_2}{\lambda^2 T} .$$

Deduce (2.84).
(2) Estimate the surface temperature of from the sensitivity of the human eye to light at 550 nm. We assume that the atmosphere is for all frequencies in the visible range similarly transparent.
For $\lambda_{\text{max,sun}} = 550$ nm, we can use Wien's displacement law to deduce that

$$T_{\text{sun}} = \frac{2.978 \times 10^6 \ \text{nm K}}{550 \ \text{nm}} = 5414 \, \text{K}$$

is the surface temperature of the sun. This estimate is close enough to the precise value of 5781 K for the effective temperature of the solar photosphere. For humans, the temperature is about 310 K and $\lambda_{\text{max}} = 9.4 \ \mu\text{m}$; and for molecular clouds, the temperature is about 20 K and $\lambda_{\text{max}} = 145 \ \mu\text{m}$.

(1) Suppose the sun emits energy at a rate of $Q = 3.8736 \times 10^{26}$ W. The solar constant is

$$S = \frac{Q}{4\pi r^2}$$

Table B.2. Ground temperatures.

Planet	T_a	T_e	N	T_g (K)
Mercury	440	441	0	440
Venus	737	240	88	737
Earth	289	255	1	303
Mars	220	217	0	217
Jupiter	165	105	5	165

where r is the distance of a given planet from the sun. Determine the solar constant of Earth, Venus, Mars, Jupiter, and Mercury.
(2) Albedos for oceans are between 0.02 and 0.1; for forests, they are between 0.06 and 0.18; for cities, they are between 0.14 and 0.18; for grasslands, they are between 0.16 and 0.20; for deserts (sand), they are between 0.35 and 0.45; for clouds (thick stratus), they are between 0.60 and 0.70; and for fresh snows, they are between 0.75 and 0.95. Provide explanations for these variations of albedo of clouds with the type of surfaces.
(1)

Planet	$r\,(10^9$ m)	S (W/m^2)
Mercury	58	9163
Venus	109	2600
Earth	150	1370
Mars	227	600
Jupiter	778	51

(2) Over the ocean the albedo is small (0.02–0.1). It is larger over the land (typically 0.35–0.45 over desert regions) and is particularly high over snow and ice. Albedo depends on the nature of the reflecting surface. Solar radiation is reflected by into space by white clouds, glaciers, and snow. Dust and volcanic ash also prevent sunlight from reaching the Earth.

Consider an atmosphere that is completely transparent to shortwave (solar) radiation but very opaque to infrared terrestrial radiation. Specifically, assume that it can be represented by n layers of atmosphere, each of which completely absorbs infrared radiation. How many layers would be required to obtain the actual average temperatures of Earth, Venus, and Mars? Explain why N is zero for some planets.
By considering the energy balance of the n-th layer, we can show that,

$$2T_n^4 = T_{n+1}^4 + T_{n-1}^4 \,,$$

where T_n is the temperature of the n-th layer, for $n > 1$. So the ground temperature is

$$T_g = (N+1)^{\frac{1}{4}} T_e \,,$$

where T_e is the planetary emission temperature and T_g is the ground temperature.
where T_a is the actual ground temperature. $N = 0$ for planets without atmosphere. Before sunlight reaches the earth it has to pass through the atmosphere. As described in Chapter 1, Earth's atmosphere has are several layers. The troposphere is where weather occurs and where climate is determined. There is also a scenario in which the atmosphere can absorb visible light rather than letting it pass through. This scenario can occur on Earth if the upper atmosphere is filled with dust resulting from nuclear accidents, nuclear wars. or any artifical design to reduce earth temperatures. The Earth will be too cold; that is why this scenario is known as nuclear winter.

As described, in Ikelle (2022a), the surface of Mercury changes in temperature are the most drastic in our entire solar system. Why?
Because the planet has no atmosphere. Temperatures on the surface of Mercury are both hot and cold. During the day, temperatures on the surface can reach up to 430 degrees Celsius. Because the planet has no atmosphere to retain that heat, nighttime temperatures on the surface can drop to -180 degrees Celsius. These changes in temperature are the most drastic in the entire Solar System.

White et al. (1997) show that the expected temperature change, ΔT, caused by the 11-year periodic change in solar activity can be described by the following equation:

$$\Delta T(t) = \frac{\Delta S}{\rho h c_P \sqrt{(K^2 + \omega_S^2)}} \cos(\omega_S t + \theta_S) , \qquad (B.5)$$

where $\Delta S \approx 0.2\,\mathrm{W/m^2}$ describes the change in the sun's constant (nearly all the energy that the Earth receives), $\omega_S = 2\pi/T_S = 2\pi/(11\,\text{years})$ is the angular version of the 11-year period, $\rho = 1.0\ \mathrm{g/cm^3}$ is the average density of ocean water, $K = 1/\text{year}$ is a dissipative timescale for energy loss to the deep ocean and to the atmosphere, $h = 50$ m is the average depth of the mixed layer of the oceans, $c_P = 4.2 \times 10^3\ \mathrm{Ws/(K\,kg)}$ is the average heat capacity of ocean water at constant pressure, and $\theta_S = \tan^{-1}(\omega_S/K)$.
(1) Determine the range of temperature change during a solar cycle. (2) Are the changes in solar activity have significant impact on the Earth'surface climate? Hint: Shaviv (2008).
(i) Using $\omega_S = 1.8 \times 10^{-8}$ 1/second, $K = 3.1 \times 10^{-8}$ 1/second, and $\rho = 1000\ \mathrm{kg/m^3}$, we can verify that the temperature change during a solar cycle ranges from -0.026 to 0.026 K. (ii) This range suggests that solar activity can have a significant impact on the Earth's climate. This result is in contrast to the official consensus from the Intergovernmental Panel on Climate Change (IPCC). The difference is due to the fact that the IPCC assumes that $\Delta S \approx 0.05\,\mathrm{W/m^2}$, which is small relative to the effect of greenhouse gases, estimated at around 2.5 W/m^2. Shaviv (2008) suggests $\Delta S \approx 0.2\,\mathrm{W/m^2}$ and maybe higher.

Is the Péclet number mainly positive or negative for glaciers? Explain your answer.
The ice velocity is mainly vertical down. Under the assumptions of only vertical advection. Thus, the Péclet number mainly positive.

(1) Define a process where a parcels temperature changes due to expansion or compression, but no heat is added or taken away from the parcel. (2) Explain why Africa and Antarctica have numerous hotspots (see Ikelle, 2022a).
Hot spots can and do occur anywhere on the Earth's surface. However, because the lithosphere is thicker in some places than others, and because lithospheric plates moves laterally relative to the mantle convection system at various rates, some plates and parts of plates have more hot spots than others. As a result, the slow moving plates that include continents such as Africa and Antarctica have numerous hot spots, as do thin plates, such as the Pacific plate, which is entirely composed of thin ocean lithosphere.

An engine transfers 2×10^3 J of energy from a hot reservoir during a cycle and transfers 1.5×10^3 J to a cold reservoir. (1) Find the efficiency of the engine. (2) How much work does the engine do in one cycle?
(1) $\eta = 1 - \frac{Q_C}{Q-H} = 1 - \frac{1.5}{2} = 0.25 = 25.0\,\%$
(2) $W = Q_H - Q_C = 5.0 \times 10^2$ J.

Chapter 3:

The microscopic and macroscopic scale distinction is not an issue in vacuum. Why?
When dealing with electromagnetics in matter, special attention must be paid to microscopic scale (or atomic scale) and macroscopic scale at which experimental measurements are made. This distinction is not important in the vacuum because we are dealing with a homogeneous (the same properties at all points) and isotropic medium (the same properties along any direction).

List two examples of conductors, two examples of insulators, and two examples of semiconductors.
(i) Conductors: metals (solids made of metallic elements), silver, cooper, tungsten (wolfram), and iron; the earth also shows conducting properties. (ii) Semiconductors: silicon (Si), germanium (Ge), gallium arsenide, and indium phosphide. (iii) Insulators: ceramics, concrete, soda-lime glass, glass, aluminum oxide, and polystyrene.

Given the polarization $P_1 = aE_1 + bE_1^2 + cE_2E_3^2$, determine $\chi_{11}^{(e;1)}$, $\chi_{12}^{(e;1)}$, $\chi_{111}^{(e;2)}$, and $\chi_{1233}^{(e;3)}$.
Taking the derivative as with respect to the field and evaluating it at zero field, we obtain

$$\chi_{11}^{(e;1)} = \left.\frac{\partial P_1}{\partial E_1}\right|_{E=0} = a + 2bE_1|_0 = a, \quad \chi_{12}^{(e;1)} = \left.\frac{\partial P_1}{\partial E_2}\right|_{E=0} = 0,$$

$$\chi_{111}^{(e;2)} = \left.\frac{1}{2}\frac{\partial^2 P_1}{\partial E_1^2}\right|_{E=0} = b, \quad \text{and} \quad \chi_{1233}^{(e;3)} = \left.\frac{1}{2}\frac{\partial P_1}{\partial E_2 \partial E_3 \partial E_3}\right|_{E=0} = c.$$

Why we are sometimes are able to hear sounds, but we are unable see emitting source?
The wavelength of sound is much greater than that of light.

Consider a hydrogen atom in which the proton (nucleus) and the electron are separated by a distance of 5.3×10^{-11} m. Determine the electrostatic and gravitational forces between the two particles.
The electrostatic force between the two particles is approximately

$$F_e = \frac{1}{4\pi\varepsilon_0}\frac{e^2}{|\boldsymbol{x}|^2} = \left(9 \times 10^9 \frac{\text{Nm}^2}{\text{C}^2}\right)\frac{(1.60 \times 10^{-19}\,\text{C})^2}{(5.3 \times 10^{-11}\,\text{m})^2} \approx 8.2 \times 10^{-8}\,\text{N} .$$

On the other hand, the gravitational force is only

$$F_g = \frac{Gm_1m_2}{|\boldsymbol{x}|^2}$$

$$- \left(6.67 \times 10^{-11} \frac{\text{Nm}^2}{\text{kg}^2}\right)\frac{(1.67 \times 10^{-27}\,\text{kg})(9.11 \times 10^{-31}\,\text{kg})}{(5.3 \times 10^{-11}\,\text{m})^2} \approx 3.6 \times 10^{-47}\,\text{N} .$$

Thus, gravitational effect can be neglected when dealing with electrostatic forces; maybe.

We cannot see dark matter with EM spectrum but it has a gravity. Remember that gravity is a property of matter, energy, and momentum. What are not see them?
Gravity is a force that acts on matter and energy by attraction; it affects the motion of everything in the universe. The fact that there is more gravity in the universe than the motion of all the visible objects in the universe may mean that there are some motionless objects.

(1) Explain what element 115 is and how it is made. Name one of its stable isotopes along with its half-life. Also describe its current industrial application, if any. (2) Physicist Robert Feynman predicted the periodic table would end at element 137. Did we reach this limit?

Moscovium is a synthetic element with the symbol Mc and the atomic number 115. Moscovium is an extremely radioactive element: its most stable known isotope, moscovium-290, has a half-life of only 0.65 seconds. It was first synthesized in 2003 by a joint team of Russian and American scientists at the Joint Institute for Nuclear Research (JINR) in Dubna, Russia. In other words, Moscovium is artificially created by colliding calcium-48 ions with americium-243 in particle accelerators. Moscovium is sometimes listed as ununpentium (symbol: Uup). So far, there has been no industrial, medical, or military use for Mc. It is very difficult to make and more importantly to preserve because of its very small half-life. Physicist Robert Feynman predicted that the periodic table would end at element 137. Currently there are 118 proven elements.

Determine the molecule in Figure 3.39 using infrared spectroscopy?

Peak #1 corresponds to about 3000 cm^{-1}. Using Figure 1.26, it can be due to C–H stretching, CH_2, or CH_3 rocking. Peak #2 corresponds to about 1700 cm^{-1}. Using Figure 1.26, it can be C=C stretch, C=O stretch, O–H, etc. Peak #3 corresponds to 1300-1400 cm^{-1}. Basically, it difficult to determine this molecule without using more peaks and the intensity of the peaks. An advanced solution is described in Chapter 4. Figure 3.39 is actually the IR spectrum of propionaldehyde (C_3H_6O).

What is a zero-point energy?

A zero-point energy is lowest possible energy that a quantum mechanical system can have. It is governed by the classical physics; $U = (1/2)hf$, where f is frequency h is the Planck's constant, f is the frequency. Note that quantum systems do not have a fixed amount of energy at any state due to the Heisenberg uncertainty principle.

Why does the cornea have the strongest refraction?

There is a large refractive index between the cornea ($n = 1.37$) and air ($n = 1$).

Suppose that you are living on a planet which is 150 million light years away from Earth. If you focus your telescope on Earth today, what species will you see?

You will be seeing dinosaurs now because the light will take this 150 million years to reach you. The inverse positions means that Hubble is looking at the past, we all are looking at the past. Through stellar travel, we can overcome this limitation and see the present. Maybe Alliens are visiting for, at least in part, similar reason.

The average temperatures of the Sun is 6000 degrees Kelvin, humans is 310 degrees Kelvin, molecular clouds is 20 degrees Kelvin, and cosmic background is 2.7 degrees Kelvin. Determine the light of the electromagnetic spectrum associated with the sun, humans, molecular clouds, and cosmic background $\lambda_{max} = 2.9/T$ in mm if T is in degrees Kelvin. The sun correspond to $\lambda_{max} = 480$ nm (ultraviolet/optical), humans correspond to $\lambda_{max} = 9.4$ μm (visible), molecular clouds correspond to $\lambda_{max} = 145$ μm (infrared), and cosmic background corresponds to $\lambda_{max} = 1.1$ mm (infrared).

What is the difference between sound waves and radio waves?

Many people confuse sound waves with radio waves. Sound and radio waves are completely different phenomena. Radio waves is one type of electromagnetic (EM) waves, just like visible light, infrared light, ultraviolet light, X-rays, and gamma rays. EM waves do not require a medium in which to propagate; they can travel through a vacuum. Sound waves describe pressure variations (waves) in liquids and gases (matter), such as air or water, and your eardrum. A radio works because sound waves at the radio station are converted into electromagnetic waves, then encoded and transmitted in the radio-frequency range. The radio in your car receives the radio waves, decodes the information, and uses a speaker to change it back into a sound wave, bringing your favorite music to your ears.

Chapter 4:

Show that $\Psi(x,t) = A\sin(kx - \omega t + \phi)$ *is not a solution to Schrödinger equation, where* k *is the wavenumber,* ω *is the angular frequency and* ϕ *is a phase constant. The wave is moving in the positive x-direction.*

Inserting $\partial \Psi(x,t)/\partial t = -\omega A\sin(kx - \omega t)$, $\partial \Psi(x,t)/\partial x = kA\cos(kx - \omega t)$, and $\partial^2 \Psi(x,t)/\partial x^2 = -k^2 \Psi(x,t)$ into the time–dependent Schrödinger equation we obtain

$$-i\hbar\omega \cos(kx - \omega t) = \left(\frac{\hbar^2 k^2}{2m} + V\right) A\sin(kx - \omega t), \tag{B.6}$$

which is not valid because $\cos(x) \neq \sin(x)$.

(1) Determine $\mathrm{H} \otimes \mathrm{H}\,|ud\rangle$ *as a function of* $|-\rangle$ *and* $|-\rangle$.

(2) We consider a system of n qubits. A number of quantum-computing algorithms use the n-bit Hadamard transformation which is the tensor product of a Hadamard transformation for each bit:

$$W_n = \underbrace{\mathrm{H} \otimes \mathrm{H} \otimes \mathrm{H} \otimes ... \otimes \mathrm{H}}_{n \text{ times}},$$

where W_n *is the Walsh-Hadamard transformation. Determine* $W_3\,|uuu\rangle$ *function of the Walsh-Hadamard transformation.*

$$\mathrm{H} \otimes \mathrm{H}\,|ud\rangle = \frac{1}{\sqrt{2}}\left(|u\rangle + |d\rangle\right) \otimes \frac{1}{\sqrt{2}}\left(|u\rangle - |d\rangle\right) + \frac{1}{2}\left(|uu\rangle - |ud\rangle + |du\rangle - |dd\rangle\right)$$

$$
\begin{aligned}
W_n\,|uuu\rangle &= \mathrm{H} \otimes \mathrm{H} \otimes \mathrm{H}\,|uuu\rangle \\[2mm]
&= \frac{1}{\sqrt{2}}\left(|u\rangle + |d\rangle\right) \otimes \frac{1}{\sqrt{2}}\left(|u\rangle + |d\rangle\right) \otimes \frac{1}{\sqrt{2}}\left(|u\rangle + |d\rangle\right) \\[2mm]
&= \frac{1}{2^{3/2}}\left(|uuu\rangle + |uud\rangle + |udu\rangle + |udd\rangle + |duu\rangle + |dud\rangle + |ddu\rangle + |ddd\rangle\right) \\[2mm]
&= \frac{1}{2^{3/2}}\left(|000\rangle + |001\rangle + |010\rangle + |011\rangle + |100\rangle + |101\rangle + |110\rangle + |111\rangle\right) \\[2mm]
&= \frac{1}{2^{3/2}} \sum_{x=000}^{111} |x\rangle .
\end{aligned}
$$

Here x is a binary string that has length 3.

Is the following quantum state

$$\left[\frac{1}{\sqrt{2}}, 0, 0, \frac{1}{\sqrt{2}}\right]^{T} \tag{B.7}$$

entangled.

Let us write this state down as follows:

$$\begin{pmatrix} \frac{1}{\sqrt{2}} \\ 0 \\ 0 \\ \frac{1}{\sqrt{2}} \end{pmatrix} = \begin{pmatrix} a \\ b \end{pmatrix} \otimes \begin{pmatrix} c \\ d \end{pmatrix} ,$$

where a, b, c, and d are unknowns. These unknowns are related as follows:

$$ac = bd = \frac{1}{\sqrt{2}} \quad \text{and} \quad ad = bc = 0 .$$

We cannot find a, b, c, and d that satisfy these relationships. Therefore, this quantum state can't be factorized, so it is entangled.

Propose five physical ways of constructing qubits.
(1) Photon vertical and horizontal polarizations; (2) two electrons going in opposite directions in space; (3) two electrons going in opposite directions in time; (4) a spin up electron and a spin down electron; (5) spins of the atomic nucleus (NMR qubits); (6) matter and antimatter; and (7) electrons and positrons.

A three-qubit quantum computer can store $2^3 = 8$ possible combinations of $|0\rangle$ and $|1\rangle$ simultaneously. As a result three-qubit quantum computer does calculations 8 times faster compared to 3-bit conventional computer; n-qubit computer has $N = 2^n$ possible states. Compare a 64-qubit quantum computer with a 64-bit our classical one.
For instance a 64-qubit quantum computer's computational power is 2^{64} times faster than our 64-bit classical computer. A 64-bit processor can work with 64-bit binary numbers at once. Because each bit participates in one process. Once the process is completed, that same bit will hold information for other processes. You see, even if you do parallel programming, you will utilize multiple processors but technically a physical bit in the processor is not shared. All bits in those processors are separate.
Simulating 3 qubits does not require $2 \times 3 = 6$ numbers. It takes $2^3 = 8$ numbers. And simulating 10 qubits does not require $2 \times 10 = 20$ numbers. It takes $2^{10} = 1024$ numbers. And so on. It is actually that multiple qubits have an advantage over several classical bits because a qubit's ability to be in multiple states simultaneously allows its performance to increase exponentially with each additional qubit whereas the increase in performance with additional classical bits is linear.

Suppose that you have a function $f(x)$. There exist $y \neq x$ with $|x - y| = kP$, where k is an integer and P is an arbitrary number, such that $f(x) = f(y)$. Characterize this function.
$f(x)$ is a periodic function with a P period. For example, $f(x) = \sin(x)$ and $f(x) = f(x + 2\pi k) = f(y)$, where $y = x + k2\pi$, $|x - y| = k2\pi$, and $P = 2\pi$.

Chapter 5:

How are neuron cells different from other cells in our bodies?
Neurons or neuron cells are considered different from other cells because the neurons communicate with each other through an electrochemical process. These neurons contain some specialized structures (synapses) and chemicals (neurotransmitters). They also have a specialized cell parts—dendrites and axons.
Neurons are very greedy in terms of energy. Brain gulps down 25 percent of glucose, 20 percent of oxygen, and 15 percent of cardiac output. Out of about 20,000 genes, 14000

are expressed in brain tissues. Mitochondria are more in number in neuronal soma and in axons. Neurons do not regenerate except in some parts in brain.

(1) What is the reversal potentials of Ca^{++} at a body temperature? (2) Determine the resting membrane potential of a cell with $g_{Na^+}/g_{K^+} = 0.03$ and $g_{Cl^-}/g_{K^+} = 0.1$. (3) Propose a method for diagnosing and identifying brain tumors.

(1) The electrical potential that exactly balances a concentration gradient is called the reversal potential. The Nernst equation reveals how the reversal potential of an ion depends on the inside and outside concentrations of this ion:

$$\Delta V = \frac{RT}{zF} \ln \left[\frac{(\text{ion})_{\text{out}}}{(\text{ion})_{\text{in}}} \right]$$

$$= \frac{27 \text{ mV}}{+2} \ln \left[\frac{(Ca^{++})_{\text{out}}}{(Ca^{++})_{\text{in}}} \right] \approx 13.5 \text{ mV} \ln \left[\frac{2 \text{ mM}}{0.0002 \text{ mM}} \right] = 124 \text{ mV} ,$$

for $T = 310$ degrees Kelvin (body temperature), $F = 96480$ Coulomb/mol, $R = 8.31$ Joule/molK, $z = 2$, $(\text{ion})_{\text{out}} = 2$ mM, $(\text{ion})_{\text{in}} = 0.0002$ mM.

(2) Let us calculate the resting potential in a concrete case:

$$\Delta V_{\text{res}} = \frac{\Delta V_{K^+} + (g_{Na^+}/g_{K^+}) \Delta V_{Na^+} + (g_{Cl^-}/g_{K^+}) \Delta V_{Cl^-}}{1 + (g_{Na^+}/g_{K^+}) + (g_{Cl^-}/g_{K^+})}$$

$$= \frac{-80 \text{ mV} + 0.03 \times 59 \text{ mV} - 0.1 \times 65 \text{ mV}}{1 + 0.03 + 0.1} = -76 \text{ mV}$$

The resting potential of a cell is a compromise between the reversal potentials of all ions to which the membrane is permeable. The influence of each ion in this compromise depends on its permeability. More permeable ions exert more influence than less permeable ones.

(1) Determine the potential level across the membrane that will exactly prevent net diffusion of Na^+, K^+, Cl^-, Ca^{++}, and HCO_3^-.

(2) Resting membrane potential: There are different kinds and concentrations of ions inside and outside cell membranes. Define them.

(3) Propose a method for diagnosing and identifying brain tumors.

(4) The SQUID magnetometer can also record the brain's magnetic field. During the alpha rhythm ,the magnetic field from the brain is about 10^{-11} Tesla. How does this magnetic field compare to that of the Earth's magnetic field?

(1) Nernst potential is equilibrium potential. The potential level across the membrane that will exactly prevent net diffusion of an ion (2) There are different kinds and concentrations of ions inside and outside cell membranes:

- Inside cell: negative ions (proteins$^-$, PO_4^{--}, SO_4^{--}); positive ions (K^+).

- Outside cell: negative ions (HCO_3^-, Cl^-); positive ions (Na^+).

there are more negative ions inside cell, especially proteins and negative ions (especially proteins) generally cannot cross membrane.

(3) The EEG confirms brain tumors since electrical activity is reduced in tumor regions.

(4) About one-billionth of the Earth's magnetic field.

Figure 5.25 shows electrocardiograms (ECGs or EKGs). Determine the ones corresponding to healthy.

None is normal. (1) Very low P-event or absent: (a), (b), (c), (d), and (e). (2) Abnormal heart rate: A heart rate that is faster or slower than normal could be a sign of atrial fibrillation: (b), (c), (d), and (e). (3) Abnormal heart rhythm: An irregular heart rhythm

Table B.3. The potential level across the membrane that will exactly prevent net diffusion of an ion.

Ion	Extracellular (mM)	Intracellular (mM)	ΔV (mV)
K^+	4	140	-92
Na^+	142	10	$+58$
Cl^-	103	4	-89
Ca^{++}	2.4	0	$+129$
HCO_3^-	28	10	-23

refers to an ECG reading with long pauses or extra beats: (b), (c), (d), and (e). (4) Abnormal waveform: Abnormalities in the waveforms of an ECG reading could mean that the electrical signals in the heart are not being transmitted or conducted properly: (b), (c), (d), and (e). (4) Abnormal intervals: A prolonged PR interval can indicate a problem with the electrical conduction system of the heart (delayed conduction of the SA node), while a prolonged QT can indicate a problem with the lower chambers of the heart(ventricles): (b), (c), (d), and (e).

Propose a method for determining formation water resistivity using SP logs.
The ideal spontaneous potential across clean bed (SSP):

$$\text{SP} = -K \log \left(\frac{R_{\text{mf}}}{R_{\text{w}}} \right) \implies \text{SSP} = -K \log \left(\frac{R_{\text{mfe}}}{R_{\text{we}}} \right) . \tag{B.8}$$

In practice the Static spontaneous potential is often written in terms of effective water resistivities (R_{mfe}) and (R_{we}) rather than actual resistivities. These are equal to R_{mf} and R_{w} except for concentrated or dilute solutions. The static spontaneous potential (SSP) is important because ratio of effective water resistivities can be obtained from is SSP. The formation temperature is also important in determining formation water resistivity. We here limit ourselves to the linear approximation the model given by the second formula highlighted in blue here.

$$T_f(z) = \frac{T_{bh} - T_0}{z_{bh}} z + T_0 , \tag{B.9}$$

T_0: surface average temperature, T_{bh}: temperature at the well bottom at z_{bh}, and z is formation depth. Arps formula: correction of R_{mf} for formation temperature

$$R_{\text{mfe}} = R_{\text{mf},0} \frac{T_0 + X}{T_f + X} \quad \text{with} \quad X = 6.77 , \tag{B.10}$$

where $R_{\text{mf},0}$ is the resistivity of mud filtrate at the surface $z = 0$. SSP is especially useful for determining formation water resistivity:

(1) Input: SP log, T_0, $R_{\text{mf},0}$
(2) Determine the formation temperature (T_f)
    and deduce $K = 65 + 0.24 \times T$.
(3) Measure $R_{\text{mf},0}$ and deduce R_{mfe}.
(4) Determine SSP and deduce $R_{\text{mfe}}/R_{\text{we}}$.
(5) Determine R_{we} and deduce R_{w}.

· Oil and gas industry has developed monograms that allow to deduce OOIP (original oil-in-place or discovery) from R_w.

Chapter 6:

(1) Find the relation between gravity, temperature, and luminosity of stars in the main sequence. (2) Find the radius of Sirius B from its surface temperature $T = 25,000$ degree Kelvin and its luminosity $L = 3.84 \times 10^{26}$ Watts.
For the main sequence (V) only

$$m \propto L^{7/2} \implies \log_{10}\left(\frac{m}{m_{sun}}\right) = 3.5 \log_{10}\left(\frac{L}{L_{sun}}\right)$$

yielding

$$\log_{10}\left(\frac{g}{g_{sun}}\right) = 4 \log_{10}\left(\frac{T}{T_{sun}}\right) - 2.5 \log_{10}\left(\frac{L}{L_{sun}}\right).$$

(2)

$$\frac{r}{r_{sun}} = \left(\frac{L}{L_{sun}}\right)^{1/2}\left(\frac{T}{T_{sun}}\right)^2 \approx 0.01$$

Perpendicular forces lead to drift of the particles; i.e.,

$$v = \underbrace{v_\| b}_{\#1} + \underbrace{v_g}_{\#2} + \underbrace{\frac{E \times B}{B^2}}_{\#3} + \underbrace{\frac{m}{qB^2}\frac{dE_\perp}{dt}}_{\#4} + \underbrace{\frac{mv_\|^2 + 0.5mv_\perp^2}{qB}\frac{B \times \nabla B}{B^2}}_{\#5},$$

where the subscripts $\|$ and $\perp$ refer to the parallel and perpendicular components with respect to the magnetic field, respectively, $b = B/B$ is the unit vector along the magnetic field, m is the particle mass, q is the particle charge, B is the magnetic field strength, and v_g is the rotating perpendicular velocity.
Define the drift terms #1, #2, #3, #4, and #5.

Term #1 is the parallel motion along the field. Term #2 represents formally the rapid rotation of the velocity connected with the gyration of the particle around the magnetic field (gyration). Terms #3 represents the $E \times B$ drift. Term #4 represents polarization drift. Term #5 is the combination of the curvature and grad-B drift.

What is the difference between a gas and a plasma?
Gas is generally electrically neutral and can't be confined magnetically, while plasma can. Plasma is electrically ionized.

Rewrite MHD equations for static equilibria (i.e., $\partial/\partial t = 0$ and $v = 0$).

$$\nabla p = -\frac{1}{\mu_0}(\nabla \times B) \times B + \rho g \quad \text{and} \quad \nabla \cdot B = 0$$

(a) Show that for solar wind, the ratio of the gas pressure gradient term and the Lorentz force known as the plasma-β is

$$\beta = \frac{3.5 \times 10^{-21} nT}{B^2} \qquad (B.11)$$

where n is in m^{-3}, T in degrees Kelvin and B in Gauss. Determine β in the solar Corona, photospheric magnetic flux tubes, and the solar wind near Earth's orbit. (b) In a configuration with a magnetic field of 5 Tesla, what is the pressure exerted on it.

(a) We simply have to assume that the plasma has an equation of $p = k_B nT$ to obtain the above equation; i.e.,

$$\beta = \frac{n k_B T}{B^2/\mu_0} = \frac{3.5 \times 10^{-21} nT}{B^2} . \qquad (B.12)$$

Use Figure 2.17. In the solar Corona, $T = 10^6$ degrees Kelvin, $n = 10^{14}$ m^{-3}, $B = 10$ Gauss, and $\beta = 3.5 \times 10^{-3}$. In photospheric magnetic flux tubes, $T = 6 \times 10^3$ degrees Kelvin, $n = 10^{23}$ m^{-3}, $B = 1000$ Gauss, and $\beta = 2$. In the solar wind near Earth's orbit, $T = 2 \times 10^5$ degrees Kelvin, $n = 10^7$ m^{-3}, $B = 6 \times 10^{-5}$ Gauss, and $\beta = 2$. (b) A magnetic field of 5 T exerts a pressure of $B^2/(2\mu_0) \approx 10^7$ N/m$^2 \approx 100$ atm.

What are the differences between Navier-Stokes and MHD equations?
Maxwell equations concerns electromagnetic fields. Navier (or Navier Stokes) equations concern fluid flow. Magnetohydrodynamics (MHD) is the a combination of the two which considers the electromagnetic and fluid behavior of a conductive fluid like a plasma. While Navier Stokes has 5 equations (density, momentum, energy), ideal MHD has 8 adding the time evolution of the magnetic field (a reduced form of Faraday's law).

Explain why these drifts contribute to a westward-directed ring current? Consider a particle at the magnetic equator of the planet's dipole field.
An ion would drift east-west while an electron would drift in the opposite direction.

Let us assume the magnetic fields are zeros, $\boldsymbol{v}_1 = -\boldsymbol{v}2 = [v_{1x}, 0, 0]^T$, $\rho_1 \neq \rho_2$, and $g = 0$. Describe the stability criteria in this case.

$$\omega_{\mathrm{RT}} = \frac{k_x v_{1x} \rho_1 \pm i \sqrt{k^2 v_{1x}^2 \rho_1 \rho_2}}{\rho_1 + \rho_2} \qquad (B.13)$$

Because ω_{RT} is a complex-valued frequency, we always have instability in this case.

Jeans instability with rotation. *To derive a more realistic instability criterion it is necessary to take into account rotation. In a rotation frame of reference it is necessary to add the Coriolis and the centrifugal accelerations to the hydrodynamic equations. We can combine the centrifugal force and the gravitavional force describe Ψ as the perturbation of the gravitational potential caused by rotation through the centrifugal force. The Coriolis acceleration can be just add to momentum equation as an additional source term. Assuming that perturbations vary only along the $z-$ axis, rewrite equations for rotating cloud and in absence of magnetic effects in time-space and frequency-wavenumber domains, (2) Deduce the Jeans wavelength for the case in which the angle between the direction of propagation of the perturbation and the rotation axis is 90 degrees.*
So equations for rotating cloud and in absence of magnetic effects reduces to

Continuity equation : $\quad \dfrac{\partial \delta \rho}{\partial t} + \boldsymbol{\nabla} \cdot (\rho_0 \delta \boldsymbol{v}) = 0$

$$\text{Momemtum equation}: \quad \frac{\partial \delta v}{\partial t} + \rho_0 \nabla (\delta \Psi) + \nabla \delta p + \underbrace{2\Omega \times \delta v}_{\text{Coriolis}} + \underbrace{\Omega \times (\Omega \times x)}_{\text{centrifugal}} = 0$$

$$\text{Poisson's equation}: \quad \nabla^2 \delta \Psi = 4\pi G_0 \delta \rho$$

where $\rho_0 = \text{constant}$, $v_0 = 0$, and $p_0 = 0$, Ω is the rotation vector. The second and third terms on the left-hand side of the latter which represent the Coriolis effect and centrifugal force, respectively, due to the rotation. $\Omega = [0, \Omega_y, \Omega_z]^T$, and $\nabla \delta p = [0, 0, \partial p / \partial p]^T$. By assuming that perturbations propagate only along the $z-$ axis. We then get the perturbation equations and that

$$\frac{\partial \delta \rho}{\partial t} + \rho_0 \frac{\partial \delta v_z}{\partial z} = 0 \iff \omega \delta \rho - \rho_0 k \delta v_z = 0$$

$$\frac{\partial \delta v_x}{\partial t} + 2\Omega_y v_z - 2\Omega_z v_y = 0 \iff \omega v_x - 2i\Omega_z v_y + 2i\Omega_y v_z = 0$$

$$\frac{\partial \delta v_y}{\partial t} + 2\Omega_z v_x \iff 2i\Omega_y v_x + \omega v_y$$

$$\frac{\partial \delta v_z}{\partial t} + \frac{1}{\rho_0} \frac{\partial \delta p}{\partial z} + \frac{\partial \delta \Psi}{\partial z} - 2\Omega_y v_x = 0 \iff -2i\Omega_y v_x + \omega v_z - \frac{c^2 k}{\rho_0} v_z + k\delta \rho = 0$$

$$\frac{\partial^2 \delta \Psi}{\partial z^2} = 4\pi G_0 \delta \rho = 0 \iff 4\pi G_0 \delta p - k^2 \delta \Psi$$

or in a matrix form as

$$\begin{pmatrix} \omega & -2i\Omega_z & 2i\Omega_y & 0 & 0 \\ 2i\Omega_z & \omega & 0 & 0 & 0 \\ 2i\Omega_y & 0 & \omega & -(c^2 k_z)/\rho_0 & k_z \\ 0 & 0 & -\rho_0 k_z & \omega & 0 \\ 0 & 0 & 0 & 4\pi G_0 & -k_z^2 \end{pmatrix} \begin{pmatrix} v_x \\ v_y \\ v_z \\ \delta \rho \\ \delta \Psi \end{pmatrix} = \begin{pmatrix} 0 \\ 0 \\ 0 \\ 0 \\ 0 \end{pmatrix}$$

Setting the determinant of the 5 x 5 coefficient matrix for the system of constants equal to zero to satisfy their nontriviality property, we obtain the secular equation We are looking for non-trivial solutions. This means that the determinant must be zero. This leads to a bi-quadratic equation in ω for the dispersion relation

$$\omega^4 - \left(4\Omega^2 + c^2 k_z^2 - 4\pi G_0 \rho_0\right) \omega^2 + 4\Omega^2 \left(c^2 k_z^2 - 4\pi G_0 \rho_0\right) \cos^2 \theta = 0 , \qquad (B.14)$$

where θ is the angle between the direction of propagation of the perturbation k and the rotation axis Ω. Using Vieta's relation, we can get the following two equations for the ω solving the equation:

$$\omega_1^2 + \omega_2^2 = 4\Omega^2 + c^2 k_z^2 - 4\pi G_0 \rho_0 \quad \text{and} \quad \omega_1^2 \omega_2^2 = 4\Omega^2 \left(c^2 k_z^2 - 4\pi G_0 \rho_0\right) \qquad (B.15)$$

The second case is the important special case where $\theta = 90°$, i.e. where the rotation axis and the perturbation are perpendicular to each other. The dispersion relation takes the simple form:

$$\omega^2 - \left(4\Omega^2 + c^2 k_z^2 - 4\pi G_0 \rho_0\right) \omega^2 = 0 , \qquad (B.16)$$

Clearly, the additional rotation term acts stabilizing. For sufficiently large rotation, in the direction perpendicular to the rotation axis, rotation can stabilize the largest scales from gravitational collapse. This is nothing else than the fact that angular momentum is an enemy of star formation, and that instead of collapsing into a point, the cloud must now collapse into a disk.

If we consider the rotating disk, the Coriolis force is protected to contraction of gas $\implies$ instability condition is changing (Toomre, 1964). To derive a more realistic instability criterion it is necessary to take into account rotation. In a rotation frame of reference it is necessary to add the Coriolis and the centrifugal accelerations to the hydrodynamic equations.

Is Jupiter a brown dwarf star?

Brown dwarfs are failed stars of mass less than $80\ m_{\text{Jupiter}}$ but more than at least 13 m_{Jupiter}—that is, the mass of Jupiter is less than $0.075\ m_{\text{Brown-dwarf}}$. Jupiter is too small to be a brown dwarf star and even too small for a gas giant planet as described in the main text. Moreover, Jupiter is too cold to be a brown dwarf. In other words, if Jupiter were a brown dwarf, it would be hotter and brighter than it is now.

Determine the amount of energy released by 1 kg of infalling matter onto (1) White dwarf with $m_{wd} = 0.85 m_{\text{sun}}$ and $r_{wd} = 6.6 \times 10^6\,\text{m} = 0.0095\,r_{\text{sun}}$ and (2) neutron star with $m_{ns} = 1.5 m_{\text{sun}}$ and $r_{ns} = 10\text{km}$.
(1) $\Delta U = \frac{Gm}{r} = 1.71 \times 10^{13}$ J (2) $\Delta U = \frac{Gm}{r} = 1.86 \times 10^{16}$ J.

References

Achten, C., and Andersson, J.T., 2015, Overview of Polycyclic Aromatic Compounds (PAC), *Polycycl Aromat Compd.*, **35**, 177–186.

Acuna, M.H. and Ness, N.F., 1976, The main magnetic field of Jupiter, *J. Geophys. Res.*, **81**, 2917–2922.

Airy, G.B., 1855, On the computations of the effect of the attraction of the mountain masses as disturbing the apparent astronomical latitude of stations in geodetic surveys, *Phil. Trans. R. Soc. London*, **145**, 101–104.

Alfvén, H., *Cosmical Electrodynamics*, Oxford Univ. Press, New York, 1950.

Alfvén, H., and Fälthammar, C.-G., *Cosmical Electrodynamics, Fundamental Principles*, Oxford Univ. Press, New York, 1963.

Altamimi et al., 2016: ITRF2014: A new release of the International Terrestrial Reference Frame modeling nonlinear station motions, *J. Geophys. Res. Solid Earth*, **121**.

Altamimi, Z. and Gross, R., 2017, *Geodesy*, In: Teunissen P, Montenbruck O (eds), Springer Handbook of Global Navigation Satellite Systems, Springer, Heidelberg, chap 36, 1039–1059.

Alvarez, L.W., Alvarez, W., Asaro, F., and Michel, H., 1980, Extraterrestrial cause for the Cretaceous/Tertiary extinction: *Science*, **208**, 1095–1108.

Aoki, I., 1991, Entropy principle for human development, growth and aging, *J. Theor. Biol.*, **150**, 215–223.

Argus, D.F., Gordon, R.G., Heflin, M.B., et al., 2010, The angular velocities of the plates and the velocity of Earth's centre from space geodesy *Geophysical Journal International*, **180**, 913–960.

Armstrong, J.A., Bloembergen, N., Ducuing, J., and Pershan, P.S., 1962, Interactions between light waves in a nonlinear dielectric, *Phys. Rev.*, **127**, 1918–1939.

Arnett, D., 1996, *Supernovae and Nucleosynthesis*, Princeton University Press.

Aspect, A., Grangier, P., and Roger, G., 1982, Experimental realization of Einstein-Podolsky-Rosen-Bohm Gedanken experiment: A new violation of Bell's inequalities, *Phys. Rev. Lett.*, **49**, 91–94.

Azevedo, F.A., Carvalho, L.R., Grinberg, L.T., Farfel, J.M., Ferretti, R.E., et al., 2009, Equal numbers of neuronal and nonneuronal cells make the human brain an isometrically scaled-up primate brain, *J Comp Neurol.*, **513**, 532–541.

Babcock, H.W., 1939, The rotation of the Andromeda Nebula, *Lick Observatory Bulletin*, **19**, 41–51.

Balbus, S., and Hawley, J., 1991, Instability, turbulence, and enhanced transport in accretion disks, *Astrophys. J.*, **376**, 214.

Ball, P., 2011, Entangled diamonds vibrate together, *Nature*, **332**.

Barnes, R., and Raymond, S.N., 2004, Predicting planets in known extrasolar planetary systems. I. Test particle simulations, *The Astrophysical Journal*, **617**, 569–574.

Baule, G., and McFee, R., 1963, Detection of the magnetic field of the heart, *American heart journal*, **66**, 95–96.

Baur, O., and Grafarend, E.W., 2006, High-Performance, GOCE Gravity Field Recovery from Gravity Gradients Tensor Invariants and Kinematic Orbit Information. In: Flury, J., Rummel, R., Reigber, C., Rothacher, M., Boedecker, G., Schreiber, U. (eds.): Observation of the Earth System from Space. Springer, New York, NY

Becker, E.D., 1969, *High Resolution NMR*, Academic Press.

Beckman, J.E., and Mahoney, T.J., 1998, The Maunder minimum and climate change: have historical records aided current research? *ASP Conference Series*, **153**, 212.

Bell, J.S., 1964, On the Einstein Podolsky Rosen paradox, *Physics*, **1**, 195–200.

Benabid, A.L., 2007, What the future holds for deep brain stimulation, *Exp. Rev. Med. Devices*, **4**, 895–903.

Bentridi, S.E., Gall, B., Hidaka, H., Amrani, N., Gauthier-Lafaye, F., Investigation of clay and neutron absorbers' roles in the genesis and evolution of Oklo natural nuclear reactors

Ben-Naim, A., 2007, *Statistical Thermodynamics Based on Information*, World Scientific.

Blakley, J., 1995, *Potential Theory in Gravity and Magentics*, Cambridge University Press, New York.

Bock, Y., and Melgar, D., 2016, Physical applications of GPS geodesy: A review, *Reports on Progress in Physics*, **79**.

Bodu, R., et al., 1972, Sur l'existence d'anomalies isotopiques rencontrées dans l'uranium du Gabon, *CR Académie des Sciences Paris*, **275**, 1731.

Bogovalov, S., and Borman, V., 2016, Separative Power of an Optimised Concurrent Gas Centrifuge, *Nuclear Engineering and Technology. Elsevier BV.*, **48**, 719–726.

Boltzmann, L.E., 1974, *The second law of thermodynamics (theoretical physics and philosophical problems)*, Springer-Verlag, NY, USA.

Bondeson, A., Iacono, R., and Bhattacharjee, A.H., 1987, Local magnetohydrodynamic instabilities of cylindrical plasma with sheared equilibrium flows, *Phys. Fluids*, **30**, 2167–2180.

Boto, E., Holmes, N., Leggett, J., Roberts, G., et al., 2018, Moving magnetoencephalography towards real-world applications with a wearable system, *Nature*, **555**, 657–6612.

Boyd, R.W., 1999, Order-of-magnitude estimates of the nonlinear optical suscepti-bility, *J. Mod. Opt.*, **46**, 367–378.

Braeutigam, S., 2013, Magnetoencephalography: Fundamentals and Established and Emerging Clinical Applications in Radiology, *ISRN Radiology*, **2013**, 529463.

Braunwald, E., 1997, Cardiovascular Medicine at the Turn of the Millennium: Tri-umphs, Concerns, and Opportunities, *The New England Journal of Medecine*, **337**, 1360–1369.

Bridgman, P.W., 1941, *The nature of thermodynamics*, Harvard University Press.

Buck, L., and Axel, R., 1991, A novel multigene family may encode odorant recep-tors: a molecular basis for odor recognition, *Cell.*, **65**, 175–187.

Bull, A.J., 1921, A hypothesis of mountain building, *Geol. Mag.*, **58**, 364–397.

Burton, D.A., and Noble, A., 2014, On the entropy of radiation reaction, *Physics Letters A*, **378**, 1031–1035. ISSN 0375-9601,

Carlson, C., 2010, Wada you do for language: fMRI and language lateralization?, *Epilepsy Curr.*, **10**, 86–88.

Carathéodory, C., 1925, 'Über die Bestimmung der Energie und der absoluten Tem-peratur mit Hilfe von reversiblen...

Carnot, S., 1824, *Reflections on the Motive Power of Fire*, translation in 2005, Dover, New York.

Carr, B., Kühnel, F. and Sandstad, M., 2016, Primordial black holes as dark matter, *Phys. Rev. D*, **94**, 083504.

Carslaw, H.S., and Jaeger, J.C., 1959, *Conduction of heat in solids*, 2nd edition, Oxford University Press, London.

Carter, R., 2019, *The human brain book*, DK Publishing

Castillo-Rogez, J.C., and Lunine, J.I., 2010, Evolution of Titan's rocky core con-strained by Cassini observations, *Geophys. Res. Lett.*, **37**, L20205.

Çengel, Y.A, 2007, *Heat and mass transfer. A practical approach*, McGraw-Hill.

Cerjan, C., Kosloff, D., Kosloff, R., and Reshef, M., 1985, A nonreflecting boundary condition for discrete acoustic and elastic wave equations, *Geophysics*, **50**, 530–743.

Chandrasekhar, S., 1960, The Stability of Non-dissipative Couette flow in hydro-magnetics, *Proceedings of the National Academy of Sciences of the United States of America*, **46**, 253–257.

Chandrasekhar, S., 2010, *An introduction to the study of stellar structure*, Dover, New York.

Cherry, E.C., 1953, Some experiments on the recognition of speech, with one and with two ears, *Journal of the Acoustical Society of America*, **25**, 975–979.

Cherry, E.C., 1957, *On human communication: A review, survey, and a criticism*, Cambridge, MA: MIT Press.

Cherry, E.C., 1961, *Two ears—but one world. In W. A. Rosenblith (Ed.), Sensory communication*, 99–117, New York: Wiley.

Cherry, E.C., and Sayers, B., 1956, Human "cross-correlation"—A technique for measuring certain parameters of speech perception, *Journal of the Acoustical Society of America*, **28**, 889–895.

Cherry, E.C., and Sayers, B., 1959, On the mechanism of binaural fusion, *Journal of the Acoustical Society of America*, **31**, 535.

Cherry, E.C., and Taylor, W.K., 1954, Some further experiments upon the recognition of speech, with one and, with two ears, *Journal of the Acoustical Society of America*, **26**, 554–559.

Clausius, R.J.E., 1850, *The mechanical theory of heat*: translation in 2008, Kessinger, Whitefish, Montana.

Clausius, R., 1865, Ueber verschiedene für die Anwendung bequeme Formen der Hauptgleichungen der mechanischen Wärmetheorie, *Ann. Phys.*, **201**, 353–400.

Coates, A.J., Wellbrock, A., Lewis, G.R., Jones, G.H., et al. 2009. Heavy Negative Ions in Titan's Ionosphere: Altitude and Latitude Dependence. *Planet. Space Sci.*, **57**, 1866–1871.

Coccarelli, A., Boileau, E., Parthimos, D., and Nithiarasu, P., 2016, An advanced computational bioheat transfer model for a human body with an embedded systemic circulation, *Biomech Model Mechanobiol*, **15**, 1173–1190.

Cohen, D., Edelsack, E.A., and Zimmerman, J.E., 1970, Magnetocardiograms taken inside a shielded room with a superconducting point-contact magnetometer, *Appl. Phys. Lett.*, **16**, 278.

Connerney, J.E.P., Acuna, M.H., and Ness, N.F., 1982, Voyager 1 assessment of Jupiter's planetary magnetic field, J. Geophys. Res. 87, 3623–3627.

Connerney, J.E.P., Acuna, M.H., Ness, N.F., and Satoh, T., 1998, New models of Jupiter's magnetic field constrained by the Io flux tube footprint, *J. Geophys. Res.*, **103**, 11,929–11,940.

Coppersmith, D., 1994, An approximate Fourier transform useful in quantum factoring, *Technical Report RC19642, IBM*.

Cornish, N.J., 1998, *The Lagrange points*, WMAP Education and Outreach.

Cox, A. and Hart, R.B., 1986, *Plate tectonics. How it works*, Blackwell Scientific Publications, Oxford.

Cuffaro, M., and Doglioni, C., 2007, Global kinematics in deep versus shallow hotspot reference, *Geological Society of America Special Paper*, **430**, 359–430.

Daniels, F., and Alberty, R.A., 1966, *Physical chemistry*, John Wiley and Sons, Inc.

Darwin, G.H., 1901, *The Tides*, John Murray, London, England.

Delrez, L., Gillon, M., Triaud, A., et al., 2018, Early 2017 observations of TRAPPIST-1 with Spitzer, *Monthly Notices*, **475**, 3577–3597.

DeMets, C., Gordon, R.G., Argus, D.F., and Stein, S., 1990, Current plate motions, *Geophys. J. Int.*, **101**, 425–478.

DeMets, C., Gordon, R.G., Argus, D.F., and Stein, S., 1994, Effect of recent revisions to the geomagnetic reversal time scale on estimates of current plate motions, *Geophys. Res. Lett.*, **21**, 2191–2194.

Deng, Z.S., and Jin, J., 2002, Analytical study on bioheat transfer problems with spatial or transient heating on skin surface or inside biological bodies, *Journal of Biomechanical Engineering*, **124**, 638–650.

de Sitter, W., 1934, On distance, magnitude, and related quantities in an expanding universe, *Bulletin of the Astronomical Institutes of the Netherlands*, **7**, 205.

Dirac, P.A.M., 1931, Quantised singularities in the electromagnetic field, *Proceedings, of The Royal Society A*, **133**, 60–72.

Doodson, A.T., 1921, The Harmonic Development of the Tide-Generating Potential, *Proceedings of the Royal Society of London Series A*, **100**, 305–329.

Dougherty, M.K., Balogh, A., Southwood, D.J., and Smith, E.J., 1996, Ulysses assessment of the jovian planetary field, *J. Geophys. Res.*, **101**, 24,929–24,942.

Drazin, P.G., and Reid, W.H, 2004, *Hydrodynamic stability*, Cambridge University Press, UK.

Dronkers, J.J., 1964, *Tidal Computations*: North-Holland Publishing Company, Amsterdam, The Netherlands.

Dunbar, R.C., 1982, Deriving the Maxwell Distribution, *J. Chem. Ed.*, **59**, 22–23.

Dür, W., Vidal, G., and J Ignacio Cirac, J.I., Three qubits can be entangled in two inequivalent ways, *Physical Review A*, 62, 43–44.

Durante, D., Hemingway, D.J., Racioppa, P., Less, L., and Stevenson, D.J., 2019, Titan's gravity field and interior structure after Cassini, *Icarus*, **326**, 123–132.

Egbert, G.D., and Erofeeva, S., 2002, Efficient inverse modeling of barotropic ocean tides, *Journal of Atmospheric and Oceanic Technology*, **19**, 183–204.

Einstein, A., Podolsky, B., and Rosen, N., 1935, Can Quantum-Mechanical Description of Physical Reality Be Considered Complete? *Phys. Rev.*, **47**, 777–780.

Ellis, D.V., Singer, J.M., 2007, *Well logging for earth scientists*, Second edition, Springer.

Ewert, H., Popov, S.V., Richter, A., Schwabe, J., Scheinert, M., and Dietrich, R., 2012, Precise analysis of ICESat altimetry data and assessment of the hydrostatic equilibrium for subglacial Lake Vostok, East Antarctica, *Geophys. J. Int.*, **191**, 557–568.

Faraday, M., 1834, Experimental researches on electricity, *7th series. Phil. Trans. R. Soc. (Lond.)*, **124**, 77–122.

Farmer, P.B., et al., 2003, Molecular epidemiology studies of carcinogenic environmental pollutants. Effects of polycyclic aromatic hydrocarbons (PAHs) in environmental pollution on exogenous and oxidative DNA damage, *Mutat Res.*, **544**, 397–402.

Farquhar, R., 1966, Station-keeping in the vicinity of collinear libration points with an application to a lunar communications problem: AAS Science and Technology Series: *Space Flight Mechanics Specialist Symposium*, **11**, 519–535.

Fiala, D., 1999, A computer model of human thermoregulation for a wide range of environmental conditions: The passive system, *J Appl Physiol*, **87**, 1957–1972.

Fiedler, C., and Burton, D.A., 2021, Analogue Hawking temperature of a laser-driven plasma, *Physics Letters A*, **403**, 127380.

Filipov, S., and Faragó, L., 2018, Implicit euler time discretization and FDM mewton in nonlinear heat transfer modeling.

Fischetti. M., and Christiansen, J., 2021, A New You in 80 Days: Cell turnover is vast and swift, *Scientific American*, **324**, 76.

Fisher, O., 1881, *Physics of the Earth's crust*, Macmillan and Co, London, UK.

Floberghagen, R., Fehringer, M., Lamarre, D., Muzi, D., Frommknecht, B., et al., 2011, Mission design, operation and exploitation of the gravity field and steady-state ocean circulation explorer mission, *J. Geod.*, **85**, 749–758.

Forsyth, D., and Uyeda, S., 1975, On the Relative Importance of the driving forces of plate motion, *Geophysical Journal International*, **43**, 163–200.

Fourier, J.B.J., 2009, *The Analytical Theory of Heat*: Cambridge Press. This reprint of the 1878 English translation of the 1822 French original, *Théorie Analytique de la Chaleur is a tour de force*.

Fowler, C.M.R., 1995, *The Solid Earth. An Introduction to Global Geophysics*, Cambridge University Press.

Francoa, M.I., Turina, L., Mershin, A., and Skoulakisa, E.M.C., 2011, Molecular vibration-sensing component in Drosophila melanogaster olfaction, *PNAS*, **108**, 3797–3802.

Frieman, E., and Rotenberg, M., 1960, On the hydromagnetic stability of stationary equilibria, *Rev. Modern Physics*, **32**, 898–902.

Gauthier-Lafaye F., Holliger, P., and Blanc, P.-L., 1996, Natural fission reactors in the Franceville basin, Gabon: A review of the conditions and results of a "critical event" in a geologic system, *Geochim. Cosmochim. Acta*, **60**, 4831–4852.

Gerlach, W., and Stern, O., 1922, Der experimentelle Nachweis der Richtungsquantelung im Magnetfeld, *Zeitschrift für Physik*, **9**, 349–352.

Geselowitz, D.B., 1970, On the Magnetic Field Generated outside an Inhomogeneous Volume Conductor by Internal Current Sources, *IEEE Transactions in Biomagnetism*, **6**, 346–347.

Gibbs, J.W., 1957, *The collected works of J. Willard Gibb*:, Yale U. Press, New Haven.

Goldman, D.E., 1943, Potential, impedance, and rectification in membranes, *The Journal of general physiology*, **27**, 37–60.

Gordon, R.G., Argus, D.F., and Heflin, M.B., 1999, Revised estimate of the angular velocity of India relative to Eurasia, *Trans. Am. Geophys. Union*, **80**, 273.

Gray, D.M., Burton-Johnson, A., and Fretwell, P.T., 2019, Evidence for a lava lake on Mt.Michael volcano, Saunders Island (South Sandwich Islands) from Landsat, Sentinel-2 and ASTER satellite imagery, *Journal of Volcanology and Geothermal Research*, **379**, 60–71.

Greenberger, D.M., Horne, M.A., and Zeilinger, A., 1989, Bell's theorem, quantum theory, and conceptions of the universe, *Kluwer, Dordrecht*, 73–76.

Gripp, A.E., and Gordon, R.G., 1990, Current plate velocities relative to the hotspots incorporating the NUVEL-1 global plate motion model, *Geophys. Res. Lett.*, **17**, 1109–1112.

Gripp, A.E., and Gordon, R.G., 2002, Young tracks of hotspots and current plate velocities, *Geophys. J. Int.*, **150**, 321–361.

Grover, L., 1996, A fast quantum mechanical algorithm for database search. *InProceedings of 28th ACM Symposium on Theory of Computing*, 212–219.

Grover, L.K., 1997, Quantum mechanics helps in searching for a needle in a haystack, *Phys. Rev. Lett.*, **79**, 325.

Gühne, O., and Tóth, G., 2009, Entanglement detection, *Physics Reports*, **474**, 1–75.

Halder, M., Beveratos, A., Gisin, N., Scarani, V., Simon, C., and Zbinden, H., 2007, Entangling independent photons by time measurement, *Nature Phys.*, **3**, 692–659.

Halliday, D., Resnick, R., and Walker, J., 2011, *Fundamentals of Physics*: 9th ed., Wiley, New York.

Hämäläinen, M.S., Hari, R., Ilmoniemi, R.J., Knuutila, J., and Lounasma, O.V., 1993, Magnetoencephalography: theory, instrumentation, and applications to non-invasive studies of the working human brain, *Rev. Mod. Phys.*, **65**, 413–497.

Hampson R., Song D., Opris I., Santos L., Shin D., Gerhardt G., et al.. (2013). Facilitation of memory encoding in primate hippocampus by a neuroprosthesis that promotes task-specific neural firing, *J. Neural Eng.*, **10(6)**, 066013.

Han, S.-C., Shum, C.K., Bevis, M., Ji, C., and Kuo, C.-Y., 2006, Crustal dilatation observed by GRACE after the 2004 Sumatra–Andaman earthquake, *Science*, **313**, 658–662.

Hartle, J.B., 2003, *Gravity: An Introduction to Einstein's General Relativity*, Benjamin Cummings.

Hatch, R., Jung, J., and Enge, P., 2000, Civilian GPS: the benefits of three frequencies, *GPS Solut.*, **3**, 1–9.

Hawley, J.F., and Balbus, S.A., 1991, A powerful local shear instability in weakly magnetized disks. II. Nonlinear evolution, *Astrophysical Journal*, **376**, 223.

Hayford, J.F., and Baldwin, A.L., 1908, The California Earthquake of April 18, 1906: *Earthquake Investigation Commission, Carnegie Inst.*, **1**, 114–115.

Heath, T.L., 2010, *The Works of Archimedes, Edited in modern notation*: Cambridge Univ. Press, Cambridge, UK.

Hellard, H., Csizmadia, S., Padovan, S., et al., 2019, Retrieval of the fluid Love number k_2 in exoplanetary transit curves, *The Astrophysical Journal*, **878**, 119.

Herculano-Houzel, S., 2009, The human brain in numbers: a linearly scaled-up primate brain, *Frontiers in Human Neuroscience*, **3**, 31.

Herculano-Houzel S., 2012, Neuronal scaling rules for primate brains: the primate advantage, *Prog. Brain Res.*, **195**, 325–340.

Hofmann-Wellenhof, B., Lichtenegger, H., and Collins, J., 1992, *Global Positioning System: Theory and Practice*, Springer-Verlag Wi en New York

Holmes, A., 1931, Radioactivity and Earth movements: Trans. geol. Soc. Glasgow 1928-1929, **18**, 559–606.

Horodecki, R., Horodecki, P., Horodecki, M., and Horodecki, K., 2009, Quantum entanglement, *Reviews of Modern Physics*, **81**, 865–942.

Howell, K.C., 1984, Three-dimensional, periodic "Halo" orbits, *Celestial Mechanics*, **32**, 53–71.

Hubbard, W.B., 1984, *Planetary interiors*, Van Nostrand Reinhold, New York.

Hubble, E., 1929, A relation between distance and radial velocity among extra-galactic nebulae, *Proceedings of the National Academy of Sciences of the United States of America*, **15**, 168–173.

Ikelle, L.T., Diet, J.P., and Tarantola, A., 1988. Linearized inversion of multioffset seismic reflection data in $\omega - k$ domain: Depth-dependent reference medium, *Geophysics*, **53**, 50–64.

Ikelle, L.T., 1995, Linearized inversion of 3-D multi-offset data: background reconstruction and AVO inversion, *Geophys. J. Int.*, **123**, 507–528.

Ikelle, L. T., and Amundsen, L., 2005, *Introduction to Petroleum Seismology*: first edition, SEG, Tulsa, Okhlahoma.

Ikelle, L. T., 2010, *Coding and Decoding: Seismic Data*, First edition, Elsevier Science.

Ikelle, L. T., 2017a, *Coding and Decoding: Seismic Data, The concept of Multishooting*, Second edition, Elsevier Science.

Ikelle, L. T., 2017b, *Introduction to Earth sciences: a physics approach*, first edition, World Scientific, Singapore.

Ikelle, L. T., and Amundsen, L., 2018, *Introduction to Petroleum Seismology*: second edition, SEG, Tulsa, Okhlahoma.

Ikelle, L.T., 2020, *Introduction to Earth sciences: a physics approach*, Second edition, World Scientific, Singapore.

Ikelle, L.T., 2023a, *Introduction to Multidisciplinary Science at Artificial-Intelligence Age: The Matter in our Universe, Biological Cells, and Plate Tectonics*, Nova Science Publishers, New York.

Ikelle, L.T., 2023b, *Introduction to Multidisciplinary Science at Artificial-Intelligence Age: Chemical, Nuclear, and Thermonuclear Reactions, and Oxygenic and Anoxygenic Photosyntheses*, Nova Science Publishers, New York.

Ikelle, L.T., 2023b, *Introduction to Multidisciplinary Science at Artificial-Intelligence Age: Elasticity, Permeability, Porosity, Viscosity, and Wettability*, Nova Science Publishers, New York.

Ikelle, L.T., 2024, *Introduction to Multidisciplinary Science at Artificial-Intelligence Age: A Binary, Cyclical, Spectroscopic, and Multiple-Time Universe*, In preparation.

Imanaka, H., Khare, B. N., Elsila, J. E., Bakes, E. L. O., et al. 2004. Laboratory experiments of Titan tholin formed in cold plasma at various Pressures:

Implications for nitrogen-containing polycyclic aromatic compounds in titan haze, *Icarus*, **168**, 344–366.

Israël, G., Szopa, C., Raulin, F., Cabane, M., et al., 2005, Complex Organic Matter in Titan's Atmospheric aerosols from in Situ pyrolysis and analysis, *Nature*, **438**, 796–799.

Jackson, J.D., 1998, *Classical electrodynamics*: third edition, John Wiley.

Jakubke, H.-D., and Jeschkeit, H., 1994, *Concise encyclopedia chemistry*, trans. rev. Eagleson, Mary. Berlin: Walter de Gruyter.

Kathuria, P., Sharma, P., Abendong, M.N., and Wetmore, S.D., 2015, Conformational preferences of DNA following damage by aristolochic acids: Structural and Energetic Insights into the Different Mutagenic Potential of the ALI and ALII-N^6-dA Adducts, *Biochemistry*, **54**, 2414–2428.

Keller, G.V., and Frischknecht, F.C., 1966, *Electrical methods in geophysical prospecting*, Pergamon Press Inc., Oxford, UK.

Khan, A.Q., Atta, M.A., and Mirza, J.A., 1986, Flow Induced Vibrations in Gas Tube Assembly of Centrifuge, *Journal of Nuclear Science and Technology*, **23**, 819–827.

Khan, A.Q., Suleman, M., Ashraf, M., and Khan, M.Z., 1987, Some Practical Aspects of Balancing an Ultra-Centrifuge Rotor, *Journal of Nuclear Science and Technology*, **24**, 951–959.

Kikuchi, M., Lackner, K., and Tran, M.Q., 2012, *Fusion physics*, IAEA International Atomic Energy Agency, Vienna, Austria.

Kippenhahn, R., and Weigert, A., 1990, *Stellar structure and evolution*, Springer-Verlag.

Knopoff, L., and Leeds, A., 1972, Lithospheric momenta and the deceleration of the earth, *Nature*, **237**, 93–95.

Kopal, Z., 1959, *Close binary systems*, Wiley, New York.

Kozak, K., Ruman, M., Kosek, K., Karasiński, G., Stachnik, L., and Polkowska, Z., 2017, Impact of volcanic eruptions on the occurrence of PAHs compounds in the aquatic ecosystem of the southern part of west spitsbergen (Hornsund Fjord, Svalbard), *Water*, **9**, 42.

Labrosse, S., and Macouin, M., 2003. The inner core and the geodynamo, *C. R. Geosciences*, **335**, 37–50.

Lagrange, J.-L., 1867, *Oeuvres de Lagrange*, Les Soins de M. J.-A. Serret, France.

Lainey, V., Jacobson, R.A., Tajeddine, R., et al., 2017, New constraints on Saturn's interior from Cassini astrometric data, *Icarus*, **281**, 286–296.

Lambeck, K., 1988, *Geophysical geodesy*: Oxford University Press.

Leick, A., 2003, *GPS Satellite Surveying*, 3rd ed/John Wiley and Sons, New York.

Le Pichon, X., 1968, Sea-floor spreading and continental drift, *J. Geophys. Res.*, **73**, 3661–3697.

Levander, A.R., 1988, Fourth-order finite-difference P-SV seismograms, *Geophysics*, **53**, 1425–1436.

Li, C., et al., 2021, Overview of the Chang'e-4 mission: opening the frontier of scientific exploration of the lunar far side, *Space Science Reviews*, **217**, 35.

Li, H.-Y., Gao, P.-P., and Ni, H.-G., 2019, Emission characteristics of parent and halogenated PAHs in simulated municipal solid waste incineration, *Sci. Total Environ.*, **665**, 11–17.

Li, H.Y., and Gregory, S., 1974, Diffusion of ions in sea water and in deep-sea sediments, *Geochimica et Cosmochimica Acta*, **38**, 703–714.

Lin, Y.C., Li, Y.C., Shangdiar, S., Chou, F.C., Sheu, Y.T., and Cheng, P.C., 2019, Assessment of PM2.5 and PAH content in PM2.5 emitted from mobile source gasoline-fueled vehicles in concomitant with the vehicle model and mileages. *Chemosphere*, **226**, 502–508.

Lindley, D., 2001, *Boltzmann's atom*, The Free Press, New York.

Liboff, L.R., 2003, *Introductory Quantum Mechanics*: Fourth Edition, Addison-Wesley.

Lorenz, R.D., Turtle, E.P, Stiles, B., Le Gall, A., et al., 2011, Hypsometry of Titan, *Icarus*, **211**, 699–706.

Love, A.E.H., 1911, *Some Problems of Geodynamics*: Cambridge Univ. Press.

Lozano, A., and Lipsman, N., 2013, Probing and regulating dysfunctional circuits using deep brain stimulation, Neuron, **77**, 406–424.

Lundquist, S., 1949, Experimental investigations of magneto-hydrodynamic waves, *Physical Review*, **107**, 1805–1809.

Madariaga, R., 1976, Dynamics of an expanding circular fault, *Bulletin of the Seismological Society of America,* **66**, 639–666.

Maeder, A., 2009, *Physics, formation and evolution of rotating Stars*: Springer-Verlag, Berlin, Germany.

Malmivuo, J., and Plonsey, R., 1995, *Bioelectromagnetism: Principles and Applications of Bioelectric and Biomagnetic Fields*, Oxford University Press.

Maniati, K., Haralambous, K.J., Turin, L., and Skoulakis, E.M.C., 2017, Vibrational detection of odorant functional groups by drosophila melanogaste, *eneuro*, **4**, 5.

Martin-Lopez, E., Laing, A., Lawson, T., Alvarez, R., Zhou, X.-Q., and O'brien, J.L., 2012, Experimental realization of shor's quantum factoring algorithm using qubit recycling, *Nature photonics*, **6**, 773–776.

Maunder, E.W., 1894, A prolonged sunspot minimum, *Knowledge*, **17**, 173–176.

Mazess, R.B., 1982, On aging bone loss, *Clin. Orthoped. Rel. Res.*, **165**, 239–252.

McConnell, R., Zhang, H., Hu, J., Ćuk, S., and Vuletić, V., 2015, Entanglement with negative Wigner function of almost 3,000 atoms heralded by one photon, *Nature*, **519**, 439–442.

McGlashan, M.L.J., 1966, The use and missuse of the law of thermodynamics, *Journal of Chemical Education*, **43**, 226.

McKay, C.P., Pollack, J.B., Courtin, R., 1991, The greenhouse and antigreenhouse effects on Titan, *Science*, **253**, 1118–1121.

McKenzie, D., Jackson, J., and Priestley, K., 2005, Thermal structure of oceanic and continental lithosphere, *Earth and Planetary Science Letters*, **233**, 337–349.

Miller, R.C., 1964, Optical second harmonic generation in piezoelectric crystals, *Applied Physics Letters*, **5**, 17.

Minster, J.B., and Jordan, T.H., 1978, Present-day plate motions, *J. Geophys. Res.*, **83**, 5331–5354.

Morgan, W.J., and Morgan, J.P., 2007, In G. R. Foulger, & D. M. Jurdy (Eds.), *Plate Velocities in the Hotspot Reference frame:* Special papers- Geological Society of America, Geological Society of America Bulletin, **430**, 65–78 Boulder, Colorado.

Morse, P.M., 1929, Diatomic molecules according to the wave mechanics. II. Vibrational levels, *Phys. Rev.*, **34**, 57.

Nagornyi, V.D., 1995, A new approach to absolute gravimeter analysis, *Metrologia*, **32**, 210–08.

Nakariakov, V.M., Ofman, L., DeLuca, E.E., Roberts, B., and Davila, J.M., 1999, TRACE observation of damped coronal loop oscillations: Implications for coronal heating, *Science*, **285**, 862–864.

Naudet, R., 1991, *OKLO: Des réacteurs nucléaires fossiles. Étude physique*, Eyrolles, Paris, France.

Nernst, W.H, 1888, Zur Kinetik der Lösung befindlichen Körper Theorie der Diffusion, *Z. Phys. Chem.*, **3**, 613-637.

Nernst, W.H., 1889, Die elektromotorische Wirksamkeit der Ionen, *Z. Phys. Chem.* **4**, 129–181.

Ni, D., 2018, Empirical models of Jupiter's interior from Juno data, Moment of inertia and tidal Love number k_2: Astronomy & Astrophysics, **613**, A32, 1–9.

Nilas, L., and Christiansen, C., 1987, Bone mass and its relationship to age and the menopause, *J. Clin. Endocrinol. Metab.*, **65**, 697–702.

Norton, I.O., 2000, Global hotspot reference frames and plate motion, *Washington DC American Geophysical Union Geophysical Monograph Series*, **121**, 339–357.

Pail, R., et al., 2011, First GOCE gravity field models derived by three different approaches, *Journal of Geodesy*, **85**, 819.

Parijskij, Y.N, 2001, The High Redshift Radio Universe In Sanchez, Norma (ed.). *Current Topics in Astrofundamental Physics*, Springer, page 223.

Parker, E.N., 1996, The alternative paradigm for magnetospheric physics: J. Geophys. Res., **101**, 10587–10625.

Pass, J., 2008, Space medicine: Medical astrosociology in the sickbay: *46th AIAA Aerospace Sciences Meeting and Exhibit*, Reno, Nevada Astrosociology.com, Huntington Beach, CA, 92647

Pennes, H.H., 1948, Analysis of tissue and arterial blood temperatures in the resting human forearm, *J. Appl. Physiol.*, **1**, 93–122.

Perera, F.P., 2000, Molecular epidemiology: on the path to prevention? *J Natl Cancer Inst*, **92**, 602–612.

Perera, F.P., Wang, S., Vishnevetsky, J., Zhang, B., Cole, K.J., Tang, D., et al., 2011, PAH/aromatic DNA adducts in cord blood and behavior scores, *in New York City children. Environ Health Perspect*, **119**, 1176–1181.

Perrin, F., 1928, *Thèse de Francis Perrin: Etude mathématique du mouvement Brownien de rotation*, Université de Paris.

Perrin, F., 1939, Calcul relatif aux conditions eventuelles de transmutation en chaine de l'uranium, *Comptes Rendus*, **208**, 1394–1396.

Planck, M., 1897, *Treatise on thermodynamics*: translation in 1990, Dover, New York.

Poirier, M.C., Weston, A., Schoket, B., Shamkhani, H., Pan, C.F., Mcdiarmid, M.A., Scott, B.G., Deeter, D.P., Heller, J.M., Jacobson-Kram, D., and Rothman, N., 1998, Biomonitoring of United States Army soldiers serving in Kuwait in 1991, *Cancer Epidemiol Biomarkers Prev*, **7**, 545–551.

Poirier, M.C., Beland, F.A., 1992, DNA adduct measurements and tumor incidence during chronic carcinogen exposure in animal models: implications for DNA adduct-based human cancer risk assessment, *Chem Res Toxicol*, **5**, 749–755.

Prialnik, D., 2009, *An introduction to the theory of stellar structure and evolution*, 2nd edition, Cambridge University Press, Cambridge, U.K.

Pratt, J.H., 1855, On the attraction of the Himalaya Mountains, and of the elevated regions beyond them, upon the plumb line in India, *Phil. Trans. R. Soc. London*, **145**, 53–100.

Raffenach, J.C., Menes, J., Devillers, C., Lucas, M., and Hagemann, R., 1976, Études chimiques et isotopiques de l'uranium, du plomb et de plusieurs produits de fission dans un échantillon de minerai du réacteur naturel d'Oklo, *Earth Planet. Sci. Lett.*, **30**, 94–108.

Randall, B.A., An improved magnetic field model for Jupiter's inner magnetosphere using a microsignature of Amalthea, *J. Geophys. Res.*, **103**, 17,535–17,542.

Rankine, W.J.M., 1908, *A Manual of the Steam Engine and Other Prime Movers*: Seventeenth Edition, Griffin, London.

Reid, H.F., 1910, The Mechanics of the Earthquake (Washington, DC: Carnegie Institution of Washington)

Reizer, E., Viskolcz, B., and Fiser, B., 2022, Formation and growth mechanisms of polycyclic aromatic hydrocarbons: A mini-review, *Chemosphere*, *291*, 132793

Reynolds, J., 1997, *An Introduction to Applied and Environmental Geophysics*, John Wiley, Chichester, New York.

Reynolds, O., 1903, *Papers on Mechanical and Physical Subjects, Volume III, The Sub-Mechanics of the Universe*, Cambridge University Press, Cambridge, UK.

Reynolds, W.C., 1968, *Thermodynamics*: Second Edition, McGraw-Hill, New York. Cambridge University Press, Cambridge, New York.

Roche, E., 1849, La figure d'une masse fluide soumise à l'attraction d'un point éloigné (The figure of a fluid mass subjected to the attraction of a distant point), part 1, *Académie des sciences de Montpellier: Mémoires de la section des sciences*, **1**, 243–262.

Roche, E., 1850, La figure d'une masse fluide soumise à l'attraction d'un point éloigné (The figure of a fluid mass subjected to the attraction of a distant point), part 2, *Académie des sciences de Montpellier: Mémoires de la section des sciences*, **1**, 333–348.

Roche, E., 1851, La figure d'une masse fluide soumise à l'attraction d'un point éloigné (The figure of a fluid mass subjected to the attraction of a distant point), part 3, *Académie des sciences de Montpellier: Mémoires de la section des sciences*, **2**, 21–32.

Rubin, V.C., and Ford, W.K. Jr., 1970, Rotation of the Andromeda Nebula from a Spectroscopic Survey of Emission Regions, *Astrophys. J.*, **159**, 379.

Rubin, V.C., Ford, Jr., W.K., and Thonnard, N., 1980, Rotational properties of 21 SC galaxies with a large range of luminosities and radii, from NGC 4605 /R = 4kpc/ to UGC 2885 /R = 122 kpc/, *Astrophys. J.*, **238**, 471.

Rummel, R., Balmino, G., Johannessen, J., Visser, P., and Woodworth, P., 2002, Dedicated gravity field missions—principles and aims, *Journal of Geodynamics*, **33**, 3–20.

Saenger, E.H., Gold, N., and Shapiro, S.A., 2000, Modeling the propagation of elastic waves using a modified finite-difference grid, *Wave motion*, **31**, 77–92.

Sahoo, B.M., Ravi Kumar, B.V.V., Banik, B.K., and Borah, P., 2020, Polyaromatic hydrocarbons (PAHs): structures, synthesis and their biological profile. *Curr. Org. Synth.*, **17**, 625–640.

Sanders, R.H., 2010, *The Dark Matter Problem: A Historical Perspective*, Cambridge University Press.

Sayers, B., and Cherry, E.C., 1957, Mechanism of binaural fusion in the hearing of speech, *Journal of the Acoustical Society of America*, **31**, 535.

Schatz, J.F., and Simmons, G., 1972, Thermal conductivity of Earth materials at high temperature, *J. Geophys. Res.*, **77**, 6966–6983.

Schettino, A., 1999, Computational methods for calculating geometric parameters of tectonic plates, *Comput. Geosci.*, **25**, 897–907.

Schrödinger, E., 1935, Die gegenwärtige situation in der quantenmechanik, *Naturwissenschaften.* **23**, 807–812.

Schrödinger, E., 1945, *What is life?*, Cambridge University Press, Cambridge, UK.

Schrödinger, E., 1952, *Statistical thermodynamics*, Cambridge University Press, Cambridge.

Schubert, G., Turcotte, D., and Olson, P., 2001, *Mantle Convection in the Earth and Planets*, Cambridge University Press, Cambridge, UK.

Segall, P., and Davis, J.L., 1998, GPS Applications for Geodynamics and Earthquake Studies, *Annual Review of Earth and Planetary Sciences*, **6**, 301–336.

Shannon, C.E., 1948, A mathematical theory of Communication, *Bell System Technical Journal*, **27**, 379–423.

Shaviv, N.J., 2008, Using the oceans as a calorimeter to quantify the solarradiative forcing, *J. Geophys. Res.*, **113**.

Shor, P.W., 1994, Algorithms for quantum computation: discrete logarithms and factoring. *In: Proceedings 35th Annual Symposium on Foundations of Computer Science*, 124–134.

Shor, P.W., 1997, Polynomial-time algorithms for prime factorization and discrete logarithms on a quantum computer, *SIAM J. Comput.*, **26**, 1484–1509.

Shyu, K.K., Wu, Y.T., Chen, T.R., Chen, H.Y., Hu, H.H., and Guo, W.Y., 2011, Measuring complexity of fetal cortical surface from mr images using 3-D modified box-counting method, *IEEE Trans. Instrum. Meas.*, **60**, 522–531.

Siegel, R., Howell, J.R., 1990, *Thermal radiation heat transfer*, McGraw-Hill.

Smith, S.M., and Zwart, S.R., 2008, *Nutritional biochemistry of spaceflight.* In: Makowsky G, ed. Adv Clin Chem, **46**, Burlington: Academic Press, 87–130.

Smythe, W.R., 1968, *Static and dynamic electricity*, 3rd ed., McGraw-Hill, New York.

Soderberg, T., 2016, *Organic chemistry with a biological emphasis*, Volumes 1 and 2, University of Minnesota Morris.

Solomon, S.C., and Sleep, N.H., 1974, Some simple physical models for absolute plate motions, *J. Geophys. Res.*, **79**, 2557–2567.

Stofan, E.R., Elachi, C., Lunine, J.I., Lorenz, R.D., Stiles, et al., 2007, The lakes of Titan, *Nature*, **445**, 61–64.

Stratton, J.A., 1941, *Electromagnetic theory*, McGraw-Hill, New York.

Strobel, D.F., 1974, The photochemistry of hydrocarbons in the atmosphere of Titan, *Icarus*, **21**, 466–470.

Sullivan, D.M., 2012, *Quantum mechanics for electrical engineers*, John Wiley & Sons, Inc.

Susskind, L., 1995, The World as a hologram, *J. Math. Phys.*, **36**, 6377–6396.

Szopa, C., Cernogora, G., Boufendi, L., Correia, J.J., et al., 2006, PAMPRE: A dusty plasma experiment for Titan's tholins production and study, *Planet. Space Sci.*, **54**, 394–404.

Taioli, E., Sram, R.J., Binkova, B., Kalina, I., et al., 2007, Biomarkers of exposure to carcinogenic PAHs and their relationship with environmental factors, *Mutation Research/Fundamental and Molecular Mechanisms of Mutagenesis*, **620**, 16–21.

Tansey, E.A., and Johnson, C.D., 2015, Recent advances in thermoregulation: Adv. Physiol. Educ. **39**, 139–148.

Tapley, B., Bettadpur, S., Ries, J., Thompson, P., and Watkins, M., 2004, GRACE measurements of mass variability in the Earth system: Science, **305**, 503–505.

Tarashi, S., Siadat, S.D., Badi, S.A., et al., 2019, Gut bacteria and their metabolites: Which one is the defendant for colorectal cancer?, *Microorganisms*, **7**, 561.

Tenzer, R., Chen, W., and Baranov, A.A., 2008, Gravity Maps of Antarctic Lithospheric Structure from Remote-Sensing and Seismic Data, Pure and Applied Geophysics, **175(1)**.

Thomson (Lord Kelvin), W., 1851, Manuscript notes for "On the dynamical theory of heat": Arch. Hist. Exact Sci., **16**, 281–282.

Thomson (Lord Kelvin), W., 1852, On a universal tendency in nature to the dissipation of mechanical energy: Proc. R. Soc. Edinb., **20**, 139–142.

Thurman, H.V., 1994, *Introductory Oceanography*, seventh edition, Macmillan, New York, NY.

Timmen, L., 2003, Precise definition of the effective measurement height of free-fall absolute gravimeters, *Metrologia*, **40**, 62–65.

Trappe, T., et al., 2009, Exercise in space: human skeletal muscle after 6 months aboard the international space station, *J Appl Physiol.*, **106**, 1159–1168.

Truesdell, C., 1969, *Rational thermodynamics*, McGraw-Hill, New York, NY, USA,

Tsurutani, B.T., and Ho, C.M., 1999. A review of discontinuities and Alfvén waves in interplanetary space: Ulysses results, *Reviews of Geophysics*, **37**, 517–541.

Turcotte, D., and Schubert, G., 2002, *Geodynamics*, Cambridge University Press, Cambridge.

Urban, B.W., and Barann, M., 2002, *Molecular and Basic Mechanisms of Anesthesia*, Pabst Science Publishers.

van der Meijde, M., Pail, R., and Bingham, R., and Floberghagen, R., GOCE data, models, and applications: A, *International Journal of Applied Earth Observation and Geoinformation*, **35**, 4–15.

VanWylen, G.J., and Sonntag, R.E., 1993, *Fundamentals of classical thermodynamics*, 4nd Edition, John Wiley and Sons, New York.

Verma, A.K., and Margot, J.L., 2016, Mercury's gravity, tides, and spin from messenger radio science data, *Journal of Geophysical Research: Planets*, **121**, 1627–1640.

Vernazza, J.E., Avrett, E.H., and Loeser, R., 1981, Structure of the solar chromosphere. III - Models of the EUV brightness components of the quiet-sun, *The Astrophysical Journal Supplement Series*, **45**, 635–725.

Velikhov, E.P., 1959, Stability of an ideally conducting liquid flowing between cylinders rotating in a magnetic field, *Sov. Phys. JETP*, **9**, 995.

Virieux, J., 1984, SH-wave propagation in heterogeneous media: Velocity-stress finite-difference method, *Geophysics*, **49**, 1933–1957.

Virieux, J., 1986, P-SV wave propagation in heterogeneous media: Velocity stress finite difference method, *Geophysics*, **51**, 889–901.

Voigt, J., 2002, Alfvén wave coupling in the auroral current circuit, *Surveys in Geophysics*, **23**, 335–377.

Waite, J. H. Jr., Young, D. T., Cravens, T. E., Coates, A. J., et al. 2007. The process of tholin formation in Titan's upper atmosphere, *Science*, **316**, 870–875. 10.1126/science.1139727.

Wald, R.M., 1997, Gravitational Collapse and Cosmic Censorship: In Iyer, B. R.; Bhawal, B.(eds.)., *Black Holes, Gravitational Radiation and the Universe*: Springer, 69–86.

Watts, A.B., 2001, *Isostasy and flexure of lithosphere*, Cambridge University Press, UK.

Wells, P.B., Thomsen, S., Jones, M.A., Baek, S., and Humphrey, J.D., 2005, Histological evidence for the role of mechanical stress in modulating thermal denaturation of collagen, *Biomech Model Mechanobiol*, **4**, 201–210.

White, W.B., Lean, J., Cayan, D., and Dettinger, M., 1997, Response of global upper ocean temperature to changing solar irradiance, *J. Geophys. Res.*, **102**, 3255–3266.

Warwick, W., and Bannister, D., 1989, *Gray's anatomy*, Churchill Livingstone, London, UK

Williams, J.G., 2007, A scheme for lunar inner core detection, *Geophys. Res. Lett.*, **34**, L03202.

Wilson, J.T., 1963, Evidence from islands on the spreading of the ocean floor, *Nature*, **197**, 536–538.

Wolf, M., 1969, Direct measurement of Earth's gravitational potential using a satellite pair, *Journal of Geophysical Research*, **74**, 5295–5300.

Wu, C.S., Shaknov, I., 1950, The angular correlation of scattered annihilation radiation, *Phys. Rev.*, **77**, 136.

Wyk de Vries, M., Bingham, R.G., and Hein, A.S., 2018, A new volcanic province: an inventory of subglacial volcanoes in West Antarctica, *Geological Society, London, Special Publications*, **461**, 231–248.

Yariv, A., and Yeh, P., 2007, *Photonics: Optical Electronics in Modern communications*, Oxford: Oxford University Press.

Yee, K., 1966, Numerical Solution of Initial Boundary Value Problems Involving Maxwell's Equations in Isotropic Media, *IEEE Transactions on Antennas and Propagation*, **14**, 302–307.

Yin, J., Cao, Y., Li, Y-H, et al., 2017, Satellite-based entanglement distribution over 1200, *Science*, **356**, 1140–1144.

Zimmerman, J. E., Thiene, P., and Harding, J. T., 1970, Design and operation of stable rf-biased superconducting point-contact quantum devices, and a note on the properties of perfectly clean metal contacts, *Journal of Applied Physics*, **41**, 1572–1580.

Zwicky, F., 1933, Die Rotverschiebung von extragalaktischen Nebeln, *Helvetica Physica Acta*, **6**, 110–127.

Zwicky, F., 1937, On the Masses of Nebulae and of Clusters of Nebulae, *Astrophys. J.*, **86**, 217.

Index

2004 Sumatra earthquake, 55

A

Abandoned oil and gas wells, 573
Absolute reference frame, 113
Accretion disk, 642, 646, 657, 687, 688–694, 696–697, 698–700
Adiabatic gradient, 220–223, 231, 236–237, 686
Advection, 180–181, 214, 218–220
Advection-diffusion equation, 218
Airy's model, 98
Alfvén speed, 641–642, 649, 651–652, 654, 655
Alfvén wave, 652–654, 656, 657
Alkanes, 352, 379, 430
Alkenes, 352, 379
Alkynes, 379
Ammonia, 434–435, 683, 698
Ampére's law, 315, 319, 343, 356–357, 370, 554, 594, 628, 634, 636
Amygdala, 511–512, 517–518
Andromeda, 418, 463, 675
Angular momentum, 17–20, 89, 107, 114, 402, 403, 404, 449, 470, 488, 489, 492, 614, 618, 648–650, 669, 672, 687, 688, 690–692
Anoxygenic photosynthesis, 202, 293
Antarctica, 59, 64–69, 93, 107, 108, 112–113, 116, 117, 137, 139–140, 214, 242, 291–294, 298, 300, 302, 372, 375, 622
Anti-greenhouse effect, 198, 279, 288–289
Anti-greenhouse gases, 288
Apollo, 36, 84, 414, 575, 624
Archimedes' principle, 94–96, 97–98
Arrhythmia, 268, 543, 551

Asteroid belt, 22–25, 63
Asteroids, 6, 22–23, 25, 59, 63, 130, 172, 584, 669, 671, 699, 717
Atom entanglements, 490
Aurora, 435–436, 583, 590, 594, 622–624, 625–626, 681, 704

B

Bandgap energy, 330
Bangui, 297
Barycenter, 70–72
Bathymetry, 42, 46–47, 64, 65, 66, 68, 78
Bay of Fundy, 69, 78, 80
Bell's inequality, 468, 484, 485, 488
Benzene, 283, 286, 352, 379, 429, 433
Bimodal distributions, 712
Bioelectromagnetism, 508, 533
Biological life, 68, 93, 203, 204, 375, 710
Biosignatures, 417, 434
Biosphere, 93
Biot-Savart law, 357–358
Blackbody, 182, 184–189, 192, 198, 201, 231, 422
Blackbody radiation, 184, 186
Blackbody spectra/spectrum, 182, 184, 187
Black hole, 1, 28–29, 83, 163, 413, 415, 422, 583, 687, 688, 690, 692, 696, 709
Bouguer anomaly, 65
Bouguer correction, 47, 67
Brain activity, 508, 538, 539
Brain-to-body ratios, 524
Brain tumors, 512, 538
Brown dwarfs, 696
Brunt-Väisälä frequency, 665–666, 685–686, 687

C

Cameroon, 7

Cancer, 138, 247, 429–433

Capacitor, 355, 544–545, 547, 548–549, 556, 580

Carnot engine, 257, 259, 261, 263

Carnot refrigerator, 266

Casimir effect, 336, 579, 580–582

Cassini-Huygens, 87–88, 279, 281, 283, 287, 290

CDC, 135

Cell membrane, 519, 526, 527, 528–530, 532, 542, 544, 547, 548, 566, 570

Center of gravity, 69, 70, 350, 352

Center of the Earth, 1, 4, 6–7, 11, 70, 72–73, 109, 112

Centrifugal acceleration, 668

Centrifugal force, 15, 60, 62, 69, 70, 72–73, 606, 609, 667

Chain reaction, 574, 575, 576

Chandrayaan-1, 139

Chang'e 4, 64, 84

Chang'e-4, 84, 140

Chromosphere, 585, 588, 597, 635, 641, 652

Cloud formations, 271–272, 274, 276, 278, 279, 291

Collisionless plasma, 538, 595, 631, 641, 705

Coma cluster, 28, 31

Compatibility equations, 314, 317

Conscious awareness, 398, 399, 508

Continuity equation, 311, 315, 318, 343, 355, 360, 368, 585, 633, 660, 662, 670

Convection, 42, 99, 149, 170–172, 173, 177–180, 204, 205, 206, 216–217, 218, 219, 220–223, 224, 226–227, 229, 231, 235, 236–237, 238, 241, 244, 246, 275, 459, 584, 585, 588, 590, 652, 657, 684, 686, 687, 725

Convection-diffusion equation, 179–180, 216, 218–219

Convection instability, 686, 687

Convective instability, 657, 666, 684, 686

Convective stability, 685–686, 687

Coriolis force, 60

Coronal loops, 588–589, 610

Coronal mass ejections, 296, 588–589, 595, 704

Coseismic, 100, 130

Coulomb's law, 344–345, 356

Cryoablation, 242

Cryobiology, 238, 241–242, 245

Cryosurgery, 242, 245

Curvature drifts, 610

Cyanobacteria, 293

Cyclones, 94

D

Dark matter, 25–32, 418, 427–428, 583, 596, 631, 692

Debye length, 632–633

Decoherence, 488, 504

Depleted oil and gas reservoirs, 572

Depolarization, 529, 530–531, 542–543

Diesel engines, 257, 260, 261, 263

Diffusion equation, 176, 177, 181, 205–206, 208–209

Dipole moment, 195, 342, 348–354, 355, 359–360, 364–365, 383, 393, 394, 402, 619

Dispersion relation, 371, 653, 655, 670, 678, 681, 684

Doppler effect, 123

Doppler shift, 123, 664, 710

DORIS, 118, 121, 122–124, 126, 146

Drake's equation, 721

Dynamo, 594–595, 647, 692

E

Earth-like exoplanets, 710, 717, 719

Earthquake, 55, 99–100, 103–104, 106, 120, 124, 126–127, 130, 139–140, 142, 145–147, 204, 293, 301, 463, 507, 721

Eccentricity, 15, 42, 44, 82, 87, 92, 123, 698

Eddington limit, 689

EDLCs, 579

EEG, 537–539, 540, 541, 543, 544

Electrical potential, 151, 347, 358–359, 533, 536, 544, 566

Electricity production, 507, 572

Electric polarization, 313, 315–316, 326, 335, 339, 355

Electrocardiogram (ECG), 541

Electroencephalogram (EEG), 537, 539

Electromotive force, 554–555, 557

Elliptical galaxies, 28–29

Endotherms, 242

Equipotential surfaces, 34, 40, 103, 105, 348

Eulerian, 658–660, 661

Euler vector, 107–109, 112, 114–115

Exoplanets, 1, 24–25, 83, 87, 88, 137, 201–203, 204, 288, 390, 416–417, 423, 425, 426, 433–435, 436–437, 463, 504–505, 524, 583–584, 591, 593, 697–698, 699, 707–710, 711–721

F

Far side of the Moon, 64, 83, 84, 93

Fifth force, 582

Fingerprint, 387, 392, 396

First law of thermodynamics, 149, 151, 154–155, 159, 175, 177, 179, 238, 257

Fission, 436, 460, 574–575, 577, 578–579, 582

Fission chain reaction, 574, 575

Fission reactors, 14, 573, 577, 578, 716

Free-air correction, 45–46, 67

Free-air-gravity anomaly, 47, 64–65

Free-air-gravity-anomaly, 64

Free-falling test, 56

Free-fall time, 12–13, 672

Frieman–Rotenberg formalism, 662, 667

Frozen flux, 706

Functional groups, 379, 381, 383, 387, 391, 396, 406

Fusion, 435–436, 459, 460, 682, 687, 695, 696–697, 698, 715–716

G

Gabon, 573, 575, 576–577

Galileo, 6, 118, 294

Gate, 478

Gaussianity, 166

Gauss's law, 315–316, 318, 343–344, 347, 356, 368, 370, 545, 552–553

Gauss's theorem, 37–38, 175, 179

GCRF, 121

Generalized Ohm's law, 700, 702, 705

Geodesy, 1, 22, 40, 58, 100, 101, 103–104, 121, 145

Geoid, 33–34, 40–42, 46, 53, 54–55, 58–59, 102–103, 105

Geotherm, 206–208, 212–213, 217, 218–220, 221–225, 226, 293

Geothermal energy, 204, 208, 572–573

Geothermics, 154, 204

Ghawar oilfield, 573

Gibbs free energy, 164

GLONASS, 118

GNSS, 58, 118, 146

GOCE, 57–59

GPR data-acquisition, 67

GPS, 57, 103–106, 110–111, 116, 118, 121, 123, 124–126, 127, 130, 590

GPS satellites, 56, 104

GRACE, 55–59

GRACE-FO, 55

Grad-B drift, 606, 608, 609, 612

Gravimeters, 48, 59

Gravitational instability, 668, 671

Gravitational lensing, 29

Gravitational microlensing, 711

Gravitational potential, 33–34, 36–37, 38, 161, 634, 667, 670

Great Oxidation Event (GOE), 293, 300

Greenhouse effect, 181, 194–195, 197–198, 201, 203

Greenhouse gases, 187, 194–196, 288–289, 344

Ground penetrating radar (GPR), 366, 371, 372

Ground state, 402, 446–447, 450, 492, 495–496

Gyrofrequency, 599–601

Gyromotion, 598, 601, 603, 616

H

Hadamard gate, 478–479, 495, 498–499

Haze, 281–282, 287–288, 289, 290, 434

Heart rhythm, 268, 551

Heat-conduction equation, 210

Heat-diffusion equation, 175–176, 180, 244, 247

Heat engine, 149, 161, 163, 237, 256, 257, 259–260, 265, 269

Heat pump, 149, 265–266

Heat transfer, 149–150, 159, 160, 170–173, 175, 179, 180, 189, 204, 205, 213, 231, 237–238, 246–247, 252, 257, 261, 274

Heat transport, 180

Heisenberg uncertainty principle, 446, 464, 466–467

Henry Cavendish, 5

Herzsprung-Russell diagram, 189–190

Hilbert space, 474, 476, 483

Hippocampus, 511–513, 517–518, 523

Homeostasis, 511, 515

Hubble, 414–415, 416, 418, 424, 625, 693, 695

Hurricanes, 94, 683

Hydrogen fusion, 459

Hydrostatic equilibrium, 29–30, 32, 234, 584, 657, 667–668, 673

Hydrothermal vents, 204, 293

Hydroxyl, 379

Hyperpolarization, 529, 531

Hyperthermia, 238, 241, 246–248, 250, 253

Hypothalamus, 239, 511, 517

Hypothermia, 238, 241, 246–248, 250

I

Ideal MHD, 637–638, 642, 661–662, 684, 700, 702, 704–706

Infrared, 171, 182, 184, 187, 190–195, 199, 203, 284, 287–289, 306, 378, 379, 381–384, 387, 388–389, 391–392, 394–395, 396, 398, 401, 409, 410–412, 413, 416–420, 423–425, 433, 434, 593, 690, 695, 696

InSAR, 58, 103, 130

InSight, 22

Interference, 78, 378, 423, 479, 494–495, 502, 540

Interferograms, 130

Interstellar, 13, 413, 417, 421, 428, 436, 460–461, 463, 632, 635, 657, 669,

672, 682, 693–694, 698, 699–700, 711

Ionization, 283, 381, 583, 596, 692

Ionosphere, 104, 366, 589, 594, 596–597, 622, 703

Ischemia, 241

Isostasy, 94, 96, 97, 99–100

Isostatic equilibrium, 96, 97–99

ITRF, 117, 120–124, 128, 129

J

James Clerk Maxwell, 161

James Van Allen, 620

James Webb Space Telescope (JWST), 416, 433

Jean Fourier, 173

Jeans instability, 657, 668–669, 671–672, 673, 674

Joseph Lagrange, 59

Journey to Mars, 22

Jupiter's atmosphere, 625, 680, 683–684

K

Kelvin-Helmholtz instability, 276, 657–658, 679–680

Keplerian disk, 691–692

Kepler's constant, 16

Kepler's law, 15–16, 20, 60, 692

Kinetic energy, 14, 30, 36–37, 264, 392, 440, 451, 452, 457–458, 495, 497, 600–601, 612, 616, 619–620, 638, 673, 695, 696

Kola Superdeep Borehole (KSDB), 293

Kronecker product, 468, 472, 476–477

Kuiper belt, 22–23, 25

L

Lagragian, 418

Lagrange points, 59, 60–63

Lagrangian, 416, 658, 660, 663

Laguerre functions, 449

Lake Vostok, 66, 68, 293, 294, 372, 374

Laplace's equation, 38, 40, 347, 545

Larmor frequency, 404, 602

Latent heat, 149, 170–173, 274

Legendre functions, 39–40, 358

Lenticular clouds, 276

Life expectancy, 137, 522–523, 539

Life on Earth, 23, 25, 92, 288, 300, 432, 508
Liquidus, 223–224, 228
Lithium, 579
LNG (liquefied natural gas), 285
Lorentz force law, 307, 309, 315
Love number, 88, 90, 91
Ludwig Boltzmann, 161
Lunik, 84

M
Magnetic bottle, 618–619, 620, 622
Magnetic diffusivity, 638, 640
Magnetic mirror, 610, 614, 616, 618, 620
Magnetic moment, 364–365, 402, 403, 470, 610, 611–614, 616
Magnetic monopole, 313, 315, 318–319, 359–360, 700
Magnetic permeability, 307, 326–327, 594
Magnetic reconnection, 638, 646, 700, 704–705
Magnetic reversal, 295, 300, 302–303, 708
Magnetocardiogram (MCG), 544
Magnetoencephalogram (MEG), 539
Magnetohydrodynamic equations, 639–640, 651
Magnetorotational instability, 657, 687, 690–692, 698
Magnetosphere, 130, 202, 204, 282–283, 295–296, 300, 303, 366, 410, 436–437, 583, 596, 622, 625, 638, 641, 657, 684, 703, 705, 716, 719
Magnetotelluric (MT) surveys, 366
Main-asteroid belt, 23
Maunder minimum, 588, 594
Maxwell relations, 164
Maxwell's equations, 306–307, 312–313, 314–316, 317, 318–319, 321, 327, 334, 339, 342–343, 347, 368–369, 373, 440, 485, 552, 554, 581, 631, 753–754
MeerKat, 428
Melting point, 67, 223–225, 280, 716
Membrane capacitor, 548–549

Membrane potential, 526, 527, 529–531, 533, 534–535, 567, 569–571
Mexico, 7, 47, 413, 717
MHD equation, 631, 633, 636, 641–642, 651, 661–662, 668, 687
Micius satellite, 488
Microgravity, 130, 133, 135, 137, 138
Microtubes, 519–520
Milankovitch Cycle, 92
Molecular cloud, 12, 416, 632, 669, 671, 675, 690, 693, 694–695, 697
Molten lava, 213
Monatomic ideal gas, 32
Monsoons, 94
Moonquakes, 139
Moon's surface gravity, 7
Multi-qubit, 476, 497
Muscle loss, 133, 138
Myocardium, 268, 541

N
Nanomaterials, 397
Navier-Stokes equation, 12, 440, 585, 641
Neodymium, 575, 578
Nernst equation, 533–534
Nervous system, 239, 508, 513, 515, 517, 529, 544, 550
Neutron stars, 189, 584, 618, 687
Newton's law, 1–3, 5, 17, 45, 99, 179, 597
NNR frame, 114, 116, 129
Noncentrosymmetric crystals, 340, 342
No-net-rotation, 114, 118
Non-Gaussian, 167, 466
Nonlinear materials, 322, 336
Non-silicate minerals, 389–391
Normal density, 167
Nuclear bombs, 436, 574
Nuclear fission, 435–436, 507, 573–575, 576, 578
Nuclear magnetic resonance (NMR), 379, 403, 407
Nuclear spins, 405

O
Oceanic trenches, 44
Oklo mine, 576
Oklo reactor, 577–578

Opacity, 231, 234–235, 689
Optical cavity, 490–492
Optical telescope, 412–414, 415, 420–421, 428
Orthometric height, 102–103, 105
Otto engines, 261–262
Oumuamua, 697, 699
Oxygenic photosynthesis, 202, 293, 435

P
PAHs, 283, 429–433
Pendulum displacement, 52
Photon entanglements, 489
Photosynthesis, 130, 202, 293
Planck's distribution, 183
Planck's law, 181, 182–183, 189
Plasma, 14, 583–722
Plate boundaries, 109, 111–112
Poisson's equation, 38, 347
Poloidal magnetic field, 629
Polycyclic aromatic hydrocarbons, 283, 429–430, 433
Post-seismic, 120, 126–128
Potential energy, 30, 33, 36–37, 340, 440–442, 444, 445, 447, 452, 454, 456, 458, 464, 630, 673–674, 688
Poynting vector, 319, 321
Pratt's hypothesis, 99
Precession, 92–93, 123, 404
Precession frequency, 404
Protoplanetary disks, 642, 687, 693
Proxima Centauri, 132, 418, 428, 437, 439, 463, 711
P-type semiconductors, 334
Putnam, 52

Q
Quantum computer, 472, 475, 476–477, 478, 488, 494–495, 496, 499, 500–501, 503–504
Quantum decoherence, 486, 496
Quantum entanglement, 439, 468, 480, 482, 486, 488–489
Quantum Fourier transform (QFT), 496, 497, 499
Quantum gate, 478, 494–495, 496, 499

Quantum measurement, 468–470, 472, 477
Quantum mechanical probability, 447
Quantum tunneling, 444, 453, 455–456, 457, 459–460, 461
Qubit, 474–475, 476–477, 478–480, 482, 486, 488, 494–496, 497–499, 501–502, 503, 521
Queqiao, 64, 84

R
Radiation pressure, 12, 186, 233, 689, 695
Radioactive waste, 574–575, 577–578
Radio telescopes, 420–421, 669, 695, 696
Raman scattering, 392–393, 394
Raman spectroscopy, 379, 383, 392, 394, 396, 398
Random variables, 165, 166, 169
Rayleigh scattering, 392–393, 394, 421
Rayleigh-Taylor instability, 655, 657, 658, 681–682
Realistic fusions, 14
Red shift, 420, 425
Reference ellipsoid, 34, 41–42, 46, 50
Repolarization, 529–531, 542–543
Resistivity, 323–324, 366–367, 555, 557, 560–561, 562–563, 564, 565–566, 568–569, 639, 701, 705
Resting potential, 529, 531, 533, 535, 549
Reversal potential, 533
Roche limit, 79, 82
Roche radius, 82
Rotational potential, 90
Rotation vector, 17
Russian Antarctic programme, 68

S
Satellite, 1, 4, 5, 7, 16, 40–42, 45, 54–58, 59, 63, 64, 82, 84–85, 103–105, 118, 122–123, 124, 129, 143, 296, 303, 415, 418, 428, 463, 488, 589, 590, 595, 597, 620, 625, 680
Satellite gravity measurements, 45
Scanning tunneling microscopy (STM), 461
Schmidt decomposition, 480, 483

Schröndinger equation, 440–442, 444, 445, 448, 450, 451, 452–453, 454–455, 475, 504
Schwarzschild criterion, 665, 685–687
Schwarzschild Radius, 29, 763
Schwarzschild stability criterion, 237, 686, 687
Seamounts, 42, 44, 143
Second Law of Thermodynamics, 159–160, 162, 256, 257, 269, 721
Self potentials, 566
Semiconductor, 323, 327, 329–331, 332–334
Semidiurnal, 76–77
Semi-major axis, 15, 16, 21, 42, 715
Semi-minor axis, 15, 42
Shor quantum algorithm, 496
Shor's algorithm, 496, 500, 501, 503
Simultaneous earthquakes, 463
SLR, 58, 118, 121, 122–124, 126, 146
SOHO, 63, 64
Solar wind, 279, 295–296, 366, 460, 583, 589, 590, 591, 594–595, 620, 622–623, 625–626, 635, 638, 641, 642, 644–646, 648–649, 654, 657, 684
Solidus, 223–225
South Atlantic Anomaly (SAA), 303
Spacecraft's engines, 22
Spherical harmonic, 38–39, 358, 449, 683
Spiral disks, 687
SP log, 566, 569, 571–572
Spontaneous generation of electricity, 507
Spontaneous potential (SP), 566
SQUID, 540, 544
Steam, 153, 256, 257, 274, 572, 575, 576, 716
Stefan-Boltzman constant, 246
Stefan-Boltzmann law, 184–185
Stellar structure, 235, 236
Stirling engine, 260–261
Stokes Raman scattering, 393, 394
Stratosphere, 282, 284, 288, 290–291
Subducting slabs, 225
Subglacial lakes, 42, 67–68, 372

Sun's photosphere, 295, 716
Supercapacitors, 579
Superchrons, 298
Supernovae, 203, 421, 423, 675, 677, 694
Superposition, 182, 345–346, 399, 424, 441, 466, 475–476, 477, 478–479, 492, 494–495, 499, 501–502, 581–582, 620
Susceptibility, 326–327, 328, 336, 340–342
Synapses, 509, 515, 529
Synthetic Aperture Radar, 103, 130

T
Tachycardia, 268
Tauri, 693, 695–696
Terahertz, 306, 337, 379, 396–397, 398–399, 401–402, 510, 519, 721
Thalamus, 511
Thermal diffusivity, 177, 178, 190, 209, 218, 725
Thermonuclear, 25, 436, 460, 695–696
Thermoregulation, 135, 239–240, 242, 244
Thermoregulatory, 240, 241
Thermosphere, 281, 282, 590, 597, 703
Thermotherapy, 149
Thorium-232, 229, 579
THz spectroscopy, 396–397
Tidal bulges, 73, 77, 78
Tide bulges, 74
Tides, 40, 44, 69, 74–75, 76–77, 78–80, 86–88, 89–90, 92–94, 110, 111
Titan, 84, 86–88, 90–91, 279–280, 281–286, 287, 288–289, 290–291, 293, 294, 300, 432–433, 680, 684
Tokamak, 610, 626–630, 635, 687–688, 704, 715
Toroidal field, 627–629
TRAPPIST-1, 23–25, 84, 190, 201, 203, 204, 417, 433, 437, 710, 719
Trojan asteroids, 22
Turbine, 263–265, 266, 572, 575, 716

U
Ultracapacitors, 579
Unfrozen flux, 706

Universal gravitational constant, 2, 5, 26, 47
Uranium-235, 239, 574–575, 576, 578–579
Uranium-enrichment, 575

V
Vacuum energy, 582
Van Allen belts, 463, 622, 624
Van der Waals equation, 159
Vasodilation, 239
Virial theorem, 29–30, 32–33, 673
VLBI, 58, 118, 120, 121, 122–124, 126, 146
Voyager, 436, 463, 624

W
Wave function, 440–442, 445–446, 447–449, 450–451, 452, 455, 458, 464–466, 468, 475, 504–505, 756
Well logging, 344, 558, 560–561, 562, 564, 565–566, 568, 571
White dwarfs, 28, 189, 584, 618
Wien's displacement law, 187, 189
Wien's law, 184, 425
WKB (Wentzel–Kramers–Brillouin) approximation, 685

Y
Yellowstone National Park, 145

For Product Safety Concerns and Information please contact our EU
representative GPSR@taylorandfrancis.com
Taylor & Francis Verlag GmbH, Kaufingerstraße 24, 80331 München, Germany